HONDA

Outboa

1978-01 R… MANUAL
2–130 HORSEPOWER, 1-4 CYLINDER

SELOC®

Managing Partners Dean F. Morgantini, S.A.E.
Barry L. Beck

Executive Editor Kevin M. G. Maher, A.S.E.

Manager-Marine/Recreation James R. Marotta, A.S.E.

Production Managers Ron Webb
Melinda Possinger

Editor Kevin M. G. Maher, A.S.E

Manufactured in USA

104 Willowbrook Lane
West Chester, PA 19382
ISBN 0-89330-048-9
4567890123 4321098765

www.selocmarine.com
1-866-SELOC55

CONTENTS

1	GENERAL INFORMATION, AND BOATING SAFETY	HOW TO USE THIS MANUAL	1-2
		BOATING SAFETY	1-3
		SAFETY IN SERVICE	1-11
2	TOOLS AND EQUIPMENT	TOOLS AND EQUIPMENT	2-2
		TOOLS	2-4
		FASTENERS, MEASUREMENTS AND CONVERSIONS	2-12
3	MAINTENANCE AND TUNE-UP	ENGINE MAINTENANCE	3-2
		BOAT MAINTENANCE	3-21
		TUNE-UP	3-25
		WINTER STORAGE CHECKLIST	3-38
		SPRING COMMISSIONING CHECKLIST	3-38
4	FUEL SYSTEM	FUEL SYSTEM BASICS	4-2
		CARBURETION	4-4
		FUEL INJECTION	4-19
5	IGNITION AND ELECTRICAL SYSTEMS	UNDERSTANDING AND TROUBLESHOOTING ELECTRICAL SYSTEMS	5-2
		BREAKER POINTS IGNITION (MAGNETO IGNITION)	5-7
		CAPACITOR DISCHARGE IGNITION (CDI) SYSTEM	5-12
		ELECTRONIC IGNITION	5-34
		CHARGING CIRCUIT	5-35
		STARTING CIRCUIT	5-43
		COMPONENT LOCATIONS	5-48
		IGNITION AND ELECTRICAL WIRING DIAGRAMS	5-52
6	LUBRICATION AND COOLING	LUBRICATION SYSTEM	6-2
		COOLING SYSTEM	6-9

CONTENTS

ENGINE MECHANICAL	7-2	POWERHEAD	7
ENGINE RECONDITIONING	7-31		
LOWER UNIT	8-2	LOWER UNIT	8
LOWER UNIT OVERHAUL	8-10		
JET DRIVE	8-68		
MANUAL TILT	9-2	TRIM AND TILT	9
GAS ASSISTED TILT	9-2		
POWER TRIM AND TILT	9-4		
REMOTE CONTROL BOX	10-2	REMOTE CONTROL	10
TILLER HANDLE	10-11		
HAND REWIND STARTER	11-2	HAND REWIND STARTER	11
GLOSSARY	11-13	GLOSSARY	
MASTER INDEX	11-17	MASTER INDEX	
SUPPLEMENT—BF2D, BF75A AND BF90A	12-1	SUPPLEMENT	12

■ **Throughout this manual you will find references to Honda models with the suffix "A" (BF#A). Procedures for later motors with a model suffix "D" (BF#D) are the same, unless noted or unless listed in the Supplement. For instance, most procedures on the BF2A apply to the BF2D, unless otherwise noted, but when working on a BF2D be sure to check the Supplement before starting a procedure. This Supplement also contains late production specifications for the BF75A/BF90A motors.**

Other titles
Brought to you by

Title	Part #
Chrysler Outboards, All Engines, 1962-84	018-7
Force Outboards, All Engines, 1984-96	024-1
Honda Outboards, All Engines 1988-98	1200
Johnson/Evinrude Outboards, 1-2 Cyl, 1956-70	007-1
Johnson/Evinrude Outboards, 1-2 Cyl, 1971-89	008-X
Johnson/Evinrude Outboards, 1-2 Cyl, 1990-95	026-8
Johnson/Evinrude Outboards, 3, 4 & 6 Cyl, 1973-91	010-1
Johnson/Evinrude Outboards, 3-4 Cyl, 1958-72	009-8
Johnson/Evinrude Outboards, 4, 6 & 8 Cyl, 1992-96	040-3
Kawasaki Personal Watercraft, 1973-91	032-2
Kawasaki Personal Watercraft, 1992-97	042-X
Marine Jet Drive, 1961-96	029-2
Mariner Outboards, 1-2 Cyl, 1977-89	015-2
Mariner Outboards, 3, 4 & 6 Cyl, 1977-89	016-0
Mercruiser Stern Drive, Type I, Alpha/MR, Bravo I & II, 1964-92	005-5
Mercruiser Stern Drive, Alpha I (Generation II), 1992-96	039-X
Mercruiser Stern Drive, Bravo I, II & III, 1992-96	046-2
Mercury Outboards, 1-2 Cyl, 1965-91	012-8
Mercury Outboards, 3-4 Cyl, 1965-92	013-6
Mercury Outboards, 6 Cyl, 1965-91	014-4
Mercury/Mariner Outboards, 1-2 Cyl, 1990-94	035-7
Mercury/Mariner Outboards, 3-4 Cyl, 1990-94	036-5
Mercury/Mariner Outboards, 6 Cyl, 1990-94	037-3
Mercury/Mariner Outboards, All Engines, 1995-99	1416
OMC Cobra Stern Drive, Cobra, King Cobra, Cobra SX, 1985-95	025-X
OMC Stern Drive, 1964-86	004-7
Polaris Personal Watercraft, 1992-97	045-4
Sea Doo/Bombardier Personal Watercraft, 1988-91	033-0
Sea Doo/Bombardier Personal Watercraft, 1992-97	043-8
Suzuki Outboards, All Engines, 1985-99	1600
Volvo/Penta Stern Drives 1968-91	011-X
Volvo/Penta Stern Drives, Volvo Engines, 1992-93	038-1
Volvo/Penta Stern Drives, GM and Ford Engines, 1992-95	041-1
Yamaha Outboards, 1-2 Cyl, 1984-91	021-7
Yamaha Outboards, 3 Cyl, 1984-91	022-5
Yamaha Outboards, 4 & 6 Cyl, 1984-91	023-3
Yamaha Outboards, All Engines, 1992-98	1706
Yamaha Personal Watercraft, 1987-91	034-9
Yamaha Personal Watercraft, 1992-97	044-6
Yanmar Inboard Diesels, 1975-98	7400

SAFETY NOTICE

Proper service and repair procedures are vital to the safe, reliable operation of all marine engines, as well as the personal safety of those performing repairs. This manual outlines procedures for servicing and repairing outboards using safe, effective methods. The procedures contain many NOTES, CAUTIONS and WARNINGS which should be followed, along with standard procedures, to eliminate the possibility of personal injury or improper service which could damage the vessel or compromise its safety.

It is important to note that repair procedures and techniques, tools and parts for servicing marine engines, as well as the skill and experience of the individual performing the work, vary widely. It is not possible to anticipate all of the conceivable ways or conditions under which these engines may be serviced, or to provide cautions as to all possible hazards that may result. Standard and accepted safety precautions and equipment should be used during cutting, grinding, chiseling, prying, or any other process that can cause material removal or projectiles.

Some procedures require the use of tools specially designed for a specific purpose. Before substituting another tool or procedure, you must be completely satisfied that neither your personal safety, nor the performance of the marine engine, will be compromised.

Although information in this manual is based on industry sources and is complete as possible at the time of publication, the possibility exists that some vehicle manufacturers made later changes which could not be included here. While striving for total accuracy, Nichols Publishing cannot assume responsibility for any errors, changes or omissions that may occur in the compilation of this data.

PART NUMBERS

Part numbers listed in this reference are not recommendations by Nichols Publishing for any product brand name. They are references that can be used with interchange manuals and aftermarket supplier catalogs to locate each brand supplier's discrete part number.

SPECIAL TOOLS

Special tools are recommended by the marine manufacturer to perform a specific task. Use has been kept to a minimum, but, where absolutely necessary, they are referred to in the text by the part number of the tool manufacturer. These tools can be purchased, under the appropriate part number, from your local dealer or regional distributor, or an equivalent tool can be purchased locally from a tool supplier or parts outlet. Before substituting any tool for the one recommended, read the SAFETY NOTICE at the top of this page.

ACKNOWLEDGMENTS

Nichols Publishing expresses appreciation to the following companies who supported the production of this manual:

- All American Manufacturing Inc.—Bronson, MI
- Belk's Marine—Holmes, PA
- Hilton Marine—Wilmington, DE
- Honda Power Corporation—Duluith, GA
- Marine Mechanics Institute—Orlando, FL

Special thanks to Mike Benfer and Robert Strickler of the Marine Mechanics Institute for providing the outboard engine and miscellaneous parts to get this project going and to Tom Truman and Sam Wright of Hilton Marine for allowing us to spend an entire day taking pictures at their marine dealership.

Nichols Publishing would like to express thanks to all of the fine companies who participate in the production of our books:

- Hand tools supplied by Craftsman are used during all phases of our vehicle teardown and photography.
- Many of the fine specialty tools used in our procedures were provided courtesy of Lisle Corporation.
- Lincoln Automotive Products (1 Lincoln Way, St. Louis, MO 63120) has provided their industrial shop equipment, including jacks (engine, transmission and floor), engine stands, fluid and lubrication tools, as well as shop presses.
- Rotary Lifts (1-800-640-5438 or www.Rotary-Lift.com), the largest automobile lift manufacturer in the world, offering the biggest variety of surface and in-ground lifts available, has fulfilled our shop's lift needs.
- Much of our shop's electronic testing equipment was supplied by Universal Enterprises Inc. (UEI).
- Safety-Kleen Systems Inc. has provided parts cleaning stations and assistance with environmentally sound disposal of residual wastes.
- United Gilsonite Laboratories (UGL), manufacturer of Drylock® concrete floor paint, has provided materials and expertise for the coating and protection of our shop floor.

HOW TO USE THIS MANUAL 1-2
CAN YOU DO IT? 1-2
WHERE TO BEGIN 1-2
AVOIDING TROUBLE 1-2
MAINTENANCE OR REPAIR? 1-2
DIRECTIONS AND LOCATIONS 1-2
PROFESSIONAL HELP 1-2
PURCHASING PARTS 1-3
AVOIDING THE MOST COMMON MISTAKES 1-3
BOATING SAFETY 1-3
REGULATIONS FOR YOUR BOAT 1-3
DOCUMENTING OF VESSELS 1-4
REGISTRATION OF BOATS 1-4
NUMBERING OF VESSELS 1-4
SALES AND TRANSFERS 1-4
HULL IDENTIFICATION NUMBER 1-4
LENGTH OF BOATS 1-4
CAPACITY INFORMATION 1-4
CERTIFICATE OF COMPLIANCE 1-4
VENTILATION 1-4
VENTILATION SYSTEMS 1-5
REQUIRED SAFETY EQUIPMENT 1-5
TYPES OF FIRES 1-5
FIRE EXTINGUISHERS 1-5
WARNING SYSTEM 1-6
PERSONAL FLOTATION DEVICES 1-6
SOUND PRODUCING DEVICES 1-8
VISUAL DISTRESS SIGNALS 1-8
EQUIPMENT NOT REQUIRED BUT RECOMMENDED 1-10
SECOND MEANS OF PROPULSION 1-10
BAILING DEVICES 1-10
FIRST AID KIT 1-10
ANCHORS 1-10
VHF-FM RADIO 1-10
TOOLS AND SPARE PARTS 1-10
COURTESY MARINE EXAMINATIONS 1-11
SAFETY IN SERVICE 1-11
DO'S 1-11
DON'TS 1-11

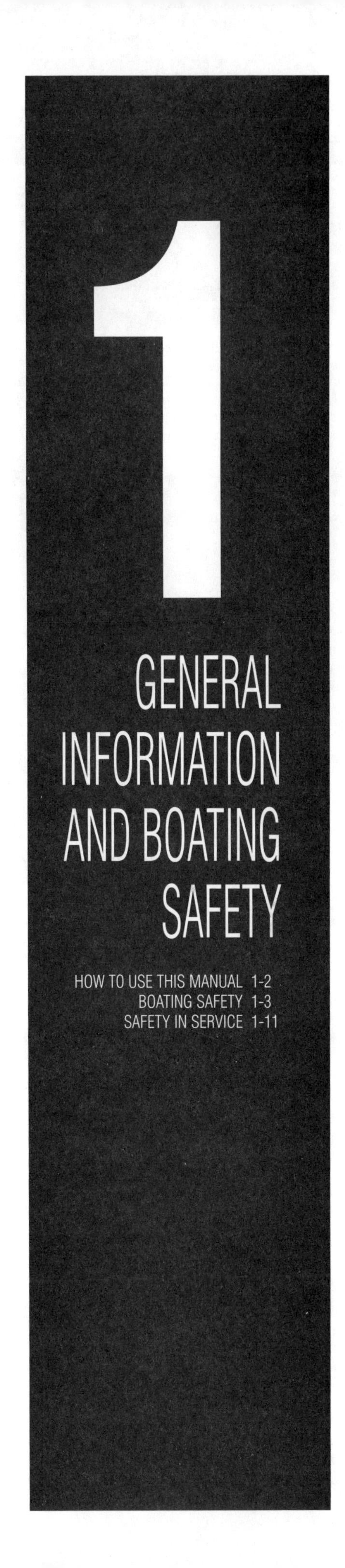

1 GENERAL INFORMATION AND BOATING SAFETY

HOW TO USE THIS MANUAL 1-2
BOATING SAFETY 1-3
SAFETY IN SERVICE 1-11

HOW TO USE THIS MANUAL

This manual is designed to be a handy reference guide to maintaining and repairing your Honda 4-stroke outboard. We strongly believe that regardless of how many or how few years experience you may have, there is something new waiting here for you.

This manual covers the topics that a factory service manual (designed for factory trained mechanics) and a manufacturer owner's manual (designed more by lawyers these days) covers. It will take you through the basics of maintaining and repairing your outboard, step-by-step, to help you understand what the factory trained mechanics already know by heart. By using the information in this manual, any boat owner should be able to make better informed decisions about what they need to do to maintain and enjoy their outboard.

Even if you never plan on touching a wrench (and if so, we hope that you will change your mind), this manual will still help you understand what a mechanic needs to do in order to maintain your engine.

Can You Do It?

If you are not the type who is prone to taking a wrench to something, NEVER FEAR. The procedures in this manual cover topics at a level virtually anyone will be able to handle. And just the fact that you purchased this manual shows your interest in better understanding your outboard.

You may find that maintaining your outboard yourself is preferable in most cases. From a monetary standpoint, it could also be beneficial. The money spent on hauling your boat to a marina and paying a tech to service the engine could buy you fuel for a whole weekend's boating. If you are unsure of your own mechanical abilities, at the very least you should fully understand what a marine mechanic does to your boat. You may decide that anything other than maintenance and adjustments should be performed by a mechanic (and that's your call), but know that every time you board your boat, you are placing faith in the mechanic's work and trusting him or her with your well-being, and maybe your life.

It should also be noted that in most areas a factory trained mechanic will command a hefty hourly rate for off site service. This hourly rate is charged from the time they leave their shop to the time they return home. The cost savings in doing the job yourself should be readily apparent at this point.

Where to Begin

Before spending any money on parts, and before removing any nuts or bolts, read through the entire procedure or topic. This will give you the overall view of what tools and supplies will be required to perform the procedure or what questions need to be answered before purchasing parts. So read ahead and plan ahead. Each operation should be approached logically and all procedures thoroughly understood before attempting any work.

Avoiding Trouble

Some procedures in this manual may require you to "label and disconnect . . ." a group of lines, hoses or wires. Don't be lulled into thinking you can remember where everything goes — you won't. If you reconnect or install a part incorrectly, things may operate poorly, if at all. If you hook up electrical wiring incorrectly, you may instantly learn a very, very expensive lesson.

A piece of masking tape, for example, placed on a hose and another on its fitting will allow you to assign your own label such as the letter "A", or a short name. As long as you remember your own code, the lines can be reconnected by matching letters or names. Do remember that tape will dissolve when saturated in fluids. If a component is to be washed or cleaned, use another method of identification. A permanent felt-tipped marker can be very handy for marking metal parts; but remember that fluids will remove permanent marker.

SAFETY is the most important thing to remember when performing maintenance or repairs. Be sure to read the information on safety in this manual.

Maintenance or Repair?

Proper maintenance is the key to long and trouble-free engine life, and the work can yield its own rewards. A properly maintained engine performs better than one that is neglected. As a conscientious boat owner, set aside a Saturday morning, at least once a month, to perform a thorough check of items which could cause problems. Keep your own personal log to jot down which services you performed, how much the parts cost you, the date, and the amount of hours on the engine at the time. Keep all receipts for parts purchased, so that they may be referred to in case of related problems or to determine operating expenses. As a do-it-yourselfer, these receipts are the only proof you have that the required maintenance was performed. In the event of a warranty problem, these receipts will be invaluable.

It's necessary to mention the difference between maintenance and repair. Maintenance includes routine inspections, adjustments, and replacement of parts that show signs of normal wear. Maintenance compensates for wear or deterioration. Repair implies that something has broken or is not working. A need for repair is often caused by lack of maintenance.

For example: draining and refilling the engine oil is maintenance recommended by all manufacturers at specific intervals. Failure to do this can allow internal corrosion or damage and impair the operation of the engine, requiring expensive repairs. While no maintenance program can prevent items from breaking or wearing out, a general rule can be stated: MAINTENANCE IS CHEAPER THAN REPAIR.

Directions and Locations

See Figure 1

Two basic rules should be mentioned here. First, whenever the Port side of the engine (or boat) is referred to, it is meant to specify the left side of the engine when you are sitting at the helm. Conversely, the Starboard means your right side. The Bow is the front of the boat and the Stern is the rear.

Most screws and bolts are removed by turning counterclockwise, and tightened by turning clockwise. An easy way to remember this is: righty-tighty; lefty-loosey. Corny, but effective. And if you are really dense (and we have all been so at one time or another), buy a ratchet that is marked ON and OFF, or mark your own.

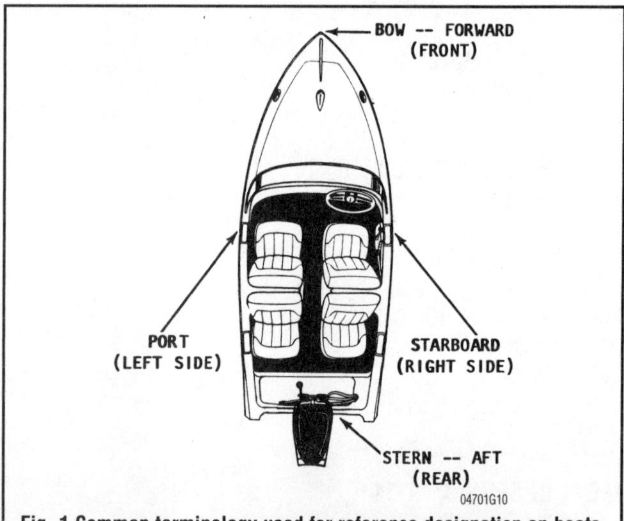

Fig. 1 Common terminology used for reference designation on boats of all size. These terms are used though out the manual

Professional Help

Occasionally, there are some things when working on an outboard that are beyond the capabilities or tools of the average Do-It-Yourselfer (DIYer). This shouldn't include most of the topics of this manual, but you will have to be the judge. Some engines require special tools or a selection of special parts, even for basic maintenance.

Talk to other boaters who use the same model of engine and speak with a trusted marina to find if there is a particular system or component on your engine that is difficult to maintain. For example, although the technique of valve adjustment on some engines may be easily understood and even performed by a DIYer, it might require a handy assortment of shims in various sizes and a few hours of disassembly to get to that point. Not having the assortment of shims handy might mean multiple trips back and forth to the parts store, and this might not be worth your time.

You will have to decide for yourself where basic maintenance ends and where professional service should begin. Take your time and do your research first (starting with the information in this manual) and then make your own decision. If you really don't feel comfortable with attempting a procedure, DON'T DO IT. If you've gotten into something that may be over your head, don't panic. Tuck your tail between your legs and call a marine mechanic. Marinas and independent shops will be able to finish a job for you. Your ego may be damaged, but your boat will be properly restored to its full running order. So, as long as you approach jobs slowly and carefully, you really have nothing to lose and everything to gain by doing it yourself.

Purchasing Parts

See Figures 2, 3 and 4

When purchasing parts there are two things to consider. The first is quality and the second is to be sure to get the correct part for your engine. To get quality parts, always deal directly with a reputable retailer. To get the proper parts always refer to the information tag on your engine prior to calling the parts counter. An incorrect part can adversely affect your engine performance and fuel economy, and will cost you more money and aggravation in the end.

Just remember, a tow back to shore will cost plenty. That charge is per hour from the time the towboat leaves their home port, to the time they return to their home port. Get the picture ....$$$?

So who should you call for parts? Well, there are many sources for the parts you will need. Where you shop for parts will be determined by what kind of parts you need, how much you want to pay, and the types of stores in your neighborhood.

Your marina can supply you with many of the common parts you require. Using a marina for as your parts supplier may be hand because of location (just walk right down the dock) or because the marina specializes in your particular brand of engine. In addition, it is always a good idea to get to know the marina staff (especially the marine mechanic).

The marine parts jobber, who is usually listed in the yellow pages or whose name can be obtained from the marina, is another excellent source for parts. In addition to supplying local marinas, they also do a sizeable business in over-the-counter parts sales for the do-it-yourselfer.

Almost every community has one or more convenient marine chain stores. These stores often offer the best retail prices and the convenience of one-stop shopping for all your needs. Since they cater to the do-it-yourselfer, these stores are almost always open weeknights, Saturdays, and Sundays, when the jobbers are usually closed.

The lowest prices for parts are most often found in discount stores or the auto department of mass merchandisers. Parts sold here are name and private brand parts bought in huge quantities, so they can offer a competitive price. Private brand parts are made by major manufacturers and sold to large chains under a store label.

Avoiding the Most Common Mistakes

There are 3 common mistakes in mechanical work:

1. Incorrect order of assembly, disassembly or adjustment. When taking something apart or putting it together, performing steps in the wrong order usually just costs you extra time; however, it CAN break something. Read the entire procedure before beginning disassembly. Perform everything in the order in which the instructions say you should, even if you can't immediately see a reason for it. When you're taking apart something that is very intricate, you might want to draw a picture of how it looks when assembled at one point in order to make sure you get everything back in its proper position. When making adjustments, perform them in the proper order; often, one adjustment affects another, and you cannot expect satisfactory results unless each adjustment is made only when it cannot be changed by another.
2. Overtorquing (or undertorquing). While it is more common for overtorquing to cause damage, undertorquing may allow a fastener to vibrate loose causing serious damage. Especially when dealing with aluminum parts, pay attention to torque specifications and utilize a torque wrench in assembly. If a torque figure is not available, remember that if you are using the right tool to perform the job, you will probably not have to strain yourself to get a fastener tight enough. The pitch of most threads is so slight that the tension you put on the wrench will be multiplied many times in actual force on what you are tightening.
3. Crossthreading. This occurs when a part such as a bolt is screwed into a nut or casting at the wrong angle and forced. Crossthreading is more likely to occur if access is difficult. It helps to clean and lubricate fasteners, then to start threading with the part to be installed positioned straight in. Always start a fastener, etc. with your fingers. If you encounter resistance, unscrew the part and start over again at a different angle until it can be inserted and turned several times without much effort. Keep in mind that some parts may have tapered threads, so that gentle turning will automatically bring the part you're threading to the proper angle, but only if you don't force it or resist a change in angle. Don't put a wrench on the part until it has been tightened a couple of turns by hand. If you suddenly encounter resistance, and the part has not seated fully, don't force it. Pull it back out to make sure it's clean and threading properly.

04891P16

Fig. 2 By far the most important asset in purchasing parts is a knowledgeable and enthusiastic parts person

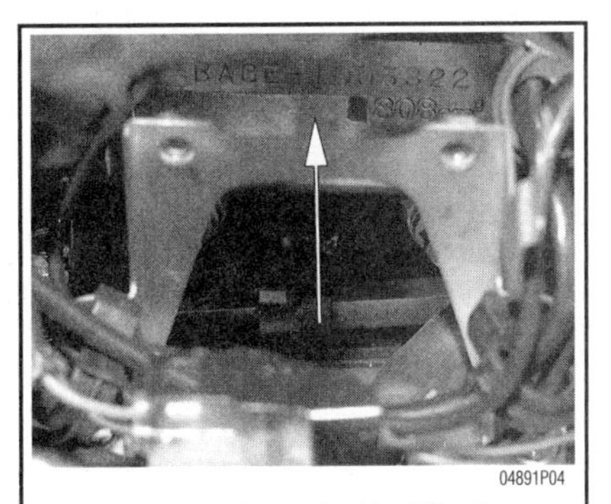

04891P04

Fig. 3 Always refer to the identification numbers on your engine when ordering parts

04971P12

Fig. 4 Parts catalogs, giving application and part number information, are provided by manufacturers for most replacement parts

BOATING SAFETY

In 1971 Congress ordered the U.S. Coast Guard to improve recreational boating safety. In response, the Coast Guard drew up a set of regulations.

Beside these federal regulations, there are state and local laws you must follow. These sometimes exceed the Coast Guard requirements. This section discusses only the federal laws. State and local laws are available from your local Coast Guard. As with other laws, "Ignorance of the boating laws is no excuse." The rules fall into two groups: regulations for your boat and required safety equipment on your boat.

Regulations For Your Boat

Most boats on waters within Federal jurisdiction must be registered or documented. These waters are those that provide a means of transportation between two or more states or to the sea. They also include the territorial waters of the United States.

DOCUMENTING OF VESSELS

A vessel of five or more net tons may be documented as a yacht. In this process, papers are issued by the U.S. Coast Guard as they are for large ships. Documentation is a form of national registration. The boat must be used solely for pleasure. Its owner must be a U.S. citizen, a partnership of U.S. citizens, or a corporation controlled by U.S. citizens. The captain and other officers must also be U.S. citizens. The crew need not be.

If you document your yacht, you have the legal authority to fly the yacht ensign. You also may record bills of sale, mortgages, and other papers of title with federal authorities. Doing so gives legal notice that such instruments exist. Documentation also permits preferred status for mortgages. This gives you additional security and aids financing and transfer of title. You must carry the original documentation papers aboard your vessel. Copies will not suffice.

REGISTRATION OF BOATS

If your boat is not documented, registration in the state of its principal use is probably required. If you use it mainly on an ocean, a gulf, or other similar water, register it in the state where you moor it.

If you use your boat solely for racing, it may be exempt from the requirement in your state. States may also exclude dinghies. Some require registration of documented vessels and non-power driven boats.

All states, except Alaska, register boats. In Alaska, the U.S. Coast Guard issues the registration numbers. If you move your vessel to a new state of principal use, a valid registration certificate is good for 60 days. You must have the registration certificate (certificate of number) aboard your vessel when it is in use. A copy will not suffice. You may be cited if you do not have the original on board.

NUMBERING OF VESSELS

A registration number is on your registration certificate. You must paint or permanently attach this number to both sides of the forward half of your boat. Do not display any other number there.

The registration number must be clearly visible. It must not be placed on the obscured underside of a flared bow. If you can't place the number on the bow, place it on the forward half of the hull. If that doesn't work, put it on the superstructure. Put the number for an inflatable boat on a bracket or fixture. Then, firmly attach it to the forward half of the boat. The letters and numbers must be plain block characters and must read from left to right. Use a space or a hyphen to separate the prefix and suffix letters from the numerals. The color of the characters must contrast with that of the background, and they must be at least three inches high.

In some states your registration is good for only one year. In others, it is good for as long as three years. Renew your registration before it expires. At that time you will receive a new decal or decals. Place them as required by state law. You should remove old decals before putting on the new ones. Some states require that you show only the current decal or decals. If your vessel is moored, it must have a current decal even if it is not in use.

If your vessel is lost, destroyed, abandoned, stolen, or transferred, you must inform the issuing authority. If you lose your certificate of number or your address changes, notify the issuing authority as soon as possible.

SALES AND TRANSFERS

Your registration number is not transferable to another boat. The number stays with the boat unless its state of principal use is changed.

HULL IDENTIFICATION NUMBER

A Hull Identification Number (HIN) is like the Vehicle Identification Number (VIN) on your car. Boats built between November 1, 1972 and July 31, 1984 have old format HINs. Since August 1, 1984 a new format has been used. Your boat's HIN must appear in two places. If it has a transom, the primary number is on its starboard side within two inches of its top. If it does not have a transom or if it was not practical to use the transom, the number is on the starboard side. In this case, it must be within one foot of the stern and within two inches of the top of the hull side. On pontoon boats, it is on the aft crossbeam within one foot of the starboard hull attachment. Your boat also has a duplicate number in an unexposed location. This is on the boat's interior or under a fitting or item of hardware.

LENGTH OF BOATS

For some purposes, boats are classed by length. Required equipment, for example, differs with boat size. Manufacturers may measure a boat's length in several ways. Officially, though, your boat is measured along a straight line from its bow to its stern. This line is parallel to its keel.

The length does not include bowsprits, boomkins, or pulpits. Nor does it include rudders, brackets, outboard motors, outdrives, diving platforms, or other attachments.

CAPACITY INFORMATION

See Figure 5

Manufacturers must put capacity plates on most recreational boats less than 20 feet long. Sailboats, canoes, kayaks, and inflatable boats are usually exempt. Outboard boats must display the maximum permitted horsepower of their engines. The plates must also show the allowable maximum weights of the people on board. And they must show the allowable maximum combined weights of people, engines, and gear. Inboards and stern drives need not show the weight of their engines on their capacity plates. The capacity plate must appear where it is clearly visible to the operator when underway. This information serves to remind you of the capacity of your boat under normal circumstances. You should ask yourself, "Is my boat loaded above its recommended capacity" and, "Is my boat overloaded for the present sea and wind conditions?" If you are stopped by a legal authority, you may be cited if you are overloaded.

Fig. 5 A U.S. Coast Guard certification plate indicates the amount of occupants and gear appropriate for safe operation of the vessel

CERTIFICATE OF COMPLIANCE

Manufacturers are required to put compliance plates on motorboats greater than 20 feet in length. The plates must say, "This boat," or "This equipment complies with the U. S. Coast Guard Safety Standards in effect on the date of certification." Letters and numbers can be no less than one-eighth of an inch high. At the manufacturer's option, the capacity and compliance plates may be combined.

VENTILATION

A cup of gasoline spilled in the bilge has the potential explosive power of 15 sticks of dynamite. This statement, commonly quoted over 20 years ago, may be an exaggeration, however, it illustrates a fact. Gasoline fumes in the bilge of a boat are highly explosive and a serious danger. They are heavier than air and will stay in the bilge until they are vented out.

Because of this danger, Coast Guard regulations require ventilation on many power boats. There are several ways to supply fresh air to engine and gasoline tank compartments and to remove dangerous vapors. Whatever the choice, it must meet Coast Guard standards.

➡The following is not intended to be a complete discussion of the regulations. It is limited to the majority of recreational vessels. Contact your local Coast Guard office for further information.

General Precautions

Ventilation systems will not remove raw gasoline that leaks from tanks or fuel lines. If you smell gasoline fumes, you need immediate repairs. The best device for sensing gasoline fumes is your nose. Use it! If you smell gasoline in an engine compartment or elsewhere, don't start your engine. The smaller the compartment, the less gasoline it takes to make an explosive mixture.

Ventilation for Open Boats

In open boats, gasoline vapors are dispersed by the air that moves through them. So they are exempt from ventilation requirements.

To be "open," a boat must meet certain conditions. Engine and fuel tank compartments and long narrow compartments that join them must be open to the atmosphere." This means they must have at least 15 square inches of open area for each cubic foot of net compartment volume. The open area must be in direct contact with the atmosphere. There must also be no long, unventilated spaces open to engine and fuel tank compartments into which flames could extend.

Ventilation for All Other Boats

Powered and natural ventilation are required in an enclosed compartment with a permanently installed gasoline engine that has a cranking motor. A compartment is exempt if its engine is open to the atmosphere. Diesel powered boats are also exempt.

VENTILATION SYSTEMS

There are two types of ventilation systems. One is "natural ventilation." In it, air circulates through closed spaces due to the boat's motion. The other type is "powered ventilation." In it, air is circulated by a motor driven fan or fans.

Natural Ventilation System Requirements

A natural ventilation system has an air supply from outside the boat. The air supply may also be from a ventilated compartment or a compartment open to the atmosphere. Intake openings are required. In addition, intake ducts may be required to direct the air to appropriate compartments.

The system must also have an exhaust duct that starts in the lower third of the compartment. The exhaust opening must be into another ventilated compartment or into the atmosphere. Each supply opening and supply duct, if there is one, must be above the usual level of water in the bilge. Exhaust openings and ducts must also be above the bilge water. Openings and ducts must be at least three square inches in area or two inches in diameter. Openings should be placed so exhaust gasses do not enter the fresh air intake. Exhaust fumes must not enter cabins or other enclosed, non-ventilated spaces. The carbon monoxide gas in them is deadly.

Intake and exhaust openings must be covered by cowls or similar devices. These registers keep out rain water and water from breaking seas. Most often, intake registers face forward and exhaust openings aft. This aids the flow of air when the boat is moving or at anchor since most boats face into the wind when anchored.

Power Ventilation System Requirements

➧ See Figure 6

Powered ventilation systems must meet the standards of a natural system. They must also have one or more exhaust blowers. The blower duct can serve as the exhaust duct for natural ventilation if fan blades do not obstruct the air flow when not powered. Openings in engine compartment, for carburetion are in addition to ventilation system requirements.

Required Safety Equipment

Coast Guard regulations require that your boat have certain equipment aboard. These requirements are minimums. Exceed them whenever you can.

TYPES OF FIRES

There are four common classes of fires:

- Class A—fires are in ordinary combustible materials such as paper or wood.
- Class B—fires involve gasoline, oil and grease.
- Class C—fires are electrical.
- Class D—fires involve ferrous metals

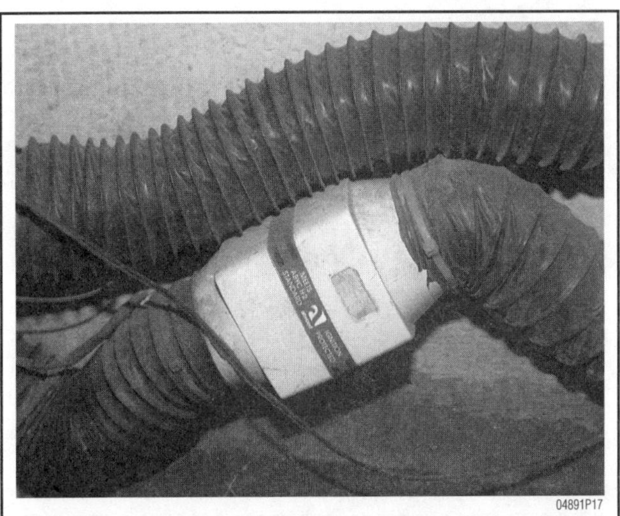

04891P17

Fig. 6 Typical blower and duct system to vent fumes from the engine compartment

One of the greatest risks to boaters is fire. This is why it is so important to carry the correct number and type of extinguishers onboard.

The best fire extinguisher for most boats is a Class B extinguisher. Never use water on Class B or Class C fires, as water spreads these types of fires. You should never use water on a Class C fire as it may cause you to be electrocuted.

FIRE EXTINGUISHERS

➧ See Figure 7

If your boat meets one or more of the following conditions, you must have at least one fire extinguisher aboard. The conditions are:

04701P29

Fig. 7 An approved fire extinguisher should be mounted close to the operator for emergency use

- Inboard or stern drive engines
- Closed compartments under seats where portable fuel tanks can be stored
- Double bottoms not sealed together or not completely filled with flotation materials
- Closed living spaces
- Closed stowage compartments in which combustible or flammable materials are stored
- Permanently installed fuel tanks
- Boat is 26 feet or more in length.

Contents of Extinguishers

Fire extinguishers use a variety of materials. Those used on boats usually contain dry chemicals, Halon, or Carbon Dioxide (CO3). Dry chemical extinguishers contain chemical powders such as Sodium Bicarbonate—baking soda.

Carbon dioxide is a colorless and odorless gas when released from an extinguisher. It is not poisonous but caution must be used in entering compartments filled with it. It will not support life and keeps oxygen from reaching your lungs. A fire-killing concentration of Carbon Dioxide is lethal. If you are in a compartment with a high concentration of CO3, you will have no difficulty breathing. But the air does not contain enough oxygen to support life. Unconsciousness or death can result.

HALON EXTINGUISHERS

Some fire extinguishers and `built-in' or `fixed' automatic fire extinguishing systems contain a gas called Halon. Like carbon dioxide it is colorless and odorless and will not support life. Some Halons may be toxic if inhaled.

To be accepted to the Coast Guard, a fixed Halon system must have an indicator light at the vessel's helm. A green light shows the system is ready. Red means it is being discharged or has been discharged. Warning horns are available to let you know the system has been activated. If your fixed Halon system discharges, ventilate the space thoroughly before you enter it. There are no residues from Halon but it will not support life.

Although Halon has excellent fire fighting properties, it is thought to deplete the earth's ozone layer and has not been manufactured since January 1, 1994. Halon extinguishers can be refilled from existing stocks of the gas until they are used up, but high federal excise taxes are being charged for the service. If you discontinue using your Halon extinguisher, take it to a recovery station rather than releasing the gas into the atmosphere. Compounds such as FE 241, designed to replace Halon, are now available.

Fire Extinguisher Approval

Fire extinguishers must be Coast Guard approved. Look for the approval number on the nameplate. Approved extinguishers have the following on their labels: "Marine Type USCG Approved, Size . . . , Type . . . , 162.208/," etc. In addition, to be acceptable by the Coast Guard, an extinguisher must be in serviceable condition and mounted in its bracket. An extinguisher not properly mounted in its bracket will not be considered serviceable during a Coast Guard inspection.

Care and Treatment

Make certain your extinguishers are in their stowage brackets and are not damaged. Replace cracked or broken hoses. Nozzles should be free of obstructions. Sometimes, wasps and other insects nest inside nozzles and make them inoperable. Check your extinguishers frequently. If they have pressure gauges, is the pressure within acceptable limits? Do the locking pins and sealing wires show they have not been used since recharging?

Don't try an extinguisher to test it. Its valves will not reseat properly and the remaining gas will leak out. When this happens, the extinguisher is useless.

Weigh and tag carbon dioxide and Halon extinguishers twice a year. If their weight loss exceeds 10 percent of the weight of the charge, recharge them. Check to see that they have not been used. They should have been inspected by a qualified person within the past six months, and they should have tags showing all inspection and service dates. The problem is that they can be partially discharged while appearing to be fully charged.

Some Halon extinguishers have pressure gauges the same as dry chemical extinguishers. Don't rely too heavily on the gauge. The extinguisher can be partially discharged and still show a good gauge reading. Weighing a Halon extinguisher is the only accurate way to assess its contents.

If your dry chemical extinguisher has a pressure indicator, check it frequently. Check the nozzle to see if there is powder in it. If there is, recharge it. Occasionally invert your dry chemical extinguisher and hit the base with the palm of your hand. The chemical in these extinguishers packs and cakes due to the boat's vibration and pounding. There is a difference of opinion about whether hitting the base helps, but it can't hurt. It is known that caking of the chemical powder is a major cause of failure of dry chemical extinguishers. Carry spares in excess of the minimum requirement. If you have guests aboard, make certain they know where the extinguishers are and how to use them.

Using a Fire Extinguisher

A fire extinguisher usually has a device to keep it from being discharged accidentally. This is a metal or plastic pin or loop. If you need to use your extinguisher, take it from its bracket. Remove the pin or the loop and point the nozzle at the base of the flames. Now, squeeze the handle, and discharge the extinguisher's contents while sweeping from side to side. Recharge a used extinguisher as soon as possible.

If you are using a Halon or carbon dioxide extinguisher, keep your hands away from the discharge. The rapidly expanding gas will freeze them. If your fire extinguisher has a horn, hold it by its handle.

Legal Requirements for Extinguishers

You must carry fire extinguishers as defined by Coast Guard regulations. They must be firmly mounted in their brackets and immediately accessible.

A motorboat less than 26 feet long must have at least one approved hand-portable, Type B-1 extinguisher. If the boat has an approved fixed fire extinguishing system, you are not required to have the Type B-1 extinguisher. Also, if your boat is less than 26 feet long, is propelled by an outboard motor, or motors, and does not have any of the first six conditions described at the beginning of this section, it is not required to have an extinguisher. Even so, it's a good idea to have one, especially if a nearby boat catches fire, or if a fire occurs at a fuel dock.

A motorboat 26 feet to under 40 feet long, must have at least two Type B-1 approved hand-portable extinguishers. It can, instead, have at least one Coast Guard approved Type B-2. If you have an approved fire extinguishing system, only one Type B-1 is required.

A motorboat 40 to 65 feet long must have at least three Type B-1 approved portable extinguishers . It may have, instead, at least one Type B-1 plus a Type B-2. If there is an approved fixed fire extinguishing system, two Type B-1 or one Type B-2 is required.

WARNING SYSTEM

Various devices are available to alert you to danger. These include fire, smoke, gasoline fumes, and carbon monoxide detectors. If your boat has a galley, it should have a smoke detector. Where possible, use wired detectors. Household batteries often corrode rapidly on a boat.

You can't see, smell, nor taste carbon monoxide gas, but it is lethal. As little as one part in 10,000 parts of air can bring on a headache. The symptoms of carbon monoxide poisoning—headaches, dizziness, and nausea—are like sea sickness. By the time you realize what is happening to you, it may be too late to take action. If you have enclosed living spaces on your boat, protect yourself with a detector. There are many ways in which carbon monoxide can enter your boat.

PERSONAL FLOTATION DEVICES

Personal Flotation Devices (PFDs) are commonly called life preservers or life jackets. You can get them in a variety of types and sizes. They vary with their intended uses. To be acceptable, they must be Coast Guard approved.

Type I PFDs

A Type I life jacket is also called an offshore life jacket. Type I life jackets will turn most unconscious people from facedown to a vertical or slightly backward position. The adult size gives a minimum of 22 pounds of buoyancy. The child size has at least 11 pounds. Type I jackets provide more protection to their wearers than any other type of life jacket. Type I life jackets are bulkier and less comfortable than other types. Furthermore, there are only two sizes, one for children and one for adults.

Type I life jackets will keep their wearers afloat for extended periods in rough water. They are recommended for offshore cruising where a delayed rescue is probable.

Type II PFDs

▸ See Figure 8

A Type II life jacket is also called a near-shore buoyant vest. It is an approved, wearable device. Type II life jackets will turn some unconscious people from facedown to vertical or slightly backward positions. The adult size gives at least 15.5 pounds of buoyancy. The medium child size has a minimum of 11 pounds. And the small child and infant sizes give seven pounds. A Type II life jacket is more comfortable than a Type I but it does not have as much buoyancy. It is not recommended for long hours in rough water. Because of this, Type IIs are recommended for inshore and inland cruising on calm water. Use them where there is a good chance of fast rescue.

Type III PFDs

Type III life jackets or marine buoyant devices are also known as flotation aids. Like Type IIs, they are designed for calm inland or close offshore water where there is a good chance of fast rescue. Their minimum buoyancy is 15.5 pounds. They will not turn their wearers face up.

Type III devices are usually worn where freedom of movement is necessary. Thus, they are used for water skiing, small boat sailing, and fishing among other activities. They are available as vests and flotation coats. Flotation coats are useful in cold weather. Type IIIs come in many sizes from small child through large adult.

Life jackets come in a variety of colors and patterns—red, blue, green, camouflage, and cartoon characters. From a safety standpoint, the best color is bright orange. It is easier to see in the water, especially if the water is rough.

Type IV PFDs

▸ See Figures 9 and 10

Type IV ring life buoys, buoyant cushions and horseshoe buoys are Coast Guard approved devices called throwables. They are made to be thrown to people in the water, and should not be worn. Type IV cushions are often used as seat cushions. Cushions are hard to hold onto in the water. Thus, they do not afford as much protection as wearable life jackets.

The straps on buoyant cushions are for you to hold onto either in the water or when throwing them. A cushion should never be worn on your back. It will turn you face down in the water.

Type IV throwables are not designed as personal flotation devices for unconscious people, non-swimmers, or children. Use them only in emergencies. They should not be used for, long periods in rough water.

Ring life buoys come in 18, 20, 24, and 30 inch diameter sizes. They have grab lines. You should attach about 60 feet of polypropylene line to the grab rope to aid in retrieving someone in the water. If you throw a ring, be careful not to hit the person. Ring buoys can knock people unconscious

Type V PFDs

Type V PFDs are of two kinds, special use devices and hybrids. Special use devices include boardsailing vests, deck suits, work vests, and others. They are approved only for the special uses or conditions indicated on their labels. Each is designed and intended for the particular application shown on its label. They do not meet legal requirements for general use aboard recreational boats.

Hybrid life jackets are inflatable devices with some built-in buoyancy provided by plastic foam or kapok. They can be inflated orally or by cylinders of compressed gas to give additional buoyancy. In some hybrids the gas is released manually. In others it is released automatically when the life jacket is immersed in water.

The inherent buoyancy of a hybrid may be insufficient to float a person unless it is inflated. The only way to find this out is for the user to try it in the water. Because of its limited buoyancy when deflated, a hybrid is recommended for use by anon-swimmer only if it is worn with enough inflation to float the wearer.

If they are to count against the legal requirement for the number of life jackets you must carry on your vessel, hybrids manufactured before February 8, 1995 must be worn whenever a boat is underway and the wearer is not below decks or in an enclosed space. To find out if your Type V hybrid must be worn to satisfy the legal requirement, read its label. If its use is restricted it will say, "REQUIRED TO BE WORN" in capital letters.

Hybrids cost more than other life jackets, but this factor must be weighed against the fact that they are more comfortable than Type I, II, or III life jackets. Because of their greater comfort, their owners are more likely to wear them than are the owners of Type I, II, or III life jackets.

The Coast Guard has determined that improved, less costly hybrids can save lives since they will be bought and used more frequently. For these reasons a new federal regulation was adopted effective February 8, 1995. The regulation increases both the deflated and inflated buoyancys of hybrids, makes them available in a greater variety of sizes and types, and reduces their costs by reducing production costs.

Even though it may not be required, the wearing of a hybrid or a life jacket is encouraged whenever a vessel is underway. Like life jackets, hybrids are now available in three types. To meet legal requirements, a Type I hybrid can be substituted for a Type I life jacket. Similarly Type II and III hybrids can be substituted for Type II and Type III life jackets. A Type I hybrid, when inflated, will turn most unconscious people from facedown to vertical or slightly backward positions just like a Type I life jacket. Type I and III hybrids function like Type II and III life jackets. If you purchase a new hybrid, it should have an owner's manual attached which describes its life jacket type and its deflated and inflated buoyancys. It warns you that it may have to be inflated to float you. The manual also tells you how to don the life jacket and how to inflate it. It also tells you how to change its inflation mechanism, recommended testing exercises, and inspection and maintenance procedures. The manual also tells you why you need a life jacket and why you should wear it. A new hybrid must be packaged with at least three gas cartridges. One of these may already be loaded into the inflation mechanism. Likewise, if it has an automatic inflation mechanism, it must be packaged with at least three of these water sensitive elements. One of these elements may be installed.

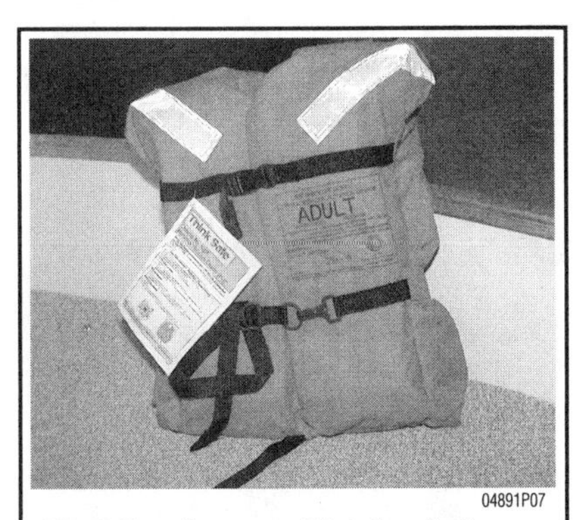

04891P07

Fig. 8 Type II approved flotation devices are recommended for inshore and inland cruising on calm water. Use them where there is a good chance of fast rescue

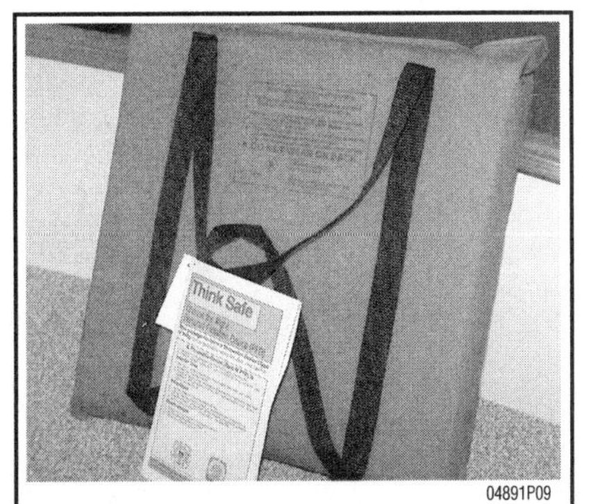

04891P09

Fig. 9 Type IV buoyant cushions are made to be thrown to people in the water. If you can squeeze air out of the cushion, it is faulty and should be replaced

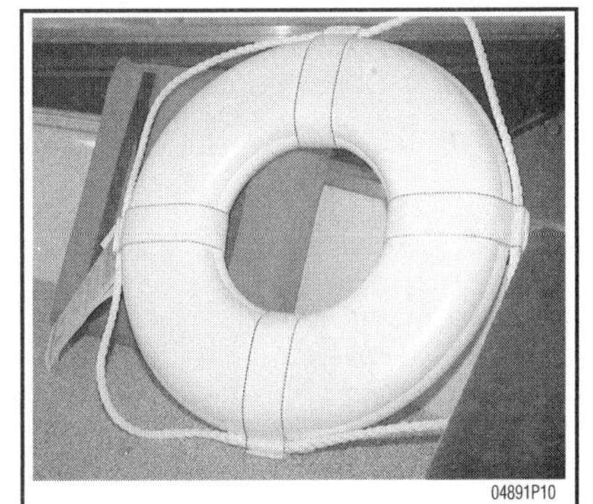

04891P10

Fig. 10 Type IV throwables, such as this ring life buoy, are not designed as personal flotation devices for unconscious people, non-swimmers, or children

Legal Requirements

A Coast Guard approved life jacket must show the manufacturer's name and approval number. Most are marked as Type I, II, III, IV, or V. All of the newer hybrids are marked for type.

You are required to carry at least one wearable life jacket or hybrid for each person on board your recreational vessel. If your vessel is 16 feet or more in length and is not a canoe or a kayak, you must also have at least one Type IV on board. These requirements apply to all recreational vessels that are propelled or controlled by machinery, sails, oars, paddles, poles, or another vessel. Sailboards are not required to carry life jackets.

You can substitute an older Type V hybrid for any required Type I, II, or III life jacket provided that its approval label shows it is approved for the activity the vessel is engaged in, approved as a substitute for a life jacket of the type required on the vessel, used as required on the labels, and used in accordance with any requirements in its owner's manual, if the approval label makes reference to such a manual.

A water skier being towed is considered to be on board the vessel when judging compliance with legal requirements.

You are required to keep your Type I, II, or III life jackets or equivalent hybrids readily accessible, which means you must be able to reach out and get them when needed. All life jackets must be in good, serviceable condition.

General Considerations

The proper use of a life jacket requires the wearer to know how it will perform. You can gain this knowledge only through experience. Each person on your boat should be assigned a life jacket. Next, it should be fitted to the person who will wear it. Only then can you be sure that it will be ready for use in an emergency.

Boats can sink fast. There may be no time to look around for a life jacket. Fitting one on you in the water is almost impossible. This advice is good even if the water is calm, and you intend to boat near shore. Most drownings occur in inland waters within a few feet of safety. Most victims had life jackets, but they weren't wearing them.

Keeping life jackets in the plastic covers they came wrapped in and in a cabin assures that they will stay clean and unfaded. But this is no way to keep them when you are on the water. When you need a life jacket it must be readily accessible and adjusted to fit you. You can't spend time hunting for it or learning how to fit it.

There is no substitute for the experience of entering the water while wearing a life jacket. Children, especially, need practice. If possible, give, your guests this experience. Tell them they should keep their arms to their sides when jumping in to keep the life jacket from riding up. Let them jump in and see how the life jacket responds. Is it adjusted so it does not ride up? Is it the proper size? Are all straps snug? Are children's life jackets the right sizes for them? Are they adjusted properly? If a child's life jacket fits correctly, you can lift the child by the jacket's shoulder straps and the child's chin and ears will not slip through. Non-swimmers, children, handicapped persons, elderly persons and even pets should always wear life jackets when they are aboard. Many states require that everyone aboard wear them in hazardous waters.

Inspect your lifesaving equipment from time to time. Leave any questionable or unsatisfactory equipment on shore. An emergency is no time for you to conduct an inspection.

Indelibly mark your life jackets with your vessel's name, number, and calling port. This can be important in a search and rescue effort. It could help concentrate effort where it will do the most good.

Care of Life Jackets

Given reasonable care, life jackets last many years. Thoroughly dry them before putting them away. Stow them in dry, well ventilated places. Avoid the bottoms of lockers and deck storage boxes where moisture may collect. Air and dry them frequently.

Life jackets should not be tossed about or used as fenders or cushions. Many contain kapok or fibrous glass material enclosed in plastic bags. The bags can rupture and are then unserviceable. Squeeze your life jacket gently. Does air leak out? If so, water can leak in and it will no longer be safe to use. Cut it Up so no one will use it, and throw it away. The covers of some life jackets are made of nylon or polyester. These materials are plastics. Like many plastics, they break down after extended exposure to the ultraviolet light in sunlight. This process may be more rapid when the materials are dyed with bright dyes such as "neon" shades.

Ripped and badly faded fabric are clues that the covering of your life jacket is deteriorating. A simple test is to pinch the fabric between your thumbs and forefingers. Now try to tear the fabric. If it can be torn, it should definitely be destroyed and discarded. Compare the colors in protected places to those exposed to the sun. If the colors have faded, the materials have been weakened. A fabric covered life jacket should ordinarily last several boating seasons with normal use. A life jacket used every day in direct sunlight should probably be replaced more often.

SOUND PRODUCING DEVICES

All boats are required to carry some means of making an efficient sound signal. Devices for making the whistle or horn noises required by the Navigation Rules must be capable of a four second blast. The blast should be audible for at least one-half mile. Athletic whistles are not acceptable on boats 12 meters or longer. Use caution with athletic whistles. When wet, some of them come apart and loose their "pea." When this happens, they are useless.

If your vessel is 12 meters long and less than 20 meters, you must have a power whistle (or power horn) and a bell on board. The bell must be in operating condition and have a minimum diameter of at least 200 mm (7.9 inches) at its mouth.

VISUAL DISTRESS SIGNALS

➧ **See Figure 11**

Visual Distress Signals (VDS) attract attention to your vessel if you need help. They also help to guide searchers in search and rescue situations. Be sure you have the right types, and learn how to use them properly.

It is illegal to fire flares improperly. In addition, they cost the Coast Guard and its Auxiliary many wasted hours in fruitless searches. If you signal a distress with flares and then someone helps you, please let the Coast Guard or the

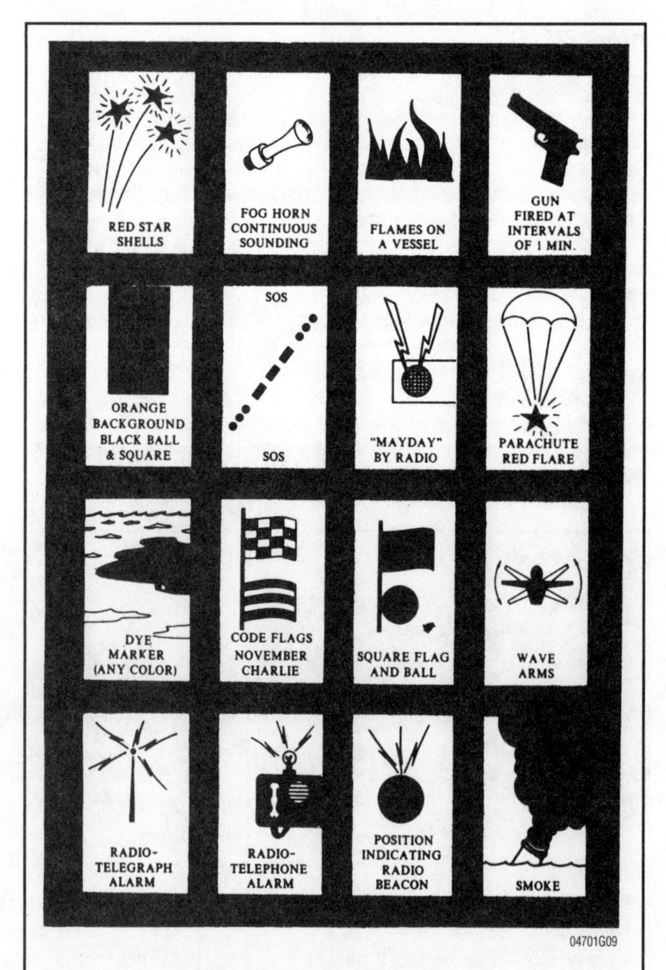

Fig. 11 Internationally accepted distress signals

appropriate Search And Rescue Agency (SAR) know so the distress report will be canceled.

Recreational boats less than 16 feet long must carry visual distress signals on coastal waters at night. Coastal waters are:

- The ocean (territorial sea)
- The Great Lakes
- Bays or sounds that empty into oceans
- Rivers over two miles across at their mouths upstream to where they narrow to two miles.

Recreational boats 16 feet or longer must carry VDS at all times on coastal waters. The same requirement applies to boats carrying six or fewer passengers for hire. Open sailboats less than 26 feet long without engines are exempt in the daytime as are manually propelled boats. Also exempt are boats in organized races, regattas, parades, etc. Boats owned in the United States and operating on the high seas must be equipped with VDS.

A wide variety of signaling devices meet Coast Guard regulations. For pyrotechnic devices, a minimum of three must be carried. Any combination can be carried as long as it adds up to at least three signals for day use and at least three signals for night use. Three day/night signals meet both requirements. If possible, carry more than the legal requirement.

➡The American flag flying upside down is a commonly recognized distress signal. It is not recognized in the Coast Guard regulations, though. In an emergency, your efforts would probably be better used in more effective signaling methods.

Types of VDS

VDS are divided into two groups; daytime and nighttime use. Each of these groups is subdivided into pyrotechnic and non-pyrotechnic devices.

DAYTIME NON-PYROTECHNIC SIGNALS

A bright orange flag with a black square over a black circle is the simplest VDS. It is usable, of course, only in daylight. It has the advantage of being a continuous signal. A mirror can be used to good advantage on sunny days. It can attract the attention of other boaters and of aircraft from great distances. Mirrors are available with holes in their centers to aid in "aiming." In the absence of a mirror, any shiny object can be used. When another boat is in sight, an effective VDS is to extend your arms from your sides and move them up and down. Do it slowly. If you do it too fast the other people may think you are just being friendly. This simple gesture is seldom misunderstood, and requires no equipment.

DAYTIME PYROTECHNIC DEVICES

Orange smoke is a useful daytime signal. Hand-held or floating smoke flares are very effective in attracting attention from aircraft. Smoke flares don't last long, and are not very effective in high wind or poor visibility. As with other pyrotechnic devices, use them only when you know there is a possibility that someone will see the display.

To be usable, smoke flares must be kept dry. Keep them in airtight containers and store them in dry places. If the "striker" is damp, dry it out before trying to ignite the device. Some pyrotechnic devices require a forceful "strike" to ignite them.

All hand-held pyrotechnic devices may produce hot ashes or slag when burning. Hold them over the side of your boat in such a way that they do not burn your hand or drip into your boat.

Nighttime Non-Pyrotechnic Signals

An electric distress light is available. This light automatically flashes the international morse code SOS distress signal (••• --- •••). Flashed four to six times a minute, it is an unmistakable distress signal. It must show that it is approved by the Coast Guard. Be sure the batteries are fresh. Dated batteries give assurance that they are current.

Under the Inland Navigation Rules, a high intensity white light flashing 50-70 times per minute is a distress signal. Therefore, use strobe lights on inland waters only for distress signals.

Nighttime Pyrotechnic Devices

➧ See Figure 12

Aerial and hand-held flares can be used at night or in the daytime. Obviously, they are more effective at night.

Fig. 12 Moisture protected flares should be carried onboard any vessel for use as a distress signal

Currently, the serviceable life of a pyrotechnic device is rated at 42 months from its date of manufacture. Pyrotechnic devices are expensive. Look at their dates before you buy them. Buy them with as much time remaining as possible.

Like smoke flares, aerial and hand-held flares may fail to work if they have been damaged or abused. They will not function if they are or have been wet. Store them in dry, airtight containers in dry places. But store them where they are readily accessible.

Aerial VDSs, depending on their type and the conditions they are used in, may not go very high. Again, use them only when there is a good chance they will be seen.

A serious disadvantage of aerial flares is that they burn for only a short time. Most burn for less than 10 seconds. Most parachute flares burn for less than 45 seconds. If you use a VDS in an emergency, do so carefully. Hold hand-held flares over the side of the boat when in use. Never use a road hazard flare on a boat, it can easily start a fire. Marine type flares are carefully designed to lessen risk, but they still must be used carefully.

Aerial flares should be given the same respect as firearms since they are firearms! Never point them at another person. Don't allow children to play with them or around them. When you fire one, face away from the wind. Aim it downwind and upward at an angle of about 60 degrees to the horizon. If there is a strong wind, aim it somewhat more vertically. Never fire it straight up. Before you discharge a flare pistol, check for overhead obstructions. These might be damaged by the flare. They might deflect the flare to where it will cause damage.

Disposal of VDS

Keep outdated flares when you get new ones. They do not meet legal requirements, but you might need them sometime, and they may work. It is illegal to fire a VDS on federal navigable waters unless an emergency exists. Many states have similar laws.

Emergency Position Indicating Radio Beacon (EPIRB)

There is no requirement for recreational boats to have EPIRBs. Some commercial and fishing vessels, though, must have them if they operate beyond the three mile limit. Vessels carrying six or fewer passengers for hire must have EPIRBs under some circumstances when operating beyond the three mile limit. If you boat in a remote area or offshore, you should have an EPIRB. An EPIRB is a small (about 6 to 20 inches high), battery-powered, radio transmitting buoy-like device. It is a radio transmitter and requires a license or an endorsement on your radio station license by the Federal Communications Commission (FCC). EPIRBs are activated by being immersed in water or by a manual switch.

Equipment Not Required But Recommended

Although not required by law, there are other pieces of equipment that are good to have onboard.

SECOND MEANS OF PROPULSION

➧ See Figure 13

All boats less than 16 feet long should carry a second means of propulsion. A paddle or oar can come in handy at times. For most small boats, a spare trolling or outboard motor is an excellent idea. If you carry a spare motor, it should have its own fuel tank and starting power. If you use an electric trolling motor, it should have its own battery.

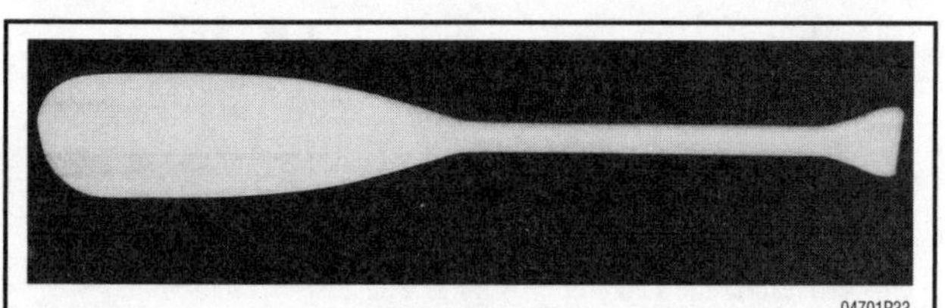

04701P33

Fig. 13 A typical wooden oar should be kept onboard as an auxiliary means of propulsion. It can also function as a grab hook for someone fallen overboard

BAILING DEVICES

All boats should carry at least one effective manual bailing device in addition to any installed electric bilge pump. This can be a bucket, can, scoop, hand operated pump, etc. If your battery "goes dead" it will not operate your electric pump.

FIRST AID KIT

➧ See Figure 14

All boats should carry a first aid kit. It should contain adhesive bandages, gauze, adhesive tape, antiseptic, aspirin, etc. Check your first aid kit from time to time. Replace anything that is outdated. It is to your advantage to know how to use your first aid kit. Another good idea would be to take a Red Cross first aid course.

ANCHORS

➧ See Figure 15

All boats should have anchors. Choose one of suitable size for your boat. Better still, have two anchors of different sizes. Use the smaller one in calm water or when anchoring for a short time to fish or eat. Use the larger one when the water is rougher or for overnight anchoring.

Carry enough anchor line of suitable size for your boat and the waters in which you will operate. If your engine fails you, the first thing you usually should do is lower your anchor. This is good advice in shallow water where

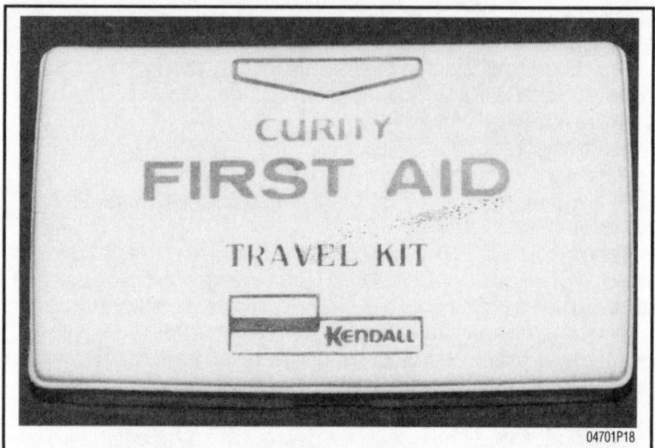

04701P18

Fig. 14 Always carry an adequately stocked first aid kit on board for the safety of the crew and guests

04891P14

Fig. 15 Choose an anchor of sufficient weight to secure the boat without dragging. In some cases separate anchors may be needed for different situations

you may be driven aground by the wind or water. It is also good advice in windy weather or rough water. The anchor will usually hold your bow into the waves.

VHF-FM RADIO

Your best means of summoning help in an emergency or in case of a breakdown is a VHF-FM radio. You can use it to get advice or assistance from the Coast Guard. In the event of a serious illness or injury aboard your boat, the Coast Guard can have emergency medical equipment meet you ashore.

TOOLS AND SPARE PARTS

➧ See Figures 16 and 17

Carry a few tools and some spare parts, and learn how to make minor repairs. Many search and rescue cases are caused by minor breakdowns that boat operators could have repaired. If your engine is an inboard or stern drive, carry spare belts and water pump impellers and the tools to change them.

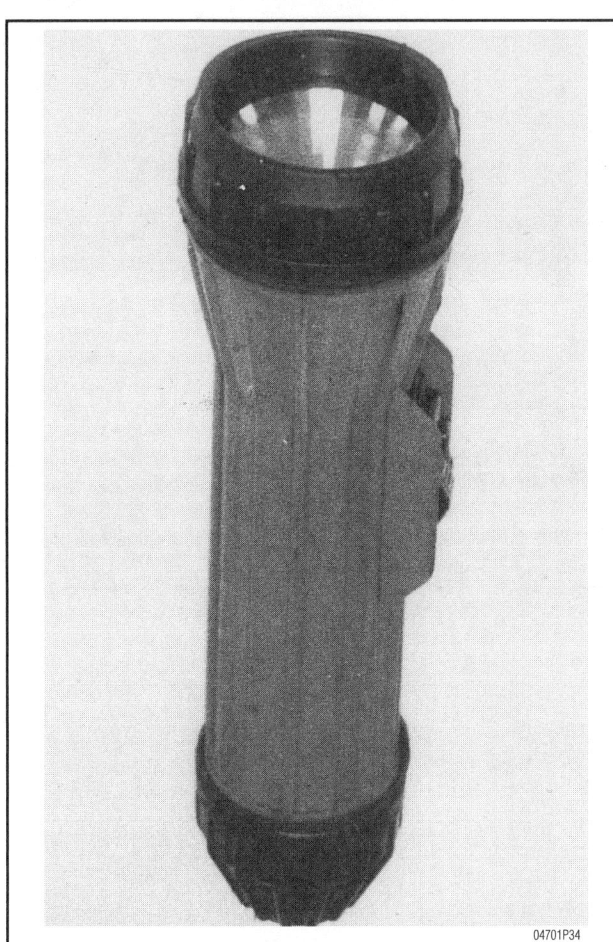

04701P34

Fig. 16 A flashlight with a fresh set of batteries is handy when repairs are needed at night. It can also double as a signaling device

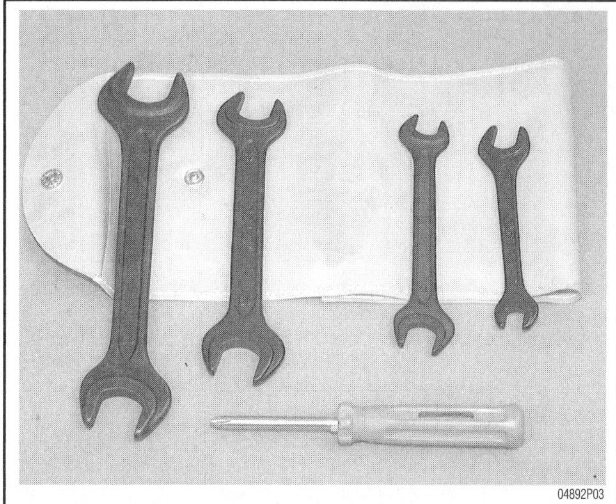

04892P03

Fig. 17 A few wrenches, a screwdriver and maybe a pair of pliers can be very helpful to make emergency repairs

Courtesy Marine Examinations

One of the roles of the Coast Guard Auxiliary is to promote recreational boating safety. This is why they conduct thousands of Courtesy Marine Examinations each year. The auxiliarists who do these examinations are well-trained and knowledgeable in the field.

These examinations are free and done only at the consent of boat owners. To pass the examination, a vessel must satisfy federal equipment requirements and certain additional requirements of the coast guard auxiliary. If your vessel does not pass the Courtesy Marine Examination, no report of the failure is made. Instead, you will be told what you need to correct the deficiencies. The examiner will return at your convenience to redo the examination.

If your vessel qualifies, you will be awarded a safety decal. The decal does not carry any special privileges, it simply attests to your interest in safe boating.

SAFETY IN SERVICE

It is virtually impossible to anticipate all of the hazards involved with maintenance and service, but care and common sense will prevent most accidents.

The rules of safety for mechanics range from "don't smoke around gasoline," to "use the proper tool(s) for the job." The trick to avoiding injuries is to develop safe work habits and to take every possible precaution. Whenever you are working on your boat, pay attention to what you are doing. The more you pay attention to details and what is going on around you, the less likely you will be to hurt yourself or damage your boat.

Do's

- Do keep a fire extinguisher and first aid kit handy.
- Do wear safety glasses or goggles when cutting, drilling, grinding or prying, even if you have 20–20 vision. If you wear glasses for the sake of vision, wear safety goggles over your regular glasses.
- Do shield your eyes whenever you work around the battery. Batteries contain sulfuric acid. In case of contact with the eyes or skin, flush the area with water or a mixture of water and baking soda, then seek immediate medical attention.
- Do use adequate ventilation when working with any chemicals or hazardous materials.
- Do disconnect the negative battery cable when working on the electrical system. The secondary ignition system contains EXTREMELY HIGH VOLTAGE. In some cases it can even exceed 50,000 volts.
- Do follow manufacturer's directions whenever working with potentially hazardous materials. Most chemicals and fluids are poisonous if taken internally.
- Do properly maintain your tools. Loose hammerheads, mushroomed punches and chisels, frayed or poorly grounded electrical cords, excessively worn screwdrivers, spread wrenches (open end), cracked sockets, or slipping ratchets can cause accidents.
- Likewise, keep your tools clean; a greasy wrench can slip off a bolt head, ruining the bolt and often harming your knuckles in the process.
- Do use the proper size and type of tool for the job at hand. Do select a wrench or socket that fits the nut or bolt. The wrench or socket should sit straight, not cocked.
- Do, when possible, pull on a wrench handle rather than push on it, and adjust your stance to prevent a fall.
- Do be sure that adjustable wrenches are tightly closed on the nut or bolt and pulled so that the force is on the side of the fixed jaw. Better yet, avoid the use of an adjustable if you have a fixed wrench that will fit.
- Do strike squarely with a hammer; avoid glancing blows. But, we REALLY hope you won't be using a hammer much in basic maintenance.
- Do use common sense whenever you work on your boat or motor. If a situation arises that doesn't seem right, sit back and have a second look. It may save an embarrassing moment or potential damage to your beloved boat.

Don'ts

- Don't run the engine in an enclosed area or anywhere else without proper ventilation—EVER! Carbon monoxide is poisonous; it takes a long time to leave the human body and you can build up a deadly supply of it in your system by simply breathing in a little every day. You may not realize you are slowly poisoning yourself.

- Don't work around moving parts while wearing loose clothing. Short sleeves are much safer than long, loose sleeves. Hard-toed shoes with neoprene soles protect your toes and give a better grip on slippery surfaces. Jewelry, watches, large belt buckles, or body adornment of any kind is not safe working around any vehicle. Long hair should be tied back under a hat.
- Don't use pockets for toolboxes. A fall or bump can drive a screwdriver deep into your body. Even a rag hanging from your back pocket can wrap around a spinning shaft.
- Don't smoke when working around gasoline, cleaning solvent or other flammable material.
- Don't smoke when working around the battery. When the battery is being charged, it gives off explosive hydrogen gas. Actually, you shouldn't smoke anyway. Save the cigarette money and put it into your boat!
- Don't use gasoline to wash your hands; there are excellent soaps available. Gasoline contains dangerous additives which can enter the body through a cut or through your pores. Gasoline also removes all the natural oils from the skin so that bone dry hands will suck up oil and grease.
- Don't use screwdrivers for anything other than driving screws! A screwdriver used as an prying tool can snap when you least expect it, causing injuries. At the very least, you'll ruin a good screwdriver.

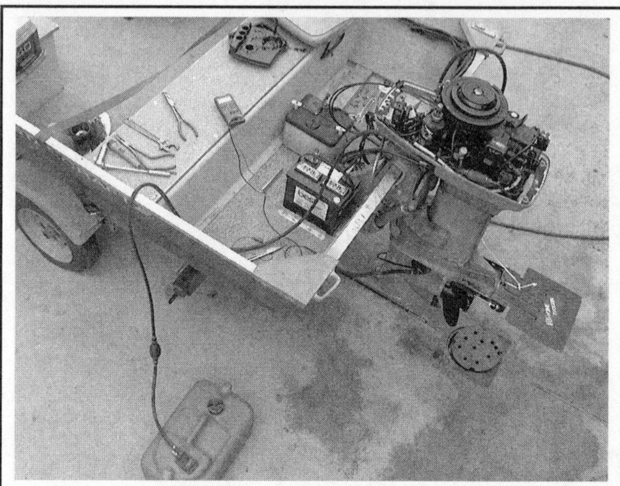

0489XP06

Combining the proper tools with the proper techniques will help to ensure a complete and satisfactory repair.

TOOLS AND EQUIPMENT 2-2
SAFETY TOOLS 2-2
WORK GLOVES 2-2
EYE AND EAR PROTECTION 2-2
WORK CLOTHES 2-3
CHEMICALS 2-3
LUBRICANTS & PENETRANTS 2-3
SEALANTS 2-3
CLEANERS 2-4
TOOLS 2-4
HAND TOOLS 2-4
SOCKET SETS 2-4
WRENCHES 2-6
PLIERS 2-7
SCREWDRIVERS 2-8
HAMMERS 2-8
OTHER COMMON TOOLS 2-8
SPECIAL TOOLS 2-8
ELECTRONIC TOOLS 2-9
GAUGES 2-9
MEASURING TOOLS 2-10
MICROMETERS & CALIPERS 2-11
DIAL INDICATORS 2-11
TELESCOPING GAUGES 2-12
DEPTH GAUGES 2-12
FASTENERS, MEASUREMENTS AND CONVERSIONS 2-12
BOLTS, NUTS AND OTHER THREADED RETAINERS 2-12
TORQUE 2-13
STANDARD AND METRIC MEASUREMENTS 2-13
SPECIFICATIONS CHARTS
USING A VACUUM GAUGE 2-10
CONVERSION FACTORS 2-14
METRIC BOLTS 2-15
SAE BOLTS 2-16

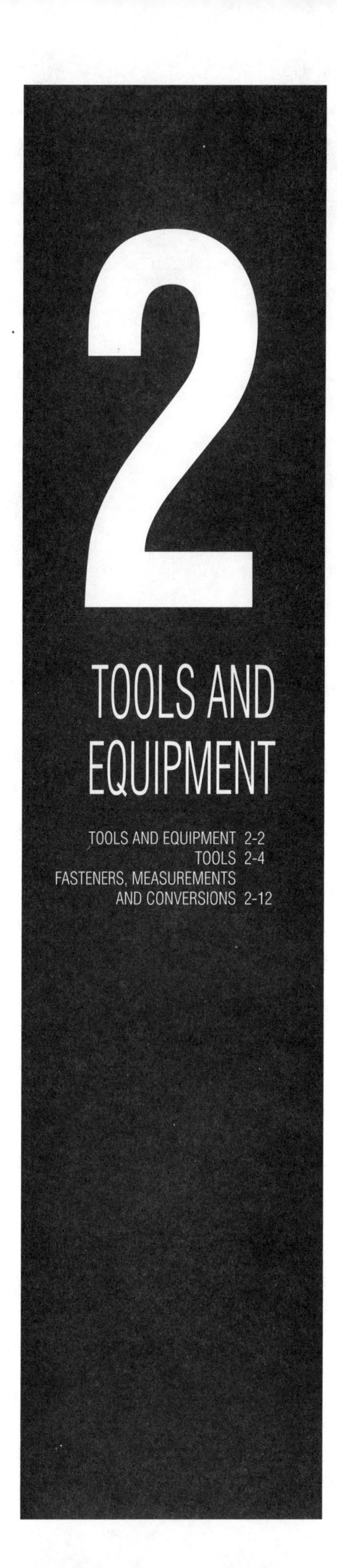

2

TOOLS AND EQUIPMENT

TOOLS AND EQUIPMENT 2-2
TOOLS 2-4
FASTENERS, MEASUREMENTS AND CONVERSIONS 2-12

TOOLS AND EQUIPMENT

Safety Tools

WORK GLOVES

See Figures 1 and 2

Unless you think scars on your hands are cool, enjoy pain and like wearing bandages, get a good pair of work gloves. Canvas or leather are the best. And yes, we realize that there are some jobs involving small parts that can't be done while wearing work gloves. These jobs are not the ones usually associated with hand injuries.

A good pair of rubber gloves (such as those usually associated with dish washing) or vinyl gloves is also a great idea. There are some liquids such as solvents and penetrants that don't belong on your skin. Avoid burns and rashes. Wear these gloves.

And lastly, an option. If you're tired of being greasy and dirty all the time, go to the drug store and buy a box of disposable latex gloves like medical professionals wear. You can handle greasy parts, perform small tasks, wash parts, etc. all without getting dirty! These gloves take a surprising amount of abuse without tearing and aren't expensive. Note however, that it has been reported that some people are allergic to the latex or the powder used inside some gloves, so pay attention to what you buy.

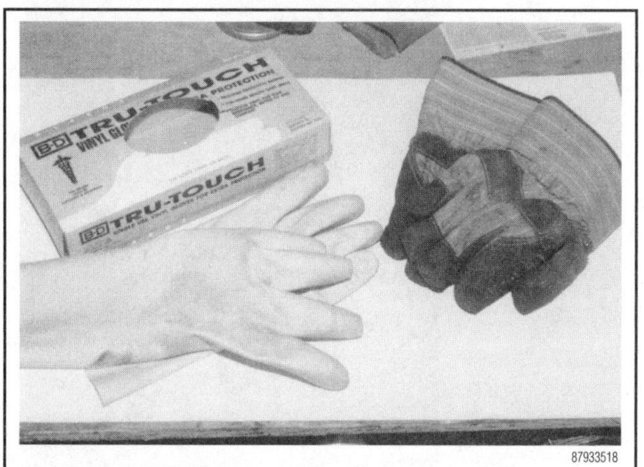

87933518

Fig. 1 Three different types of work gloves. The box contains latex gloves

04892P14

Fig. 2 Latex gloves come in handy when you are doing those messy jobs, like handling an oil soaked filter

EYE AND EAR PROTECTION

See Figures 3 and 4

Don't begin any job without a good pair of work goggles or impact resistant glasses! When doing any kind of work, it's all too easy to avoid eye injury through this simple precaution. And don't just buy eye protection and leave it on the shelf. Wear it all the time! Things have a habit of breaking, chipping, splashing, spraying, splintering and flying around. And, for some reason, your eye is always in the way!

If you wear vision correcting glasses as a matter of routine, get a pair made with polycarbonate lenses. These lenses are impact resistant and are available at any optometrist.

Often overlooked is hearing protection. Power equipment is noisy! Loud noises damage your ears. It's as simple as that! The simplest and cheapest form of ear protection is a pair of noise-reducing ear plugs. Cheap insurance for your ears. And, they may even come with their own, cute little carrying case.

More substantial, more protection and more money is a good pair of noise reducing earmuffs. They protect from all but the loudest sounds. Hopefully those are sounds that you'll never encounter since they're usually associated with disasters.

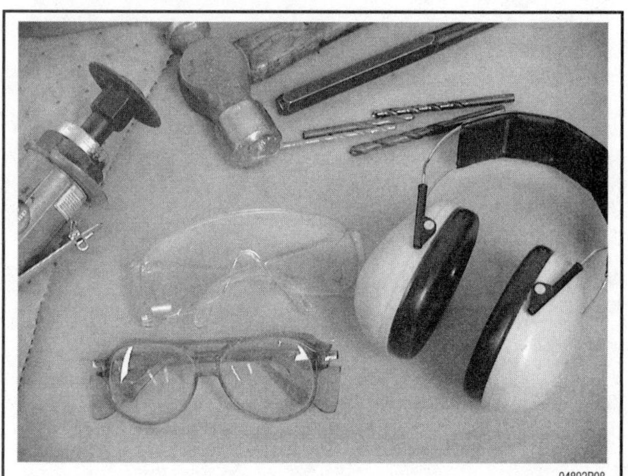

04892P08

Fig. 3 Don't begin any job without a good pair of work goggles or impact resistant glasses. Also good noise reducing earmuffs are cheap insurance to protect your hearing

04892P28

Fig. 4 Things have a habit of breaking, chipping, splashing, spraying, splintering and flying around. And, for some reason, your eye is always in the way

WORK CLOTHES

Everyone has "work clothes." Usually these consist of old jeans and a shirt that has seen better days. That's fine. In addition, a denim work apron is a nice accessory. It's rugged, can hold some spare bolts, and you don't feel bad wiping your hands or tools on it. That's what it's for.

When working in cold weather, a one-piece, thermal work outfit is invaluable. Most are rated to below zero (Fahrenheit) temperatures and are ruggedly constructed. Just look at what the marine mechanics are wearing and that should give you a clue as to what type of clothing is good.

Chemicals

There is a whole range of chemicals that you'll find handy for maintenance work. The most common types are, lubricants, penetrants and sealers. Keep these handy onboard. There are also many chemicals that are used for detailing or cleaning.

When a particular chemical is not being used, keep it capped, upright and in a safe place. These substances may be flammable, may be irritants or might even be caustic and should always be stored properly, used properly and handled with care. Always read and follow all label directions and be sure to wear hand and eye protection!

LUBRICANTS & PENETRANTS

See Figure 5

Anti-seize is used to coat certain fasteners prior to installation. This can be especially helpful when two dissimilar metals are in contact (to help prevent corrosion that might lock the fastener in place). This is a good practice on a lot of different fasteners, BUT, NOT on any fastener which might vibrate loose causing a problem. If anti-seize is used on a fastener, it should be checked periodically for proper tightness.

Lithium grease, chassis lube, silicone grease or a synthetic brake caliper grease can all be used pretty much interchangeably. All can be used for coating rust-prone fasteners and for facilitating the assembly of parts that are a tight fit. Silicone and synthetic greases are the most versatile.

➡Silicone dielectric grease is a non-conductor that is often used to coat the terminals of wiring connectors before fastening them. It may sound odd to coat metal portions of a terminal with something that won't conduct electricity, but here is it how it works. When the connector is fastened the metal-to-metal contact between the terminals will displace the grease (allowing the circuit to be completed). The grease that is displaced will then coat the non-contacted surface and the cavity around the terminals, SEALING them from atmospheric moisture that could cause corrosion.

Silicone spray is a good lubricant for hard-to-reach places and parts that shouldn't be gooped up with grease.

Penetrating oil may turn out to be one of your best friends when taking something apart that has corroded fasteners. Not only can they make a job easier, they can really help to avoid broken and stripped fasteners. The most familiar penetrating oils are Liquid Wrench® and WD-40®. A newer penetrant, PB Blaster® also works well. These products have hundreds of uses. For your purposes, they are vital!

Before disassembling any part (especially on an exhaust system), check the fasteners. If any appear rusted, soak them thoroughly with the penetrant and let them stand while you do something else (for particularly rusted or frozen parts you may need to soak them a few days in advance). This simple act can save you hours of tedious work trying to extract a broken bolt or stud.

SEALANTS

See Figures 6 and 7

Sealants are an indispensable part for certain tasks, especially if you are trying to avoid leaks. The purpose of sealants is to establish a leak-proof bond between or around assembled parts. Most sealers are used in conjunction with gaskets, but some are used instead of conventional gasket material.

The most common sealers are the non-hardening types such as Permatex®No.2 or its equivalents. These sealers are applied to the mating surfaces of each part to be joined, then a gasket is put in place and the parts are assembled.

➡A sometimes overlooked use for sealants like RTV is on the threads of vibration prone fasteners.

One very helpful type of non-hardening sealer is the "high tack" type. This type is a very sticky material that holds the gasket in place while the parts are being assembled. This stuff is really a good idea when you don't have enough hands or fingers to keep everything where it should be.

The stand-alone sealers are the Room Temperature Vulcanizing (RTV) silicone gasket makers. On some engines, this material is used instead of a gasket. In those instances, a gasket may not be available or, because of the shape of the mating surfaces, a gasket shouldn't be used. This stuff, when used in conjunction with a conventional gasket, produces the surest bonds.

RTV does have its limitations though. When using this material, you will have a time limit. It starts to set-up within 15 minutes or so, so you have to assemble the parts without delay. In addition, when squeezing the material out of the tube, don't drop any glops into the engine. The stuff will form and set and travel around the oil gallery, possibly plugging up a passage. Also, most types are not fuel-proof. Check the tube for all cautions.

04892P09

Fig. 5 Antiseize, penetrating oil, lithium grease, electronic cleaner and silicone spray. These products have hundreds of uses and should be a part of your chemical tool collection

87933507

Fig. 6 Sealants are essential for preventing leaks

04892P10

Fig. 7 On some engines, RTV is used instead of gasket material to seal components

87933018

Fig. 8 The new citrus hand cleaners not only work well, but they smell pretty good too. Choose one with pumice for added cleaning power

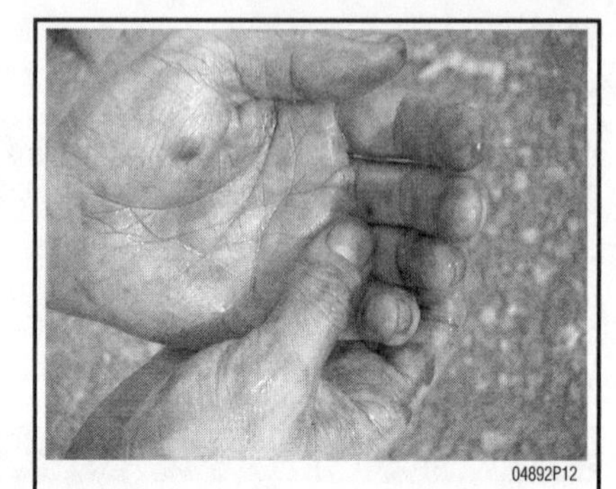

04892P12

Fig. 9 The use of hand lotion seals your hands and keeps dirt and grease from sticking to your skin

CLEANERS

➧ See Figures 8 and 9

There are two types of cleaners on the market today: parts cleaners and hand cleaners. The parts cleaners are for the parts; the hand cleaners are for you. They are not interchangeable.

There are many good, non-flammable, biodegradable parts cleaners on the market. These cleaning agents are safe for you, the parts and the environment. Therefore, there is no reason to use flammable, caustic or toxic substances to clean your parts or tools.

As far as hand cleaners go, the waterless types are the best. They have always been efficient at cleaning, but leave a pretty smelly odor. Recently though, just about all of them have eliminated the odor and added stuff that actually smells good. Make sure that you pick one that contains lanolin or some other moisture-replenishing additive. Cleaners not only remove grease and oil but also skin oil.

➡Most women will tell you to use a hand lotion when you're all cleaned up. It's okay. Real men DO use hand lotion! Believe it or not, using hand lotion before your hands are dirty will actually make them easier to clean when you're finished with a dirty job. Lotion seals your hands, and keeps dirt and grease from sticking to your skin.

TOOLS

➧ See Figure 10

Tools; this subject could fill a completely separate manual. The first thing you will need to ask yourself, is just how involved do you plan to get. If you are serious about your maintenance you will want to gather a quality set of tools to make the job easier, and more enjoyable. BESIDES, TOOLS ARE FUN!!!

Almost every do-it-yourselfer loves to accumulate tools. Though most find a way to perform jobs with only a few common tools, they tend to buy more over time, as money allows. So gathering the tools necessary for maintenance does not have to be an expensive, overnight proposition.

When buying tools, the saying "You get what you pay for ..." is absolutely true! Don't go cheap! Any hand tool that you buy should be drop forged and/or chrome vanadium. These two qualities tell you that the tool is strong enough for the job. With any tool, go with a name that you've heard of before, or, that is recommended buy your local professional retailer. Let's go over a list of tools that you'll need.

Most of the world uses the metric system. However, some American-built engines and aftermarket accessories use standard fasteners. So, accumulate your tools accordingly. Any good DIYer should have a decent set of both U.S. and metric measure tools.

➡Don't be confused by terminology. Most advertising refers to "SAE and metric", or "standard and metric." Both are misnomers. The Society of Automotive Engineers (SAE) did not invent the English system of measurement; the English did. The SAE likes metrics just fine. Both English (U.S.) and metric measurements are SAE approved. Also, the current "standard" measurement IS metric. So, if it's not metric, it's U.S. measurement.

Hand Tools

SOCKET SETS

➧ See Figures 11 thru 17

Socket sets are the most basic hand tools necessary for repair and maintenance work. For our purposes, socket sets come in three drive sizes: ¼ inch, ⅜ inch and ½ inch. Drive size refers to the size of the drive lug on the ratchet, breaker bar or speed handle.

A ⅜ inch set is probably the most versatile set in any mechanic's tool box. It

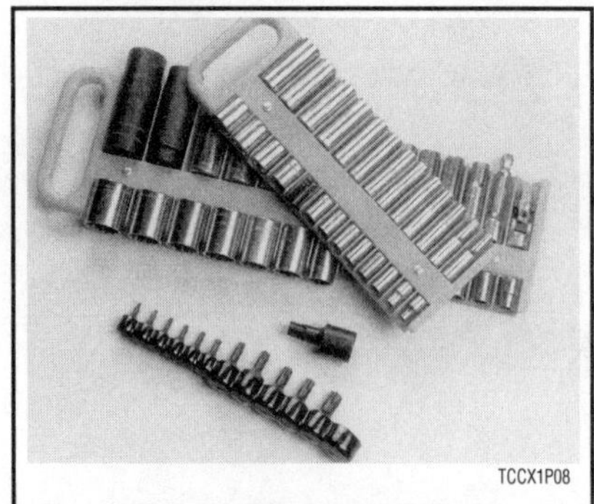

TCCX1P08

Fig. 10 Socket holders, especially the magnetic type, are handy items to keep tools in order

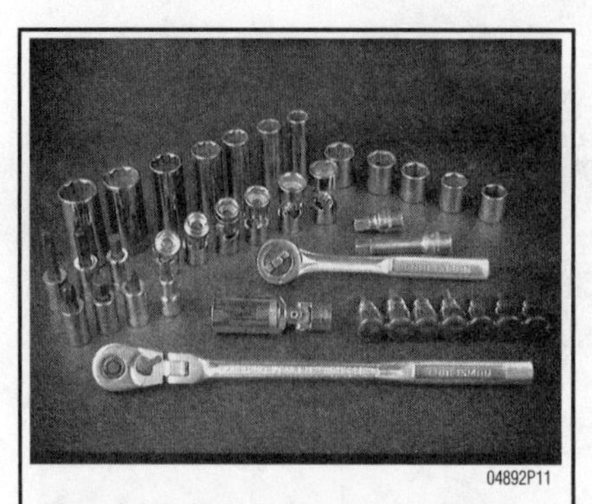

04892P11

Fig. 11 A ⅜ inch socket set is probably the most versatile tool in any mechanic's tool box

90991P38

Fig. 12 A swivel (U-joint) adapter (left), a ¼ inch-to-⅜ inch adapter (center) and a ⅜ inch-to-¼ inch adapter (right)

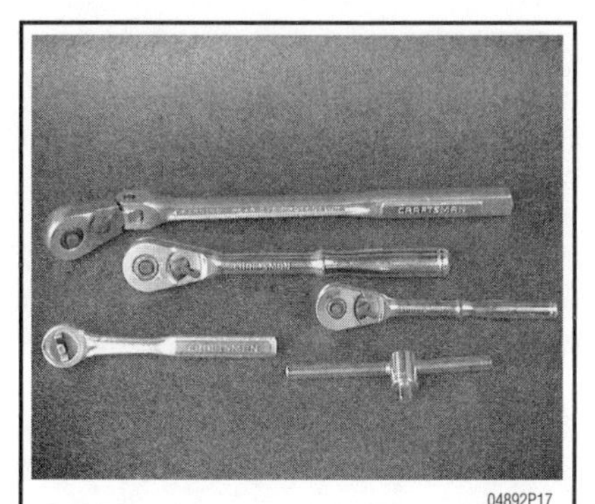

04892P17

Fig. 13 Ratchets come in all sizes and configurations from rigid to swivel-headed

04892P15

Fig. 14 Standard length sockets (top) are good for just about all jobs. However, some bolts may require deep sockets (bottom)

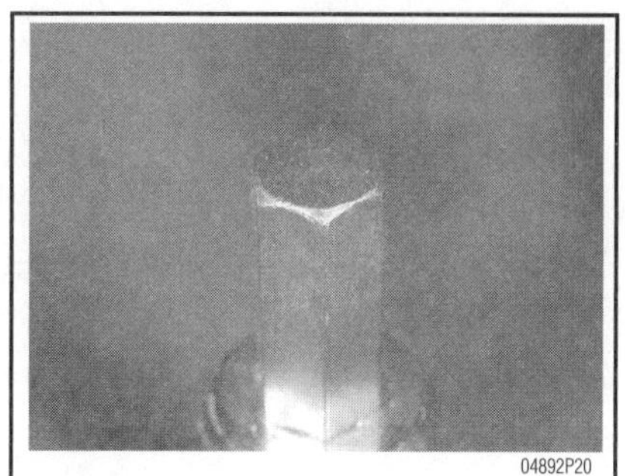

04892P20

Fig. 15 Hex-head fasteners retain many components on modern powerheads. These fasteners require a socket with a hex shaped driver

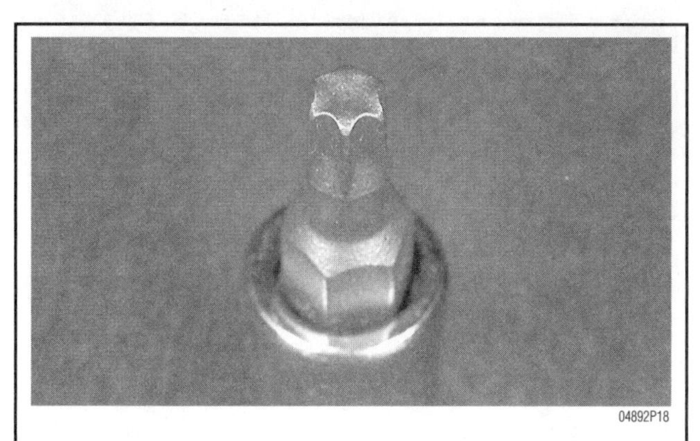

04892P18

Fig. 16 Torx® drivers . . .

04892P19

Fig. 17 . . . and tamper resistant drivers are required to remove special fasteners installed by the manufacturers

allows you to get into tight places that the larger drive ratchets can't and gives you a range of larger sockets that are still strong enough for heavy duty work. The socket set that you'll need should range in sizes from ⅜ inch through 1 inch for standard fasteners, and a 6mm through 19mm for metric fasteners.

You'll need a good ½ inch set since this size drive lug assures that you won't break a ratchet or socket on large or heavy fasteners. Also, torque wrenches with a torque scale high enough for larger fasteners are usually ½ inch drive.

¼ inch drive sets can be very handy in tight places. Though they usually duplicate functions of the ⅜ inch set, ¼ inch drive sets are easier to use for smaller bolts and nuts.

As for the sockets themselves, they come in standard and deep lengths as well as 6 or 12 point. 6 and 12 points refers to how many sides are in the socket itself. Each has advantages. The 6 point socket is stronger and less prone to slipping which would strip a bolt head or nut. 12 point sockets are more common, usually less expensive and can operate better in tight places where the ratchet handle can't swing far.

Standard length sockets are good for just about all jobs, however, some stud-head bolts, hard-to-reach bolts, nuts on long studs, etc., require the deep sockets.

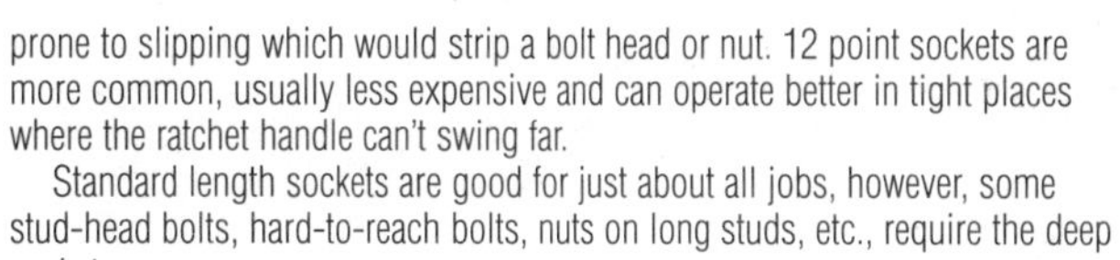

Most manufacturers use recessed hex-head fasteners to retain many of the engine parts. These fasteners require a socket with a hex shaped driver or a large sturdy hex key. To help prevent torn knuckles, we would recommend that you stick to the sockets on any tight fastener and leave the hex keys for lighter applications. Hex driver sockets are available individually or in sets just like conventional sockets.

More and more, manufacturers are using Torx® head fasteners, which were once known as tamper resistant fasteners (because many people did not have tools with the necessary odd driver shape). They are still used where the manufacturer would prefer only knowledgeable mechanics or advanced Do-It-Yourselfers (DIYers) to work.

Torque Wrenches

➧ See Figure 18

In most applications, a torque wrench can be used to assure proper installation of a fastener. Torque wrenches come in various designs and most stores will carry a variety to suit your needs. A torque wrench should be used any time you have a specific torque value for a fastener. Keep in mind that because there is no worldwide standardization of fasteners, the charts at the end of this section are a general guideline and should be used with caution. If you are using the right tool for the job, you should not have to strain to tighten a fastener.

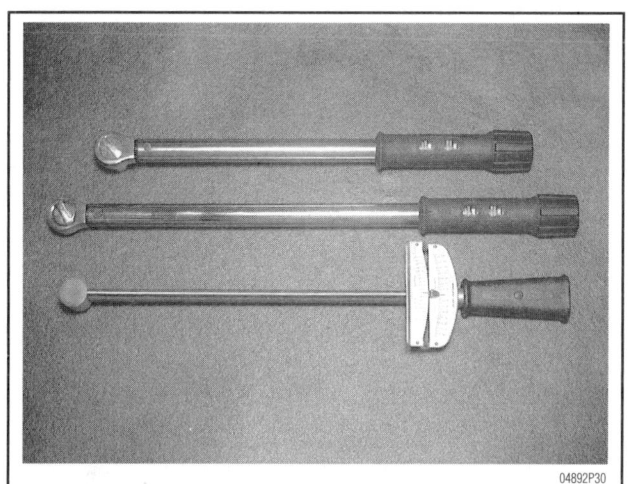

04892P30

Fig. 18 Three types of torque wrenches. Top to bottom: a ⅜ inch drive clicker type that reads in inch lbs., a ½ inch drive clicker type and a ½ inch drive beam type

BEAM TYPE

➧ See Figures 19 and 20

The beam type torque wrench is one of the most popular styles in use. If used properly, it can be the most accurate also. It consists of a pointer attached to the head that runs the length of the flexible beam (shaft) to a scale located near the handle. As the wrench is pulled, the beam bends and the pointer indicates the torque using the scale.

CLICK (BREAKAWAY) TYPE

➧ See Figures 21 and 22

Another popular torque wrench design is the click type. The clicking mechanism makes achieving the proper torque easy and most use ratcheting head for ease of bolt installation. To use the click type wrench you pre-adjust it to a torque setting. Once the torque is reached, the wrench has a reflex signaling feature that causes a momentary breakaway of the torque wrench body, sending an impulse to the operator's hand.

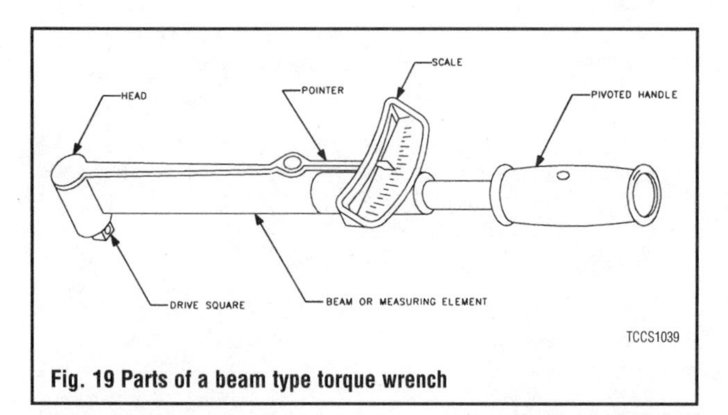

Fig. 19 Parts of a beam type torque wrench

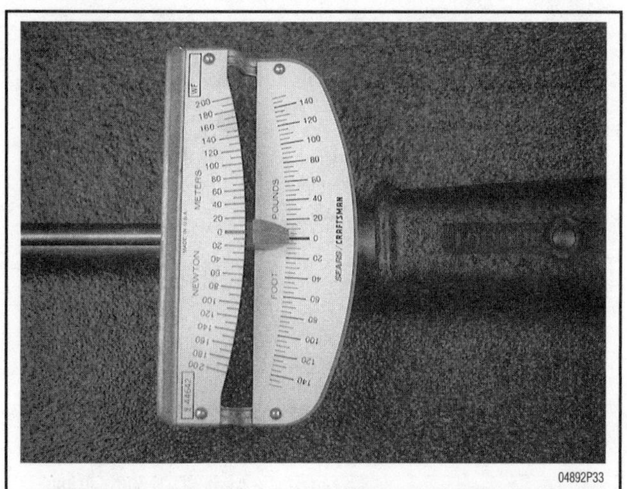

Fig. 20 A beam type torque wrench consists of a pointer attached to the head that runs the length of the flexible beam (shaft) to a scale located near the handle

Breaker Bars

➧ See Figure 23

Breaker bars are long handles with a drive lug. Their main purpose is to provide extra turning force when breaking loose tight bolts or nuts. They come in all drive sizes and lengths. Always take extra precautions and use proper technique when using a breaker bar.

WRENCHES

➧ See Figures 24, 25, 26 and 27

Basically, there are 3 kinds of fixed wrenches: open end, box end, and combination.

Open end wrenches have 2-jawed openings at each end of the wrench. These wrenches are able to fit onto just about any nut or bolt. They are extremely versatile but have one major drawback. They can slip on a worn or rounded bolt head or nut, causing bleeding knuckles and a useless fastener.

Box-end wrenches have a 360° circular jaw at each end of the wrench. They come in both 6 and 12 point versions just like sockets and each type has the same advantages and disadvantages as sockets.

Combination wrenches have the best of both. They have a 2-jawed open end and a box end. These wrenches are probably the most versatile.

As for sizes, you'll probably need a range similar to that of the sockets, about 1/4 inch through 1 inch for standard fasteners, or 6mm through 19mm for metric fasteners. As for numbers, you'll need 2 of each size, since, in many instances, one wrench holds the nut while the other turns the bolt. On most fasteners, the nut and bolt are the same size so having two wrenches of the same size comes in handy.

➡Although you will typically just need the sizes we specified, there are some exceptions. Occasionally you will find a nut which is larger. For these, you will need to buy ONE expensive wrench or a very large adjustable. Or you can always just convince the spouse that we are talking about safety here and buy a whole (read expensive) large wrench set.

One extremely valuable type of wrench is the adjustable wrench. An adjustable wrench has a fixed upper jaw and a moveable lower jaw. The lower jaw is moved by turning a threaded drum. The advantage of an adjustable wrench is its ability to be adjusted to just about any size fastener.

The main drawback of an adjustable wrench is the lower jaw's tendency to move slightly under heavy pressure. This can cause the wrench to slip if it is not facing the right way. Pulling on an adjustable wrench in the proper direction will cause the jaws to lock in place. Adjustable wrenches come in a large range of sizes, measured by the wrench length.

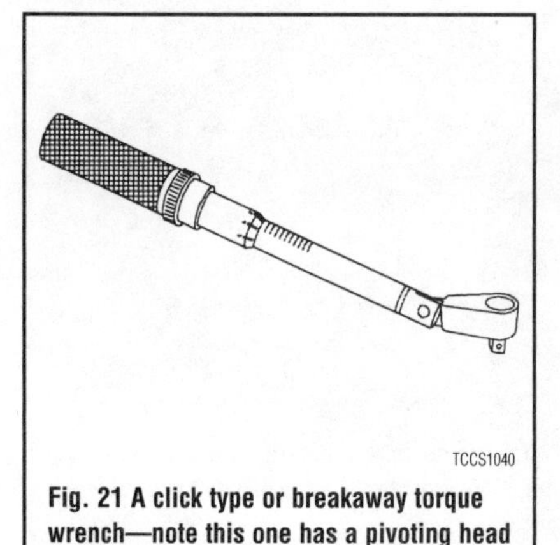

Fig. 21 A click type or breakaway torque wrench—note this one has a pivoting head

Fig. 22 Setting the proper torque on a click type torque wrench involves turning the handle until the proper torque specification appears on the dial

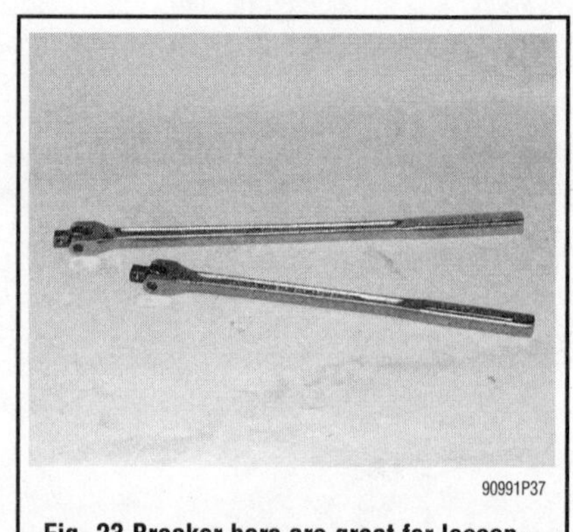

Fig. 23 Breaker bars are great for loosening large or stuck fasteners

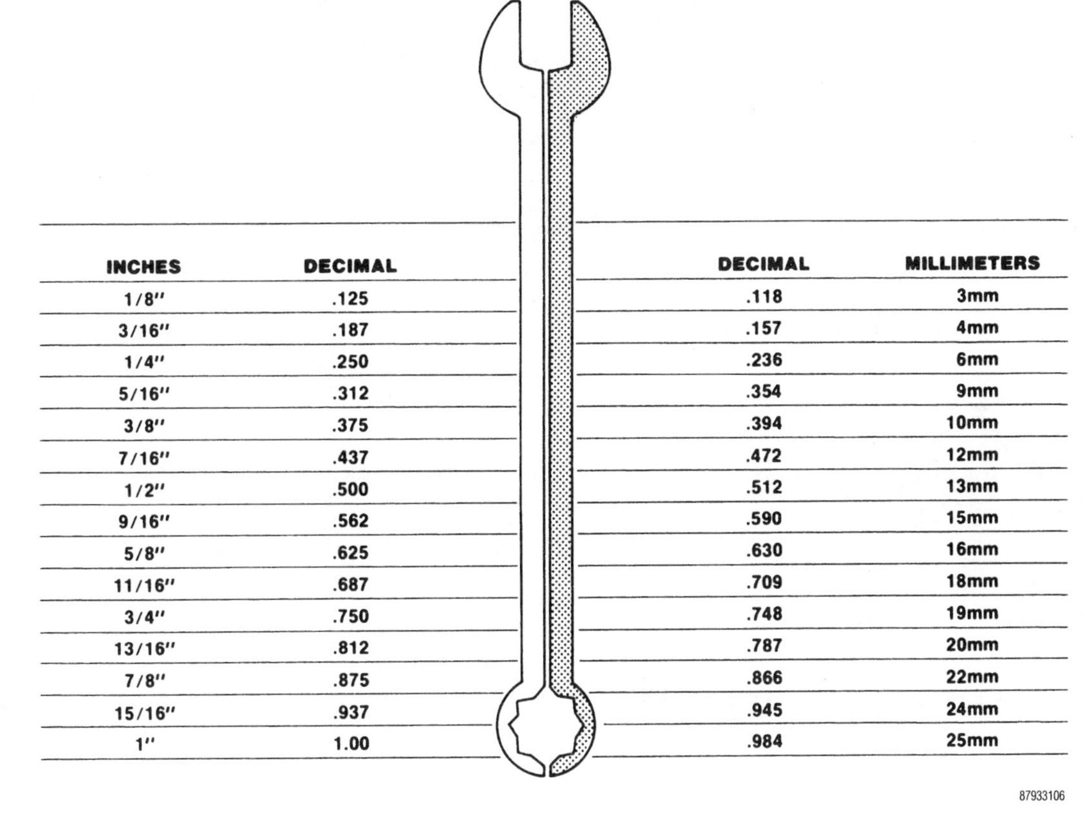

INCHES	DECIMAL	DECIMAL	MILLIMETERS
1/8"	.125	.118	3mm
3/16"	.187	.157	4mm
1/4"	.250	.236	6mm
5/16"	.312	.354	9mm
3/8"	.375	.394	10mm
7/16"	.437	.472	12mm
1/2"	.500	.512	13mm
9/16"	.562	.590	15mm
5/8"	.625	.630	16mm
11/16"	.687	.709	18mm
3/4"	.750	.748	19mm
13/16"	.812	.787	20mm
7/8"	.875	.866	22mm
15/16"	.937	.945	24mm
1"	1.00	.984	25mm

87933106

Fig. 24 Comparison of U.S. measure and metric wrench sizes

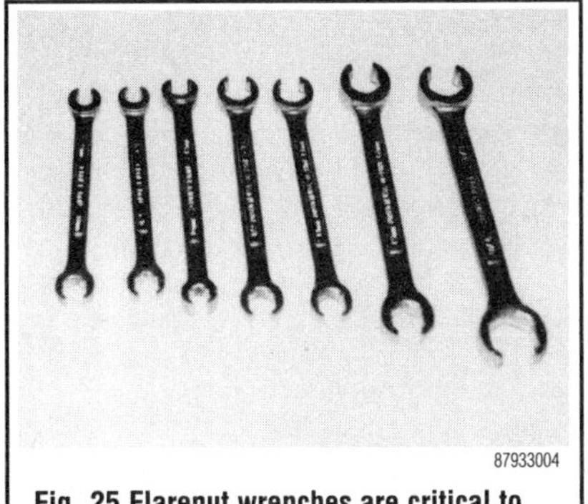

87933004

Fig. 25 Flarenut wrenches are critical to ensure tube fittings do not become rounded

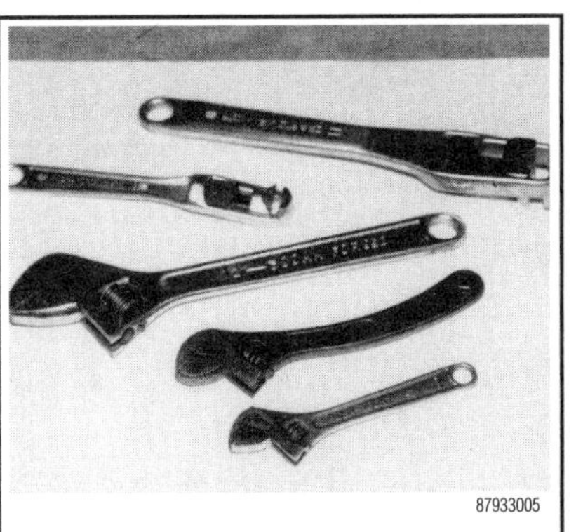

87933005

Fig. 26 Several types and sizes of adjustable wrenches

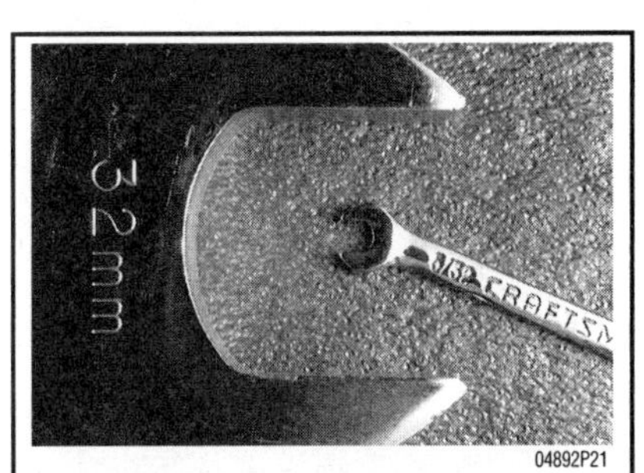

04892P21

Fig. 27 Occasionally you will find a nut which requires a particularly large or particularly small wrench. Rest assured that the proper wrench to fit is available at your local tool store

PLIERS

See Figure 28

Pliers are simply mechanical fingers. They are, more than anything, an extension of your hand. At least 3 pair of pliers are an absolute necessity—standard, needle nose and channel lock.

In addition to standard pliers there are the slip-joint, multi-position pliers such as ChannelLock® pliers and locking pliers, such as Vise Grips®.

Slip joint pliers are extremely valuable in grasping oddly sized parts and fasteners. Just make sure that you don't use them instead of a wrench too often since they can easily round off a bolt head or nut.

Locking pliers are usually used for gripping bolts or studs that can't be removed conventionally. You can get locking pliers in square jawed, needle-nosed and pipe-jawed. Locking pliers can rank right up behind duct tape as the handy-man's best friend.

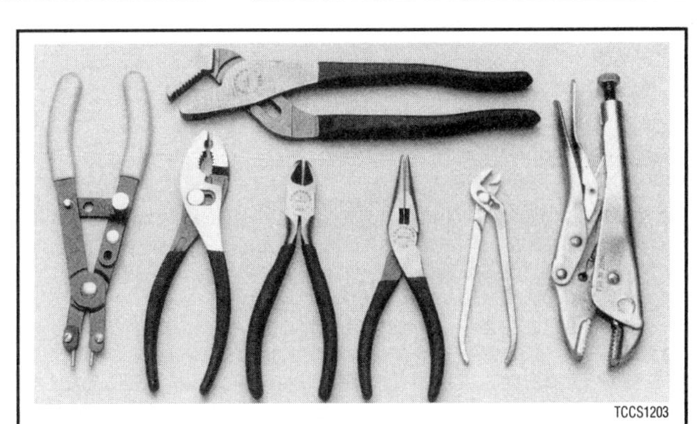

TCCS1203

Fig. 28 Pliers and cutters come in many shapes and sizes. You should have an assortment on hand

SCREWDRIVERS

You can't have too many screwdrivers. They come in 2 basic flavors, either standard or Phillips. Standard blades come in various sizes and thicknesses for all types of slotted fasteners. Phillips screwdrivers come in sizes with number designations from 1 on up, with the lower number designating the smaller size. Screwdrivers can be purchased separately or in sets.

HAMMERS

See Figure 29

You always need a hammer for just about any kind of work. You need a ball-peen hammer for most metal work when using drivers and other like tools. A plastic hammer comes in handy for hitting things safely. A soft-faced dead-blow hammer is used for hitting things safely and hard. Hammers are also VERY useful with non air-powered impact drivers.

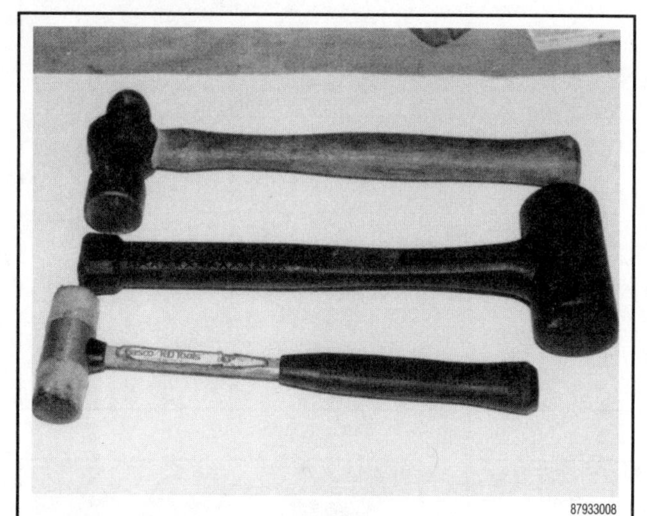

87933008

Fig. 29 Three types of hammers. Top to bottom: ball peen, rubber dead-blow, and plastic

OTHER COMMON TOOLS

There are a lot of other tools that every DIYer will eventually need (though not all for basic maintenance). They include:

- Funnels (for adding fluid)
- Chisels
- Punches
- Files
- Hacksaw
- Portable Bench Vise
- Tap and Die Set
- Flashlight
- Magnetic Bolt Retriever
- Gasket scraper
- Putty Knife
- Screw/Bolt Extractors
- Prybar

Hacksaws have just one use—cutting things off. You may wonder why you'd need one for something as simple as maintenance, but you never know. Among other things, guide studs to ease parts installation can be made from old bolts with their heads cut off.

A tap and die set might be something you've never needed, but you will eventually. It's a good rule, when everything is apart, to clean-up all threads, on bolts, screws and threaded holes. Also, you'll likely run across a situation in which stripped threads will be encountered. The tap and die set will handle that for you.

Gasket scrapers are just what you'd think, tools made for scraping old gasket material off of parts. You don't absolutely need one. Old gasket material can be removed with a putty knife or single edge razor blade. However, putty knives may not be sharp enough for some really stubborn gaskets and razor blades have a knack of breaking just when you don't want them to, inevitably slicing the nearest body part! As the old saying goes, "always use the proper tool for the job". If you're going to use a razor to scrape a gasket, be sure to always use a blade holder.

Putty knives really do have a use in a repair shop. Just because you remove all the bolts from a component sealed with a gasket doesn't mean it's going to come off. Most of the time, the gasket and sealer will hold it tightly. Lightly driving a putty knife at various points between the two parts will break the seal without damage to the parts.

A small — 8-10 inches (20–25 centimeters) long — prybar is extremely useful for removing stuck parts.

Never use a screwdriver as a prybar! Screwdrivers are not meant for prying. Screwdrivers, used for prying, can break, sending the broken shaft flying!

Screw/bolt extractors are used for removing broken bolts or studs that have broke off flush with the surface of the part.

SPECIAL TOOLS

See Figure 30

Almost every marine engine around today requires at least one special tool to perform a certain task. In most cases, these tools are specially designed to overcome some unique problem or to fit on some oddly sized component.

When manufacturers go through the trouble of making a special tool, it is usually necessary to use it to assure that the job will be done right. A special tool might be designed to make a job easier, or it might be used to keep you from damaging or breaking a part.

Don't worry, MOST basic maintenance procedures can either be performed without any special tools OR, because the tools must be used for such basic things, they are commonly available for a reasonable price. It is usually just the low production, highly specialized tools (like a super thin 7-point star-shaped socket capable of 150 ft. lbs. (203 Nm) of torque that is used only on the crankshaft nut of the limited production what-dya-callit engine) that tend to be outrageously expensive and hard to find. Luckily, you will probably never need such a tool.

Special tools can be as inexpensive and simple as an adjustable strap wrench or as complicated as an ignition tester. A few common specialty tools are listed here, but check with your dealer or with other boaters for help in determining if there are any special tools for YOUR particular engine. There is an added advantage in seeking advice from others, chances are they may have already found the special tool you will need, and know how to get it cheaper.

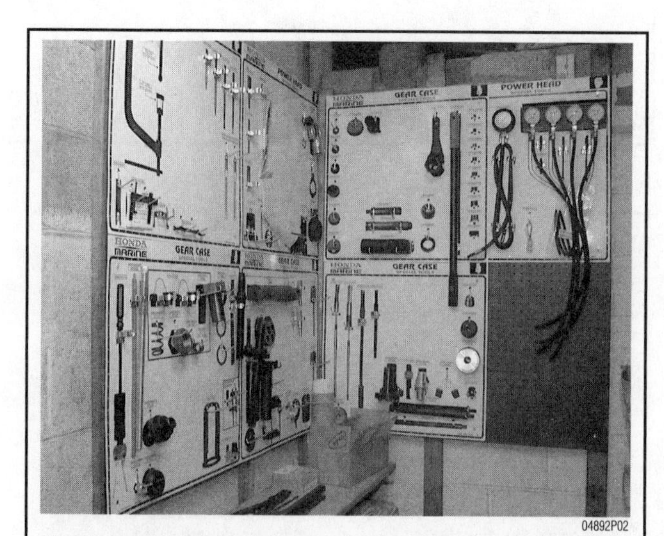

04892P02

Fig. 30 Almost every marine engine requires at least one special tool to perform a certain task

ELECTRONIC TOOLS

Battery Testers

The best way to test a non-sealed battery is using a hydrometer to check the specific gravity of the acid. Luckily, these are usually inexpensive and are available at most parts stores. Just be careful because the larger testers are usually designed for larger batteries and may require more acid than you will be able to draw from the battery cell. Smaller testers (usually a short, squeeze bulb type) will require less acid and should work on most batteries.

Electronic testers are available and are often necessary to tell if a sealed battery is usable. Luckily, many parts stores have them on hand and are willing to test your battery for you.

Battery Chargers

See Figure 31

If you are a weekend boater and take your boat out every week, then you will most likely want to buy a battery charger to keep your battery fresh. There are many types available, from low amperage trickle chargers to electronically controlled battery maintenance tools which monitor the battery voltage to prevent over or undercharging. This last type is especially useful if you store your boat for any length of time (such as during the severe winter months found in many Northern climates).

Even if you use your boat on a regular basis, you will eventually need a battery charger. Remember that most batteries are shipped dry and in a partial charged state. Before a new battery can be put into service it must be filled and properly charged. Failure to properly charge a battery (which was shipped dry) before it is put into service will prevent it from ever reaching a fully charged state.

Digital Volt/Ohm Meter (DVOM)

See Figure 32

Multimeters are an extremely useful tool for troubleshooting electrical problems. They can be purchased in either analog or digital form and have a price range to suit any budget. A multimeter is a voltmeter, ammeter and ohmmeter (along with other features) combined into one instrument. It is often used when testing solid state circuits because of its high input impedance (usually 10 megaohms or more). A brief description of the multimeter main test functions follows:

- Voltmeter—the voltmeter is used to measure voltage at any point in a circuit, or to measure the voltage drop across any part of a circuit. Voltmeters usually have various scales and a selector switch to allow the reading of different voltage ranges. The voltmeter has a positive and a negative lead. To avoid damage to the meter, always connect the negative lead to the negative (-) side of the circuit (to ground or nearest the ground side of the circuit) and connect the positive lead to the positive (+) side of the circuit (to the power source or the nearest power source). Note that the negative voltmeter lead will always be black and that the positive voltmeter will always be some color other than black (usually red).
- Ohmmeter—the ohmmeter is designed to read resistance (measured in ohms) in a circuit or component. Most ohmmeters will have a selector switch which permits the measurement of different ranges of resistance (usually the selector switch allows the multiplication of the meter reading by 10, 100, 1,000 and 10,000). Some ohmmeters are "auto-ranging" which means the meter itself will determine which scale to use. Since the meters are powered by an internal battery, the ohmmeter can be used like a self-powered test light. When the ohmmeter is connected, current from the ohmmeter flows through the circuit or component being tested. Since the ohmmeter's internal resistance and voltage are known values, the amount of current flow through the meter depends on the resistance of the circuit or component being tested. The ohmmeter can also be used to perform a continuity test for suspected open circuits. In using the meter for making continuity checks, do not be concerned with the actual resistance readings. Zero resistance, or any ohm reading, indicates continuity in the circuit. Infinite resistance indicates an opening in the circuit. A high resistance reading where there should be none indicates a problem in the circuit. Checks for short circuits are made in the same manner as checks for open circuits, except that the circuit must be isolated from both power and normal ground. Infinite resistance indicates no continuity, while zero resistance indicates a dead short.

WARNING

Never use an ohmmeter to check the resistance of a component or wire while there is voltage applied to the circuit.

- Ammeter—an ammeter measures the amount of current flowing through a circuit in units called amperes or amps. At normal operating voltage, most circuits have a characteristic amount of amperes, called "current draw" which can be measured using an ammeter. By referring to a specified current draw rating, then measuring the amperes and comparing the two values, one can determine what is happening within the circuit to aid in diagnosis. An open circuit, for example, will not allow any current to flow, so the ammeter reading will be zero. A damaged component or circuit will have an increased current draw, so the reading will be high. The ammeter is always connected in series with the circuit being tested. All of the current that normally flows through the circuit must also flow through the ammeter; if there is any other path for the current to follow, the ammeter reading will not be accurate. The ammeter itself has very little resistance to current flow and, therefore, will not affect the circuit, but it will measure current draw only when the circuit is closed and electricity is flowing. Excessive current draw can blow fuses and drain the battery, while a reduced current draw can cause motors to run slowly, lights to dim and other components to not operate properly.

GAUGES

Compression Gauge

See Figure 33

An important element in checking the overall condition of your engine is to check compression. This becomes increasingly more important on outboards with high hours. Compression gauges are available as screw-in types and hold-in types. The screw-in type is slower to use, but eliminates the possibility of a faulty reading due to escaping pressure. A compression reading will uncover many problems that can cause rough running. Normally, these are not the sort of problems that can be cured by a tune-up.

90991P34

Fig. 31 The Battery Tender® is more than just a battery charger, when left connected, it keeps your battery fully charged

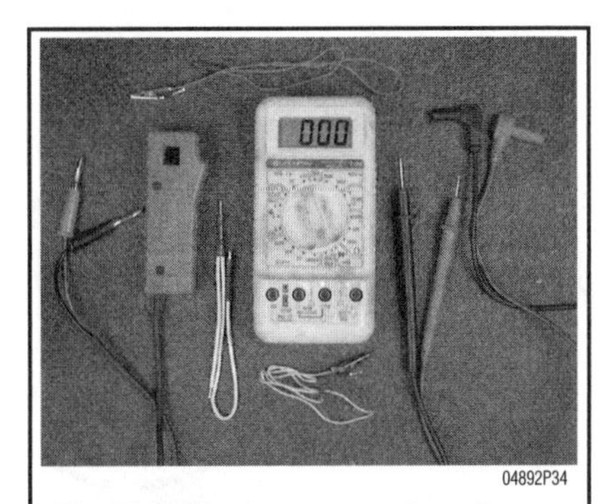
04892P34

Fig. 32 Multimeters are an extremely useful tool for troubleshooting electrical problems

04892P37

Fig. 33 Cylinder compression test results are extremely valuable indicators of internal engine condition

Vacuum Gauge

➧ See Figures 34 thru 35

Vacuum gauges are handy for discovering air leaks, late ignition or valve timing, and a number of other problems.

Measuring Tools

Eventually, you are going to have to measure something. To do this, you will need at least a few precision tools in addition to the special tools mentioned earlier.

USING A VACUUM GAUGE

White needle = steady needle ***Dark needle = drifting needle***

The vacuum gauge is one of the most useful and easy-to-use diagnostic tools. It is inexpensive, easy to hook up, and provides valuable information about the condition of your engine.

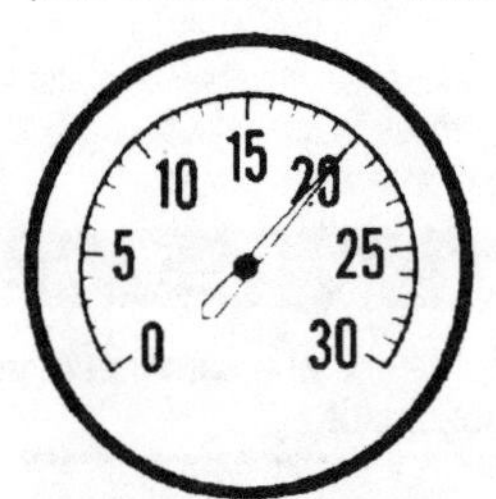

Indication: Normal engine in good condition

Gauge reading: Steady, from 17-22 in./Hg.

Indication: Sticking valve or ignition miss

Gauge reading: Needle fluctuates from 15-20 in./Hg. at idle

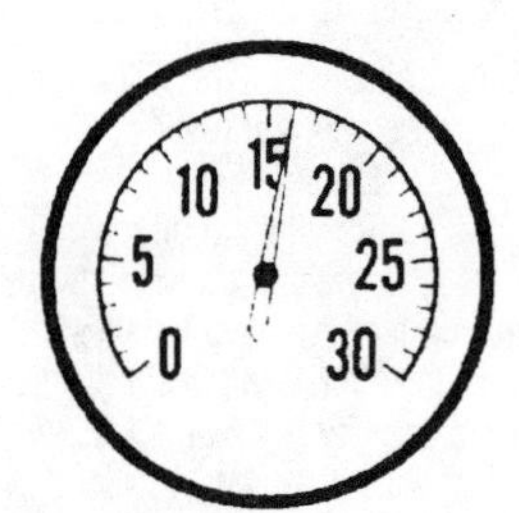

Indication: Late ignition or valve timing, low compression, stuck throttle valve, leaking carburetor or manifold gasket.

Gauge reading: Low (15-20 in./Hg.) but steady

Indication: Improper carburetor adjustment, or minor intake leak at carburetor or manifold

NOTE: Bad fuel injector O-rings may also cause this reading.

Gauge reading: Drifting needle

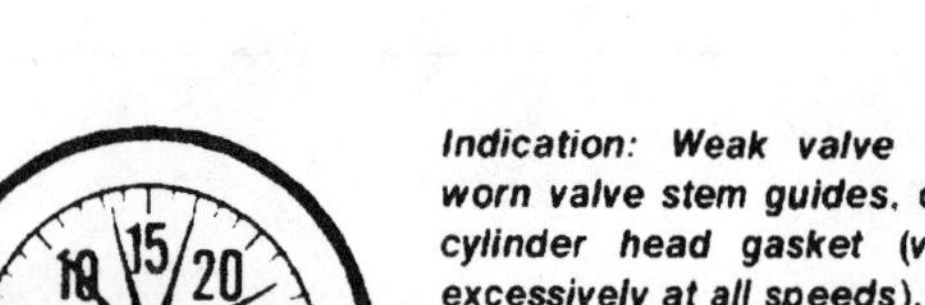

Indication: Weak valve springs, worn valve stem guides, or leaky cylinder head gasket (vibrating excessively at all speeds).

Gauge reading: Needle fluctuates as engine speed increases

Indication: Burnt valve or improper valve clearance. The needle will drop when the defective valve operates.

Gauge reading: Steady needle, but drops regularly

Indication: Choked muffler or obstruction in system. Speed up the engine. Choked muffler will exhibit a slow drop of vacuum to zero.

Gauge reading: Gradual drop in reading at idle

Indication: Worn valve guides

Gauge reading: Needle vibrates excessively at idle, but steadies as engine speed increases

TCCS3C01

04892P23

Fig. 34 Vacuum gauges are useful for many diagnostic tasks including testing of some fuel pumps

04892P25

Fig. 35 In a pinch, you can also use the vacuum gauge on a hand operated vacuum pump

MICROMETERS & CALIPERS

Micrometers and calipers are devices used to make extremely precise measurements. The simple truth is that you really won't have the need for many of these items just for simple maintenance. You will probably want to have at least one precision tool such as an outside caliper to measure rotors or brake pads, but that should be sufficient to most basic maintenance procedures.

Should you decide on becoming more involved in boat engine mechanics, such as repair or rebuilding, then these tools will become very important. The success of any rebuild is dependent, to a great extent on the ability to check the size and fit of components as specified by the manufacturer. These measurements are made in thousandths and ten-thousandths of an inch.

Micrometers

➧ **See Figure 36**

A micrometer is an instrument made up of a precisely machined spindle which is rotated in a fixed nut, opening and closing the distance between the end of the spindle and a fixed anvil.

Outside micrometers can be used to check the thickness parts such shims or the outside diameter of components like the crankshaft journals. They are also used during many rebuild and repair procedures to measure the diameter of components such as the pistons. The most common type of micrometer reads in 1/1000 of an inch. Micrometers that use a vernier scale can estimate to 1/10 of an inch.

Inside micrometers are used to measure the distance between two parallel surfaces. For example, in powerhead rebuilding work, the inside mike measures cylinder bore wear and taper. Inside mikes are graduated the same way as outside mikes and are read the same way as well.

Remember that an inside mike must be absolutely perpendicular to the work being measured. When you measure with an inside mike, rock the mike gently from side to side and tip it back and forth slightly so that you span the widest part of the bore. Just to be on the safe side, take several readings. It takes a certain amount of experience to work any mike with confidence.

Metric micrometers are read in the same way as inch micrometers, except that the measurements are in millimeters. Each line on the main scale equals 1 mm. Each fifth line is stamped 5, 10, 15, and so on. Each line on the thimble scale equals 0.01 mm. It will take a little practice, but if you can read an inch mike, you can read a metric mike.

Calipers

➧ **See Figure 37**

Inside and outside calipers are useful devices to have if you need to measure something quickly and precise measurement is not necessary. Simply take the reading and then hold the calipers on an accurate steel rule.

DIAL INDICATORS

➧ **See Figure 38**

A dial indicator is a gauge that utilizes a dial face and a needle to register measurements. There is a movable contact arm on the dial indicator. When the arms moves, the needle rotates on the dial. Dial indicators are calibrated to show readings in thousandths of an inch and typically, are used to measure end-play and runout on various parts.

Dial indicators are quite easy to use, although they are relatively expensive. A variety of mounting devices are available so that the indicator can be used in a

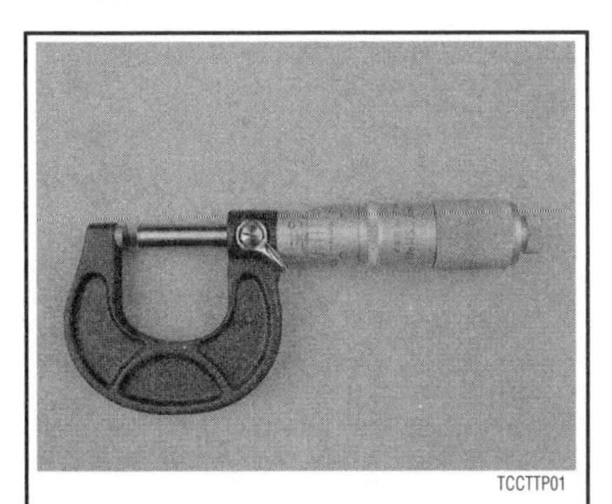

TCCTTP01

Fig. 36 Outside micrometers can be used to measure the thickness of shims or the outside diameter of a crankshaft journal

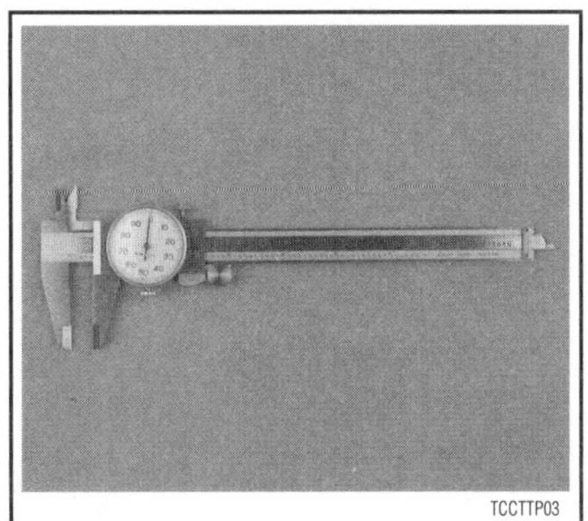

TCCTTP03

Fig. 37 Calipers are the fast and easy way to make precise measurements

TCCS3805

Fig. 38 Here, a dial indicator is used to measure the axial clearance (end play) of a crankshaft during a powerhead rebuilding procedure

number of situations. Make certain that the contact arm is always parallel to the movement of the work being measured.

TELESCOPING GAUGES

See Figure 39

A telescope gauge is used during rebuilding procedures (NOT usually basic maintenance) to measure the inside of bores. It can take the place of an inside mike for some of these jobs. Simply insert the gauge in the hole to be measured and lock the plungers after they have contacted the walls. Remove the tool and measure across the plungers with an outside micrometer.

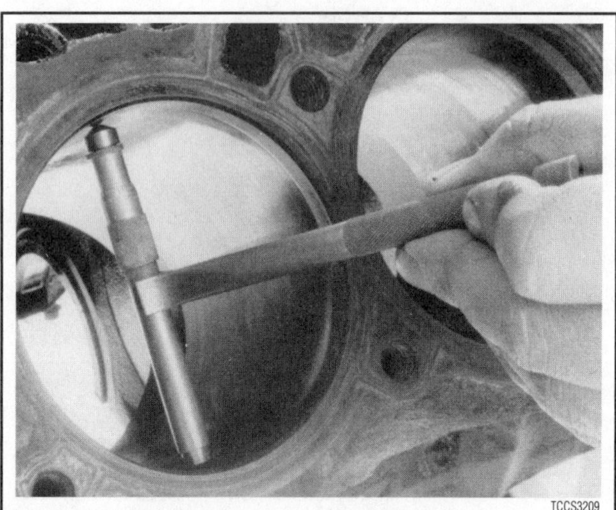

TCCS3209

Fig. 39 Telescoping gauges are used during powerhead rebuilding procedures to measure the inside diameter of bores

DEPTH GAUGES

See Figure 40

A depth gauge can be inserted into a bore or other small hole to determine exactly how deep it is. One common use for a depth gauge is measuring the distance the piston sits below the deck of the block at top dead center. Some outside calipers contain a built-in depth gauge so money can be saved by just buying one tool.

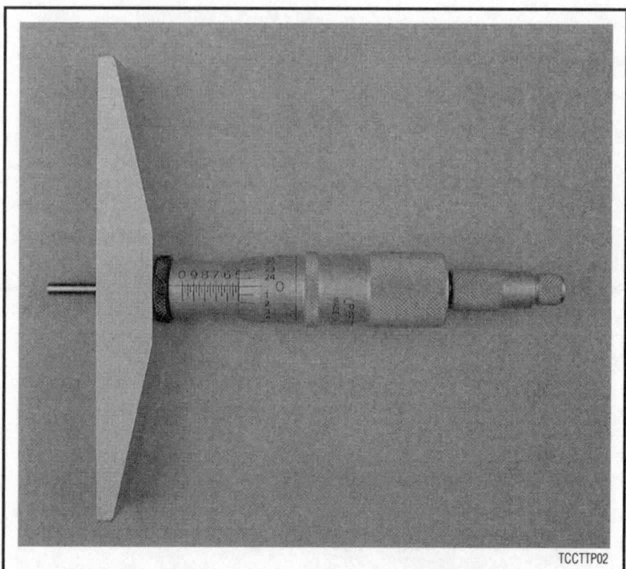

TCCTTP02

Fig. 40 Depth gauges are used to measure the depth of bore or other small holes

FASTENERS, MEASUREMENTS AND CONVERSIONS

Bolts, Nuts and Other Threaded Retainers

See Figures 41, 42, 43, 44 and 45

Although there are a great variety of fasteners found in the modern boat engine, the most commonly used retainer is the threaded fastener (nuts, bolts, screws, studs, etc). Most threaded retainers may be reused, provided that they are not damaged in use or during the repair.

Some retainers (such as stretch bolts or torque prevailing nuts) are designed to deform when tightened or in use and should not be reused.

Whenever possible, we will note any special retainers which should be replaced during a procedure. But you should always inspect the condition of a retainer when it is removed and you should replace any that show signs of damage. Check all threads for rust or corrosion which can increase the torque necessary to achieve the desired clamp load for which that fastener was originally selected. Additionally, be

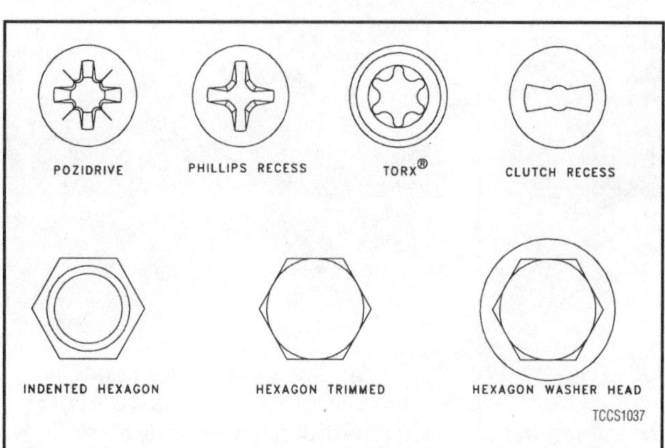

Fig. 41 Here are a few of the most common screw/bolt driver styles

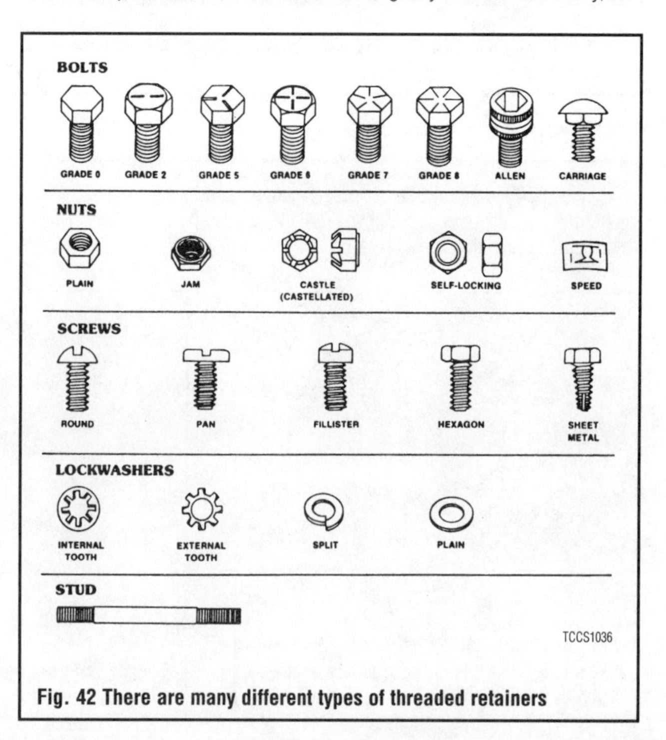

Fig. 42 There are many different types of threaded retainers

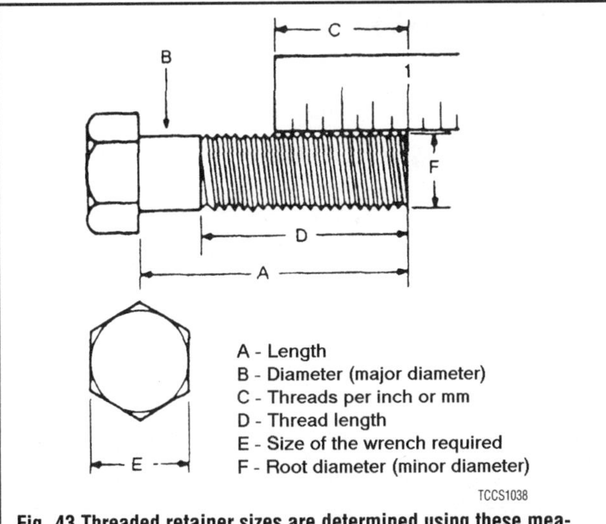

Fig. 43 Threaded retainer sizes are determined using these measurements

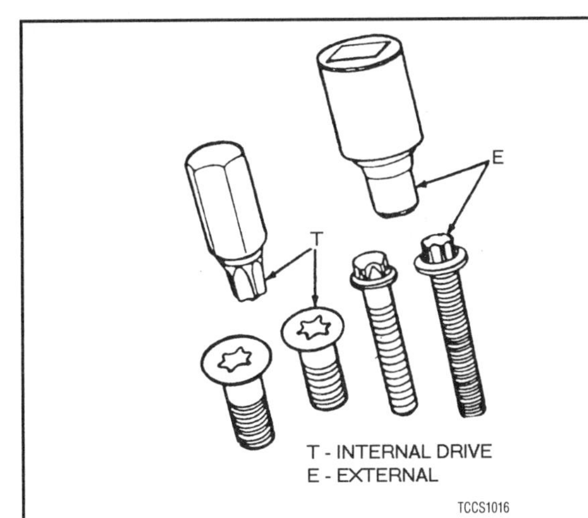

Fig. 44 Special fasteners such as these Torx® head bolts are used by manufacturers to discourage people from working on vehicles without the proper tools (and knowledge)

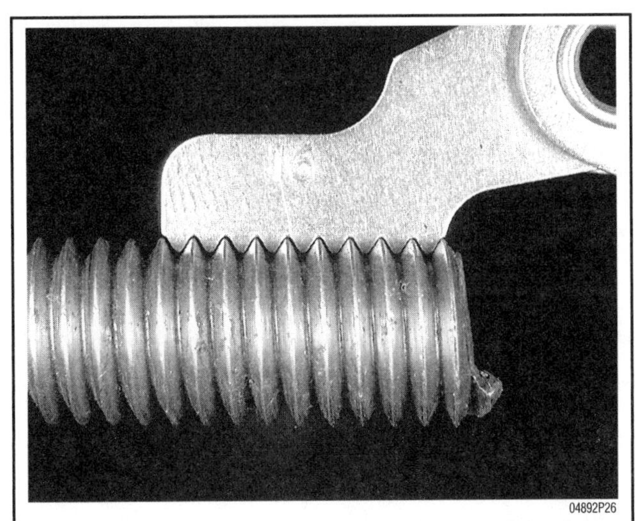

Fig. 45 Thread gauges measure the threads-per-inch and the pitch of a bolt or stud's threads

sure that the driver sur-face of the fastener has not been compromised by rounding or other damage. In some cases a driver surface may become only partially rounded, allowing the driver to catch in only one direction. In many of these occurrences, a fastener may be installed and tightened, but the driver would not be able to grip and loosen the fastener again. (This could lead to frustration down the line should that component ever need to be disassembled again).

If you must replace a fastener, whether due to design or damage, you must always be sure to use the proper replacement. In all cases, a retainer of the same design, material and strength should be used. Markings on the heads of most bolts will help determine the proper strength of the fastener. The same material, thread and pitch must be selected to assure proper installation and safe operation of the vehicle afterwards.

Thread gauges are available to help measure a bolt or stud's thread. Most part or hardware stores keep gauges available to help you select the proper size. In a pinch, you can use another nut or bolt for a thread gauge. If the bolt you are replacing is not too badly damaged, you can select a match by finding another bolt which will thread in its place. If you find a nut which threads properly onto the damaged bolt, then use that nut to help select the replacement bolt. If however, the bolt you are replacing is so badly damaged (broken or drilled out) that its threads cannot be used as a gauge, you might start by looking for another bolt (from the same assembly or a similar location) which will thread into the damaged bolt's mounting. If so, the other bolt can be used to select a nut; the nut can then be used to select the replacement bolt.

In all cases, be absolutely sure you have selected the proper replacement. Don't be shy, you can always ask the store clerk for help.

**** WARNING**

Be aware that when you find a bolt with damaged threads, you may also find the nut or drilled hole it was threaded into has also been damaged. If this is the case, you may have to drill and tap the hole, replace the nut or otherwise repair the threads. NEVER try to force a replacement bolt to fit into the damaged threads.

Torque

Torque is defined as the measurement of resistance to turning or rotating. It tends to twist a body about an axis of rotation. A common example of this would be tightening a threaded retainer such as a nut, bolt or screw. Measuring torque is one of the most common ways to help assure that a threaded retainer has been properly fastened.

When tightening a threaded fastener, torque is applied in three distinct areas, the head, the bearing surface and the clamp load. About 50 percent of the measured torque is used in overcoming bearing friction. This is the friction between the bearing surface of the bolt head, screw head or nut face and the base material or washer (the surface on which the fastener is rotating). Approximately 40 percent of the applied torque is used in overcoming thread friction. This leaves only about 10 percent of the applied torque to develop a useful clamp load (the force which holds a joint together). This means that friction can account for as much as 90 percent of the applied torque on a fastener.

Standard and Metric Measurements

Specifications are often used to help you determine the condition of various components, or to assist you in their installation. Some of the most common measurements include length (in. or cm/mm), torque (ft. lbs., inch lbs. or Nm) and pressure (psi, in. Hg, kPa or mm Hg).

In some cases, that value may not be conveniently measured with what is available in your toolbox. Luckily, many of the measuring devices which are available today will have two scales so Standard or Metric measurements may easily be taken. If any of the various measuring tools which are available to you do not contain the same scale as listed in your specifications, use the accompanying conversion factors to determine the proper value.

The conversion factor chart is used by taking the given specification and multiplying it by the necessary conversion factor. For instance, looking at the first line, if you have a measurement in inches such as "free-play should be 2 in." but your ruler reads only in millimeters, multiply 2 in. by the conversion factor of 25.4 to get the metric equivalent of 50.8mm. Likewise, if the specification was given only in a Metric measurement, for example in Newton Meters (Nm), then look at the center column first. If the measurement is 100 Nm, multiply it by the conversion factor of 0.738 to get 73.8 ft. lbs.

CONVERSION FACTORS

LENGTH–DISTANCE

Inches (in.)	x 25.4	= Millimeters (mm)	x .0394	= Inches
Feet (ft.)	x .305	= Meters (m)	x 3.281	= Feet
Miles	x 1.609	= Kilometers (km)	x .0621	= Miles

VOLUME

Cubic Inches (in3)	x 16.387	= Cubic Centimeters	x .061	= in3
IMP Pints (IMP pt.)	x .568	= Liters (L)	x 1.76	= IMP pt.
IMP Quarts (IMP qt.)	x 1.137	= Liters (L)	x .88	= IMP qt.
IMP Gallons (IMP gal.)	x 4.546	= Liters (L)	x .22	= IMP gal.
IMP Quarts (IMP qt.)	x 1.201	= US Quarts (US qt.)	x .833	= IMP qt.
IMP Gallons (IMP gal.)	x 1.201	= US Gallons (US gal.)	x .833	= IMP gal.
Fl. Ounces	x 29.573	= Milliliters	x .034	= Ounces
US Pints (US pt.)	x .473	= Liters (L)	x 2.113	= Pints
US Quarts (US qt.)	x .946	= Liters (L)	x 1.057	= Quarts
US Gallons (US gal.)	x 3.785	= Liters (L)	x .264	= Gallons

MASS–WEIGHT

Ounces (oz.)	x 28.35	= Grams (g)	x .035	= Ounces
Pounds (lb.)	x .454	= Kilograms (kg)	x 2.205	= Pounds

PRESSURE

Pounds Per Sq. In. (psi)	x 6.895	= Kilopascals (kPa)	x .145	= psi
Inches of Mercury (Hg)	x .4912	= psi	x 2.036	= Hg
Inches of Mercury (Hg)	x 3.377	= Kilopascals (kPa)	x .2961	= Hg
Inches of Water (H_2O)	x .07355	= Inches of Mercury	x 13.783	= H_2O
Inches of Water (H_2O)	x .03613	= psi	x 27.684	= H_2O
Inches of Water (H_2O)	x .248	= Kilopascals (kPa)	x 4.026	= H_2O

TORQUE

Pounds–Force Inches (in–lb)	x .113	= Newton Meters (N·m)	x 8.85	= in–lb
Pounds–Force Feet (ft–lb)	x 1.356	= Newton Meters (N·m)	x .738	= ft–lb

VELOCITY

Miles Per Hour (MPH)	x 1.609	= Kilometers Per Hour (KPH)	x .621	= MPH

POWER

Horsepower (Hp)	x .745	= Kilowatts	x 1.34	= Horsepower

FUEL CONSUMPTION*

Miles Per Gallon IMP (MPG)	x .354	= Kilometers Per Liter (Km/L)
Kilometers Per Liter (Km/L)	x 2.352	= IMP MPG
Miles Per Gallon US (MPG)	x .425	= Kilometers Per Liter (Km/L)
Kilometers Per Liter (Km/L)	x 2.352	= US MPG

*It is common to covert from miles per gallon (mpg) to liters/100 kilometers (1/100 km), where mpg (IMP) x 1/100 km = 282 and mpg (US) x 1/100 km = 235.

TEMPERATURE

Degree Fahrenheit (°F)	= (°C x 1.8) + 32
Degree Celsius (°C)	= (°F – 32) x .56

TCCS1044

Metric Bolts

Relative Strength Marking	4.6, 4.8			8.8		
Bolt Markings	4.6 / 4.8			8.8		
Usage	Frequent			Infrequent		
Bolt Size	Maximum Torque			Maximum Torque		
Thread Size x Pitch (mm)	Ft-Lb	Kgm	Nm	Ft-Lb	Kgm	Nm
6 x 1.0	2–3	.2–.4	3–4	3–6	.4–.8	5–8
8 x 1.25	6–8	.8–1	8–12	9–14	1.2–1.9	13–19
10 x 1.25	12–17	1.5–2.3	16–23	20–29	2.7–4.0	27–39
12 x 1.25	21–32	2.9–4.4	29–43	35–53	4.8–7.3	47–72
14 x 1.5	35–52	4.8–7.1	48–70	57–85	7.8–11.7	77–110
16 x 1.5	51–77	7.0–10.6	67–100	90–120	12.4–16.5	130–160
18 x 1.5	74–110	10.2–15.1	100–150	130–170	17.9–23.4	180–230
20 x 1.5	110–140	15.1–19.3	150–190	190–240	26.2–46.9	160–320
22 x 1.5	150–190	22.0–26.2	200–260	250–320	34.5–44.1	340–430
24 x 1.5	190–240	26.2–46.9	260–320	310–410	42.7–56.5	420–550

88523G12

SAE Bolts

SAE Grade Number	1 or 2			5			6 or 7		
Bolt Markings Manufacturers' marks may vary—number of lines always two less than the grade number.									
Usage	Frequent			Frequent			Infrequent		
Bolt Size (inches)—(Thread)	Maximum Torque			Maximum Torque			Maximum Torque		
	Ft-Lb	kgm	Nm	Ft-Lb	kgm	Nm	Ft-Lb	kgm	Nm
$^1/_4$—20	5	0.7	6.8	8	1.1	10.8	10	1.4	13.5
—28	6	0.8	8.1	10	1.4	13.6			
$^5/_{16}$—18	11	1.5	14.9	17	2.3	23.0	19	2.6	25.8
—24	13	1.8	17.6	19	2.6	25.7			
$^3/_8$—16	18	2.5	24.4	31	4.3	42.0	34	4.7	46.0
—24	20	2.75	27.1	35	4.8	47.5			
$^7/_{16}$—14	28	3.8	37.0	49	6.8	66.4	55	7.6	74.5
—20	30	4.2	40.7	55	7.6	74.5			
$^1/_2$—13	39	5.4	52.8	75	10.4	101.7	85	11.75	115.2
—20	41	5.7	55.6	85	11.7	115.2			
$^9/_{16}$—12	51	7.0	69.2	110	15.2	149.1	120	16.6	162.7
—18	55	7.6	74.5	120	16.6	162.7			
$^5/_8$—11	83	11.5	112.5	150	20.7	203.3	167	23.0	226.5
—18	95	13.1	128.8	170	23.5	230.5			
$^3/_4$—10	105	14.5	142.3	270	37.3	366.0	280	38.7	379.6
—16	115	15.9	155.9	295	40.8	400.0			
$^7/_8$— 9	160	22.1	216.9	395	54.6	535.5	440	60.9	596.5
—14	175	24.2	237.2	435	60.1	589.7			
1— 8	236	32.5	318.6	590	81.6	799.9	660	91.3	894.8
—14	250	34.6	338.9	660	91.3	849.8			

88523G10

ENGINE MAINTENANCE 3-2
SERIAL NUMBER IDENTIFICATION 3-2
ENGINE COVER 3-7
REMOVAL & INSTALLATION 3-7
ENGINE OIL 3-7
ENGINE OIL RECOMMENDATIONS 3-8
OIL LEVEL CHECK 3-8
OIL & FILTER CHANGE 3-10
LOWER UNIT 3-12
DRAINING AND FILLING 3-12
AIR CLEANER 3-13
FUEL FILTER 3-13
RELIEVING FUEL SYSTEM PRESSURE 3-14
REMOVAL & INSTALLATION 3-14
FUEL/WATER SEPARATOR 3-15
TIMING BELT 3-15
INSPECTION 3-15
ADJUSTMENT 3-16
TRIM, TILT & PIVOT POINTS 3-19
INSPECTION AND LUBRICATION 3-19
PROPELLER 3-19
BOAT MAINTENANCE 3-21
INSIDE THE BOAT 3-21
FIBERGLASS HULLS 3-21
TRIM TABS, ANODES AND LEAD WIRES 3-22
BATTERY 3-23
CLEANING 3-23
CHECKING SPECIFIC GRAVITY 3-23
BATTERY TERMINALS 3-24
BATTERY AND CHARGING SAFETY PRECAUTIONS 3-24
BATTERY CHARGERS 3-25
REPLACING BATTERY CABLES 3-25
TUNE-UP 3-25
INTRODUCTION 3-25
TUNE-UP SEQUENCE 3-25
COMPRESSION CHECK 3-26
CHECKING COMPRESSION 3-26
LOW COMPRESSION 3-26
SPARK PLUGS 3-27
SPARK PLUG HEAT RANGE 3-27
SPARK PLUG SERVICE 3-27
REMOVAL & INSTALLATION 3-27
READING SPARK PLUGS 3-28
INSPECTION & GAPPING 3-28
SPARK PLUG WIRES 3-30
TESTING 3-30
CHECKING RESISTANCE 3-30
REMOVAL & INSTALLATION 3-30
VALVE CLEARANCE 3-30
ADJUSTMENT 3-30
IGNITION SYSTEM 3-36
TIMING AND SYNCHRONIZATION 3-36
TIMING 3-36
SYNCHRONIZATION 3-36
PREPARATION 3-36
IDLE SPEED ADJUSTMENT 3-37
CARBURETOR SYNCHRONIZATION 3-37
WINTER STORAGE CHECKLIST 3-38
SPRING COMMISSIONING CHECKLIST 3-38
SPECIFICATIONS CHARTS
GENERAL ENGINE SPECIFICATIONS 3-3
MAINTENANCE INTERVALS 3-6
CAPACITIES 3-11
TUNEUP SPECIFICATIONS 3-26

3 MAINTENANCE AND TUNE-UP

ENGINE MAINTENANCE 3-2
BOAT MAINTENANCE 3-21
TUNE-UP 3-25
WINTER STORAGE CHECKLIST 3-38
SPRING COMMISSIONING CHECKLIST 3-38

ENGINE MAINTENANCE

➧ See Figures 1 and 2

Most engine repair work can be directly or indirectly attributed to lack of proper care for the engine. This is especially true of care during the off-season period. There is no way a mechanical engine, particularly an outboard engine, can be left sitting idle for an extended period of time, say for six months, and then be ready for instant satisfactory service.

Imagine if you will, leaving your car for six months, and then expecting to turn the key, have it roar to life, and be able to drive off in the same manner as a daily occurrence.

It is critical for an outboard engine to be run at least once a month, preferably, in the water, but if this is not possible, then a flush attachment must be connected to the lower unit to allow the engine to be operated while it is out of the water.

Water must circulate through the lower unit to the powerhead anytime the engine is operating to prevent damage to the water pump in the lower unit. Running the engine without water for only five seconds will damage the water pump impeller.

** CAUTION

Never operate the engine at high rpm with a flush device attached. Without a load on the propeller, the engine can be seriously damaged.

04893P42

Fig. 1 One of the most basic but important maintenance items on a four stroke is checking the engine oil level

04893P71

Fig. 2 Unlike two-stroke outboards, Honda uses an automotive-style spin-on oil filter on its four-stroke outboards

At the same time, the shift mechanism should be operated through the full range several times and the steering operated from hard-over to hard-over.

Only through a regular maintenance program can the owner expect to receive long life and satisfactory performance at minimum cost.

Serial Number Identification

➧ See Figures 3 thru 17

The frame serial number and the engine serial number are the manufacturer's key to engine changes. These numbers usually identify the year of manufacture, the qualified horsepower rating and other important items necessary when researching components in the parts book. If any correspondence or parts are

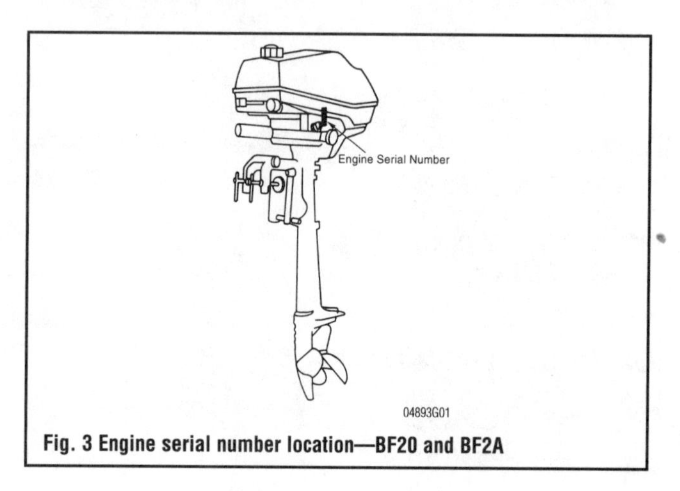

04893G01

Fig. 3 Engine serial number location—BF20 and BF2A

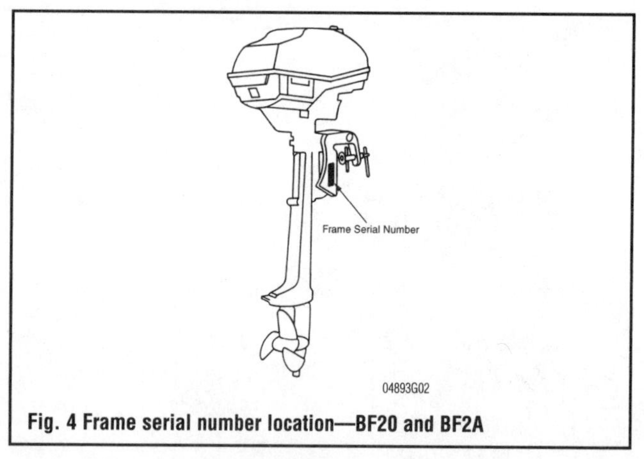

04893G02

Fig. 4 Frame serial number location—BF20 and BF2A

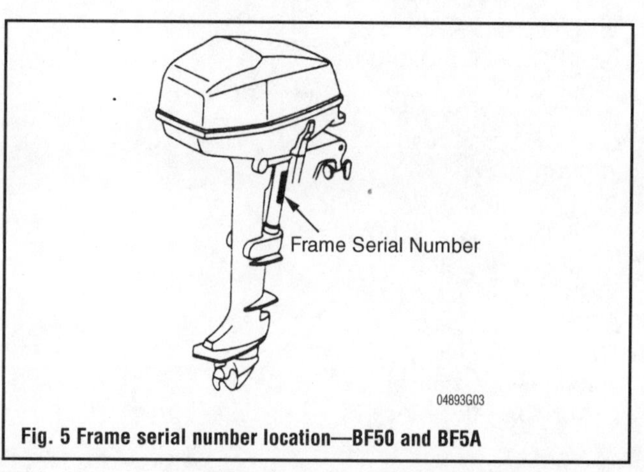

04893G03

Fig. 5 Frame serial number location—BF50 and BF5A

General Engine Specifications

Year	Model	Horsepower	Engine Type	Displace cu.in. (cc)	Bore and Stroke	Compression Ratio	Fuel System	Ignition System	Starting System	Cooling System
1999-	BF130A	130	4cyl SOHC	137 (2254)	3.4 x 3.8	N/A	Fuel Injection	Capacitor Discharge (CDI)	Electric	Impeller Pump / Thermostat Controlled
2001	BF115A	115	4cyl SOHC	137 (2254)	3.4 x 3.8	N/A	Fuel Injection	Capacitor Discharge (CDI)	Electric	Impeller Pump / Thermostat Controlled
	BF90A	90	4cyl SOHC	97 (1590)	3.0 x 3.5	8.8 to 1	Four Carburetors	Capacitor Discharge (CDI)	Electric	Impeller Pump / Thermostat Controlled
	BF75A	75	4cyl SOHC	97 (1590)	3.0 x 3.5	8.8 to 1	Four Carburetors	Capacitor Discharge (CDI)	Electric	Impeller Pump / Thermostat Controlled
	BF50A	50	3cyl SOHC	49.3 (808)	2.8 x 2.8	9.2 to 1	Three Carburetors	Capacitor Discharge (CDI)	Electric	Impeller Pump / Thermostat Controlled
	BF40A	40	3cyl SOHC	49.3 (808)	2.8 x 2.8	9.2 to 1	Three Carburetors	Capacitor Discharge (CDI)	Electric	Impeller Pump / Thermostat Controlled
	BF30A	30	3cyl SOHC	30.5 (499)	2.3 x 2.5	9.2 to 1	Three Carburetors	Capacitor Discharge (CDI)	Electric / Recoil	Impeller Pump / Thermostat Controlled
	BF25A	25	3cyl SOHC	30.5 (499)	2.3 x 2.5	9.2 to 1	Three Carburetors	Capacitor Discharge (CDI)	Electric / Recoil	Impeller Pump / Thermostat Controlled
	BF15A/BF15D	15	2cyl SOHC	17.1 (280)	2.3 x 2.1	8.6 to 1	One Carburetor	Capacitor Discharge (CDI)	Electric / Recoil	Impeller Pump / Thermostat Controlled
	BF9.9A/BF9.9D	9.9	2cyl SOHC	17.1 (280)	2.3 x 2.1	8.6 to 1	One Carburetor	Capacitor Discharge (CDI)	Electric / Recoil	Impeller Pump / Thermostat Controlled
	BF8A/BF8D	8	2cyl SOHC	12 (197)	2.2 x 1.6	8.6 to 1	One Carburetor	Capacitor Discharge (CDI)	Recoil	Impeller Pump / Thermostat Controlled
	BF5A/BF5D	5	1cyl OHV	7.8 (127)	2.4 x 1.8	8.7 to 1	One Carburetor	Transistorized Pointless	Recoil	Impeller Pump / Thermostat Controlled
	BF2A/BF2D	2	1cyl OHV	3.4 (57)	1.8 x 1.4	6.5 to 1 / 8.0:1	One Carburetor	Transistorized Pointless	Recoil	Forced Air
1998	BF130A	130	4cyl SOHC	137 (2254)	3.4 x 3.8	N/A	Fuel Injection	Capacitor Discharge (CDI)	Electric	Impeller Pump / Thermostat Controlled
	BF115A	115	4cyl SOHC	137 (2254)	3.4 x 3.8	N/A	Fuel Injection	Capacitor Discharge (CDI)	Electric	Impeller Pump / Thermostat Controlled
	BF90A	90	4cyl SOHC	97 (1590)	3.0 x 3.5	8.8 to 1	Four Carburetors	Capacitor Discharge (CDI)	Electric	Impeller Pump / Thermostat Controlled
	BF75A	75	4cyl SOHC	97 (1590)	3.0 x 3.5	8.8 to 1	Four Carburetors	Capacitor Discharge (CDI)	Electric	Impeller Pump / Thermostat Controlled
	BF50A	50	3cyl SOHC	49.3 (808)	2.8 x 2.8	9.2 to 1	Three Carburetors	Capacitor Discharge (CDI)	Electric	Impeller Pump / Thermostat Controlled
	BF40A	40	3cyl SOHC	49.3 (808)	2.8 x 2.8	9.2 to 1	Three Carburetors	Capacitor Discharge (CDI)	Electric	Impeller Pump / Thermostat Controlled
	BF30A	30	3cyl SOHC	30.5 (499)	2.3 x 2.5	9.2 to 1	Three Carburetors	Capacitor Discharge (CDI)	Electric / Recoil	Impeller Pump / Thermostat Controlled
	BF25A	25	3cyl SOHC	30.5 (499)	2.3 x 2.5	9.2 to 1	Three Carburetors	Capacitor Discharge (CDI)	Electric / Recoil	Impeller Pump / Thermostat Controlled
	BF15A	15	2cyl SOHC	17.1 (280)	2.3 x 2.1	8.6 to 1	One Carburetor	Capacitor Discharge (CDI)	Electric / Recoil	Impeller Pump / Thermostat Controlled
	BF9.9A	9.9	2cyl SOHC	17.1 (280)	2.3 x 2.1	8.6 to 1	One Carburetor	Capacitor Discharge (CDI)	Electric / Recoil	Impeller Pump / Thermostat Controlled
	BF8A	8	2cyl SOHC	12 (197)	2.2 x 1.6	8.6 to 1	One Carburetor	Capacitor Discharge (CDI)	Recoil	Impeller Pump / Thermostat Controlled
	BF5A	5	1cyl OHV	7.8 (127)	2.4 x 1.8	8.7 to 1	One Carburetor	Transistorized Pointless	Recoil	Impeller Pump / Thermostat Controlled
	BF2A	2	1cyl OHV	3.4 (57)	1.8 x 1.4	6.5 to 1	One Carburetor	Transistorized Pointless	Recoil	Forced Air
1997	BF90A	90	4cyl SOHC	97 (1590)	3.0 x 3.5	8.8 to 1	Four Carburetors	Capacitor Discharge (CDI)	Electric	Impeller Pump / Thermostat Controlled
	BF75A	75	4cyl SOHC	97 (1590)	3.0 x 3.5	8.8 to 1	Four Carburetors	Capacitor Discharge (CDI)	Electric	Impeller Pump / Thermostat Controlled
	BF50A	50	3cyl SOHC	49.3 (808)	2.8 x 2.8	9.2 to 1	Three Carburetors	Capacitor Discharge (CDI)	Electric	Impeller Pump / Thermostat Controlled
	BF40A	40	3cyl SOHC	49.3 (808)	2.8 x 2.8	9.2 to 1	Three Carburetors	Capacitor Discharge (CDI)	Electric	Impeller Pump / Thermostat Controlled
	BF30A	30	3cyl SOHC	30.5 (499)	2.3 x 2.5	9.2 to 1	Three Carburetors	Capacitor Discharge (CDI)	Electric / Recoil	Impeller Pump / Thermostat Controlled
	BF25A	25	3cyl SOHC	30.5 (499)	2.3 x 2.5	9.2 to 1	Three Carburetors	Capacitor Discharge (CDI)	Electric / Recoil	Impeller Pump / Thermostat Controlled
	BF15A	15	2cyl SOHC	17.1 (280)	2.3 x 2.1	8.6 to 1	One Carburetor	Capacitor Discharge (CDI)	Electric / Recoil	Impeller Pump / Thermostat Controlled
	BF9.9A	9.9	2cyl SOHC	17.1 (280)	2.3 x 2.1	8.6 to 1	One Carburetor	Capacitor Discharge (CDI)	Electric / Recoil	Impeller Pump / Thermostat Controlled
	BF8A	8	2cyl SOHC	12 (197)	2.2 x 1.6	8.6 to 1	One Carburetor	Capacitor Discharge (CDI)	Recoil	Impeller Pump / Thermostat Controlled
	BF5A	5	1cyl OHV	7.8 (127)	2.4 x 1.8	8.7 to 1	One Carburetor	Transistorized Pointless	Recoil	Impeller Pump / Thermostat Controlled
	BF2A	2	1cyl OHV	4.6 (76)	1.8 x 1.8	6.5 to 1	One Carburetor	Transistorized Pointless	Recoil	Forced Air
1996	BF90A	90	4cyl SOHC	97 (1590)	3.0 x 3.5	8.8 to 1	Four Carburetors	Capacitor Discharge (CDI)	Electric	Impeller Pump / Thermostat Controlled
	BF75A	75	4cyl SOHC	97 (1590)	3.0 x 3.5	8.8 to 1	Four Carburetors	Capacitor Discharge (CDI)	Electric	Impeller Pump / Thermostat Controlled
	BF50A	50	3cyl SOHC	49.3 (808)	2.8 x 2.8	9.2 to 1	Three Carburetors	Capacitor Discharge (CDI)	Electric	Impeller Pump / Thermostat Controlled
	BF45A	45	3cyl SOHC	49.3 (808)	2.8 x 2.8	9.2 to 1	Three Carburetors	Capacitor Discharge (CDI)	Electric	Impeller Pump / Thermostat Controlled
	BF40A	40	3cyl SOHC	49.3 (808)	2.8 x 2.8	9.2 to 1	Three Carburetors	Capacitor Discharge (CDI)	Electric / Recoil	Impeller Pump / Thermostat Controlled
	BF35A	35	3cyl SOHC	30.5 (499)	2.3 x 2.5	9.2 to 1	Three Carburetors	Capacitor Discharge (CDI)	Electric / Recoil	Impeller Pump / Thermostat Controlled
	BF30A	30	3cyl SOHC	30.5 (499)	2.3 x 2.5	9.2 to 1	Three Carburetors	Capacitor Discharge (CDI)	Electric / Recoil	Impeller Pump / Thermostat Controlled
	BF25A	25	3cyl SOHC	30.5 (499)	2.3 x 2.5	9.2 to 1	Three Carburetors	Capacitor Discharge (CDI)	Electric / Recoil	Impeller Pump / Thermostat Controlled
	BF15A	15	2cyl SOHC	17.1 (280)	2.3 x 2.1	8.6 to 1	One Carburetor	Capacitor Discharge (CDI)	Electric / Recoil	Impeller Pump / Thermostat Controlled

General Engine Specifications

Year	Model	Horsepower	Engine Type	Displace cu.in. (cc)	Bore and Stroke	Compression Ratio	Fuel System	Ignition System	Starting System	Cooling System
1996	BF9.9A	9.9	2cyl SOHC	17.1 (280)	2.3 x 2.1	8.6 to 1	One Carburetor	Capacitor Discharge (CDI)	Electric / Recoil	Impeller Pump / Thermostat Controlled
	BF8A	8	2cyl SOHC	12 (197)	2.2 x 1.6	8.6 to 1	One Carburetor	Capacitor Discharge (CDI)	Recoil	Impeller Pump / Thermostat Controlled
	BF5A	5	1cyl OHV	7.75 (127)	2.4 x 1.8	8.7 to 1	One Carburetor	Transistorized Pointless	Recoil	Impeller Pump / Thermostat Controlled
	BF2A	2	1cyl OHV	4.6 (76)	1.8 x 1.8	6.5 to 1	One Carburetor	Transistorized Pointless	Recoil	Forced Air
1995	BF50A	50	3cyl SOHC	49.5 (808)	2.8 x 2.8	9.2 to 1	Three Carburetors	Capacitor Discharge (CDI)	Electric	Impeller Pump / Thermostat Controlled
	BF40A	40	3cyl SOHC	49.5 (808)	2.8 x 2.8	9.2 to 1	Three Carburetors	Capacitor Discharge (CDI)	Electric	Impeller Pump / Thermostat Controlled
	BF30A	30	3cyl SOHC	30.5 (499)	2.3 x 2.5	9.2 to 1	Three Carburetors	Capacitor Discharge (CDI)	Electric / Recoil	Impeller Pump / Thermostat Controlled
	BF25A	25	3cyl SOHC	30.5 (499)	2.3 x 2.5	9.2 to 1	Three Carburetors	Capacitor Discharge (CDI)	Electric / Recoil	Impeller Pump / Thermostat Controlled
	BF15A	15	2cyl SOHC	17.1 (280)	2.3 x 2.1	8.6 to 1	One Carburetor	Capacitor Discharge (CDI)	Electric / Recoil	Impeller Pump / Thermostat Controlled
	BF9.9A	9.9	2cyl SOHC	17.1 (280)	2.3 x 2.1	8.6 to 1	One Carburetor	Capacitor Discharge (CDI)	Electric / Recoil	Impeller Pump / Thermostat Controlled
	BF8A	8	2cyl SOHC	12 (197)	2.2 x 1.6	8.6 to 1	One Carburetor	Capacitor Discharge (CDI)	Recoil	Impeller Pump / Thermostat Controlled
	BF5A	5	1cyl OHV	7.8 (127)	2.4 x 1.8	8.7 to 1	One Carburetor	Transistorized Pointless	Recoil	Impeller Pump / Thermostat Controlled
	BF2A	2	1cyl OHV	4.6 (76)	1.8 x 1.8	6.5 to 1	One Carburetor	Transistorized Pointless	Recoil	Forced Air
1994	BF45A	45	3cyl SOHC	49.3 (808)	2.8 x 2.8	9.2 to 1	Three Carburetors	Capacitor Discharge (CDI)	Electric	Impeller Pump / Thermostat Controlled
	BF35A	35	3cyl SOHC	49.3 (808)	2.8 x 2.8	9.2 to 1	Three Carburetors	Capacitor Discharge (CDI)	Electric / Recoil	Impeller Pump / Thermostat Controlled
	BF25A	25	3cyl SOHC	30.5 (499)	2.3 x 2.5	9.2 to 1	Three Carburetors	Capacitor Discharge (CDI)	Electric / Recoil	Impeller Pump / Thermostat Controlled
	BF15A	15	2cyl SOHC	17.1 (280)	2.3 x 2.1	8.6 to 1	One Carburetor	Capacitor Discharge (CDI)	Electric / Recoil	Impeller Pump / Thermostat Controlled
	BF9.9A	9.9	2cyl SOHC	17.1 (280)	2.3 x 2.1	8.6 to 1	One Carburetor	Capacitor Discharge (CDI)	Electric / Recoil	Impeller Pump / Thermostat Controlled
	BF8A	8	2cyl SOHC	12 (197)	2.2 x 1.6	8.6 to 1	One Carburetor	Capacitor Discharge (CDI)	Recoil	Impeller Pump / Thermostat Controlled
	BF5A	5	1cyl OHV	7.75 (127)	2.36 x 1.8	8.7 to 1	One Carburetor	Transistorized Pointless	Recoil	Impeller Pump / Thermostat Controlled
	BF2A	2	1cyl OHV	4.6 (76)	1.8 x 1.8	6.5 to 1	One Carburetor	Transistorized Pointless	Recoil	Forced Air
1993	BF45A	45	3cyl SOHC	49.5 (808)	2.8 x 2.8	9.2 to 1	Three Carburetors	Capacitor Discharge (CDI)	Electric	Impeller Pump / Thermostat Controlled
	BF35A	35	3cyl SOHC	30.5 (499)	2.8 x 2.8	9.2 to 1	Three Carburetors	Capacitor Discharge (CDI)	Electric / Recoil	Impeller Pump / Thermostat Controlled
	BF15A	15	2cyl SOHC	17.1 (280)	2.3 x 2.1	8.6 to 1	One Carburetor	Capacitor Discharge (CDI)	Electric / Recoil	Impeller Pump / Thermostat Controlled
	BF9.9A	9.9	2cyl SOHC	17.1 (280)	2.3 x 2.1	8.6 to 1	One Carburetor	Capacitor Discharge (CDI)	Electric / Recoil	Impeller Pump / Thermostat Controlled
	BF8A	8	2cyl SOHC	12 (197)	2.2 x 1.6	8.6 to 1	One Carburetor	Capacitor Discharge (CDI)	Recoil	Impeller Pump / Thermostat Controlled
	BF5A	5	1cyl OHV	7.8 (127)	2.36 x 1.8	8.7 to 1	One Carburetor	Transistorized Pointless	Recoil	Impeller Pump / Thermostat Controlled
	BF2A	2	1cyl OHV	4.6 (76)	1.8 x 1.8	6.5 to 1	One Carburetor	Transistorized Pointless	Recoil	Forced Air
1992	BF45A	45	3cyl SOHC	49.3 (808)	2.8 x 2.8	9.2 to 1	Three Carburetors	Capacitor Discharge (CDI)	Electric	Impeller Pump / Thermostat Controlled
	BF35A	35	3cyl SOHC	49.3 (808)	2.8 x 2.8	9.2 to 1	Three Carburetors	Capacitor Discharge (CDI)	Electric / Recoil	Impeller Pump / Thermostat Controlled
	BF15A	15	2cyl SOHC	17.1 (280)	2.3 x 2.1	8.6 to 1	One Carburetor	Capacitor Discharge (CDI)	Electric / Recoil	Impeller Pump / Thermostat Controlled
	BF9.9A	9.9	2cyl SOHC	17.1 (280)	2.3 x 2.1	8.6 to 1	One Carburetor	Capacitor Discharge (CDI)	Electric / Recoil	Impeller Pump / Thermostat Controlled
	BF8A	8	2cyl SOHC	12 (197)	2.2 x 1.6	8.6 to 1	One Carburetor	Capacitor Discharge (CDI)	Recoil	Impeller Pump / Thermostat Controlled
	BF5A	5	1cyl OHV	7.8 (127)	2.36 x 1.8	8.7 to 1	One Carburetor	Transistorized Pointless	Recoil	Impeller Pump / Thermostat Controlled
	BF2A	2	1cyl OHV	4.6 (76)	1.8 x 1.8	6.5 to 1	One Carburetor	Transistorized Pointless	Recoil	Forced Air
1991	BF45A	45	3cyl SOHC	49.3 (808)	2.8 x 2.8	9.2 to 1	Three Carburetors	Capacitor Discharge (CDI)	Electric	Impeller Pump / Thermostat Controlled
	BF35A	35	3cyl SOHC	49.3 (808)	2.8 x 2.8	9.2 to 1	Three Carburetors	Capacitor Discharge (CDI)	Electric / Recoil	Impeller Pump / Thermostat Controlled
	BF15A	15	2cyl SOHC	17.1 (280)	2.3 x 2.1	8.6 to 1	One Carburetor	Capacitor Discharge (CDI)	Electric / Recoil	Impeller Pump / Thermostat Controlled
	BF9.9A	9.9	2cyl SOHC	17.1 (280)	2.3 x 2.1	8.6 to 1	One Carburetor	Capacitor Discharge (CDI)	Electric / Recoil	Impeller Pump / Thermostat Controlled
	BF8A	8	2cyl SOHC	12 (197)	2.2 x 1.6	8.6 to 1	One Carburetor	Capacitor Discharge (CDI)	Recoil	Impeller Pump / Thermostat Controlled
	BF5A	5	1cyl OHV	7.75(127)	2.36 x 1.8	8.7 to 1	One Carburetor	Transistorized Pointless	Recoil	Impeller Pump / Thermostat Controlled
	BF2A	2	1cyl OHV	4.6 (76)	1.8 x 1.8	6.5 to 1	One Carburetor	Transistorized Pointless	Recoil	Forced Air
1990	BF15A	15	2cyl SOHC	17.1 (280)	2.3 x 2.1	8.6 to 1	One Carburetor	Capacitor Discharge (CDI)	Electric / Recoil	Impeller Pump / Thermostat Controlled
	BF9.9A	9.9	2cyl SOHC	17.1 (280)	2.3 x 2.1	8.6 to 1	One Carburetor	Capacitor Discharge (CDI)	Electric / Recoil	Impeller Pump / Thermostat Controlled
	BF8A	8	2cyl SOHC	12 (197)	2.2 x 1.6	8.6 to 1	One Carburetor	Capacitor Discharge (CDI)	Recoil	Impeller Pump / Thermostat Controlled
	BF5A	5	1cyl OHV	7.8 (127)	2.4 x 1.8	8.7 to 1	One Carburetor	Transistorized Pointless	Recoil	Impeller Pump / Thermostat Controlled
	BF2A	2	1cyl OHV	4.6 (76)	1.8 x 1.8	6.5 to 1	One Carburetor	Transistorized Pointless	Recoil	Forced Air

04893C05

General Engine Specifications

Year	Model	Horsepower	Engine Type	Displace cu.in. (cc)	Bore and Stroke	Compression Ratio	Fuel System	Ignition System	Starting System	Cooling System
1989	BF15A	15	2cyl SOHC	17.1 (280)	2.3 x 2.1	8.6 to 1	One Carburetor	Capacitor Discharge (CDI)	Electric / Recoil	Impeller Pump / Thermostat Controlled
	BF9.9A	9.9	2cyl SOHC	12 (197)	2.2 x 1.57	8.6 to 1	One Carburetor	Capacitor Discharge (CDI)	Electric / Recoil	Impeller Pump / Thermostat Controlled
	BF8A	8	2cyl SOHC	12 (197)	2.2 x 1.57	8.6 to 1	One Carburetor	Capacitor Discharge (CDI)	Recoil	Impeller Pump / Thermostat Controlled
	BF50/BF5A	5	1cyl OHV	7.75 (127)	2.36 x 1.77	8.7 to 1	One Carburetor	Transistorized Pointless	Recoil	Impeller Pump / Thermostat Controlled
	BF20/BF2A	2	1cyl OHV	4.6 (76)	1.8 x 1.8	6.5 to 1	One Carburetor	Transistorized Pointless	Recoil	Forced Air
1988	BF15	15	2cyl SOHC	17.1 (280)	2.3 x 2.1	8.6 to 1	One Carburetor	Capacitor Discharge (CDI)	Electric / Recoil	Impeller Pump / Thermostat Controlled
	BF100	9.9	2cyl SOHC	12 (197)	2.2 x 1.57	8.6 to 1	One Carburetor	Capacitor Discharge (CDI)	Electric / Recoil	Impeller Pump / Thermostat Controlled
	BF80	8	2cyl SOHC	12 (197)	2.2 x 1.57	8.6 to 1	One Carburetor	Capacitor Discharge (CDI)	Recoil	Impeller Pump / Thermostat Controlled
	BF50	5	1cyl OHV	7.75 (127)	2.36 x 1.77	8.7 to 1	One Carburetor	Transistorized Pointless	Recoil	Impeller Pump / Thermostat Controlled
	BF20	2	1cyl OHV	4.6 (76)	1.8 x 1.8	6.5 to 1	One Carburetor	Transistorized Pointless	Recoil	Forced Air
1987	BF100	9.9	2cyl SOHC	12 (197)	2.2 x 1.57	8.6 to 1	One Carburetor	Transistorized Pointless	Recoil	Impeller Pump / Thermostat Controlled
	BF75	7.5	2cyl SOHC	12 (197)	2.2 x 1.57	8.6 to 1	One Carburetor	Transistorized Pointless	Recoil	Impeller Pump / Thermostat Controlled
	BF50	5	1cyl OHV	7.75 (127)	2.36 x 1.77	8.7 to 1	One Carburetor	Transistorized Pointless	Recoil	Impeller Pump / Thermostat Controlled
	BF20	2	1cyl OHV	4.6 (76)	1.8 x 1.8	6.5 to 1	One Carburetor	Transistorized Pointless	Recoil	Forced Air
1986	BF100	9.9	2cyl SOHC	12 (197)	2.2 x 1.57	8.6 to 1	One Carburetor	Transistorized Pointless	Recoil	Impeller Pump / Thermostat Controlled
	BF75	7.5	2cyl SOHC	12 (197)	2.2 x 1.57	8.6 to 1	One Carburetor	Transistorized Pointless	Recoil	Impeller Pump / Thermostat Controlled
	BF50	5	1cyl OHV	7.75 (127)	2.36 x 1.77	8.7 to 1	One Carburetor	Transistorized Pointless	Recoil	Impeller Pump / Thermostat Controlled
	BF20	2	1cyl OHV	4.6 (76)	1.8 x 1.8	6.5 to 1	One Carburetor	Transistorized Pointless	Recoil	Forced Air
1985	BF100	9.9	2cyl SOHC	12 (197)	2.2 x 1.57	8.6 to 1	One Carburetor	Transistorized Pointless	Recoil	Impeller Pump / Thermostat Controlled
	BF75	7.5	2cyl SOHC	12 (197)	2.2 x 1.57	8.6 to 1	One Carburetor	Transistorized Pointless	Recoil	Impeller Pump / Thermostat Controlled
	BF20	2	1cyl OHV	4.6 (76)	1.8 x 1.8	6.5 to 1	One Carburetor	Transistorized Pointless	Recoil	Forced Air
1984	BF100	9.9	2cyl SOHC	12 (197)	2.2 x 1.57	8.6 to 1	One Carburetor	Transistorized Pointless	Recoil	Impeller Pump / Thermostat Controlled
	BF75	7.5	2cyl SOHC	12 (197)	2.2 x 1.57	8.6 to 1	One Carburetor	Transistorized Pointless	Recoil	Impeller Pump / Thermostat Controlled
	BF20	2	1cyl OHV	4.6 (76)	1.8 x 1.8	6.5 to 1	One Carburetor	Transistorized Pointless	Recoil	Forced Air
1983	BF100	9.9	2cyl SOHC	12 (197)	2.2 x 1.57	8.6 to 1	One Carburetor	Transistorized Pointless	Recoil	Impeller Pump / Thermostat Controlled
	BF75	7.5	2cyl SOHC	12 (197)	2.2 x 1.57	8.6 to 1	One Carburetor	Transistorized Pointless	Recoil	Impeller Pump / Thermostat Controlled
	BF20	2	1cyl OHV	4.6 (76)	1.8 x 1.8	6.5 to 1	One Carburetor	Transistorized Pointless	Recoil	Forced Air
1982	BF100	9.9	2cyl SOHC	12 (197)	2.2 x 1.57	8.6 to 1	One Carburetor	Transistorized Pointless	Recoil	Impeller Pump / Thermostat Controlled
	BF75	7.5	2cyl SOHC	12 (197)	2.2 x 1.57	8.6 to 1	One Carburetor	Transistorized Pointless	Recoil	Impeller Pump / Thermostat Controlled
	BF20	2	1cyl OHV	4.6 (76)	1.8 x 1.8	6.5 to 1	One Carburetor	Transistorized Pointless	Recoil	Forced Air
1981	BF100	9.9	2cyl SOHC	12 (197)	2.2 x 1.57	8.6 to 1	One Carburetor	Transistorized Pointless	Recoil	Impeller Pump / Thermostat Controlled
	BF75	7.5	2cyl SOHC	12 (197)	2.2 x 1.57	8.6 to 1	One Carburetor	Transistorized Pointless	Recoil	Impeller Pump / Thermostat Controlled
	BF20	2	1cyl OHV	4.6 (76)	1.8 x 1.8	6.5 to 1	One Carburetor	Transistorized Pointless	Recoil	Forced Air
1980	BF100	9.9	2cyl SOHC	12 (197)	2.2 x 1.57	8.6 to 1	One Carburetor	Transistorized Pointless	Recoil	Impeller Pump / Thermostat Controlled
	BF75	7.5	2cyl SOHC	12 (197)	2.2 x 1.57	8.6 to 1	One Carburetor	Transistorized Pointless	Recoil	Impeller Pump / Thermostat Controlled
	BF20	2	1cyl OHV	4.6 (76)	1.8 x 1.8	6.5 to 1	One Carburetor	Transistorized Pointless	Recoil	Forced Air
1979	BF100	9.9	2cyl SOHC	12 (197)	2.2 x 1.57	8.6 to 1	One Carburetor	Transistorized Pointless	Recoil	Impeller Pump / Thermostat Controlled
	BF75	7.5	2cyl SOHC	12 (197)	2.2 x 1.57	8.6 to 1	One Carburetor	Transistorized Pointless	Recoil	Impeller Pump / Thermostat Controlled
	BF20	2	1cyl OHV	4.6 (76)	1.8 x 1.8	6.5 to 1	One Carburetor	Transistorized Pointless	Recoil	Forced Air
1978	BF100	9.9	2cyl SOHC	12 (197)	2.2 x 1.57	8.6 to 1	One Carburetor	Transistorized Pointless	Recoil	Impeller Pump / Thermostat Controlled
	BF75	7.5	2cyl SOHC	12 (197)	2.2 x 1.57	8.6 to 1	One Carburetor	Transistorized Pointless	Recoil	Impeller Pump / Thermostat Controlled
	BF20	2	1cyl OHV	4.6 (76)	1.8 x 1.8	6.5 to 1	One Carburetor	Transistorized Pointless	Recoil	Forced Air

N/A - Not Available

04893C06

Maintenance Interval Chart

Component	Each	First 1mth/20hrs	Every 6mths/100hrs	Every 1year/200 hrs
Anode(s)	*Check*			*Check*
Bolts/Fasteners (all accessible)	*Check*			*Check*
Caburetor Linkage			*Check/Adjust*	
Combustion Chamber and Valves (BF 2A, BF20 only)	*Check/Replace Every 300 hrs*			
Engine Oil	*Check*	*Replace **	*Replace*	
Fuel Filter			*Check/Replace*	
Fuel Line				
All except BF25A, BF30A	*Check/Replace Every 2yr/400 hrs*			
BF25A, BF30A	*Check*			*Replace*
Fuel Tank and Strainer				*Check/Clean*
Lower Unit Oil		*Replace **	*Check*	*Replace*
Lubrication (perform more often for salt water use)		*Lubricate **	*Lubricate*	
Propeller and Retainer				
Tab Washer Retained	*Check*		*Replace*	
Cotter Pin Retained	*Check*		*Check/Replace*	
Shear Pin	*Check*		*Replace*	
Spark Plug				
All except BF80, BF8A BF9.9A, BF50, BF5A, BF75, BF100, BF15A		*Check/Adjust **		*Check/Adjust*
BF80, BF8A BF9.9A, BF50, BF5A, BF75, BF100, BF15A			*Check/Adjust*	
Starter Rope			*Check/Replace*	
Swivel Case Liner, Bushing and Water Seal (BF2D only)	*Replace Every 3 years*			
Thermostat				*Check/Replace*
Throttle linkage and idle speed		*Check/Adjust **	*Check/Adjust*	
Timing Belt				
BF25A, BF30A				*Replace*
BF35A, BF40A, BF45A, BF50A				*Replace*
BF75A, BF90A				*Check/Adjust*
BF115A, BF130A				*Check/Adjust*
Valve Clearance		*Check/Adjust **		*Check/Adjust*

** First service is performed at 10 hours for some 1999 and later models. Check your owner's manual for more details*

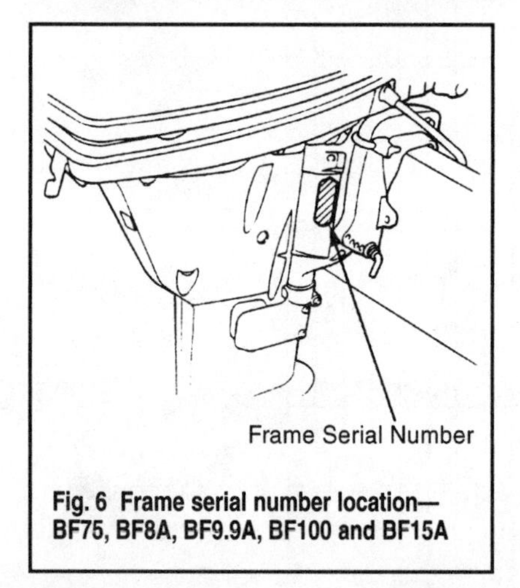

Fig. 6 Frame serial number location—BF75, BF8A, BF9.9A, BF100 and BF15A

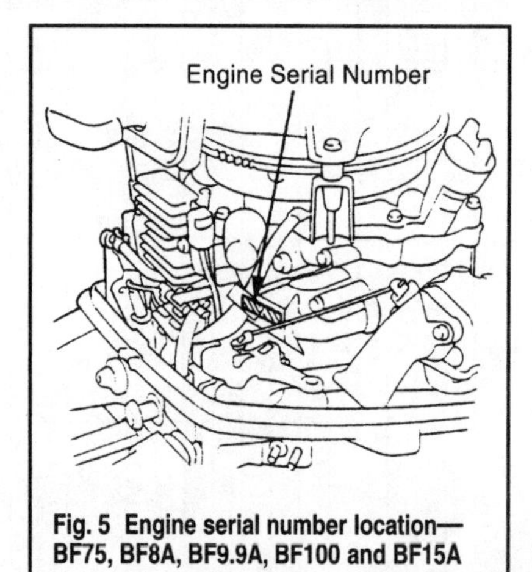

Fig. 5 Engine serial number location—BF75, BF8A, BF9.9A, BF100 and BF15A

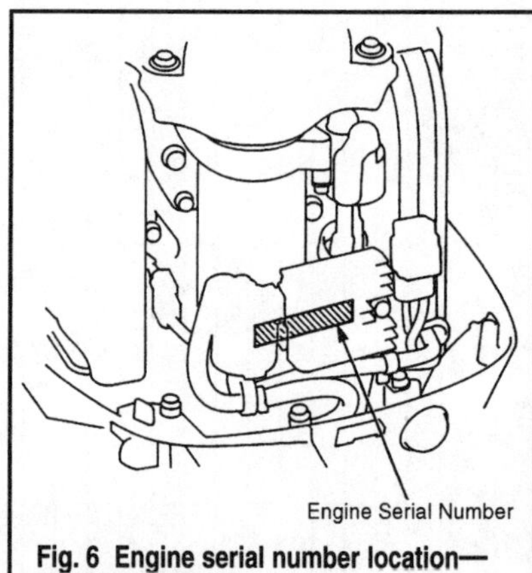

Fig. 6 Engine serial number location—BF25A and BF30A

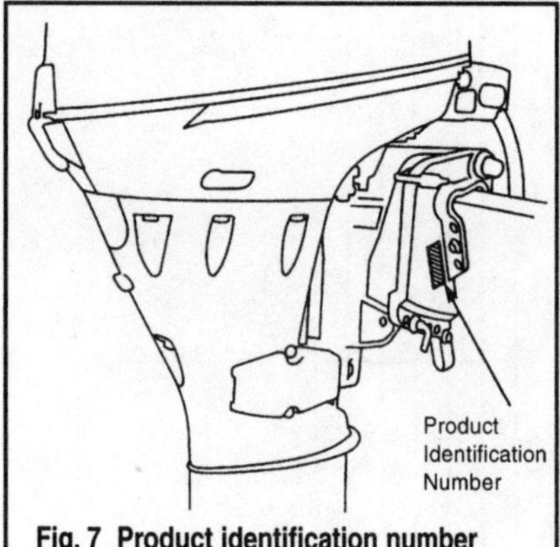

Fig. 7 Product identification number location—BF25A and BF30A

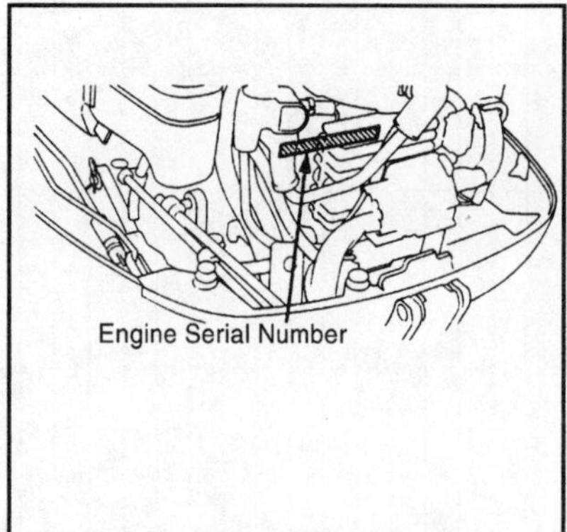

Fig. 8 Engine serial number location—BF35A and BF45A

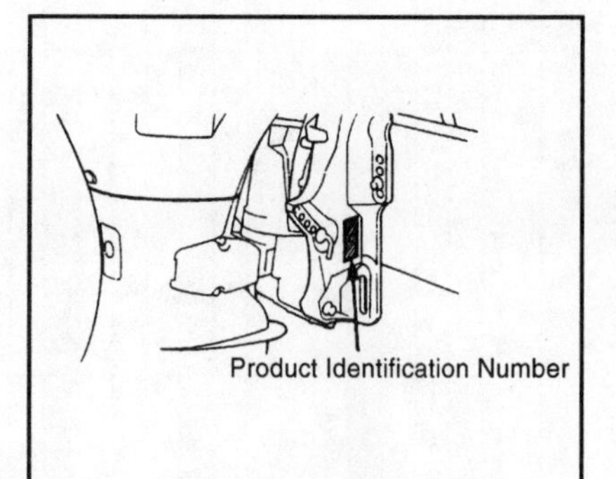

Fig. 9 Product identification number location—BF35A and BF45A

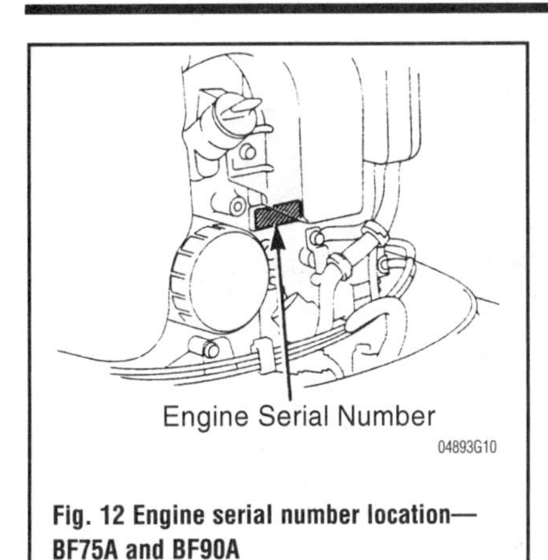

Fig. 12 Engine serial number location—BF75A and BF90A

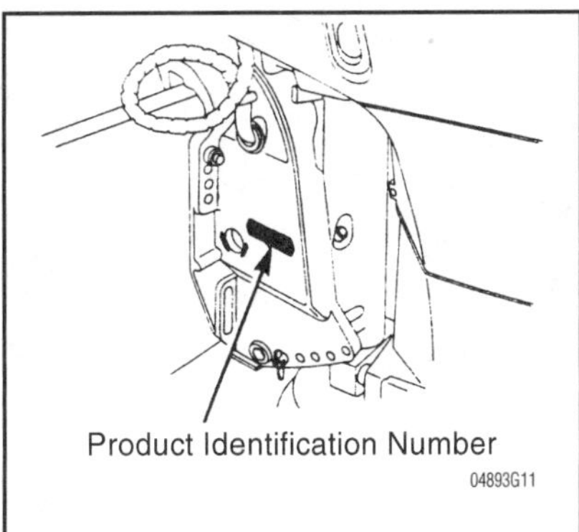

Fig. 13 Product identification number—BF75A and BF90A

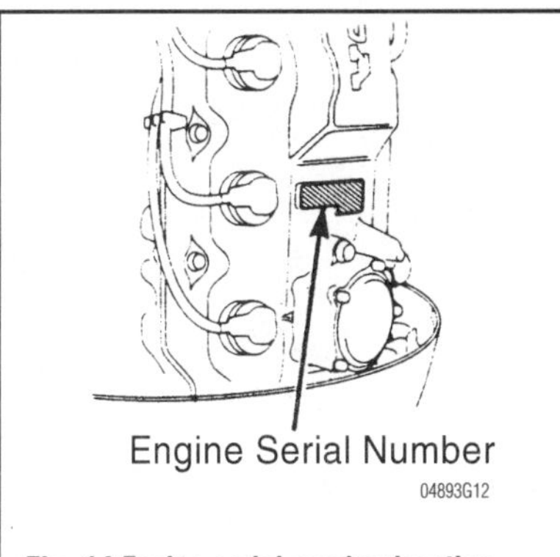

Fig. 14 Engine serial number location—BF115A and BF130A

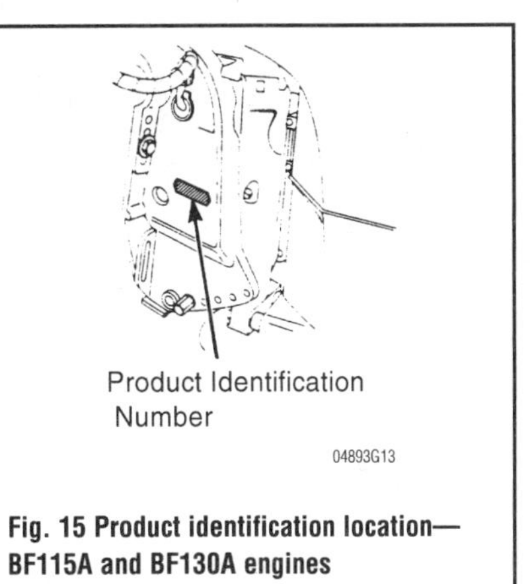

Fig. 15 Product identification location—BF115A and BF130A engines

Fig. 16 Typically, Honda product identification number plates . . .

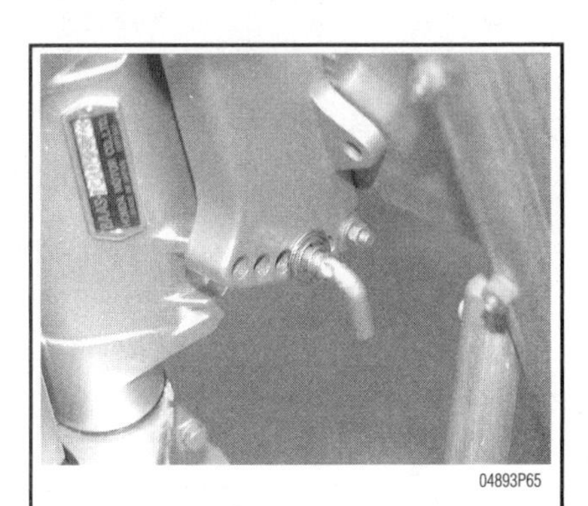

Fig. 17 . . . are located on the transom bracket

required, the engine serial number must be used or proper identification is not possible. The accompanying illustrations will be very helpful in locating the engine identification tag for the various models.

The product identification number for all engines is located on a plate attached to the stern bracket.

The engine serial number is stamped on the cylinder head or engine block in various locations as shown in the illustrations.

Engine Cover

REMOVAL & INSTALLATION

See Figures 18, 19 and 20

On all models, the engine cover is attached by some type of lever or latch. No tools are necessary to remove the cover.

Two types of mechanisms are used. The first is a simple over center latch. Lift the latch to disengage the cover. The second is a single lever that is turned approximately 45 degrees to release the cover. Once the retaining mechanisms are free, the cover can be tilted to disengage any locating pins or hooks.

When installing the cover, make sure that any hooks or pins are in their proper positions before engaging the latch or lever.

After the latch or lever is disengaged, tilt the cover to disengage any hooks or tabs, if equipped and lift cover off the engine.

Engine Oil

See Figures 21, 22, 23 and 24

Engine oil is the life blood of a 4-stroke engine and if it is not maintained at the correct level and changed on a regular basis, major engine damage can occur. It is not uncommon to see outboard units well over 20 years of age moving a boat through the water as if the unit had recently been purchased from the

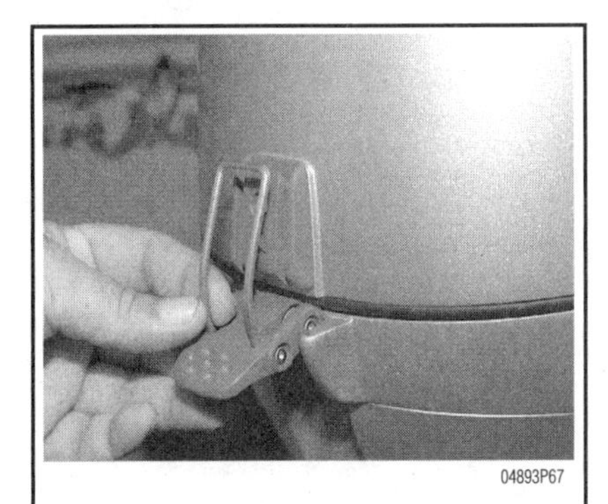

Fig. 18 To remove the engine cover, release any latches . . .

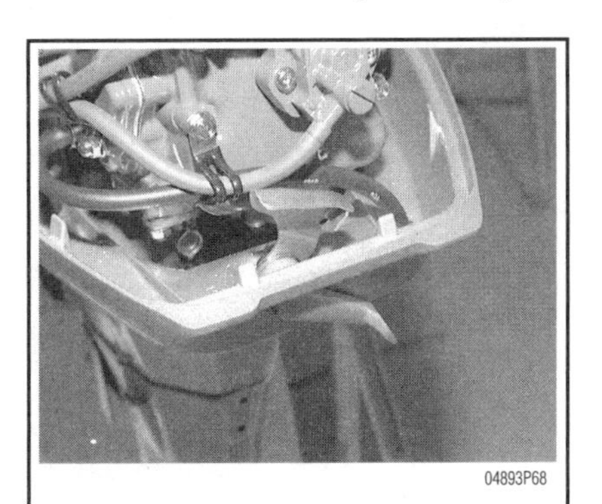

Fig. 19 . . . or unlock any levers that hold the cover in place . . .

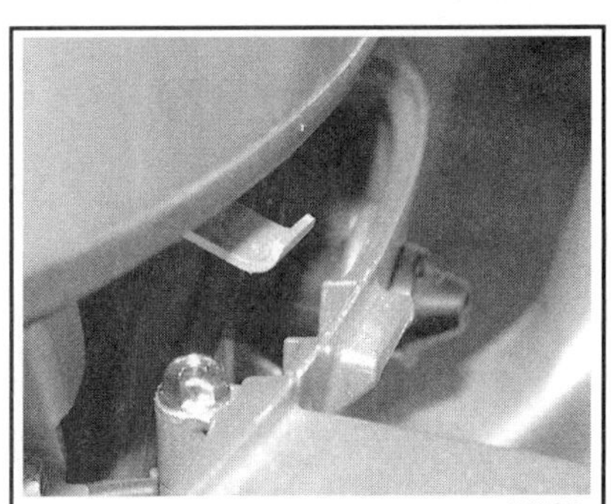

Fig. 20 . . . and lift the cover off any alignment tabs

current line of models. An inquiry with the proud owner will undoubtedly reveal his main credit for its performance to be regular periodic maintenance

Maintaining the correct engine oil level is one of the most basic (and essential) forms of engine maintenance. Get into the habit of checking your oil level every time you start your engine. All engines naturally consume small amounts of oil, and if left neglected, can consume enough oil to damage the internal components of the engine. Assuming the oil level is correct because you "checked it the last time" can be a costly mistake.

ENGINE OIL RECOMMENDATIONS

Every bottle of engine oil for sale in the U.S. should have a label describing what standards it meets. Engine oil service classifications are designated by the American Petroleum Institute (API), based on the chemical composition of a given type of oil and testing of samples. The ratings include "S" (normal gasoline engine use) and "C" (commercial and fleet) applications. Over the years, the S rating has been supplemented with various letters, each one representing the latest and greatest rating available at the time of its introduction. During recent years these ratings have included SF, SG, SH and most recently (at the time of this book's publication), SJ. Each successive rating usually meets all of the standards of the previous alpha designation, but also meets some new criteria, meets higher standards and/or contains newer or different additives. Since oil is so important to the life of your engine, you should obviously NEVER use an oil of questionable quality. Oils that are labeled with modern API ratings, including the "energy conserving" donut symbol, have been proven to meet the API quality standards. Always use the highest grade of oil available. The better quality of the oil, the better it will lubricate the internals of your engine.

In addition to meeting the classification of the American Petroleum Institute,

88525G02

Fig. 21 The API donut symbol

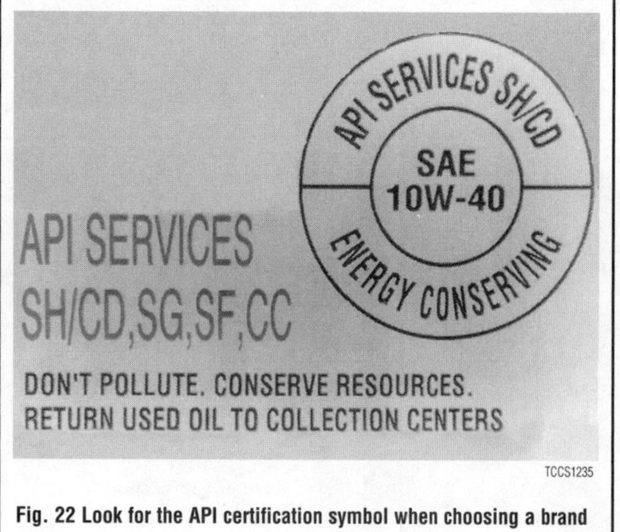

TCCS1235

Fig. 22 Look for the API certification symbol when choosing a brand of oil

88525G05

Fig. 23 A lot of information can be found on your average bottle of oil

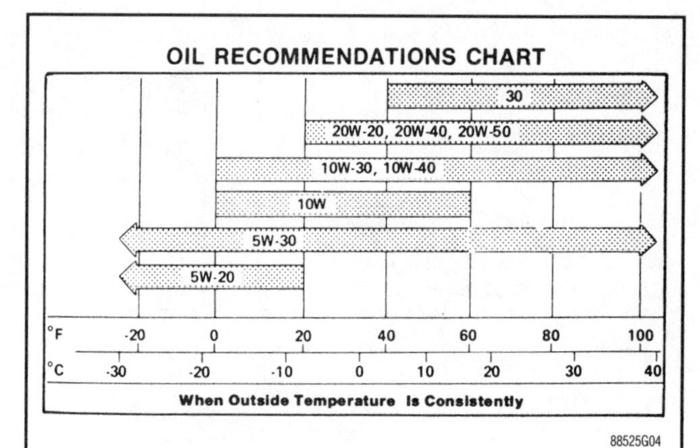

88525G04

Fig. 24 Typical oil recommendation chart—check your owners manual for specific recommendations

your oil should be of a viscosity suitable for the outside temperature in which your engine will be operating. Oil must be thin enough to get between the close-tolerance moving parts it must lubricate. Once there, it must be thick enough to separate them with a slippery oil film. If the oil is too thin, it won't separate the parts; if it's too thick, it can't squeeze between them in the first place—either way, excess friction and wear takes place. To complicate matters, cold-morning starts require a thin oil to reduce engine resistance, while high speed driving requires a thick oil which can lubricate vital engine parts at temperatures.

According to the Society of Automotive Engineers' (SAE) viscosity classification system, an oil with a high viscosity number (such as SAE 40 or SAE 50) will be thicker than one with a lower number (SAE 10W). The "W" in 10W indicates that the oil is desirable for use in winter driving, and does not stand for "weight". Through the use of special additives, multiple-viscosity oils are available to combine easy starting at cold temperatures with engine protection at high speeds. For example, a 10W40 oil is said to have the viscosity of a 10W oil when the engine is cold and that of a 40W oil when the engine is warm. The use of such an oil will decrease engine resistance and improve efficiency.

OIL LEVEL CHECK

**** CAUTION**

The EPA warns that prolonged contact with used engine oil may cause a number of skin disorders, including cancer! You should make every effort to minimize your exposure to used engine oil.

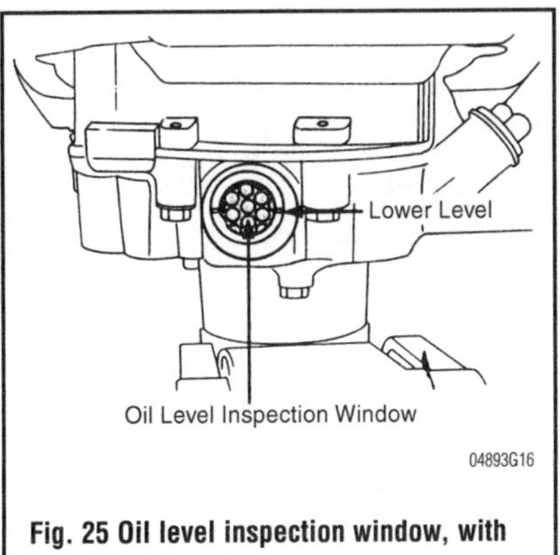

Fig. 25 Oil level inspection window, with low oil level mark

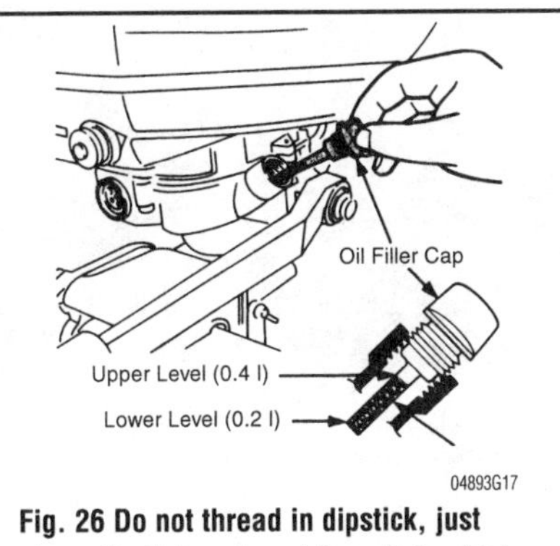

Fig. 26 Do not thread in dipstick, just place dipstick on top of threads to obtain the correct reading

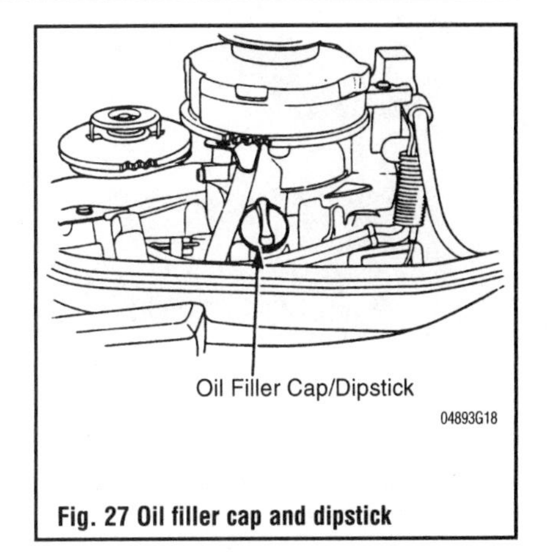

Fig. 27 Oil filler cap and dipstick

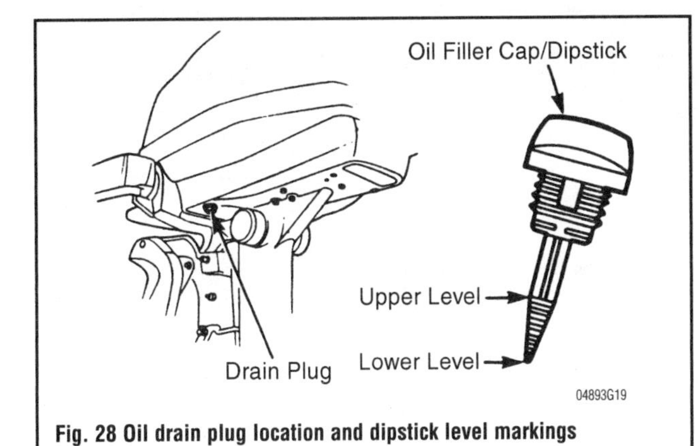

Fig. 28 Oil drain plug location and dipstick level markings

Protective gloves should be worn when changing the oil. Wash your hands and any other exposed skin areas as soon as possible after exposure to used engine oil. Soap and water, or waterless hand cleaner should be used.

BF2A

➧ See Figures 25 and 26

1. Position the outboard upright so it is as close to a 90 degree angle to the water as possible, If the outboard is tilted at an angle (for proper planing of the boat) it will cause a false reading in the oil level inspection window.
2. Remove the engine cover by disengaging the latches or lever(s).
3. Remove the oil level dipstick, and wipe it clean with a rag.
4. Insert the dipstick. Do not thread in dipstick, just place dipstick on top of threads to obtain the correct reading.
5. Pull out the dipstick again, and read the level.

➡An alternate method is to check the oil level in the oil level inspection window. If the oil level is down toward the lower level mark on the window, fill up to the upper level mark on the dipstick

6. If the oil level is low, add oil in small increments through the dipstick hole until the oil level is correct.
7. Once the oil level is correct, install the dipstick and tighten it securely.

➡Do not overfill the engine with oil. Excessive oil can cause engine damage.

BF5A, BF50 and BF8A

➧ See Figures 27 and 28

1. Position the outboard upright so it is as close to a 90 degree angle to the water as possible. If the outboard is tilted at an angle (for proper planing of the boat) it will cause a false reading on the dipstick.
2. Unscrew the dipstick from the engine, and wipe it clean with a rag.
3. Insert the dipstick back into the engine, without screwing it back in the threads.
4. Remove the dipstick and read the level.

➡Do not overfill the engine with oil. Excessive oil can cause engine damage.

5. If the oil level is low, add oil in small increments through the dipstick hole until the oil level is correct.
6. Once the oil level is correct, install the dipstick and tighten it securely.

All Other Models

➧ See Figures 29, 30, 31, 32 and 33

1. Position the outboard upright, so it is as close to a 90 degree angle to the water as possible. If the outboard is tilted at an angle (for proper planing of the boat) it will cause a false dipstick reading.
2. Remove the engine cover by disengaging the latches or lever(s).
3. Remove the oil level dipstick, and wipe it clean with a rag.

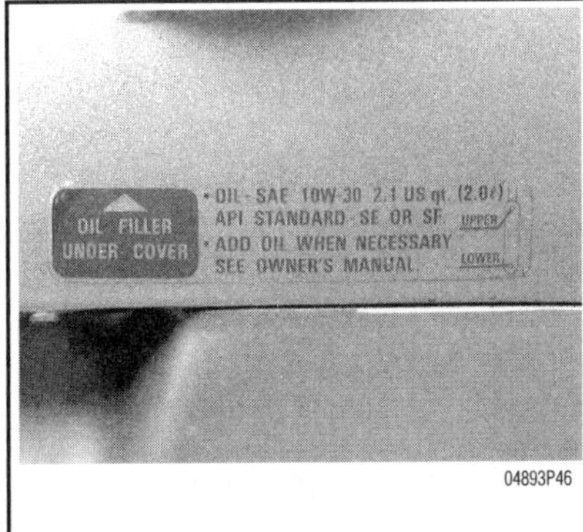

Fig. 29 Oil filler location and model specific information on engine cover

Fig. 30 Make sure to insert the dipstick fully into the tube to obtain an accurate reading

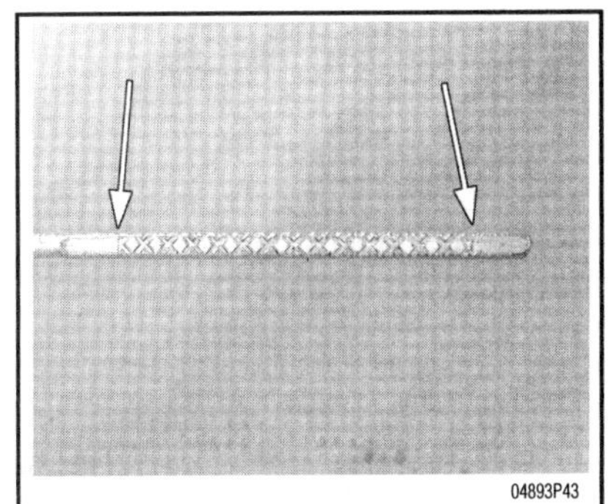

Fig. 31 The oil level should always be maintained to read at the top of the crosshatch marks on the dipstick

Fig. 32 If the level on the dipstick is low, add oil as necessary

Fig. 33 Always use a funnel to prevent spills when adding oil

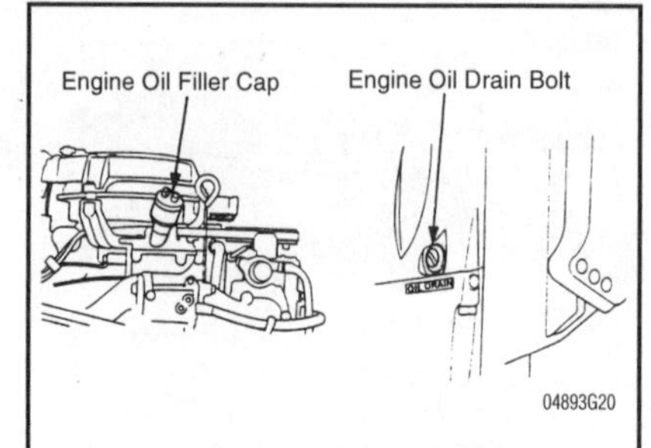

Fig. 34 The engine oil drain plug is usually located on the upper half of the casing, and is marked for identification. BF15A shown, others similar

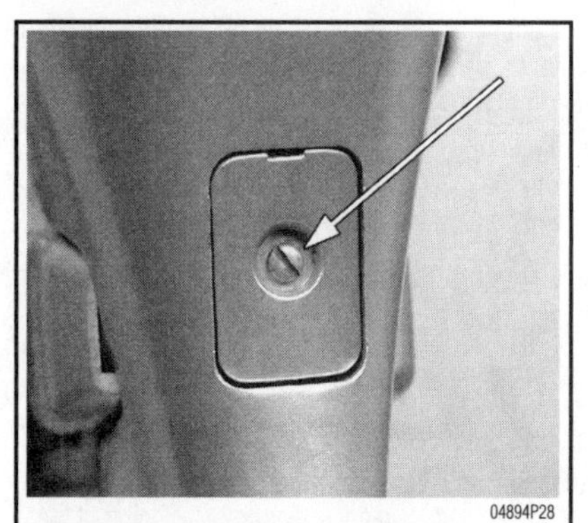

Fig. 35 To access the engine oil drain plug, remove the cover

Fig. 36 Once the cover is removed, the drain plug can be accessed

Fig. 37 After a suitable container has been placed under the drain plug, it can be loosened, and unscrewed, allowing the oil to drain

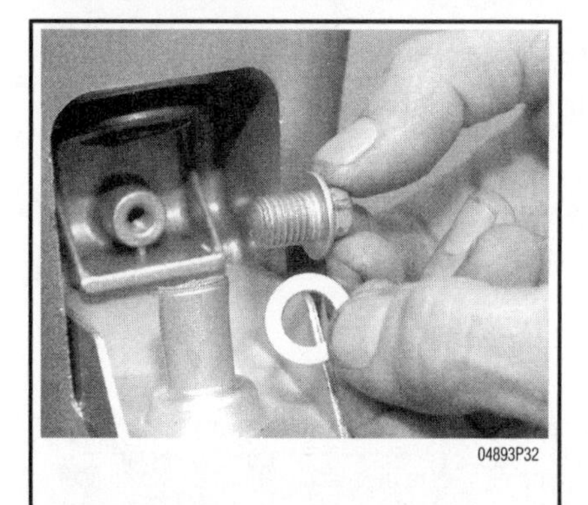

Fig. 38 The washer on the drain plug should be replaced to prevent leaks

Fig. 39 Loosen the oil filter by turning it counterclockwise. Cap style oil filter tools work well in this application

Fig. 40 A plastic water bottle trimmed to size can be useful for catching the oil in the filter when it is removed from the engine

4. Insert the dipstick, making sure that it is all the way seated in the tube.
5. Pull out the dipstick again, and read the level.

➡Do not overfill the engine with oil. Excessive oil can cause engine damage.

6. If the oil level is low, add oil through the filler cap on the top of the engine until the level is correct.

OIL & FILTER CHANGE

➧ See Figures 34 thru 44

Most outboard manufacturers recommend changing the filter at every other oil change. We recommend a filter change with EVERY oil change. The cost involved in purchasing an oil filter for every change is worth the additional cleanliness of the engine oil, which lengthens the life of the internal parts of the engine.

Honda recommends changing the oil after the first 20 hours or one month of operation and thereafter every 100 hours of operation, or 6 months, whichever comes first. If you choose, you can follow the manufacturer recommended interval for changing the engine oil and filter. But no harm can be done by changing the oil and filter at shorter intervals than the manufacturer recommends.

✱✱ CAUTION

The EPA warns that prolonged contact with used engine oil may cause a number of skin disorders, including cancer! You should make every effort to minimize your exposure to used engine oil. Protective gloves should be worn when changing the oil. Wash your hands and any other exposed skin areas as soon as possible after exposure to used engine oil. Soap and water, or waterless hand cleaner should be used.

Fig. 41 Simply slip the cut edge of the bottle under the lip cast under the oil filter mounting point to catch any oil drips

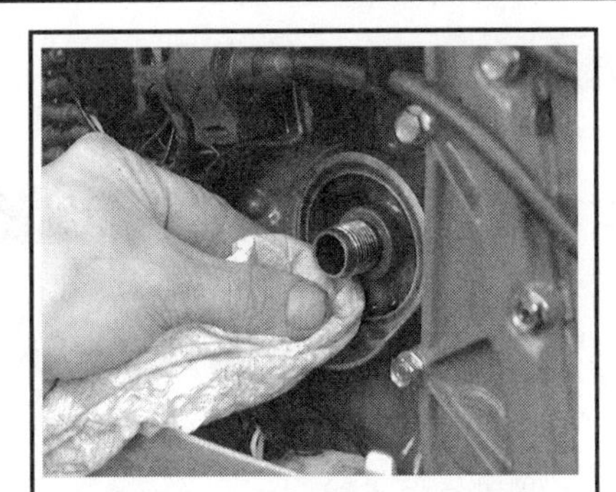

Fig. 42 Before installing the new oil filter, wipe the mating surface on the engine clean

Fig. 43 Make sure to coat the gasket of the new oil filter with clean engine oil

Fig. 44 Using your hand, tighten the filter approximately 3/4 to 7/8 of a turn by hand

Capacities

Horsepower	Engine Quart (Liter)		Lower Unit Quart (Liter)		Fuel Tank Gallon (Liter)
BF130A	5.9 (5.6)		1.1 (1.0)		-
BF115A	5.9 (5.6)		1.1 (1.0)		-
BF90A	4.8 (4.5)	①	0.70 (0.66)	②	-
BF75A	4.8 (4.5)	①	0.70 (0.66)	②	-
BF50A	2.5 (2.4)		0.53 (0.50)		6.6 (25)
BF45A	2.5 (2.4)		0.53 (0.50)		6.6 (25)
BF40A	2.5 (2.4)		0.53 (0.50)		6.6 (25)
BF35A	2.5 (2.4)		0.53 (0.50)		6.6 (25)
BF30A	2.0 (1.9)	①	0.30 (0.28)		6.6 (25)
BF25A	2.0 (1.9)	①	0.30 (0.28)		6.6 (25)
BF15A / BF15D	1.1 (1.0)		0.254 (.240)		3.4 (13)
BF9.9A / BF9.9D	1.1 (1.0)		0.254 (.240)		3.4 (13)
BF75 / BF100	0.80 (0.76)		0.23 (0.22)		3.4 (13)
BF8A / BF80 / BF8D	0.80 (0.76)		0.23 (0.22)		3.4 (13)
BF5A / BF50 / BF5D	0.58 (0.55)		0.10 (0.09)		3.4 (13)
BF2A / BF20	0.42 (0.40)		0.05 (0.04)		0.26 (1.0)
BF2D	0.26 (0.25)		0.05 (0.04)		0.26 (1.0)

① *With oil filter replaced*

② *Specification is for models through 1998. For 1999 and later models spec is 1.1 (1.0)*

1. If possible, run the engine to heat the oil to allow for quicker and more thorough draining.

**** WARNING**

Do NOT run the engine without having a flushing device attached to the engine. The water pump can be seriously damaged, and the engine may overheat.

2. Position the outboard upright so it is as close to a 90 degree angle to the water as possible to facilitate proper draining of the oil.
3. Locate and remove the drain plug cover (if equipped) to access the drain plug.
4. On BF35A—BF50A , position drain plug cover to act as a guide for the draining oil.
5. Position a suitable container under the drain plug to contain the used oil.
6. Loosen the drain plug until it able to be unscrewed by hand. While pushing in on the drain plug, unscrew it until it is free of the threads, and then quickly pull it away and allow the oil to drain.
7. Once the oil has completely drained, install the drain plug. Replace the washer if necessary.
8. Using an appropriately sized wrench, remove the oil filter from the side of the engine. BF25A—BF130A only. BF2A—BF15A are not equipped with disposable oil filters.

■ **On some models, a plastic bottle can be trimmed to fit under the filter when it is removed to keep oil from draining inside the engine pan.**

9. Wipe the filter mating surface clean with a rag or lint-free towel.
10. Coat the gasket of the new filter with a small amount of fresh oil.
11. Screw the filter onto the housing until the gasket just touches the mating surface.
12. Turn the filter an additional 3/4 to 7/8 turn by hand. Do not use a wrench to tighten the filter.
13. Add oil (of the proper, quantity, type and viscosity) to the engine.
14. Once the oil is added to the engine, use the dipstick to verify the proper level. It is normal for the oil level to be a little high, since the oil filter has not been filled with oil.
15. Start the engine (with a flushing device attached to prevent engine damage) and allow it to run for at least one minute.

**** WARNING**

Do NOT run the engine without having a flush device attached to the engine. The water pump can be seriously damaged, and the engine may overheat.

16. Shut the engine **OFF**, and let the oil settle back into the oil pan.
17. Check the level of the oil. If the level is low, add oil in small increments until the oil level is correct.
18. Dispose of the used oil by bringing it to a service station or auto parts store, where it can be collected for recycling.

■ **Do not pour used oil into the ground or into the water. Improper disposal of used engine oil not only pollutes the environment, it is a violation of federal law.**

Lower Unit

➧ See Figures 45 and 46

Regular maintenance and inspection of the lower unit is critical for proper operation and reliability. A lower unit can quickly fail if it becomes heavily contaminated with water, or excessively low on oil. The most common cause of a lower unit failure is water contamination.

Water in the lower unit is usually caused by fishing line, or other foreign material, becoming entangled around the propeller shaft and damaging the seal. If the line is not removed, it will eventually cut the propeller shaft seal and allow water to enter the lower unit. Fishing line has also been known to cut a groove in the propeller shaft if left neglected over time. This area should be checked frequently.

DRAINING AND FILLING

➧ See Figures 47 thru 54

**** CAUTION**

The EPA warns that prolonged contact with used engine oil may cause a number of skin disorders, including cancer! You should make every effort to minimize your exposure to used engine oil. Protective gloves should be worn when changing the oil. Wash your hands and any other exposed skin areas as soon as possible after exposure to used engine oil. Soap and water, or waterless hand cleaner should be used.

04703P10

Fig. 45 This lower unit was destroyed because the bearing carrier was frozen due to lack of lubrication

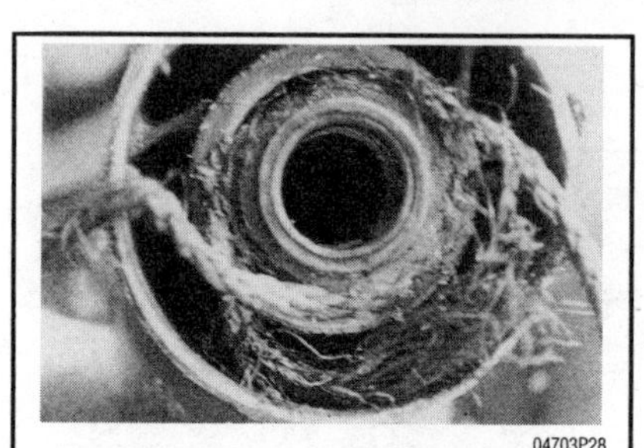

04703P28

Fig. 46 Excellent view of rope and fishing line entangled behind the propeller. Entangled fishing line can actually cut through the seal, allowing water to enter the lower unit and lubricant to escape

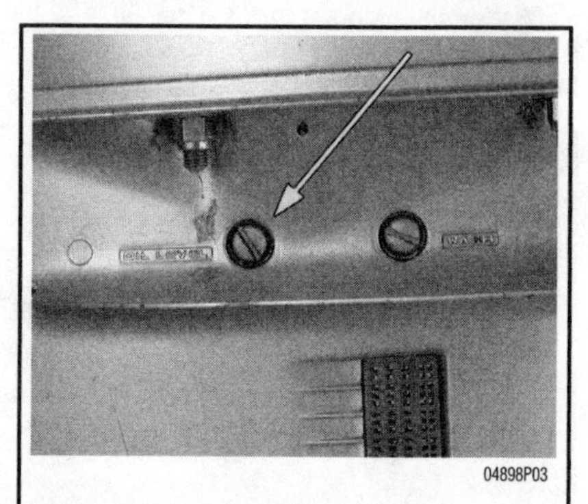

04898P03

Fig. 47 The oil level plug is almost always located at the top of the lower unit

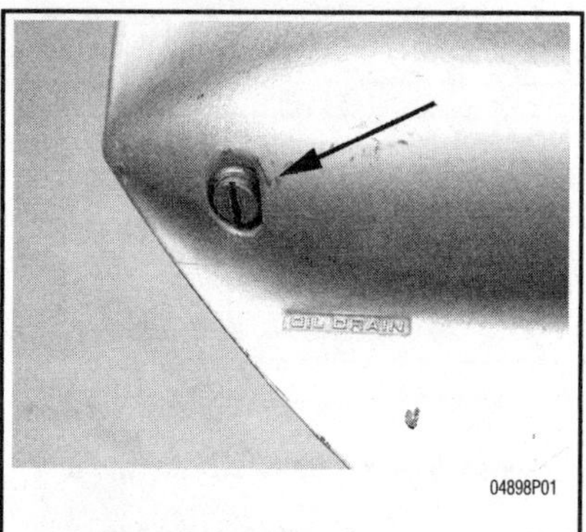

04898P01

Fig. 48 To drain the lower unit, remove the drain plug . . .

04898P06

Fig. 49 . . . with a large flathead screwdriver to prevent rounding the screw

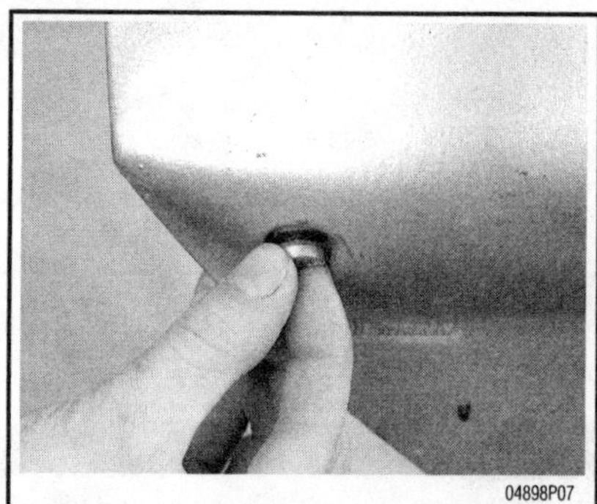

04898P07

Fig. 50 Once the drain plug is loosened, unscrew it by hand, while pushing inward until it is free of the threads . . .

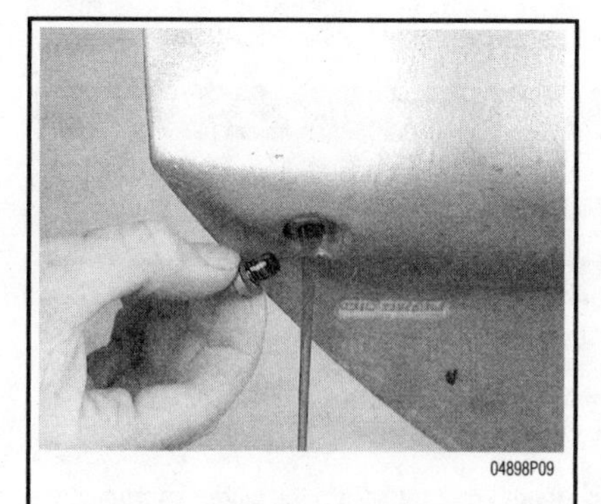

04898P09

Fig. 51 . . . then quickly pull it away to drain the oil

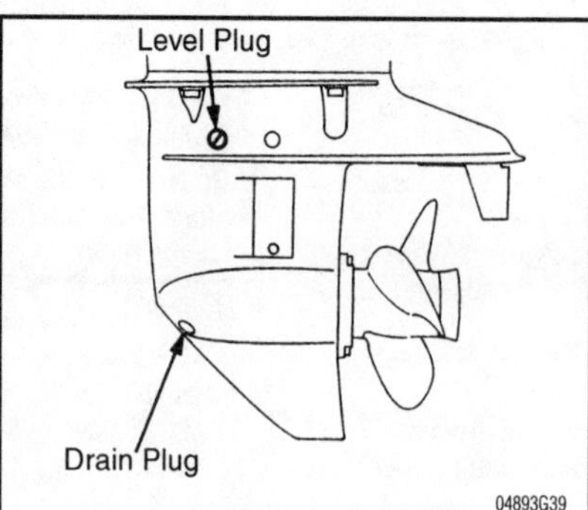

04893G39

Fig. 52 The oil in the lower unit is contained between the oil level plug and the oil drain plug

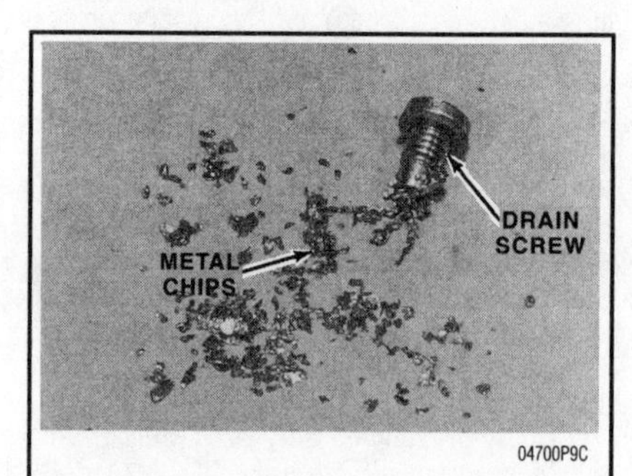

04700P9C

Fig. 53 This is what you don't want to see. Metal particles like these mean the lower unit must be torn down to find the cause of the problem

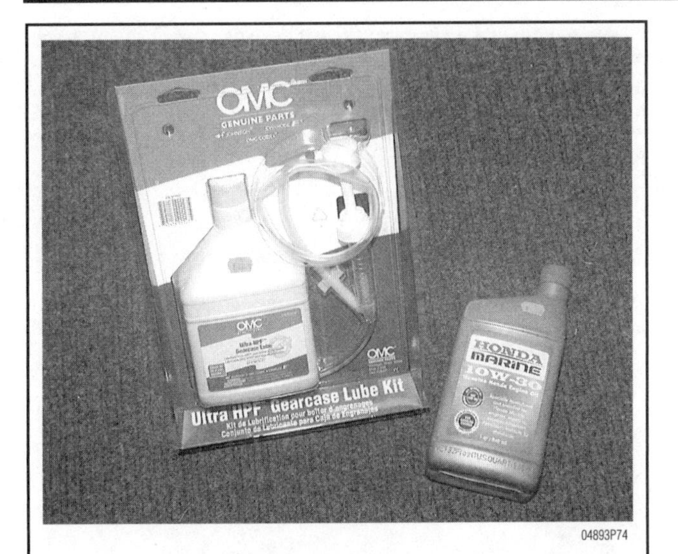

Fig. 54 Oil should be squeezed, or on larger outboards, pumped into the drain plug of the lower unit using a gear case lube kit

1. Place a suitable container under the lower unit.
2. Loosen the oil level plug on the lower unit. This step is important! If the oil level plug cannot be loosened or removed, the complete lower unit lubricant service cannot be performed.

➡Never remove the vent or filler plugs when the lower unit is hot. Expanded lubricant will be released through the plug hole.

3. Remove the fill plug from the lower end of the gear housing followed by the oil level plug.
4. Allow the lubricant to completely drain from the lower unit.

➡If applicable, check the magnet end of the drain screw for metal particles. Some normal wear is to be expected, but if there are signs of metal chips or excessive metal particles, the gear case needs to be disassembled and inspected.

5. Inspect the lubricant for the presence of a milky white substance, water or metallic particles. If any of these conditions are present, the lower unit should be serviced immediately.
6. Place the outboard in the proper position for filling the lower unit. The lower unit should not list to either port or starboard, and should be completely vertical.
7. Insert the lubricant tube into the oil drain hole at the bottom of the lower unit, and inject lubricant until the excess begins to come out the oil level hole.
8. Using new gaskets, (washers) install the oil level and vent plugs (if applicable) first, then install the oil fill plug.
9. Place the used lubricant in a suitable container for transportation to an authorized recycling station.

Air Cleaner

➧ See Figures 55, 56, 57 and 58

Honda outboard engines do not use an air cleaner like most engines. Instead, there is an assembly which fits over the carburetor(s) known as a "silencer cover" or an "air guide." This unit, while helping to cut down on engine noise, also (on most models) has a coarse-mesh screen which keeps leaves, bugs and other large debris from being drawn through the carburetor(s) and into the engine.

The only maintenance required for the silencer cover or air guide is periodic inspection for any accumulated debris within the inlet ductwork, and the silencer itself.

Fuel Filter

The fuel filter is designed to keep particles of dirt and debris from entering the carburetor(s) and clogging the internal passages. A small speck of dirt or sand can drastically affect the ability of the carburetor(s) to deliver the proper amount of air and fuel to enter the engine. If a filter becomes clogged, the flow of gasoline will be impeded. This could cause lean fuel mixtures, hesitation and stumbling, and idle problems.

Regular replacement of the fuel filter will decrease the risk of blocking the flow of fuel to the engine, which could leave you stranded on the water. Fuel filters are usually inexpensive, and replacement is a simple task. Change your fuel filter on a regular basis to avoid fuel delivery problems to the carburetor.

In addition to the fuel filter mounted on the engine, a filter is usually found inside or near the fuel tank (with the exception of BF2A and BF20). Because of the large variety of differences in both portable and fixed fuel tanks, it is impossible to give a detailed procedure for removal and installation. Most in-tank filters are sim-

Fig. 55 Air inlet detail—BF2A and BF20

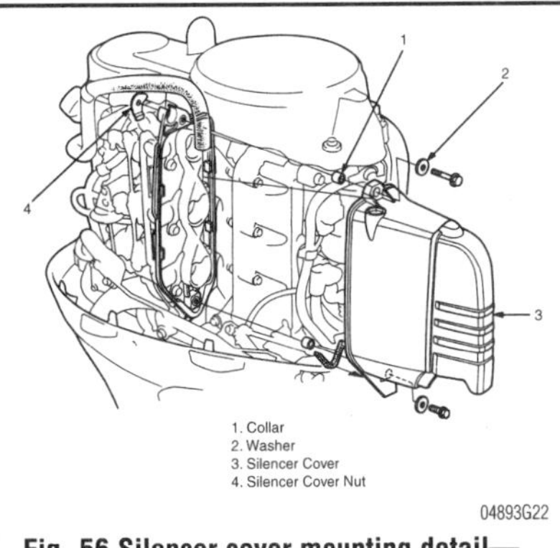

Fig. 56 Silencer cover mounting detail—BF35A shown, other models similar

Fig. 57 Silencer cover mounted on a BF35A powerhead

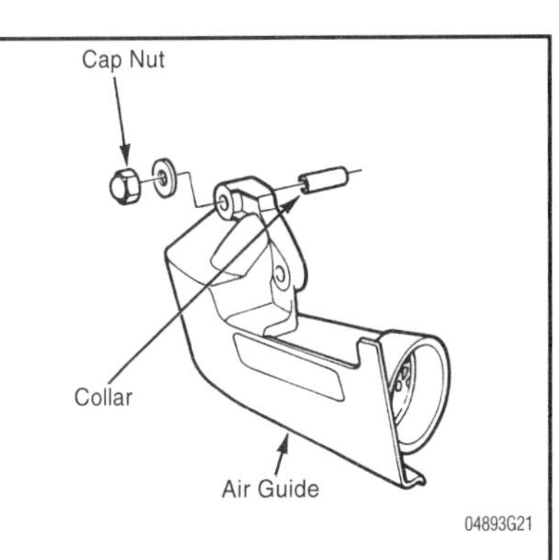

Fig. 58 Air guide used on the BF20 and BF2A powerheads

ply a screen on the pick-up line inside the fuel tank. Filters of this type usually only need to be cleaned and returned to service. Fuel filters on the outside of the tank are typically of the inline type, and are replaced by simply removing the clamps, disconnecting the hoses, and installing a new filter. When installing the new filter, make sure the arrow on the filter points in the direction of fuel flow.

RELIEVING FUEL SYSTEM PRESSURE

On fuel injected engines, always relieve system pressure prior to disconnecting any fuel system component, fitting or fuel line.

**** CAUTION**

Exercise extreme caution whenever relieving fuel system pressure to avoid fuel spray and potential serious bodily injury. Please be advised that fuel under pressure may penetrate the skin or any part of the body it contacts.

To avoid the possibility of fire and personal injury, always disconnect the negative battery cable.

Always place a shop towel or cloth around the fitting or connection prior to loosening to absorb any excess fuel due to spillage. Ensure that all fuel spillage is removed from engine surfaces. Ensure that all fuel soaked clothes or towels in suitable waste container.

1. Remove the engine cover.
2. Place a wrench on both the service check bolt and fitting nut to prevent the fitting from twisting and breaking off.
3. Holding the service check bolt and fuel pressure check nut with both wrenches, place a shop towel or equivalent material over the service check bolt.
4. Loosen the service check bolt approximately one turn slowly to relieve the fuel pressure.
5. After relieving the fuel pressure, remove the service check bolt and replace the 6mm sealing washer with a new one. Tighten the service check bolt to 9 ft. lbs. (12 Nm).

REMOVAL & INSTALLATION

Integral Fuel Tank Filter

▶ See Figure 59

**** CAUTION**

Observe all applicable safety precautions when working around fuel. Whenever servicing the fuel system, always work in a well ventilated area. Do not allow fuel spray or vapors to come in contact with a spark or open flame. Do not smoke while working around gasoline. Keep a dry chemical fire extinguisher near the work area. Always keep fuel in a container specifically designed for fuel storage; also, always properly seal fuel containers to avoid the possibility of fire or explosion.

1. Remove the engine cover.
2. Drain the fuel in the tank into a suitable container.
3. Remove the fuel tank.
4. Remove the clamp on the fuel tank hose.
5. Remove the hose from the tank, followed by removing the filter from the tank.
6. Clean the filter assembly in solvent and blow it dry with compressed air. If excessively dirty or contaminated with water, replace the filter.
7. Install the filter into the tank, followed by the fuel hose.
8. Secure the hose to the fuel tank with the clamp. The clamp should be positioned 5mm from the end of the hose.
9. Install the fuel tank onto the engine.
10. Pour a small amount of fuel into the fuel tank and check for leaks.
11. If there are no signs of leakage, the fuel tank can be filled to full capacity.
12. Install the engine cover.

Inline Filter

▶ See Figures 60, 61, 62, 63 and 64

**** CAUTION**

Observe all applicable safety precautions when working around fuel. Whenever servicing the fuel system, always work in a well ventilated area. Do not allow fuel spray or vapors to come in contact with a spark or open flame. Do not smoke while working around gasoline. Keep a dry chemical fire extinguisher near the work area. Always keep fuel in a container specifically designed for fuel storage; also, always properly seal fuel containers to avoid the possibility of fire or explosion.

1. Remove the engine cover.
2. Locate the fuel filter in the engine pan.
3. Lift the fuel filter from the engine pan, and place a pan or clean rag underneath it to absorb any spilled fuel.
4. Slide the hose retaining clips off the filter nipple with a pair of pliers and disconnect the hoses from the filter.
5. Reinstall the hoses on the filter nipples of the new filter. Make sure the embossed arrow on the filter points in the direction of fuel flow.
6. Slide the clips on each hose over the filter nipples.
7. Check the fuel filter installation for leakage by priming the fuel system with the fuel line primer bulb.

Fuel Tank
Fuel Filter
04893G23

Fig. 59 On BF2A and BF20, the fuel filter is integrated with the fuel tank

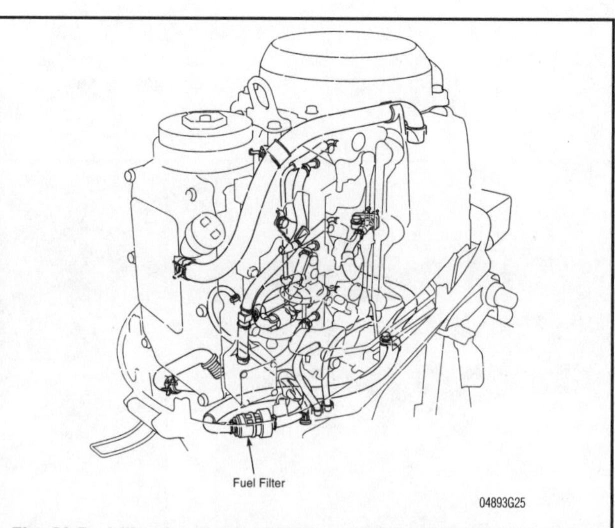

Fig. 60 Fuel filter location on BF25A and BF30A; BF5A–BF15A similar

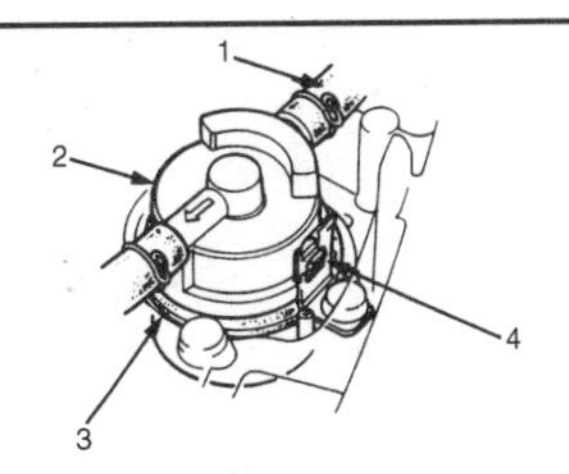

Fig. 61 On BF35A–BF50A, the fuel filter is held in place by a metal retainer. Hold the clip back when lifting the filter

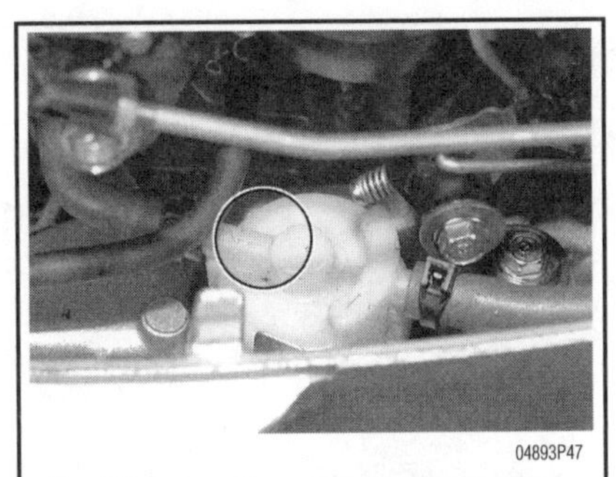

Fig. 62 Be sure to mount the filter in the proper direction. The arrow on the filter (circled) indicates the direction of the flow of fuel

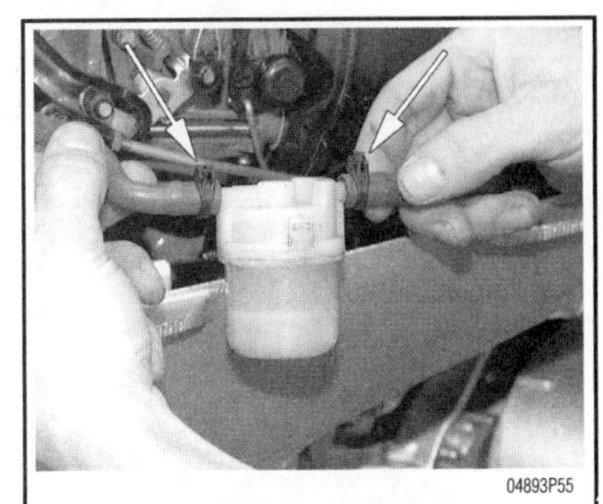

Fig. 63 If the clamps on the filter become damaged, nylon tie wraps can be used to secure the hoses to the filter

Fig. 64 Most fuel connections use spring-type hose clamps that can loose tension over time. It is wise to replace this type of clamp during servicing to prevent fuel leaks

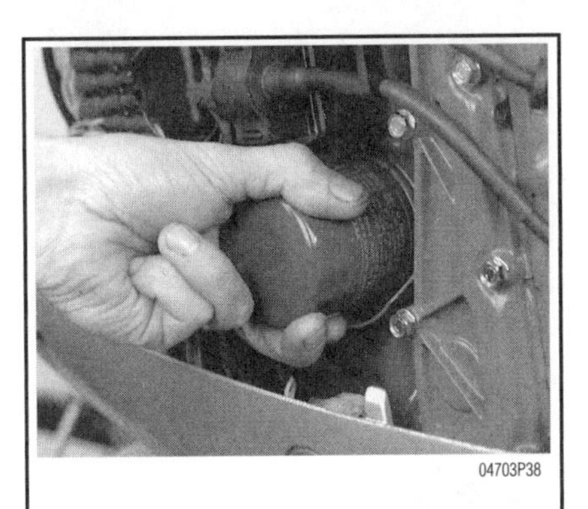

Fig. 65 A water separating fuel filter installed inside the boat on the transom

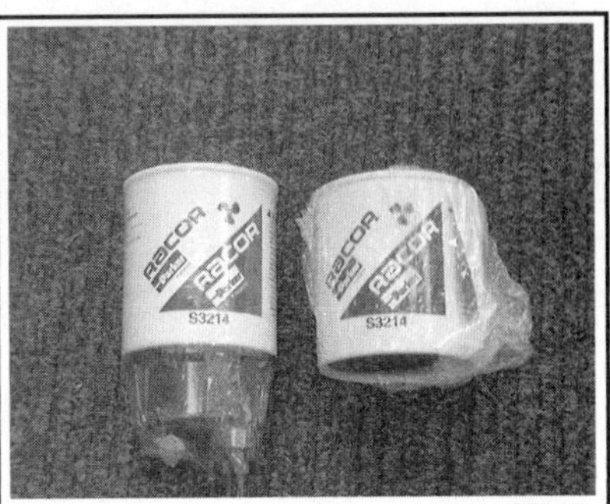

Fig. 66 A typical water separating fuel filter assembly ready to be installed on the boat

8. Once it is confirmed that there is no leakage from the connections, place the filter back to its proper position in the engine pan.
9. Replace the engine cover.

Fuel/Water Separator

See Figures 65 and 66

In addition to the engine and inline fuel filters, there is usually another filter located in the fuel supply line. This is the fuel/water separator. It is used to remove water particles from the fuel prior to entering the engine or inline filter. Water can enter the fuel supply from a variety of sources and can lead to poor engine performance and ultimately, serious engine damage.

Because of the large variety of differences in both portable and fixed fuel tanks, it is impossible to give a single procedure to cover all applications. Check with the boat manufacturer or the marina who rigged the boat to get the specifics of your particular fuel filtration system.

Timing Belt

The BF8A–BF130A utilize a timing belt to drive the camshaft from the crankshaft's turning motion and to maintain proper valve timing. Some manufacturers schedule periodic timing belt replacement to assure optimum engine performance, to make sure the operator is never stranded in the water should the belt break (as the engine will stop instantly) and for some (manufacturers with interference engines) to prevent the possibility of severe internal engine damage should the belt break.

INSPECTION

See Figures 67, 68 and 69

Inspection of the timing belt is (in most cases) a simple matter of removing the belt cover (usually held on by a few bolts or screws) and inspecting the belt. Keeping a close eye on the condition of your engine's timing belt can help avoid an enjoyable outing turning into a miserable one.

A severely worn or damaged belt may not give any warning that it is about to fail. In general, any time the engine timing cover(s) is (are) removed you should inspect the belt for premature parting, severe cracks or missing teeth.

➡The BF8A–BF130A are listed as interference engines (meaning that the valves will contact the pistons if the camshaft was rotated separately from the crankshaft). This places increased emphasis on timing belt inspection and maintenance, as the results of a timing belt failure can be catastrophic.

Fig. 67 Inspect the full length of the timing belt for wear and damage. Replace the belt if necessary

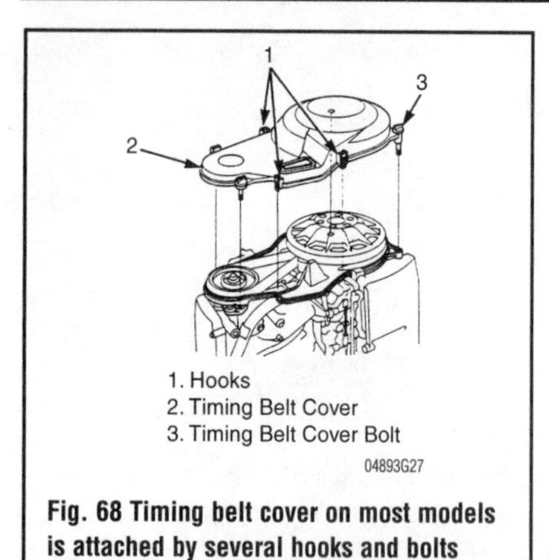

Fig. 68 Timing belt cover on most models is attached by several hooks and bolts

Fig. 69 You have to remove the timing belt cover to access the timing belt below

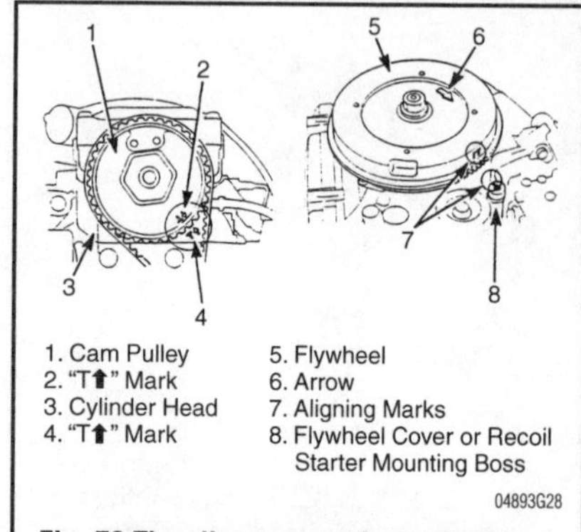

Fig. 70 The alignment marks on the flywheel for TDC—BF25A and BF30A

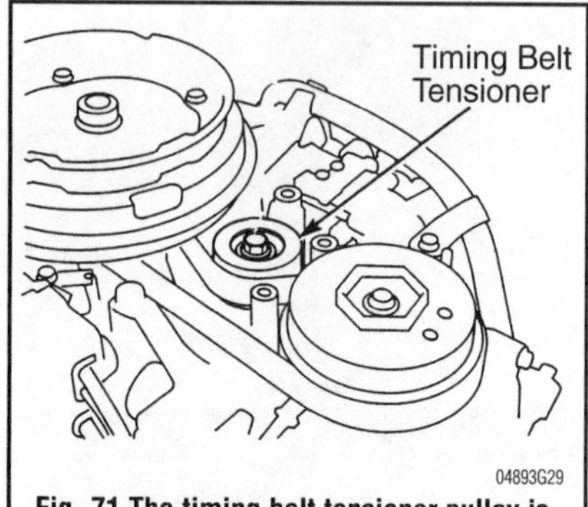

Fig. 71 The timing belt tensioner pulley is held in place by one bolt—BF25A and BF30A

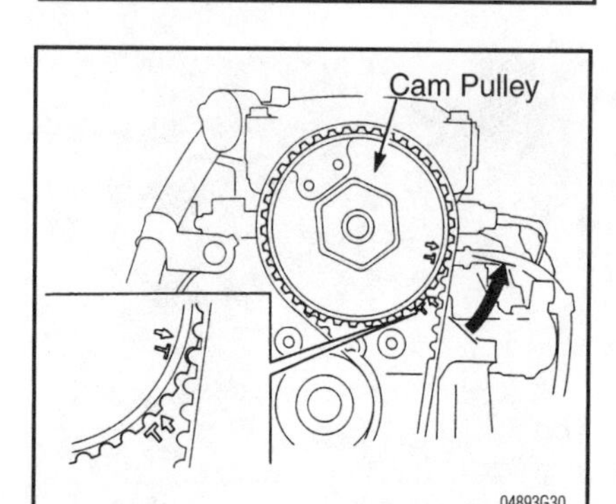

Fig. 72 The cam pulley should be three teeth before TDC when tightening the tensioner pulley bolt—BF25A and BF30A

Fig. 73 Tighten the timing belt tensioner bolt to the proper torque

1. Remove the engine cover.
2. On BF35A–BF130A , the timing belt cover is secured by four bolts; loosen the bolts and remove the cover.
3. Remove the spark plugs. This will make it easier to rotate the engine, allowing inspection of the full length of the belt.
4. Using the flywheel, rotate the engine **clockwise** by hand, and inspect the full length of the belt, looking for fraying, cracks, and missing teeth. If the belt shows any of wear, it should be replaced. Refer to the "Powerhead" section for replacement of the belt.
5. If the belt is in suitable condition, check for proper adjustment. Refer to the Adjustment procedure below.
6. Once the belt is adjusted properly, install the timing belt cover (if equipped) and the spark plugs.
7. Install the engine cover.

ADJUSTMENT

BF8A–BF15A

The timing belt on BF8A–BF15A is not adjustable.

BF25A and BF30A

➧ See Figures 70, 71, 72 and 73

1. Remove the spark plugs. This will make it easier to rotate the engine.
2. Turn the flywheel to align the timing marks. The number one cylinder should be at Top Dead Center (TDC).
3. Remove the engine hanger bracket and timing belt tensioner cap.
4. Loosen the timing belt tensioner flange bolt.
5. Starting from the number one cylinder timing mark, rotate the flywheel counterclockwise by at least 5 gear teeth of the camshaft pulley, then turn it clockwise until the camshaft pulley timing mark "T" is 3 teeth before the "T" mark on the cylinder head (3 teeth before TDC of the number one cylinder.)
6. Tighten the timing belt tensioner flange bolt to the specified torque. Do not apply additional pressure on the tensioner pulley; the spring on the pulley bracket spring provides adequate tension for the belt.
7. Install the engine hanger bracket and timing belt tensioner cap.
8. Install the spark plugs.

BF35A–BF50A

➧ See Figures 74 thru 81

1. Remove the engine cover.
2. Remove the spark plugs. This will make it easier to rotate the engine.

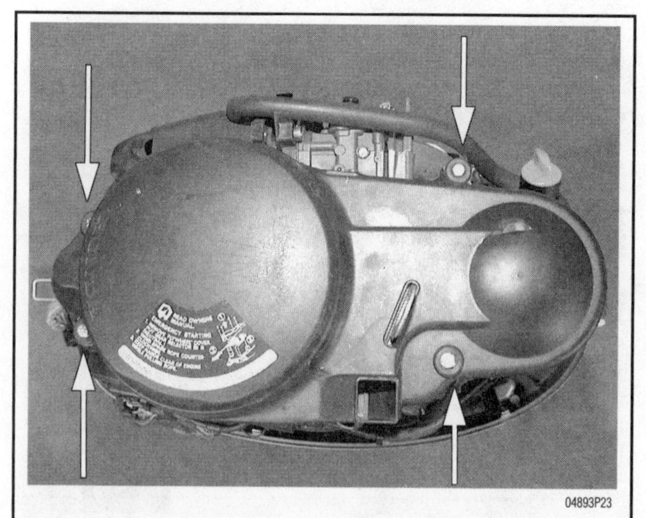

Fig. 74 The timing belt cover is held in place by four bolts—BF35A–BF50A

Fig. 75 Loosen the tensioner pulley bolt to allow the spring on the pulley to provide tension on the belt

Fig. 76 Using the flywheel, turn the engine by hand (note the counterclockwise directional arrow on the flywheel)

Fig. 77 Flywheel timing mark for BF35A–BF50A

Fig. 78 With the flywheel timing marks aligned, the timing marks on the camshaft pulley should be aligned—BF35A–BF50A

Fig. 79 Turn the flywheel to align the timing marks as shown; the tensioner pulley bolt can now be tightened—BF35A–BF50A

Fig. 80 Notice that the tensioner hole is slotted. This is to take up any slack in the timing belt. Do not apply more than the specified torque to the bolt or the tensioner may bind

Fig. 81 Use a torque wrench on the tensioner pulley bolt to ensure it is properly tightened

3. Remove the timing belt cover.

➡**BF35A–BF50A operate in a counterclockwise direction, which is the opposite of most engines.**

4. Turn the flywheel to align the timing marks. The number one cylinder should be at Top Dead Center (TDC).

⁂ CAUTION

The gear teeth on the flywheel can become sharp from constant engagement of the starter. Use caution when turning the flywheel by hand.

5. Remove the engine hanger bracket and timing belt tensioner cap.
6. Loosen the timing belt tensioner bolt.

⁂ WARNING

Do not turn the engine clockwise; the water pump impeller can be damaged.

7. Starting from the number one cylinder timing mark, rotate the flywheel **counterclockwise** at least two full turns.
8. After realigning the timing marks for the number one cylinder, slowly turn the flywheel counterclockwise until the timing mark "T" is three teeth past the "T" mark on the cylinder head.
9. Tighten the timing belt tensioner flange bolt to the specified torque. Do not apply additional pressure on the tensioner pulley; the spring on the pulley bracket spring provides adequate tension for the belt.
10. Install the engine hanger bracket and timing belt tensioner cap.
11. Install the spark plugs.
12. Install the timing belt cover.
13. Install the engine cover.

BF75A and BF90A

➧ See Figures 82, 83, 84 and 85

➡**Make sure that all intake and exhaust valve clearances are within specification before adjusting timing belt tension.**

1. Remove the timing belt cover.
2. Remove the spark plugs. This will make it easier to rotate the engine.
3. Turn the starter pulley clockwise to align the timing marks on the camshaft pulley and the timing belt lower cover. Also align the timing mark on the starter pulley with the mark on the timing belt lower cover. The number one cylinder should be at Top Dead Center (TDC).
4. Remove the engine hanger bracket and timing belt tensioner cap.

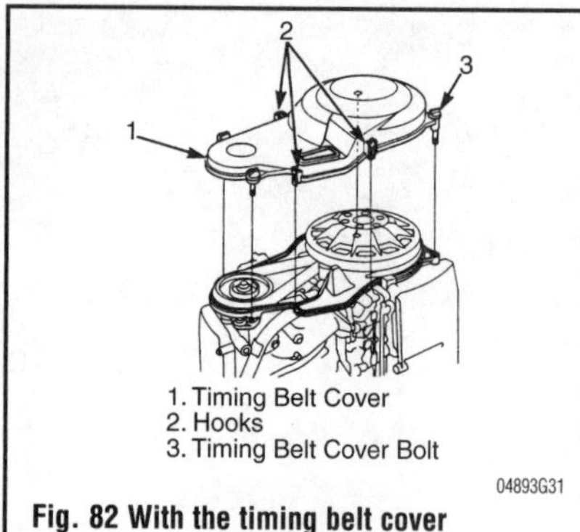

Fig. 82 With the timing belt cover removed carefully inspect the timing belt ribs for damage—BF75A and BF90A

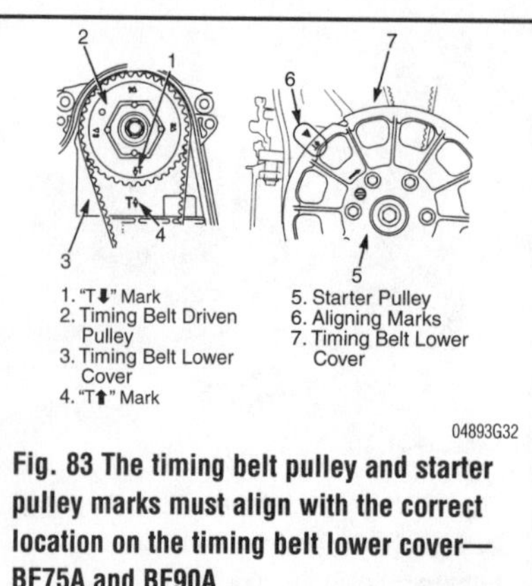

Fig. 83 The timing belt pulley and starter pulley marks must align with the correct location on the timing belt lower cover—BF75A and BF90A

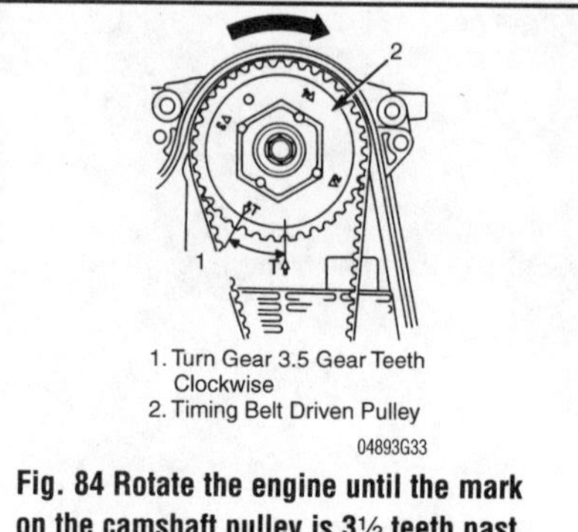

Fig. 84 Rotate the engine until the mark on the camshaft pulley is 3½ teeth past the mark on the cylinder head—BF75A and BF90A

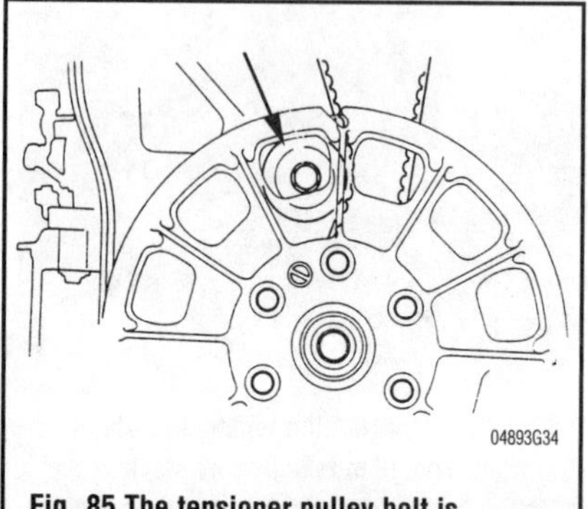

Fig. 85 The tensioner pulley bolt is accessed through an opening in the starter pulley—BF75A and BF90A

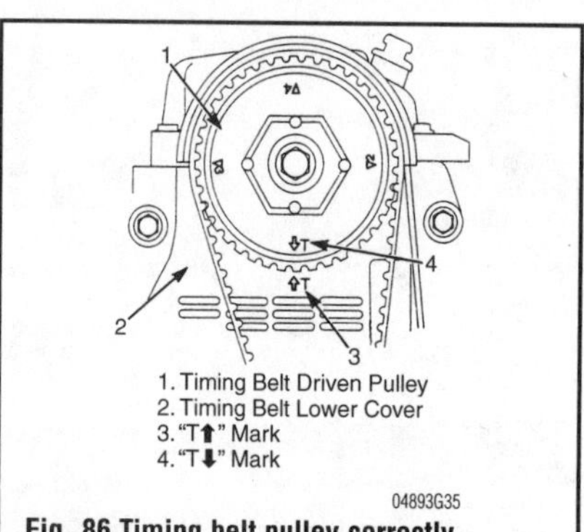

Fig. 86 Timing belt pulley correctly aligned with the mark on the timing belt lower cover—BF115A and BF130A

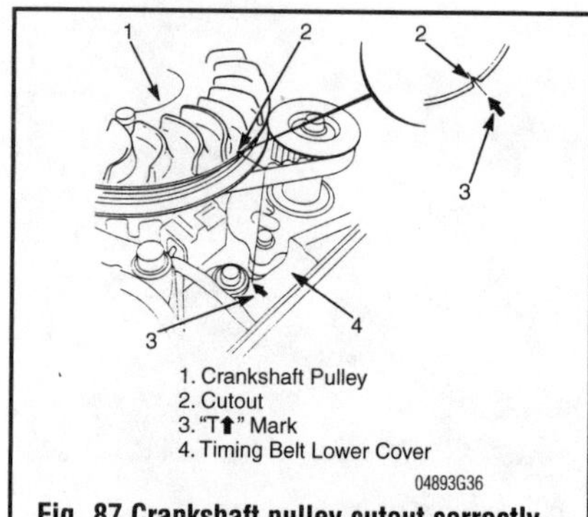

Fig. 87 Crankshaft pulley cutout correctly aligned with mark on the timing belt lower cover—BF115A and BF130A

WARNING

Do not turn the engine counterclockwise; the water pump impeller can be damaged.

5. Starting from the number one cylinder timing mark, rotate the flywheel **clockwise** at least two full turns.
6. After realigning the timing marks for the number one cylinder, slowly turn the starter pulley clockwise until the "T" on the camshaft pulley is at least 3 ½ teeth past the "T" mark on the cylinder head.
7. Loosen the timing belt tensioner bolt.
8. Tighten the timing belt tensioner flange bolt to the specified torque. Do not apply additional pressure on the tensioner pulley; the spring on the pulley bracket spring provides adequate tension for the belt.
9. Install the engine hanger bracket and timing belt tensioner cap.
10. Install the spark plugs.
11. Install the timing belt cover.

BF115A and BF130A

See Figures 86, 87, 88 and 89

Make sure the valve clearance is correct before adjusting the belt tension. The balancer belt tension and the timing belt timing should be adjusted simultaneously.

1. Turning the crankshaft pulley clockwise, align the "T" mark on the camshaft pulley with the "T" mark on the timing belt lower cover. Also make sure that the cutout in the crankshaft pulley aligns with the "T" mark on the timing belt lower cover.
2. Loosen the adjusting nut and the adjusting bracket bolt $FR2/3to one turn and be sure that the balancer belt tensioner moves freely to the opposite side from the belt (i.e. belt loosening direction).

Do not loosen the adjusting nut and adjusting bracket more than one turn.

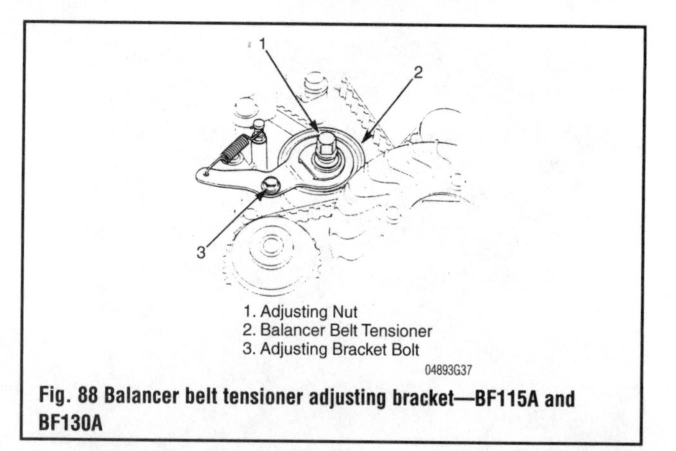

Fig. 88 Balancer belt tensioner adjusting bracket—BF115A and BF130A

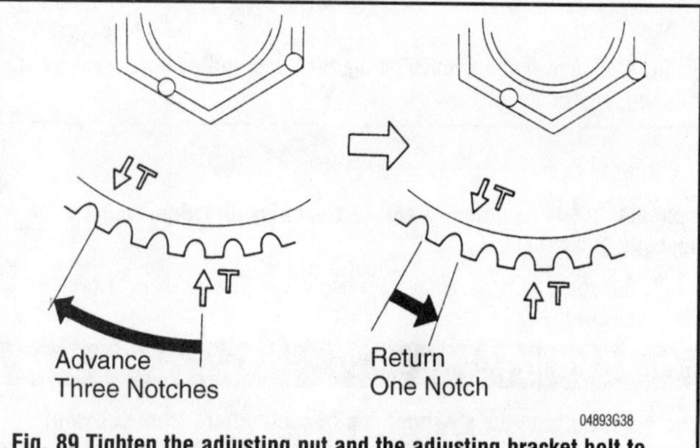

Fig. 89 Tighten the adjusting nut and the adjusting bracket bolt to the proper torque specifications—BF115A and BF130A

3. Make sure that each alignment mark is in proper alignment, then turn the crankshaft pulley 2 turns **CLOCKWISE**. Then slowly turn it a further three notches of the timing belt driven gear. Now, turn the crankshaft pulley **COUNTERCLOCKWISE** by 1 notch of the timing belt driven pulley gear.

➡If the crankshaft pulley was turned more than three notches of the timing belt driven pulley gear, turn the crankshaft pulley clockwise to bring No.1 piston to it compression stroke, then repeat the belt tension adjustment procedure.

4. Tighten the adjusting nut to 33 ft. lbs. (44Nm) and the adjusting bracket bolt to 9 ft. lbs. (12Nm).

➡Each belt tensioner is of the automatically adjusting type. Do not push the tensioners in the direction the timing belt with your hand.

Trim, Tilt & Pivot Points

INSPECTION AND LUBRICATION

▸ See Figures 90, 91, 92 and 93

The steering head and other pivot points of the outboard-to-engine mounting components need periodic lubrication with marine grade grease to provide smooth operation and prevent corrosion. Usually, these pivot points are easily lubricated by simply attaching a grease gun to the fittings.

If the engine is used in salt water, the frequency of applying lubricant is usually doubled in comparison to operation in fresh water. Due to the very corrosive nature of salt water, some sort or anti-seize type thread compound should be used on all exposed fasteners outside of the cowling to reduce the chance of them seizing in place and breaking off when you try to remove them.

➡Rinsing off the engine after each use is a very good habit to get into, not only does it help preserve the appearance of the engine, it virtually eliminates the corrosive effects of operating in salt water.

Propeller

▸ See Figures 94, 95 and 96

The propeller should be inspected regularly to be sure the blades are in good condition. If any of the blades become bent or nicked, this condition will set up vibrations in the motor. Remove and inspect the propeller. Use a file to trim nicks and burrs. Take care not to remove any more material than is absolutely necessary.

Also, check the rubber and splines **INSIDE** the propeller hub for damage. If there is damage to either of these, take the propeller to your local marine dealer or a "prop shop". They can evaluate the damaged propeller and determine if it can be saved by rehubbing.

Additionally, the propeller should be removed each time the boat is hauled from the water at the end of an outing. Any material entangled behind the propeller should be removed before any damage to the shaft and seals can occur. This may seem like a waste of time, but the small amount of time involved in removing the propeller is returned many times by reduced maintenance and repair, including the replacement of expensive parts.

04893P04

Fig. 90 The steering head pivot on this BF35A has a fitting for lubricating the bushings

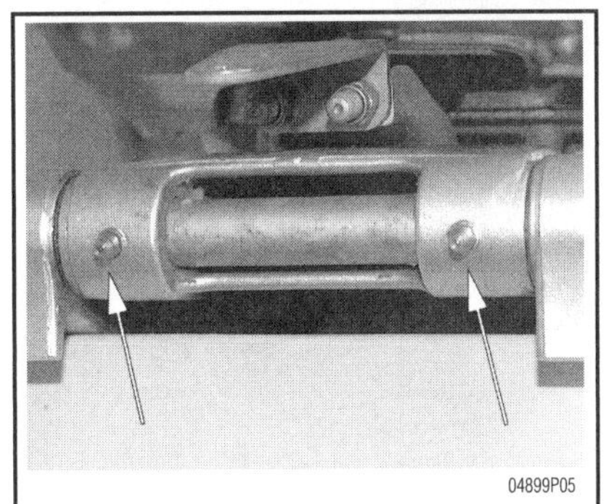
04899P05

Fig. 91 The tilt pivot on the mounting bracket also has fittings for lubricating the bushings

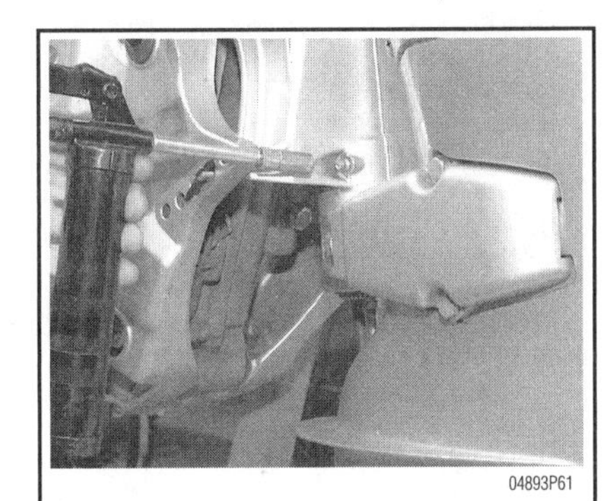
04893P61

Fig. 92 Some lubrication fittings may be hidden so look carefully . . .

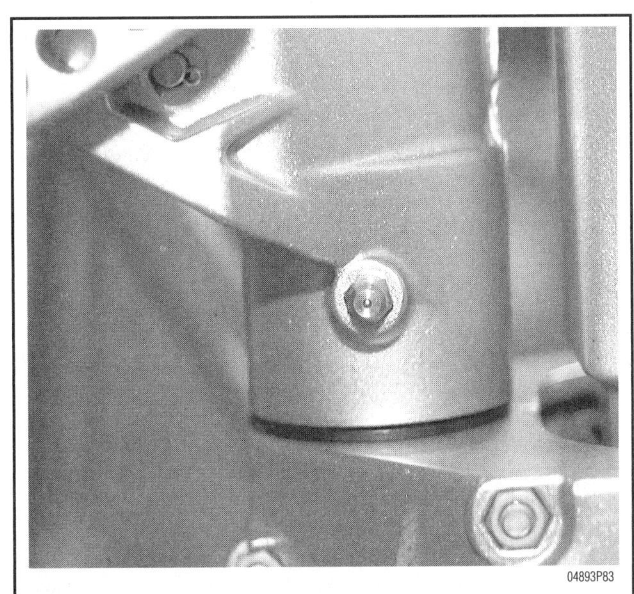
04893P83

Fig. 93 . . . like this grease fitting on the steering tube

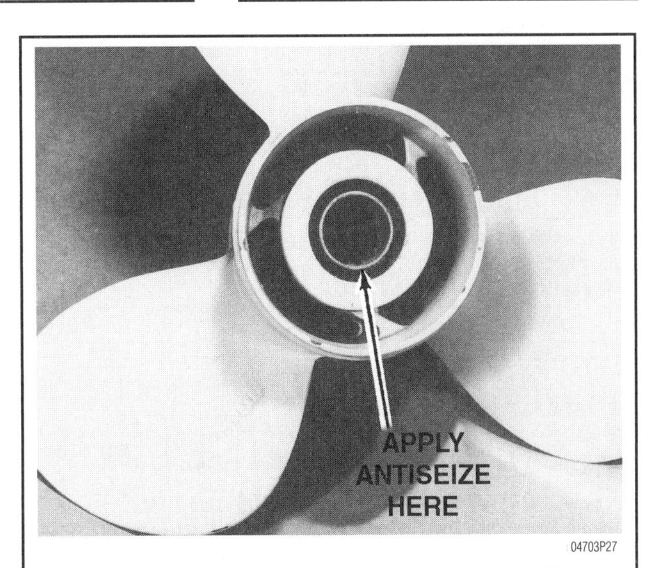

04703P27

Fig. 94 An application of anti-seize on the propeller shaft splines will prevent the propeller from seizing on the shaft and facilitate easier removal for the next service

Fig. 95 A block of wood will prevent the propeller from turning as the nut is being turned

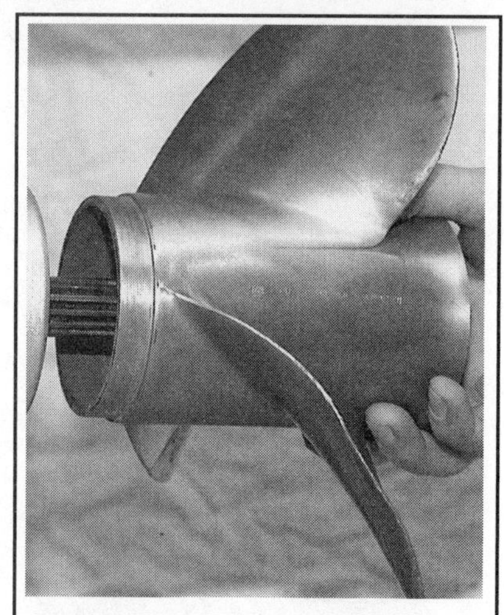

Fig. 75 Once the propeller nut and washer is removed, the propeller can be slid off the shaft

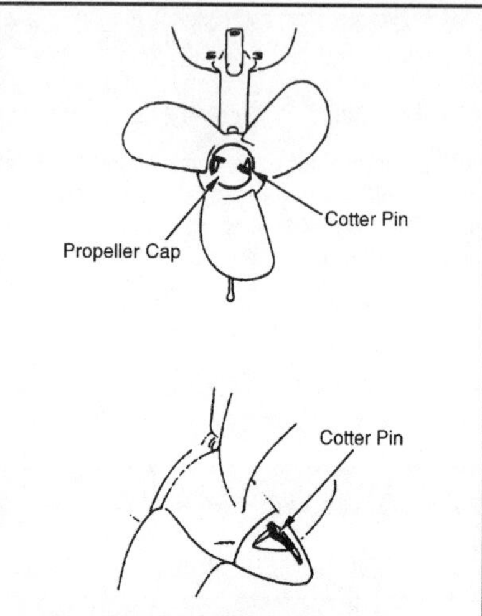

Fig.97 On BF2A—BF8A, remove the cotter pin and unscrew the propeller cap to remove the propeller

BF2A—BF8A

◆ **See Figures 97 and 98**

The propeller on BF2A<en dash>BF8A is held on by a propeller cap, secured with a stainless steel cotter pin. A shear pin (a small pin mounted in a cross-drilled hole in the propeller shaft) is used to keep the propeller from spinning on its shaft when power is applied. The shear pin also protects the drivetrain and propeller. If the propeller happens to hit a foreign object in the water, the shear pin breaks, instead of damaging the lower unit or the propeller itself. If this happens while you're on the water, a new shear pin and cotter pin will be needed to get you back to port. Most outboards of this type have extra shear and cotter pins stored on or inside the engine cover, or on the stern bracket. If you do break a shear pin, be sure to replace the standby pins with new ones as soon as you return to port.

1. Using pliers, straighten the cotter pin, and pull it out of the propeller cap.
2. Unscrew the propeller cap, and slide the propeller off the shaft. Be prepared to catch the shear pin, which is behind the propeller.
3. If the shear pin is broken, it may have to be tapped out with a small punch.

To install:

4. Install the shear pin into the propeller shaft.
5. If the shear pin was broken, make sure the remains of the pin are not lodged in the slot on the mating surface of the propeller.
6. Coat the surface of the propeller shaft with anti-seize or marine lubricant.
7. Slide the propeller on the shaft until it is against the shear pin, then turn the propeller until the slot in the propeller aligns with the shear pin.

✻✻ WARNING

Make certain that the shear pin is engaged with the slot in the propeller. The propeller will spin freely on the shaft if the shear pin is not aligned.

8. Install the propeller cap.
9. Install a new stainless steel cotter pin, and bend the ends away from each other to secure it in the propeller cap.

✻✻ WARNING

Always use a stainless steel cotter pin on the propeller shaft. Standard steel pins will rust quickly, and can completely disintegrate, allowing the propeller nut to vibrate free of the shaft. If this were to happen, the propeller can fall off the shaft, which can cause the engine to spin without a load (causing major engine damage) and you stranded on the water.

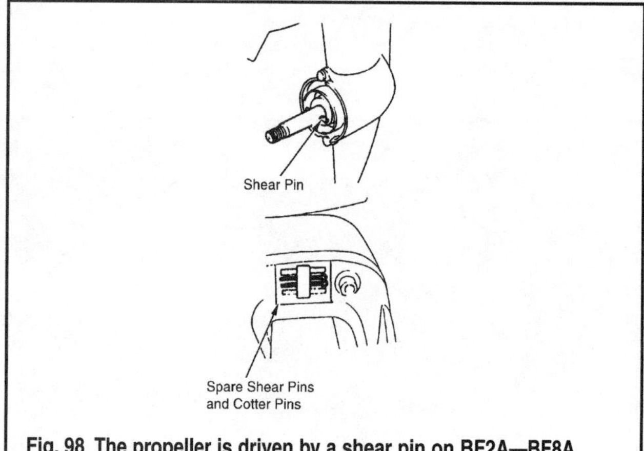

Fig. 98 The propeller is driven by a shear pin on BF2A—BF8A

BF75A and BF90A

◆ **See Figure 98**

These engines are unique in the way the propeller is secured to the shaft. Instead of a castellated nut and cotter pin, a special star-shaped lockwasher is used that secures the propeller nut when the ends of the washer are folded into place.

1. Bend the tabs of the special lockwasher away from the propeller nut.
2. Remove the propeller nut from the shaft. If necessary, use a block of wood between the anti-cavitation plate and the propeller to keep it from turning.
3. Remove the special lockwasher, followed by the propeller.
4. If necessary, remove the thrust washer from the propeller shaft. (Note the chamfered edge of the thrust washer; it should face the lower unit.)

To install:

5. Install the thrust washer on the propeller shaft.
6. Lightly coat the splines of the propeller shaft with anti-seize compound.
7. Slide the propeller onto the shaft, followed by the special lockwasher.

■ **A new lockwasher should be used whenever the propeller is removed.**

8. Install the propeller nut, and torque it to 55 ft. lbs. (75 Nm) for all except 1999 and later BF75A/BF90A which should be tightened to 35 ft. lbs. (44 Nm). If necessary, hold the propeller stationary by using a block of wood placed between the propeller and the anti-cavitation plate.

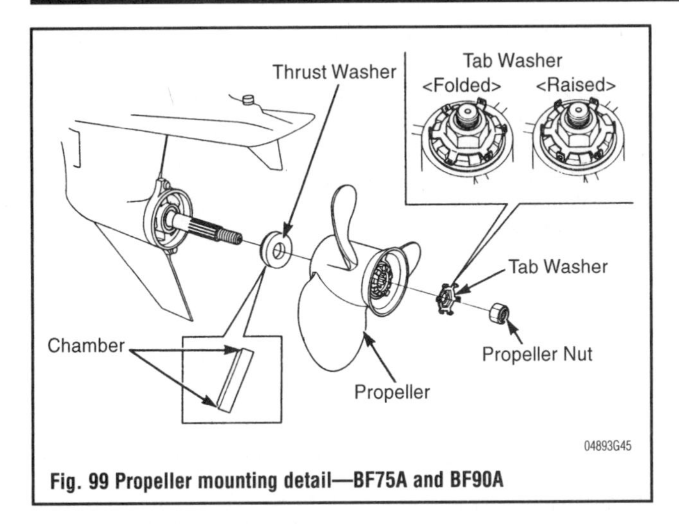

Fig. 99 Propeller mounting detail—BF75A and BF90A

9. Using a medium-sized punch, push the tabs on the lockwasher against the propeller nut to hold it in place.

Except BF2A, BF20, BF5A, BF50, BF8A, BF75A and BF90A

See Figure 100

The propeller is held in place on these engines by a castellated nut and a stainless steel cotter pin. This is the most common type of method employed by outboard manufacturers.

1. Using needle-nose pliers, straighten the cotter pin on the propeller shaft and push it out from the nut.
2. Remove the castellated propeller nut, followed by any washers or spacers. Keep track of their order for installation.
3. Slide the propeller from the shaft, followed by any washers or spacers mounted behind the propeller.

To install:

4. Lightly coat the splines of the propeller shaft with anti-seize compound to prevent corrosion.
5. Install the thrust washer (in the proper direction) on the propeller shaft.
6. Slide the propeller on the shaft, and install any washers or spacers.

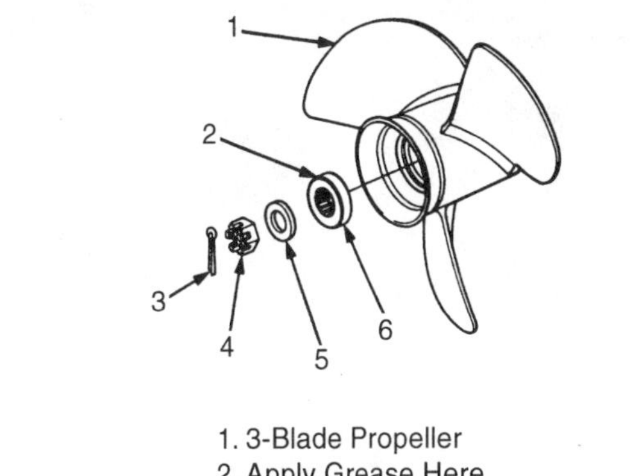

1. 3-Blade Propeller
2. Apply Grease Here
3. Cotter Pin
4. Castle Nut
5. Washer
6. Spacer

04893G44

Fig. 100 Propeller mounting detail—except BF75A and BF90A

7. Install the castellated propeller nut. Tighten it only by hand. If the nut has to be turned to allow installation of the cotter pin, turn it clockwise with a wrench until the cotter pin can be installed. The torque on the nut should not exceed 25 ft. lbs. (35 Nm) for the BF9.9A–BF50A and 33 ft. lbs. (44Nm) for the BF115A and BF130A .

**** WARNING**

Always use a stainless steel cotter pin on the propeller shaft. Standard steel pins will rust quickly, and can completely disintegrate, allowing the propeller nut to vibrate free of the shaft. If this were to happen, the propeller can fall off the shaft, which can cause the engine to spin without a load (causing major engine damage) and leaving you stranded in the water.

8. Using a new stainless steel cotter pin, place it through the propeller shaft and bend the ends around the nut.

BOAT MAINTENANCE

Inside The Boat

See Figure 101

The following points may be lubricated with an all purpose marine lubricant:

- Remote control cable ends next to the hand nut. DO NOT over-lubricate the cable
- Steering arm pivot socket
- Exposed shaft of the cable passing through the cable guide tube
- Steering link rod to steering cable

Fiberglass Hulls

See Figures 102, 103 and 104

Fiberglass reinforced plastic hulls are tough, durable, and highly resistant to impact. However, like any other material they can be damaged. One of the advantages of this type of construction is the relative ease with which it may be repaired. Because of its break characteristics, and the simple techniques used in restoration, these hulls have gained popularity throughout the world. From the most congested urban marina, to isolated lakes in wilderness areas, to the

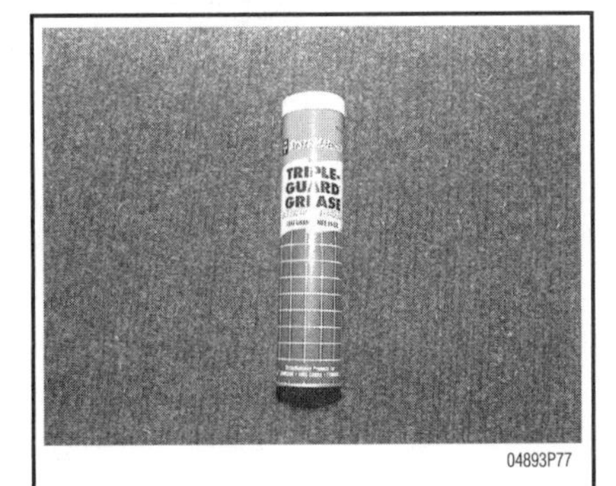

04893P77

Fig. 101 Use only a good quality marine grade grease for lubrication

SELSTK00

Fig. 102 In areas where marine growth is a problem, a coating of anti-foul bottom paint should be applied

SELSTK55

Fig. 103 The best way to care for a fiberglass hull is to wash it thoroughly

04892P04

Fig. 104 Fiberglass, vinyl and rubber care products, such as those available from Meguiar's are available to protect every part of your boat

severe cold of far off northern seas, and in sunny tropic remote rivers of primitive islands or continents, fiberglass boats can be found performing their daily task with a minimum of maintenance.

A fiberglass hull has almost no internal stresses. Therefore, when the hull is broken or stove-in, it retains its true form. It will not dent to take an out-of-shape set. When the hull sustains a severe blow, the impact will be either absorbed by deflection of the laminated panel or the blow will result in a definite, localized break. In addition to hull damage, bulkheads, stringers, and other stiffening structures attached to the hull may also be affected and therefore, should be checked. Repairs are usually confined to the general area of the rupture.

➡The best way to care for a fiberglass hull is to wash it thoroughly, immediately after hauling the boat while the hull is still wet.

A foul bottom can seriously affect boat performance. This is one reason why racers, large and small, both powerboat and sail, are constantly giving attention to the condition of the hull below the waterline.

In areas where marine growth is prevalent, a coating of vinyl, anti-fouling bottom paint should be applied. If growth has developed on the bottom, it can be removed with a solution of Muriatic acid applied with a brush or swab and then rinsed with clear water. Always use rubber gloves when working with Muriatic acid and take extra care to keep it away from your face and hands. The fumes are toxic. Therefore, work in a well-ventilated area, or if outside, keep your face on the windward side of the work.

Barnacles have a nasty habit of making their home on the bottom of boats which have not been treated with anti-fouling paint. Actually they will not harm the fiberglass hull, but can develop into a major nuisance.

If barnacles or other crustaceans have attached themselves to the hull, extra work will be required to bring the bottom back to a satisfactory condition. First, if practical, put the boat into a body of fresh water and allow it to remain for a few days. A large percentage of the growth can be removed in this manner. If this remedy is not possible, wash the bottom thoroughly with a high-pressure fresh water source and use a scraper. Small particles of hard shell may still hold fast. These can be removed with sandpaper.

Trim Tabs, Anodes and Lead Wires

See Figures 105 thru 112

Check the trim tabs and the anodes (zinc). Replace them, if necessary. The trim tab must make a good ground inside the lower unit. Therefore, the trim tab and the cavity must not be painted. In addition to trimming the boat, the trim tab acts as a zinc electrode to prevent electrolysis from acting on more expensive parts. It is normal for the tab to show signs of erosion. The tabs are inexpensive and should be replaced frequently.

Clean the exterior surface of the unit thoroughly. Inspect the finish for dam-

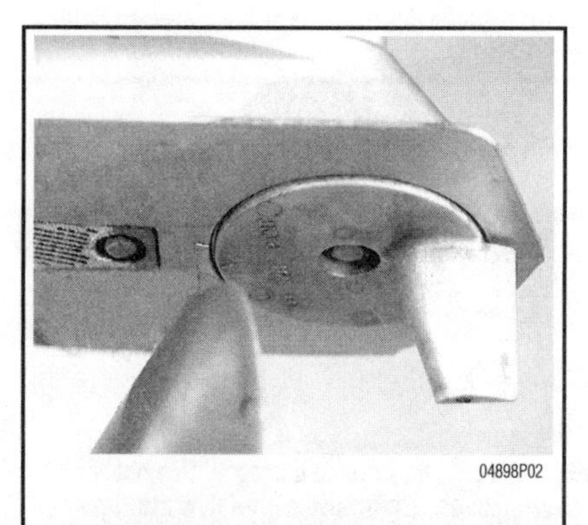

04898P02

Fig. 105 What a trim tab should look like when it's in good condition

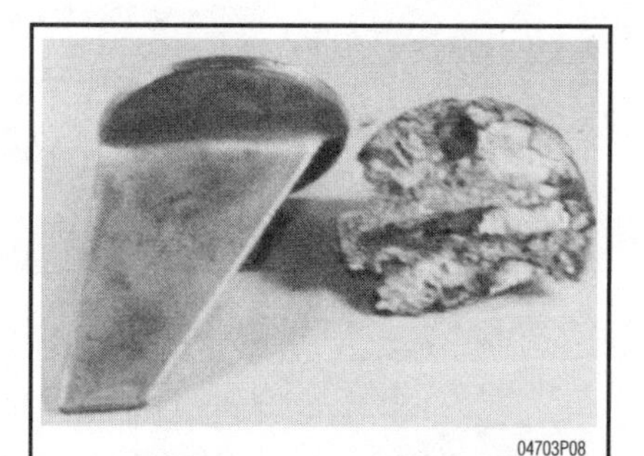

04703P08

Fig. 106 Such extensive erosion of a trim tab compared with a new tab suggests an electrolysis problem or complete disregard for periodic maintenance

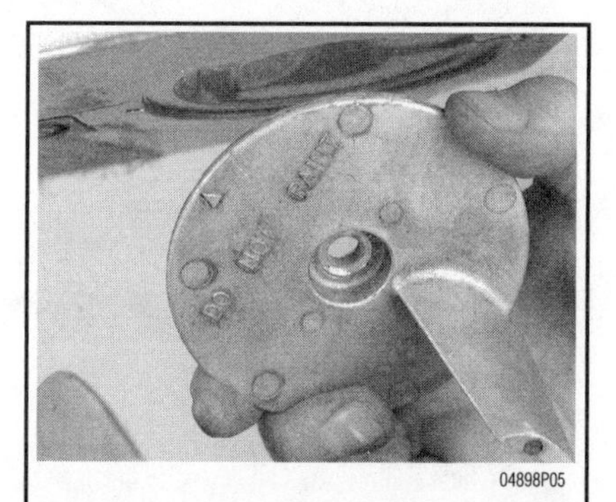

04898P05

Fig. 107 Although many outboards use the trim tab as an anode . . .

04899P21

Fig. 108 . . . other types of anodes are also used throughout the outboard, like this one on the stern bracket . . .

04893P80

Fig. 109 . . . and this one on the lower unit

04891P02

Fig. 110 Anodes installed in the water jacket of a powerhead provide added protection against corrosion

Fig. 111 Most anodes are easily removed by loosening and removing their attaching fasteners

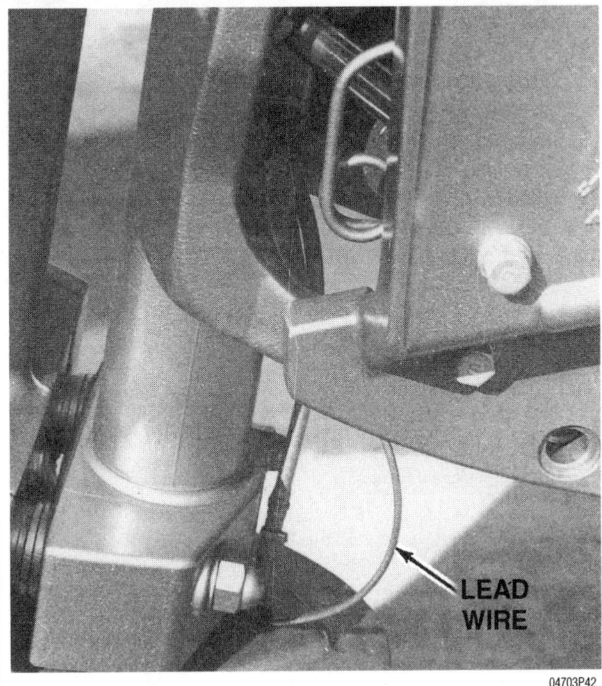

Fig. 112 One of the many lead wires used to connect bracketed parts. Lead wires are used as an assist in reducing corrosion

age or corrosion. Clean any damaged or corroded areas, and then apply primer and matching paint.

Check the entire unit for loose, damaged, or missing parts.

An anode is attached across both clamp brackets. It also serves as protection for the coil of hydraulic hoses beneath the trim/tilt unit between the brackets.

Lead wires provide good electrical continuity between various brackets which might be isolated from the trim tab by a coating of lubricant between moving parts.

Battery

Difficulty in starting accounts for almost half of the service required on boats each year. A survey by Champion Spark Plug Company indicated that roughly one third of all boat owners experienced a "won't start" condition in a given year. When an engine won't start, most people blame the battery when, in fact, it may be that the battery has run down in a futile attempt to start an engine with other problems.

Maintaining your battery in peak condition may be though of as either tune-up or maintenance material. Most wise boaters will consider it to be both. A complete check up of the electrical system in your boat at the beginning of the boating season is a wise move. Continued regular maintenance of the battery will ensure trouble free starting on the water.

A complete battery service procedure is included in the "Maintenance" section of this manual. The following are a list of basic electrical system service procedures that should be performed as part of any tune-up.

- Check the battery for solid cable connections
- Check the battery and cables for signs of corrosion damage
- Check the battery case for damage or electrolyte leakage
- Check the electrolyte level in each cell
- Check to be sure the battery is fastened securely in position
- Check the battery's state of charge and charge as necessary
- Check battery voltage while cranking the starter. Voltage should remain above 9.5 volts
- Clean the battery, terminals and cables
- Coat the battery terminals with dielectric grease or terminal protector

Batteries which are not maintained on a regular basis can fall victim to parasitic loads (small current drains which are constantly drawing current from the battery). Normal parasitic loads may drain a battery on boat that is in storage and not used frequently. Boats that have additional accessories with increased parasitic load may discharge a battery sooner. Storing a boat with the negative battery cable disconnected or battery switch turned off will minimize discharge due to parasitic loads.

CLEANING

Keep the battery clean, as a film of dirt can help discharge a battery that is not used for long periods. A solution of baking soda and water mixed into a paste may be used for cleaning, but be careful to flush this off with clear water.

➡Do not let any of the solution into the filler holes on non-sealed batteries. Baking soda neutralizes battery acid and will de-activate a battery cell.

CHECKING SPECIFIC GRAVITY

The electrolyte fluid (sulfuric acid solution) contained in the battery cells will tell you many things about the condition of the battery. Because the cell plates must be kept submerged below the fluid level in order to operate, maintaining the fluid level is extremely important. In addition, because the specific gravity of the acid is an indication of electrical charge, testing the fluid can be an aid in determining if the battery must be replaced. A battery in a boat with a properly operating charging system should require little maintenance, but careful, periodic inspection should reveal problems before they leave you stranded.

**** CAUTION**

Battery electrolyte contains sulfuric acid. If you should splash any on your skin or in your eyes, flush the affected area with plenty of clear water. If it lands in your eyes, get medical help immediately.

As stated earlier, the specific gravity of a battery's electrolyte level can be used as an indication of battery charge. At least once a year, check the specific gravity of the battery. It should be between 1.20 and 1.26 on the gravity scale. Most parts stores carry a variety of inexpensive battery testing hydrometers. These can be used on any non-sealed battery to test the specific gravity in each cell.

Conventinal Battery

➧ See Figures 113 and 114

A hydrometer is required to check the specific gravity on all batteries that are not maintenance-free. The hydrometer has a squeeze bulb at one end and a nozzle at the other. Battery electrolyte is sucked into the hydrometer until the float or pointer is lifted from its seat. The specific gravity is then read by noting the position of the float/pointer. If gravity is low in one or more cells, the battery should be slowly charged and checked again to see if the gravity has come up. Generally, if after charging, the specific gravity of any two cells varies more than 50 points (0.50), the battery should be replaced, as it can no longer produce sufficient voltage to guarantee proper operation.

Check the battery electrolyte level at least once a month, or more often in hot

Fig. 113 On non-maintenance free batteries with translucent cases, the electrolyte level can be seen through the case; on other types (such as the one shown), the cell cap must be removed

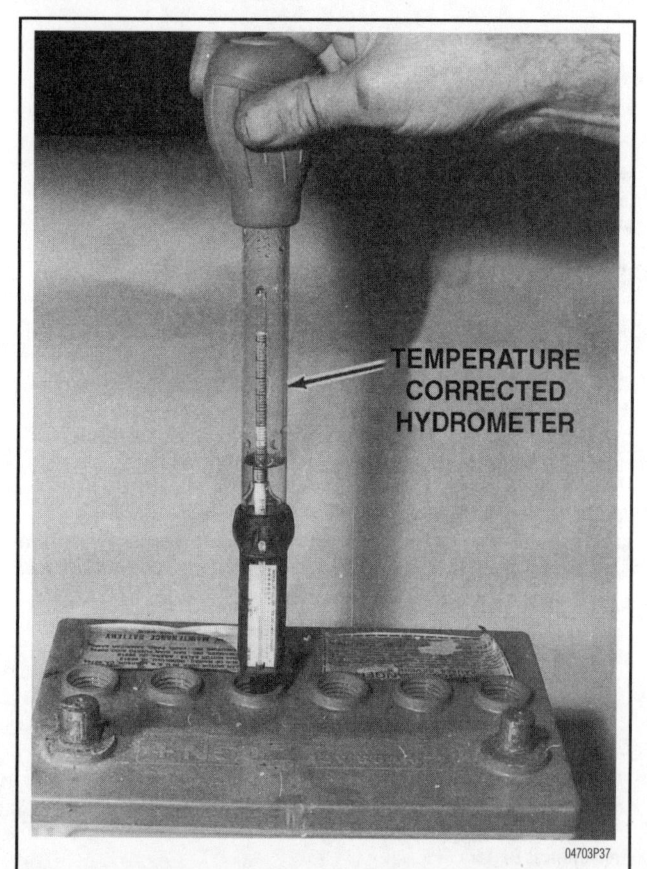

Fig. 114 The best way to determine the condition of a battery is to test the specific gravity of the electrolyte with a battery tester

weather or during periods of extended operation. Electrolyte level can be checked either through the case on translucent batteries or by removing the cell caps on opaque-case types. The electrolyte level in each cell should be kept filled to the split ring inside each cell, or the line marked on the outside of the case.

If the level is low, add only distilled water through the opening until the level is correct. Each cell is separate from the others, so each must be checked and filled individually. Distilled water should be used, because the chemicals and minerals found in most drinking water are harmful to the battery and could significantly shorten its life.

If water is added in freezing weather, the battery should be warmed to allow the water to mix with the electrolyte. Otherwise, the battery could freeze.

Maintenance-Free Batteries

➧ See Figure 115

Although some maintenance-free batteries have removable cell caps for access to the electrolyte, the electrolyte condition and level is usually checked using the built-in hydrometer "eye". The exact type of eye varies between battery manufacturers, but most apply a sticker to the battery itself explaining the possible readings. When in doubt, refer to the battery manufacturer's instructions to interpret battery condition using the built-in hydrometer.

The readings from built-in hydrometers may vary, however a green eye usually indicates a properly charged battery with sufficient fluid level. A dark eye is normally an indicator of a battery with sufficient fluid, but one that may be low in charge. In addition, a light or yellow eye is usually an indication that electrolyte supply has dropped below the necessary level for battery (and hydrometer) operation. In this last case, sealed batteries with an insufficient electrolyte level must usually be discarded.

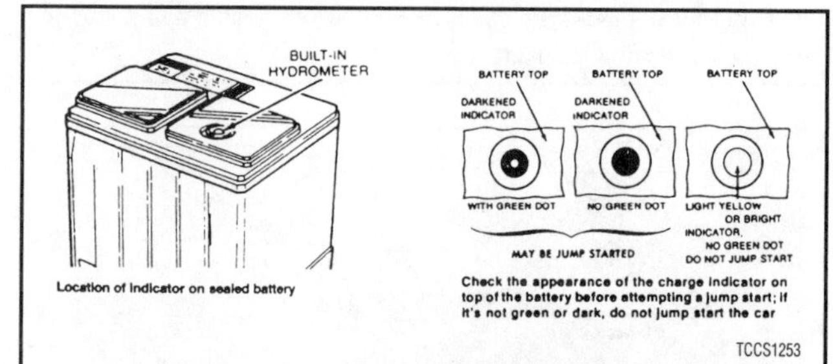

Fig. 115 A typical sealed (maintenance-free) battery with a built-in hydrometer—note that the hydrometer eye may vary between manufacturers; always refer to the battery's label

BATTERY TERMINALS

At least once a season, the battery terminals and cable clamps should be cleaned. Loosen the clamps and remove the cables, negative cable first. On batteries with top mounted posts, the use of a puller specially made for this purpose is recommended. These are inexpensive and available from most auto parts stores.

Clean the cable clamps and the battery terminal with a wire brush, until all corrosion, grease, etc., is removed and the metal is shiny. It is especially important to clean the inside of the clamp thoroughly (a wire brush is useful here), since a small deposit of foreign material or oxidation there will prevent a sound electrical connection and inhibit either starting or charging. It is also a good idea to apply some dielectric grease to the terminal, as this will aid in the prevention of corrosion.

After the clamps and terminals are clean, reinstall the cables, negative cable last; Do not hammer the clamps onto battery posts. Tighten the clamps securely, but do not distort them. Give the clamps and terminals a thin external coating of grease after installation, to retard corrosion.

Check the cables at the same time that the terminals are cleaned. If the insulation is cracked or broken, or if its end is frayed, that cable should be replaced with a new one of the same length and gauge.

BATTERY AND CHARGING SAFETY PRECAUTIONS

Always follow these safety precautions when charging or handling a battery.

1. Wear eye protection when working around batteries. Batteries contain corrosive acid and produce explosive gas a byproduct of their operation. Acid on the skin should be neutralized with a solution of baking soda and water made into a paste. In case acid contacts the eyes, flush with clear water and seek medical attention immediately.
2. Avoid flame or sparks that could ignite the hydrogen gas produced by the battery and cause an explosion. Connection and disconnection of cables to battery terminals is one of the most common causes of sparks.
3. Always turn a battery charger OFF, before connecting or disconnecting the leads. When connecting the leads, connect the positive lead first, then the negative lead, to avoid sparks.

4. When lifting a battery, use a battery carrier or lift at opposite corners of the base.
5. Ensure there is good ventilation in a room where the battery is being charged.
6. Do not attempt to charge or load-test a maintenance-free battery when the charge indicator dot is indicating insufficient electrolyte.
7. Disconnect the negative battery cable if the battery is to remain in the boat during the charging process.
8. Be sure the ignition switch is OFF before connecting or turning the charger ON. Sudden power surges can destroy electronic components.
9. Use proper adapters to connect charger leads to batteries with non-conventional terminals.

BATTERY CHARGERS

See Figure 116

Before using any battery charger, consult the manufacturer's instructions for its use. Battery chargers are electrical devices that change Alternating Current

90991P34

Fig. 116 Automatic battery chargers, like the Battery Tender® from Deltran, have an important advantage—they can stay connected to your battery for extended periods without the possibility of overcharging

(AC) to a lower voltage of Direct Current (DC) ... marine battery. There are two types of battery ch...
A manual battery charger must be physically disc... has come to a full charge. If not, the battery can be ove... Excess charging current at the end of the charging cycle ... resulting in loss of water and active material, substantially r...

➡As a rule, on manual chargers, when the ammeter on ... registers half the rated amperage of the charger, the battery i... charged. This can vary, and it is recommended to use a hydrometer to accurately measure state of charge.

Automatic battery chargers have an important advantage—they can be left connected (for instance, overnight) without the possibility of overcharging the battery. Automatic chargers are equipped with a sensing device to allow the battery charge to taper off to near zero as the battery becomes fully charged. When charging a low or completely discharged battery, the meter will read close to full rated output. If only partially discharged, the initial reading may be less than full rated output, as the charger responds to the condition of the battery. As the battery continues to charge, the sensing device monitors the state of charge and reduces the charging rate. As the rate of charge tapers to zero amps, the charger will continue to supply a few milliamps of current—just enough to maintain a charged condition.

REPLACING BATTERY CABLES

Battery cables don't go bad very often, but like anything else, they can wear out. If the cables on your boat are cracked, frayed or broken, they should be replaced.
When working on any electrical component, it is always a good idea to disconnect the negative (-) battery cable. This will prevent potential damage to many sensitive electrical components
Always replace the battery cables with one of the same length, or you will increase resistance and possibly cause hard starting. Coat the battery posts with a light film of dielectric grease, or a battery terminal protectant spray once you've installed the new cables. If you replace the cables one at a time, you won't mix them up.

➡Any time you disconnect the battery cables, it is recommended that you disconnect the negative (-) battery cable first. This will prevent you from accidentally grounding the positive (+) terminal when disconnecting it, thereby preventing damage to the electrical system.

Before you disconnect the cable(s), first turn the ignition to the OFF position. This will prevent a draw on the battery which could cause arcing. When the battery cable(s) are reconnected (negative cable last), be sure to check all electrical accessories are all working correctly.

TUNE-UP

Introduction

A proper tune-up is the key to long and trouble-free engine life, and the work can yield its own rewards. As a conscientious boater, set aside a Saturday morning, say once a month, to check or replace items which could cause major problems later. Keep your own personal log to jot down which services you performed, how much the parts cost you, the date, and the number of hours on the engine at that time. Keep all receipts for such items as engine oil and filters, so that they may be referred to in case of related problems or to determine operating expenses. As a do-it-yourselfer, these receipts are the only proof you have that the required maintenance was performed. In the event of a warranty problem, these receipts will be invaluable.
The efficiency, reliability, fuel economy and enjoyment available from engine performance are directly dependent on having your outboard tuned properly. The importance of performing service work in the proper sequence cannot be over emphasized. Before making any adjustments, check the specifications. Never rely on memory when making critical adjustments.
A practical maintenance program that is followed throughout the year, is one of the best methods of ensuring the engine will give satisfactory performance. As they say, you can spend a little time now or a lot of time later.
The extent of the engine tune-up is usually dependent on the time lapse since the last service. A complete tune-up of the entire engine would entail almost all of the work outlined in this manual. However, this is usually not necessary in most cases.
In this section, a logical sequence of tune-up steps will be presented in general terms. If additional information or detailed service work is required, refer to the section containing the appropriate instructions.
Each year higher compression ratios are built into modern outboard engines and the electrical systems become more complex. Therefore, the need for reliable, authoritative, and detailed instructions becomes more critical. The information in this section fulfill that requirement.

Tune-Up Sequence

During a major tune-up, a definite sequence of service work should be followed to return the engine its maximum performance level. This type of work should not be confused with troubleshooting (attempting to locate a problem when the engine is not performing satisfactorily). In many cases, these two areas will overlap, because many times a minor or major tune-up will correct the malfunction and return the system to normal operation.
The following list is a suggested sequence of tasks to perform during a tune-up.

- Perform a compression check of each cylinder
- Inspect the spark plugs to determine their condition. Test for adequate spark at the plug
- Start the engine in a body of water and check the water flow through the engine
- Check the gear oil in the lower unit

Tuneup Specifications Chart

Horsepower	Spark Plug NGK	Spark Plug Champion	Spark Plug Gap Inch(mm)	Ignition Timing ° BTDC	Carb Pilot Screw Turns Out	Idle Speed RPM (Neutral)	Valve Lash Intake Inch (mm)	Valve Lash Exhaust Inch (mm)
BF130A	ZFR7F	KJ22CR-L8	0.028-0.031 (0.7-0.8)	10-24°	-	700-800	0.009-0.011 (0.24-0.28)	0.011-0.012 (0.28-0.32)
BF115A	ZFR7F	KJ22CR-L8	0.028-0.031 (0.7-0.8)	10-24°	-	700-800	0.009-0.011 (0.24-0.28)	0.011-0.012 (0.28-0.32)
BF90A	DR7EA	X22ESR-U	0.024-0.028 (0.6-0.7)	5-29°	2 1/4	900-1050	0.007-0.009 (0.18-0.22)	0.010-0.012 (0.26-0.30)
BF75A	DR7EA	X22ESR-U	0.024-0.028 (0.6-0.7)	5-29°	1 7/8	900-1050	0.007-0.009 (0.18-0.22)	0.010-0.012 (0.26-0.30)
BF50A	DR7EA	X22ESR-U	0.024-0.028 (0.6-0.7)	5-32°	1	900-1050	0.005-0.007 (0.13-0.17)	0.008-0.010 (0.21-0.25)
BF45A	DR7EA	X22ESR-U	0.024-0.028 (0.6-0.7)	5-32°	2 1/8	900-1050	0.005-0.007 (0.13-0.17)	0.008-0.010 (0.21-0.25)
BF40A	DR7EA	X22ESR-U	0.024-0.028 (0.6-0.7)	5-28°	2 1/4	900-1050	0.005-0.007 (0.13-0.17)	0.008-0.010 (0.21-0.25)
BF35A	DR7EA	X22ESR-U	0.024-0.028 (0.6-0.7)	5-28°	2 1/8	900-1050	0.005-0.007 (0.13-0.17)	0.008-0.010 (0.21-0.25)
BF30A	DR7EA	X22ESR-U	0.024-0.028 (0.6-0.7)	5-32°	3	850-950	0.004-0.006 (0.10-0.14)	0.007-0.009 (0.18-0.22)
BF25A	DR7EA	X22ESR-U	0.024-0.028 (0.6-0.7)	5-26°	2	850-950	0.004-0.006 (0.10-0.14)	0.007-0.009 (0.18-0.22)
BF15A/BF15D	DR-6HS	X20FSR-U	0.024-0.028 (0.6-0.7)	5-35°	1 5/8	1050-1150	0.004-0.006 (0.10-0.14)	0.007-0.009 (0.18-0.22)
BF9.9A/BF9.9D	DR-5HS	X16FSR-U	0.024-0.028 (0.6-0.7)	5-35°	2 3/4	1050-1150	0.004-0.006 (0.10-0.14)	0.007-0.009 (0.18-0.22)
BF75/BF100	DR-5HS	X16FSR-U	0.024-0.028 (0.6-0.7)	5-35°	2 1/2	1050-1150	0.002-0.004 (0.06-0.10)	0.002-0.004 (0.06-0.10)
BF8A/BF80/BF8D	DR-5HS	X16FSR-U	0.024-0.028 (0.6-0.7)	5-35°	2 1/2	1050-1150	0.002-0.004 (0.06-0.10)	0.002-0.004 (0.06-0.10)
BF5A/BF50/BF5D	BPR5ES	W16EPR-U	0.028-0.031 (0.7-0.8)	15-35°	2 3/8	1050-1150	0.002-0.006 (0.06-0.14)	0.004-0.007 (0.11-0.19)
BF2A/BF20	BMR-4A	W14MR-U	0.028-0.031 (0.7-0.8)	20°	2 1/8	1050-1150	0.003-0.006 (0.08-0.16)	0.003-0.006 (0.08-0.16)
BF2D	CR5HSB	①	0.024-0.028 (0.6-0.7)	27°	2	1900-2100	0.002-0.004 (0.06-0.10)	0.004-0.005 (0.09-0.13)

① No Champion Number Provided, however Denso: U16FSR-UB

- Check the carburetor adjustments and the need for an overhaul.
- Check the fuel pump for adequate performance and delivery
- Make a general inspection of the ignition system
- Test the starter motor and the solenoid, if so equipped
- Check the internal wiring
- Check the timing and synchronization

Compression Check

Before beginning to tune any engine, ensure the engine has satisfactory compression. An engine with worn or broken piston rings, burned pistons, or scored cylinder walls, will not perform properly no matter how much time and expense is spent on the tune-up. Poor compression must be corrected or the tune-up will not give the desired results.

Cylinder compression test results are extremely valuable indicators of internal engine condition. The best marine mechanics automatically check an engine's compression as the first step in a comprehensive tune-up. Obviously, it is useless to try to tune an engine with extremely low or erratic compression readings, since a simple tune-up will not cure the problem.

The pressure created in the combustion chamber may be measured with a gauge that remains at the highest reading it measures during the action of a one-way valve. This gauge is inserted into the spark plug hole.
A compression test will uncover many mechanical problems that can cause rough running or poor performance.

CHECKING COMPRESSION

◆ **See Figure 117**

Prepare the engine for a compression test as follows:

1. Run the engine until it reaches operating temperature. The engine is at operating temperature a few minutes. If the test is performed on a cold engine, the readings will be considerably lower than normal, even if the engine is in perfect mechanical condition.
2. Label and disconnect the spark plug wires. Always grasp the molded cap and pull it loose with a twisting motion to prevent damage to the connection.
3. Clean all dirt and foreign material from around the spark plugs, and then remove all the plugs. Keep them in order by cylinder for later evaluation.
4. Ground the spark plug leads to the engine to render the ignition system inoperative while performing the compression check.
5. Insert a compression gauge into the No. 1, top, spark plug opening.
6. Crank the engine with the starter through at least 4 complete strokes with the throttle at the wide-open position, to obtain the highest possible reading. Record the reading.
7. Repeat the test and record the compression for each cylinder.

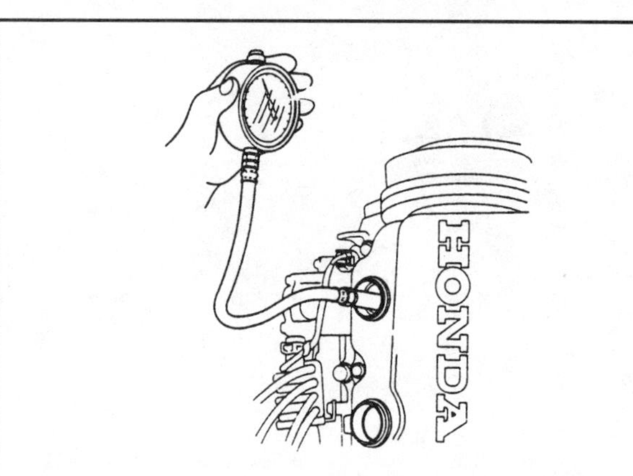

Fig. 117 Insert a compression gauge into the spark plug hole and crank the engine with the starter through at least 4 complete strokes with the throttle wide open to obtain the highest possible reading

8. On single cylinder engines, look for a steady build in pressure during cranking. The first compression stroke should produce the largest needle movement, followed by a slightly less movement from the first reading, and so on. By the fifth compression stroke, the jump of the needle on the pressure gauge should be minimal.
9. A variation between cylinders is far more important than the actual readings. A variation of more than 15 psi (103 kPa), between cylinders indicates the lower compression cylinder is defective. Not all engines will exhibit the same compression readings. In fact, two identical engines may not have the same compression. Generally, the rule of thumb is that the lowest cylinder should be within 25% of the highest (difference between the two readings).
10. If compression is low in one or more cylinders, the problem may be worn, broken, or sticking piston rings, worn valves, seats or guides, scored pistons or worn cylinders.

LOW COMPRESSION

Compression readings that are generally low indicate worn, broken, or sticking piston rings, scored pistons or worn cylinders, or problems with the valves. Low compression in two adjacent cylinders (with normal compression in the other cylinders) indicates a blown head gasket between the low-reading cylinders. Other problems are possible (broken ring, hole burned in a piston), but a blown head gasket is most likely.

Spark Plugs

➧ See Figures 118 and 119

Spark plug life and efficiency depend upon the condition of the engine and the combustion chamber temperatures to which the plug is exposed. These temperatures are affected by many factors, such as compression ratio of the engine, air/fuel mixtures and the type of normally placed on your engine.

Factory installed plugs are, in a way, compromise plugs, since the factory has no way of knowing what typical loads your engine will see. However, most people never have reason to change their plugs from the factory recommended heat range.

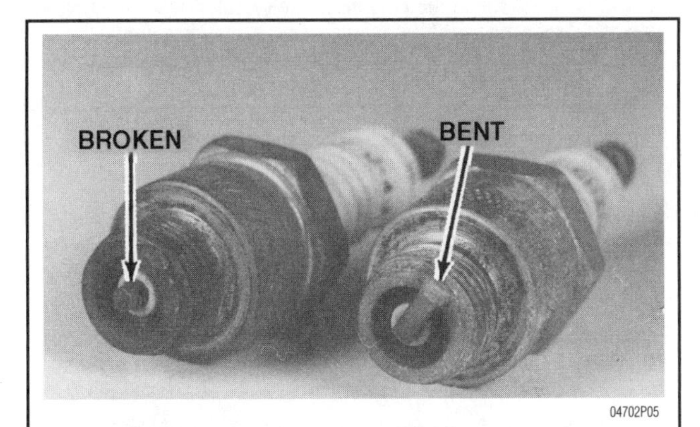

Fig. 118 These are severely damaged spark plugs. Notice the broken electrode on the left plug. The electrode must be found and retrieved prior to returning the powerhead to service

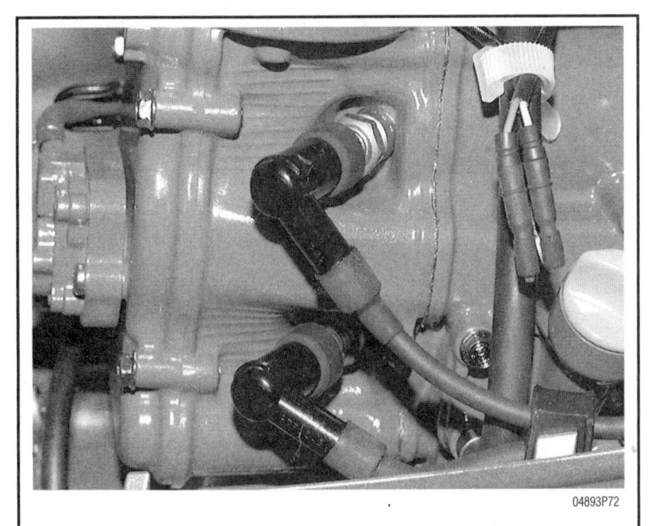

Fig. 119 Typically, spark plugs are screwed into the side of the cylinder head and intersect the cylinder at an angle to provide better combustion

SPARK PLUG HEAT RANGE

➧ See Figure 120

Spark plug heat range is the ability of the plug to dissipate heat. The longer the insulator (or the farther it extends into the engine), the hotter the plug will operate; the shorter the insulator (the closer the electrode is to the block's cooling passages) the cooler it will operate. A plug that absorbs little heat and remains too cool will quickly accumulate deposits of oil and carbon since it is not hot enough to burn them off. This leads to plug fouling and consequently to misfiring. A plug that absorbs too much heat will have no deposits but, due to the excessive heat, the electrodes will burn away quickly and might possibly lead to pre-ignition or other ignition problems. Pre-ignition takes place when plug tips get so hot that they glow sufficiently to ignite the air/fuel mixture before the actual spark occurs. This early ignition will usually cause a pinging during heavy loads.

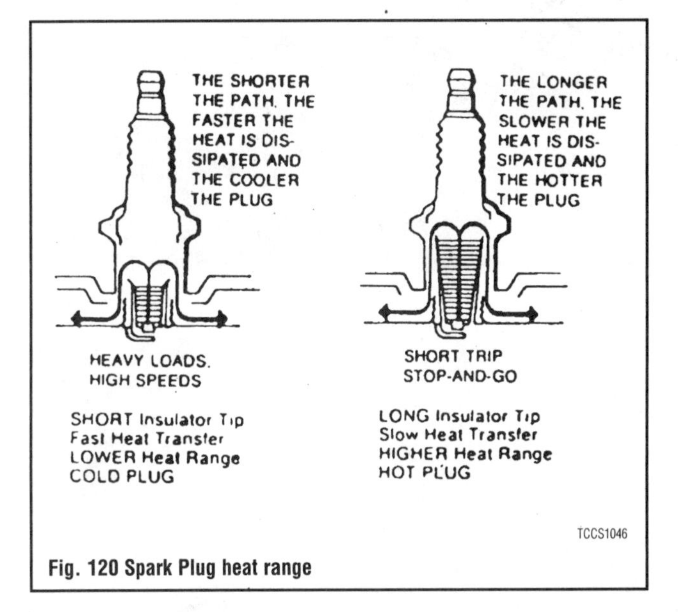

Fig. 120 Spark Plug heat range

SPARK PLUG SERVICE

➡New technologies in spark plug and ignition system design have pushed the recommended replacement interval to every 100 hours of operation (6 months). However, this depends on usage and conditions. This holds true unless internal engine wear or damage cause plug fouling. If you suspect this, you may wish to remove and inspect the plugs before the recommended time.

Spark plugs should only require replacement once a season. The electrode on a new spark plug has a sharp edge, but with use, this edge becomes rounded by wear, causing the plug gap to increase. As the gap increases, the plug's voltage requirement also increases. It requires a greater voltage to jump the wider gap and about two to three times as much voltage to fire a plug at high speeds than at idle.

Tools needed for spark plug replacement include a ratchet, short extension, spark plug socket (there are two types; either $FR13/16 inch or ⅝ inch, depending upon the type of plug), a combination spark plug gauge and gapping tool, and a can of penetrating oil or anti-seize type grease for engines with aluminum heads.

When removing spark plugs, work on one at a time. Don't start by removing the plug wires all at once, because unless you number them, they may become mixed up. Take a minute before you begin and number the wires with tape.

REMOVAL & INSTALLATION

➧ See Figures 121, 122 and 123

1. Disconnect the negative battery cable, and if the engine has been run recently, allow the engine to thoroughly cool. Attempting to remove plugs from a hot cylinder head could cause the plugs to seize and damage the threads in the cylinder head. Especially on aluminum heads!
2. Carefully twist the spark plug wire cap to loosen it, then pull the cap using a twisting motion and remove it from the plug. Be sure to pull on the boot and not on the wire, otherwise the connector located inside the boot may become separated.
3. Using compressed air (and safety glasses), blow debris from the spark plug well to assure that no harmful contaminants are allowed to enter the combustion chamber when the spark plug is removed. If compressed air is not available, use a rag or a brush to clean the area. Compressed air is available from both an air compressor or from compressed air in cans available at photography and computer stores.

04893P39

Fig. 121 First remove the spark plug cap using a twisting motion

04893P40

Fig. 122 Using a spark plug socket that is equipped with a rubber insert to properly hold the plug insulator, turn the spark plug counterclockwise to loosen it . . .

04893P41

Fig. 123 . . . and remove the spark plug from the bore

➡Remove the spark plugs when the engine is cold, if possible, to prevent damage to the threads. If plug removal is difficult, apply a few drops of penetrating oil to the area around the base of the plug, and allow it a few minutes to work.

4. Using a spark plug socket that is equipped with a rubber insert to properly hold the plug insulator, turn the spark plug counterclockwise to loosen and remove the spark plug from the bore.

***** WARNING**

Avoid the use of a flexible extension on the socket. Use of a flexible extension may allow a shear force to be applied to the plug. A shear force could break the plug off in the cylinder head, leading to costly and frustrating repairs. In addition, be sure to support the ratchet with your other hand—this will also help prevent the socket from damaging the plug.

To install:

5. Inspect the spark plug cap for tears or damage.
6. Apply a thin coating of anti-seize on the thread of the plug. This is extremely important on aluminum head engines; it will keep the spark plug from "seizing" in the cylinder head.
7. Carefully thread the plug into the bore by hand. If resistance is felt before the plug completely bottomed, back the plug out and begin threading again.

➡Whenever possible, thread the spark plug into the cylinder head by hand. Using a socket to thread the spark plugs may cause crossthreading.

8. Carefully tighten the spark plug. If the plug you are installing is equipped with a crush washer, seat the plug, then tighten to 10–15 ft. lbs. (14–20 Nm) or about ¼ turn to crush the washer. Whenever possible, spark plugs should be tightened to the factory torque specification.
9. Apply a small amount of silicone dielectric compound to the end of the spark plug lead or inside the spark plug boot to prevent sticking, then install the boot to the spark plug and push until it clicks into place. The click may be felt or heard. Gently pull back on the boot to assure proper contact.

READING SPARK PLUGS

See Figure 124

Your spark plugs are the single most valuable indicator of your engine's internal condition. Study your spark plugs carefully every time you remove them. Compare them to illustrations shown to identify the most common plug conditions.

INSPECTION & GAPPING

See Figures 125, 126 and 127

Evaluate each cylinder's performance by comparing the spark condition. Check each spark plug to be sure they are all of the same manufacturer and have the same heat range rating. Inspect the threads in the spark plug opening of the block, and clean the threads before installing the plug.

When purchasing new spark plugs, always ask the dealer if there has been a spark plug change for the engine being serviced. Sometimes manufacturers "update" the recommended type of plug (or possibly change the gap) for a particular engine.

Always check the gap on new plugs as they are not always set correctly at the factory. Do not use a flat feeler gauge when measuring the gap on a used plug, because the reading may be inaccurate. A round-wire type gapping tool is the best way to check the gap. The correct gauge should pass through the electrode gap with a slight drag. If you're in doubt, try a wire that is one size smaller and one larger. The smaller gauge should go through easily, while the larger one shouldn't go through at all.

Wire gapping tools usually have a bending tool attached. Use this tool to adjust the side electrode until the proper distance is obtained. Never attempt to bend the center electrode. Also, be careful not to bend the side electrode too far or too often as it may weaken and break off within the engine, requiring removal of the cylinder head to retrieve it. The ground electrode (the L-shaped one connected to the body of the plug) must be parallel to the center electrode and the specified size wire gauge must pass between the electrodes with a slight drag.

A normally worn spark plug should have light tan or gray deposits on the firing tip.

A carbon fouled plug, identified by soft, sooty, black deposits, may indicate an improperly tuned vehicle. Check the air cleaner, ignition components and engine control system.

This spark plug has been **left in the engine too long**, as evidenced by the extreme gap- Plugs with such an extreme gap can cause misfiring and stumbling accompanied by a noticeable lack of power.

An oil fouled spark plug indicates an engine with worn poston rings and/or bad valve seals allowing excessive oil to enter the chamber.

A physically damaged spark plug may be evidence of severe detonation in that cylinder. Watch that cylinder carefully between services, as a continued detonation will not only damage the plug, but could also damage the engine.

A bridged or almost bridged spark plug, identified by a build-up between the electrodes caused by excessive carbon or oil build-up on the plug.

TCCA1P40

Fig. 124 Spark plug condition can be a good indicator of how well your outboard is running

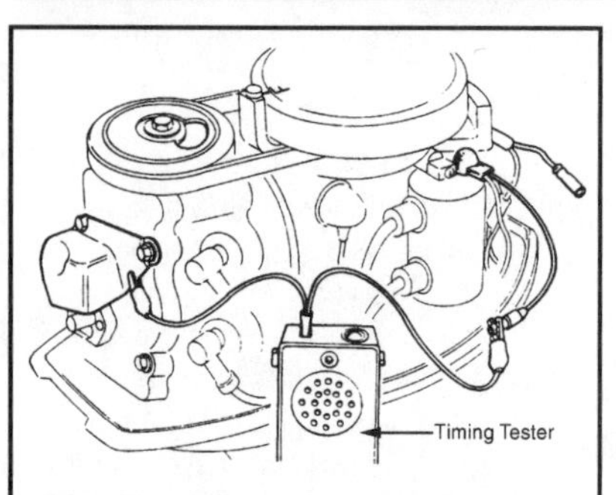

Fig. 125 Use a wire brush to clean debris off the end of the spark plug and measure the gap between the center electrode and the ground electrode

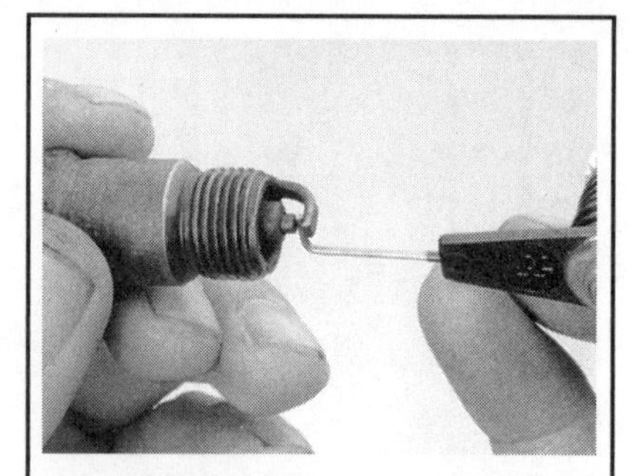

Fig. 126 Use a wire type plug gauge to check the distance between the center and ground electrodes

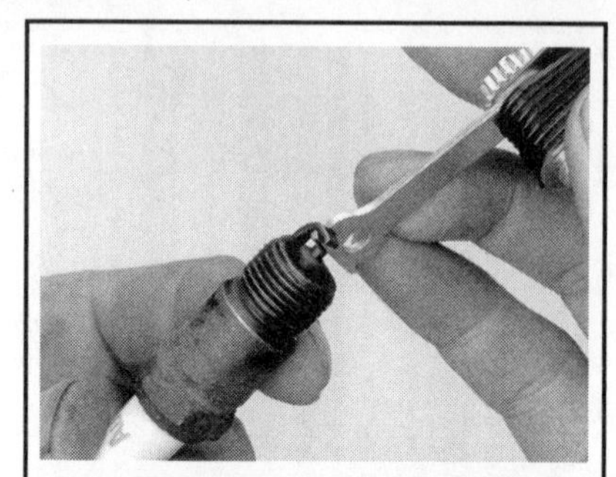

Fig. 127 Most spark plug gappers have an adjusting tool to bend the ground electrode, thus increasing or decreasing the distance to the center electrode

Spark Plug Wires

TESTING

◆ **See Figures 128, 129 and 130**

At every tune-up/inspection, visually check the spark plug cables for burns cuts, or breaks in the insulation. Check the boots, nipples and wire connections on the individual coils. Replace any damaged wiring.

Although in recent years the reliability of ignition components has increased dramatically, the fact that these components are used in a harsh environment can lead to breakdowns due to corrosion, vibration, etc. The resistance of the wires and individual coils should be checked periodically with an ohmmeter. Wires with excessive resistance will cause misfiring, and may make the engine difficult to start in damp weather.

To check primary side resistance,

1. Disconnect negative battery cable.
2. Disconnect wiring harness from the coil.
3. Measure resistance between the terminals on the coil.

If a spark plug wire is cut, frayed, or otherwise damaged, replacement of the wire requires that the entire coil and wire be replaced as a unit.

4. Remove the spark plug cap from the plug.
5. Holding the wire steady, twist the plug cap counterclockwise while gently pulling it away from the wire.
6. Using an ohmmeter, check the resistance between the ends of the cap.

CHECKING RESISTANCE

Once a year, usually when you change your spark plugs, check the resistance of the spark plug wire(s) and cap(s) with an ohmmeter. Wires with excessive resistance will cause misfiring and may make the engine difficult to start. Additionally, worn wires will allow arcing and misfiring in humid conditions.

Remove the spark plug wire from the engine. Test the wires by connecting one lead of the ohmmeter to the coil end of the wire and the other lead to the spark plug end of the wire. Resistance should measure approximately less than 7000 ohms per foot of wire.

REMOVAL & INSTALLATION

When installing a new set of spark plug wires, replace the wires one at a time so there will be no confusion. Coat the inside of the boots with dielectric grease to prevent sticking. Install the boot firmly over the spark plug until it clicks into place. The click may be felt or heard. Gently pull back on the boot to assure proper contact. Route the wire the same as the original and install it in a similar manner on the engine. Repeat the process for each wire.

Valve Clearance

As the valve tip and the valve actuating components wear, excessive clearance develops between these components leading to reduced engine performance and valve noise, commonly called "valve tap". Valve actuation on these engines is performed by the rocker arms acting directly on the valve stem.

Adjustment is done by rotating the engine until Top Dead Center of the compression stroke. This is when both the intake and exhaust valves closed and a clearance reading can be taken with a feeler gauge. Although the basic procedure for all Honda four-stroke outboard engines are similar, follow the procedure below for your specific application.

ADJUSTMENT

BF20 and BF2A

◆ **See Figure 131**

■ **For BF2D models please refer to the Supplement at the back of the manual.**

1. Remove the fan cover.
2. Unscrew the two 5mm nut and remove the carburetor.
3. Unscrew the two 5 x 10mm flange bolts and the valve cover.
4. With the engine cold, line up the triangle mark on the starter pulley with

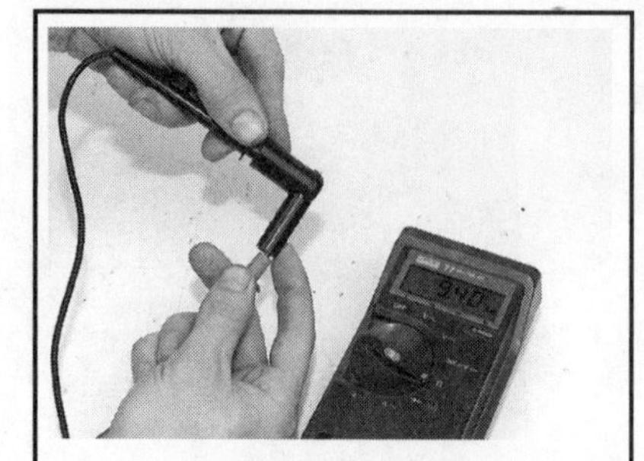

Fig. 128 Spark plug caps are an important part of the ignition system and should be checked for high resistance

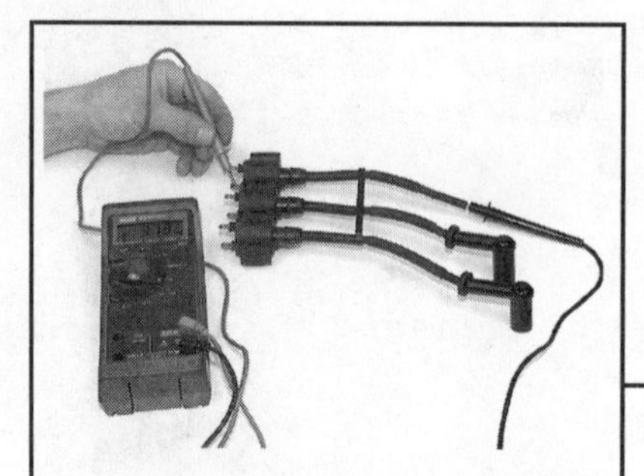

Fig. 129 Checking secondary side spark plug wires for resistance with a digital ohmmeter

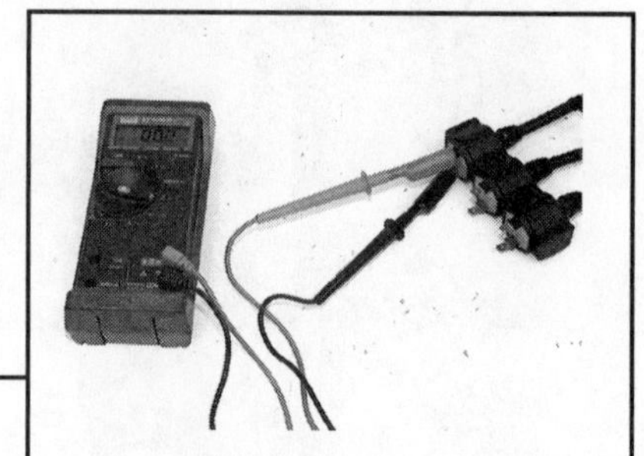

Fig. 130 Checking primary side ignition coil resistance with digital ohmmeter

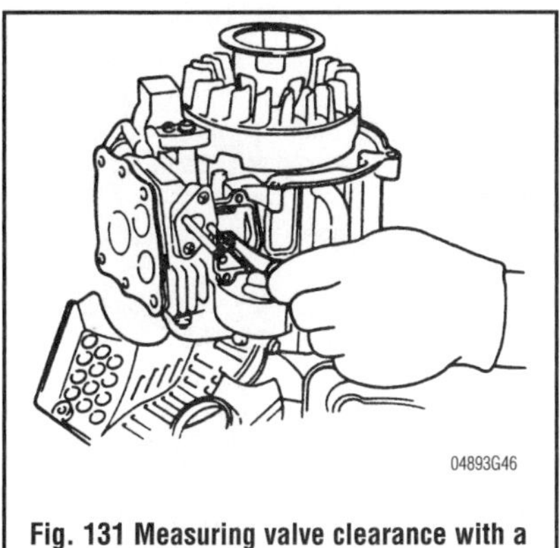

Fig. 131 Measuring valve clearance with a feeler gauge—BF20 and BF2A

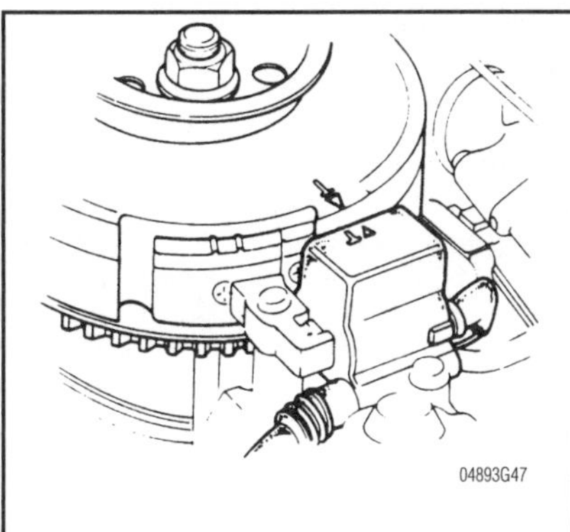

Fig. 132 Correct alignment of the timing marks—BF50 and BF5A

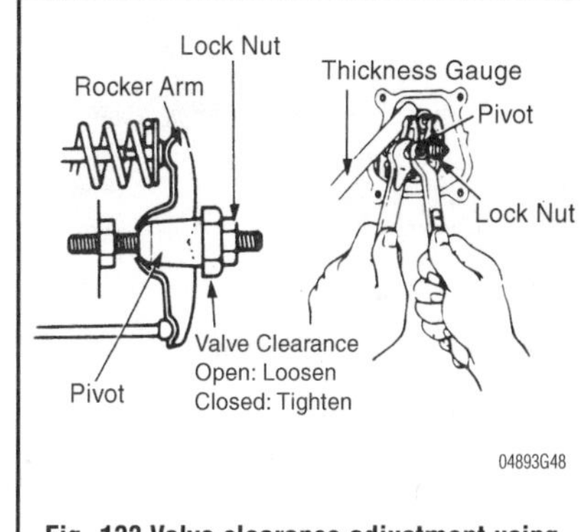

Fig. 133 Valve clearance adjustment using a feeler gauge—BF50 and BF5A

the mark on the ignition coil (with the piston at TDC with both valves closed), and measure the valve clearance with a feeler gauge.

➡Cylinder head removal is not required for valve clearance inspection.

5. Standard valve clearance is 0.003–0.006 in. (0.08–0.16mm) for both intake and exhaust.
6. If the valve clearance is not within specification, you must replace the valve adjuster to obtain the correct clearance. Valve adjusters are available from your local dealer in increments of 0.004 in. from 0.124 in. (3.15mm) to 0.150 in (0.150mm). Measure the used adjuster thickness and then select the replacement adjuster that will result in the correct clearance.

➡If the correct clearance cannot be obtained with a replacement adjuster, you then have the bottom of the adjuster lapped. This procedure requires special equipment and it is recommended that you have a dealer or machine shop perform this procedure.

BF50 and BF5A

▸ See Figures 132 and 133

1. Remove the cylinder head cover and align the "T triangle" marks on both the flywheel and the ignition coil. The piston must be at TDC of the compression stroke.
2. Measure the clearance between valve and the rocker arm using a feeler gauge. Standard valve clearance is 0.002 in.–0.006 in. (0.06mm–0.14mm) for the intake valve and 0.004 in.–0.007 in. (0.11–0.19mm) for the exhaust valve.
3. If the valve clearance is not within the specified range, adjust the clearance by loosening the lock nut and holding the pivot nut.
4. Turn down the pivot nut to achieve the correct clearance as measured with a feeler gauge.
5. After obtaining the correct clearance, hold the pivot nut and tighten the lock nut.
6. Recheck the clearance.

BF75, BF8A and BF100

▸ See Figure 134

1. With the engine cold and the shifter in "Neutral", remove the engine cover.
2. Remove the spark plugs.
3. Rotate the flywheel so that the "T" mark on the flywheel and the index mark on the starter case line up.
4. With No.1 cylinder at TDC, check the exhaust and intake valves with a feeler gauge. Standard clearance is 0.002 in.–0.004 in. (0.06mm–0.10mm) for both intake and exhaust.
5. If adjustment is necessary, loosen the lock nut and turn the adjusting screw in or out until a slight drag is felt on the gauge.
6. Rotate the flywheel 360°and repeat the above procedure.
7. Hold the adjusting screw and tighten the lock nut to 4.3–7.2 ft. lbs. (60–100kg-cm).
8. Recheck the clearance and readjust the valve clearance if necessary.

BF9.9A and BF15A

▸ See Figures 135 and 136

1. With the engine cold, and the shift lever in "Neutral", remove the recoil starter assembly.
2. Turn the engine over by hand, aligning the "T" mark on the flywheel and the "I" mark on the starter case.

➡ Make sure the cylinder to be checked is at TDC of the compression stroke. The compression stroke between the two cylinders is 360° different. Upon aligning the marks, adjust the cylinder with the valves closed, then turn the flywheel another 360° and adjust the other cylinder.

3. Remove the cylinder head cover.
4. Measure the intake and exhaust valves with a feeler gauge. Standard

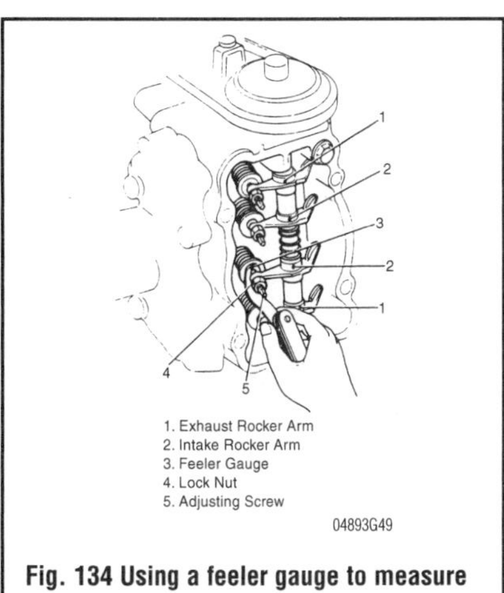

Fig. 134 Using a feeler gauge to measure valve clearance— BF75, BF8A and BF100

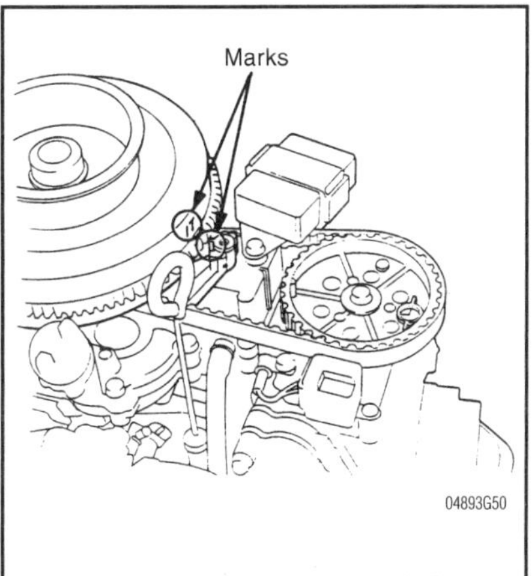

Fig. 135 Proper alignment of the timing marks BF9.9A and BF15A

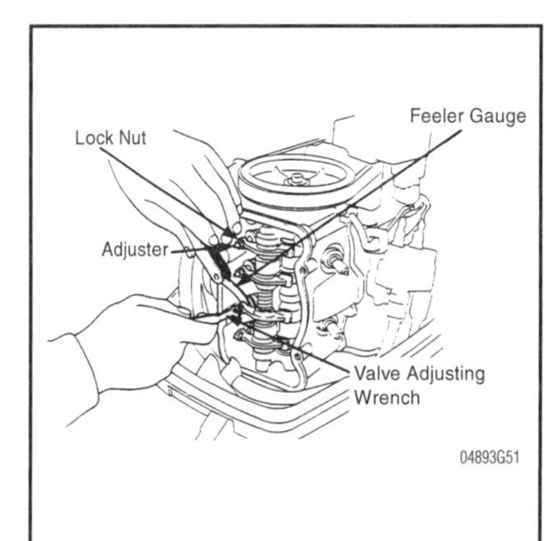

Fig. 136 Using a feeler gauge and wrench to adjust the valve clearance BF9.9A and BF15A

valve clearance is 0.005 in.–0.001 in. (0.10mm–0.14mm) for the intake valve and 0.008 in.–0.001 in. (0.18mm–0.22mm) for the exhaust valve.

5. If the clearance is not within the specified range, adjust the clearance by loosening the lock nut and turning the adjuster. There is a special tool for turning the adjuster (No. 07908–KE90000, but a suitably modified box end wrench will also work for this procedure.

6. After adjustment, tighten the lock nut and recheck the clearances.

BF25A and BF30A

➧ See Figures 137, 138, 139, 140 and 141

1. With the engine cold and the shifter in neutral, remove the engine cover.
2. Remove the recoil starter or the flywheel cover.
3. Disconnect the oil return tube and the breather tube from the cylinder head cover and remove the cylinder head cover.
4. Remove the spark plugs and engine hanger bracket.

**** CAUTION**

To prevent damaging the water pump impeller, always turn the crankshaft in a clockwise direction.

5. Manually turn the flywheel in the same direction as the embossed arrow on the flywheel (clockwise). Align the "TI" mark on the flywheel with the "I" mark on the top of the flywheel cover or recoil starter mounting boss. The "T arrow" mark on the camshaft pulley will align with the "T arrow" mark on the cylinder head. In this position the No.1 piston is at TDC of its compression stroke.
6. With the engine in this position, you can check the intake and exhaust valve clearances by inserting a feeler gauge between the valve stem and the rocker arm adjusting screw. Standard valve clearance is 0.004 in.–0.006 in. (0.10mm–0.14mm) for the intake valve and 0.007 in.–0.009 in. (0.007mm–0.009mm) for the exhaust valve.
7. Turn the flywheel clockwise until the cam pulley turns an additional 240° and the "triangle 2" mark aligns with the "T arrow" mark on the cylinder head. This will put the No.2 piston at TDC of its compression stroke. Check the intake and exhaust valve clearances of No.2 cylinder.
8. Turn the flywheel clockwise until the cam pulley turns an additional 240° and the "triangle 3" mark aligns with the "T arrow" mark on the cylinder head. This will put the No.3 piston at TDC of its compression stroke. Check the intake and exhaust valve clearances on the No.3 cylinder.
9. If any of the clearances are out of specification the valve must be adjusted.
10. With the cylinder at TDC of it compression stroke, loosen the adjusting screw lock nut, and turn the adjusting screw to obtain the correct clearance for the intake and exhaust valves with a feeler gauge.
11. Hold the adjusting screw using a special tool (Valve Adjusting Screw No.07908-KE9000) and tighten the lock nut to 5.8 ft. lbs. (8.0 Nm).

BF35A and BF45A

➧ See Figures 142 thru 151

1. With the engine cold, and the shifter in neutral, remove the engine cover.
2. Remove the flywheel cover.
3. Disconnect the oil return and breather tubes from the valve cover, then remove all the bolts and lift off the cover.
4. Remove the spark plugs and engine lifting bracket.

**** CAUTION**

The starter stopper collar is close to the flywheel and the starter ring can become sharp. Avoid hand contact with the starter stopper collar or flywheel ring gear. Hold the upper portion of the flywheel when rotating the engine by hand.

**** CAUTION**

To avoid damaging the water pump impeller, always rotate the engine in a counterclockwise direction.

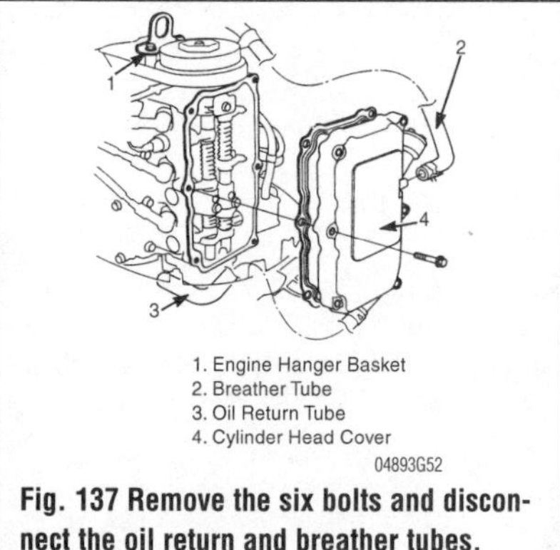

Fig. 137 Remove the six bolts and disconnect the oil return and breather tubes, then remove the cylinder head cover

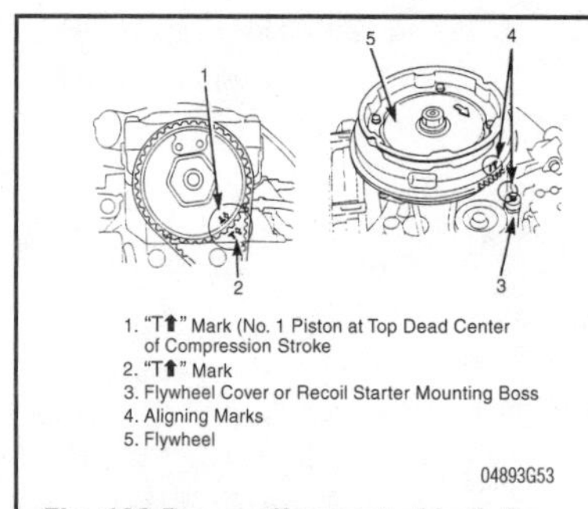

Fig. 138 Proper alignment of both the camshaft pulley and flywheel timing marks for adjusting No.1 cylinder at TDC—BF25A and BF30A

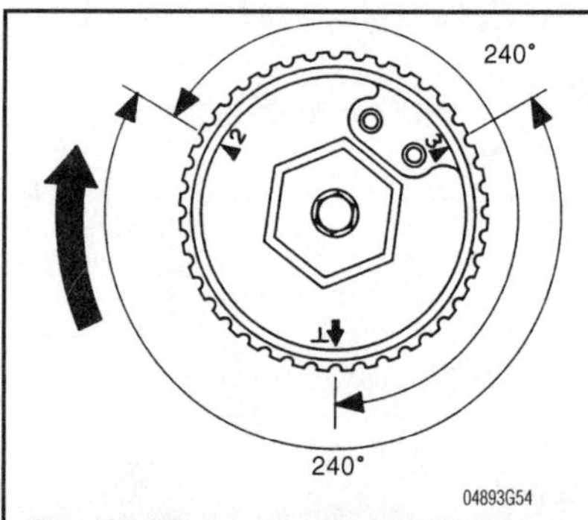

Fig. 139 Alignment marks for adjusting No.2 and No.3 cylinders at TDC—BF25A and BF30A

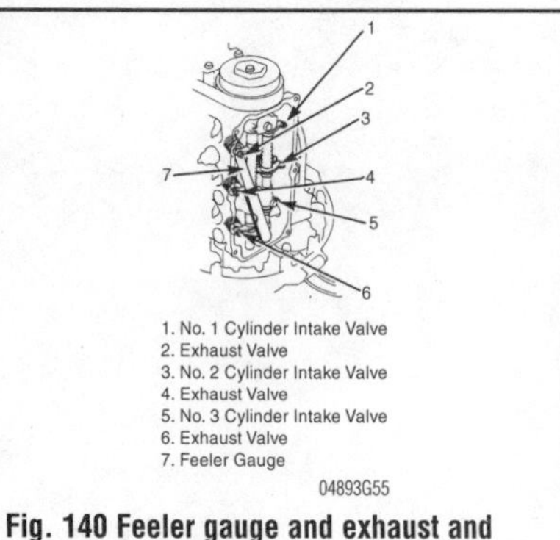

Fig. 140 Feeler gauge and exhaust and intake valve placement on the cylinder head—BF25A and BF30A

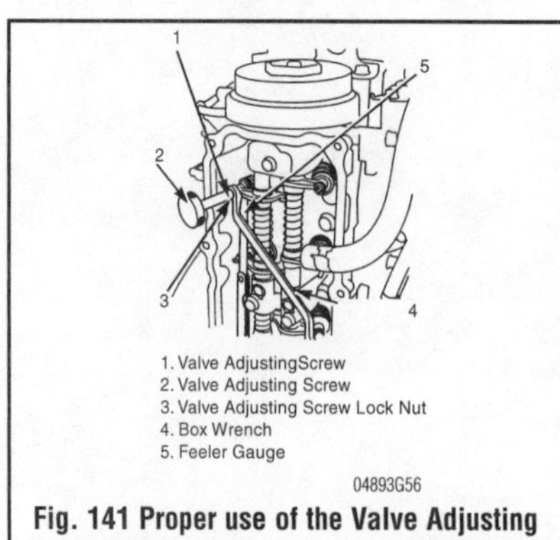

Fig. 141 Proper use of the Valve Adjusting Screw tool to adjust the valves—BF25A and BF30A

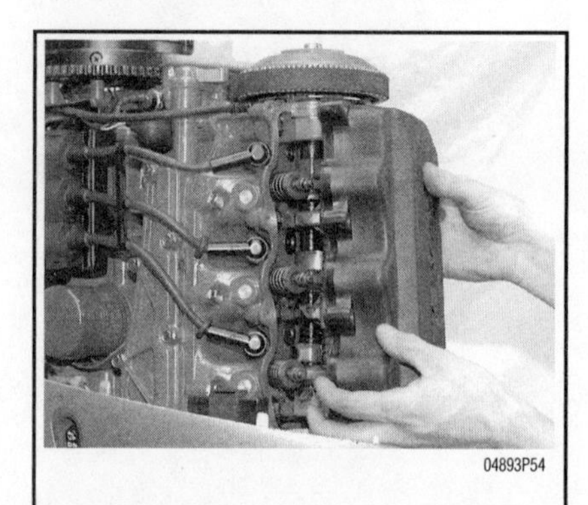

Fig. 142 After removing the bolts, lift off the cylinder head cover

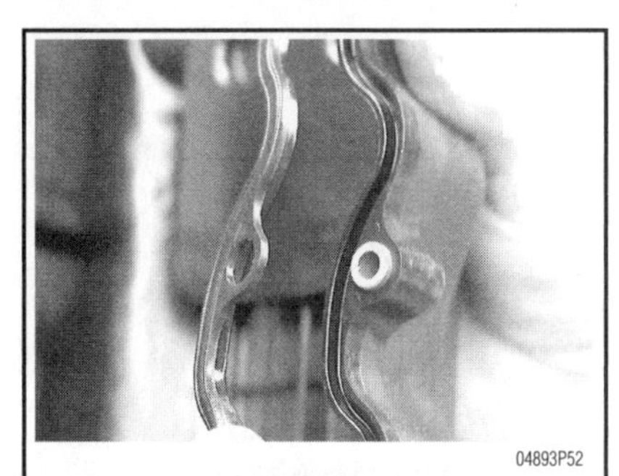

04893P52

Fig. 143 Now is the time to inspect the gasket condition. Replace it if there are any problems such as tears, cracks or stretching

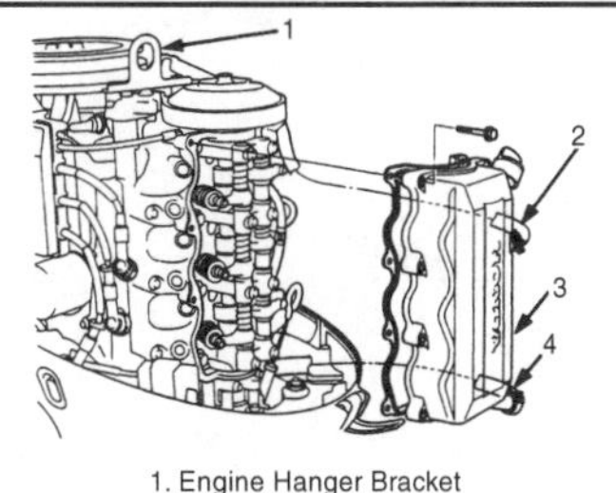

1. Engine Hanger Bracket
2. Breather Tube
3. Cylinder Head Cover
4. Oil Return Tube

04893G57

Fig. 144 Cylinder head cover removal, including oil return and breather tube

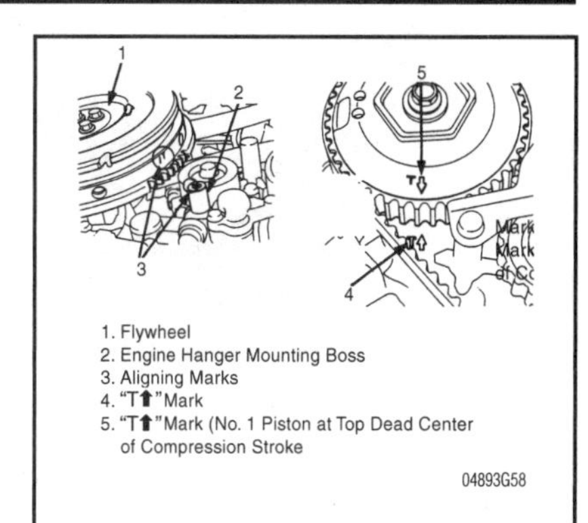

1. Flywheel
2. Engine Hanger Mounting Boss
3. Aligning Marks
4. "T↑"Mark
5. "T↑"Mark (No. 1 Piston at Top Dead Center of Compression Stroke

04893G58

Fig. 145 Flywheel and camshaft pulley timing marks—BF35A and BF45A

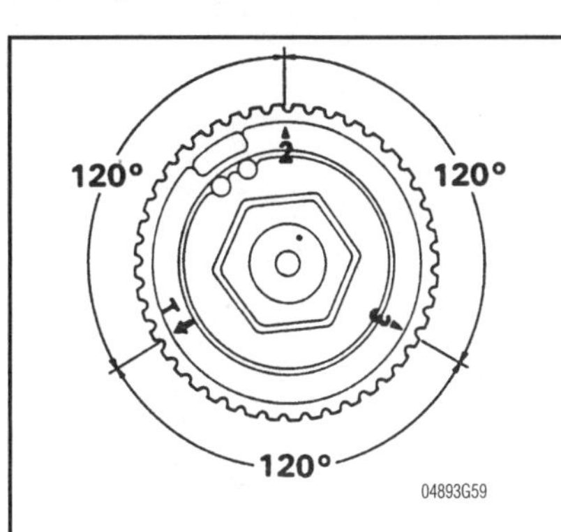

04893G59

Fig. 146 Timing marks for No.1, No.2 and No.3 cylinders—BF35A and BF45A

04893P48

Fig. 147 First, slide the feeler gauge between the adjuster on the rocker arm and the valve. There should be slight drag on the feeler gauge if the clearance is correct . . .

04893P49

Fig. 148 . . . if the clearance is out of specification, loosen the lock nut on the rocker arm . . .

04893P50

Fig. 149 . . . and reinsert the feeler gauge. Then tighten or loosen the adjuster with a screwdriver to obtain the correct clearance . . .

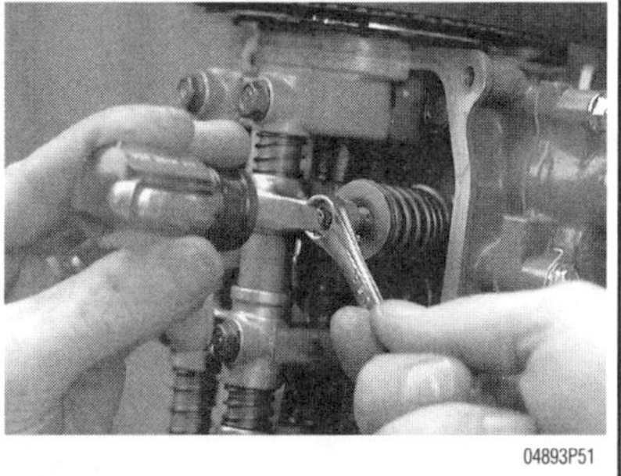

04893P51

Fig. 150 . . . after getting the correct clearance, hold the adjuster in position with the screwdriver, and tighten the lock nut

1. No. 1 Cylinder Intake Valve
2. Exhaust Valve
3. No. 2 Cylinder Intake Valve
4. Exhaust Valve
5. No. 3 Cylinder Intake Valve
6. Exhaust Valve
7. Feeler Gauge

04893G60

Fig. 151 Feeler gauge and intake and exhaust valve placement on the cylinder head—BF35A and BF45A

5. Turn the flywheel by hand, counterclockwise, in the same direction as the embossed arrow on the flywheel. Then align the "T" mark on the flywheel with the "I" mark on the top of the engine hanger mounting boss. The "T arrow" mark on the cam pulley will then align with the "T arrow" mark on the cylinder head. In this position, the No.1 piston is at TDC of the compression stroke.

6. Check the valve clearance for the No.1 cylinder. Standard valve clearance is 0.005 in.–0.007 in. (0.13mm–0.007mm) for the intake valves and 0.008 in.–0.010 in. (0.21mm–0.25mm) for the exhaust valves.

7. Turn the flywheel counterclockwise until the cam pulley turns an additional 120° and the "triangle 2" mark aligns with "T arrow" mark on the cylinder head. This will put the No.2 piston at TDC of the compression stroke. You can now check the intake and exhaust clearances on No.2 cylinder.

8. Turn the flywheel counterclockwise until the cam pulley turns an additional 120° and the "triangle 3" mark aligns with the "T arrow" mark on the cylinder head. This will put the No.3 piston at TDC of the compression stroke. You can now check the intake and exhaust clearances on No.3 cylinder.

9. If any of the clearances are out of specification the valves must be adjusted.

10. With the cylinder at TDC of it compression stroke, loosen the adjusting screw lock nut, and turn the adjusting screw to obtain the correct clearance for the intake and exhaust valves with a feeler gauge.

11. Hold the adjusting screw using a screwdriver, tighten the lock nut to 16.6 ft. lbs. (23Nm).

12. Recheck the valve clearance after tightening the lock nut.

BF75A and BF90A

➧ See Figures 152 thru 158

1. With the engine cold, and the shifter in neutral, remove the engine cover.

2. Loosen the bolts and release the hooks on the timing belt cover and remove the cover.

3. Remove the two grommets on the engine undercase.

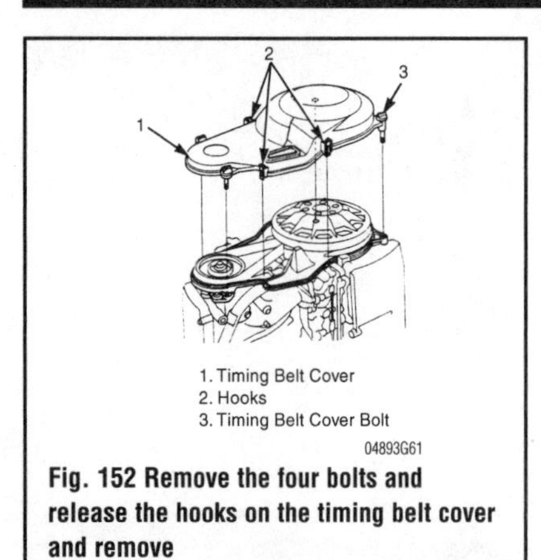

1. Timing Belt Cover
2. Hooks
3. Timing Belt Cover Bolt

04893G61

Fig. 152 Remove the four bolts and release the hooks on the timing belt cover and remove

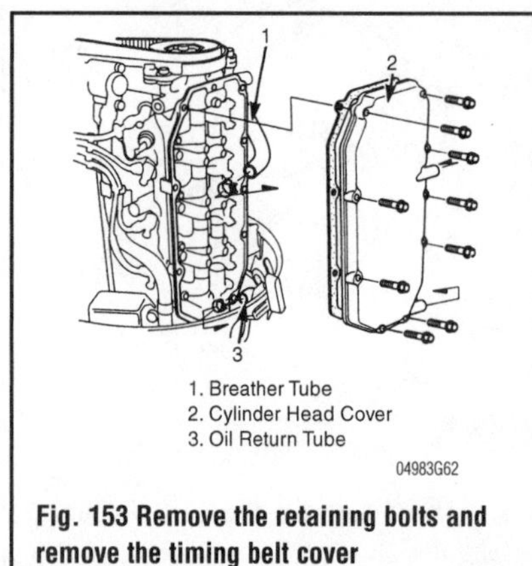

1. Breather Tube
2. Cylinder Head Cover
3. Oil Return Tube

04983G62

Fig. 153 Remove the retaining bolts and remove the timing belt cover

1. Timing Belt Driven Pulley
2. "T↑" Mark (No. 1 Piston at Top Dead Center of Its Compression Stroke
3. Timing Belt Lower Cover
4. "T↑" Mark

04983G63

Fig. 154 Alignment marks on the camshaft pulley . . .

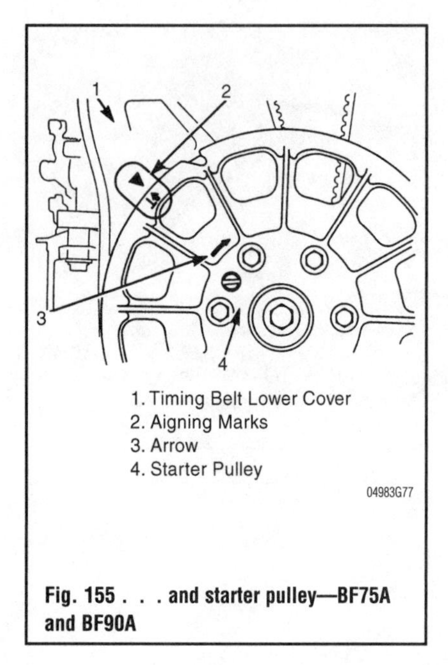

1. Timing Belt Lower Cover
2. Aigning Marks
3. Arrow
4. Starter Pulley

04983G77

Fig. 155 . . . and starter pulley—BF75A and BF90A

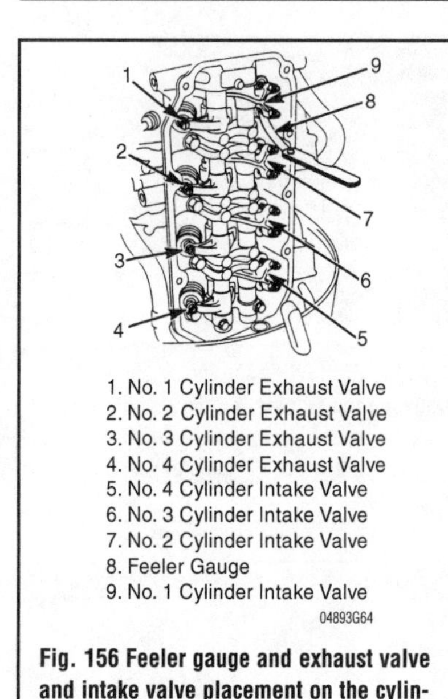

1. No. 1 Cylinder Exhaust Valve
2. No. 2 Cylinder Exhaust Valve
3. No. 3 Cylinder Exhaust Valve
4. No. 4 Cylinder Exhaust Valve
5. No. 4 Cylinder Intake Valve
6. No. 3 Cylinder Intake Valve
7. No. 2 Cylinder Intake Valve
8. Feeler Gauge
9. No. 1 Cylinder Intake Valve

04893G64

Fig. 156 Feeler gauge and exhaust valve and intake valve placement on the cylinder head—BF75A and BF90A

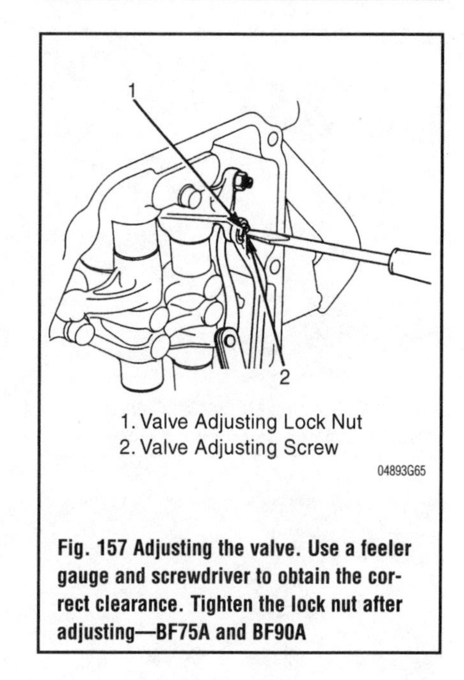

1. Valve Adjusting Lock Nut
2. Valve Adjusting Screw

04893G65

Fig. 157 Adjusting the valve. Use a feeler gauge and screwdriver to obtain the correct clearance. Tighten the lock nut after adjusting—BF75A and BF90A

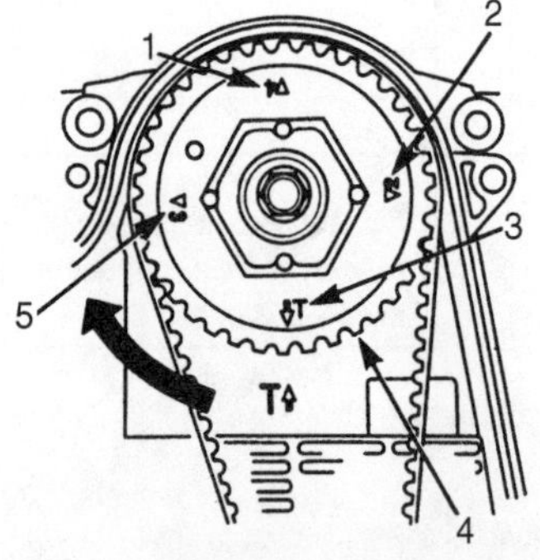

1. No. 4 Piston at Top Dead Center of Compression Stroke
2. No. 2 Piston at Top Dead Center of Compression Stroke
3. No. 1 Piston at Top Dead Center of Compression Stroke
4. Timing Belt Driven Pulley
5. No. 3 Piston at Top Dead Center of Compression Stroke

04893G66

Fig. 158 Timing marks and correct pulley rotation for each cylinder at TDC—BF75A and BF90A

4. Disconnect the breather and oil return tubes from the cylinder head cover and remove the cover.
5. Remove the spark plugs.
6. Manually turn starter pulley clockwise, in the same direction as the arrow on the starter pulley, and align the "T arrow" mark on the starter pulley with the "triangle" mark on the timing belt lower cover. Make sure that the "T arrow" mark on the timing belt driven pulley aligns with the "T arrow" mark on the timing belt lower cover at this time. In this position the No.1 is at TDC of the compression stroke.

➡Do not turn the starter pulley counterclockwise or water pump damage can occur.

7. With the No.1 cylinder at TDC, check the intake and exhaust valves with a feeler gauge. Standard clearance is 0.007 in.–0.009 in. (0.18mm–0.22mm) for the intake valves and 0.010 in.–0.012 in. (0.26mm–0.30mm) for the exhaust valves.
8. If any of the clearances are out of specification the valves must be adjusted.
9. With the cylinder at TDC of it compression stroke, loosen the adjusting screw lock nut, and turn the adjusting screw to obtain the correct clearance for the intake and exhaust valves with a feeler gauge.
10. After adjusting the intake and exhaust valves on No.1 cylinder, turn the driven pulley an additional 90° to align the "triangle 2" mark with the "T arrow" mark on the timing belt lower cover. This will put the No.2 piston at TDC of its compression stroke. Check the clearances of the intake and exhaust valves of the No.2 cylinder.
11. After adjusting the intake and exhaust valves of No.2 cylinder, adjust the

intake and exhaust valve clearances on No.4 and No.3 cylinder in that order by repeating the above steps.

12. After adjusting each valve, tighten down the adjusting screw and loosely tighten the lock nut.

13. Recheck the clearances and tighten the lock nut to 17 ft. lbs. (23Nm).

BF115A and BF130A

See Figures 159, 160, 161, 162 and 163

1. With the engine cold, and the shifter in neutral, remove the engine cover.
2. Release the breather tube and control cable from the clamps.
3. Remove the bolts and release the locking tabs of the upper timing belt cover from the lower cover and remove the upper cover.
4. Remove the spark plugs and spark plug caps.
5. Remove the bolts from the rear case and remove.
6. Remove the fuel tube clip and release the fuel tube.
7. Disconnect the breather tube from the engine.
8. Remove the harness holder from the connector bracket.
9. Disconnect both fuel tubes from the fuel pump (low pressure side).
10. Remove the bolts and remove the cylinder head cover and oil return tube together.

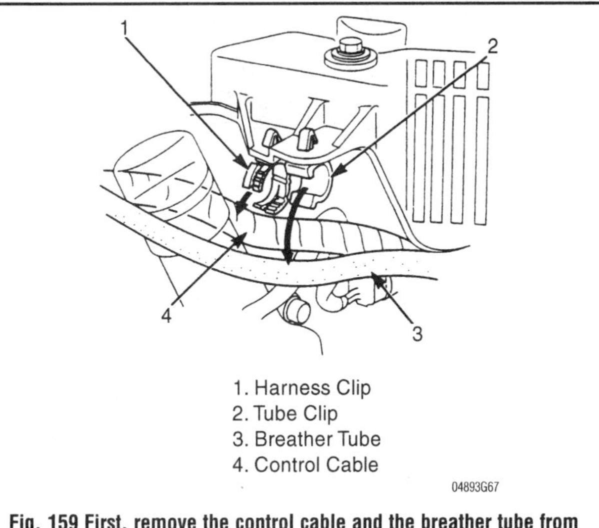

Fig. 159 First, remove the control cable and the breather tube from the hold down clamps

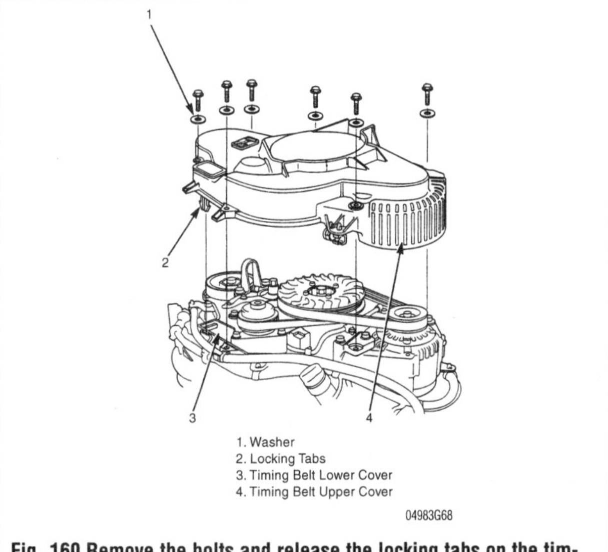

Fig. 160 Remove the bolts and release the locking tabs on the timing belt upper cover and remove it from the lower cover

CAUTION

Gasoline is highly flammable and explosive. You can be burned or seriously injured when handling fuel. Keep heat, sparks and open flames away from gasoline. Use only in a well ventilated space. Wipe up fuel spills immediately.

11. Before removing fuel tube, clamp with a hose clamp to prevent fuel from leaking.

12. Manually turn the crankshaft pulley clockwise and align the "T arrow" mark on the camshaft pulley with the "T arrow" mark on the lower timing belt cover. In this position, the No.1 piston is TDC at its compression stroke.

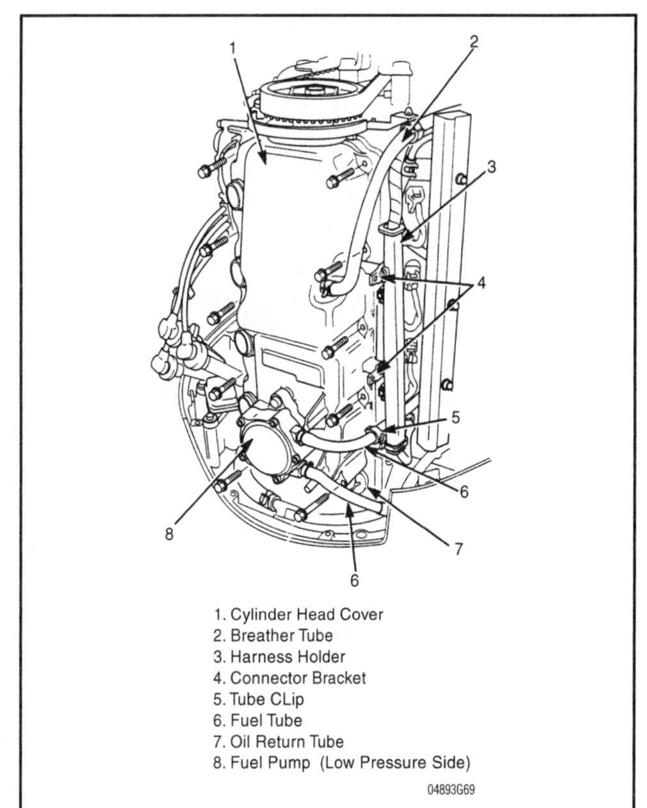

Fig. 161 Make sure that all hoses, tubes and other obstructions are disconnected before removing the cylinder head cover

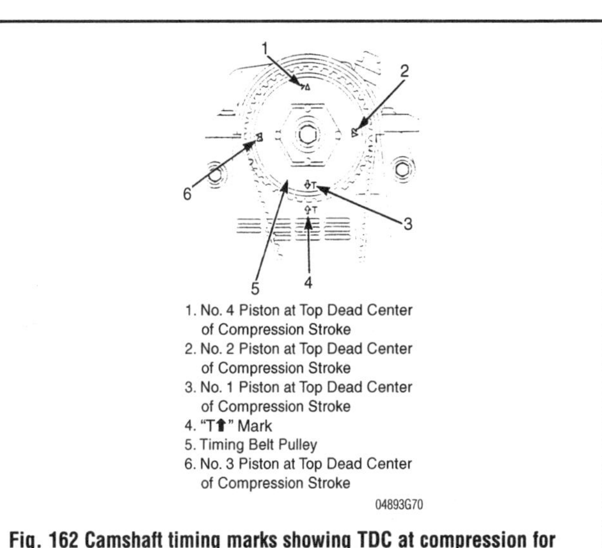

Fig. 162 Camshaft timing marks showing TDC at compression for each cylinder—BF115A and BF130A

**** WARNING**

Do not rotate the crankshaft pulley in a counterclockwise direction. Damage to the water pump impeller will occur.

13. With the No.1 cylinder at TDC, check the intake and exhaust valves with a feeler gauge. Standard clearance is .0094 in.–.0110 in. (0.24mm–0.28mm) for the intake valve and .0110 in.–.0125 in. (0.28mm–0.32mm) for the exhaust valve.
14. If any of the clearances are out of specification the valves must be adjusted.
15. With the cylinder at TDC of it compression stroke, loosen the adjusting screw lock nut, and turn the adjusting screw until a slight resistance is felt on the gauge to obtain the correct clearance for the intake and exhaust valves.
16. After adjusting the intake and exhaust valve clearances of the No.1 cylinder, turn the crankshaft pulley clockwise and align the "2 triangle" mark on the camshaft pulley with the "T arrow" mark on the timing belt lower cover. In this position, the No.2 piston is at the TDC of its compression stroke. Adjust the intake and exhaust valves of this cylinder while the engine is in this position.
17. After adjusting the valves of No.2 cylinder, adjust the valve clearances of No.4 and No.3 cylinders, in that order, in the same steps as above.
18. After each final valve adjustment, snug down the adjusting screw and tighten the lock nut to 14 ft. lbs. (20Nm).

Ignition System

See Figure 164

The electronic CDI ignition system has become one of the most reliable components on the modern outboard engine. There is very little maintenance involved in the operation of the ignition and even less to repair if the component fails. Most systems are sealed and there is no option other than to replace the failed component.

It is very important to narrow down the ignition problem and replace the correct component rather than just replace parts hoping to solve the problem. Electronic components can be very expensive and are usually not returnable. Please refer to the "Ignition and Electrical" Section for more information on troubleshooting and repairing the CDI ignition system.

Timing And Synchronization

Timing and synchronization on an outboard engine is extremely important to obtain maximum efficiency. The powerhead cannot perform properly and produce its designed horsepower output if the fuel and ignition systems have not been precisely adjusted.

TIMING

See Figure 165

Timing on all Honda outboards with the exception of early model BF75 and BF100 is either fixed or electronically advanced through the Electronic Control Module. The timing of these systems is not adjustable.

The BF75 and BF100 use a contact breaker type of ignition system which can be adjusted as follows.

1. Connect a timing tester (buzz box) or ohmmeter to the ignition coil secondary side cable and to a good engine ground.
2. Turn the flywheel to align the "F" mark on the flywheel to the index mark on the starter case boss. The contact breaker points should just start to open (15°BTDC).
3. To adjust the timing, loosen the breaker plate locking screw and move the contact breaker plate to achieve the correct timing. Retighten the locking screw.
4. Rotate the flywheel one full turn, and check the ignition timing for the other cylinder. Readjust the contact breaker points if necessary.

SYNCHRONIZATION

On the multi-carburetted models, the carburetors must be synchronized in order to ensure that each cylinder is performing the same as the others. In order to do this you must use a vacuum gauge set on the intake manifold of each carburetor.

Before making adjustments with the timing or synchronizing, the ignition system should be thoroughly checked and the fuel system verified to be in good working order.

PREPARATION

The Synchronizing of the fuel systems on an outboard motor are critical adjustments. The following equipment is essential and is called out repeatedly in this section. This equipment must be used as described, unless otherwise instructed by the equipment manufacturer. Naturally, the equipment is removed following completion of the adjustments.

Tachometer

A tachometer connected to the powerhead must be used to accurately determine engine speed during idle and high-speed adjustment. Engine speed readings range from 0 to 6,000 rpm in increments of 100 rpm. Choose a tachometers with solid state electronic circuits which eliminates the need for relays or batteries and contribute to their accuracy.

A tachometer is installed as standard equipment on most powerheads covered in this manual. Due to local conditions, it may be necessary to adjust the carburetor while the outboard unit is running in a test tank or with the boat in a body of water. For maximum performance, the idle rpm should be adjusted under actual operating conditions. Under such conditions it might be necessary to attach a tachometer closer to the powerhead than the one installed on the control panel.

Flywheel Rotation

The instructions may call for rotating the flywheel until certain marks are aligned with the timing pointer. When the flywheel must be rotated, always move the flywheel in the indicated direction. If the flywheel should be rotated in the opposite direction, the water pump impeller vanes would be twisted.

Should the powerhead be started with the pump tangs bent back in the wrong direction, the tangs may not have time to bend in the correct direction before

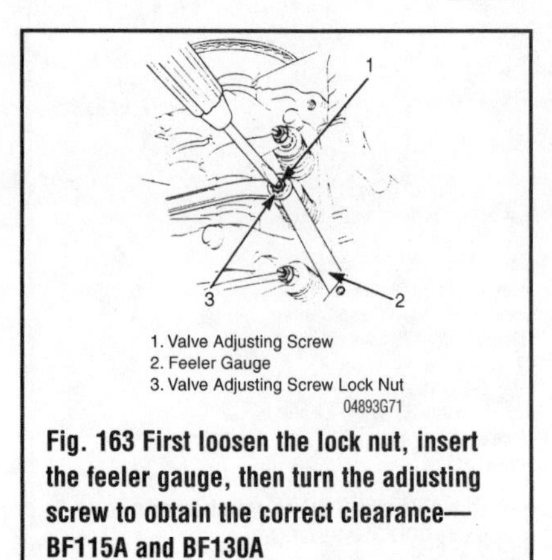

Fig. 163 First loosen the lock nut, insert the feeler gauge, then turn the adjusting screw to obtain the correct clearance—BF115A and BF130A

Fig. 164 Honda Electronic Control Module

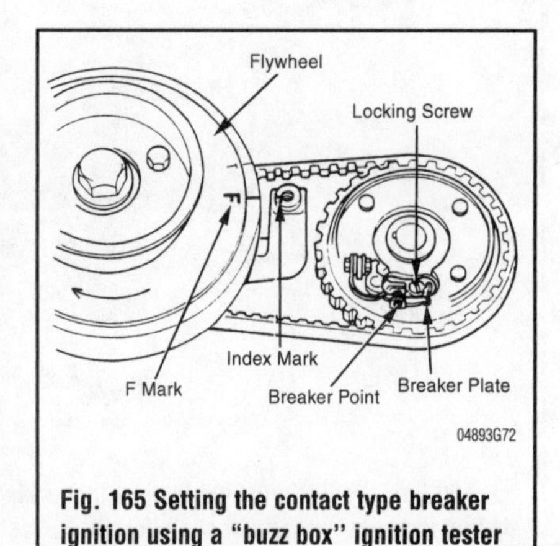

Fig. 165 Setting the contact type breaker ignition using a "buzz box" ignition tester

they are damaged. The least amount of damage to the water pump will affect cooling of the powerhead.

Test Tank

Since the engine must be operated at various times and engine speeds during some procedures, a test tank or moving the boat into a body of water, is necessary. If installing the engine in a test tank, outfit the engine with an appropriate test propeller.

**** CAUTION**

Water must circulate through the lower unit to the powerhead anytime the powerhead is operating to prevent damage to the water pump in the lower unit. Just five seconds without water will damage the water pump impeller.

■ **Remember the powerhead will not start without the emergency tether in place behind the kill switch knob.**

**** CAUTION**

Never operate the powerhead above a fast idle with a flush attachment connected to the lower unit. Operating the powerhead at a high rpm with no load on the propeller shaft could cause the powerhead to runaway causing extensive damage to the unit.

IDLE SPEED ADJUSTMENT

◆ **See Figure 166**

**** CAUTION**

Water must circulate through the lower unit to the engine any time the engine is run to prevent damage to the water pump in the lower unit. Just five seconds without water will damage the water pump.

1. On the BF75A and BF90A you must first remove the propeller before doing this procedure.
2. Run the engine, in neutral, in an outboard test tank with at least 4 inches of water above the anti-ventilation plate or use a flush attachment (optional equipment) and connect a supply of fresh water to the WASH plug hole. Always allow the engine to warm up to proper operating temperature (approximately 10 minutes).
3. Stop the engine and remove the engine cover. On tiller handle type engines, turn the throttle grip to the SLOW position.
4. Attach a tachometer to the engine and restart the engine.
5. After the speed has settled, turn the throttle stop screw to obtain the specified idle speed. Idle speed is 900-1000 rpm for most models, except for the BF25A, which is 850-950 rpm or the BF2D which is 1900-2100 rpm. If the engine idle speed will not stabilize, you must perform the carburetor synchronization adjustment.

CARBURETOR SYNCHRONIZATION

◆ **See Figure 167**

1. Remove the engine cover.
2. Remove the plugs and sealing washers from the intake manifolds of each engine.
3. Attach the vacuum gauge adapters to each intake manifold plug hole, and connect the vacuum gauge hose to the adapters.

■ **All vacuum gauge hoses must be of equal length and all gauges must be calibrated the same.**

4. Attatch a tachometer to the engine.
5. Have the engine in an upright position with the gear selector or control lever in the "N" (Neutral) position.
6. Start the engine in an outboard test tank, making sure that the water level is at least 4 inches above the anti-ventilation plate. Allow the engine to warm up to normal operating temperature.
7. Check that the standard idle speed is correct.
8. Check the difference in intake manifold vacuum between the cylinders. The maximum vacuum difference between all cylinders should be 0.75 in. (20mm) Hg or less.
9. If the vacuum difference between cylinders is not within specification adjust the carburetors to specification.

■ **The No.4 carburetor on 4 cylinder engines and the No.3 carburetor on 3 cylinder engines are the synchronization base carburetors. These carburetors will not have a synchronization adjusting screw.**

10. Adjust the carburetors to get the least amount of vacuum difference between the cylinders. As the manifold vacuum difference decreases, the idle speed will become more stable. Each increment on the vacuum gauge indicates 1 in. (25.4mm) Hg.
11. Turn the No.1, No.2, or No.3 carburetor adjusting screws so that the vacuum difference between all cylinders is 0.75 in. (20mm) Hg or less.
12. After each adjustment, check the idle speed and adjust if necessary by turning the base carburetor throttle stop screw.
13. Rev the throttle several times and allow the engine to return to idle. Check the vacuum difference between cylinders and if they are not within 0.75 in. (20mm) Hg or less, readjust as necessary.
14. If the adjustments are correct, stop the engine, remove the tachometer and gauge adapters and install the plugs and sealing washers and reinstall the engine cover.

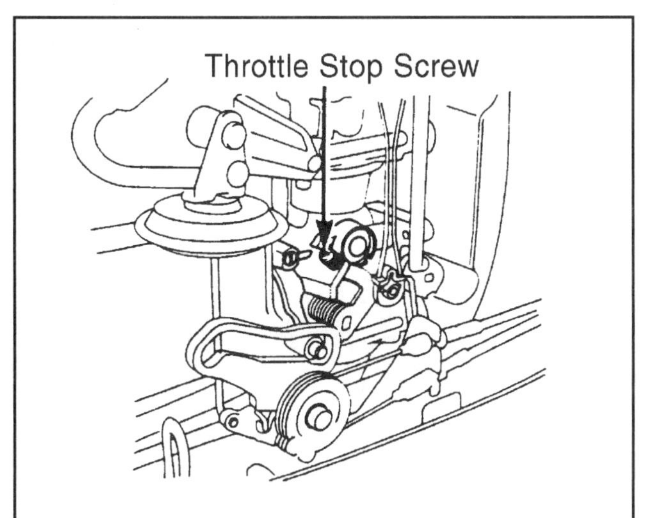

Fig. 166 Turn the throttle stop screw to get the correct idle—BF75A shown, most models similar)

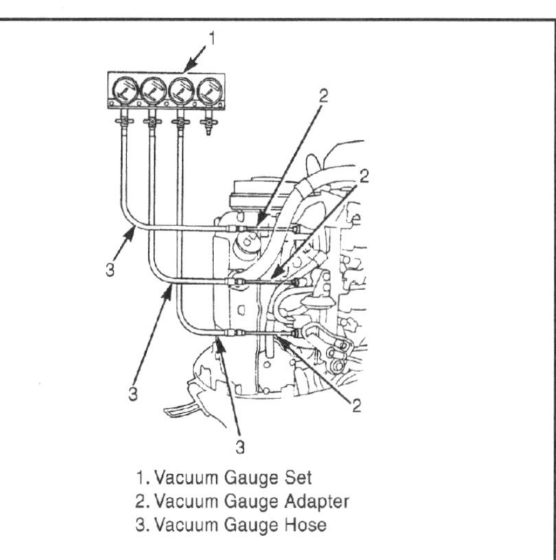

Fig. 167 Carburetor synchronization with a vacuum gauge set on the BF25A. Other engine application are similar

WINTER STORAGE CHECKLIST

Taking extra time to store the boat properly at the end of each season will increase the chances of satisfactory service at the next season. Remember, storage is the greatest enemy of an outboard motor. The unit should be run on a monthly basis. The boat steering and shifting mechanism should also be worked through complete cycles several times each month. If a small amount of time is spent in such maintenance, the reward will be satisfactory performance, increased longevity and greatly reduced maintenance expenses.

For many years there has been the widespread belief simply shutting off the fuel at the tank and then running the powerhead until it stops is the proper procedure before storing the engine for any length of time. Right? WRONG!

First, it is not possible to remove all fuel in the carburetor by operating the powerhead until it stops. Considerable fuel is trapped in the float chamber and other passages and in the line leading to the carburetor. The only guaranteed method of removing all fuel is to take the time to remove the carburetors, and drain the fuel.

Proper storage involves adequate protection of the unit from physical damage, rust, corrosion, and dirt. The following steps provide an adequate maintenance program for storing the unit at the end of a season.

1. On four-stroke outboards it is important drain the engine oil and replace with new oil and filter.
2. Squirt a small quantity of engine oil into each spark plug hole and crank the engine over to distribute the oil around the engine internals. Reinstall the old spark plugs (you will install new spark plugs in the spring).
3. Drain all fuel from the carburetor float bowls. On fuel injected models, drain the fuel from the vapor separator.
4. Drain the fuel tank and the fuel lines Store the fuel tank in a cool dry area with the vent OPEN to allow air to circulate through the tank. Do not store the fuel tank on bare concrete. Place the tank to allow air to circulate around it.
5. Change the fuel filter.
6. Drain, and then fill the lower unit with new lower unit gear oil.
7. Lubricate the throttle and shift linkage and the steering pivot shaft.
8. Clean the outboard unit thoroughly. Coat the powerhead with a commercial corrosion and rust preventative spray. Install the cowling, and then apply a thin film of fresh engine oil to all painted surfaces.
9. Remove the propeller. Apply Perfect Seal® or a waterproof sealer to the propeller shaft splines, and then install the propeller back in position.
10. Be sure all drain holes in the gear housing are open and free of obstructions. Check to be sure the flush plug has been removed to allow all the water to drain. Trapped water could freeze, expand, and cause expensive castings to crack.
11. Always store the outboard unit off the boat with the lower unit below the powerhead to prevent any water from being trapped inside.
12. Be sure to consult your owners manual for any particular storage procedures applicable to your specific model.

SPRING COMMISSIONING CHECKLIST

See Figures 168 thru 175

A spring tune-up is essential to getting the most out of your engine. If the engine has been properly winterized, it is usually no problem to get it in top running condition again in the springtime. If the engine has just been put in the garage and forgotten for the winter, then it is doubly important to do a complete tune up before putting the engine back into service. If you have ever been stranded out on the water because your engine has died, and you had to suffer the embarrassment of having to be towed back to the marina, now is the time to prevent that from occurring.

Satisfactory performance and maximum enjoyment can be realized if a little time is spent in preparing the outboard unit for service at the beginning of the season. Assuming the unit has been properly stored, a minimum amount of work is required to prepare the unit for use. The following steps outline an adequate and logical sequence of tasks to be performed before using the outboard the first time in a new season.

1. Lubricate the outboard according to the manufacturer's recommendations.
2. Perform a tune-up on the engine. This should include replacing the spark plugs and making a thorough check of the ignition system. The ignition

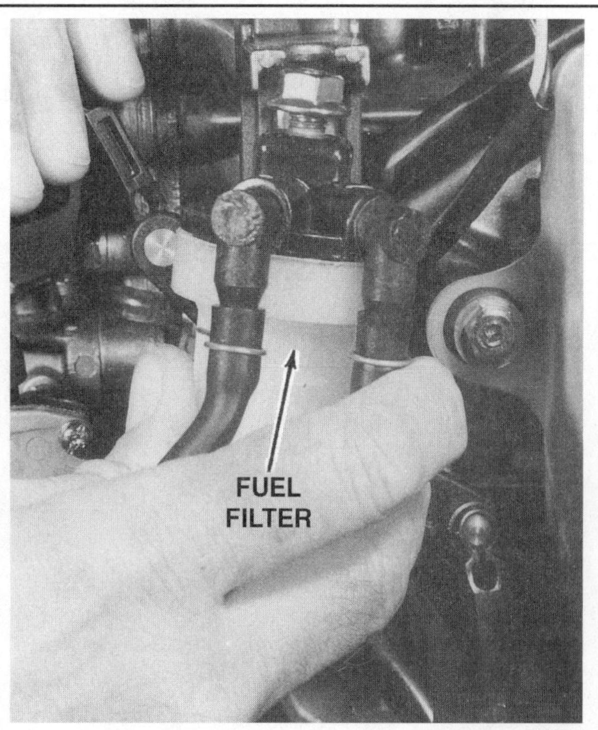

Fig. 168 Removing the fuel filter for inspection and possible replacement

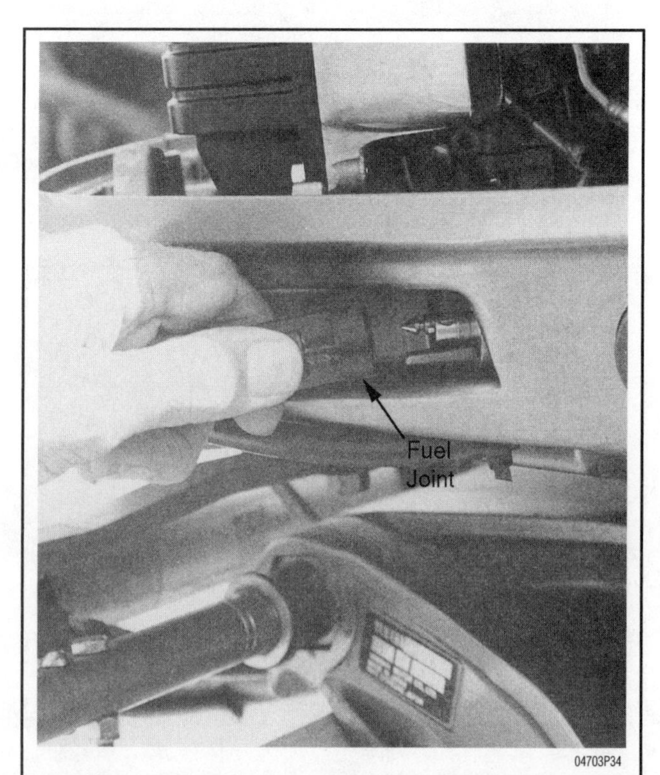

Fig. 169 Make a pre-season check of the fuel line coupling at the fuel joint to ensure a proper and clean connection

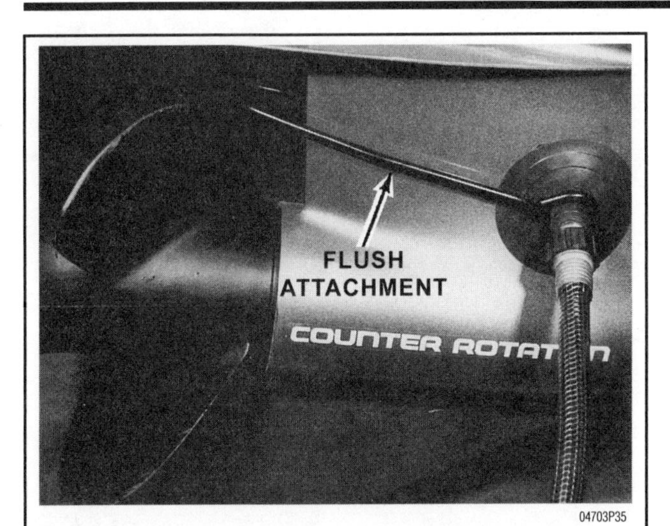

04703P35

Fig. 170 This popular and inexpensive flushing device should be included in every boat owner's maintenance kit

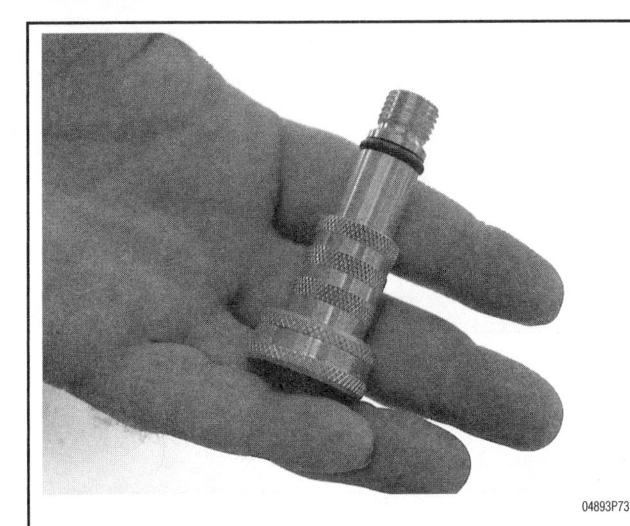

04893P73

Fig. 171 Honda outboards come with a self contained flushing port on the lower unit that uses a special flush kit adapter available from your dealer

system check should include the ignition coils, stator assembly, condition of the wiring and the battery.

3. If a built-in fuel tank is installed, take time to check the gasoline tank and all fuel lines, fittings, couplings, valves, including the flexible tank fill and vent. Turn on the fuel supply valve at the tank. If the fuel was not drained at the end of the previous season, make a careful inspection for gum formation. If a six-gallon fuel tank is used, take the same action. When gasoline is allowed to stand for long periods of time, particularly in the presence of copper, gummy deposits form. This gum can clog the filters, lines, and passageways in the carburetor.

4. Replace the oil in the lower unit.

5. Replace the fuel filter.

6. Replace the engine oil and filter. Make sure to use only a quality four stroke engine oil and NEVER use two stroke oil in a four stroke engine.

7. Close all water drains. Check and replace any defective water hoses. Check to be sure the connections do not leak. Replace any spring-type hose clamps with band-type clamps, if they have lost their tension or if they have distorted the water hose.

8. The engine can be run with the lower unit in water to flush it. If this is not practical, a flush attachment may be used. This unit is attached to the water pick-up in the lower unit. Attach a garden hose, turn on the water, allow the water to flow into the engine for awhile, and then run the engine.

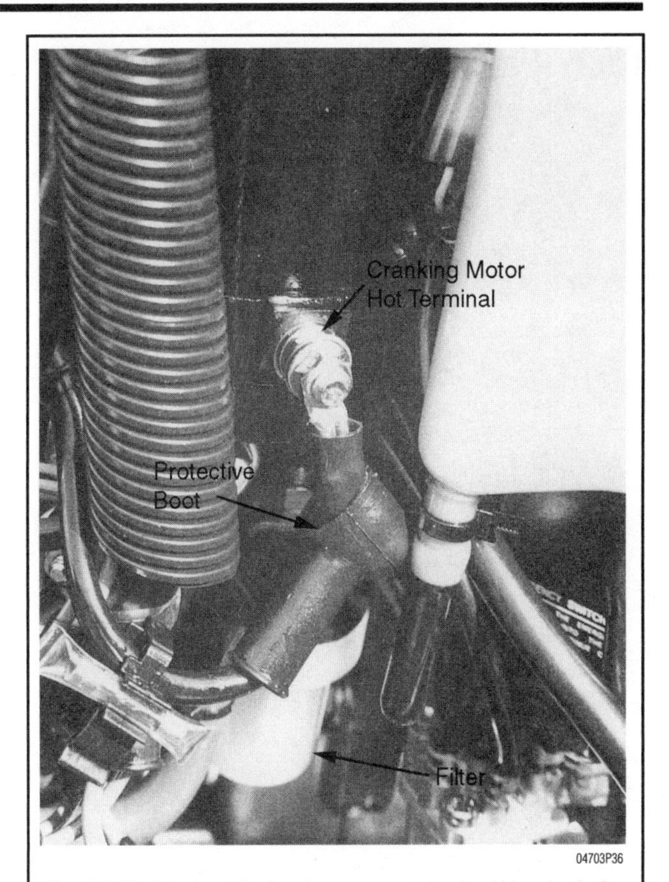

04703P36

Fig. 172 Electrical and fuel system components should be checked on a regular basis

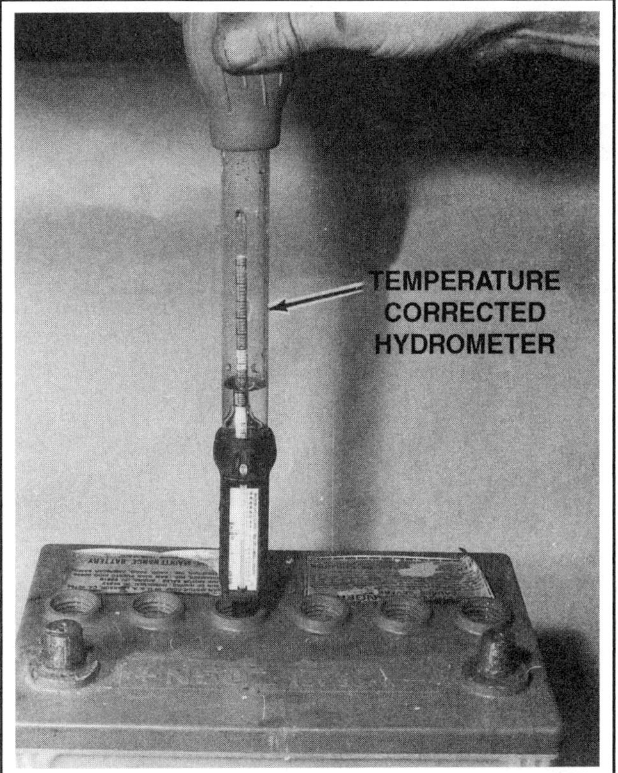

04703P37

Fig. 173 Checking the condition of the battery electrolyte using a hydrometer

04703P38

Fig. 174 A water separating fuel filter installed inside the boat on the transom

CAUTION

Water must circulate through the lower unit to the powerhead anytime the powerhead is operating to prevent the engine from overheating and damage to the water pump in the lower unit. Just five seconds without water will damage the water pump impeller.

04897P69

Fig. 175 The thermostat is usually located in an accessible place for easy maintenance or replacement

9. Check the exhaust outlet for water discharge. Check for leaks. Check operation of the thermostat.
10. Check the electrolyte level in the battery and the voltage for a full charge. Clean and inspect the battery terminals and cable connections. Take time to check the polarity, if a new battery is being installed. Cover the cable connections with grease or special protective compound as a prevention to corrosion formation. Check all electrical wiring and grounding circuits.
11. Check all electrical parts on the engine and lower portions of the hull. Rubber boots help keep electrical connections clean and reduce the possibility of arcing.

➡Electric cranking motors and high tension wiring harnesses should be of a marine type that cannot cause an explosive mixture to ignite.

12. If a water separating filter is installed between the fuel tank and the powerhead fuel filter, replace the element at least once each season. This filter removes water and fuel system contaminants such as dirt, rust, and other solids, thus reducing potential problems.
13. As a last step in spring commissioning, perform a full engine tune-up.

CAUTION

Before putting the boat in the water, take time to verify the drain plugs are installed. Countless number of boating excursions have had a very sad beginning because the boat was eased into the water only to have the boat begin to fill with the water.

FUEL SYSTEM 4-2
FUEL SYSTEM BASICS 4-2
FUEL 4-2
RECOMMENDATIONS 4-2
OCTANE RATING 4-2
VAPOR PRESSURE AND ADDITIVES 4-2
THE BOTTOM LINE WITH FUELS 4-3
HIGH ALTITUDE OPERATION 4-3
ALCOHOL-BLENDED FUELS 4-3
COMBUSTION 4-3
ABNORMAL COMBUSTION 4-3
FACTORS AFFECTING COMBUSTION 4-3
FUEL SYSTEM SERVICE 4-3
CARBURETION 4-4
CARBURETOR IDENTIFICATION 4-4
DRAINING THE FUEL SYSTEM 4-4
TROUBLESHOOTING THE CARBURETED FUEL SYSTEM 4-4
LOGICAL TROUBLESHOOTING 4-4
COMMON PROBLEMS 4-5
COMMON SYMPTOMS 4-5
CARBURETORS 4-6
DESCRIPTION AND OPERATION 4-6
REMOVAL & INSTALLATION 4-8
DISASSEMBLY 4-9
CLEANING & INSPECTION 4-11
ASSEMBLY 4-11
FUEL PUMP 4-14
DESCRIPTION & OPERATION 4-14
FUEL PRESSURE CHECK 4-14
REMOVAL & INSTALLATION 4-14
DISASSEMBLY 4-14
CLEANING & INSPECTION 4-15
ASSEMBLY 4-15
FUEL LINES 4-15
GENERAL INFORMATION 4-15
REPLACEMENT & ROUTING 4-15
FUEL INJECTION 4-19
FUEL INJECTION BASICS 4-19
TROUBLESHOOTING THE FUEL INJECTION SYSTEM 4-20
LOGICAL TROUBLESHOOTING 4-20
PRELIMINARY INSPECTION 4-20
DIAGNOSIS BY SYMPTOM 4-21
COMMON PROBLEMS 4-22
DIAGNOSTIC TROUBLE CODES 4-22
SERVICE PRECAUTIONS 4-22
READING 4-23
CLEARING 4-24
THROTTLE POSITION SENSOR (TPS) 4-26
DESCRIPTION & OPERATION 4-26
TESTING 4-26
REMOVAL & INSTALLATION 4-26
INTAKE AIR TEMPERATURE SENSOR (IAT) 4-27
DESCRIPTION & OPERATION 4-27
TESTING 4-27
REMOVAL & INSTALLATION 4-27
ENGINE COOLANT TEMPERATURE (ECT) 4-27
DESCRIPTION & INSTALLATION 4-27
TESTING 4-27
REMOVAL & INSTALLATION 4-27
OVERHEAT SENSOR 4-28
DESCRIPTION & OPERATION 4-28
TESTING 4-28
REMOVAL & INSTALLATION 4-28
FUEL INJECTORS 4-28
DESCRIPTION & OPERATION 4-28
TESTING 4-28
REMOVAL & INSTALLATION 4-28
VAPOR SEPARATOR 4-29
DESCRIPTION & OPERATION 4-29
REMOVAL & INSTALLATION 4-30
CLEANING & INSPECTION 4-30
FUEL PUMP 4-31
DESCRIPTION & OPERATION 4-31
FUEL PRESSURE CHECK 4-31
REMOVAL & INSTALLATION 4-31
OVERHAUL 4-32
FUEL PRESSURE REGULATOR 4-32
DESCRIPTION & OPERATION 4-32
REMOVAL & INSTALLATION 4-32
INSPECTION 4-32
FUEL LINES 4-32
GENERAL INFORMATION 4-32
REPLACEMENT & ROUTING 4-33
COMMON PROBLEMS 4-34
SPECIFICATIONS CHARTS
CARBURETOR FLOAT HEIGHTS 4-13
CARBURETOR PILOT SCREW SPECIFICATIONS 4-13
DIAGNOSTIC TROUBLE CODE CHART 4-23

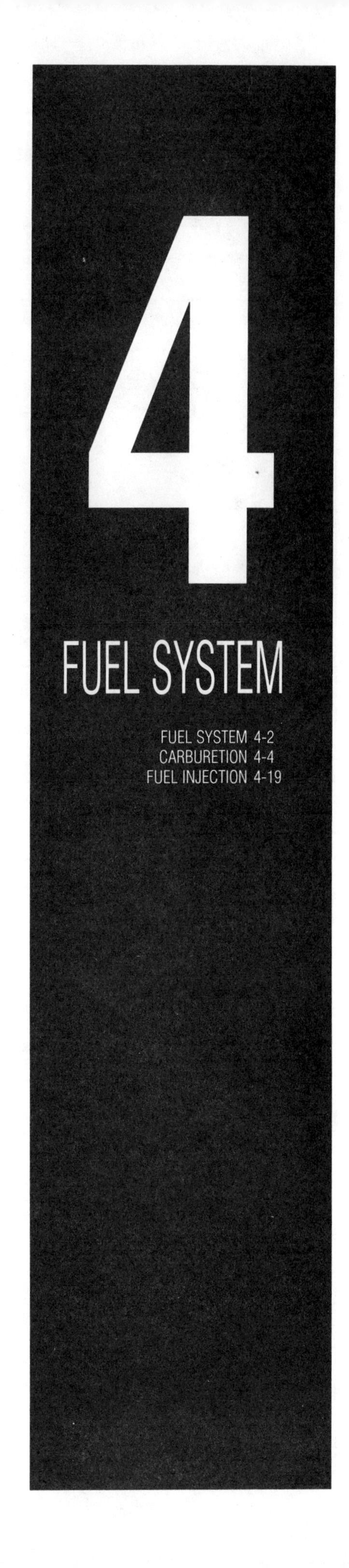

FUEL SYSTEM BASICS

◆ See Figure 1

The fuel delivery system consists of all the components that supply the engine with fuel. This includes the tank itself, all the lines, one or more fuel filters, a fuel pump (mechanical or electric), and the fuel metering components (carburetor or fuel injection system).

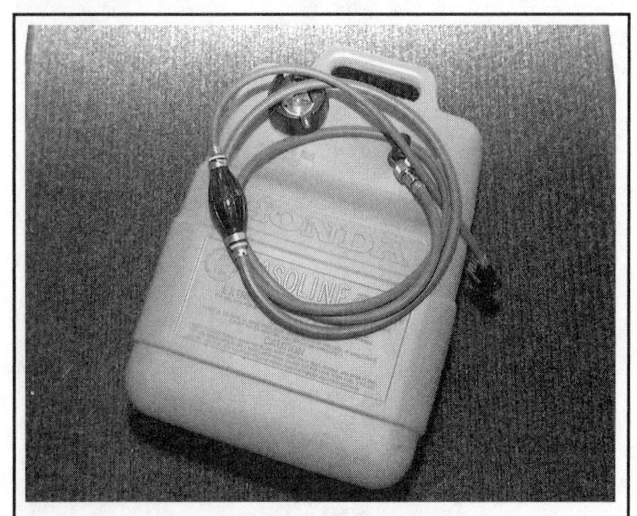

Fig. 1 Typical portable fuel tank used with small horsepower outboards

Fuel

◆ See Figure 2

RECOMMENDATIONS

Reformulated gasoline fuels are now found in many market areas. Current testing indicates no particular problems with using this fuel. Shelf life is shorter and, because of the oxygenates, a slight leaning out at idle may be experienced. This slightly lean condition can be compensated for by adjusting idle mixture screws.

Fuel recommendations have become more complex as the chemistry of modern gasoline changes. The major driving force behind the changes in gasoline chemistry is the search for additives to replace lead as an octane booster and lubricant.

Fig. 2 Always obtain your fuel from a reputable fuel dock

These new additives are governed by the types of emissions they produce in the combustion process. Also, the replacement additives do not always provide the same level of combustion stability, making a fuel's octane rating less meaningful.

In the search for new fuel additives, automobiles are used as the test medium. Not one high performance two cycle engine was tested in the process of determining the chemistry of today's gasoline.

In the 1960's and 1970's, leaded fuel was common. The lead served two functions. The lead served as an octane booster (combustion stabilizer) and, in four cycle engines, served as a valve seat lubricant. For two cycle engines, the primary benefit of lead was to serve as a combustion stabilizer. Lead served very well for this purpose, even in high heat applications.

Today, all lead has been removed from the gasoline process. This means that the benefit of lead as an octane booster has been eliminated. Several substitute octane boosters have been introduced in the place of lead. While many are adequate in an automobile, most do not perform nearly as well as lead did, even though the octane rating of the fuel is the same.

OCTANE RATING

A fuel's octane rating is a measurement of how stable the fuel is when heat is introduced. Octane rating is a major consideration when deciding whether a fuel is suitable for a particular application. For example, in an engine, we want the fuel to ignite when the spark plug fires and not before, even under high pressure and temperatures. Once the fuel is ignited, it must burn slowly and smoothly, even though heat and pressure are building up while the burn occurs. The unburned fuel should be ignited by the traveling flame front, not by some other source of ignition, such as carbon deposits or the heat from the expanding gasses. A fuel's octane rating is known as a measurement of the fuel's anti-knock properties (ability to burn without exploding).

Usually a fuel with a higher octane rating can be subjected to a more severe combustion environment before spontaneous or abnormal combustion occurs. To understand how two gasoline samples can be different, even though they have the same octane rating, we need to know how octane rating is determined.

The American Society of Testing and Materials (ASTM) has developed a universal method of determining the octane rating of a fuel sample. The octane rating you see on the pump at a gasoline station is known as the pump octane number. Look at the small print on the pump. The rating has a formula. The rating is determined by the R+M/2 method.

Therefore, the number you see on the pump is the average of two other octane ratings.

The Research Octane Reading is a measure of a fuel's anti-knock properties under a light load, or part throttle conditions. During this test, combustion heat is easily dissipated.

The Motor Octane Rating is a measure of a fuel's anti-knock properties under a heavy load, or full throttle conditions, when heat buildup is at maximum.

VAPOR PRESSURE AND ADDITIVES

Two other factors besides octane rating affect how suitable the fuel is for a particular application.

Fuel vapor pressure is a measure of how easily a fuel sample evaporates. Many additives used in gasoline contain aromatics. Aromatics are light hydrocarbons distilled off the top of a crude oil sample. They are effective at increasing the research octane of a fuel sample, but can cause vapor lock on a very hot day. If you have an inconsistent running engine and you suspect vapor lock, use a piece of clear fuel line to look for bubbles, indicating that the fuel is vaporizing.

One negative side effect of aromatics is that they create additional combustion products such as carbon and varnish. If your engine requires high octane fuel to prevent detonation, de-carbon the engine more frequently with an internal engine cleaner to prevent ring sticking due to excessive varnish buildup.

Besides aromatics, two types of alcohol are used in fuel today as octane boosters, ethanol and methanol. Again, alcohol tends to raise the research octane of the fuel. This usually means they will have limited benefit in an outboard motor. Also, alcohol contains oxygen, which means that since it is replacing gasoline without oxygen content, alcohol fuel blends cause the fuel-air mixture to be leaner.

THE BOTTOM LINE WITH FUELS

If we could buy fuel of the correct octane rating, free of alcohol and aromatics, this would be our first choice.

Honda continues to recommend unleaded fuel. This is almost a redundant recommendation due to the near universal unavailability of any other type fuel.

According to the fuel recommendations that come with your outboard, there is no engine in the product line that requires more than 86 octane. Unleaded fuel produces fewer engine and spark plug deposits and extends exhaust system life.

Regardless of the fuel octane rating you choose, try to stay with a name brand fuel. You never know for sure what kinds of additives or how much is in off brand fuel.

HIGH ALTITUDE OPERATION

At elevated altitudes there is less oxygen in the atmosphere than at sea level. Less oxygen means lower combustion efficiency and less power output. Power output is reduced three percent for every thousand feet above sea level. At ten thousand feet, power is reduced 30 percent from that available at sea level.

Re-jetting for high altitude does not restore this lost power. Re-jetting simply corrects the air-fuel ratio for the reduced air density, and makes the most of the remaining available power. If you re-jet an engine, you are locked into the higher elevation. You cannot operate at sea level until you re-jet for sea level. Understand that going below the elevation jetted for your motor will damage the engine. As a general rule, jet for the lowest elevation anticipated. Spark plug insulator tip color is the best guide for high altitude jetting.

If you are in an area of known poor fuel quality, you may want to use fuel additives. Today's additives are mostly alcohol and aromatics, and their effectiveness may be limited. It is difficult to find additives without ethanol, methanol, or aromatics. If you use octane boosters frequent de-carboning may be necessary. If possible, the best policy is to use name brand pump fuel with no additional additives except Honda fuel conditioners and Ring-Free.

ALCOHOL-BLENDED FUELS

The Environmental Protection Agency mandated a phase-out of all leaded fuels. Lead was used to boost the octane of fuel. By January of 1986, the maximum allowable amount of lead was 0.1 gm/gal, down from 1.1 gm/gal.

Gasoline suppliers, in general, feel that the 0.1 gm/gal limit is too low to make lead of any real use to improve octane. Therefore, alternate octane enhancers are being used. There are multiple methods currently employed to improve octane but the most inexpensive additive seems to be alcohol.

There are, however, some special considerations due to the effects of alcohol in fuel. You should know about them and what steps to take when using alcohol-blended fuels commonly called gasohol.

Alcohol in fuel is either methanol (wood alcohol) or ethanol (grain alcohol). Either type can have serious effects when applied to outboard motor applications.

The leaching affect of alcohol will, in time, cause fuel lines and plastic components to become brittle to the point of cracking. Unless replaced, these cracked lines could leak fuel, increasing the potential for hazardous situations.

➡Honda fuel lines and plastic fuel system components have been specially formulated to resist alcohol leaching effects.

When gasohol becomes contaminated with water, the water combines with the alcohol then settles to the bottom. This leaves the gasoline and the oil for models using premix, on a top layer. With alcohol-blended fuels, the amount of water necessary for this phase separation to occur is 0.5% by volume.

All fuels have chemical compounds added to reduce the tendency towards phase separation. If phase separation occurs, however, there is a possibility of a lean oil/fuel mixture with the potential for engine damage.

Combustion

The four-stroke engine requires four complete strokes of the piston to complete one power cycle. During the intake stroke, the intake valve opens and the fuel/air mixture is drawn into the combustion chamber as a result of the of the sudden drop in pressure in the cylinder. As the piston moves upward on it's compression stroke, both the exhaust and intake valves are closed and the air/fuel mixture is compressed. The ignition system then triggers the spark plug and ignites the mixture. This forces the piston downward in the cylinder bore and is called the power stroke.

ABNORMAL COMBUSTION

There are two types of abnormal combustion:

- Pre-ignition—Occurs when the air-fuel mixture is ignited by some other incandescent source other than the correctly timed spark from the spark plug.
- Detonation—Occurs when excessive heat and or pressure ignites the air/fuel mixture rather than the spark plug. The burn becomes explosive.

FACTORS AFFECTING COMBUSTION

The combustion process is affected by several interrelated factors. This means that when one factor is changed, the other factors also must be changed to maintain the same controlled burn and level of combustion stability.

- Compression—Determines the level of heat buildup in the cylinder when the air-fuel mixture is compressed. As compression increases, so does the potential for heat buildup.
- Ignition Timing—Determines when the gasses will start to expand in relation to the motion of the piston. If the gasses begin to expand too soon, such as they would during pre-ignition or in an overly advanced ignition timing, the motion of the piston opposes the expansion of the gasses, resulting in extremely high combustion chamber pressures and heat.

As ignition timing is retarded, the burn occurs later in relation to piston position. This means that the piston has less distance to travel under power to the bottom of the cylinder, resulting in less usable power.

- Fuel Mixture—Determines how efficient the burn will be. A rich mixture burns slower than a lean one. If the mixture is too lean, it can't become explosive. The slower the burn, the cooler the combustion chamber, because pressure buildup is gradual.
- Fuel Quality (Octane rating)—Determines how much heat is necessary to ignite the mixture. Once the burn is in progress, heat is on the rise. The unburned poor quality fuel is ignited all at once by the rising heat instead of burning gradually as a flame front of the burn passing by. This action results in detonation (pinging).
- Other Factors—In general, anything that can cause abnormal heat buildup can be enough to push an engine over the edge to abnormal combustion, if any of the four basic factors previously discussed are already near the danger point, for example, excessive carbon buildup raises the compression and retains heat as glowing embers.

Fuel System Service

Safety is the most important factor when performing fuel system service. Failure to conduct repairs in a safe manner may result in serious personal injury or death. Work on an outboard's fuel system components can be accomplished safely and effectively by adhering to the following rules and guidelines.

- To avoid the possibility of fire and personal injury, always disconnect the negative battery cable unless the repair or test procedure requires that battery voltage be applied.
- Always relieve the fuel system pressure prior to disconnecting any fuel system component (injector, fuel rail, pressure regulator, etc.) fitting or fuel line connection. Exercise extreme caution whenever relieving fuel system pressure to avoid exposing skin, face and eyes to fuel spray. Please be advised that fuel under pressure may penetrate the skin or any part of the body that it contacts.
- Always place a shop towel or cloth around the fitting or connection prior to loosening to absorb any excess fuel due to spillage. Ensure that all fuel spillage is quickly remove from engine surfaces. Ensure that all fuel-soaked cloths or towels are deposited into a flame-proof waste container with a lid.
- Always keep a dry chemical (Class B) fire extinguisher near the work area.
- Do not allow fuel spray or fuel vapors to come into contact with a spark or open flame.
- Always use a second wrench when loosening or tightening fuel line con-

nections fittings. This will prevent unnecessary stress and torsion to fuel piping. Always follow the proper torque specifications.

- Always replace worn fuel fitting O-rings with new ones. Do not substitute fuel hose where rigid pipe is installed.
- Use unleaded gasoline with a pump octane rating of 86 or higher
- Try to only use a known, quality brand of fuel. Off brand fuels can often affect engine performance
- If you use an oxygenated fuel, make sure that it is unleaded and meets the minimum octane rating requirement
- Use a fuel/water separating filter in the fuel line between the tank and the engine
- Never use stale or contaminated gasoline
- These are four-stroke engines. Never, ever run these engines with a gasoline/oil mixture like a two-stroke engine

CARBURETION

**** CAUTION**

Observe all applicable safety precautions when working around fuel. Whenever servicing the fuel system, always work in a well ventilated area. Do not allow fuel spray or vapors to come in contact with a spark or open flame. Keep a dry chemical fire extinguisher near the work area. Always keep fuel in a container specifically designed for fuel storage; also, always properly seal fuel containers to avoid the possibility of fire or explosion.

The carburetor is merely a metering device for mixing fuel and air in the proper proportions for efficient engine operation. Too much fuel, or not enough air, will alter this ratio and cause performance problems.

Carburetor Identification

See Figure 3

All carbureted Honda outboard engines have float bowl-type, horizontal butterfly carburetors. Currently, there are eight distinct carburetor code groups. The code is cast into the carburetor as shown in the illustration.

The code groups differ in body shape, control location and jet types. But they all share similar inspection and adjustment procedures. Refer to the correct shop manual for more specific information on individual models.

Draining The Fuel System

See Figure 4

For many years there has been the widespread belief that simply shutting off the fuel at the tank and then running the engine until it stops is the proper procedure before storing the engine for any length of time. Right? Wrong!

It is not possible to remove all of the fuel in the carburetor by operating the engine until it stops. Some fuel is trapped in the float chamber and other passages and in the line leading to the carburetor. The only guaranteed method of removing ALL of the fuel is to take the time to remove the carburetor, and drain the fuel.

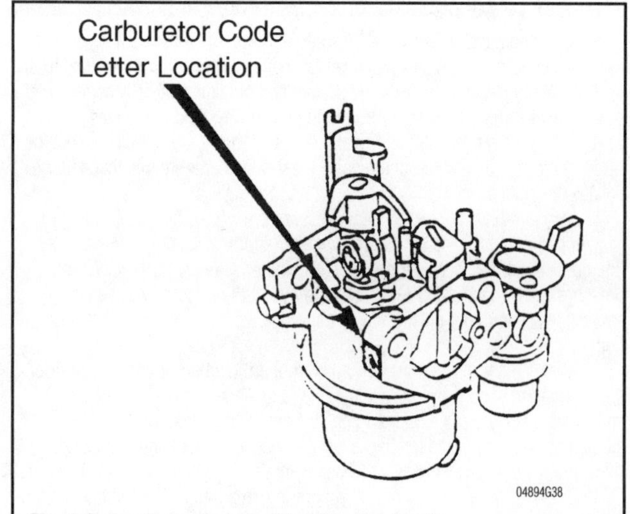

Fig. 3 Carburetor code letter identification location

Fig. 4 Typical fuel line quick disconnect fittings

Proper procedure involves:

1. Shutting off the fuel supply at the tank.
2. Disconnecting the fuel line at the tank.
3. Operating the engine until it begins to run rough, then stopping the engine, which will leave some fuel inside.
4. Removing and draining the carburetor.

By disconnecting the fuel supply, all small passages are cleared of fuel even though some fuel is left in the carburetor. A light oil should be put in the combustion chamber as instructed in the owner's manual. On most carburetors there is a drain plug on the float bowl to drain the fuel from the carburetor.

For short periods of storage, simply running the carburetor dry may help prevent severe gum and varnish from forming in the carburetor. This is especially true during hot weather.

Troubleshooting The Carbureted Fuel System

LOGICAL TROUBLESHOOTING

Troubleshooting the fuel system requires the same techniques used in other areas on the engine. A through, step-by-step process to troubleshooting will almost always pay-off in the discovery of the problem.

The following paragraphs provide an orderly sequence of tests to pinpoint problems in the fuel system.

1. Gather as much information as you can.
2. Duplicate the condition. Take the boat out and verify the complaint.
3. If the problem cannot be duplicated, you cannot fix it. This could be a product operation problem.
4. Once the problem has been duplicated, you can begin troubleshooting. Give the entire unit a careful visual inspection. You can tell a lot about the engine from the care and condition of the entire rig. What's the condition of the propeller and the lower unit? Remove the engine cover and look for any visible signs of failure. Are there any signs of head gasket leakage. Is the engine paint discolored from high temperature or are there any holes or cracks in the engine

block? Perform a compression and leak down test. While cranking the engine during the compression test, listen for any abnormal sounds. If the engine passes these simple tests we can assume that the mechanical condition of the engine is good. All other engine mechanical inspection would be too time consuming at this point.

5. Your next step is to isolate the fuel system into two sub-systems. Separate the fuel delivery components from the carburetors. To do this, substitute the boat's fuel supply with a known good supply. Use a 6 gallon portable tank and fuel line. Connect the portable fuel supply directly to the engine fuel pump, bypassing the boat fuel delivery system. Now test the engine. If the problem is no longer present, you know where to look. If the problem is still present, further troubleshooting is required.

6. When testing the engine, observe the throttle position when the problem occurs. This will help you pinpoint the circuit that is malfunctioning. Carburetor troubleshooting and repair is very demanding. You must pay close attention to the location, position and sometimes the numbering on each part removed. The ability to identify a circuit by the operating RPM it affects is important. Often your best troubleshooting tool is a can of cleaner. This can be used to trace those mystery circuits and find that last speck of dirt. Be careful and wear safety glasses when using this method.

COMMON PROBLEMS

Fuel Delivery

▶ See Figure 5

Many times fuel system troubles are caused by a plugged fuel filter, a defective fuel pump, or by a leak in the line from the fuel tank to the fuel pump. A defective choke may also cause problems. Would you believe, a majority of starting troubles which are traced to the fuel system are the result of an empty fuel tank or aged sour fuel.

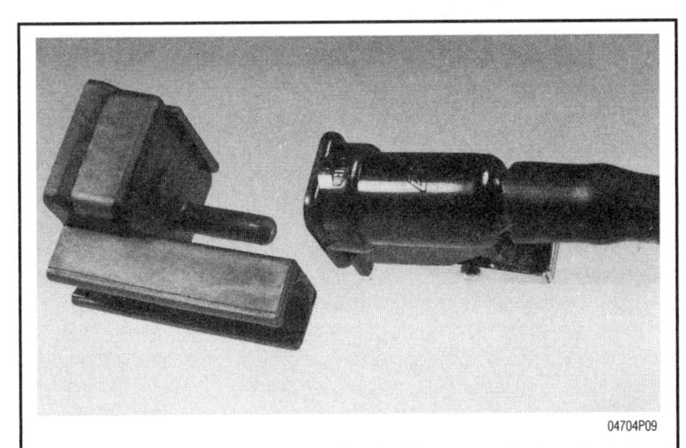

04704P09

Fig. 5 An excellent way of protecting fuel hoses against contamination is an end cap filter

Sour Fuel

▶ See Figure 6

Under average conditions (temperate climates), fuel will begin to break down in about four months. A gummy substance forms in the bottom of the fuel tank and in other areas. The filter screen between the tank and the carburetor and small passages in the carburetor will become clogged. The gasoline will begin to give off an odor similar to rotten eggs. Such a condition can cause the owner much frustration, time in cleaning components, and the expense of replacement or overhaul parts for the carburetor.

Even with the high price of fuel, removing gasoline that has been standing unused over a long period of time is still the easiest and least expensive preventative maintenance possible. In most cases, this old gas can be used without harmful effects in an automobile using regular gasoline.

Choke Problems

When the engine is hot, the fuel system can cause starting problems. After a hot engine is shut down, the temperature inside the fuel bowl may rise to 200 degrees F and cause the fuel to actually boil. All carburetors are vented to allow this pressure to escape to the atmosphere. However, some of the fuel may percolate over the high-speed nozzle.

04894P47

Fig. 6 The use of an approved fuel additive, such as Stabil® fuel conditioner and stabilizer, will prevent fuel from souring for up to twelve months

If the choke should stick in the open position, the engine will be hard to start. If the choke should stick in the closed position, the engine will flood, making it very difficult to start.

In order for this raw fuel to vaporize enough to burn, considerable air must be added to lean out the mixture. Therefore, the only remedy is to remove the spark plugs, ground the leads, crank the powerhead through about ten revolutions, clean the plugs, reinstall the plugs, and start the engine.

If the needle valve and seat assembly is leaking, an excessive amount of fuel may enter the intake manifold and create a flooded condition.

COMMON SYMPTOMS

Rough Engine Idle

If an engine does not idle smoothly, the most reasonable approach to the problem is to perform a tune-up to eliminate such areas as:

- Defective points (if applicable)
- Faulty spark plugs
- Timing out of adjustment

Other problems that can prevent an engine from running smoothly include:

- An air leak in the intake manifold
- Uneven compression between the cylinders

Of course any problem in the carburetor affecting the air/fuel mixture will also prevent the engine from operating smoothly at idle speed. These problems usually include:

- Too high a fuel level in the bowl
- A heavy float
- Leaking needle valve and seat
- Defective automatic choke
- Improper adjustments for idle mixture or idle speed

Excessive Fuel Consumption

Excessive fuel consumption can be the result of any one of four conditions, or a combination of all.

- Inefficient engine operation.
- Faulty condition of the hull, including excessive marine growth.
- Poor boating habits of the operator.
- Leaking or out of tune carburetor.

If the fuel consumption suddenly increases over what could be considered normal, then the cause can probably be attributed to the engine or boat and not the operator.

Marine growth on the hull can have a very marked effect on boat performance. This is why sail boats always try to have a haul-out as close to race time as possible.

While you are checking the bottom, take note of the propeller condition. A bent blade or other damage will definitely cause poor boat performance.

If the hull and propeller are in good shape, then check the fuel system for possible leaks. Check the line between the fuel pump and the carburetor while the engine is running and the line between the fuel tank and the pump when the engine is not running. A leak between the tank and the pump many times will not appear when the engine is operating, because the suction created by the pump drawing fuel will not allow the fuel to leak. Once the engine is turned off and the suction no longer exists, fuel may begin to leak.

If a minor tune-up has been performed and the spark plugs, points, and timing are properly adjusted, then the problem most likely is in the carburetor and an overhaul is in order.

Check the needle valve and seat for leaking. Use extra care when making any adjustments affecting the fuel consumption, such as the float level or automatic choke.

Engine Surge

If the engine operates as if the load on the boat is being constantly increased and decreased, even though an attempt is being made to hold a constant engine speed, the problem can most likely be attributed to the carburetor pilot screw out of adjustment, clogged carburetor passages, air leakage from carburetor insulator or insufficient fuel in the fuel tank.

Carburetors

See Figures 7 and 8

The BF20 and BF2A uses a horizontal-type butterfly valve carburetor with a manual choke. This is the only engine with fuel supplied by gravity from an engine mounted fuel tank.

The BF50, BF5A, BF75, BF8A and BF100 also use horizontal-type, butterfly valve carburetors with a manual choke. Fuel is supplied by a diaphragm-type fuel pump from a remote fuel tank.

The BF9.9A and BF15A use horizontal-type, butterfly valve carburetors with manual choke. Fuel is supplied by a mechanical plunger-type fuel pump from a remote fuel tank.

The BF25A and BF30A use horizontal-type, butterfly valve carburetors (3 per engine) with a mechanical plunger-type fuel pump.

The BF35A, BF45A, BF75A and BF90A use horizontal-type, butterfly valve carburetors (4 per engine) with a mechanical plunger-type fuel pump.

DESCRIPTION AND OPERATION

Float Systems

See Figures 9 and 10

A small chamber in the carburetor serves as a fuel reservoir. A float valve admits fuel into the reservoir to replace the fuel consumed by the engine. If the carburetor has more than one reservoir, the fuel level in each reservoir (chamber) is controlled by identical float systems.

Fuel level in each chamber is extremely critical and must be maintained accurately. Accuracy is obtained through proper adjustment of the floats. This adjustment will provide a balanced metering of fuel to each cylinder at all speeds.

Following the fuel through its course, from the fuel tank to the combustion chamber of the cylinder, will provide an appreciation of exactly what is taking place. In order to start the engine, the fuel must be moved from the tank to the carburetor by a squeeze bulb installed in the fuel line. This action is necessary because the fuel pump does not have sufficient pressure to draw fuel from the tank during cranking before the engine starts.

The fuel for some small horsepower units is gravity fed from a tank mounted at the rear of the powerhead. After the engine starts, the fuel passes through the pump to the carburetor. All systems have some type of filter installed somewhere in the line between the tank and the carburetor. Many units have a filter as an integral part of the carburetor.

At the carburetor, the fuel passes through the inlet passage to the needle and seat, and then into the float chamber (reservoir). A float in the chamber rides up and down on the surface of the fuel. After fuel enters the chamber and the level rises to a predetermined point, a tang on the float closes the inlet needle and the flow entering the chamber is cut off. When fuel leaves the chamber as the engine operates, the fuel level drops and the float tang allows the inlet needle to move off its seat and fuel once again enters the chamber. In this manner, a constant reservoir of fuel is maintained in the chamber to satisfy the demands of the engine at all speeds.

A fuel chamber vent hole is located near the top of the carburetor body to permit atmospheric pressure to act against the fuel in each chamber. This pressure assures an adequate fuel supply to the various operating systems of the powerhead.

Air/Fuel Mixture

See Figure 11

A suction effect is created each time the piston moves upward in the cylinder. This suction draws air through the throat of the carburetor. A restriction in the throat, called a venturi, controls air velocity and has the effect of reducing air pressure at this point.

The difference in air pressures at the throat and in the fuel chamber, causes the fuel to be pushed out of metering jets extending down into the fuel chamber. When the fuel leaves the jets, it mixes with the air passing through the venturi. This fuel/air mixture should then be in the proper proportion for burning in the cylinders for maximum engine performance.

In order to obtain the proper air/fuel mixture for all engine speeds, some models have high and low speed jets. These jets have adjustable needle valves which are used to compensate for changing atmospheric conditions. In almost all cases, the high-speed circuit has fixed high-speed jets and are not adjustable.

A throttle valve controls the flow of air/fuel mixture drawn into the combustion chambers. A cold powerhead requires a richer fuel mixture to start and during the brief period it is warming to normal operating temperature. A choke valve is placed ahead of the metering jets and venturi. As this valve begins to close, the volume of air intake is reduced, thus enriching the mixture entering the cylinders. When this choke valve is fully closed, a very rich fuel mixture is drawn into the cylinders.

The throat of the carburetor is usually referred to as the barrel. Carburetors

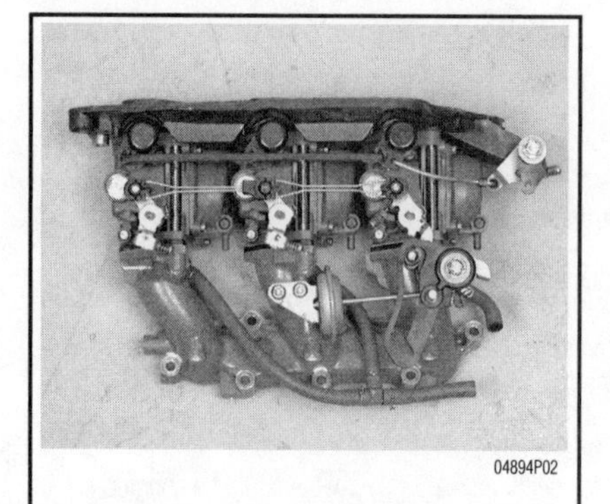

04894P02

Fig. 7 Horizontal-type, butterfly valve carburetors are used on all Honda outboards

04894P07

Fig. 8 Typical multiple carburetor assembly showing individual carburetors and the common manifold

04894P28

Fig. 9 The float chamber consists of two pieces on this carburetor

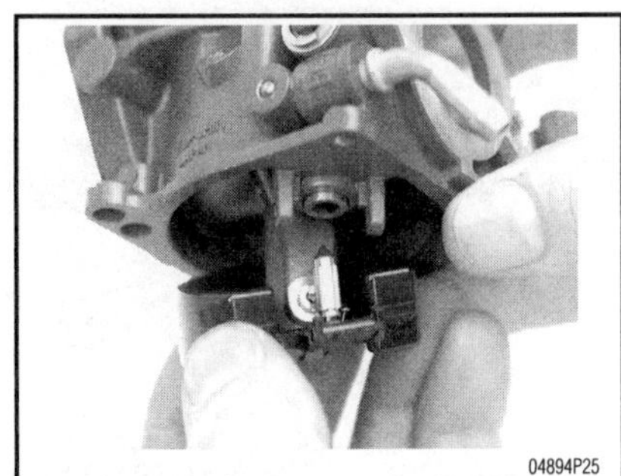

Fig. 10 The needle and seat are controlled by float level and admit fuel into the reservoir to replace the fuel consumed by the engine

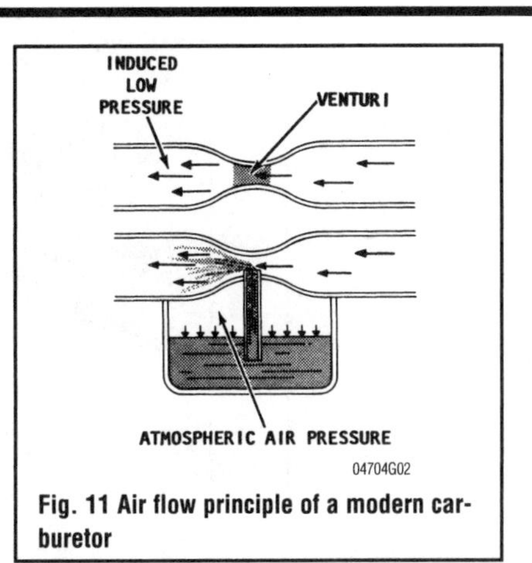

Fig. 11 Air flow principle of a modern carburetor

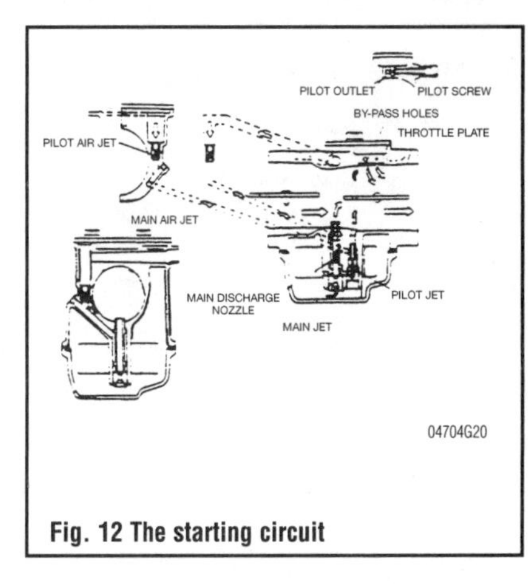

Fig. 12 The starting circuit

with single, double, or four barrels have individual metering jets, needle valves, throttle and choke plates for each barrel. Single and two barrel carburetors are fed by a single float and chamber.

Carburetor Circuits

The following section illustrates the circuit functions and locations of a typical marine carburetor.

STARTING CIRCUIT

See Figure 12

The choke plate is closed, creating a partial vacuum in the venturi. As the piston rises, negative pressure in the crankcase draws the rich air-fuel mixture from the float bowl into the venturi and on into the engine.

LOW SPEED CIRCUIT

See Figure 13

Zero–one-eighth throttle, when the pressure in the crankcase is lowered, the air-fuel mixture is discharged into the venturi through the pilot outlet because the throttle plate is closed. No other outlets are exposed to low venturi pressure. The fuel is metered by the pilot jet. The air is metered by the pilot air jet. The combined air-fuel mixture is regulated by the pilot air screw.

MID-RANGE CIRCUIT

See Figure 14

One-eighth–three-eighths throttle, as the throttle plate continues to open, the air-fuel mixture is discharged into the venturi through the bypass holes. As the throttle plate uncovers more bypass holes, increased fuel flow results because of the low pressure in the venturi. Depending on the model, there could be two, three or four bypass holes.

HIGH SPEED CIRCUIT

See Figure 15

Three-eighths–wide-open throttle, as the throttle plate moves toward wide open, we have maximum air flow and very low pressure. The fuel is metered through the main jet, and is drawn into the main discharge nozzle. Air is metered by the main air jet and enters the discharge nozzle, where it combines with fuel. The mixture atomizes, enters the venturi, and is drawn into the engine.

Basic Functions

See Figure 16

The carburetor systems on in line engines require careful cleaning and adjustment if problems occur. These carburetors are complicated but not too complex to understand. All carburetors operate on the same principles.

Traditional carburetor theory often involves a number of laws and principles. To troubleshoot carburetors learn the basic principles, watch how the carburetor comes apart, trace the circuits, see what they do and make sure they are clean. These are the basic steps for troubleshooting and successful repair.

The diagram illustrates several carburetor basics. If you blow through the straw an atomized mixture (air and fuel droplets) comes out. When you blow through the straw a pressure drop is created in the straw column inserted in the liquid. In a carburetor this is mostly air and a little fuel. The actual ratio of air to fuel differs with engine conditions but is usually from 15 parts air to one part fuel at optimum cruise to as little as 7 parts air to one part fuel at full choke.

If the top of the container is covered and sealed around the straw what will happen? No flow. This is typical of a clogged carburetor bowl vent. If the base of the straw is clogged or restricted what will happen? No flow or low flow. This represents a clogged main jet. If the liquid in the glass is lowered and you blow through the straw with the same force what will happen? Not as much fuel will flow. A lean condition occurs. If the fuel level is raised and you blow again at the same velocity what happens? The result is a richer mixture.

Honda carburetors control air flow semi-independently of RPM. This is done

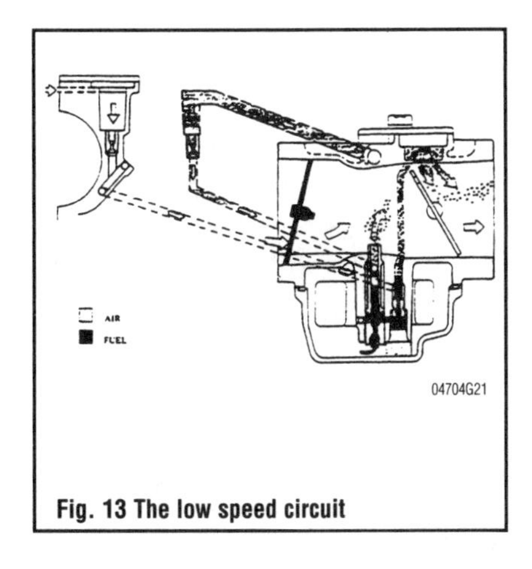

Fig. 13 The low speed circuit

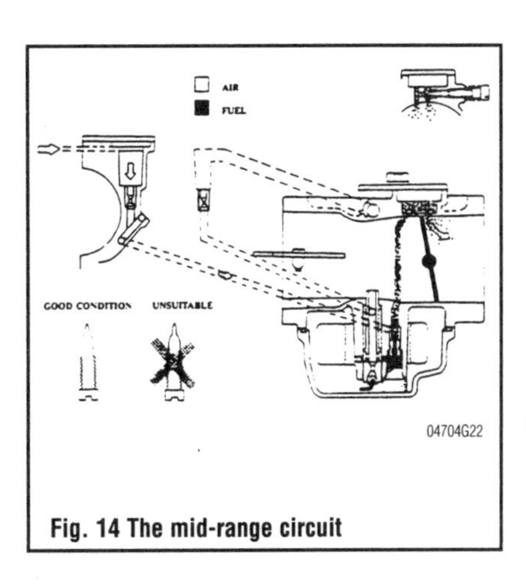

Fig. 14 The mid-range circuit

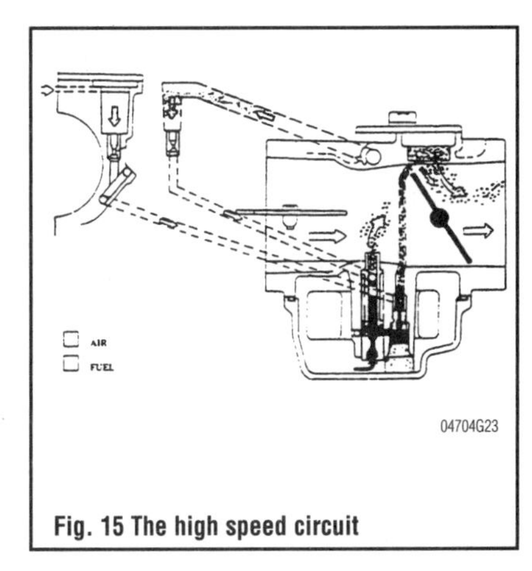

Fig. 15 The high speed circuit

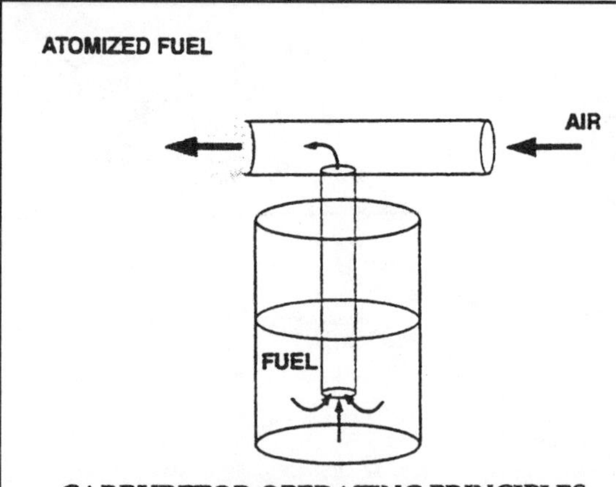

Fig. 16 If you blow through the straw, an atomized mixture (air and fuel droplets) comes out

with a throttle plate. The throttle plate works in conjunction with other systems or circuits to deliver correct mixtures within certain RPM bands. The idle circuit pilot outlet controls from 0-1/8 throttle. The series of small holes in the carburetor throat called transition holes control the 1/8-3/8 throttle range. At wide open throttle the main jet handles most of the fuel metering chores, but the low and mid-range circuits continue to supply part of the fuel.

Enrichment is necessary to start a cold engine. Fuel and air mix does not want to vaporize in a cold engine. In order to get a little fuel to vaporize, a lot of fuel is dumped into the engine. On many engines a choke plate is used for cold starts. This plate restricts air entering the engine and increases the fuel to air ratio.

REMOVAL & INSTALLATION

◆ **See Figures 17 thru 23**

Good shop practice dictates a carburetor repair kit be purchased and new parts be installed any time the carburetor is disassembled.

Make an attempt to keep the work area organized and to cover parts after they have been cleaned. This practice will prevent foreign objects from entering the passageways or adhering to critical parts of the carburetor.

** WARNING

Remove the drain screw and completely drain the float bowl or bowls, of all fuel before disassembling the carburetor. Fuel vapor or spilled fuel may ignite. Wipe up all spilled fuel at once.

1. Remove the engine cover for access.
2. For models so equipped, remove the three cap nuts on the recoil starter and remove the recoil starter.

** WARNING

Do not allow the spring to jump out of the recoil starter assembly.

3. If necessary on models equipped with integral fuel tanks, remove the cap nuts on the fuel tank mounting rubber, disconnect the fuel hose from the fuel tank and lift the tank off the engine.

■ **Now is a good time to inspect the fuel filter and replace it if necessary.**

4. Disconnect the throttle cable and choke rod at the carburetor or carburetor assembly (as applicable).
5. For single carburetor motors, proceed as follows:
   a. Loosen the two cap nuts on the carburetor studs or the 2 retaining bolts and remove the air guide.
   b. Carefully pull the carburetor and gasket off the powerhead.
   c. If necessary remove the two screws on the carburetor joint plate, then remove the joint plate, plate gasket, carburetor insulator and insulator gasket.

Fig. 17 Drain each carburetor float bowl into an appropriate container before removing

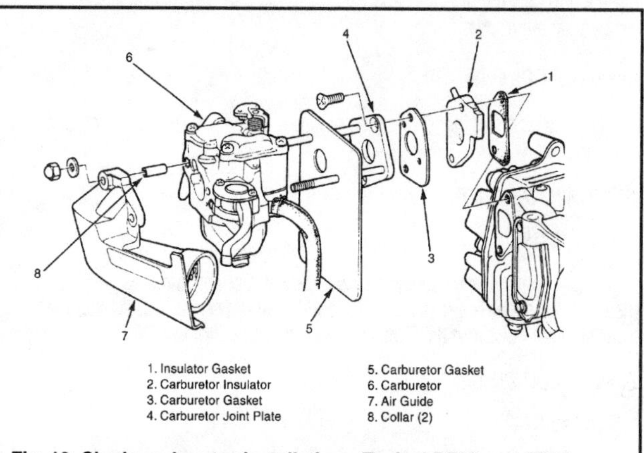

Fig. 18 Single carburetor installation—Typical BF20 and BF2A carburetor shown (BF2D similar, but not the same)

Fig. 19 Disconnect the throttle linkage before carb removal

■ **Now is the time to check all the carburetor gasket surfaces for tears and other damage. Replace if the gasket is in any way damaged.**

6. For multiple carburetor motors, proceed as follows:
   a. If not done already, tag and disconnect the fuel line(s) from the carburetors.
   b. Make sure all choke linkages and throttle cables are disconnected.
   c. Remove the air guide or air silencer.
   d. Loosen the bolts at the silencer plate until the carburetors separate from the intake manifold. Support the carburetors at this time, the bolts through the silencer plate will hold the carburetors together.
   e. Remove carburetors from the powerhead as a single unit and separate them, as necessary.

Fig. 20 The choke linkage rod and the . . .

Fig. 21 . . . throttle rod pivot must be disconnected on multi-carb models...

Fig. 22 . . . before removing the carburetor assembly from the engine

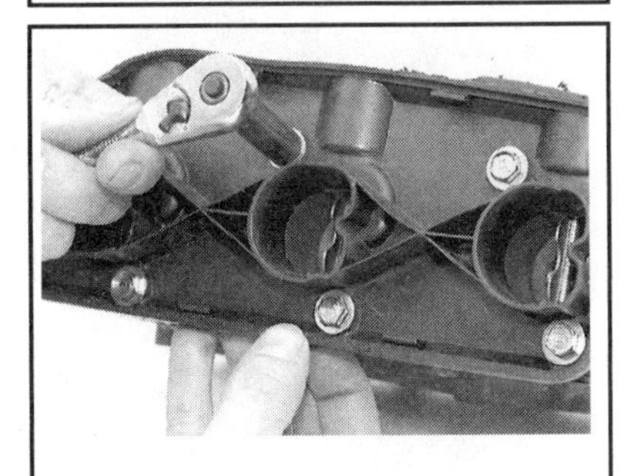

Fig. 23 Remove the carburetors from the silencer plate

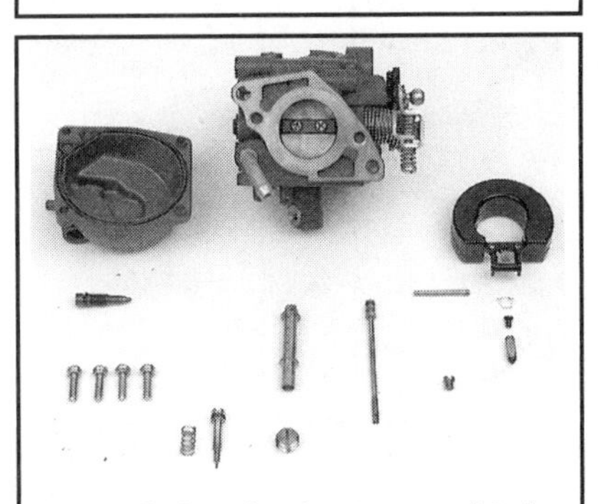

Fig. 24 Outboard carburetors consist of many small and delicate parts which must be reassembled in the proper sequence

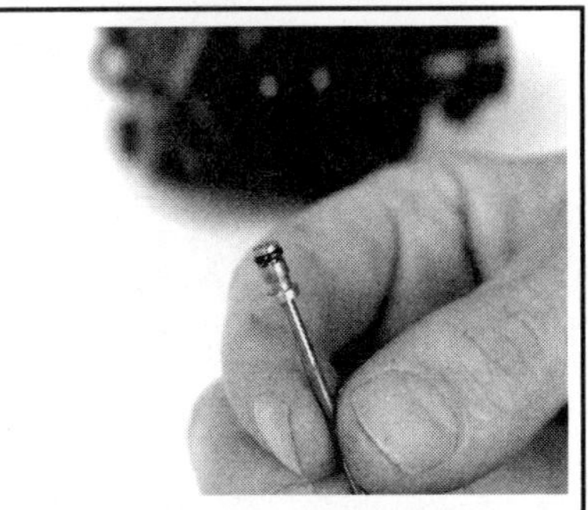

Fig. 25 A component as small as this O-ring can cause major fuel control problems if it fails or is damaged during installation

To install:

7. Inspect the carburetor insulators and gaskets for wear and tear and reinstall them on the engine.
8. For multiple carburetor motors, install the carbs as follows:
   a. Sandwich the carburetors between the silencer plate and the intake manifold and install the insulators and gaskets.
   b. Install the bolts through the silencer plate, carburetors and into the intake manifold.
   c. When everything lines up, tighten the bolts to the correct torque
9. For single carburetor motors. Install the carbs as follows:
   a. If removed install the insulator (using a new insulator gasket) and the joint plate (also using a new gasket). Tighten the retaining screws securely.
   b. Install the carburetor to the powerhead along with the air horn and secure using the cap nuts or the retaining bolts.
10. After the carburetors have been installed, reattach the choke and throttle linkages and the fuel hoses.
11. If applicable install the fuel tank and/or the hand-rewind starter assembly.
12. Pressurize the fuel system using the primer bulb and check for leaks.

**** CAUTION**

NEVER operate the engine, even for a moment, without a source of cooling water for the motor otherwise impeller or possibly even powerhead damage may occur.

13. Operate the engine for a few minutes with the engine cover still removed in order to observe for fuel leaks. Shut the powerhead down and install the engine cover.

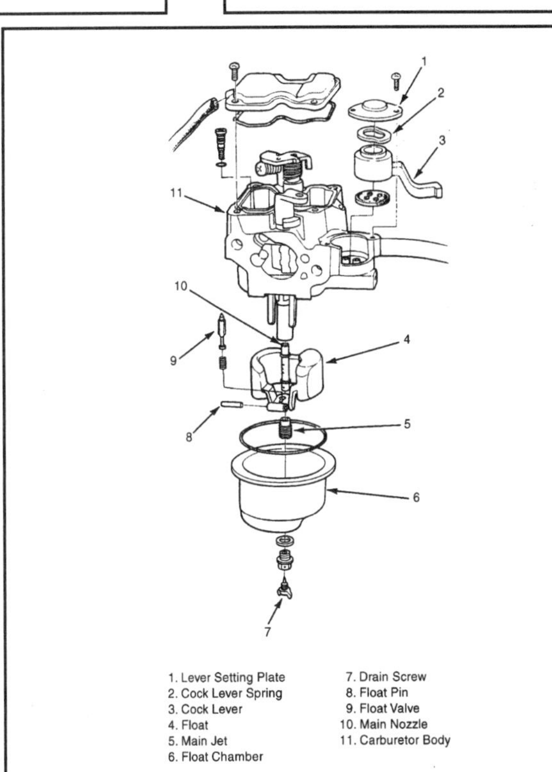

Fig. 26 Exploded view of the BF20 and BF2A carburetor

DISASSEMBLY

◆ **See Figures 24 thru 30**

■ **Prior to disassembling your carburetor(s), make absolutely sure that the carburetor is the cause of the problem. More outboard engines fail or are subject to poor performance because the carburetors have been incorrectly adjusted than any other reason.**

■ **In a well lit and ventilated space, lay the carburetors on a clean work surface, on a clean shop towel for example. Use containers for all the small parts that will be coming out of the carburetor. On multiple carbu-**

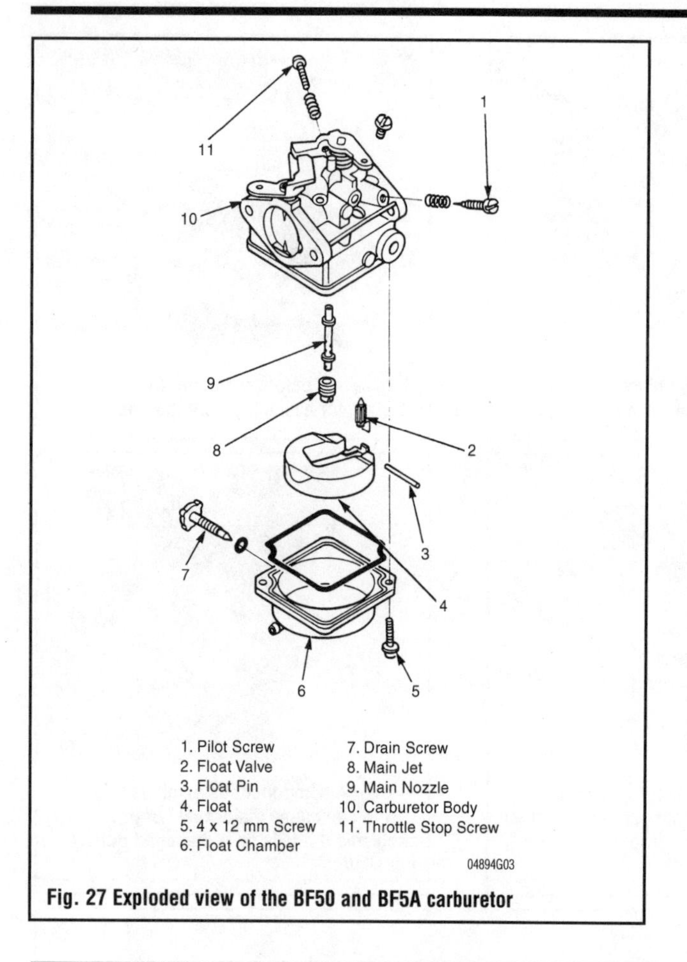

Fig. 27 Exploded view of the BF50 and BF5A carburetor

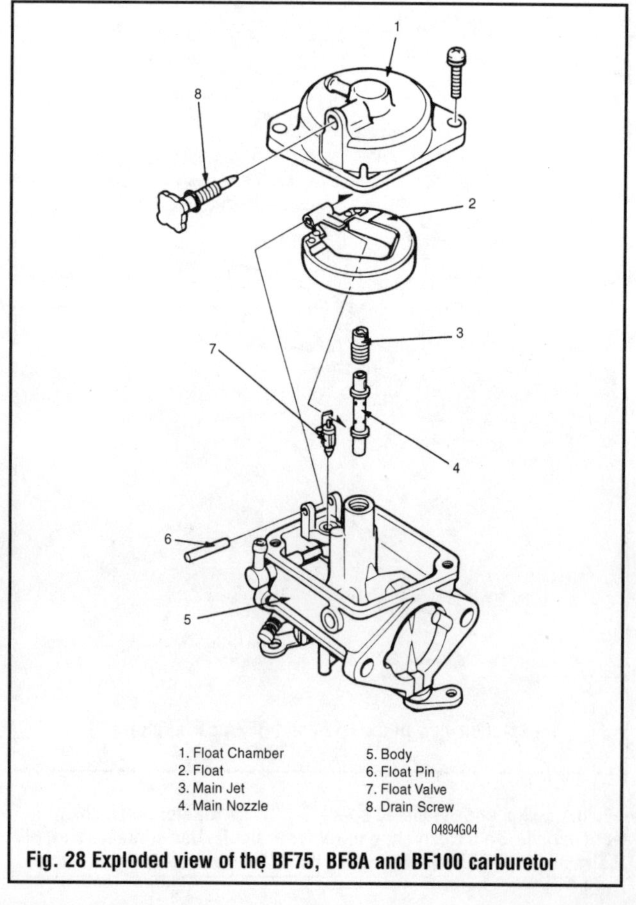

Fig. 28 Exploded view of the BF75, BF8A and BF100 carburetor

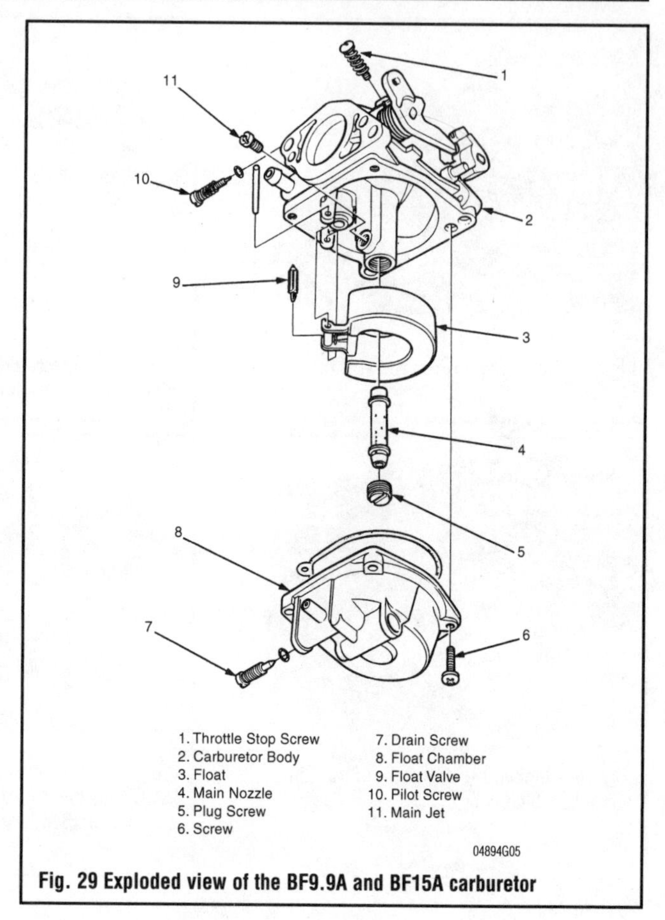

Fig. 29 Exploded view of the BF9.9A and BF15A carburetor

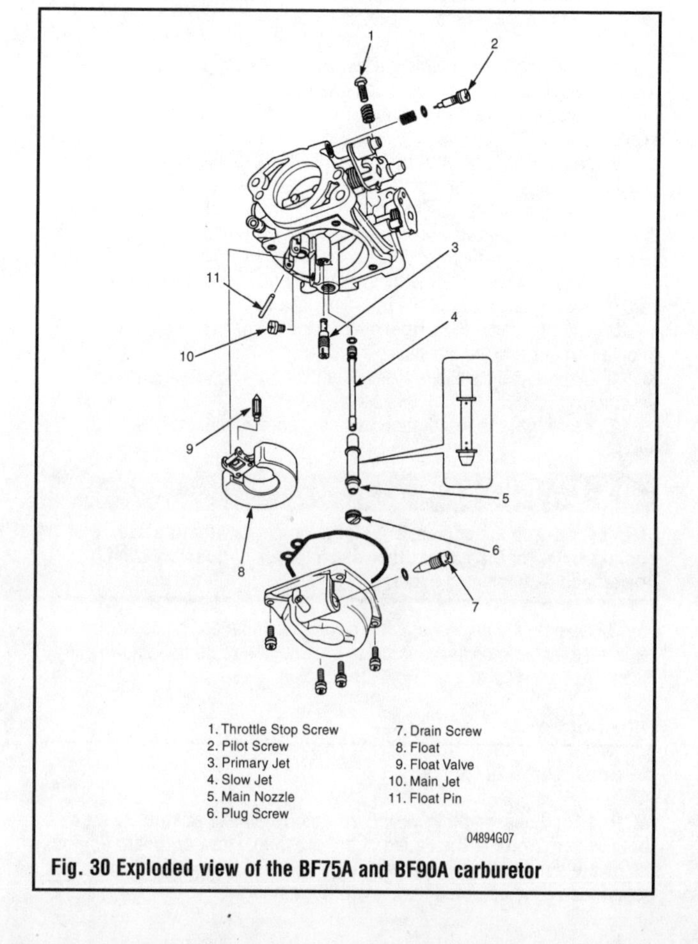

Fig. 30 Exploded view of the BF75A and BF90A carburetor

retor sets, be sure to not mix parts from different carburetors, keep them separate.

1. Pull out the float pin and remove the float assembly.
2. Carefully remove the float valve.
3. Remove the main jet and main nozzle assembly.
4. Remove all air screws (note how many turns in the screw is before removing)

CLEANING & INSPECTION

See Figures 31 and 32

If you don't have access to compressed air, many computer and camera shops sell cans of clean, dry, compressed air which are perfect for cleaning small outboard engine carburetors.

CAUTION

Never dip rubber parts, plastic parts, diaphragms, or pump plungers in carburetor cleaner. These parts should be cleaned only in solvent, and then blown dry with compressed air.

Place all metal parts in a screen-type tray and dip them in carburetor cleaner until they appear completely clean, then blow them dry with compressed air.

Blow out all passages in the castings with compressed air. Check all parts and passages to be sure they are not clogged or contain any deposits. Never use a piece of wire or any type of pointed instrument to clean drilled passages or calibrated holes in a carburetor.

Move the throttle shaft back and forth to check for wear. If the shaft appears to be too loose, replace the complete throttle body because individual replacement parts are not available.

Inspect the main body, air horn, and venturi cluster gasket surfaces for cracks and burrs which might cause a leak. Check the float for deterioration. Check to be sure the float spring has not been stretched. If any part of the float is damaged, the unit must be replaced. Check the float arm needle contacting surface and replace the float if this surface has a groove worn in it.

Inspect the tapered section of the idle adjusting needles and replace any that have developed a groove. As previously mentioned, most of the parts which should be replaced during a carburetor overhaul are included in overhaul kits available from your local marine dealer. One of these kits will contain a matched fuel inlet needle and seat. This combination should be replaced each time the carburetor is disassembled as a precaution against leakage.

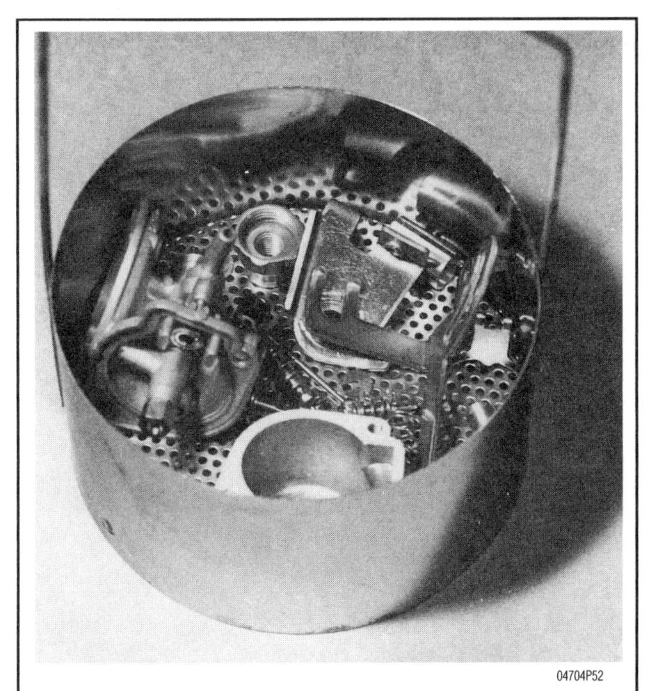

04704P52

Fig. 31 Metal parts from a disassembled carburetor in a basket ready to be immersed in carburetor cleaner

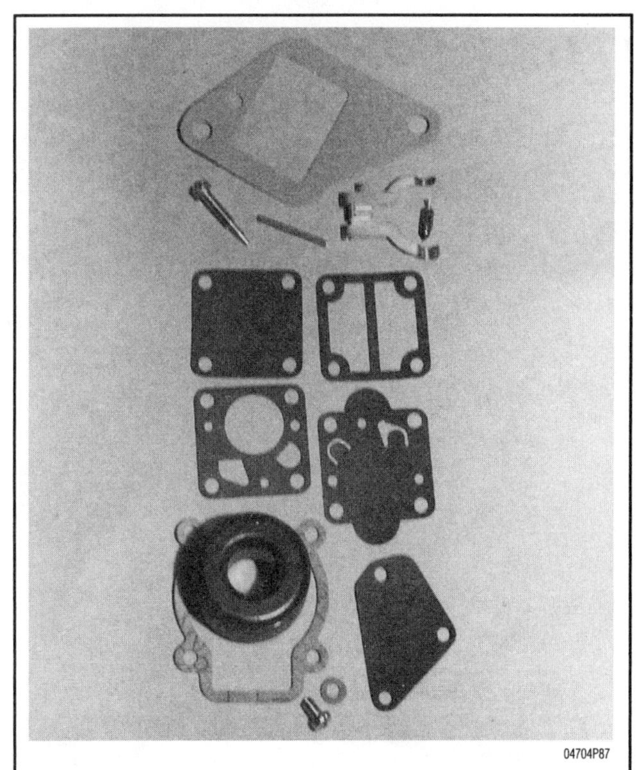

04704P87

Fig. 32 A carburetor repair kit, similar to this one, are available at your local service dealer. They contain the necessary replacement parts for a complete carburetor overhaul

ASSEMBLY

See Figures 33 thru 52

1. Install a new carburetor O-ring or gasket into the carburetor body.
2. Install a new O-ring on the new pilot and throttle stop screws. Install the pilot and throttle stop screw into the carburetor body.
3. Install the main nozzle into the carburetor body.
4. Install the main jet into the main nozzle and tighten it just snug with a screwdriver.

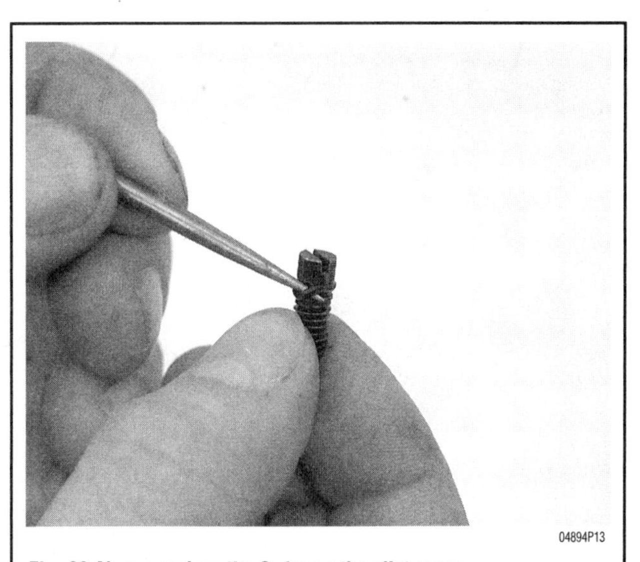

04894P13

Fig. 33 Always replace the O-ring on the pilot screw . . .

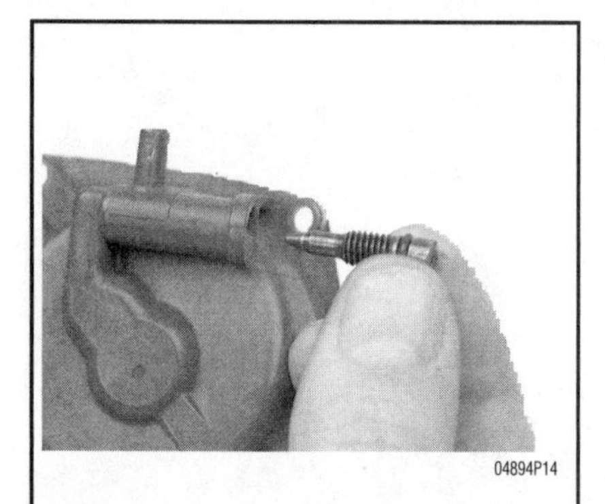

Fig. 34 . . . and reinstall into the carburetor body

Fig. 35 Install the Jet Set Assembly into the main well area . . .

Fig. 36 . . . take a look at the condition of the threads and make sure that there is no dirt or other debris inside . . .

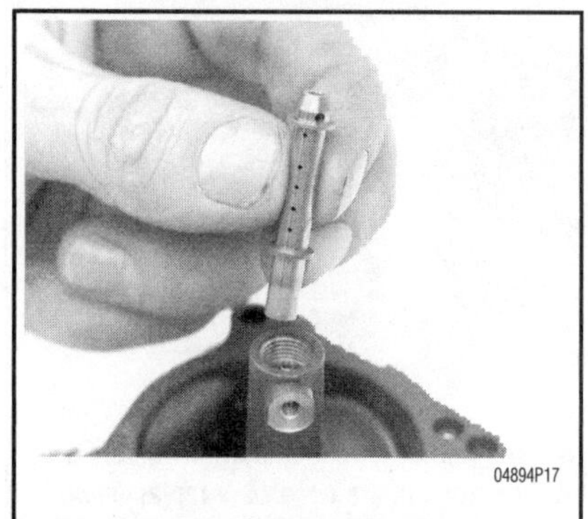

Fig. 37 . . . then install the main nozzle . . .

Fig. 38 . . . don't forget to install the plug screw . . .

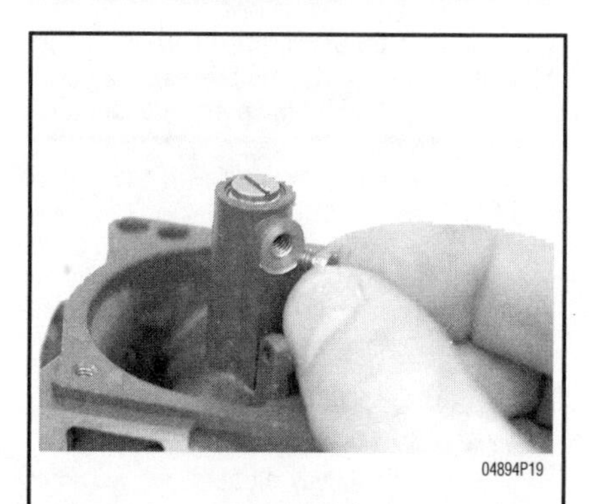

Fig. 39 . . . and then install the main jet itself.

Fig. 40 Check the condition of the rubber O-ring and spring, then install the pilot screw into the carburetor body

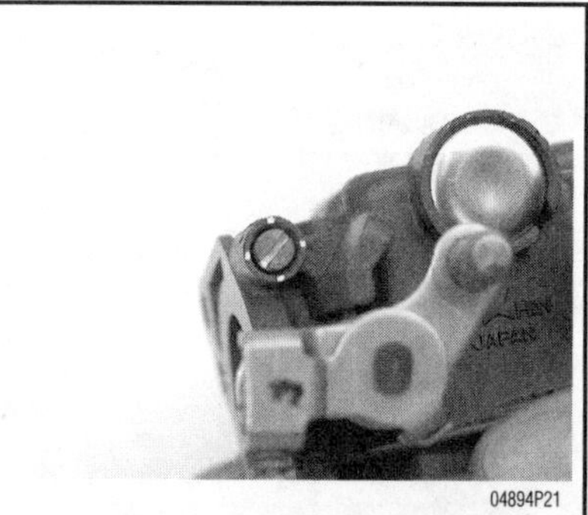

Fig. 41 Lightly seat the pilot screw. Damage to the pilot screw will occur if it is tightened against the seat

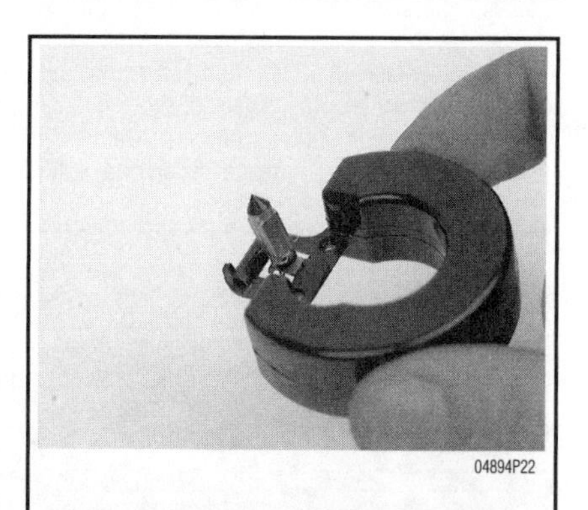

Fig. 42 Thoroughly inspect the float and float needle for damage . . .

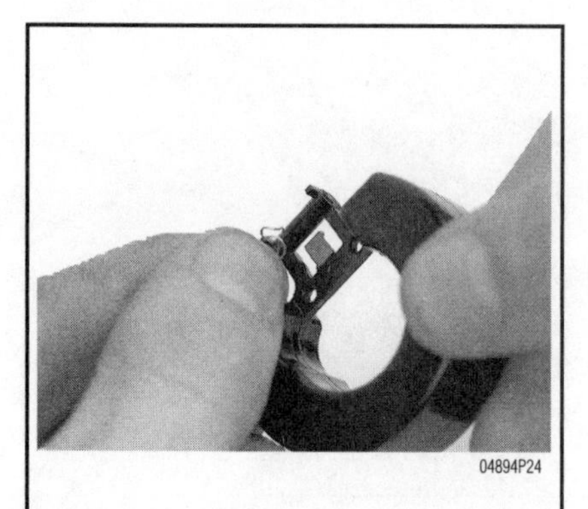

Fig. 43 . . . also inspect the float spring and needle retainer for damage or wear

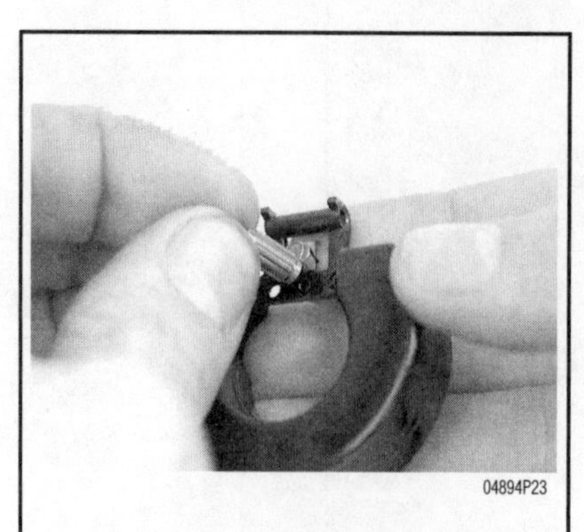

Fig. 44 Slide the new needle valve onto the float arm

Fig. 45 Push the pin into the holes in the carburetor body and float hinge.

Carburetor Float Heights	
Model	**Float Height**
BF20 / BF2A / BF2D	*0.413 - 0.531 in. (10.5 - 13.5 mm) non-adjustable*
BF50 / BF5A / BF5D	*0.35 - 0.43 in. (9.0 -11.0 mm) non-adjustable*
BF75 / BF100 / BF8A / BF8D	*0.39 in. (10.0 mm) adjustable*
BF9.9 / BF15	*0.51 - 0.59 in. (13 - 15 mm) non-adjustable*
BF25A	*0.6 in. (14 mm) adjustable*
BF30A	*0.5 in. (13 mm) adjustable*
BF35A / BF45A	*0.6 in. (14 mm) adjustable*
BF75A / BF90A	*0.45 IN. (11.5 mm) adjustable*

5. Slide a new needle valve into the groove of the float arm.
6. Lower the float arm into position with the needle valve sliding into the needle valve seat. Now, push the float pin through the holes in the carburetor body and hinge using a small awl or similar tool if needed.
7. Hold the carburetor body in a perfect upright position. Measure the distance from the float top to the carburetor body. If the measurement is out of specification, carefully bend the hinge, to achieve the required measurement

■ **BF 20A, BF2A, BF50, BF5A, BF9.9 and BF15 are non-adjustable and the entire float assembly must be replaced).**

This is a very important procedure, as it will determine how much fuel is maintained in the float bowl and will prevent fuel starvation and flooding.

8. Install the float valve assembly and check for free movement.

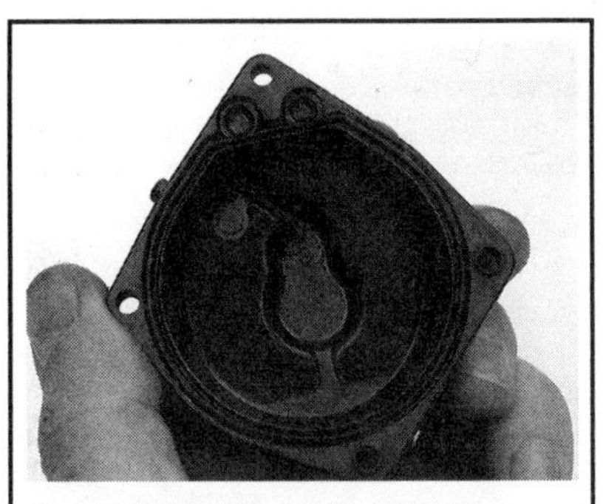

Fig. 46 Inspect the inside of the float bowl for dirt and debris. Also check the condition of the rubber O-ring

Fig. 47 Do not over tighten the float bowl screws. If you don't break the screw off in the carburetor body, it is possible to crack the float bowl itself

Fig. 48 Clean the carburetor gasket surface, making sure there is no old gasket material left to create a potential air leak

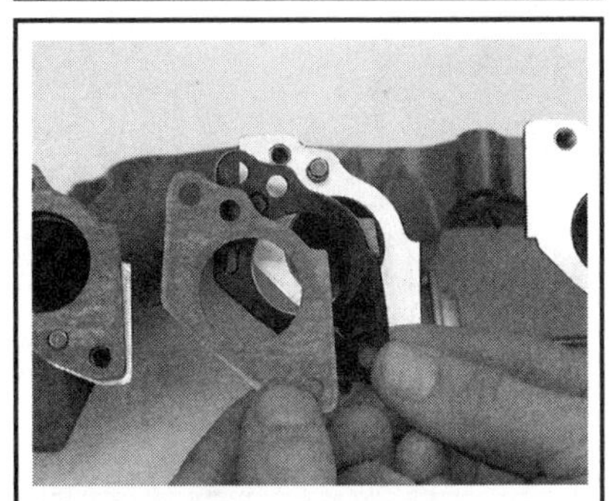

Fig. 49 Make sure the gaskets are installed in their correct positions prior to installation

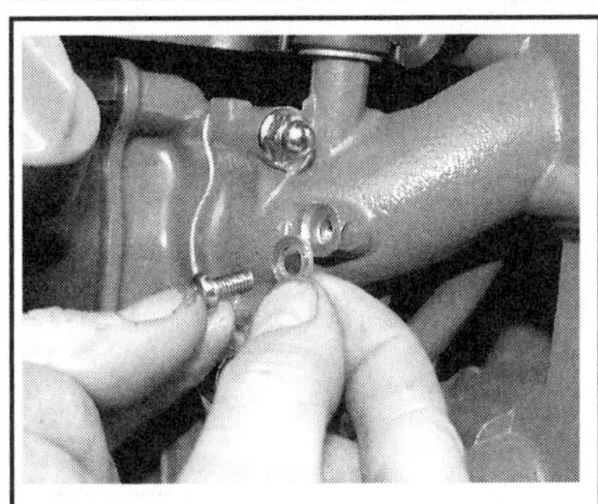

Fig. 50 To synchronize the carburetors, remove the screw and sealing washer on each intake manifold . . .

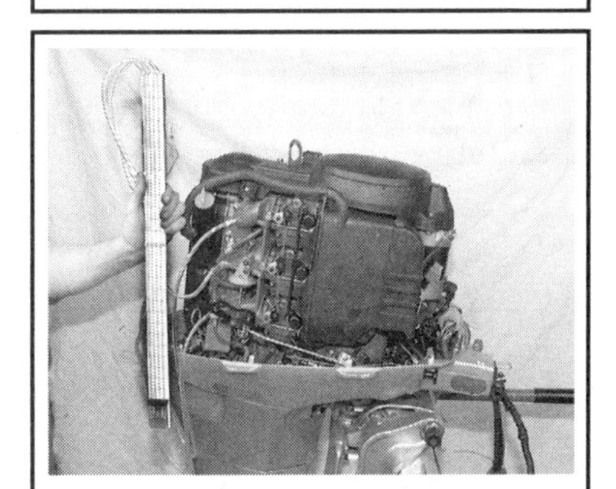

Fig. 51 . . . and install the vacuum gauges to each cylinder intake manifold with the correct adapters

Fig. 52 Final installation on one model of the 3-cylinder on the engines

Carburetor Pilot Screw Specifications

Horsepower	Carb Pilot Screw Turns Out
BF130A	-
BF115A	-
BF90A	*2 1/4*
BF75A	*1 7/8*
BF50A	*1*
BF45A	*2 1/8*
BF40A	*2 1/4*
BF35A	*2 1/8*
BF30A	*3*
BF25A	*2*
BF15A / BF15D	*1 5/8*
BF9.9A / BF9.9D	*2 3/4*
BF75 / BF100	*2 1/2*
BF8A / BF80 / BF8D	*2 1/2*
BF5A / BF50 / BF5D	*2 3/8*
BF2A / BF20	*2 1/8*
BF2D	*2*

9. Reinstall the float bowl, making sure that the o-ring has not slipped and is not being pinched.
10. Check and adjust the pilot screw setting to the specification on the chart.
11. Once the carburetor is reassembled, check the new gaskets to make sure of their direction when installing.

Fuel Pump

DESCRIPTION & OPERATION

See Figure 53

Honda uses diaphragm pumps, mechanical plunger pumps and on the fuel injected engines a combination of electric and mechanical plunger pumps to operate the fuel injection system. The pumps are similar from model to model, with the major differences being the BF20 and BF2A engines which rely on gravity feed fuel from the engine mounted fuel tanks and the BF75A and BF90A which use two mechanical plunger pumps instead of a single pump. The fuel pumps are automotive-style and mechanically driven and do not rely on crankcase vacuum like two-stroke engine fuel pumps.

Diaphragm Pumps are operated by a push rod activated by the engine camshaft. The push rod is connected to a flexible diaphragm in the pump body. A spring or springs inside the body maintains tension against the diaphragm. As the push rod is actuated, it moves the diaphragm in one direction and spring tension moves it in the other direction.

This continual motion of the diaphragm creates a partial vacuum and pressure in the space above the diaphragm. The vacuum draws the fuel from the fuel tank and the pressure pushes it toward the carburetor. A one-way check valve is used in the pump to prevent fuel from being pumped back into the fuel tank.

Mechanical plunger pumps are similar in operation to the diaphragm pumps, but instead of using a flexible diaphragm to create the pressure and vacuum needed to pump the fuel, it uses a plunger or piston in a cylinder to create the pressure and vacuum. Like a diaphragm pump, there is also a one-way check valve in the pump body to prevent fuel from being pumped back into the fuel tank.

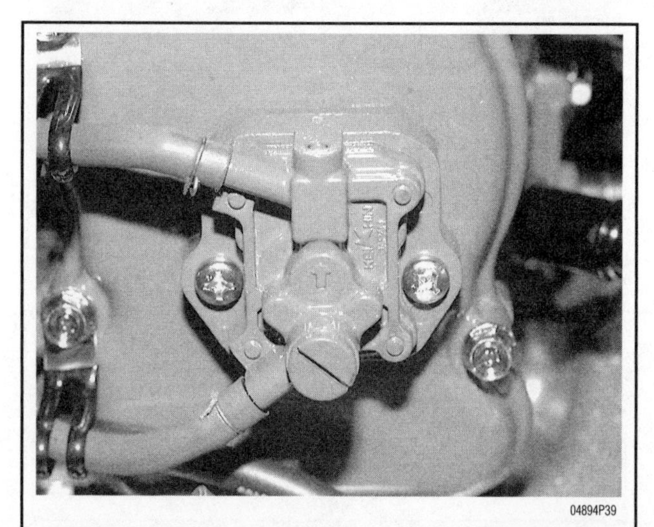

04894P39

Fig. 53 The mechanical plunger pumps are mechanically driven and do not rely on crankcase vacuum like two-stroke engine fuel pumps

FUEL PRESSURE CHECK

1. Disconnect the fuel hose at the carburetor.
2. Install a pressure gauge with a T-fitting in the fuel hose between the fuel tank and the carburetor.
3. Start the engine and record the pressure. Standard fuel pressure for this engine is 4.7 psi at 5,000 rpm.
4. If the fuel pressure is not within specifications, check the following related items before taking the pump apart:

- Check for obstructions in the fuel hose going both into and out of the pump.
- Make sure the vent line on the fuel tank is open.
- Make sure there is sufficient fuel in the fuel tank.
- Check the fuel filter for blockage.

If these checks do not reveal the fuel pressure problem, then it is probable that the pump is the cause of the problem. Remove and inspect the fuel pump.

➡Make sure to have the correct fuel pump rebuild kit on hand before taking the pump apart. It is good practice to go ahead and rebuild the pump any time it is taken apart.

REMOVAL & INSTALLATION

See Figure 54

1. Remove the engine cover.
2. Mark each hose with a piece of tape to indicate it's correct location.
3. Loosen the hose clips and remove the fuel hoses from the pump body.
4. Unbolt and remove the fuel pump from the engine block.

➡When reinstalling the fuel pump, make the embossed directional arrows face in the correct direction.

5. With the fuel pump facing in the right direction, insert the bolts through the pump body and into the engine block.
6. Tighten the bolts to the specified torque for your model.
7. Reinstall the fuel hoses onto the fuel pump, making sure that they are in the correct position.

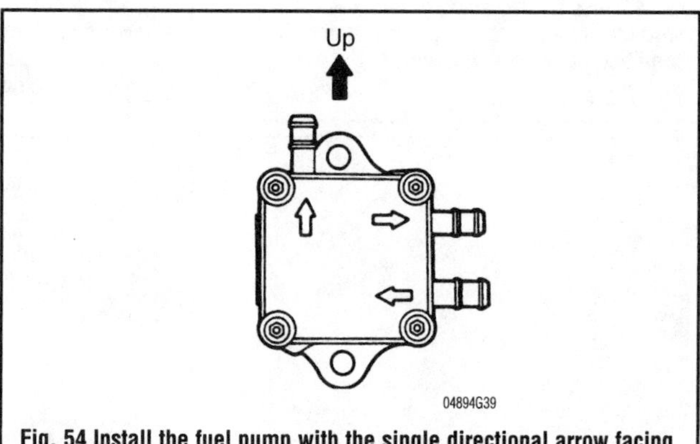

04894G39

Fig. 54 Install the fuel pump with the single directional arrow facing up

DISASSEMBLY

See Figure 55

Only the diaphragm-type pumps are rebuildable. The mechanical plunger-type pumps are not and must be replaced as an entire unit if found to be faulty.

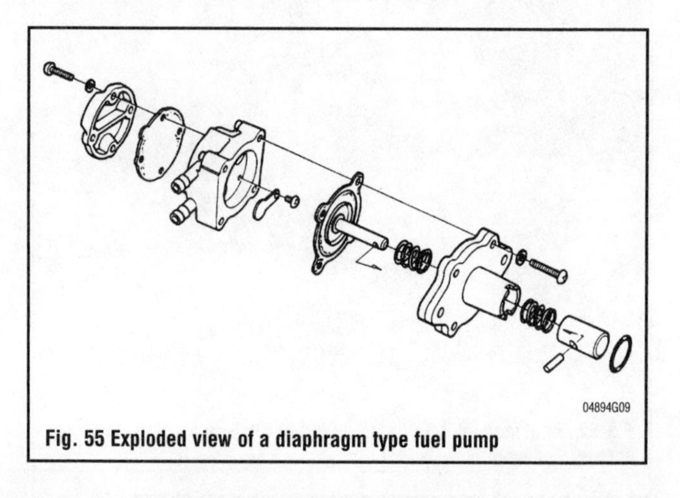

04894G09

Fig. 55 Exploded view of a diaphragm type fuel pump

➡Lay out all the fuel pump parts on clean work surface and keep the small parts in a container top prevent loosing any.

1. Open the pump body by removing the bolts securing the two halves together.
2. Push the diaphragm downward and then rotate until the foot is parallel with the groove in the push rod.

**** CAUTION**

The diaphragm spring is under tension. When removing the diaphragm, use caution when releasing the diaphragm from the lower pump body.

3. Keeping pressure on the diaphragm, slowly let the diaphragm spring push the diaphragm up from the lower body.
4. Lay out the separate components for inspection.

CLEANING & INSPECTION

1. Carefully inspect the diaphragm for holes, tears, deterioration. If there are any problems at all, replace the diaphragm.
2. Inspect the one-way check valve for cracks and wear.
3. Check the diaphragm and push rod springs for wear, replace if necessary.
4. Check the filter for debris or damage. It is a good idea to replace the filter any time the pump is taken apart for repairs or inspection.
5. Check all O-rings for tears, cuts, stretching or distortion. Again it is good idea to replace the O-rings any time the pump is taken apart.

ASSEMBLY

1. Lubricate the push rod with a small dab of grease and insert into the pump body and push it back and forth several time to make sure moves freely.
2. Insert the diaphragm spring into the lower body.
3. Insert the diaphragm and press down and twist to engage the diaphragm foot in the push rod groove.
4. Reassemble the one-way check valve and install it back into the upper body.
5. Install the upper body and bolt the entire assembly together. Make sure that the O-rings and diaphragm are seated correctly.

**** WARNING**

Fuel leaking past the diaphragm can cause fuel dilution of the engine oil and eventually cause serious engine damage.

Fuel Lines

See Figure 56

GENERAL INFORMATION

The fuel system should be routinely inspected for leaks. Check all the fuel lines for cracks, leaks and deformation. Fittings are usually the most common points for leaks to develop. Leaks sometimes also develop at the fuel injectors as the sealing O-rings age. Any type of damage or leaks should be fixed immediately.

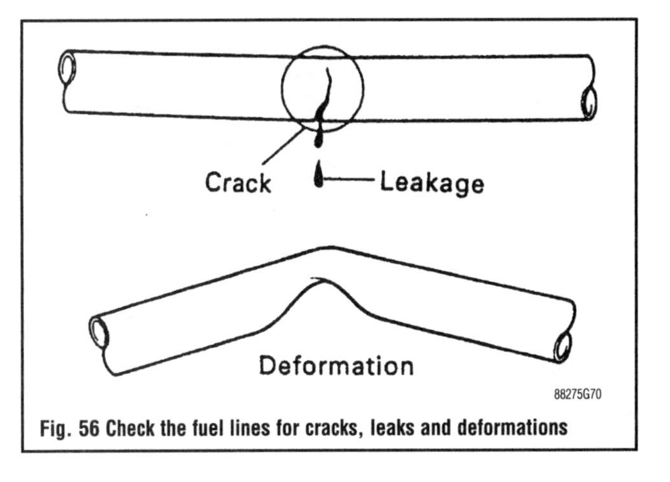

Fig. 56 Check the fuel lines for cracks, leaks and deformations

REPLACEMENT & ROUTING

See Figures 57 thru 64

If any hoses, lines or fittings are found to be damaged or defective, its time for replacement. Always replace hoses one at a time rather than pulling a bunch hoses off at once and then having to try to remember when they all go when its time to reassemble. Mark each individual hose as to its direction and location and it will make reassembly much easier.

Most hose clamps are the band or wire type tension clamps, and they need to be checked for tension. These clamps are normally reliable, but they can cor-

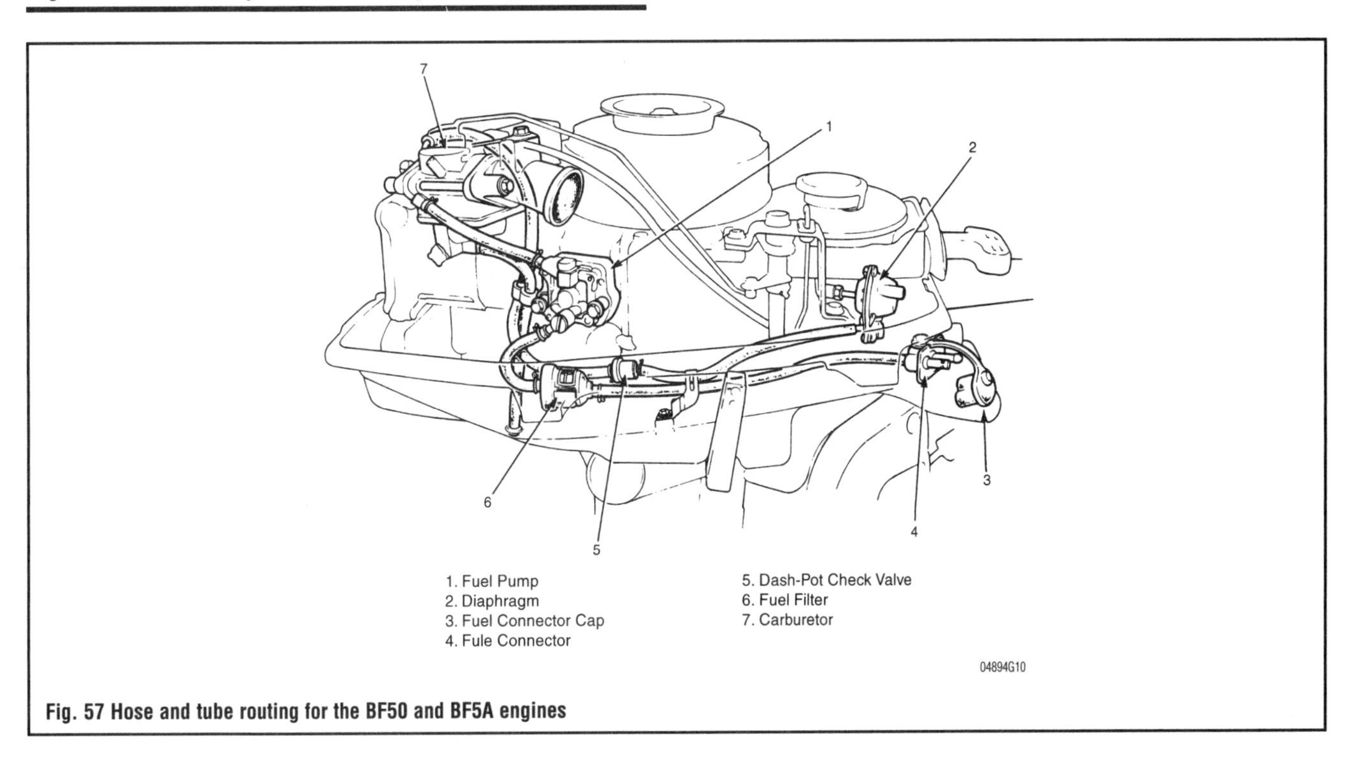

Fig. 57 Hose and tube routing for the BF50 and BF5A engines

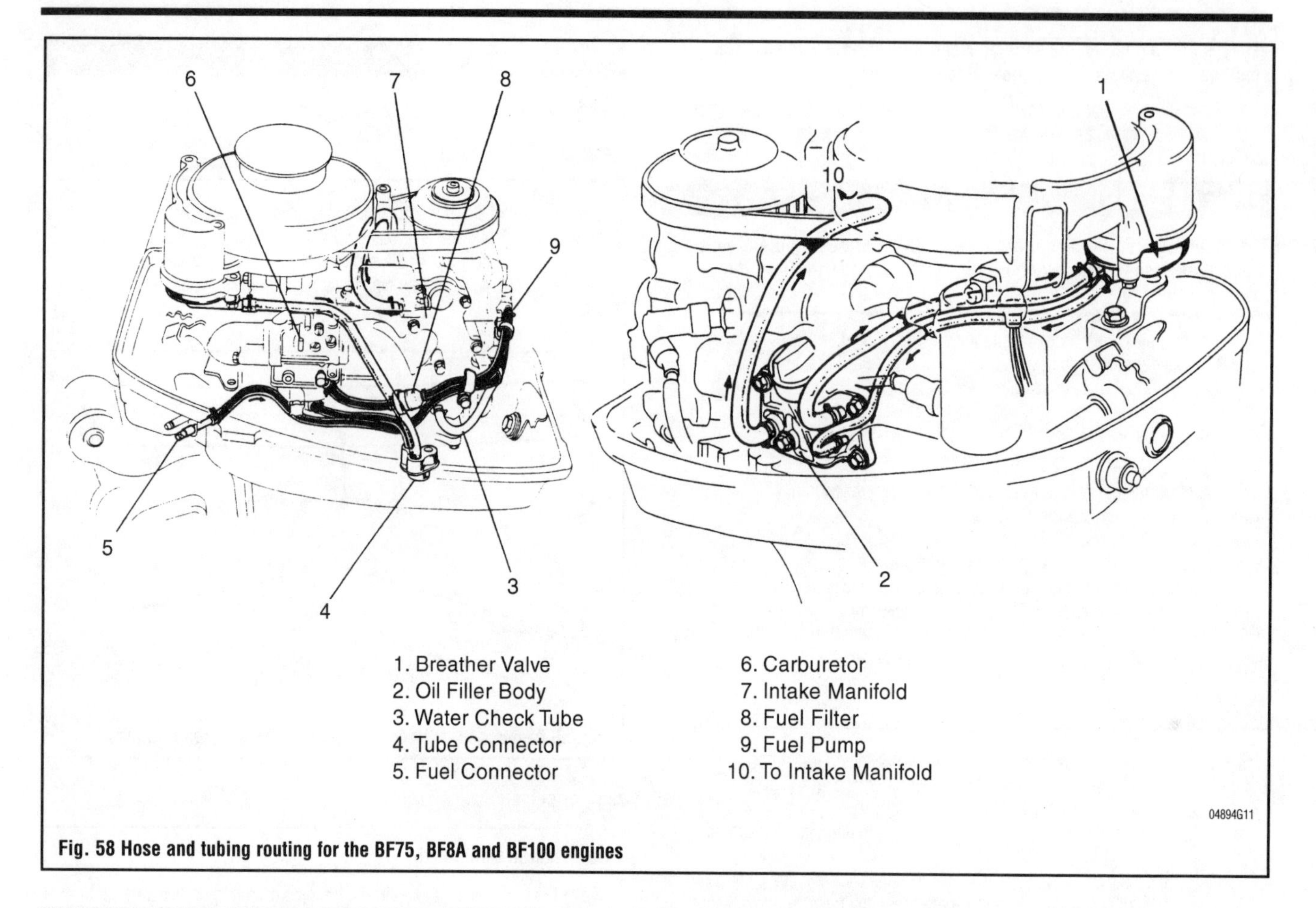

Fig. 58 Hose and tubing routing for the BF75, BF8A and BF100 engines

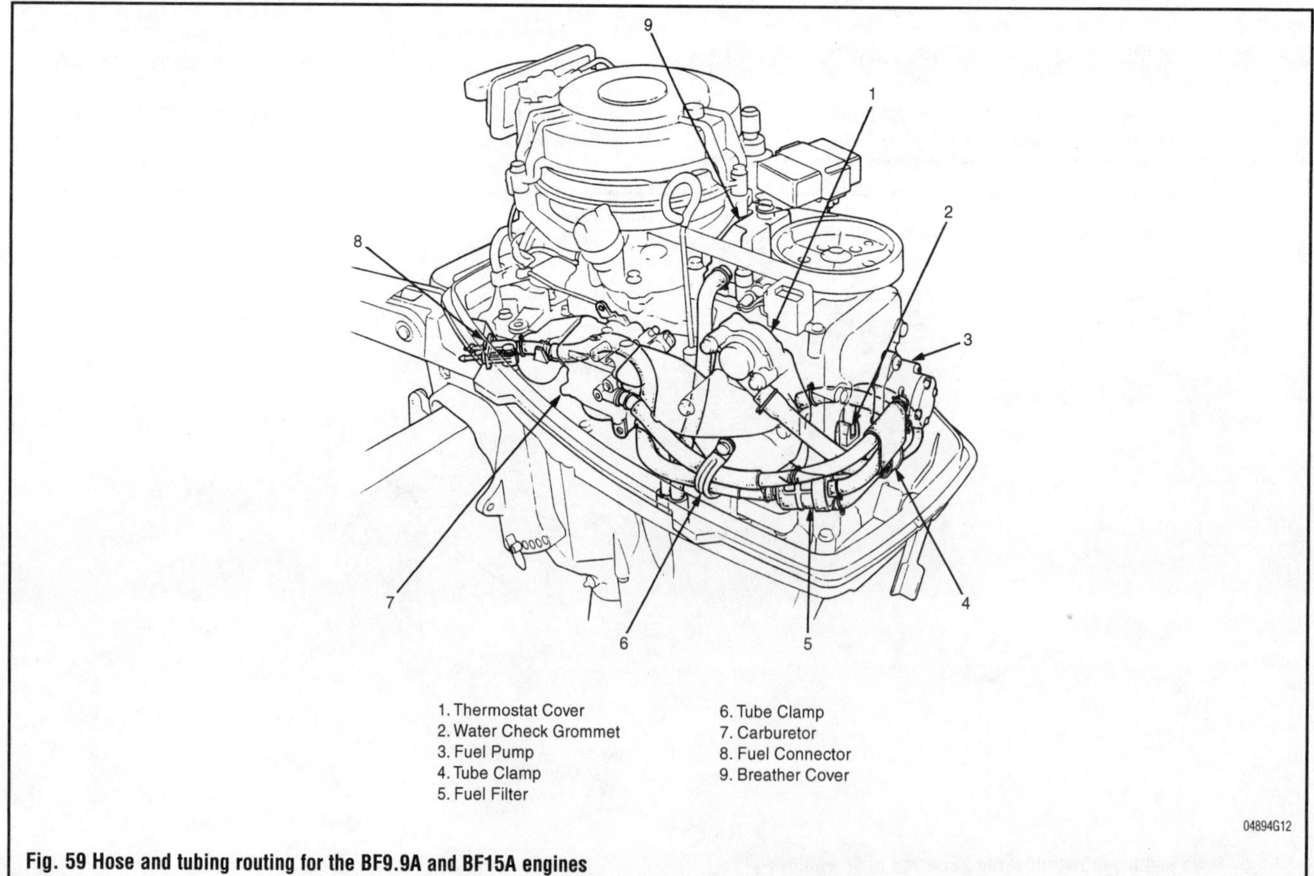

Fig. 59 Hose and tubing routing for the BF9.9A and BF15A engines

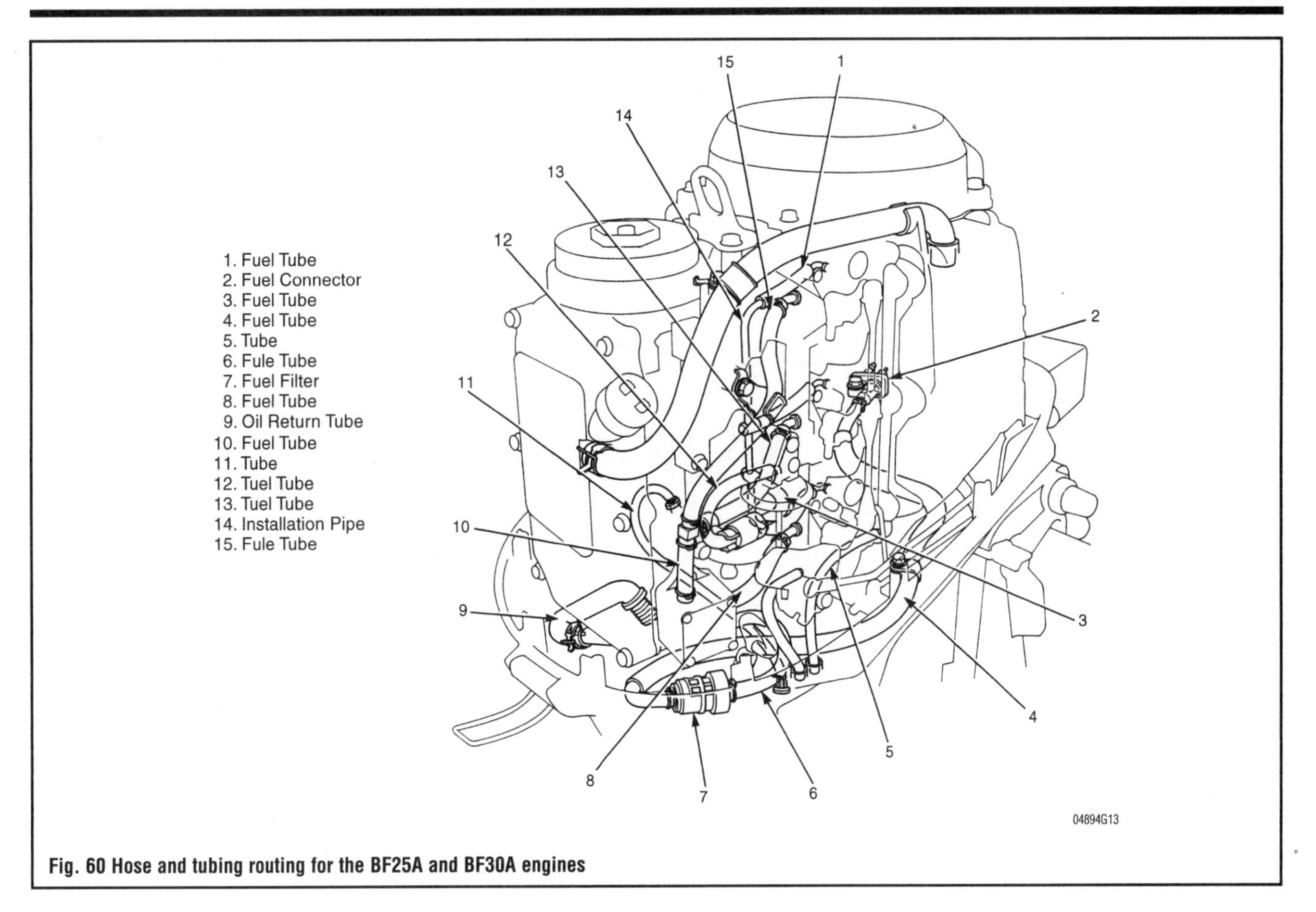

Fig. 60 Hose and tubing routing for the BF25A and BF30A engines

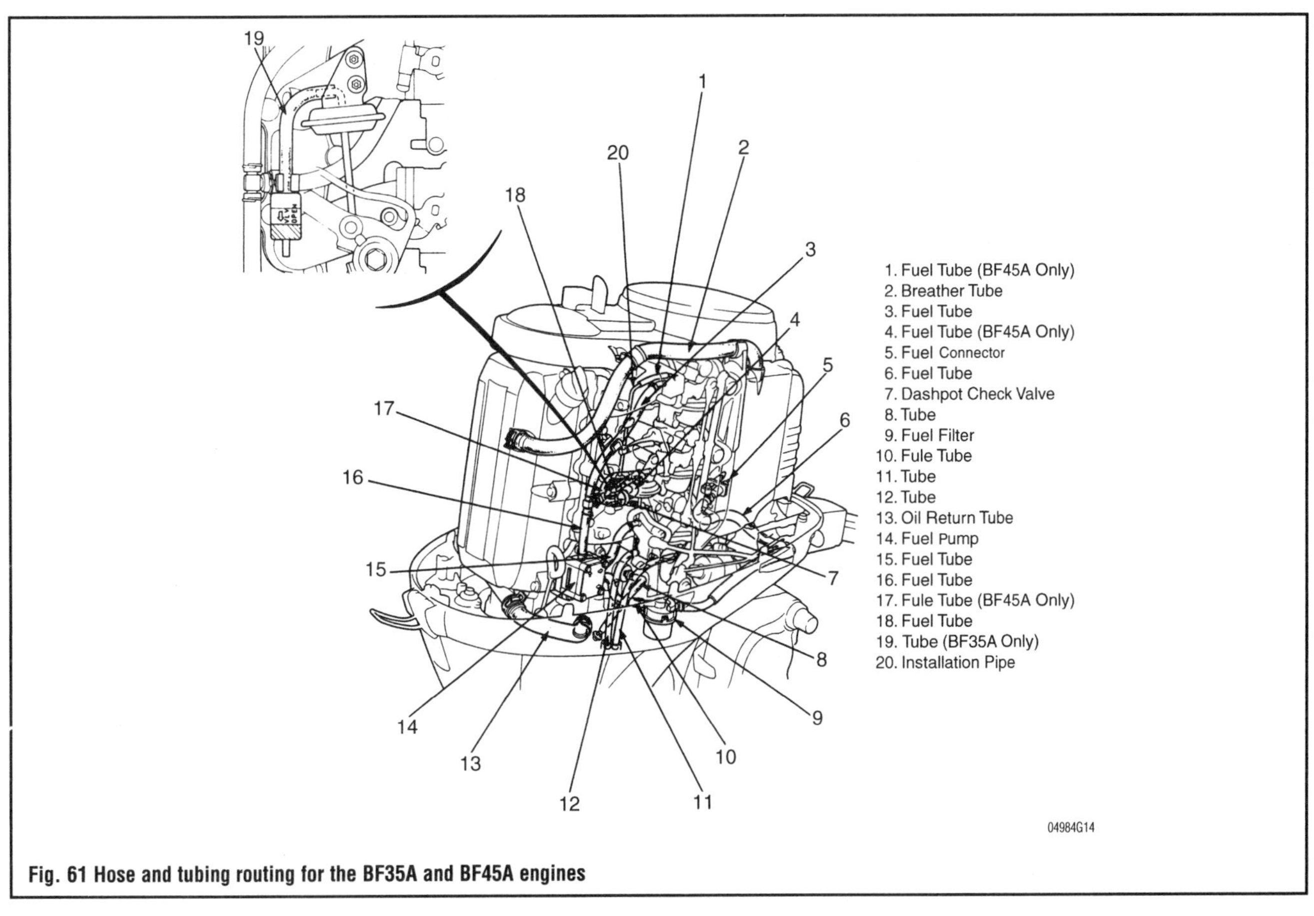

Fig. 61 Hose and tubing routing for the BF35A and BF45A engines

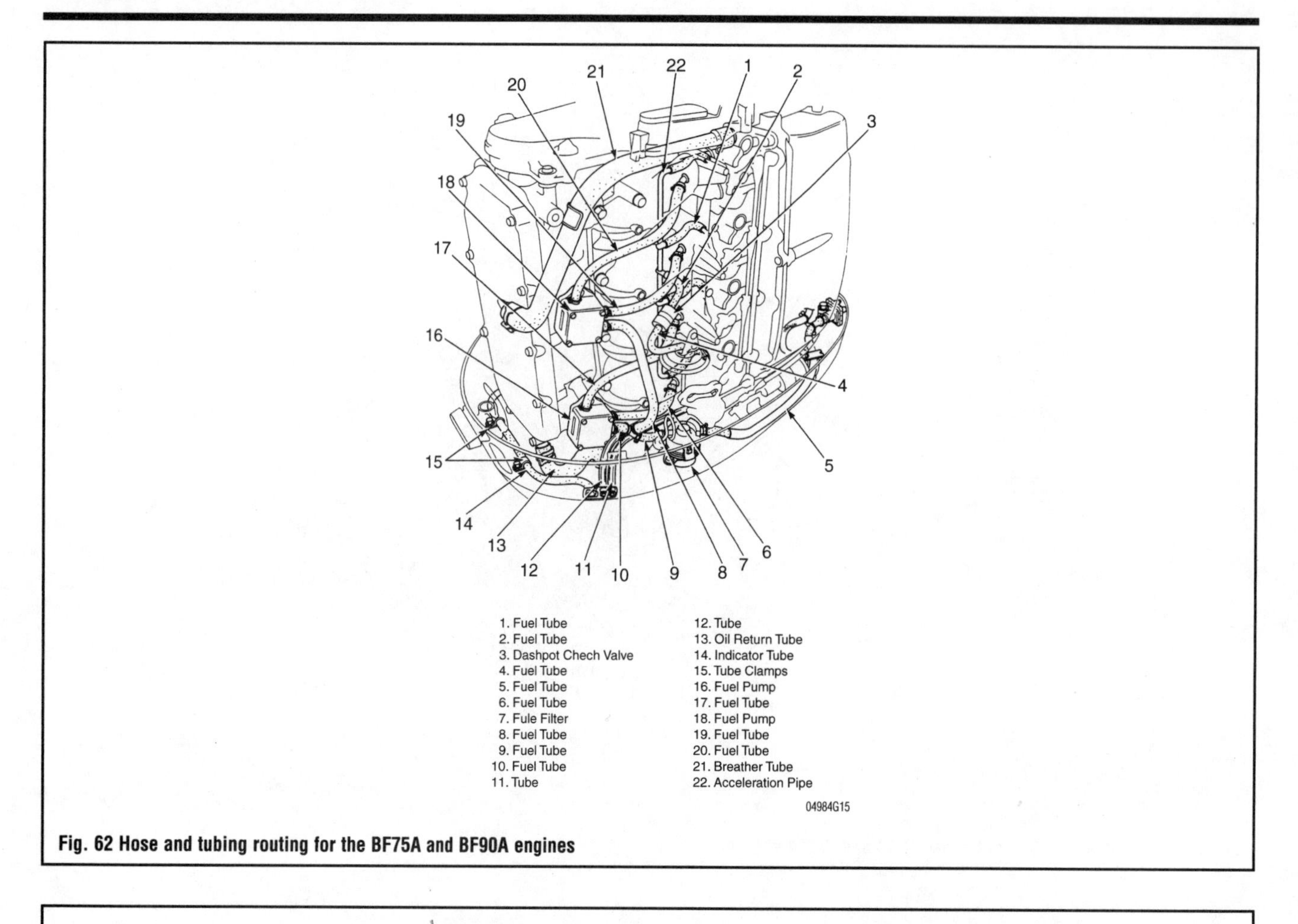

Fig. 62 Hose and tubing routing for the BF75A and BF90A engines

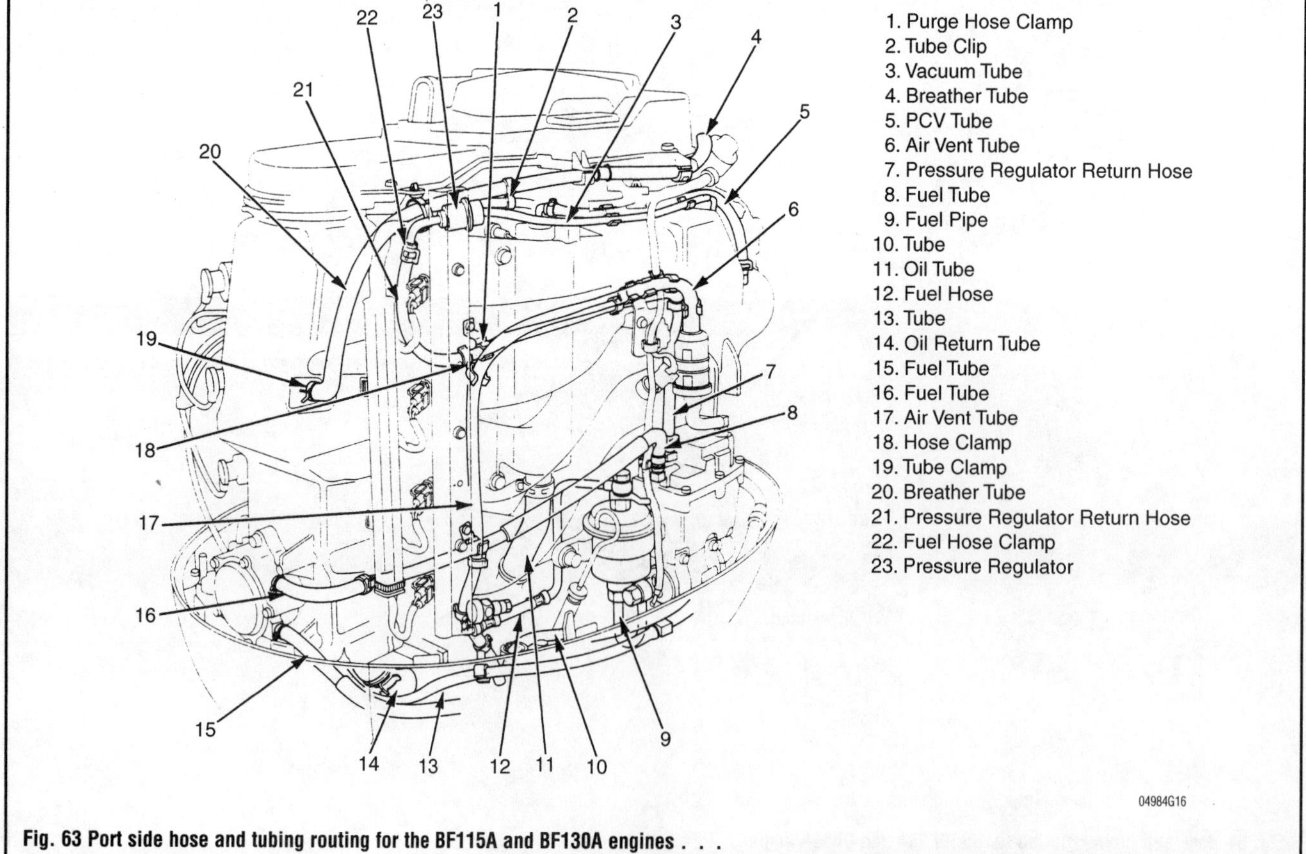

Fig. 63 Port side hose and tubing routing for the BF115A and BF130A engines . . .

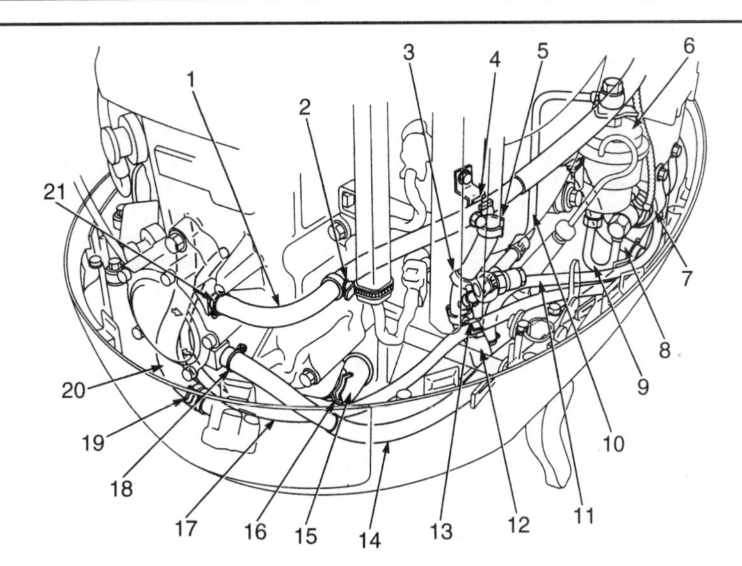

1. Fuel Tube
2. Tube Clip
3. Tube Clamp
4. Tube Clip
5. Sensor Tube Clip
6. Fuel Filter (High Pressure Side)
7. Fuel ump Unit
8. Tube
9. Fuel Pipe
10. Fuel Hose
11. Tube
12. Air Vent Tube
13. Hose Clamp
14. Fuel Tube
15. Oil Return Tube
16. Tube Clamp
17. Tube
18. Tube Clip
19. Hose Clamp
20. Flushing Water Hose
21. Tube Clip

04894G17

Fig. 64 . . . and Starboard side hose and tubing for the BF115A and BF130A engines

rode and break after time so now is the time to replace them with new clamps. These clamps are loosened and adjusted with a pair of common pliers or clamp pliers specific for each of these clamps. They are available at tool and automotive parts stores.

If there are screw type clamps being used, do not overtighten the clamp. This will cause the clamp to dig into the hose and lead to cuts in the end of hose and the possibility of the hose coming off the fitting. leaks

When replacing fittings, be extra careful to not put any extra pressure on the fitting itself. Most fittings are fragile when twisted and can break off clean. This means having to somehow remove what's left of the broken fitting with a special tool such as a screw extractor, then hopefully not damaging the threads. When this happens, the hole may need to be re-threaded with a tap or in the worse case, drilled out and the threads repaired with a thread insert.

➡Due to the corrosive environment that outboard engines operate in, when replacing hardware use marine specific equipment. These clamps are usually manufactured out of stainless steel and are much more corrosion resistant than the more common hardware store variety of clamp.

Follow the hose routing guides in this section for the proper routing and clamping of hoses and lines. If your application is not shown here, use common sense and remember not to route fuel lines near anything hot enough to melt the hose or place hoses near rotating parts such as the flywheel. Clamp hoses together with wire-ties to keep them from chafing and also detail up the engine itself.

FUEL INJECTION

Beginning with the 1998 model year, Honda introduced the BF115 and BF130 engines, equipped with Programmed Fuel Injection. This system, which uses a 16-bit digital computer, allows very accurate fuel control and as a result, achieves high performance and lower fuel consumption. These engines have the following features:

- The system is of the speed density type. This system is excellent in response to changes in running conditions and provides accurate control of the amount of fuel injected. It determines the required fuel injection amount based on engine speed and intake manifold pressure.
- Two-group injection is used. This method separates the four cylinders into groups of two. Each cylinder, which has its own injector, is fired separately from the other group.
- The difference between the intake manifold vacuum and the fuel pressure is handled by the fuel pressure regulator. The regulator maintains a constant fuel pressure and results in the proper amount of fuel being injected.

The features give these engines better performance and better fuel economy than a two-stroke outboard engine while insuring compliance with strict environmental regulations now being imposed on the marine industry.

Fuel Injection Basics

➧ See Figure 65

Fuel injection systems have been used on vehicles for many years. The earliest ones were purely mechanical. As technology advanced, electronic fuel injection systems became more popular. Early mechanical and electronic fuel injection systems did not use feedback controls. As emissions became more of a concern, feedback controls were adapted to both types of fuel injection systems.

Both mechanical and electronic fuel injection systems can be found on gasoline engines. Diesel engines are most commonly found with mechanical type systems, although the newest generations of these engines have been using electronic fuel injection.

The fuel delivery system consists of all the components which supply the engine with fuel. This includes the tank itself, all the lines, one or more fuel filters, a fuel pump (mechanical or electric), and the fuel metering components (carburetor or fuel injection system).

Fuel injection is not a new invention. Even as early as the 1950s, various automobile manufacturers experimented with mechanical-type injection systems. There was even a vacuum tube equipped control unit offered for one system! This might have been the first "electronic fuel injection system." Early problems with fuel injection revolved around the control components. The electronics were not very smart or reliable. These systems have steadily improved since. Today's fuel injection technology, responding to the need for better economy and emission control, has become amazingly reliable and efficient. Computerized engine management, the brain of fuel injection, continues to get more reliable and more precise.

Components needed for a basic computer-controlled system are as follows:

- A computer-controlled engine manager, which is the Engine Control Module (or ECM), with a set of internal maps to follow.
- A set of input devices to inform the ECM of engine performance parameters.
- A set of output devices. Each device is controlled by the ECM. These devices modify fuel delivery and timing. Changes to fuel and timing are based on input information matched to the map programs.

This list gets a little more complicated when you start to look at specific components. Some fuel injection systems may have twenty or more input devices. On many systems, output control can extend beyond fuel and timing. The Honda Programmed Fuel Injection System provides more than just the basic functions, but is still straight forward in its layout. There are six input devices and three output controls. The diagram on the following page shows the input and output devices with their functions.

There are several fuel injection delivery methods. Throttle body injection is relatively inexpensive and was used widely in early automotive systems. This is usually a low pressure system running at 15 PSI or less. Often an engine with a single carburetor was selected for throttle body injection. The carburetor was recast to hold a single injector. and the original manifold was retained. Throttle body injection is not as precise or efficient as port injection.

The second type is more precise and follows the firing order of the engine. Each cylinder gets a squirt of fuel precisely when needed.

Troubleshooting The Fuel Injection System

Programmed Fuel Injection (PGM–FI) System is a fully electronic microprocessor based engine management system. The Electronic Control Unit (ECU) is given responsibility for control of injector timing and duration, intake air control, ignition timing, cold start enrichment, fuel pump control, fuel cutoff.

The ECU receives electric signals from many sensors and sources on and around the engine. The signals are processed against pre-programmed values; correct output signals from the ECU are determined by these calculations. The ECU contains additional memories, back-up and fail-safe functions as well as self diagnostic capabilities.

Diagnosis is generally based on symptom diagnosis and stored fault codes, if any. Testing always requires the use of the diagnostic charts in conjunction with the Honda test harness (pin-out box), a voltmeter and an ohmmeter.

LOGICAL TROUBLESHOOTING

Diagnosis of a driveability problem requires attention to detail and following the diagnostic procedures in the correct order. Resist the temptation to begin extensive testing before completing the preliminary diagnostic steps. The preliminary or visual inspection must be completed in detail before diagnosis begins. In many cases this will shorten diagnostic time and often cure the problem without the need for involved electronic testing.

There are two basic ways to check your vehicle fuel system for problems. They are by symptom diagnosis and by the on-board computer self-diagnostic system. The first place to start is always the preliminary inspection. Intermittent problems are the most difficult to locate. If the problem is not present at the time you are testing you may not be able to locate the fault.

PRELIMINARY INSPECTION

See Figure 66

The visual inspection of all components is possibly the most critical step of diagnosis. A detailed examination of connectors, wiring and vacuum hoses can often lead to a repair without further diagnosis. Also, take into consideration if the engine has been serviced recently. Sometimes things get reconnected in the wrong place, or not at all. A careful inspector will check the undersides of hoses as well as the integrity of hard-to-reach hoses blocked by the air silencer or other components. Correct routing for vacuum hoses (if applicable to your model) can be obtained from your specific vehicle service manual. Wiring

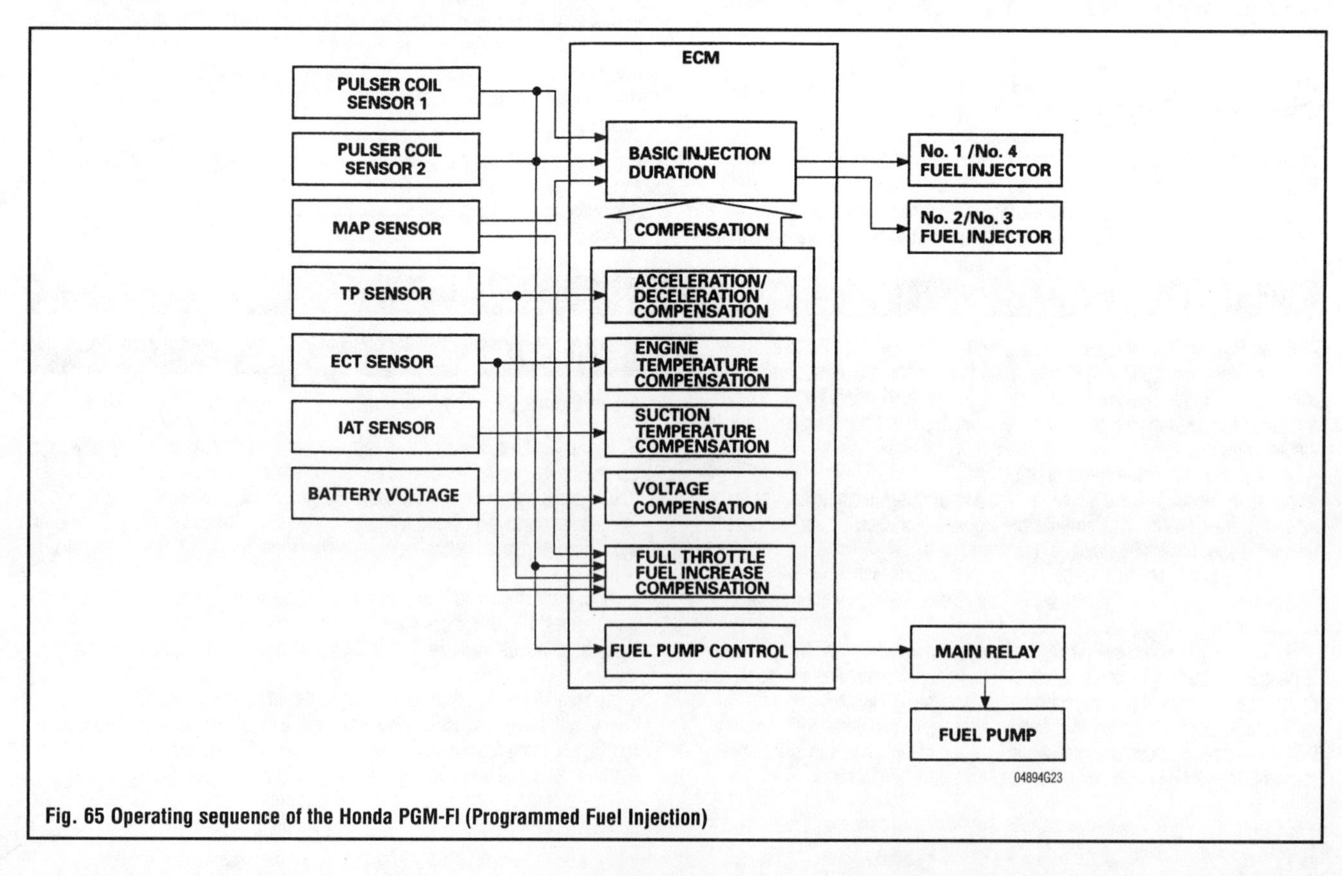

Fig. 65 Operating sequence of the Honda PGM-FI (Programmed Fuel Injection)

Fig. 66 A visual inspection of all fuel injection components is possibly the most critical step of diagnosis

should be checked carefully for any sign of strain, burning, crimping or terminals pulled out from a connector.

Checking connectors at components or in harnesses is required; usually, pushing them together will reveal a loose fit. Also, check electrical connectors for corroded, bent, damaged, improperly seated pins, and bad wire crimps to terminals. Pay particular attention to ground circuits, making sure they are not loose or corroded. Any component or wiring in the vicinity of a fluid leak or spillage should be given extra attention during inspection. Remember how corrosive an environment it is that an outboard engine operates in. Salt spray is especially corrosive to electrical connectors.

Additionally, inspect maintenance items such as belt condition and tension, and the battery charge and condition. Any of these very simple items may affect the system enough to set a fault code.

DIAGNOSIS BY SYMPTOM

When diagnosing by symptom, the first step is to find out if the problem really exists. This may sound like a waste of time but you must be able to recreate the problem before you begin testing. This is called an "operational check". Each operational check will give either a positive or negative answer (symptom). A positive answer is found when the check gives a positive result (the instruments function when you turn the key). A negative answer is found when the check gives a negative result (the instruments do not function when you turn the key). After performing several operational checks, a pattern may develop. This pattern is used in the next step of diagnosis to determine related symptoms.

In order to determine related symptoms, perform operational checks on circuits related to the problem circuit (the instruments and the tilt and trim system do not work). These checks can be made without the use of any test equipment. Simply follow the wires in the wiring harness or, if available, obtain a copy of your engine's specific wiring diagram. If you see that the instruments and the tilt and trim system are on the same circuit, first check the instruments to see if they work. Then check the tilt and trim system. If neither the instruments or tilt and trim system work, this tells you that there is a problem in that circuit. Perform additional operational checks on that circuit and compile a list of symptoms.

When analyzing your answers, a defect will always lie between a check which gave a positive answer and one which gave a negative answer. Look at your list of symptoms and try to determine probable areas to test. If you get negative answers on related circuits, then maybe the problem is at the common junction. After you have determined what the symptoms are and where you are going to look for defects, develop a plan for isolating the trouble. Ask a knowledgeable person which components frequently fail on your powerhead. Also notice which parts or components are easiest to reach and how can you accomplish the most by doing the least amount of checks.

A common way of diagnosis is to use the split-in-half technique. Each test that is made essentially splits the trouble area in half. By performing this technique several times the area where a problem is located becomes smaller and smaller until the problem can be isolated in a single wire or component. This area is most commonly between the two closest checkpoints that produced a negative answer and a positive answer.

After the problem is located, perform the repair procedure. This may involve replacing a component, repairing a component or damaged wire, or making an adjustment.

➡Never assume a component is defective until you have thoroughly tested it.

The final step is to make sure the complaint is corrected. Remember that the symptoms that you uncover may lead to several problems that require separate repairs. Repeat the diagnosis and test procedures repeatedly until all negative symptoms are corrected.

Two simple tests-pressure and volume-can easily verify the mechanical integrity of the vehicle's fuel supply and return systems. Problems within the fuel supply or return system can cause many driveability problems because the control system can't compensate for a mechanical problem. Nor can the system adjust the air/fuel mixture when the system is in open loop or back-up mode. Moreover, a control system's parameters are programmed based on a set flow rate and system pressure. Therefore, even if the computer were adjusting the air/fuel mixture, the computer's calculations would be wrong whenever the system pressure or volume was incorrect.

➡When testing fuel system pressure, always use a pressure gauge that is capable of handling the system pressure. Old style vacuum/pressure gauges will not work with most fuel injection systems

The fuel system's pressure can be thoroughly verified by performing both static and dynamic tests. Dynamic tests are used to determine the system's overall operating condition while the engine is running. For example, if the pressure is low, the problem is most likely to be in the supply or high side of the system. Likewise, if the fuel pressure is high, the problem is going to be on the return or low side of the fuel system. All pressure tests should be taken on the high or supply side of the fuel system.

If a dynamic test reveals that the pressure is below manufacturer specifications, the problem is caused by one or more of the following items: weak pumps, clogged or restricted fuel filter or lines, external leaks bad check valves, or a faulty fuel pressure regulator. But if a test reveals pressures above the desired specification, then the problem is caused by one or more of these items: restricted or clogged return lines, vacuum leak at the pressure regulator, or a defective fuel pressure regulator.

However, if the pressure is high on a returnless fuel system, the cause, most likely, is a defective pressure regulator or clogged/faulty injectors. The causes are slightly different on returnless systems because the regulator is located in or near the fuel tank and there is no fuel return line used on these systems. However, the systems which use a remotely mounted fuel pressure regulator require a short return line. Thus, there is no excess fuel beyond the injectors and the measured pressure is already the regulated pressure.

Many manufacturers are switching to returnless fuel systems because of the advantages they provide. One advantage is lower evaporative emissions which allows less frequent purge cycles and the use of smaller canisters. The evaporative emissions are lower on a return-less fuel system because the fuel doesn't absorb any heat from the engine compartment. This prevents it from turning into a vapor as it returns to the fuel tank.

Another example of how returnless fuel systems are different from conventional fuel injection systems is that they may not use a frame/engine compartment mounted fuel filter. Most return-less systems use a combination fuel filter/pressure regulator which is mounted on top of the fuel pump module. The fuel pump module is located in the tank. Other return-less systems use an in-line filter on the frame that incorporates an integral fuel pressure regulator. These systems also utilize a short return line from the fuel filter to the fuel pump module in the tank. Yet, others use a regular frame/engine mounted filter in conjunction with a fuel pressure regulator and filter assembly in the tank.

Previously we mentioned that a static pressure test should also be performed. This test can be performed before or after a dynamic test with the engine off. These tests are useful for determining the sealing capacity of the system. If the fuel system leaks down after the engine is shut off, the vehicle will experience extended crank times while the system builds up enough fuel pressure to start. If the system is prone to losing residual pressure too quickly, it also usually suffers from low fuel pressure during the dynamic test. In addition to verifying that the system is holding residual pressure, a static test can also verify that the system is receiving the initial priming pulse at startup.

However, don't assume that the correct amount of fuel is being delivered just because the system pressure is within specifications. The only way to measure fuel volume is to open the fuel system into a graduated container and measure the fuel output while cranking the engine. Some manufacturers specify that a certain amount of fuel be dispersed over a short period of time, but generally the pump should be able to supply at least 1.5 oz. (45 ml.) for approximately 2 seconds. If the fuel output is below the recommended specification, there is a restriction in the fuel supply or the fuel pump is weak.

In summary, pressure and volume tests provide information about the overall condition of the fuel system. The information obtained from these tests will set the direction and path of your diagnosis.

When troubleshooting the fuel injection system, always remember to make absolutely sure that the problem is with the fuel injection system and not another component. This particular system has very sophisticated electronic controls, which in most cases, can only be diagnosed and repaired by an authorized dealer who has the expensive diagnostic equipment.

There are some trouble shooting procedures that can be done outside a professional shop. By follow the following steps:

COMMON PROBLEMS

Fuel Delivery

➧ See Figure 67

Many times fuel system troubles are caused by a plugged fuel filter, a defective fuel pump, or by a leak in the line from the fuel tank to the fuel pump. A defective choke may also cause problems. would you believe, a majority of starting troubles which are traced to the fuel system are the result of an empty fuel tank or aged sour fuel.

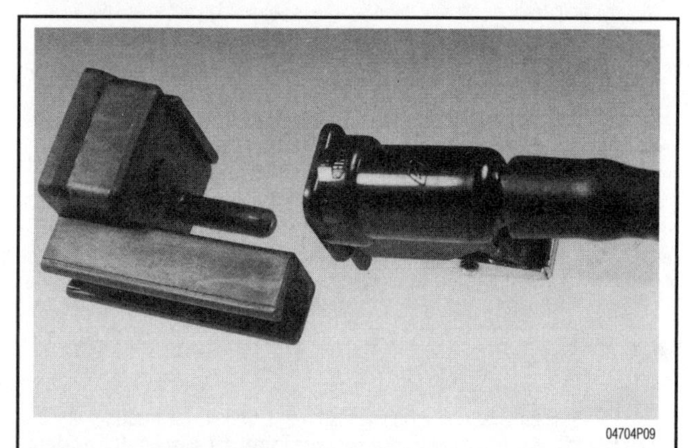

04704P09

Fig. 67 An excellent way of protecting fuel hoses against contamination is an end cap filter

Sour Fuel

➧ See Figure 68

Under average conditions (temperate climates), fuel will begin to break down in about four months. A gummy substance forms in the bottom of the fuel tank and in other areas. The filter screen between the tank and the carburetor and small passages in the carburetor will become clogged. The gasoline will begin to give off an odor similar to rotten eggs. Such a condition can cause the owner much frustration, time in cleaning components, and the expense of replacement or overhaul parts for the carburetor.

Even with the high price of fuel, removing gasoline that has been standing unused over a long period of time is still the easiest and least expensive preventative maintenance possible. In most cases, this old gas can be used without harmful effects in an automobile using regular gasoline.

The gasoline preservative will keep the fuel fresh for up to twelve months. If this particular product is not available in your area, other similar additives are produced under various trade names.

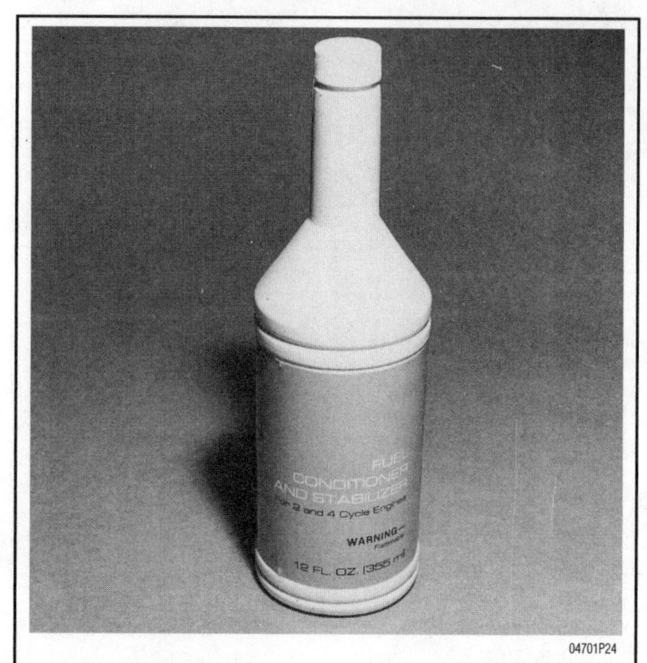

04701P24

Fig. 68 The use of a brand name fuel additive, will prevent fuel from souring for up to twelve months

Choke Problems

Fuel injected engines don't actually have a choke circuit in the fuel system. These engines rely on the ECU to control the air/fuel mixture and injector function according to the electronic signals sent by a host of sensors and monitors. If a fuel injected engine suffers a driveability problem, and it can be traced to the fuel injection system, then the sensors and computer-controlled output components must be tested and repaired or replaced.

Diagnostic Trouble Codes

The ECM (Engine Control Module) has the self-diagnosis function which turns the MIL (Malfunction Indicator Lamp) ON when it detects an out of specification sensor reading with the input/output system. You need to short-circuit the service check connector when the MIL is ON. The MIL should blink indicating the probable cause of the problem by the number and length of the blinks. When multiple problems occur simultaneously, the MIL will indicate these by blinking separate codes, one after another.

Codes 10 and after are indicated by a series of long and short blinks. The number of long blinks equals the first digit of the code and the number of short blinks equals the second digit of the code. When the overheat sensor is faulty, the overheat light as well as the MIL should blink simultaneously.

If the ECM body is faulty, the MIL turns ON and does not go out when the service check connector is short-circuited. If the MIL blinks by short-circuiting the service check connector, detect the problem part by referring to the troubleshooting Guide and following the troubleshooting flow chart.

SERVICE PRECAUTIONS

- Do not operate the fuel pump when the fuel lines are empty.
- Do not operate the fuel pump when removed from the fuel tank.
- Do not reuse fuel hose clamps.
- The washer(s) below any fuel system bolt (banjo fittings, service bolt, fuel filter, etc.) must be replaced whenever the bolt is loosened. Do not reuse the washers; a high-pressure fuel leak may result.
- Make sure all ECU harness connectors are fastened securely. A poor connection can cause an extremely high voltage surge and result in damage to integrated circuits.
- Keep ECU all parts and harnesses dry during service. Protect the ECU and all solid-state components from rough handling or extremes of temperature.

- Before attempting to remove any parts, turn the ignition switch **OFF** and disconnect the battery ground cable.
- Always use a 12 volt battery as a power source, never a booster or high-voltage charging unit.
- Do not disconnect the battery cables with the engine running.
- Do not disconnect any wiring connector with the engine running or the ignition **ON** unless specifically instructed to do so.
- Do not depress the accelerator pedal when starting.
- Do not rev up the engine immediately after starting or just prior to shutdown.
- Do not apply battery power directly to injectors.
- Whenever possible, use a flashlight instead of a drop light.
- Keep all open flame and smoking material out of the area.
- Use a shop cloth or similar to catch fuel when opening a fuel system.
- Relieve fuel system pressure before servicing.
- Always use eye or full-face protection when working around fuel lines, fittings or components.
- Always keep a dry chemical (class B-C) fire extinguisher near the area.

READING

➧ **See Figure 69**

1. Allow the Malfunction Indicator Lamp (MIL) to continue blinking. When the MIL is on, short-circuit the RED 2P service check connector located inside the ECM case using a special tool (SCS Connector) available from your dealer.

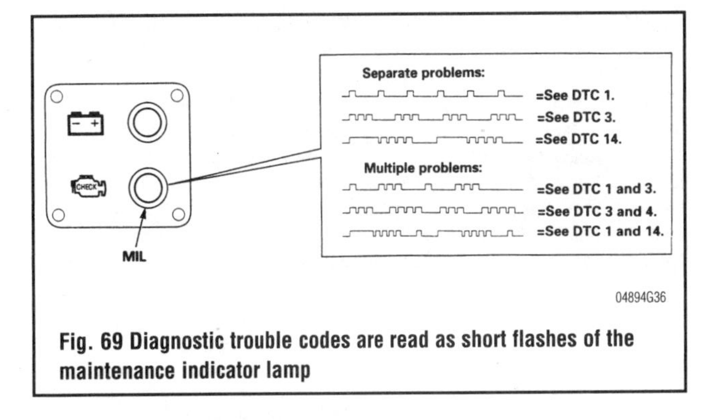

Fig. 69 Diagnostic trouble codes are read as short flashes of the maintenance indicator lamp

2. Count the number of blinks when the MIL starts blinking. The diagnostic trouble code (DTC) is blinked over and over. The long blinks equal the first digit and the number of short blinks equal the second digit of the DTC.

When a fault code is detected, it appears as a flash of the Malfunction Indicator Lamp (MIL) on the instrument panel. This indicates that an abnormal signal in the system has been recognized by the control module.

When diagnosing by code, the first step is to read any fault codes as flashes of the MIL from the control module.. The fault codes will identify the area to perform more in-depth testing. After the fault codes have been read, proceed to test each of the components and component circuits indicated. Continue per-

Diagnostic Trouble Code Chart

MIL Flashes	Component
MIL does not turn on or blink	MIL open circuit in wire
	MIL bulb blown
	ECM ground wire open circuit
	ECM faulty
MIL stays ON	Service Check connector wire short circuit
	MIL wire short circuit
	Sensor system power supply line short circuit
	ECM power supply line open circuit
	ECM faulty
3	MAP sensor connector disconnected
	MAP sensor wire short or open circuit
	MAP sensor faulty
4	Pulser coil sensor connection loose or disconnected
	Pulser coil sensor 1 wire short or open circuit
	Pulser coil sensor 1 faulty
6	ECT sensor connector disconnected
	ECT sensor wire short or open circuit
	ECT sensor faulty
7	TP sensor connector disconnected
	TP sensor wire short or open circuit
	TP sensor faulty
8	Pulser coil sensor connector loose or disconnected
	Pulser coil sensor 2 wire short or open circuit
	Pulser coil sensor 2 faulty
10	IAT sensor connector disconnected
	IAT sensor wire short or open circuit
	IAT sensor faulty
14	IAC valve connector disconnected
	IAC valve faulty
	IAC valve wire short or open circuit
24	Overheat sensor connector disconnected
	Overheat sensor wire short or open circuit
	Overheat sensor faulty

04894C03

forming individual component tests until the failed component is located, Remember, fault codes do indicate the presence of a failure, but they do not identify the failed component directly.

Codes 1–9 are indicated by a series of short flashes; two-digit codes use a number of long flashes for the first digit followed by the appropriate number of short flashes. For example, Code 14 would be indicated by 1 long flash followed by 4 short flashes. Codes are separated by a pause between transmissions. The position of the codes during output can be helpful in diagnostic work. Multiple codes transmitted in isolated order indicate unique occurrences; 3 flashes on the MIL, followed by 7 flashes, would indicate an open circuit in the sensor output voltage line. 3, 6, 7, 10 and 24 would read as 3 flashes followed by 6 flashes, then 7 flashes, then 1 long flash and finally 2 short and 4 long flashes. This would indicate an open circuit in the sensor ground line.

➡In the event that a code is encountered which is not on the chart, re-count the number of flashes. If the code is truly wrong, it will be necessary to rep!ace the ECM with a known good unit and re-test. Since this can be very expensive, you may wish to bring the engine to an authorized repair center to have the suspected unit checked and a good unit tried on the engine before buying a new ECM.

CLEARING

▶ See Figure 70

➡Perform the following steps within 20-seconds in order to reset the ECM.

1. Turn the ignition switch to the OFF position.
2. Short-circuit the RED 2P service check connector which is located inside the ECM case by using the SCS connector special tool.
3. Turn the ignition switch to the ON position.
4. With the emergency stop lanyard clip inserted into the emergency stop switch, press the emergency stop switch for one half second or more, then release the switch for one tenth of a second or more. Repeat this procedure 5 times.

Proceed to the next steps immediately. Steps 4 through 6 must be performed within 20 seconds.

5. Make sure that the buzzer sounds twice. At this time the MIL should stay ON.
6. Turn the ignition switch to the OFF position. This completes the ECM reset procedure.

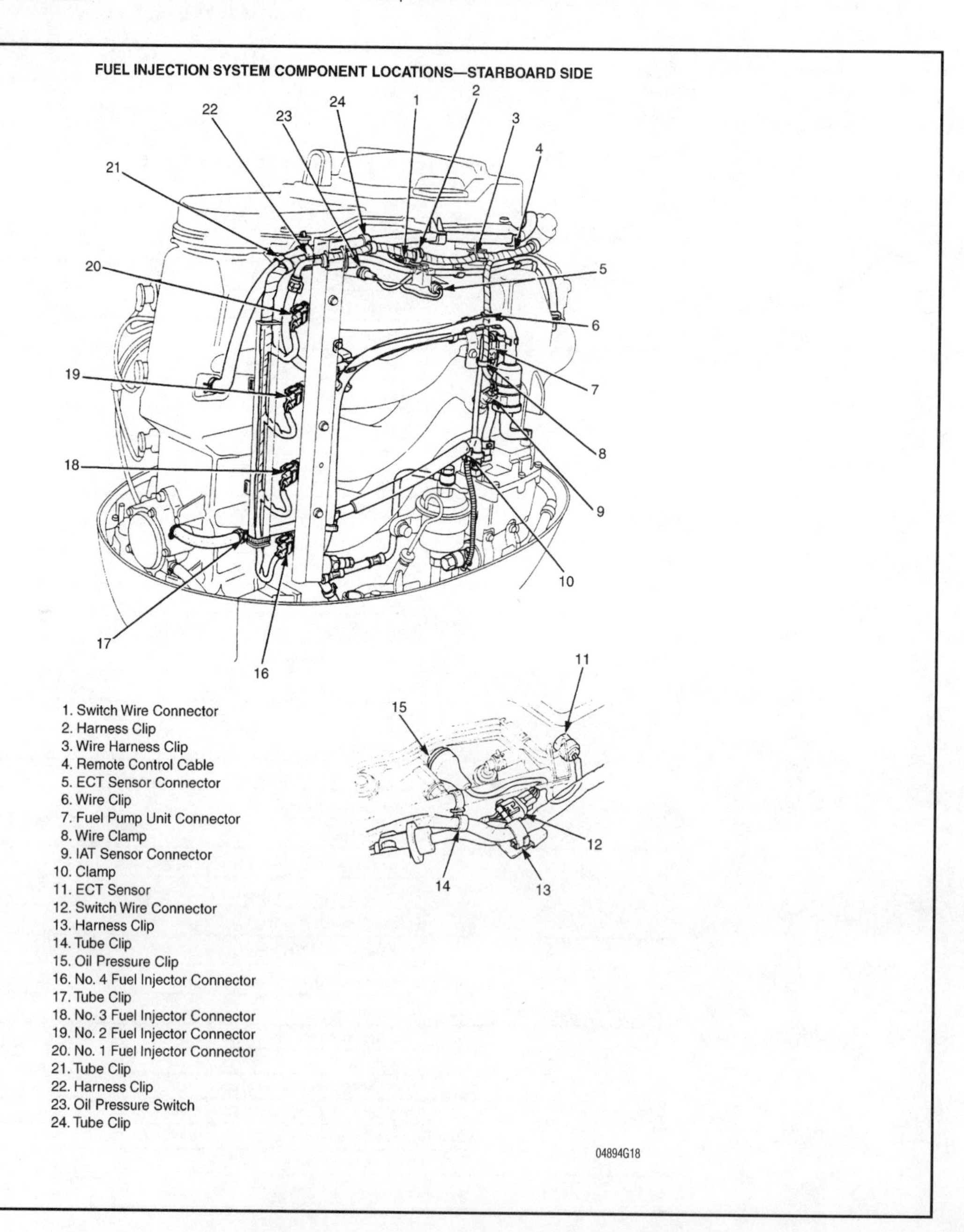

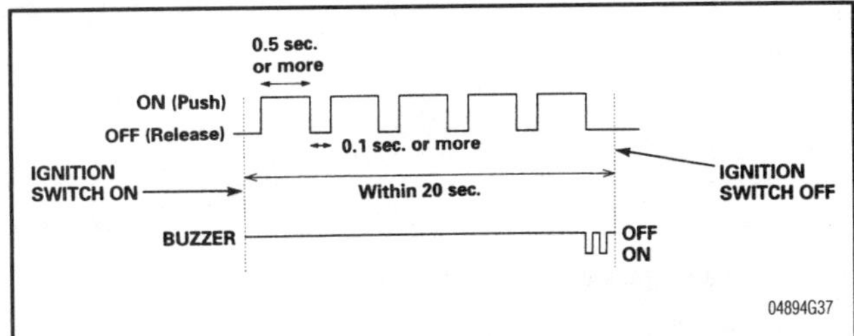

Fig. 70 Press the emergency stop switch for one half second or more, then release the switch for one tenth of a second or more. Repeat this procedure 5 times

After troubleshooting, disconnect the SCS connector from the service check connector. The MIL will stay illuminated while the SCS is connected to the service check connector.

Be sure to reset the ECM after troubleshooting. The MIL can turn ON or blink when the ECM has detected a poor or loose connector contact in the ECM and that the connector of the probable problem part needs to be cleaned or repaired immediately.

If the problem cannot be identified by cleaning or repairing the connector, check the throttle cable, idle adjustment and the fuel supply system.

➡The MIL goes OFF when the ignition switch is turned OFF. It does not turn back on when the ignition switch is turned ON unless the problem is detected again. It is a good idea to note the fault code before turning off the ignition so it can be looked up in the diagnostic trouble code (DTC) chart and investigated.

The MIL blinks when the SCS connector is connected to the RED 2P service check connector and the ignition switch is turned ON. This is because the memory of the problem is stored in the ECM. Turn the ignition switch ON again and, even though MIL does not turn ON again, connect the SCS connector to the RED 2P service check connector to short and check the DTC.

When the ignition switch is turned on with the special tool (Test Harness) disconnected during the inspection as per the troubleshooting flow chart, the MIL can turn ON. Be sure to reset the ECM.

FUEL INJECTION SYSTEM COMPONENT LOCATIONS—PORT SIDE

1. Wire Harness Clip
2. Alternator Connector
3. Tube Clip
4. Alternator Terminal
5. Wire Harness Clip
6. Wire Harness Band
7. Map Sensor Connector
8. TP Sensor
9. TP Sensor Connector
10. Clamp
11. IAT Sensor Connector
12. Wire Clamp
13. Fuel Pump Unit Connector
14. Wire Clip
15. Wire Harness Clip
16. Harness Clip
17. Remote Control Cable A
18. IAC Valve Connector

04894G19

Throttle Position Sensor (TPS)

DESCRIPTION & OPERATION

▸ See Figures 71

This sensor is a potentiometer which translates the position of the throttle plate to an electrical signal. The signal is near zero at idle and increases to just under 5 volts at wide open throttle. The sensor is mounted to the side of the throttle body.

TESTING

▸ See Figure 72

1. Make sure the ignition switch is in the off position.
2. Disconnect the throttle position sensor 3P connector.
3. Turn the ignition switch to the on position.
4. Connect the positive (+) tester lead to the to the No.1 terminal of the harness side of the TPS 3P connector , and the negative (-) tester lead to the No. 3 terminal of the 3P connector.
5. Then measure the voltage. The measurement should be between 4.75 and 5.25 volts.

➡If the TPS is found to be faulty, the entire throttle body must be replaced as an entire assembly.

REMOVAL & INSTALLATION

▸ See Figures 73 thru 78

1. Remove the silencer casing.
2. Disconnect the TPS connector at the harness.
3. Disconnect the throttle side remote control cable.
4. Remove the throttle arm from the shift link bracket. Disconnect the two throttle cables from the throttle arm.
5. Unscrew the cap nuts and bolts and remove the throttle body assembly.

To install:

6. Install the new throttle body gasket. Never reuse this gasket.
7. Line throttle body up on the two studs and slide it in place.
8. Torque down the cap nuts and bolts to 20 ft. lbs. (26 Nm).
9. Reconnect the two throttle cables to the throttle arm and adjust them to the correct specifications.

➡Now is a good time to grease the cable ends and throttle cam grooves.

10. Check the throttle cable length of the sections shown at the throttle cam open side and close side. The cable length at the throttle cam open side and close side should be 0.67 in. (17mm) from the point shown.
11. If the measurement is above or below the specification, adjust the cables by loosing the adjusting nut on either the open side or close side as needed to obtain the correct measurement.
12. Measure the throttle cable length from the end of the threaded part to the lock nut end. Measure at the throttle arm open and close sides respectively, they should measure 0.71 in. (18mm) on the open side and 0.63 in. (16mm) on the close side.
13. If the measurement is above or below the correct specification, adjust it by loosing the adjusting nut on either the open or close side as is required.
14. Move the remote control lever to the forward position (wide open throttle).

If is hard to move the remote control lever to the forward position, with the engine stopped, move the lever while turning the propeller or propeller shaft.

Do not use force to move the lever or damage to the gear shifting system may occur.

15. Check to see if the throttle arm is in contact with the fully open position stopper at this time. If it is not, further adjustment will be necessary.
16. Move the remote control lever to the neutral position.
17. Then loosen the shift pivot lock nut and remove the lock pin and wash-

04894P45

Fig. 71 The TPS is located on the side of the throttle body and senses the rotation of the throttle

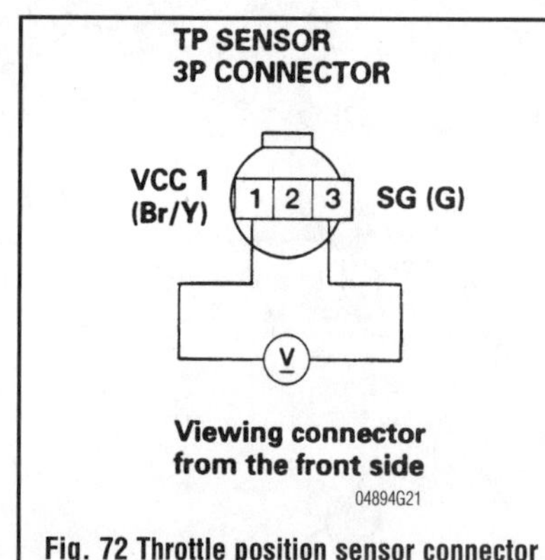

04894G21

Fig. 72 Throttle position sensor connector terminal identification

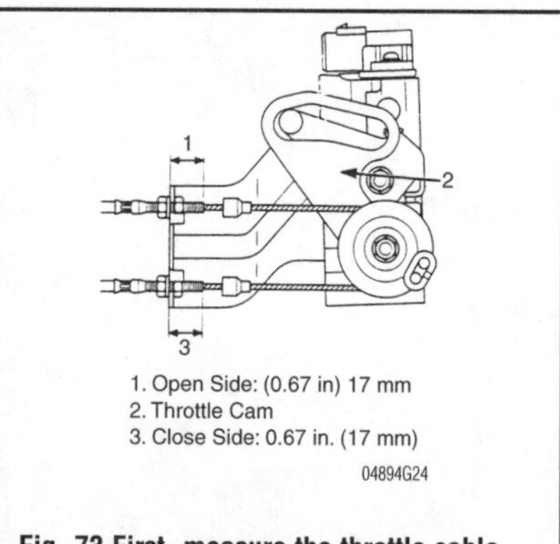

04894G24

Fig. 73 First, measure the throttle cable length on the throttle cam . . .

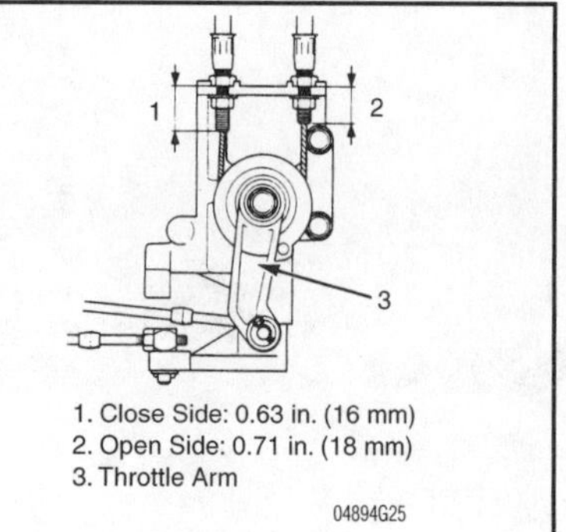

04894G25

Fig. 74 . . . next, measure the length on the throttle arm . . .

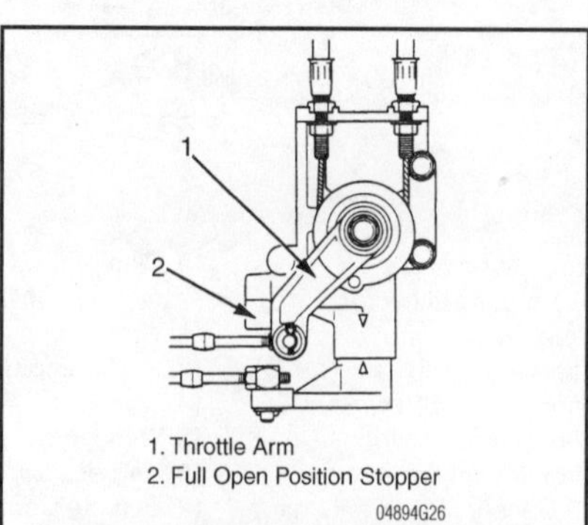

04894G26

Fig. 75 . . . then make sure the throttle arm is in contact with the full open position stopper

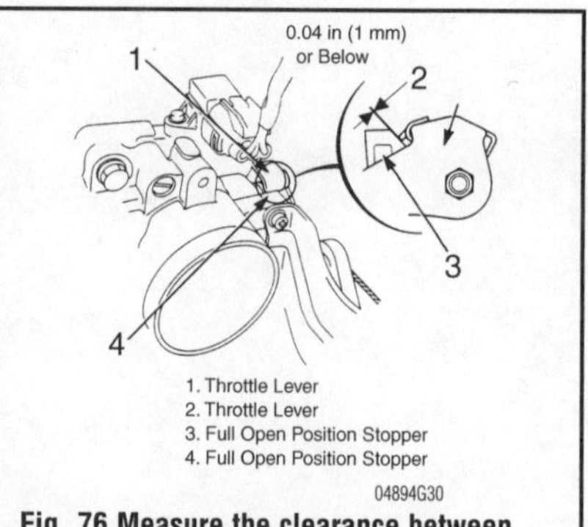

04894G30

Fig. 76 Measure the clearance between the throttle body throttle lever and the full open stopper . . .

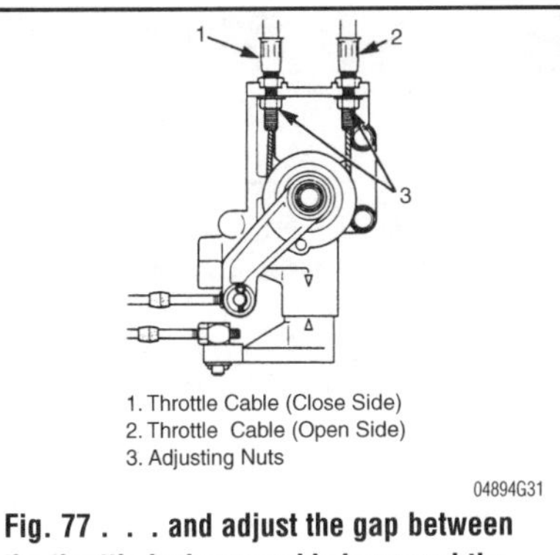

Fig. 77 . . . and adjust the gap between the throttle body assembly lever and the full open stopper

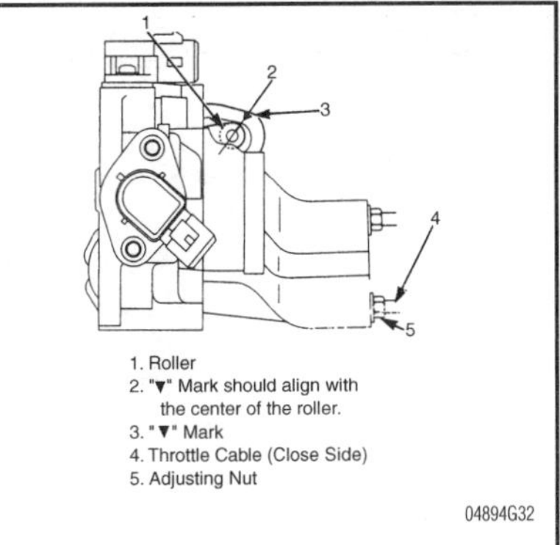

Fig. 78 Finally, check the alignment marks and adjust by turning the adjusting nut

Fig. 79 The air temperature sensor is located in the intake manifold and reads the temperature of the incoming air

ers from the shift pivot. Disconnect the shift pivot from the throttle arm and adjust it by turning the shift pivot.

18. Install the shift pivot on the throttle arm. Move the remote control lever to the forward full open position and check to see if the throttle arm is in contact with the full open position stopper. Then, move the lever to the neutral position, and tighten the shift pivot lock nut.
19. With the throttle arm in contact with the full open position stopper, measure the clearance between the throttle lever of the throttle body assembly and the full open position stopper. It should be 0.04 in. (1mm) or less.

Adjustment is needed if the clearance is more than 0.04 in. (1mm).

20. The adjustment must be made as follows. Move the remote control lever to the neutral position and loosen the adjusting nuts at the open and close sides of the throttle cable of the throttle arm side.
21. Move the remote control lever to the forward full open position.
22. Turn the adjusting nut at the open side of the throttle cable in or out as needed to adjust the gap between the throttle lever of the throttle body assembly and the full open position stopper to 0.04 in. (1mm) or below.
23. Loosely tighten the adjusting nuts at the open side And close side of the throttle cable, and return the remote control lever to the neutral position.
24. Check whether or not the triangle mark on the throttle cam aligns with the center of the throttle cam roller. If the mark is out of alignment, adjust by turning the adjusting nut at the close side of the throttle cable.
25. After adjustment, tighten the adjusting nuts at the open and close sides of the throttle cable securely.
26. Install the silencer case and connect the TPS connector.

➡Check to be sure that the power tilt motor wire is routed between the close side and the open side of the throttle cables.

Intake Air Temperature Sensor (IAT)

DESCRIPTION & OPERATION

The Intake Air Temperature Sensor (IAT) compensates the fuel injection amount according to the intake air temperature

This device is a temperature-dependent resistor (thermistor) and is placed in the intake manifold. It acts much like the coolant temperature sensor, but with a reduced thermal capacity for quicker response. The injector duration determined by the ECM is altered for different operating conditions by the signals sent by this sensor.

TESTING

1. Turn the ignition to the OFF position.
2. Disconnect the IAT sensor 2P connector and measure the resistance between the two terminals of the IAT sensor.
3. Check whether the adequate resistance for the IAT is available. If there is no resistance, replace the sensor.
4. Turn the ignition switch back to the ON position and connect the (+) positive test lead to the to the No. 2 terminal (right side of connector as viewed from the front side) of the harness side of the IAT sensor 2P connector and the (-) negative tester lead to a good engine ground. Measure the voltage. Voltage should measure between 4.30–5.25 volts.

REMOVAL & INSTALLATION

➧ **See Figure 79**

1. Disconnect the wire harness from the sensor at the intake manifold.
2. Remove the sensor from the intake manifold, and discard the rubber O-ring.
3. Inspect the sensor, checking for cracks, broken connectors, etc. If the sensor is suspect, do the voltage and resistance test as explained above.

To install:

4. Place a new, lubricated O-ring on the sensor.
5. Thread the sensor into the intake manifold to 13 ft. lbs. (18Nm).
6. Reconnect the wiring harness to the sensor.

Engine Coolant Temperature (ECT)

DESCRIPTION & INSTALLATION

When the engine temperature is low, the ECT signals the ECM to increase the fuel injection amount according to the intake manifold vacuum. After the engine starts , and starts to warm up, the ECT gradually decreases the amount of fuel being injected.

TESTING

1. Turn the ignition switch OFF.
2. Disconnect the ECT sensor connector and measure the resistance between the two terminals of the ECT sensor.
3. If there is not adequate resistance, replace the sensor.
4. Turn the ignition switch to the ON position.
5. Connect the (+) positive test lead to the No. 2 terminal on the connector harness side and the (-) negative lead to a good engine ground and measure the voltage. It should measure between 4.30–5.25 volts.
6. If the voltage is not to specification, replace the sensor.

REMOVAL & INSTALLATION

1. Remove the intake manifold assembly.
2. Disconnect the ECT sensor from the wiring harness.
3. Unscrew the sensor from the engine.
4. Discard the O-ring. Do not reuse.

To install:

5. After testing, install a new O-ring on the sensor.
6. Thread the sensor back into the engine and torque to 13 ft. lbs. (18Nm).
7. Reconnect the wiring harness to the sensor

Overheat Sensor

DESCRIPTION & OPERATION

The overheat sensor monitors cooling water temperature. This is essential information during cold starts and warm-up. Should the temperature become excessive, the overheat sensor triggers a visual and audible alarm to alert the operation of the overheat condition.

TESTING

1. Turn the ignition switch OFF.
2. Disconnect the overheat sensor from the wiring harness.
3. Measure the resistance between the two terminals of the overheat sensor.
4. If there is not adequate resistance, replace the sensor.
5. Turn the ignition switch to the ON position.
6. Connect the (+) positive lead to the No. 2 terminal on the harness side and the (-) negative lead to a good engine ground. The voltage should measure 4.30–5.25 volts.
7. If the voltage is not to specification, replace the sensor.

REMOVAL & INSTALLATION

See Figure 80

1. Disconnect the wiring harness from the sensor.
2. Unscrew the sensor from the engine. Discard the O-ring.

To install:

3. After testing, replace the O-ring and screw in the overheat sensor.
4. Reconnect the wiring harness.

Fuel Injectors

DESCRIPTION & OPERATION

An electronic fuel injector is an electronic solenoid driven by the control module. The control module, based on sensor inputs, determines when and how long the fuel injector should operate. When an electric current is supplied to the injector, a spring loaded ball is lifted from its seat. This allows fuel to flow through spray orifices and deflect off the sharp edge of the injector nozzle. This action causes the fuel to form an angled, cone shaped spray pattern before entering the air stream in the intake manifold.

TESTING

1. Disconnect the fuel injector connector from the wiring harness.
2. Measure the resistance between the fuel injector terminals. The resistance should measure 11.1–12.3 ohms.
3. When the measurement does not meet specification, replace the injector.

REMOVAL & INSTALLATION

1. Disconnect the negative battery cable.
2. Relieve the fuel system pressure.

**** CAUTION**

Observe all applicable safety precautions when working around fuel. Whenever servicing the fuel system, always work in a well ventilated area. Do not allow fuel spray or vapors to come in con-

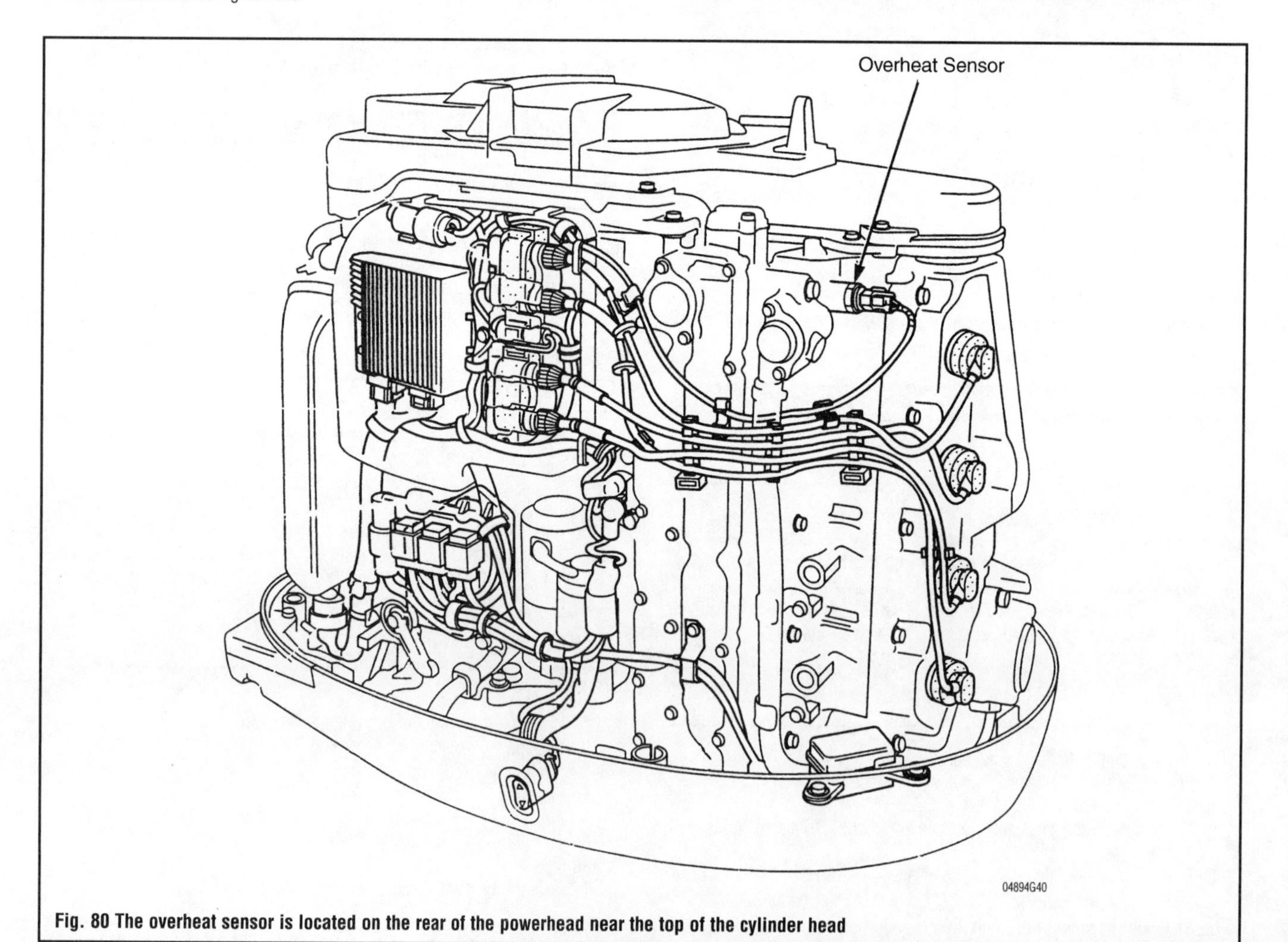

Fig. 80 The overheat sensor is located on the rear of the powerhead near the top of the cylinder head

tact with a spark or open flame. Keep a dry chemical fire extinguisher near the work area. Always keep fuel in a container specifically designed for fuel storage; also, always properly seal fuel containers to avoid the possibility of fire or explosion.

3. Disconnect the fuel injector electrical connection from the wiring harness.
4. Remove the wiring harness holder and brackets.
5. Loosen and remove the service check bolt and move the fuel supply line out of the way.
6. Loosen the pressure regulator bolts so it can be removed from the fuel rail.
7. Remove the cap nuts, and remove the fuel rail.
8. Remove the fuel injector/s from the intake manifold.

To install:

9. Replace the O-ring, cushion ring and the injector seal with new components.
10. Install a new cushion rings on the injectors.
11. Lightly coat the new O-rings with engine oil and install them in the groove on the injectors.
12. Install the injectors on the fuel rail.

➡Make sure that the injectors are installed into the fuel rail straight. If they are installed off-center, damage can occur to the O-ring and a fuel leak will occur.

13. Lightly coat the inner wall of the new injector seal rings and install them into the intake manifold.
14. Install the fuel pipe insulators on the intake manifold.
15. Install the fuel rail, with the injectors installed, on the intake manifold.

➡To prevent damaging the O-rings, have the injectors already installed on the fuel rail and install the entire assembly on the intake manifold.

16. Tighten the fuel rail to the intake manifold with the cap nuts.
17. Connect the vacuum tube, pressure regulator return hose, injector connectors.
18. With new sealing washers, tighten the fuel hose.
19. Reconnect the negative battery cable and turn the ignition switch to the ON position. The fuel pump should cycle for about 2 seconds, causing the fuel pressure in the high pressure fuel line to rise. Do this procedure several time and check all sealing areas for fuel leakage.
20. Install the seal rings into the intake manifold.

➡Do not reuse the seal rings. Apply engine oil to the inner wall of the rings before installation.

21. Insert the fuel injector into the intake manifold.
22. Insert new cushion rings and O-rings on the injectors.

➡Do not reuse cushion and O-rings. Lightly coat the new O-rings with engine oil.

23. Install the fuel rail, with the fuel rail insulators installed behind, and tighten the cap nuts.
24. Install the pressure regulator on the fuel rail using a new O-ring and tighten the bolts.
25. Install the fuel hose with new sealing washers and torque to 16 ft. lbs. (22 Nm).
26. Reconnect the wiring harness holder and brackets.
27. Reconnect the individual fuel injection connectors to the wiring harness.

➡After the fuel system has been pressurized, check all sealing areas for leaks.

Vapor Separator

DESCRIPTION & OPERATION

➧ See Figure 82

The supply side of this system is similar to Honda carbureted engines. Fuel is pulled from the six gallon or boat tank through a primer bulb and hose to a

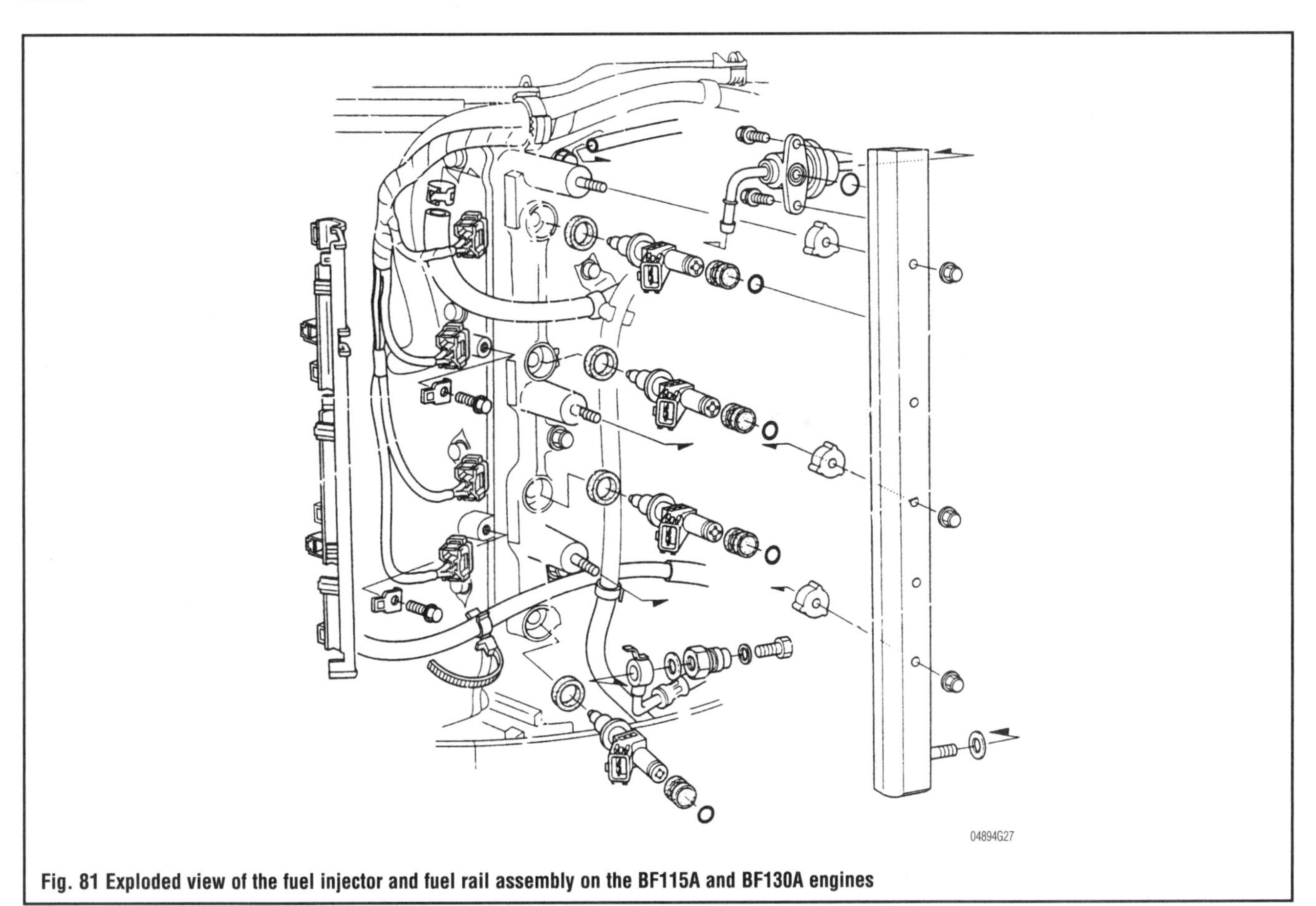

Fig. 81 Exploded view of the fuel injector and fuel rail assembly on the BF115A and BF130A engines

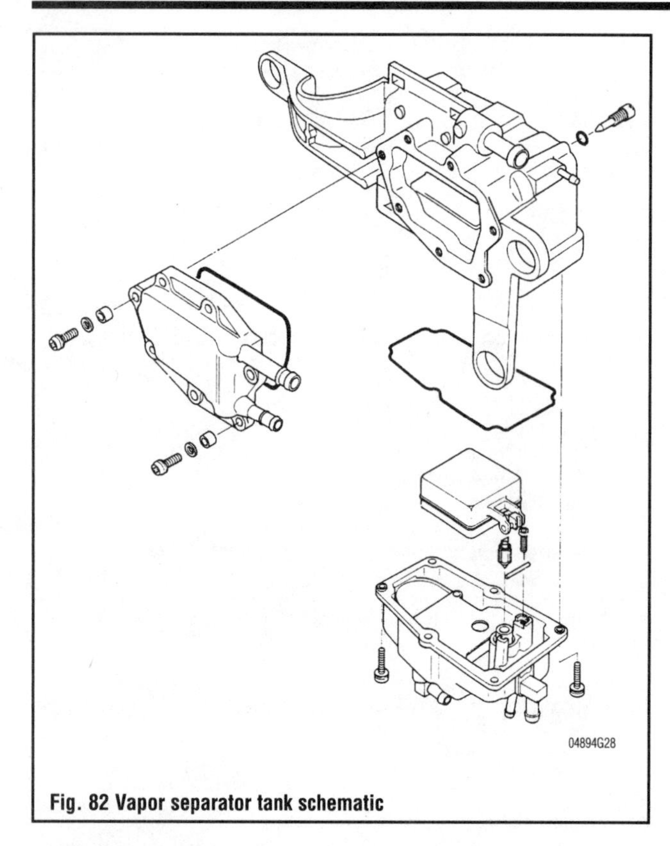

Fig. 82 Vapor separator tank schematic

low-pressure fuel filter. Next the low-pressure fuel pump sends the fuel on to the vapor separator tank.

The vapor separator tank is a multipurpose reservoir. It acts as a "burp" tank for any air in the supply side of the system. This ensures that the high-pressure pump does not pump any air. The separator tank is vented into the air intake to recirculate any vapor that accumulates. Fuel is pumped from the separator by an externally mounted high-pressure electric fuel pump. Fuel drawn in to this pump passes through the high-pressure fuel filter and then to the injectors. Pressure is maintained by the fuel pressure regulator.

The pump is designed to supply more fuel to the injectors than maximum demand requires. The regulator controls this over supply by diverting excess high pressure fuel back to the vapor separator. This loop (vapor separator tank to the injectors through the regulator and back to the separator tank) is typical of most fuel injection systems.

REMOVAL & INSTALLATION

See Figure 83

CAUTION

Observe all applicable safety precautions when working around fuel. Whenever servicing the fuel system, always work in a well ventilated area. Do not allow fuel spray or vapors to come in contact with a spark or open flame. Keep a dry chemical fire extinguisher near the work area. Always keep fuel in a container specifically designed for fuel storage; also, always properly seal fuel containers to avoid the possibility of fire or explosion.

1. Loosen the drain screw and drain the vapor separator before disassembly.
2. Remove the water jacket screws and remove the water jacket cover.
3. Remove the screws and separate the float assembly cover.
4. Remove the float pin retaining screw and remove the float pin.
5. Remove the float and the float valve.

To install:

6. Install new O-rings on the water cover jacket, float cover and drain screw.
7. Install the float, float valve and float pin. Secure the float pin with the pin screw.

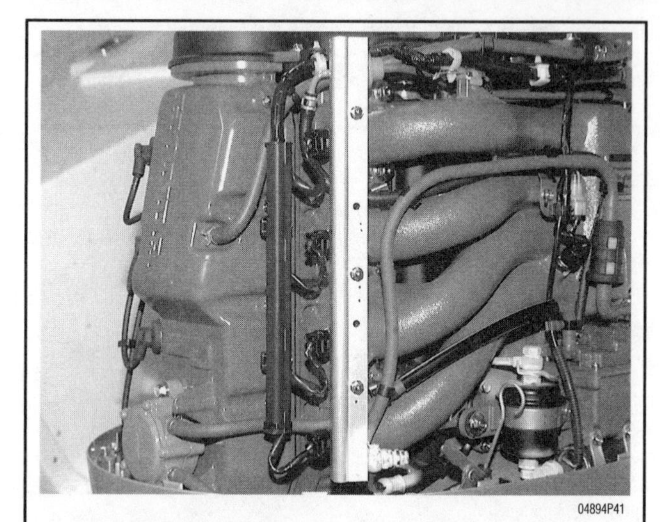

Fig. 83 The vapor separator tank is located on the starboard side of the powerhead near the fuel filter

After you install the float assembly, check the movement of the float by touching it with your finger.

8. Install the covers back on the chamber body and tighten the screws.

CLEANING & INSPECTION

See Figure 84

WARNING

If there is gasoline overflowing from the vapor separator, you must perform the following steps to correct the problem:

1. Remove the float assembly from the vapor separator.
2. Remove the float valve from the float.
3. Check the tip of the float valve and the valve seat for wear or damage. If damaged, replace the float valve or float cover as needed.

If the float valve and seat are undamaged, check the float height by the following procedure:

4. Install the float valve and float in the cover.
5. Hold the cover as shown in the illustration and measure the distance between the float and the cover at the edge of the cover. Float height should measure between 1.1–1.3 in. (29–34mm).
6. If the float height is not within this specification, adjust the float height by gently bending the brass float tab.
7. Reinstall assembly and check measurement again.

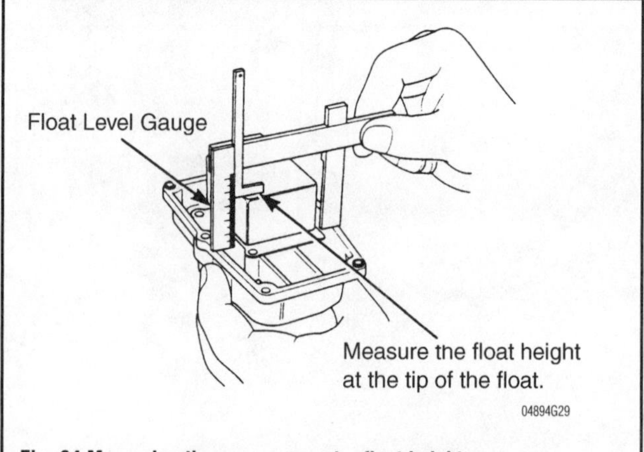

Fig. 84 Measuring the vapor separator float height

Fuel Pump

DESCRIPTION & OPERATION

The fuel injected Honda system consists of two separate fuel pumps. An engine driven plunger-type mechanical pump for the low-pressure side of the system, which takes a suction from the fuel tank and sends it to the vapor separator. Then the externally-mounted, electrically-driven, high-pressure pump takes its suction from the vapor separator and sends the fuel to the fuel injectors at

FUEL PRESSURE CHECK

1. Relieve fuel pressure as per the instructions in Chapter 3, Engine Maintenance.
2. Remove the service check bolt and insert the Honda special tool No. 07406-004000A or equivalent in the hole.
3. Disconnect the vacuum hose from the pressure regulator and clamp or plug the hose.

➡Golf tee's make excellent vacuum hose plugs.

4. Remove the propeller and set the engine in a test tank.
5. Start the engine and with the engine idling, measure the fuel pressure. Standard fuel pressure at idle (700–800 rpm) should be 38–46 psi (265–314 kPa)

When the fuel pressure is reading is not within specification, perform the following checks:

- When the fuel pressure is higher than normal, check the pressure regulator hose for kinking or obstructions. Also inspect the fuel pressure regulator for proper operation.
- When the fuel pressure is lower than normal, check the fuel pressure regulator for proper operation, the high-pressure fuel filter for clogging and the fuel pump itself for proper operation.

6. After the check is completed, always replace the sealing washers with new ones and retighten the service check bolt to 9 ft. lbs. (12Nm).

REMOVAL & INSTALLATION

Mechanical Type

1. Remove the engine cover.
2. Remove the rear separate case.

CAUTION

Observe all applicable safety precautions when working around fuel. Whenever servicing the fuel system, always work in a well ventilated area. Do not allow fuel spray or vapors to come in contact with a spark or open flame. Keep a dry chemical fire extinguisher near the work area. Always keep fuel in a container specifically designed for fuel storage; also, always properly seal fuel containers to avoid the possibility of fire or explosion.

3. Remove the hose clips and pull off the fuel hoses.
4. Remove the two retaining bolts and remove the pump from the engine.

To install:

5. Discard the rubber O-ring and replace with a new one. Inspect the fuel hoses and hose clamps for cracks and other damage and replace as needed.
6. Lightly coat the new O-ring with engine oil and install it on the pump.
7. Install the two retaining bolts and tighten snugly.
8. Install the fuel hoses. Make sure that the fuel hose from the low-pressure filter goes onto the lower nipple (arrow facing inwards) and the fuel hose going to the vapor separator goes on the top nipple (arrow facing outwards).
9. Make sure all the claps are tight and check for fuel leaks after running the engine.

Electric Type

➧ See Figures 85 and 86

1. Disconnect the negative battery cable.
2. Make sure that the fuel pressure is relieved.

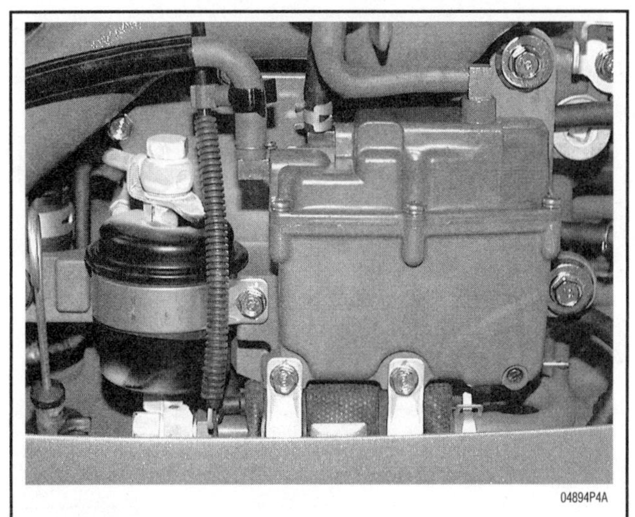

Fig. 85 Vapor separator (right), fuel filter (left) and fuel pump (bottom) assembly

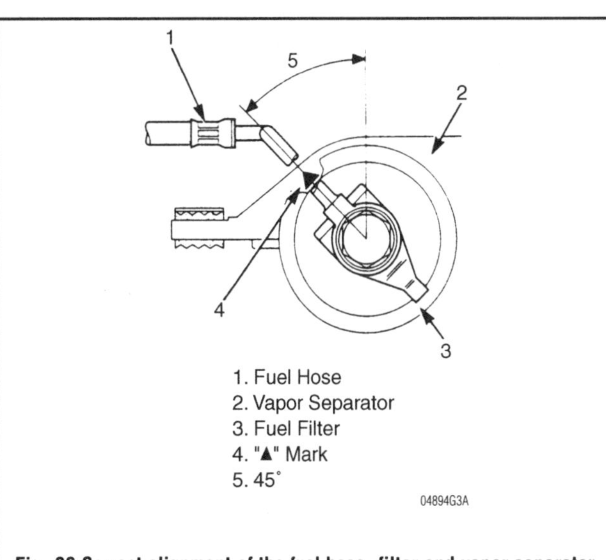

Fig. 86 Correct alignment of the fuel hose, filter and vapor separator

3. Disconnect the wiring harness connector.
4. Loosen the flare nut at the fuel filter and remove the sealing nut on the Banjo fitting.
5. Remove the fuel pipe and discard the sealing washers.
6. Loosen and remove the sealing bolt on the fuel filter Banjo fitting. Discard the two sealing washers.
7. Loosen the fuel filter band and slide out the fuel filter.
8. Loosen the clamp bands and slid out the fuel pump.

To install:

9. Discard the rubber O-ring and replace with a new one. Inspect the fuel hoses and hose clamps for cracks and other damage and replace as needed.
10. Install the grommet on the fuel filter.
11. Install new sealing washers on either side of the fuel hose Banjo fitting at the top of the filter and tighten the bolt to 25 ft. lbs. (33Nm).
12. Set the fuel filter on the vapor separator by aligning the mark on the separator and the filter as per the illustration.
13. Loosely tighten the fuel pipe flare nut.
14. Install the fuel filter band and loosely tighten the flange bolts.
15. Clean all the fuel line connections and install the two new sealing washers
16. Set the two insulators over the fuel pump and secure with the two grommets.
17. Hold the pump with the positive (+) terminal up when viewing the pump from the front.

18. Install the insulator material so that the mating parts of the insulators do not come in contact with
the bottom of the vapor separator.
19. Hold the pump securely in position, making sure it does not move out of position, install the two fuel pump bands and loosely tighten the bolts.
20. Making sure all the fuel lines are aligned and tighten the clamp nuts.
21. Tighten the fuel pipe flare nut tightly, making sure not to move the alignment marks on the fuel filter and vapor separator. Torque the fitting to 27 ft. lbs. (37Nm).

OVERHAUL

These particular fuel pumps are not rebuildable and must be replaced as a complete assembly.

Fuel Pressure Regulator

DESCRIPTION & OPERATION

The pressure regulator is a vacuum operated device located downstream of the fuel injector. Its function is to maintain a constant pressure across the fuel injector tip. The regulator uses a spring loaded rubber diaphragm to uncover a fuel return port. When the fuel pump becomes operational, fuel flows past the injector into the regulator, and is restricted from flowing any further by the blocked return port. When fuel pressure reaches the predetermined setting, it pushes on the diaphragm, compressing the spring, and uncovers the fuel return port. The diaphragm and spring will constantly move from an open to closed position to keep the fuel pressure constant.

REMOVAL & INSTALLATION

See Figure 87

CAUTION

Observe all applicable safety precautions when working around fuel. Whenever servicing the fuel system, always work in a well ventilated area. Do not allow fuel spray or vapors to come in contact with a spark or open flame. Keep a dry chemical fire extinguisher near the work area. Always keep fuel in a container specifically designed for fuel storage; also, always properly seal fuel containers to avoid the possibility of fire or explosion.

1. Disconnect the battery negative cable.
2. Relieve the fuel pressure at the service check bolt.
3. Remeove the vacuum hose and the pressure regulator return hose.
4. Remove the two bolts and lift off the regulator.

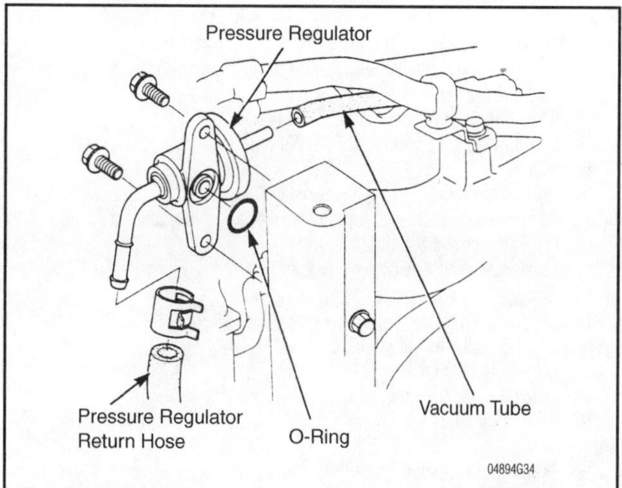

Fig. 87 Fuel pressure regulator placement on the BF115A and BF130A engines

To install:

➡Always replace the sealing washers and O-rings with new ones.

5. Lightly coat the new O-ring with engine oil and place it on the regulator.
6. Install the two bolts and tighten them lightly.
7. Install the vacuum and return hoses.
8. Tighten the bolts snugly.
9. Tighten the service check bolt (after installing new sealing washers).

INSPECTION

1. Check the pressure regulator vacuum tube. Make sure that the hoses are connected and that there are no kinks or obstructions.
2. Remove the propeller and start the engine in a test tank.
3. With the engine idling in neutral, disconnect the vacuum hose from the regulator and plug or clamp it closed.
4. Check the fuel pressure, it should measure higher than with the hose connected. Standard fuel pressure at idle (700–800 rpm) should be 38–46 psi (265–314 kPa).
5. When the fuel pressure does not rise, connect the vacuum tube back to the regulator. Lightly pinch the regulator return hose that goes between the regulator and vapor separator 2 or 3 times, and then measure the fuel pressure again. If the pressure is not within specification, replace the pressure regulator.

➡When pinching the return hose, protect it by wrapping a rag around it and then lightly pinching it closed with a pair of pliers.

Fuel Lines

GENERAL INFORMATION

See Figures 88, 89, 90 and 91

All marine hose must meet special standards, which means that not only must the hose be compatible with the fuel itself, but must meet strict standards for fire resistance. The use of automotive-grade hoses in a marine application violates US Coast Guard regulations making it illegal and is unsafe in any event. Only use hose with "Approved, USCG type" embossed on it.

Always use fuel-hose clamps made entirely of stainless steel. Do not use automotive-grade or hardware store clamps on the fuel line. They may only be partly stainless and the non-stainless part of the clamp will rust in a marine

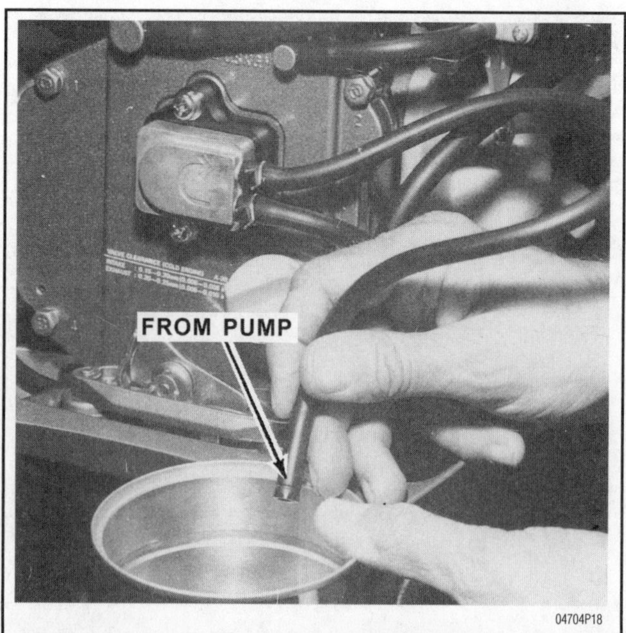

Fig. 88 To test the fuel pickup in the fuel tank, operate the squeeze bulb and observe fuel flowing from the disconnected line at the fuel pump. Discharge fuel into an approved container.

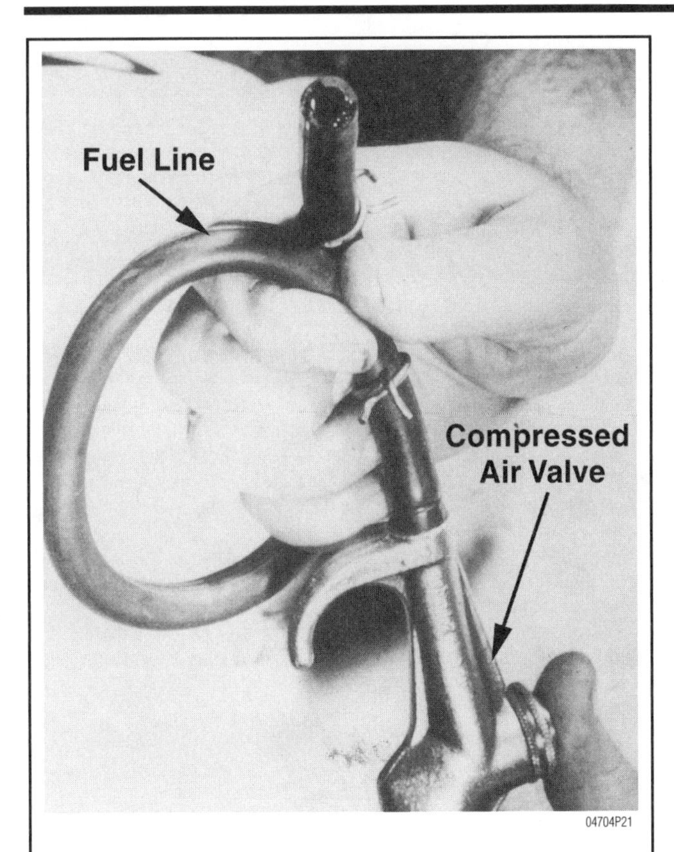

Fig. 89 Many times restrictions suck as foreign material may be cleared from the fuel lines using compressed air. Ensure the open end of the hose is pointing in a clear direction to avoid personal injury

environment and eventually fail. Always check your fuel hose and fittings at least annually and even better, before each outing, and replace any components that show signs of corrosion, wear and tear or leakage.

On most installations, the fuel line is provided with quick-disconnect fittings at the tank and at the engine. If there is reason to believe the problem is at the quick-disconnects, the hose ends should be replaced as an assembly. For a small additional expense, the entire fuel line can be replaced and thus eliminate this entire area as a problem source for many future seasons.

The primer squeeze bulb can be replaced in a short time. First, cut the hose line as close to the old bulb as possible. Slide a small clamp over the end of the

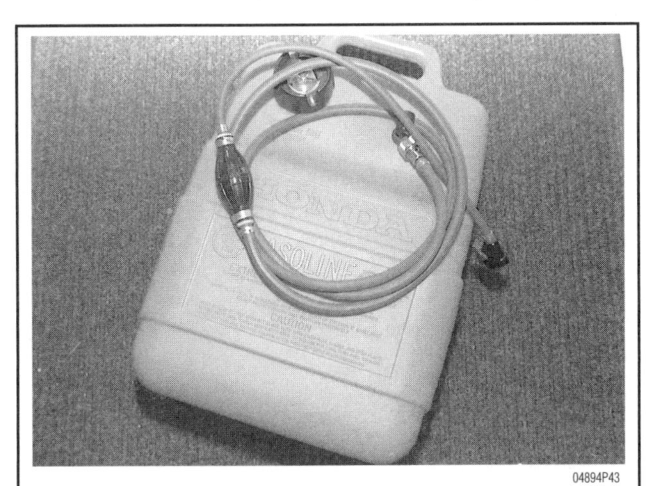

Fig. 90 Factory approved fuel tank and hose. Always check the entire hose and bulb assembly for wear and tear and replace if necessary

Fig. 91 Quick connect hose fittings. Both the tank and engine fitting must be in good working order and free of any debris to prevent a fuel leak

fuel line from the tank. Next, install the small end of the check valve assembly into this side of the fuel line. Always install the primer bulb in the direction which the embossed arrow points. Place a large clamp over the end of the check valve assembly. Use Primer Bulb Adhesive when the connections are made. Tighten the clamps. Repeat the procedure with the other side of the bulb assembly and the line leading to the engine.

REPLACEMENT & ROUTING

See Figure 92

Make sure that all fuel lines are routed properly and are placed away from any rotating parts, heat sources or where they can be pinched or cut. Check all fittings and clamps and replace or repair any that are broken or worn out.

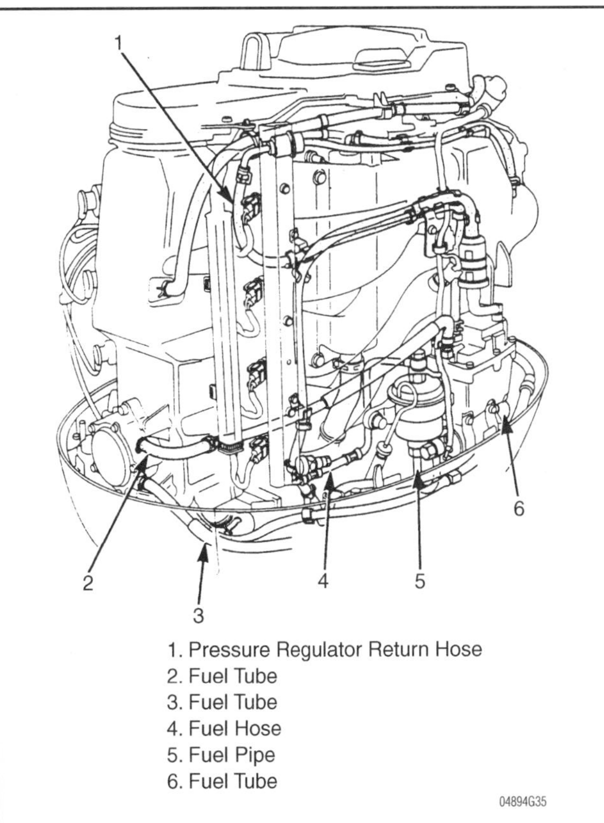

Fig. 92 Fuel hose routing on the BF115A and BF130A engines

COMMON PROBLEMS

Rough Engine Idle

If an engine does not idle smoothly, the most reasonable approach to the problem is to perform a tune-up to eliminate such areas as:

- Defective points
- Faulty spark plugs
- Timing out of adjustment

Other problems that can prevent an engine from running smoothly include:

- An air leak in the intake manifold
- Uneven compression between the cylinders

Of course any problem in the carburetor or fuel injection system that affects the air/fuel mixture will also prevent the engine from operating smoothly at idle speed. These problems usually include:

- Too high a fuel level in the bowl
- A heavy float
- Leaking needle valve and seat
- Defective automatic choke
- Improper adjustments for idle mixture or idle speed
- Malfunctioning sensors or ECM

Excessive Fuel Consumption

Excessive fuel consumption can be the result of any one of four conditions, or a combination of all.

- Inefficient engine operation.
- Faulty condition of the hull, including excessive marine growth.
- Poor boating habits of the operator.
- Leaking or out of tune carburetor.

If the fuel consumption suddenly increases over what could be considered normal, then the cause can probably be attributed to the engine or boat and not the operator.

Marine growth on the hull can have a very marked effect on boat performance. This is why sail boats always try to have a haul-out as close to race time as possible.

While you are checking the bottom, take note of the propeller condition. A bent blade or other damage will definitely cause poor boat performance.

If the hull and propeller are in good shape, then check the fuel system for possible leaks. Check the line between the fuel pump and the carburetor while the engine is running and the line between the fuel tank and the pump when the engine is not running. A leak between the tank and the pump many times will not appear when the engine is operating, because the suction created by the pump drawing fuel will not allow the fuel to leak. Once the engine is turned off and the suction no longer exists, fuel may begin to leak.

If a minor tune-up has been performed and the spark plugs, points, and timing are properly adjusted, then the problem most likely is in the carburetor and an overhaul is in order.

Check the needle valve and seat for leaking. Use extra care when making any adjustments affecting the fuel consumption, such as the float level or automatic choke.

UNDERSTANDING AND TROUBLESHOOTING ELECTRICAL SYSTEMS 5-2
BASIC ELECTRICAL THEORY 5-2
HOW DOES ELECTRICITY WORK: THE WATER ANALOGY 5-2
OHM'S LAW 5-2
ELECTRICAL COMPONENTS 5-2
POWER SOURCE 5-2
GROUND 5-3
PROTECTIVE DEVICES 5-3
SWITCHES & RELAYS 5-3
LOAD 5-3
WIRING & HARNESSES 5-4
CONNECTORS 5-4
TEST EQUIPMENT 5-4
JUMPER WIRES 5-4
TEST LIGHTS 5-5
MULTIMETERS 5-5
TROUBLESHOOTING ELECTRICAL SYSTEMS 5-6
TESTING 5-6
VOLTAGE 5-6
VOLTAGE DROP 5-6
RESISTANCE 5-6
OPEN CIRCUITS 5-7
SHORT CIRCUITS 5-7
WIRE AND CONNECTOR REPAIR 5-7
ELECTRICAL SYSTEM PRECAUTIONS 5-7
BREAKER POINTS IGNITION (MAGNETO IGNITION) 5-7
SYSTEM TESTING 5-8
SPARK CHECK 5-8
BREAKER POINTS 5-9
GAP CHECK 5-9
REMOVAL & INSTALLATION 5-9
CONDENSER 5-10
DESCRIPTION & OPERATION 5-10
TESTING 5-10
REMOVAL & INSTALLATION 5-10
CHARGE COIL 5-10
DESCRIPTION & OPERATION 5-10
TESTING 5-10
REMOVAL & INSTALLATION 5-10
PRIMARY COIL 5-11
DESCRIPTION & OPERATION 5-11
TESTING 5-11
REMOVAL & INSTALLATION 5-11
IGNITION COIL 5-11
DESCRIPTION & OPERATION 5-11
TESTING 5-11
REMOVAL & INSTALLATION 5-11
BREAKER POINTS IGNITION OVERHAUL 5-11
DISASSEMBLY 5-11
CLEANING & INSPECTION 5-11
ASSEMBLY 5-12
CAPACITOR DISCHARGE IGNITION (CDI) SYSTEM 5-12
DESCRIPTION AND OPERATION 5-12
SINGLE-CYLINDER IGNITION 5-12
TWO & THREE-CYLINDER IGNITION 5-13
FOUR-CYLINDER CARBURETED IGNITION 5-13
FOUR-CYLINDER EFI IGNITION 5-13
SYSTEM TESTING 5-16
PROCEDURE 5-16
PULSAR COIL 5-16
DESCRIPTION & OPERATION 5-16
TESTING 5-16
REMOVAL & INSTALLATION 5-17
CHARGE COIL 5-18
DESCRIPTION & OPERATION 5-18
TESTING 5-18
REMOVAL & INSTALLATION 5-19
LIGHTING COIL 5-20
DESCRIPTION & OPERATION 5-20
REMOVAL & INSTALLATION 5-20
TESTING 5-20
IGNITION COILS 5-20
DESCRIPTION & OPERATION 5-20
TESTING 5-20
REMOVAL & INSTALLATION 5-22
CDI UNIT 5-25
DESCRIPTION & OPERATION 5-25
TESTING 5-26
REMOVAL & INSTALLATION 5-30
RECTIFIER 5-31
TESTING 5-31
REMOVAL & INSTALLATION 5-32
REGULATOR 5-33
DESCRIPTION & OPERATION 5-33
TESTING 5-33
REMOVAL & INSTALLATION 5-33
REGULATOR/RECTIFIER 5-33
TESTING 5-33
REMOVAL & INSTALLATION 5-34
ELECTRONIC IGNITION 5-34
DESCRIPTION AND OPERATION 5-34
SELF DIAGNOSIS 5-34
CHARGING CIRCUIT 5-35
DESCRIPTION AND OPERATION 5-35
SINGLE PHASE CHARGING SYSTEM 5-36
DESCRIPTION AND OPERATION 5-36
THREE-PHASE CHARGING SYSTEM 5-36
DESCRIPTION AND OPERATION 5-36
TROUBLESHOOTING 5-37
SERVICING 5-37
PRECAUTIONS 5-37
TROUBLESHOOTING THE CHARGING SYSTEM 5-38
ALTERNATOR (STATOR) 5-38
TESTING 5-38
REMOVING & INSTALLATION 5-39
BATTERY 5-40
MARINE BATTERIES 5-40
BATTERY CONSTRUCTION 5-40
BATTERY RATINGS 5-40
BATTERY LOCATION 5-41
BATTERY SERVICE 5-41
BATTERY TERMINALS 5-42
BATTERY & CHARGING SAFETY PRECAUTIONS 5-42
BATTERY CHARGERS 5-43
BATTERY CABLES 5-43
BATTERY STORAGE 5-43
STARTING CIRCUIT 5-43
DESCRIPTION AND OPERATION 5-43
TROUBLESHOOTING THE STARING SYSTEM 5-44
STARTER MOTOR 5-44
DESCRIPTION & OPERATION 5-44
TESTING 5-44
REMOVAL & INSTALLATION 5-45
STARTER MOTOR SOLENOID SWITCH 5-46
DESCRIPTION AND OPERATION 5-46
TESTING 5-46
REMOVAL & INSTALLATION 5-47
COMPONENT LOCATIONS 5-48
IGNITION AND ELECTRICAL WIRING DIAGRAMS 5-52
SPECIFICATIONS CHARTS
CDI TEST CHARTS 5-26
VOLTAGE REGULATOR RECTIFER 5-32

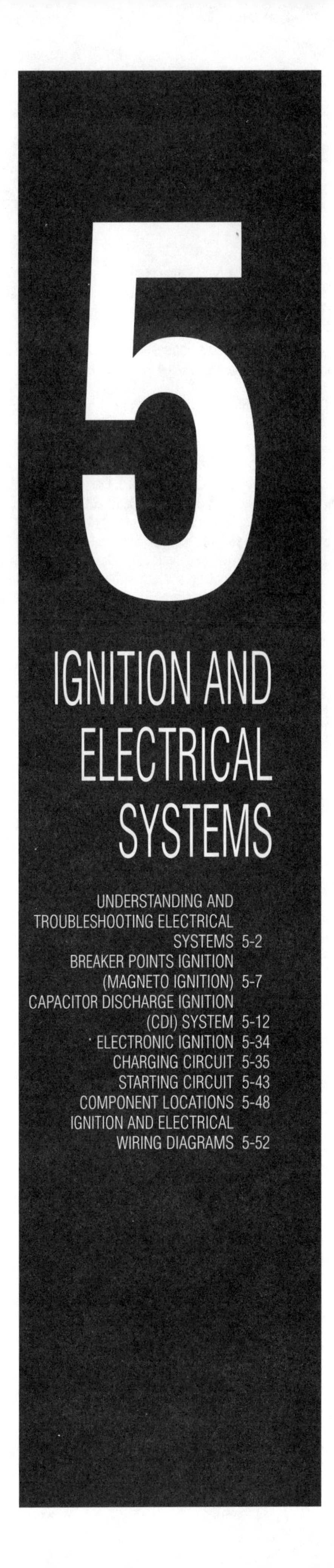

UNDERSTANDING AND TROUBLESHOOTING ELECTRICAL SYSTEMS 5-2
BREAKER POINTS IGNITION (MAGNETO IGNITION) 5-7
CAPACITOR DISCHARGE IGNITION (CDI) SYSTEM 5-12
ELECTRONIC IGNITION 5-34
CHARGING CIRCUIT 5-35
STARTING CIRCUIT 5-43
COMPONENT LOCATIONS 5-48
IGNITION AND ELECTRICAL WIRING DIAGRAMS 5-52

UNDERSTANDING AND TROUBLESHOOTING ELECTRICAL SYSTEMS

Basic Electrical Theory

▶ See Figure 1

For any 12 volt, negative ground, electrical system to operate, the electricity must travel in a complete circuit. This simply means that current (power) from the positive terminal (+) of the battery must eventually return to the negative terminal (-) of the battery. Along the way, this current will travel through wires, fuses, switches and components. If, for any reason, the flow of current through the circuit is interrupted, the component fed by that circuit will cease to function properly.

Perhaps the easiest way to visualize a circuit is to think of connecting a light bulb (with two wires attached to it) to the battery—one wire attached to the negative (-) terminal of the battery and the other wire to the positive (+) terminal. With the two wires touching the battery terminals, the circuit would be complete and the light bulb would illuminate. Electricity would follow a path from the battery to the bulb and back to the battery. It's easy to see that with longer wires on our light bulb, it could be mounted anywhere. Further, one wire could be fitted with a switch so that the light could be turned on and off.

The normal marine circuit differs from this simple example in two ways. First, instead of having a return wire from each bulb to the battery, the current travels through a single ground wire which handles all the grounds for a specific circuit. Secondly, most marine circuits contain multiple components which receive power from a single circuit. This lessens the amount of wire needed to power components.

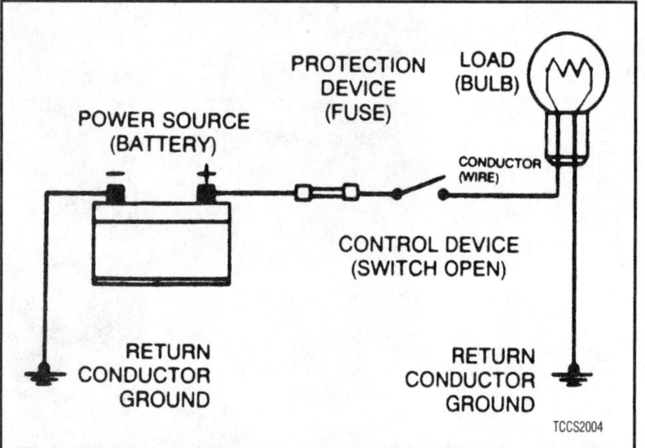

Fig. 1 This example illustrates a simple circuit. When the switch is closed, power from the positive (+) battery terminal flows through the fuse and the switch, and then to the light bulb. The light illuminates and the circuit is completed through the ground wire back to the negative (-) battery terminal.

HOW DOES ELECTRICITY WORK: THE WATER ANALOGY

Electricity is the flow of electrons—the sub-atomic particles that constitute the outer shell of an atom. Electrons spin in an orbit around the center core of an atom. The center core is comprised of protons (positive charge) and neutrons (neutral charge). Electrons have a negative charge and balance out the positive charge of the protons. When an outside force causes the number of electrons to unbalance the charge of the protons, the electrons will split off the atom and look for another atom to balance out. If this imbalance is kept up, electrons will continue to move and an electrical flow will exist.

Many people have been taught electrical theory using an analogy with water. In a comparison with water flowing through a pipe, the electrons would be the water and the wire is the pipe.

The flow of electricity can be measured much like the flow of water through a pipe. The unit of measurement used is amperes, frequently abbreviated as amps (a). You can compare amperage to the volume of water flowing through a pipe. When connected to a circuit, an ammeter will measure the actual amount of current flowing through the circuit. When relatively few electrons flow through a circuit, the amperage is low. When many electrons flow, the amperage is high.

Water pressure is measured in units such as pounds per square inch (psi); The electrical pressure is measured in units called volts (v). When a voltmeter is connected to a circuit, it is measuring the electrical pressure.

The actual flow of electricity depends not only on voltage and amperage, but also on the resistance of the circuit. The higher the resistance, the higher the force necessary to push the current through the circuit. The standard unit for measuring resistance is an ohm (Ω). Resistance in a circuit varies depending on the amount and type of components used in the circuit. The main factors which determine resistance are:

- Material—some materials have more resistance than others. Those with high resistance are said to be insulators. Rubber materials (or rubber-like plastics) are some of the most common insulators used, as they have a very high resistance to electricity. Very low resistance materials are said to be conductors. Copper wire is among the best conductors. Silver is actually a superior conductor to copper and is used in some relay contacts, but its high cost prohibits its use as common wiring. Most marine wiring is made of copper.
- Size—the larger the wire size being used, the less resistance the wire will have. This is why components which use large amounts of electricity usually have large wires supplying current to them.
- Length—for a given thickness of wire, the longer the wire, the greater the resistance. The shorter the wire, the less the resistance. When determining the proper wire for a circuit, both size and length must be considered to design a circuit that can handle the current needs of the component.
- Temperature—with many materials, the higher the temperature, the greater the resistance (positive temperature coefficient). Some materials exhibit the opposite trait of lower resistance with higher temperatures (negative temperature coefficient). These principles are used in many of the sensors on the engine.

OHM'S LAW

There is a direct relationship between current, voltage and resistance. The relationship between current, voltage and resistance can be summed up by a statement known as Ohm's law.

Voltage (E) is equal to amperage (I) times resistance (R): $E=I \times R$

Other forms of the formula are $R=E/I$ and $I=E/R$

In each of these formulas, E is the voltage in volts, I is the current in amps and R is the resistance in ohms. The basic point to remember is that as the resistance of a circuit goes up, the amount of current that flows in the circuit will go down, if voltage remains the same.

The amount of work that the electricity can perform is expressed as power. The unit of power is the watt (w). The relationship between power, voltage and current is expressed as:

Power (W) is equal to amperage (I) times voltage (E): $W=I \times E$

This is only true for direct current (DC) circuits; The alternating current formula is a tad different, but since the electrical circuits in most vessels are DC type, we need not get into AC circuit theory.

Electrical Components

POWER SOURCE

Power is supplied to the vessel by two devices: The battery and the alternator. The battery supplies electrical power during starting or during periods when the current demand of the vessel's electrical system exceeds the output capacity of the alternator. The alternator supplies electrical current when the engine is running. The alternator does not just supply the current needs of the vessel, but it recharges the battery.

The Battery

In most modern vessels, the battery is a lead/acid electrochemical device consisting of six 2 volt subsections (cells) connected in series, so that the unit is capable of producing approximately 12 volts of electrical pressure. Each subsection consists of a series of positive and negative plates held a short distance apart in a solution of sulfuric acid and water.

The two types of plates are of dissimilar metals. This sets up a chemical reaction, and it is this reaction which produces current flow from the battery when its positive and negative terminals are connected to an electrical load. The power removed from the battery is replaced by the alternator, restoring the battery to its original chemical state.

The Alternator

On some vessels there isn't an alternator, but a generator. The difference is that an alternator supplies alternating current which is then changed to direct current for use on the vessel, while a generator produces direct current. Alternators tend to be more efficient and that is why they are used on almost all modern engines.

Alternators and generators are devices that consist of coils of wires wound together making big electromagnets. One group of coils spins within another set and the interaction of the magnetic fields causes a current to flow. This current is then drawn off the coils and fed into the vessel's electrical system.

GROUND

Two types of grounds are used in marine electric circuits. Direct ground components are grounded to the electrically conductive metal through their mounting points. All other components use some sort of ground wire which leads back to the battery. The electrical current runs through the ground wire and returns to the battery through the ground (-) cable; if you look, you'll see that the battery ground cable connects between the battery and a heavy gauge ground wire.

➡It should be noted that a good percentage of electrical problems can be traced to bad grounds.

PROTECTIVE DEVICES

See Figure 2

It is possible for large surges of current to pass through the electrical system of your vessel. If this surge of current were to reach the load in the circuit, the surge could burn it out or severely damage it. It can also overload the wiring, causing the harness to get hot and melt the insulation. To prevent this, fuses, circuit breakers and/or fusible links are connected into the supply wires of the electrical system. These items are nothing more than a built-in weak spot in the system. When an abnormal amount of current flows through the system, these protective devices work as follows to protect the circuit:

- Fuse—when an excessive electrical current passes through a fuse, the fuse "blows" (the conductor melts) and opens the circuit, preventing the passage of current.
- Circuit Breaker—a circuit breaker is basically a self-repairing fuse. It will open the circuit in the same fashion as a fuse, but when the surge subsides, the circuit breaker can be reset and does not need replacement.
- Fusible Link—a fusible link (fuse link or main link) is a short length of special, high temperature insulated wire that acts as a fuse. When an excessive electrical current passes through a fusible link, the thin gauge wire inside the link melts, creating an intentional open to protect the circuit. To repair the circuit, the link must be replaced. Some newer type fusible links are housed in plug-in modules, which are simply replaced like a fuse, while older type fusible links must be cut and spliced if they melt. Since this link is very early in the electrical path, it's the first place to look if nothing on the vessel works, yet the battery seems to be charged and is properly connected.

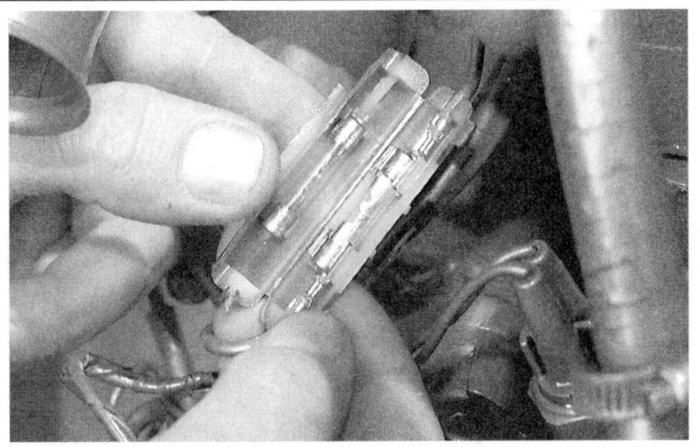

04975P11

Fig. 2 Fuses protect the vessel's electrical system from abnormally high amounts of current flow

CAUTION

Always replace fuses, circuit breakers and fusible links with identically rated components. Under no circumstances should a component of higher or lower amperage rating be substituted.

SWITCHES & RELAYS

See Figure 3

Switches are used in electrical circuits to control the passage of current. The most common use is to open and close circuits between the battery and the various electric devices in the system. Switches are rated according to the amount of amperage they can handle. If a sufficient amperage rated switch is not used in a circuit, the switch could overload and cause damage.

Some electrical components which require a large amount of current to operate use a special switch called a relay. Since these circuits carry a large amount of current, the thickness of the wire in the circuit is also greater. If this large wire were connected from the load to the control switch, the switch would have to carry the high amperage load and the space needed for wiring in the vessel would be twice as big to accommodate the increased size of the wiring harness. To prevent these problems, a relay is used.

Relays are composed of a coil and a set of contacts. When the coil has a current passed though it, a magnetic field is formed and this field causes the contacts to move together, completing the circuit. Most relays are normally open, preventing current from passing through the circuit, but they can take any electrical form depending on the job they are intended to do. Relays can be considered "remote control switches." They allow a smaller current to operate devices that require higher amperages. When a small current operates the coil, a larger current is allowed to pass by the contacts. Some common circuits which may use relays are horns, lights, starter, electric fuel pumps and other high draw circuits.

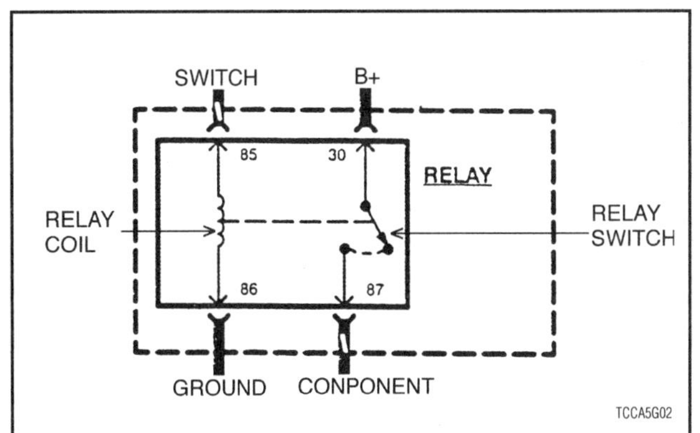

Fig. 3 Relays are composed of a coil and a switch. These two components are linked together so that when one operates, the other operates at the same time. The large wires in the circuit are connected from the battery to one side of the relay switch (B+) and from the opposite side of the relay switch to the load (component). Smaller wires are connected from the relay coil to the control switch for the circuit and from the opposite side of the relay coil to ground

LOAD

Every electrical circuit must include a "load" (something to use the electricity coming from the source). Without this load, the battery would attempt to deliver its entire power supply from one pole to another. This is called a "short circuit". All this electricity would take a short cut to ground and cause a great amount of damage to other components in the circuit by developing a tremendous amount of heat. This condition could develop sufficient heat to melt the insulation on all the surrounding wires and reduce a multiple wire cable to a lump of plastic and copper.

WIRING & HARNESSES

The average vessel contains miles of wiring, with hundreds of individual connections. To protect the many wires from damage and to keep them from becoming a confusing tangle, they are organized into bundles, enclosed in plastic or taped together and called wiring harnesses. Different harnesses serve different parts of the vessel. Individual wires are color coded to help trace them through a harness where sections are hidden from view.

Marine wiring or circuit conductors can be either single strand wire, multi-strand wire or printed circuitry. Single strand wire has a solid metal core and is usually used inside such components as alternators, motors, relays and other devices. Multi-strand wire has a core made of many small strands of wire twisted together into a single conductor. Most of the wiring in a marine electrical system is made up of multi-strand wire, either as a single conductor or grouped together in a harness. All wiring is color coded on the insulator, either as a solid color or as a colored wire with an identification stripe. A printed circuit is a thin film of copper or other conductor that is printed on an insulator backing. Occasionally, a printed circuit is sandwiched between two sheets of plastic for more protection and flexibility. A complete printed circuit, consisting of conductors, insulating material and connectors is called a printed circuit board. Printed circuitry is used in place of individual wires or harnesses in places where space is limited, such as behind instrument panels.

Since marine electrical systems are very sensitive to changes in resistance, the selection of properly sized wires is critical when systems are repaired. A loose or corroded connection or a replacement wire that is too small for the circuit will add extra resistance and an additional voltage drop to the circuit.

The wire gauge number is an expression of the cross-section area of the conductor. Vessels from countries that use the metric system will typically describe the wire size as its cross-sectional area in square millimeters. In this method, the larger the wire, the greater the number. Another common system for expressing wire size is the American Wire Gauge (AWG) system. As gauge number increases, area decreases and the wire becomes smaller. An 18 gauge wire is smaller than a 4 gauge wire. A wire with a higher gauge number will carry less current than a wire with a lower gauge number. Gauge wire size refers to the size of the strands of the conductor, not the size of the complete wire with insulator. It is possible, therefore, to have two wires of the same gauge with different diameters because one may have thicker insulation than the other.

It is essential to understand how a circuit works before trying to figure out why it doesn't. An electrical schematic shows the electrical current paths when a circuit is operating properly. Schematics break the entire electrical system down into individual circuits. In a schematic, usually no attempt is made to represent wiring and components as they physically appear on the vessel; switches and other components are shown as simply as possible. Face views of harness connectors show the cavity or terminal locations in all multi-pin connectors to help locate test points.

CONNECTORS

➧ **See Figures 4, 5 and 6**

Three types of connectors are commonly used in marine applications—weatherproof, molded and hard shell.

- Weatherproof—these connectors are most commonly used where the connector is exposed to the elements. Terminals are protected against moisture and dirt by sealing rings which provide a weather tight seal. All repairs require the use of a special terminal and the tool required to service it. Unlike standard blade type terminals, these weatherproof terminals cannot be straightened once they are bent. Make certain that the connectors are properly seated and all of the sealing rings are in place when connecting leads.
- Molded—these connectors require complete replacement of the connector if found to be defective. This means splicing a new connector assembly into the harness. All splices should be soldered to insure proper contact. Use care when probing the connections or replacing terminals in them, as it is possible to create a short circuit between opposite terminals. If this happens to the wrong terminal pair, it is possible to damage certain components. Always use jumper wires between connectors for circuit checking and NEVER probe through weatherproof seals.
- Hard Shell—unlike molded connectors, the terminal contacts in hard-shell connectors can be replaced. Replacement usually involves the use of a special terminal removal tool that depresses the locking tangs (barbs) on the connector terminal and allows the connector to be removed from the rear of the shell. The connector shell should be replaced if it shows any evidence of burning, melting, cracks, or breaks. Replace individual terminals that are burnt, corroded, distorted or loose.

Test Equipment

Pinpointing the exact cause of trouble in an electrical circuit is most times accomplished by the use of special test equipment. The following sections describe different types of commonly used test equipment and briefly explain how to use them in diagnosis. In addition to the information covered below, the tool manufacturer's instruction manual (provided with most tools) should be read and clearly understood before attempting any test procedures.

JUMPER WIRES

**** CAUTION**

Never use jumper wires made from a thinner gauge wire than the circuit being tested. If the jumper wire is of too small a gauge, it may overheat and possibly melt. Never use jumpers to bypass high resistance loads in a circuit. Bypassing resistances, in effect, creates a short circuit. This may, in turn, cause damage and fire. Jumper wires should only be used to bypass lengths of wire or to simulate switches.

TCCA6P03

Fig. 4 Hard shell (left) and weatherproof (right) connectors have replaceable terminals

TCCA6P04

Fig. 5 Weatherproof connectors are most commonly used in the engine compartment or where the connector is exposed to the elements

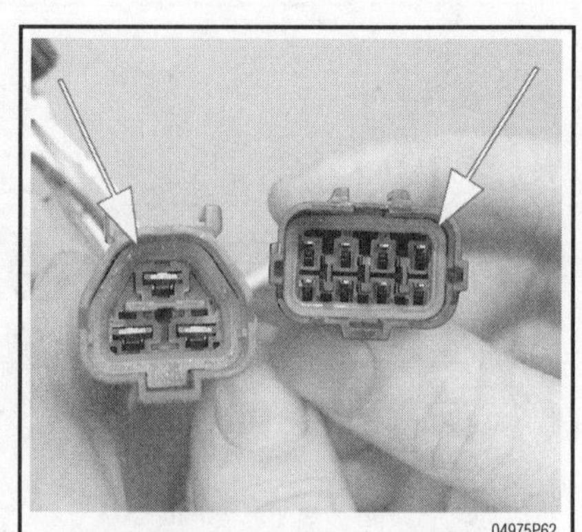

04975P62

Fig. 6 The seals on weatherproof connectors must be kept in good condition to prevent the terminals from corroding

Jumper wires are simple, yet extremely valuable, pieces of test equipment. They are basically test wires which are used to bypass sections of a circuit. Although jumper wires can be purchased, they are usually fabricated from lengths of standard marine wire and whatever type of connector (alligator clip, spade connector or pin connector) that is required for the particular application being tested. In cramped, hard-to-reach areas, it is advisable to have insulated boots over the jumper wire terminals in order to prevent accidental grounding. It is also advisable to include a standard marine fuse in any jumper wire. This is commonly referred to as a "fused jumper". By inserting an in-line fuse holder between a set of test leads, a fused jumper wire can be used for bypassing open circuits. Use a 5 amp fuse to provide protection against voltage spikes.

Jumper wires are used primarily to locate open electrical circuits, on either the ground (-) side of the circuit or on the power (+) side. If an electrical component fails to operate, connect the jumper wire between the component and a good ground. If the component operates only with the jumper installed, the ground circuit is open. If the ground circuit is good, but the component does not operate, the circuit between the power feed and component may be open. By moving the jumper wire successively back from the component toward the power source, you can isolate the area of the circuit where the open is located. When the component stops functioning, or the power is cut off, the open is in the segment of wire between the jumper and the point previously tested.

You can sometimes connect the jumper wire directly from the battery to the "hot" terminal of the component, but first make sure the component uses 12 volts in operation. Some electrical components, such as fuel injectors or sensors, are designed to operate on about 4 to 5 volts, and running 12 volts directly to these components will cause damage.

TEST LIGHTS

See Figure 7

The test light is used to check circuits and components while electrical current is flowing through them. It is used for voltage and ground tests. To use a 12 volt test light, connect the ground clip to a good ground and probe wherever necessary with the pick. The test light will illuminate when voltage is detected. This does not necessarily mean that 12 volts (or any particular amount of voltage) is present; it only means that some voltage is present. It is advisable before using the test light to touch its ground clip and probe across the battery posts or terminals to make sure the light is operating properly.

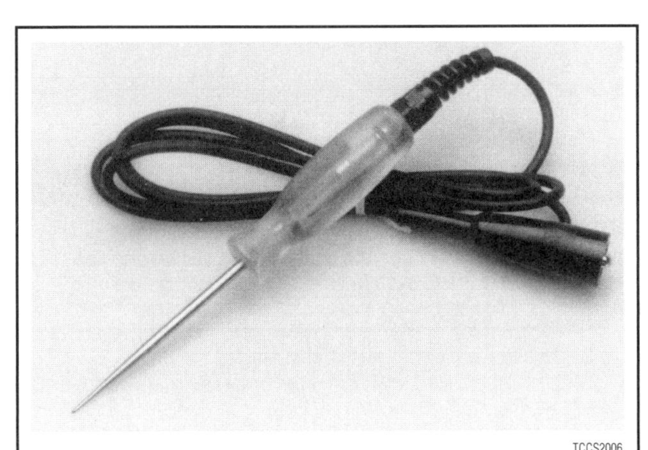

TCCS2006

Fig. 7 A 12 volt test light is used to detect the presence of voltage in a circuit

WARNING

Do not use a test light to probe electronic ignition, spark plug or coil wires. Never use a pick-type test light to probe wiring on electronically controlled systems unless specifically instructed to do so. Any wire insulation that is pierced by the test light probe should be taped and sealed with silicone after testing.

Like the jumper wire, the 12 volt test light is used to isolate opens in circuits. But, whereas the jumper wire is used to bypass the open to operate the load, the 12 volt test light is used to locate the presence of voltage in a circuit. If the test light illuminates, there is power up to that point in the circuit; if the test light does not illuminate, there is an open circuit (no power). Move the test light in successive steps back toward the power source until the light in the handle illuminates. The open is between the probe and a point which was previously probed.

The self-powered test light is similar in design to the 12 volt test light, but contains a 1.5 volt penlight battery in the handle. It is most often used in place of a multimeter to check for open or short circuits when power is isolated from the circuit (continuity test).

The battery in a self-powered test light does not provide much current. A weak battery may not provide enough power to illuminate the test light even when a complete circuit is made (especially if there is high resistance in the circuit). Always make sure that the test battery is strong. To check the battery, briefly touch the ground clip to the probe; if the light glows brightly, the battery is strong enough for testing.

A self-powered test light should not be used on any electronically controlled system or component. The small amount of electricity transmitted by the test light is enough to damage many electronic marine components.

MULTIMETERS

See Figure 8

Multimeters are an extremely useful tool for troubleshooting electrical problems. They can be purchased in either analog or digital form and have a price range to suit any budget. A multimeter is a voltmeter, ammeter and ohmmeter (along with other features) combined into one instrument. It is often used when testing solid state circuits because of its high input impedance (usually 10 megaohms or more). A brief description of the multimeter main test functions follows:

- Voltmeter—the voltmeter is used to measure voltage at any point in a circuit, or to measure the voltage drop across any part of a circuit. Voltmeters usually have various scales and a selector switch to allow the reading of different voltage ranges. The voltmeter has a positive and a negative lead. To avoid damage to the meter, always connect the negative lead to the negative (-) side of the circuit (to ground or nearest the ground side of the circuit) and connect the positive lead to the positive (+) side of the circuit (to the power source or the nearest power source). Note that the negative voltmeter lead will always be black and that the positive voltmeter will always be some color other than black (usually red).
- Ohmmeter—the ohmmeter is designed to read resistance (measured in ohms) in a circuit or component. Most ohmmeters will have a selector switch which permits the measurement of different ranges of resistance (usually the selector switch allows the multiplication of the meter reading by 10, 100, 1,000 and 10,000). Some ohmmeters are "auto-ranging" which means the meter itself will determine which scale to use. Since the meters are powered by an internal battery, the ohmmeter can be used like a self-powered test light. When the ohm-

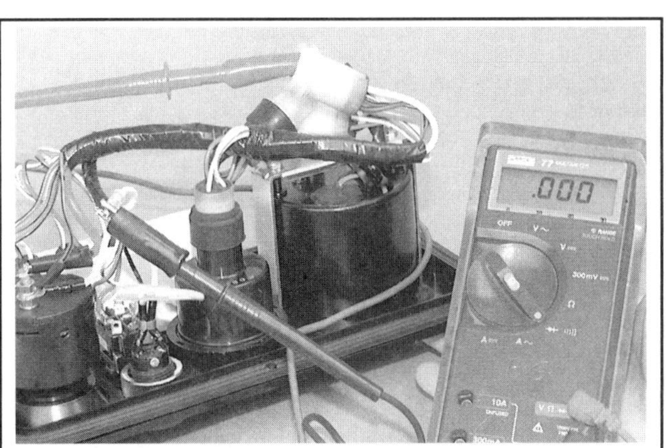

04975P60

Fig. 8 Multimeters are essential for diagnosing faulty wires, switches and other electrical components

meter is connected, current from the ohmmeter flows through the circuit or component being tested. Since the ohmmeter's internal resistance and voltage are known values, the amount of current flow through the meter depends on the resistance of the circuit or component being tested. The ohmmeter can also be used to perform a continuity test for suspected open circuits. In using the meter for making continuity checks, do not be concerned with the actual resistance readings. Zero resistance, or any ohm reading, indicates continuity in the circuit. Infinite resistance indicates an opening in the circuit. A high resistance reading where there should be none indicates a problem in the circuit. Checks for short circuits are made in the same manner as checks for open circuits, except that the circuit must be isolated from both power and normal ground. Infinite resistance indicates no continuity, while zero resistance indicates a dead short.

✱✱ WARNING

Never use an ohmmeter to check the resistance of a component or wire while there is voltage applied to the circuit.

- Ammeter—an ammeter measures the amount of current flowing through a circuit in units called amperes or amps. At normal operating voltage, most circuits have a characteristic amount of amperes, called "current draw" which can be measured using an ammeter. By referring to a specified current draw rating, then measuring the amperes and comparing the two values, one can determine what is happening within the circuit to aid in diagnosis. An open circuit, for example, will not allow any current to flow, so the ammeter reading will be zero. A damaged component or circuit will have an increased current draw, so the reading will be high. The ammeter is always connected in series with the circuit being tested. All of the current that normally flows through the circuit must also flow through the ammeter; if there is any other path for the current to follow, the ammeter reading will not be accurate. The ammeter itself has very little resistance to current flow and, therefore, will not affect the circuit, but it will measure current draw only when the circuit is closed and electricity is flowing. Excessive current draw can blow fuses and drain the battery, while a reduced current draw can cause motors to run slowly, lights to dim and other components to not operate properly.

Troubleshooting Electrical Systems

When diagnosing a specific problem, organized troubleshooting is a must. The complexity of a modern marine vessel demands that you approach any problem in a logical, organized manner. There are certain troubleshooting techniques, however, which are standard:

- **Establish when the problem occurs**. Does the problem appear only under certain conditions? Were there any noises, odors or other unusual symptoms? Isolate the problem area. To do this, make some simple tests and observations, then eliminate the systems that are working properly. Check for obvious problems, such as broken wires and loose or dirty connections. Always check the obvious before assuming something complicated is the cause.
- **Test for problems systematically to determine the cause once the problem area is isolated**. Are all the components functioning properly? Is there power going to electrical switches and motors. Performing careful, systematic checks will often turn up most causes on the first inspection, without wasting time checking components that have little or no relationship to the problem.
- **Test all repairs after the work is done to make sure that the problem is fixed**. Some causes can be traced to more than one component, so a careful verification of repair work is important in order to pick up additional malfunctions that may cause a problem to reappear or a different problem to arise. A blown fuse, for example, is a simple problem that may require more than another fuse to repair. If you don't look for a problem that caused a fuse to blow, a shorted wire (for example) may go undetected.

Experience has shown that most problems tend to be the result of a fairly simple and obvious cause, such as loose or corroded connectors, bad grounds or damaged wire insulation which causes a short. This makes careful visual inspection of components during testing essential to quick and accurate troubleshooting.

Testing

VOLTAGE

This test determines voltage available from the battery and should be the first step in any electrical troubleshooting procedure after visual inspection. Many electrical problems, especially on electronically controlled systems, can be caused by a low state of charge in the battery. Excessive corrosion at the battery cable terminals can cause poor contact that will prevent proper charging and full battery current flow.

1. Set the voltmeter selector switch to the 20V position.
2. Connect the multimeter negative lead to the battery's negative (-) post or terminal and the positive lead to the battery's positive (+) post or terminal.
3. Turn the ignition switch **ON** to provide a load.
4. A well charged battery should register over 12 volts. If the meter reads below 11.5 volts, the battery power may be insufficient to operate the electrical system properly.

VOLTAGE DROP

When current flows through a load, the voltage beyond the load drops. This voltage drop is due to the resistance created by the load and also by small resistances created by corrosion at the connectors and damaged insulation on the wires. The maximum allowable voltage drop under load is critical, especially if there is more than one load in the circuit, since all voltage drops are cumulative.

1. Set the voltmeter selector switch to the 20 volt position.
2. Connect the multimeter negative lead to a good ground.
3. Operate the circuit and check the voltage prior to the first component (load).
4. There should be little or no voltage drop in the circuit prior to the first component. If a voltage drop exists, the wire or connectors in the circuit are suspect.
5. While operating the first component in the circuit, probe the ground side of the component with the positive meter lead and observe the voltage readings. A small voltage drop should be noticed. This voltage drop is caused by the resistance of the component.
6. Repeat the test for each component (load) down the circuit.
7. If a large voltage drop is noticed, the preceding component, wire or connector is suspect.

RESISTANCE

✱✱ WARNING

Never use an ohmmeter with power applied to the circuit. The ohmmeter is designed to operate on its own power supply. The normal 12 volt electrical system voltage could damage the meter!

1. Isolate the circuit from the vessel's power source.
2. Ensure that the ignition key is **OFF** when disconnecting any components or the battery.
3. Where necessary, also isolate at least one side of the circuit to be checked, in order to avoid reading parallel resistances. Parallel circuit resistances will always give a lower reading than the actual resistance of either of the branches.
4. Connect the meter leads to both sides of the circuit (wire or component) and read the actual measured ohms on the meter scale. Make sure the selector switch is set to the proper ohm scale for the circuit being tested, to avoid misreading the ohmmeter test value.

OPEN CIRCUITS

See Figure 9

This test already assumes the existence of an open in the circuit and it is used to help locate the open portion.

1. Isolate the circuit from power and ground.
2. Connect the self-powered test light or ohmmeter ground clip to the ground side of the circuit and probe sections of the circuit sequentially.
3. If the light is out or there is infinite resistance, the open is between the probe and the circuit ground.
4. If the light is on or the meter shows continuity, the open is between the probe and the end of the circuit toward the power source.

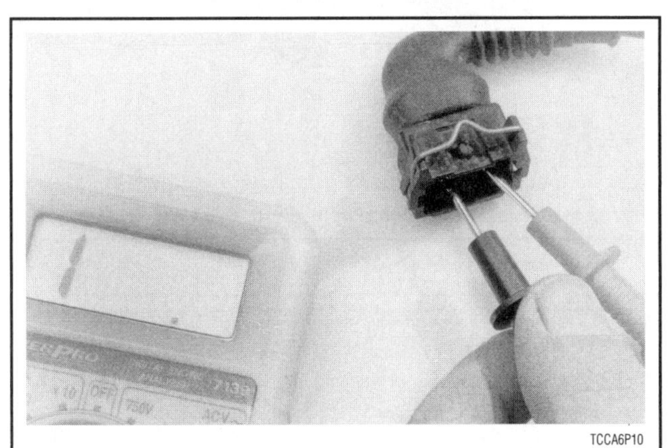

TCCA6P10

Fig. 9 The infinite reading on this multimeter (1 .) indicates that the circuit is open

SHORT CIRCUITS

Never use a self-powered test light to perform checks for opens or shorts when power is applied to the circuit under test. The test light can be damaged by outside power.

1. Isolate the circuit from power and ground.
2. Connect the self-powered test light or ohmmeter ground clip to a good ground and probe any easy-to-reach point in the circuit.
3. If the light comes on or there is continuity, there is a short somewhere in the circuit.
4. To isolate the short, probe a test point at either end of the isolated circuit (the light should be on or the meter should indicate continuity).
5. Leave the test light probe engaged and sequentially open connectors or switches, remove parts, etc. until the light goes out or continuity is broken.
6. When the light goes out, the short is between the last two circuit components which were opened.

Wire And Connector Repair

Almost anyone can replace damaged wires, as long as the proper tools and parts are available. Wire and terminals are available to fit almost any need. Even the specialized weatherproof, molded and hard shell connectors are now available from aftermarket suppliers.

Be sure the ends of all the wires are fitted with the proper terminal hardware and connectors. Wrapping a wire around a stud is never a permanent solution and will only cause trouble later. Replace wires one at a time to avoid confusion. Always route wires in the same manner of the manufacturer.

When replacing connections, make absolutely certain that the connectors are certified for marine use. Automotive wire connectors may not meet United States Coast Guard (USCG) specifications.

If connector repair is necessary, only attempt it if you have the proper tools. Weatherproof and hard shell connectors require special tools to release the pins inside the connector. Attempting to repair these connectors with conventional hand tools will damage them.

Electrical System Precautions

- Wear safety glasses when working on or near the battery.
- Don't wear a watch with a metal band when servicing the battery or starter. Serious burns can result if the band completes the circuit between the positive battery terminal and ground.
- Be absolutely sure of the polarity of a booster battery before making connections. Connect the cables positive-to-positive, and negative-to-negative. Connect positive cables first, and then make the last connection to ground on the body of the booster vessel so that arcing cannot ignite hydrogen gas that may have accumulated near the battery. Even momentary connection of a booster battery with the polarity reversed will damage alternator diodes.
- Disconnect both vessel battery cables before attempting to charge a battery.
- Never ground the alternator or generator output or battery terminal. Be cautious when using metal tools around a battery to avoid creating a short circuit between the terminals.
- When installing a battery, make sure that the positive and negative cables are not reversed.
- Always disconnect the battery (negative cable first) when charging.
- Never smoke or expose an open flame around the battery . Hydrogen gas accumulates near the battery and is highly explosive.

BREAKER POINTS IGNITION (MAGNETO IGNITION)

See Figures 10, 11, 12 and 13

All Honda outboard engines use a pointless electronic ignition system with the exception of early model BF75, BF8A and BF100 which use a breaker point type magneto.

This ignition system uses a mechanically switched, collapsing field to induce spark at the plug. A magnet moving by a coil produces current in the primary coil winding. The current in the primary winding creates a magnetic field. When the points are closed the current goes to ground. As the breaker points open the primary magnetic field collapses across the secondary field. This induces (transforms) a high voltage potential in the secondary coil winding. This high voltage current travels to the spark plug and jumps the gap.

The breaker point ignition system contains a condenser that works like a sponge in the circuit. Current that is flowing through the primary circuit tries to keep going. When the breaker point switch opens the current will arc over the widening gap. The condenser is wired in parallel with the points. The condenser absorbs some of the current flow as the points open. This reduces arc over and extends the life of the points.

The breaker point ignition consists of the point cover, camshaft pulley, spark advancer, contact breaker and condenser. All these components are installed underneath the camshaft pulley.

As the pole pieces of the magnet pass over the heels of the coil, a magnetic field is built up about the coil, causing a current to flow through the primary winding.

At the proper time, the breaker points are separated by action of a cam designed into the collar of the camshaft and the primary circuit is broken. When the circuit is broken, the flow of primary current stops and causes the magnetic field about the coil to break down instantly. At this precise moment, an electrical current of extremely high voltage is induced in the fine secondary windings of the coil. This high voltage is conducted to the spark plug where it jumps the gap between the points of the plug to ignite the compressed charge of air-fuel mixture in the cylinder.

The breaker points must be aligned accurately to provide the best contact surface. This is the only way to assure maximum contact area between the point surfaces; accurate setting of the point gap; proper synchronization; and satisfactory point life. If the points are not aligned properly, the result will be premature wear or pitting. This type of damage may change the cam angle, although the actual distance will remain the same.

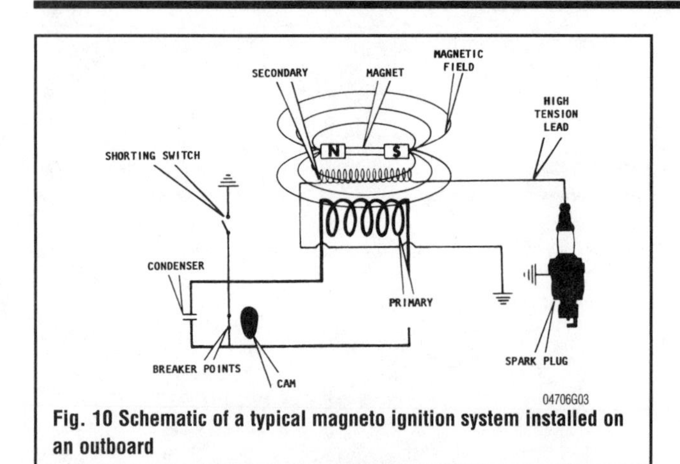

Fig. 10 Schematic of a typical magneto ignition system installed on an outboard

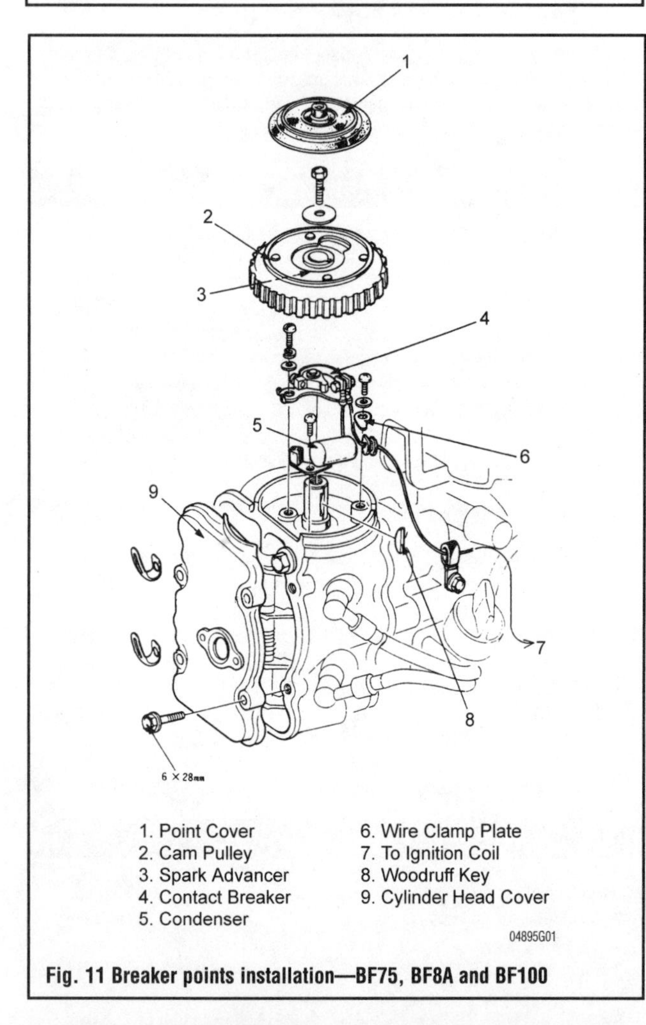

Fig. 11 Breaker points installation—BF75, BF8A and BF100

Magnetos installed on outboard engines will usually operate over extremely long periods of time without requiring adjustment or repair. However, if ignition system problems are encountered, and the usual corrective actions such as replacement of spark plugs does not correct the problem, the magneto output should be checked to determine if the unit is functioning properly.

Unfortunately, the breaker point set of the contact breaker point ignition system is located under the camshaft pulley. This location requires the hand rewind starter to be removed, and the camshaft pulley to be pulled in order to replace the point set.

However, the manufacturer made provisions for the point gap to be checked with a feeler gauge after the point cover has been removed.

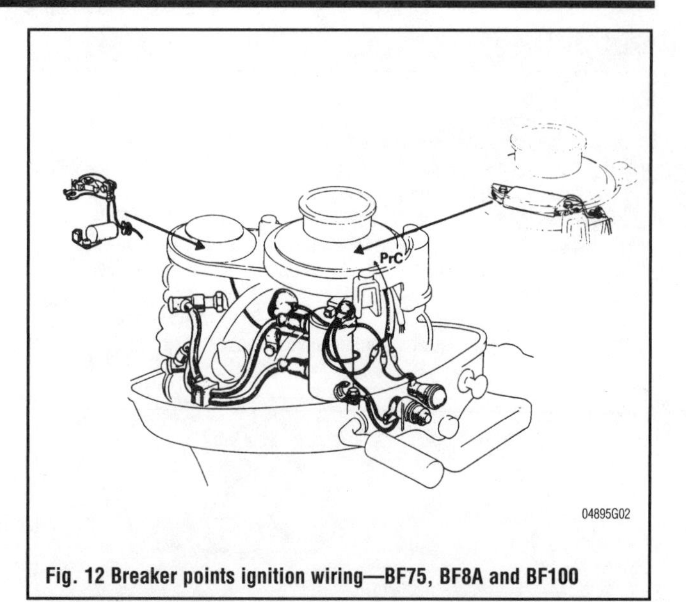

Fig. 12 Breaker points ignition wiring—BF75, BF8A and BF100

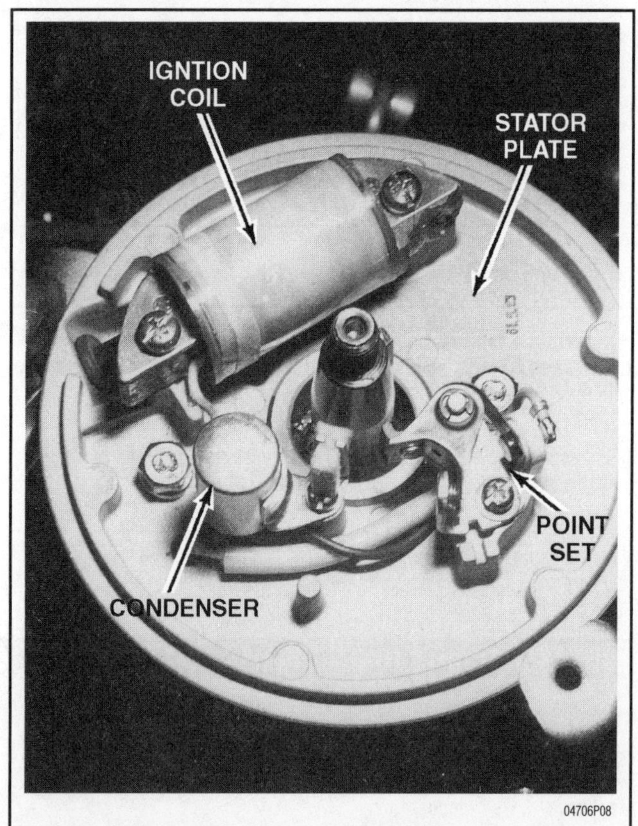

Fig. 13 After the flywheel is removed, components on the stator plate may be easily serviced

System Testing

SPARK CHECK

Perform a spark test if you suspect the ignition system of not working properly.

⁂ WARNING

When checking the spark, make sure there is no fuel on either the engine or the spark plug. Also keep your hands away from high voltage electrical components.

1. Remove the spark plug and ground the plug electrode to the engine.
2. Pull the recoil starter and check for spark at the plug.

If there is a good spark at the plug, the ignition system should be performing properly. If there is no spark, precede to the next step in troubleshooting the ignition system problem.

Breaker Points

GAP CHECK

See Figure 14

➡The point gap may be checked using a ohmmeter or timing tester (buzz box) without removing the hand rewind starter or the camshaft pulley. This procedure is not necessary if the point gap has already been checked using a feeler gauge.

1. Connect a timing tester or ohmmeter to the primary side wire of the ignition coil and a good engine ground.
2. Remove the points cover on the camshaft pulley.
3. Make sure to align the "F" mark on the flywheel with the index mark on the starter case. The contact breaker points should just be starting to open at 15°BTDC.
4. Loosen the breaker plate locking screw and move the breaker plate to achieve the correct timing. Retighten the locking screw.
5. Rotate the flywheel one full turn and recheck the ignition timing for the other cylinder. If the timing is not correct, readjust so that the timing is correct for both cylinders.

When the points are open, creating an open in the circuit, the meter will register an infinite resistance—an air gap. Therefore, the meter needle will swing either to the far right or to the far left, depending on the scale of the ohmmeter. If you are using a digital multimeter, read your owners manual to find out what type of infinity reading it will show on the display.

6. Set the point gap with a feeler gauge. Standard point gap is 0.012–1.016 in. (0.3–0.4 mm).
7. If the point gap is incorrect, the gap must be adjusted or the point set replaced.

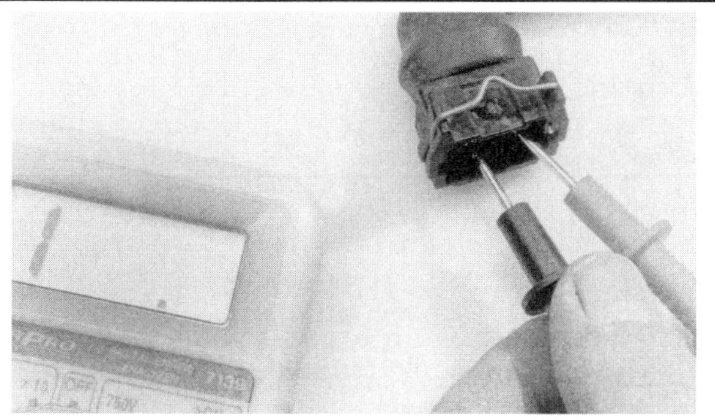

TCCA6P10

Fig. 14 The infinite reading on this multimeter (1 .) indicates an open circuit

REMOVAL & INSTALLATION

See Figures 15 and 16

1. Remove the engine cover.
2. With a flywheel pulley holder (No. 07925–893000) or a commonly available strap wrench, hold the flywheel and loosen the retaining nut.
3. With the flywheel puller (No. 07935–8050002) remove the flywheel.

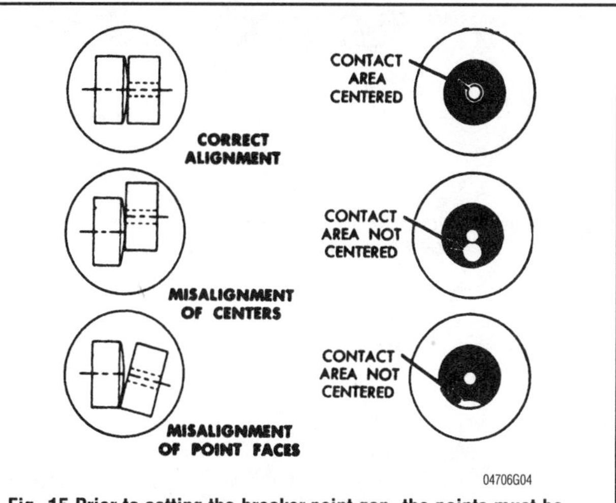

04706G04

Fig. 15 Prior to setting the breaker point gap, the points must be properly aligned

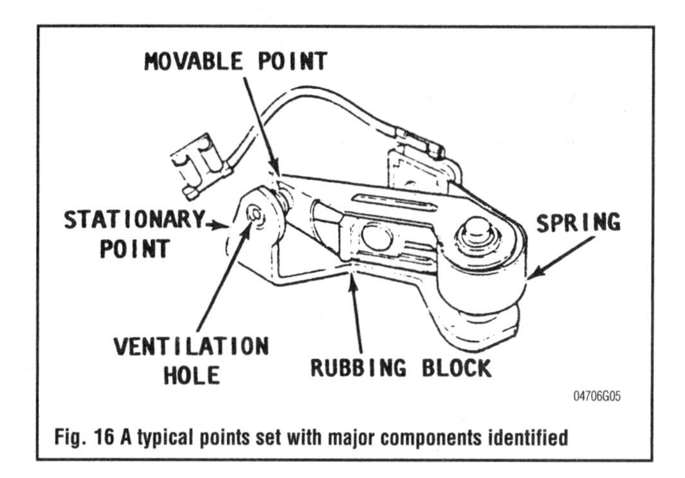

04706G05

Fig. 16 A typical points set with major components identified

4. Remove the timing belt from the cam pulley with your fingers. DO NOT pry the belt off with a screwdriver or other type of tool.
5. Remove the point cover.
6. Loosen the camshaft pulley bolt and remove the camshaft pulley. DO NOT hammer or bang on the pulley when removing.
7. Disconnet the wire at the primary side of the ignition coil.
8. Remove the screws and lift off the breaker point assembly and condenser.

To install:

9. Install the breaker point assembly and condenser back on the camshaft. Make sure that the Woodruff key is in place and seated properly on the shaft.

➡Make sure to note the installation direction on the wire clamp plate. If it is not installed properly, the breaker will be grounded and the engine will not start.

10. Align the keyway on the cam pulley onto the camshaft and install the camshaft pulley.
11. Install the retaining bolt and washer, and it to 7 ft. lbs. (100 kg-cm).
12. Align the "T" mark on the flywheel and the punch mark on the pulley to their respective alignment marks.

➡Prior to it's installation, check the timing belt for cracks or other damage and replace the belt if necessary.

13. Install the timing belt. Do not use force to install the belt, and always make sure that you can read the "HONDA" printed on the belt.
14. Align the keyway on the points cover with the camshaft and install the cover, making sure that it's seated correctly.
15. Connect the wire from the breaker points assembly to the ignition coil primary side.
16. Reinstall the engine cover.

Condenser

DESCRIPTION & OPERATION

See Figure 17

In simple terms, a condenser is composed of two sheets of tin or aluminum foil laid one on top of the other, but separated by a sheet of insulating material such as waxed paper, etc. The sheets are rolled into a cylinder to conserve space and then inserted into a metal case for protection and to permit easy assembling.

The purpose of the condenser is to prevent excessive arcing across the points and to extend their useful life. When the flow of primary current is brought to a sudden stop by the opening of the points, the magnetic field in the primary windings collapses instantly, and is not allowed to fade away, which would happen if the points were allowed to arc.

The condenser stores the electricity that would have arced across the points and discharges that electricity when the points close again. This discharge is in the opposite direction to the original flow, and tends to smooth out the current. The more quickly the primary field collapses, the higher the voltage produced in the secondary windings and delivered to the spark plugs. In this way, the condenser (in the primary circuit), affects the voltage (in the secondary circuit) at the spark plugs.

Modern condensers seldom cause problems, therefore, it is not necessary to install a new one each time the points are replaced. However, if the points show evidence of arcing, the condenser may be at fault and should be replaced. A faulty condenser may not be detected without the use of special test equipment. Testing will reveal any defects in the condenser, but will not predict the useful life left in the unit.

The modest cost of a new condenser justifies its purchase and installation to eliminate this item as a source of trouble.

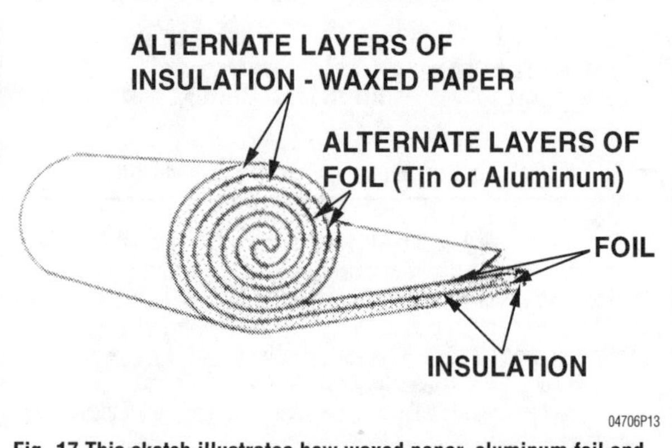

Fig. 17 This sketch illustrates how waxed paper, aluminum foil and insulation are rolled in the manufacture of a typical condenser

TESTING

Discharge the condenser by grounding it prior to testing the capacity.

1. Place a tester lead on the condenser output wire and another on the condenser body. The meter should read 0.24 microfarads.
2. Replace the condenser if it is not within specification.

REMOVAL & INSTALLATION

1. Remove the engine cover.
2. With a flywheel pulley holder (No. 07925–893000) or a commonly available strap wrench, hold the flywheel and loosen the retaining nut.
3. With the flywheel puller (No. 07935–8050002) remove the flywheel.
4. Remove the timing belt from the cam pulley with your fingers. DO NOT pry the belt off with a screwdriver or other type of tool.
5. Remove the point cover.
6. Loosen the camshaft pulley bolt and remove the camshaft pulley. DO NOT hammer or bang on the pulley when removing.
7. Disconnet the wire at the primary side of the ignition coil.
8. Remove the screws and lift off the condenser.

To install:

9. Install the condenser back onto the breaker point assembly and install both back on the camshaft. Make sure that the Woodruff key is in place and seated properly on the shaft.

Make sure to note the installation direction on the wire clamp plate. If it is not installed properly, the breaker will be grounded and the engine will not start.

10. Align the keyway on the cam pulley onto the camshaft and install the camshaft pulley.
11. Install the retaining bolt and washer, and it to 7 ft. lbs. (100 kg-cm).
12. Align the "T" mark on the flywheel and the punch mark on the pulley to their respective alignment marks.

Prior to it's installation, check the timing belt for cracks or other damage and replace the belt if necessary.

13. Install the timing belt. Do not use force to install the belt, and always make sure that you can read the "HONDA" printed on the belt.
14. Align the keyway on the points cover with the camshaft and install the cover, making sure that it's seated correctly.
15. Connect the wire from the breaker points assembly to the ignition coil primary side.
16. Clean the crankshaft taper and insert the Woodruff key into the keyway, making sure that the key is seated properly.
17. Install the flywheel onto the crankshaft taper. DO NOT hit the flywheel while installing it.
18. Torque down the flywheel retaining bolt to 47 ft. lbs. (650 kg-cm).
19. Reinstall the engine cover.

Charge Coil

DESCRIPTION & OPERATION

The main function of the charging system is to recharge the battery after starting the engine. The charging system also helps support the accessory loads once the engine is running. On many outboard engines, the charging system is a dealer installed option.

The charging system consists of the charge coil, magnets permanently installed in the flywheel, a voltage regulator/rectifier, a 5a fuse/fuse holder assembly, a battery and the wiring system connecting the entire system together.

The rotation of the flywheel magnets past the coil creates alternating current (A/C). This current is sent to the voltage regulator/rectifier and is converted in to the direct current that will charge the battery or run the appropriate accessories.

TESTING

The basic test for a charge coil is continuity. This measures the actual resistance from one end of the charge coil to the other. The correct specification for each coil can be found in the Ignition Testing Specifications chart.

1. Adjust the meter to ohms.
2. Connect the tester across the pulser leads and note the reading.
3. Compare the coil reading to the specifications. Resistance should measure 0.12–0.15 ohms.
4. If the reading is above or below the correct value and beyond the allowable deviation then it must be replaced.

Remember that temperature has an affect on resistance. Most resistance specifications are given assuming a temperature of 70° F.

REMOVAL & INSTALLATION

1. Remove the engine cover.
2. With a flywheel pulley holder (No. 07925–893000) or a commonly available strap wrench, hold the flywheel and loosen the retaining nut.
3. With the flywheel puller (No. 07935–8050002) remove the flywheel.

4. Disconnet the charge coil wires at the rectifier.
5. Remove the screws and lift off the charge coil.

To install:

6. Install the charge coil and tighten the screws.
7. Clean the crankshaft taper and insert the Woodruff key into the keyway, making sure that the key is seated in the keyway properly.
8. Install the flywheel onto the crankshaft taper. DO NOT hit the flywheel while installing it.
9. Torque down the flywheel retaining bolt to 47 ft. lbs. (650 kg-cm).
10. Reconnect the charge coil wires at the rectifier
11. Reinstall the engine cover.

Primary Coil

DESCRIPTION & OPERATION

As the flywheel rotates, magnets attached to the edge of the flywheel create a current that will flow through the closed breaker points into the ignition coil primary windings. This flow of current through the coil primary windings builds a very strong magnetic field. When the breaker cam opens the points, the field collapses, inducing high voltage in the coil secondary winding. This voltage is sent to the spark plug. The condenser will then absorb any residual current remaining in the primary windings while the breaker points are open. This eliminates arcing at the breaker points and in turn produces a stronger spark at the spark plug. The breaker points then close and the flywheel continues to rotate.

TESTING

Primary coil does not have to be removed from the engine to test.

1. Unplug the primary coil wires at the connectors from the ignition coil primary side and the D/C output plug.
2. Measure the resistance between the two wires with a ohmmeter. Resistance should measure 2.0 ohms.
3. If the primary coil does not measure to specification, replace the coil.

REMOVAL & INSTALLATION

1. Remove the engine cover.
2. With a flywheel pulley holder (No. 07925–893000) or a commonly available strap wrench, hold the flywheel and loosen the retaining nut.
3. With the flywheel puller (No. 07935–8050002) remove the flywheel.
4. Disconnet the primary coil wires at the ignition coil and D/C output plug.
5. Remove the screws and lift off the primary coil.

To install:

6. Install the primary coil and tighten the screws.

➡Make sure that the primary coil and wires do not interfere with flywheel movement.

7. Clean the crankshaft taper and insert the Woodruff key into the keyway, making sure that the key is seated in the keyway properly.
8. Install the flywheel onto the crankshaft taper. DO NOT hit the flywheel while installing it.
9. Torque down the flywheel retaining bolt to 47 ft. lbs. (650 kg-cm).
10. Reconnect the primary coil wires at the at the ignition coil and D/C output plug.
11. Reinstall the engine cover.

Ignition Coil

DESCRIPTION & OPERATION

The coil is the heart of the ignition system. Essentially, it is nothing more than a transformer which takes the relatively low voltage (12 volts) available from the primary coil and increases it to a point where it will fire the spark plug as much as 20,000 volts.

Once the voltage is discharged from the ignition coil the secondary circuit begins and only stretches from the ignition coil to the spark plugs via extremely large high tension leads. At the spark plug end, the voltage arcs in the form of a spark, across from the center electrode to the outer electrode, and then to ground via the spark plug threads. This completes the ignition circuit.

TESTING

1. Remove the spark plug caps
2. With an ohmmeter, measure the resistance between the spark plug wires. Resistance should measure 6.4K–9.6K ohms.
3. Measure the resistance between the primary terminal and the coil mounting lug (for ground).
4. Resistance should measure 0.46–0.66 ohms.

REMOVAL & INSTALLATION

1. Remove the engine cover.
2. Pull the spark plug wires off the plugs by the caps.
3. Unscrew the nut and remove the primary side leads on the ignition coil.
4. Unscrew the ignition coil retaining screws from the engine.

➡Make sure to not drop the nuts and the bottom spacer into the engine case.

5. Remove the ignition coil from the engine.

To install:

6. Making sure that the bottom spacer is in place, line up the ignition coil with the engine and insert the screws into the coil and tighten the screws securely.
7. Reinstall the primary side leads on the coil and tighten the nut securely.
8. Reinstall the spark plug wires on the plugs.
9. Reinstall the engine cover.

Breaker Points Ignition Overhaul

DISASSEMBLY

1. Disconnect the negative battery cable lead.
2. Remove the engine cover.
3. Disconnect the breaker point lead wire from the ignition coil.
4. Remove the recoil starter.
5. Remove the three retaining bolts holding the recoil starter to the flywheel.
6. Remove the recoil starter from the starter pulley.
7. Using pulley holder (No. 07925-8930000) or a commercially available strap wrench, hold the flywheel steady and loosen the flywheel retaining bolt and carefully remove the flywheel, making sure not to damage the magnets inside.
8. Remove the timing belt by hand. Do not use a screwdriver or other prying tool to remove the belt.
9. Remove the points cover.
10. Remove the camshaft pulley retaining bolt and lift off the camshaft pulley. Do not hammer on the pulley to remove it.
11. Remove the two retaining screws and lift off the breaker points and condenser.

CLEANING & INSPECTION

1. Check the surface of the point contacts. Pitting, corrosion or premature wear may indicate that the point contacts are not aligned.
2. Check the resistance across the contacts.

➡If the test indicates zero resistance, the points are serviceable. A slight resistance across the points will affect idle operation. A high resistance may cause the ignition to malfunction and loose spark. Therefore, if any resistance across the points is indicated, the point set should be replaced.

3. Discharge the capacitor before testing.
4. Check capacitor capacity: Standard reading is: 0.24 microfarads.
5. If the points are slightly pitted, fine emery cloth or a point file can be used to polish the point surfaces.

ASSEMBLY

1. Reinstall the breaker point assembly on the engine using the two retaining screws. Make sure the breaker plate is installed in the correct direction. If the plate is not installed properly, the breaker points will be grounded, making it impossible to start the engine.
2. Install the camshaft pulley on the camshaft.

■ **Make sure that the Woodruff key is installed and seated properly on the camshaft.**

3. Holding the pulley securely, install the pulley retaining bolt and it to 7 ft. lbs. (100 kg—cm).
4. Inspect the timing belt for wear or cracks and then install the timing belt. Align the "T" mark on the flywheel with the punch mark on the pulley. Always install the belt so that the word "HONDA" can be read.
5. Install the flywheel onto the crankshaft, making sure that the Woodruff key is installed and seated properly on the crankshaft.
6. Using the flywheel holder or strap wrench, tighten the flywheel retaining bolt to 47 ft. lbs. (650 kg—cm.

■ **Align the breaker points by bending or twisting the fixed contact point only.**

7. Reconnect the negative battery cable.
8. Set the breaker points.

CAPACITOR DISCHARGE IGNITION (CDI) SYSTEM

Description and Operation

◆ **See Figures 18 thru 26**

SINGLE-CYLINDER IGNITION

In its simplest form, a CDI ignition is composed of the following elements:

- Magneto
- Pulser coil
- Charge, or source coil
- Igniter (CDI) box
- Ignition coil
- Spark plug

Other components such as main switches, stop switches, or computer systems may be included, though, these items are not necessary for basic CDI operation.

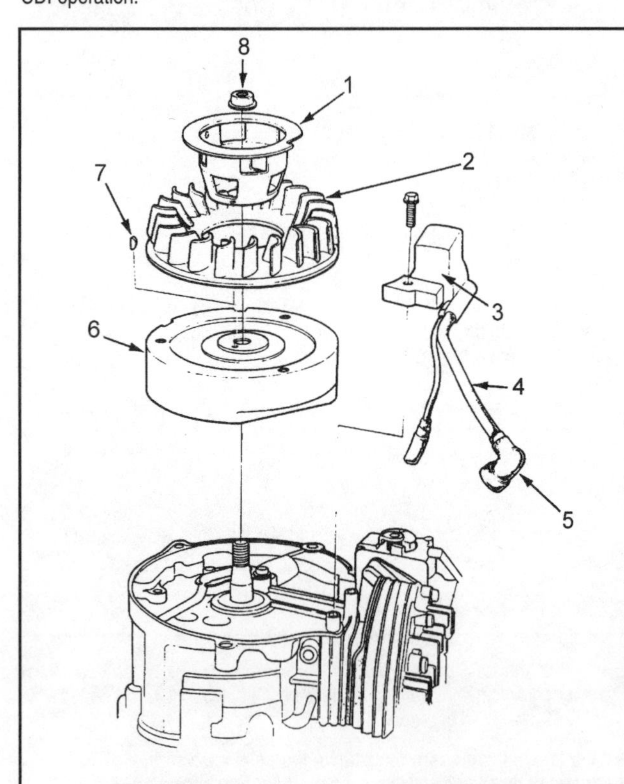

Fig. 18 The very simple CDI system used on BF2A, BF2D and BF20 models (Note the cooling fan is integrated with the flywheel on some models)

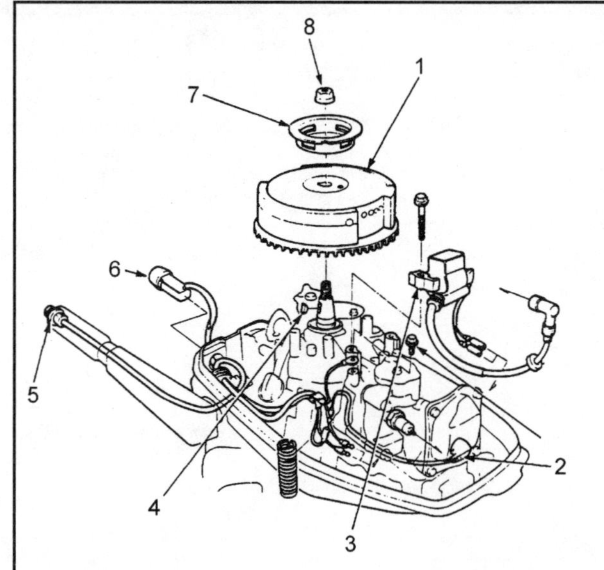

Fig. 19 The BF5A and BF50 CDI ignition configuration

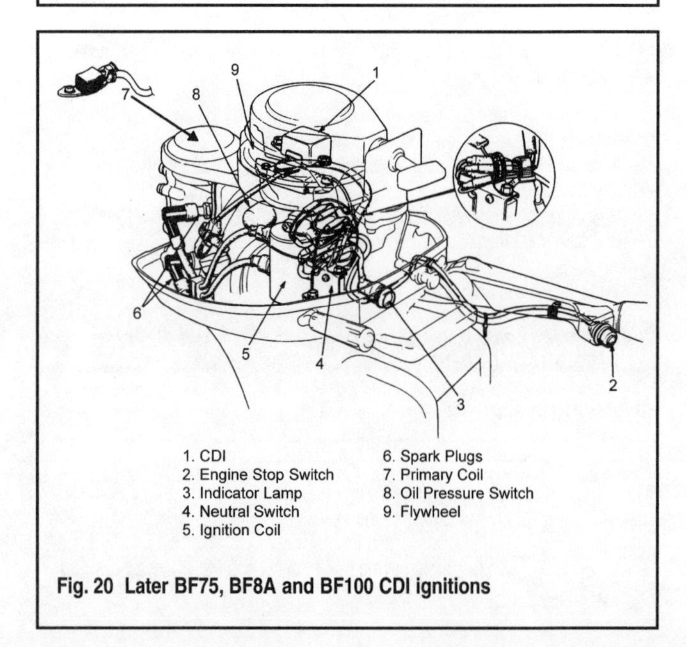

Fig. 20 Later BF75, BF8A and BF100 CDI ignitions

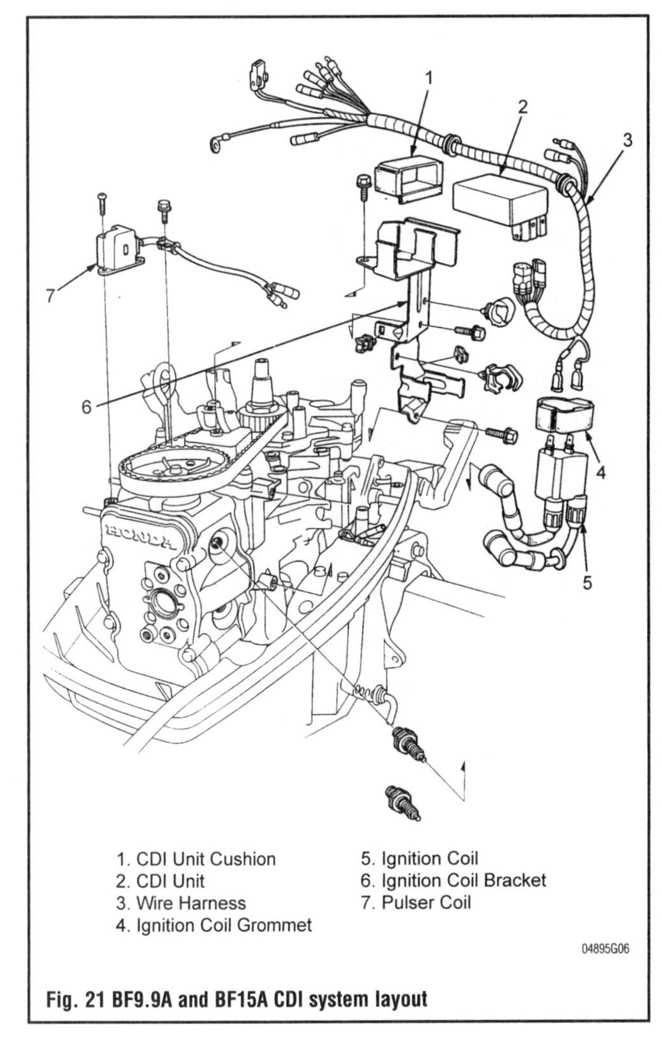

Fig. 21 BF9.9A and BF15A CDI system layout

To understand basic CDI operation, it is important to understand the basic theory of induction. Induction theory states that if we move a magnet (magnetic field) past a coil of wire(or the coil by the magnet), AC current will be generated in the coil.

The amount of current produced depends on several factors:

- How fast the magnet moves past the coil
- The size of the magnet(strength)
- How close the magnet is to the coil
- Number of turns of wire and the size of the windings

When the flywheel rotates, the electrical power generated at the exciter coil is rectified by the diode and charged into the ignition condenser. The thyristor is off at this time.

When the magnet on the camshaft pulley passes the pulser coil, the electric pulser coil signal is emitted by the magnetic force. This signal passes the gate circuit, turns on the thyristor, and discharges the electric charge from the condenser. When the discharged current flows through the ignition coil primary circuit, high voltage is generated in the secondary circuit and the spark plug sparks.

The spark advance is handled by electronic advance spark system, which advances the ignition timing when the gate circuit turns on the thyristor according to the engine speed to obtain high speed power.

The current produced in the charge coil goes to the CDI box. On the way in, it is converted to DC current by a diode. This DC current is stored in the capacitor located inside the box. As the charge coil produces current, the capacitor stores it.

At a specific time in the magneto's revolution, the magnets go past the pulser coil. The pulser coil is smaller than the charge coil so it has less current output. The current from the pulser also goes into the CDI box. This current signals the CDI box when to fire the capacitor (the pulser may be called a trigger coil for obvious reasons). The current from the capacitor flows out to the ignition coil and spark plug. The pulser acts much like the points in older ignitions systems.

When the pulser signal reaches the CDI box, all the electricity stored in the capacitor is released at once. This current flows through the ignition coil's primary windings.

The ignition coil is a step-up transformer. It turns the relatively low voltage entering the primary windings into high voltage at the secondary windings. This occurs due to a phenomena known as induction.

The high voltage generated in the secondary windings leaves the ignition coil and goes to the spark plug. The spark in turn ignites the air-fuel charge in the combustion chamber.

Once the complete cycle has occurred, the spinning magneto immediately starts the process over again.

Main switches, engine stop switches, and the like are usually connected on the wire in between the CDI box and the ignition coil. When the main switch or stop switch is turned to the OFF position, the switch is closed. This closed switch short-circuits the charge coil current to ground rather than sending it through the CDI box. With no charge coil current through the CDI box, there is no spark and the engine stops or, if the engine is not running, no spark is produced.

TWO & THREE-CYLINDER IGNITION

The Capacitor Discharge Ignition (CDI) system installed on the twin-cylinder Honda models use a semi-conductor switching element called an "SCR" (Silicone Controlled Rectifier) or more commonly as thyristor.

➡The three-cylinder engines are equipped with three ignition condensers, one voltage doubler condenser, and an independent ignition system is provided for each cylinder.

1. With the engine running, the flywheel magneto induces alternating current in the exciter coil. Positive half-cycle current flows through the diode an into the CDI unit to charge the capacitor.
2. When the advance rotor in the camshaft pulley turn to the ignition position, alternating (AC) current is induced in the pulser coil. The current flows through another diode and triggers the thyristor into a conductive state. This causes the capacitor to discharge into the ignition coil.
3. This momentary rush of discharging current induces a high voltage surge in the ignition coil secondary winding and causes the spark plugs to spark.

FOUR-CYLINDER CARBURETED IGNITION

The BF75A and BF90A use an electronic-advance, DC-CDI system which is operated by battery voltage. These are very reliable ignition systems and have very good spark from low speeds and are maintenance-free.

1. When the ignition switch is turned on, the current flows from the battery to the capacitor through a DC-DC converter which then increases voltage.
2. To start and run the engine, an electrical signal is generated at the pulser coil, that signal then makes the thyristor turn on through the gate circuit. This results in the capacitor discharging. When the discharged electrical current flows through the ignition coil primary circuit, high voltage is generated in the secondary circuit which then produces a spark at the spark plugs.
3. For high speed operation, an electronic spark advance system advances the ignition timing when the gate circuit turns on the thyristor.
4. When the ignition is turned off, or the engine stop switch is turned on, no spark is produced at the spark plugs and the engine stops.

➡If the engine is shut down with the engine stop switch, the ignition switch must be turned off to prevent the battery from discharging.

FOUR-CYLINDER EFI IGNITION

The BF115A and BF130A use a completely transistorized ignition with the igniter unit being part of the Engine Control Module (ECM). The ignition timing control system first determines basic ignition timing based on an engine speed signal sent by the pulser coil and intake manifold vacuum signal send by the MAP sensor, then alters the basic timing with inputs from the throttle position sensor (TPS), the engine temperature compensation sensor (ECT) and the intake air temperature compensation sensor (IAT).

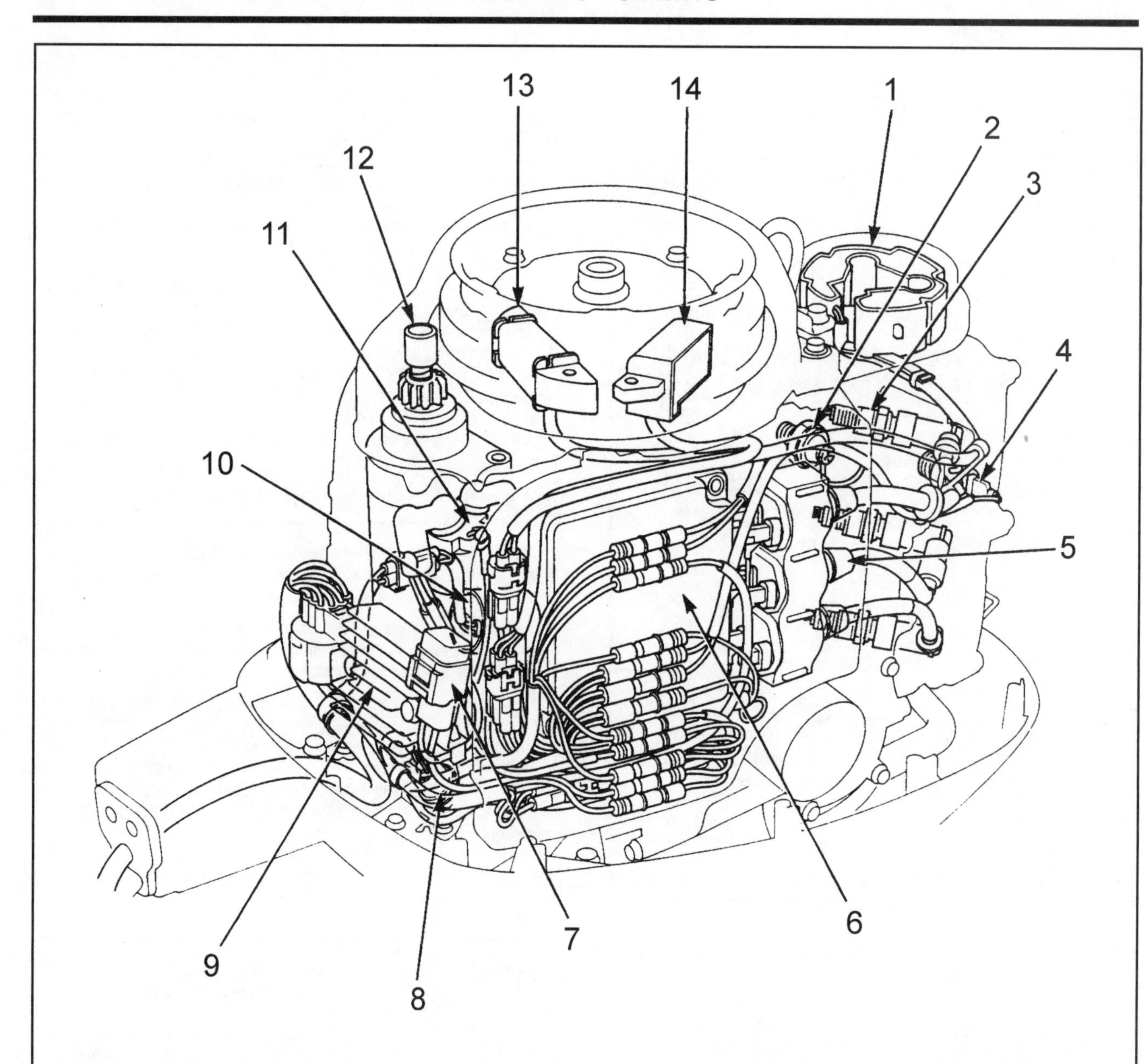

1. Pulser Coil
2. Oil Pressure Switch
3. Spark Plug
4. Thermo Switch
5. Ignition Coil
6. CDI Unit
7. Fuse Case [Electric Starter Type Only]
8. Neutral Switch [Electric Starter Type Only]
9. Regulator/Rectifier [Electric Starter Type Only]
10. Choke Solenoid [Remote Control Type Only]
11. Starter Solenoid [Electric Starter Type Only]
12. Electric Starter [Electric Starter Type Only]
13. Charge Coil [Electric Starter Type Only]
14. Exciter Coil

04895G07

Fig. 22 The BF25A and BF30A use a more densely packaged CDI system

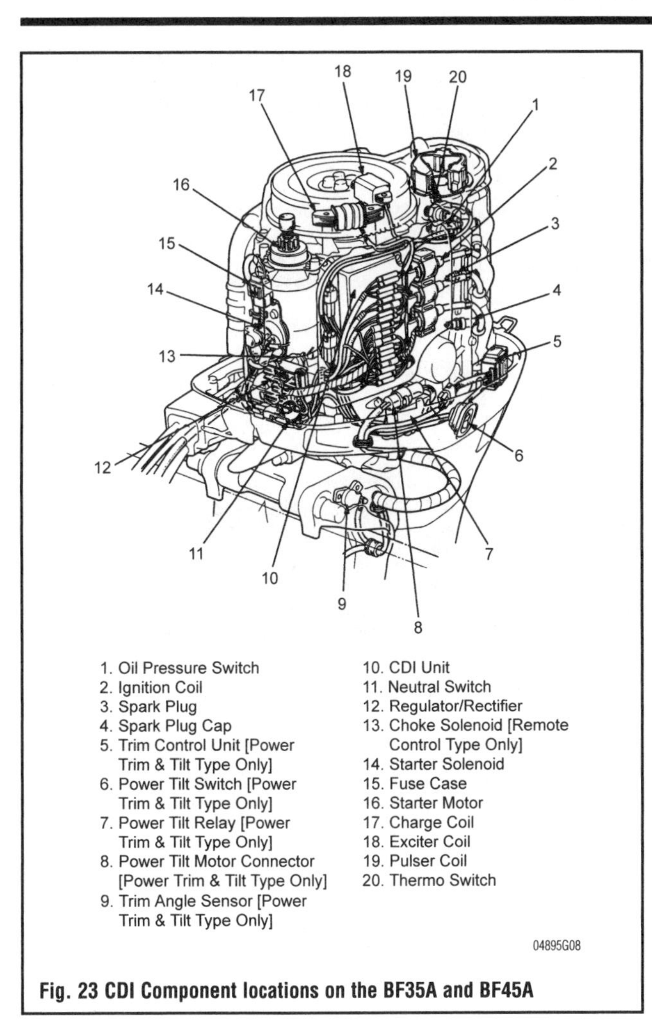

1. Oil Pressure Switch
2. Ignition Coil
3. Spark Plug
4. Spark Plug Cap
5. Trim Control Unit [Power Trim & Tilt Type Only]
6. Power Tilt Switch [Power Trim & Tilt Type Only]
7. Power Tilt Relay [Power Trim & Tilt Type Only]
8. Power Tilt Motor Connector [Power Trim & Tilt Type Only]
9. Trim Angle Sensor [Power Trim & Tilt Type Only]
10. CDI Unit
11. Neutral Switch
12. Regulator/Rectifier
13. Choke Solenoid [Remote Control Type Only]
14. Starter Solenoid
15. Fuse Case
16. Starter Motor
17. Charge Coil
18. Exciter Coil
19. Pulser Coil
20. Thermo Switch

04895G08

Fig. 23 CDI Component locations on the BF35A and BF45A

1. Overheat Sensor
2. 90A Fuse Box
3. Starter Solenoid
4. Starter Motor
5. Power Tilt Switch
6. 30A Fuse Boxes
7. 10A Fuse Box
8. Power Tilt Relay
9. No. 2/No. 3 Ignition Coil
10. ECM
11. No. 1/No. 4 Ignition Coil

04895G10

Fig. 25 BF115A and BF130A port side ignition components . . .

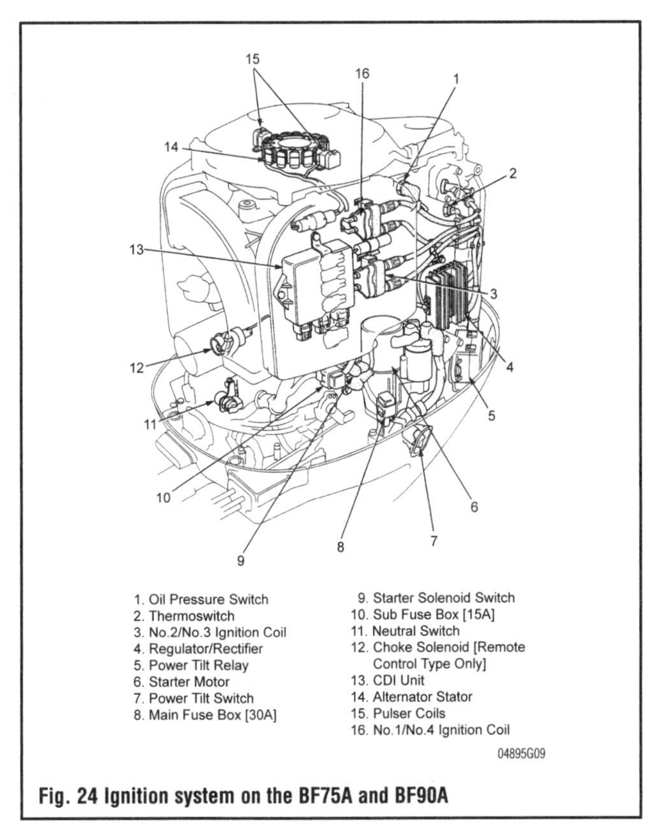

1. Oil Pressure Switch
2. Thermoswitch
3. No.2/No.3 Ignition Coil
4. Regulator/Rectifier
5. Power Tilt Relay
6. Starter Motor
7. Power Tilt Switch
8. Main Fuse Box [30A]
9. Starter Solenoid Switch
10. Sub Fuse Box [15A]
11. Neutral Switch
12. Choke Solenoid [Remote Control Type Only]
13. CDI Unit
14. Alternator Stator
15. Pulser Coils
16. No.1/No.4 Ignition Coil

04895G09

Fig. 24 Ignition system on the BF75A and BF90A

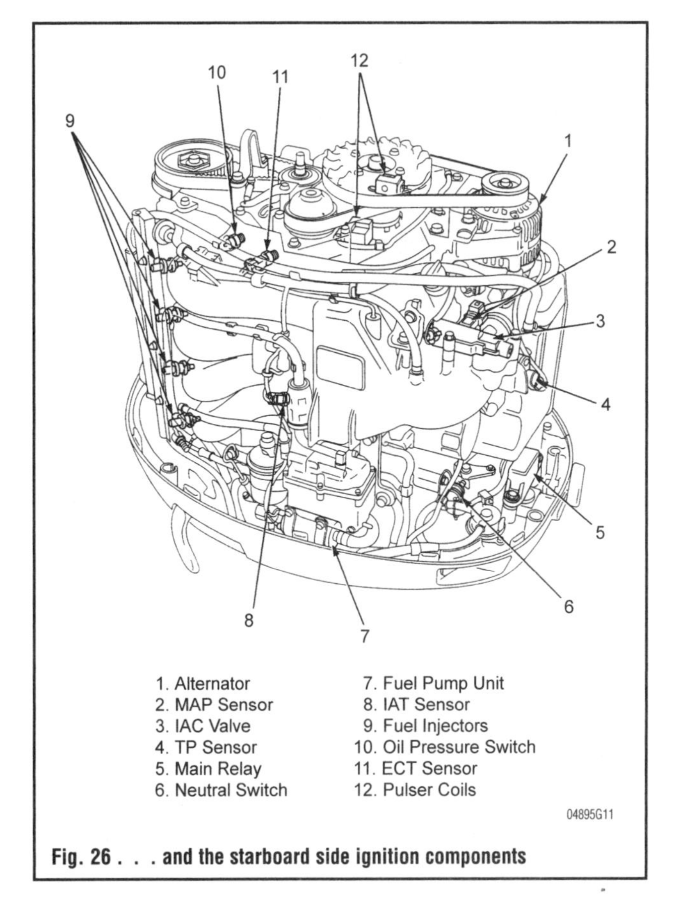

1. Alternator
2. MAP Sensor
3. IAC Valve
4. TP Sensor
5. Main Relay
6. Neutral Switch
7. Fuel Pump Unit
8. IAT Sensor
9. Fuel Injectors
10. Oil Pressure Switch
11. ECT Sensor
12. Pulser Coils

04895G11

Fig. 26 . . . and the starboard side ignition components

System Testing

PROCEDURE

Perform a visual inspection of the wiring connections and grounds. Determine if the problem affects all or just certain cylinders and perform a spark check using a spark gap tool ("spark checker") and then check ignition timing with a timing light.

If the problem affects all the cylinders, check the capacitor charge coil output, engine stop switch, CDI unit output and pulse coil output.

If the problem affects individual cylinders only, check the components whose failure would affect that particular cylinder such as the pulse coil and ignition coil performance.

If the problem is timing related, check the mechanical part of the system, such as the pulse coil or CDI box itself and then check the electronic timing advance components of the system, the throttle position sensor (if applicable), the pulse coil/s and the CDI module.

CDI troubleshooting can be performed with a peak reading voltmeter. This will check the CDI voltage to the ignition coils.

- If CDI voltage is good, isolate individual ignition coils or spark plugs and check output voltage.
- If the CDI voltage is bad, check all CDI input voltages.
- Check the pulse coil output to the CDI unit.
- Check the capacitor charge coil output to the CDI unit.
- Check the pulser coil output to the CDI unit.

If all the input voltages are normal, the problem has now been isolated to the CDI unit itself. If any input voltage is abnormal, check the appropriate coil for winding resistance and insulation breakdown. If the problem is timing related, check all the timing inputs to the CDI unit, such as the throttle position sensor. If the timing inputs are good, the problem is isolated to the CDI unit.

Pulsar Coil

DESCRIPTION & OPERATION

The second circuit used in CDI systems is the pulsar circuit. The pulsar circuit has its own flywheel magnet, a pulsar coil, a diode, and a thyristor. A thyristor is a solid state electronic switching device which permits voltage to flow only after it is triggered by another voltage source.

At the point in time when the ignition timing marks align, an alternating current is induced in the pulsar coil, in the same manner as previously described for the charge coil. This current is then passed to a second diode located in the CDI unit where it becomes DC current and flows on to the thyristor. This voltage triggers the thyristor to permit the voltage stored in the capacitor to be discharged. The capacitor voltage passes through the thyristor and on to the primary windings of the ignition coil.

In this manner, a spark at the plug may be accurately timed by the timing marks on the flywheel relative to the magnets in the flywheel and to provide as many as 100 sparks per second for a powerhead operating at 6000 rpm.

On the BF75A and BF90A , the two pulsar coils are mounted to produce a pulse every 180° revolution of the crankshaft. Therefore two pulses are produced each crankshaft revolution.

The BF115A and BF130A use a different type of signaling system. Two pulsar coils are evenly spaced at 180° around the timing belt guide plate. The pulses sent by the two coils determine engine speed and are used by the ECM to determine ignition timing.

TESTING

BF115A and BF130A

See Figure 27

Pulser coil does not need to be removed to test.

1. Disconnect the pulser coil at the wiring harness connector.
2. Measure the resistance with an ohmmeter between the terminals of the pulser coil wiring harness.
3. Resistance between the Gray/White and Brown/Red wires, and between Orange/Brown and Blue wires should be 970–1,170 ohms.
4. If resistance is not as specified, the pulser coil may be faulty.
5. If resistance is within specification, inspect the wiring harness.

BF75A and 90A

See Figure 28

Pulser coil does not need to be removed for testing

1. Disconnect the pulser coil wiring harness connector.
2. Measure the resistance between the BL2 and BU/G, and the BL1 and BU terminals with an ohmmeter.
3. Resistance should measure between 168–252 ohms.
4. If resistance is not as specified, the pulser coil may be faulty.
5. If resistance is within specification, inspect the wiring harness.

BF35A, BF40A and BF45A

See Figure 29

Pulser coil does not need to be removed for testing.

1. Disconnect the pulser coil connector from the side of the engine.
2. Measure the resistance with an ohmmeter between the BL terminal and the other terminals.
3. Resistance should measure between 288–352 ohms.
4. If resistance is not as specified, the pulser coil may be faulty.
5. If resistance is within specification, inspect the wiring harness.

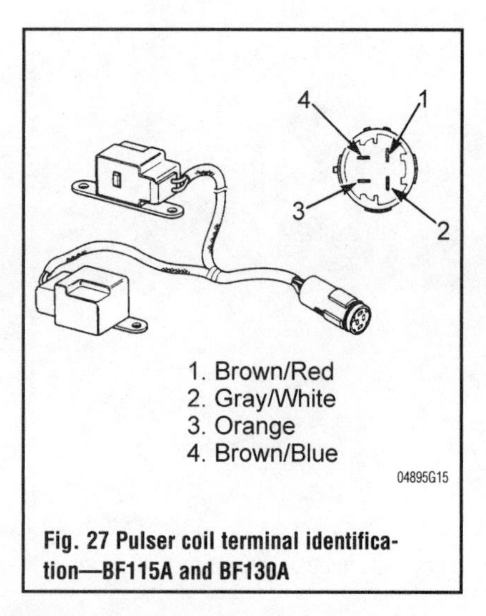

Fig. 27 Pulser coil terminal identification—BF115A and BF130A

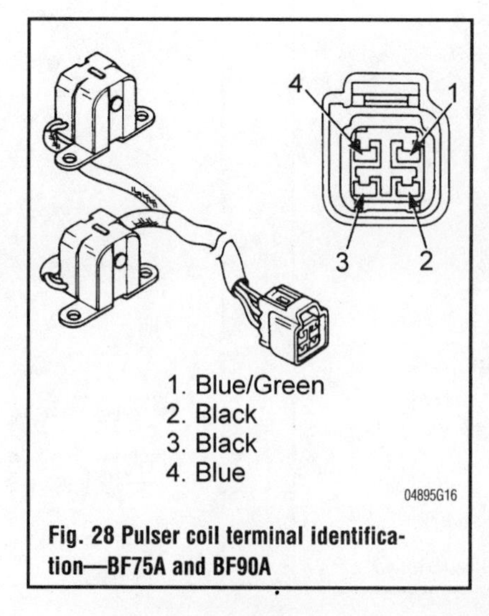

Fig. 28 Pulser coil terminal identification—BF75A and BF90A

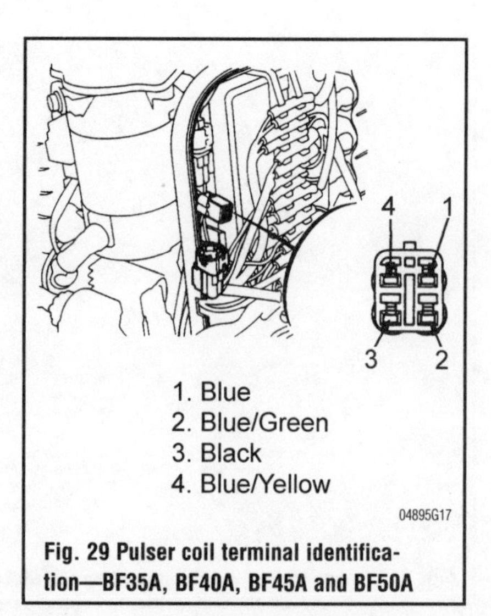

Fig. 29 Pulser coil terminal identification—BF35A, BF40A, BF45A and BF50A

BF25A and BF30A

➧ See Figure 30

Pulser coil does not need to be removed for testing.

1. Disconnect the pulser coil connector located on the right side of the engine.
2. Measure the resistance with an ohmmeter between the BL terminal and the other terminals.
3. Resistance should measure between 290–355 ohms.
4. If resistance is not as specified, the pulser coil may be faulty.
5. If resistance is within specification, inspect the wiring harness.

BF9.9A and BF15A

➧ See Figure 31

Pulser coil does not have to be removed for testing.

1. Disconnect the pulser coil wires from the wiring harness.
2. Measure the resistance with an ohmmeter between the two pulser wires.
3. Resistance should measure between 351–429 ohms.
4. If resistance is not as specified, the pulser coil may be faulty.
5. If resistance is within specification, inspect the wiring harness.

REMOVAL & INSTALLATION

BF115A and BF130A

The pulser coils are mounted below the timing belt drive pulley.

1. Remove the engine cover.
2. Remove the timing belt upper cover.
3. Remove the alternator, balance and timing belts.
4. Disconnect the pulser coil wire connector.
5. Remove the pulser coil retaining screws, wire clamp screws, and remove the pulser coils from either side of the crankshaft.

To install:

6. Install the pulser coils and wire clamps.
7. Install the timing balancer and alternator belts.
8. Install the timing belt upper cover.
9. Reconnect the pulser coil wire connector.

BF75A and BF90A

➧ See Figures 32 and 33

The pulser coils are mounted on either side of the alternator stator which is mounted below the starter pulley.

1. Remove the engine cover.
2. Remove the timing belt cover.
3. Remove the starter pulley, alternator rotor, and alternator stator.
4. Disconnect the pulser coil wire connector and remove the pulser coils from the engine.

To install:

5. Pass the pulser coil wire underneath each pulser coil and install the coils on the base and screw down.
6. Make sure the pulser coil wire is inserted in the groove of the base and secured with the two clamps. Remove any slack at the wire pickups.
7. Make sure that the pulser coil wire passes on the right side of the guide rib on the base.
8. Turn the alternator rotor clockwise until the center of the projection on the pulser coil aligns with the center of the projection on the alternator rotor.

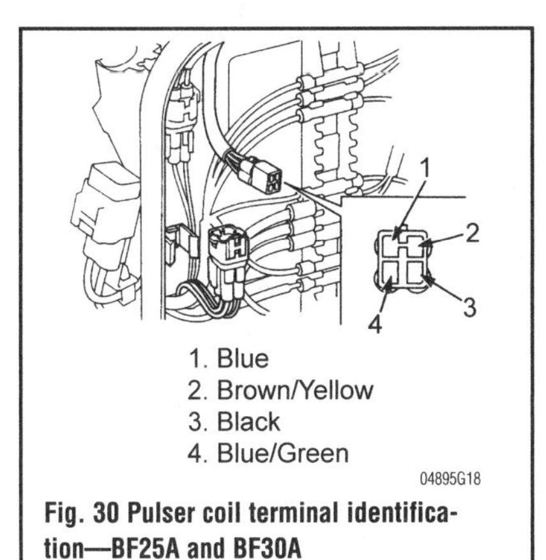

Fig. 30 Pulser coil terminal identification—BF25A and BF30A

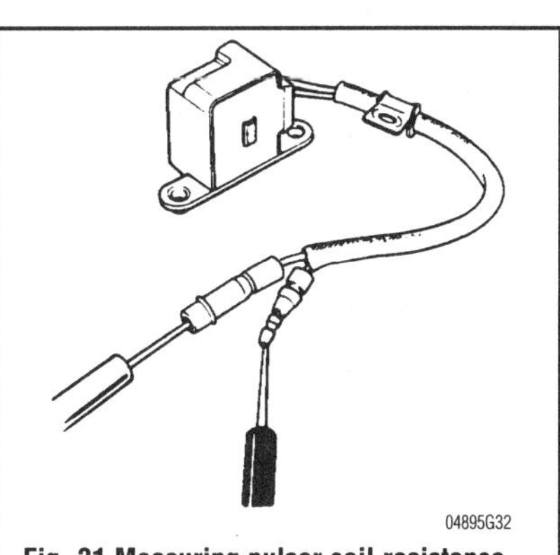
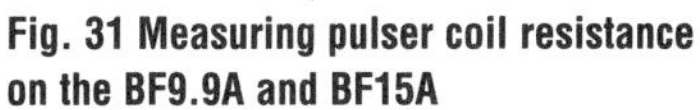

Fig. 31 Measuring pulser coil resistance on the BF9.9A and BF15A

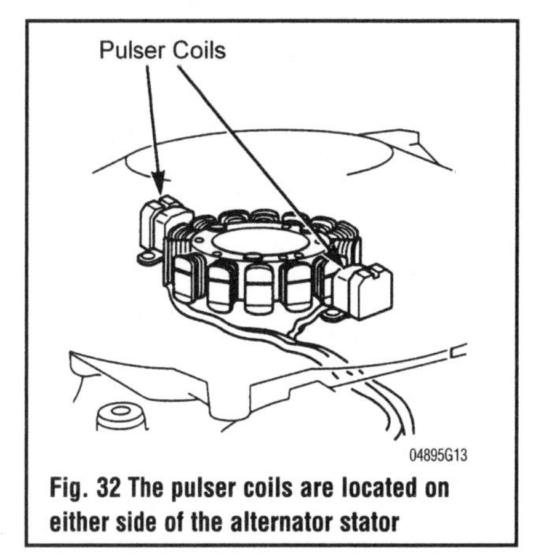

Fig. 32 The pulser coils are located on either side of the alternator stator

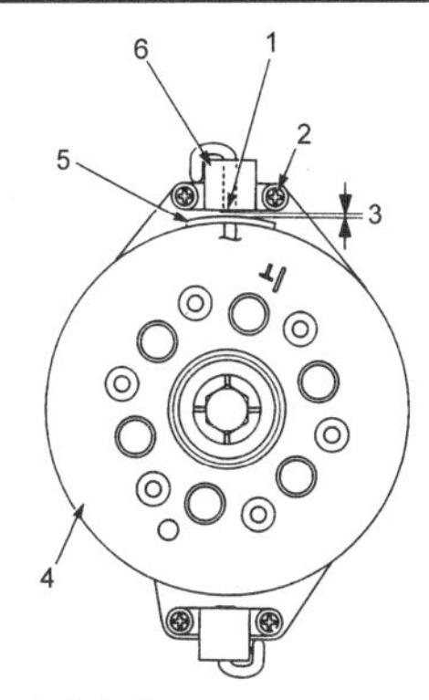

Fig. 33 Adjust the pulser coil clearance from the alternator rotor

9. Adjust the clearance between the pulser coil projection and the alternator rotor projection to 0.028–0.031 in. (0.7–0.8 mm).
10. To adjust the pulser coil, loosen the retaining screw and move the pulser coil. Adjust the opposite coil the same way.

➡When adjusting the pulser coil, do not move the center of the pulser coil projection and the alternator rotor projection out of alignment.

BF25A, BF30A, BF35A, BF40A, BF45A and BF50A

➧ See Figures 34, 35, 36 and 37

The pulser coil is mounted below the timing pulley on these models.

1. Disconnect the pulser coil connector inside the CDI unit.
2. Remove the engine and flywheel covers.
3. Remove the flywheel, timing belt and camshaft pulley.
4. Unscrew the retaining screws and remove the pulser coil assembly.

To install:

5. Install the pulser coil assembly onto the engine and tighten the retaining screws.
6. Install the camshaft pulley, timing belt and flywheel assembly.
7. Install the engine and flywheel covers.

Fig. 34 BF35A Pulser coil located below the timing pulley

Fig. 35 Removing the BF35A pulser coil

Fig. 36 Make sure to line up the dowel pin on the camshaft pulley with the notch on the end of the camshaft

Fig. 37 The flywheel is held in place by a single bolt, tighten it securely

Charge Coil

DESCRIPTION & OPERATION

The main function of the charging system is to recharge the battery after starting the engine. The charging system also helps support the accessory loads once the engine is running. On many outboard engines, the charging system is a dealer installed option.

The charging system consists of the charge coil, magnets permanently installed in the flywheel, a voltage regulator/rectifier, a 5a fuse/fuse holder assembly, a battery and the wiring system connecting the entire system together.

The rotation of the flywheel magnets past the coil creates alternating current (A/C). This current is sent to the voltage regulator/rectifier and is converted in to the direct current that will charge the battery or run the appropriate accessories.

TESTING

The charge coil checks are the same as those for the pulser coil. There are two types of charge coils. One type has a charge coil lead connected to ground. The other type charge coil has both leads go directly to the CDI box. Both charge coil types require the continuity test. This checks the coil's internal resistance. On charge coils that do not have a grounded lead, the short-to-ground test must also be done

The basic test for a charge coil is continuity. This measures the actual resistance from one end of the charge coil to the other. The correct specification for each coil can be found in the Ignition Testing Specifications chart.

1. Adjust the meter to ohms.
2. Connect the tester across the pulser leads and note the reading.
3. Compare the coil reading to the specifications.
4. If the reading is above or below the correct value and beyond the allowable deviation then it must be replaced.

➡Remember that temperature has an affect on resistance. Most resistance specifications are given assuming a temperature of 70° F.

BF25A and BF30A

➧ See Figures 38 and 39

➡It is not necessary to remove the coil from the engine for testing..

1. Remove the engine cover.
2. Remove the CDI unit cover and unplug the charge coil wires.
3. Measure the resistance with an ohmmeter between the each wire terminal of the charge coil.
4. The resistance should read 0.45–0.54 ohms for the standard 12V–6A charge coil and 0.27–0.33 ohms for the optional 12V–10A charge coil.

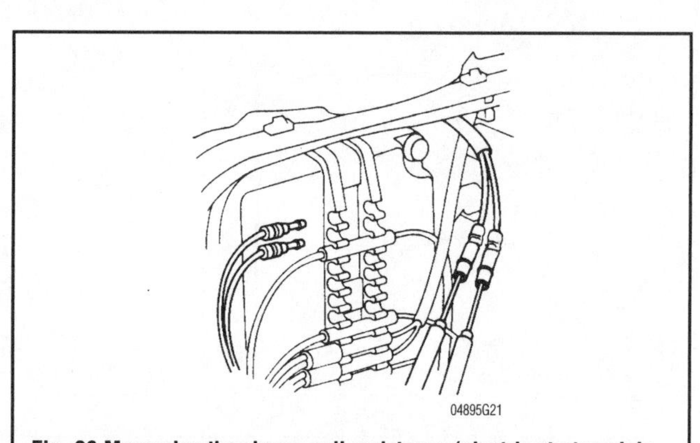

Fig. 38 Measuring the charge coil resistance (electric start models only)—BF25A and BF30A

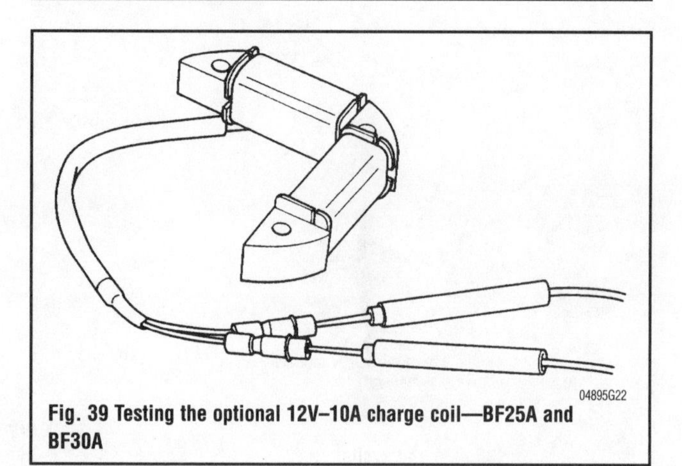

Fig. 39 Testing the optional 12V–10A charge coil—BF25A and BF30A

BF35, BF40A, BF45A and BF50A

➧ See Figures 40 and 41

➡It is not necessary to remove the coil from the engine for testing..

1. Remove the engine cover.
2. Remove the CDI unit cover and unplug the charge coil wires.
3. Measure the resistance with an ohmmeter between the each wire terminal of the charge coil.
4. The resistance should read 0.20–0.26 ohms for the standard 12V–6A charge coil and 0.17–0.23 ohms for the optional 12V–10A charge coil.

BF75, BF8A and BF100

➧ See Figure 42

➡It is not necessary to remove the charge coil from the engine for testing.

1. Remove the engine cover.
2. Disconnect the charge coil wiring from the rectifier connector.
3. Measure the resistance with an ohmmeter between each wire terminal.
4. The resistance should measure 0.03–0.23 ohms on the BF75 and BF100 and 0.12–0.15 ohms for the BF8A model.

REMOVAL & INSTALLATION

BF25A, BF30A, BF35A, BF40A, BF45A and BF50A

➧ See Figure 43

1. Disconnect the negative battery cable.
2. Remove the CDI unit cover and unplug the charge coil wires.
3. Remove the flywheel. Make sure to keep track of the dowel pin.
4. Unscrew the two retaining screws, the cord holder and remove the charge coil.

To install:

5. Install the charge coil with the two retaining screws onto the engine.
6. Install the cord holder so that the wires are not pinched.
7. Install the flywheel, making sure that the dowel pin is in place and aligned.
8. Install the flywheel cover.
9. Install the engine cover and reattach the battery negative cable.

BF75, BF8A and BF100

➧ See Figure 44

Charge coil is located below the flywheel.

1. Disconnect the charge coil wiring harness at the connector.
2. Remove the flywheel to access the charge coil.
3. Unscrew the retaining bolts and remove the charge coil.

To install:

4. Install the coil and retaining bolts. Be sure that the coil and wiring does not interfere with flywheel rotation or timing belt.
5. Install the flywheel.
6. Connect the charge coil wiring harness at the connector.

BF50 and BF5A

➧ See Figure 45

These engines use a pair of optional, dealer installed, combination lighting and charge coils.

The combination charging and lighting coils are located below the flywheel.

1. Disconnect the coil at the rectifier connection.
2. Remove the flywheel to gain access to the charge coil.

To install:

3. Install the coil and retaining bolts. Make sure to route the wire inside the holder so that they do not interfere with flywheel rotation and then secure them with the wire clamp.
4. Install the flywheel.

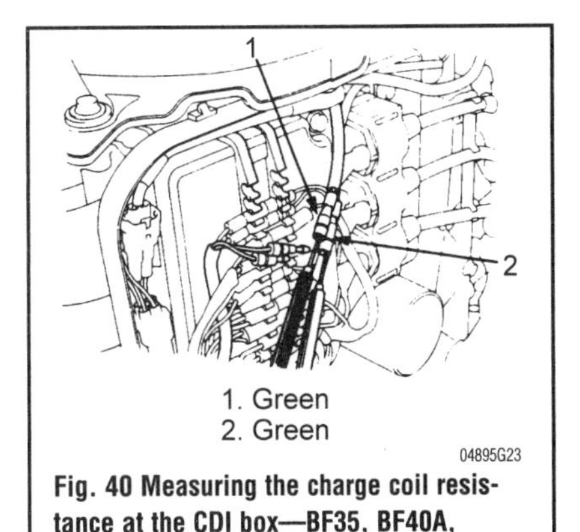

Fig. 40 Measuring the charge coil resistance at the CDI box—BF35, BF40A, BF45A and BF50A

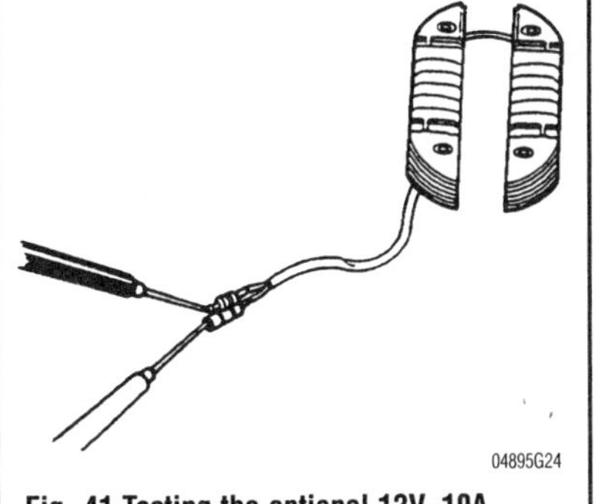

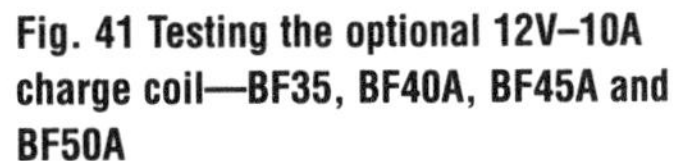

Fig. 41 Testing the optional 12V–10A charge coil—BF35, BF40A, BF45A and BF50A

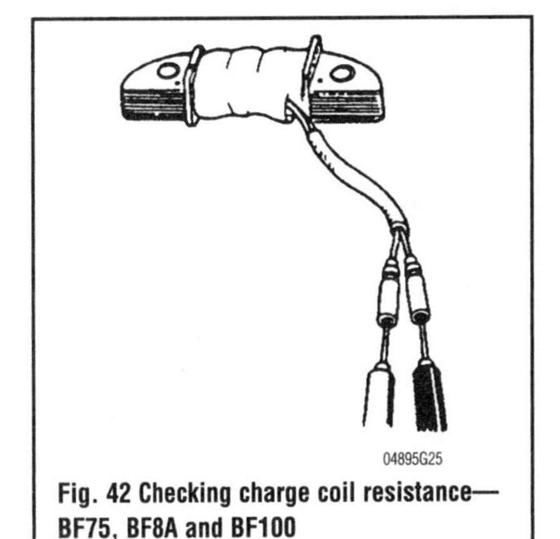

Fig. 42 Checking charge coil resistance—BF75, BF8A and BF100

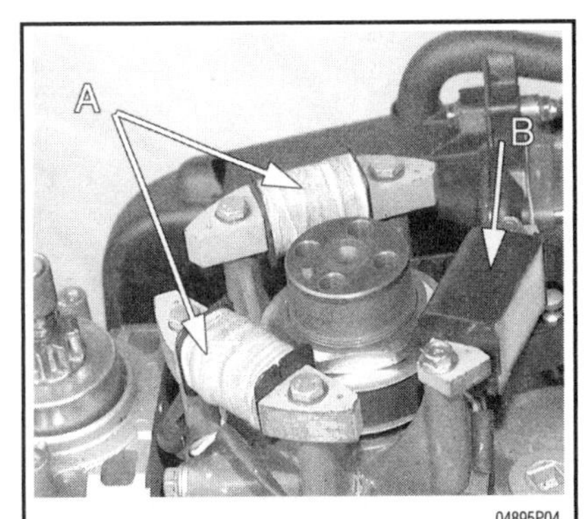

Fig. 43 Charge coils (A) and exciter coil (B) mounted on the BF35A

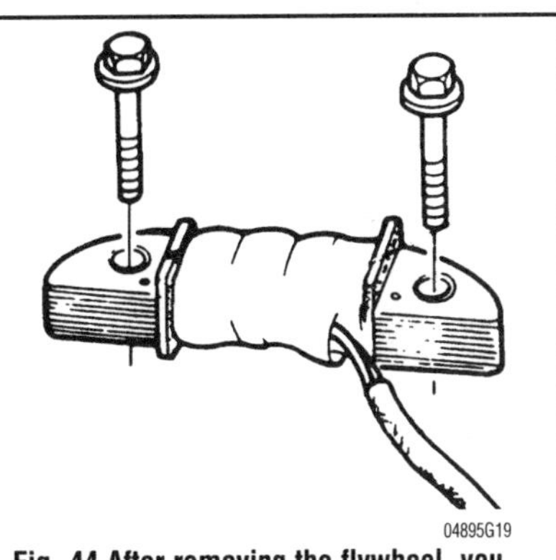

Fig. 44 After removing the flywheel, you can remove the charge coil

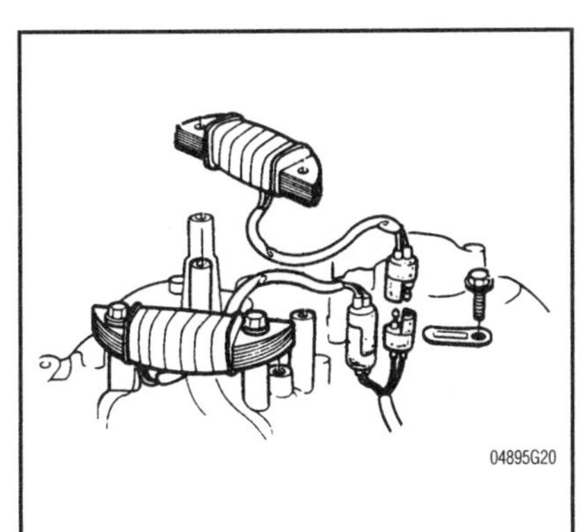

Fig. 45 Unplug the harness leads and remove the charging and lighting coils

Lighting Coil

DESCRIPTION & OPERATION

➡The BF50 and BF5A are the only Honda models to use a lighting coil. In this case, they use an optional, dealer installed, combination lighting/charge coil.

The A/C current from the coil is changed to D/C current in the silicone rectifier then it's wired to the D/C receptacle on the engine housing.

REMOVAL & INSTALLATION

Removal and installation procedures for the lighting coil are the same as for the charge coil. See "Charge Coil" in this section for more information.

TESTING

Testing procedures for the lighting coil are the same as for the charge coil. See "Charge Coil" in this section for more information.

Ignition Coils

DESCRIPTION & OPERATION

See Figure 46

Ignition coils, one per cylinder, boost the DC voltage instantly to approximately 20,000 volts to the spark. This completes the primary side of the ignition circuit.

Once the voltage is discharged from the ignition coil the secondary circuit begins and only stretches from the ignition coil to the spark plugs via extremely large high tension leads. At the spark plug end, the voltage arcs in the form of a spark, across from the center electrode to the outer electrode, and then to ground via the spark plug threads. This completes the ignition circuit.

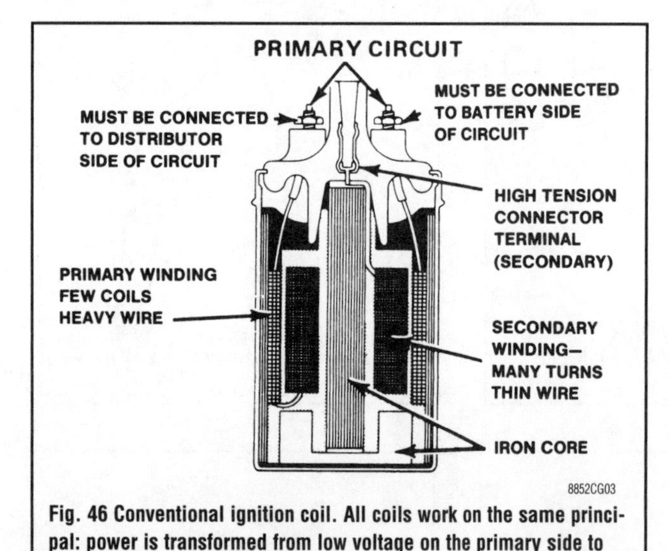

Fig. 46 Conventional ignition coil. All coils work on the same principal: power is transformed from low voltage on the primary side to high voltage on the secondary side

TESTING

Although the best test for an ignition coil is on a dynamic ignition coil tester, resistance checks can also be done.

There are two circuits in an ignition coil, the primary winding circuit and the secondary winding circuit. Both need to be checked.

The tester connection procedure for a continuity check will depend on how the coil is constructed. Generally, the primary circuit is the small gauge wire or wires, while the secondary circuit contains the high tension or plug lead.

Some ignition coils have the primary and/or secondary circuits grounded on one end. On these type coils, only the continuity check is done. On ignition coils that are not grounded on one end, the short-to-ground test must also be done. Regardless of the coil type, compare the resistance with the Electrical Testing Specification chart.

➡When checking the secondary side, make sure to read the procedure for your particular engine. Some models require the spark plug caps be removed and other require them to remain on. In some cases the cap is bad, not the coil. Bad resistor caps can be the cause of high-speed misfire. Unscrew the cap and check the resistance (5 kilo-ohms).

The other method used to test ignition coils is with a Dynamic Ignition Coil Tester. Since the output side of the ignition coil has very high voltage, a regular voltmeter can not be used. While resistance reading can be valuable, the best tool for checking dynamic coil performance is a dynamic ignition coil tester.

1. Connect the coil to the tester according to the manufacturer's instructions.
2. Set the spark gap according to the specifications.
3. Operate the coil for about 5 minutes.
4. If the spark jumps the gap with the correct spark color, the coil is probably good.

BF115A and BF130A

See Figures 47 and 48

1. Measure the primary side resistance between the two terminals with an ohmmeter. Resistance should measure 0.60–0.72 ohms.
2. Measure the secondary side resistance. With the spark plug caps still attached, place the ohmmeter test leads into the inside of the spark plug caps and measure the resistance between the two wires. Resistance should measure 25–38 kilo-ohms.
3. If either resistance is not within specification, the coil may be defective.

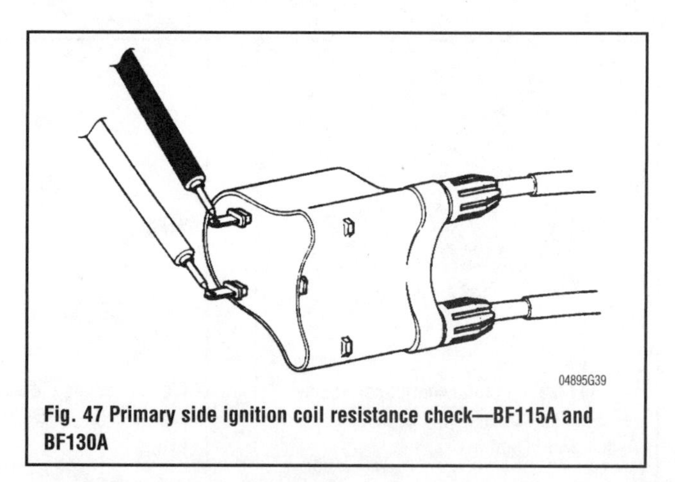

Fig. 47 Primary side ignition coil resistance check—BF115A and BF130A

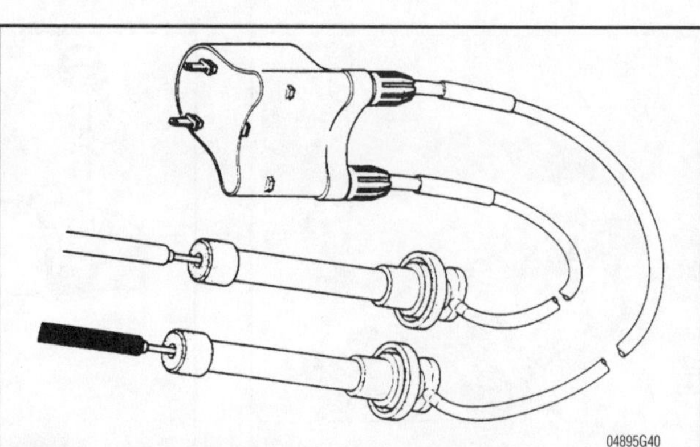

Fig. 48 Secondary side ignition coil resistance check—BF115A and BF130A

BF75A and BF90A

◆ See Figures 49 and 50

1. Measure the primary side resistance between the two terminals with an ohmmeter. Resistance should measure 0.35-0.43 ohms.
2. Measure the secondary side resistance. With the spark plug caps still attached, place the ohmmeter test leads into the inside of the spark plug caps and measure the resistance between the two wires. Resistance should measure 23.1-34.7 kilo-ohms.
3. If either resistance is not within specification, the coil may be defective.

BF25A, BF30A, BF35A, BF40A, BF45A and BF50A

◆ See Figures 51, 52 and 53

1. Measure the primary side resistance between the two terminals with an ohmmeter. Resistance should measure 0.19-0.23 ohms.

2. Measure the secondary side resistance. On the BF25A and BF30A , with the spark plug caps still attached, place one test lead into the inside of the spark plug cap and the other on the green terminal and measure the resistance between the two. Resistance should measure 10.3-15.9 kilo-ohms.

3. On the BF35A, BF450, BF45A and BF50A, remove the spark plug cap and measure the resistance of the secondary coil between the plug wire and the green terminal. Resistance should measure 2.8-3.4 kilo-ohms.

4. Measure the resistance of the detached spark plug cap by touching one test lead to the wire end of the cap and the other test lead at the spark plug end of the cap. Resistance should measure 7.5-12.5 kilo-ohms.

5. If either resistance is not within specification, the coil may be defective.

BF9.9A and BF15A

◆ See Figures 54 and 55

1. Measure the primary side resistance between the two terminals with an ohmmeter. Resistance should measure 0.35-0.43 ohms.
2. Measure the secondary resistance with the spark plug caps removed. Attach the ohmmeter test leads to both spark plug wires and measure the resistance. Resistance should be 8.01-9.79 ohms.
3. If either resistance is not within specification, the coil may be defective.

BF75, BF8A and BF100

1. Measure the primary side resistance between the two terminals with an ohmmeter. Resistance should measure 0.46-0.66 ohms.
2. Measure the secondary resistance with the spark plug caps removed. Attach the ohmmeter test leads to both spark plug wires and measure the resistance, 6.4-9.6 kilo-ohms.
3. If either resistance is not within specification, the coil may be defective.

BF50, BF5A, BF20, BF2A and BF2D

◆ See Figures 56 and 57

■ **Please refer to the Supplement for art depicting the BF2D coil.**

1. Measure the primary resistance between the wire connector and the coil core. Resistance should measure 0.7-0.9 ohms for all except the BF2D for which resistance should be 0.98-1.2 ohms.
2. Measure secondary resistance with the spark plug cap removed and measure the resistance between the spark plug wire and the coil core. Resistance should read 6.3-7.7 kilo-ohms for all except the BF2D for which resistance should be 11-15 kilo-ohms.
3. If either resistance is not within specification, the coil may be defective.

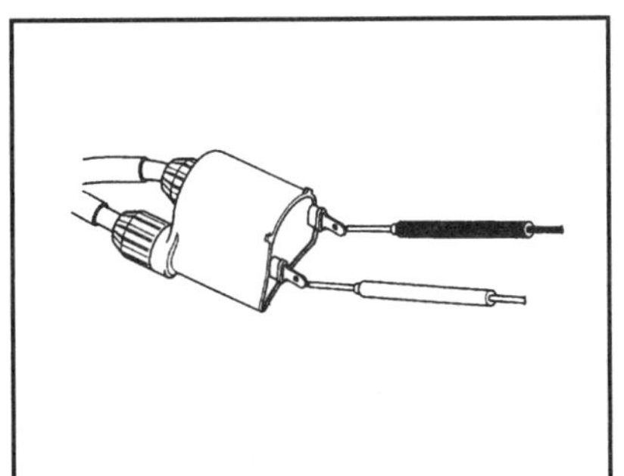

Fig. 49 Primary side ignition coil resistance check—BF75A and BF90A

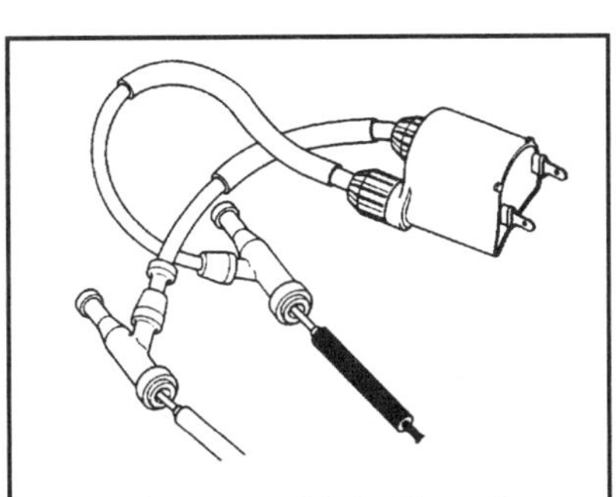

Fig. 50 Secondary side ignition coil resistance check—BF75A and BF90A

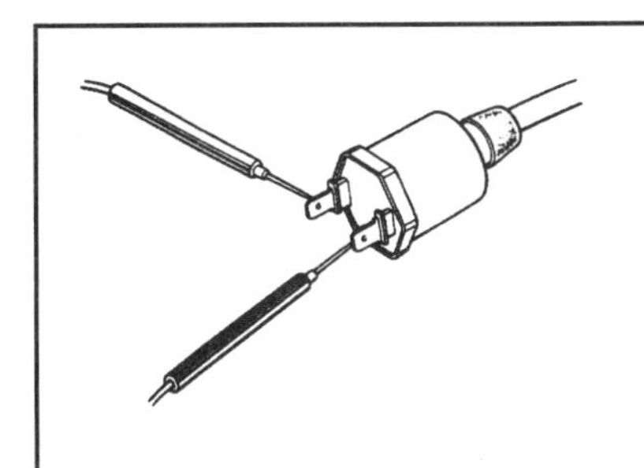

Fig. 51 BF25A, BF30A, BF35A, BF40A, BF45A and BF50A

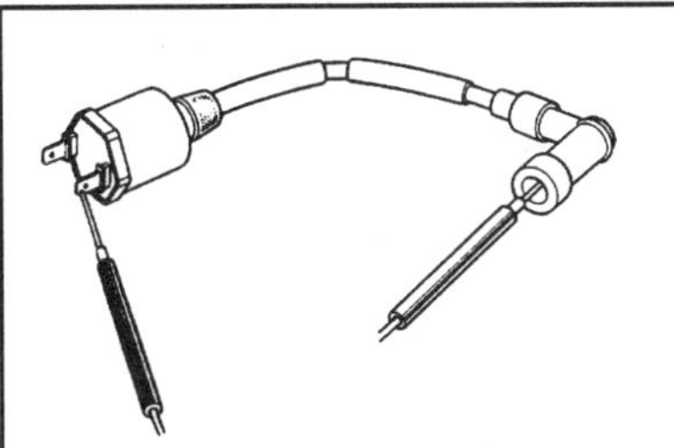

Fig. 52 Secondary side ignition coil resistance check—BF25A, BF30A, BF35A, BF40A, BF45A and BF50A

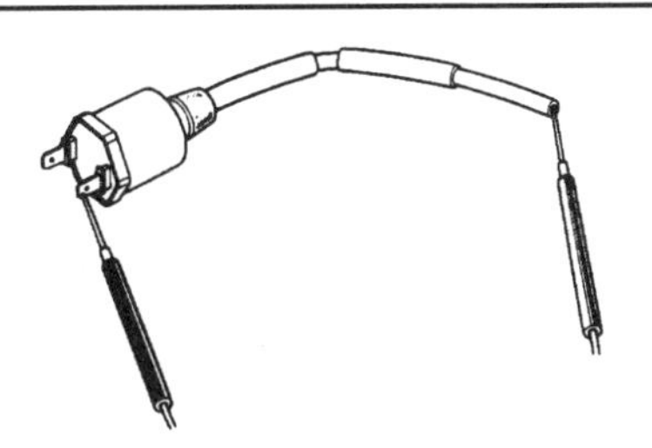

Fig. 53 Secondary side resistance check with the plug cap removed

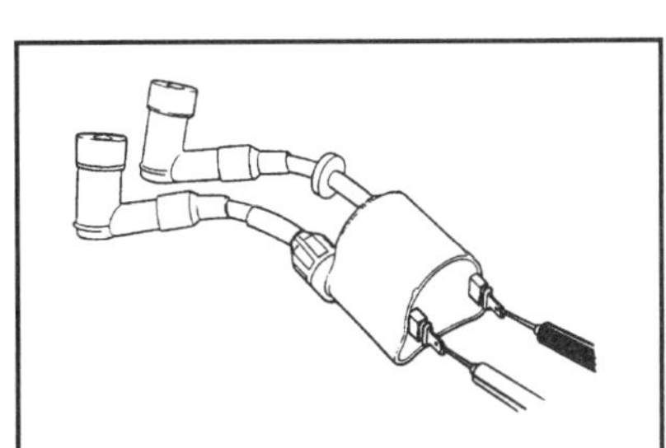

Fig. 54 Primary side ignition coil resistance check—BF9.9A and BF15A

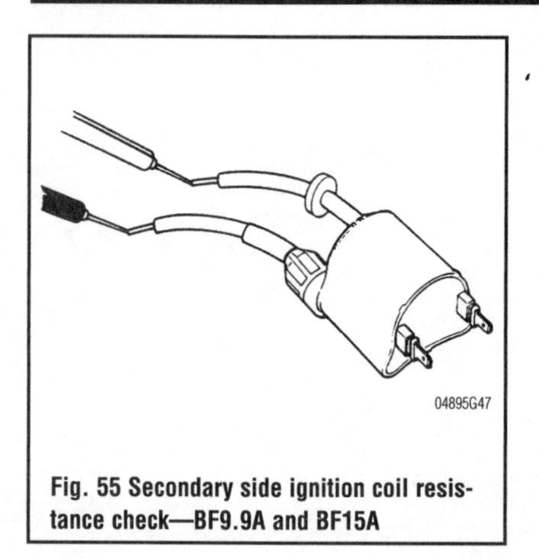

Fig. 55 Secondary side ignition coil resistance check—BF9.9A and BF15A

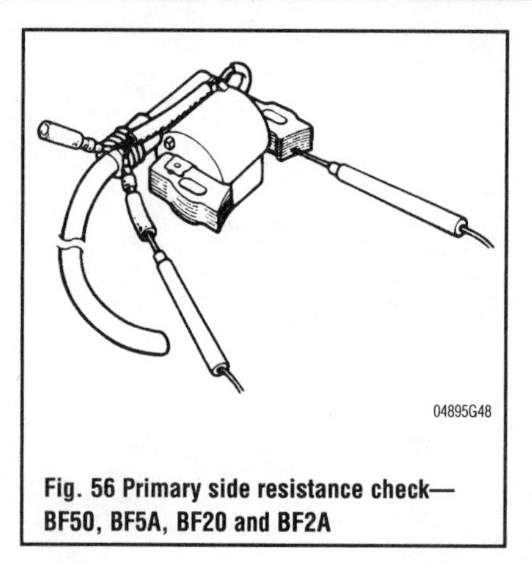

Fig. 56 Primary side resistance check—BF50, BF5A, BF20 and BF2A

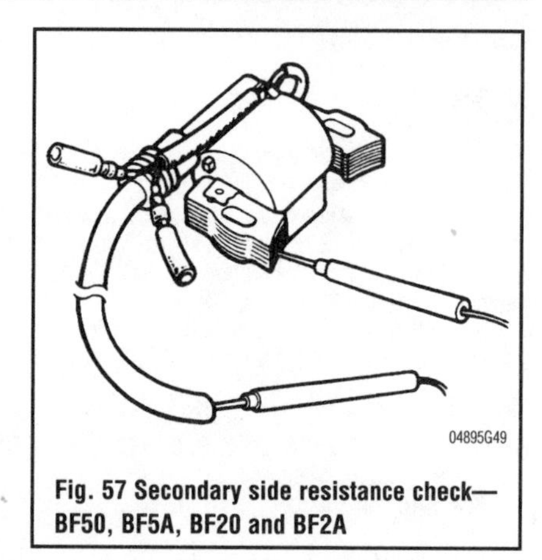

Fig. 57 Secondary side resistance check—BF50, BF5A, BF20 and BF2A

REMOVAL & INSTALLATION

BF115A and BF130A

➧ See Figures 58 and 59

1. Disconnect the positive and negative cables from the battery.
2. Remove the engine cover.
3. Remove the ECM cover.
4. Disconnect the ignition coil primary side wire terminal connectors.
5. Pull the spark plug wires off the plugs. Always pull the plug wires at the caps and not by the wires.
6. Unhook the ignition coil grommets from the ECM case and remove the ignition coils.

To install:

7. Slide the ignition coil into the grommet and attach the grommet and ignition coil to the ECM case.
8. Connect the ignition coil primary wires to the proper ignition coil terminals.
9. Make sure all the spark plug wire clamps and clips are in place.
10. Put the spark plug wires back on the spark plugs.
11. Install the ECM cover and the engine cover.
12. Connect the positive and negative cables to the battery.

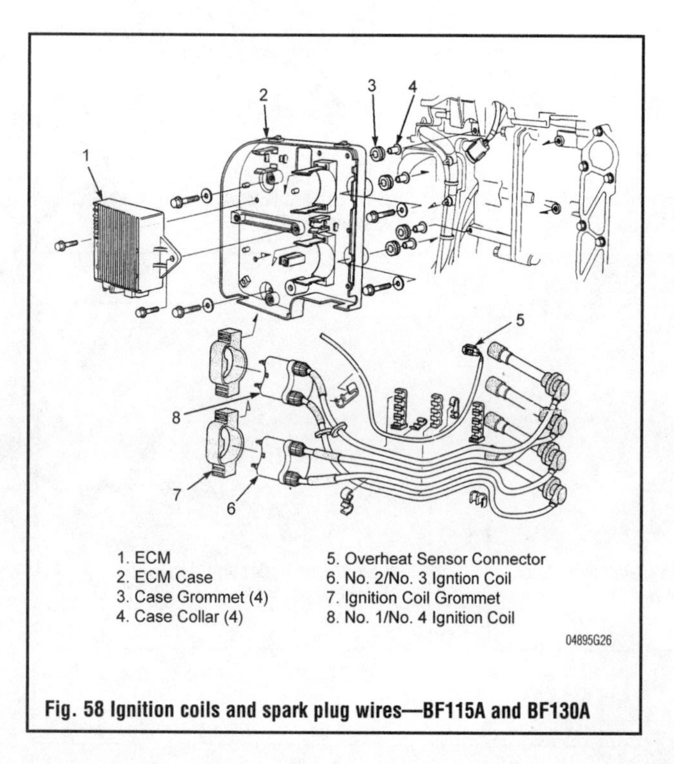

Fig. 58 Ignition coils and spark plug wires—BF115A and BF130A

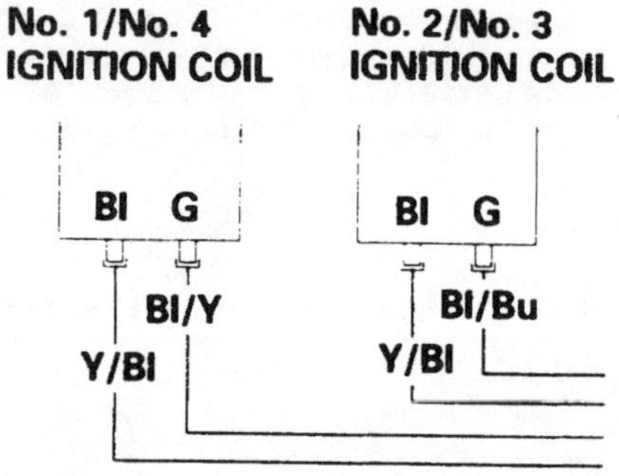

No. 1/No. 4 IGNITION COIL
No. 2/No. 3 IGNITION COIL
Bl G
Bl G
Bl/Y
Y/Bl
Bl/Bu
Y/Bl
04895G27

Fig. 59 Make sure the primary side wires are attached to the proper coils

BF75A and BF90A

➧ See Figures 60 and 61

1. Disconnect the negative and positive battery cables.
2. Remove the engine cover.
3. Remove the plug wires from the spark plugs.
4. Remove the CDI unit cover.
5. Disconnect the ignition coil primary leads from the ignition coils.
6. Unclip the spark plug wires from the wire clamps and clips.
7. Pull the ignition coils and away from their attachment points in the CDI unit case.

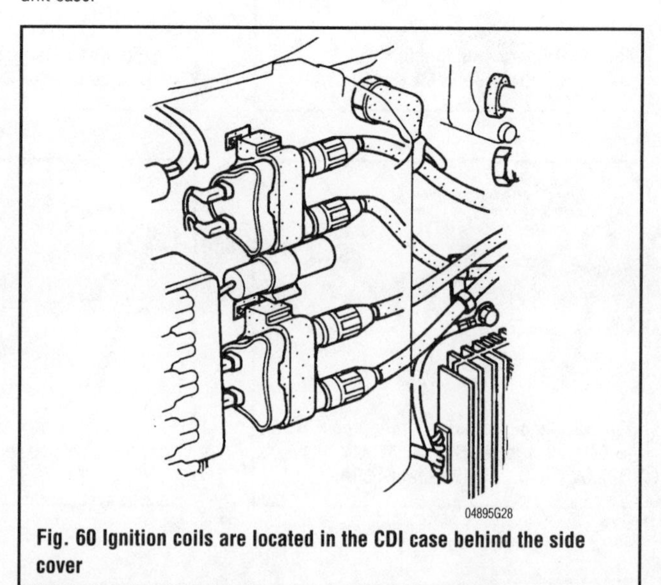

Fig. 60 Ignition coils are located in the CDI case behind the side cover

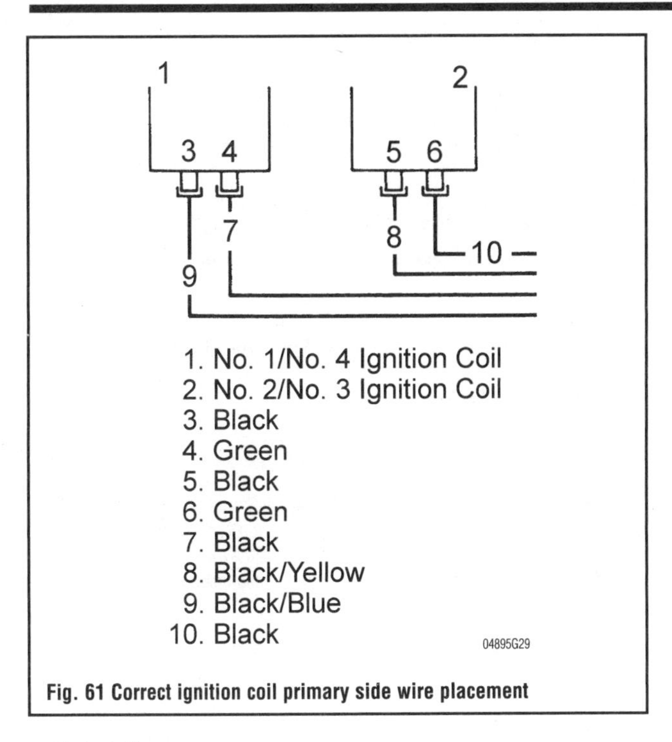

Fig. 61 Correct ignition coil primary side wire placement

To install:

8. Insert the ignition coil into its respective grommet.
9. Install the ignition coil and grommet onto the proper attachment point in the CDI case.
10. Connect the ignition coil primary side leads to the proper ignition coil.
11. Put the spark plug wires and caps back on the spark plugs.

➡Make sure to put the spark plug wires back into the proper clamps and clips.

12. Install the CDI unit cover.
13. Install the engine cover.
14. Reconnect the negative and positive battery cables.

BF30A, BF35A BF40A, BF45A, BF50A and

➧ See Figures 62, 63, 64 and 65

1. Disconnect the negative and positive battery cables.
2. Remove the engine cover.
3. Remove the CDI unit cover.
4. Disconnect the ignition coil primary leads from the ignition coils.
5. Unplug the spark plug wires from the spark plugs.
6. Unclip the spark plug wires from the wire clamps and clips.
7. Pull the ignition coils and away from their attachment points in the CDI unit case.

To install:

8. Insert the ignition coil into its respective grommet.
9. Install the ignition coil and grommet onto the proper attachment point in the CDI case.
10. Connect the ignition coil primary side leads to the proper ignition coil.
11. Install the spark plug wires and caps back on the spark plugs.

➡Make sure to put the spark plug wires back into the proper clamps and clips.

12. Install the CDI unit cover.
13. Install the engine cover.
14. Reconnect the negative and positive battery cables.

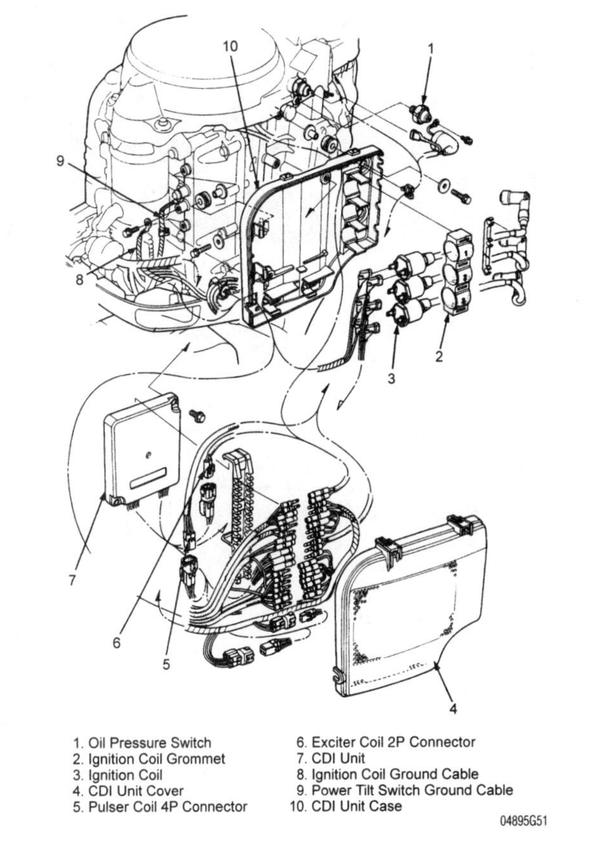

Fig. 62 CDI unit and ignition system component locations—BF35A, BF40A, BF45A, BF50A

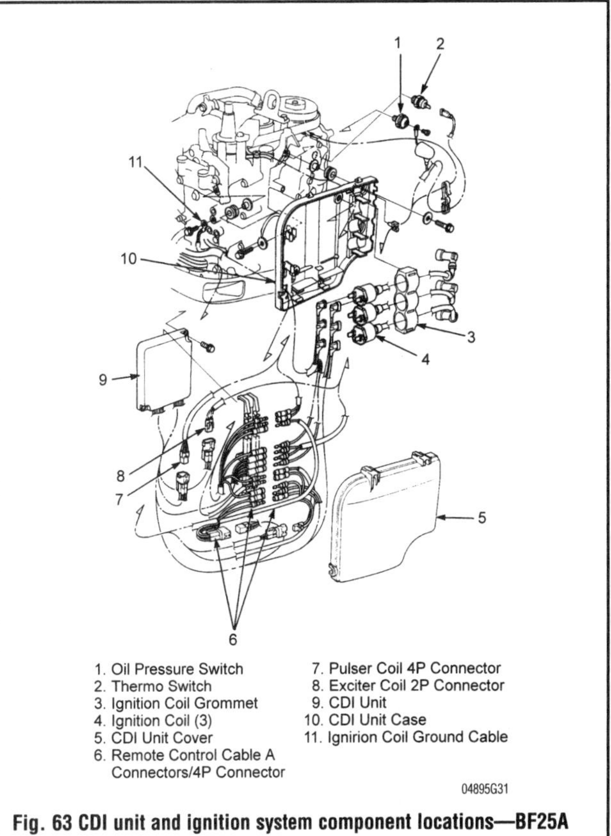

Fig. 63 CDI unit and ignition system component locations—BF25A and BF30A

Fig. 64 Ignition coils removed from the mounts in the CDI case

Fig. 65 Number the primary side wires prior to disconnecting them, this will avoid confusion during installation

BF9.9A and BF15A

➧ See Figure 66

1. Disconnect the negative and positive battery cables.
2. Remove the engine cover.
3. Disconnect the ignition coil primary leads from the ignition coils.
4. Unplug the spark plug wires and caps from the spark plugs
5. Unclip the spark plug wires from the wire clamps and clips.
6. Pull the ignition coils and away from their attachment points in the CDI unit case.

To install:

7. Insert the ignition coil into its respective grommet.
8. Install the ignition coil and grommet onto the proper attachment point on the engine.
9. Connect the ignition coil primary side leads to the proper ignition coil.
10. Install the spark plug wires back on the spark plugs.

➡Make sure to put the spark plug wires back into the proper clamps and clips.

11. Install the engine cover.
12. Reconnect the negative and positive battery cables.

BF75, BF8A and BF100

➧ See Figures 67 and 68

1. Remove the engine cover.
2. Disconnect the ignition coil primary leads from the ignition coils.
3. Unplug the spark plug wires and caps from the spark plugs
4. Unclip the spark plug wires from the wire clamps and clips.
5. Pull the ignition coils and away from their attachment points on the engine.

To install:

6. Insert the ignition coil into its respective grommet.
7. Install the ignition coil and grommet onto the proper attachment point on the engine.
8. Connect the ignition coil primary side leads to the proper ignition coil.
9. Install the spark plug wires back on the spark plugs.

➡Make sure to put the spark plug wires back into the proper clamps and clips.

10. Install the engine cover.
11. Reconnect the negative and positive battery cables.

BF50 and BF5A

➧ See Figure 69

1. Disconnect the ignition coil primary leads from the ignition coils.
2. Unplug the spark plug wire and cap from the spark plug.
3. Unclip the spark plug wire from the grommet.
4. Unscrew the bolts, and pull the ignition coil and away from the attachment point on the engine.

To install:

5. Install the ignition coil onto the proper attachment point on the engine.
6. Connect the ignition coil primary side leads to the proper ignition coil.

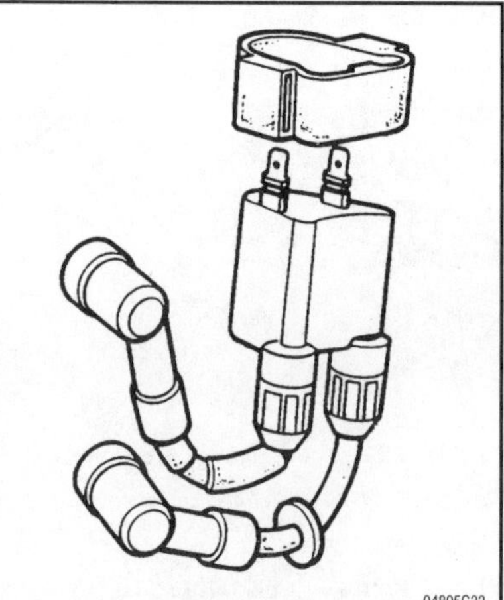

Fig. 66 The ignition coil and spark plug wires are located on the starboard side of the engine near the front of the case—BF9.9A and BF15A

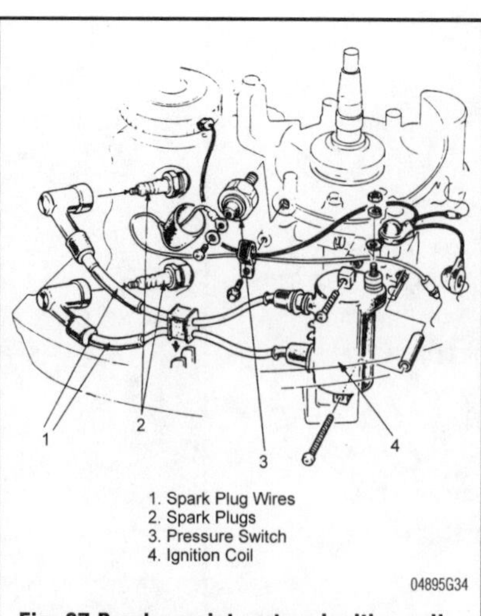

Fig. 67 Breaker point system ignition coil and spark plug wires—BF75, BF8A and BF100

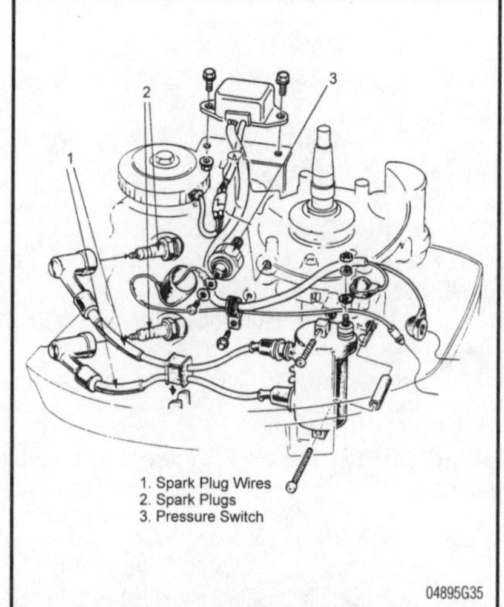

Fig. 68 CDI system ignition coil and spark plug wires—BF75, BF8A and BF100

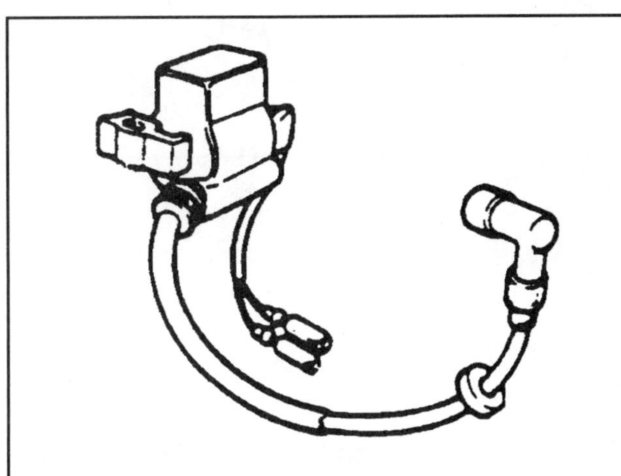

Fig. 69 **Ignition coil and spark plug wire removed from the engine—BF50 and BF5A**

7. Insert a long feeler gauge between the coil and the flywheel, then push the coil toward the flywheel while at the same time tightening the bolts. The clearance at both ends of the coil must be adjusted simultaneously. If a long feeler gauge is not available, use same thickness feeler gauges at both ends. Avoid the magnet area of the flywheel when making the ignition coil adjustment. Specified clearance between the flywheel and the ignition coil is: 0.008-0.024 in. (0.2-0.6 mm).
8. Install the spark plug wire back on the spark plug.

■ **Make sure to put the spark plug wire back into the proper clamp, clip or grommet.**

9. Install the engine cover.

BF20 and BF2A

◆ **See Figures 70 and 71**

1. Disconnect the ignition coil primary lead from the ignition coils.
2. Unplug the spark plug wire and cap from the spark plug.
3. Unclip the spark plug wire from the grommet.
4. Unscrew the bolts, and pull the ignition coil and away from the attachment point on the engine.

To install:

5. Install the ignition coil onto the proper attachment point on the engine.
6. Connect the ignition coil primary side leads to the proper ignition coil.
7. Insert a long feeler gauge between the coil and the flywheel, then push the coil toward the flywheel while at the same time tightening the bolts.

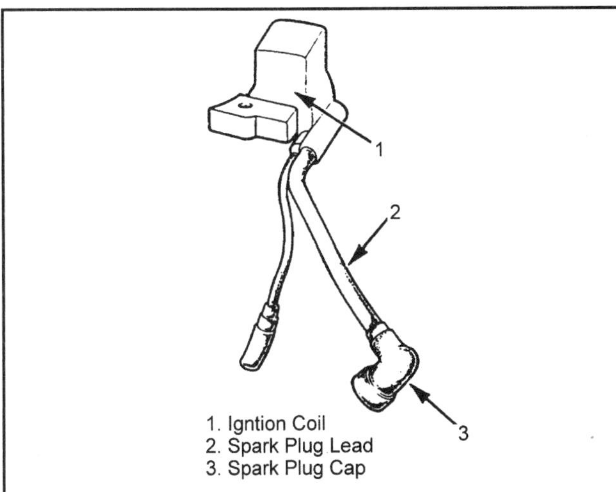

Fig. 70 **Ignition coil and spark plug wire are mounted next to the flywheel—BF20 and BF2A**

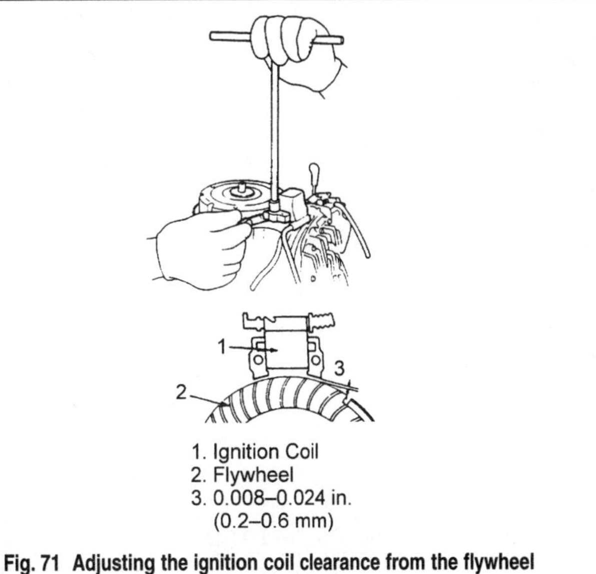

Fig. 71 **Adjusting the ignition coil clearance from the flywheel**

The clearance at both ends of the coil must be adjusted simultaneously. If a long feeler gauge is not available, use same thickness feeler gauges at both ends. Avoid the magnet area of the flywheel when making the ignition coil adjustment. Specified clearance between the flywheel and the ignition coil is: 0.008-0.024 in. (0.2-0.6 mm) for all except the BF2D on which clearance should be 0.012-0.020 in. (0.3-0.5mm).

8. Install the spark plug wire back on the spark plug.

■ **Make sure to put the spark plug wire back into the proper clamp, clip or grommet.**

9. Install the engine cover.

CDI Unit

DESCRIPTION AND OPERATION

In its simplest form, a CDI ignition is composed of the following elements:

- Magneto
- Pulser coil
- Charge, or source coil
- Igniter (CDI) box
- Ignition coil
- Spark plug

Other components such as main switches, stop switches, or computer systems may be included, though, these items are not necessary for basic CDI operation.

To understand basic CDI operation, it is important to understand the basic theory of induction. Induction theory states that if we move a magnet (magnetic field) past a coil of wire (or the coil by the magnet), AC current will be generated in the coil.

The amount of current produced depends on several factors:

- How fast the magnet moves past the coil
- The size of the magnet (strength)
- How close the magnet is to the coil
- Number of turns of wire and the size of the windings

The current produced in the charge coil goes to the CDI box. On the way in, it is converted to DC current by a diode. This DC current is stored in the capacitor located inside the box. As the charge coil produces current, the capacitor stores it.

At a specific time in the magneto's revolution, the magnets go past the pulser coil. The pulser coil is smaller than the charge coil so it has less current output. The current from the pulser also goes into the CDI box. This current signals the CDI box when to fire the capacitor (the pulser may be called a trigger coil for obvious reasons). The current from the capacitor flows out to the ignition coil and spark plug. The pulser acts much like the points in older ignitions systems.

When the pulser signal reaches the CDI box, all the electricity stored in the capacitor is released at once. This current flows through the ignition coil's primary windings.

The ignition coil is a step-up transformer. It turns the relatively low voltage entering the primary windings into high voltage at the secondary windings. This occurs due to a phenomena known as induction.

The high voltage generated in the secondary windings leaves the ignition coil and goes to the spark plug. The spark in turn ignites the air-fuel charge in the combustion chamber.

Once the complete cycle has occurred, the spinning magneto immediately starts the process over again.

Main switches, engine stop switches, and the like are usually connected on the wire in between the CDI box and the ignition coil. When the main switch or stop switch is turned to the OFF position, the switch is closed. This closed switch short-circuits the charge coil current to ground rather than sending it through the CDI box. With no charge coil current through the CDI box, there is no spark and the engine stops or, if the engine is not running, no spark is produced.

TESTING

The charts outline testing procedures for the CDI unit. The unit may remain installed on the powerhead, or it may be removed for testing. In either case, the testing procedures are identical.

Measure the continuity between the CDI unit terminals. Use the accompanying charts for correct resistance specifications. If the any of the readings are not within specifications, the CDI unit must be replaced.

CDI Test Chart—BF75A and BF90A

Unit : kΩ

TESTER (−) \ TESTER (+)	1	2	3	4	5	6	7	8	9	10	11	12	13	14	15	16	17	18
1		10-30	10-30	10-30	11-32	18-50	18-50	10-30	∞	18-50	6-17	10-30	9-25	∞	10-30	6-17	∞	6-17
2	10-30		10-30	10-30	11-32	18-50	18-50	10-30	∞	18-50	6-17	10-30	10-30	∞	9-25	6-17	∞	6-17
3	10-30	10-30		2-8	1.5-5	8-22	8-22	6-17	∞	7-20	4-12	8-22	9-25	∞	9-25	4-12	∞	4-12
4	10-30	10-30	2-8		1.5-5	8-22	8-22	6-17	∞	7-20	4-12	8-22	9-25	∞	9-25	4-12	∞	4-12
5	10-30	10-30	1.5-5	1.5-5		5-19	5-19	6-23	∞	5-19	3-10	8-22	9-25	∞	9-25	3-10	∞	3-10
6	∞	∞	∞	∞	∞		∞	∞	∞	∞	∞	∞	∞	∞	∞	∞	∞	∞
7	∞	∞	∞	∞	∞	∞		∞	∞	∞	∞	∞	∞	∞	∞	∞	∞	∞
8	∞	∞	∞	∞	∞	∞	∞		∞	∞	∞	∞	∞	∞	∞	∞	∞	∞
9	14-40	14-40	4-11	4-11	2-9	12-33	10-30	10-30		11-32	7-18	12-33	15-42	∞	15-42	7-18	∞	7-18
10	∞	∞	∞	∞	∞	∞	∞	∞	∞		∞	∞	∞	∞	∞	∞	∞	∞
11	5-13	5-13	4-10	4-10	4-10	10-30	10-30	2-7	∞	10-30		3-10	2-7	∞	2-7	0	∞	0
12	9-25	9-25	5-13	5-13	3-12	13-26	14-40	7-20	∞	13-36	3-10		8-22	∞	8-22	3-10	∞	3-10
13	18-50	18-50	17-47	17-47	19-53	36-100	36-100	20-57	∞	30-80	10-30	16-43		∞	22-60	10-30	∞	10-30
14	∞	∞	∞	∞	∞	∞	∞	∞	∞	∞	∞	∞	∞		∞	∞	∞	∞
15	18-50	18-50	17-47	17-47	19-53	36-100	36-100	20-57	∞	30-80	10-30	16-43	22-60	∞		10-30	∞	10-30
16	5-13	5-13	3-10	3-10	3-10	10-30	10-30	2-7	∞	10-30	0	3-10	2-7	∞	2-7		∞	0
17	∞	∞	∞	∞	∞	∞	∞	∞	∞	∞	∞	∞	∞	∞	∞	∞		∞
18	5-13	5-13	3-10	3-10	3-10	10-30	10-30	2-7	∞	10-30	0	3-10	2-7	∞	2-7	0	∞	

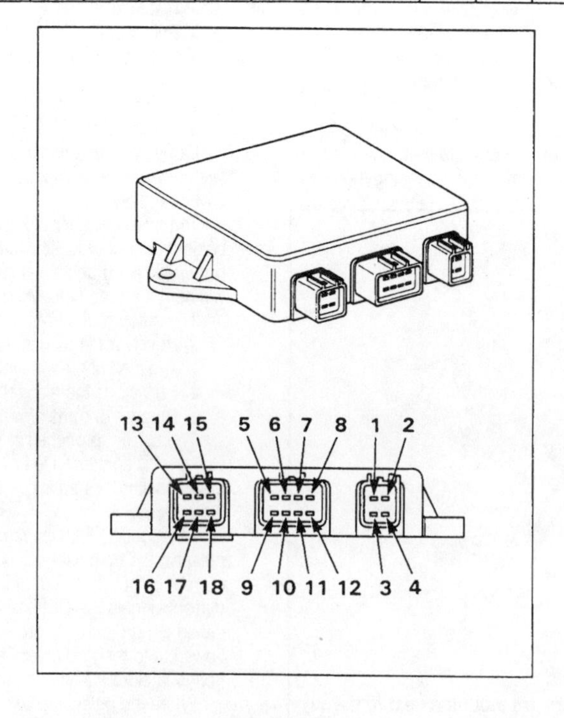

04895C01

CDI Test Chart: BF35A–BF45A (Tiller Handle)

CDI UNIT

Bl/G 8, R 10, Y 9, Bl 11, Bl/Y 7, Bl/Bu 6, R 10, Bl 11, Y 9, Bu 1, Bu/G 3, Bl 11, Bu/Y 2, G 4, Bl 11, Bl/R 5, Bl 11

Bl	BLACK	Br	BROWN
Y	YELLOW	O	ORANGE
Bu	BLUE	Lb	LIGHT BLUE
G	GREEN	Lg	LIGHT GREEN
R	RED	P	PINK
W	WHITE	Gr	GRAY

(KΩ)

COLOR	Tester (−) / Tester (+)	1	2	3	4	5	6	7	8	9	10	11
Bu	1		10–500	10–500	10–500	∞	∞	∞	∞	∞	∞	3–300
Bu/Y	2	10–500		10–500	10–500	∞	∞	∞	∞	∞	∞	3–300
Bu/G	3	10–500	10–500		10–500	∞	∞	∞	∞	∞	∞	3–300
G	4	10–500	10–500	10–500		∞	∞	∞	∞	∞	∞	0.5–50
Bl/R	5	20–500	20–500	20–500	0.5–50		∞	∞	∞	∞	∞	0.5–50
Bl/Bu	6	∞	∞	∞	∞	∞		∞	∞	∞	∞	∞
Bl/Y	7	∞	∞	∞	∞	∞	∞		∞	∞	∞	∞
Bl/G	8	∞	∞	∞	∞	∞	∞	∞		∞	∞	∞
Y	9	10–500	10–500	10–500	10–500	∞	∞	∞	∞		∞	0.5–50
R	10	3–300	3–300	3–300	3–300	∞	∞	∞	∞	∞		0.5–50
Bl	11	3–300	3–300	3–300	3–300	∞	∞	∞	∞	∞	∞	

TESTER (−) / TESTER (+)	PC.E	EXC.E.	IND.E.	SW.E.
GND	Continuity	Continuity	Continuity	Continuity

04895C02

CDI Test Chart: BF35A–BF45A (Non-Tiller Handle)

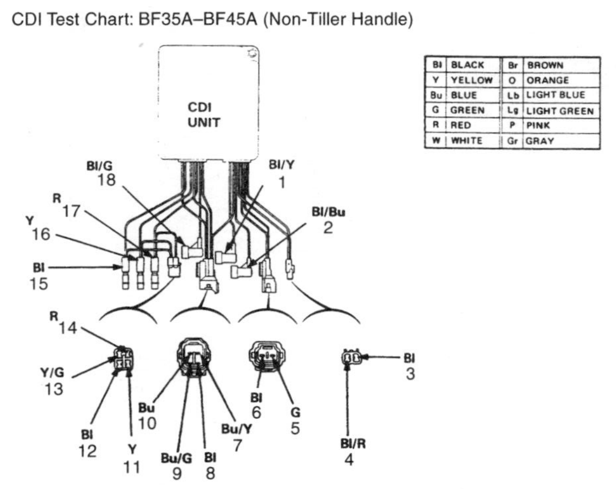

Bl	BLACK	Br	BROWN
Y	YELLOW	O	ORANGE
Bu	BLUE	Lb	LIGHT BLUE
G	GREEN	Lg	LIGHT GREEN
R	RED	P	PINK
W	WHITE	Gr	GRAY

(KΩ)

COLOR	Tester (−) / Tester (+)	10	7	9	5	4	2	1	18	11, 16	14, 17	3, 6, 8, 12, 15	13
Bu	10		10–500	10–500	10–500	∞	∞	∞	∞	∞	∞	3–300	∞
Bu/Y	7	10–500		10–500	10–500	∞	∞	∞	∞	∞	∞	3–300	∞
Bu/G	9	10–500	10–500		10–500	∞	∞	∞	∞	∞	∞	3–300	∞
G	5	10–500	10–500	10–500		∞	∞	∞	∞	∞	∞	0.5–50	∞
Bl/R	4	20–500	20–500	20–500	0.5–50		∞	∞	∞	∞	∞	0.5–50	∞
Bl/Bu	2	∞	∞	∞	∞	∞		∞	∞	∞	∞	∞	∞
Bl/Y	1	∞	∞	∞	∞	∞	∞		∞	∞	∞	∞	∞
Bl/G	18	∞	∞	∞	∞	∞	∞	∞		∞	∞	∞	∞
Y	11, 16	10–500	10–500	10–500	10–500	∞	∞	∞	∞		∞	0.5–50	∞
R	14, 17	3–300	3–300	3–300	3–300	∞	∞	∞	∞	∞		0.5–50	∞
Bl	3, 6, 8, 12, 15	3–300	3–300	3–300	3–300	∞	∞	∞	∞	∞	∞		∞
Y/G	13	10–500	10–500	10–500	10–500	∞	∞	∞	∞	∞	∞	0.5–500	

TESTER (−) / TESTER (+)	PC.E	EXC.E.	IND.E.	SW.E.
GND	Continuity	Continuity	Continuity	Continuity

04895C03

CDI Test Chart: BF25A and BF30A (Tiller Handle)

CDI UNIT

Bl	BLACK	Br	BROWN
Y	YELLOW	O	ORANGE
Bu	BLUE	Lb	LIGHT BLUE
G	GREEN	Lg	LIGHT GREEN
R	RED	P	PINK
W	WHITE	Gr	GRAY

Bl/Y 1, Bl/Bu 2, Bl 3, Bl/R 4, G 5, Bl 6, Bu/Y 7, Bl 8, Bl/G 9, Bu 10, Bl/G 11, Bl 12, Y 13, Bl 14, R/Bu 15, Y 16

(KΩ)

COLOR	Tester(–) / Tester(+)	10	7	9	5	4	2	1	11	13, 16	15	3, 6, 8, 12, 14
Bu	10		10–500	10–500	10–500	∞	∞	∞	∞	∞	∞	3–300
Bu/Y	7	10–500		10–500	10–500	∞	∞	∞	∞	∞	∞	3–300
Bu/G	9	10–500	10–500		10–500	∞	∞	∞	∞	∞	∞	3–300
G	5	∞	∞	∞		∞	∞	∞	∞	∞	∞	0.5–50
Bl/R	4	20–500	20–500	20–500	0.5–50		∞	∞	∞	∞	∞	0.5–50
Bl/Bu	2	∞	∞	∞	∞	∞		∞	∞	∞	∞	∞
Bl/Y	1	∞	∞	∞	∞	∞	∞		∞	∞	∞	∞
Bl/G	11	∞	∞	∞	∞	∞	∞	∞		∞	∞	∞
Y	13, 16	10–500	10–500	10–500	10–500	∞	∞	∞	∞		∞	0.5–50
R/Bu	15	3–300	3–300	3–300	3–300	∞	∞	∞	∞	∞		0.5–50
Bl	3, 6, 8, 12, 14	3–300	3–300	3–300	3–300	∞	∞	∞	∞	∞	∞	

TESTER (–) / TESTER (+)	PC.E	EXC.E	IND.E
GND	Continuity	Continuity	Continuity

04895C04

CDI Test Chart: BF25A–BF30A (Non-Tiller Handle)

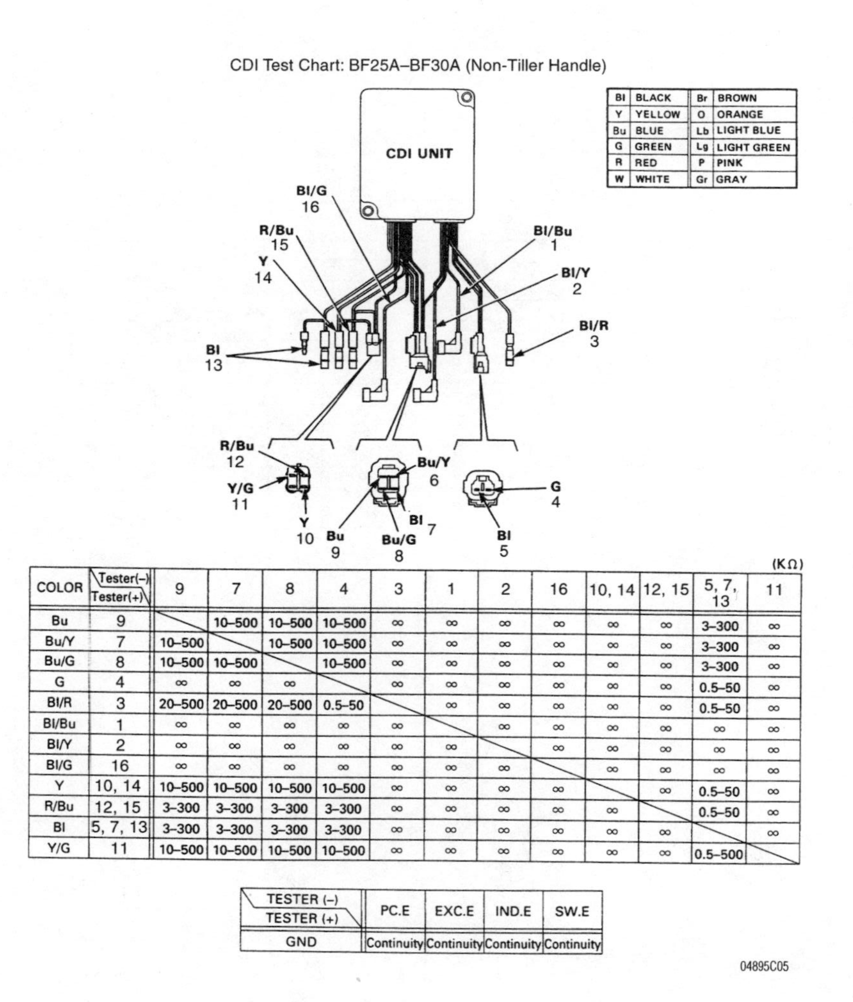

(KΩ)

COLOR	Tester(–) / Tester(+)	9	7	8	4	3	1	2	16	10, 14	12, 15	5, 7, 13	11
Bu	9		10–500	10–500	10–500	∞	∞	∞	∞	∞	∞	3–300	∞
Bu/Y	7	10–500		10–500	10–500	∞	∞	∞	∞	∞	∞	3–300	∞
Bu/G	8	10–500	10–500		10–500	∞	∞	∞	∞	∞	∞	3–300	∞
G	4	∞	∞	∞		∞	∞	∞	∞	∞	∞	0.5–50	∞
Bl/R	3	20–500	20–500	20–500	0.5–50		∞	∞	∞	∞	∞	0.5–50	∞
Bl/Bu	1	∞	∞	∞	∞	∞		∞	∞	∞	∞	∞	∞
Bl/Y	2	∞	∞	∞	∞	∞	∞		∞	∞	∞	∞	∞
Bl/G	16	∞	∞	∞	∞	∞	∞	∞		∞	∞	∞	∞
Y	10, 14	10–500	10–500	10–500	10–500	∞	∞	∞	∞		∞	0.5–50	∞
R/Bu	12, 15	3–300	3–300	3–300	3–300	∞	∞	∞	∞	∞		0.5–50	∞
Bl	5, 7, 13	3–300	3–300	3–300	3–300	∞	∞	∞	∞	∞	∞		∞
Y/G	11	10–500	10–500	10–500	10–500	∞	∞	∞	∞	∞	∞	0.5–500	

TESTER (–) / TESTER (+)	PC.E	EXC.E	IND.E	SW.E
GND	Continuity	Continuity	Continuity	Continuity

04895C05

CDI TEST CHART—BF9.9A AND BF15A

Red tester lead \ Black tester read	1	2	3	4	5	6
1		50–500	50–500	∞	100–∞	100–∞
2	20–200		0–5	∞	50–500	50–500
3	20–200	0–5		∞	50–500	50–500
4	∞	∞	∞		∞	∞
5	50–500	5–50	5–50	∞		∞
6	50–500	0.5–20	0.5–20	∞	5–50	

04895C06

CDI TEST CHART—BF75 AND BF100

Engine serial numbers 1200001 – 1299999

[kΩ]

(–) Probe \ (+) Probe	Brown	Green	Black	Yellow	Orange	White
Brown		6 – 10	0 – 0.5	10 – 20	∞	∞
Green	0.5 – 10		0.5 – 10	15 – 30	∞	∞
Black	0 – 0.5	6 – 10		10 – 20	∞	∞
Yellow	10 – 20	15 – 30	10 – 20		∞	∞
Orange	100 – 400	100 – 400	100 – 400	100 – 400		∞
White	10 – 100	10 – 100	10 – 100	10 – 100	∞	

Engine serial number 1300001 – 1300700:

[kΩ]

(–) Probe \ (+) Probe	Brown	Green	Black	Yellow	Orange	Blue	Black/White	White
Brown		6 – 10	0 – 0.5	10 – 20	∞	6 – 10	8 – 30	∞
Green	0.5 – 10		0.5 – 10	15 – 30	∞	0.05 – 5	1 – 20	∞
Black	0 – 0.5	6 – 10		10 – 20	∞	6 – 10	8 – 30	∞
Yellow	10 – 20	15 – 30	10 – 20		∞	15 – 30	15 – 50	∞
Orange	100 – 400	100 – 400	100 – 400	100 – 400		100 – 400	100 – 400	∞
Blue	0.5 – 10	0.05 – 5	0.5 – 10	15 – 30	∞		1 – 20	∞
Black/White	20 – 40	15 – 25	20 – 40	35 – 70	∞	15 – 25		∞
White	10 – 100	10 – 100	10 – 100	10 – 100	∞	20–	0.5 – 10	

Engine serial number 1300701/1300703 (BF75/BF100) and subsequent:

[kΩ]

(–) Probe \ (+) Probe	Brown	Green	Black	Yellow	Orange	Blue	Black/White	White
Brown		35 – 100	0 – 0.5	10 – 20	∞	30 – 100	30 – 100	∞
Green	0.5 – 10		0.5 – 10	15 – 30	∞	10 – 40	0	∞
Black	0 – 0.5	20 – 100		10 – 20	∞	30 – 100	30 – 100	∞
Yellow	10 – 20	30 – 200	10 – 20		∞	30 – 200	30 – 200	∞
Orange	∞	∞	∞	∞		∞	∞	∞
Blue	3 – 30	0.5 – 10	3 – 30	20 – 200	∞		0.5 – 10	∞
Black/White	0.5 – 10	0	0.5 – 10	15 – 30	∞	10 – 40		∞
White	2 – 20	0.5 – 10	2 – 20	20 – 200	∞	20 – 200	0.5 – 10	

04895C07

REMOVAL & INSTALLATION

BF75A and BF90A

See Figure 72

1. Disconnect the negative and positive battery cables.
2. Remove the engine cover.
3. Remove the CDI unit cover.

CDI unit is mounted behind the wiring harness.

4. Disconnect or gently move the necessary wiring to access the CDI unit.
5. Unplug the three connectors on the CDI unit.
6. Unscrew the two mounting bolts and remove the CDI unit.

To install:

7. Install the CDI unit and screw in the two mounting bolts.
8. Plug in the three connectors on the CDI unit.
9. Reconnect and secure the wiring harness in front of the CDI unit.
10. Replace the CDI unit cover.
11. Install the engine cover.
12. Reconnect the negative and positive battery cables.

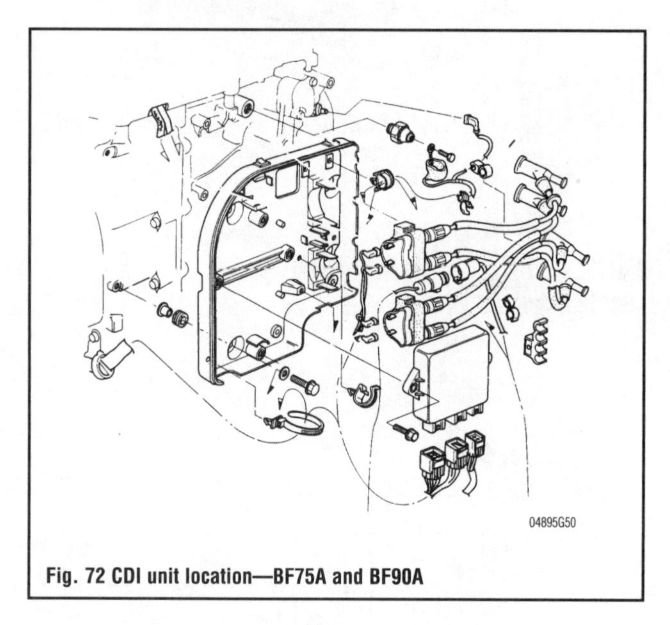

Fig. 72 CDI unit location—BF75A and BF90A

BF50A, BF45A, BF40A, BF35A, BF30A and BF25A

See Figures 73 and 74

1. Disconnect the negative and positive battery cables.
2. Remove the engine cover.
3. Remove the CDI unit cover.

CDI unit is mounted behind the wiring harness.

4. Disconnect or gently move the necessary wiring to access the CDI unit.
5. Unplug the two connectors on the CDI unit.
6. Unscrew the two mounting bolts and remove the CDI unit.

To install:

7. Install the CDI unit and screw in the two mounting bolts.
8. Plug in the two connectors on the CDI unit.
9. Reconnect and secure the wiring harness in front of the CDI unit.
10. Replace the CDI unit cover.
11. Install the engine cover.
12. Reconnect the negative and positive battery cables.

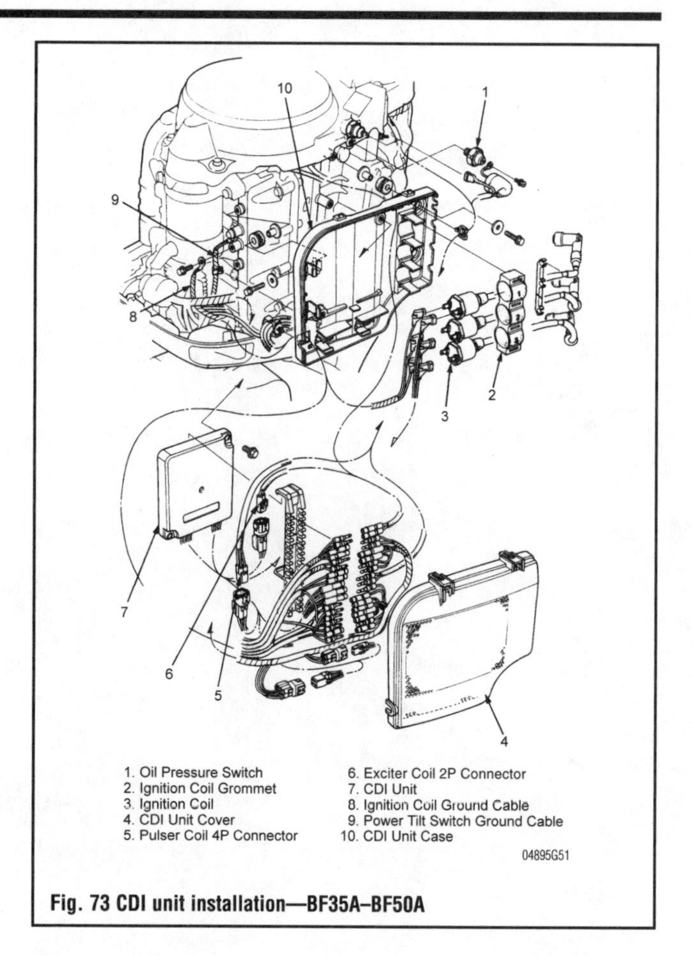

Fig. 73 CDI unit installation—BF35A-BF50A

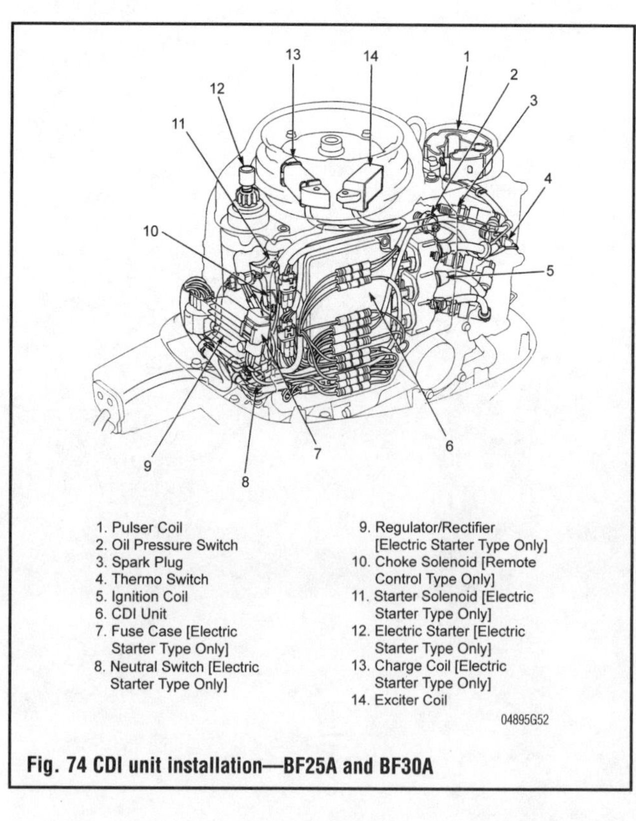

Fig. 74 CDI unit installation—BF25A and BF30A

BF9.9A and BF15A

➧ See Figure 75

1. Disconnect the negative and positive battery cables.
2. Remove the engine cover.
3. Unplug the wire harness from the CDI unit.
4. Slide the CDI unit out of the unit cushion on the ignition coil bracket.

To install:

5. Slide the CDI unit back into the CDI unit cushion and insert the entire assembly back into the ignition coil bracket.
6. Plug the wire harness back into the CDI unit.
7. Install the engine cover.
8. Reconnect the negative and positive battery cables.

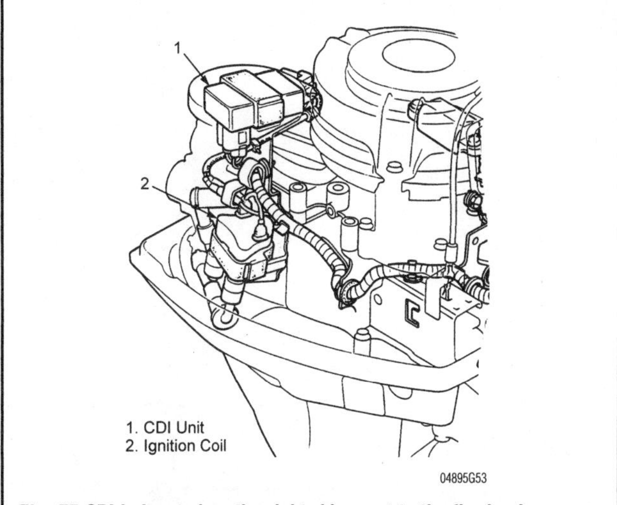

Fig. 75 CDI is located on the right side, next to the flywheel—BF9.9A and BF15A

BF75, BF8A and BF100

1. Remove the engine cover.
2. Disconnect the CDI unit connectors at the magneto pulser coil, ignition coil and engine switch.
3. Unscrew the CDI unit bolts and remove the CDI unit from the engine

To install:

4. Install the CDI unit on the engine and tighten the bolts.
5. Carefully route the wires, making sure not to interfere with the flywheel or camshaft pulley rotation. Always use the wire clamps.
6. Connect the wire connectors at the pulser coil, ignition coil, and engine switch
7. Install the engine cover.

Rectifier

TESTING

The rectifier uses a diode that acts as a one-way check valve; it can be checked out using an ohmmeter. Hook up the ohmmeter leads, one to a common ground and one to a lead (terminal) coming from the rectifier. Using the high ohm scale, look for a reading, then reverse the leads. This time there should be no reading (infinity). This means the diode is good. Repeat the test on the other leads. Also test between the red lead and the other two leads (terminals). A normal diode will show a reading

1. The charts outline testing procedures for the rectifier. The unit may remain installed on the powerhead, or it may be removed for testing. In either case, the testing procedures are identical.
2. Select the appropriate scale on the ohmmeter. Make contact with the Red meter lead to the leads called out as (+) and the Black meter lead to the leads called out as (-). Proceed slowly and carefully in the order given.
3. If resistance is not as specified, the rectifier is faulty and should be replaced.

Rectifier Test Chart — Except V4 and V6 Powerheads

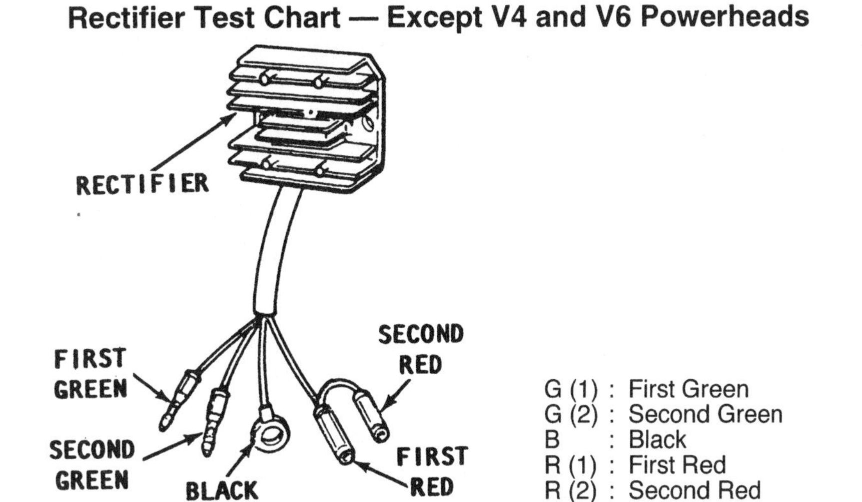

	G (1)	G (2)	B	R (1)	R (2)
G (1)		NC	NC	C	C
G (2)	NC		NC	C	C
B	C	C		C	C
R (1)	NC	NC	NC		
R (2)	NC	NC	NC		

C: Continuity
NC: No Continuity

04706C16

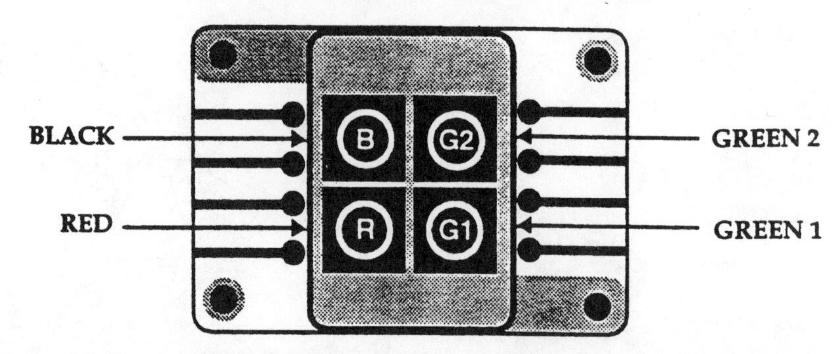

		Positive Test Lead Position			
		(1) G1	(2) G2	(3) R	(4) B
Negative Test Lead Position	(1) G1		NO	YES	NO
	(2) G2	YES		YES	YES
	(3) R	NO	NO		
	(4) B	YES	YES	YES	

YES = Continuity (Needle moves) NO = No continuity (Needle does not move)

04706C17

BF115A and BF130A

See Figure 76

➡The rectifier on this engine is incorporated into the alternator itself. The rectifier can be checked for continuity only after removing the alternator from the engine.

1. Check the rectifier for continuity between the "B" terminal and the respective "P" terminals (P1, P2, P3, P4), and between the "E" terminal and the "P" terminals (P1, P2, P3, P4). Check for continuity in two directions by reversing the test meter leads.

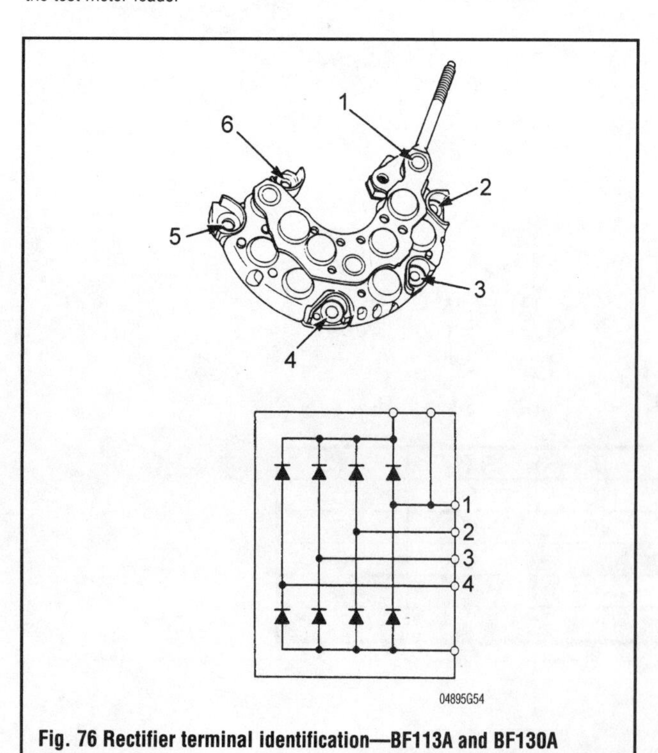

Fig. 76 Rectifier terminal identification—BF113A and BF130A

2. There should be continuity in one direction only. Replace the rectifier assembly if there is continuity in both directions.

BF50 and BF5A

See Figure 77

1. Using the R x 100 ohms range of the multimeter, check the resistance of the silicone rectifier as specified in the rectifier test chart.
2. If the readings are out of specification, replace the rectifier.

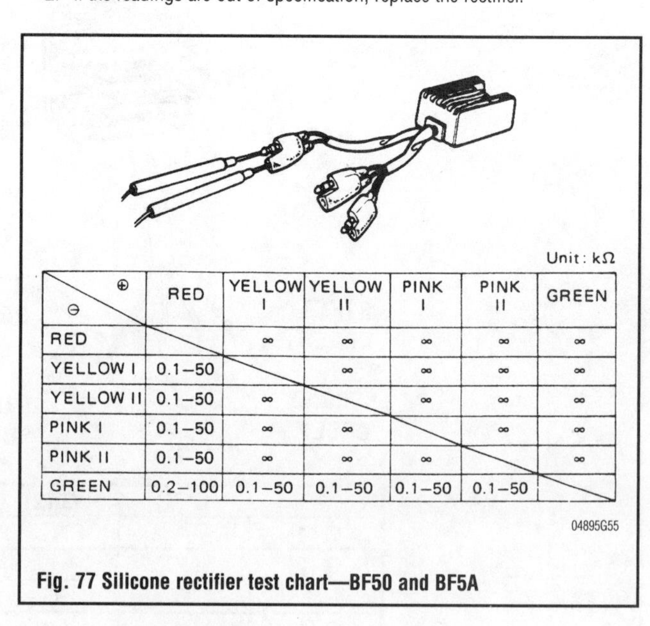

Unit: kΩ

⊖ \ ⊕	RED	YELLOW I	YELLOW II	PINK I	PINK II	GREEN
RED		∞	∞	∞	∞	∞
YELLOW I	0.1–50		∞	∞	∞	∞
YELLOW II	0.1–50	∞		∞	∞	∞
PINK I	0.1–50	∞	∞		∞	∞
PINK II	0.1–50	∞	∞	∞		∞
GREEN	0.2–100	0.1–50	0.1–50	0.1–50	0.1–50	

04895G55

Fig. 77 Silicone rectifier test chart—BF50 and BF5A

REMOVAL & INSTALLATION

BF115A and BF130A

The rectifier on this engine is incorporated into the alternator. For more information on removal and installation of the rectifier, refer to the "Alternator" section.

BF50 and BF5A

The silicone rectifier is located on the left, rear of the engine.

1. Remove the engine cover.
2. Unplug the rectifier from the fuse holder and the charging and lighting coils.
3. Unscrew the retaining bolt and remove the rectifier from the engine.

To install:

4. Install the rectifier on the engine and tighten the retaining bolt.

➡Make sure to route the silicone rectifier wires between the mounting boss and the inside of the engine housing wall and clamp them together with the lighting coil wires.

5. Plug the rectifier into the connections on the fuse holder and charging and lighting coils.
6. Replace the engine cover.

Regulator

DESCRIPTION & OPERATION

The voltage regulator controls the alternators field voltage by grounding one end of the field windings very rapidly. The frequency varies according to current demand the more the field is grounded, the more voltage and current the alternator produces. Voltage is maintained at about 13.5–15 volts. During high engine speeds and low current demands, the regulator will adjust the voltage of the alternator field to lower the alternator output voltage. Conversely, when the engine is idling and the current demands may be high, the regulator will increase field voltage, increasing the output of the alternator.

➡The BF115A and BF130A are the only ones to use a regulator.

TESTING

Honda does not provide a specific voltage regulator test procedure for this model, however, on some models it is possible to isolate the alternator from the regulator by grounding the **F** (field) terminal. Grounding the **F** terminal removes the regulator from the circuit and forces full alternator output. On alternators equipped with internal regulators, we recommend replacing the complete assembly if either the alternator or regulator is defective.

The voltage regulator (IC Regulator) on this engine is incorporated into the alternator itself. The voltage regulator can be checked for continuity only after removing the alternator from the engine. See the alternator overhaul procedure for removing the alternator.

REMOVAL & INSTALLATION

The regulator on this engine is incorporated into the alternator. For more information on removal and installation of the rectifier, refer to the "Alternator" section.

Regulator/Rectifier

The BF9.9A to BF90A use a regulator and rectifier built into one unit. Both components function separately, but share a common housing.

TESTING

BF75A and BF90A

➧ See Figure 78

1. Measure the resistance between the terminals shown in the diagram.
2. Replace the regulator/rectifier if any of the readings are out of specification.

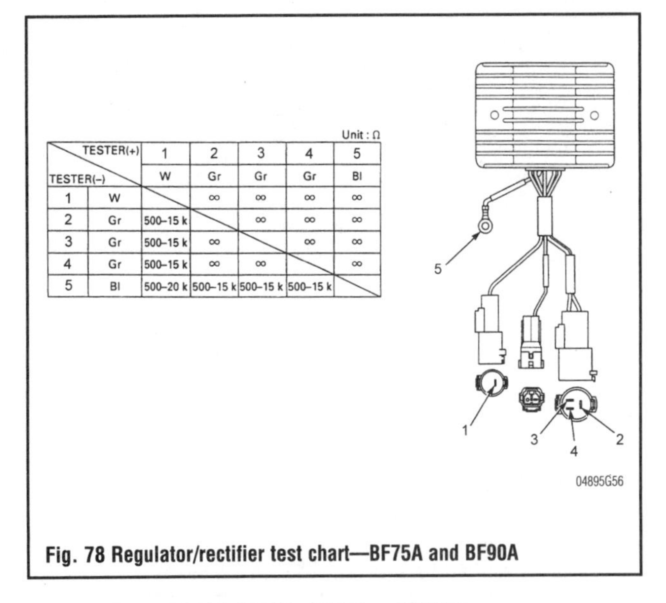

Unit : Ω

TESTER(−) \ TESTER(+)		1	2	3	4	5
		W	Gr	Gr	Gr	Bl
1	W		∞	∞	∞	∞
2	Gr	500–15 k		∞	∞	∞
3	Gr	500–15 k	∞		∞	∞
4	Gr	500–15 k	∞	∞		∞
5	Bl	500–20 k	500–15 k	500–15 k	500–15 k	

Fig. 78 Regulator/rectifier test chart—BF75A and BF90A

BF50A, BF45A, BF40A, BF35A, BF30A and BF25A

➧ See Figures 79 and 80

1. Measure the resistance between the terminals shown in the diagram. Note that there are different model regulator/rectifiers for both remote control/electric start and tiller handle/electric start engines.

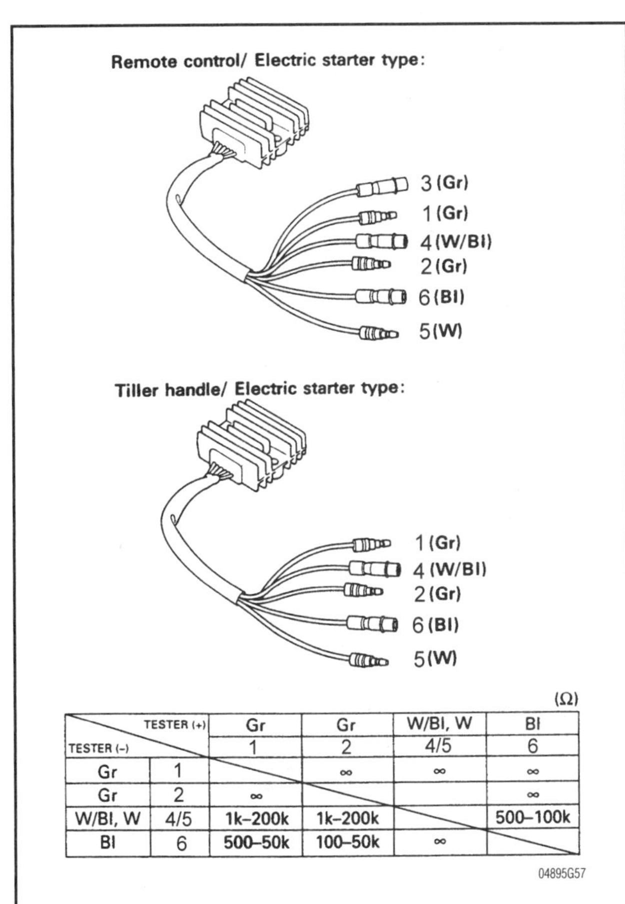

(Ω)

TESTER (−) \ TESTER (+)		Gr	Gr	W/Bl, W	Bl
		1	2	4/5	6
Gr	1		∞	∞	∞
Gr	2	∞		∞	∞
W/Bl, W	4/5	1k–200k	1k–200k		500–100k
Bl	6	500–50k	100–50k	∞	

Fig. 79 Regulator/rectifier test chart—BF25A and BF30A

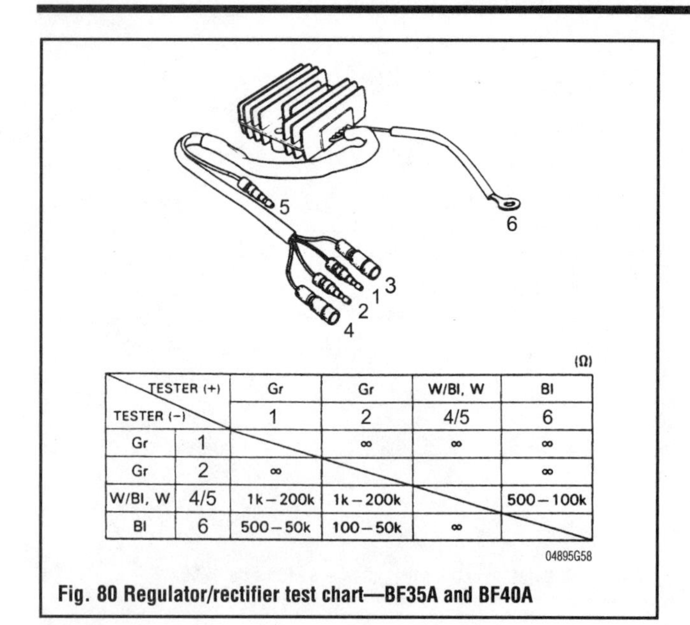

(Ω)

TESTER (+) / TESTER (−)		Gr 1	Gr 2	W/Bl, W 4/5	Bl 6
Gr	1		∞	∞	∞
Gr	2	∞			∞
W/Bl, W	4/5	1k – 200k	1k – 200k		500 – 100k
Bl	6	500 – 50k	100 – 50k	∞	

04895G58

Fig. 80 Regulator/rectifier test chart—BF35A and BF40A

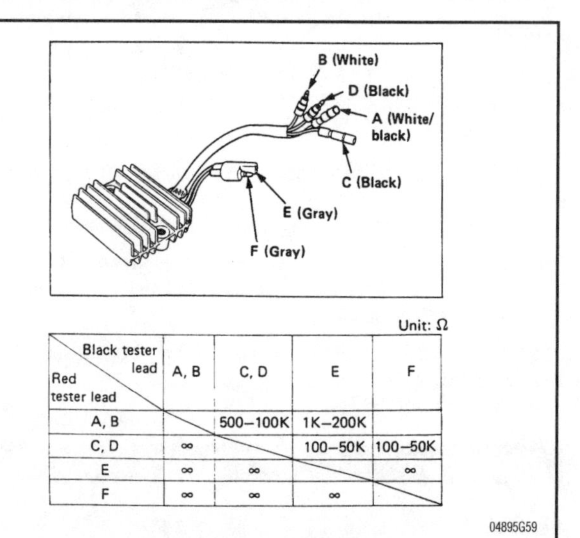

Unit: Ω

Red tester lead / Black tester lead	A, B	C, D	E	F
A, B		500–100K	1K–200K	
C, D	∞		100–50K	100–50K
E	∞	∞		∞
F	∞	∞	∞	

04895G59

Fig. 81 Regulator/rectifier test chart—BF9.9 and BF15A

BF9.9 and BF15A

➧ See Figure 81

1. Measure the resistance between the terminals.
2. Replace the regulator/rectifier if any of the measurements are out of specification.

REMOVAL & INSTALLATION

BF75A and BF90A

The regulator/rectifier is located on the port side of the engine below the spark plug wires.

1. Disconnect the negative and positive battery cables.
2. Remove the engine cover.
3. Unplug the three wiring connectors and disconnect the ground wire.
4. Remove the retaining bolts and remove the regulator/rectifier from the engine.

To install:

5. Reinstall the regulator/rectifier on the engine and tighten the retaining bolts.
6. Plug in the three wiring connectors and reconnect the ground wire.
7. Install the engine cover.
8. Connect the negative and positive battery cables.

BF25A, BF30A, BF35A, BF40A, BF45A AND BF50A

The location of the regulator/rectifier on these is on the forward part of the engine, in front of the starter motor. The removal and installation is similar on all the engines in this series.

1. Disconnect the negative and positive battery cables.
2. Remove the engine cover.
3. Unplug the regulator/rectifier wire terminals and disconnect the ground wire.
4. Remove the retaining screws and lift out the rectifier/regulator from the engine.

To install:

5. Install the regulator/rectifier on the engine and tighten the retaining screws.
6. Plug in the wire terminals and tighten the ground wire connector.
7. Install the engine cover.
8. Reconnect the battery positive and negative cables.

BF9.9 and BF15A

The regulator/rectifier is easily accessible, located on the front of the engine below the flywheel.

1. Remove the engine cover.
2. Unplug the regulator/rectifier leads from the wiring harness.
3. Unscrew the retaining bolts and remove the regulator/rectifier from the engine.

To install:

4. Install the regulator/rectifier on the engine and tighten the retaining bolts.
5. Plug in the wire leads to the wiring harness.
6. Replace the engine cover.

ELECTRONIC IGNITION

Description And Operation

The BF113A and BF130A fuel injected engines use an electronic ignition system which, when combined with the fuel injection system, becomes an integrated electronic engine management system that improves fuel consumption, performance and exhaust emissions.

The ignition timing control system controls the ignition timing by first determining the basic ignition timing based on an engine speed signal sent by the pulser coil and the intake manifold vacuum signal sent by the MAP sensor.

The ignition timing control system, with inputs from the Throttle Position Sensor (TPS), the Engine Temperature Compensation Sensor (ECT), and the Suction Air Temperature Sensor (IAT), compensates the base timing for optimum ignition timing

Self Diagnosis

➧ See Figures 82 and 83

The ECM (Engine Control Module) has the self-diagnosis function which turns the MIL (Malfunction Indicator Lamp) ON when it detects an out of specification sensor reading within the input/output system. You need to short-circuit the service check connector when the MIL is ON. The MIL should blink indicating the probable cause of the problem by the number and length of the blinks. When multiple problems occur simultaneously, the MIL will indicate these by blinking separate codes, one after another.

Codes 10 and after are indicated by a series of long and short blinks. The number of long blinks equals the first digit of the code and the number of short

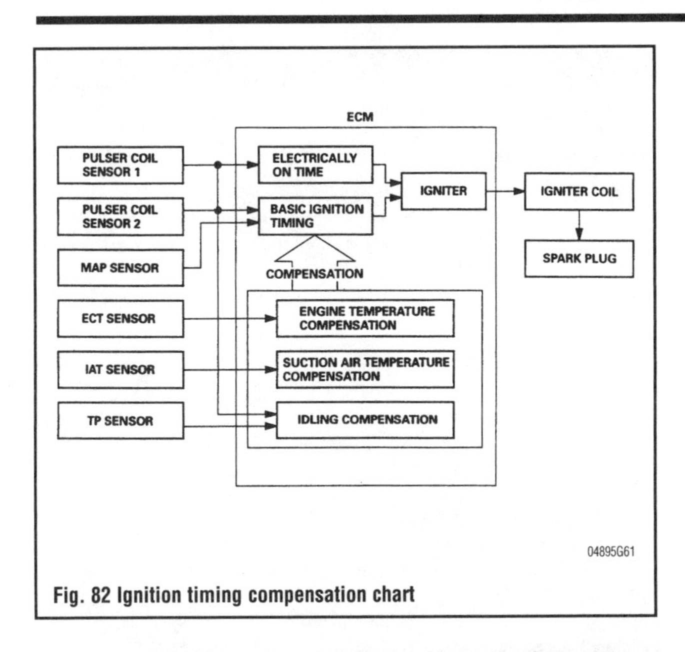

Fig. 82 Ignition timing compensation chart

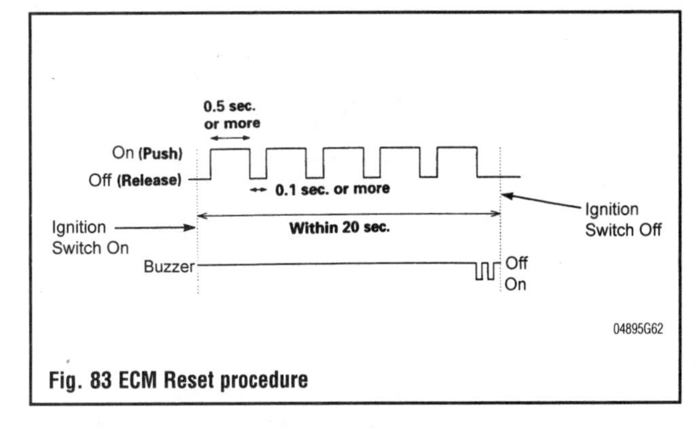

Fig. 83 ECM Reset procedure

blinks equals the second digit of the code. When the overheat sensor is faulty, the overheat light as well as the MIL should blink simultaneously.

If the ECM body is faulty, the MIL turns ON and does not go out when the service check connector is short-circuited.

For more information on self diagnosis and how to read and clear codes, refer to "Diagnostic Trouble Codes".

CHARGING CIRCUIT

Description and Operation

See Figures 84, 85, 86 and 87

The BF113A and BF130A use an automotive-style, belt-driven alternator, while the BF75A and BF90A use a stator-type alternator, driven off the crankshaft and mounted below the starter pulley.

A typical charging system contains an alternator (generator), drive belt, battery, voltage regulator and the associated wiring. The charging system, like the starting system is a series circuit with the battery wired in parallel. After the engine is started and running, the alternator takes over as the source of power and the battery then becomes part of the load on the charging system.

Some manufacturers use the term generator instead of alternator. Many years ago there used to be a difference, now they are one and the same. The alternator, which is driven by the belt, consists of a rotating coil of laminated wire called the rotor. Surrounding the rotor are more coils of laminated wire that remain stationary (which is how we get the term stator) just inside the alternator case. When current is passed through the rotor via the slip rings and brushes, the rotor becomes a rotating magnet with, of course, a magnetic field. When a magnetic field passes through a conductor (the stator), alternating current (A/C) is generated. This A/C current is rectified, turned into direct current (D/C), by the diodes located within the alternator.

The voltage regulator controls the alternator's field voltage by grounding one end of the field windings very rapidly. The frequency varies according to current demand. The more the field is grounded, the more voltage and current the alternator produces. Voltage is maintained at about 13.5–15 volts. During high engine speeds and low current demands, the regulator will adjust the voltage of the alternator field to lower the alternator output voltage. Conversely, when the vehicle is idling and the current demands may be high, the regulator will increase the field voltage, increasing the output of the alternator. Some vehicles actually turn the alternator off during periods of no load and/or wide open throttle. This was designed to reduce fuel consumption and increase power. Depend-

STATOR
SIX DIODES
AC
AC
DC
DC
+ −
8852BG59

Fig. 84 Negative and positive diodes convert AC current into DC current-note that the AC current reverses direction while the DC current flows in only one direction

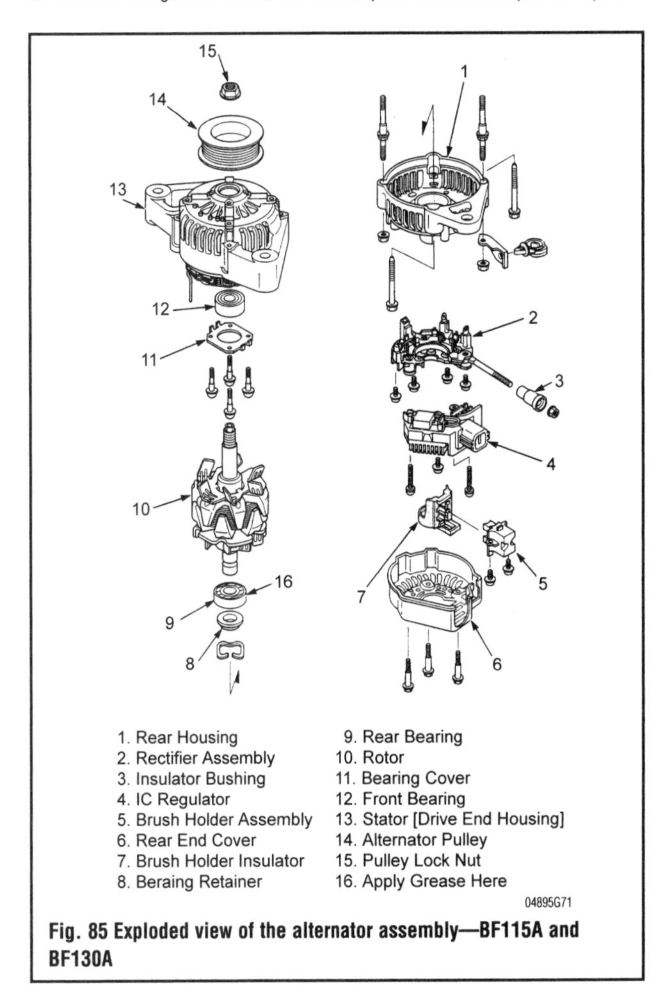

Fig. 85 Exploded view of the alternator assembly—BF115A and BF130A

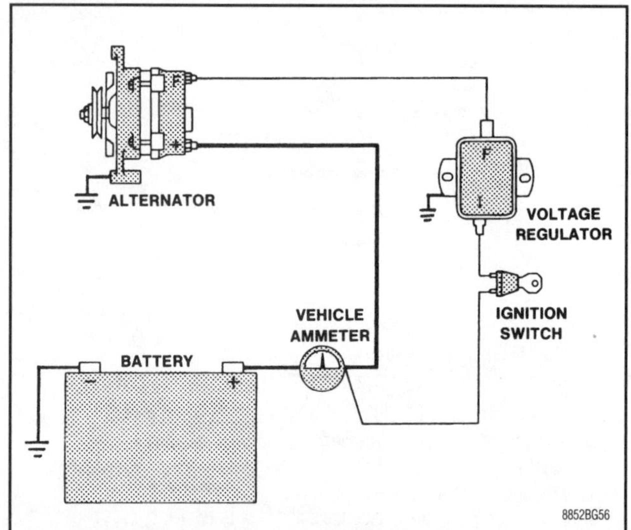

Fig. 86 Common charging system schematic. Note that not all vehicles are equipped with an ammeter in the circuit

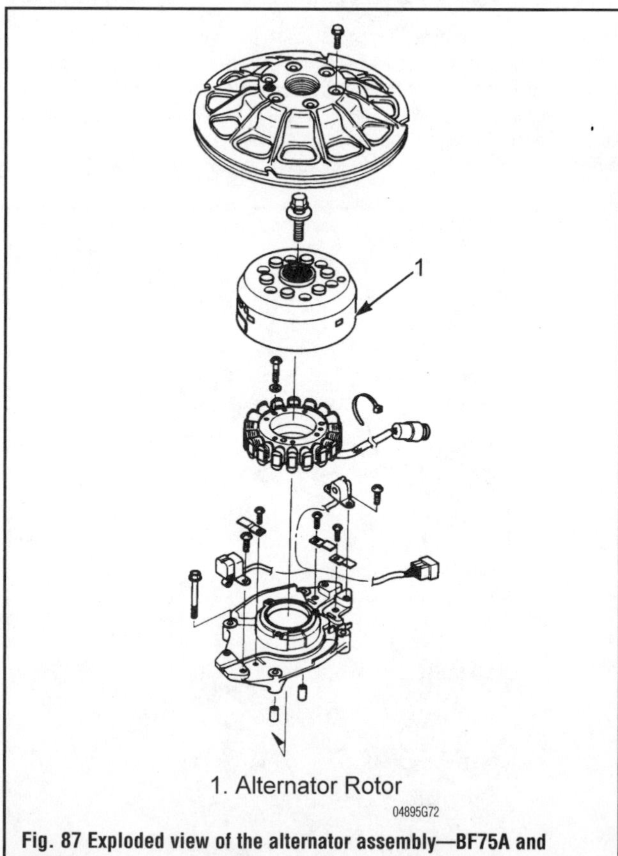

Fig. 87 Exploded view of the alternator assembly—BF75A and BF90A

ing on the manufacturer, voltage regulators can be found in different locations, including inside or on the alternator, and even inside the PCM.

Drive belts are often overlooked when diagnosing a charging system failure. Check the belt tension on the alternator pulley and replace/adjust the belt. A loose belt will result in an undercharged battery and a no-start condition. This is especially true in wet weather conditions when the moisture causes the belt to become more slippery.

Single Phase Charging System

DESCRIPTION & OPERATION

The single-phase charging system found on inline engines provides basic battery maintenance. Single-phase, full wave systems like these are found on a variety of products. Many outboard engines, water vehicles, motorcycles, golf cars and snowmobiles use similar systems.

This charging system produces electricity by moving a magnet past a fixed coil. Alternating current is produced by this method. Since a battery cannot be charged by AC (alternating current), the AC current produced by the lighting coil is rectified or changed into DC (direct current) to charge the battery.

To control the charging rate an additional device called a regulator is used. When the battery voltage reaches approximately 14.6 volts the regulator sends the excess current to ground. This prevents the battery from overcharging and boiling away the electrolyte.

The charging system consists of the following components:

- A flywheel containing magnets
- The lighting coil or alternator coil
- The battery, fuse assembly and wiring
- A regulator/rectifier

The lighting coil is usually a bright exposed copper wire with a lacquer-type coating. Lighting coils are built in with the ignition charge coils on some models. If the charge coil or lighting coil fails the whole stator assembly must be replaced.

The flywheel contains the magnets. The number of magnets determines the number of poles. Each magnet has two poles, so a 4-pole system has two magnets. You need to know the number of poles in order to set the tachometer correctly.

Servicing charging systems is not difficult if you follow a few basic rules. Always start by verifying the problem. If the complaint is that the battery will not stay charged do not automatically assume that the charging system is at fault. Something as simple as an accessory that draws current with the key off will convince anyone they have a bad charging system. Another culprit is the battery. Remember to clean and service your battery regularly. Battery abuse is the number one charging system problem.

The regulator/rectifier is the brains of the charging system. This assembly controls current flow in the charging system. If battery voltage is below 14.6 volts the regulator sends the available current from the rectifier to the battery. If the battery is fully charged (about 15 volts) the regulator diverts most of the current from the rectifier back to the lighting coil through ground.

Do not expect the regulator/rectifier to send current to a fully charged battery. You may find that you must pull down the battery voltage below 12.5 volts to test charging system output. Running the power trim and tilt will reduce the battery voltage. Even a pair of 12 volt sealed beam lamps hooked to the battery will reduce the battery voltage quickly.

In the charging system the regulator/rectifier is the most complex item to troubleshoot. You can avoid troubleshooting the regulator/rectifier by checking around it. Check the battery and charge or replace it as needed. Check the AC voltage output of the lighting coil. If AC voltage is low check the charge coil for proper resistance and insulation to ground. If these check OK measure the resistance of the Black wire from the rectifier/regulator to ground and for proper voltage output on the Red lead coming from the rectifier/regulator going to the battery. If all the above check within specification replace the rectifier/regulator and verify the repair by performing a charge rate test. This same check around method is used on other components like the CDI unit.

Three-Phase Charging System

DESCRIPTION & OPERATION

Three-phase systems have two more coils in the stator and one more wire than single-phase charging systems. They create higher amperage output than single-phase in nearly the same space.

➡If you do not have a solid grasp of single-phase charging systems, please read the description and operation for single-phase systems before continuing.

AC current is generated identically in both three-phase and single-phase systems. These charging systems produce AC (alternating current) by moving magnets past a fixed set of coils. Since a battery cannot be charged by AC, the AC produced by the lighting coils is rectified or changed into DC (direct current). The rate at which the battery receives this rectified current is controlled by the regulator.

The two additional lighting coils found in a three-phase charging system add complexity to circuit tracing and troubleshooting. Some systems also incorporate a battery isolator. These additional components can make these systems intimidating.

When attempting to troubleshoot these systems, apply a divide and conquer method to demystify this system. Once you have separated the components and circuitry into digestible blocks the system will be much easier to understand.

The charging system consists of the following components:

- A flywheel containing magnets
- The stator assembly, consisting of three individual lighting coils tied together in a "Y" configuration
- The battery, fuse assemblies and wiring
- A battery isolator and wiring, if so equipped

Servicing this system requires a consistent approach using a reliable check-list. If you are not systematic you may forget to check a critical component.

TROUBLESHOOTING

Servicing three-phase charging systems is not difficult if you follow a few basic rules. Always start by verifying the problem.

- Do not automatically assume that the charging system is at fault.
- A small draw with the key off, a battery with a low electrolyte level, or an overdrawn system can cause the same symptoms.
- It has become common practice on outboard engines to overload the electrical system with accessories. This places an excessive demand on the charging system. If the system is "overdrawn at the amp bank" then no amount of parts changing will fix it.

When troubleshooting a three-phase charging system use the following procedure:

1. The regulator/rectifier assembly is the brains of the charging system. The regulator controls current flow in the charging system. If battery voltage is below about 14.6 volts the regulator sends the available current to the battery. If the battery is fully charged (about 14.5 to 15 volts) the regulator diverts the current/amps to ground.
2. Do not expect the regulator to send current to a fully charged battery. Check the battery for a possible draw with the key off. This draw may be the cumulative effect of several radio and/or clock memories. If these accessories are wired to the cranking battery then a complaint of charging system failure may really be excessive draw. Draw over about 25 milliamps should arouse your suspicions. The fuel management gauge memory and speedometer clock draw about 10 milliamps each. Remember that a milliamp is 1⁄1000 of an amp. Check battery condition thoroughly because it is the #1 culprit in charging system failures.
3. Do not forget to check through the fuse and Red wire back to the battery. It can be embarrassing to overlook a blown fuse or open Red wire.
4. You must pull the battery voltage down below 12.5 volts to test charging system output. Running the power trim and tilt will reduce the battery voltage. A load bank or even a pair of 12-volt sealed-beam headlamps hooked to the battery can also be used to reduce the battery voltage.
5. Once the battery's good condition is verified and it has been reduced to below 12.5 volts you can test further.
6. Install an ammeter to check actual amp output. Several tool manufacturers produce a shunt adapter that will attach to your multi-meter and allow you to read the amp output. Verify that the system is delivering sufficient amperage. Too much amperage and a battery that goes dry very quickly indicates that the rectifier/regulator should be replaced.
7. If the system does not put out enough amperage, then test the lighting coil. Isolate the coil and test for correct resistance and short to ground.
8. During these test procedures the regulator/rectifier has not been bench checked. Usually it is advisable to avoid troubleshooting the regulator/rectifier directly. The procedures listed so far have focused on checking around the rectifier/regulator. If you verify that the battery, Black lead, Red lead, and stator are good then what is left in the system to cause the verified problem? The process of elimination has declared the rectifier/regulator bad.
9. This check around method is also useful on other components that can not be checked directly or involve time-consuming test procedures. This is the same method suggested for checking the capacitor discharge ignition box.

SERVICING

The charging system is an integral part of the ignition system. For information on service procedures, please refer to the "Ignition" section of this manual.

Charging System Checks

EXCESSIVE CHARGING

There is really only one cause for this type of failure, the regulator is not working. It isn't controlling charging output to the battery. Since there is no repair of this part, replace it.

UNDERCHARGING

If there is an undercharge condition after running the DC amperage check at the fuse assembly, then disconnect the stator coupling from the harness and perform AC voltage checks between the three stator leads. Check between two stator leads at a time. There are three volt checks done to cover all possible combinations.

At idle, there is typically 14+ volts on each test. It can be higher if the idle is higher. All three readings should be equal, within a volt or two. Stator shorts to ground can be checked by doing a voltage test between one stator lead and ground, engine running. There should be roughly half the normal stator voltage check reading.

If the readings are all within specification, the stator is working correctly. Proceed to the Red wire and Black wire checks.

If any or all readings are below normal, turn the engine **OFF** and check the stator windings using an ohmmeter. An isolated continuity check and a short to ground check should be done. If the stator is bad, replace it since it can't be repaired.

CHARGING SYSTEM DIAGNOSIS CHART

As the chart shows, this is a complete check of the whole system. A problem in the system can be found if the chart is followed. Don't jump around the chart! With some practice, this troubleshooting procedure will become second nature.

PRECAUTIONS

To prevent damage to the on-board computer, alternator and regulator, the following precautionary measures must be taken when working with the electrical system:

- Wear safety glasses when working on or near the battery.
- Don't wear a watch with a metal band when servicing the battery. Serious burns can result if the band completes the circuit between the positive battery terminal and ground.
- Be absolutely sure of the polarity of a booster battery before making connections. Connect the cables positive-to-positive, and negative-to-negative. Connect positive cables first, and then make the last connection to ground on the body of the booster vehicle so that arcing cannot ignite hydrogen gas that may have accumulated near the battery. Even momentary connection of a booster battery with the polarity reversed will damage alternator diodes.
- Never ground the alternator or generator output or battery terminal. Be cautious when using metal tools around a battery to avoid creating a short circuit between the terminals.
- Never ground the field circuit between the alternator and regulator.
- Never run an alternator or generator without load unless the field circuit is disconnected.
- Never attempt to polarize an alternator.
- When installing a battery, make sure that the positive and negative cables are not reversed.

- When jump-starting the boat, be sure that like terminals are connected. This also applies to using a battery charger. Reversed polarity will burn out the alternator and regulator in a matter of seconds.
- Never operate the alternator with the battery disconnected or on an otherwise uncontrolled open circuit.
- Do not short across or ground any alternator or regulator terminals.
- Do not try to polarize the alternator.
- Do not apply full battery voltage to the field (brown) connector.
- Always disconnect the battery ground cable before disconnecting the alternator lead.
- Always disconnect the battery (negative cable first) when charging it.
- Never subject the alternator to excessive heat or dampness. If you are steam cleaning the engine, cover the alternator.
- Never use arc-welding equipment on the car with the alternator connected.

Troubleshooting The Charging System

The charging system should be inspected if:

- A Diagnostic Trouble Code (DTC) is set relating to the charging system
- The charging system warning light is illuminated
- The voltmeter on the instrument panel indicates improper charging (either high or low) voltage
- The battery is overcharged (electrolyte level is low and/or boiling out)
- The battery is undercharged (insufficient power to crank the starter)

The starting point for all charging system problems begins with the inspection of the battery, related wiring and the alternator drive belt (if equipped). The battery must be in good condition and fully charged before system testing. If a Diagnostic Trouble Code (DTC) is set, diagnose and repair the cause of the trouble code first.

If equipped, the charging system warning light will illuminate if the charging voltage is either too high or too low. The warning light should light when the key is turned to the **ON** position as a bulb check. When voltage is produced due to the engine starting, the light should go out. A good sign of voltage that is too high are lights that burn out and/or burn very brightly. Over-charging can also cause damage to the battery and electronic circuits.

A thorough, systematic approach to troubleshooting will pay big rewards. Build your troubleshooting check list with the most likely offenders at the top. Do not be tempted to throw parts at a problem without systematically troubleshooting the system first.

1. Do a visual check of the battery, wiring and fuses. Are there any new additions to the wiring? An excellent clue might be, "Everything was working OK until I added that live well pump." With a comment like this you would know where to check first.
2. Test the battery thoroughly. Check the electrolyte level, the wiring connections and perform a load test to verify condition.
3. Perform a fuse and Red wire check with the voltmeter. Verify the ground at the rectifier. Do you have 12 volts and a good fuse? While you are at the Red wire, check alternator output with an ammeter. Be sure the battery is down around 12 volts.
4. Do a draw test if it fits the symptoms. Many times a battery that will not charge overnight or week-to-week. Put a test lamp or ammeter in the line with everything off and look for a draw.
5. A similar problem can be a system that is simply overdrawn. The electrical system cannot keep up with the demand. Do a consumption survey. More amps out than the alternator can return requires a different strategy.
6. Next, go to the source. Check the lighting coil for correct resistance and shorts to ground.
7. If all these tests fail to pinpoint the problem and you have verified low or no output to the battery then replace the rectifier.

Alternator (Stator)

TESTING

➡Before testing, make sure all connections and mounting bolts are clean and tight. Many charging system problems are related to loose and corroded terminals or bad grounds. Don't overlook the engine ground connection to the body, or the tension of the alternator drive belt.

Voltage Drop Test

1. Make sure the battery is in good condition and fully charged.
2. Perform a voltage drop test of the positive side of the circuit as follows:
   - Start the engine and allow it to reach normal operating temperature.
   - Turn the headlamps, heater blower motor and interior lights on.
   - Bring the engine to about 2,500 rpm and hold it there.
   - Connect the negative (-) voltmeter lead directly to the battery positive (+) terminal.
   - Touch the positive voltmeter lead directly to the alternator B+ output stud, not the nut. The meter should read no higher than about 0.5 volts. If it does, then there is higher than normal resistance between the positive side of the battery and the B+ output at the alternator.
   - Move the positive (+) meter lead to the nut and compare the voltage reading with the previous measurement. If the voltage reading drops substantially, then there is resistance between the stud and the nut.

➡The theory is to keep moving closer to the battery terminal one connection at a time in order to find the area of high resistance (bad connection).

3. Perform a voltage drop test of the negative side of the circuit as follows:
   a. Start the engine and allow it to reach normal operating temperature.
   b. Turn the headlamps, heater blower motor and interior lights ON.
   c. Bring the engine to about 2,500 rpm and hold it there.
   d. Connect the negative (-) voltmeter lead directly to the negative battery terminal.
   e. Touch the positive (+) voltmeter lead directly to the alternator case or ground connection. The meter should read no higher than about 0.3 volts. If it does, then there is higher than normal resistance between the battery ground terminal and the alternator ground.
   f. Move the positive (+) meter lead to the alternator mounting bracket, if the voltage reading drops substantially then you know that there is a bad electrical connection between the alternator and the mounting bracket.

➡The theory is to keep moving closer to the battery terminal one connection at a time in order to find the area of high resistance (bad connection).

Current Output Test

▶ See Figure 88

1. Perform a current output test as follows:

➡The current output test requires the use of a volt/amp tester with battery load control and an inductive amperage pick-up. Follow the manufacturer's instructions on the use of the equipment.

   a. Start the engine and allow it to reach normal operating temperature.
   b. Turn **OFF** all electrical accessories.

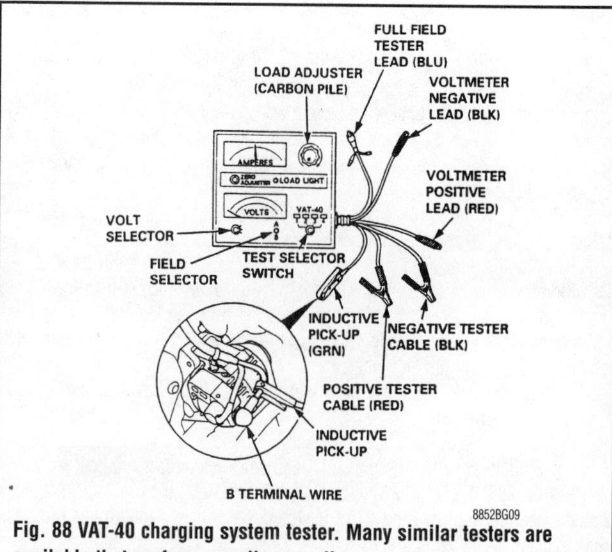

Fig. 88 VAT-40 charging system tester. Many similar testers are available that perform equally as well

c. Connect the tester to the battery terminals and cable according to the instructions.

d. Bring the engine to about 2,500 rpm and hold it there.

e. Apply a load to the charging system with the rheostat on the tester. Do not let the voltage drop below 12 volts.

f. The alternator should deliver to within 10% of the rated output. If the amperage is not within 10% and all other components test good, replace the alternator.

REMOVING & INSTALLATION

BF115A and BF130A

See Figure 89

1. Disconnect the battery negative cable.
2. Remove the engine cover and upper timing belt cover.
3. Disconnect the alternator wiring connector and the 90A fuse cable.
4. Loosen the alternator bolts and move the alternator to the side.
5. Remove the alternator belt.
6. Remove the alternator from the bracket.

To install:

7. Set the alternator into the bracket and install the bolts.
8. Install the alternator belt.

Adjust the belt by the deflection method. When pushed by 22 lbs. (98 Nm) of force, belt deflection should measure:

- Used belt: 0.30–0.35 in. (7.7–9.0 mm).
- New belt: 0.26–0.30 in. (6.7–7.7 mm).

9. If belt tension is still out of specification, replace the alternator belt.
10. After belt is adjusted, tighten the flange bolt to 33 ft. lbs. (44 Nm) and the flange nut to 17 ft. lbs. (24 Nm.).

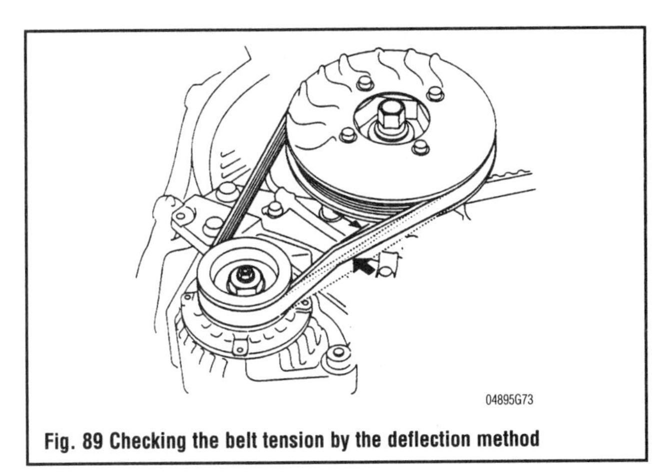

Fig. 89 Checking the belt tension by the deflection method

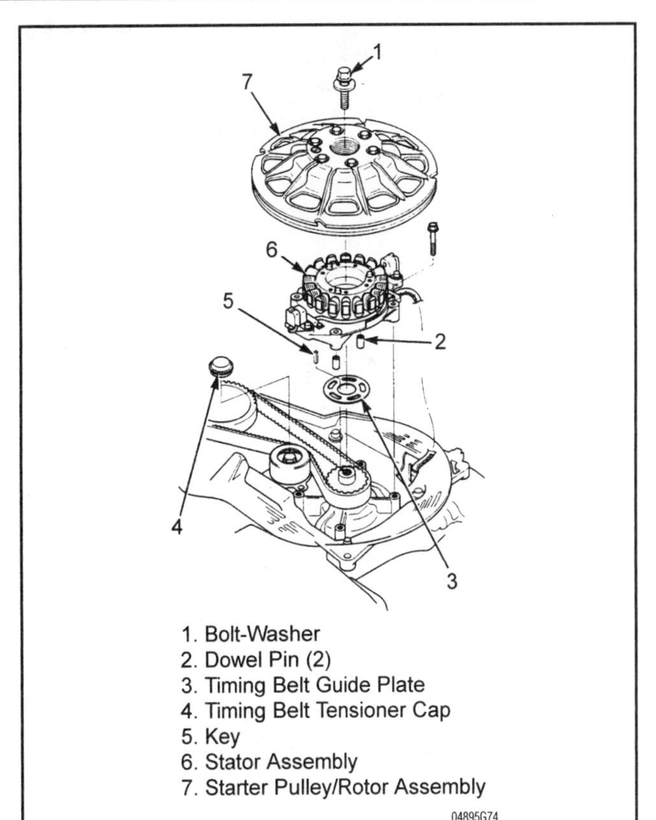

Fig. 90 Alternator/stator pulley assembly

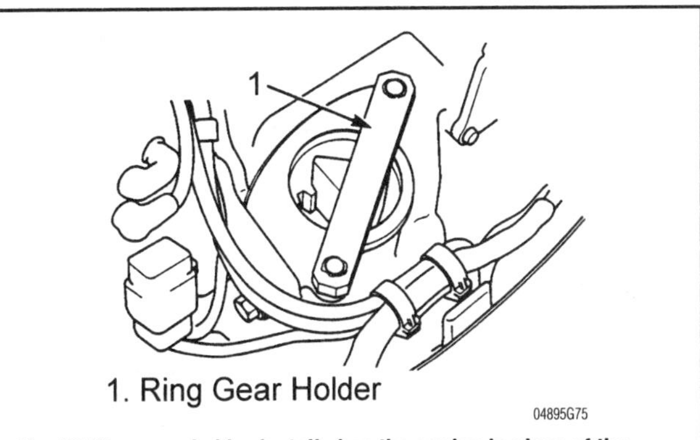

Fig. 91 Ring gear holder installed on the engine in place of the starter motor

BF75A and BF90A

See Figures 90 and 91

1. Remove the timing belt cover.
2. Remove the starter motor and install the special tool (ring gear holder 07SPB-ZW10100) in its place.
3. Disconnect the any alternator wiring connections.

Do not use a screwdriver or other prying type tool to hold the starter pulley. Use only the special tool or fabricate an equalivant ring gear holder.

4. Loosen the alternator bolt-washer and remove the starter pulley/alternator rotor assembly.
5. Remove the four flange bolts on the alternator stator assembly and remove the stator.

CAUTION

Make sure to keep track of the Woodruff key, the two dowel pins and the timing belt guide plate.

To install:

6. Clean all the surfaces with a degreasing agent.
7. With the ring gear holder in place, install the timing belt guide plate with the flanged parts facing out.
8. Set the two dowel pins in place and install the alternator stator assembly.
9. Set the Woodruff key in place on the crankshaft and install the alternator rotor.

If the starter pulley has been separated from the alternator rotor, make sure to align the projection on the back side of the pulley to the hole on the alternator rotor. Tighten the bolts to 9 ft. lbs. (12 Nm).

10. After applying oil to the flange, washer and threads of the bolt, tighten the bolt to 134 ft. lbs. (181 Nm).
11. Reconnect the alternator wiring connectors.
12. Remove the special tool and replace the starter motor.

Battery

The battery is one of the most important parts of the electrical system. In addition to providing electrical power to start the engine, it also provides power for operation of the running lights, radio, and electrical accessories.

Because of its job and the consequences (failure to perform in an emergency), the best advice is to purchase a well-known brand, with an extended warranty period, from a reputable dealer.

The usual warranty covers a pro-rated replacement policy, which means the purchaser would be entitled to a consideration for the time left on the warranty period if the battery should prove defective before its time.

Do not consider a battery of less than 70- amp/hour or 100-minute reserve capacity. If in doubt as to how large the boat requires, make a liberal estimate and then purchase the one with the next higher amp rating. Outboards equipped with an onboard computer, should be equipped with a battery of at least 100 to 105 amp/hour capacity.

MARINE BATTERIES

See Figure 92

Because marine batteries are required to perform under much more rigorous conditions than automotive batteries, they are constructed differently than those used in automobiles or trucks. Therefore, a marine battery should always be the No. 1 unit for the boat and other types of batteries used only in an emergency.

Marine batteries have a much heavier exterior case to withstand the violent pounding and shocks imposed on it as the boat moves through rough water and in extremely tight turns. The plates are thicker and each plate is securely anchored within the battery case to ensure extended life. The caps are spill proof to prevent acid from spilling into the bilges when the boat heels to one side in a tight turn, or is moving through rough water. Because of these features, the marine battery will recover from a low charge condition and give satisfactory service over a much longer period of time than any type intended for automotive use.

Fig. 92 A fully charged battery, filled to the proper level with electrolyte, is the heart of the ignition and electrical systems. Engine cranking and efficient performance of electrical items depend on a full rated battery

**** WARNING**

Never use a Maintenance-free battery with an outboard engine that is not voltage regulated. The charging system will continues to charge as long as the engine is running and it is possible that the electrolyte could boil out if periodic checks of the cell electrolyte level are not done.

BATTERY CONSTRUCTION

See Figure 93

A battery consists of a number of positive and negative plates immersed in a solution of diluted sulfuric acid. The plates contain dissimilar active materials and are kept apart by separators. The plates are grouped into elements. Plate straps on top of each element connect all of the positive plates and all of the negative plates into groups.

The battery is divided into cells holding a number of the elements apart from the others. The entire arrangement is contained within a hard plastic case. The top is a one-piece cover and contains the filler caps for each cell. The terminal posts protrude through the top where the battery connections for the boat are made. Each of the cells is connected to its neighbor in a positive-to-negative manner with a heavy strap called the cell connector.

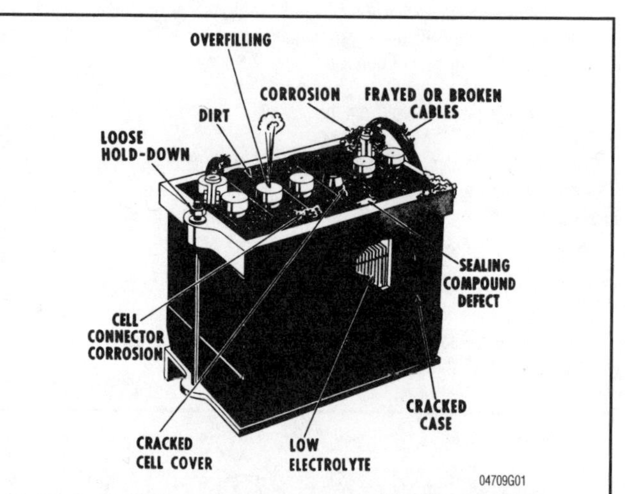

Fig. 93 A visual inspection of the battery should be made each time the boat is used. Such a quick check may reveal a potential problem in its early stages. A dead battery in a busy waterway or far from assistance could have serious consequences

BATTERY RATINGS

See Figure 94

Three different methods are used to measure and indicate battery electrical capacity:

- Amp/hour rating
- Cold cranking performance
- Reserve capacity

The amp/hour rating of a battery refers to the battery's ability to provide a set amount of amps for a given amount of time under test conditions at a constant temperature. Therefore, if the battery is capable of supplying 4 amps of current for 20 consecutive hours, the battery is rated as an 80 amp/hour battery. The amp/hour rating is useful for some service operations, such as slow charging or battery testing.

Cold cranking performance is measured by cooling a fully charged battery to 0°F (-17°C) and then testing it for 30 seconds to determine the maximum current flow. In this manner the cold cranking amp rating is the number of amps available to be drawn from the battery before the voltage drops below 7.2 volts.

The illustration depicts the amount of power in watts available from a battery at different temperatures and the amount of power in watts required of the engine at the same temperature. It becomes quite obvious—the colder the climate, the more necessary for the battery to be fully charged.

Reserve capacity of a battery is considered the length of time, in minutes, at 80°F (27°C), a 25 amp current can be maintained before the voltage drops below 10.5 volts. This test is intended to provide an approximation of how long the engine, including electrical accessories, could operate satisfactorily if the stator assembly or lighting coil did not produce sufficient current. A typical rating is 100 minutes.

If possible, the new battery should have a power rating equal to or higher than the unit it is replacing.

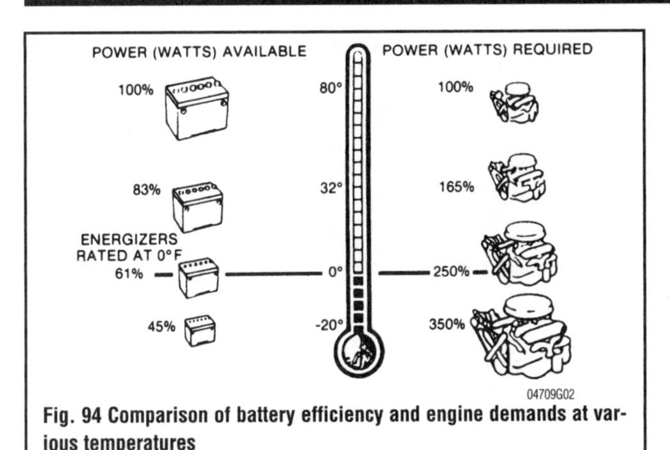

04709G02

Fig. 94 Comparison of battery efficiency and engine demands at various temperatures

BATTERY LOCATION

Every battery installed in a boat must be secured in a well protected, ventilated area. If the battery area lacks adequate ventilation, hydrogen gas, which is given off during charging, is very explosive. This is especially true if the gas is concentrated and confined.

BATTERY SERVICE

See Figures 95, 96 and 97

Batteries require periodic servicing and a definite maintenance program will ensure extended life. If the battery should test satisfactorily but still fails to perform properly, one of four problems could be the cause.

1. An accessory might have accidentally been left on overnight or for a long period during the day. Such an oversight would result in a discharged battery.
2. Using more electrical power than the stator assembly or lighting coil can replace would result in an undercharged condition.
3. A defect in the charging system. A faulty stator assembly or lighting coil, defective rectifier, or high resistance somewhere in the system could cause the battery to become undercharged.
4. Failure to maintain the battery in good order. This might include a low level of electrolyte in the cells, loose or dirty cable connections at the battery terminals or possibly an excessively dirty battery top.

Electrolyte Level

See Figures 98 and 99

The most common procedure for checking the electrolyte level in a battery is to remove the cell caps and visually observe the level in the cells. The bottom of each cell has a split vent which will cause the surface of the electrolyte to appear distorted when it makes contact. When the distortion first appears at the bottom of the split vent, the electrolyte level is correct.

During hot weather and periods of heavy use, the electrolyte level should be checked more often than during normal operation. Add distilled water to bring the level of electrolyte in each cell to the proper level. Take care not to overfill, because adding an excessive amount of water will cause loss of electrolyte and any loss will result in poor performance, short battery life, and will contribute quickly to corrosion.

Never add electrolyte from another battery. Use only distilled water.

Battery Testing

A hydrometer is a device to measure the percentage of sulfuric acid in the battery electrolyte in terms of specific gravity. When the condition of the battery drops from fully charged to discharged, the acid leaves the solution and enters the plates, causing the specific gravity of the electrolyte to drop.

It may not be common knowledge, but hydrometer floats are calibrated for use at 80°F (27°C). If the hydrometer is used at any other temperature, hotter or colder, a correction factor must be applied.

Remember, a liquid will expand if it is heated and will contract if cooled. Such expansion and contraction will cause a definite change in the specific gravity of the liquid, in this case the electrolyte.

A quality hydrometer will have a thermometer/temperature correction table in the lower portion, as shown in the accompanying illustration. By knowing the air temperature around the battery and from the table, a correction factor may be applied to the specific gravity reading of the hydrometer float. In this manner, an accurate determination may be made as to the condition of the battery.

When using a hydrometer, pay careful attention to the following points:

1. Never attempt to take a reading immediately after adding water to the battery. Allow at least ¼ hour of charging at a high rate to thoroughly mix the electrolyte with the new water. This time will also allow for the necessary gases to be created.
2. Always be sure the hydrometer is clean inside and out as a precaution against contaminating the electrolyte.
3. If a thermometer is an integral part of the hydrometer, draw liquid into it several times to ensure the correct temperature before taking a reading.

04709P02

Fig. 95 Explosive hydrogen gas is normally released from the cells under a wide range of circumstances. This battery exploded when the gas ignited from someone smoking in the area when the caps were removed. Such an explosion could also be caused by a spark from the battery terminals

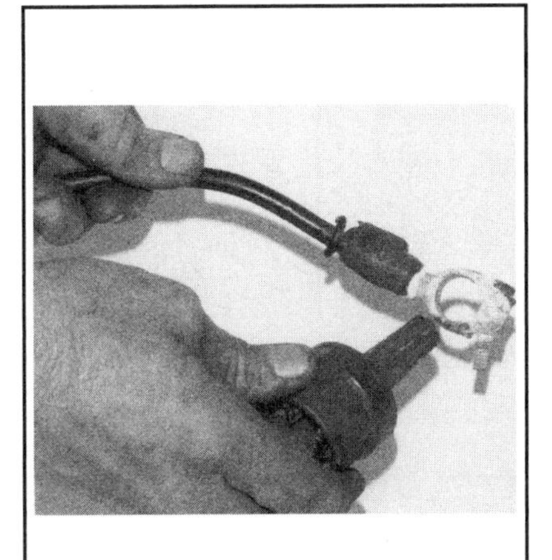

04709P04

Fig. 96 A two part battery cable cleaning tool will do an excellent job of cleaning the inside of the cable connectors

04709P05

Fig. 97 The second part of the battery cable cleaning tool contains a brush for cleaning the battery terminals

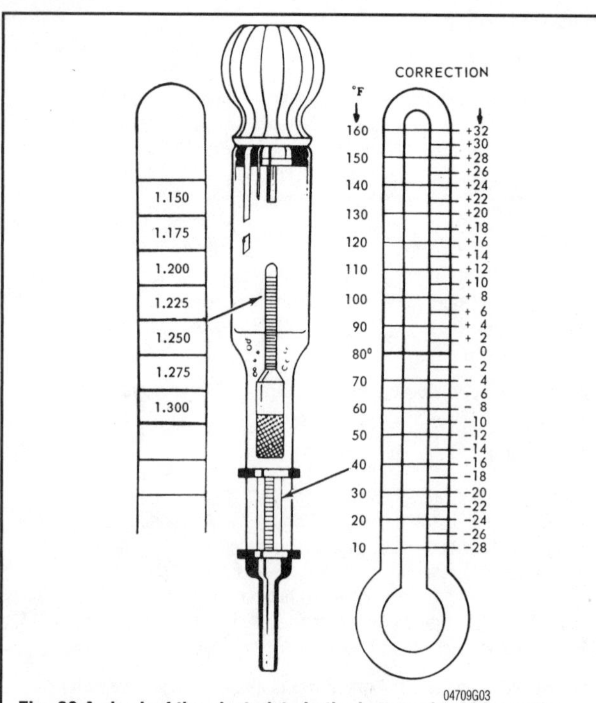

Fig. 98 A check of the electrolyte in the battery should be on the maintenance schedule for any boat

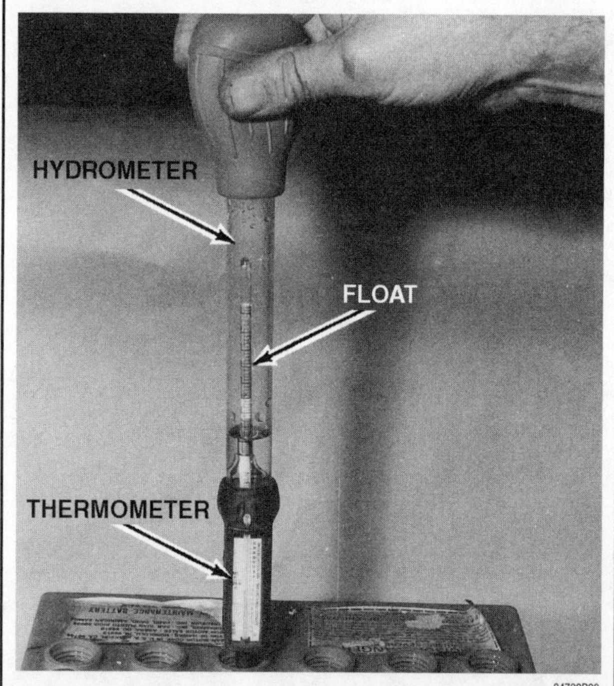

Fig. 99 Testing the electrolyte's specific gravity using a temperature corrected hydrometer

4. Be sure to hold the hydrometer vertically and suck up liquid only until the float is free and floating.
5. Always hold the hydrometer at eye level and take the reading at the surface of the liquid with the float free and floating.
6. Disregard the slight curvature appearing where the liquid rises against the float stem. This phenomenon is due to surface tension.
7. Do not drop any of the battery fluid on the boat or on your clothing, because it is extremely caustic. Use water and baking soda to neutralize any battery liquid that does accidentally drop.
8. After drawing electrolyte from the battery cell until the float is barely free, note the level of the liquid inside the hydrometer. If the level is within the Green band range for all cells, the condition of the battery is satisfactory. If the level is within the white band for all cells, the battery is in fair condition.
9. If the level is within the Green or white band for all cells except one, which registers in the red, the cell is shorted internally. No amount of charging will bring the battery back to satisfactory condition.
10. If the level in all cells is about the same, even if it falls in the Red band, the battery may be recharged and returned to service. If the level fails to rise above the Red band after charging, the only solution is to replace the battery.

Battery Cleaning

Dirt and corrosion should be cleaned from the battery as soon as it is discovered. Any accumulation of acid film or dirt will permit current to flow between the terminals. Such a current flow will drain the battery over a period of time.

Clean the exterior of the battery with a solution of diluted ammonia or a paste made from baking soda to neutralize any acid which may be present. Flush the cleaning solution off with clean water.

➡Take care to prevent any of the neutralizing solution from entering the cells, by keeping the caps tight.

A poor contact at the terminals will add resistance to the charging circuit. This resistance will cause the voltage regulator to register a fully charged battery, and thus cut down on the stator assembly or lighting coil output adding to the low battery charge problem.

Scrape the battery posts clean with a suitable tool or with a stiff wire brush. Clean the inside of the cable clamps to be sure they do not cause any resistance in the circuit.

BATTERY TERMINALS

At least once a season, the battery terminals and cable clamps should be cleaned. Loosen the clamps and remove the cables, negative cable first. On batteries with top mounted posts, the use of a puller specially made for this purpose is recommended. These are inexpensive and available in most parts stores.

Clean the cable clamps and the battery terminal with a wire brush, until all corrosion, grease, etc., is removed and the metal is shiny. It is especially important to clean the inside of the clamp thoroughly (a wire brush is useful here), since a small deposit of foreign material or oxidation there will prevent a sound electrical connection and inhibit either starting or charging. It is also a good idea to apply some dielectric grease to the terminal, as this will aid in the prevention of corrosion.

After the clamps and terminals are clean, reinstall the cables, negative cable last, do not hammer the clamps onto battery posts. Tighten the clamps securely, but do not distort them. Give the clamps and terminals a thin external coating of grease after installation, to retard corrosion.

Check the cables at the same time that the terminals are cleaned. If the insulation is cracked or broken, or if its end is frayed, that cable should be replaced with a new one of the same length and gauge.

BATTERY & CHARGING SAFETY PRECAUTIONS

Always follow these safety precautions when charging or handling a battery:

- Wear eye protection when working around batteries. Batteries contain corrosive acid and produce explosive gas a byproduct of their operation. Acid on the skin should be neutralized with a solution of baking soda and water made into a paste. In case acid contacts the eyes, flush with clear water and seek medical attention immediately.
- Avoid flame or sparks that could ignite the hydrogen gas produced by the battery and cause an explosion. Connection and disconnection of cables to battery terminals is one of the most common causes of sparks.
- Always turn a battery charger **OFF**, before connecting or disconnecting the leads. When connecting the leads, connect the positive lead first, then the negative lead, to avoid sparks.
- When lifting a battery, use a battery carrier or lift at opposite corners of the base.

- Ensure there is good ventilation in a room where the battery is being charged.
- Do not attempt to charge or load-test a maintenance-free battery when the charge indicator dot is indicating insufficient electrolyte.
- Disconnect the negative battery cable if the battery is to remain in the boat during the charging process.
- Be sure the ignition switch is **OFF**before connecting or turning the charger **ON**. Sudden power surges can destroy electronic components.
- Use proper adapters to connect charger leads to batteries with non-conventional terminals.

BATTERY CHARGERS

Before using any battery charger, consult the manufacturer's instructions for its use. Battery chargers are electrical devices that change Alternating Current (AC) to a lower voltage of Direct Current (DC) that can be used to charge a marine battery. There are two types of battery chargers—manual and automatic.

A manual battery charger must be physically disconnected when the battery has come to a full charge. If not, the battery can be overcharged, and possibly fail. Excess charging current at the end of the charging cycle will heat the electrolyte, resulting in loss of water and active material, substantially reducing battery life.

➡As a rule, on manual chargers, when the ammeter on the charger registers half the rated amperage of the charger, the battery is fully charged. This can vary, and it is recommended to use a hydrometer to accurately measure state of charge.

Automatic battery chargers have an important advantage—they can be left connected (for instance, overnight) without the possibility of overcharging the battery. Automatic chargers are equipped with a sensing device to allow the battery charge to taper off to near zero as the battery becomes fully charged. When charging a low or completely discharged battery, the meter will read close to full rated output. If only partially discharged, the initial reading may be less than full rated output, as the charger responds to the condition of the battery. As the battery continues to charge, the sensing device monitors the state of charge and reduces the charging rate. As the rate of charge tapers to zero amps, the charger will continue to supply a few milliamps of current—just enough to maintain a charged condition.

BATTERY CABLES

Battery cables don't go bad very often, but like anything else, they can wear out. If the cables on your boat are cracked, frayed or broken, they should be replaced.

When working on any electrical component, it is always a good idea to disconnect the negative (-) battery cable. This will prevent potential damage to many sensitive electrical components

Always replace the battery cables with one of the same length, or you will increase resistance and possibly cause hard starting. Smear the battery posts with a light film of dielectric grease, or a battery terminal protectant spray once you've installed the new cables. If you replace the cables one at a time, you won't mix them up.

➡Any time you disconnect the battery cables, it is recommended that you disconnect the negative (-) battery cable first. This will prevent you from accidentally grounding the positive (+) terminal when disconnecting it, thereby preventing damage to the electrical system.

Before you disconnect the cable(s), first turn the ignition to the **OFF**position. This will prevent a draw on the battery which could cause arcing. When the battery cable(s) are reconnected (negative cable last), be sure to check all electrical accessories are all working correctly.

BATTERY STORAGE

If the boat is to be laid up for the winter or for more than a few weeks, special attention must be given to the battery to prevent complete discharge or possible damage to the terminals and wiring. Before putting the boat in storage, disconnect and remove the batteries. Clean them thoroughly of any dirt or corrosion, and then charge them to full specific gravity reading. After they are fully charged, store them in a clean cool dry place where they will not be damaged or knocked over, preferably on a couple blocks of wood. Storing the battery up off the deck, will permit air to circulate freely around and under the battery and will help to prevent condensation.

Never store the battery with anything on top of it or cover the battery in such a manner as to prevent air from circulating around the filler caps. All batteries, both new and old, will discharge during periods of storage, more so if they are hot than if they remain cool. Therefore, the electrolyte level and the specific gravity should be checked at regular intervals. A drop in the specific gravity reading is cause to charge them back to a full reading.

In cold climates, care should be exercised in selecting the battery storage area. A fully-charged battery will freeze at about 60 degrees below zero. A discharged battery, almost dead, will have ice forming at about 19 degrees above zero.

STARTING CIRCUIT

Description and Operation

➧ See Figure 100

In the early days, all outboard engines were started by simply pulling on a rope wound around the flywheel. As time passed and owners were reluctant to use muscle power, it was necessary to replace the rope starter with some form of power cranking system. Today, many small engines are still started by pulling on a rope, but others have a powered starter motor installed.

The system utilized to replace the rope method was an electric starter motor coupled with a mechanical gear mesh between the starter motor and the powerhead flywheel, similar to the method used to crank an automobile engine.

As the name implies, the sole purpose of the starter motor circuit is to control operation of the starter motor to crank the powerhead until the engine is operating. The circuit includes a solenoid or magnetic switch to connect or disconnect the motor from the battery. The operator controls the switch with a key switch.

A neutral safety switch is installed into the circuit to permit operation of the starter motor only if the shift control lever is in neutral. This switch is a safety device to prevent accidental engine start when the engine is in gear.

The starter motor is a series wound electric motor which draws a heavy cur-

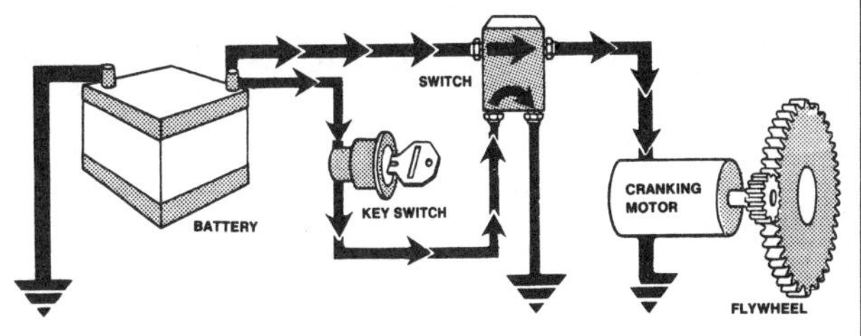

Fig. 100 A typical starting system converts electrical energy into mechanical energy to turn the engine. The components are: Battery, to provide electricity to operate the starter; Ignition switch, to control the energizing of the starter relay or solenoid; Starter relay or solenoid, to make and break the circuit between the battery and starter; Starter, to convert electrical energy into mechanical energy to rotate the engine; Starter drive gear, to transmit the starter rotation to the engine flywheel

rent from the battery. It is designed to be used only for short periods of time to crank the engine for starting. To prevent overheating the motor, cranking should not be continued for more than 30-seconds without allowing the motor to cool for at least three minutes. Actually, this time can be spent in making preliminary checks to determine why the engine fails to start.

Power is transmitted from the starter motor to the powerhead flywheel through a Bendix drive. This drive has a pinion gear mounted on screw threads. When the motor is operated, the pinion gear moves upward and meshes with the teeth on the flywheel ring gear.

When the powerhead starts, the pinion gear is driven faster than the shaft, and as a result, it screws out of mesh with the flywheel. A rubber cushion is built into the Bendix drive to absorb the shock when the pinion meshes with the flywheel ring gear. The parts of the drive must be properly assembled for efficient operation. If the drive is removed for cleaning, take care to assemble the parts as shown in the accompanying illustrations in this section. If the screw shaft assembly is reversed, it will strike the splines and the rubber cushion will not absorb the shock.

The sound of the motor during cranking is a good indication of whether the starter motor is operating properly or not. Naturally, temperature conditions will affect the speed at which the starter motor is able to crank the engine. The speed of cranking a cold engine will be much slower than when cranking a warm engine. An experienced operator will learn to recognize the favorable sounds of the powerhead cranking under various conditions.

Troubleshooting the Starting System

If the starter motor spins, but fails to crank the engine, the cause is usually a corroded or gummy Bendix drive. The drive should be removed, cleaned, and given an inspection.

If the starter motor cranks the engine too slowly, the following are possible causes and the corrective actions that may be taken:

- Battery charge is low. Charge the battery to full capacity.
- High resistance connections at the battery, solenoid, or motor. Clean and tighten all connections.
- Undersize battery cables. Replace cables with sufficient size.
- Battery cables too long. Relocate the battery to shorten the run to the solenoid.

Before wasting too much time troubleshooting the starter motor circuit, the following checks should be made. Many times, the problem will be corrected.

- Battery fully charged.
- Shift control lever in neutral.
- Main 20-amp fuse located at the base of the fuse cover is good (not blown).
- All electrical connections clean and tight.
- Wiring in good condition, insulation not worn or frayed.

Two more areas may cause the powerhead to crank slowly even though the starter motor circuit is in excellent condition

- A tight or frozen powerhead
- Water in the lower unit.

Starter Motor

DESCRIPTION & OPERATION

See Figure 101

As the name implies, the sole purpose of the cranking motor circuit is to control operation of the cranking motor to crank the powerhead until the engine is operating. The circuit includes a solenoid or magnetic switch to connect or disconnect the motor from the battery. The operator controls the switch with a key switch.

A neutral safety switch is installed into the circuit to permit operation of the cranking motor only if the shift control lever is in neutral. This switch is a safety device to prevent accidental engine start when the engine is in gear.

The cranking motor is a series wound electric motor which draws a heavy current from the battery. It is designed to be used only for short periods of time to crank the engine for starting. To prevent overheating the motor, cranking should not be continued for more than 30-seconds without allowing the motor to cool for at least three minutes. Actually, this time can be spent in making preliminary checks to determine why the engine fails to start.

Power is transmitted from the cranking motor to the powerhead flywheel through a Bendix drive. This drive has a pinion gear mounted on screw threads. When the motor is operated, the pinion gear moves upward and meshes with the teeth on the flywheel ring gear.

When the powerhead starts, the pinion gear is driven faster than the shaft, and as a result, it screws out of mesh with the flywheel. A rubber cushion is built into the Bendix drive to absorb the shock when the pinion meshes with the flywheel ring gear. The parts of the drive must be properly assembled for efficient operation. If the drive is removed for cleaning, take care to assemble the parts as shown in the accompanying illustrations in this section. If the screw shaft assembly is reversed, it will strike the splines and the rubber cushion will not absorb the shock.

The sound of the motor during cranking is a good indication of whether the cranking motor is operating properly or not. Naturally, temperature conditions will affect the speed at which the cranking motor is able to crank the engine. The speed of cranking a cold engine will be much slower than when cranking a warm engine. An experienced operator will learn to recognize the favorable sounds of the powerhead cranking under various conditions.

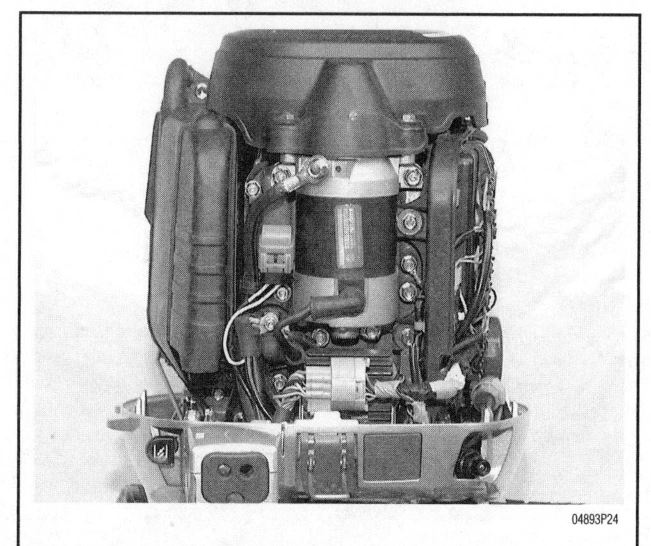

04893P24

Fig. 101 Typical location of the starter motor

TESTING

See Figure 102

1. Starter motor Rotates Slowly
   a. Battery charge is low. Charge the battery to full capacity.
   b. Electrical connections corroded or loose. Clean and tighten.
   c. Defective starter motor. Perform an amp draw test. Lay an amp draw-gauge on the cable leading to the starter motor. Turn the key on and attempt to crank the engine. If the gauge indicates an excessive amperage draw, the starter motor must be replaced or rebuilt.
2. Starter motor Fails To Crank Powerhead
   a. Disconnect the starter motor lead from the solenoid to prevent the powerhead from starting during the testing process.

This lead is to remain disconnected from the solenoid during tests No. 2–6.

   b. Disconnect the Black ground wire from the No. 2 terminal.
   c. Connect a voltmeter between the No. 2 terminal and a common engine ground.
   d. Turn the key switch to the start position.
   e. Observe the voltmeter reading. If there is the slightest amount of reading, check the Black ground wire connection or check for an open circuit.
3. Test Starter motor Solenoid
   a. Connect a voltmeter between the engine common ground and the No. 3 terminal.
   b. Turn the ignition key switch to the start position.
   c. Observe the voltmeter reading. If the meter indicates more than 0.3 volt, the solenoid is defective and must be replaced.

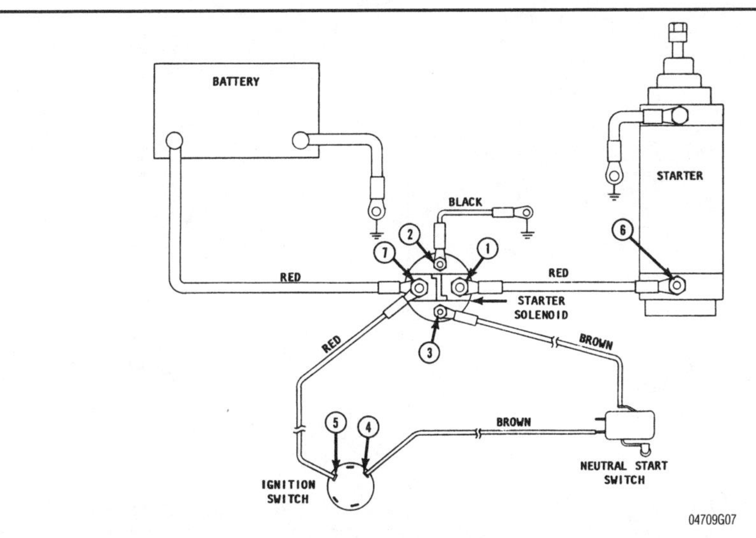

Fig. 102 Perform the troubleshooting steps using this typical starter motor system diagram. Step numbers correspond to circled numbers in the diagram

4. Test Neutral Start Switch
   a. Connect a voltmeter between the common engine ground and the No.
4. Turn the ignition key switch to the start position.
   b. Observe the voltmeter. If there is any indication of a reading, the neutral start switch is open or the brown wire lead is open between the No. 3 and No. 4.
5. Test for Open Wire
   a. Connect a voltmeter between the common engine ground and No. 5.
   b. The voltmeter should indicate 12-volts. If the meter needle flickers (fails to hold steady), check the circuit between No. 5 and common engine ground. If meter fails to indicate voltage, replace the positive battery cable.
6. Further Tests for Solenoid
   a. Connect the voltmeter between the common engine ground and No. 1.
   b. Turn the ignition key switch to the start position.
   c. Observe the voltmeter. If there is no reading, the starter motor solenoid is defective and must be replaced.
7. Test Large Red Cable
   a. Connect the Red cable to the starter motor solenoid.
   b. Connect the voltmeter between the engine common ground and No. 6.
   c. Turn the ignition key switch to the start position, or depress the start button.
   d. Observe the voltmeter. If there is no reading, check the Red cable for a poor connection or an open circuit. If there is any indication of a reading, and the starter motor does not rotate, the starter motor must be replaced.

REMOVAL & INSTALLATION

BF90A, BF75A , BF115A and BF130A

See Figure 103

Starter motor is located on the port side of the engine below the ECM and just behind the fuse boxes.

1. Disconnect the positive and negative battery cables.
2. Remove the positive starter cable from starter motor solenoid.
3. Disconnect the negative lead from the starter motor solenoid.
4. Remove the two retaining bolts on the starter and lift out the starter motor.
5. Discard the old O-ring and replace with a new one.

To install:

6. Install the starter motor onto the engine and tighten the retaining bolts to 29 ft. lbs. (39 Nm).
7. Replace the positive starter cable on the starter motor solenoid and tighten to 8 ft. lbs. (10.8 Nm).
8. Slide the negative lead back onto the starter motor solenoid terminal.
9. Reconnect the positive and negative battery cables.

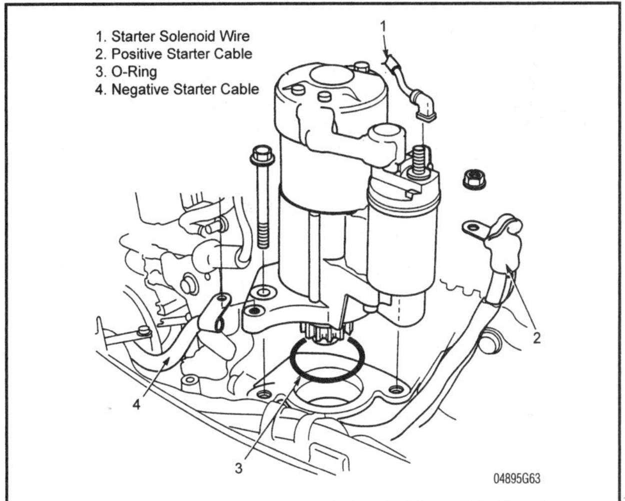

Fig. 103 The starter motor on the BF90A, BF75A , BF115A and BF130A is installed with the drive end facing down

BF50A, BF45A, BF40A and BF35A

See Figure 104

The starter motor is located on the front of the engine.

1. Remove the engine cover and flywheel cover.
2. Disconnect the positive and negative battery cables.

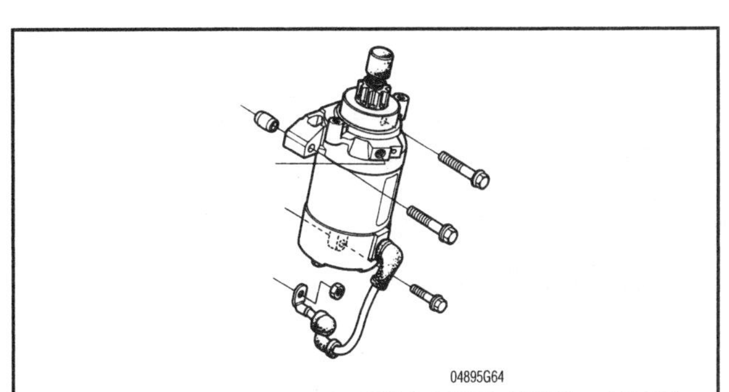

Fig. 104 The starter motor on the BF50A, BF45A, BF40A and BF35A is installed with the drive end facing up to contact the underside of the flywheel

3. Disconnect the positive and negative starter motor cables. Disconnect the positive at the solenoid switch.
4. Remove the two retaining bolts, being careful to not drop the dowel pin, and lift out the starter motor.

To install:

5. Insert the two retaining bolts, slipping the dowel pin onto the left bolt, into the starter motor.
6. Install the starter motor and tighten the retaining bolts to 15.2 ft. lbs. (21 Nm).
7. Connect the positive and negative starter motor cables keeping the upper cable 90°to the starter housing and the positive 90°to the solenoid switch terminal.
8. Connect the positive and negative battery cable.
9. Replace the flywheel and engine covers.

BF30A and BF25A

➧ See Figure 105

Starter is located on the front of the engine.

1. Disconnect the positive and negative battery cables.
2. Remove the engine cover.
3. Remove the flywheel.
4. Remove the carburetor assembly.
5. Remove or disconnect electrical components as needed to access the starter motor bracket.
6. Disconnect the positive starter motor cable and the negative ground cable.
7. Remove the two retaining bolts holding the starter to the electric starter bracket, being careful of the two dowels, and lift the starter off the bracket.

To install:

8. Install the starter onto the bracket with the two bolts and dowels and tighten to 15.2 ft. lbs. (21 Nm).
9. Connect the positive starter motor cable to the starter and tighten to 3.6 ft. lbs. (5 Nm). Install the negative ground cable from the starter to the engine ground.
10. Reconnect or reinstall all electrical components.
11. Install the carburetor assembly.

➡Perform all carburetor assembly procedures prior to operating the engine.

12. Install the flywheel.
13. Install the engine cover.

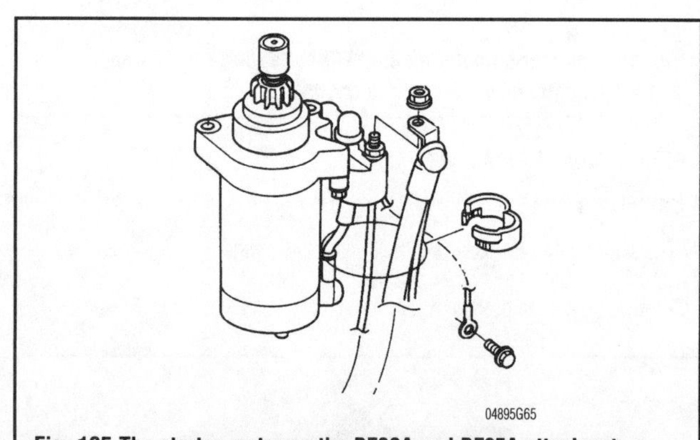

Fig. 105 The starter motor on the BF30A and BF25A attaches to a bracket which is then installed on the powerhead

BF9.9 and BF15A

➧ See Figure 106

Starter motor is located on the starboard side of the engine.

1. Disconnect the positive and negative battery cables.
2. Remove the engine and flywheel covers.
3. Remove the flywheel assembly.
4. Disconnect the starter motor positive cable and negative ground wire.
5. Remove the two retaining bolts and lift starter motor off the engine.

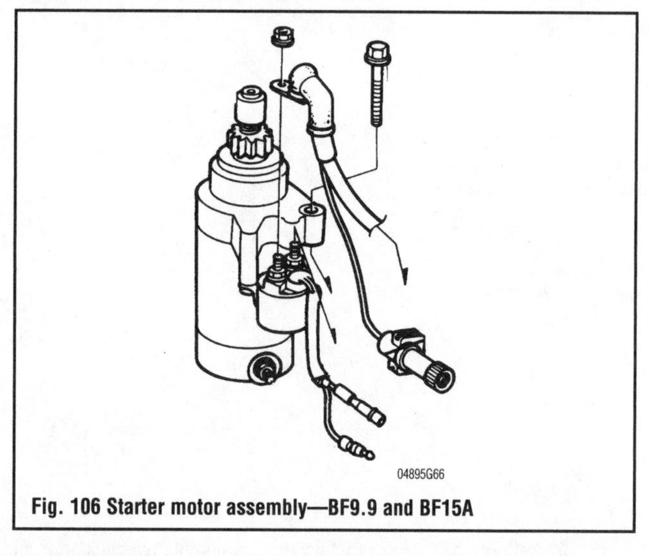

Fig. 106 Starter motor assembly—BF9.9 and BF15A

To install:

6. Install the starter motor on the engine and insert the retaining bolts.
7. Tighten the bolts to 15.2 ft. lbs. (21 Nm).
8. Connect the starter motor positive cable and negative ground wire.
9. Install the flywheel.
10. Install the flywheel and engine covers.
11. Connect the positive and negative battery cables.

Starter Motor Solenoid Switch

DESCRIPTION & OPERATION

When the starter button on the instrument panel is depressed, current flows and energizes the starter's solenoid coil. The energized coil becomes an electromagnet, which pulls the plunger into the coil, and closes a set of contacts which allow high current to reach the starter motor. At the same time, the plunger also serves to push the starter pinion to mesh with the teeth on the flywheel.

TESTING

➧ See Figures 107, 108, 109 and 110

1. Make sure that the battery is in good shape before performing this test.
2. Connect the leads of the starter solenoid switch to a 12V power source using jumper wires.
3. With voltage applied, check continuity with a multimeter between the terminals of the solenoid. Multimeter should show continuity.
4. Disconnect the jumper wires from the battery terminals. With voltage turned off, there should not be any continuity between the terminals.
5. If the solenoid fails the continuity test, replace the solenoid.

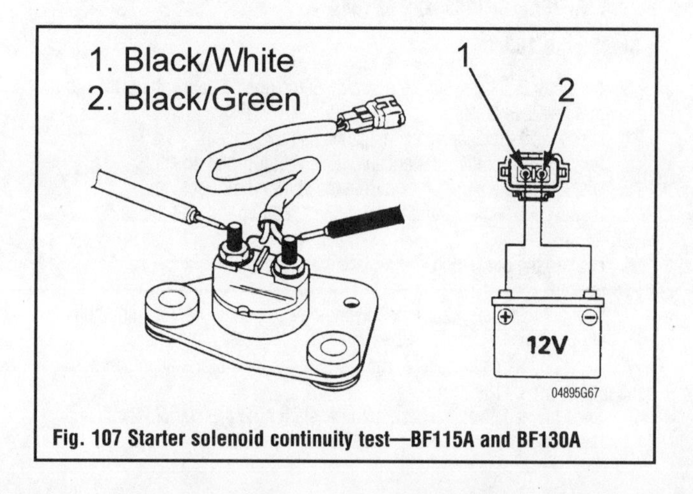

Fig. 107 Starter solenoid continuity test—BF115A and BF130A

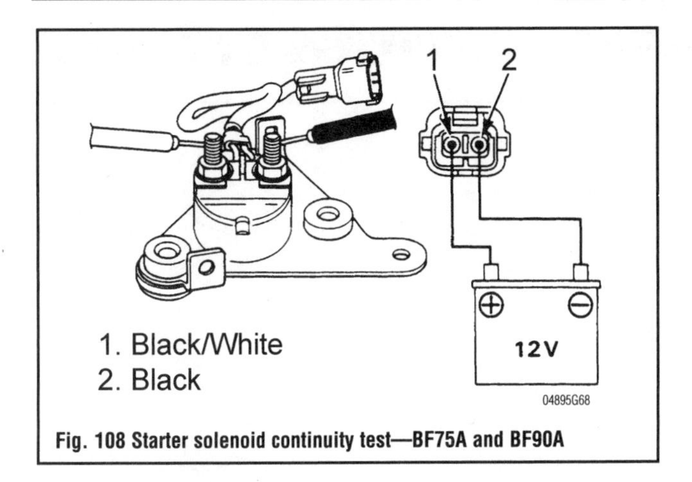

Fig. 108 Starter solenoid continuity test—BF75A and BF90A

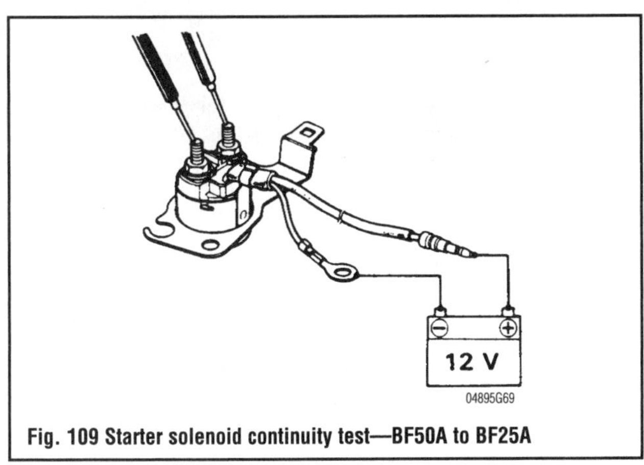

Fig. 109 Starter solenoid continuity test—BF50A to BF25A

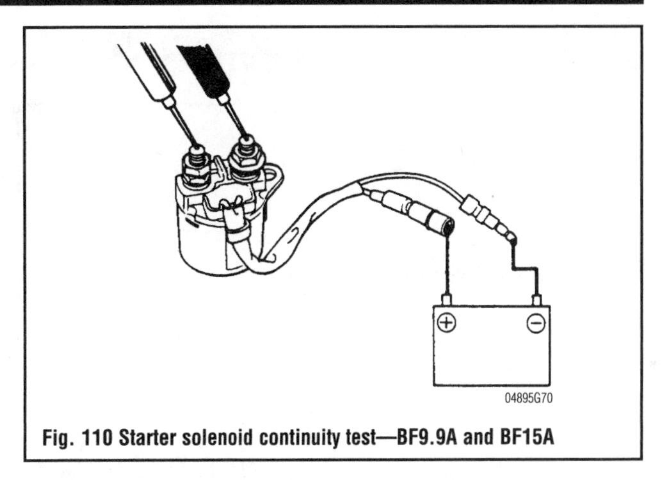

Fig. 110 Starter solenoid continuity test—BF9.9A and BF15A

REMOVAL & INSTALLATION

1. Disconnect the positive and negative battery cables before removing the solenoid.
2. Disconnect the positive and negative starter cables at the solenoid terminals.
3. Loosen the retaining bolts remove the solenoid from the engine.
4. Make sure to keep track of all wire clamps, clips and rubber bushings when removing the solenoid.

To install:

5. Install the solenoid and any rubber bushings on the engine and tighten the retaining bolts.
6. Connect the starter positive and negative cables to the solenoid terminals.
7. Clamp all wires and cables.
8. Connect the positive and negative battery cables.

COMPONENT LOCATIONS

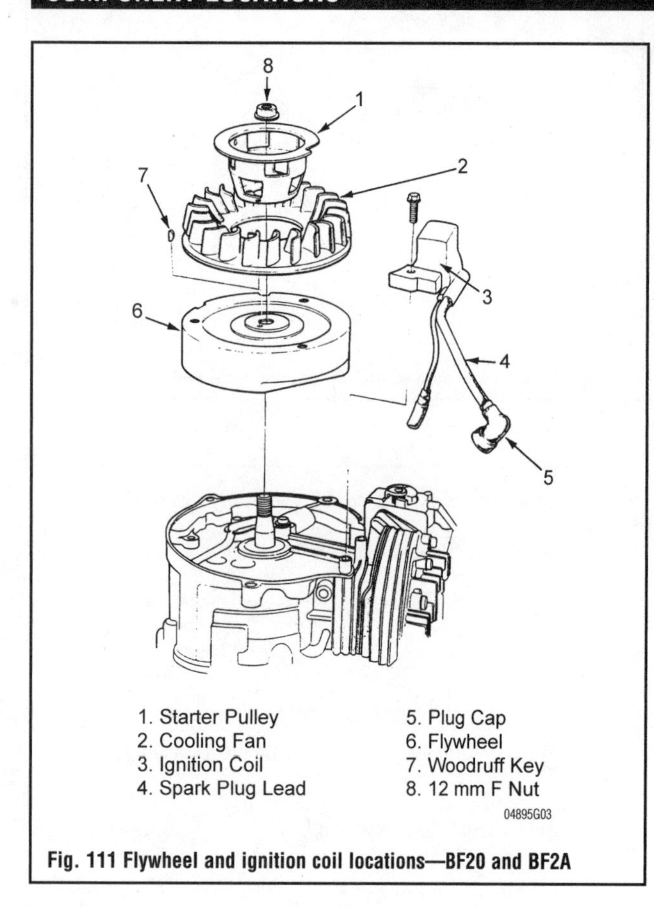

Fig. 111 Flywheel and ignition coil locations—BF20 and BF2A

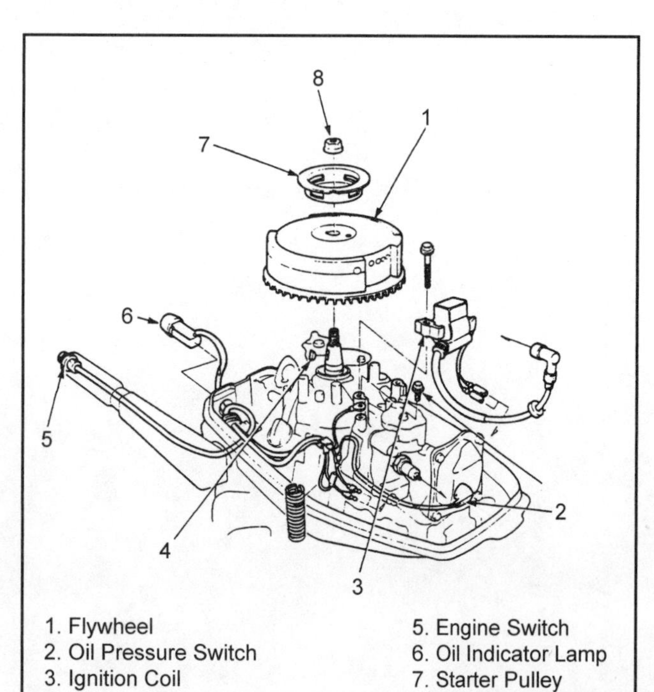

Fig. 112 Flywheel and electrical component locations—BF50 and BF5A

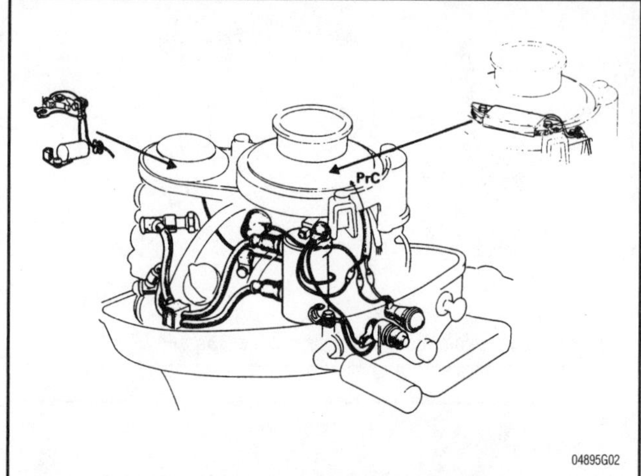

Fig. 113 Electrical and breaker point ignition component locations—BF75 and BF100 Early model (Serial No. 1000004–1199999)

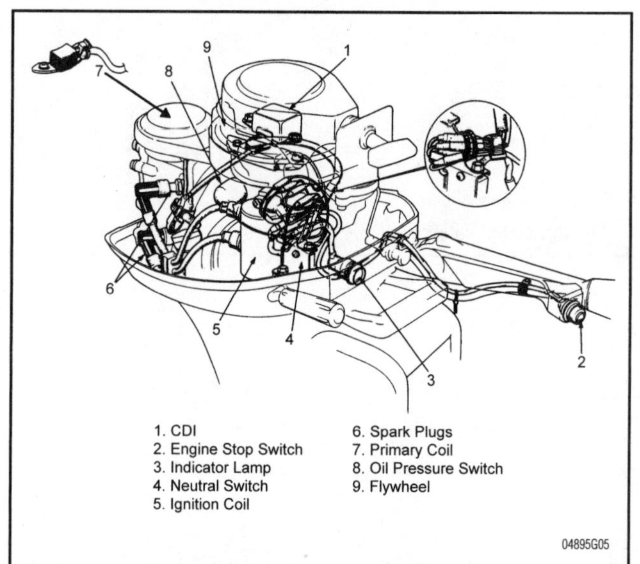

Fig. 114 Electrical and CDI ignition component locations—BF75, BF8A and BF100 Later model (serial no. 1300001 and later)

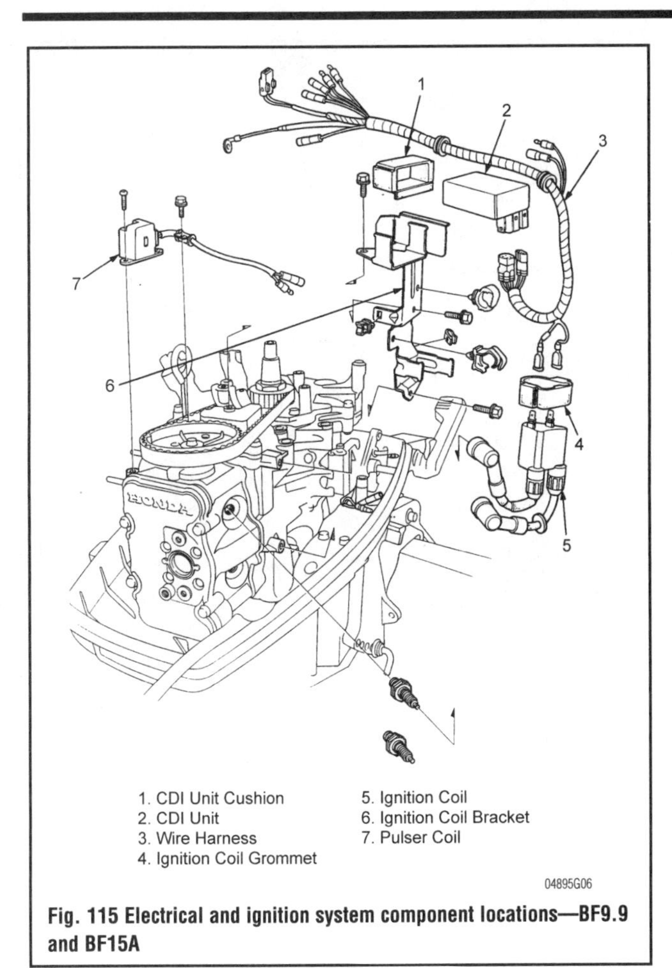

1. CDI Unit Cushion
2. CDI Unit
3. Wire Harness
4. Ignition Coil Grommet
5. Ignition Coil
6. Ignition Coil Bracket
7. Pulser Coil

04895G06

Fig. 115 Electrical and ignition system component locations—BF9.9 and BF15A

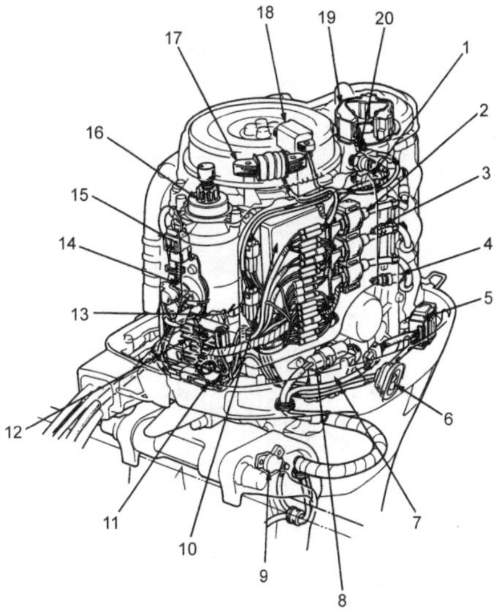

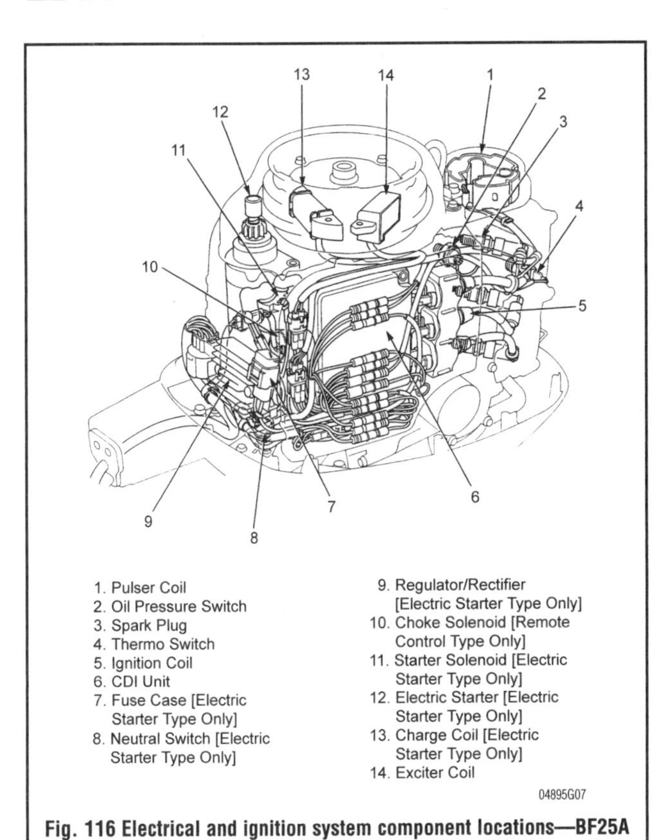

1. Pulser Coil
2. Oil Pressure Switch
3. Spark Plug
4. Thermo Switch
5. Ignition Coil
6. CDI Unit
7. Fuse Case [Electric Starter Type Only]
8. Neutral Switch [Electric Starter Type Only]
9. Regulator/Rectifier [Electric Starter Type Only]
10. Choke Solenoid [Remote Control Type Only]
11. Starter Solenoid [Electric Starter Type Only]
12. Electric Starter [Electric Starter Type Only]
13. Charge Coil [Electric Starter Type Only]
14. Exciter Coil

04895G07

Fig. 116 Electrical and ignition system component locations—BF25A and BF30A

1. Oil Pressure Switch
2. Ignition Coil
3. Spark Plug
4. Spark Plug Cap
5. Trim Control Unit [Power Trim & Tilt Type Only]
6. Power Tilt Switch [Power Trim & Tilt Type Only]
7. Power Tilt Relay [Power Trim & Tilt Type Only]
8. Power Tilt Motor Connector [Power Trim & Tilt Type Only]
9. Trim Angle Sensor [Power Trim & Tilt Type Only]
10. CDI Unit
11. Neutral Switch
12. Regulator/Rectifier
13. Choke Solenoid [Remote Control Type Only]
14. Starter Solenoid
15. Fuse Case
16. Starter Motor
17. Charge Coil
18. Exciter Coil
19. Pulser Coil
20. Thermo Switch

04895G08

Fig. 117 Electrical and ignition system component locations—BF35A, BF40A, BF45A and BF50A

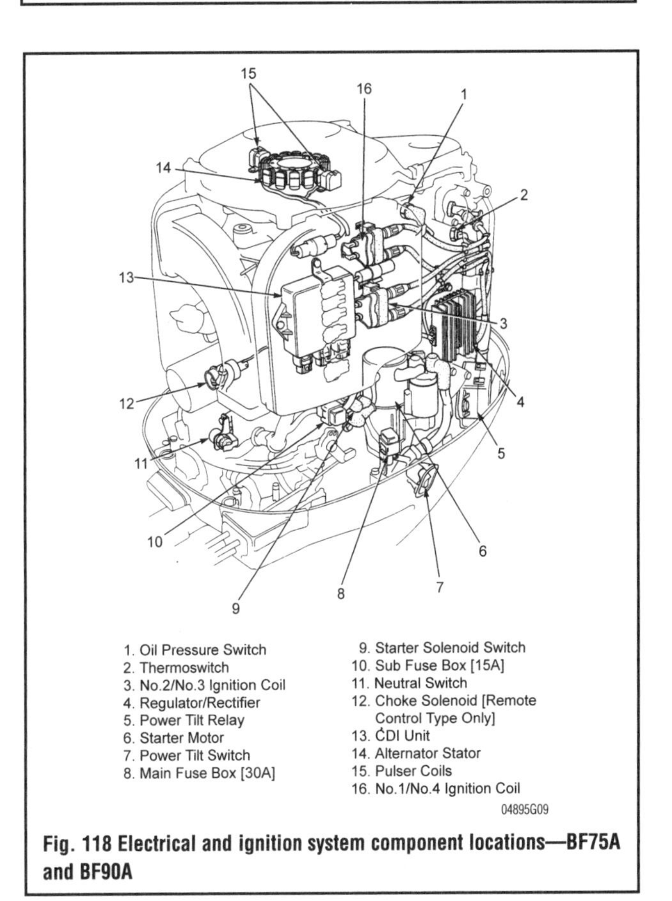

1. Oil Pressure Switch
2. Thermoswitch
3. No.2/No.3 Ignition Coil
4. Regulator/Rectifier
5. Power Tilt Relay
6. Starter Motor
7. Power Tilt Switch
8. Main Fuse Box [30A]
9. Starter Solenoid Switch
10. Sub Fuse Box [15A]
11. Neutral Switch
12. Choke Solenoid [Remote Control Type Only]
13. CDI Unit
14. Alternator Stator
15. Pulser Coils
16. No.1/No.4 Ignition Coil

04895G09

Fig. 118 Electrical and ignition system component locations—BF75A and BF90A

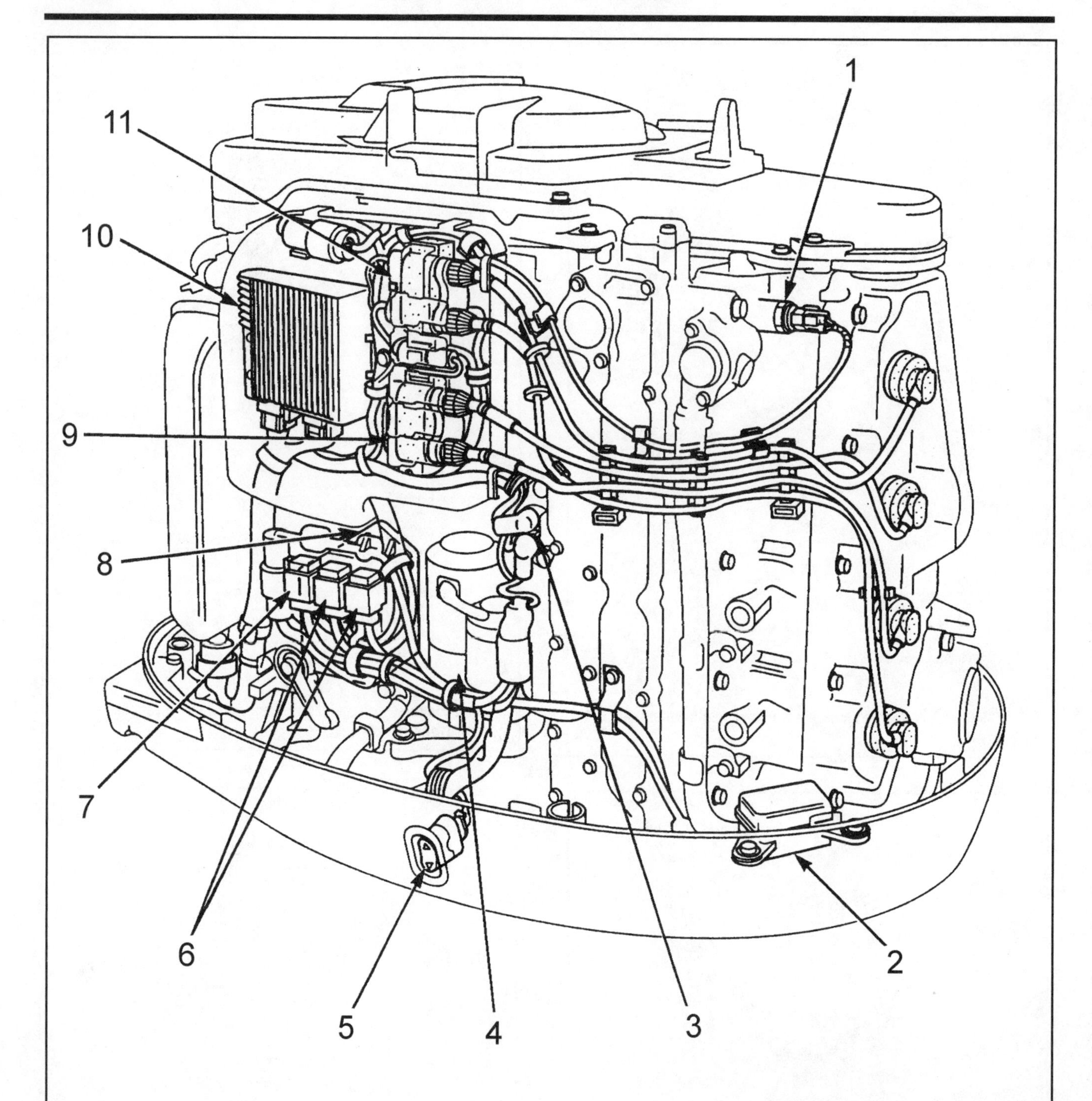

1. Overheat Sensor
2. 90A Fuse Box
3. Starter Solenoid
4. Starter Motor
5. Power Tilt Switch
6. 30A Fuse Boxes
7. 10A Fuse Box
8. Power Tilt Relay
9. No. 2/No. 3 Ignition Coil
10. ECM
11. No. 1/No. 4 Ignition Coil

04895G10

Fig. 119 Port side electrical and ignition system component locations—BF115A and BF130A

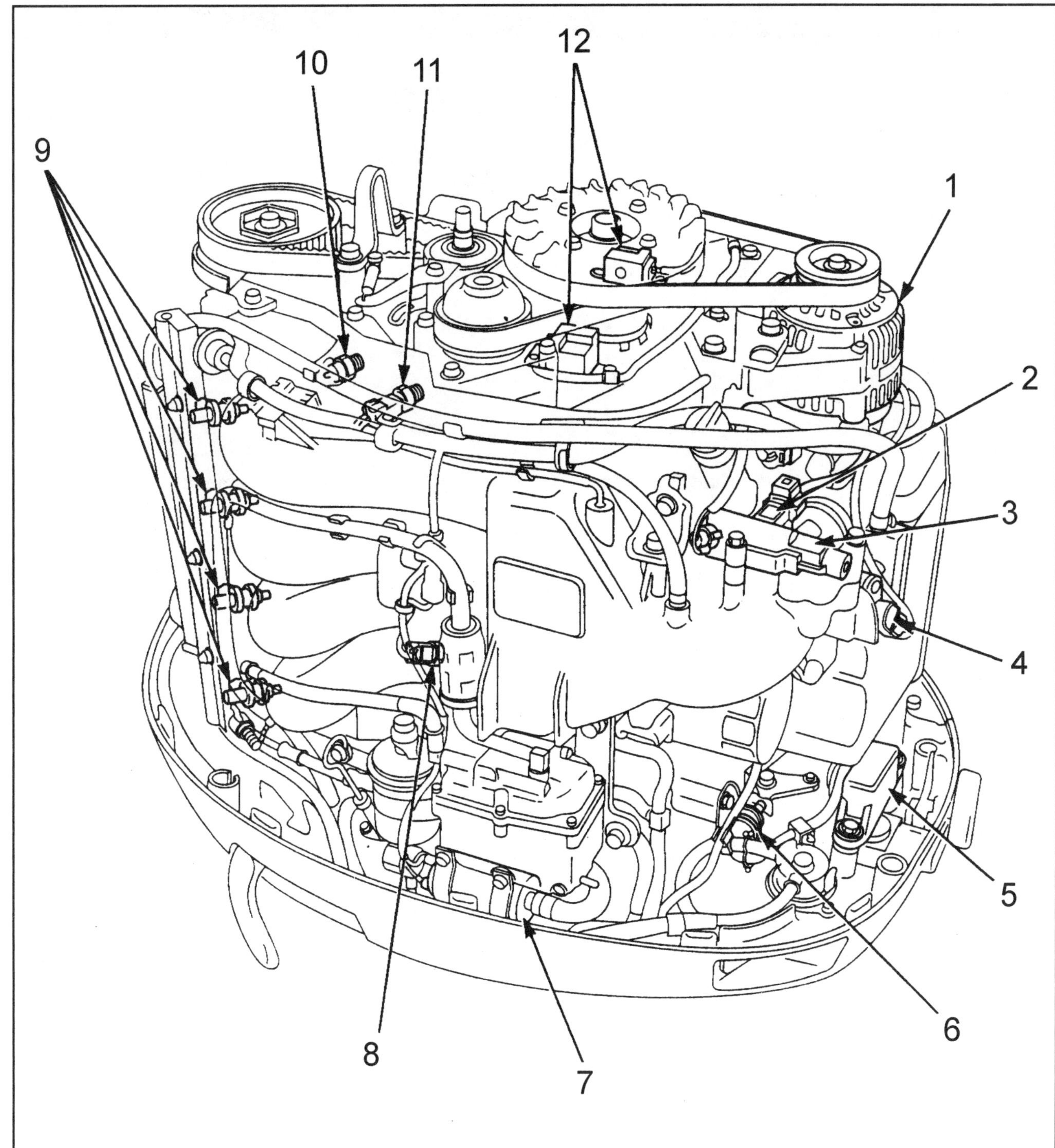

1. Alternator
2. MAP Sensor
3. IAC Valve
4. TP Sensor
5. Main Relay
6. Neutral Switch
7. Fuel Pump Unit
8. IAT Sensor
9. Fuel Injectors
10. Oil Pressure Switch
11. ECT Sensor
12. Pulser Coils

04895G11

Fig. 120 Starboard side electrical and ignition system component locations—BF115A and BF130A

IGNITION AND ELECTRICAL WIRING DIAGRAMS

INDEX OF WIRING DIAGRAMS

DIAGRAM 1	Sample Diagram: How To Read & Interpret Wiring Diagrams
DIAGRAM 2	Sample Diagram: Wiring Diagram Symbols
DIAGRAM 3	BF2A/20 Ignition Schematic
DIAGRAM 4	BF5A/50 Ignition, Charging, Oil Pressure Schematic
DIAGRAM 5	BF75/100 Power Tilt, Engine Schematic
DIAGRAM 6	BF8A Ignition, Charging, Oil Pressure Schematic
DIAGRAM 7	BF9.9A/15A Starting, Charging, Oil Pressure, Ignition System Schematic
DIAGRAM 8	BF25A/30A Starting, Charging, Choke, Oil Pressure, Thermo Switch Schematic
DIAGRAM 9	BF25A/30A Ignition System Schematic
DIAGRAM 10	BF35A/40A/45A/50A Ignition, Trim Control, Emergency Stop Schematic
DIAGRAM 11	BF35A/40A/45A/50A (w/o Tiller Handle) Starting, Charging, Choke, Oil Pressure, Thermo Switch Schematic
DIAGRAM 12	BF35A/40A/45A/50A (w/ Tiller Handle) Starting, Charging, Emergency Stop, Oil Pressure, Thermo Switch Schematic
DIAGRAM 13	BF75A/90A Starting, Charging, Choke, Oil Pressure, Thermo Switch Schematic
DIAGRAM 14	BF75A/90A Ignition, Heater, Trim Control, Buzzer, Emergency Stop Schematic
DIAGRAM 15	BF115/130 Engine Control Unit Schematic
DIAGRAM 16	BF115/130 Starting, Fuel Pump, Trim Angle Sensor, Tilt Control Schematic

04895W01

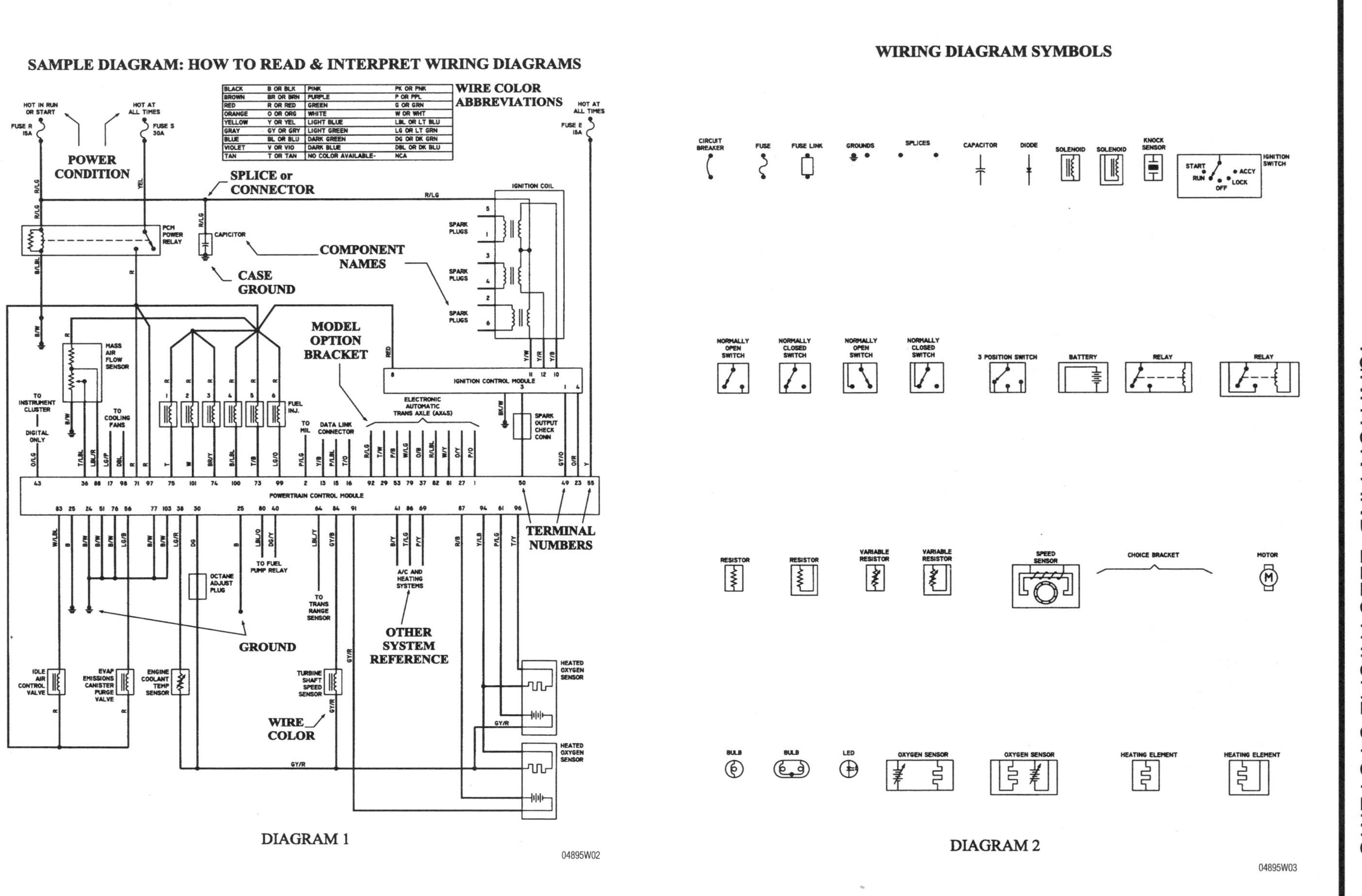
SAMPLE DIAGRAM: HOW TO READ & INTERPRET WIRING DIAGRAMS
WIRE COLOR ABBREVIATIONS
BLACK B OR BLK
BROWN BR OR BRN
RED R OR RED
ORANGE O OR ORG
YELLOW Y OR YEL
GRAY GY OR GRY
BLUE BL OR BLU
VIOLET V OR VIO
TAN T OR TAN
PINK PK OR PNK
PURPLE P OR PPL
GREEN G OR GRN
WHITE W OR WHT
LIGHT BLUE LBL OR LT BLU
LIGHT GREEN LG OR LT GRN
DARK GREEN DG OR DK GRN
DARK BLUE DBL OR DK BLU
NO COLOR AVAILABLE- NCA
HOT IN RUN OR START
FUSE R 15A
HOT AT ALL TIMES
FUSE S 30A
HOT AT ALL TIMES
FUSE E 15A
POWER CONDITION
SPLICE or CONNECTOR
PCM POWER RELAY
CAPICITOR
COMPONENT NAMES
CASE GROUND
IGNITION COIL
SPARK PLUGS
SPARK PLUGS
SPARK PLUGS
MODEL OPTION BRACKET
MASS AIR FLOW SENSOR
IGNITION CONTROL MODULE
TO INSTRUMENT CLUSTER
DIGITAL ONLY
TO COOLING FANS
FUEL INJ.
TO MIL
DATA LINK CONNECTOR
ELECTRONIC AUTOMATIC TRANS AXLE (AX4S)
SPARK OUTPUT CHECK CONN
POWERTRAIN CONTROL MODULE
TERMINAL NUMBERS
TO FUEL PUMP RELAY
OCTANE ADJUST PLUG
TO TRANS RANGE SENSOR
A/C AND HEATING SYSTEMS
OTHER SYSTEM REFERENCE
GROUND
IDLE AIR CONTROL VALVE
EVAP EMISSIONS CANISTER PURGE VALVE
ENGINE COOLANT TEMP SENSOR
TURBINE SHAFT SPEED SENSOR
WIRE COLOR
HEATED OXYGEN SENSOR
HEATED OXYGEN SENSOR
DIAGRAM 1
04895W02
WIRING DIAGRAM SYMBOLS
CIRCUIT BREAKER
FUSE
FUSE LINK
GROUNDS
SPLICES
CAPACITOR
DIODE
SOLENOID
SOLENOID
KNOCK SENSOR
IGNITION SWITCH
START
RUN
OFF
LOCK
ACCY
NORMALLY OPEN SWITCH
NORMALLY CLOSED SWITCH
NORMALLY OPEN SWITCH
NORMALLY CLOSED SWITCH
3 POSITION SWITCH
BATTERY
RELAY
RELAY
RESISTOR
RESISTOR
VARIABLE RESISTOR
VARIABLE RESISTOR
SPEED SENSOR
CHOICE BRACKET
MOTOR
BULB
BULB
LED
OXYGEN SENSOR
OXYGEN SENSOR
HEATING ELEMENT
HEATING ELEMENT
DIAGRAM 2
04895W03

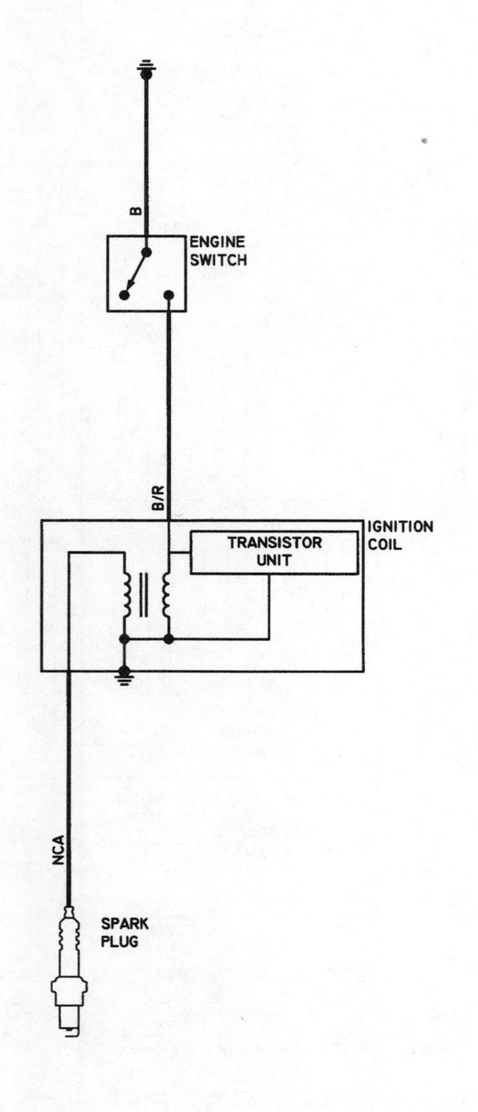
HONDA BF20/2A WIRING SCHEMATICS
B
ENGINE SWITCH
B/R
IGNITION COIL
TRANSISTOR UNIT
NCA
SPARK PLUG
DIAGRAM 3
04895W04

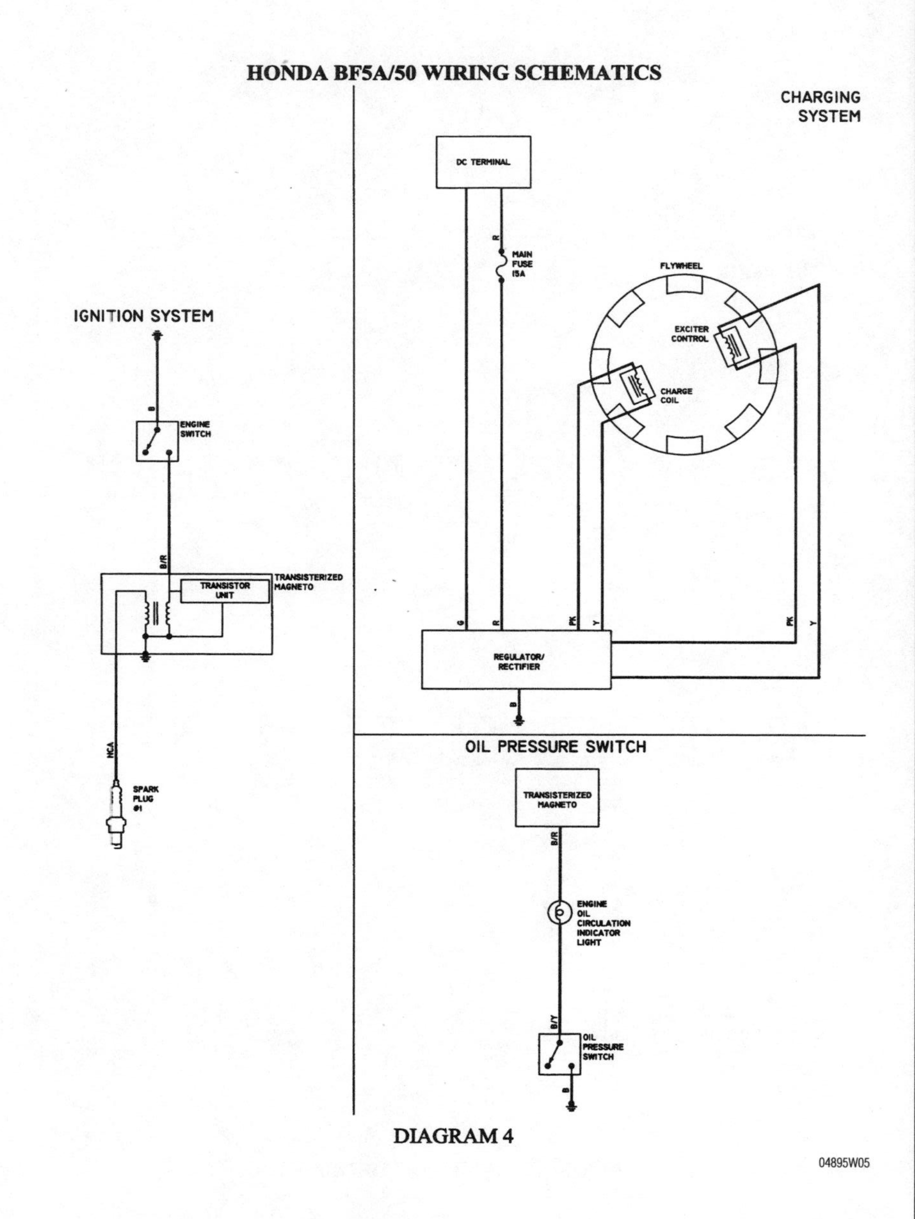
HONDA BF5A/50 WIRING SCHEMATICS
CHARGING SYSTEM
DC TERMINAL
MAIN FUSE 15A
FLYWHEEL
EXCITER CONTROL
CHARGE COIL
REGULATOR/ RECTIFIER
IGNITION SYSTEM
ENGINE SWITCH
TRANSISTOR UNIT
TRANSISTERIZED MAGNETO
NCA
SPARK PLUG #1
OIL PRESSURE SWITCH
TRANSISTERIZED MAGNETO
ENGINE OIL CIRCULATION INDICATOR LIGHT
OIL PRESSURE SWITCH
DIAGRAM 4
04895W05

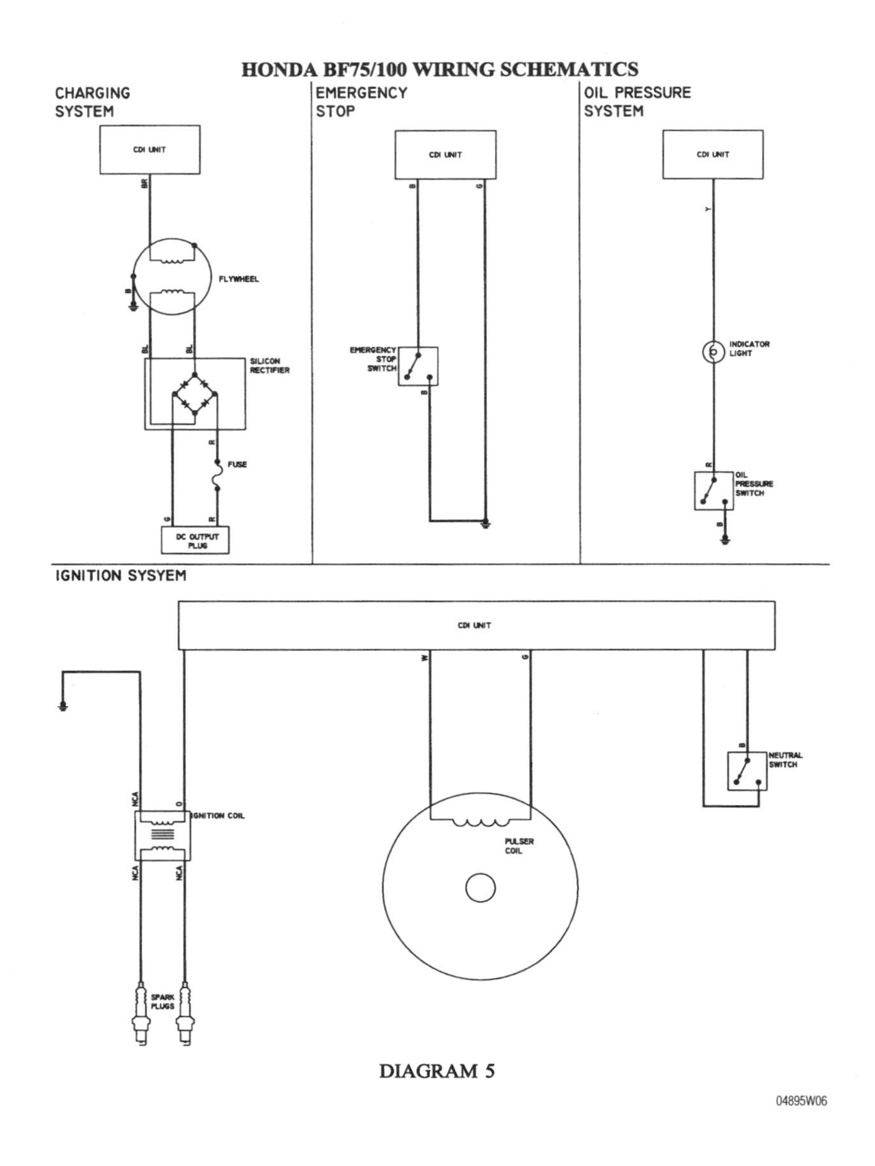

DIAGRAM 5

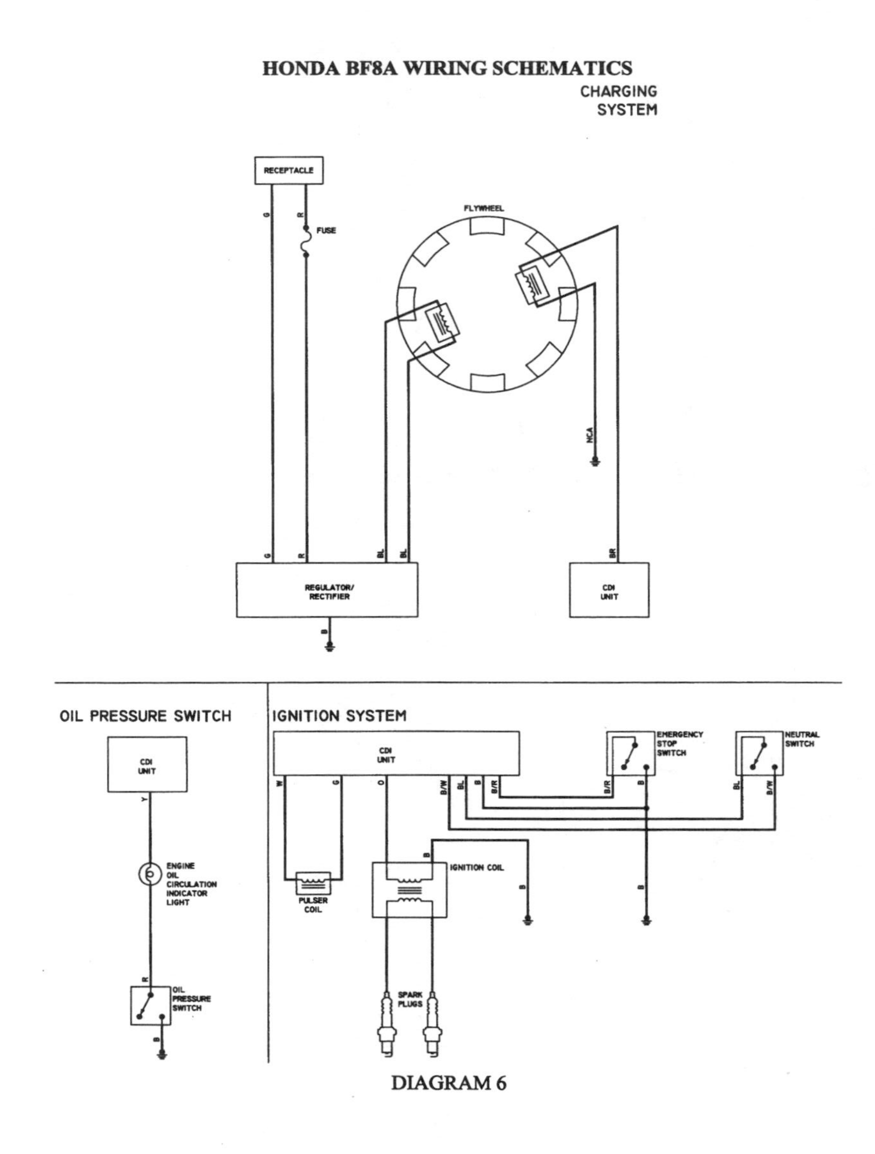

DIAGRAM 6

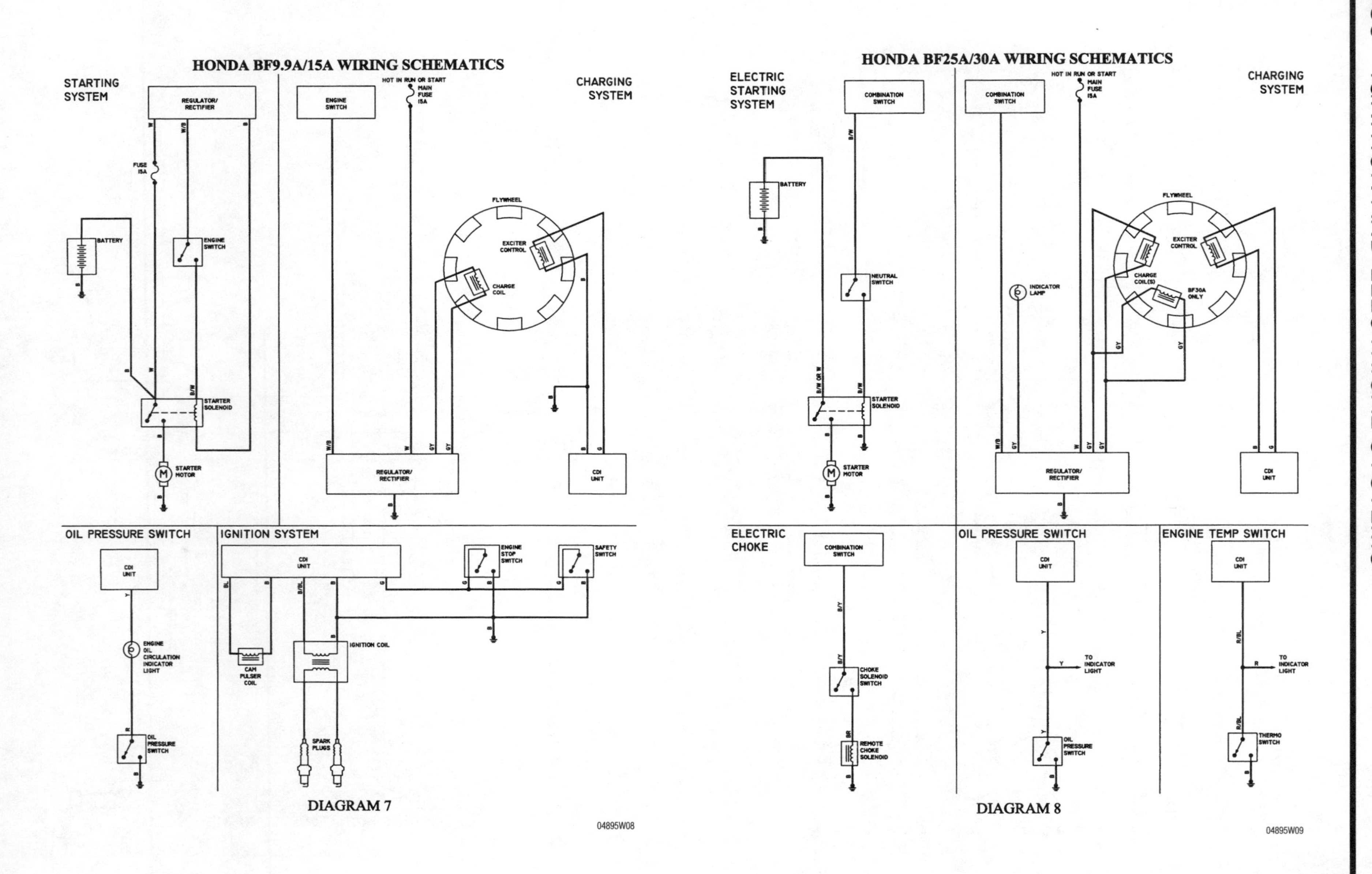
HONDA BF9.9A/15A WIRING SCHEMATICS
STARTING SYSTEM
CHARGING SYSTEM
REGULATOR/ RECTIFIER
ENGINE SWITCH
HOT IN RUN OR START
MAIN FUSE 15A
FUSE 15A
BATTERY
ENGINE SWITCH
FLYWHEEL
EXCITER CONTROL
CHARGE COIL
STARTER SOLENOID
STARTER MOTOR
REGULATOR/ RECTIFIER
CDI UNIT
OIL PRESSURE SWITCH
IGNITION SYSTEM
CDI UNIT
CDI UNIT
ENGINE STOP SWITCH
SAFETY SWITCH
ENGINE OIL CIRCULATION INDICATOR LIGHT
CAM PULSER COIL
IGNITION COIL
OIL PRESSURE SWITCH
SPARK PLUGS
DIAGRAM 7
04895W08
HONDA BF25A/30A WIRING SCHEMATICS
ELECTRIC STARTING SYSTEM
CHARGING SYSTEM
COMBINATION SWITCH
COMBINATION SWITCH
HOT IN RUN OR START
MAIN FUSE 15A
BATTERY
FLYWHEEL
EXCITER CONTROL
CHARGE COIL(S)
BF30A ONLY
NEUTRAL SWITCH
INDICATOR LAMP
STARTER SOLENOID
STARTER MOTOR
REGULATOR/ RECTIFIER
CDI UNIT
ELECTRIC CHOKE
COMBINATION SWITCH
OIL PRESSURE SWITCH
CDI UNIT
ENGINE TEMP SWITCH
CDI UNIT
TO INDICATOR LIGHT
TO INDICATOR LIGHT
CHOKE SOLENOID SWITCH
REMOTE CHOKE SOLENOID
OIL PRESSURE SWITCH
THERMO SWITCH
DIAGRAM 8
04895W09

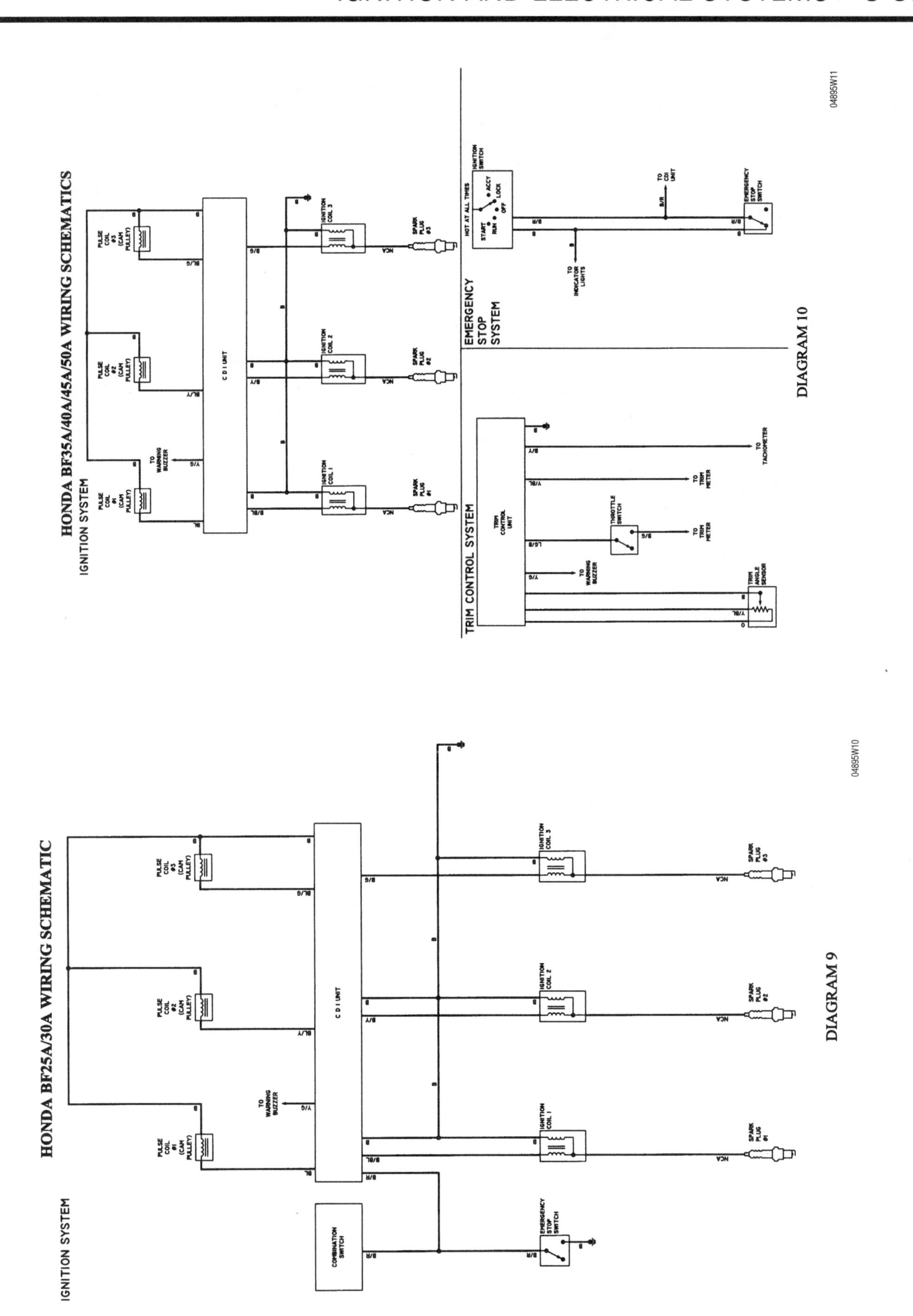
HONDA BF35A/40A/45A/50A WIRING SCHEMATICS
IGNITION SYSTEM
EMERGENCY STOP SYSTEM
TRIM CONTROL SYSTEM
DIAGRAM 10
04895W11
HONDA BF25A/30A WIRING SCHEMATIC
IGNITION SYSTEM
DIAGRAM 9
04895W10

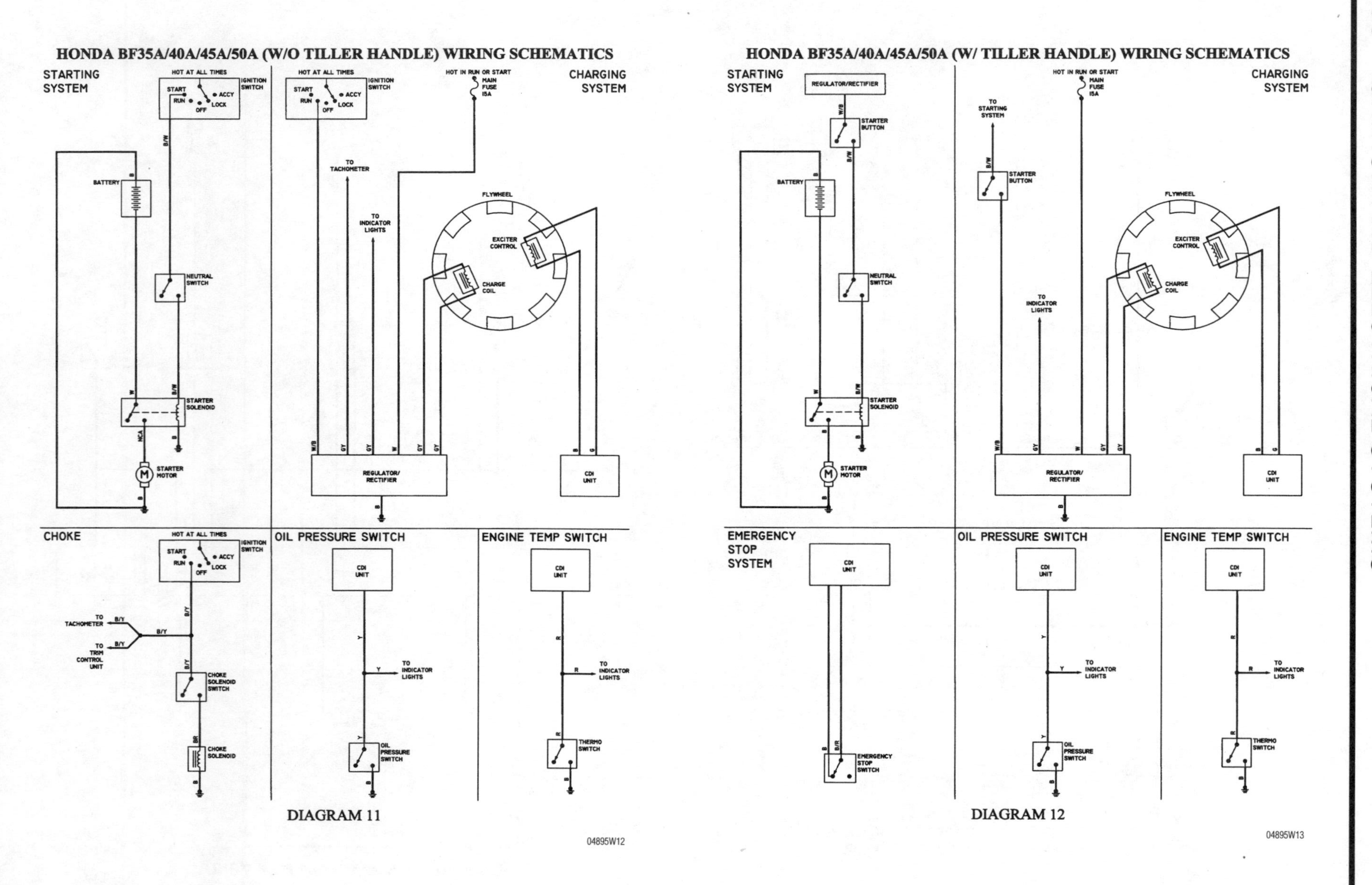
HONDA BF35A/40A/45A/50A (W/O TILLER HANDLE) WIRING SCHEMATICS
STARTING SYSTEM
HOT AT ALL TIMES
IGNITION SWITCH
START
RUN
ACCY
LOCK
OFF
BATTERY
NEUTRAL SWITCH
STARTER SOLENOID
STARTER MOTOR
TO TACHOMETER
TO INDICATOR LIGHTS
HOT IN RUN OR START
MAIN FUSE 15A
CHARGING SYSTEM
FLYWHEEL
EXCITER CONTROL
CHARGE COIL
REGULATOR/RECTIFIER
CDI UNIT
CHOKE
TO TACHOMETER
TO TRIM CONTROL UNIT
CHOKE SOLENOID SWITCH
CHOKE SOLENOID
OIL PRESSURE SWITCH
OIL PRESSURE SWITCH
ENGINE TEMP SWITCH
THERMO SWITCH
DIAGRAM 11
04895W12
HONDA BF35A/40A/45A/50A (W/ TILLER HANDLE) WIRING SCHEMATICS
STARTING SYSTEM
REGULATOR/RECTIFIER
STARTER BUTTON
TO STARTING SYSTEM
EMERGENCY STOP SYSTEM
EMERGENCY STOP SWITCH
DIAGRAM 12
04895W13

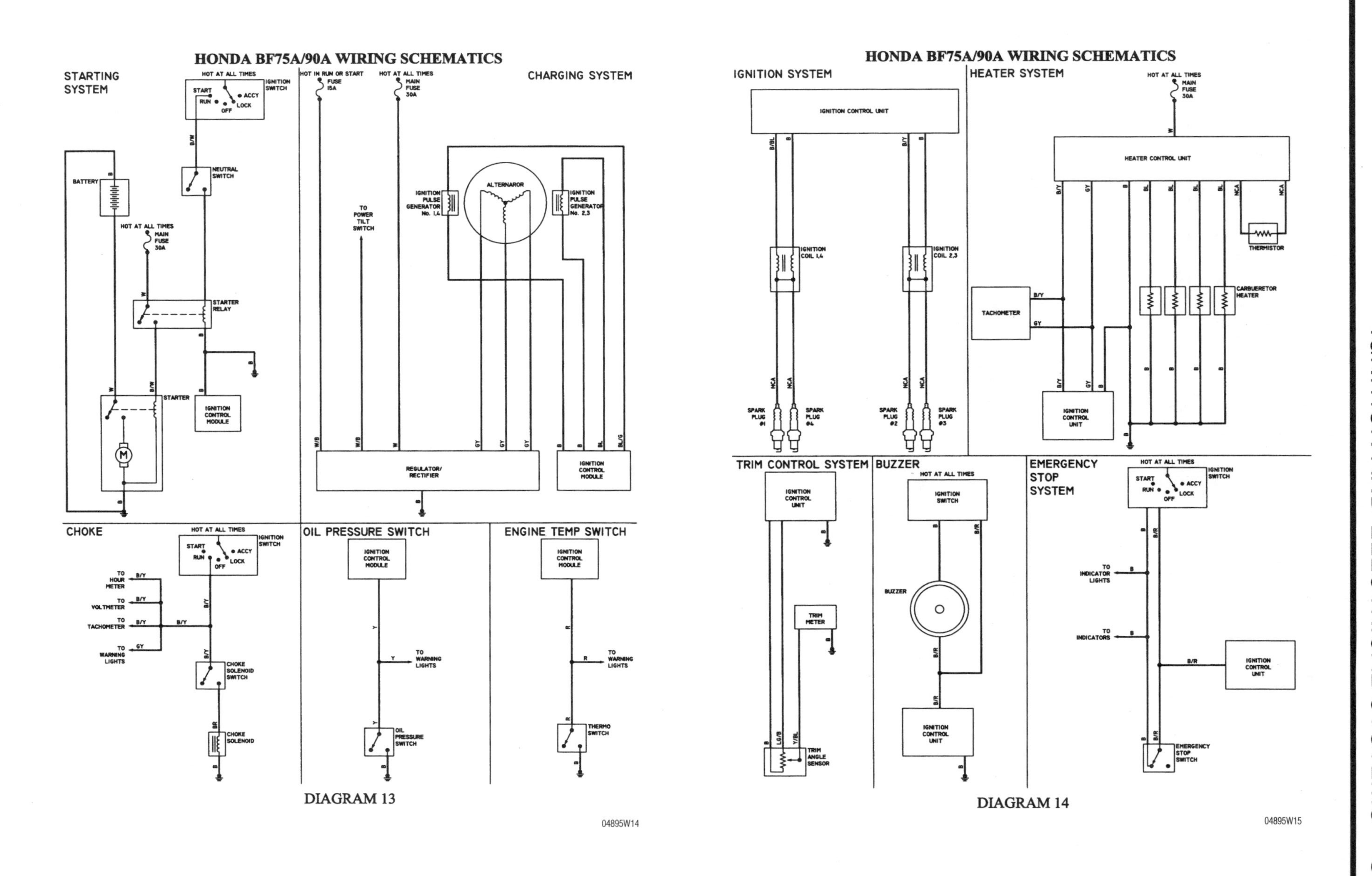
HONDA BF75A/90A WIRING SCHEMATICS
STARTING SYSTEM
CHARGING SYSTEM
CHOKE
OIL PRESSURE SWITCH
ENGINE TEMP SWITCH
DIAGRAM 13
04895W14
HONDA BF75A/90A WIRING SCHEMATICS
IGNITION SYSTEM
HEATER SYSTEM
TRIM CONTROL SYSTEM
BUZZER
EMERGENCY STOP SYSTEM
DIAGRAM 14
04895W15

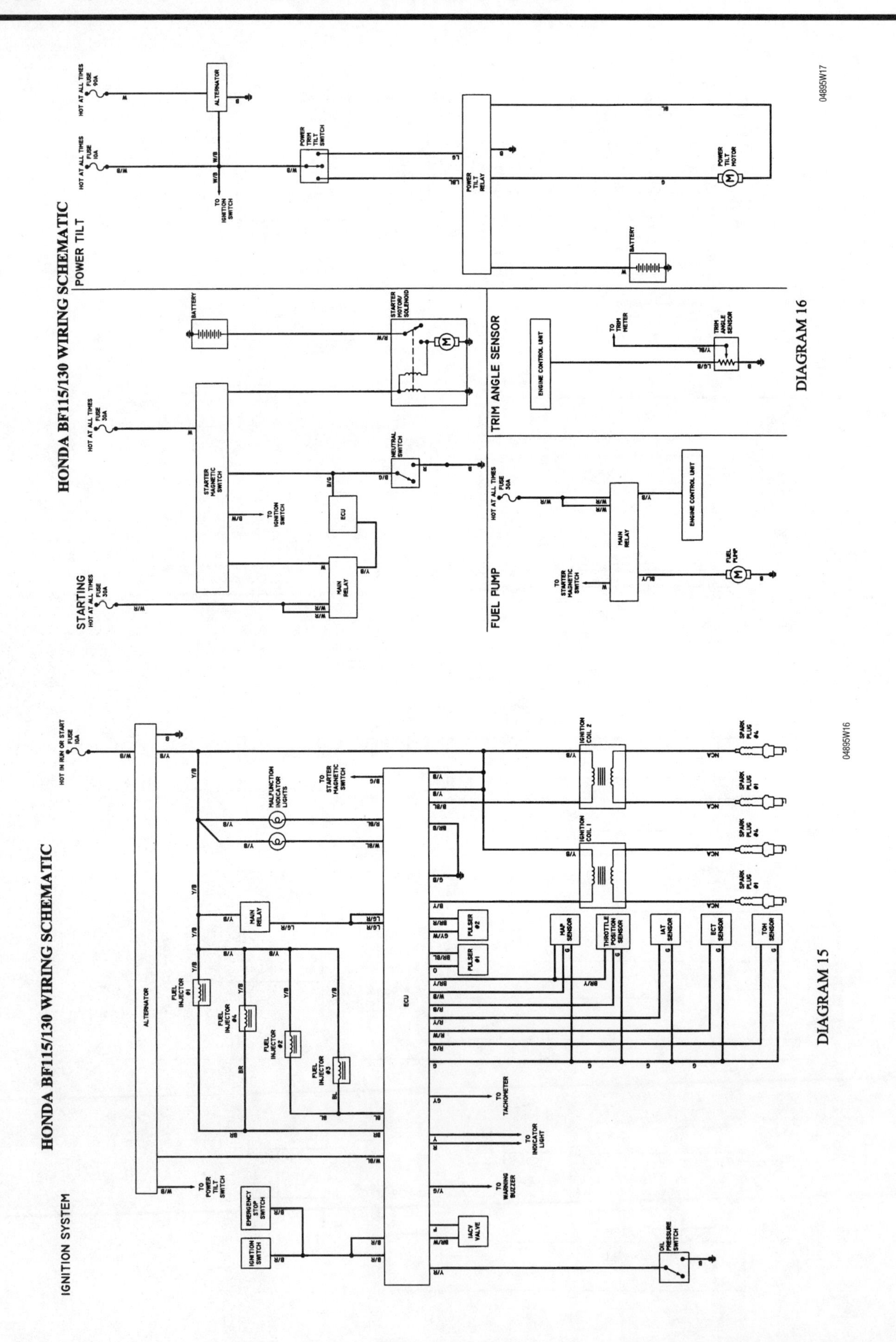
HONDA BF115/130 WIRING SCHEMATIC
POWER TILT
STARTING
TRIM ANGLE SENSOR
FUEL PUMP
DIAGRAM 16
04895W17
HONDA BF115/130 WIRING SCHEMATIC
IGNITION SYSTEM
DIAGRAM 15
04895W16
ALTERNATOR
ECU

LUBRICATION SYSTEM 6-2
DESCRIPTION AND OPERATION 6-2
SPLASH LUBRICATION 6-2
FORCED LUBRICATION 6-2
TROUBLESHOOTING THE LUBRICATION SYSTEM 6-2
OIL PUMP 6-3
REMOVAL & INSTALLATION 6-3
OVERHAUL 6-4
OIL PRESSURE SWITCH 6-6
TESTING 6-6
REMOVAL & INSTALLATION 6-7
OIL PRESSURE WARNING LAMP 6-7
TESTING 6-7
REMOVAL & INSTALLATION 6-7
WARNING BUZZER 6-8
TESTING 6-8
REMOVAL & INSTALLATION 6-8
COOLING SYSTEM 6-9
DESCRIPTION AND OPERATION 6-9
WATER PUMP 6-9
THERMOSTAT & PRESSURE RELIEF 6-9
OVERHEAT WARNING SYSTEM 6-9
TROUBLESHOOTING THE COOLING SYSTEM 6-10
WATER PUMP 6-10
REMOVAL & INSTALLATION 6-10
OVERHEAT WARNING LAMP 6-15
TESTING 6-15
REMOVAL & INSTALLATION 6-16
OVERHEAT THERMOSWITCH 6-16
TESTING 6-16
REMOVAL & INSTALLATION 6-16
WARNING BUZZER 6-16
TESTING 6-16
REMOVAL & INSTALLATION 6-16

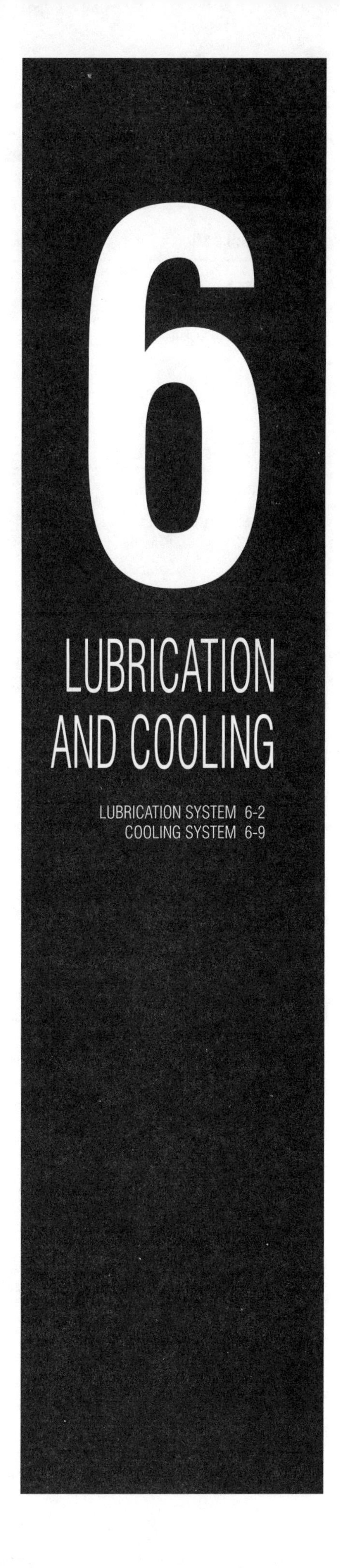

LUBRICATION SYSTEM

Description and Operation

SPLASH LUBRICATION

See Figure 1

The BF20 and BF2A are lubricated by the simplest of lubrication methods called the splash lubrication system. All vital moving parts are splashed with lubricating oil stored in the crankcase. An oil slinger attached to a shaft extends down into the area which the oil lies. As the crankshaft turns, the slinger, splashes oil up onto the cylinder walls, crankshaft bearings and camshaft bearings.

FORCED LUBRICATION

See Figure 2

The oil system used on forced lubrication outboards closely resembles the system used on other 4-stroke engines like the one in your car or truck. Its primary components consist of an oil sump at the bottom of the engine, an oil pickup tube submerged in the sump, an oil pump, an oil filter and various oil passage ways in the engine.

Oil is drawn from the sump by the pump through a filter, where dirt and metal particles are removed. The oil is then delivered under pressure to various parts of the engine via oil galleys machined into the engine. Oil from the oil pump flows through drilled passages in the camshaft to the valves and rocker arms. The oil also travels through the oil gallery in the cylinder block. The oil gallery consists of two passages which feed the upper and lower main journals of the crankshaft. As oil enters the hole in the upper journal, it travels down through the drilled passage in the crankshaft to lubricate the connecting rod journal. The main journal is lubricated by oil coming up from the lower passage via the crankshaft. Oil splash from the crankshaft lubricates the connecting rods, pistons and the cylinders.

The oil pump is a Trochoid design and consists of an inner rotor, outer rotor and pump body. The inner rotor is driven by the camshaft. The outer rotor is free in the pump body and is driven by the inner rotor. As the rotors spin, the volume of the oil between them changes and provides the force to push oil out into the oil gallery. A check valve prevents over pressurization of the system and allows directs oil back to the oil pan at a predetermined pressure.

A breather chamber allows crankcase ventilation but prevents airborne oil mist from leaving the engine.

Since oil pressure in a 4-stroke engine is so essential, the lubrication system contains two warning devices to notify the operator of problems. The first is an oil pressure warning lamp mounted on the engine, or, in the case of larger engines, mounted on the remote control panel. The second is an emergency warning buzzer is used on some engines to provide an audible notification of problems. The warning system is activated by an oil pressure switch mounted to the engine. If oil pressure drops to a predetermined level, the switch grounds completing the lamp and buzzer circuit.

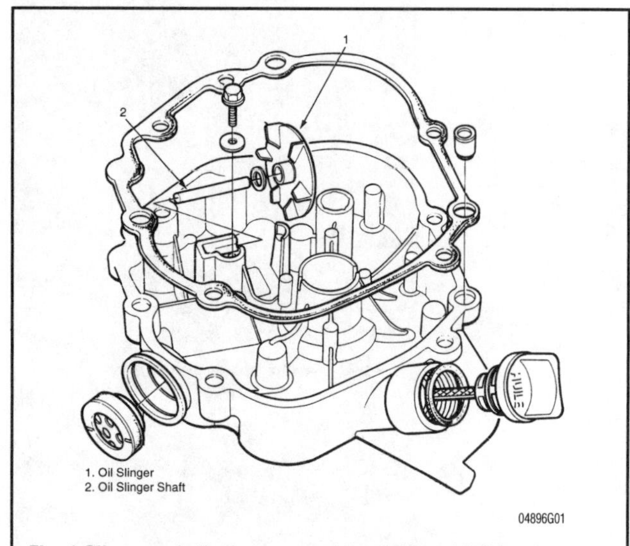

Fig. 1 Oil pan and oil slinger assembly—BF20 and BF2A only

Troubleshooting the Lubrication System

See Figure 3

Check for proper oil pressure at the oil pressure switch passage with an externally mounted mechanical oil pressure gauge (as opposed to relying on a factory installed dash-mounted lamp).

1. With the engine cold, locate and remove the oil pressure switch.
2. Following the manufacturer's instructions, connect a mechanical oil pressure gauge.
3. Start the engine and allow it to idle.
4. Check the oil pressure reading when cold and record the number.
5. Run the engine until normal operating temperature is reached. This may require you to take the boat out for a short cruise or run the outboard in a test tank.

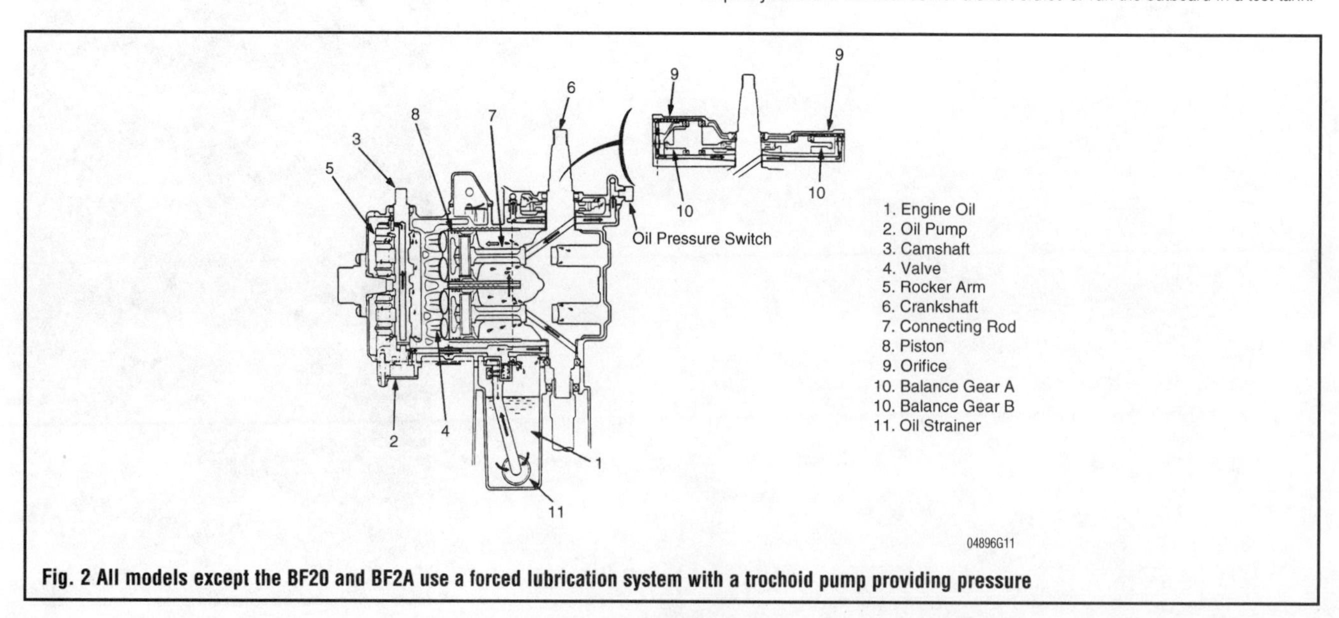

Fig. 2 All models except the BF20 and BF2A use a forced lubrication system with a trochoid pump providing pressure

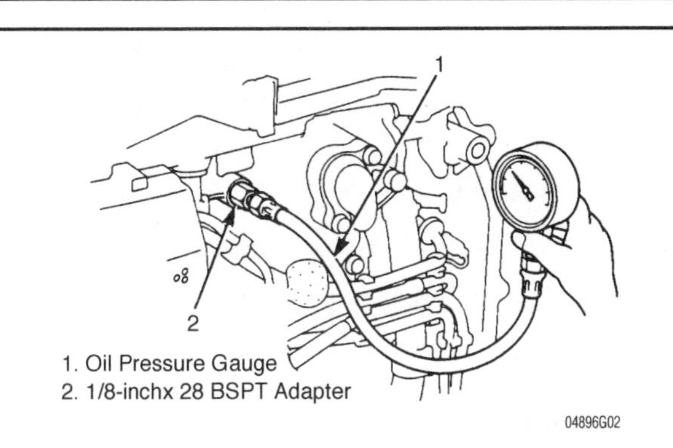

Fig. 3 Using a commonly available oil pressure gauge, measure oil pressure at the oil pressure switch port

6. Check the oil pressure reading again with the engine hot and record the number. Turn the engine **OFF**.
7. Compare your oil pressure reading to the specification. If the cold pressure is well above the specification, and the hot reading was lower than the specification, you may have the wrong viscosity oil in the engine. Change the oil, making sure to use the proper grade and quantity, then repeat the test.

Low oil pressure readings could be attributed to internal component wear, pump related problems, a low oil level, or oil viscosity that is too low. High oil pressure readings could be caused by an overfilled crankcase, too high of an oil viscosity or a faulty pressure relief valve.

Oil Pump

REMOVAL & INSTALLATION

BF115A and BF130A

See Figure 4

1. Remove the engine.
2. Remove the oil pump body from the crankcase.

Do not reuse the gasket.

3. Remove the two dowel pins from the oil pump body.

To install:

4. Install the two dowel pins into the oil pump body.

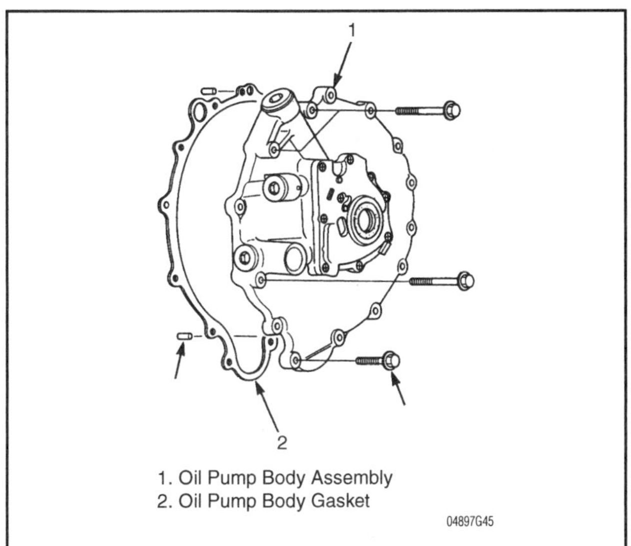

Fig. 4 Oil pump and oil pump body assembly—BF115A and BF130A

5. Install the oil pump to the crankcase using a new gasket and tighten the bolts securely.
6. Reinstall the engine.

BF75A and BF90A

1. Remove the cylinder head.
2. Remove the oil pump from the cylinder head.

To install:

3. Fill the pump body with fresh engine oil.
4. Inspect the oil pump O-ring for damage or wear. Replace if needed.
5. Align the projection on the oil pump shaft with the indentation on the camshaft. Turn the oil pump shaft to align.
6. Install the pump body on the cylinder head.
7. Tighten the 6 x 18 mm bolts to 9 ft. lbs. (12 Nm) and the 8 x 65 mm bolts to 20 ft. lbs. (26 Nm).

BF50A, BF45A, BF40 and BF35A

1. Remove the cylinder head.
2. Remove the oil pump from the cylinder head.

To install:

3. Fill the pump body with fresh engine oil.
4. Inspect the oil pump O-rings for damage or wear. Replace if needed.
5. Align the projection on the oil pump shaft with the indentation on the camshaft. Turn the oil pump shaft to align.
6. Insatll the pump body to the cylinder head and tighten the bolts to 9.4 ft. lbs. (13 Nm).

BF25A and BF30A

1. Remove the cylinder head.
2. Remove the oil pump from the cylinder head.

To install:

3. Fill the pump body with fresh engine oil.
4. Inspect the oil pump O-rings for damage or wear. Replace if needed.
5. Align the projection on the oil pump shaft with the indentation on the camshaft. Turn the oil pump shaft to align.
6. Install the pump body on the cylinder head. Tighten the bolts to 9.4 ft. lbs. (13 Nm).

BF9.9A and BF15A

1. Remove the cylinder head.
2. Remove the oil pump from the cylinder head.

To install:

3. Fill the pump body with fresh engine oil.
4. Inspect the oil pump O-rings for damage or wear. Replace if needed.
5. Align the oil pump shaft with the camshaft. Turn the oil pump shaft to align.
6. Install the pump body on the cylinder head and tighten the bolts securely.

BF75, BF8A and BF100

1. Remove the cylinder head.
2. Remove the oil pump from the cylinder head.

To install:

3. Fill the pump body with fresh engine oil.
4. Inspect the oil pump O-rings for damage or wear. Replace if needed.
5. Align the oil pump shaft with the camshaft. Turn the oil pump shaft to align.
6. Bolt the pump body to the cylinder head and tighten the bolts securely.

BF50 and BF5A

1. Remove the engine.
2. The oil pump is located on the bottom of the engine, on the oil pan.

To install:

3. Inspect the pump body, shaft and rotors for any signs of damage, wear or other abnormality. Replace any parts which are damaged.
4. Insert the shaft into the pump body, making sure to align the flats on the shaft to the rotor and camshaft.
5. Fill the pump body with clean engine oil.
6. Install the new O-ring, pump cover and tighten the screws securely.

BF2A and BF20

➡These engines are splash lubricated and do not use a mechanical oil pump for lubrication.

OVERHAUL

BF130A and BF115A

➧ See Figure 5

1. Remove the oil pump cover. Don't drop the dowel pins.
2. Remove the inner and outer rotors. Clean both thoroughly, looking for signs of wear, scratches or any other type of damage. Replace any part not meeting specifications or that is damaged.
3. Measure the oil pump inner diameter. Standard measurement should be 3.3071–3.3083 in. (84.000–84.030 mm).
4. Measure the pump body depth. Standard measurement should be 0.4929–0.4941 in. (12.520–12.550 mm).
5. Measure the outer rotor height. Standard measurement should be 0.4913–0.4921 in. (12.480–12.500 mm).
6. Measure the inner rotor-to-outer rotor clearance. Standard measurement should be 0.002–0.006 in. (0.04–0.16 mm). If the clearance is greater than 0.008 in (0.20 mm) replace the component.
7. Measure the rotor-to-pump body clearance. Standard measurement should be 0.004–0.007 in. (0.10–0.18 mm). If the clearance is greater than 0.008 in. (0.20 mm) replace the component.
8. Measure the outer rotor-to-oil pump body side clearance. Standard measurement should be 0.001–0.003 in. (0.02–0.07 mm). If the clearance is greater than 0.005 in (0.12 mm) replace the component.

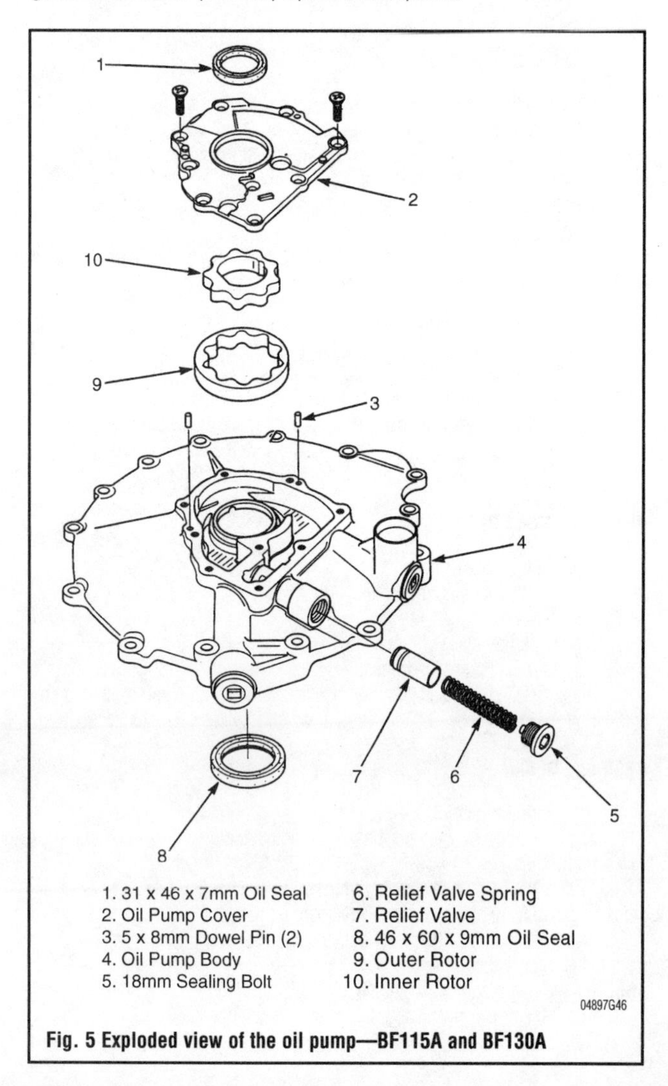

Fig. 5 Exploded view of the oil pump—BF115A and BF130A

9. Replace both oil seals, apply grease to the seal lips before assembling the pump.
10. Remove the sealing bolt and remove the relief valve and relief valve spring.
11. Check both for damage, then oil the spring before reassembling.
12. Add a locking agent, such as Loctite®, to the sealing bolt and torque to 29 ft. lbs. (39 Nm).
13. After cleaning the rotors, oil them both thoroughly and install in the oil pump body.
14. Install the oil pump cover and torque the screws to 5.1 ft. lbs. (7 Nm).
15. Install the oil pump body onto the crankcase and torque the bolts to 20 ft. lbs. (26 Nm).

BF75A and BF90A

➧ See Figure 6

1. Remove the oil pump cover and O-ring.
2. Remove the outer and inner rotors.
3. Remove the thrust washer.
4. Remove the sealing bolt and remove the sealing washer, relief valve spring and the relief valve.
5. Inspect the sealing washer, relief valve spring and the relief valve for scratches, wear or any other type of damage. Replace any damaged part.
6. Coat the spring and valve with fresh engine oil before assembly.
7. Apply a locking agent, such as Loctite®, to the sealing bolt and tighten to 29 ft. lbs. (39 Nm).
8. Measure the oil pump body inner diameter. Standard measurement should be 3.150–3.151 in. (80.00–80.04 mm). If the clearance is greater than 3.152 in. (80.06 mm) replace the components.
9. Measure the oil pump depth. Standard measurement should be 0.709–0.711 in. (18.02–18.05 mm). If the clearance is greater than 0.712 in. (18.09 mm) replace the components.
10. Measure the outer rotor height. Standard measurement should be 0.708–0.709 in. (17.98–18.00 mm). If the clearance is greater than 0.707 in. (17.96 mm) replace the components.
11. Measure the inner rotor-to-outer rotor clearance. Standard measurement should be 0.0008–0.0063 in. (0.02–0.16 mm). If the clearance is greater than 0.008 in. (0.2 mm) replace the components.
12. Measure the outer rotor-to-pump body clearance. Standard measurement should be 0.004–0.007 in. (0.10–0.19 mm). If the clearance is greater than 0.009 in. (0.23 mm) replace the components.
13. Measure the pump end clearance. Standard measurement should be 0.0008–0.0028 in. (0.02–0.07 mm). If the clearance is greater than 0.004 in. (0.10 mm) replace the components.
14. Replace all parts not meeting specifications or that are damaged in any way.

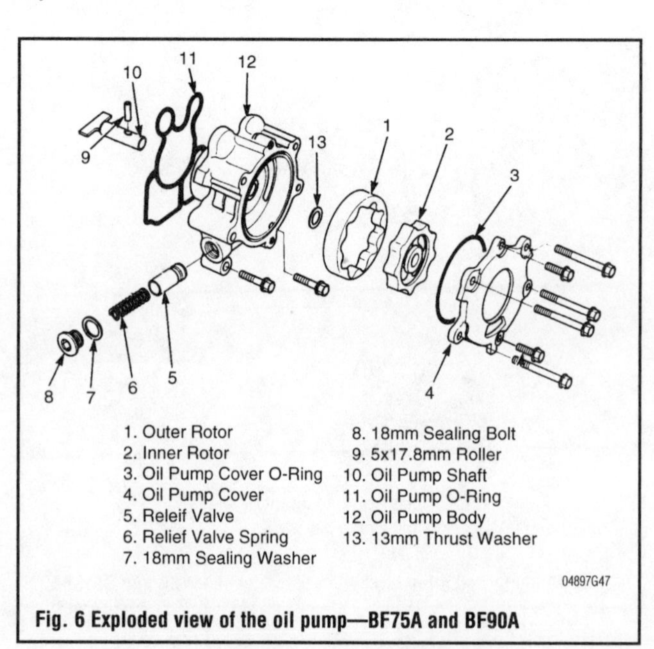

Fig. 6 Exploded view of the oil pump—BF75A and BF90A

BF50A, BF45A, BF40A and BF35A

➧ See Figure 7

1. Remove the oil pump cover and O-ring.
2. Remove the outer and inner rotors.
3. Remove the thrust washer.
4. Remove the oil pump shaft and roller.
5. Remove the sealing bolt and remove the sealing washer, relief valve spring and the relief valve.
6. Inspect the sealing washer, relief valve spring and the relief valve for scratches, wear or any other type of damage. Replace any damaged part.
7. Coat the spring and valve with fresh engine oil before assembly.
8. Apply a locking agent, such as Loctite®, to the sealing bolt.
9. Measure the oil pump body inner diameter. Standard measurement should be 1.974–1.975 in. (50.15–50.18 mm). If the clearance is greater than 3.152 in. (80.06 mm) replace the components.
10. Measure the oil pump depth. Standard measurement should be 0.709–0.711 in. (18.02–18.05 mm). If the clearance is greater than 0.712 in. (18.09 mm) replace the components.
11. Measure the outer rotor height. Standard measurement should be 0.708–0.709 in. (17.98–18.00 mm). If the clearance is greater than 0.707 in. (17.96 mm) replace the components.
12. Measure the inner rotor-to-outer rotor clearance. Standard measurement should be 0.0008–0.0063 in. (0.02–0.16 mm) If the clearance is greater than 0.008 in. (0.2 mm) replace the components.
13. Measure the outer rotor-to-pump body clearance. Standard measurement should be 0.004–0.007 in. (0.10–0.19 mm). If the clearance is greater than 0.009 in. (0.23 mm) replace the components.
14. Measure the pump end clearance. Standard measurement should be 0.0008–0.0028 in. (0.02–0.07 mm). If the clearance is greater than 0.004 in. (0.10 mm) replace the components.
15. Replace all parts not meeting specifications or that are damaged in any way.

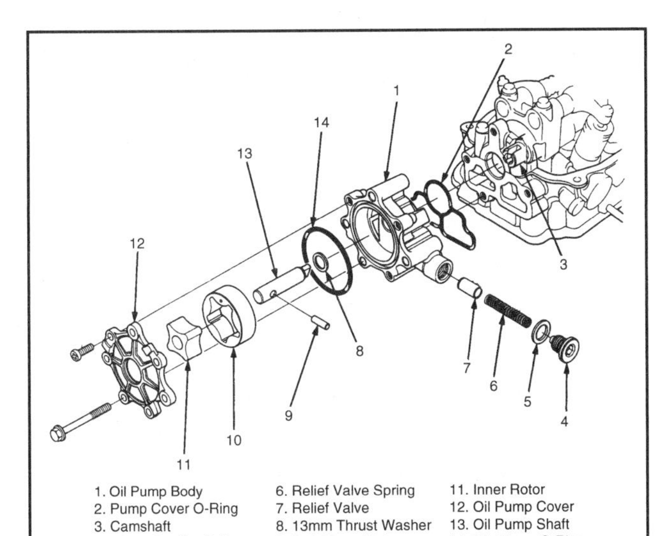

Fig. 7 Exploded view drawing of the oil pump assembly—BF35A–BF50A

BF25A and BF30A

➧ See Figure 8

1. Remove the oil pump cover and O-ring.
2. Remove the outer and inner rotors.
3. Remove the thrust washer.
4. Remove the oil pump shaft and roller.
5. Remove the sealing bolt and remove the sealing washer, relief valve spring and the relief valve.

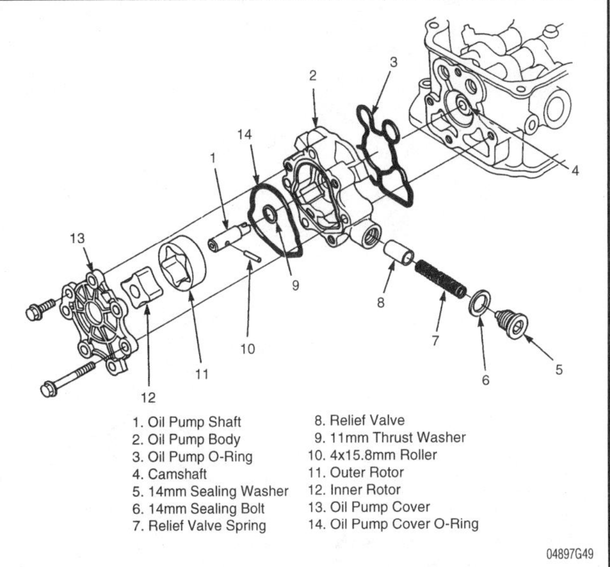

Fig. 8 Exploded view drawing of the model oil pump assembly—BF25A and BF30A

6. Inspect the sealing washer, relief valve spring and the relief valve for scratches, wear or any other type of damage. Replace any damaged part.
7. Coat the spring and valve with fresh engine oil before assembly.
8. Apply a locking agent, such as Loctite®, to the sealing bolt.
9. Measure the oil pump body inner diameter. Standard measurement should be 1.603–1.604 in. (40.71–40.74 mm). If the clearance is greater than 1.605 in. (40.76 mm) replace the components.
10. Measure the oil pump depth. Standard measurement should be 0.591–0.593 in. (15.02–15.05 mm). If the clearance is greater than 0.594 in. (15.09 mm) replace the components.
11. Measure the outer rotor height. Standard measurement should be 0.590–0.591 in. (14.98–15.00 mm). If the clearance is greater than 0.589 in. (14.96 mm) replace the components.
12. Measure the inner rotor-to-outer rotor clearance. Standard measurement should be 0.006in. MAX (0.15 mm). If the clearance is greater than 0.01 in. (0.2 mm) replace the components..
13. Measure the outer rotor-to-pump body clearance. Standard measurement should be 0.006–0.008 in. (0.15–0.21 mm). If the clearance is greater than 0.010 in. (0.26 mm) replace the components.
14. Measure the pump end clearance. Standard measurement should be 0.001–0.003 in. (0.02–0.07 mm). If the clearance is greater than 0.004 in. (0.1 mm) replace the components.
15. Replace all parts not meeting specifications or that are damaged in any way.

BF9.9A and BF15A

➧ See Figure 9

1. Remove the oil pump cover and O-ring.
2. Remove the outer and inner rotors.
3. Remove the oil pump shaft.
4. Measure the oil pump body inner diameter. Standard measurement should be 1.146 in. (29.10 mm).

   If the clearance is greater than 1.154 in. (29.30 mm).
5. Measure the oil pump depth. Standard measurement should be 0.513in. (13.02 mm). If the clearance is greater than 0.515 in. (13.08 mm) replace the components.
6. Measure the outer rotor height. Standard measurement should be 0.51in. (13.0 mm). If the clearance is greater than 0.509 in. (12.95 mm) replace the components.
7. Measure the inner rotor-to-outer rotor clearance. Standard measurement should be 0.0059in. (0.15 mm). If the clearance is greater than 0.0079 in. (0.20 mm) replace the components.
8. Measure the outer rotor-to-pump body clearance. Standard measurement

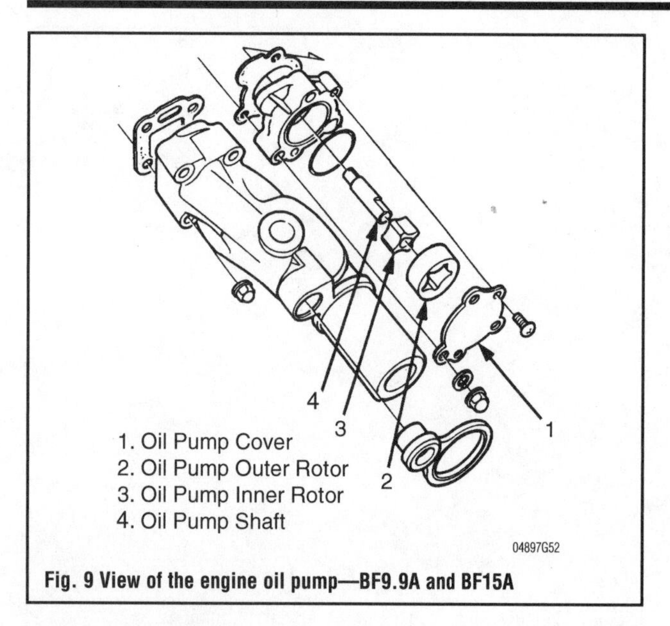

Fig. 9 View of the engine oil pump—BF9.9A and BF15A

should be 0.0039–0.0083 in. (0.10–0.21 mm). If the clearance is greater than 0.0102 in. (0.26 mm) replace the components.

9. Replace all parts not meeting specifications or that are damaged in any way.

BF75, BF8A and BF100

➧ See Figure 10

1. Remove the oil pump cover and O-ring.
2. Remove the outer and inner rotors.
3. Remove the oil pump shaft.
4. Measure the oil pump body inner diameter. Standard measurement should be 0.911–0.913 in. (23.15–23.18 mm). If the clearance is greater than 0.915 in. (23.23 mm) replace the components.
5. Measure the inner rotor-to-outer rotor clearance. Standard measurement should be 0.006in. (0.15 mm). If the clearance is greater than 0.008 in. (0.20 mm) replace the components.
6. Measure the outer rotor-to-pump body clearance. Standard measurement should be 0.006 in. (0.15 mm). If the clearance is greater than 0.010 in. (0.26 mm) replace the components.
7. Replace all parts not meeting specifications or that are damaged in any way.

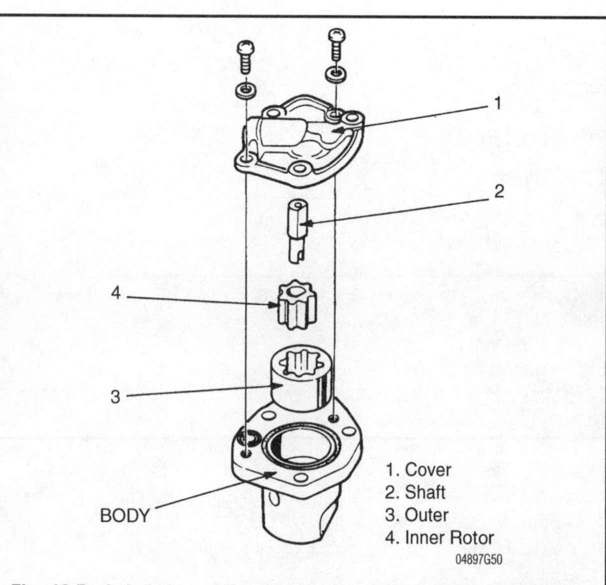

Fig. 10 Exploded view of the oil pump assembly—BF75 and BF100

BF50 and BF5A

➧ See Figure 11

1. Remove the oil pump cover and O-ring.
2. Remove the outer and inner rotors.
3. Remove the oil pump shaft.
4. Measure the oil pump body inner diameter. Standard measurement should be 0.911in. (23.15 mm). If the clearance is greater than 0.913 in. (23.20 mm) replace the components.
5. Measure the inner rotor-to-outer rotor clearance. Standard measurement should be 0.006in. (0.15 mm). If the clearance is greater than 0.008 in. (0.20 mm) replace the components.
6. Measure the outer rotor-to-pump body clearance. Standard measurement should be 0.006–0.008 in. (0.15–0.21 mm). If the clearance is greater than 0.010 in. (0.26 mm) replace the components.
7. Measure the outer rotor height. Standard measurement should be 0.472 in. (11.98 mm). If the clearance is greater than 0.470 in. (11.95 mm) replace the components.
8. Measure the rotor-to-side body clearance. Standard measurement should be 0.0008–0.0035 in. (0.02–0.09 mm). If the clearance is greater than 0.004 in. (0.11 mm) replace the components.
9. Measure the pump body depth. Standard measurement should be 0.472 in. (12.0 mm). If the clearance is greater than 0.475 in. (12.06 mm) replace the components.
10. Replace all parts not meeting specifications or that are damaged in any way.

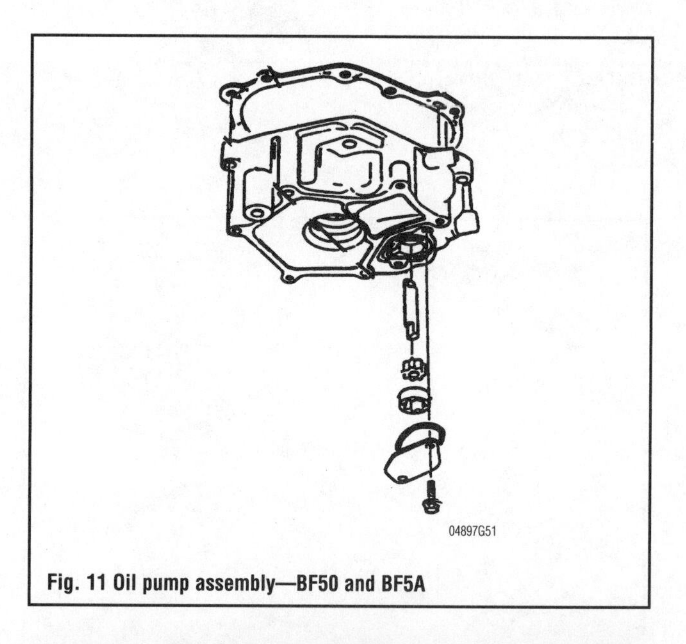

Fig. 11 Oil pump assembly—BF50 and BF5A

Oil Pressure Switch

TESTING

➧ See Figure 12

1. Disconnect the oil pressure switch wiring harness.
2. Check for continuity between the switch terminal and ground.
3. On outboards up to the BF30A, continuity should exist when the outboard is operating and oil pressure is present. Continuity should not exist when the engine is turned off.
4. On BF35A outboards and higher, continuity should not exist when the outboard is operating and oil pressure is present. Continuity should exist when the engine is turned off.
5. If the switch functions as stated, remove the switch and check for proper oil pressure using a mechanical pressure gauge.
6. If the switch does not function as stated, it may be faulty.

Fig. 12 The oil pressure switch is tested by checking for continuity between the switch terminal and ground

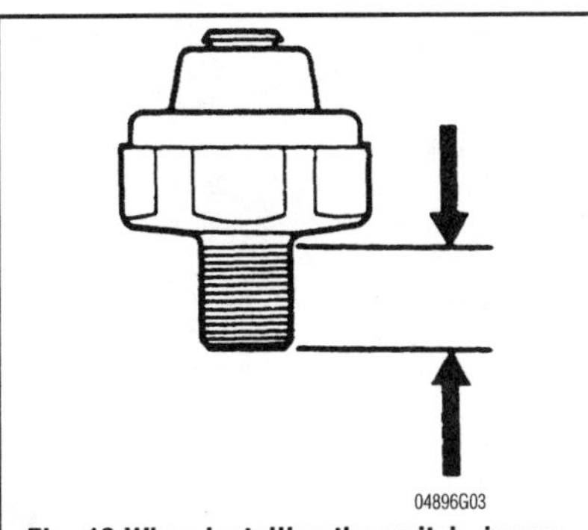

Fig. 13 When installing the switch, insure the threads are coated with sealant to prevent oil leaks

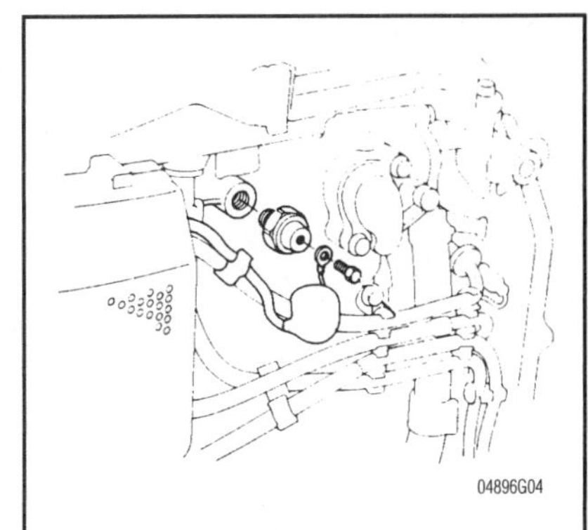

Fig. 14 The oil pressure switch is screwed into a port on the engine block

REMOVAL & INSTALLATION

See Figures 13 and 14

1. Remove the engine cowling and note the oil pressure switch location.
2. Label and disconnect the oil pressure switch wiring harness.
3. Using an appropriate size wrench, remove the switch from the engine.

A small amount of oil may spill from the switch hole when it is removed. Place a drain pan under the switch hole prior to removal.

To install:

4. If not already done, coat the oil pressure switch with sealant or Teflon® tape to prevent leakage.
5. Install the switch and tighten securely.
6. Connect the switch electrical harness.
7. Test the switch for proper operation.
8. Install the engine cowling.

Oil Pressure Warning Lamp

TESTING

Single Lamp

The oil pressure warning lamp illuminates when the engine is running and adequate oil pressure is present. If the lamp should go out or start to flash, an oil system problem exists and the engine should be shut down immediately.

1. Disconnect the oil pressure warning lamp wiring harness.
2. Connect the lamp wires to a 9 volt battery. The lamp should illuminate. If the lamp does not illuminate, it may be faulty.

Some lamps contain a diode and may only function with the wires placed on the battery in a certain way. If the lamp does not illuminate when the wires are touched to the battery, reverse the wires on the battery terminals and retest.

3. If the lamp functions as stated, check the overheat switch for proper operation.

Dual Lamp

See Figure 15

1. Disconnect the indicator lamp wiring harness.
2. Connect a 12 volt switched power source as noted in the illustration.
3. With switch 1 (SW1) **ON**, the green lamp should illuminate.
4. With switches 1 and 3(SW1, 3) **ON**, the green lamp should not illuminate.
5. If the switch functions properly, the problem may be in the wiring harness
6. With switch 2 (SW2) **ON**, the red lamp should illuminate.
7. If the lights function as stated, there may be a problem in the wiring harness.

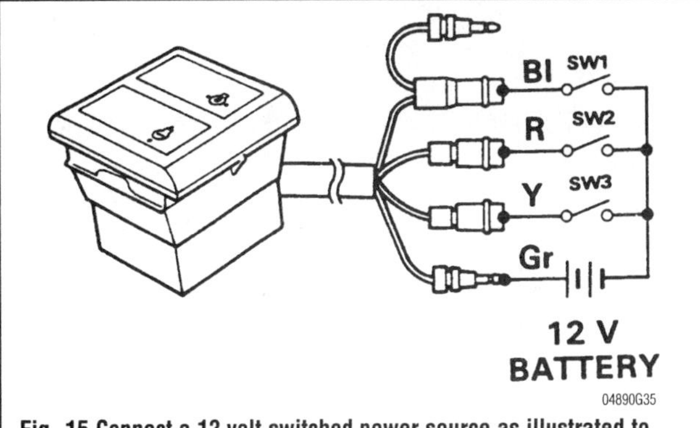

Fig. 15 Connect a 12 volt switched power source as illustrated to test the indicator lamps

8. If the lights do not function as stated, the indicator lamp assembly may be faulty.

REMOVAL & INSTALLATION

Engine Cowl Mounted

See Figure 16

1. Remove the engine cowling and note the oil pressure warning lamp location.
2. Disconnect the oil pressure warning lamp wiring harness.
3. Press in on the tabs and slide the lamp through the engine cowling to remove.
4. When installing, insure the switch tabs are fully seated and the wiring harness is properly routed.

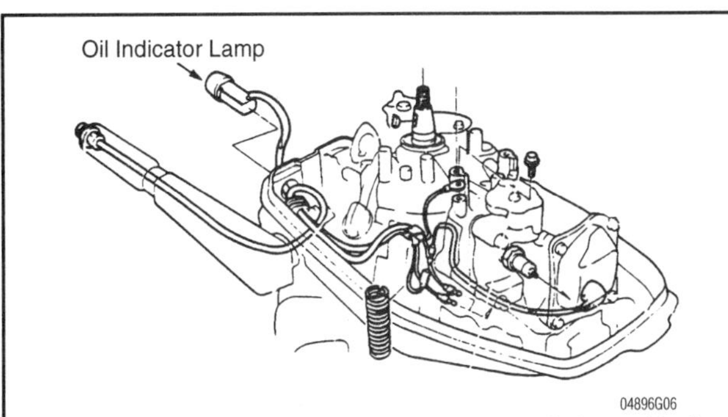

Fig. 16 Single oil pressure warning lamps are usually located in the engine cowling and snap into place using clips on the back side of the lamp

Remote Control Box Mounted

See Figures 17, 18 and 19

1. Remove the control box from the side of the boat and open the side covers to allow access to the internal components.
2. Disconnect the oil pressure warning lamp wiring harness.
3. On lamps installed using tabs, press in on the tabs and slide the lamp through the control box to remove.
4. On lamps installed using locknuts, remove the locknut and slide the lamp through the control box to remove.

To install:

5. On lamps installed using tabs, insure the switch tabs are fully seated
6. On lamps installed using locknuts, tighten the locknut securely.
7. Always insure the wiring harness is properly routed.

Warning Buzzer

TESTING

See Figure 20

On some engines, a warning buzzer will sound any time an oil pressure problem is encountered and the oil pressure warning lamp is lit.

1. The warning buzzer is tested by simply connecting it to a 12 volt power source.
2. The buzzer should sound when properly connected to power.
3. If the buzzer functions properly, there may be a problem in the wiring harness.
4. If the buzzer does not perform as stated, it may be faulty.

REMOVAL & INSTALLATION

See Figure 21

1. Remove the control box from the side of the boat and open the side covers to allow access to the internal components.
2. Disconnect the buzzer wiring harness.
3. Remove any wire straps that connect the buzzer wiring harness to the box.
4. Remove the buzzer from the control box.

To install:

5. Install the buzzer in the control box.
6. Install any wire straps that connect the buzzer harness to the control box.
7. Connect the buzzer wiring harness.
8. Test the buzzer for proper operation.
9. Install the control box side covers and mount the control box in the boat.

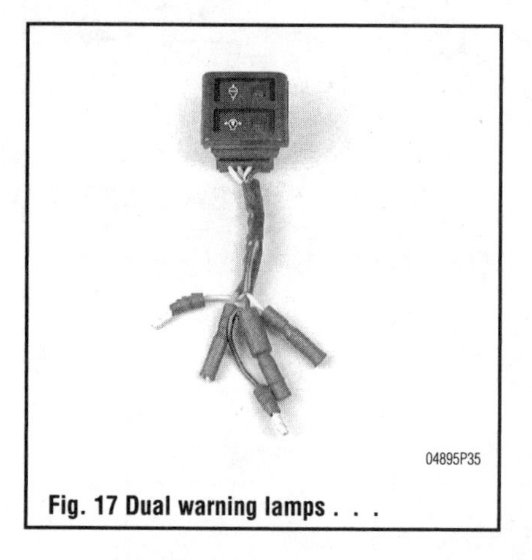

Fig. 17 Dual warning lamps . . .

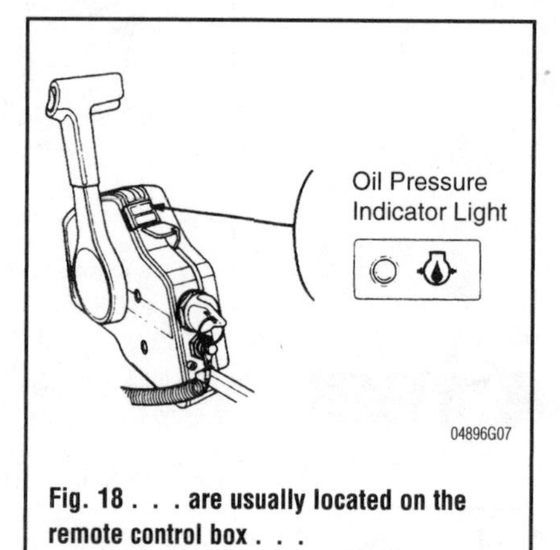

Fig. 18 . . . are usually located on the remote control box . . .

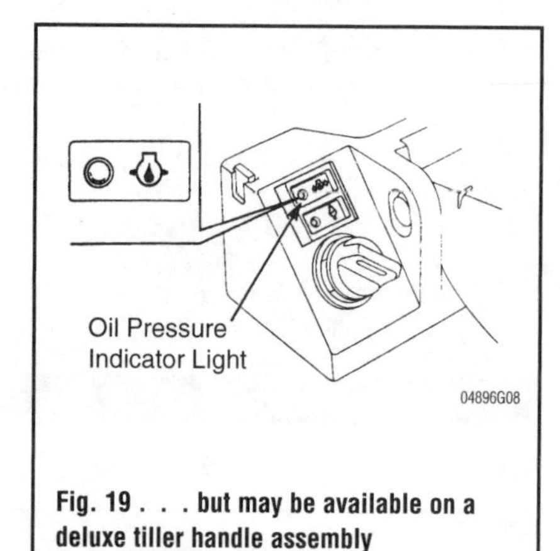

Fig. 19 . . . but may be available on a deluxe tiller handle assembly

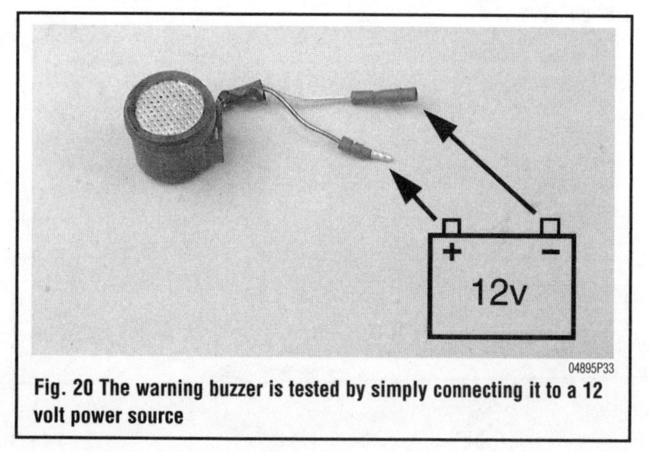

Fig. 20 The warning buzzer is tested by simply connecting it to a 12 volt power source

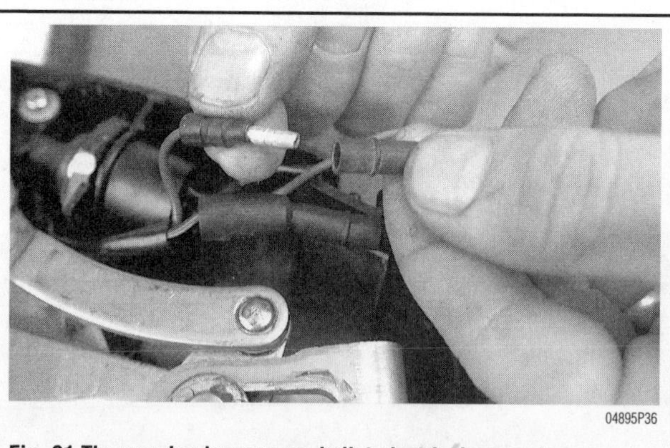

Fig. 21 The warning buzzer uses bullet electrical connectors

COOLING SYSTEM

Description and Operation

WATER PUMP

➧ See Figures 22 and 23

The water pump uses an impeller driven by the drive shaft, sealing between an offset housing and lower plate to create a flexing of the impeller blades. The rubber impeller inside the pump maintains an equal volume of water flow at most operating RPMs.

At low speeds the pump acts like a full displacement pump with the longer impeller blades following the contour of the pump housing. As pump RPMs increase, and because of resistance to the flow of water, the impellers bend back away from the pump housing and the pump acts like a centrifugal pump. If the impeller blades are short, they remain in contact throughout the full RPM range, supplying full pressure.

➡The outboard should never be run without water, not even for a moment. As the dry impeller tips come in contact with the pump housing, the impeller will be damaged. On later models, the impeller comes in contact with a stainless steel cup inside a nylon housing, and damage occurs to the impeller in seconds.

On most powerheads, if the powerhead overheats, a warning is signaled through the ignition circuit and a temperature unit signals the operator of an overheat condition. This should happen before major damage can occur. Reasons for overheating can be as simple as a plastic bag over the water inlet, or as serious as a leaking head gasket.

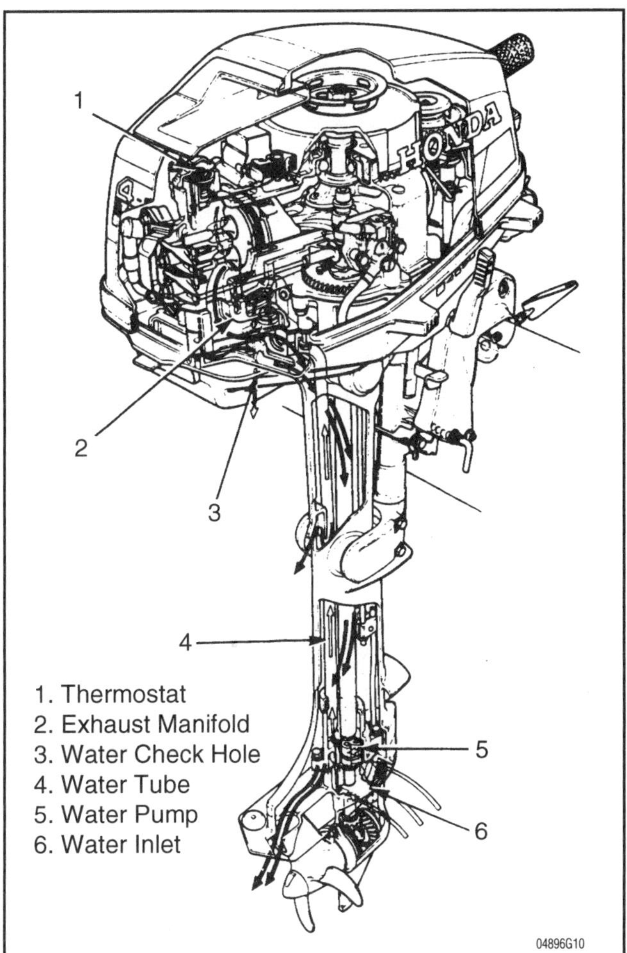

Fig. 22 Typical water flow and component locations for an outboard cooling system

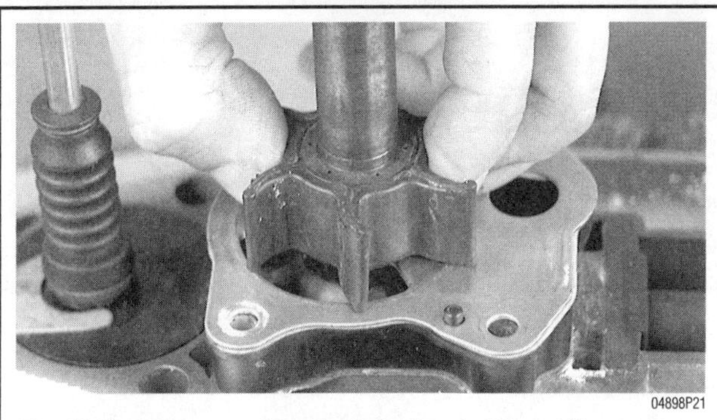

Fig. 23 The water pump impeller is the heart of the cooling system and should always be kept in top condition

THERMOSTAT & PRESSURE RELIEF

➧ See Figure 24

The thermostat controls the water temperature by regulating the flow of water through the water jacket passages at idle. At high RPM water pressure may be great enough to force the thermostat off its seat, compressing a spring and causing a release of pressure. There may be a separate pressure relief valve that opens at higher RPM when the thermostat is either closed or opened that allows water to return to the water pump or to be exhausted. Routing of the water through a jacket may be accomplished by the use of water deflectors placed in the water jacket. These neoprene deflectors must be properly positioned. They deflect the water and therefore prevent it from going between lower cylinders and route the water upward and around the upper cylinders. If left out or improperly installed, overheating will occur. The heat developed in the combustion process, which is conducted through the cylinder wall, is picked up and carried away by the water, and flows out with the exhaust gases.

Fig. 24 A thermostat, located in a water passage on the powerhead, controls water temperature by regulating water flow through the powerhead

OVERHEAT WARNING SYSTEM

The water-cooled system is designed to maintain powerhead temperatures controlled by a thermostat, pressure relief valve, and water pump RPM. There are variations of this system, using a by sounding device. The horn or buzzer is hooked up through cold water return to the pump. On most outboards the water inlet is located in the lower unit. From the inlet water is pulled through a screen into the pump, and pushed up a water tube to the exhaust side of the powerhead. The incoming water passing near the exhaust is warmed and then circulated past the thermostat, or through the thermostat, controlling water temperature at idle in the water jacket passages. It then exits at the bottom of the powerhead into the exhaust. A tell-tale outlet is used on most models indicating water pump operation. Some units will not show water pump operation until the thermostat opens. Water dropped into the exhaust stream cools the exhaust and the housing preventing damage to the casting.

➡Some units will not show water pump operation until the thermostat opens.

Troubleshooting the Cooling System

The water-cooled powerhead has a lot of problems to consider when talking about overheating. The most obvious is that the thermostat may be damaged or corroded, or the thermostat cover gaskets may be leaking or damaged.

The water pump is the heart of the cooling system, and requires a periodic inspection of the pump body, bottom plate, and impeller for scoring, which would prevent a seal. Check for grooves in the driveshaft where the seal rides. Any damage in these areas may cause air or exhaust gases to be drawn into the pump, putting bubbles into the water. In this case, air does not aid in cooling.

- Is the pump inlet clear and clean of foreign material or marine growth? Check that the inlet screen is totally open. How about the impeller?
- Try and separate the impeller hub from the rubber. If it shows signs of loosening or cracking away from the hub, replace the impeller.
- Has the impeller taken a set, and are the blade tips worn down or do they look burned? Are the side sealing rings on the impeller worn away? If so, replace the impeller.

The life of the powerhead depends on this pump, so don't reuse any parts that look damaged. Are any parts of the impeller missing! If so, they must be found. Broken pieces will migrate up the water tube into the water jacket passages and cause a restriction which might block a water passage. It can be expensive to locate the broken pieces in the water passages, but they must be found, or major damage could occur. The best insurance against breaking the impeller is to replace it at the beginning of each boating season, and don't run it out of the water. When installing a metal-bodied pump housing, coat all screws with non-hardening sealing compound to retard galvanic corrosion. The water tube carries the water from the pump to the powerhead. Grommets seal the water tube to the water pump and exhaust housing at each end of the tube, and can deteriorate. Also, the water tube(s) should be checked for holes through the side of the tube, for restrictions, dents, or kinks.

Overheating at high RPMs, but not under light load, may indicate a leaking head gasket. If a head gasket it leaking, water can go into the cylinder, or hot exhaust gases may go into the water jacket, creating exhaust bubbles and excessive heat. The aluminum heads have a tendency to warp, and need to be surfaced each time they are removed. Resurface them by using emery paper and a surface block moving in a eight motion. Also, inspect the cylinders and pistons for damage. Other areas to consider are the exhaust cover gaskets and plate. Look for corrosion pin holes. This is rare, but if the outboard has been operated in salt water over the years, there may just be a problem.

If the outboard is mounted too high, air may be drawn into the water inlet or sufficient water may not be available at the water inlet. When underway the outboard anti-ventilation plate should be running at or near the bottom of the boat and parallel to the surface of the water. This will allow undisturbed water to come to the lower unit, and the water pick-up should be able to draw sufficient water for proper cooling. When the outboard has been run in brackish or saltwater, the cooling system should be flushed. The outboard should be run at idle speed on the flushing tool for at least five minutes. This will wash the salt from the castings and reduce internal corrosion. If the outboard is small and there is no flushing tool that will fit, run the outboard in a tank, drum, or bucket.

There is no need to run in gear during the flushing operation. After the flushing job is done, rinse the external parts of the outboard off to remove the salt spray.

When service work is done on the water pump or lower unit, all the bolts that attach the lower unit to the exhaust housing, and bolts that hold the water pump housing, should be coated with nonhardening gasket sealing compound to guard against corrosion. If this is not done, the bolts will be difficult to remove or will be seized by galvanic corrosion next time service work is performed.

Last but not least, check to be sure the overheat warning system is working properly. By grounding the wire at the sending unit, the horn should sound, and/or a light should turn on. To proper test this circuit for a specific year, make, and model, refer to the appropriate service manual.

Water Pump

REMOVAL & INSTALLATION

BF20 and BF2A

➧ See Figures 25 and 26

1. Remove the bolts securing the extension plate to the gear case and remove the extension plate.

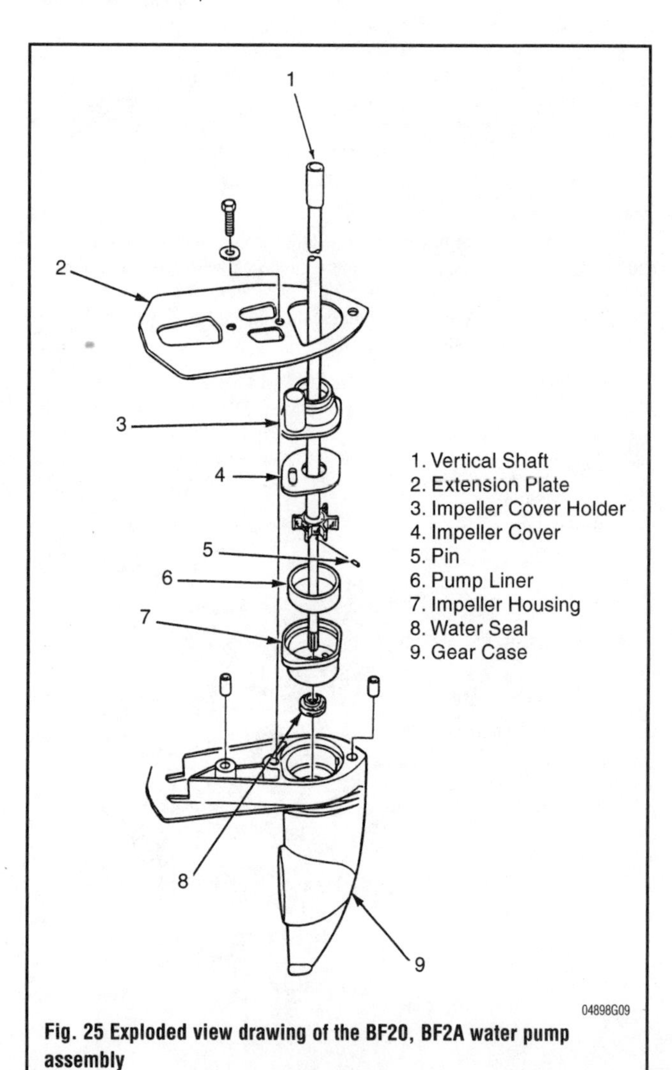

Fig. 25 Exploded view drawing of the BF20, BF2A water pump assembly

1. Pump Liner
2. Water Pump Impeller

04898G11

Fig. 26 Make sure the impeller vanes are bending in this direction during installation

2. Raise the water pump cover a bit to remove the indexing pin. Leave the water pump cover in this position at this time.

3. Pry the circlip free from the end of the Pinion Shaft with a thin screwdriver. This clip holds the pinion gear onto the Pinion Shaft. The clip may not come free on the first try, but have patience and it will come free. With one hand, remove Pinion Shaft up and out of the lower unit housing and at the same time, with the other hand, catch the pinion gear and thrust washer from behind the gear.

With the Pinion Shaft on the workbench, remove the impeller cover holder, the impeller cover and the water pump impeller. Check the condition of the impeller carefully and replace with a new one if there is any question as to its condition for satisfactory service. Never turn the impeller over in an attempt to gain further life from the impeller. Now is the time to replace the impeller. It's cheap insurance against burning up an engine due to impeller failure.

Inspect the pump liner and the impeller housing.

To install:

Obtain the following special tools:

- Seal driver tool—p/n 07749-0010000
- Pilot (22 mm)—p/n 07746-0041000.

➡After installation of the water pump housing cover, the oil seal lip faces downward to prevent water in the water pump from contaminating the lubricant around the Pinion Shaft. However, to install the seal, the water pump cover is turned upside down on the work surface. In this position, the lip of the seal must face upward.

Using the seal installer and handle, install the seal. After installation, pack the seal with water resistant lubricant.

4. Install the impeller housing and pump liner into the gear case.
5. Install the impeller on the pinion shaft with the keyway facing downward.
6. Slide the impeller up the shaft and insert the pin below the impeller.
7. Align the impeller keyway with the pin and slide the impeller down over the pin.
8. Seat the impeller in the pump liner while rotating the impeller and shaft clockwise. The impeller vanes must bend away from the direction of shaft rotation or the impeller will be damaged.
9. Lower the assembled Pinion Shaft into the lower unit but do not mate the cover with the lower unit surface at this time. Leave some space, as shown. The splines on the lower end of the Pinion Shaft will protrude into the lower unit cavity.
10. Slide the thrust washer onto the lower end of the Pinion Shaft. Slide the pinion gear up onto the end of the Pinion Shaft. The splines of the pinion gear will index with the splines of the Pinion Shaft and the gear teeth will mesh with the teeth of the forward gear. Rotating the pinion gear slightly will permit the splines to index and the gears to mesh. Now, comes the hard part. Snap the circlip into the groove on the end of the Pinion Shaft to secure the pinion gear in place. If the first attempt is not successful, try again. Take a break, have a cup of coffee, tea, whatever, then give it another go. With patience, the task can be accomplished.
11. Apply some Loctite, or equivalent, to the extension plate retaining bolt. Tighten the bolt to a torque value of 4.4–8.7 in. lbs. (60–120 kg-cm).

BF50 and BF5A

➧ See Figures 27, 28 and 29

1. Remove the lower unit.
2. Remove the retaining bolts.
3. Remove the impeller housing.
4. Remove the impeller gaskets and the impeller cover.
5. Remove the distance spacers.
6. Slide the impeller off the pinion shaft.

To install:

7. Install the impeller cover, with the new impeller gaskets on either side of it, onto the gear case. Be sure that the cover is not bent or damaged.
8. Set the distance spacers in place. Make sure the collars are installed in their proper locations.
9. Install the impeller into the impeller housing while turning it counterclockwise.

⁂ WARNING

Do not install the impeller any other way or damage to both the impeller and engine will occur.

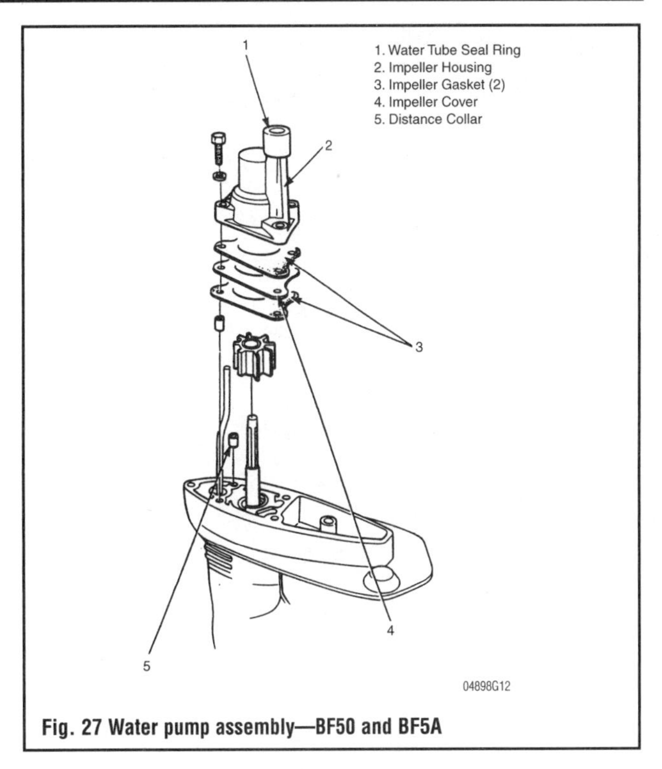

Fig. 27 Water pump assembly—BF50 and BF5A

Fig. 28 Install the impeller in the pump housing by turning it counterclockwise

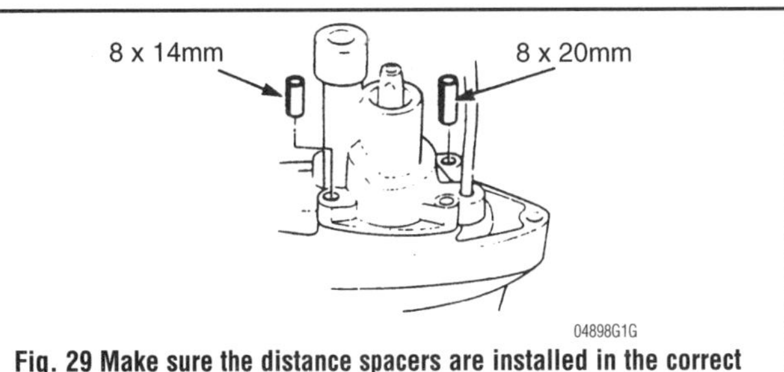

Fig. 29 Make sure the distance spacers are installed in the correct locations. They are not interchangeable.

10. Line up the flats on the impeller and pinion shaft and install the assembly onto the pinion shaft
11. Install a thread locking compound on the retaining bolts and tighten the impeller housing to the gear case.
12. Install the seal ring on the pump discharge port.
13. Install the lower unit.

BF75, BF8A and BF100

➧ See Figures 30, 31 and 32

1. Remove the lower unit.
2. Remove the retaining bolts on the water pump housing and the two retaining bolts on the cover holder.

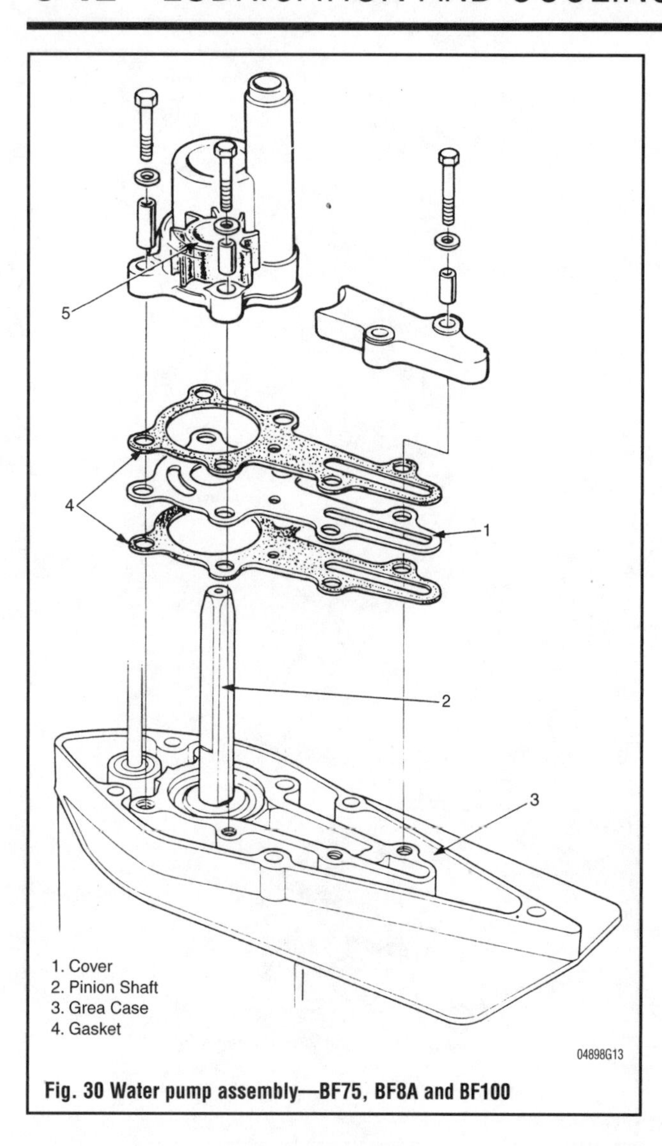

Fig. 30 Water pump assembly—BF75, BF8A and BF100

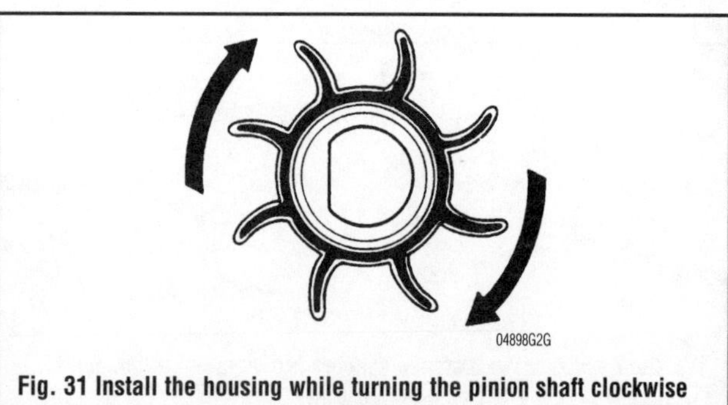

Fig. 31 Install the housing while turning the pinion shaft clockwise

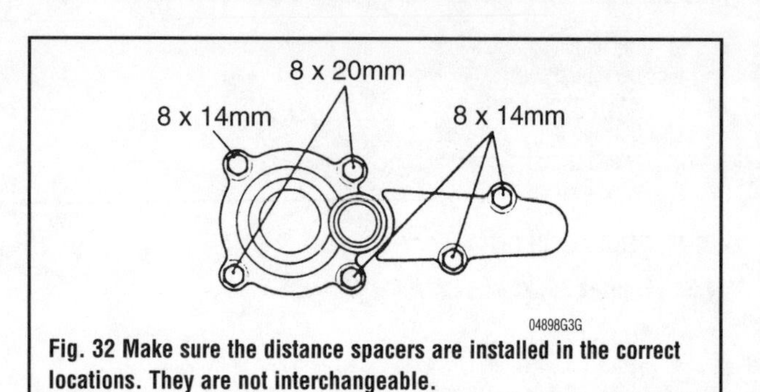

Fig. 32 Make sure the distance spacers are installed in the correct locations. They are not interchangeable.

3. Remove the impeller gaskets and the impeller cover.
4. Remove the distance spacers.
5. Slide the impeller off the pinion shaft.

To install:

6. Install the impeller cover, with the new impeller gaskets on either side of it, onto the gear case. Be sure that the cover is not bent or damaged.
7. Slide the impeller over the shaft, aligning the flats on the pinion shaft and the impeller.
8. While turning the pinion shaft clockwise, install the impeller housing over the impeller.
9. Install the distance spacers in their proper locations and tighten the retaining bolts.
10. Install the lower unit.

BF9.9A and BF15A

See Figures 33 and 34

1. Remove the lower unit.
2. Remove the retaining bolts.
3. Remove the impeller housing.
4. Remove the impeller gaskets and the impeller cover.
5. Remove the distance spacers. These collars are not interchangeable and must be installed as shown in the drawing.
6. Slide the impeller off the pinion shaft.

To install:

7. Install the impeller cover, with the 2 new impeller gaskets on either side of it, onto the gear case. Be sure that the cover is not bent or damaged.
8. While turning the impeller counterclockwise, install it into the impeller housing.

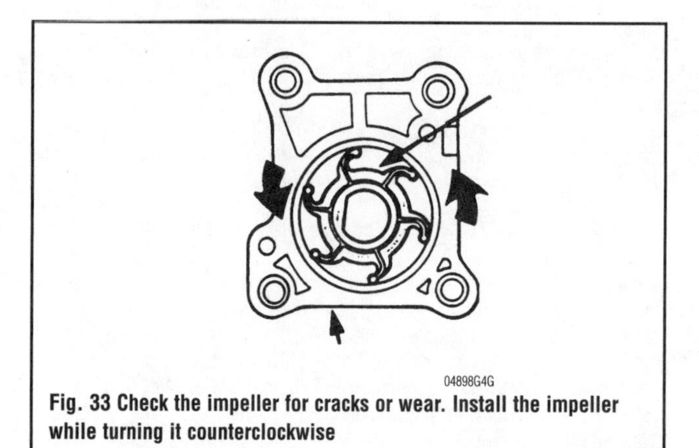

Fig. 33 Check the impeller for cracks or wear. Install the impeller while turning it counterclockwise

Fig. 34 BF9.9A, BF15A water pump assembly

9. Align up the flats on the pinion shaft and the impeller, and install the assembly onto the pinion shaft.
10. Install the distance spacers, then tighten the retaining bolts.
11. Install the water seal on the discharge port.
12. Install the lower unit.

BF25A, BF30A, BF35A, BF40A, BF45A and BF50A

See Figures 35, 36 and 37

1. Remove the lower unit.
2. Remove the retaining bolts on the impeller housing.
3. Remove the impeller housing and water pump O-ring. Discard the O-ring.
4. Remove the water tube seal ring.
5. Remove the pump liner.

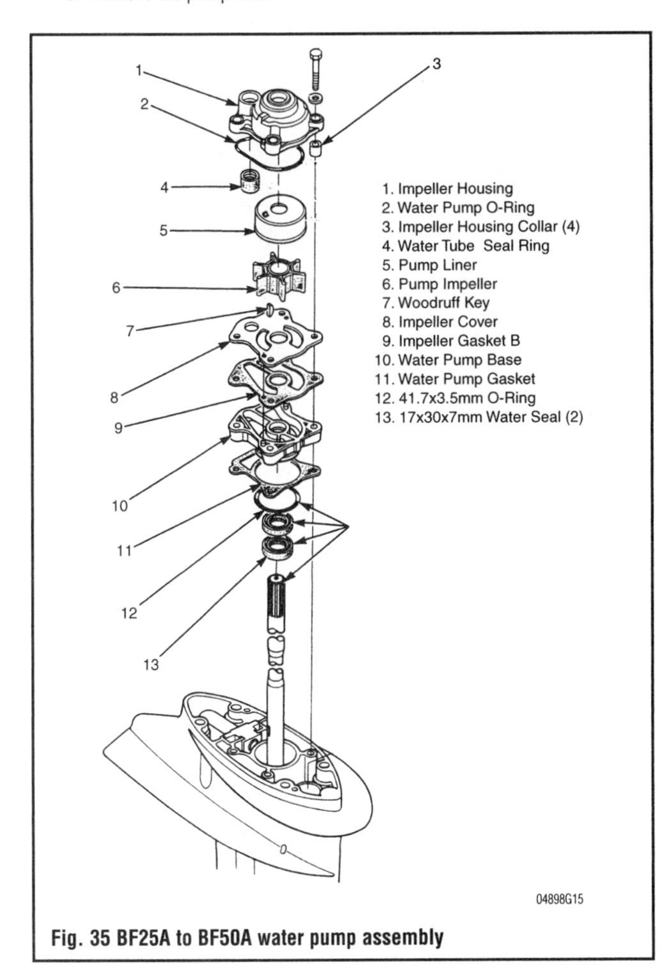

Fig. 35 BF25A to BF50A water pump assembly

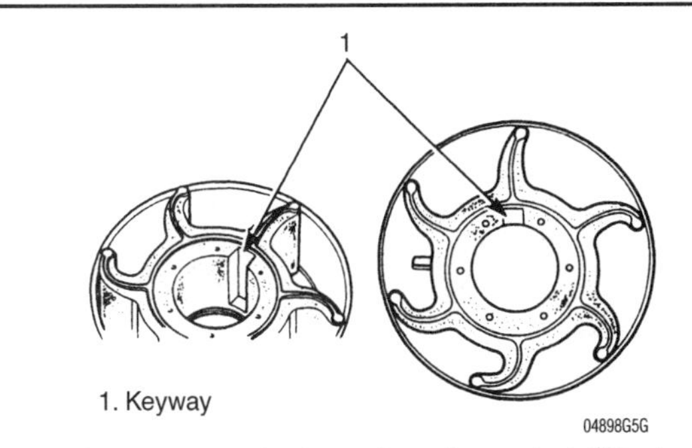

Fig. 36 Check the impeller for cracks and wear. Install the impeller by turning it counterclockwise with the open end of the keyway facing the bottom of the engine

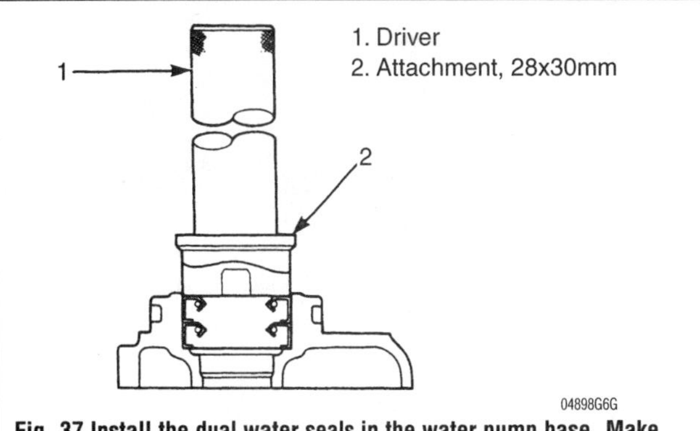

Fig. 37 Install the dual water seals in the water pump base. Make sure the seal are installed in the direction shown in the drawing

6. Remove the impeller. Keep track of the Woodruff key.
7. Remove the impeller cover, impeller gasket "B", water pump base, water pump gasket "A" and the bottom O-ring. Discard the O-ring.
8. Using a commercially available seal remover, pry out the water seals from the water pump base.

To install:

9. To install the new upper and lower water seals, obtain the following special tools:
   - Driver—07749-0010000
   - Attachment (28 x 30 mm)—07946-1870100
10. Install the 2 new water seals in the direction shown in the drawing.
11. Install a new bottom O-ring and lower impeller gasket.
12. Install the water pump base.
13. Install the upper impeller gasket.
14. Install the impeller cover.
15. Install the impeller into the pump liner by turning it counterclockwise. Keep the open end of the keyway toward the bottom of the engine.
16. Install the Woodruff key onto the pinion shaft.
17. With the impeller keyway aligned to the Woodruff key, install the pump liner onto the pinion shaft.
18. Install the water tube seal ring, new O-ring, impeller housing collars and impeller housing.
19. Install the retaining bolts and washers and tighten the assembly.
20. Install the lower unit.

BF75A and BF90A

See Figures 38, 39, 40 and 41

1. Remove the lower unit.
2. Remove the retaining bolts on the impeller housing.
3. Remove the impeller housing.
4. Remove the upper impeller gasket and discard. Do not reuse the gasket.
5. Remove the impeller. Keep track of the Woodruff key.
6. Remove the impeller cover and the lower impeller gasket. Discard the gasket.
7. Remove the water pump housing retaining bolts and remove the water pump housing and water pump gasket.
8. Using a commercially available seal remover, pry out the upper and lower water seals from the water pump base. Discard the seals. Do not reuse.

To install:

9. To install the new upper and lower water seals, obtain the following special tools:
   - Driver—07749-0010000.
   - Attachment (37 x 40 mm)—07946-0010200.
10. Install the new upper water seal first in the water pump housing. Note the direction of installation, the upper seal lip faces down and the lower seal lip faces up.
11. Apply Loctite®271 or equivalent to the circumference of the seal and grease to the seal lips.
12. Using the special tools, drive the seal into the water pump housing until it seats firmly.
13. Install the water pump housing and new water pump gasket. Apply a thread-locking agent and tighten the retaining bolts to 5.1 ft. lbs. (7 Nm)

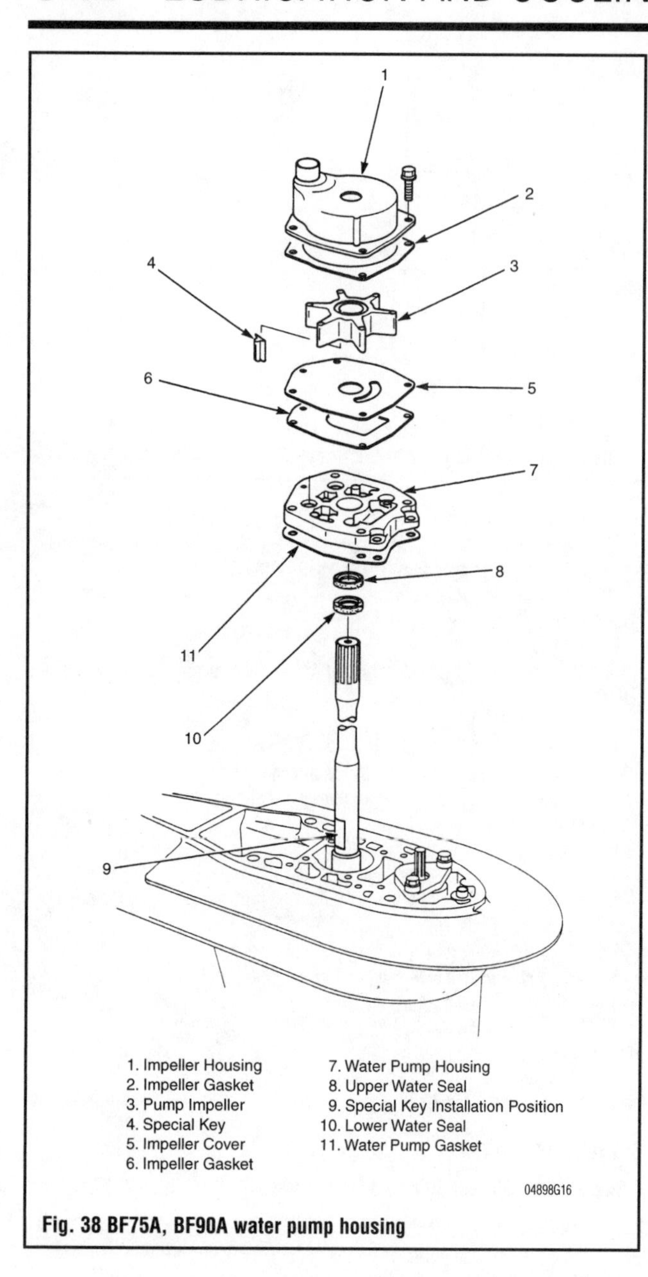

Fig. 38 BF75A, BF90A water pump housing

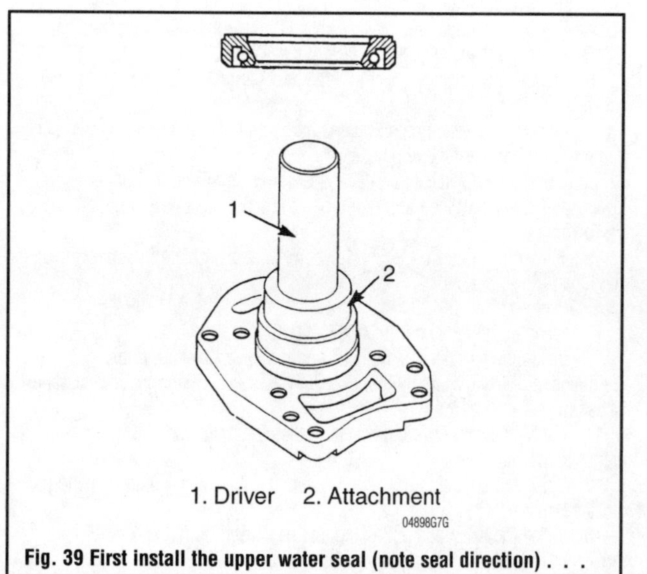

Fig. 39 First install the upper water seal (note seal direction) . . .

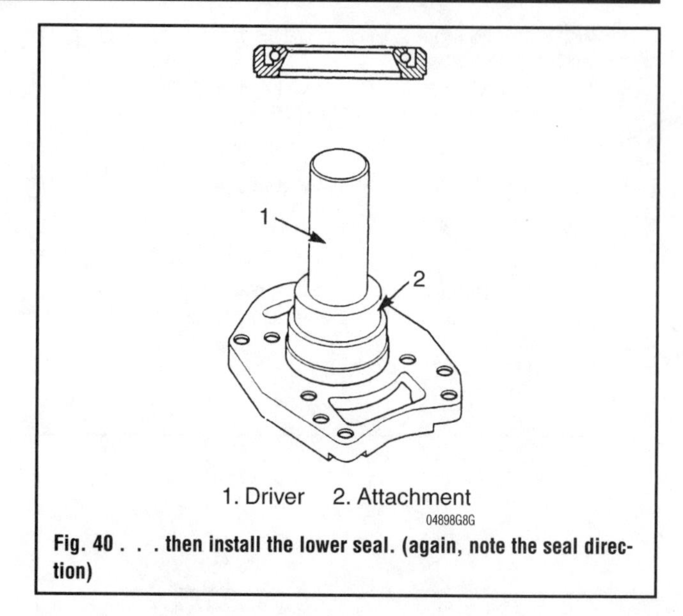

Fig. 40 . . . then install the lower seal. (again, note the seal direction)

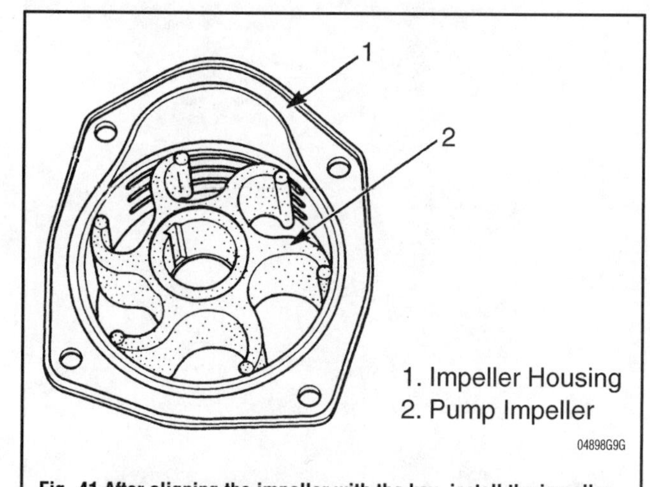

Fig. 41 After aligning the impeller with the key, install the impeller housing on the pump housing by turning the impeller housing clockwise as viewed from the top of the housing

14. Install a new impeller gasket and the impeller cover.
15. Apply grease to the inside of the impeller housing. While turning the impeller counterclockwise, install the impeller into the housing with the open end of the keyway facing up. Line up the hole in the impeller with the hole in the housing.
16. Slide the impeller housing down the pinion shaft and place the special key into position.
17. Align the keyway in the impeller with the key on the pinion shaft and install the impeller housing by turning the housing clockwise as viewed from the top of the housing.

➡After installation, double-check the position of the impeller gasket.

18. Apply a thread-locking agent and tighten the retaining bolts to 5.1 ft. lbs. (7 Nm)
19. Apply beads of RTV sealant to the mating surface of the water pump housing and the gear case.
20. Install the lower unit.

BF115A and BF130A

➧ See Figures 42, 43, 44 and 45

1. Remove the lower unit.
2. Remove the retaining bolts and distance spacers on the impeller housing.
3. Remove the impeller housing and water pump O-ring. Discard the O-ring.
4. Remove the pump liner.

5. Remove the impeller. Keep track of the Woodruff key.
6. Remove the impeller cover and impeller gasket. Discard the gasket.

To install:

7. To install the new upper and lower water seals, obtain the following special tools:
- Driver—07749-0010000.
- Attachment (32 x 35 mm)—07746-0010100.

8. Apply grease to the outer circumference and lips of the new seals.
9. Using the special tools, drive the seals into the water pump housing with the lips on both seals facing down.
10. Apply grease to the inside of the pump liner. While turning the impeller counterclockwise, install the impeller into the liner with the open end of the keyway facing up. Line up the hole in the impeller with the hole in the liner.
11. Install the pump liner in the impeller housing by aligning the two projections on the line with the two grooves in the impeller housing. Be sure that the cutout section of the liner is aligned with the open end of the impeller housing.
12. Install the new O-ring and new water pump gasket.
13. Install the water pump housing.
14. Install the new impeller gasket and impeller cover over the water pump housing. Make sure to install the impeller cover with the "ZW5" mark toward the impeller.
15. Slide the impeller housing down the pinion shaft and place the special key into position.

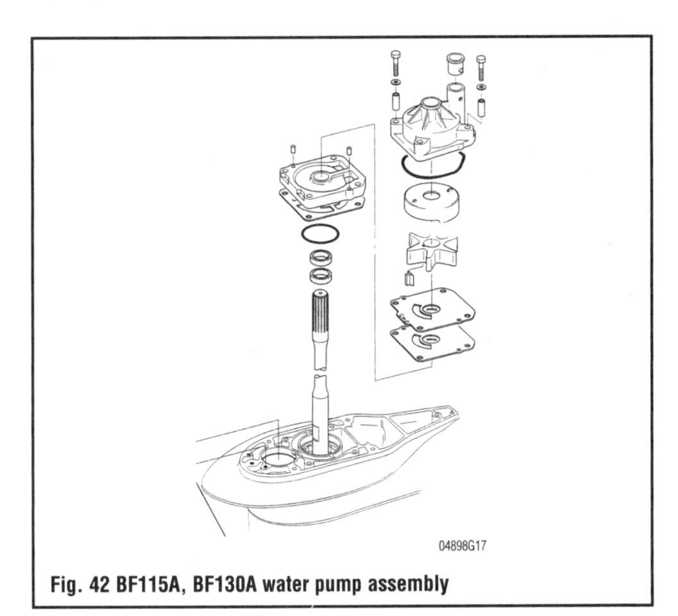

Fig. 42 BF115A, BF130A water pump assembly

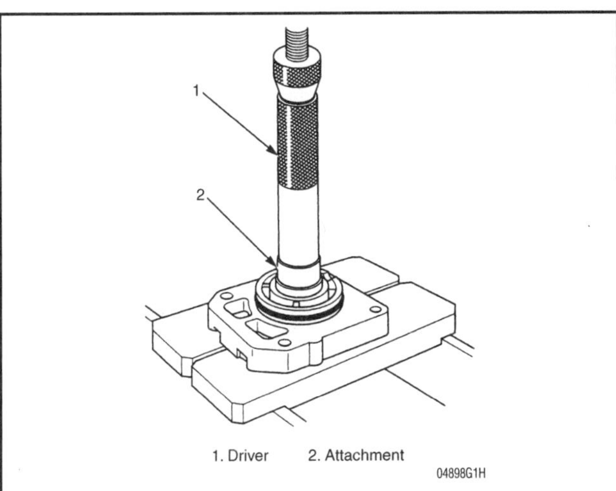

Fig. 43 Drive the water seals one at a time into the housing using the driver. Make sure they are installed in the right direction as shown

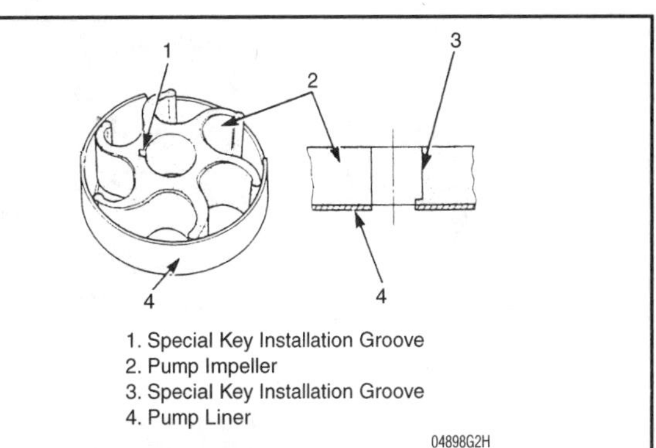

Fig. 44 Check the impeller for cracks and wear. Install the impeller in the pump liner as shown in the drawing

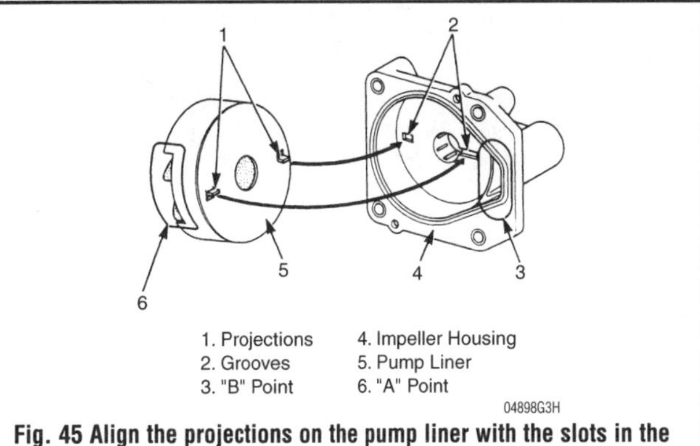

Fig. 45 Align the projections on the pump liner with the slots in the impeller housing as shown

16. Align the keyway in the impeller with the key on the pinion shaft and install the impeller housing by turning the housing clockwise as viewed from the top of the housing.

➡After installation, double-check the position of the impeller gasket.

17. Install the distance spacers and tighten the retaining bolts to 14 ft. lbs. (19.7 Nm).
18. Apply grease to the inside of the seal ring on the pump discharge port.
19. Install the lower unit.

Overheat Warning Lamp

TESTING

Single Lamp

The engine overheat warning lamp illuminates during an engine overheat condition only.

1. Disconnect the overheat warning lamp wiring harness.
2. Connect the lamp wires to a 9 volt battery. The lamp should illuminate. If the lamp does not illuminate, it may be faulty.

➡Some lamps contain a diode and may only function with the wires placed on the battery in a certain way. If the lamp does not illuminate when the wires are touched to the battery, reverse the wires on the battery terminals and retest.

3. If the switch functions as stated, check the overheat switch for proper operation.

Dual Lamp

▶ See Figure 46

1. Disconnect the lamp wiring harness.
2. Connect a 12 volt switched power source as noted in the illustration.

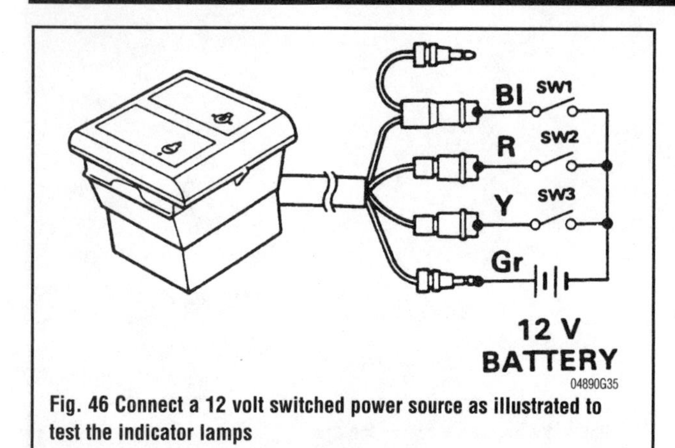

Fig. 46 Connect a 12 volt switched power source as illustrated to test the indicator lamps

3. With switch 1 (SW1) **ON**, the green lamp should illuminate.
4. With switches 1 and 3(SW1, 3) **ON**, the green lamp should not illuminate.
5. If the switch functions properly, the problem may be in the wiring harness
6. With switch 2 (SW2) **ON**, the red lamp should illuminate.
7. If the lights function as stated, there may be a problem in the wiring harness.
8. If the lights do not function as stated, the lamp assembly may be faulty.

REMOVAL & INSTALLATION

Engine Cowl Mounted

1. Remove the engine cowling and note the oil pressure warning lamp location.
2. Disconnect the oil pressure warning lamp wiring harness.
3. Press in on the tabs and slide the lamp through the engine cowling to remove.
4. When installing, insure the switch tabs are fully seated and the wiring harness is properly routed.

Remote Control Box Mounted

1. Remove the control box from the side of the boat and open the side covers to allow access to the internal components.
2. Disconnect the oil pressure warning lamp wiring harness.
3. On lamps installed using tabs, press in on the tabs and slide the lamp through the control box to remove.
4. On lamps installed using locknuts, remove the locknut and slide the lamp through the control box to remove.

To install:

5. On lamps installed using tabs, insure the switch tabs are fully seated
6. On lamps installed using locknuts, tighten the locknut securely.
7. Always insure the wiring harness is properly routed.

Overheat Thermoswitch

TESTING

1. Disconnect the overheat thermoswitch wiring harness.
2. Check the switch for continuity between the terminal and ground. Continuity should exist with coolant temperature above 176–248°F (98–102°C). Continuity should not exist when coolant temperature falls below 27–45°F (3–7°C).

➡Trying to control engine temperature to check thermoswitch continuity may be tricky. An alternate method is to remove the thermoswitch and suspend it in a container of water. Heat the water to raise the switch temperature as needed.

3. If the thermoswitch functions as stated, check the wiring harness for damage or a short.
4. If the thermoswitch does not function as stated, it may be faulty.

REMOVAL & INSTALLATION

1. Remove the engine cowling and note the thermoswitch location.
2. Drain and recycle the engine coolant.
3. Label and disconnect the thermoswitch wiring harness.
4. Using an appropriate size wrench, remove the switch from the engine.

To install:

5. If not already done, coat the switch with sealant or Teflon® tape to prevent leakage.
6. Install the switch and tighten securely.
7. Connect the switch electrical harness.
8. Fill the engine with coolant and bleed the system of air.
9. Test the switch for proper operation.
10. Install the engine cowling.

Warning Buzzer

TESTING

On some engines, a warning buzzer will sound any time an overheat problem is encountered and the overheat warning lamp is lit.

1. The warning buzzer is tested by simply connecting it to a 12 volt power source.
2. The buzzer should sound when properly connected to power.
3. If the buzzer functions properly, there may be a problem in the wiring harness.
4. If the buzzer does not perform as stated, it may be faulty.

REMOVAL & INSTALLATION

1. Remove the control box from the side of the boat and open the side covers to allow access to the internal components.
2. Disconnect the buzzer wiring harness.
3. Remove any wire straps that connect the buzzer wiring harness to the box.
4. Remove the buzzer from the control box.

To install:

5. Install the buzzer in the control box.
6. Install any wire straps that connect the buzzer harness to the control box.
7. Connect the buzzer wiring harness.
8. Test the buzzer for proper operation.
9. Install the control box side covers and mount the control box in the boat.

7

POWERHEAD

ENGINE MECHANICAL 7-2
ENGINE RECONDITIONING 7-31

ENGINE MECHANICAL 7-2
THE FOUR-STROKE CYCLE 7-3
INTAKE STROKE 7-3
COMPRESSION STROKE 7-3
POWER STROKE 7-3
EXHAUST STROKE 7-3
VALVE (ROCKER) COVER 7-3
REMOVAL & INSTALLATION 7-3
ROCKER ARM/SHAFTS 7-5
REMOVAL & INSTALLATION 7-5
INTAKE MANIFOLD 7-9
REMOVAL & INSTALLATION 7-9
CYLINDER HEAD 7-11
REMOVAL & INSTALLATION 7-11
OIL PAN 7-16
REMOVAL & INSTALLATION 7-16
TIMING BELT COVER 7-19
REMOVAL & INSTALLATION 7-19
TIMING BELT 7-19
INSPECTION 7-19
REMOVAL & INSTALLATION 7-20
ADJUSTMENT 7-22
CAMSHAFT, BEARINGS AND LIFTERS 7-25
REMOVAL & INSTALLATION 7-25
FLYWHEEL 7-29
REMOVAL & INSTALLATION 7-29
ENGINE RECONDITIONING 7-31
DETERMINING ENGINE CONDITION 7-31
COMPRESSION TEST 7-31
OIL PRESSURE TEST 7-32
BUY OR REBUILD? 7-32
ENGINE OVERHAUL TIPS 7-32
TOOLS 7-32
OVERHAUL TIPS 7-32
CLEANING 7-33
REPAIRING DAMAGED THREADS 7-33
ENGINE PREPARATION 7-34
CYLINDER HEAD 7-34
DISASSEMBLY 7-34
INSPECTION 7-35
REFINISHING & REPAIRING 7-37
ASSEMBLY 7-38
ENGINE BLOCK 7-39
GENERAL INFORMATION 7-39
DISASSEMBLY 7-39
INSPECTION 7-40
REFINISHING 7-42
ASSEMBLY 7-42
ENGINE START-UP AND BREAK-IN 7-45
STARTING THE ENGINE 7-45
BREAKING IT IN 7-45
KEEP IT MAINTAINED 7-45
SPECIFICATIONS CHARTS
ENGINE REBUILDING SPECIFICATIONS 7-46

ENGINE MECHANICAL

See Figures 1, 2 and 3

The basic piston engine is a metal block containing a series of chambers. The upper engine block is usually an iron or aluminum alloy casting, consisting of outer walls that form hollow jackets around the cylinder walls. The lower block, which provides a number of rigid mounting points for the bearings that hold the crankshaft in place, is known as the crankcase. The hollow jackets of the upper block add rigidity to the engine and contain the liquid coolant that carries heat away from the cylinders and other engine parts.

An air-cooled engine block consists of a crankcase that provides a rigid mounting for the crankshaft and has studs to hold the cylinders in place. The cylinders are individual, single-wall castings, finned for cooling, and they are usually bolted to the crankcase, rather than cast integrally with the block.

In a water-cooled engine, only the cylinder head is bolted to the block (usually on top).

The crankshaft is a long iron or steel shaft (and sometimes aluminum in more high-tech or high performance applications) mounted rigidly at a number of points in the bottom of the crankcase. The crankshaft is free to turn and contains several counterweighted crankpins (one centered under each cylinder) that are offset several inches from the center of the crankshaft and turn in a circle as the crankshaft turns. Pistons are connected to the crankpins by steel connecting rods. The rods connect the pistons at their upper ends with the crankpins at their lower ends. Circular rings seal the small space between the pistons and wall of the cylinders.

88528G01

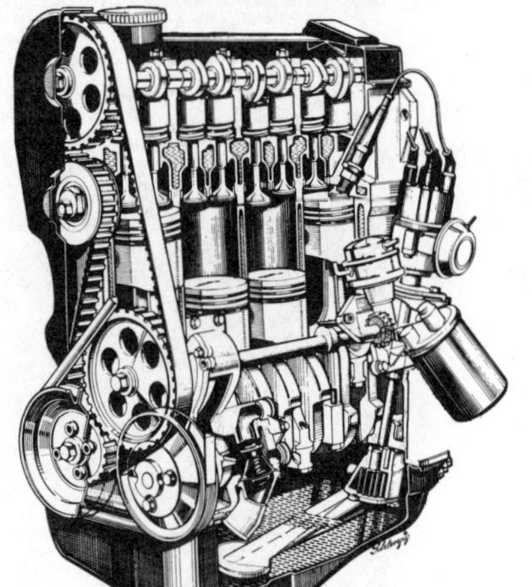

Fig. 1 Cutaway view of an in-line overhead cam four-cylinder engine

When the crankshaft spins, the pistons move up and down in the cylinders, varying the volume of each cylinder, depending on the position of the piston. At least two openings in each cylinder head (above the cylinders) allow the intake of the air/fuel mixture and the exhaust of burned gasses. After intake, the pistons compress the fuel mixture at the top of the cylinder, the fuel is ignited, and, as the pistons are forced downward by the expansion of burning fuel, the connecting rods convert the up and down motion of the pistons into rotary (turning) motion of the crankshaft. A round flywheel at the rear of the crankshaft provides a large, stable mass to smooth out the rotation.

The cylinder heads form tight covers for the tops of the cylinders and contain chambers into which the fuel mixture is forced as it is compressed by the pistons reaching the upper limit of their travel. Each combustion chamber contains at least one intake valve, one exhaust valve, and one spark plug per cylinder (depending on the design). The tips of the spark plugs protrude into the combustion chambers.

The valve in each opening of the cylinder head is opened and closed by the action of the camshaft. The camshaft is driven by the crankshaft through a gear, chain, or belt at 1/2 crankshaft speed (the camshaft gear is twice the size of the crankshaft gear). The valves are operated either through rocker arms and pushrods (overhead valve and some overhead cam engines) or directly by the camshaft using cam followers which usually contain shims for adjustment (overhead cam engine).

Lubricating oil is stored in a pan at the bottom of the engine and is force-fed to all parts of the engine by a gear-type pump, driven from the camshaft. The oil lubricates the entire engine and seals the piston rings, giving good compression.

The four-stroke engine, with its somewhat more complex mechanicals, operates just like an automobile engine. The fuel economy is better than a two-stroke, and mixing the fuel isn't required. Possibly the only drawback of the four-stroke engine is the weight-to-horsepower ratio.

As a workhorse, the four-stroke engine shines. Four-strokes are inherently torquey, regardless of displacement. This characteristic makes towing and hauling a strong point of the four-stroke. Also, if you plan to ride in areas with a lot of steep hills, rocks or tight, twisty trails, a four-stroke would be a prime candidate.

From a maintenance standpoint, a four-stroke might require slightly more maintenance than a two-stroke. The valve lash must be kept in adjustment for maximum engine performance. Also, the camshaft drive chain needs to be periodically adjusted on some engines. Adjustment intervals vary from one manufacturer to another, and some manufacturers equip their four-stroke engines with automatic cam chain adjusters. However. this periodic maintenance in no way

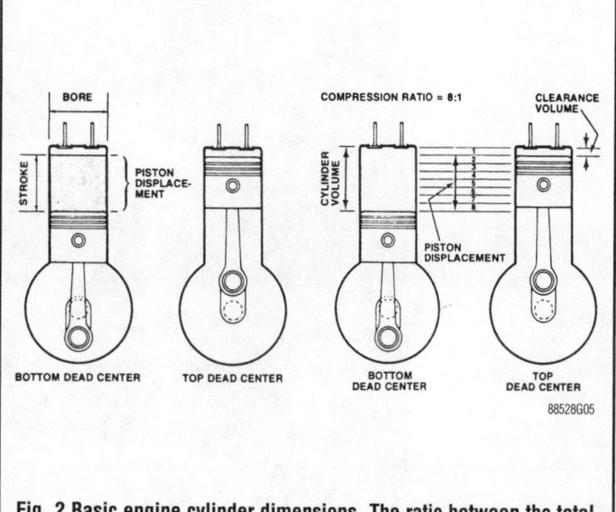

Fig. 2 Basic engine cylinder dimensions. The ratio between the total cylinder and clearance volume is the compression ratio

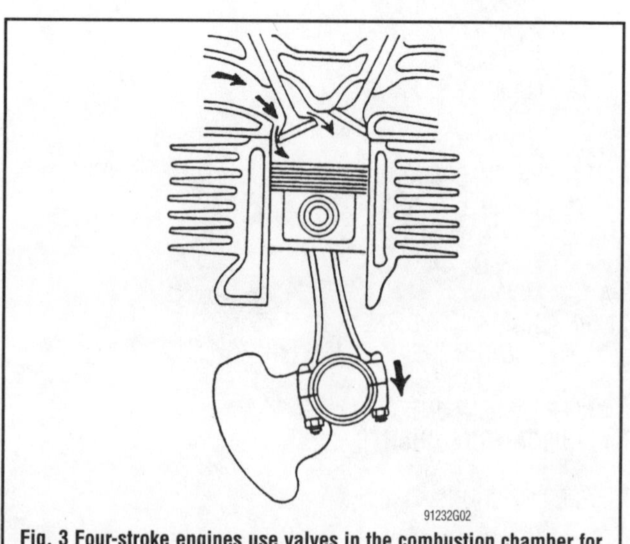

Fig. 3 Four-stroke engines use valves in the combustion chamber for entry and exit of the air and fuel mixture

designates a four-stroke as a "high maintenance" engine. Four-stroke engines are typically very reliable and will provide many years of service before a major overhaul is required, making them an excellent choice for most boats.

One final thing to consider when deciding between a two-stroke and a four-stroke is fuel economy. A two-stroke engine is an inherently inefficient engine, even though more power is produced per cubic centimeter than a four-stroke. Because of the inefficient (put powerful) operation of the two-stroke, significant amounts of raw fuel exit directly through the exhaust system. A two-stroke will consume more fuel than a four stroke, no matter how you look at it. This will limit your overall range of your boat, unless a larger fuel tank is fitted.

The Four-Stroke Cycle

◆ See Figure 4

In order for the four stroke cycle diesel engine to function properly, valves and injectors must act in direct relation to each other and to the four strokes of the engine. The intake and exhaust valves are camshaft operated, linked by tappets or cam followers, pushrods and rocker arms. The injectors are operated by either hydraulic or mechanical means, timed to the crankshaft and/or the camshaft rotation to provide the spray of fuel into the combustion chamber at the precise moment for efficient combustion.

INTAKE STROKE

During the intake stroke, the piston travels downward with the intake valve open and the exhaust valve closed. The downward travel of the piston allows and draws atmospheric air into the cylinder from the induction system. The intake charge consists of air only and contains no fuel mixture.

COMPRESSION STROKE

At the bottom of the intake stroke with the piston at Bottom Dead Center (BDC), the intake valve closes and the piston starts upwards on its compression stroke. The exhaust valve remains closed. At the end of the compression stroke, air in the combustion chamber has been forced by the piston to occupy a smaller space than it occupied at the beginning of the stroke. Thus, compression ratio is the direct proportion of the amount of space the air occupied in the combustion chamber before and after being compressed.

POWER STROKE

During the beginning of the power stroke, the piston is pushed downward by the burning and expanding gases. Both the intake and exhaust valves remain closed. As more fuel is added to the cylinder and burns, the gases become hotter and expand more rapidly, forcing the piston downward with much driving force and causing the crankshaft to rotate, in a power delivering action.

EXHAUST STROKE

As the piston reaches its Bottom Dead Center (BDC), the exhaust valve opens and the piston moves upward. The intake valve remains closed. The upward travel of the piston forces the burned gases from the combustion chamber through the open exhaust port and into the exhaust manifold. As the piston reaches the Top Dead Center (TDC) and starts its downward movement, the intake stroke is repeated and the cycling stroke continue in their proper sequence.

Valve (Rocker) Cover

REMOVAL & INSTALLATION

BF113A and BF130A

◆ See Figures 5, 6 and 7

1. Disconnect the breather hose from the cylinder head cover and remove the wiring harness holder from the connector bracket.
2. Disconnect the fuel hoses at the low pressure fuel pump.

THE FOUR STROKE CYCLE

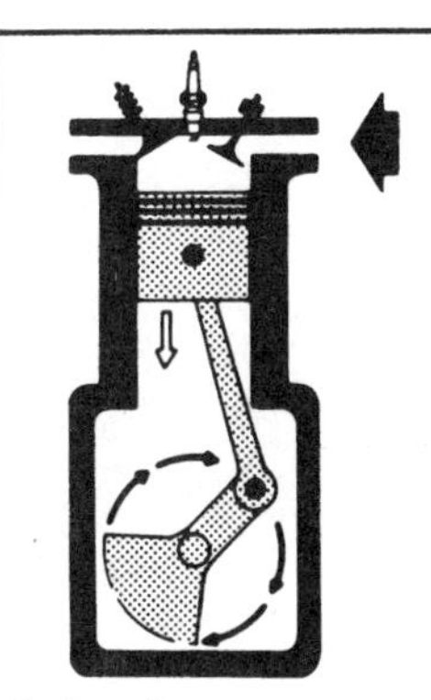

1. Intake

The intake stroke begins with the piston near the top of its travel. As the piston begins its descent, the exhaust valve closes fully, the intake valve opens and the volume of the combustion chamber begins to increase, creating a vacuum. As the piston descends, an air/fuel mixture is drawn from the carburetor into the cylinder through the intake manifold. The intake stroke ends with the intake valve closed just after the piston has begun its upstroke.

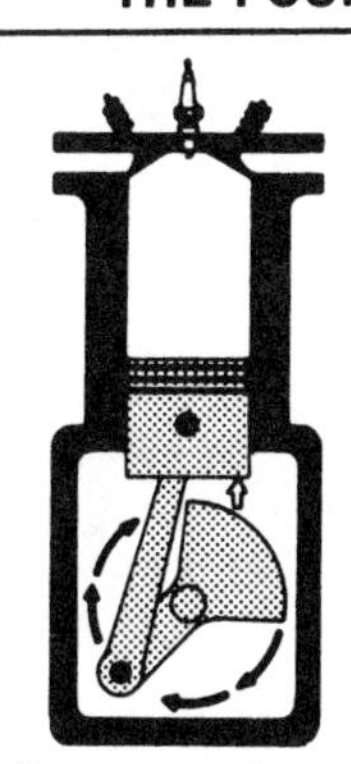

2. Compression

As the piston ascends, the fuel/air mixture is forced into the small chamber machined into the cylinder head. This compresses the mixture until it occupies 1/8th to 1/11th of the volume that it did at the time the piston began its ascent. This compression raises the temperature of the mixture and increases its pressure, increasing the force generated by the expansion of gases during the power stroke.

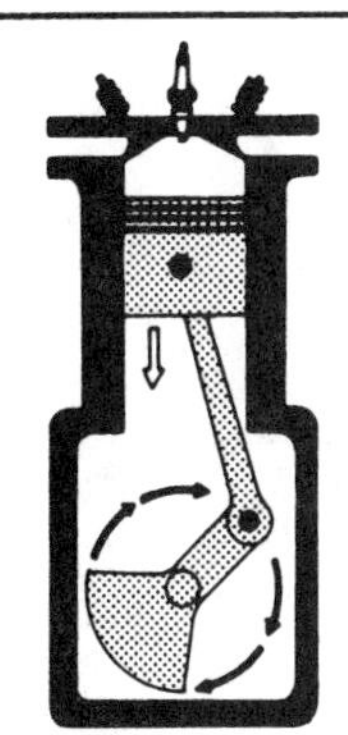

3. Ignition

The fuel/air mixture is ignited by the spark plug just before the piston reaches the top if its stroke so that a very large portion of the fuel will have burned by the time the piston begins descending again. The heat produced by combustion increases the pressure in the cylinder, forcing the piston down with great force.

4. Exhaust

As the piston approaches the bottom of its stroke, the exhaust valve begins opening and the pressure in the cylinder begins to force the gases out around the valve. The ascent of the piston then forces nearly all the rest of the unburned gases from the cylinder. The cycle begins again as the exhaust valve closes, the intake valve opens and the piston begins descending and bringing a fresh charge of fuel and air into the combustion chamber.

88528G03

Fig. 4 The four-stroke cycle of a basic two-valve, carburetted, gasoline, spark ignition engine (multi-valve and fuel injected engines operate the same way)

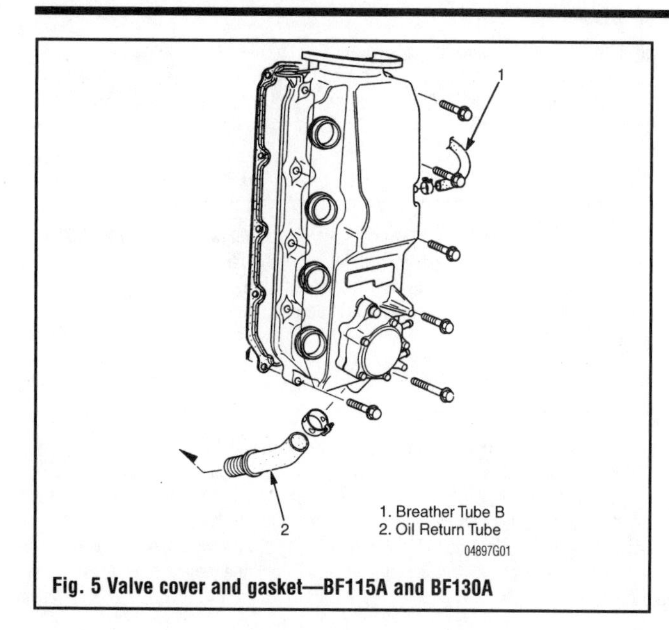

Fig. 5 Valve cover and gasket—BF115A and BF130A

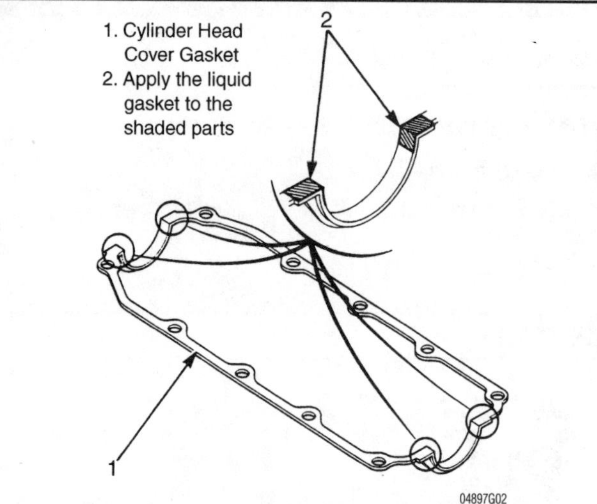

Fig. 6 Apply liquid gasket to the shaded areas to prevent a potentially messy oil leak

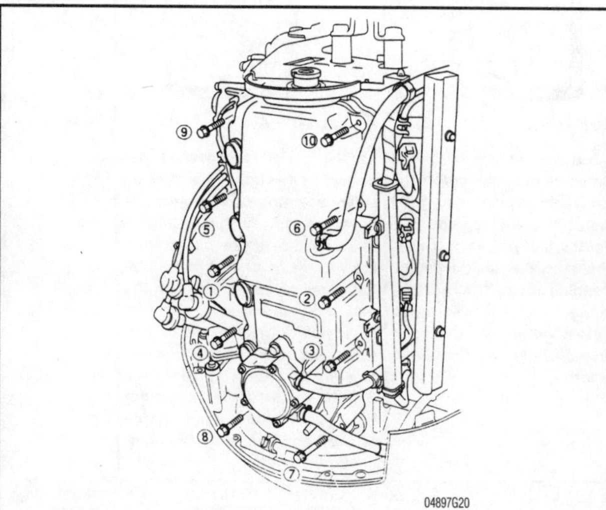

Fig. 7 Tighten the valve cover bolts to specification in the order shown—BF113A and BF130A

➡To prevent fuel spillage, clamp the hoses with hose clips.

3. Remove the cover bolts and lift off the cover and oil return hose as a set.
4. Remove the 90A fuse box.

To install:

5. Replace the 90A fuse box.
6. Apply liquid gasket sealant (Three Bond #1201, #1215 or equalivant) to the gasket at the areas indicated.

➡Make sure to follow the manufactures instructions for the liquid gasket and install it according to those instructions. If the gasket does not seal, a very messy oil leak into the engine pan will result.

7. Install the cover and oil return hose on the engine.
8. Make sure the cover is centered and tighten the cover bolts to 9 ft. lbs. (12 Nm) in two steps in the order shown.
9. Unclip the fuel hoses and reinstall them on the low pressure fuel pump.
10. Reconnect the breather hose and install the wiring harness holder.

BF75A and BF90A

See Figure 8

1. Remove the engine cover.
2. Disconnect the oil return hose from the cylinder head cover.
3. Remove the nine flange bolts and lift off the cylinder head cover.

To install:

4. Inspect the gasket for any cuts, wear or other defects. Replace the gasket if any problems are found.
5. Making sure that the gasket is centered, install the cylinder head cover and tighten the bolts snugly.

➡Do not over torque the bolts or the gasket will distort and an oil leak will occur.

6. Push the oil return hose approximately 0.8 in. (20 mm) onto the return pipe on the cylinder head cover and tighten the clamp.

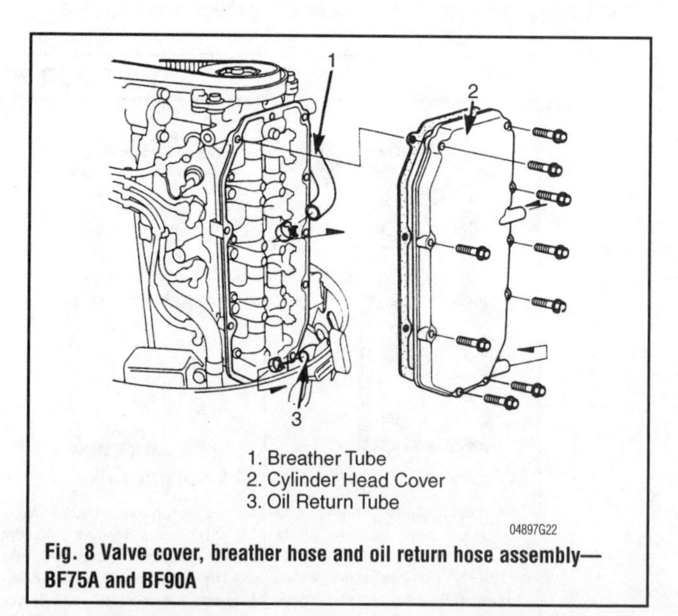

Fig. 8 Valve cover, breather hose and oil return hose assembly—BF75A and BF90A

BF50A, BF45A, BF40A and BF35A

See Figures 9 and 10

1. Remove the engine cover.
2. Disconnect the oil breather and oil return tubes.
3. Remove the bolts and lift off the cylinder head cover.

To install:

4. Inspect the cover gasket for tears, cuts, etc. Replace the gasket if necessary.
5. Install the cover on the engine, making sure its centered, and tighten the bolts snugly.

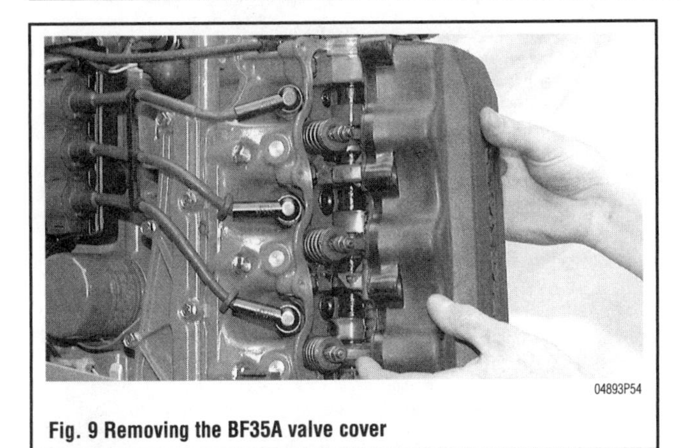

Fig. 9 Removing the BF35A valve cover

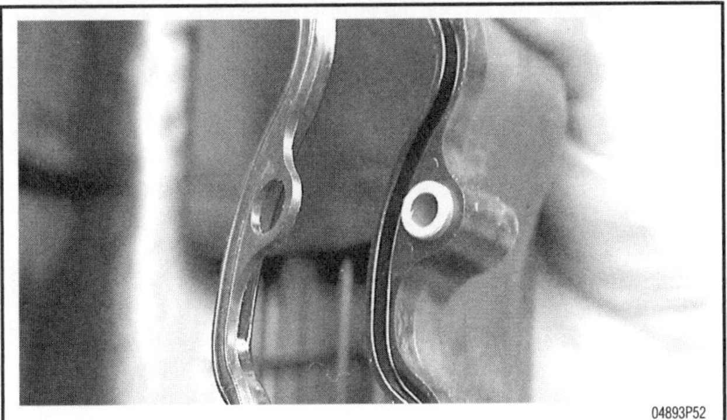

Fig. 10 Inspect the valve cover gasket for wear, tears, or any type of deformity that will prevent the gasket from sealing

6. Install the oil breather and oil return tubes.
7. Install the engine cover.

BF30A and BF25A

➧ See Figures 11 and 12

1. Remove the engine cover.
2. Remove the flywheel cover or recoil starter (if equipped).
3. Disconnect the oil return and breather hoses from the cylinder head cover.
4. Remove the bolts and lift off the valve cover.

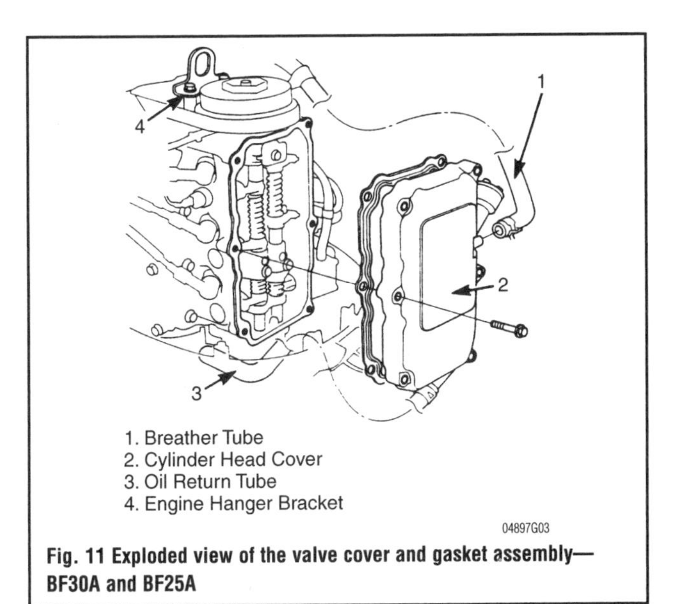

Fig. 11 Exploded view of the valve cover and gasket assembly—BF30A and BF25A

Fig. 12 Valve cover and breather tube assembly—BF25A

To install:

5. Inspect the cover gasket for wear, tear, etc. Replace the gasket if necessary.
6. Install the cover and tighten the bolts snugly. Do not overtighten the bolts or the gasket will deform and leak.
7. Connect the oil return and breather hoses to the cover.
8. Replace the recoil starter or flywheel cover.
9. Replace the engine cover.

BF75, BF8.8A, BF9.9A, BF100 and BF15A

1. After removing the fuel pump, loosen the four bolts and remove the valve cover.

To install:

2. Inspect the gasket for wear and tear and replace if necessary.
3. Install the gasket on the engine and tighten the bolts snugly.
4. Install the fuel pump.

Rocker Arm/Shafts

REMOVAL & INSTALLATION

BF115A and BF130A

➧ See Figures 13, 14 and 15

1. Remove the cylinder head, cylinder head gasket and the two 14 x 20 mm dowel pins.
2. Loosen the intake and exhaust valve adjuster lock nuts completely.

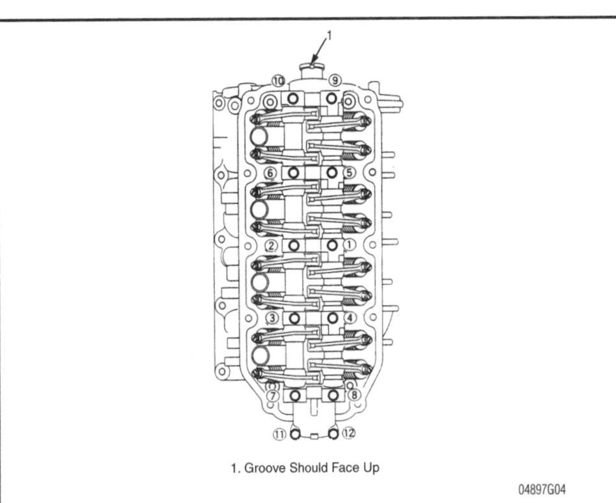

Fig. 13 Tighten the bolts in the order shown in two or three torque increments. Notice the notch in the camshaft is facing up— BF115A and BF130A

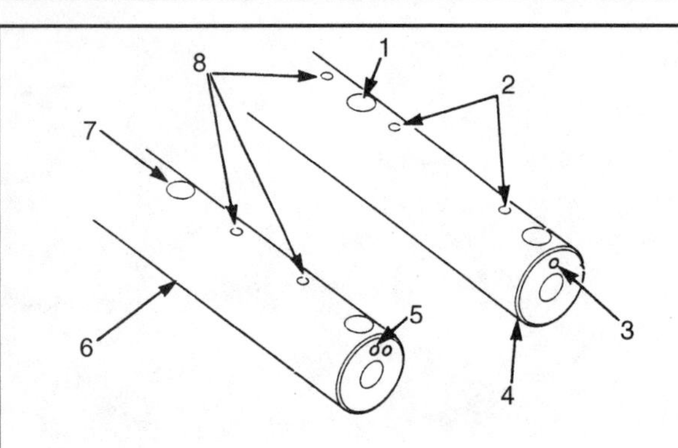

Fig. 14 Make sure the holes at the end of the rocker shafts are facing upward—BF115A and BF130A

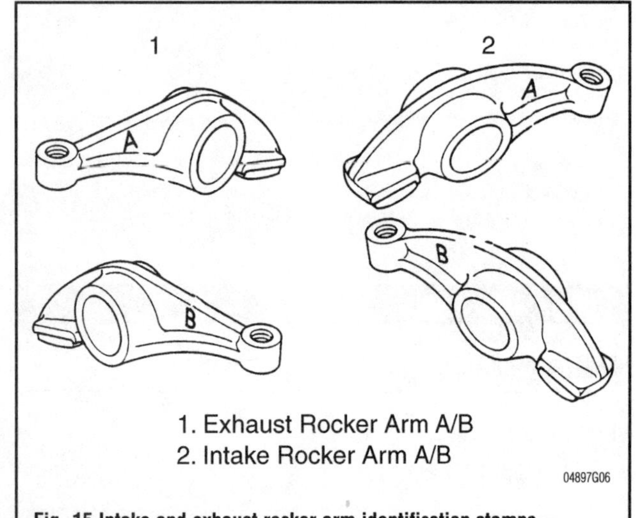

Fig. 15 Intake and exhaust rocker arm identification stamps—BF115A and BF130A

3. Loosen the camshaft holders evenly from the center holder out each end and lift the rocker shafts and arms off the engine.
4. Slide the camshaft holders, rocker arms, wave washers and rocker collars off the rocker shafts.

To install:

➡Check the installation position of each part before disassembly. Make sure to install the parts in the correct position during reassembly.

➡Check the rocker arms and fuel pump rocker arm surfaces that contact the camshaft for wear or damage.

➡Note that the intake rocker shaft has a larger shaft diameter than the exhaust rocker shaft.

➡Each camshaft holder has an identification number stamped on it. Check each holder and make sure the correct holder is used during installation.

5. Apply liquid gasket (Three Bond #1201, #1215 or something equivalent) to the cylinder head mating surfaces of the two end camshaft holders.
6. Both the intake and exhaust rocker arms are stamped with "A" or "B" for identification, make sure that they are installed in the proper order.
7. Install the wave washers. The exhaust and intake washers can be identified by the different diameters (intake washer larger, exhaust smaller).
8. Install the rocker collars. Keep them separated in the following order:
   - Rocker collar "A" has the highest collar.
   - Rocker collar "B" is higher than the intake collar "A".
   - Intake rocker collar "A" is lower than rocker collar "B".
9. Install the fuel pump rocker arm on the intake rocker shaft with the "arrow" mark pointed toward the cylinder head cover.
10. After all the components have been installed on each rocker shaft, install the camshaft holders onto the shafts. Make sure that the No.1 camshaft holder is mounted so that the hole at the end of the rocker arm shaft faces up.
11. Install the rocker arm assembly with the groove in the camshaft flange facing upwards toward the cylinder head.
12. Tighten the bolts in the order shown in two or three increments. Tighten the ten 8 x 76 mm flange bolts to 16 ft. lbs. (22 Nm) and the two 6 x 40 mm flange bolts to 9 ft. lbs. (12 Nm).

BF75A and BF90A

➧ See Figures 16 and 17

1. Remove the cylinder head bolts and lift off the cylinder head.
2. Remove the oil pump.
3. Loosen the intake and exhaust adjusting lock nuts and loosen the adjusting screws until they are even with the rocker arm.
4. Remove the two flange bolts on the cam thrust holder.
5. Remove the two 20 mm sealing bolts behind the oil pump.
6. To remove the intake and exhaust rocker arms, align the camshaft keyway on the timing belt driven side tot he position shown in the drawing.
   - To remove the intake rocker arms, set the camshaft at 10°clockwise.
   - To remove the exhaust rocker arms, set the camshaft at 10°counterclockwise.
7. Insert a 12 mm bolt (1.25 mm pitch) into the oil pump mounting side of the rocker shafts. As you pull out the shaft using the bolt, remove the rocker arm collars, wave washers and rocker arms.

➡Make sure to keep the intake and exhaust components separated and reassemble everything in the reverse order of disassembly.

8. Remove the cam thrust holder.
9. Remove the fuel pump lifter and remove the camshaft from the oil pump side of the cylinder head.

To install:

10. Apply assembly lube (molybdenum disulfide) to the camshaft journal and camshaft surface and install the cam into the head.
11. Set the two dowel pins on the cam thrust holder and then loosely install into the head.
12. Apply assembly lube to the rocker arm face and the outer surface of the rocker arm shaft.
13. Set the camshaft keyway on the timing belt side of the to 10°clockwise.
14. Using the 12 mm bolt, insert the intake rocker arm shaft (the one with

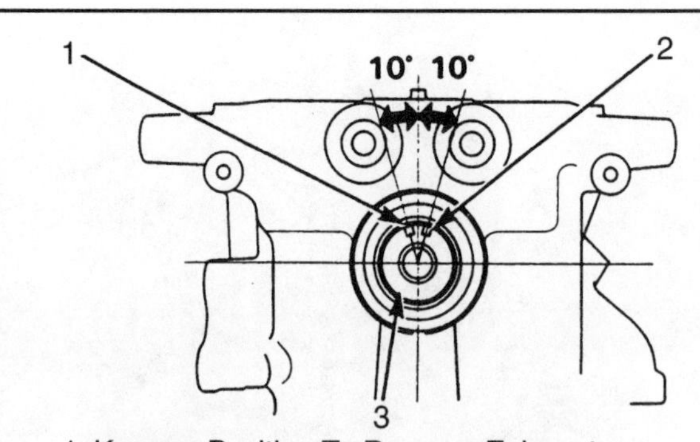

Fig. 16 Camshaft keyway position for removing the intake and exhaust rocker arms—BF75A and BF90A

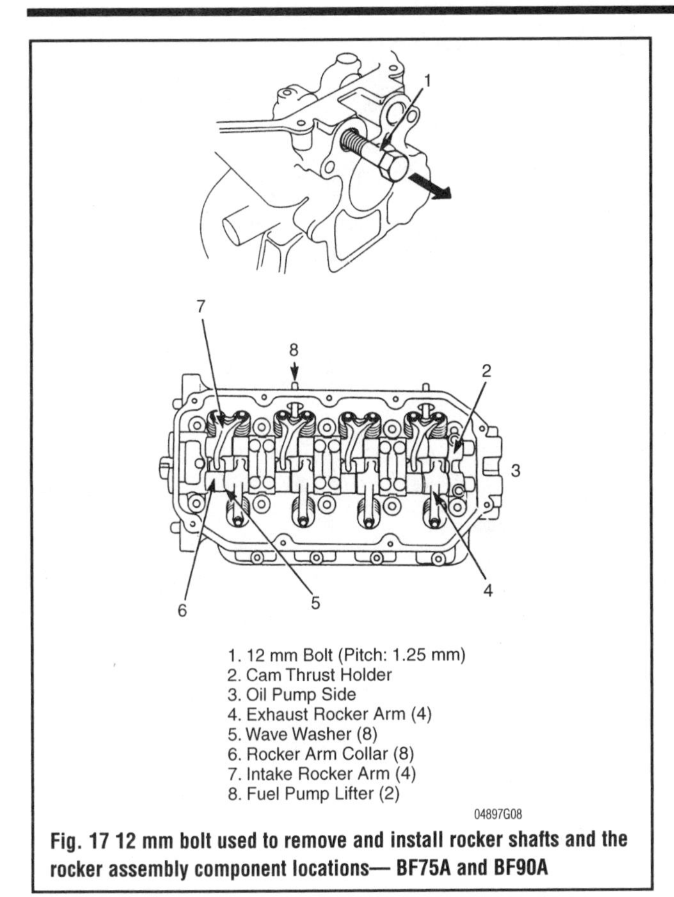

Fig. 17 12 mm bolt used to remove and install rocker shafts and the rocker assembly component locations— BF75A and BF90A

seven oil holes) into the cylinder head from the oil pump side. Starting with the No.4 cylinder, install the rocker arm collars, wave washers, and intake rocker arms in the reverse order that they were taken apart.

15. Install the exhaust rocker arm (the one with four oil holes) components in the same order as the intake components.

➡It is very important to install the intake and exhaust rocker shafts with the keyways in the correct positions where they are inclined 10°clockwise or counterclockwise respectively. If this is not done properly, the rocker arm will hit the camshaft lobe and prevent the smooth installation of the rocker arm shaft.

16. Set the flange bolt into the camshaft thrust holder. If it is hard to start the bolt, turn the 12 mm bolt clockwise until the groove in the rocker shaft aligns with the hole in the cam thrust holder and the flange bolt will thread in.

17. Tighten the camshaft thrust holder flange bolts to 6.5 ft. lbs. (9 Nm) and the 20 mm sealing bolts to 36 ft. lbs. (49 Nm).

BF50A, BF45A, BF40A and BF35A

See Figures 18, 19 and 20

1. Remove the oil pump.
2. Loosen the intake and exhaust valve adjuster lock nuts and loosen the adjusting screws completely.
3. Loosen the camshaft holders evenly, from the center holder, out to each end and lift the rocker shafts and arms off the engine.
4. Slide the rocker arms, rocker arm springs and rocker arm collars from the rocker arm shafts.

To install:

5. Cleaning all the bearing surfaces prior to installation.
6. Inspect the rocker shaft for any wear marks or scratches. Measure the outer shaft diameter. If the O.D. is larger than the service limit, the shaft must be replaced.

- Standard: 0.5502–0.5509 in. (13.976–13.994 mm).
- Service limit: 0.549 in. (13.95 mm).

7. Inspect the rocker arms for wear or scratches on both the cam contact

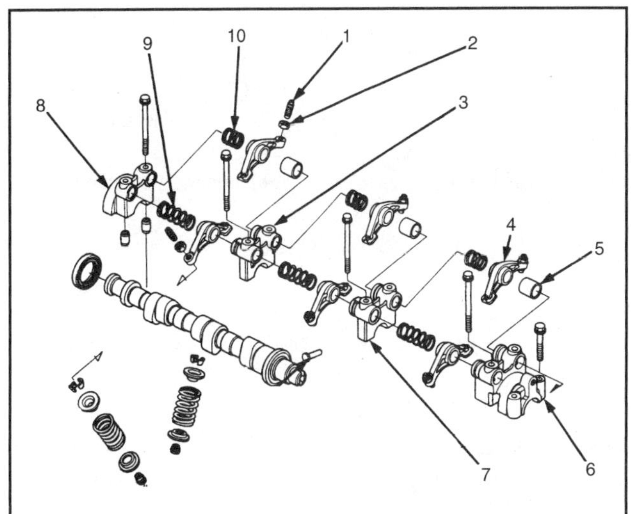

Fig. 18 Exploded view drawing of the rocker component locations— BF50A, BF45A, BF40A and BF35A

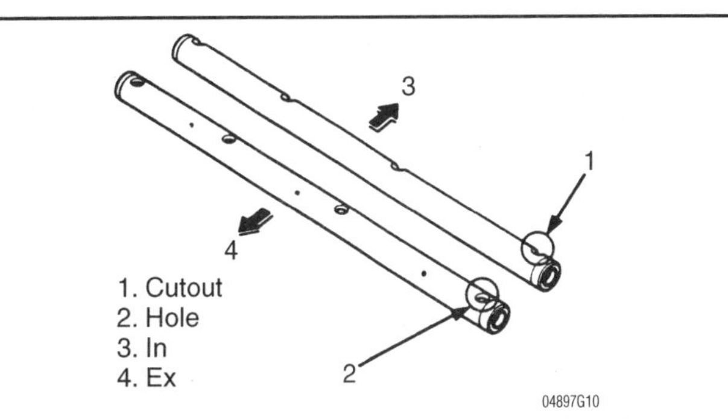

Fig. 19 The rocker shafts must be installed with the hole and cutout in the proper position on each shaft— BF50A, BF45A, BF40A and BF35A

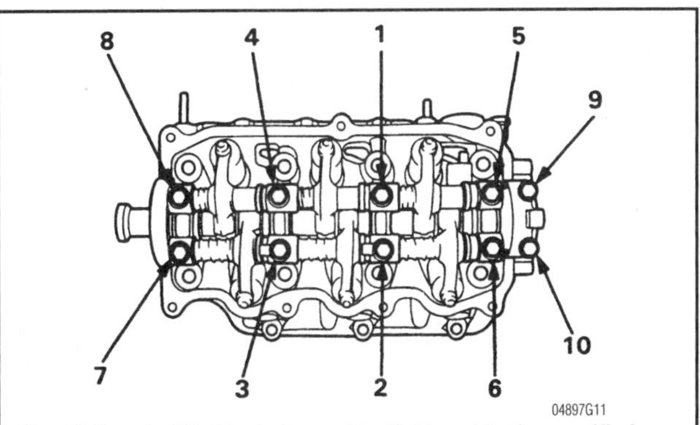

Fig. 20 Camshaft holder bolts must be tightened to the specified torque in the sequence shown— BF50A, BF45A, BF40A and BF35A

surface and the inner bore. Measure the rocker arm inner diameter and replace the rocker if the I.D. is larger than the service limit.

- Standard: 0.5516–0.5523 in. (14.010–14.028 mm)
- Service limit: 0.553 in. (14.05)

8. Measure the rocker arm shaft to rocker arm clearance.
- Standard: 0.0006–0.0020 in. (0.016–0.052 mm)
- Service limit: 0.003 in. (0.07 mm)

9. Reversing the disassembly procedure, install the rocker arms, rocker arm springs, rocker arm collars and camshaft holders onto the rocker arm shafts.

➡Do not interchange the exhaust side rocker arm shaft with the intake rocker shaft.

10. Install the rocker arm shaft with the hole in it on the exhaust side and the rocker arm shaft with the cutout in it on the intake side.
11. Install the No. 2 and No.3 camshaft holders facing the cam pulley side of the rocker shaft.
12. Torque the bolts to specification in the proper sequence.
- Eight 6 x 70 mm cam holder bolts 10.1 ft. lbs. (14 Nm).
- Two 6 x 35 mm cam holder bolts to 8.7 ft. lbs. (12 Nm).

BF25A and BF30A

➧ See Figures 21 and 22

1. Remove the oil pump.
2. Loosen the valve adjusting screw lock nuts and loosen the adjusting screws completely.
3. Remove the two rocker shaft retaining bolts.
4. Slide the rocker shafts out through the oil pump end of the cylinder head.
5. Keep the exhaust and intake components separated.

To install:

6. Inspect the rocker shaft for any wear marks or scratches. Measure the outer shaft diameter. If the O.D. is larger than the service limit, the shaft must be replaced.
- Standard: 0.5103–0.5110 in. (12.962–12.980 mm).
- Service limit: 0.509 in. (12.92 mm).

7. Inspect the rocker arms for wear or scratches on both the cam contact surface and the inner bore. Measure the rocker arm inner diameter and replace the rocker if the I.D. is larger than the service limit.
- Standard: 0.5118–0.5125 in. (13.000–13.018mm)
- Service limit. 0.513 in. (13.04 mm)

8. Measure the rocker arm shaft to rocker arm clearance.
- Standard: 0.0007–0.0022 (0.020–0.056 mm)
- Service limit: 0.003 in. (0.07 mm)

9. Slide the rocker shafts into the cylinder head from the oil pump side.
10. As the shaft enters the cylinder head, install the rocker arm, rocker arm spring and rocker arm collar in the proper order for each rocker shaft.
11. Make sure that the cut outs in the rocker shafts face outward.
12. After the shaft is in place, install the shaft retaining bolts and torque to 6.5 ft. lbs. (9 Nm).
13. Replace the oil pump.

BF9.9A and BF15A

➧ See Figure 23

1. Remove the oil pump.
2. Loosen all the valve adjusting screws.
3. Screw a 10 x 1.25 mm bolt into the threaded end of the rocker shaft and pull out the shaft.
4. As the shaft is being pulled out, remove the rocker arms and the rocker arm spring.

To install:

5. Inspect the rocker shaft for wear and scratches, then measure the shaft outer diameter. If the shaft O.D. measures out of specification, replace the shaft.
- Standard: 0.5106 in. (12.968 mm).
- Service limit: 0.5087 in. 12.92 mm).

6. Inspect the rocker arms for wear and scratches, then measure the rocker arm inner diameter. If the I.D. out of specification, replace the rocker arm.
- Standard: 0.5118 in. (13.0 mm).
- Service limit: 0.5133 in. (13.04 mm).

7. Lubricate the rocker arm bushings, rocker shaft and rocker shaft bushing before assembly.
8. Slide the rocker shaft through the cylinder head with the threaded end toward you and install the rocker arms and rocker arm spring as the shaft enters. Unless the part has been replaced, install the parts in the order they came out.
9. Replace the oil pump.

BF75, BF8A and BF100

➧ See Figure 24

1. Remove the oil pump.
2. Loosen the lock nuts and adjusting screws to unload the valve springs before removing the rocker arm shaft.
3. Screw in a 10 x 1.25 mm bolt into the threaded hole on the rocker shaft and pull out the shaft.
4. As the shaft leaves the cylinder head, remove the rocker arms, rocker collars and rocker arm spring.
5. After the rocker shaft has been removed, you can now remove the camshaft and thrust washers.

To install:

6. Inspect the rocker shaft for wear and scratches, then measure the shaft outer diameter. If the shaft O.D. measures out of specification, replace the shaft.
- Standard: 0.510–0.511 in. (12.947–12.968 mm).
- Service limit: 0.509 in. 12.92 mm).

7. Inspect the rocker arms for wear and scratches, then measure the rocker arm inner diameter. If the I.D. is out of specification, replace the rocker arm.
- Standard: 0.512–0.513 in. (13.0–13.03 mm).
- Service limit: 0.514 in. (13.06 mm).

8. Lubricate the rocker arm bushings, rocker shaft and rocker shaft bushing before assembly.
9. Slide the rocker shaft through the cylinder head with the threaded end toward you and install the rocker arms and rocker arm spring as the shaft enters. Unless the part has been replaced, install the parts in the order they came out.
10. Install the oil pump back on the cylinder head.

➡When reassembling, apply a coat of grease on the end of the exhaust pipe and insert it through the seal. Be careful to not dislodge the seal spring.

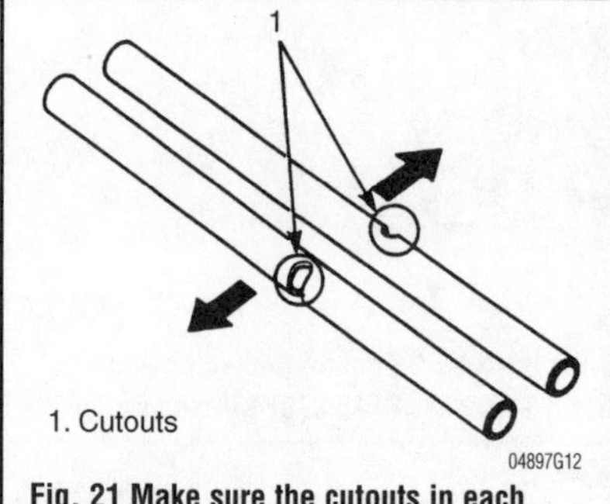

Fig. 21 Make sure the cutouts in each shaft are facing outward so the rocker shaft retaining bolts clear the shafts— BF50A, BF45A, BF40A and BF35A

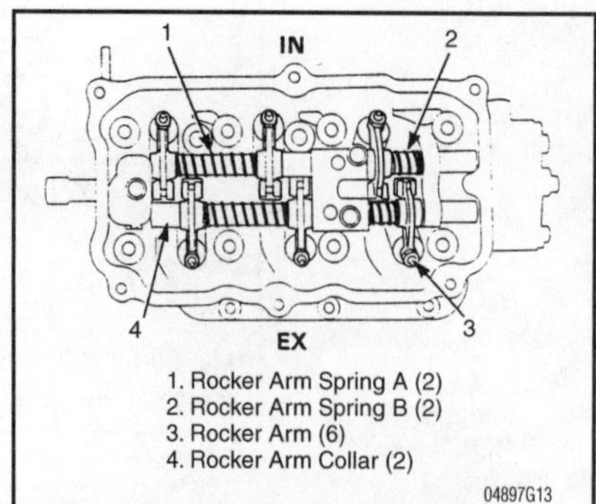

Fig. 22 Correct assembly of the rocker arms and shafts— BF50A, BF45A, BF40A and BF35A

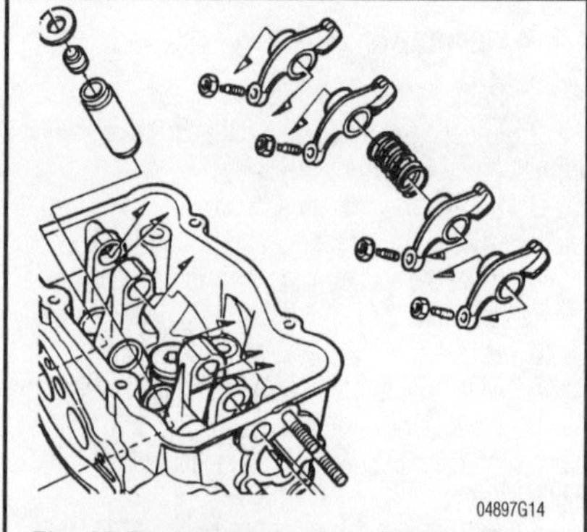

Fig. 23 Rocker arm and shaft assembly— BF9.9A and BF15A

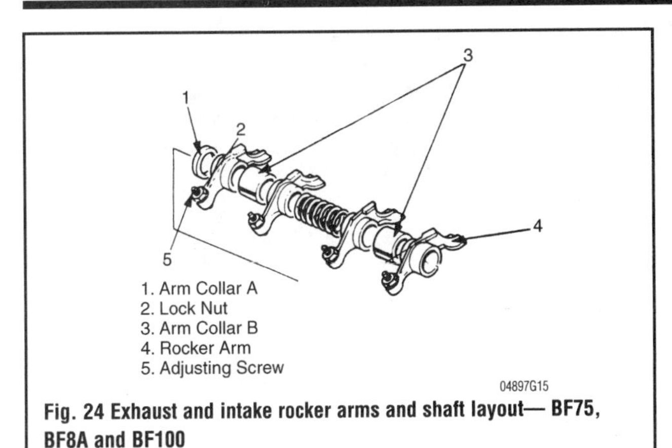

Fig. 24 Exhaust and intake rocker arms and shaft layout— BF75, BF8A and BF100

BF50 and BF5A

➧ See Figure 25

➡These models use a stud type rocker arm assembly and have no rocker shafts.

1. Remove the engine cover.
2. Remove the valve cover.
3. Loosen the lock nut and remove the nut and the rocker arm pivot.
4. Lift off the rocker arm and pull out the pushrod.
5. Inspect the center hole, pivot and pushrod contact surfaces for wear and scratches. Replace any component if needed.

To install:

6. Install the pushrod and pushrod guide.
7. Install the pivot stud and torque to 20–22 ft. lbs. (28–30 Nm).
8. Install the rocker arm, rocker arm pivot and pivot lock nut onto the pivot stud.

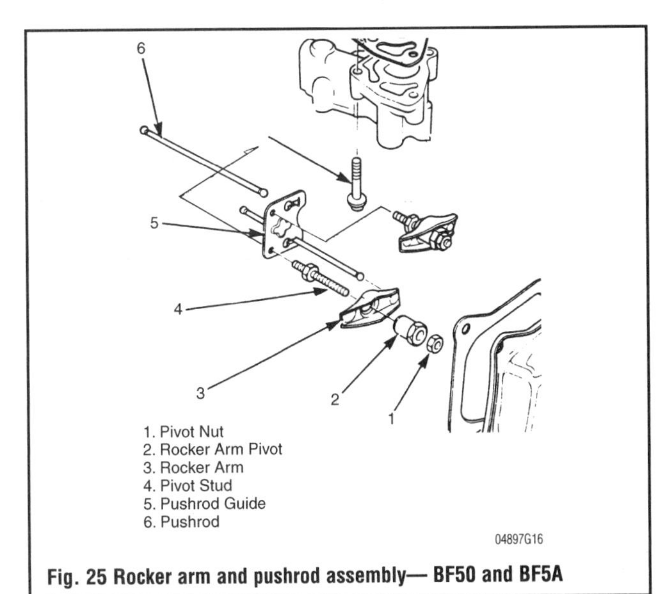

Fig. 25 Rocker arm and pushrod assembly— BF50 and BF5A

Intake Manifold

REMOVAL & INSTALLATION

BF115A and BF130A

➧ See Figure 26

1. Disconnect the negative battery cable form the battery terminal.
2. Remove the engine cover.

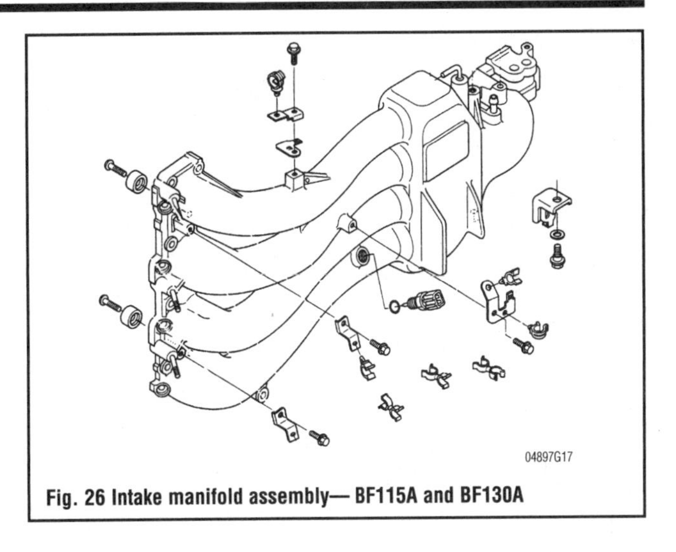

Fig. 26 Intake manifold assembly— BF115A and BF130A

3. Relieve fuel pressure per the instructions in the fuel system chapter.
4. Remove the separate rear case.
5. Remove the upper fuel hose on the low pressure fuel pump.
6. Remove the air vent tube from the mounting case.
7. Disconnect the fuel injector connectors from the fuel injectors.
8. Disconnect the upper and lower water tubes from their "T" fittings.
9. Disconnect the 3P connector
10. Disconnect the fuel pump 2P connector, disconnect the IAT sensor connector and release the remote control cable from the wire clip.
11. Disconnect the MAP sensor connector.
12. Remove the silencer.
13. Remove the throttle body from the intake manifold and discard the gasket.
14. Disconnect he IAC valve connector and remove the PCV hose.
15. Remove the dipstick.
16. Remove the two bolts securing the vapor separator assembly and the one bolt securing the throttle body.
17. Remove the nine bolts securing the intake manifold.
18. When the manifold has cleared the mounting studs, move the manifold toward the cylinder head and lift upwards to remove. Discard the intake manifold gasket.

To install:

➡When installing the assembly, make sure the projection on the cylinder block goes between the two hose joints. Take care not to damage the back of the vapor separator with the projection at this time.

19. Loosely install the nine bolts securing the intake manifold.
20. Mount the upper intake manifold stay in contact with the left side of the crankcase boss and mount the lower intake manifold stay in contact with the crankcase sealing bolt and loosely tighten the bolts.
21. After aligning the intake manifold and its brackets, tighten the cap nuts to 29 ft. lbs. (39 Nm)
22. Install and tighten securely the two bolts securing the vapor separator assembly and the one bolt securing the throttle body.
23. Install the dipstick.
24. Connect he IAC valve connector and install the PCV hose.
25. Install the throttle body to the intake manifold with a new gasket.
26. Install the silencer.
27. Connect the MAP sensor connector.
28. Connect the fuel pump 2P connector, connect the IAT sensor connector and secure the remote control cable to the wire clip.
29. Connect the 3P connector
30. Connect the upper and lower water tubes from their "T" fittings.
31. Connect the fuel injector connectors from the fuel injectors.
32. Install the air vent tube from the mounting case.
33. Install the upper fuel hose on the low pressure fuel pump.
34. Install the separate rear case.
35. Relieve fuel pressure per the instructions in the fuel system chapter.
36. Install the engine cover.
37. Connect the negative battery cable to the battery terminal.

BF75A and BF90A

➧ See Figure 27

1. Remove the engine cover and timing belt cover.
2. Disconnect the choke knob and choke rod.
3. Disconnect the two throttle cables at the throttle cam side.
4. Drain the carburetor float bowls.
5. Disconnect the fuel lines.
6. Disconnect the two hose from the water check grommet.
7. Disconnect the breather tube from the valve cover.
8. Remove the two bolts and the silencer duct.
9. Remove the carburetor assembly.
10. Remove the throttle cam.
11. Separate the carburetor components to remove the intake manifold. Discard the carburetor gaskets.

To install:

➡Make sure the rollers are installed between the intake manifold, gaskets, insulator and carburetor bodies.

12. Reassemble the carburetor assembly.
13. Install the throttle cam.
14. Install the carburetor assembly.
15. Install the two bolts and the silencer duct.
16. Connect the breather tube from the valve cover.
17. Connect the two hose from the water check grommet.
18. Connect the fuel lines.
19. Connect the two throttle cables at the throttle cam side.
20. Connect the choke knob and choke rod.
21. Install the engine cover and timing belt cover.

BF50A, BF45A, BF40A and BF35A

➧ See Figures 28 and 29

1. Remove the engine cover.
2. Remove the silencer cover.
3. Remove choke knob and choke rod.
4. Remove the throttle pivot from the throttle cam.
5. On the remote control model only, slide the choke solenoid rearward and disconnect the rod joint from the plunger rod and remove the rod from the choke arm.
6. Disconnect he breather tube and fuel hoses.
7. Remove the bolts and cap nuts and then remove the carburetor assembly.
8. Separate the carburetor sections, Discard the gaskets and remove the intake manifold.

To install:

9. Assemble the carburetor sections, install the new gaskets and install the intake manifold on the carburetor assembly.
10. Install the carburetor assembly onto the engine.
11. Connect the breather tube and fuel hoses.
12. Reassemble the choke solenoid assembly
13. Reattach the throttle pivot from the throttle cam.
14. Install the choke knob and choke rod.
15. Install the silencer cover.
16. Install the engine cover.

BF25A and BF30A

➧ See Figure 30

1. Remove the engine cover.
2. Remove choke knob and choke rod.
3. Remove the throttle pivot from the throttle cam.
4. On the remote control model only, slide the choke solenoid rearward and disconnect the rod joint from the plunger rod and remove the rod from the choke arm.
5. Disconnect he breather tube and fuel hoses.
6. Remove the bolts and cap nuts and then remove the carburetor assembly.
7. Separate the carburetor sections, Discard the gaskets and remove the intake manifold.

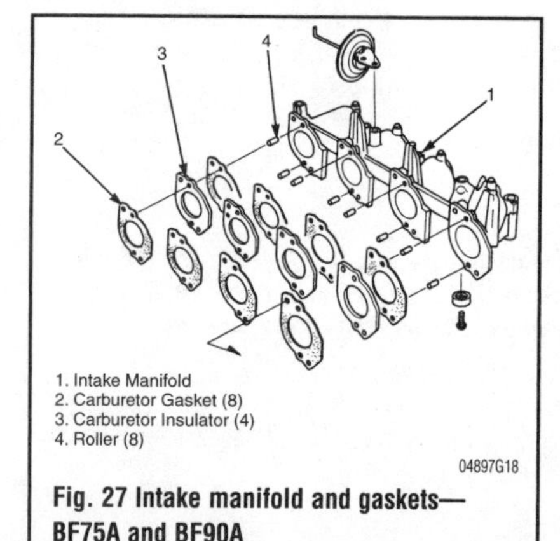

Fig. 27 Intake manifold and gaskets— BF75A and BF90A

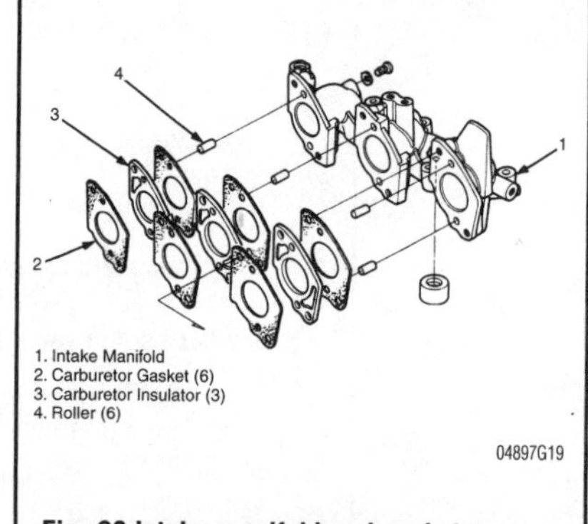

Fig. 28 Intake manifold and gasket assembly— BF50A, BF45A, BF40A and BF35A

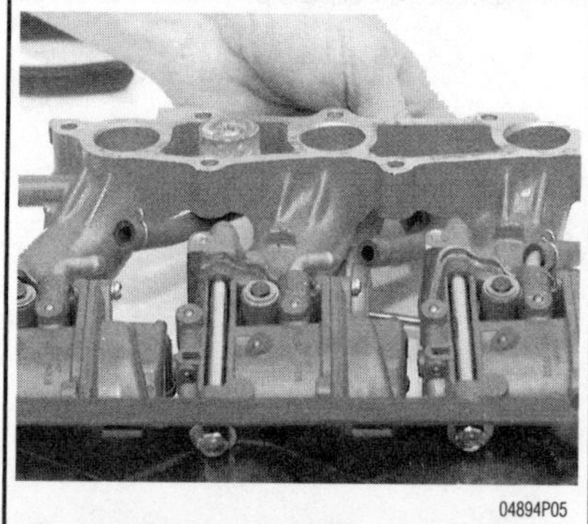

Fig. 29 Removing the intake manifold from the cylinder head

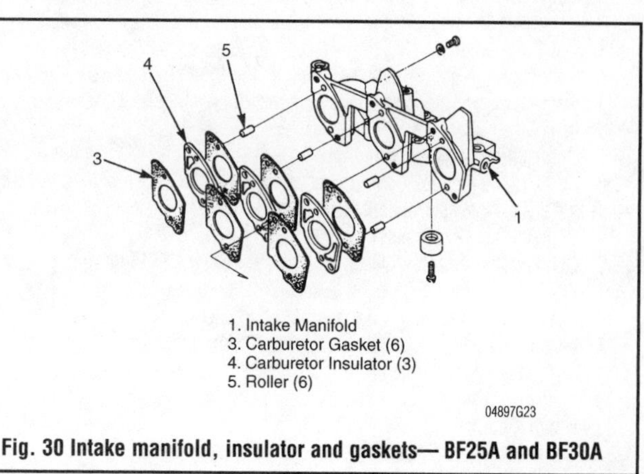

Fig. 30 Intake manifold, insulator and gaskets— BF25A and BF30A

To install:

8. Assemble the carburetor sections, install the new gaskets and install the intake manifold on the carburetor assembly.
9. Install the carburetor assembly onto the engine.
10. Connect the breather tube and fuel hoses.
11. Reassemble the choke solenoid assembly
12. Reattach the throttle pivot from the throttle cam.
13. Install the choke knob and choke rod.
14. Install the silencer cover.
15. Install the engine cover.

BF9.9A and BF15A

➧ See Figure 31

1. Remove the engine cover.
2. Disconnect the choke rod and throttle rod.

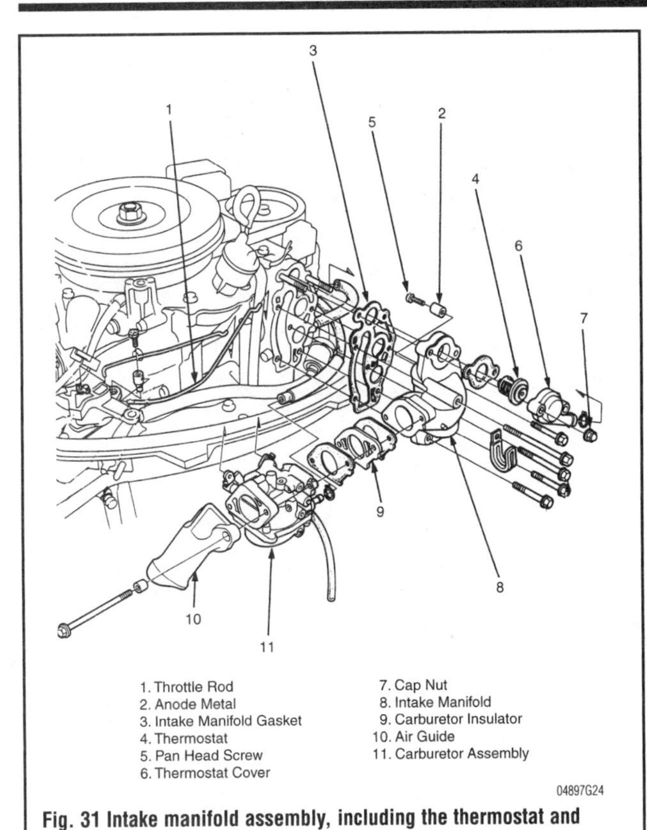

Fig. 31 Intake manifold assembly, including the thermostat and cover— BF9.9A and BF15A

3. Disconnect the fuel hose.
4. Unbolt the intake manifold from the engine.
5. Separate the carburetor from the intake manifold. Discard the old gaskets.
6. Remove the thermostat from the intake manifold.

To install:

7. Install the thermostat in the intake manifold.
8. With the new gaskets, attach the carburetor to the intake manifold.
9. Install the intake manifold on the engine.
10. Connect the fuel hose.
11. Connect the choke rod and throttle rod.
12. Install the engine cover.

BF75, BF8A and BF100

See Figure 32

1. Remove the engine cover.
2. Disconnect the choke rod and throttle rod.
3. Disconnect the fuel hose.
4. Unbolt carburetor from the intake manifold.

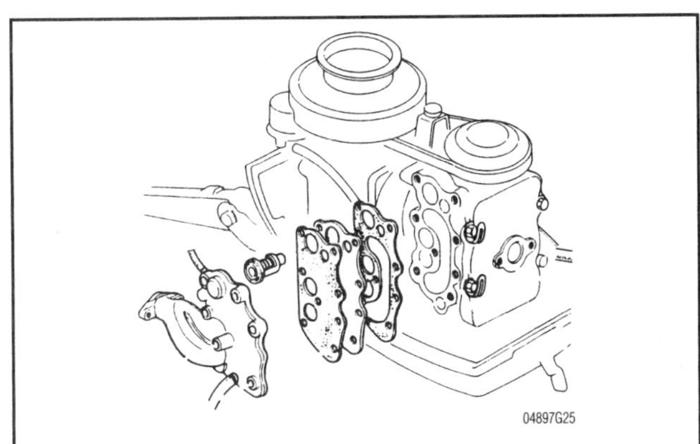

Fig. 32 Intake manifold assembly— BF75, BF8A and BF100

5. Remove the intake manifold from the engine and Discard the gaskets.
6. Remove the thermostat from the intake manifold.
7. Clean the inside of the intake manifold with compressed air.

To install:

8. Install the thermostat in the intake manifold.
9. With the new gaskets, attach the intake manifold to the engine.
10. Install the carburetor onto the intake manifold.
11. Connect the fuel hose.
12. Connect the choke rod and throttle rod.
13. Install the engine cover.

BF50 and BF5A

See Figure 33

1. Remove the engine cover.
2. Disconnect the choke rod and throttle rod.
3. Disconnect the fuel hose.
4. Unbolt carburetor from the intake manifold.
5. Remove the intake manifold from the engine and Discard the gaskets.
6. Remove the thermostat from the intake manifold.
7. Clean the inside of the intake manifold with compressed air.

To install:

8. Install the thermostat in the intake manifold.
9. With the new gaskets, and making sure to install the carburetor insulator in the proper direction, attach the intake manifold to the engine.
10. Install the carburetor onto the intake manifold.
11. Connect the fuel hose.
12. Connect the choke rod and throttle rod.
13. Install the engine cover.

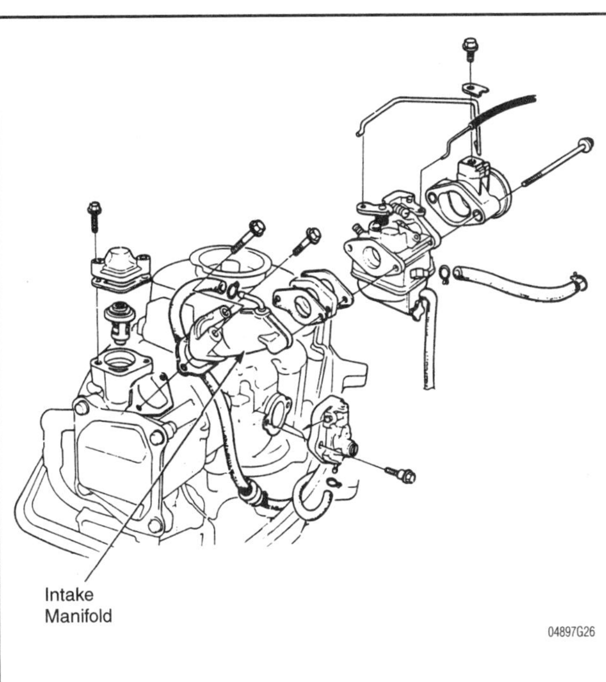

Fig. 33 Intake manifold and gasket assembly—BF50 and BF5A

Cylinder Head

REMOVAL & INSTALLATION

BF115A and BF130A

See Figures 34, 35 and 36

1. Remove the timing belt upper cover.
2. Remove the flushing water hose.

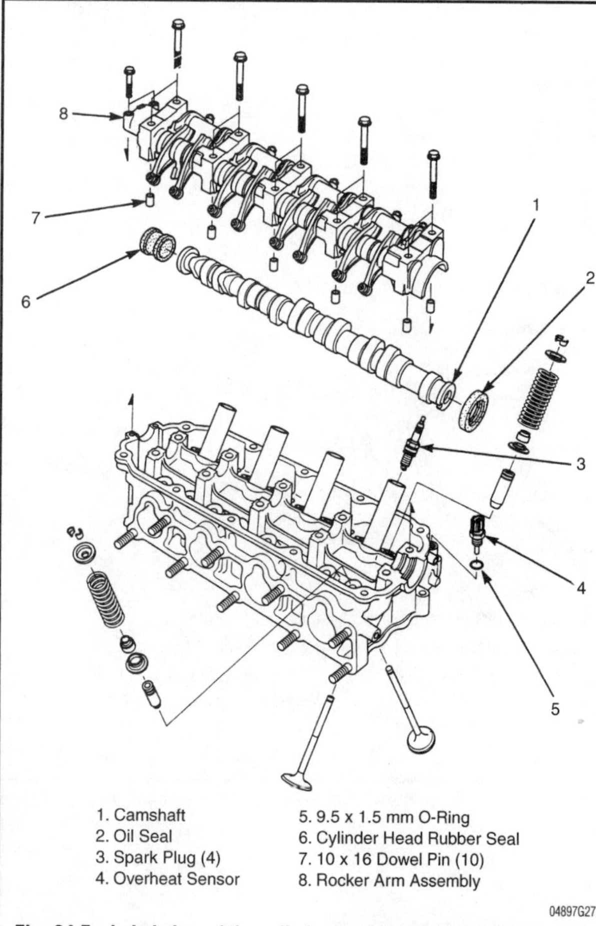

Fig. 34 Exploded view of the cylinder head assembly—BF115A and BF130A

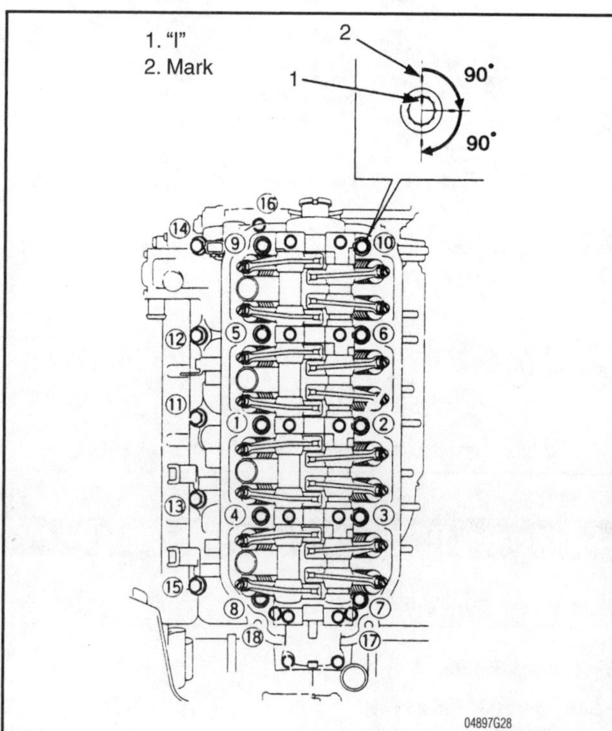

Fig. 35 Cylinder head torque sequence and bolt torque procedure—BF115A and BF130A

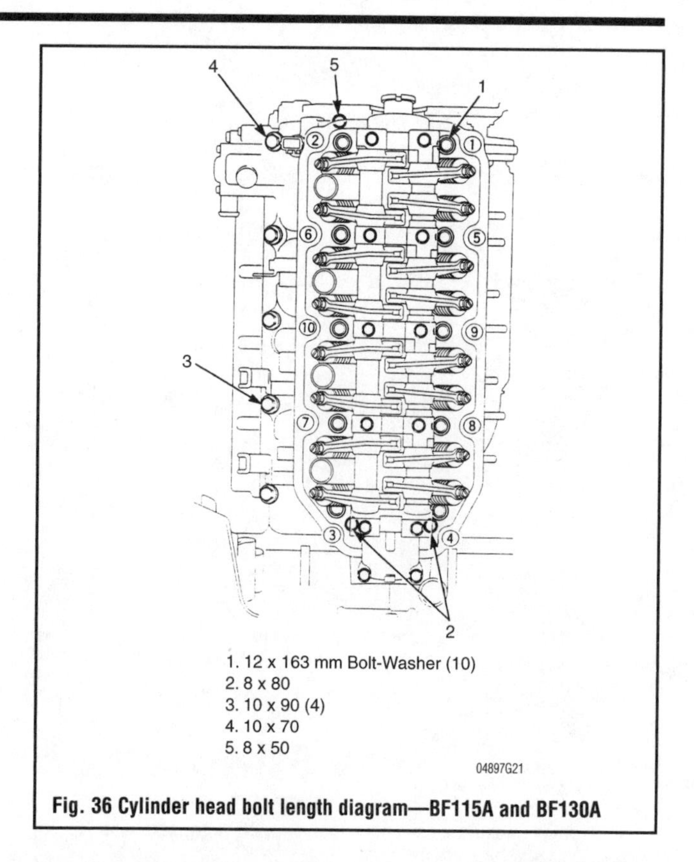

Fig. 36 Cylinder head bolt length diagram—BF115A and BF130A

3. Remove the intake manifold assembly.
4. Remove the spark plug caps and disconnect the connector from the overheat sensor.
5. Remove the two spark plug wire clamps.
6. Be sure that the timing belt driven pulley is at the top dead center of the compression stroke of the No. 1 cylinder.
7. Remove the timing belt.
8. Remove the timing belt driven pulley.
9. Disconnect the breather hose from the valve cover.
10. Remove the harness holder from the bracket.
11. Disconnect the low pressure fuel hoses and clamp them off.
12. Remove the valve cover and oil return hose as a set.
13. Remove the 90A fuse box.
14. Remove the cylinder head, cylinder head gasket and the two 14 x 20 mm dowel pins.

To install:

15. Check to make sure that the oil control orifice is mounted on the cylinder block. Install the two 14 x 20 mm dowel pins and a new head gasket.
16. Install the cylinder head on the cylinder block.
17. Apply engine oil to the threaded and seating sections of the bolts and loosely thread each bolt into the cylinder head.
18. Tighten the ten 12 x 163 mm flange bolts to 33 ft. lbs. (44 Nm).
19. With the "I" mark on the flange part of each bolt-washer as a reference point, make a mark on the cylinder head at right angles to each reference mark. Tighten the bolt-washers to 90° in two steps in the numbered order.
20. After tightening the 12 x 163 mm bolt-washers, tighten the other bolts in the numbered order in two or three steps to the standard torque. 10 mm flange bolt: 29 ft. lbs. (39 Nm); 8 mm flange bolt: 20 ft. lbs. (26 Nm).
21. Replace the 90A fuse box.
22. Replace the valve cover and oil return hose as a set.
23. Connect the low pressure fuel hoses and clamp them off.
24. Replace the harness holder from the bracket.
25. Connect the breather hose from the valve cover.
26. Replace the timing belt driven pulley.
27. Replace the timing belt.
28. Replace the two spark plug wire clamps.
29. Replace the spark plug caps and connect the connector from the overheat sensor.
30. Replace the intake manifold assembly.

31. Replace the timing belt upper cover.
32. Reconnect the flushing water hose.
33. Adjust the valve timing.
34. Replace the 90A fuse box.

BF75A and BF90A

See Figures 37 and 38

1. Remove the engine cover and timing belt cover.
2. Remove the timing belt and timing pulley.
3. Remove the timing belt lower cover.
4. Disconnect the fuel hoses.
5. Remove the carburetor assembly.
6. Remove the spark plugs.
7. Disconnect the thermoswitich wire from the thermoswitch.
8. Remove the flushing water hose clamps.
9. Move the power tilt relay assembly to the relief valve side without disconnecting the wire.
10. Tilt the engine until the cylinder head faces slightly upward.
11. Disconnect the oil return hose from the cylinder head cover.
12. Remove the valve cover.
13. Remove the cylinder head bolts and lift off the cylinder head and Discard the used head gasket.

To install:

14. Clean all cylinder head mating surfaces.
15. Install the cylinder head and head gasket and torque the bolts to specification in sequence.
16. Replace the oil pump.
17. Replace the timing belt lower cover.
18. Replace the timing belt and timing pulley.
19. Replace the engine cover and timing belt cover.
20. Connect the oil return hose from the cylinder head cover.
21. Move the power tilt relay assembly back to its correct place.
22. Replace the flushing water hose clamps.
23. Connect the thermoswitch wire to the thermoswitich.
24. Replace the spark plugs.
25. Replace the carburetor assembly.
26. Connect the fuel hoses.
27. Replace the timing belt lower cover.
28. Replace the timing belt and timing pulley.
29. Adjust the valve timing.
30. Replace the valve cover.
31. Replace the engine cover and timing belt cover.

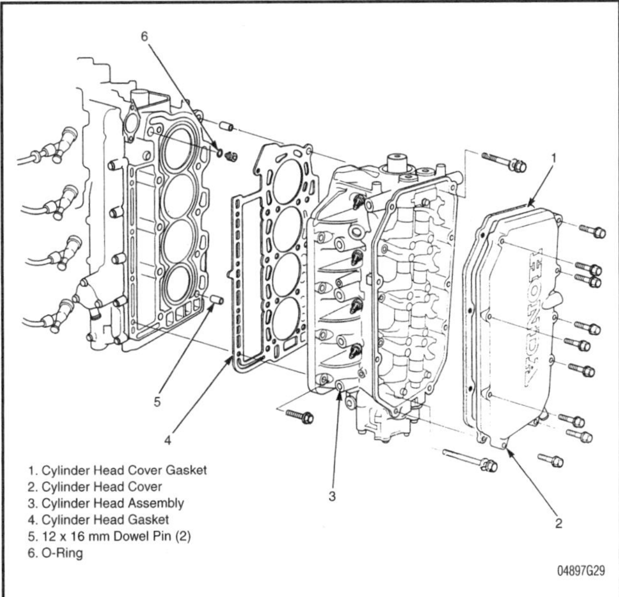

Fig. 37 Cylinder head and head gasket assembly— BF75A and BF90A

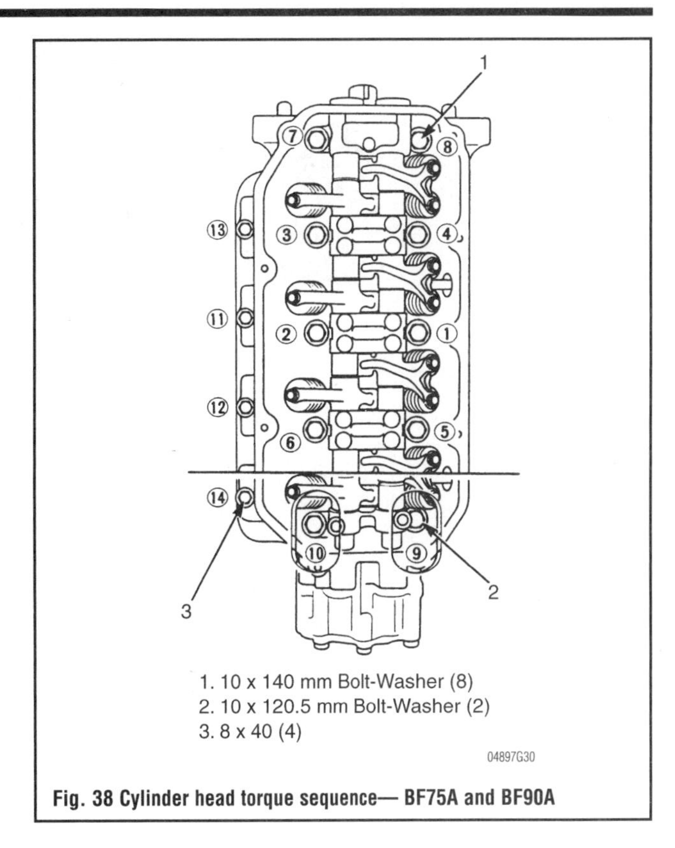

Fig. 38 Cylinder head torque sequence— BF75A and BF90A

BF50A, BF45A, BF40A and BF35A

See Figures 39 and 40

1. Remove the engine cover and flywheel cover.
2. Remove the cam pulley and pulser coil.
3. Disconnect the breather hose from the silencer cover and the fuel hose from the fuel pump.
4. Remove the choke knob, throttle rod pivot and the carburetor assembly.
5. Remove the three 8 x 40 mm flange bolts and the eight 10 x 85 mm flange bolts.
6. Remove the cylinder head assembly and Discard the used head gasket.

To install:

7. Clean all the mating surfaces before reassembling the cylinder head assembly.
8. Install the cylinder head and gasket assembly on the engine.

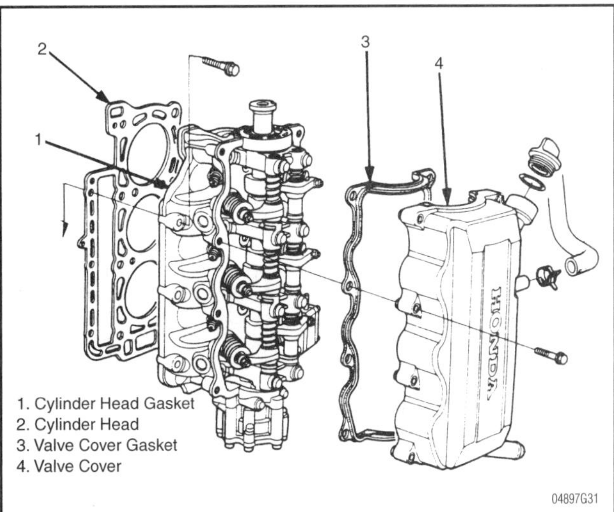

Fig. 39 Exploded view of the cylinder head and head gasket assembly— BF50A, BF45A, BF40A and BF35A

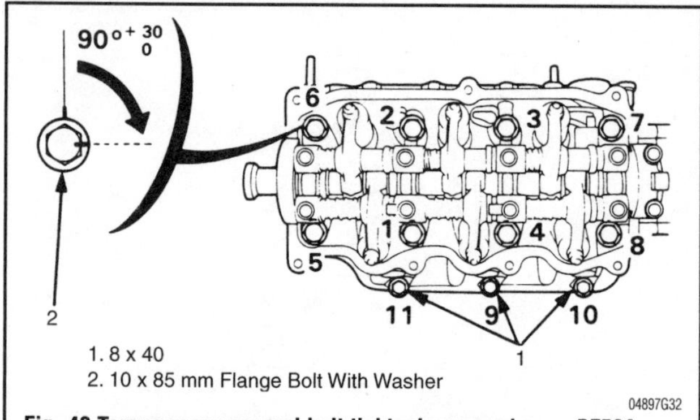

Fig. 40 Torque sequence and bolt tightening procedure— BF50A, BF45A, BF40A and BF35A

9. Tighten the cylinder head to the proper torque specification in the proper sequence.

- 10 x 85 mm flange bolt with washer: 27.5 ft. lbs. (38 Nm) plus an additional 90°.
- 8 x 40 mm flange bolt: 19.5 ft. lbs. (27 Nm).

10. Install the choke knob, throttle rod pivot and the carburetor assembly.
11. Connect the breather hose to the silencer cover and the fuel hose to the fuel pump.
12. Install the cam pulley and pulser coil.
13. Install the engine cover and flywheel cover.

BF25A and BF30A

➧ See Figures 41 and 42

1. Remove the engine cover and the flywheel cover.
2. Remove the spark plug caps from the spark plugs.
3. Remove the camshaft pulley and the pulser coil.
4. Remove the choke knob rod and throttle rod pivot.
5. Remove the carburetor assembly.
6. Disconnect the oil breather and oil return hoses.
7. Remove the valve cover.
8. Remove the eight 8 x 83 mm flange bolts and plain washers.
9. Remove the cylinder head assembly and Discard the head gasket.

To install:

10. Clean all the mating surfaces before assembly.
11. Install the cylinder head assembly.

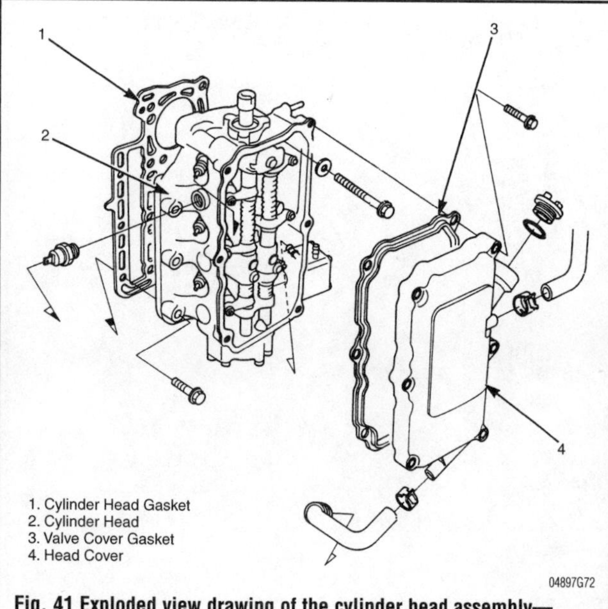

Fig. 41 Exploded view drawing of the cylinder head assembly— BF25A and BF30A

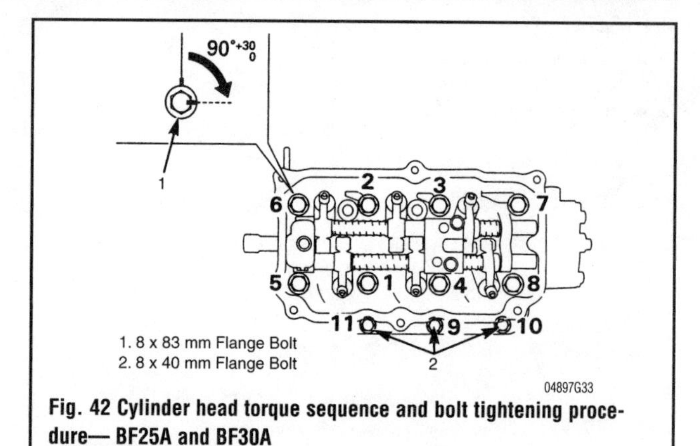

Fig. 42 Cylinder head torque sequence and bolt tightening procedure— BF25A and BF30A

12. Tighten the bolts to the proper torque specification and in the proper sequence.

- 8 x 83 mm flange bolt: 21.0 ft. lbs. (29 Nm) plus an additional 90°
- 8 x 40 mm flange bolt; 19.5 ft. lbs. (27 Nm).

13. Install the cylinder head cover.
14. Reconnect the oil breather and oil return hoses. Make sure to align the oil return hose with the alignment mark on the valve cover.
15. Install the carburetor assembly.
16. Install the choke knob rod and throttle rod pivot.
17. Install the camshaft pulley and the pulser coil.
18. Install the spark plug caps from the spark plugs.
19. Install the engine cover and the flywheel cover.

BF9.9A and BF15A

➧ See Figure 43

1. Remove the engine cover.
2. Remove the fuel pump.
3. Remove the valve cover.
4. Remove the cam pulley and timing belt.
5. Remove the cylinder head assembly and Discard the head gasket.

To install:

6. Clean all mating surfaces before installation.
7. Install the cylinder head assembly and torque the bolts to the proper specification.

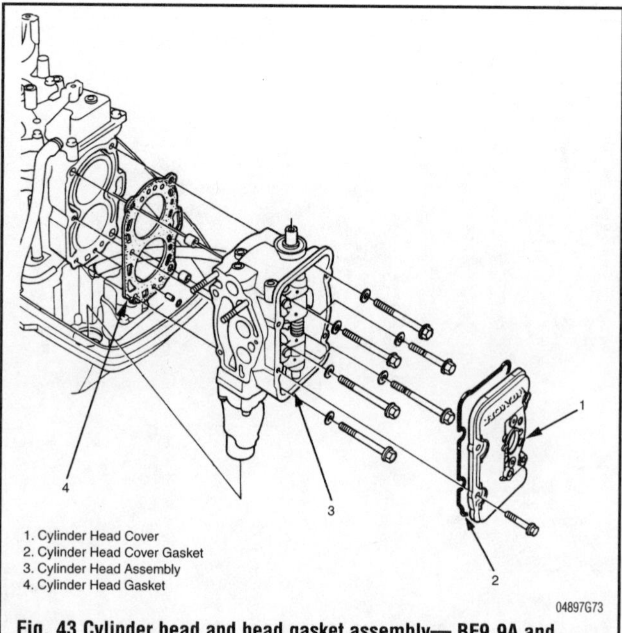

Fig. 43 Cylinder head and head gasket assembly— BF9.9A and BF15A

- Tighten the bolts in a criss-cross fashion from the middle out to the sides.
- Torque the bolts to: 18.8 ft. lbs. (26 Nm).

8. Install the camshaft pulley and timing belt.
9. Install the valve cover.
10. Install the fuel pump.
11. Install the engine cover.

BF75, BF8A, BF100

See Figure 44

1. Remove the engine cover.
2. Remove the valve cover and fuel hoses.
3. Remove the carburetor assembly.
4. Remove the camshaft pulley and timing belt.
5. Remove ignition (CDI or contact breaker) components.
6. Remove the cylinder head with the intake manifold, exhaust pipe, oil pump and valve train installed. Discard the old head gasket.

To install:

7. Throughly clean all mating surfaces before assembly
8. Install the new head gasket. Do not forget to install the O-ring below the gasket.
9. Install the cylinder head on the engine and tighten the bolts to specification.
   - Make sure that the proper length bolts are installed in each hole.
   - Tighten the bolts in a criss-cross fashion from the middle out to the sides to 16–20 ft. lbs. (22–28 Nm).
10. Replace the ignition components.
11. Replace the timing pulley and timing belt.
12. Replace the carburetor assembly.
13. Adjust the valves.
14. Replace the valve cover and fuel hoses.
15. Replace the engine cover.

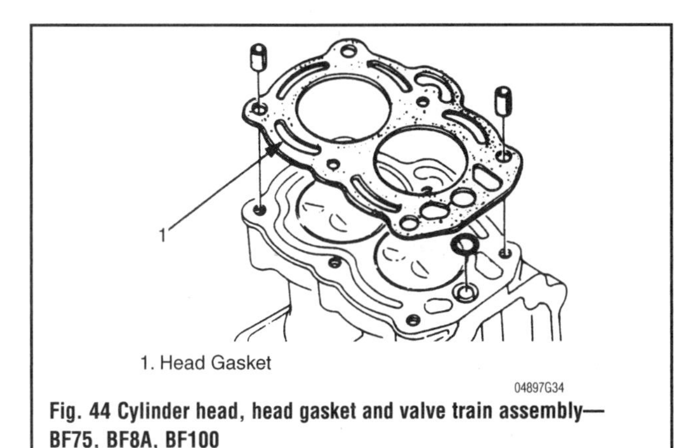

Fig. 44 Cylinder head, head gasket and valve train assembly— BF75, BF8A, BF100

BF50 and BF5A

See Figure 45

1. Remove the engine cover.
2. Remove the valve cover and gasket.
3. Remove the ignition coil and recoil starter.
4. Remove the fuel pump.
5. Remove the flywheel assembly.
6. Remove the carburetor and thermostat.
7. Remove the exhaust manifold.
8. Remove the cylinder head.

To install:

9. After inspecting the gasket for tears or wear, install the gasket and cylinder head on the engine.
10. Torque the bolts to 16–20 ft. lbs. (22–28 Nm).
11. Install the exhaust manifold.
12. Install the carburetor and thermostat.
13. Install the flywheel assembly.

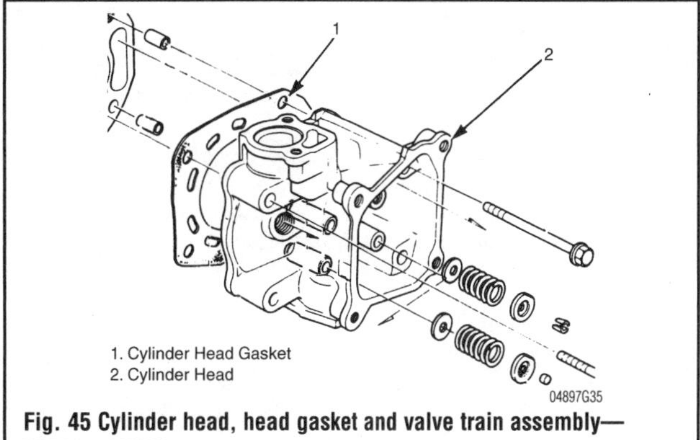

Fig. 45 Cylinder head, head gasket and valve train assembly— BF50 and BF5A

14. Install the fuel pump.
15. Install the recoil starter and ignition coil.
16. Install the valve cover and gasket.
17. Install the engine cover.

BF20 and BF2A

See Figures 46 and 47

1. Remove the engine cover, recoil starter, 2 cap nuts and the fuel tank.
2. Remove the choke rod, throttle cable and air chamber.
3. Remove the spark plug

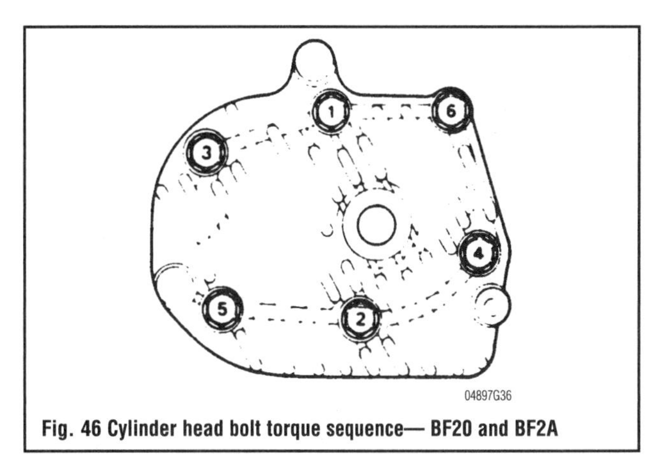

Fig. 46 Cylinder head bolt torque sequence— BF20 and BF2A

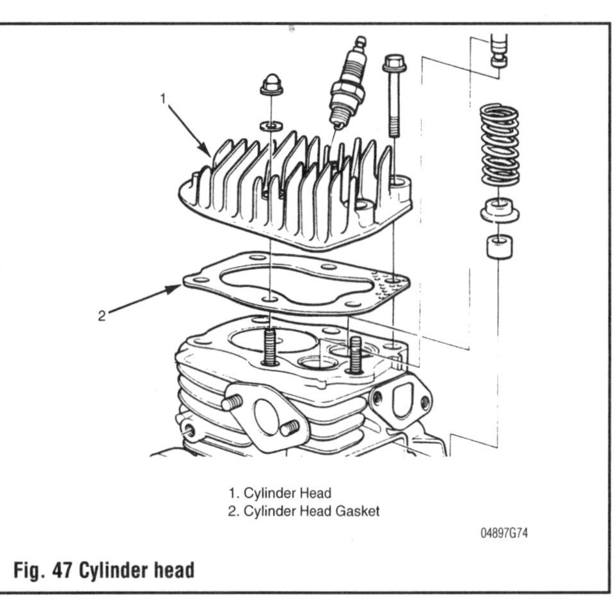

Fig. 47 Cylinder head

4. Disconnect the engine switch wires, spark plug cap and the 6 x 20 mm bolt.
5. Remove the four self-tapping screws, two bolts and the fan cover.
6. Remove the four cylinder head bolts, two cap nuts and lift off the cylinder head.
7. Remove accumulated carbon deposits from inside the combustion chamber before installing the cylinder head.
8. Inspect the cylinder head for warpage with a flat edge and feeler gauge, and the gasket for wear and tears. Replace if necessary.

To install:

9. Clean all mating surfaces thoroughly before assembly
10. Install the cylinder head and gasket on the engine.
11. Torque the bolts in the pattern shown to 5.8–8.7 ft. lbs. (80–120 kg-cm)
12. Install the fan cover, bolts and self-tapping screws.
13. Connect the engine switch wires, spark plug cap and install the 6 x 20 mm bolt.
14. Install the spark plug.
15. Connect the choke rod, throttle cable and the air chamber.
16. Install the fuel tank, two cap nuts, recoil starter and the engine cover.

Oil Pan

REMOVAL & INSTALLATION

BF25A to BF130A

See Figures 48 thru 57

1. Remove the lower unit.
2. Remove the extension case and the undercover.
3. Drain the engine oil.
4. Unbolt the oil pan from the engine crankcase. Discard the oil pan gasket.
5. Remove the exhaust pipe and the water tube from the oil pan assembly. Discard the exhaust pipe gasket.

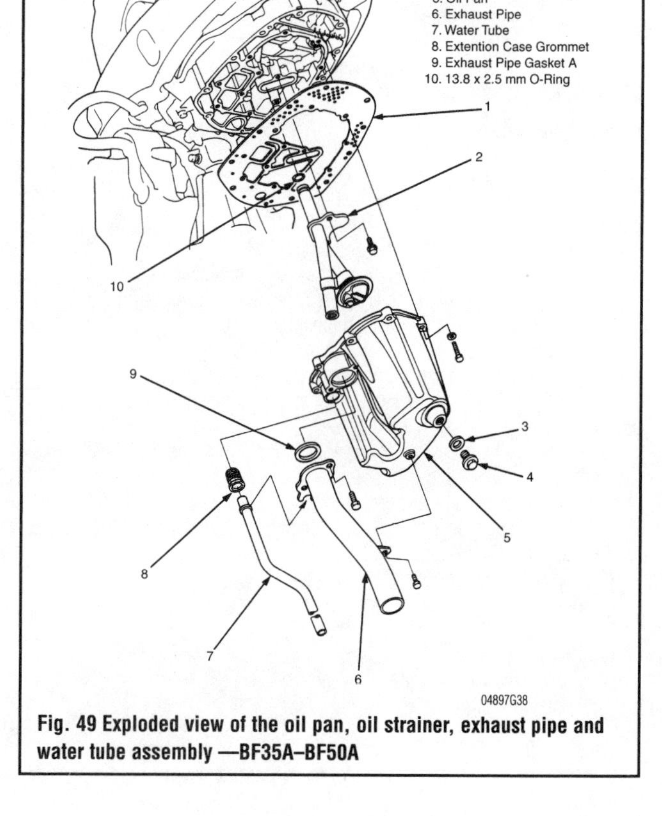

Fig. 49 Exploded view of the oil pan, oil strainer, exhaust pipe and water tube assembly —BF35A–BF50A

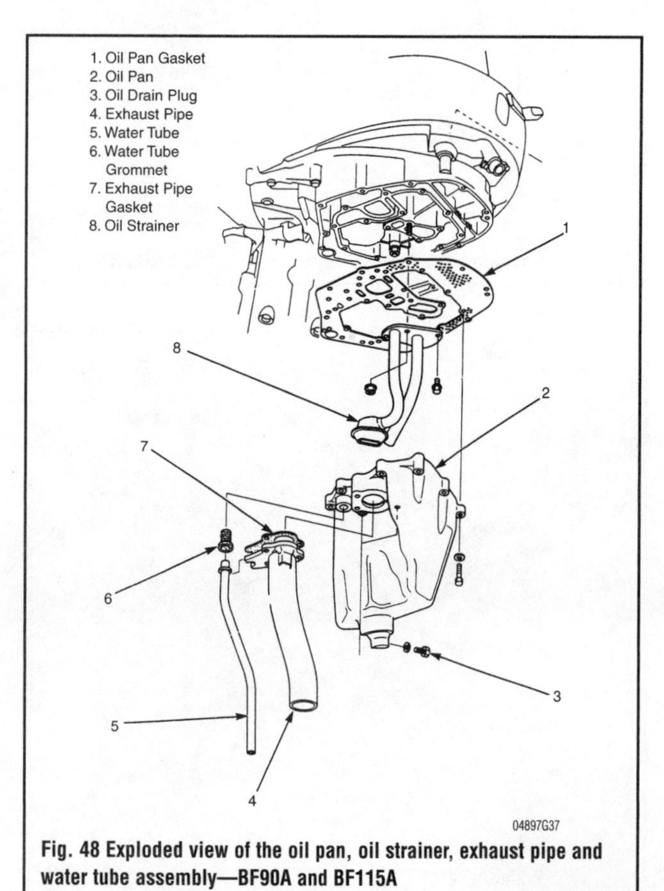

Fig. 48 Exploded view of the oil pan, oil strainer, exhaust pipe and water tube assembly—BF90A and BF115A

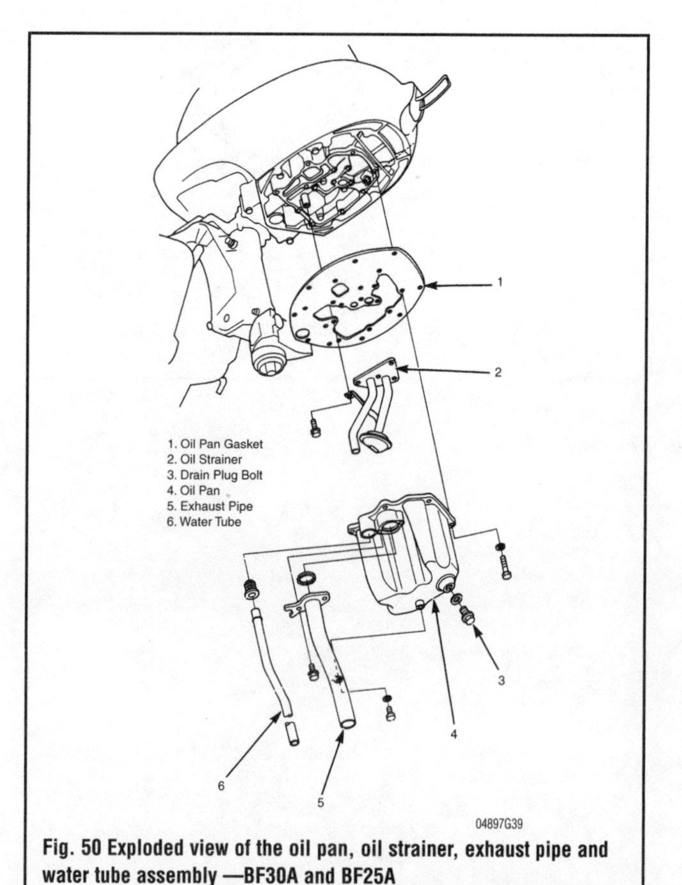

Fig. 50 Exploded view of the oil pan, oil strainer, exhaust pipe and water tube assembly —BF30A and BF25A

Fig. 51 Remove the lower unit, extension case and the undercover . . .

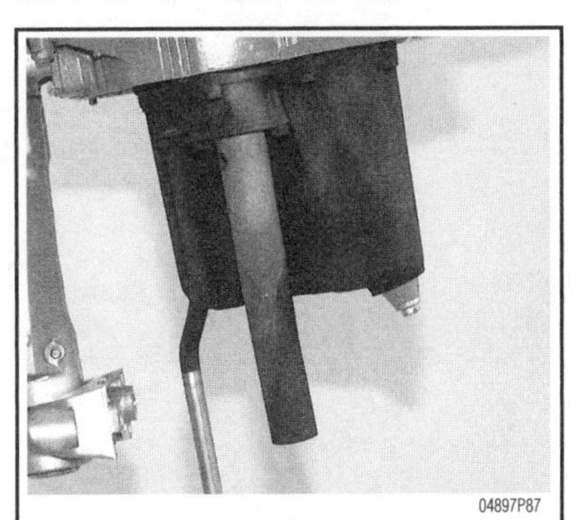

Fig. 52 . . . this will expose the oil pan, exhaust pipe and water tube . . .

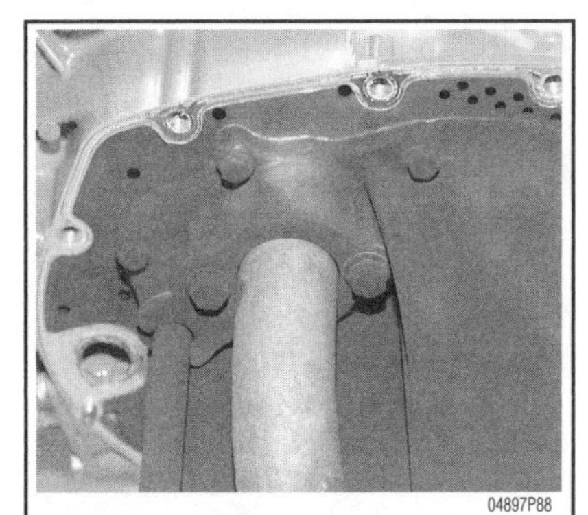

Fig. 53 . . . now, unbolt the oil pan from the engine crankcase . . .

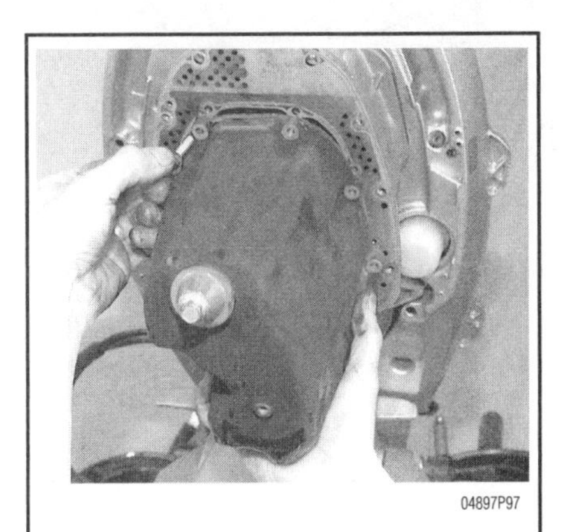

Fig. 54 . . . and remove the oil pan . . .

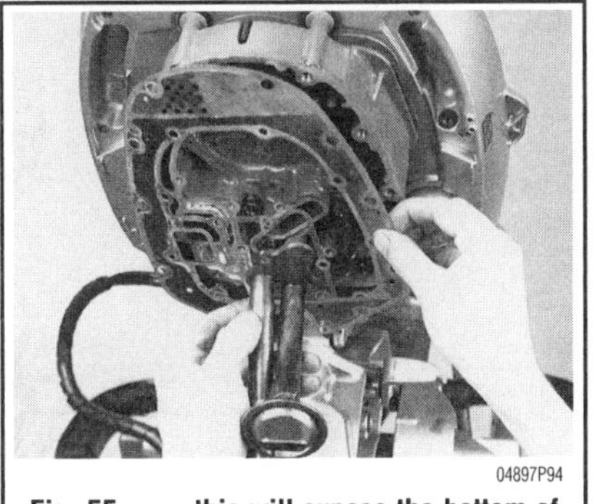

Fig. 55 . . . this will expose the bottom of the engine crankcase, the oil pan gasket, the oil strainer and oil return line . . .

Fig. 56 Install the exhaust seal . . .

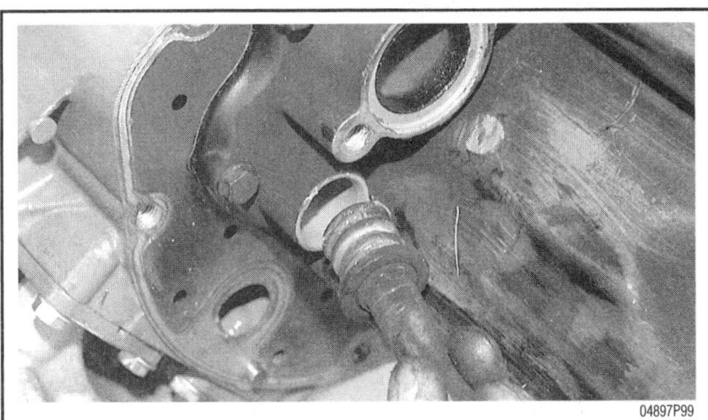

Fig. 57 . . . apply grease to the water tube grommet and install it in the oil pan

To install:

6. Now is a good time to clean the oil strainer in solvent and inspect the screen for damage. Replace the O-ring on the oil drain tube before assembly.
7. Clean the oil pan and crankcase mating surfaces prior to assembly.
8. Install the exhaust pipe and water tube onto the oil pan. Apply grease to the entire circumference of the water tube grommet before installation.
9. Install the new gasket and bolt the oil pan onto the crankcase.
10. Install the lower unit, extension case and undercover.

BF9.9A and BF15A

See Figure 58

1. Drain the engine oil.
2. Remove the engine. The oil pick-up tube and strainer are mounted on the engine.

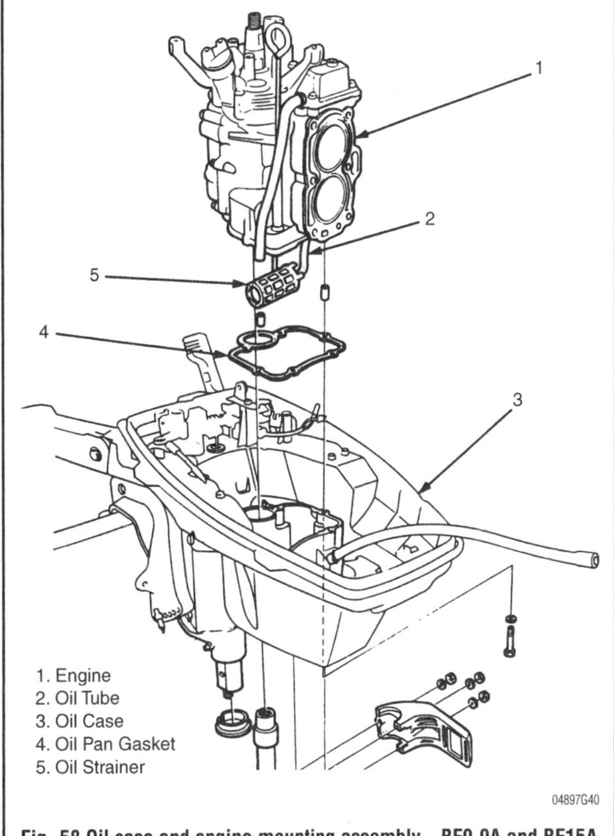

Fig. 58 Oil case and engine mounting assembly—BF9.9A and BF15A

To install:

3. Replace the gasket and clean both mating surfaces.
4. Install the engine back onto the oil case.

BF75, BF8A and BF100

➧ See Figures 59 and 60

➡The oil pan can be removed from the swivel case without removing the engine. Remove the two nuts fastening the lower mount housing and oil case, and the thrust mount rubber.

If the engine is being removed, follow the following procedure:

1. Remove the engine and Discard the gasket.

To install:

2. Check the oil filter for damage. Install it in the correct direction.
3. Install a new gasket and install the engine.

➡Make sure the engine mounting bolts are installed their correct positions.

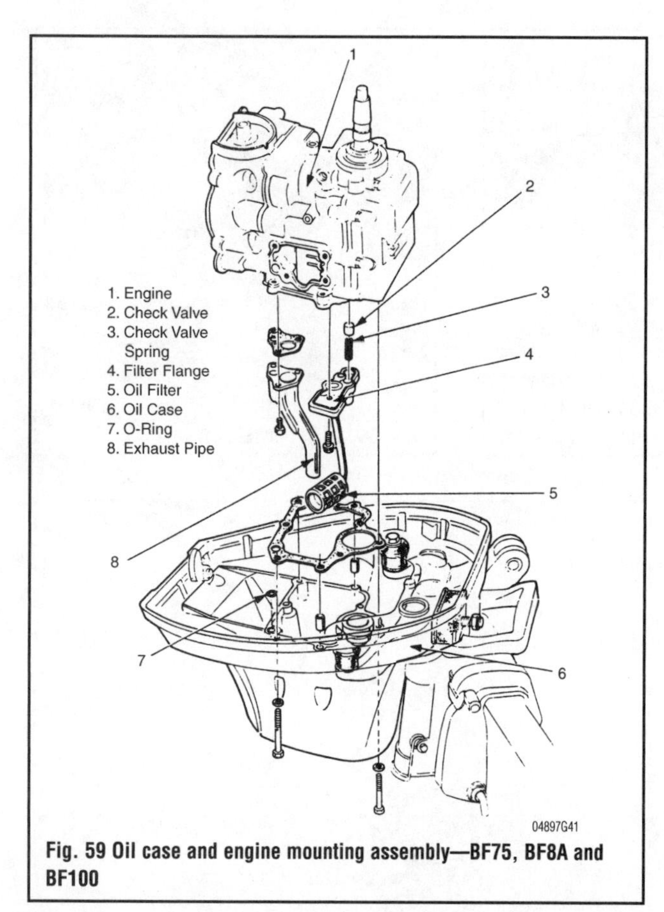

Fig. 59 Oil case and engine mounting assembly—BF75, BF8A and BF100

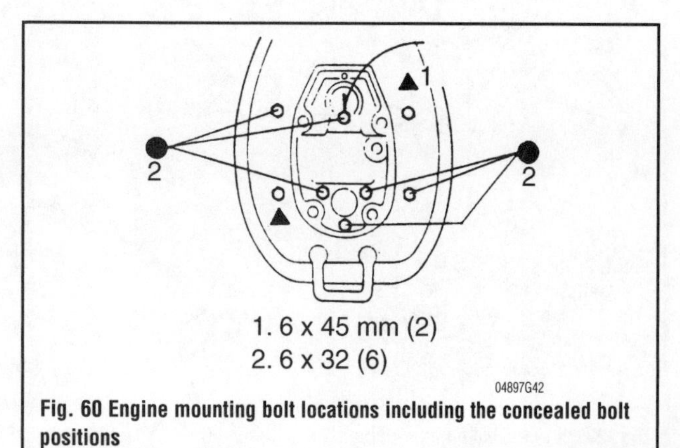

Fig. 60 Engine mounting bolt locations including the concealed bolt positions

BF50 and BF5A

➧ See Figure 61

1. After removing the engine, unbolt the oil pan from the bottom of the engine and Discard the gasket.

To install:

2. Clean all the mating surfaces before assembly.
3. After installing the oil strainer, pour engine oil into the suction hole at the oil pump.
4. Replace the gasket and install the oil pan on the crankcase.
5. Install the engine.

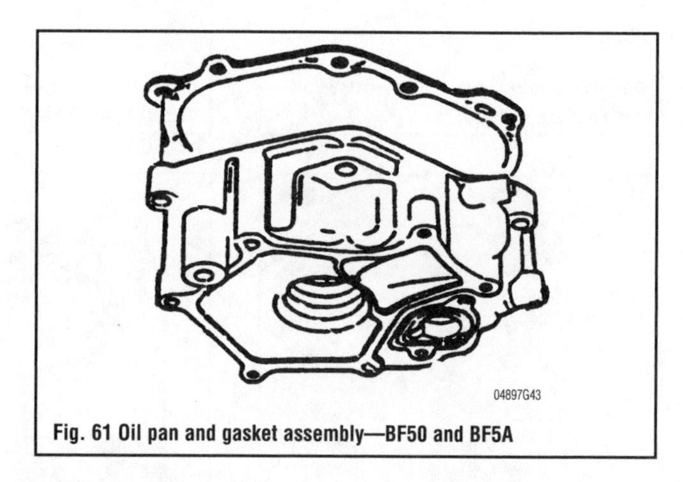

Fig. 61 Oil pan and gasket assembly—BF50 and BF5A

BF20 and BF2A

➧ See Figure 62

1. Remove the engine.
2. Remove the oil pan and Discard the oil pan gasket
3. If needed, replace the oil and water seals.

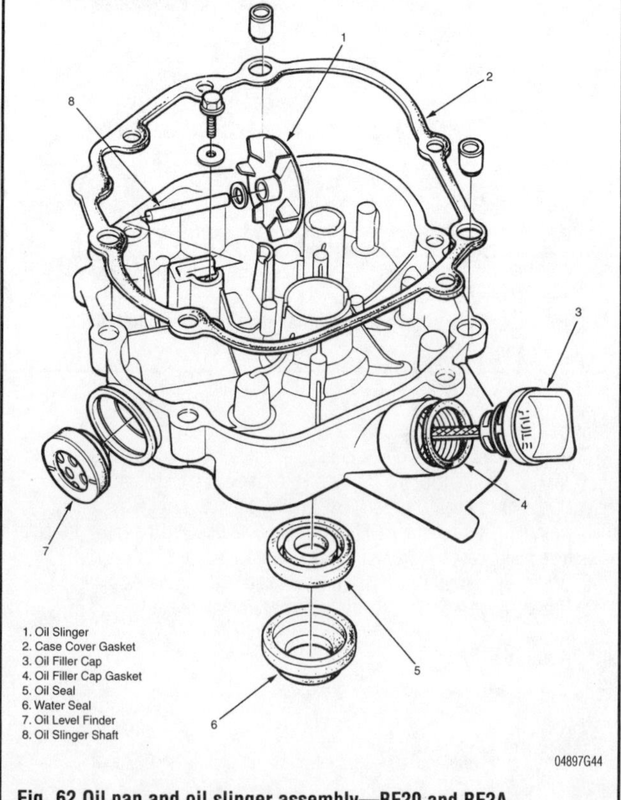

Fig. 62 Oil pan and oil slinger assembly—BF20 and BF2A

To install:

4. Throughly clean both mating surfaces before installing the new gasket.
5. Bolt together the crankcase and the oil pan. Tighten the bolts to 5.8–8.7 ft. lbs. (80–120 kg-cm)

Timing Belt Cover

REMOVAL & INSTALLATION

See Figures 63, 64 and 65

The timing belt covers are secured to the engines using a variety of fasteners; some use bolts and washers, some flange bolts, some use hooks and others use sliding locks. What ever the fastening system used, for the most part removing the timing belt cover is a very easy and straightforward procedure.

After removing the engine cover, take a look at the fasteners and make sure that you only remove the ones needed to remove the timing belt cover and not some other component.

After removing the cover, inspect it for cracks and general wear and tear. Make sure the tabs are not broken and that it will be secure when reinstalled and not flap around and rub up against a timing belt or pulley. Turn the cover upside down and look at the underside, and clean it if necessary. Remember, you want to keep the timing belt underneath clean to avoid belt problems which can lead to very serious, very expensive engine problems.

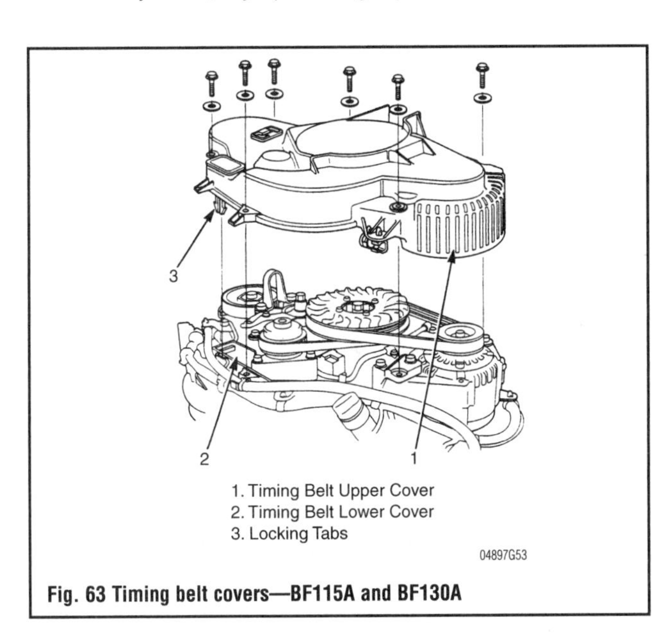

Fig. 63 Timing belt covers—BF115A and BF130A

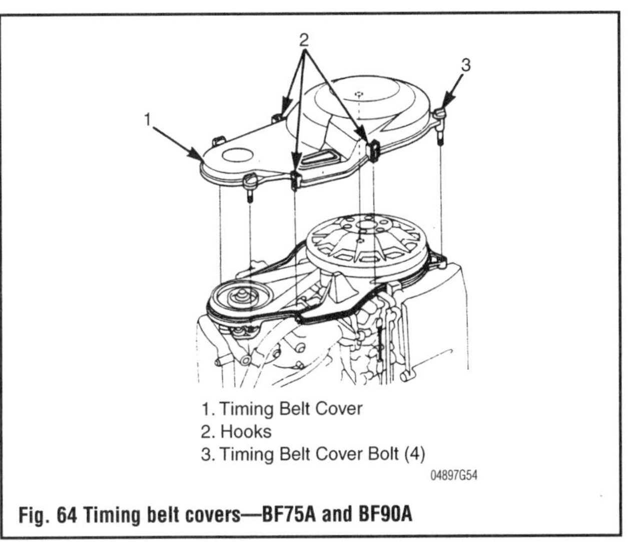

Fig. 64 Timing belt covers—BF75A and BF90A

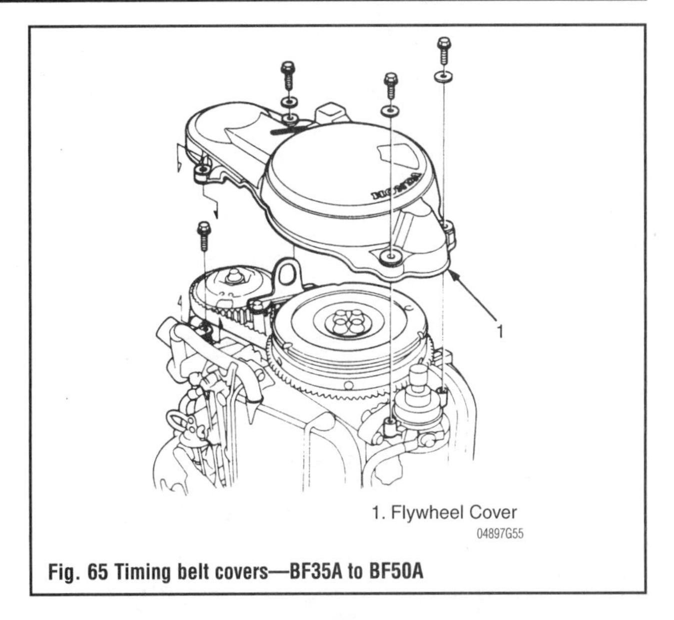

Fig. 65 Timing belt covers—BF35A to BF50A

Timing Belt

The BF8A and up models utilize a timing belt to drive the camshaft from the crankshaft's turning motion and to maintain proper valve timing. Some manufacturers schedule periodic timing belt replacement to assure optimum engine performance, to make sure the operator is never stranded in the water should the belt break (as the engine will stop instantly) and for some (manufacturers with interference engines) to prevent the possibility of severe internal engine damage should the belt break.

The BF8A and up models are listed as interference engines (meaning that valves will contact the pistons if the camshaft was rotated separately from the crankshaft). You will have to decide for yourself if the peace of mind offered by a new belt is worth it on high hour engines.

INSPECTION

See Figure 66

Inspection of the timing belt is (in most cases) a simple matter of removing the belt cover (usually held on by a few bolts or screws) and inspecting the belt. Keeping a close eye on the condition of your engine's timing belt can help avoid an enjoyable outing turning into a miserable one.

A severely worn or damaged belt may not give any warning that it is about to fail. In general, any time the engine timing cover(s) is (are) removed you should inspect the belt for premature parting, severe cracks or missing teeth.

Fig. 66 Inspect the full length of the timing belt for wear and damage. Replace the belt if necessary

REMOVAL & INSTALLATION

BF115A and BF130A

See Figures 67, 68 and 69

1. Remove the balancer belt drive pulley.
2. Remove the timing belt from the timing belt driven pulley, then remove the belt from the timing belt drive pulley.
   - Do not pry the belt off.
   - Do not let the belt get contaminated with dirt, grease, etc.

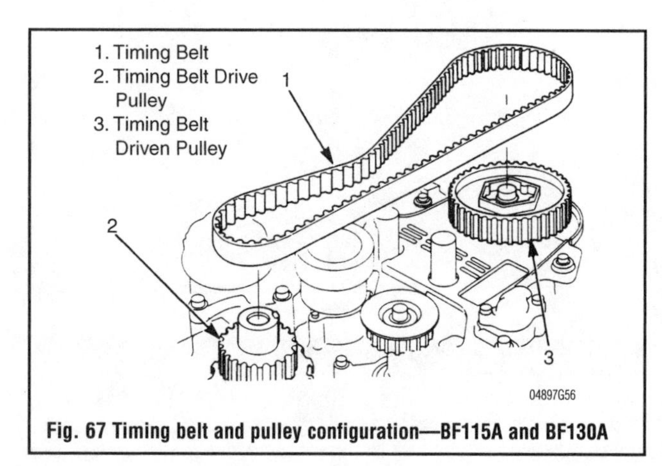

Fig. 67 Timing belt and pulley configuration—BF115A and BF130A

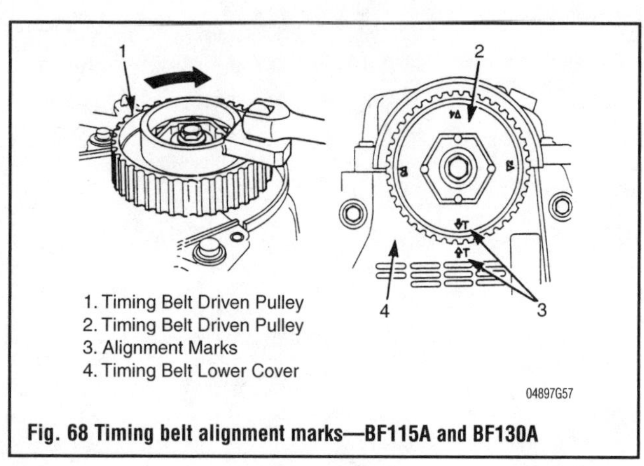

Fig. 68 Timing belt alignment marks—BF115A and BF130A

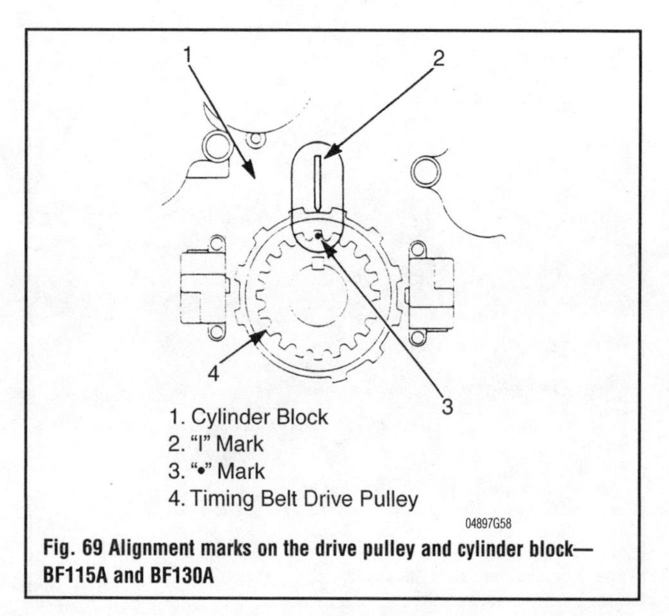

Fig. 69 Alignment marks on the drive pulley and cylinder block—BF115A and BF130A

   - Do not bend the timing belt. Always store the belt by hanging it up on a wall.

To install:

3. Remove the spark plugs and plug wire caps.
4. Set the special tool (lock nut wrench, 56 mm: 07LPA-ZV30200) on the timing belt driven pulley as shown. Align the timing mark on the driven pulley with the mark on the timing belt lower cover by turning the driven pulley.
5. Align the mark on the timing belt drive pulley with the mark on the cylinder block.

Be sure that the drive pulley and the timing belt guide are installed properly when the alignment marks are aligned. If there is a gap between the pulley and the plate, they are not in correct alignment.

6. Install the timing belt on the drive and driven pulleys in that order. Be careful to not allow the timing marks to move out of alignment.
7. After the belt installation, double check the alignment marks for proper alignment.

BF75A and BF90A

1. Remove the engine cover.
2. Make sure the gear case is in neutral.
3. Loosen the timing cover bolts and remove the timing belt cover.
4. Remove the starter motor assembly and install the special tool (ring gear holder P/N 07SPB-ZW10100) to hold the flywheel in place.
5. Remove the bolt from the alternator rotor and remove the starter pulley/alternator rotor from the engine.
6. Remove the bolts and remove the alternator stator assembly.
7. Remove the dowel pins, key, timing belt guide plate and the timing belt tensioner cap.
8. Remove the special tool from the starter hole.
9. Loosen the flange bolts on the timing belt tensioner. Push the tensioner to the port side of the engine and tighten the flange bolt to hold it out of the way.
10. Remove the timing belt from the timing belt camshaft pulley first, then off the crankshaft pulley.

Do not pry the belt off with a screwdriver or any other type of prying tool. Use only your hands to remove the belt.

BF50A, BF45A, BF40A and BF35A

See Figures 70, 71, 72 and 73

1. Remove the engine cover, timing belt cover and spark plugs.

CAUTION

The starter stopper collar is close to the flywheel and the starter ring gear can become sharp. Avoid hand contact with the stopper collar

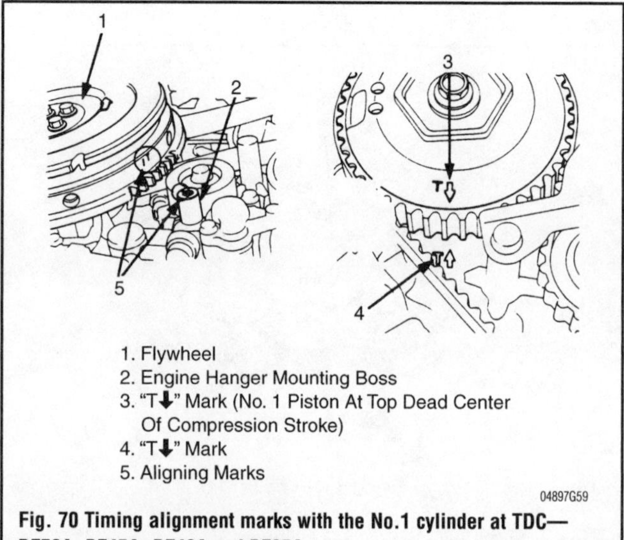

Fig. 70 Timing alignment marks with the No.1 cylinder at TDC—BF50A, BF45A, BF40A and BF35A

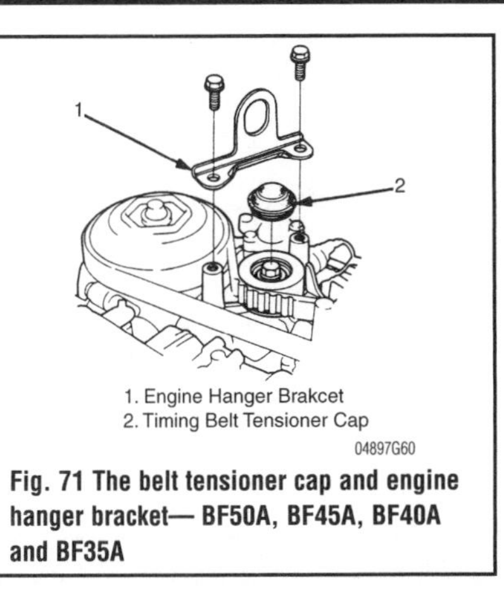

Fig. 71 The belt tensioner cap and engine hanger bracket— BF50A, BF45A, BF40A and BF35A

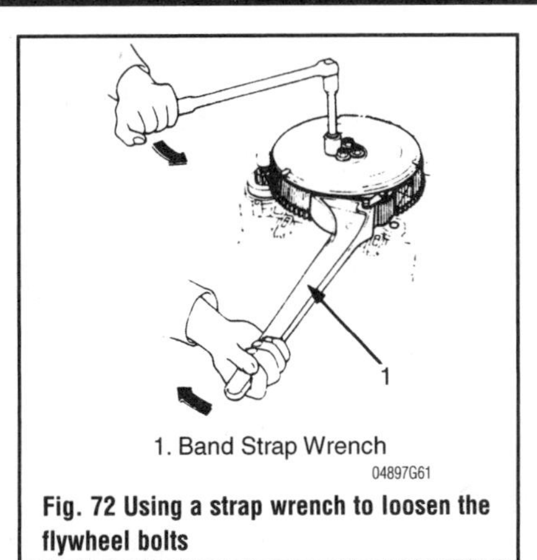

Fig. 72 Using a strap wrench to loosen the flywheel bolts

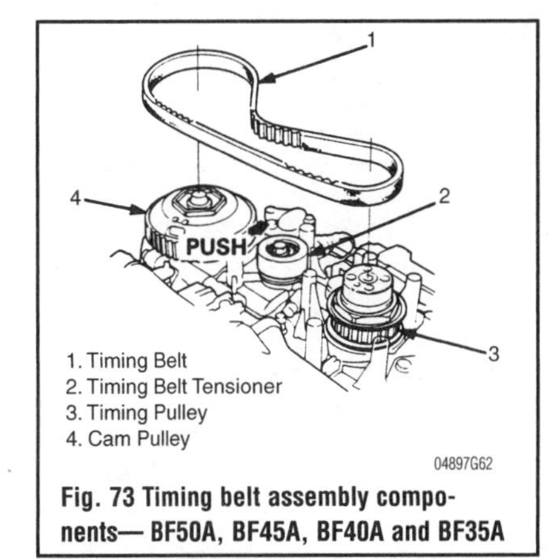

Fig. 73 Timing belt assembly components— BF50A, BF45A, BF40A and BF35A

or flywheel ring gear. Hold the upper portion of the flywheel when rotating the engine. To avoid damaging the water pump impeller, always turn the crankshaft in a counterclockwise direction only.

2. Remove the two flange bolts and the engine hanger bracket.
3. Rotate the engine counterclockwise, using the flywheel, until the No. 1 cylinder timing marks align.
4. Remove the timing belt tensioner cap.
5. Attach a strap wrench to the flywheel and remove the four flange bolts from the flywheel.
6. Remove the flywheel.
7. Disconnect and remove the exciter and charge coils.
8. Loosen the flange bolt on the timing belt tensioner. Then while pushing the belt tensioner in the direction of the arrow, tighten the flange bolt.

➡Do not use excessive force or pry on the timing belt while removing it.

9. Lift the timing belt off the belt tensioner first, then from the camshaft pulley and finally from the timing pulley.

To install:

⁂ CAUTION

The valves may become bent if the cam pulley or crankshaft are rotated with the timing belt removed or improperly installed. If the timing marks have become substantially misaligned, it may be necessary to loosen the valve adjusting screw lock nuts and back off the valve adjusting screws before aligning the No.1 cylinder timing marks.

10. Check and align the cam pulley timing mark with the timing mark on the cylinder head.
11. Install the flywheel on the crankshaft by aligning the dowel pin.
12. Install and hand tighten the flywheel flange bolts.
13. Check and align the flywheel timing mark with the timing mark on the top of the engine hanger mounting boss.
14. Remove the flywheel carefully. Do not change the timing mark alignment.
15. Install the timing belt on the timing pulley first, then over the camshaft pulley, then over the tensioner pulley.

BF25A and BF30A

➧ See Figure 74

1. Remove the spark plugs.
2. Align the timing marks so that the No.1 piston is a t TDC of the compression stroke.
3. Remove the flywheel, charge coil and the exciter coil.
4. Disconnect he pulser coil wire connector.

➡Do not use excessive force or pry the belt off the pulley.

To install:

⁂ CAUTION

The valves may become bent if the cam pulley or crankshaft are rotated with the timing belt removed or improperly installed. If the timing marks have become substantially misaligned, it may be necessary to loosen the valve adjusting screw lock nuts and back off the valve adjusting screws before aligning the No. 1 cylinder timing marks.

5. Install the camshaft pulley.
6. Align the camshaft pulley timing marks as shown in the drawing
7. Install the timing belt guide plate, Woodruff key and the timing pulley onto the crankshaft.
8. Align the timing mark on the timing pulley with the mark on the cylinder block.
9. Iinstall the timing belt.

⁂ WARNING

When installing the belt, do not move the timing marks out of alignment.

10. After installing the belt, recheck each mark for proper positioning.

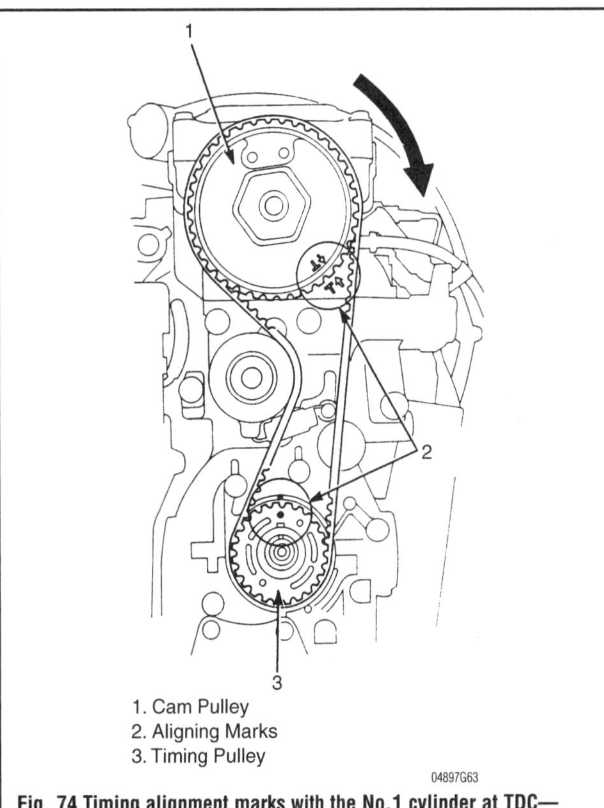

Fig. 74 Timing alignment marks with the No.1 cylinder at TDC— BF25A and BF30A

BF9.9A and BF15A

See Figure 75

1. Remove the engine cover.
2. Remove the recoil starter.
3. Remove the flywheel.
4. Disconnect and remove the charging and exciter coils.
5. Lift the belt off the camshaft pulley by hand. Do not use excessive for or pry the belt off the pulley.

On some occasions it may be necessary to remove the timing pulley flange in order to remove the timing belt. This requires removing the large nut at the base of the crankshaft with a special tool (crankshaft holder: No. 07923-ZA00000) and a commercially available 1⅜in. (34 mm) crowfoot wrench. Perform the following procedure to remove the timing pulley flange:

- Flatten the sides of the lock washer that secures the lock nut.
- Position the special tool on the crankshaft by aligning the keyway groove in the tool with the Woodruff key.
- Install the flyway nut on the crankshaft and tighten it hand-tight to prevent the special tool from slipping.
- Hold the special tool with a 30 mm box-end wrench and loosen the lock nut with the 1⅜in. (34 mm) crowfoot wrench.
- Tighten the lock nut to 18 ft. lbs. (25 Nm) on reassembly and bend up the sides of the washer.

To install:

6. Install the crankshaft holding tool on the crankshaft.
7. Make a reference mark on the special tool to show the keyway location.
8. Put the flywheel nut on the crankshaft and hand-tighten it to prevent the tool from slipping.
9. Turn the camshaft pulley by hand until the timing mark on the cam pulley aligns with the timing mark on the breather cover.
10. Rotate the crankshaft until the reference mark on the special tool aligns with the timing mark on the starter cover and install the belt.

BF75, BF8A and BF100

1. Remove the engine cover.
2. Remove the recoil starter and starter pulley.
3. Remove the flywheel
4. Remove the timing belt from the cam pulley by hand. Do not pry the belt off the pulley.

To install:

5. Align the timing mark on the flywheel and the punch mark on the pulley to their respective aligning marks.
6. Install the belt onto the pulleys.

Always install the belt so that the word Honda can be read on the outside of the belt.

ADJUSTMENT

BF115A and BF130A

See Figures 76, 77, 78 and 79

Make sure the valve clearance is correct before adjusting the belt tension. The balancer belt tension and the timing belt timing should be adjusted simultaneously.

1. Turning the crankshaft pulley clockwise, align the "T" mark on the camshaft pulley with the "T" mark on the timing belt lower cover. Also make sure that the cutout in the crankshaft pulley aligns with the "T" mark on the timing belt lower cover.
2. Loosen the adjusting nut and the adjusting bracket bolt ⅔ to one turn and be sure that the balancer belt tensioner moves freely to the opposite side from the belt (i.e. belt loosening direction).

Do not loosen the adjusting nut and adjusting bracket more than one turn.

3. Make sure that each alignment mark is in proper alignment, then turn the crankshaft pulley 2 turns **CLOCKWISE**. Then slowly turn it a further three

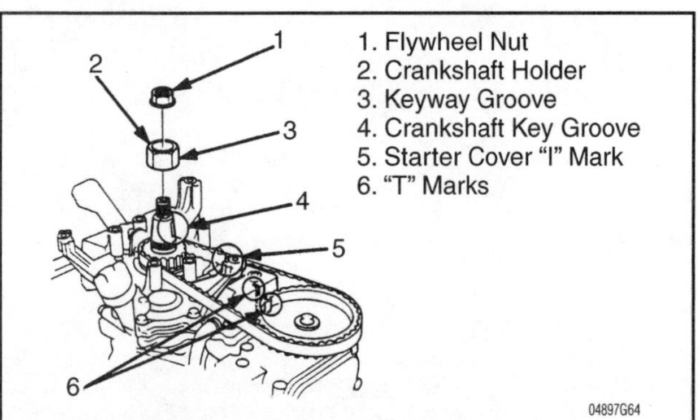

Fig. 75 Timing alignment marks with the No.1 cylinder at TDC—BF9.9A and BF15A

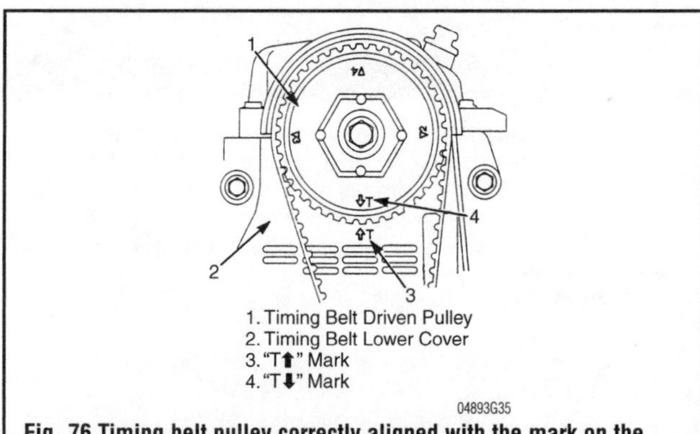

Fig. 76 Timing belt pulley correctly aligned with the mark on the timing belt lower cover—BF115A and BF130A

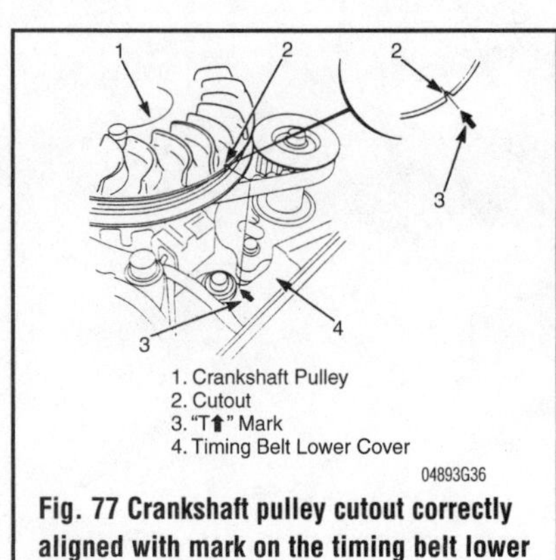

Fig. 77 Crankshaft pulley cutout correctly aligned with mark on the timing belt lower cover—BF115A and BF130A

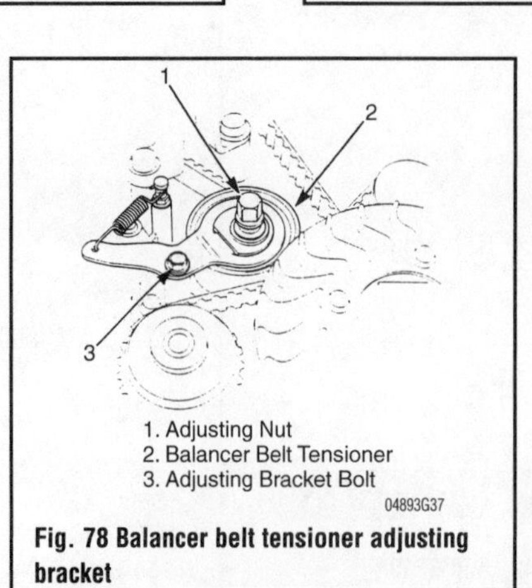

Fig. 78 Balancer belt tensioner adjusting bracket

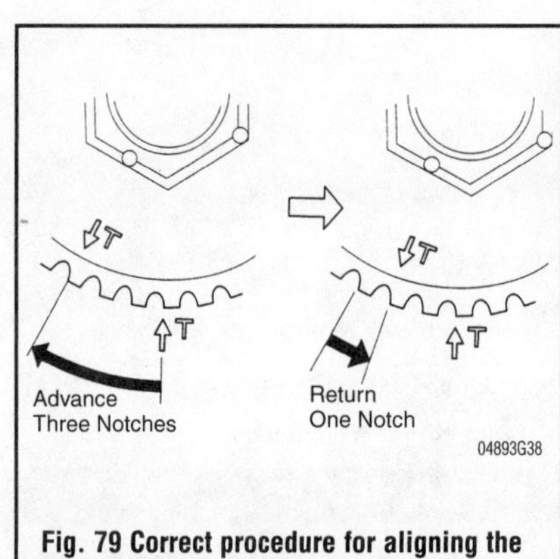

Fig. 79 Correct procedure for aligning the timing marks—BF115A and BF130A

notches of the timing belt driven gear. Now, turn the crankshaft pulley **COUNTERCLOCKWISE** by 1 notch of the timing belt driven pulley gear.

➡If the crankshaft pulley was turned more than three notches of the timing belt driven pulley gear, turn the crankshaft pulley clockwise to bring No.1 piston to it compression stroke, then repeat the belt tension adjustment procedure.

4. Tighten the adjusting nut to 33 ft. lbs. (44Nm) and the adjusting bracket bolt to 9 ft. lbs. (12Nm).

➡Each belt tensioner is of the automatically adjusting type. Do not push the tensioners in the direction the timing belt with your hand.

As the valve tip and the valve actuating components wear, excessive clearance develops between these components leading to reduced engine performance and valve noise, commonly called "valve tap". Valve actuation on these engines is performed by the rocker arms acting directly on the valve stem.

Adjustment is done by rotating the engine until Top Dead Center of the compression stroke. This is when both the intake and exhaust valves closed and a clearance reading can be taken with a feeler gauge. Although the basic procedure for all Honda four-stroke outboard engines are similar, follow the procedure below for your specific application.

BF75A and BF90A

See Figures 80, 81, 82 and 83

➡Make sure that all intake and exhaust valve clearances are within specification before adjusting timing belt tension.

1. Remove the timing belt cover.
2. Remove the spark plugs. This will make it easier to rotate the engine.
3. Turn the starter pulley clockwise to align the timing marks on the camshaft pulley and the timing belt lower cover. Also align the timing mark on the starter pulley with the mark on the timing belt lower cover. The number one cylinder should be at Top Dead Center (TDC).
4. Remove the engine hanger bracket and timing belt tensioner cap.

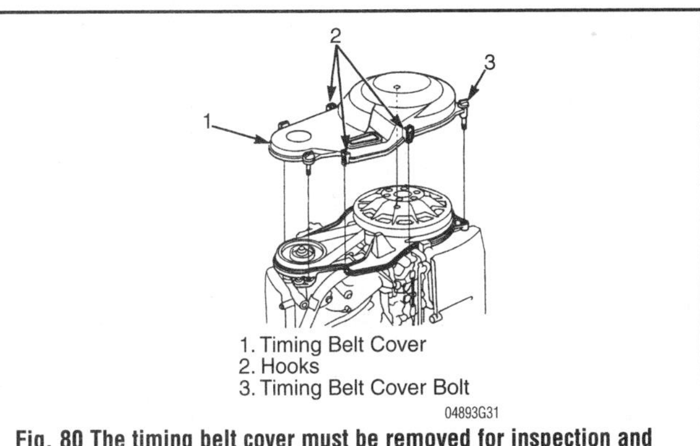

Fig. 80 The timing belt cover must be removed for inspection and adjustment—BF75A and BF90A

⁂ WARNING

Do not turn the engine counterclockwise; the water pump impeller can be damaged.

5. Starting from the number one cylinder timing mark, rotate the flywheel **clockwise** at least two full turns.
6. After realigning the timing marks for the number one cylinder, slowly turn the starter pulley clockwise until the "T" on the camshaft pulley is at least 3 ½ teeth past the "T" mark on the cylinder head.
7. Loosen the timing belt tensioner bolt.
8. Tighten the timing belt tensioner flange bolt to the specified torque. Do not apply additional pressure on the tensioner pulley; the spring on the pulley bracket spring provides adequate tension for the belt.
9. Install the engine hanger bracket and timing belt tensioner cap.
10. Install the spark plugs.
11. Install the timing belt cover.

BF50A, BF45A, BF40A and BF35A

See Figures 84 thru 91

1. Remove the engine cover.
2. Remove the spark plugs. This will make it easier to rotate the engine.
3. Remove the timing belt cover.

➡BF35A–BF50A operate in a counterclockwise direction, which is the opposite of most engines.

4. Turn the flywheel to align the timing marks. The number one cylinder should be at Top Dead Center (TDC).

⁂ CAUTION

The gear teeth on the flywheel can become sharp from constant engagement of the starter. Use caution when turning the flywheel by hand.

5. Remove the engine hanger bracket and timing belt tensioner cap.
6. Loosen the timing belt tensioner bolt.

⁂ WARNING

Do not turn the engine clockwise; the water pump impeller can be damaged.

7. Starting from the number one cylinder timing mark, rotate the flywheel **counterclockwise** at least two full turns.
8. After realigning the timing marks for the number one cylinder, slowly turn the flywheel counterclockwise until the timing mark "T" is three teeth past the "T" mark on the cylinder head.
9. Tighten the timing belt tensioner flange bolt to the specified torque. Do not apply additional pressure on the tensioner pulley; the spring on the pulley bracket spring provides adequate tension for the belt.
10. Install the engine hanger bracket and timing belt tensioner cap.

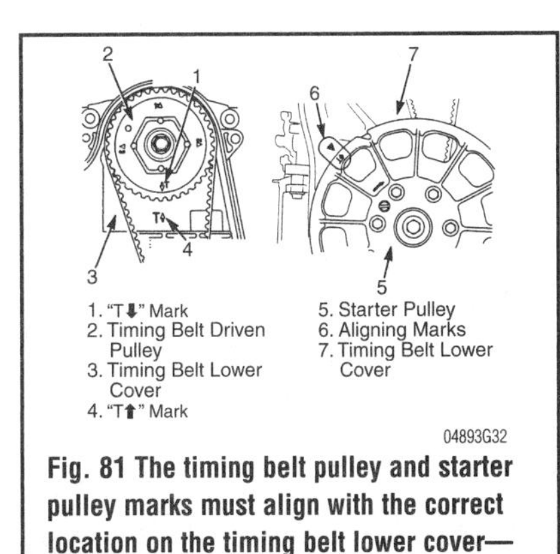

Fig. 81 The timing belt pulley and starter pulley marks must align with the correct location on the timing belt lower cover—BF75A and BF90A

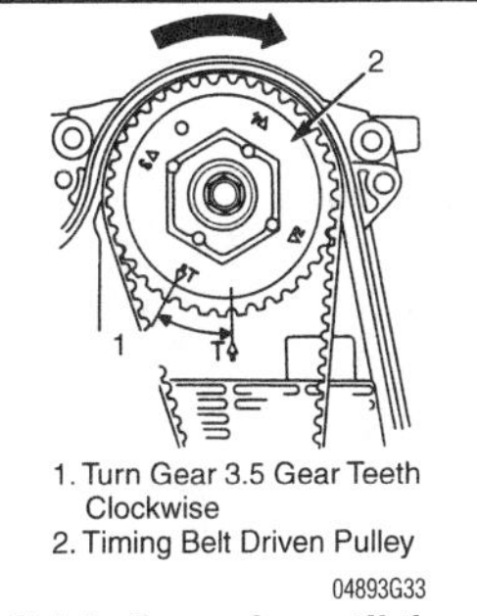

Fig. 82 Rotate the engine until the mark on the camshaft pulley is 3½ teeth past the mark on the cylinder head—BF75A and BF90A

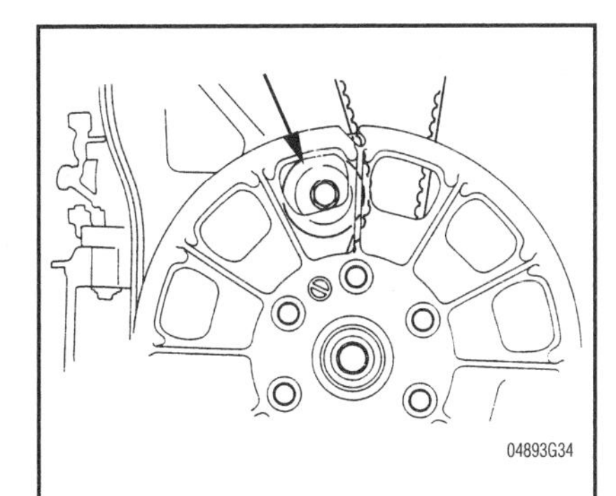

Fig. 83 The tensioner pulley bolt is accessed through an opening in the starter pulley—BF75A and BF90A

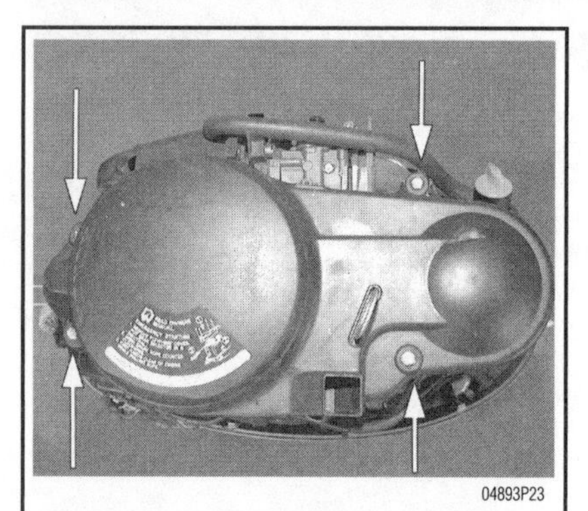

Fig. 84 The timing belt cover is held in place by four bolts—BF35A–BF50A

Fig. 85 Loosen the tensioner pulley bolt to allow the spring on the pulley to provide tension on the belt—BF35A–BF50A

Fig. 86 Using the flywheel, turn the engine by hand (note the counterclockwise directional arrow on the flywheel)—BF35A–BF50A

Fig. 87 Flywheel timing mark—BF35A–BF50A

Fig. 88 With the flywheel timing marks aligned, the timing marks on the camshaft pulley should be aligned —BF35A–BF50A

Fig. 89 Turn the flywheel to align the timing marks as shown; the tensioner pulley bolt can now be tightened

Fig. 90 Notice that the tensioner hole is slotted. This is to take up any slack in the timing belt. Do not apply more than the specified torque to the bolt or tensioner may bind

Fig. 91 Use a torque wrench on the tensioner pulley bolt to ensure it is properly tightened

1. Cam Pulley
2. "T↑" Mark
3. Cylinder Head
4. "T↑" Mark
5. Flywheel
6. Arrow
7. Aligning Marks
8. Flywheel Cover or Recoil Starter Mounting Boss

Fig. 92 The aligning marks on the flywheel for TDC—BF25A and BF30A

11. Install the spark plugs.
12. Install the timing belt cover.
13. Install the engine cover.

BF25A and BF30A

See Figures 92, 93 and 94

1. Remove the spark plugs. This will make it easier to rotate the engine.
2. Turn the flywheel to align the timing marks. The number one cylinder should be at Top Dead Center (TDC).
3. Remove the engine hanger bracket and timing belt tensioner cap.
4. Loosen the timing belt tensioner flange bolt.
5. Starting from the number one cylinder timing mark, rotate the flywheel counterclockwise by at least 5 gear teeth of the camshaft pulley, then turn it clockwise until the camshaft pulley timing mark "T" is 3 teeth before the "T" mark on the cylinder head (3 teeth before TDC of the number one cylinder.)

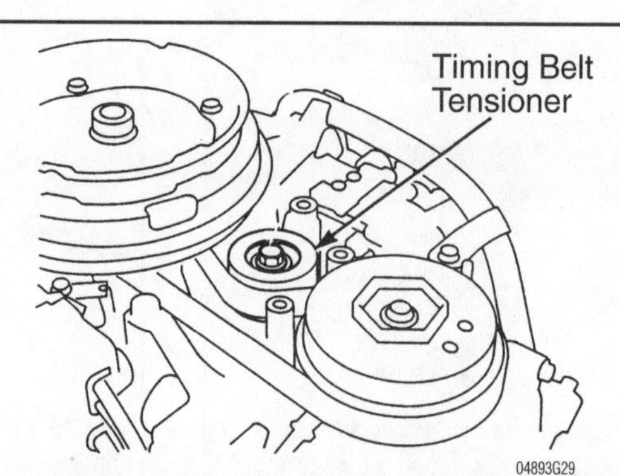

Fig. 93 The timing belt tensioner pulley is held in place by one bolt —BF25A–BF30A

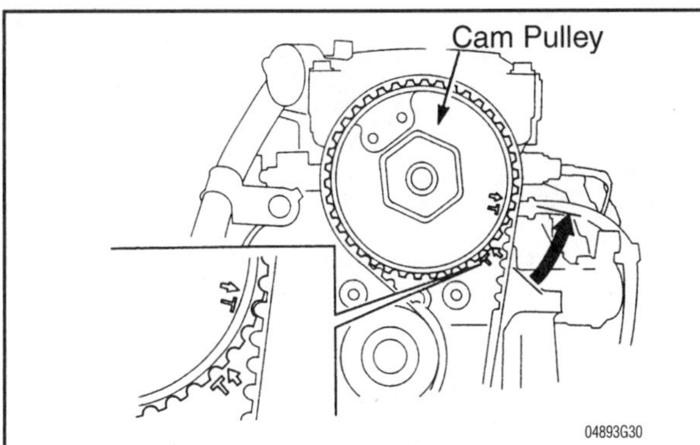

Fig. 94 The cam pulley should be three teeth before TDC when tightening the tensioner pulley bolt —BF25A–BF30A

6. Tighten the timing belt tensioner flange bolt to the specified torque. Do not apply additional pressure on the tensioner pulley; the spring on the pulley bracket spring provides adequate tension for the belt.
7. Install the engine hanger bracket and timing belt tensioner cap.
8. Install the spark plugs.

BF8A to BF15A

The timing belts on BF8A–BF15A are not adjustable.

Camshaft, Bearings And Lifters

REMOVAL & INSTALLATION

BF115A and BF130A

See Figure 95

Overhead camshafts are very sensitive to proper lubrication and need clean, fresh oil to perform at their best. A worn camshaft may be your report card for poor maintenance intervals and late oil changes.

1. Remove the cylinder head assembly.

Make sure to store all the parts separately, sorted according to location (intake side or exhaust side).

2. Remove the rocker arm assembly. Loosen the lock nuts of all the rocker arms and loosen the adjusting screws. When removing the rocker arm assembly, do not remove the mounting bolts.
3. Lift out the camshaft and discard the oil seals.
4. Measure the camshaft height:
   - Intake—Standard: 1.5068–1.5102 in. (38.274–38.359 mm)
   - Exhaust—Standard: 1.4823–1.4865 in. (37.651–37.765 mm)
5. Measure the camshaft journal outer diameter (O.D.) No.1 to No.5.
   - Standard: 1.0998–1.004 in. (27.935–27.950 mm)

   Check the camshaft oil clearance.
6. Clean off the camshaft and check the cam lobes and journals for any damage.
7. Throughly clean the cylinder head and the camshaft holder bearings and install the camshaft on the head.
8. Set Plastigage® in the axial direction on each journal.
9. Install the camshaft holder assembly and tighten the bolts in the numbered sequence in two or three increments. Make sure to tighten the 8 x 76 mm flange bolts tot he specified torque and tighten the 6 x 40 flange bolts to the standard torque.
   - 8 x 76 mm flange bolt: 16 ft. lbs. (22 Nm).
   - 6 x 40 mm flange bolt: 9 ft. lbs. (12 Nm).
10. Measure the camshaft holder journal inner diameter (I.D.) No.1 to No.5.
    - Standard: 1.1024–1.1033 in. (28.000–28.024 mm).
11. Remove the camshaft holders and measure the width of the pressed Plastigage® using the scale printed on the container.
    - Standard: 0.0020–0.0035 in. (0.050–0.089 mm).
    - Service limit: 0.006 in. (0.15 mm).
12. If the camshaft oil clearance exceeds the service limit, replace the camshaft. Don't forget to remove the Plastigage® from the journals before assembly.

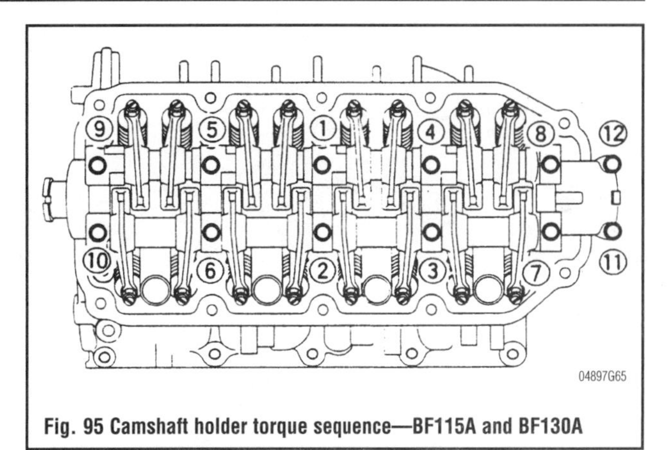

Fig. 95 Camshaft holder torque sequence—BF115A and BF130A

Measure the height of the new replacement camshaft. If the oil clearance now exceeds the service limit, you will need to replace the cylinder head or camshaft holders.

13. Measure the camshaft axial clearance.
14. Put the camshaft at TDC of the compression stroke of the No.1 piston, (with the groove in the camshaft flange pointing up). Check the camshaft axial play by pulling and pushing from the opposite side of the dial indicator.
    - Standard: 0.002–0.006 in. (0.05–0.15 mm).
    - Serice limit: 0.020 in. (0.5 mm).
15. If the measurement exceeds the service limit, replace the camshaft.

Install a new replacement camshaft and re-measure the axial clearance. If the measurement still exceeds the service limit, replace the cylinder head and all the camshaft holders.

16. Measure the camshaft run out.
    - Standard: 0.001 in. (0.03 mm) MAX
    - Service limit: 0.002 in. (0.04 mm).

To install:

17. Apply molybdenum disulfide (moly assembly lube) to each journal and cam lobe.
18. Apply grease to the lip of each oil seal.
19. Install the camshaft and seals into the cylinder head.
20. Install the rocker arm assembly and tighten the bolts in the numbered sequence and to the correct torque.

BF75A and BF90A

See Figures 96 and 97

1. Remove the cylinder head bolts and lift off the cylinder head.
2. Remove the oil pump.
3. Loosen the intake and exhaust adjusting lock nuts and loosen the adjusting screws until they are even with the rocker arm.

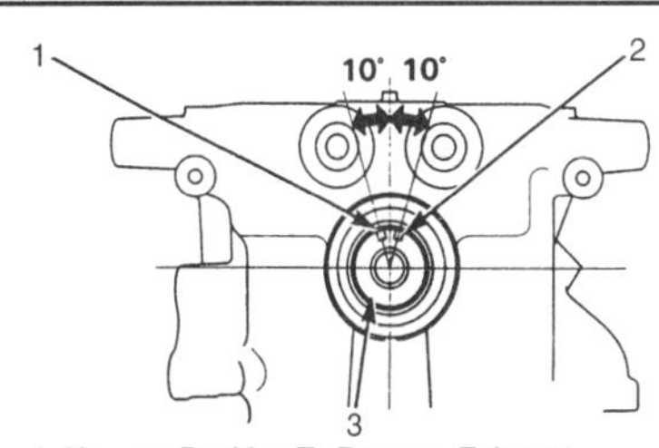

1. Keyway Position To Remove Exhaust Rocker Arm
2. Keyway Position To Remove Intake Rocker Arm
3. Camshaft

Fig. 96 Camshaft keyway position for removing the intake and exhaust rocker arms—BF75A and BF90A

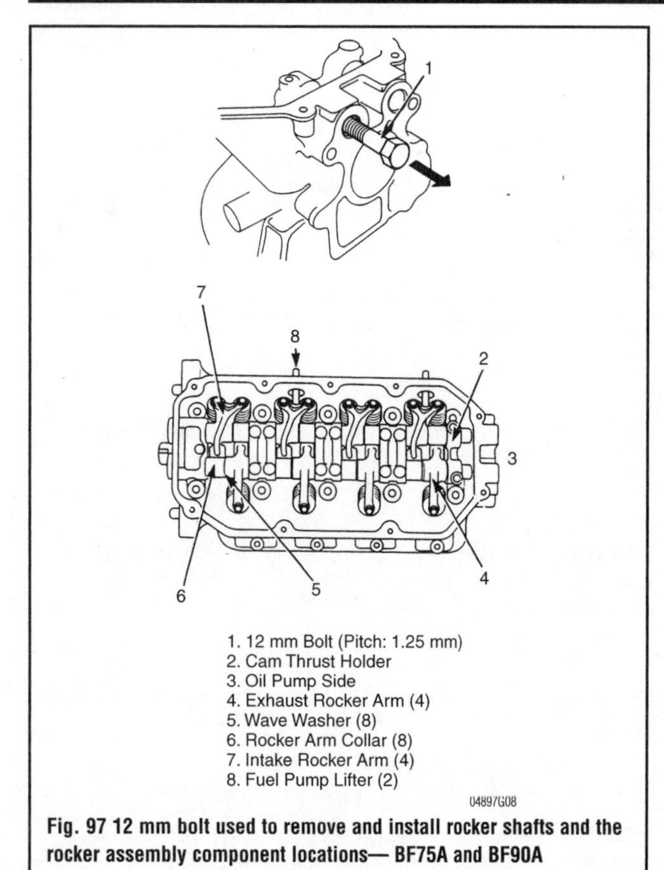

Fig. 97 12 mm bolt used to remove and install rocker shafts and the rocker assembly component locations— BF75A and BF90A

4. Remove the two flange bolts on the cam thrust holder.
5. Remove the two 20 mm sealing bolts behind the oil pump.
6. To remove the intake and exhaust rocker arms, align the camshaft keyway on the timing belt driven side tot he position shown in the drawing.
   - To remove the intake rocker arms, set the camshaft at 10°clockwise.
   - To remove the exhaust rocker arms, set the camshaft at 10°counterclockwise.
7. Insert a 12 mm bolt (1.25 mm pitch) into the oil pump mounting side of the rocker shafts. As you pull out the shaft using the bolt, remove the rocker arm collars, wave washers and rocker arms.

➡Make sure to keep the intake and exhaust components separated and reassemble everything in the reverse order of disassembly.

8. Remove the cam thrust holder.
9. Remove the fuel pump lifter and remove the camshaft from the oil pump side of the cylinder head.

To install:

10. Apply assembly lube (molybdenum disulfide) to the camshaft journal and camshaft surface and install the cam into the head.
11. Set the two dowel pins on the cam thrust holder and then loosely install into the head.
12. Apply assembly lube to the rocker arm face and the outer surface of the rocker arm shaft.
13. Set the camshaft keyway on the timing belt side of the to 10°clockwise.
14. Using the 12 mm bolt, insert the intake rocker arm shaft (the one with seven oil holes) into the cylinder head from the oil pump side. Starting with the No.4 cylinder, install the rocker arm collars, wave washers, and intake rocker arms in the reverse order that they were taken apart.
15. Install the exhaust rocker arm (the one with four oil holes) components in the same order as the intake components.

➡It is very important to install the intake and exhaust rocker shafts with the keyways in the correct positions where they are inclined 10°clockwise or counterclockwise respectively. If this is not done properly, the rocker arm will hit the camshaft lobe and prevent the smooth installation of the rocker arm shaft.

16. Set the flange bolt into the camshaft thrust holder. If it is hard to start the bolt, turn the 12 mm bolt clockwise until the groove in the rocker shaft aligns with the hole in the cam thrust holder and the flange bolt will thread in.
17. Tighten the camshaft thrust holder flange bolts to 6.5 ft. lbs. (9 Nm) and the 20 mm sealing bolts to 36 ft. lbs. (49 Nm).

BF50A, BF45A, BF40A and BF35A

➧ See Figures 98, 99 and 100

1. Remove the oil pump.
2. Loosen the intake and exhaust valve adjuster lock nuts and loosen the adjusting screws completely.
3. Loosen the camshaft holders evenly, from the center holder, out to each end and lift the rocker shafts and arms off the engine.
4. Remove the camshaft from the cylinder head and perform the following measurements. If any of the measurements are not within specification, replace the camshaft.

To install:

5. Measure the camshaft height.
   - Intake: Standard 1.3751–1.3877 in. (34.928–35.248 mm). Service limit: 1.3665 in. (34.708 mm).
   - Exhaust: Standard 1.3769–1.3895 in. (34.973–35.293 mm). Service limit: 1.3682 in. (34.753 mm).

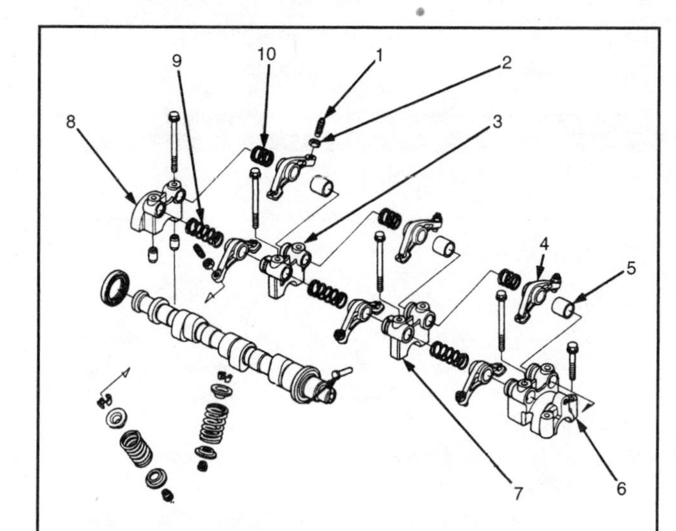

Fig. 98 Exploded view drawing of the rocker component locations— BF50A, BF45A, BF40A and BF35A

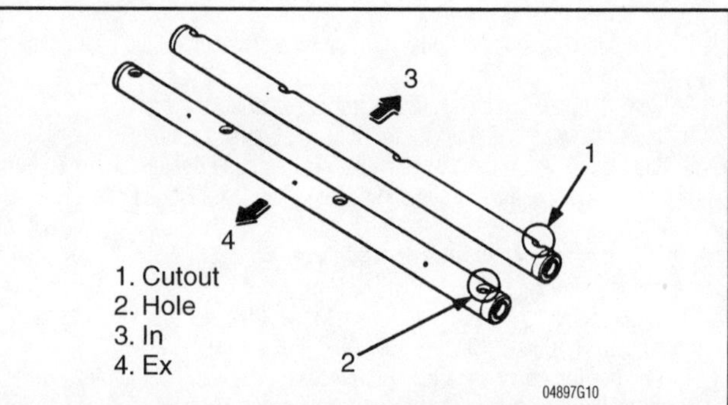

Fig. 99 The rocker shafts must be installed with the hole and cutout in the proper position on each shaft— BF50A, BF45A, BF40A and BF35A

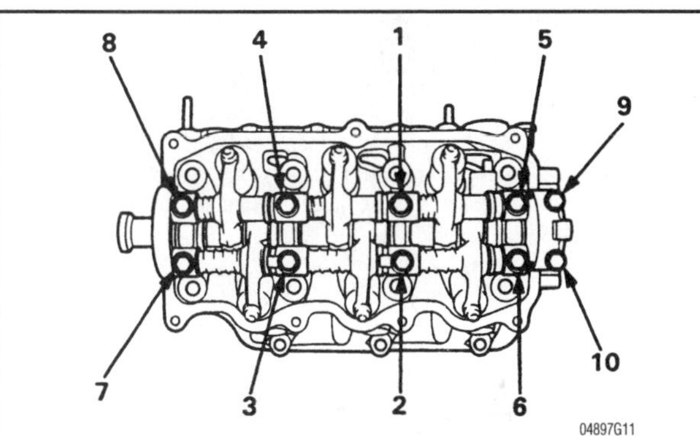

Fig. 100 Camshaft holder bolts must be tightened to the specified torque in the sequence shown— BF50A, BF45A, BF40A and BF35A

6. Measure the camshaft journal outer diameter (O.D.).
- Standard: 0.9039–0.9047 in. (22.959–22.980 mm).
- Service limit: 0.903 in. (22.93 mm).

7. Measure the camshaft runout.
- Standard: 0.0012 in. Max. (0.03 mm).
- Service limit: 0.0019 in.

8. Measure the camshaft axial clearance.

➡Measure the axial runout before removing the cylinder head from the block. Also, loosen the valve adjusting lock nuts and loosen the adjusting screws before performing this procedure.

- Standard: 0.0012–0.0043 in. (0.03–0.11 mm).
- Service limit: 0.012 in. (0.3 mm).

9. Measure the camshaft oil clearance.
10. Use Plastigage®, set axially on each camshaft journal. Install the camshaft holders, No.1 to No.4 and the dowel pins.
11. Tighten the camshaft holders to the specified torque.
- 6 x 70 mm flange bolts: 10.1 ft. lbs. (14 Nm).
- 6 x 35 mm flange bolts: 8.7 ft. lbs. (12 Nm).

➡Do not rotate the camshaft while tightening the bolts. Tighten the bolts from the inside to the outside.

12. Remove the camshaft holders and check the width of the Plastigage®using the scale on the package.
- Standard: 0.0008–0.0026 in. (0.020–0.065 mm).
- Service limit: 0.003 in. (0.08 mm).
- If the clearance exceeds the service limit, replace the camshaft. If the clearance still exceeds the service limit with a new camshaft in place, replace the cylinder head.

13. Clean all the bearing surfaces prior to installation.
14. Reversing the disassembly procedure, install the rocker arms, rocker arm springs, rocker arm collars and camshaft holders onto the rocker arm shafts.

➡Do not interchange the exhaust side rocker arm shaft with the intake rocker shaft.

15. Install the rocker arm shaft with the hole in it on the exhaust side and the rocker arm shaft with the cutout in it on the intake side.
16. Install the No. 2 and No.3 camshaft holders facing the cam pulley side of the rocker shaft.
17. Torque the bolts to specification in the proper sequence.
- Eight 6 x 70 mm cam holder bolts 10.1 ft. lbs. (14 Nm).
- Two 6 x 35 mm cam holder bolts to 8.7 ft. lbs. (12 Nm).

BF25A and BF30A

See Figures 101, 102 and 103

1. Remove the oil pump.
2. Loosen the valve adjusting screw lock nuts and loosen the adjusting screws completely.
3. Remove the two rocker shaft retaining bolts.
4. Slide the rocker shafts out through the oil pump end of the cylinder head.
5. Keep the exhaust and intake components separated.
6. Remove the fuel pump lifter.
7. Remove the camshaft out through the oil pump end of the cylinder head. Discard the oil seals.

Perform the following camshaft measurements.

8. Measure the camshaft height.
- Intake: Standard 0.9478–0.9572 in. (24.073–24.313 mm). Service limit: 0.9387 in. (23.843 mm).
- Exhaust: Standard 0.9490–0.9585 in. (24.105–24.345 mm). Service limit: 0.9411 in. (23.905 mm).

9. Measure the camshaft journal outer diameter (O.D.).
- Standard No.1: 0.7858–0.7866 in. (19.959–19.980 mm).
- Service limit No.1: 0.785 in. (19.93 mm).
- Standard No.2: 1.1787–1.1795 in. (29.939–29.960 mm).
- Service limit No.2: 1.18 in. (29.9 mm).
- Standard No.3: 0.7073–0.7080 in. (17.996–17.984 mm).
- Service limit No.3: 0.706 in. (17.94 mm).

10. Measure the camshaft runout.
- Standard: 0.001 in. Max. (0.03 mm).
- Service limit: 0.002 in. (0.05 mm).

11. Measure the camshaft axial clearance.

➡Measure the axial runout before removing the cylinder head from the block. Also, loosen the valve adjusting lock nuts and loosen the adjusting screws before performing this procedure.

- Standard: 0.006–0.0016 in. (0.15–0.40 mm).
- Service limit: 0.06 in. (0.02 mm).

12. Measure the camshaft oil clearance.

Cylinder head I.D. of camshaft journal.
- Standard No.1: 0.7874–0.7882 in. (20.000–20.021 mm).
- Service limit No.1: 0.789 in. (20.05 mm).
- Standard No.2: 1.1811–1.1821 in. (30.000–30.025 mm).
- Service limit No.2: 1.183 in. (18.06 mm).
- Standard No.3: 0.7087–0.7097 in. (18.000–18.027 mm)
- Service limit: 0.711 in. (18.06 mm).

Camshaft oil clearance.
- Standard No.1: 0.0008–0.0024 in. (0.020–0.062 mm).

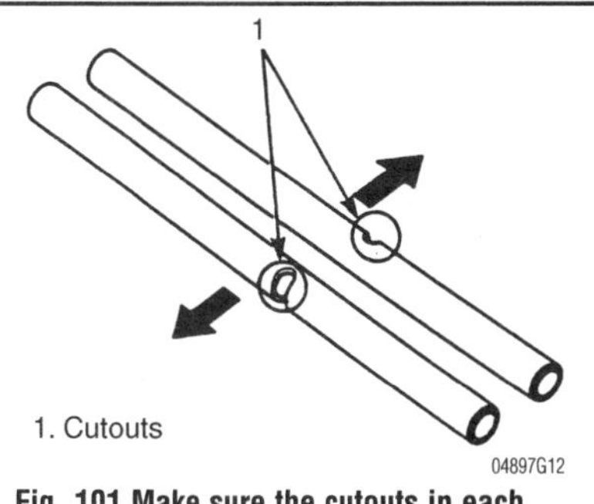

Fig. 101 Make sure the cutouts in each shaft are facing outward so the rocker shaft retaining bolts clear the shafts— BF50A, BF45A, BF40A and BF35A

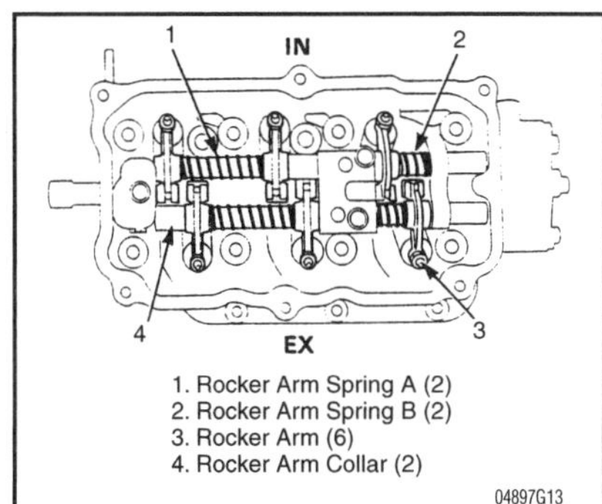

Fig. 102 Correct assembly of the rocker arms and shafts— BF50A, BF45A, BF40A and BF35A

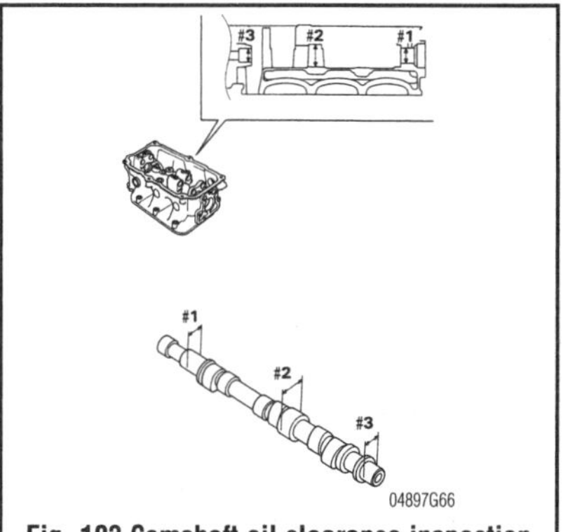

Fig. 103 Camshaft oil clearance inspection points— BF50A, BF45A, BF40A and BF35A

- Service limit No.1: 0.003 in. (0.08 mm).
- Standard No.2: 0.0016–0.0034 in. (0.040–0.086 mm).
- Service limit No.2: 0.004 in. (0.11 mm).
- Standard No.3: 0.0006–0.0024 (0.016–0.061 mm).
- Service limit No.3: 0.003 in. (0.08 mm).
- If the clearance exceeds the service limit, replace the camshaft. If the clearance still exceeds the service limit with a new camshaft in place, replace the cylinder head.

To install:

13. Clean all the bearing surfaces prior to installation.
14. Reversing the disassembly procedure, install the camshaft, rocker arms, rocker arm springs and rocker arm collars back into the cylinder head, from the oil pump side.

➡Do not interchange the exhaust side rocker arm shaft with the intake rocker shaft.

15. Install the rocker arm shafts with the facing outward.
16. Torque the bolts to specification.
17. Install the oil pump.

BF9.9A and BF15A

See Figure 104

1. Remove the oil pump.
2. Loosen all the valve adjusting screws.
3. Screw a 10 x 1.25 mm bolt into the threaded end of the rocker shaft and pull out the shaft.
4. As the shaft is being pulled out, remove the rocker arms and the rocker arm spring.
5. Remove the camshaft and thrust washer.

To install:

6. Inspect the camshaft for wear and scratches, then perform the following measurements.
7. Measure the camshaft height.

- BF15A
- Intake standard: 0.9643 in. (24.47 mm).
- Service limit: 0.9535 in. (24.22 mm).
- Exhaust standard: 0.9654 (24.52 mm).
- Service limit: 0.9555 in. (24.27 mm).
- BF9.9A
- Intake standard: 0.9405 in. (23.89 mm).
- Service limit: 0.9307 in. (23.64 mm).
- Exhaust standard: 0.9425 in. (23.94 mm).
- Service limit: 0.9326 in. (23.69 mm)

8. Measure the camshaft journal O.D.

- Oil pump side: Standard: 0.6292 in. (15.984 mm); Service limit: 0.6279 in. (15.95 mm)
- Pulley side: Standard: 0.7080 in. (17.984 mm); Service limit: 0.7066 in. (17.95 mm).

➡If any of the measurements do not meet specifications, replace the camshaft

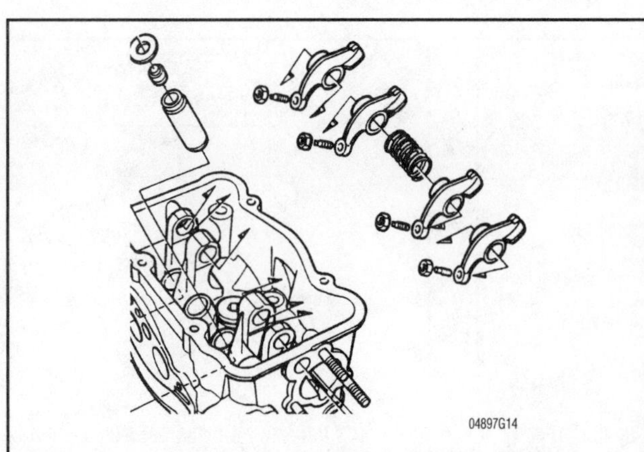

Fig. 104 Rocker arm and shaft assembly—BF9.9A and BF15A

9. Lubricate the rocker arm bushings, rocker shaft, rocker shaft bushing and camshaft journals before assembly.
10. Slide the camshaft into the cylinder head from the oil pump side. Do not forget to install the thrust washer.
11. Slide the rocker shaft through the cylinder head with the threaded end toward you and install the rocker arms and rocker arm spring as the shaft enters. Unless the part has been replaced, install the parts in the order they came out.
12. Replace the oil pump.

BF75, BF8A and BF100

See Figure 105

1. Remove the cylinder head.
2. Remove the oil pump.
3. Loosen the lock nuts and adjusting screws to unload the valve springs before removing the rocker arm shaft.
4. Screw in a 10 x 1.25 mm bolt into the threaded hole on the rocker shaft and pull out the shaft.
5. As the shaft leaves the cylinder head, remove the rocker arms, rocker collars and rocker arm spring.
6. After removing the rocker assembly, remove the camshaft also.

To install:

7. Inspect the camshaft for wear and scratches, then perform the following measurements.
8. Measure the cam lobe height.
9. CDI type ignition.

- BF100: Standard: 0.99 in. (25.2 mm); Service limit: 0.982 in. (24.95 mm)
- BF75/8.8A: Intake standard: 1.04 in. (26.5 mm); Service limit: 1.033 in. (26.25 mm).

Exhaust standard: 0.91 in. (23.2 mm); Service limit: 0.904 in. (22.95 mm).

10. Contact breaker type ignition.

- BF100: Standard: 0.98 in. (25.0 mm); Service limit: 0.974 in. (24.75 mm).
- BF75/8.8A: Standard: 0.91 in. (23.0 mm); Service limit: 0.896 in. (22.75 mm).

11. Measure the camshaft O.D. (oil pump side).

- Standard: 0.6285–0.6292 in. (15.966–15.984 mm).
- Service limit: 0.627 in. (15.916 mm).

12. Lubricate the rocker arm bushings, rocker shaft, rocker shaft bushing and camshaft journals before assembly.
13. Install the camshaft and thrust washers into the cylinder head.
14. Slide the rocker shaft through the cylinder head with the threaded end toward you and install the rocker arms and rocker arm spring as the shaft enters. Unless the part has been replaced, install the parts in the order they came out.
15. Install the oil pump back on the cylinder head.

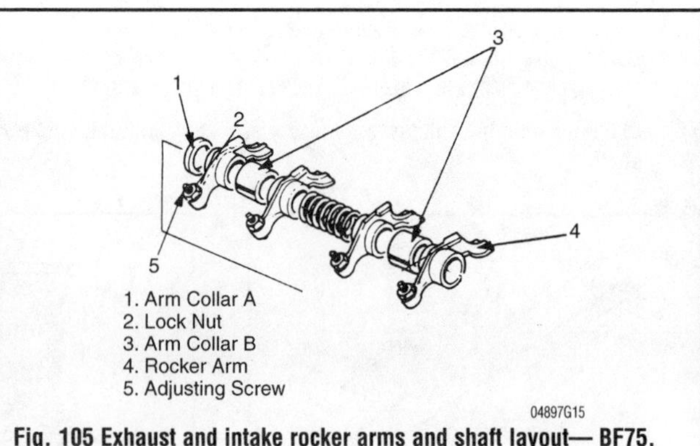

Fig. 105 Exhaust and intake rocker arms and shaft layout— BF75, BF8A and BF100

BF50 and BF5A

See Figure 106

1. Remove the engine cover.
2. Remove the recoil starter assembly.

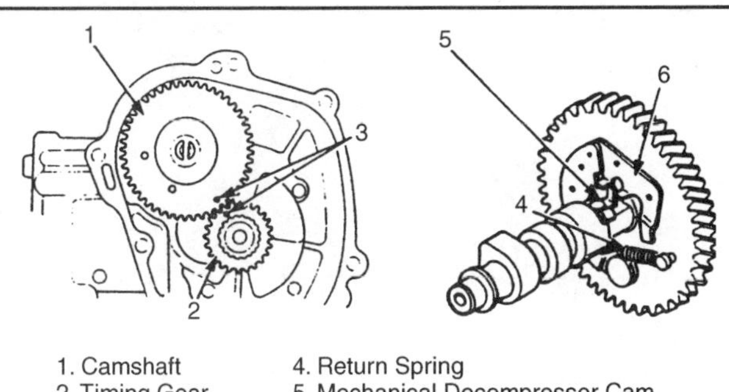

Fig. 106 Camshaft-to-timing gear alignment marks—BF50 and BF5A

3. Remove the engine.
4. Remove the oil pump assembly.
5. Remove the oil pan assembly.
6. Remove the camshaft assembly out of the crankcase.

To install:

7. Inspect the mechanical decompressor assembly for worn or weak springs and check the decompressor weight for smooth movement.
8. Align the timing marks on the camshaft and crankshaft and install the camshaft assembly into the crankcase.
9. Install the oil pan assembly.
10. Install the oil pump assembly.
11. Install the engine.

BF20 and BF2A

See Figure 107

1. Remove the engine cover.
2. Remove the engine.
3. Remove the recoil starter assembly.
4. Remove the starter pulley, cooling fan and flywheel assemblies.
5. Remove the cylinder head.
6. Remove the intake and exhaust valve assemblies.
7. Remove the valve lifters. Mark each lifter so its reassembled in the correct position.
8. Remove the oil pan assembly.
9. Remove the camshaft/timing gear assembly.

To install:

Perform the following measurements before installing the camshaft.

10. Measure the cam height.
- Standard: 0.8197 in. (20.82 mm).
- Service limit: 0.8059 in. (20.47 mm).
11. Measure the camshaft O.D.
- Standard: 0.4797 in. (12.184 mm).
- Service limit: 0.4783 in. (12.15 mm).
12. Measure the camshaft journal I.D.

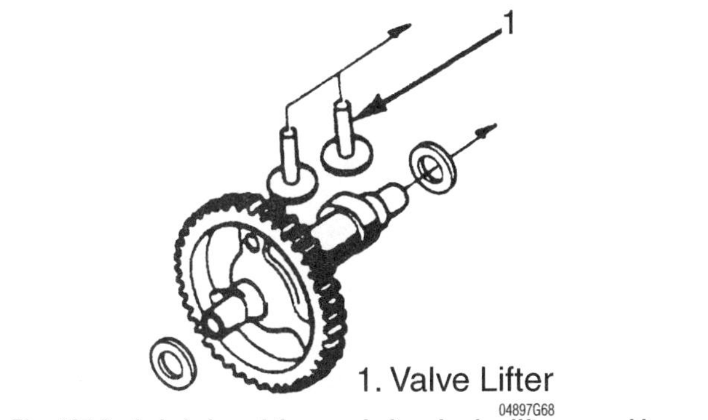

Fig. 107 Exploded view of the camshaft and valve lifter assembly—BF20 and BF2A

- Standard: 0.4083 in. (12.20 mm).
- Service limit: 0.4823 in. (12.25 mm).
13. If the camshaft does not meet specifications, replace it.
14. Before installing the camshaft, install the valve lifters. Tilt the engine to one side to prevent the lifters from falling out during assembly.
15. Install the camshaft when the timing marks are aligned on both the camshaft and crankshaft.
16. Install the intake and exhaust valve assemblies.
17. Install the cylinder head.
18. Install the starter pulley, cooling fan and flywheel assemblies.
19. Install the recoil starter assembly.
20. Install the engine.
21. Install the engine cover.

Flywheel

REMOVAL & INSTALLATION

BF115A and BF130A

1. Remove the engine.
2. Remove the oil pump body assembly. Discard the used gasket.
3. Remove the flywheel boss.
4. Remove the flywheel.

To install:

5. Check the ring gear for wear and damage. If the ring gear is damaged, make sure to check the starter motor pinion gear also.

Before the installing the flywheel, use a degreaser to clean all the mating surfaces between the flywheel and crankshaft.

6. Install the flywheel. Oil the threads and seat of the 12 mm bolts and tighten to 87 ft. lbs. (118 Nm).
7. Install the flywheel boss. Again, oil the threads and seat on the bolts and tighten to 24 ft. lbs. (32 Nm).
8. Install the oil pump body and gasket. Tighten the bolts to standard torque.

BF75A and BF90A

See Figure 108

1. Remove the engine
2. Remove the starter motor.

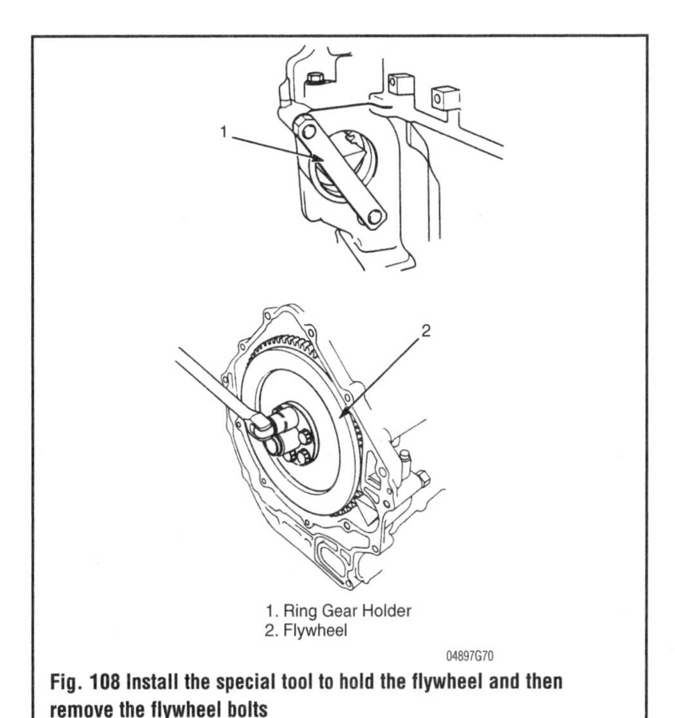

Fig. 108 Install the special tool to hold the flywheel and then remove the flywheel bolts

3. Install the special tool (ring gear holder 07SPB-ZW10100) into the starter motor mounting base.
4. Remove the six bolts and lift off the flywheel.

To install:

5. Throughly clean all the mating surfaces with a degreaser and install the flywheel onto the crankshaft.
6. Oil the threads and seats on the bolts and loosely tighten the bolts.
7. Attach the ring gear holder and tighten the flywheel bolts to 76 ft. lbs. (103 Nm).
8. Remove the ring gear holder and replace the starter motor.
9. Install the engine.

BF50A, BF45A, BF40A and BF35A

See Figures 109, 110 and 111

1. Remove the engine cover.
2. Open the CDI unit cover and disconnect the charge and exciter coils.
3. Remove the flywheel cover.
4. Remove the charge and exciter coils.

Do not hit the flywheel with a hammer during removal, damage to the embedded magnets can occur.

5. Use a strap wrench to hold the flywheel and loosen the bolts. Remove the flywheel.

To install:

6. Install the charge and exciter coils and connect them at the CDI unit.
7. Install the flywheel onto the crankshaft.

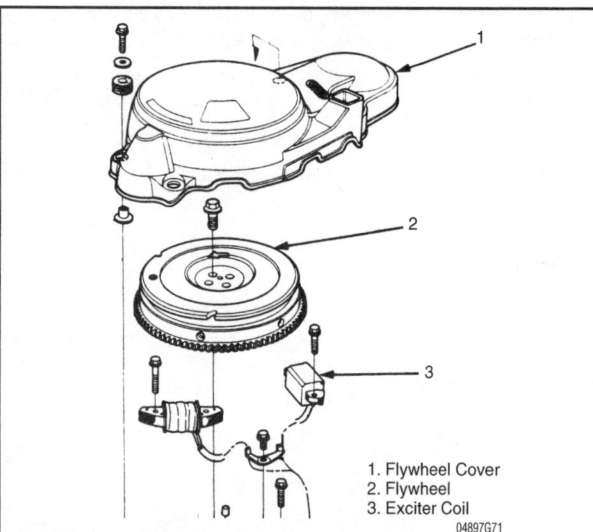

Fig. 109 Exploded view drawing of the flywheel assembly (BF35A and BF45A models)

Fig. 110 Loosen the flywheel bolts . . .

Fig. 111 . . . and remove the flywheel

8. Oil the threads and seats of the bolts before installation and tighten the bolts to 47.7 ft. lbs. (66 Nm).
9. Install the flywheel cover.
10. Install the engine cover.

BF25A and BF30A

1. Remove the engine cover.
2. Open up the CDI unit cover and disconnect the exciter coil 2P connector (on electric start models, disconnect the charge coil connectors).
3. On electric start models, remove the flywheel cover. On recoil start models, remove the recoil starter.
4. Using a strap wrench, loosen the 16 mm nut.
5. Using a commonly available flywheel puller, remove the flywheel.

CAUTION

Make sure that the puller is positioned equally around the flywheel. Use caution to not damage the ring gear teeth.

Do not use a hammer to remove the flywheel.

To install:

6. Throughly clean both the taper on the crankshaft and the tapered hole in the flywheel.
7. Make sure that the Woodruff key is in place and install the flywheel.
8. Hold the flywheel with strap wrench and tighten the 16 mm bolt to 83.2 ft. lbs. (115 Nm).
9. Install the flywheel cover or recoil starter as applicable to the model.
10. Connect the exciter or charge coil connectors and close up the CDI unit cover.
11. Install the engine cover.

BF15A and BF9.9A

1. Remove the engine cover.
2. Hold the flywheel with a commonly available strap wrench and loosen the nut.
3. If needed for clearance, disconnect and remove the exciter and charge coils.
4. Using the special tool (flywheel puller No. 07935-8050002 or No. 07935-8050003) remove the flywheel.

To install:

5. Make sure that the Woodruff key is in place and install the flywheel.
6. Hold the flywheel with strap wrench and tighten the 16 mm bolt to 83.2 ft. lbs. (115 Nm).
7. Install the flywheel cover or recoil starter as applicable to the model.
8. Connect the exciter or charge coil connectors.
9. Install the engine cover.

BF75, BF8A and BF100

1. Remove the engine cover.
2. Remove the recoil starter.
3. Using a commonly available strap wrench or the special tool (pulley holder No. 07925-8930000), hold the flywheel and loosen the nut.
4. Using the special tool (flywheel puller 07935-8050002), remove the flywheel.

To install:

5. Make sure that the Woodruff key is in place and install the flywheel.
6. Hold the flywheel with strap wrench or special tool and tighten the nut to 47 ft. lbs. (650 kg-cm).
7. Install the flywheel cover or recoil starter as applicable to the model.
8. Install the engine cover.

BF50 and BF5A

1. Remove the engine cover.
2. Remove the recoil starter.
3. Use a strap wrench or insert a screwdriver or equalivant tool into the hole in the starter pulley and loosen the flywheel nut.
4. Using the special tool (flywheel puller 07935-8050002), remove the flywheel.

To install:

5. Make sure that the Woodruff key is in place and install the flywheel.
6. Hold the flywheel with strap wrench or special tool and tighten the nut to 51–58 ft. lbs. (700–800 kg-cm).
7. Install the flywheel cover or recoil starter as applicable to the model.
8. Install the engine cover.

BF20 and BF2A

1. Remove the engine cover.
2. Remove the recoil starter.
3. Use a strap wrench or insert a screwdriver or equalivant tool into the hole in the starter pulley and loosen the flywheel nut.
4. Using a four-inch common puller, remove the flywheel.

**** CAUTION**

Do not place the puller arms on the flywheel magnet.

To install:

5. Make sure that the Woodruff key is in place and install the flywheel.
6. Hold the flywheel with strap wrench and tighten the nut to 32.5–39.8 ft. lbs. (450–550 kg-cm).
7. Install the flywheel cover or recoil starter as applicable to the model.
8. Install the engine cover.

ENGINE RECONDITIONING

Determining Engine Condition

Anything that generates heat and/or friction will eventually burn or wear out (for example, a light bulb generates heat, therefore its life span is limited). With this in mind, a running engine generates tremendous amounts of both; friction is encountered by the moving and rotating parts inside the engine and heat is created by friction and combustion of the fuel. However, the engine has systems designed to help reduce the effects of heat and friction and provide added longevity. The oiling system reduces the amount of friction encountered by the moving parts inside the engine, while the cooling system reduces heat created by friction and combustion. If either system is not maintained, a break-down will be inevitable. Therefore, you can see how regular maintenance can affect the service life of your marine engine. This applies to your oil and filter; if it is not changed often enough it becomes laden with contaminates and is unable to properly lubricate the engine. This increases friction and wear.

There are a number of methods for evaluating the condition of your engine. A compression test can reveal the condition of your pistons, piston rings, cylinder bores, head gasket(s), valves and valve seats. An oil pressure test can warn you of possible engine bearing, or oil pump failures. Excessive oil consumption, evidence of oil in the engine air intake area and/or bluish smoke from the tailpipe may indicate worn piston rings, worn valve guides and/or valve seals. As a general rule, an engine that uses no more than one quart of oil every 100 hours or six-months is in good condition. Engines that use one quart of oil or more in less than 100 hours or six-months should first be checked for oil leaks. If any oil leaks are present, have them fixed before determining how much oil is consumed by the engine, especially if blue smoke is not visible at the tailpipe.

COMPRESSION TEST

See Figure 112

A noticeable lack of engine power, excessive oil consumption and/or poor fuel mileage measured over an extended period are all indicators of internal engine wear. Worn piston rings, scored or worn cylinder bores, blown head gaskets, sticking or burnt valves, and worn valve seats are all possible culprits. A check of each cylinder's compression will help locate the problem.

A screw-in type compression gauge is more accurate than the type you simply hold against the spark plug hole. Although it takes slightly longer to use, it's worth the effort to obtain a more accurate reading.

1. Make sure that the proper amount and viscosity of engine oil is in the crankcase, then ensure the battery is fully charged.
2. Warm-up the engine to normal operating temperature, then shut the engine **OFF**.
3. Disable the ignition system.
4. Label and disconnect all of the spark plug wires from the plugs.
5. Thoroughly clean the cylinder head area around the spark plug ports, then remove the spark plugs.
6. Set the throttle plate to the fully open (wide-open throttle) position. You can block the accelerator linkage open for this, or you can have an assistant fully open the throttle.
7. Install a screw-in type compression gauge into the No. 1 spark plug hole until the fitting is snug.

**** WARNING**

Be careful not to crossthread the spark plug hole.

8. According to the tool manufacturer's instructions, connect a remote starting switch to the starting circuit.
9. With the ignition switch in the **OFF** position, use the remote starting

04892P37

Fig. 112 A screw in compression gauge is more accurate and easier to use without an assistant

switch to crank the engine through at least five compression strokes (approximately 5 seconds of cranking) and record the highest reading on the gauge.

10. Repeat the test on each cylinder, cranking the engine approximately the same number of compression strokes and/or time as the first.

11. Compare the highest readings from each cylinder to that of the others. The indicated compression pressures are considered within specifications if the lowest reading cylinder is within 75 percent of the pressure recorded for the highest reading cylinder. For example, if your highest reading cylinder pressure was 150 psi (1034 kPa), then 75 percent of that would be 113 psi (779 kPa). So the lowest reading cylinder should be no less than 113 psi (779 kPa).

12. If a cylinder exhibits an unusually low compression reading, pour a tablespoon of clean engine oil into the cylinder through the spark plug hole and repeat the compression test. If the compression rises after adding oil, it means that the cylinder's piston rings and/or cylinder bore are damaged or worn. If the pressure remains low, the valves may not be seating properly (a valve job is needed), or the head gasket may be blown near that cylinder. If compression in any two adjacent cylinders is low, and if the addition of oil doesn't help raise compression, there is leakage past the head gasket. Oil and coolant in the combustion chamber, combined with blue or constant white smoke from the tailpipe, are symptoms of this problem. However, don't be alarmed by the normal white smoke emitted from the tailpipe during engine warm-up or from cold weather driving. There may be evidence of water droplets on the engine dipstick and/or oil droplets in the cooling system if a head gasket is blown.

OIL PRESSURE TEST

Check for proper oil pressure at the sending unit passage with an externally mounted mechanical oil pressure gauge (as opposed to relying on a factory installed dash-mounted gauge). A tachometer may also be needed, as some specifications may require running the engine at a specific rpm.

1. With the engine cold, locate and remove the oil pressure sending unit.
2. Following the manufacturer's instructions, connect a mechanical oil pressure gauge and, if necessary, a tachometer to the engine.
3. Start the engine and allow it to idle.
4. Check the oil pressure reading when cold and record the number. You may need to run the engine at a specified rpm, so check the specifications.
5. Run the engine until normal operating temperature is reached.
6. Check the oil pressure reading again with the engine hot and record the number. Turn the engine **OFF**.
7. Compare your hot oil pressure reading to that given in the chart. If the reading is low, check the cold pressure reading against the chart. If the cold pressure is well above the specification, and the hot reading was lower than the specification, you may have the wrong viscosity oil in the engine. Change the oil, making sure to use the proper grade and quantity, then repeat the test.

Low oil pressure readings could be attributed to internal component wear, pump related problems, a low oil level, or oil viscosity that is too low. High oil pressure readings could be caused by an overfilled crankcase, too high of an oil viscosity or a faulty pressure relief valve.

Buy or Rebuild?

Now that you have determined that your engine is worn out, you must make some decisions. The question of whether or not an engine is worth rebuilding is largely a subjective matter and one of personal worth. Is the engine a popular one, or is it an obsolete model? Are parts available? Will it get acceptable gas mileage once it is rebuilt? Is the boat it's being put into worth keeping? Would it be less expensive to buy a new engine, have your engine rebuilt by a pro, rebuild it yourself or buy a used engine from a salvage yard? Or would it be simpler and less expensive to buy another boat? If you have considered all these matters and more, and have still decided to rebuild the engine, then it is time to decide how you will rebuild it.

➡The editors at Chilton feel that most engine machining should be performed by a professional machine shop. Don't think of it as wasting money, rather, as an assurance that the job has been done right the first time. There are many expensive and specialized tools required to perform such tasks as boring and honing an engine block or having a valve job done on a cylinder head. Even inspecting the parts requires expensive micrometers and gauges to properly measure wear and clearances. Also, a machine shop can deliver to you clean, and ready to assemble parts, saving you time and aggravation. Your maximum savings will come from performing the removal, disassembly, assembly and installation of the engine and purchasing or renting only the tools required to perform the above tasks. Depending on the particular circumstances, you may save 40 to 60 percent of the cost doing these yourself.

A complete rebuild or overhaul of an engine involves replacing all of the moving parts (pistons, rods, crankshaft, camshaft, etc.) with new ones and machining the non-moving wearing surfaces of the block and heads. Unfortunately, this may not be cost effective. For instance, your crankshaft may have been damaged or worn, but it can be machined undersize for a minimal fee.

So, as you can see, you can replace everything inside the engine, but, it is wiser to replace only those parts which are really needed, and, if possible, repair the more expensive ones. Later in this section, we will break the engine down into its two main components: the cylinder head and the engine block. We will discuss each component, and the recommended parts to replace during a rebuild on each.

Engine Overhaul Tips

Most engine overhaul procedures are fairly standard. In addition to specific parts replacement procedures and specifications for your individual engine, this section is also a guide to acceptable rebuilding procedures. Examples of standard rebuilding practice are given and should be used along with specific details concerning your particular engine.

Competent and accurate machine shop services will ensure maximum performance, reliability and engine life. In most instances it is more profitable for the do-it-yourself mechanic to remove, clean and inspect the component, buy the necessary parts and deliver these to a shop for actual machine work.

Much of the assembly work (crankshaft, bearings, piston rods, and other components) is well within the scope of the do-it-yourself mechanic's tools and abilities. You will have to decide for yourself the depth of involvement you desire in an engine repair or rebuild.

TOOLS

The tools required for an engine overhaul or parts replacement will depend on the depth of your involvement. With a few exceptions, they will be the tools found in a mechanic's tool kit. More in-depth work will require some or all of the following:

- A dial indicator (reading in thousandths) mounted on a universal base
- Micrometers and telescope gauges
- Jaw and screw-type pullers
- Scraper
- Valve spring compressor
- Ring groove cleaner
- Piston ring expander and compressor
- Ridge reamer
- Cylinder hone or glaze breaker
- Plastigage®
- Engine stand

The use of most of these tools is illustrated in this section. Many can be rented for a one-time use from a local parts jobber or tool supply house.

Occasionally, the use of special tools is called for. See the information on Special Tools and the Safety Notice in the front of this book before substituting another tool.

OVERHAUL TIPS

Aluminum has become extremely popular for use in engines, due to its low weight. Observe the following precautions when handling aluminum parts:

- Never hot tank aluminum parts (the caustic hot tank solution will eat the aluminum.
- Remove all aluminum parts (identification tag, etc.) from engine parts prior to the tanking.
- Always coat threads lightly with engine oil or anti-seize compounds before installation, to prevent seizure.
- Never overtighten bolts or spark plugs especially in aluminum threads.

When assembling the engine, any parts that will be exposed to frictional contact must be prelubed to provide lubrication at initial start-up. Any product specifically formulated for this purpose can be used, but engine oil is not recommended as a prelube in most cases.

When semi-permanent (locked, but removable) installation of bolts or nuts is desired, threads should be cleaned and coated with Loctite® or another similar, commercial non-hardening sealant.

CLEANING

See Figures 113, 114 and 115

Before the engine and its components are inspected, they must be thoroughly cleaned. You will need to remove any engine varnish, oil sludge and/or carbon deposits from all of the components to insure an accurate inspection. A crack in the engine block or cylinder head can easily become overlooked if hidden by a layer of sludge or carbon.

Most of the cleaning process can be carried out with common hand tools and readily available solvents or solutions. carbon deposits can be chipped away using a hammer and a hard wooden chisel. Old gasket material and varnish or sludge can usually be removed using a scraper and/or cleaning solvent. Extremely stubborn deposits may require the use of a power drill with a wire brush. If using a wire brush, use extreme care around any critical machined surfaces (such as the gasket surfaces, bearing saddles, cylinder bores, etc.). USE OF A WIRE BRUSH IS NOT RECOMMENDED ON ANY ALUMINUM COMPONENTS. Always follow any safety recommendations given by the manufacturer of the tool and/or solvent. You should always wear eye protection during any cleaning process involving scraping, chipping or spraying of solvents.

An alternative to the mess and hassle of cleaning the parts yourself is to drop them off at a local garage or machine shop. They will, more than likely, have the necessary equipment to properly clean all of the parts for a nominal fee.

**** CAUTION**

Always wear eye protection during any cleaning process involving scraping, chipping or spraying of solvents.

Remove any oil galley plugs, freeze plugs and/or pressed-in bearings and carefully wash and degrease all of the engine components including the fasteners and bolts. Small parts such as the valves, springs, etc., should be placed in a metal basket and allowed to soak. Use pipe cleaner type brushes, and clean all passageways in the components. Use a ring expander and remove the rings from the pistons. Clean the piston ring grooves with a special tool or a piece of broken ring. Scrape the carbon off of the top of the piston. You should never use a wire brush on the pistons. After preparing all of the piston assemblies in this manner, wash and degrease them again.

**** WARNING**

Use extreme care when cleaning around the cylinder head valve seats. A mistake or slip may cost you a new seat.

When cleaning the cylinder head, remove carbon from the combustion chamber with the valves installed. This will avoid damaging the valve seats.

REPAIRING DAMAGED THREADS

See Figures 116, 117, 118, 119 and 120

Several methods of repairing damaged threads are available. Heli-Coil® (shown here), Keenserts® and Microdot® are among the most widely used. All involve basically the same principle—drilling out stripped threads, tapping the hole and installing a prewound insert—making welding, plugging and oversize fasteners unnecessary.

Two types of thread repair inserts are usually supplied: a standard type for most inch coarse, inch fine, metric course and metric fine thread sizes and a spark lug type to fit most spark plug port sizes. Consult the individual tool manufacturer's catalog to determine exact applications. Typical thread repair kits will contain a selection of prewound threaded inserts, a tap (corresponding to the outside diameter threads of the insert) and an installation tool. Spark plug inserts usually differ because they require a tap equipped with pilot threads and a combined reamer/tap section. Most manufacturers also supply blister-packed thread repair inserts separately in addition to a master kit containing a variety of taps and inserts plus installation tools.

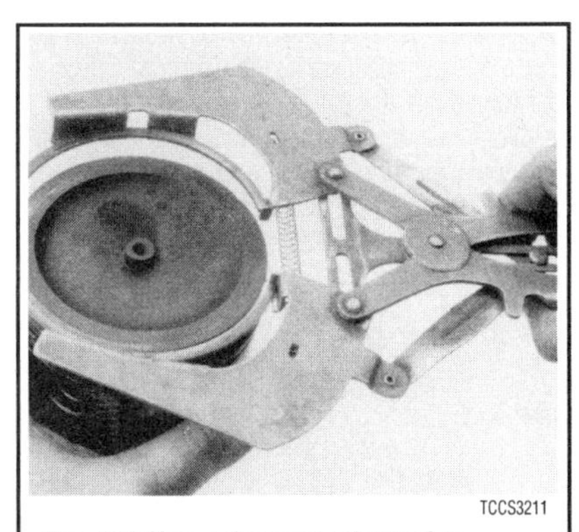

TCCS3211

Fig. 113 Use a ring expander tool to remove the piston rings

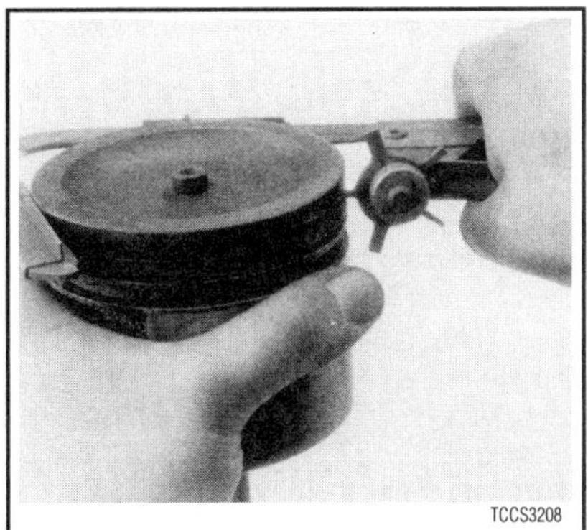

TCCS3208

Fig. 114 Clean the piston ring grooves using a ring groove cleaner tool, or . . .

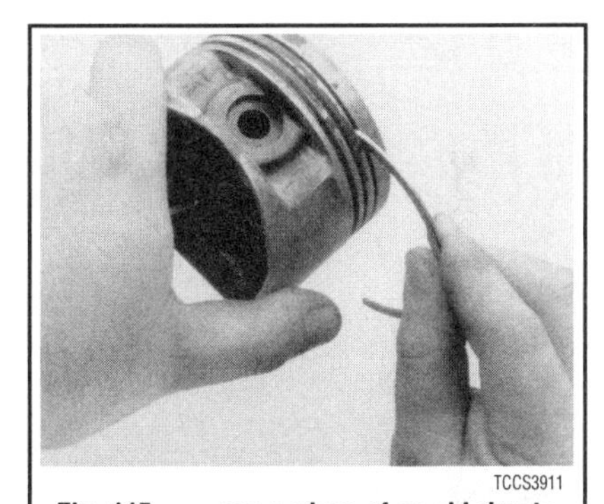

TCCS3911

Fig. 115 . . . use a piece of an old ring to clean the grooves. Be careful, the ring can be quite sharp

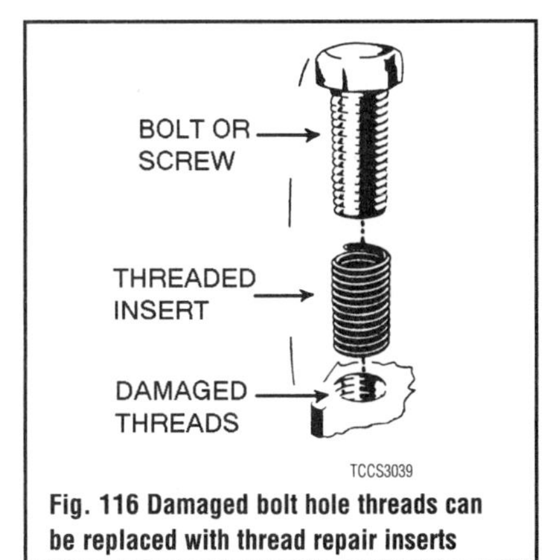

TCCS3039

Fig. 116 Damaged bolt hole threads can be replaced with thread repair inserts

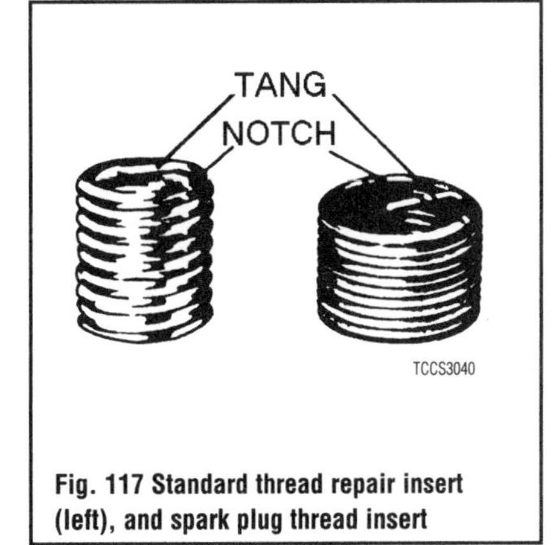

TCCS3040

Fig. 117 Standard thread repair insert (left), and spark plug thread insert

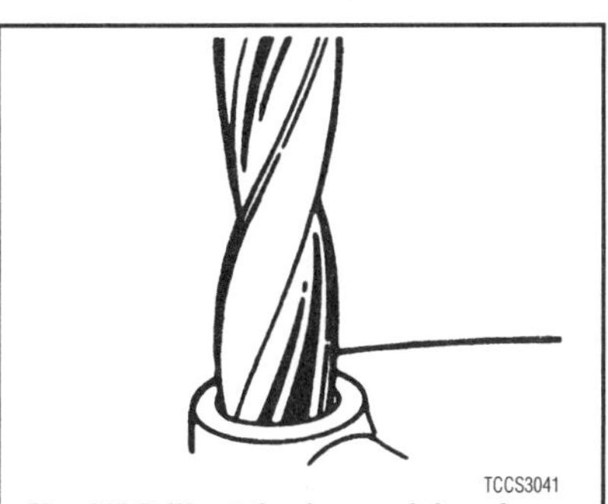

TCCS3041

Fig. 118 Drill out the damaged threads with the specified size bit. Be sure to drill completely through the hole or to the bottom of a blind hole

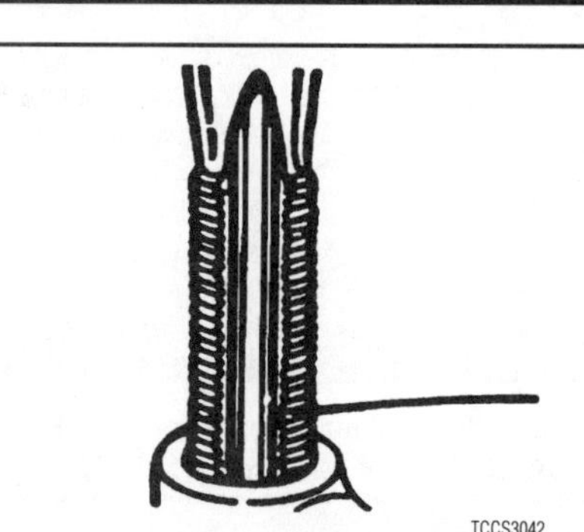

Fig. 119 Using the kit, tap the hole in order to receive the thread insert. Keep the tap well oiled and back it out frequently to avoid clogging the threads

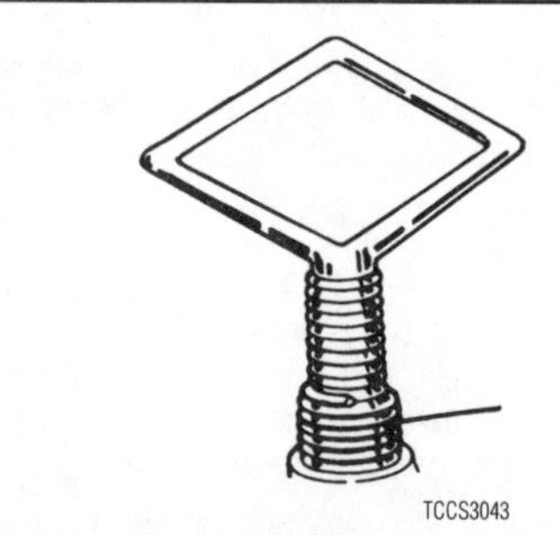

Fig. 120 Screw the insert onto the installer tool until the tang engages the slot. Thread the insert into the hole until it is ¼–½ turn below the top surface, then remove the tool and break off the tang using a punch

Before attempting to repair a threaded hole, remove any snapped, broken or damaged bolts or studs. Penetrating oil can be used to free frozen threads. The offending item can usually be removed with locking pliers or using a screw/stud extractor. After the hole is clear, the thread can be repaired, as shown in the series of accompanying illustrations and in the kit manufacturer's instructions.

Engine Preparation

To properly rebuild an engine, you must first remove it from the vessel, then disassemble and diagnose it. Ideally you should place your engine on an engine stand. This affords you the best access to the engine components. Follow the manufacturer's directions for using the stand with your particular engine. Remove the flywheel or flexplate before installing the engine to the stand.

Now that you have the engine on a stand, and assuming that you have drained the oil and coolant from the engine, it's time to strip it of all but the necessary components. Before you start disassembling the engine, you may want to take a moment to draw some pictures, or fabricate some labels or containers to mark the locations of various components and the bolts and/or studs which fasten them. Modern day engines use a lot of little brackets and clips which hold wiring harnesses and such, and these holders are often mounted on studs and/or bolts that can be easily mixed up. The manufacturer spent a lot of time and money designing your marine engine, and they wouldn't have wasted any of it by haphazardly placing brackets, clips or fasteners. If it's present when you disassemble it, put it back when you assemble, you will regret not remembering that little bracket which holds a wire harness out of the path of a rotating part.

You should begin by unbolting any accessories still attached to the engine, such as the alternator, etc. Then, unfasten any manifolds (intake or exhaust) which were not removed during the engine removal procedure. Finally, remove any covers remaining on the engine such as the rocker arm, timing cover and oil pan. The idea is to reduce the engine to the bare necessities (cylinder head(s), valve train, engine block, crankshaft, pistons and connecting rods), plus any other `in block' components such as oil pumps, balance shafts and auxiliary shafts.

Finally, remove the cylinder head(s) from the engine block and carefully place on a bench. Disassembly instructions for each component follow later in this section.

Cylinder Head

There are two basic types of cylinder heads used on today's four-stroke outboards: the Overhead Valve (OHV) and the Overhead Camshaft (OHC). The latter can also be broken down into two subgroups: the Single Overhead Camshaft (SOHC) and the Dual Overhead Camshaft (DOHC). Generally, if there is only a single camshaft on a head, it is just referred to as an OHC head. Also, an engine with an OHV cylinder head is also known as a pushrod engine.

Most cylinder heads these days are made of an aluminum alloy due to its light weight, durability and heat transfer qualities. However, cast iron was the material of choice in the past, and is still used on many engines today. Whether made from aluminum or iron, all cylinder heads have valves and seats. Some use two valves per cylinder, while the more hi-tech engines will utilize a multi-valve configuration using 3, 4 and even 5 valves per cylinder. When the valve contacts the seat, it does so on precision machined surfaces, which seals the combustion chamber. All cylinder heads have a valve guide for each valve. The guide centers the valve to the seat and allows it to move up and down within it. The clearance between the valve and guide can be critical. Too much clearance and the engine may consume oil, lose vacuum and/or damage the seat. Too little, and the valve can stick in the guide causing the engine to run poorly if at all, and possibly causing severe damage. The last component all cylinder heads have are valve springs. The spring holds the valve against its seat. It also returns the valve to this position when the valve has been opened by the valve train or camshaft. The spring is fastened to the valve by a retainer and valve locks (sometimes called keepers). Aluminum heads will also have a valve spring shim to keep the spring from wearing away the aluminum.

An ideal method of rebuilding the cylinder head would involve replacing all of the valves, guides, seats, springs, etc. with new ones. However, depending on how the engine was maintained, often this is not necessary. A major cause of valve, guide and seat wear is an improperly tuned engine. An engine that is running too rich, will often wash the lubricating oil out of the guide with gasoline, causing it to wear rapidly. Conversely, an engine which is running too lean will place higher combustion temperatures on the valves and seats allowing them to wear or even burn. Springs fall victim to the operating habits of the individual. An operator who often runs the engine rpm to the redline will wear out or break the springs faster then one that stays well below it. Unfortunately, operating time takes a toll on all of the parts. Generally, the valves, guides, springs and seats in a cylinder head can be machined and re-used, saving you money. However, if a valve is burnt, it may be wise to replace all of the valves, since they were all operating in the same environment. The same goes for any other component on the cylinder head. Think of it as an insurance policy against future problems related to that component.

Unfortunately, the only way to find out which components need replacing, is to disassemble and carefully check each piece. After the cylinder head(s) are disassembled, thoroughly clean all of the components.

DISASSEMBLY

OHV Heads

Before disassembling the cylinder head, you may want to fabricate some containers to hold the various parts, as some of them can be quite small (such as keepers) and easily lost. Also keeping yourself and the components organized will aid in assembly and reduce confusion. Where possible, try to maintain a components original location; this is especially important if there is not going to be any machine work performed on the components.

1. If you haven't already removed the rocker arms and/or shafts, do so now.
2. Position the head so that the springs are easily accessed.
3. Use a valve spring compressor tool, and relieve spring tension from the retainer.

➡Due to engine varnish, the retainer may stick to the valve locks. A gentle tap with a hammer may help to break it loose.

4. Remove the valve locks from the valve tip and/or retainer. A small magnet may help in removing the locks.
5. Lift the valve spring, tool and all, off of the valve stem.

6. If equipped, remove the valve seal. If the seal is difficult to remove with the valve in place, try removing the valve first, then the seal. Follow the steps below for valve removal.
7. Position the head to allow access for withdrawing the valve.

➡Cylinder heads that have seen a lot of hours and/or abuse may have mushroomed the valve lock grove and/or tip, causing difficulty in removal of the valve. If this has happened, use a metal file to carefully remove the high spots around the lock grooves and/or tip. Only file it enough to allow removal.

8. Remove the valve from the cylinder head.
9. If equipped, remove the valve spring shim. A small magnetic tool or screwdriver will aid in removal.
10. Repeat Steps 3 though 9 until all of the valves have been removed.

OHC Heads

➧ See Figures 121 and 122

Whether it is a single or dual overhead camshaft cylinder head, the disassembly procedure is relatively unchanged. One aspect to pay attention to is careful labeling of the parts on the dual camshaft cylinder head. There will be an intake camshaft and followers as well as an exhaust camshaft and followers and they must be labeled as such. In some cases, the components are identical and could easily be installed incorrectly. DO NOT MIX THEM UP! Determining which is which is very simple; the intake camshaft and components are on the same side of the head as was the intake manifold. Conversely, the exhaust camshaft and components are on the same side of the head as was the exhaust manifold.

ROCKER ARM TYPE CAMSHAFT FOLLOWERS

➧ See Figures 123 thru 131

Most cylinder heads with rocker arm-type camshaft followers are easily disassembled using a standard valve spring compressor. However, certain may not have enough open space around the spring for the standard tool and may require you to use a C-clamp style compressor tool instead.

1. If not already removed, remove the rocker arms and/or shafts and the camshaft. Mark their positions for assembly.
2. Position the cylinder head to allow access to the valve spring.
3. Use a valve spring compressor tool to relieve the spring tension from the retainer.

➡Due to engine varnish, the retainer may stick to the valve locks. A gentle tap with a hammer may help to break it loose.

4. Remove the valve locks from the valve tip and/or retainer. A small magnet may help in removing the small locks.
5. Lift the valve spring, tool and all, off of the valve stem.
6. If equipped, remove the valve seal. If the seal is difficult to remove with the valve in place, try removing the valve first, then the seal. Follow the steps below for valve removal.
7. Position the head to allow access for withdrawing the valve.

➡Cylinder heads that have seen a lot of miles and/or abuse may have mushroomed the valve lock grove and/or tip, causing difficulty in removal of the valve. If this has happened, use a metal file to carefully remove the high spots around the lock grooves and/or tip. Only file it enough to allow removal.

8. Remove the valve from the cylinder head.
9. If equipped, remove the valve spring seat. A small magnetic tool or screwdriver will aid in removal.
10. Repeat Steps 3 though 9 until all of the valves have been removed.

INSPECTION

Now that all of the cylinder head components are clean, it's time to inspect them for wear and/or damage. To accurately inspect them, you will need some specialized tools:

- A 0–1 in. micrometer for the valves
- A dial indicator or inside diameter gauge for the valve guides
- A spring pressure test gauge

04897P36

Fig. 121 Example of a two-valve cylinder head, the top view . . .

04897P35

Fig. 122 . . . and the bottom view

04897P25

Fig. 123 Example of the shaft mounted rocker arms on a BF35A cylinder head

04897P26

Fig. 124 Before the camshaft can be removed, all of the followers must first be removed

04897P28

Fig. 125 Install a valve spring compressor on the cylinder head . . .

04897P29

Fig. 126 . . . and compress the valve spring in order to remove the keepers . . .

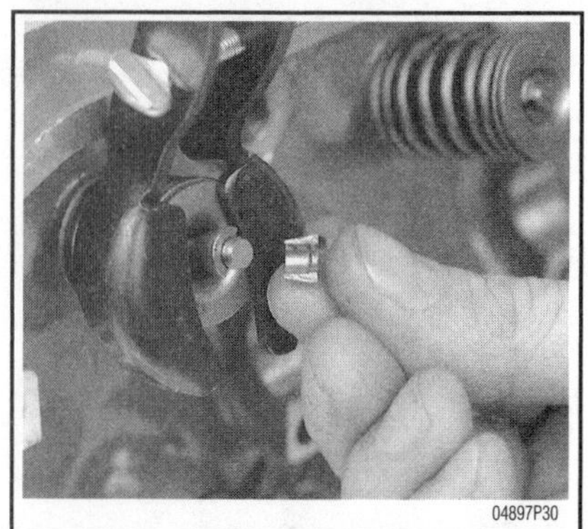
04897P30

Fig. 127 . . . then remove the valve locks from the valve stem and spring retainer

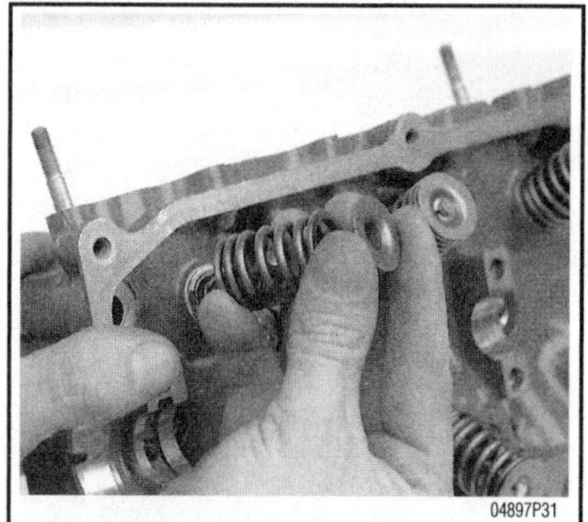
04897P31

Fig. 128 Remove the valve spring and retainer from the cylinder head

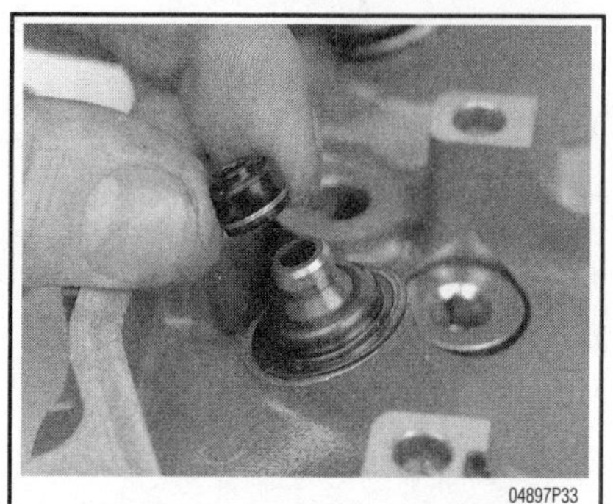
04897P33

Fig. 129 Remove the valve seal from the guide. Some gentle prying or pliers may help to remove stubborn ones

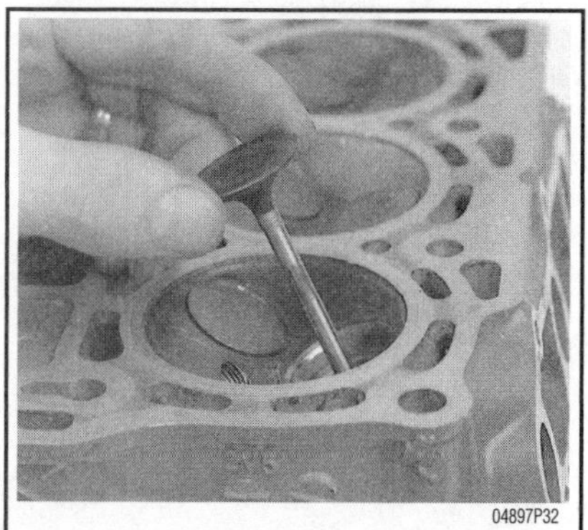
04897P32

Fig. 130 Invert the cylinder head and withdraw the valve from the valve guide bore

04897P34

Fig. 131 All aluminum and some cast iron heads will have these valve spring seats. Remove all of them as well

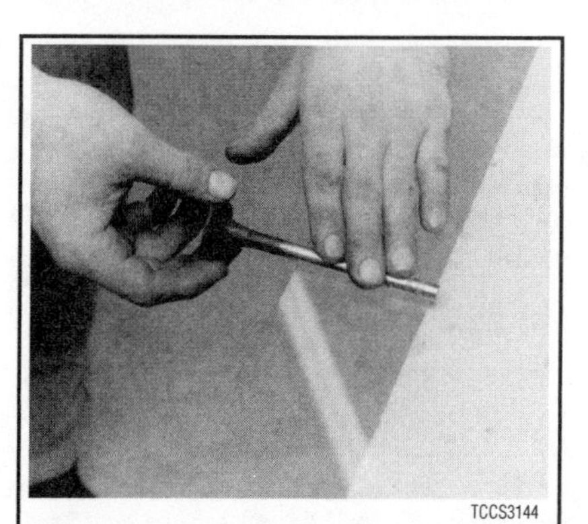
TCCS3144

Fig. 132 Valve stems may be rolled on a flat surface to check for bends

If you do not have access to the proper tools, you may want to bring the components to a shop that does.

Valves

➧ See Figures 132 and 133

The first thing to inspect are the valve heads. Look closely at the head, margin and face for any cracks, excessive wear or burning. The margin is the best place to look for burning. It should have a squared edge with an even width all around the diameter. When a valve burns, the margin will look melted and the edges rounded. Also inspect the valve head for any signs of tuliping. This will show as a lifting of the edges or dishing in the center of the head and will usually not occur to all of the valves. All of the heads should look the same, any that seem dished more than others are probably bad. Next, inspect the valve lock grooves and valve tips. Check for any burrs around the lock grooves, especially if you had to file them to remove the valve. Valve tips should appear flat, although slight rounding with high mileage engines is normal. Slightly worn valve tips will need to be machined flat. Last, measure the valve stem diameter with the micrometer. Measure the area that rides within the guide, especially towards the tip where most of the wear occurs. Take several measurements along its length and compare them to each other. Wear should be even along the length with little to no taper. If no minimum diameter is given in the specifications, then the stem should not read more than 0.001 in. (0.025mm) below the unworn area of the valve stem. Any valves that fail these inspections should be replaced.

Springs, Retainers and Valve Locks

➧ See Figures 134 and 135

The first thing to check is the most obvious, broken springs. Next check the free length and squareness of each spring. If applicable, insure to distinguish between intake and exhaust springs. Use a ruler and/or carpenter's square to measure the length. A carpenter's square should be used to check the springs for squareness. If a spring pressure test gauge is available, check each springs

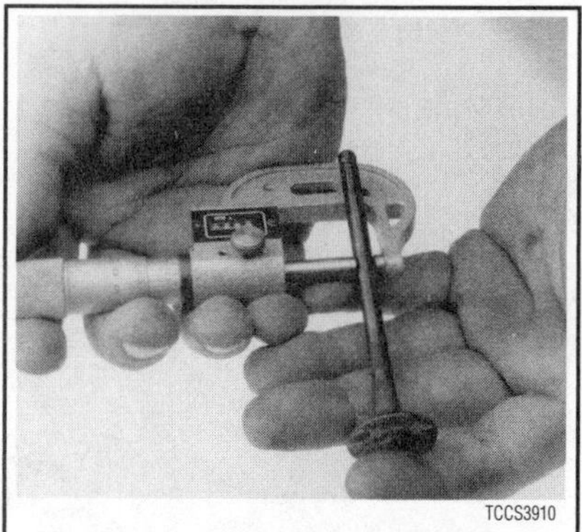
TCCS3910

Fig. 133 Use a micrometer to check the valve stem diameter

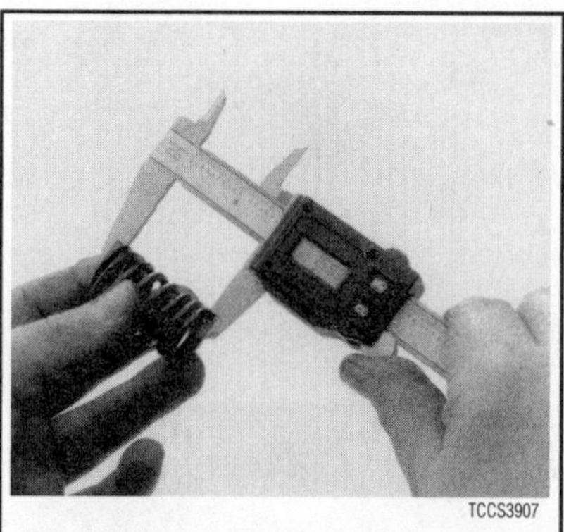
TCCS3907

Fig. 134 Use a caliper to check the valve spring free-length

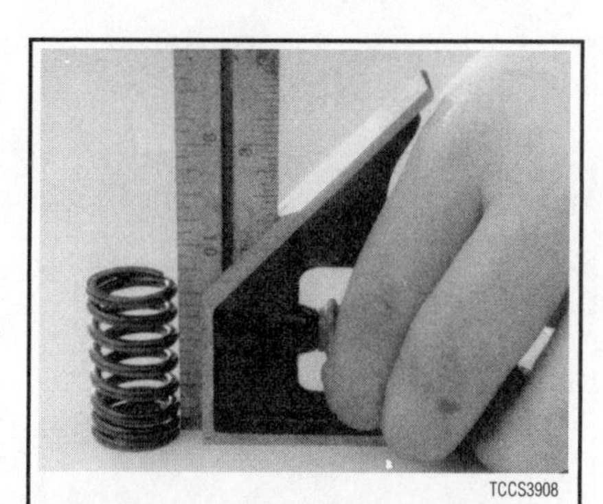
TCCS3908

Fig. 135 Check the valve spring for squareness on a flat surface; a carpenter's square can be used

rating and compare to the specifications chart. Check the readings against the specifications given. Any springs that fail these inspections should be replaced.

The spring retainers rarely need replacing, however they should still be checked as a precaution. Inspect the spring mating surface and the valve lock retention area for any signs of excessive wear. Also check for any signs of cracking. Replace any retainers that are questionable.

Valve locks should be inspected for excessive wear on the outside contact area as well as on the inner notched surface. Any locks which appear worn or broken and its respective valve should be replaced.

Cylinder Head

There are several things to check on the cylinder head: valve guides, seats, cylinder head surface flatness, cracks and physical damage.

VALVE GUIDES

Now that you know the valves are good, you can use them to check the guides, although a new valve, if available, is preferred. Before you measure anything, look at the guides carefully and inspect them for any cracks, chips or breakage. Also if the guide is a removable style (as in most aluminum heads), check them for any looseness or evidence of movement. All of the guides should appear to be at the same height from the spring seat. If any seem lower (or higher) from another, the guide has moved. Mount a dial indicator onto the spring side of the cylinder head. Lightly oil the valve stem and insert it into the cylinder head. Position the dial indicator against the valve stem near the tip and zero the gauge. Grasp the valve stem and wiggle towards and away from the dial indicator and observe the readings. Mount the dial indicator 90 degrees from the initial point and zero the gauge and again take a reading. Compare the two readings for a out of round condition. Check the readings against the specifications given. An Inside Diameter (I.D.) gauge designed for valve guides will give you an accurate valve guide bore measurement. If the I.D. gauge is used, compare the readings with the specifications given. Any guides that fail these inspections should be replaced or machined.

VALVE SEATS

A visual inspection of the valve seats should show a slightly worn and pitted surface where the valve face contacts the seat. Inspect the seat carefully for severe pitting or cracks. Also, a seat that is badly worn will be recessed into the cylinder head. A severely worn or recessed seat may need to be replaced. All cracked seats must be replaced. A seat concentricity gauge, if available, should be used to check the seat run-out. If run-out exceeds specifications the seat must be machined (if no specification is given use 0.002 in. or 0.051mm).

CYLINDER HEAD SURFACE FLATNESS

See Figures 136 and 137

After you have cleaned the gasket surface of the cylinder head of any old gasket material, check the head for flatness.

Place a straightedge across the gasket surface. Using feeler gauges, determine the clearance at the center of the straightedge and across the cylinder head at several points. Check along the centerline and diagonally on the head surface. If the warpage exceeds 0.003 in. (0.076mm) within a 6.0 in. (15.2cm) span, or 0.006 in. (0.152mm) over the total length of the head, the cylinder head must be resurfaced. After resurfacing the heads of a V-type engine, the intake manifold flange surface should be checked, and if necessary, milled proportionally to allow for the change in its mounting position.

CRACKS AND PHYSICAL DAMAGE

Generally, cracks are limited to the combustion chamber, however, it is not uncommon for the head to crack in a spark plug hole, port, outside of the head or in the valve spring/rocker arm area. The first area to inspect is always the hottest: the exhaust seat/port area.

A visual inspection should be performed, but just because you don't see a crack does not mean it is not there. Some more reliable methods for inspecting for cracks include Magnaflux®, a magnetic process or Zyglo®, a dye penetrant. Magnaflux® is used only on ferrous metal (cast iron) heads. Zyglo® uses a spray-on fluorescent mixture along with a black light to reveal the cracks. It is strongly recommended to have your cylinder head checked professionally for cracks, especially if the engine was known to have overheated and/or leaked or consumed coolant. Contact a local shop for availability and pricing of these services.

Physical damage is usually very evident. For example, a broken mounting ear from dropping the head or a bent or broken stud and/or bolt. All of these defects should be fixed or, if unrepairable, the head should be replaced.

Camshaft and Followers

See Figure 138

Inspect the camshaft(s) and followers as described under "Camshaft, Bearings and Lifters".

REFINISHING & REPAIRING

Many of the procedures given for refinishing and repairing the cylinder head components must be performed by a machine shop. Certain steps, if the inspected part is not worn, can be performed yourself inexpensively. However, you spent a lot of time and effort so far, why risk trying to save a couple bucks if you might have to do it all over again?

Valves

Any valves that were not replaced should be refaced and the tips ground flat. Unless you have access to a valve grinding machine, this should be done by a machine shop. If the valves are in extremely good condition, as well as the valve seats and guides, they may be lapped in without performing machine work.

It is a recommended practice to lap the valves even after machine work has been performed and/or new valves have been purchased. This insures a positive seal between the valve and seat.

LAPPING THE VALVES

➡Before lapping the valves to the seats, read the rest of the cylinder head section to insure that any related parts are in acceptable enough condition to continue.

➡Before any valve seat machining and/or lapping can be performed, the guides must be within factory recommended specifications.

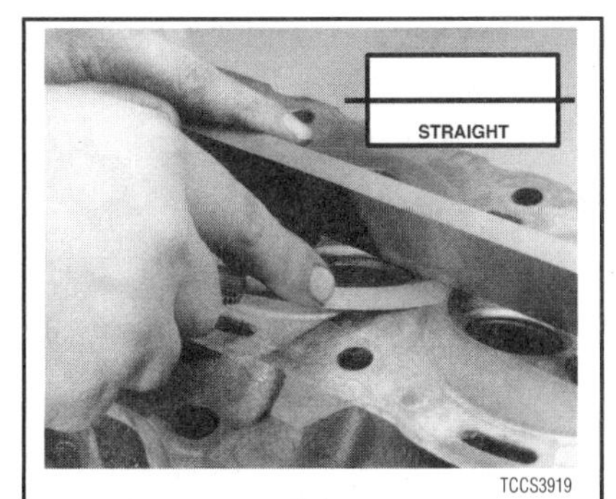

Fig. 136 Check the head for flatness across the center of the head surface using a straightedge and feeler gauge

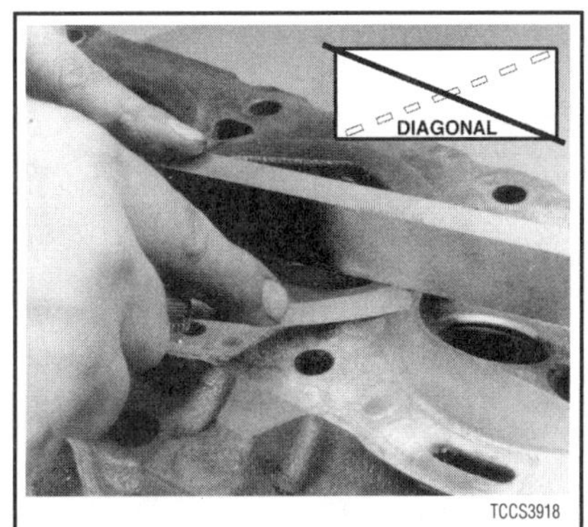

Fig. 137 Checks should also be made along both diagonals of the head surface

Fig. 138 Inspect the cam follower faces for wear or damage

1. Invert the cylinder head.
2. Lightly lubricate the valve stems and insert them into the cylinder head in their numbered order.
3. Raise the valve from the seat and apply a small amount of fine lapping compound to the seat.
4. Moisten the suction head of a hand-lapping tool and attach it to the head of the valve.
5. Rotate the tool between the palms of both hands, changing the position of the valve on the valve seat and lifting the tool often to prevent grooving.
6. Lap the valve until a smooth, polished circle is evident on the valve and seat.
7. Remove the tool and the valve. Wipe away all traces of the grinding compound and store the valve to maintain its lapped location.

**** WARNING**

Do not get the valves out of order after they have been lapped. They must be put back with the same valve seat with which they were lapped.

Springs, Retainers and Valve Locks

There is no repair or refinishing possible with the springs, retainers and valve locks. If they are found to be worn or defective, they must be replaced with new (or known good) parts.

Cylinder Head

Most refinishing procedures dealing with the cylinder head must be performed by a machine shop. Read the sections below and review your inspection data to determine whether or not machining is necessary.

VALVE GUIDE

➡If any machining or replacements are made to the valve guides, the seats must be machined.

Unless the valve guides need machining or replacing, the only service to perform is to thoroughly clean them of any dirt or oil residue.

There are only two types of valve guides used on marine outboard engines: the replaceable-type (all aluminum heads) and the cast-in integral-type (most cast iron heads). There are four recommended methods for repairing worn guides.

- Knurling
- Inserts
- Reaming oversize
- Replacing

Knurling is a process in which metal is displaced and raised, thereby reducing clearance, giving a true center, and providing oil control. It is the least expensive way of repairing the valve guides. However, it is not necessarily the best, and in some cases, a knurled valve guide will not stand up for more than a short time. It requires a special knurlizer and precision reaming tools to obtain proper clearances. It would not be cost effective to purchase these tools, unless you plan on rebuilding several of the same cylinder head.

Installing a guide insert involves machining the guide to accept a bronze insert. One style is the coil-type which is installed into a threaded guide. Another is the thin-walled insert where the guide is reamed oversize to accept a split-sleeve insert. After the insert is installed, a special tool is then run through the guide to expand the insert, locking it to the guide. The insert is then reamed to the standard size for proper valve clearance.

Reaming for oversize valves restores normal clearances and provides a true valve seat. Most cast-in type guides can be reamed to accept an valve with an oversize stem. The cost factor for this can become quite high as you will need to purchase the reamer and new, oversize stem valves for all guides which were reamed. Oversizes are generally 0.003 to 0.030 in. (0.076 to 0.762mm), with 0.015 in. (0.381mm) being the most common.

To replace cast-in type valve guides, they must be drilled out, then reamed to accept replacement guides. This must be done on a fixture which will allow centering and leveling off of the original valve seat or guide, otherwise a serious guide-to-seat misalignment may occur making it impossible to properly machine the seat.

Replaceable-type guides are pressed into the cylinder head. A hammer and a stepped drift or punch may be used to install and remove the guides. Before removing the guides, measure the protrusion on the spring side of the head and record it for installation. Use the stepped drift to hammer out the old guide from the combustion chamber side of the head. When installing, determine whether or not the guide also seals a water jacket in the head, and if it does, use the recommended sealing agent. If there is no water jacket, grease the valve guide and its bore. Use the stepped drift, and hammer the new guide into the cylinder head from the spring side of the cylinder head. A stack of washers the same thickness as the measured protrusion may help the installation process.

VALVE SEATS

➡Before any valve seat machining can be performed, the guides must be within factory recommended specifications.

➡If any machining or replacements were made to the valve guides, the seats must be machined.

If the seats are in good condition, the valves can be lapped to the seats, and the cylinder head assembled. See the valves section for instructions on lapping.

If the valve seats are worn, cracked or damaged, they must be serviced by a machine shop. The valve seat must be perfectly centered to the valve guide, which requires very accurate machining.

CYLINDER HEAD SURFACE

If the cylinder head is warped, it must be machined flat. If the warpage is extremely severe, the head may need to be replaced. In some instances, it may be possible to straighten a warped head enough to allow machining. In either case, contact a professional machine shop for service.

➡Any OHC cylinder head that shows excessive warpage should have the camshaft bearing journals align bored after the cylinder head has been resurfaced.

**** WARNING**

Failure to align bore the camshaft bearing journals could result in severe engine damage including but not limited to: valve and piston damage, connecting rod damage, camshaft and/or crankshaft breakage.

CRACKS AND PHYSICAL DAMAGE

Certain cracks can be repaired in both cast iron and aluminum heads. For cast iron, a tapered threaded insert is installed along the length of the crack. Aluminum can also use the tapered inserts, however welding is the preferred method. Some physical damage can be repaired through brazing or welding. Contact a machine shop to get expert advice for your particular dilemma.

ASSEMBLY

The first step for any assembly job is to have a clean area in which to work. Next, thoroughly clean all of the parts and components that are to be assembled. Finally, place all of the components onto a suitable work space and, if necessary, arrange the parts to their respective positions.

OHV Engines

1. Lightly lubricate the valve stems and insert all of the valves into the cylinder head. If possible, maintain their original locations.
2. If equipped, install any valve spring shims which were removed.
3. If equipped, install the new valve seals, keeping the following in mind:
   - If the valve seal presses over the guide, lightly lubricate the outer guide surfaces.
   - If the seal is an O-ring type, it is installed just after compressing the spring but before the valve locks.
4. Place the valve spring and retainer over the stem.
5. Position the spring compressor tool and compress the spring.
6. Assemble the valve locks to the stem.
7. Relieve the spring pressure slowly and insure that neither valve lock becomes dislodged by the retainer.
8. Remove the spring compressor tool.
9. Repeat Steps 2 through 8 until all of the springs have been installed.

OHC Engines

ROCKER ARM TYPE CAMSHAFT FOLLOWERS

See Figure 139

1. Lightly lubricate the valve stems and insert all of the valves into the cylinder head. If possible, maintain their original locations.
2. If equipped, install any valve spring shims which were removed.
3. If equipped, install the new valve seals, keeping the following in mind:
   - If the valve seal presses over the guide, lightly lubricate the outer guide surfaces.
   - If the seal is an O-ring type, it is installed just after compressing the spring but before the valve locks.
4. Place the valve spring and retainer over the stem.
5. Position the spring compressor tool and compress the spring.
6. Assemble the valve locks to the stem.
7. Relieve the spring pressure slowly and insure that neither valve lock becomes dislodged by the retainer.
8. Remove the spring compressor tool.
9. Repeat Steps 2 through 8 until all of the springs have been installed.
10. Install the camshaft(s), rockers, shafts and any other components that were removed for disassembly.

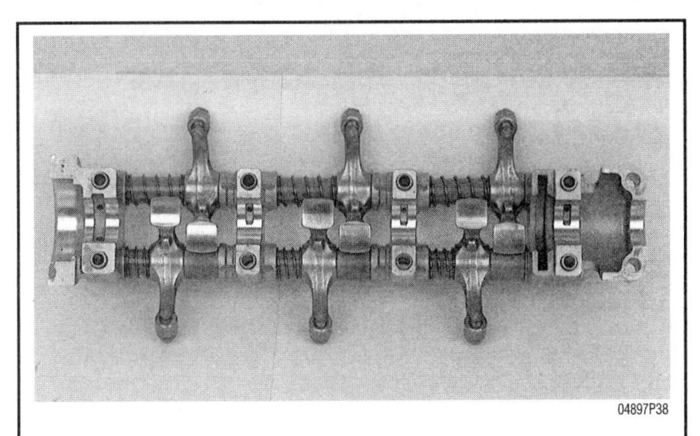
04897P38

Fig. 139 Rocker arm type cam followers

Engine Block

GENERAL INFORMATION

A thorough overhaul or rebuild of an engine block would include replacing the pistons, rings, bearings, timing belt/chain assembly and oil pump. For OHV engines also include a new camshaft and lifters. The block would then have the cylinders bored and honed oversize (or if using removable cylinder sleeves, new sleeves installed) and the crankshaft would be cut undersize to provide new wearing surfaces and perfect clearances. However, your particular engine may not have everything worn out. What if only the piston rings have worn out and the clearances on everything else are still within factory specifications? Well, you could just replace the rings and put it back together, but this would be a very rare example. Chances are, if one component in your engine is worn, other components are sure to follow, and soon. At the very least, you should always replace the rings, bearings and oil pump. This is what is commonly called a "freshen up".

Cylinder Ridge Removal

Because the top piston ring does not travel to the very top of the cylinder, a ridge is built up between the end of the travel and the top of the cylinder bore.

Pushing the piston and connecting rod assembly past the ridge can be difficult, and damage to the piston ring lands could occur. If the ridge is not removed before installing a new piston or not removed at all, piston ring breakage and piston damage may occur.

➡It is always recommended that you remove any cylinder ridges before removing the piston and connecting rod assemblies. If you know that new pistons are going to be installed and the engine block will be bored oversize, you may be able to forego this step. However, some ridges may actually prevent the assemblies from being removed, necessitating its removal.

There are several different types of ridge reamers on the market, none of which are inexpensive. Unless a great deal of engine rebuilding is anticipated, borrow or rent a reamer.

1. Turn the crankshaft until the piston is at the bottom of its travel.
2. Cover the head of the piston with a rag.
3. Follow the tool manufacturers instructions and cut away the ridge, exercising extreme care to avoid cutting too deeply.
4. Remove the ridge reamer, the rag and as many of the cuttings as possible. Continue until all of the cylinder ridges have been removed.

DISASSEMBLY

See Figures 140, 141 and 142

The engine disassembly instructions following assume that you have the engine mounted on an engine stand. If not, it is easiest to disassemble the engine on a bench or the floor. You must be able to access the connecting rod fasteners and turn the crankshaft during disassembly. Also, all engine covers (timing, front, side, oil pan, whatever) should have already been removed. Engines which are seized or locked up may not be able to be completely disassembled, and a core (salvage yard) engine should be purchased.

If not done during the cylinder head removal on pushrod engines, remove the pushrods and lifters, keeping them in order for assembly. Remove the timing gears and/or timing chain assembly, then remove the oil pump drive assembly and withdraw the camshaft from the engine block. Remove the oil pick-up and pump assembly. If equipped, remove any balance or auxiliary shafts. If necessary, remove the cylinder ridge from the top of the bore. See the cylinder ridge removal procedure earlier in this section.

If not done during the cylinder head removal on OHC Engines, remove the timing belt and or gear assembly. Remove the oil pick-up and pump assembly and, if necessary, the pump drive. If equipped, remove any balance or auxiliary

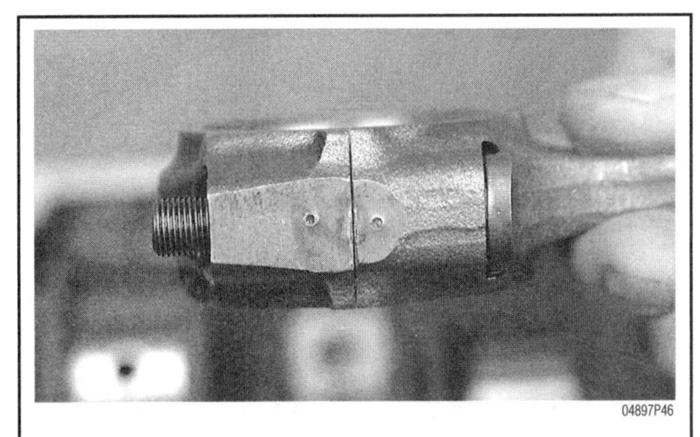
04897P46

Fig. 140 Punch marks indicating a particular cylinder number

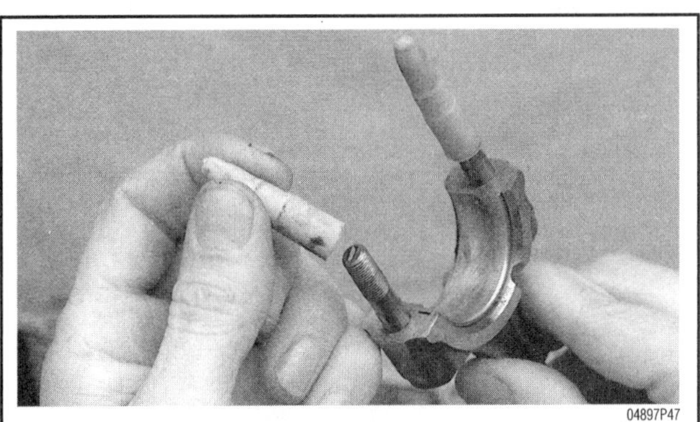
04897P47

Fig. 141 Place rubber hose over the connecting rod studs to protect the crankshaft and cylinder bores from damage

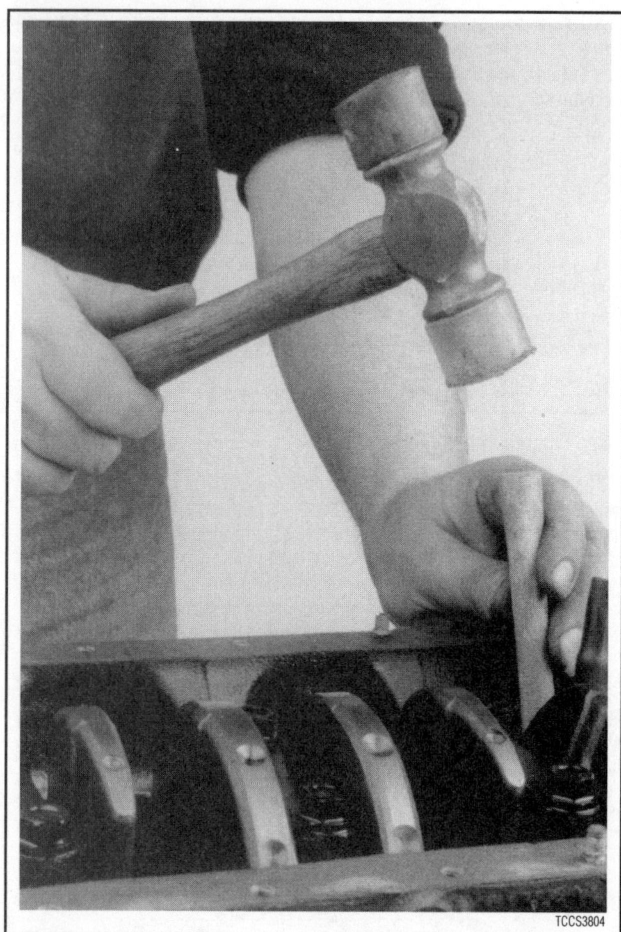

Fig. 142 Carefully tap the piston out of the bore using a wooden dowel

shafts. If necessary, remove the cylinder ridge from the top of the bore. See the "Cylinder Ridge Removal" for more information.

Rotate the engine over so that the crankshaft is exposed. Use a number punch or scribe and mark each connecting rod with its respective cylinder number. The cylinder closest to the front of the engine is always number 1. However, depending on the engine placement, the front of the engine could either be the flywheel or damper/pulley end. Use a number punch or scribe and also mark the main bearing caps from front to rear with the front most cap being number 1 (if there are five caps, mark them 1 through 5, front to rear).

WARNING

Take special care when pushing the connecting rod up from the crankshaft because the sharp threads of the rod bolts/studs will score the crankshaft journal. Insure that special plastic caps are installed over them, or cut two pieces of rubber hose to do the same.

Again, rotate the engine, this time to position the number one cylinder bore (head surface) up. Turn the crankshaft until the number one piston is at the bottom of its travel, this should allow the maximum access to its connecting rod. Remove the number one connecting rods fasteners and cap and place two lengths of rubber hose over the rod bolts/studs to protect the crankshaft from damage. Using a sturdy wooden dowel and a hammer, push the connecting rod up about 1 in. (25mm) from the crankshaft and remove the upper bearing insert. Continue pushing or tapping the connecting rod up until the piston rings are out of the cylinder bore. Remove the piston and rod by hand, put the upper half of the bearing insert back into the rod, install the cap with its bearing insert installed, and hand-tighten the cap fasteners. If the parts are kept in order in this manner, they will not get lost and you will be able to tell which bearings came form what cylinder if any problems are discovered and diagnosis is necessary. Remove all the other piston assemblies in the same manner.

The only remaining component in the engine block should now be the crankshaft. Loosen the main bearing caps evenly until the fasteners can be turned by hand, then remove them and the caps. Remove the crankshaft from the engine block. Thoroughly clean all of the components.

INSPECTION

Now that the engine block and all of its components are clean, it's time to inspect them for wear and/or damage. To accurately inspect them, you will need some specialized tools:

- Two or three separate micrometers to measure the pistons and crankshaft journals
- A dial indicator
- Telescoping gauges for the cylinder bores
- A rod alignment fixture to check for bent connecting rods

If you do not have access to the proper tools, you may want to bring the components to a shop that does.

Generally, you shouldn't expect cracks in the engine block or its components unless it was known to leak, consume or mix engine fluids, it was severely overheated, or there was evidence of bad bearings and/or crankshaft damage. A visual inspection should be performed on all of the components, but just because you don't see a crack does not mean it is not there. Some more reliable methods for inspecting for cracks include Magnaflux®, a magnetic process or Zyglo®, a dye penetrant. Magnaflux® is used only on ferrous metal (cast iron). Zyglo® uses a spray on fluorescent mixture along with a black light to reveal the cracks. It is strongly recommended to have your engine block checked professionally for cracks, especially if the engine was known to have overheated and/or leaked or consumed coolant. Contact a local shop for availability and pricing of these services.

Engine Block

ENGINE BLOCK BEARING ALIGNMENT

The Honda four stroke engines, BF15A–BF130A use a split crankcase. The two halves of the crankcase provide the crankshaft bearing saddles and are bolted together along the crank centerline. There is no way to align-bore this type of block. If after measuring the oil clearances on the crankshaft and connecting rods, they do not meet specification, even after installing undersize bearings on the crankshaft and connecting rods, the crankcase must be replaced.

Separate the crankcase halves and inspect all of the main bearing saddles for damage, burrs or high spots. If damage is found, and it is caused from a spun main bearing, the block will need to have the crankshaft oil clearances checked. Any burrs or high spots should be carefully removed with a metal file.

Place a straightedge on the bearing saddles, in the engine block, along the centerline of the crankshaft. If any clearance exists between the straightedge and the saddles, the block must be replaced.

DECK FLATNESS

The top of the engine block where the cylinder head mounts is called the deck. Insure that the deck surface is clean of dirt, carbon deposits and old gasket material. Place a straightedge across the surface of the deck along its centerline and, using feeler gauges, check the clearance along several points. Repeat the checking procedure with the straightedge placed along both diagonals of the deck surface. If the reading exceeds 0.003 in. (0.076mm) within a 6.0 in. (15.2cm) span, or 0.006 in. (0.152mm) over the total length of the deck, it must be machined.

CYLINDER BORES

See Figure 143

The cylinder bores house the pistons and are slightly larger than the pistons themselves. A common piston-to-bore clearance is 0.0015–0.0025 in. (0.0381mm–0.0635mm). Inspect and measure the cylinder bores. The bore should be checked for out-of-roundness, taper and size. The results of this inspection will determine whether the cylinder can be used in its existing size and condition, or a rebore to the next oversize is required (or in the case of removable sleeves, have replacements installed).

The amount of cylinder wall wear is always greater at the top of the cylinder than at the bottom. This wear is known as taper. Any cylinder that has a taper of 0.0012 in. (0.305mm) or more, must be rebored. Measurements are taken at a

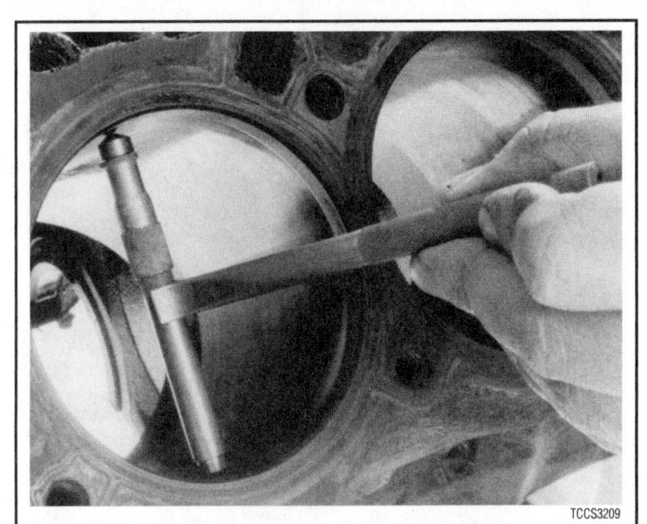

Fig. 143 Use a telescoping gauge to measure the cylinder bore diameter—take several readings within the same bore

number of positions in each cylinder: at the top, middle and bottom and at two points at each position; that is, at a point 90 degrees from the crankshaft centerline, as well as a point parallel to the crankshaft centerline. The measurements are made with either a special dial indicator or a telescopic gauge and micrometer. If the necessary precision tools to check the bore are not available, take the block to a machine shop and have them mike it. Also if you don't have the tools to check the cylinder bores, chances are you will not have the necessary devices to check the pistons, connecting rods and crankshaft. Take these components with you and save yourself an extra trip.

For our procedures, we will use a telescopic gauge and a micrometer. You will need one of each, with a measuring range which covers your cylinder bore size.

1. Position the telescopic gauge in the cylinder bore, loosen the gauges lock and allow it to expand.

➡Your first two readings will be at the top of the cylinder bore, then proceed to the middle and finally the bottom, making a total of six measurements.

2. Hold the gauge square in the bore, 90 degrees from the crankshaft centerline, and gently tighten the lock. Tilt the gauge back to remove it from the bore.
3. Measure the gauge with the micrometer and record the reading.
4. Again, hold the gauge square in the bore, this time parallel to the crankshaft centerline, and gently tighten the lock. Again, you will tilt the gauge back to remove it from the bore.
5. Measure the gauge with the micrometer and record this reading. The difference between these two readings is the out-of-round measurement of the cylinder.
6. Repeat steps 1 through 5, each time going to the next lower position, until you reach the bottom of the cylinder. Then go to the next cylinder, and continue until all of the cylinders have been measured.

The difference between these measurements will tell you all about the wear in your cylinders. The measurements which were taken 90 degrees from the crankshaft centerline will always reflect the most wear. That is because at this position is where the engine power presses the piston against the cylinder bore the hardest. This is known as thrust wear. Take your top, 90 degree measurement and compare it to your bottom, 90 degree measurement. The difference between them is the taper. When you measure your pistons, you will compare these readings to your piston sizes and determine piston-to-wall clearance.

Crankshaft

Inspect the crankshaft for visible signs of wear or damage. All of the journals should be perfectly round and smooth. Slight scores are normal for a used crankshaft, but you should hardly feel them with your fingernail. When measuring the crankshaft with a micrometer, you will take readings at the front and rear of each journal, then turn the micrometer 90 degrees and take two more readings, front and rear. The difference between the front-to-rear readings is the journal taper and the first-to-90 degree reading is the out-of-round measurement. Generally, there should be no taper or out-of-roundness found, however, up to 0.0005 in. (0.0127mm) for either can be overlooked. Also, the readings should fall within the factory specifications for journal diameters.

If the crankshaft journals fall within specifications, it is recommended that it be polished before being returned to service. Polishing the crankshaft insures that any minor burrs or high spots are smoothed, thereby reducing the chance of scoring the new bearings.

Pistons and Connecting Rods

PISTONS

➧ See Figure 144

The piston should be visually inspected for any signs of cracking or burning (caused by hot spots or detonation), and scuffing or excessive wear on the skirts. The wrist pin attaches the piston to the connecting rod. The piston should move freely on the wrist pin, both sliding and pivoting. Grasp the connecting rod securely, or mount it in a vise, and try to rock the piston back and forth along the centerline of the wrist pin. There should not be any excessive play evident between the piston and the pin. If there are C-clips retaining the pin in the piston then you have wrist pin bushings in the rods. There should not be any excessive play between the wrist pin and the rod bushing. Normal clearance for the wrist pin is approx. 0.001–0.002 in. (0.025mm–0.051mm).

Use a micrometer and measure the diameter of the piston, perpendicular to the wrist pin, on the skirt. Compare the reading to its original cylinder measurement obtained earlier. The difference between the two readings is the piston-to-wall clearance. If the clearance is within specifications, the piston may be used as is. If the piston is out of specification, but the bore is not, you will need a new piston. If both are out of specification, you will need the cylinder rebored and oversize pistons installed. Generally if two or more pistons/bores are out of specification, it is best to rebore the entire block and purchase a complete set of oversize pistons.

Fig. 144 Measure the piston's outer diameter, perpendicular to the wrist pin, with a micrometer

CONNECTING ROD

You should have the connecting rod checked for straightness at a machine shop. If the connecting rod is bent, it will unevenly wear the bearing and piston, as well as place greater stress on these components. Any bent or twisted connecting rods must be replaced. If the rods are straight and the wrist pin clearance is within specifications, then only the bearing end of the rod need be checked. Place the connecting rod into a vice, with the bearing inserts in place, install the cap to the rod and torque the fasteners to specifications. Use a telescoping gauge and carefully measure the inside diameter of the bearings. Compare this reading to the rods original crankshaft journal diameter measurement. The difference is the oil clearance. If the oil clearance is not within specifications, install new bearings in the rod and take another measurement. If the clearance is still out of specifications, and the crankshaft is not, the rod will need to be reconditioned by a machine shop.

➡You can also use Plastigage® to check the bearing clearances. The assembling section has complete instructions on its use.

Camshaft

Inspect the camshaft and lifters/followers as described in "Camshaft, Bearings and Lifters" earlier in this section.

Bearings

All of the engine bearings should be visually inspected for wear and/or damage. The bearing should look evenly worn all around with no deep scores or pits. If the bearing is severely worn, scored, pitted or heat blued, then the bearing, and the components that use it, should be brought to a machine shop for inspection. Full-circle bearings (used on most auxiliary shafts, balance shafts, etc.) require specialized tools for removal and installation, and should be brought to a machine shop for service.

Oil Pump

➡The oil pump is responsible for providing constant lubrication to the whole engine and so it is recommended that a new oil pump be installed when rebuilding the engine.

Completely disassemble the oil pump and thoroughly clean all of the components. Inspect the oil pump gears and housing for wear and/or damage. Insure that the pressure relief valve operates properly and there is no binding or sticking due to varnish or debris. If all of the parts are in proper working condition, lubricate the gears and relief valve, and assemble the pump.

REFINISHING

▶ See Figure 145

Almost all engine block refinishing must be performed by a machine shop. If the cylinders are not to be rebored, then the cylinder glaze can be removed with a ball hone. When removing cylinder glaze with a ball hone, use a light or penetrating type oil to lubricate the hone. Do not allow the hone to run dry as this may cause excessive scoring of the cylinder bores and wear on the hone. If new pistons are required, they will need to be installed to the connecting rods. This should be performed by a machine shop as the pistons must be installed in the correct relationship to the rod or engine damage can occur.

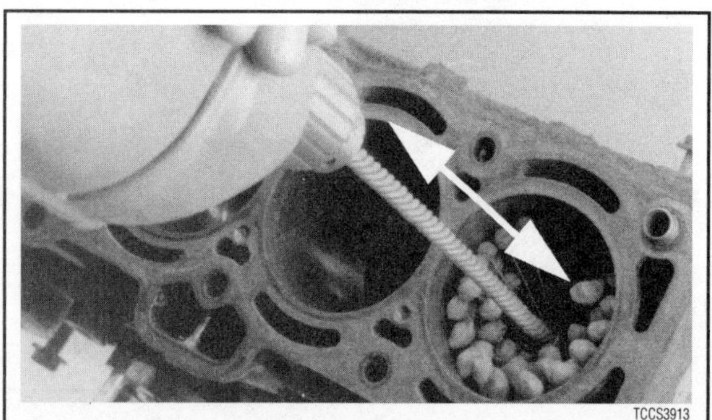

Fig. 145 Use a ball type cylinder hone to remove any glaze and provide a new surface for seating the piston rings

Pistons and Connecting Rods

▶ See Figure 146

Only pistons with the wrist pin retained by C-clips are serviceable by the home-mechanic. Press fit pistons require special presses and/or heaters to remove/install the connecting rod and should only be performed by a machine shop.

All pistons will have a mark indicating the direction to the front of the engine and the must be installed into the engine in that manner. Usually it is a notch or arrow on the top of the piston, or it may be the letter F cast or stamped into the piston.

Fig. 146 Most pistons are marked to indicate positioning in the engine (usually a mark means the side facing the front)

C-CLIP TYPE PISTONS

If piston pin removal requires special tools or jigs, Chilton recommends taking the piston/connecting rod assembly to a qualified machine shop or dealer. The expense of these tool is usually prohibitive unless lot's of pistons are going to be repaired

1. Note the location of the forward mark on the piston and mark the connecting rod in relation.
2. Remove the C-clips from the piston and withdraw the wrist pin.

➡Varnish build-up or C-clip groove burrs may increase the difficulty of removing the wrist pin. If necessary, use a punch or drift to carefully tap the wrist pin out.

3. Insure that the wrist pin bushing in the connecting rod is usable, and lubricate it with assembly lube.
4. Remove the wrist pin from the new piston and lubricate the pin bores on the piston.
5. Align the forward marks on the piston and the connecting rod and install the wrist pin.
6. The new C-clips will have a flat and a rounded side to them. Install both C-clips with the flat side facing out.
7. Repeat all of the steps for each piston being replaced.

ASSEMBLY

Before you begin assembling the engine, first give yourself a clean, dirt free work area. Next, clean every engine component again. The key to a good assembly is cleanliness.

Mount the engine block into the engine stand and wash it one last time using water and detergent (dishwashing detergent works well). While washing it, scrub the cylinder bores with a soft bristle brush and thoroughly clean all of the oil passages. Completely dry the engine and spray the entire assembly down with an anti-rust solution such as WD-40® or similar product. Take a clean lint-free rag and wipe up any excess anti-rust solution from the bores, bearing saddles, etc. Repeat the final cleaning process on the crankshaft. Replace any freeze or oil galley plugs which were removed during disassembly.

Crankshaft

▶ See Figures 147, 148, 149, 150 and 151

1. Remove the main bearing inserts from the crankcase halves.
2. Be sure that the bearing inserts and bearing saddles are clean. Foreign material under inserts will distort bearing and cause failure.
3. Place the upper main bearing inserts in the bearing saddles with the tang in the slot.

➡The oil holes in the bearing inserts must be aligned with the oil holes in the cylinder block.

4. Install the lower main bearing inserts into the bearing saddles.
5. Clean the mating surfaces of block and rear main bearing cap.
6. Carefully lower the crankshaft into place. Be careful not to damage bearing surfaces.
7. Check the clearance of each main bearing by using the following procedure:
   a. Place a piece of Plastigage® or its equivalent, on bearing surface across full width of bearing cap and about 1/4 in. off center.
   b. Install the crankcase half and tighten bolts to specifications. Do not turn crankshaft while Plastigage® is in place.
   c. Remove the crankcase half. Using the supplied Plastigage® scale, check width of Plastigage® at widest point to get maximum clearance. Difference between readings is taper of journal.
   d. If clearance exceeds specified limits, try an undersize bearing in combination with the standard bearing. Bearing clearance must be within specified limits. If standard and undersize bearing do not bring clearance within desired limits, refinish crankshaft journal, then install undersize bearings.
8. Install the rear main seal.
9. After the bearings have been fitted, apply a light coat of engine oil to the journals and bearings. Install the rear main bearing cap. Install all bearing caps except the thrust bearing cap. Be sure that main bearing inserts are installed in original locations. Tighten the crankcase bolts to specifications.
10. Install the thrust bearing cap with bolts finger-tight.
11. Pry the crankshaft forward against the thrust surface of upper half of bearing.
12. Hold the crankshaft forward and pry the thrust bearing cap to the rear. This aligns the thrust surfaces of both halves of the bearing.
13. Retain the forward pressure on the crankshaft. Tighten the cap bolts to specifications.
14. Measure the crankshaft end-play as follows:
   a. Mount a dial gauge to the engine block and position the tip of the gauge to read from the crankshaft end.
   b. Carefully pry the crankshaft toward the rear of the engine and hold it there while you zero the gauge.
   c. Carefully pry the crankshaft toward the front of the engine and read the gauge.

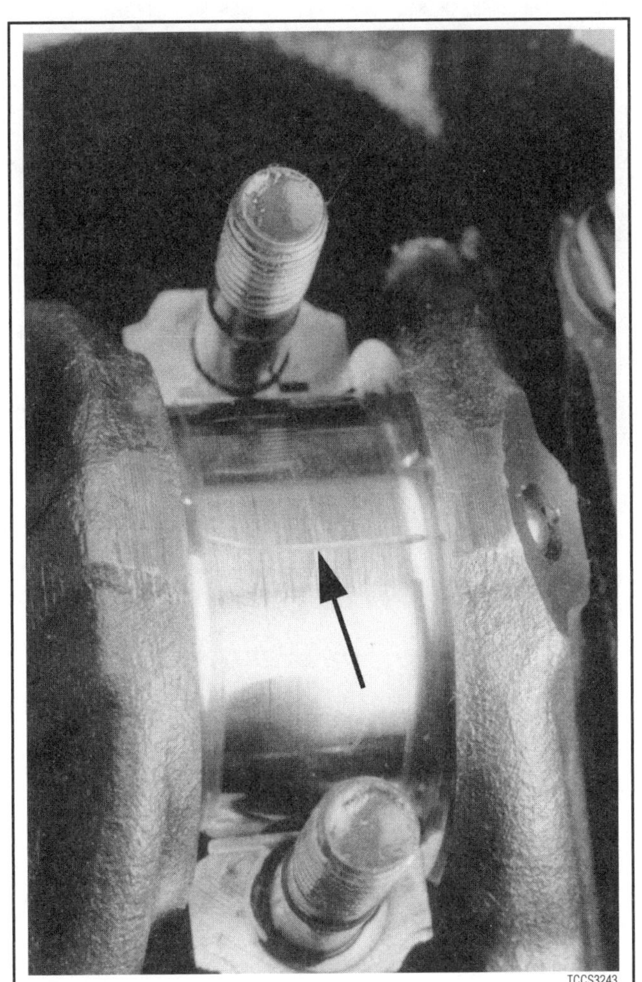

TCCS3243

Fig. 147 Apply a strip of gauging material to the bearing journal, then install

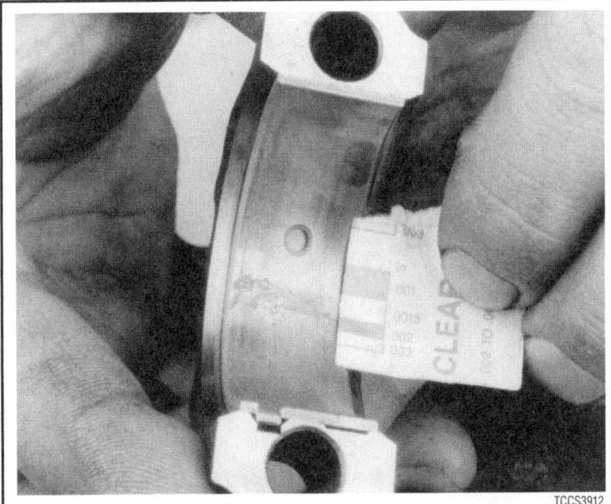

TCCS3912

Fig. 148 After the cap is removed again, use the scale supplied with the gauging material to check the clearance

TCCS3805

Fig. 149 A dial gauge may be used to check crankshaft end-play . . .

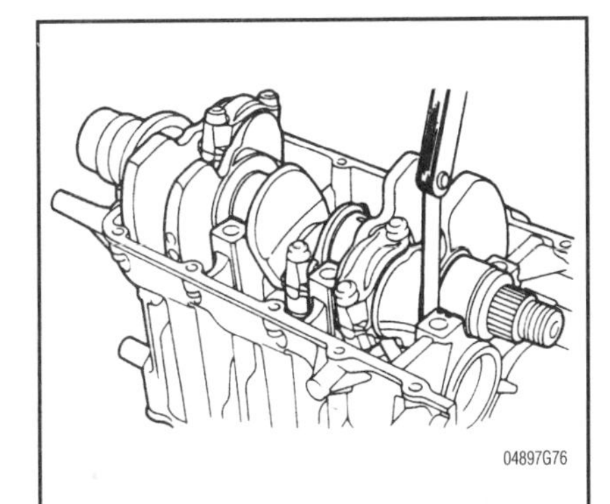

04897G76

Fig. 150 . . . or a feeler gauge is used to measure the crankshaft side play

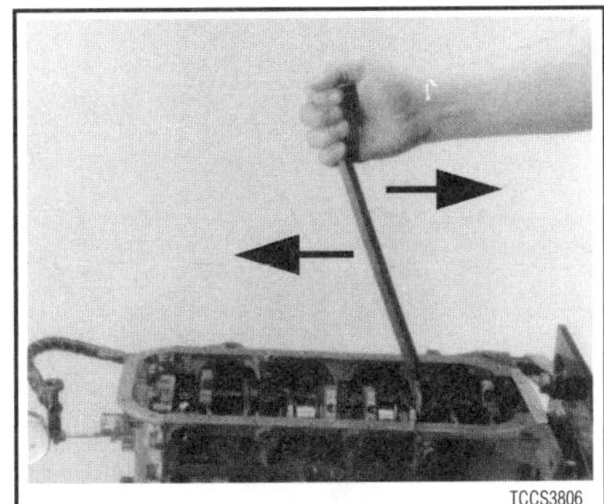

TCCS3806

Fig. 151 Carefully pry the crankshaft back and forth while reading the dial gauge for end-play

d. Confirm that the reading is within specifications. If not, install a new thrust bearing and repeat the procedure. If the reading is still out of specifications with a new bearing, have a machine shop inspect the thrust surfaces of the crankshaft, and if possible, repair it.

15. Rotate the crankshaft so as to position the first rod journal to the bottom of its stroke.

Pistons and Connecting Rods

See Figures 152, 153, 154 and 155

1. Before installing the piston/connecting rod assembly, oil the pistons, piston rings and the cylinder walls with light engine oil. Install connecting rod bolt protectors or rubber hose onto the connecting rod bolts/studs. Also perform the following:
   a. Select the proper ring set for the size cylinder bore.
   b. Position the ring in the bore in which it is going to be used.
   c. Push the ring down into the bore area where normal ring wear is not encountered.
   d. Use the head of the piston to position the ring in the bore so that the ring is square with the cylinder wall. Use caution to avoid damage to the ring or cylinder bore.
   e. Measure the gap between the ends of the ring with a feeler gauge. Ring gap in a worn cylinder is normally greater than specification. If the ring gap is greater than the specified limits, try an oversize ring set.
   f. Check the ring side clearance of the compression rings with a feeler gauge inserted between the ring and its lower land according to specification. The gauge should slide freely around the entire ring circumference without binding. Any wear that occurs will form a step at the inner portion of the lower land. If the lower lands have high steps, the piston should be replaced.

Fig. 152 Checking the piston ring-to-ring groove side clearance using the ring and a feeler gauge

2. Unless new pistons are installed, be sure to install the pistons in the cylinders from which they were removed. The numbers on the connecting rod and bearing cap must be on the same side when installed in the cylinder bore. If a connecting rod is ever transposed from one engine or cylinder to another, new bearings should be fitted and the connecting rod should be numbered to correspond with the new cylinder number. The notch on the piston head goes toward the front of the engine.
3. Install all of the rod bearing inserts into the rods and caps.
4. Install the rings to the pistons. Install the oil control ring first, then the second compression ring and finally the top compression ring. Use a piston ring expander tool to aid in installation and to help reduce the chance of breakage.
5. Make sure the ring gaps are properly spaced around the circumference of the piston. Fit a piston ring compressor around the piston and slide the piston and connecting rod assembly down into the cylinder bore, pushing it in with the wooden hammer handle. Push the piston down until it is only slightly below the top of the cylinder bore. Guide the connecting rod onto the crankshaft bearing journal carefully, to avoid damaging the crankshaft.
6. Check the bearing clearance of all the rod bearings, fitting them to the crankshaft bearing journals. Follow the procedure in the crankshaft installation above.
7. After the bearings have been fitted, apply a light coating of assembly oil to the journals and bearings.
8. Turn the crankshaft until the appropriate bearing journal is at the bottom of its stroke, then push the piston assembly all the way down until the connecting rod bearing seats on the crankshaft journal. Be careful not to allow the bearing cap screws to strike the crankshaft bearing journals and damage them.
9. After the piston and connecting rod assemblies have been installed, check the connecting rod side clearance on each crankshaft journal.
10. Install the auxiliary/balance shaft(s)/assembly(ies).

OHV Engines

CAMSHAFT, LIFTERS AND TIMING ASSEMBLY

1. Install the camshaft.
2. Install the lifters/pushrods into their bores.
3. Install the pivot bolt and pushrod guide assembly.

CYLINDER HEAD(S)

1. Install the cylinder head(s) using new gaskets.
2. Assemble the rest of the valve train (pushrods and rocker arms and/or shafts).

OHC Engines

CYLINDER HEAD(S)

1. Install the cylinder head(s) using new gaskets.
2. Install the timing belt assemblies.

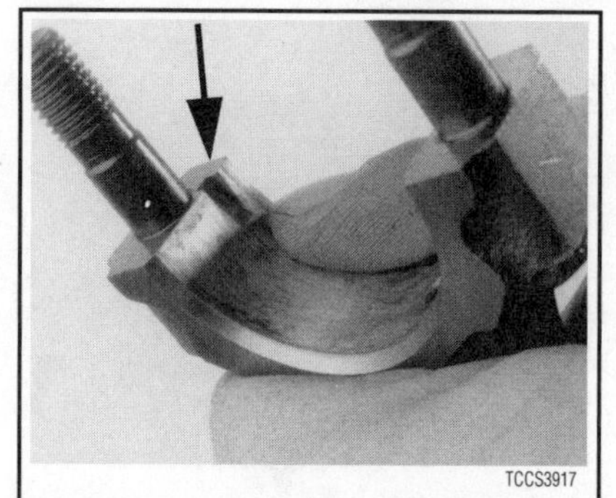

Fig. 153 The notch on the side of the bearing cap matches the tang on the bearing insert

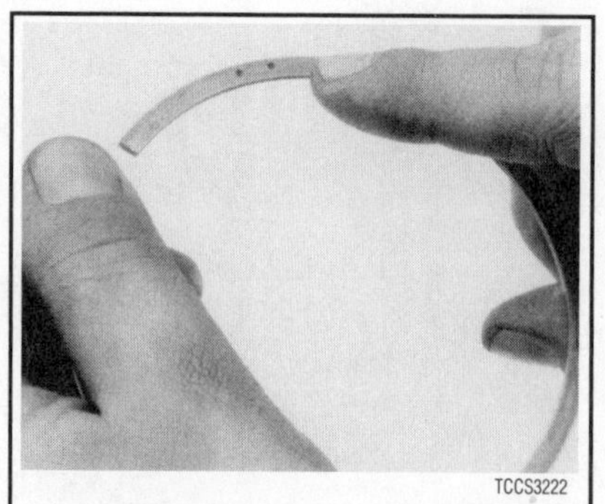

Fig. 154 Most rings are marked to show which side of the ring should face up when installed to the piston

Fig. 155 Install the piston and rod assembly into the block using a ring compressor and the handle of a hammer

Engine Start-up and Break-in

STARTING THE ENGINE

Now that the engine is installed and every wire and hose is properly connected, go back and double check that all hoses are properly connected. Check that your oil drain plug is installed and properly tightened. If not already done, install a new oil filter onto the engine. Fill the crankcase with the proper amount and grade of engine oil.

1. Connect the battery.
2. Start the engine. Keep your eye on your oil pressure indicator; if it does not indicate oil pressure within 10 seconds of starting, turn it off.

**** WARNING**

Damage to the engine can result if it is allowed to run with no oil pressure. Check the engine oil level to make sure that it is full. Check for any leaks and if found, repair the leaks before continuing. If there is still no indication of oil pressure, you may need to prime the system.

3. Confirm that there are no fluid leaks (oil or other).
4. Allow the engine to reach normal operating temperature
5. At this point you can perform any necessary checks or adjustments, such as checking the ignition timing.
6. Install any remaining components or engine covers which were removed.

BREAKING IT IN

Properly breaking in an engine is essential to ensuring long life and engine performance.

On the BF2A–BF15A run the engine at low speeds for the first ten hours and avoid running at full speed.

On the BF20A–BF130A the break-in procedure is as follows:

- First 15 minutes—Run the engine at trolling speed. Use only the minimum amount of throttle opening to operate the boat safely at trolling speed.
- Next 45 minutes—Run the engine up to a maximum of 2,000–3,000 rpm or 10% to 30% of the throttle opening.
- Next 60 minutes—Run the engine up to a maximum of 4,000–5,000 rpm or 50% to 80% of the throttle opening. Short bursts of full throttle are acceptable but do not run the engine at full throttle continuously.
- Next 8 hours—Avoid using full throttle continuously. Do not run the engine at full throttle for more than 5 minutes at a time.

For boats that come up on plane easily, bring the boat up on plane, then reduce the throttle setting to the break-in settings above.

KEEP IT MAINTAINED

Now that you have just gone through all of that hard work, keep yourself from doing it all over again by thoroughly maintaining it. Not that you may not have maintained it before, heck you could have had thousands of hours on it before doing this. However, you may have bought the engine used, and the previous owner did not keep up on maintenance. Which is why you just went through all of that hard work. See?

Engine Rebuilding Specifications—BF2A and BF20

Component	Standard		Service Limit	
	Standard (in.)	Metric (mm)	Standard (in.)	Metric (mm)
Engine				
Type	4-Stroke, Side Valve, 1-Cylinder			
Fuel Consumption	400 g/Psh			
Cooling System	Forced Air Cooling (Water Cooled Exhaust)			
Ignition System	Transistorized Magneto			
Carburetor	Horizontal-Type Butterfly Valve			
Lubrication System	Spalsh-Type			
Starting System	Recoil Starter			
Stopping System	Ground Primary Circuit			
Fuel	Regular Grade Automobile Gasoline			
Exhaust System	Underwater Exhaust System			
Lower Unit				
Gear Ratio	13:28			
Propeller				
Number of Blades	3			
Diameter	7.2	184		
Pitch	4.7	120		
Propeller Rotation	Clockwise (Viewed From Rear)			
Valve				
Stem OD				
Intake	0.216	5.49	0.215	5.45
Exhaust	0.217	5.50	0.213	5.40
Guide ID	0.217	5.50	0.219	5.56
Stem-to-Guide Clearance				
Intake	0.0004	0.010	0.004	0.11
Exhaust	0.0022	0.055	0.006	0.16
Seat Width	0.028	0.70	0.04	1
Spring Free Length	1.07	27.1	0.98	25
Cylinder				
Sleeve ID	1.8110	46	1.813	46.05
Piston				
Skirt OD	1.8108	45.995	1.808	45.92
Piston-to-Cylinder Clearance	0-0.0012	0-0.03	0.0051	0.13
Pin Bore ID	0.3938	10.002	0.3957	10.05
Piston Pin				
Pin OD	0.3937	10.00	0.3917	9.95
Pin-to-Bore Clearance	0.0006	0.015	0.0039	0.10
Piston Ring				
Width				
Top/Second	0.0591	1.5	0.0539	1.37
Side Clearance				
Top	0.0022-0.0035	0.055-0.090	0.0059	0.15
Second	0.0022-0.0033	0.055-0.085	0.0059	0.15

Engine Rebuilding Specifications—BF2A and BF20

Component	Standard		Service Limit	
	Standard (in.)	Metric (mm)	Standard (in.)	Metric (mm)
Piston Ring				
End Gap				
Top/Second	0.0059-0.014	0.15-0.035	0.039	1
Oil	0.0079-0.031	0.2-0.8	0.039	1
Connecting Rod				
Small End ID	0.3939	10.006	0.3957	10.05
Rod-to-Pin Clearance	0.00024-0.00091	0.006-0.023	0.0039	0.10
Big End Oil Clearance	0.00063-0.0015	0.016-0.038	0.0039	0.10
Big End Axial Clarance	0.0079-0.035	0.20-0.90	0.043	1.10
Big End ID	0.7087	18	0.7102	18.04
Crankshaft				
Crank Pin OD	0.7080	17.984	0.7063	17.94
Camshaft				
Cam Lift	0.8197	20.82	0.8059	20.47
Journal OD	0.4797	12.184	0.4783	12.15
Crankcase				
Journal ID	0.4803	12.20	0.4823	12.25
Propeller Shaft				
OD at Bevel Gear	0.4320-0.4324	10.973-10.984	0.4303	10.93
Propeller Shaft Holder				
Shaft Bore ID	0.4331-0.4338	11-11.018	0.4354	11.06
Shaft-to-Bore Clearance	0.0006-0.0018	0.016-0.045	-	-
Vertical Shaft				
OD at Gear Case	0.4320-0.4324	10.973-10.984	0.4303	10.93
Gear Case				
Vertical Shaft Bore ID	0.4331-0.4338	11-11.018	0.4354	11.06
Vertical Shaft Bore Clearance	0.0006-0.0018	0.016-0.045	-	-

■ **Please refer to the Supplement for BF2D Engine Rebuilding Specifications**

Engine Rebuilding Specifications—BF5A and BF50

Component	Standard		Service Limit	
	Standard (in.)	Metric (mm)	Standard (in.)	Metric (mm)
Engine				
Type	4-Stroke, Vertical OHV, 1-Cylinder			
Fuel Consumption	310 g/PSh			
Cooling System	Forced Water Circulation by Impeller Pump with Thermostat			
Ignition System	Transistorized Magneto			
Carburetor	Horizontal-Type Butterfly Valve			
Lubrication System	Pressure Lubrication by Trochoid Pump			
Starting System	Recoil Starter			
Stopping System	Ground Primary Circuit			
Fuel	Regular Grade Automobile Gasoline			
Fuel Pump	Diaphragm type			
Exhaust System	Underwater Exhaust System			
Lower Unit				
Gear Ratio	13:27			
Propeller				
Number of Blades	3			
Diameter	7.875	200		
Pitch	7.5	190		
Propeller Rotation	Clockwise (Viewed From Rear)			
Clutch	Dog Clutch (Forward-Neutral-Reverse)			
Valve				
Stem OD				
Intake	0.216	5.49	0.215	5.45
Exhaust	0.214	5.44	0.213	5.40
Guide ID				
Intake	0.217	5.50	0.218	5.54
Exhaust	0.217	5.50	0.219	5.57
Stem-to-Guide Clearance				
Intake	0.0008-0.0017	0.020-0.044	0.003	0.07
Exhaust	0.0024-0.0034	0.060-0.087	0.006	0.15
Seat Width	0.03	0.80	0.070	1.80
Spring Free Length	1.165	29.60	1.106	28.10
Cylinder				
Sleeve ID	2.3622	60	2.365	60.07
Piston				
Skirt OD	2.3616	59.985	2.359	59.92
Piston-to-Cylinder Clearance	0.0006-0.0020	0.015-0.050	0.004	0.10
Pin Bore ID	0.7087	18.002	0.709	18.02
Piston Pin				
Pin OD	0.7087	18.0	0.707	17.97
Pin-to-Bore Clearance	0.0001-0.0006	0.002-0.014	0.002	0.04
Piston Ring				
Side Clearance	0.0006-0.0018	0.015-0.045	0.004	0.10
End Gap	0.006-0.014	0.15-0.35	0.020	0.50

04987C03

Engine Rebuilding Specifications—BF5A and BF50

Component	Standard		Service Limit	
	Standard (in.)	Metric (mm)	Standard (in.)	Metric (mm)
Connecting Rod				
Small End ID	0.7089	18.005	0.710	18.04
Big End Oil Clearance	0.0016-0.0025	0.04-0.063	0.003	0.08
Big End Axial Clarance	0.004-0.028	0.1-0.7	0.039	1
Crankshaft				
Crank Pin OD	1.180	29.98	1.179	29.95
Journal OD	0.9840	24.993	0.983	24.97
Camshaft				
Cam Lift	0.8197	20.82	0.8059	20.47
Journal OD	0.4797	12.184	0.4783	12.15
Crankcase				
Journal ID	0.9848	25.013	0.986	25.04
Crankcase-to-crankshaft clearance	0.0008-0.0018	0.020-0.046	0.003	0.08
Propeller Shaft				
OD at Bevel Gear				
Forward	0.4718	11.984	0.470	11.95
Reverse	0.5112	12.984	0.510	12.95
Bevel Gear ID				
Forward	0.472	12.00	0.474	12.04
Reverse	0.512	13.00	0.513	13.04

04987C04

Engine Rebuilding Specifications—BF8A, BF75 and BF100

Component	Standard: Standard (in.)	Standard: Metric (mm)	Service Limit: Standard (in.)	Service Limit: Metric (mm)
Engine				
Type	4-Stroke, OHC, 2-Cylinder			
Fuel Consumption	270 g/PSh			
Cooling System	Forced Water Circulation by Impeller Pump with Thermostat			
Ignition System	Transistorized Magneto			
Carburetor	Horizontal-Type Butterfly Valve			
Lubrication System	Pressure Lubrication by Trochoid Pump			
Starting System	Recoil Starter			
Stopping System	Ground Primary Circuit			
Fuel	Regular Grade Automobile Gasoline			
Fuel Pump	Diaphragm type			
Exhaust System	Underwater Exhaust System			
Lower Unit				
Gear Ratio	12:29			
Propeller				
Number of Blades	3			
Diameter	9.5	240		
Pitch	8.75	220		
Propeller Rotation	Clockwise (Viewed From Rear)			
Clutch	Dog Clutch (Forward-Neutral-Reverse)			
Valve				
Stem OD				
Intake	0.220	5.50	0.200	5.08
Exhaust	0.220	5.50	0.187	4.75
Guide ID	0.220	5.50	0.218	5.54
Seat Width	0.03	0.70	0.079	2.0
Spring Free Length	1.138	28.9	1.079	27.4
Rock Arm				
Rocker Arm ID	0.51	13.0	0.514	13.06
Rocker Arm Shaft OD	0.51	13.0	0.509	12.92
Cylinder				
Sleeve ID	2.20	56	2.20	55.880
Piston				
Skirt OD	2.20	56.00	2.20	55.880
Piston-to-Cylinder Clearance	0.0004-0.0022	0.010-0.055	-	-
Pin Bore ID	0.55	14	0.553	14.048
Piston Pin				
Pin OD	0.55	14	0.549	13.954
Piston Ring				
Ring Width				
Top/Second	0.06	1.5	0.054	1.36
Oil	0.1	2.5	0.093	2.37

04987C05

Engine Rebuilding Specifications—BF8A, BF75 and BF100

Component	Standard: Standard (in.)	Standard: Metric (mm)	Service Limit: Standard (in.)	Service Limit: Metric (mm)
Piston Ring				
Side Clearance				
Top/Second	0.001	0.025	0.004	0.10
Oil	0.0006	0.015	0.004	0.10
End Gap	0.006	0.15	0.039	1.0
Connecting Rod				
Small End ID	0.55	14	0.554	14.070
Big End Oil Clearance	0.002	0.04	0.003	0.083
Big End Axial Clarance	0.02	0.6	0.051	1.3
Crankshaft				
Crank Pin OD	1.10	28	1.10	27.952
Camshaft				
Cam Lift				
BF75				
Serial # 1000004–1199999	0.91	23	0.896	22.75
Except Serial # 1000004–1199999	1.04	26.5	1.033	26.25
BF100				
Serial # 1000004–1199999	0.98	25	0.974	24.75
Except Serial # 1000004–1199999	0.99	25.2	0.982	24.95
Journal OD	0.63	16	0.627	15.916
Oil Pump				
Body ID	0.91	23	0.915	23.23
Inner Rotor-to-Outer Rotor Clearance	0.006	0.15	0.008	0.15
Outer Rotor-to-Body Clearance	0.006	0.15	0.010	0.26
Propeller Shaft				
OD at Bevel Gear	0.67	17	0.667	16.930
Bevel Gear ID	0.67	17	0.672	17.06

04987C06

Engine Rebuilding Specifications—BF9.9A and BF15A

Component	Standard: Standard (in.)	Standard: Metric (mm)	Service Limit: Standard (in.)	Service Limit: Metric (mm)
Engine				
Type	4-Stroke, OHC, Inline, 2-Cylinder, Water Cooled			
Fuel Consumption				
BF9.9A	270 g/PSh			
BF15A	264 g/PSh			
Cooling System	Forced Water Circulation by Impeller Pump with Thermostat			
Ignition System	CDI			
Carburetor	Horizontal-Type Butterfly Valve			
Lubrication System	Pressure Lubrication by Trochoid Pump			
Starting System	Recoil Starter (Electric Starter)			
Stopping System	Ground Primary Circuit			
Fuel	Regular Grade Automobile Gasoline			
Fuel Pump	Mechanical Plunger Type			
Exhaust System	Underwater Exhaust System			
Lower Unit				
Gear Ratio	13:27			
Propeller				
Number of Blades	3			
Diameter	9.5	240		
Pitch	8.75	220		
Propeller Rotation	Clockwise (Viewed From Rear)			
Clutch	Dog Clutch (Forward-Neutral-Reverse)			
Valve				
Stem OD				
Intake	0.2161	5.49	0.2153	5.47
Exhaust	0.2153	5.47	0.2145	5.45
Guide ID	0.2165	5.5	0.2181	5.54
Stem-to-Guide Clearance				
Intake	0.0039-0.00145	0.010-0.037	0.00275	0.07
Exhaust	0.00118-0.00224	0.030-0.057	0.00472	0.12
Seat Width	0.04	1.0	0.078	2.0
Spring Free Length	1.45	36.8	1.39	35.3
Rock Arm				
Rocker Arm ID	0.5118	13	0.5133	13.04
Rocker Arm Shaft OD	0.5106	12.968	0.5087	12.92
Cylinder				
Sleeve ID	2.28	58	2.2856	58.055
Piston				
Skirt OD	2.2829	573985	2.280	57392
Piston-to-Cylinder Clearance	0.006-0.0020	0.015-0.050	0.0039	0.1
Pin Bore ID	0.5513	14.002	0.552	14.02
Piston Pin				
Pin OD	0.550	14	0.550	13.97
Pin-to-Pin Bore Clearance	0.00007-0.00055	0.002-0.014	0.002	0.04

04987C07

Engine Rebuilding Specifications—BF9.9A and BF15A

Component	Standard: Standard (in.)	Standard: Metric (mm)	Service Limit: Standard (in.)	Service Limit: Metric (mm)
Piston Ring				
Ring Width				
Top/Second	0.047	1.2	0.043	1.08
Side Clearance				
Top/Second	0.00098-0.00217	0.025-0.055	0.0039	0.1
Oil	0.00217-0.00551	0.055-0.140	0.0079	0.2
End Gap				
Top	0.0059-0.0118	0.15-0.30	0.0197	0.5
Second	0.0138-0.0197	0.35-0.50	0.0276	0.7
Oil	0.0079-0.0315	0.2-0.8	0.039	1.0
Connecting Rod				
Small End ID	0.5514	14.005	0.5528	14.04
Big End Oil Clearance	0.0016-0.0026	0.040-0.066	0.0031	0.08
Big End Axial Clarance	0.0059-0.0138	0.15-0.35	0.0276	0.7
Crankshaft				
Crank Pin OD	1.180	29.98	1.179	29.95
Main Journal OD	1.299	33.0	1.298	32.98
Side Clearance	0.0039-0.0118	0.10-0.30	0.0236	0.6
Oil Clearance	0.0008-0.0016	0.021-0.040	0.0020	0.05
Camshaft				
Cam Lobe Height				
BF9.9A				
Intake	0.9405	23.89	0.9307	23.64
Exhaust	0.9425	23.94	0.9326	23.69
BF15A				
Intake	0.9634	24.47	0.9535	24.22
Exhaust	0.9654	24.52	0.9555	24.27
Journal OD				
Oil Pump Side	0.6292	15.984	0.6279	15.95
Pulley Side	0.7080	17.984	0.7066	17.95
Oil Pump				
Body ID	1.146	29.10	1.154	29.30
Inner Rotor-to-Outer Rotor Clearance	0.0059	.015	0.0079	0.20
Outer Rotor-to-Body Clearance	0.0039-0.0083	0.10-0.21	0.0102	0.26
Rotor-to-Body Side Clearance	0.0079-0.00354	0.02-0.09	0.0043	0.11
Propeller Shaft				
OD at Bevel Gear	0.66866	16.984	0.66732	16.95
Bevel Gear ID	0.66929	17.00	0.67086	17.04

04987C08

Engine Rebuilding Specifications—BF25A and BF30A

Component	Standard: Standard (in.)	Standard: Metric (mm)	Service Limit: Standard (in.)	Service Limit: Metric (mm)
Engine				
Type	4-Stroke, OHC, Inline, 3-Cylinder, Water Cooled			
Fuel Consumption	238 g/PSh			
Cooling System	Forced Water Circulation by Impeller Pump with Thermostat			
Ignition System	CDI			
Carburetor	Horizontal-Type Butterfly Valve			
Lubrication System	Pressure Lubrication by Trochoid Pump			
Starting System	Recoil Starter (Electric Starter)			
Stopping System	Ground Primary Circuit			
Fuel	Regular Grade Automobile Gasoline			
Fuel Pump	Mechanical Plunger Type			
Exhaust System	Underwater Exhaust System			
Lower Unit				
Gear Ratio	13:27			
Propeller				
Number of Blades	3			
Diameter	9.25	235		
Pitch	12	305		
Propeller Rotation	Clockwise (Viewed From Rear)			
Clutch	Dog Clutch (Forward-Neutral-Reverse)			
Valve				
Stem OD				
Intake	0.2156-0.2161	5.475-5.490	0.215	5.45
Exhaust	0.2148-0.2154	5.455-5.470	0.214	5.43
Guide ID	0.2165-0.2170	5.500-5.512	0.218	5.54
Stem-to-Guide Clearance				
Intake	0.0004-0.0015	0.010-0.037	0.003	0.07
Exhaust	0.0012-0.0022	0.030-0.057	0.005	0.12
Seat Width	0.035-0.043	0.9-1.1	0.08	2.0
Spring Free Length	1.45	36.8	1.39	35.3
Rock Arm				
Rocker Arm ID	0.5118-0.5125	13-13.018	0.513	13.04
Rocker Arm Shaft OD	0.5103-0.5110	12.962-12.980	0.509	12.92
Shaft-to-Rocker Arm Clearance	0.007-0.0022	0.020-0.056	0.003	0.07
Cylinder				
Sleeve ID	2.2835-2.2841	58-58.015	2.2856	58.055
Cylinder Head				
Distortion	0.002	0.05	0.004	0.10
Piston				
Skirt OD	2.2823-2.2831	57.970-57.990	2.280	57.92
Piston-to-Cylinder Clearance	0.004-0.0018	0.010-0.045	0.004	0.1
Pin Bore ID	0.5513-0.5515	14.002-14.008	0.552	14.02

04987C09

Engine Rebuilding Specifications—BF25A and BF30A

Component	Standard: Standard (in.)	Standard: Metric (mm)	Service Limit: Standard (in.)	Service Limit: Metric (mm)
Piston Pin				
Pin OD	0.5509-0.5512	13.994-14.000	0.550	13.97
Pin-to-Pin Bore Clearance	0.00001-0.00006	0.002-0.014	0.002	0.04
Piston Ring				
Ring Width				
Top/Second	0.0463-0.0469	1.175-1.190	0.043	1.080
Side Clearance				
Top/Second	0.0010-0.0022	0.025-0.055	0.004	0.1
Oil	0.0022-0.0055	0.055-0.140	0.008	0.2
End Gap				
Top	0.006-0.012	0.15-0.30	0.020	0.5
Second	0.014-0.020	0.35-0.50	0.030	0.7
Oil	0.008-0.031	0.2-0.8	0.04	1.0
Connecting Rod				
Small End ID	0.5516-0.5520	14.010-14.022	0.553	14.05
Big End Oil Clearance	0.004-0.0011	0.010-0.028	0.010	0.40
Big End Axial Clarance	0.005-0.011	0.12-0.27	0.010	0.40
Connecting Rod Bearing Oil Clearance	0.0007-0.0017	0.018-0.042	0.030	0.80
Crankshaft				
Crank Pin OD	1.2593-1.2603	31.987-32.011	1.258	31.96
Main Journal OD	1.4168-1.4174	35.986-36.002	1.416	35.96
Side Clearance	0.002-0.012	0.05-0.30	0.018	0.45
Oil Clearance	0.0008-0.0017	0.020-0.044	0.0020	0.06
Camshaft				
Cam Lobe Height				
Intake	0.9478-0.9572	24.073-24.313	0.9387	23.843
Exhaust	0.9490-0.9585	24.105-24.345	0.9411	23.905
Camshaft Axial Clearance	0.006-0.016	0.15-0.40	0.02	0.60
Journal OD				
No. 1	0.7858-0.7866	19.959-19.980	0.785	19.93
No. 2	1.1787-1.1795	29.939-29.960	1.18	29.90
No. 3	0.7073-0.7080	17.966-17.984	0.706	17.94
Shaft Oil Clearance				
No. 1	0.0008-0.0024	0.020-0.062	0.003	0.08
No. 2	0.0016-0.0034	0.040-0.086	0.004	0.11
No. 3	0.0006-0.0024	0.016-0.061	0.003	0.08
Oil Pump				
Body ID	1.603-1.604	40.71-40.74	1.605	40.76
Inner Rotor-to-Outer Rotor Clearance	0.006	0.15	0.01	0.20
Outer Rotor-to-Body Clearance	0.006-0.008	0.15-0.21	0.010	0.26
Outer Rotor Height	0.590-0.591	14.98-15.00	0.589	14.96
Pump Body Depth	0.591-0.593	15.02-15.05	0.594	15.09
Rotor-to-Body Side Clearance	0.001-0.003	0.02-0.07	0.004	0.10

04987C10

Engine Rebuilding Specifications—BF25A and BF30A

Component	Standard: Standard (in.)	Standard: Metric (mm)	Service Limit: Standard (in.)	Service Limit: Metric (mm)
Vertical Shaft				
Shaft OD	0.8747-0.8752	22.217-22.230	0.8739	22.196
Bevel Gear ID				
Forward	0.7480-0.7489	19.000-19.021	0.750	19.04
Reverse	0.868-0.878	22.050-22.30	0.880	22.35
Propeller Shaft				
Shaft OD				
Front	0.7467-0.7472	18.967-18.980	0.7459	18.946
Rear	0.82-0.83	20.90-21.20	0.821	20.85
At Needle Bearing	0.8664-0.8669	22.007-22.020	0.866	21.99
Bevel Gear ID	0.66929	17.00	0.67086	17.04

04987C11

Engine Rebuilding Specifications—BF35A and BF45A

Component	Standard: Standard (in.)	Standard: Metric (mm)	Service Limit: Standard (in.)	Service Limit: Metric (mm)
Engine				
Type	4-Stroke, OHC, Inline, 3-Cylinder, Water Cooled			
Fuel Consumption	210 g/hp-h			
Cooling System	Forced Water Circulation by Impeller Pump with Thermostat			
Ignition System	CDI			
Carburetor	Horizontal-Type Butterfly Valve			
Lubrication System	Pressure Lubrication by Trochoid Pump			
Starting System	Recoil Starter (Electric Starter)			
Stopping System	Ground Primary Circuit			
Fuel	Regular Grade Automobile Gasoline			
Fuel Pump	Mechanical Plunger Type			
Exhaust System	Underwater Exhaust System			
Lower Unit				
Gear Ratio	14:23			
Propeller				
Number of Blades	3			
Diameter	11.25	286		
Pitch	13	330		
Propeller Rotation	Clockwise (Viewed From Rear)			
Clutch	Dog Clutch (Forward-Neutral-Reverse)			
Valve				
Stem OD				
Intake	0.2157-0.2161	5.448-5.490	0.215	5.45
Exhaust	0.2150-0.2154	5.446-5.470	0.213	5.42
Guide ID	0.2165-0.2170	5.500-5.512	0.218	5.54
Stem-to-Guide Clearance				
Intake	0.0004-0.0013	0.010-0.032	0.002	0.06
Exhaust	0.0012-0.0020	0.030-0.052	0.004	0.10
Seat Width	0.049-0.061	1.25-1.55	0.08	2.0
Spring Free Length	1.45	36.8	1.39	35.3
Rock Arm				
Rocker Arm ID	0.5516-0.5523	14.010-14.028	0.553	14.05
Rocker Arm Shaft OD	0.5502-0.5509	13.976-13.994	0.549	13.95
Shaft-to-Rocker Arm Clearance	0.006-0.0020	0.016-0.052	0.003	0.07
Cylinder				
Sleeve ID	2.7559-2.7565	70-70.015	2.2758	70.06
Cylinder Head				
Distortion	0.0019	0.05	0.004	0.10
Piston				
Skirt OD	2.7547-2.7555	69.970-69.990	2.7524	69.910
Piston-to-Cylinder Clearance	0.004-0.0018	0.010-0.045	0.0035	0.9
Pin Bore ID	0.7087-0.7090	18.002-18.008	0.709	18.02

04987C12

Engine Rebuilding Specifications—BF35A and BF45A

Component	Standard		Service Limit	
	Standard (in.)	Metric (mm)	Standard (in.)	Metric (mm)
Piston Pin				
Pin OD	0.7084-0.7086	17.994-18.000	0.7068	17.954
Pin-to-Pin Bore Clearance	0.00001-0.00005	0.002-0.014	0.0016	0.04
Piston Ring				
Ring Width				
Top	0.0390-0.0404	0.990-1.025	0.038	0.960
Second	0.0469-0.0482	1.190-1.1225	0.0457	1.160
Side Clearance				
Top/Second	0.0006-0.0026	0.040-0.065	0.004	0.1
Oil	0.00059-0.0018	0.015-0.045	0.006	0.15
End Gap				
Top	0.006-0.012	0.15-0.30	0.030	0.5
Second	0.002-0.018	0.30-0.45	0.037	0.95
Oil	0.0079-0.028	0.2-0.7	0.04	1.0
Connecting Rod				
Small End ID	0.7093-0.7100	18.016-18.034	0.711	18.05
Big End Oil Clearance	0.006-0.0016	0.016-0.040	0.019	0.05
Big End Axial Clarance	0.0019-0.079	0.05-0.20	0.002	0.20
Connecting Rod Bearing Oil Clearance	0.0008-0.0015	0.020-0.038	0.003	0.08
Crankshaft				
Crank Pin OD	1.4951-1.4961	37.976-38.00	1.494	37.94
Main Journal OD	1.5741-1.5750	39.982-40.006	1.572	39.95
Side Clearance	0.0019-0.012	0.05-0.30	0.018	0.45
Oil Clearance	0.0008-0.0015	0.020-0.038	0.0019	0.05
Camshaft				
Cam Lobe Height				
Intake	1.3751-1.3877	34.928-35.248	1.3665	34.708
Exhaust	1.3769-1.3895	34.973-35.293	1.3682	34.753
Camshaft Axial Clearance	0.0012-0.0043	0.03-0.11	0.012	0.30
Journal OD	0.9039-0.9047	22.959-22.980	0.903	22.93
Shaft Oil Clearance	0.0008-0.0026	0.020-0.065	0.003	0.08
Shaft Runout	0.012	0.03	0.0019	0.05
Oil Pump				
Body ID	1.974-1.975	50.15-50.18	1.976	50.20
Inner Rotor-to-Outer Rotor Clearance	0.006	0.15	0.0079	0.20
Outer Rotor-to-Body Clearance	0.006-0.009	0.15-0.22	0.0102	0.26
Outer Rotor Height	0.6685-0.6693	16.98-17.0	0.667	16.93
Pump Body Depth	0.670-0.671	17.02-01705	0.673	17.09
Pump End Clearance	0.008-0.0028	0.02-0.07	0.004	0.10
Vertical Shaft				
Shaft OD	0.8747-0.8752	22.217-22.230	0.8739	22.196

04987C13

Engine Rebuilding Specifications—BF35A and BF45A

Component	Standard		Service Limit	
	Standard (in.)	Metric (mm)	Standard (in.)	Metric (mm)
Bevel Gear ID				
Forward	0.7480-0.7489	19.000-19.021	0.750	19.04
Reverse	0.868-0.878	22.050-22.30	0.880	22.35
Propeller Shaft				
Shaft OD				
Front	0.7467-0.7472	18.967-18.980	0.7459	18.946
Rear	0.82-0.83	20.90-21.20	0.821	20.85
At Needle Bearing	0.8664-0.8669	22.007-22.020	0.866	21.99

04987C14

Engine Rebuilding Specifications—BF40A and BF50A

Component	Standard		Service Limit	
	Standard (in.)	Metric (mm)	Standard (in.)	Metric (mm)
Engine				
Type	4-Stroke, OHC, Inline, 3-Cylinder, Water Cooled			
Fuel Consumption				
BF40A	203 g/hp-h			
BF50A	195 g/hp-h			
Cooling System	Forced Water Circulation by Impeller Pump with Thermostat			
Ignition System	CDI			
Carburetor	Horizontal-Type Butterfly Valve			
Lubrication System	Pressure Lubrication by Trochoid Pump			
Starting System	Recoil Starter (Electric Starter)			
Stopping System	Ground Primary Circuit			
Fuel	Regular Grade Automobile Gasoline			
Fuel Pump	Mechanical Plunger Type			
Exhaust System	Underwater Exhaust System			
Lower Unit				
Gear Ratio	26:33 (14:23)			
Propeller				
Number of Blades	3			
Diameter	11.25	286		
Pitch	13	330		
Propeller Rotation	Clockwise (Viewed From Rear)			
Clutch	Dog Clutch (Forward-Neutral-Rev. rse)			
Valve				
Stem OD				
Intake	0.2157-0.2161	5.448-5.490	0.215	5.45
Exhaust	0.2150-0.2154	5.446-5.470	0.213	5.42
Guide ID	0.2165-0.2170	5.500-5.512	0.218	5.54
Stem-to-Guide Clearance				
Intake	0.0004-0.0013	0.010-0.032	0.002	0.06
Exhaust	0.0012-0.0020	0.030-0.052	0.004	0.10
Seat Width	0.049-0.061	1.25-1.55	0.08	2.0
Spring Free Length	1.45	36.8	1.39	35.3
Rock Arm				
Rocker Arm ID	0.5516-0.5523	14.010-14.028	0.553	14.05
Rocker Arm Shaft OD	0.5502-0.5509	13.976-13.994	0.549	13.95
Shaft-to-Rocker Arm Clearance	0.006-0.0020	0.016-0.052	0.003	0.07
Cylinder				
Sleeve ID	2.7559-2.7565	70-70.015	2.2758	70.06
Cylinder Head				
Distortion	0.0019	0.05	0.004	0.10
Piston				
Skirt OD	2.7547-2.7555	69.970-69.990	2.7524	69.910
Piston-to-Cylinder Clearance	0.004-0.0018	0.010-0.045	0.0035	0.9
Pin Bore ID	0.7087-0.7090	18.002-18.008	0.709	18.02

04987C15

Engine Rebuilding Specifications—BF40A and BF50A

Component	Standard		Service Limit	
	Standard (in.)	Metric (mm)	Standard (in.)	Metric (mm)
Piston Pin				
Pin OD	0.7084-0.7086	17.994-18.000	0.7068	17.954
Pin-to-Pin Bore Clearance	0.00001-0.00005	0.002-0.014	0.0016	0.04
Piston Ring				
Ring Width				
Top	0.0390-0.0404	0.990-1.025	0.038	0.960
Second	0.0469-0.0482	1.190-1.1225	0.0457	1.160
Side Clearance				
Top/Second	0.0006-0.0026	0.040-0.065	0.004	0.1
Oil	0.00059-0.0018	0.015-0.045	0.006	0.15
End Gap				
Top	0.006-0.012	0.15-0.30	0.030	0.5
Second	0.002-0.018	0.30-0.45	0.037	0.95
Oil	0.0079-0.028	0.2-0.7	0.04	1.0
Connecting Rod				
Small End ID	0.7093-0.7100	18.016-18.034	0.711	18.05
Big End Oil Clearance	0.006-0.0016	0.016-0.040	0.019	0.05
Big End Axial Clarar.ce	0.0019-0.079	0.05-0.20	0.002	0.20
Connecting Rod Bearing Oil Clearance	0.0008-0.0015	0.020-0.038	0.003	0.08
Crankshaft				
Crank Pin OD	1.4951-1.4961	37.976-38.00	1.494	37.94
Main Journal OD	1.5741-1.5750	39.982-40.006	1.572	39.95
Side Clearance	0.0019-0.012	0.05-0.30	0.018	0.45
Oil Clearance	0.0008-0.0015	0.020-0.038	0.0019	0.05
Camshaft				
Cam Lobe Height				
Intake	1.3751-1.3877	34.928-35.248	1.3665	34.708
Exhaust	1.3769-1.3895	34.973-35.293	1.3682	34.753
Camshaft Axial Clearance	0.0012-0.0043	0.03-0.11	0.012	0.30
Journal OD	0.9039-0.9047	22.959-22.980	0.903	22.93
Shaft Oil Clearance	0.0008-0.0026	0.020-0.065	0.003	0.08
Shaft Runout	0.012	0.03	0.0019	0.05
Oil Pump				
Body ID	1.974-1.975	50.15-50.18	1.976	50.20
Inner Rotor-to-Outer Rotor Clearance	0.006	0.15	0.0079	0.20
Outer Rotor-to-Body Clearance	0.006-0.009	0.15-0.22	0.0102	0.26
Outer Rotor Height	0.6685-0.6693	16.98-17.0	0.667	16.93
Pump Body Depth	0.670-0.671	17.02-01705	0.673	17.09
Pump End Clearance	0.008-0.0028	0.02-0.07	0.004	0.10
Vertical Shaft				
Shaft OD	0.8747-0.8752	22.217-22.230	0.8739	22.196

04987C16

Engine Rebuilding Specifications—BF40A and BF50A

Component	Standard		Service Limit	
	Standard (in.)	Metric (mm)	Standard (in.)	Metric (mm)
Bevel Gear ID				
Forward	0.7480-0.7489	19.000-19.021	0.750	19.04
Reverse	0.868-0.878	22.050-22.30	0.880	22.35
Propeller Shaft				
Shaft OD				
Front	0.7467-0.7472	18.967-18.980	0.7459	18.946
Rear	0.82-0.83	20.90-21.20	0.821	20.85
At Needle Bearing	0.8664-0.8669	22.007-22.020	0.866	21.99

04987C17

Engine Rebuilding Specifications—BF75A and BF90A

Component	Standard: Standard (in.)	Standard: Metric (mm)	Service Limit: Standard (in.)	Service Limit: Metric (mm)
Engine				
Type	4-Stroke, OHC, Inline, 3-Cylinder, Water Cooled			
Fuel Consumption				
BF40A	203 g/hp-h			
BF50A	195 g/hp-h			
Cooling System	Forced Water Circulation by Impeller Pump with Thermostat			
Ignition System	CDI			
Carburetor	Horizontal-Type Butterfly Valve			
Lubrication System	Pressure Lubrication by Trochoid Pump			
Starting System	Recoil Starter (Electric Starter)			
Stopping System	Ground Primary Circuit			
Fuel	Regular Grade Automobile Gasoline			
Fuel Pump	Mechanical Plunger Type			
Exhaust System	Underwater Exhaust System			
Lower Unit				
Gear Ratio	26:33 (14:23)			
Propeller				
Number of Blades	3			
Diameter	11.25	286		
Pitch	13	330		
Propeller Rotation	Clockwise (Viewed From Rear)			
Clutch	Dog Clutch (Forward-Neutral-Reverse)			
Valve				
Stem OD				
Intake	0.2157-0.2161	5.448-5.490	0.215	5.45
Exhaust	0.2581-0.2587	6.555-6.570	0.258	6.53
Guide ID				
Intake	0.2165-0.2170	5.500-5.512	0.218	5.54
Exhaust	0.2598-0.2604	6.600-6.615	0.261	6.64
Stem-to-Guide Clearance				
Intake	0.0004-0.0014	0.010-0.037	0.0028	0.07
Exhaust	0.0012-0.0024	0.030-0.060	0.005	0.12
Seat Width	0.049-0.061	1.25-1.55	0.08	2.0
Spring Free Length				
Intake	1.81	46.0	-	-
Exhaust				
Inner	1.64	41.7	-	-
Outer	1.74	44.1	-	-
Rock Arm				
Rocker Arm ID	0.6693-0.6700	17.00-17.018	0.671	17.04
Rocker Arm Shaft OD	0.6678-0.6685	16.962-16.980	0.666	16.92
Shaft-to-Rocker Arm Clearance	0.0008-0.0022	0.020-0.056	0.0028	0.07
Cylinder				
Sleeve ID	2.9528-2.9533	75.000-75.015	2.9549	75.055

04987C18

Engine Rebuilding Specifications—BF75A and BF90A

Component	Standard: Standard (in.)	Standard: Metric (mm)	Service Limit: Standard (in.)	Service Limit: Metric (mm)
Cylinder Head				
Distortion	0.0019	0.05	0.004	0.10
Piston				
Skirt OD	2.9520-2.9524	74.980-74.990	2.950	74.92
Piston-to-Cylinder Clearance	0.0004-0.0013	0.010-0.035	0.004	0.10
Pin Bore ID	0.7484-0.7487	19.010-19.016	-	-
Piston Pin				
Pin OD	0.7479-0.7480	18.996-19.000	0.747	18.97
Pin-to-Pin Bore Clearance	0.0004-0.0008	0.010-0.020	-	-
Piston Ring				
Ring Width				
Top	0.0461-0.0469	1.170-1.190	0.043	1.08
Second	0.0581-0.0587	1.475-1.490	0.054	1.38
Side Clearance				
Top	0.0012-0.0024	0.030-0.060	0.004	0.10
Second	0.0012-0.0022	0.030-0.055	0.004	0.10
Oil	0.0018-0.0057	0.045-0.145	0.008	0.20
End Gap				
Top/Second	0.006-0.012	0.15-0.30	0.030	0.50
Oil	0.008-0.031	0.20-0.80	0.035	0.90
Connecting Rod				
Small End ID	0.7465-0.7472	18.960-18.980	-	-
Big End Axial Clearance	0.006-0.012	0.15-0.30	0.020	0.40
Connecting Rod Bearing Oil Clearance	0.0008-0.0015	0.020-0.038	0.0019	0.050
Crankshaft				
Crank Pin OD	1.7707-1.7717	44.976-45.000	1.770	44.96
Main Journal OD	2.1644-2.1654	54.976-55.000	2.164	54.96
Side Clearance	0.004-0.014	0.10-0.35	0.018	0.450
Oil Clearance	0.0010-0.0017	0.025-0.043	0.0024	0.060
Camshaft				
Cam Lobe Height				
Intake	1.5751-1.5846	40.008-40.248	1.5672	39.808
Exhaust	1.5691-1.5786	39.857-40.097	1.5613	39.657
Camshaft Axial Clearance	0.0012-0.0043	0.030-0.110	0.012	0.30
Journal ID				
#1	1.2579-1.2589	31.950-31.975	1.257	31.93
#2	1.8870-1.8880	47.930-47.955	1.886	47.90
#3	1.9067-1.9077	48.430-48.455	1.906	48.40
#4	1.9264-1.9274	48.930-48.955	1.925	48.90
#5 (Oil Pump Journal)	1.4154-1.4163	35.950-35.975	1.415	35.93
Shaft Oil Clearance				
#1and #5	0.0010-0.0030	0.025-0.075	0.004	0.10
#2, #3 and #4	0.00180.0037	0.045-0.095	0.005	0.12
Shaft Runout	0.0012	0.03	0.0019	0.05

04987C19

Engine Rebuilding Specifications—BF75A and BF90A

Component	Standard: Standard (in.)	Standard: Metric (mm)	Service Limit: Standard (in.)	Service Limit: Metric (mm)
Oil Pump				
Body ID	3.150-3.151	80.00-80.04	3.152	80.06
Inner Rotor-to-Outer Rotor Clearance	0.0008-0.0063	0.02-0.16	0.008	0.20
Outer Rotor-to-Body Clearance	0.004-0.007	0.10-0.19	0.0009	0.23
Outer Rotor Height	0.708-0.709	17.98-18.00	0.707	17.96
Pump Body Depth	0.709-0.711	18.02-18.05	0.712	18.09
Pump End Clearance	0.0008-0.0028	0.02-0.07	0.004	0.10
Vertical Shaft				
Shaft OD	1.1245-1.1250	28.562-28.575	1.1237	28.541
Propeller Shaft				
Shaft OD				
Forward Bevel Gear	0.9996-1.0010	25.390-25.425	0.9988	25.369
Holder	1.1870-1.1875	30.149-30.162	1.1861	30.128

04987C20

Engine Rebuilding Specifications—BF115A and BF130A

Component	Standard: Standard (in.)	Standard: Metric (mm)	Service Limit: Standard (in.)	Service Limit: Metric (mm)
Engine				
Type	4-Stroke, OHC, Inline, 3-Cylinder, Water Cooled			
Fuel Consumption				
BF115A	228 g/hp-h			
BF130A	254 g/hp-h			
Cooling System	Forced Water Circulation by Impeller Pump with Thermostat			
Ignition System	Fully Transistorized			
Fuel System	Programmed Fuel Injection			
Lubrication System	Pressure Lubrication by Trochoid Pump			
Starting System	Recoil Starter (Electric Starter)			
Stopping System	Ground Primary Circuit			
Fuel	Regular Grade Automobile Gasoline			
Fuel Pump	Electric and Mechanical Plunger Types			
Exhaust System	Underwater Exhaust System			
Lower Unit				
Gear Ratio	14:28			
Propeller	Per Application			
Propeller Rotation	Clockwise (Viewed From Rear) LA and XA Types			
	Counterclockwise (Viewed From Rear) LCA and XCA Types			
Clutch	Dog Clutch (Forward-Neutral-Reverse)			
Valve				
Stem OD				
Intake	0.2159-0.2163	5.485-5.495	0.2148	5.455
Exhaust	0.2146-0.2150	5.450-5.460	0.2134	5.420
Guide ID	0.2171-0.2177	5.515-5.530	0.219	5.55
Stem-to-Guide Clearance				
Intake	0.0008-0.0018	0.020-0.045	0.0031	0.080
Exhaust	0.0022-0.0031	0.055-0.080	0.0047	0.120
Seat Width	0.049-0.061	1.28-1.55	0.080	2.00
Seat Installation Height				
Intake	1.841-1.872	46.75-47.55	1.882	47.80
Exhaust	1.838-1.869	46.68-47.48	1.879	47.73
Spring Free Length				
Intake	2.113	53.66	-	-
Exhaust	2.188	55.58	-	-
Rock Arm				
Rocker Arm ID				
Intake	0.7879-0.7886	20.012-20.030	-	-
Exhaust	0.7091-0.7098	18.012-18.030	-	-
Rocker Arm Shaft OD				
Intake	0.7863-0.7871	19.972-19.993	-	-
Exhaust	0.7077-0.7084	17.976-17.994	-	-
Shaft-to-Rocker Arm Clearance				
Intake	0.0007-0.0023	0.019-0.0580	0.003	0.08
Exhaust	0.0007-0.0021	0.018-0.0540	0.003	0.08

04987C21

Engine Rebuilding Specifications—BF115A and BF130A

Component	Standard		Service Limit	
	Standard (in.)	Metric (mm)	Standard (in.)	Metric (mm)
Cylinder Block				
Sleeve ID	3.3858-3.3864	86.00-86.015	3.389	86.07
Warpage	0.003	0.07	0.004	0.10
Cylinder Head				
Distortion	-	-	0.002	0.05
Piston				
Skirt OD	3.3846-3.3850	85.97-85.98	3.384	85.96
Piston-to-Cylinder Clearance	0.0008-0.0018	0.020-0.045	0.0020	0.050
Pin Bore ID	0.8645-0.8647	21.960-21.963	-	-
Ring Groove Width				
Top/Second	0.0480-0.0484	1.220-1.230	0.049	1.25
Oil	0.1104-0.1112	2.805-2.825	0.112	2.85
Piston Pin				
Pin OD	0.8646-0.8648	21.961-21.965	-	-
Pin-to-Pin Bore Clearance	-0.0002- +0.0001	-0.005- +0.002	-	-
Piston Ring				
Ring Width				
Top	0.0461-0.0467	1.170-1.185	-	-
Second	0.0462-0.0469	1.175-1.190	-	-
Side Clearance				
Top	0.0014-0.0024	0.035-0.060	0.005	0.13
Second	0.0012-0.0022	0.030-0.055	0.004	0.10
End Gap				
Top	0.008-0.014	0.20-0.35	0.024	0.60
Second	0.016-0.022	0.40-0.55	0.028	0.70
Oil	0.008-0.028	0.20-0.70	0.031	0.80
Connecting Rod				
Small End ID	0.8650-0.8652	21.970-21.976	-	-
Small End-To-Piston Clearance	0.0002-0.0006	0.005-0.015	-	-
Big End Axial Clearance	0.006-0.012	0.15-0.30	0.016	0.40
Connecting Rod Bearing Oil Clearance	0.0010-0.0017	0.026-0.044	-	-
Crankshaft				
Crank Pin OD	1.7707-1.7717	44.976-45.000	1.770	44.96
Main Journal OD				
#1 and #2	2.1646-2.1655	54.980-55.004	-	-
#3	2.1644-2.1654	54.976-55.000	-	-
#4	2.1646-2.1655	54.980-55.004	-	-
#5	2.1650-2.1660	54.992-55.016	-	-
Journal Roundness	0.0002	0.005	-	-
Crankshaft Runout	0.0012	0.030	0.0016	0.040
Oil Clearance				
#1, #2 and #4	0.0011-0.0018	0.027-0.045	0.0020	0.050
#3	0.0012-0.0019	0.031-0.049	0.0022	0.055
#5	0.0007-0.0014	0.017-0.035	0.0016	0.040

04987C22

Engine Rebuilding Specifications—BF115A and BF130A

Component	Standard		Service Limit	
	Standard (in.)	Metric (mm)	Standard (in.)	Metric (mm)
Crankshaft				
Crankshaft End-play	0.004-0.014	0.10-0.35	0.018	0.45
Camshaft				
Cam Lobe Height				
Intake	1.5068-1.5102	38.274-38.359	-	-
Exhaust	1.4823-1.4865	37.651-37.756	-	-
Journal OD	1.0998-1.1004	27.935-27.950		
Shaft Oil Clearance	0.0020-0.0035	0.050-0.089	0.006	0.15
Shaft Runout	0.002-0.006	0.05-0.15	0.020	0.50
Camshaft Axial Clearance	0.002-0.006	0.05-0.15	0.020	0.50
Balancer Shaft				
Journal OD				
#1				
Intake	1.6820-1.6824	42.722-42.734	1.681	42.71
Exhaust	0.8243-0.8248	20.938-20.950	0.824	20.92
#2	1.5241-1.5246	38.712-38.724	1.524	38.70
#3	1.3670-1.3675	34.722-34.734	1.367	34.71
Bearing ID				
#1				
Intake	1.6850-1.6858	42.800-42.820	1.686	42.83
Exhaust	0.8268-0.8273	21.000-21.013	0.828	21.02
#2	1.5276-1.5283	38.800-38.820	1.529	38.83
#3	1.3701-1.3709	34.800-34.820	1.371	34.83
Journal Oil Clearance				
#1				
Intake	0.0020-0.0030	0.050-0.075	0.004	0.09
Exhaust	0.0026-0.0040	0.066-0.099	0.005	0.12
#2	0.0030-0.0043	0.076-0.108	0.005	0.13
#3	0.0026-0.0039	0.066-0.098	0.005	0.12
Journal Roundness	0.0002	0.005	0.0002	0.005
Shaft End-Play				
Intake	0.002-0.006	0.40-0.15	-	-
Exhaust	0.004-0.016	0.10-0.40	-	-
Shaft Runout	0.001	0.02	0.001	0.03
Oil Pump				
Body ID	3.3071-3.3083	84.000-84.030	-	-
Inner Rotor-to-Outer Rotor Clearance	0.002-0.006	0.04-0.16	0.008	0.20
Outer Rotor-to-Body Clearance	0.004-0.007	0.10-0.19	0.0009	0.23
Outer Rotor Height	0.4913-0.4921	12.480-12.500	-	-
Pump Body Depth	0.4929-0.4941	12.520-12.550	-	-
Pump End Clearance	0.001-0.003	0.02-0.07	0.005	0.12

04987C23

LOWER UNIT 8-2
GENERAL INFORMATION 8-2
SHIFTING PRINCIPLES 8-2
STANDARD ROTATING UNIT 8-2
COUNTERROTATING UNIT 8-3
TROUBLESHOOTING THE LOWER UNIT 8-4
PROPELLER 8-4
REMOVAL & INSTALLATION 8-4
LOWER UNIT—NO REVERSE GEAR 8-5
REMOVAL & INSTALLATION 8-5
LOWER UNIT—WITH REVERSE GEAR 8-8
REMOVAL & INSTALLATION 8-8
LOWER UNIT OVERHAUL 8-10
BF20 AND BF2A (FORWARD ONLY) 8-10
DISASSEMBLY 8-10
CLEANING & INSPECTION 8-10
ASSEMBLY 8-11
EXCEPT BF20 AND BF2A (REVERSE GEAR) 8-12
DISASSEMBLY 8-12
CLEANING & INSPECTING 8-29
ASSEMBLY 8-33
SHIMMING PROCEDURE 8-50
JET DRIVE 8-68
DESCRIPTION AND OPERATION 8-68
MODEL IDENTIFICATION AND SERIAL NUMBERS 8-68
JET DRIVE ASSEMBLY 8-69
REMOVAL & INSTALLATION 8-69
ADJUSTMENT 8-72
DISASSEMBLY 8-74
CLEANING & INSPECTING 8-74
ASSEMBLING 8-75

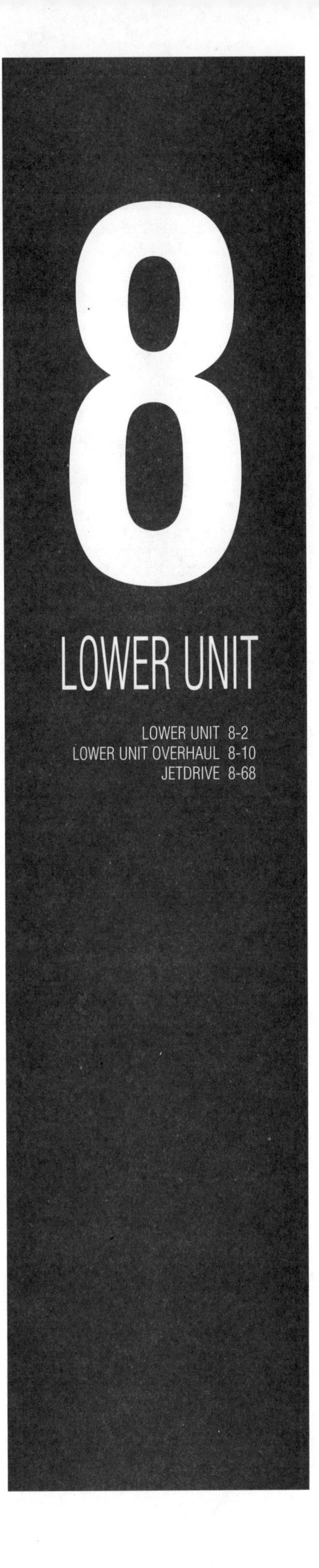

LOWER UNIT 8-2
LOWER UNIT OVERHAUL 8-10
JETDRIVE 8-68

LOWER UNIT

General Information

The lower unit is considered to be the part of the outboard below the exhaust housing. The unit contains the propeller shaft, the driven and pinion gears, the pinion shaft from the powerhead and the water pump. Torque is transferred from the powerhead crankshaft to the gearcase by a driveshaft. A pinion gear on the pinion shaft meshes with a drive gear in the gearcase to change the vertical power flow into a horizontal flow through the propeller shaft. The power head pinion shaft rotates clockwise continuously when the engine is running, but propeller rotation is controlled by the gear train shifting mechanism.

The lower units on all but the BF20 and BF2A are equipped with shifting capabilities. The forward and reverse gears together with the clutch, shift assembly and related linkage are all housed within the lower unit.

On Honda outboards with a reverse gear, a sliding clutch engages the appropriate gear in the gearcase when the shift mechanism is placed in forward or reverse. This creates a direct coupling that transfers the power flow from the pinion to the propeller shaft.

Two types of lower units are available. The first type is a conventional propeller driven lower unit which includes both the standard model and the counterrotating model installed on some outboards.

Counterrotating models are designated with as "LCA and XCA". On these models, the propeller rotates in the opposite direction than on regular models and is used when dual engines are mounted in the boat to equalize the propeller's directional churning force on the water. This allows the operator to maintain a true course instead of being pulled off toward one side while underway.

The second type is a jet drive propulsion system. Water is drawn in from the forward edge of the lower unit and forced out under pressure to propel the boat forward.

The lower unit can be removed without removing the entire outboard from the boat. Each part of this section is presented with complete detailed instructions for removal, disassembly, cleaning and inspecting, assembling, adjusting and installation of only one unit. Each part is complete from removal of the first item to final test operation.

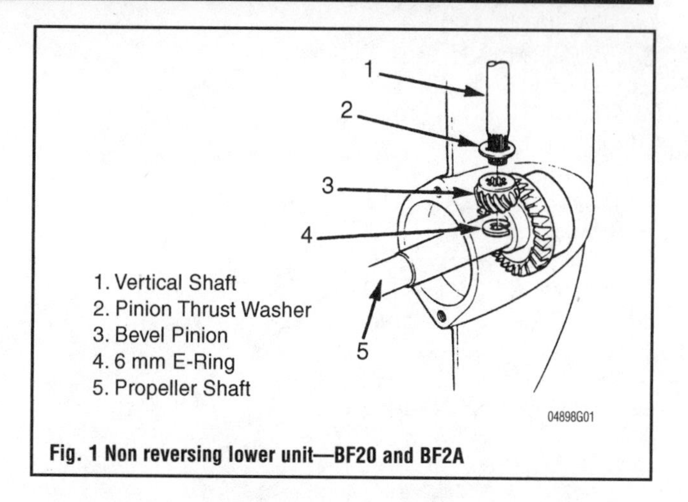

Fig. 1 Non reversing lower unit—BF20 and BF2A

Shifting Principles

STANDARD ROTATING UNIT

Non-Reverse Type

See Figure 1

A non-reversing type lower unit is used only on the BF20 and BF2A. It is a direct drive unit—the pinion gear on the lower end of the driveshaft is in constant mesh with the forward gear. Reverse action of the propeller is accomplished by the operator swinging the engine with the tiller handle 180° and holding it in this position while the boat moves sternward. When the operator is ready to move forward again, they simply swing the tiller handle back to the normal forward position.

Reverse Type

See Figures 2, 3 and 4

A typical lower unit is equipped with a clutch shifter permitting operation in neutral, forward and reverse. When the unit is shifted from neutral to reverse, the shift rod is raised and lowered. This action moves the plunger and clutch shifter toward the reverse gear. When the unit is shifted from reverse to neutral and then to forward gear, the movement of the shift rod is reversed and the clutch shifter is moved toward the forward gear.

The shift mechanism design consists of a cam on the shift rod, a shift slider, steel ball, cross pin ring, shifter pin, clutch shifter, shifter pin holder and a shift spring. This assembly fits into and around the hollow end of the propeller shaft.

The ball inside the shift slide is held under tension by the compression spring and holds the shift slider in position against the shift cam, holding the lower unit in the selected gear.

In the neutral position, the cam is centered in the shifter.

When the lower unit is shifted into forward gear, the shift cam is moved to the lowest position against the shift slider. This allows the spring tension to

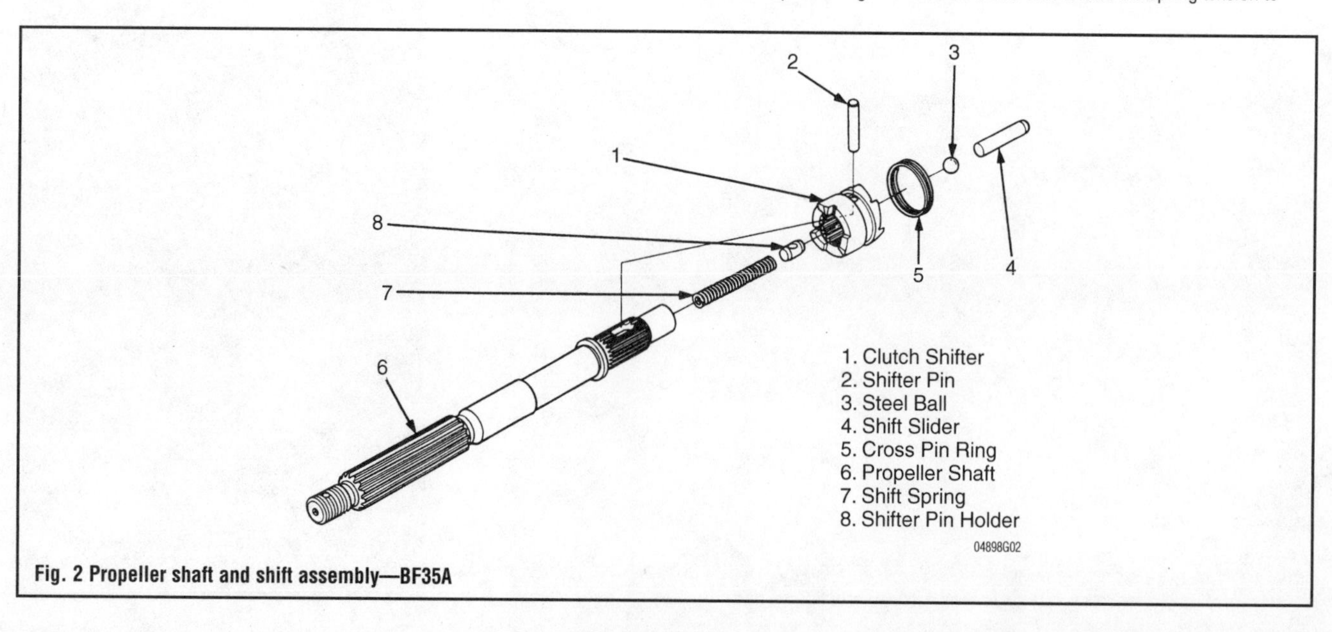

Fig. 2 Propeller shaft and shift assembly—BF35A

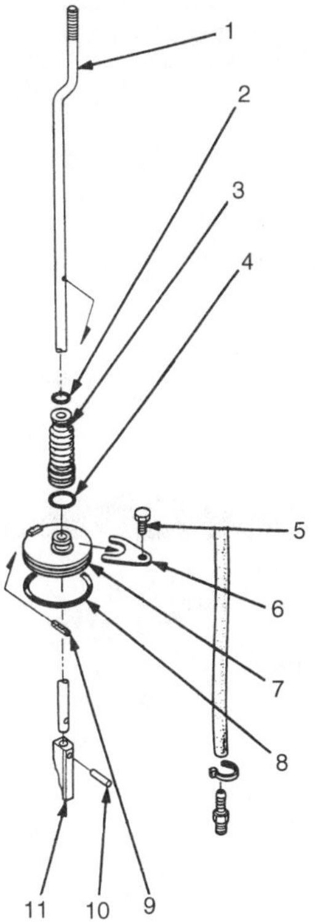

1. Shift Rod B
2. Grommet Band A
3. Shift Rod Grommet
4. Grommet Band B
5. 6 x 12 mm Bolt
6. Seal Holder Set Plate
7. Seal Holder
8. 47 x 3.1 mm O-Ring
9. 2.5 x 12 mm Spring Pin
10. Roller
11. Push Rod

04898G03

Fig. 3 Shift shaft assembly—BF35A

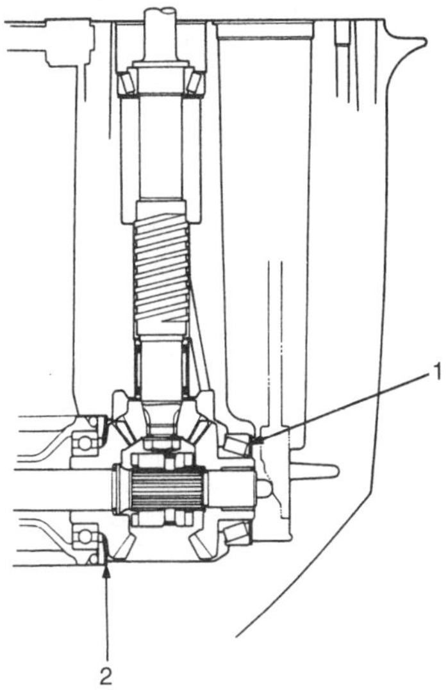

1. Forward Gear Shim
2. Reverse Gear Shim

04898G04

Fig. 4 Cutaway view of the gear case—BF35A

move the shift clutch forward and engage the forward bevel gear. The shift into forward gear is complete.

When the lower unit is shifted into reverse gear, the unit first moves into the neutral position. The shift cam is raised to the upper position again allowing spring tension to move the shift clutch into contact with the reverse bevel gear. The shift movement to reverse gear is complete.

The BF75A to BF130A models use a rotary type shifting system. The shift shaft itself is turned, moving the shift shaft back and forth in the propeller shaft, either engaging or disengaging the forward and reverse gears with the clutch shifter.

The pinion gear on the lower end of the driveshaft is in constant mesh with both the forward and reverse gear. These three gears constantly rotate anytime the powerhead is operating.

A sliding clutch shifter is mounted on the propeller shaft. A shifting motion at the control box will translate into a back and forth motion at the clutch shifter via a series of shift mechanism components. When the clutch shifter is moved forward, it engages with the forward gear. Because the clutch is secured to the propeller shaft with a pin, the shaft rotates at the same speed as the clutch. The propeller is thereby rotated to move the boat forward.

When the clutch shifter is moved aft, the clutch engages only the reverse gear. The propeller shaft and the propeller are thus moved in the opposite direction to move the boat sternward.

When the clutch shifter is in the neutral position, neither the forward nor reverse gear is engaged with the clutch and the propeller shaft does not rotate.

From this explanation, an understanding of wear characteristics can be appreciated. The pinion gear and the clutch shifter receive the most wear, followed by the forward gear, with the reverse gear receiving the least wear. All three gears, the forward, reverse and pinion, are spiral bevel type gears.

A mixture of ball bearings, tapered roller bearings and caged or loose needle bearings is used in each unit. The type bearing used is clearly indicated in the procedures.

COUNTERROTATING UNIT

See Figure 5

As mentioned earlier in this section a single design shifting mechanism is employed on both the standard and counter-rotating units, with the counter-rotating shift mechanism having the shift rod being turned 180°from the standard shift mechanism.

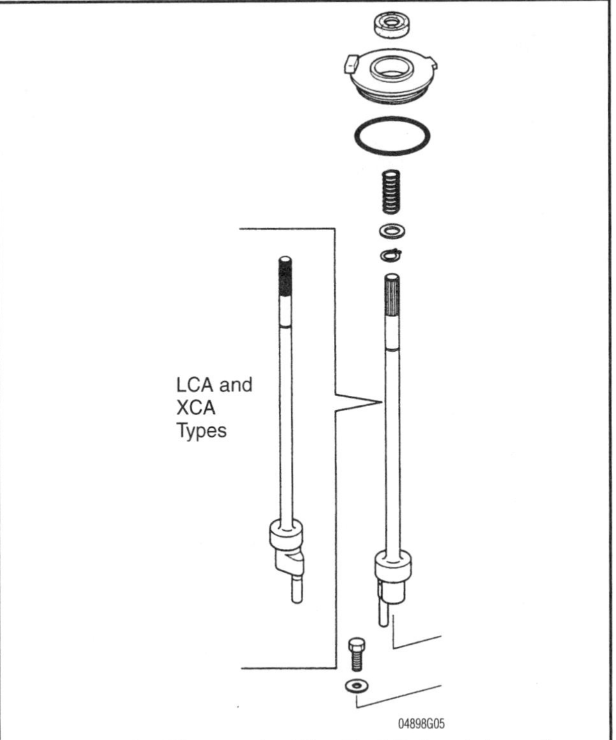

Fig. 5 Notice the differences in shift rod installations between the counter-rotating and standard types

The main physical differences lie in the shifting mechanism (mirror image) and the propeller shaft. The counterrotating unit has a shoulder machined into it for the forward bevel gear tapered roller bearing. Another difference regards nomenclature: what would be the forward gear on a standard unit becomes the reverse gear on a counterrotating unit and what would normally be the reverse gear on a standard unit, becomes the forward gear on a counterrotating unit. The pinion gear remains the same and driveshaft rotation remains the same as on a standard lower unit.

Mirror image shifting mechanisms produce counter rotation of the propeller shaft. This type lower unit consists of the same major identical components as the standard unit.

On a standard lower unit, the cam on the shift shaft is located on the starboard side of the shifter. Therefore, when the rod is rotated counterclockwise, the clutch shifter is pulled forward and the forward gear is engaged.

On a counterrotating lower unit, the cam on the shift rod is located on the port side of the shifter. Therefore, when the rod is rotated counterclockwise, the clutch shifter is pushed back and the gear in the aft end of the housing (which normally is the reverse gear) is engaged. In this manner, the rotation of the propeller shaft is reversed. The same logic applies to the selection of reverse gear.

➡Counter-rotational shifting is accomplished without modification to the shift cable at the shift box. The normal setup is essential for correct shifting. The only special equipment the counterrotating unit requires is the installation of a left-hand propeller.

Troubleshooting the Lower Unit

Troubleshooting must be done before the unit is removed from the powerhead to permit isolating the problem to one area. Always attempt to proceed with troubleshooting in an orderly manner. The shot-in the-dark approach will only result in wasted time, incorrect diagnosis, frustration and unnecessary replacement of parts.

The following procedures are presented in a logical sequence with the most prevalent, easiest and less costly items to be checked listed first.

Check the propeller and the rubber hub. See if the hub is shredded. If the propeller has been subjected to many strikes against underwater objects, it could slip on its hub. If the hub appears to be damaged, replace it with a new hub. Replacement of the hub must be done by a propeller rebuilding shop equipped with the proper tools and experience for such work.

1. Verify the ignition switch is **OFF**, to prevent possible personal injury, should the engine start. Shift the unit into reverse gear and at the same time have an assistant turn the propeller shaft to ensure the clutch is fully engaged.
2. If the shift handle is hard to move, the trouble may be in the lower unit shift rod, requiring an adjustment, or in the shift box.

Disconnect the remote control cable at the engine and then remove the remote control shift cable. Operate the shift lever. If shifting is still hard, the problem is in the shift cable or control box.

3. If the shifting feels normal with the remote control cable disconnected, the problem must be in the lower unit. To verify the problem is in the lower unit, have an assistant turn the propeller and at the same time move the shift cable back and forth. Determine if the clutch engages properly.

Propeller

REMOVAL & INSTALLATION

➧ See Accompanying Illustrations

1. Straighten the cotter pin and pull it free of the propeller shaft. Remove the castle nut, washer and outer spacer. Never pry on the edge of the propeller. Any small distortion will affect propeller performance.
2. If the nut is frozen, place a block of wood between one blade of the propeller and the anti-cavitation plate to keep the shaft from turning. Use a socket and breaker bar to loosen the castle nut. Remove the nut, washer, outer spacer and then the propeller and inner spacer.
3. If the propeller is frozen to the shaft, heat must be applied to the shaft to melt out the rubber inside the hub. Using heat will destroy the hub, but there is no other way. As heat is applied, the rubber will expand and the propeller will actually be blown from the shaft. Therefore, stand clear to avoid personal injury.
4. Use a knife and cut the hub off the inner sleeve.

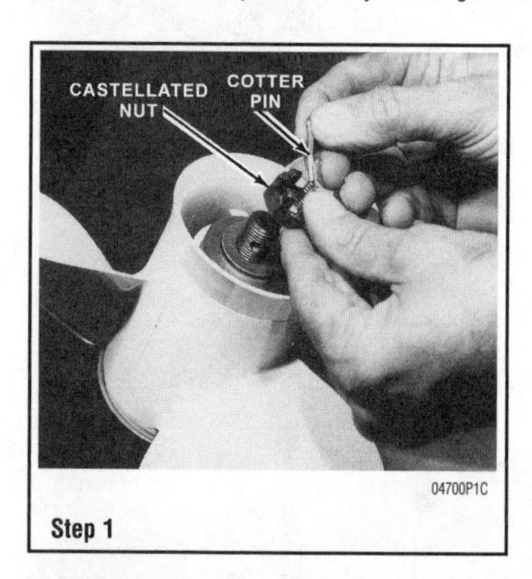

Step 1

Step 2

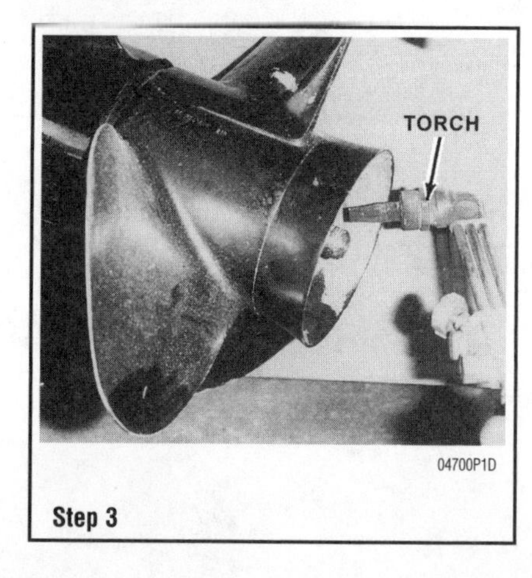

Step 3

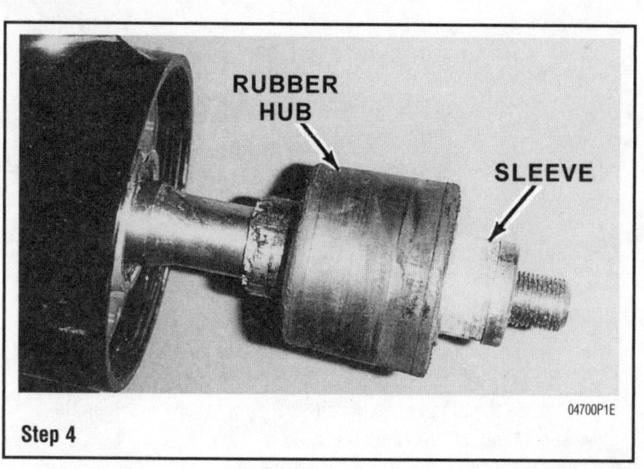

Step 4

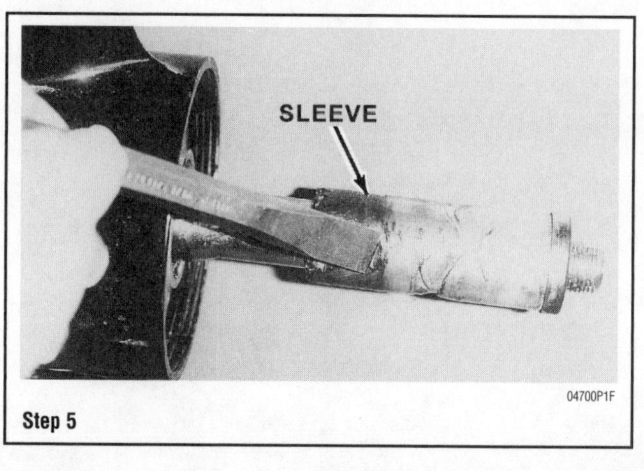

Step 5

5. The sleeve can be removed by cutting it with a hacksaw, or it can be removed with a puller. Again, if the sleeve is frozen, it may be necessary to apply heat. Remove the thrust hub from the propeller shaft.

To install:

➡An anti-seize compound will prevent the propeller from freezing to the shaft and permit propeller removal, without difficulty, the next time the propeller needs to be pulled.

Apply Hondaline Grease, or equivalent anti-seize compound, to the propeller shaft.

6. Install the inner spacer onto the propeller shaft
7. Install the propeller.
8. Install the outer spacer, washer and the castle nut. Place a block of wood between one of the propeller blades and the anti-cavitation plate to prevent the propeller from rotating. Tighten the nut to specification, then back off the nut until the cotter pin may be inserted through the nut and the hole in the propeller shaft. Bend the arms of the cotter pin around the nut to secure it in place.

Remove the block of wood. Connect the spark plug wires to the spark plugs. Connect the electrical lead to the battery terminal.

9. Position the trim tab so the helmsperson is able to handle the boat with equal ease to starboard and port at normal cruising speed. If the boat seems to turn more easily to starboard, loosen the mounting bolt and move the trim tab trailing edge to the starboard. Move the trailing edge of the trim tab to the left if the boat tends to turn more easily to port.

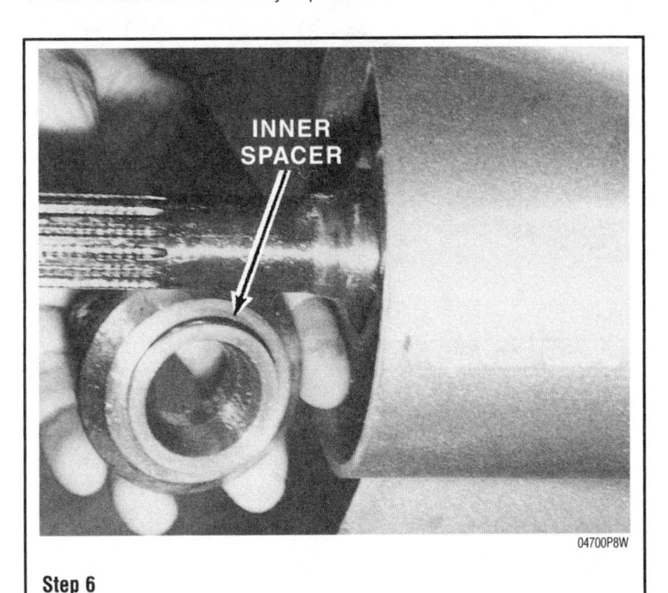

Step 6

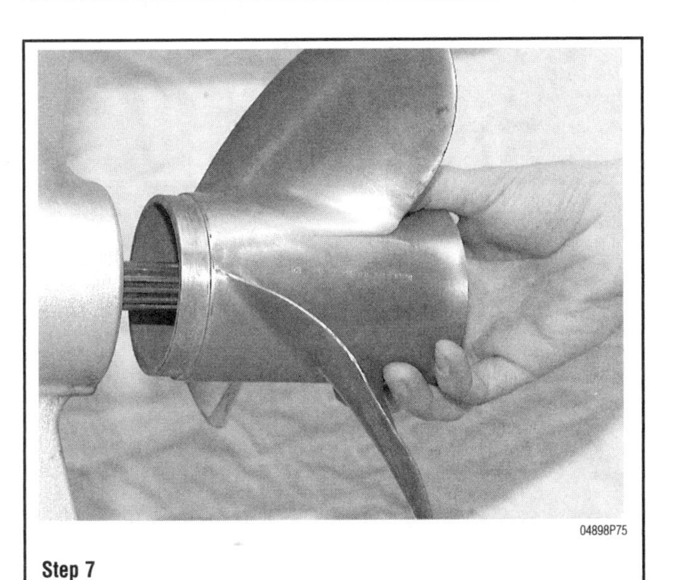

Step 7

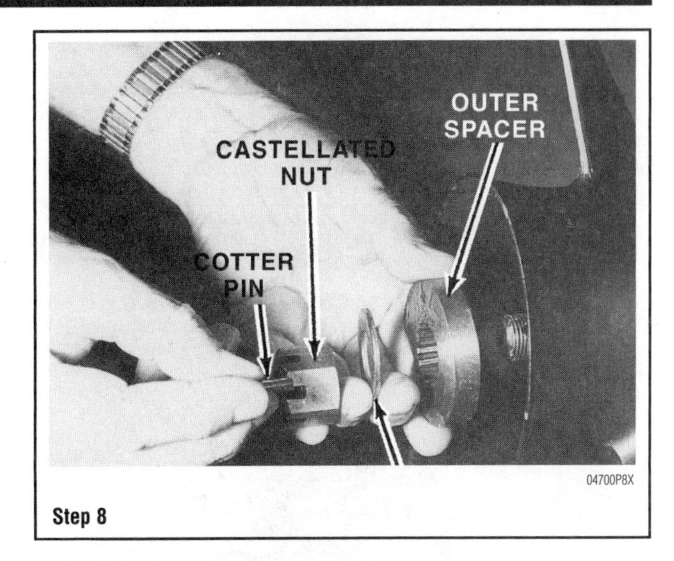

Step 8

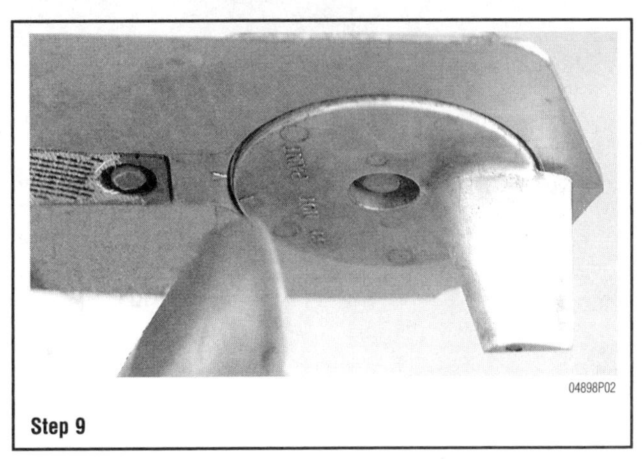

Step 9

Lower Unit—No Reverse Gear

REMOVAL & INSTALLATION

◗ See Accompanying Illustrations

The following procedures present complete instructions to remove, disassemble, assemble and adjust the lower unit of the BF2A and BF20 units.

➡In order to remove the water pump impeller, the driveshaft must be removed from the lower unit. The impeller and pump housing can only be removed from the lower end of the driveshaft. Therefore, the circlip securing the pinion gear to the driveshaft must first be removed. This can only be accomplished by removing the lower unit bearing carrier and removing the propeller shaft.

Before purchasing a lower unit gasket replacement kit, take time to establish the year of manufacture for the unit being serviced. On some units the cap gasket has been replaced with an O-ring.

1. Position a suitable container under the lower unit and then remove the oil plug and the oil level screw. Allow the gear lubricant to drain into the container. As the lubricant drains, catch some with your fingers from time to time and rub it between your thumb and finger to determine if any metal particles are present. If metal is detected in the lubricant, the unit must be completely disassembled, inspected and the damaged parts replaced. Check the color of the lubricant as it drains. A whitish or creamy color indicates the presence of water in the lubricant. Check the drain pan for signs of water separation from the lubricant. The presence of any water in the gear lubricant is bad news. The unit must be completely disassembled, inspected and the cause of the problem determined and corrected.

After the lubricant has drained, temporarily install both the drain and oil level screws.

2. Straighten the cotter pin and then pull it free of the propeller with a pair of pliers or a cotter pin removal tool. Remove the propeller and then push the shear pin out of the propeller shaft.

3. Remove the two bolts securing the bearing carrier.

4. Rotate the holder through 90° and then gently tap the holder to break it free from the lower unit.

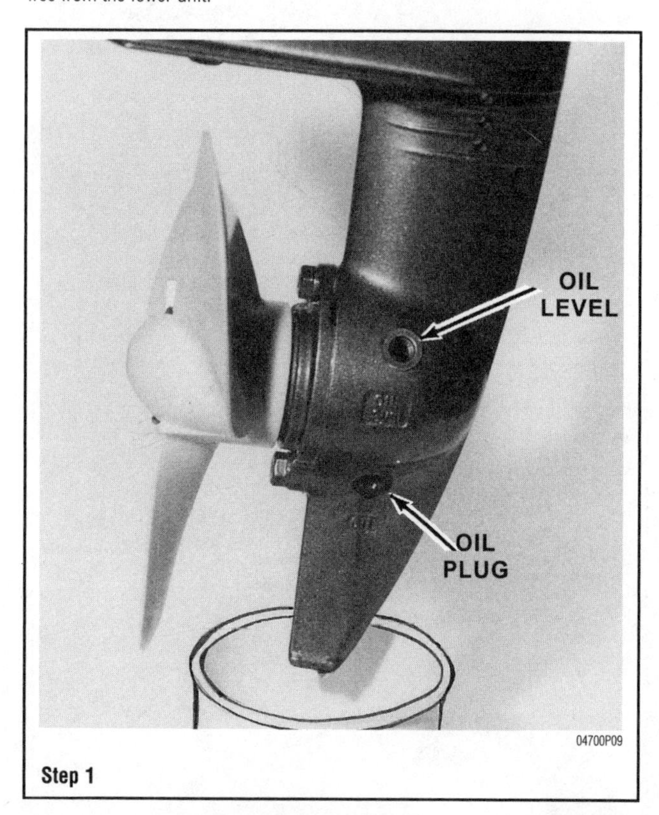

Step 1

5. Remove the holder from the lower unit. Remove and discard the gasket, as required. Remove and discard the O-ring, if equipped.

➡Perform the next step only if the seal has been damaged and is no longer fit for service. Removal of the seal destroys its sealing qualities and it cannot be installed a second time. Therefore, be absolutely sure a new seal is available before removing the old seal in the next step.

6. Use a screwdriver or a bearing/seal removal tool to remove the water seal from the bearing carrier.

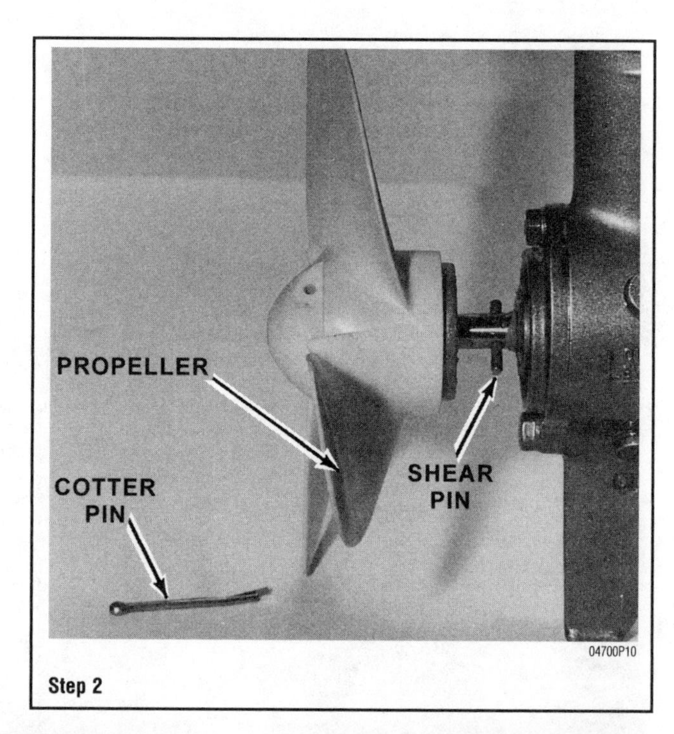

Step 2

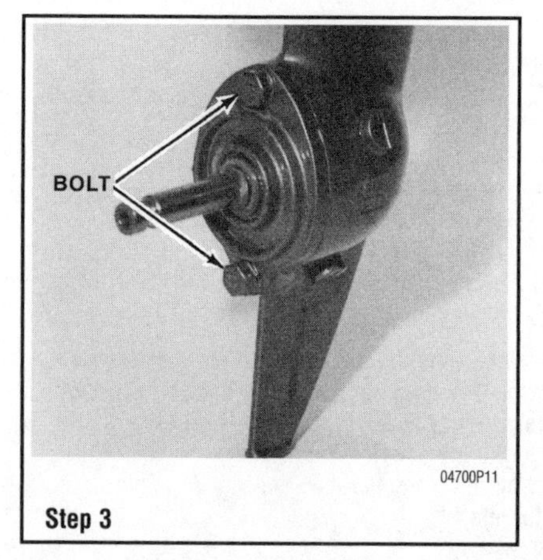

Step 3

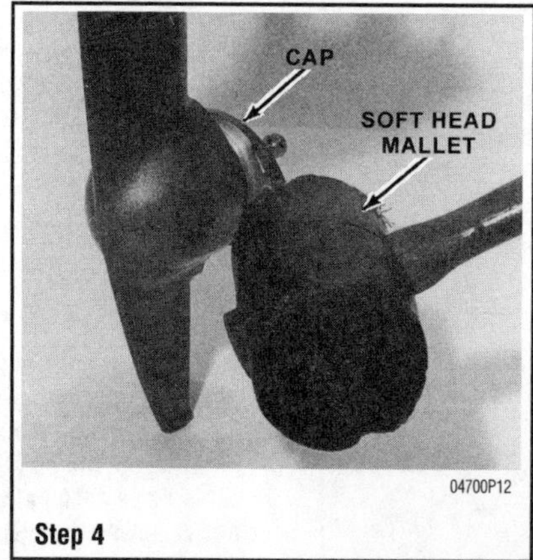

Step 4

Step 5

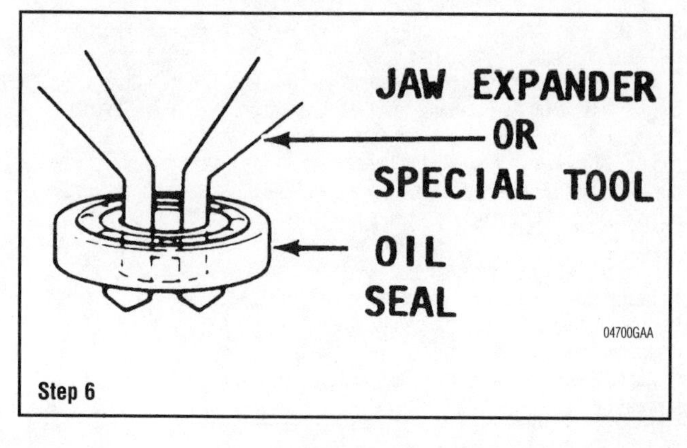

Step 6

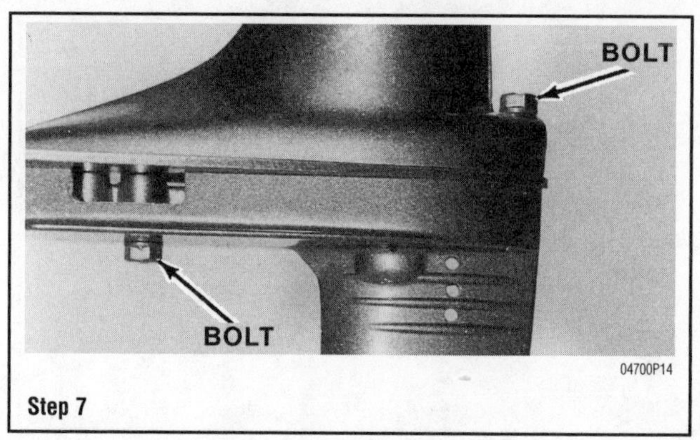

Step 7

7. Remove the two attaching bolts securing the lower unit to the intermediate housing.

8. Separate the lower unit from the intermediate housing. Watch for and save the two dowel pins when the two units are separated. The water tube will come out of the grommet and remain with the intermediate housing. The driveshaft will remain with the lower unit.

To install:

9. Apply just a dab of Hondaline Grease or an equivalent water resistant lubricant, to the indexing pin on the mating surface of the lower unit. Install the anti-cavitation plate onto the lower surface of the intermediate housing. Begin to bring the intermediate housing and the lower gear housing together. As the two units come closer, rotate the propeller shaft slightly to index the upper end of the lower driveshaft tube with the upper driveshaft. At the same time, feed the water tube into the water tube seal. Push the lower gear housing and the intermediate housing together. The pin on the upper surface of the lower unit will index with a matching hole in the lower surface of the anti-cavitation plate.

10. Apply Loctite® to the threads of the two bolts used to secure the lower unit to the intermediate housing. Install and tighten the two bolts.

Install the bearing carrier and gasket. Tighten the two bolts to 5.8–8.7 in. lbs. (80–120 kg-cm).

Apply Honda All Purpose Grease, or equivalent anti-seize compound to the propeller shaft.

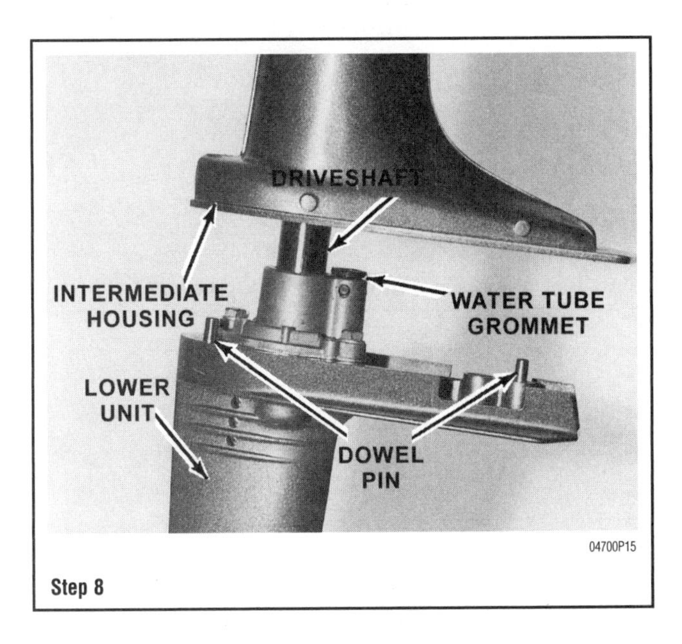

Step 8

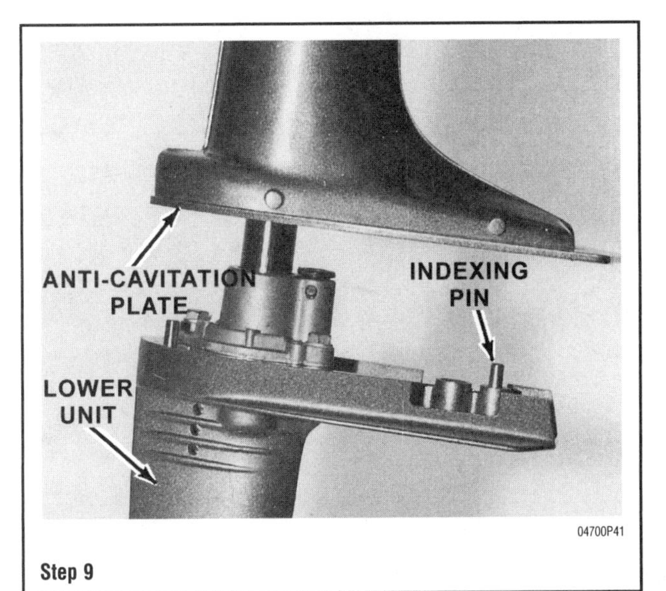

Step 9

➡The compound will prevent the propeller from freezing to the shaft and permit the propeller to be removed, without difficulty, the next time removal is required.

11. Install a new shear pin through the propeller shaft. Guide the propeller onto the pin. Insert a new cotter pin into the hole and bend the ends of the cotter pin in opposite directions.

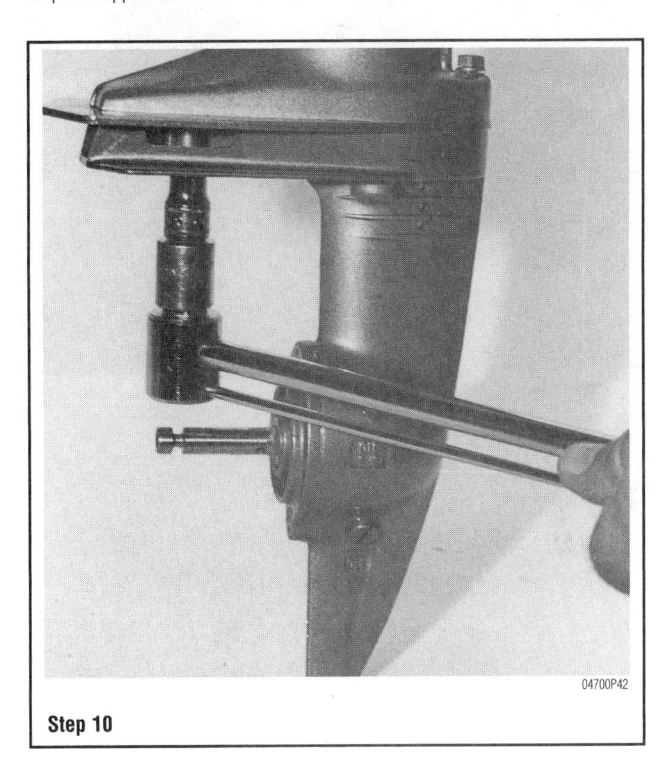

Step 10

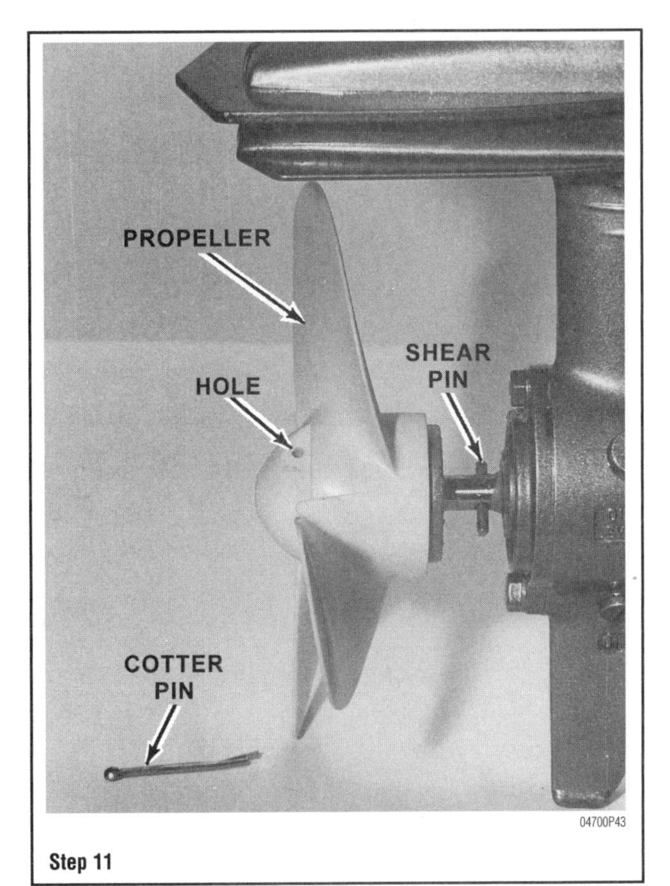

Step 11

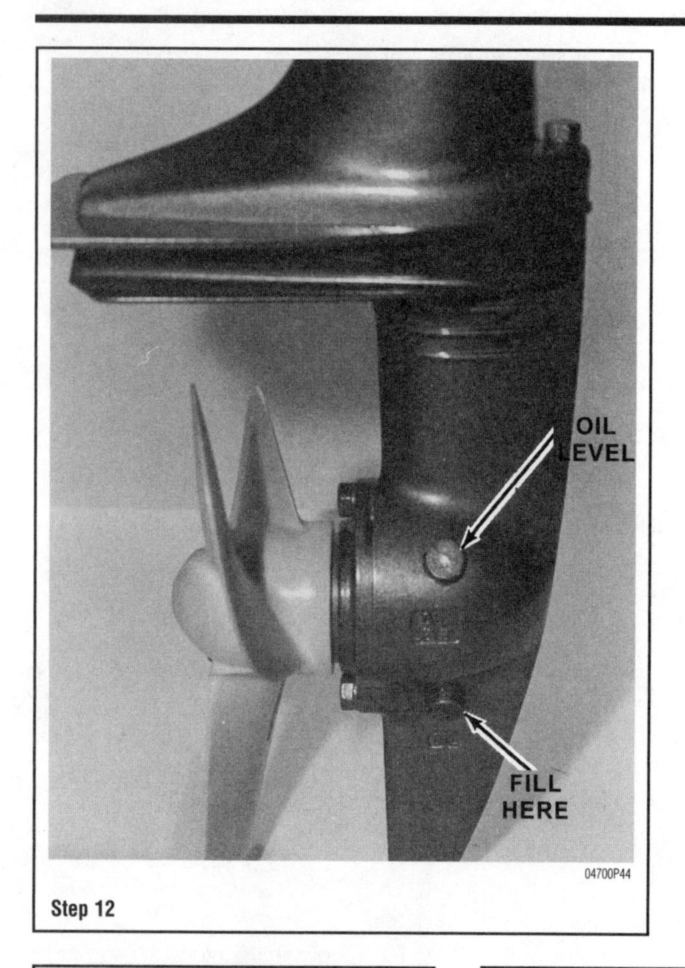

04700P44

Step 12

12. Remove the oil level plug and the oil plug. Fill the lower unit with Hondaline gearcase lubricant until the lubricant flows from the top hole. Install both plugs and clean any excess fluid from the lower unit.

Lower Unit—With Reverse Gear

REMOVAL & INSTALLATION

See Figures 6 thru 16

The following procedures present complete instructions to remove, disassemble, assemble and adjust the lower unit. When the standard and counterrotating units differ, special instructions are given for each unit.

Because so many different models are covered in this manual, it would not be feasible to provide an illustration of each and every unit. Therefore, the accompanying illustrations are of a typical unit. The unit being serviced may

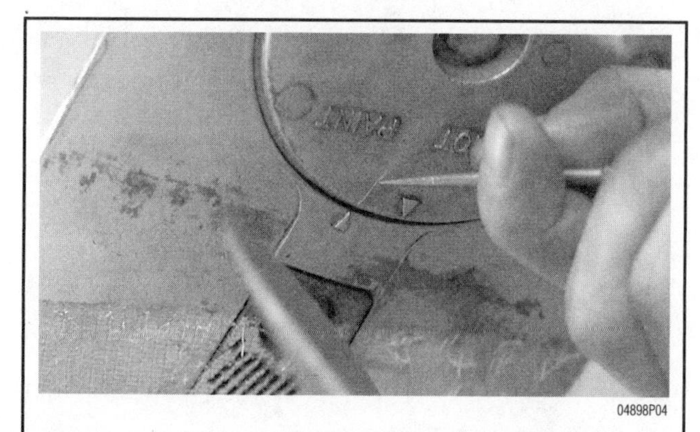
04898P04

Fig. 6 Mark the position of the trim tab before removing it . . .

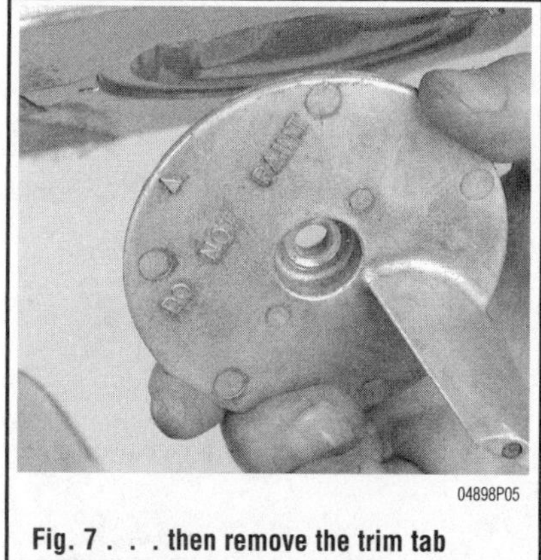
04898P05

Fig. 7 . . . then remove the trim tab

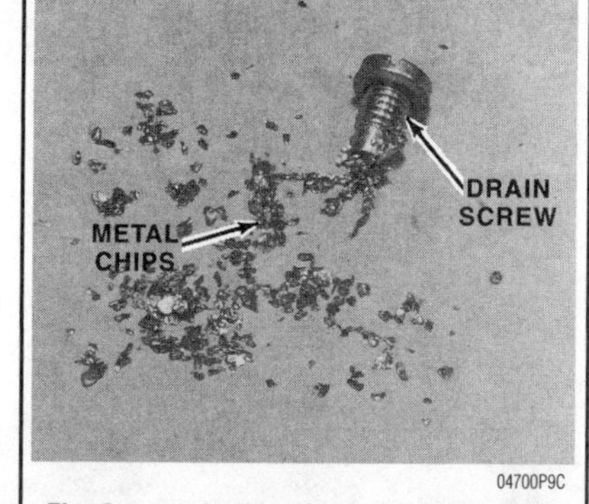

04898P09

Fig. 8 Drain the gear case of all lubricants . . .

04700P9C

Fig. 9 . . . checking for metal chips while draining the gear case

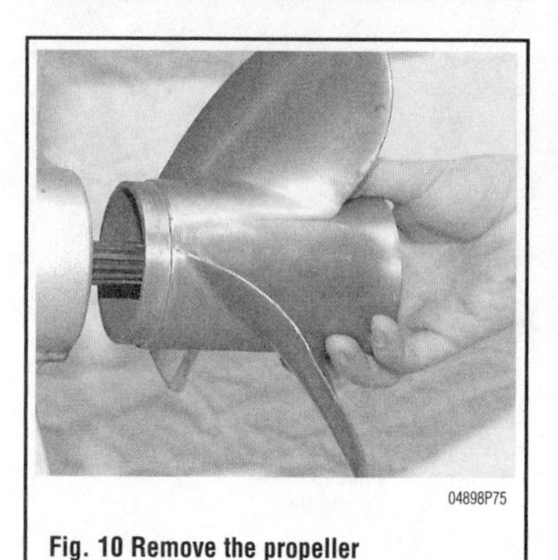
04898P75

Fig. 10 Remove the propeller

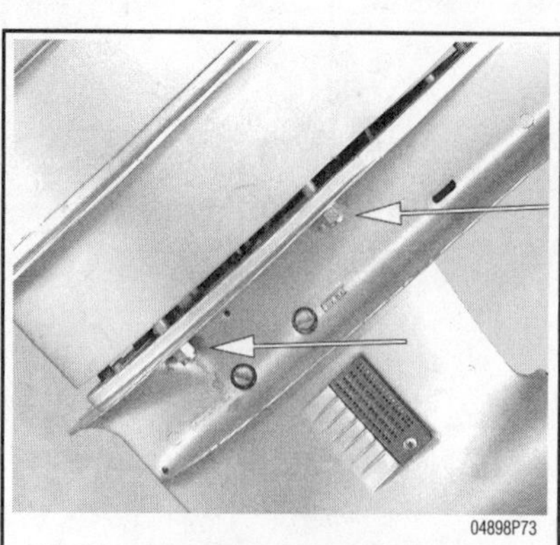
04898P73

Fig. 11 Loosen the bolts holding the gear case to the extension case . . .

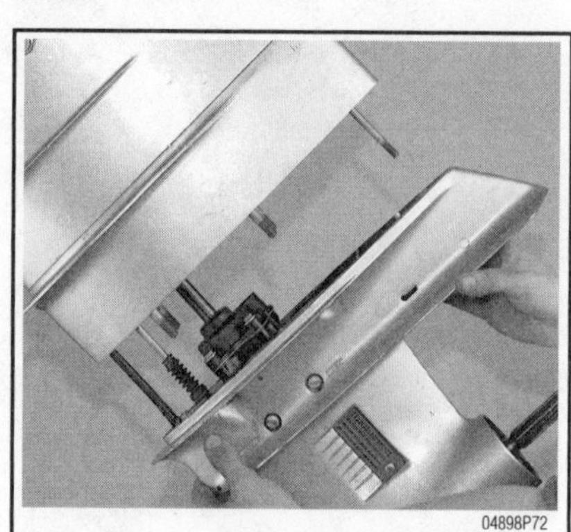
04898P72

Fig. 12 . . . remove the gear case from the extension case . . .

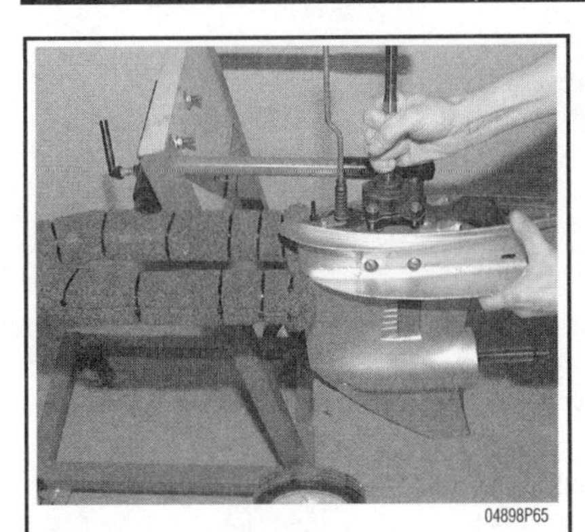
Fig. 13 . . . and place the gear case in a suitable holding fixture

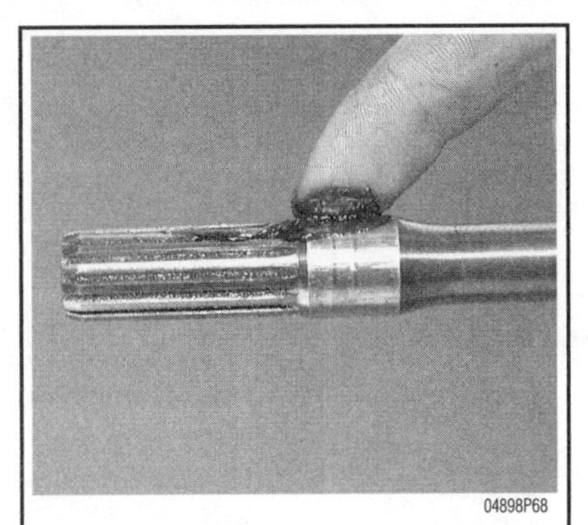
Fig. 14 Apply a dab of grease to the driveshaft splines . . .

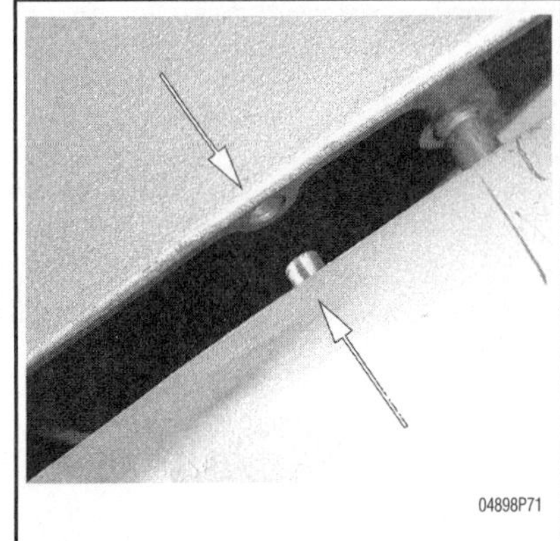
Fig. 15 . . . and to the indexing pins

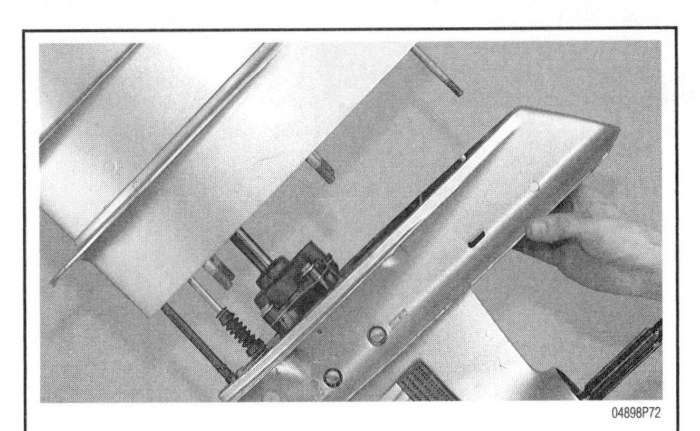
Fig. 16 Line up the components and bring them together

differ slightly in appearance due to engineering or cosmetic changes but the procedures are valid. If a difference should occur, the models affected will be clearly identified.

1. Mark the position of the trim tab, as an aid to installing it back in its original location.
2. Using the correct size socket, remove the bolt and the trim tab.
3. Position a suitable container under the lower unit and then remove the oil screw and the oil level screw. Allow the gear lubricant to drain into the container. As the lubricant drains, catch some with your fingers from time to time and rub it between your thumb and finger to determine if any metal particles are present. If metal is detected in the lubricant, the unit must be completely disassembled, inspected and the damaged parts replaced.

Check the color of the lubricant as it drains. A whitish or creamy color indicates the presence of water in the lubricant. Check the drain pan for signs of water separation from the lubricant. The presence of any water in the gear lubricant is bad news. The unit must be completely disassembled, inspected, the cause of the problem determined and then corrected.

After the lubricant has drained, temporarily install both the drain and oil level screws.

On BF2A , BF20 and BF5A models, straighten the cotter pin and then pull it free of the propeller with a pair of pliers or a cotter pin removal tool. Remove the propeller and then push the shear pin out of the propeller shaft.

4. Straighten the cotter pin and then pull it free of the castle nut with a pair of pliers or a cotter pin removal tool. Remove the castle propeller nut by first placing a block of wood between one of the propeller blades and the anti-cavitation plate to prevent the propeller from rotating and then remove the nut. Remove the washer and the outer spacer. Remove the outer thrust hub from the propeller shaft. If the thrust hub is stubborn and refuses to budge, use two padded pry bars on opposite sides of the hub and work the hub loose. Take care not to damage the lower unit.

Remove the propeller. All models have a spacer between the bearing carrier and the propeller. Slide this spacer free of the propeller shaft.

Remove the inner thrust hub. If this hub is also stubborn, use padded pry bars and work the hub loose. Again, take care not to damage the lower unit.

Move the shift lever to the "N" position, loosen the lock nut and the adjusting nut that connect shift rod "A" and "B", and disconnect the two shift rods.

➡On BF75A to BF130A models, the shift rod is splined and will just pull apart, with shift rod "A" separating from the gear case.

Remove the water hose.

Tilt and lock the outboard unit in the raised position.

5. Remove the external bolts securing the lower unit to the intermediate housing.
6. One bolt securing the lower unit to the intermediate housing is located under the trim tab.
7. Separate the lower unit from the intermediate housing. Watch for and save the dowel pins when the two units are separated. The water tube will come out of the grommet and remain with the intermediate housing. The driveshaft will remain with the lower unit.

The upper and lower driveshafts will automatically disengage as the unit separates.

To install:

8. Apply a dab of Hondaline Grease or equivalent water-resistant lubricant to the splines at the upper end of the driveshaft and on any shift rod that is splined.

➡An excessive amount of lubricant on top of the driveshaft to crankshaft splines will be trapped in the clearance space. This trapped lubricant will not allow the driveshaft to fully engage with the crankshaft.

Apply some of the same lubricant to the end of the water tube in the intermediate housing.

Apply a dab of Hondaline Grease, or equivalent water-resistant lubricant, to the indexing pin on the mating surface of the lower unit. Check to see if the lower unit is in neutral gear position.

Begin to bring the intermediate housing and lower gear housing together.

➡The next step takes time and patience. Success will probably not be achieved on the first attempt. Three items must mate at the same time before the lower unit can be seated against the intermediate housing.

- The top of the driveshaft on the lower unit indexes with the lower end of the crankshaft.
- The water tube in the intermediate housing slides into the grommet on the water pump housing.
- The top splines of the lower shift rod in the lower unit slide into the internal splines of the upper shift rod in the intermediate housing if applicable.

9. As the two units come closer, rotate the propeller shaft slightly to index the upper end of the lower driveshaft tube with the crankshaft. At the same time, feed the water tube into the water tube grommet and feed the lower shift rod into the upper shift rod.

Push the lower unit housing and the intermediate housing together. The dowel pin on the upper surface of the lower unit will index with a matching hole in the lower surface of the extension case.

Tighten the adjusting nut on shift rod "A" to secure the two shift rods together and tighten the lock nut.

10. Apply Loctite® to the threads of the six bolts used to secure the lower unit to the intermediate housing. Install and tighten the bolts to the correct torque specification for that particular model.

11. Apply Loctite® to the threads of the trim tab bolt and tighten this bolt to the same torque specification as given in the previous step.
12. Make sure the water hose is reattached and clamped tightly.
13. Operate the shift lever through all gears. The shifting should be smooth and the propeller should rotate in the proper direction when the flywheel is rotated by hand in a clockwise direction. Naturally the propeller should not rotate when the unit is in neutral.

➡An anti-seize compound will prevent the propeller from freezing to the shaft and permit propeller removal, without difficulty, the next time the propeller needs to be pulled.

14. Apply Hondaline Grease, or equivalent anti-seize compound, to the propeller shaft. Install the propeller.

Remove the block of wood. Connect the spark plug wires to the spark plugs. Connect the electrical lead to the battery terminal.

15. The trim tab should be positioned to enable the helmsperson to handle the boat with equal ease to starboard and port at normal cruising speed. If the boat seems to turn more easily to starboard, loosen the socket head screw and move the trim tab trailing edge to the right. Move the trailing edge of the trim tab to the left if the boat tends to turn more easily to port.

Shift the unit into forward gear, release the tilt lock lever and lower the outboard to the normal operating position.

16. Remove the oil level plug and the drain plug. Fill the lower unit with Hondaline gearcase lubricant or Hypoid gear oil 90 weight until the lubricant flows from the top hole. Install both plugs and clean any excess lubricant from the lower unit.

Mount the engine in a test tank or body of water.

✽✽ CAUTION

Water must circulate through the lower unit to the powerhead anytime the powerhead is operating to prevent damage to the water pump in the lower unit. Just five seconds without water will damage the water pump impeller.

Start the engine and check the completed work for satisfactory operation, shifting and no leaks.

LOWER UNIT OVERHAUL

BF20 and BF2A (Forward Only)

DISASSEMBLY

See Figures 17 and 18

1. Pull the Pinion Shaft tube free of the Pinion Shaft.
2. Remove the bolts securing the extension plate to the gear case and remove the extension plate.
3. Raise the water pump cover a bit to remove the indexing pin. Leave the water pump cover in this position at this time.
4. Pry the circlip free from the end of the Pinion Shaft with a thin screwdriver. This clip holds the pinion gear onto the Pinion Shaft. The clip may not come free on the first try, but have patience and it will come free. With one hand, remove Pinion Shaft up and out of the lower unit housing and at the same time, with the other hand, catch the pinion gear and thrust washer from behind the gear.

With the Pinion Shaft on the workbench, remove the impeller cover holder, the impeller cover and the water pump impeller. Check the condition of the impeller carefully and replace with a new one if there is any question as to its condition for satisfactory service. Never turn the impeller over in an attempt to gain further life from the impeller. Now is the time to replace the impeller. It's cheap insurance against burning up an engine due to impeller failure.

Inspect the pump liner and the impeller housing.

5. Inspect the condition of the upper seal in the lower unit housing. If replacement is required, use the same slide hammer and jaw attachment to remove the seal.

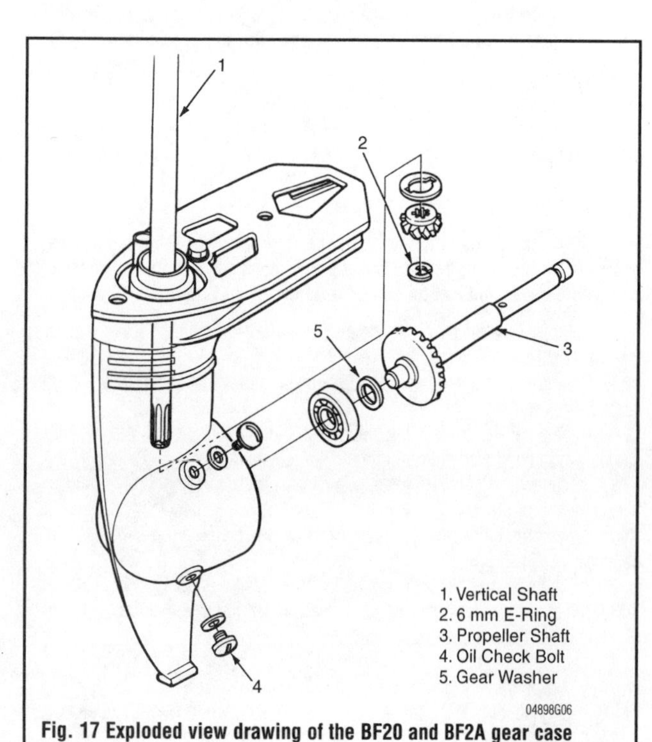

Fig. 17 Exploded view drawing of the BF20 and BF2A gear case assembly

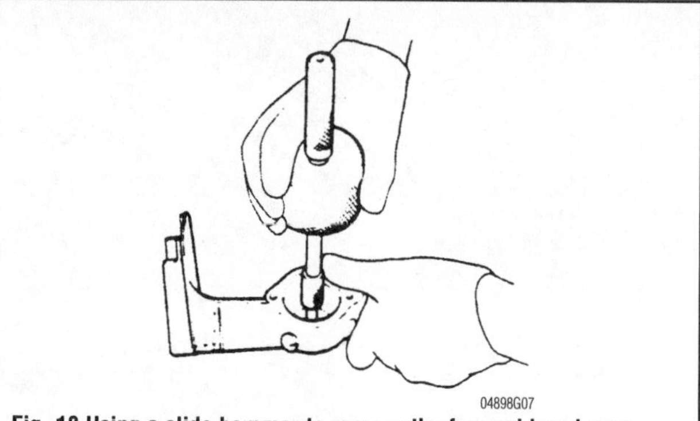

Fig. 18 Using a slide hammer to remove the forward bevel gear roller bearing

➡Removal of the seal destroys its sealing qualities and it cannot be installed a second time. Therefore, remove the seal only if it is unfit for further service and be absolutely sure a new seal is available before removing the old seal.

Obtain a slide hammer with jaw expander attachment and use it remove the water seal from gear case.

6. Remove the propeller shaft from the lower unit housing. The forward gear will come out with the shaft.

➡Inspect the forward propeller shaft bearing, if any roughness is felt while turning the bearing, or any other damage such as rust from water in the gear case is noticed, the bearing must be replaced. Removing the bearing will damage it beyond further use so don't remove it unless you need to.

Using the same slide hammer, now remove the forward propeller shaft bearing.

CLEANING & INSPECTION

Clean all water pump parts with solvent and then dry them with compressed air. Inspect the water pump cover and base for cracks and distortion. If possible, always install a new water pump impeller while the lower unit is disassembled. A new impeller will ensure extended satisfactory service and give peace of mind to the owner. If the old impeller must be returned to service, never install it in reverse to the original direction of rotation. Installation in reverse will cause premature impeller failure.

Inspect the ends of the impeller blades for cracks, tears and wear. Check for a glazed or melted appearance, caused from operating without sufficient water. If any question exists, as previously stated, install a new impeller if at all possible.

➡If an old impeller is installed be sure the impeller is installed in the same manner from which it was removed—the blades will rotate in the same direction. Never turn the impeller over thinking it will extend its life. On the contrary, the blades would crack and break after just a short time of operation.

Inspect the bearing surface of the propeller shaft. Check the shaft surface for pitting, scoring, grooving, imbedded particles, uneven wear and discoloration.

Check the straightness of the propeller shaft with a set of V-blocks. Rotate the propeller on the blocks.

Good shop practice dictates installation of new O-rings and oil seals regardless of their appearance.

Clean the pinion gear and the propeller shaft with solvent. Dry the cleaned parts with compressed air.

Check the pinion gear and the drive gear for abnormal wear.

ASSEMBLY

➧ See Figure 19

Forward Propeller Shaft Bearing

This first section applies only if the forward propeller shaft bearing was removed from the housing.

Place the propeller shaft forward bearing squarely into the housing with the side embossed with the bearing size facing outward. Drive the bearing into the housing until it is fully seated. The bearing must be properly installed to receive the forward end of the propeller shaft.

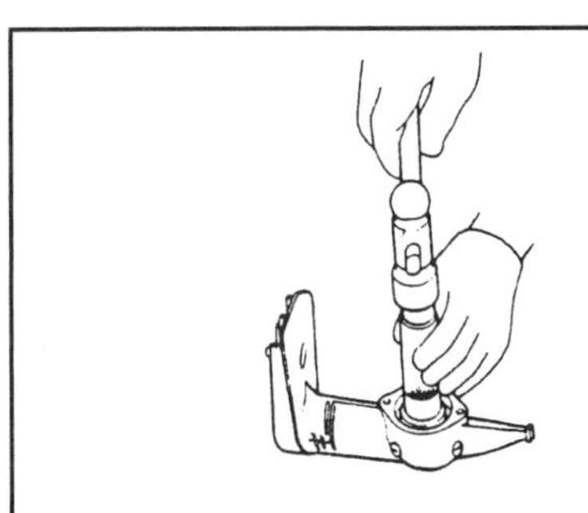

04898G10

Fig. 19 Using the bearing driver, install the forward gear bearing squarely into the gear case

Water Seal

1. Obtain a seal installer. Place the water seal over the end of the installer with the lip facing upward.

➡The seal must be installed with the lip facing upward to prevent water from entering and contaminating the lubricant in the lower unit.

Lower the seal installer and the seal squarely into the seal recess. Tap the end of the handle with a hammer until the seal is fully seated. After installation, pack the seal with Honda Grease A or equivalent water resistant lubricant.

Propeller Shaft

1. make sure the gear washer is in place and insert the propeller shaft into the lower unit housing. Push the propeller shaft into the forward propeller shaft bearing as far as possible.

Water Pump

➧ See Figures 20 and 21

Obtain the following special tools:

- Seal driver tool—p/n 07749-0010000
- Pilot (22 mm)—p/n 07746-0041000.

➡After installation of the water pump housing cover, the oil seal lip faces downward to prevent water in the water pump from contaminating

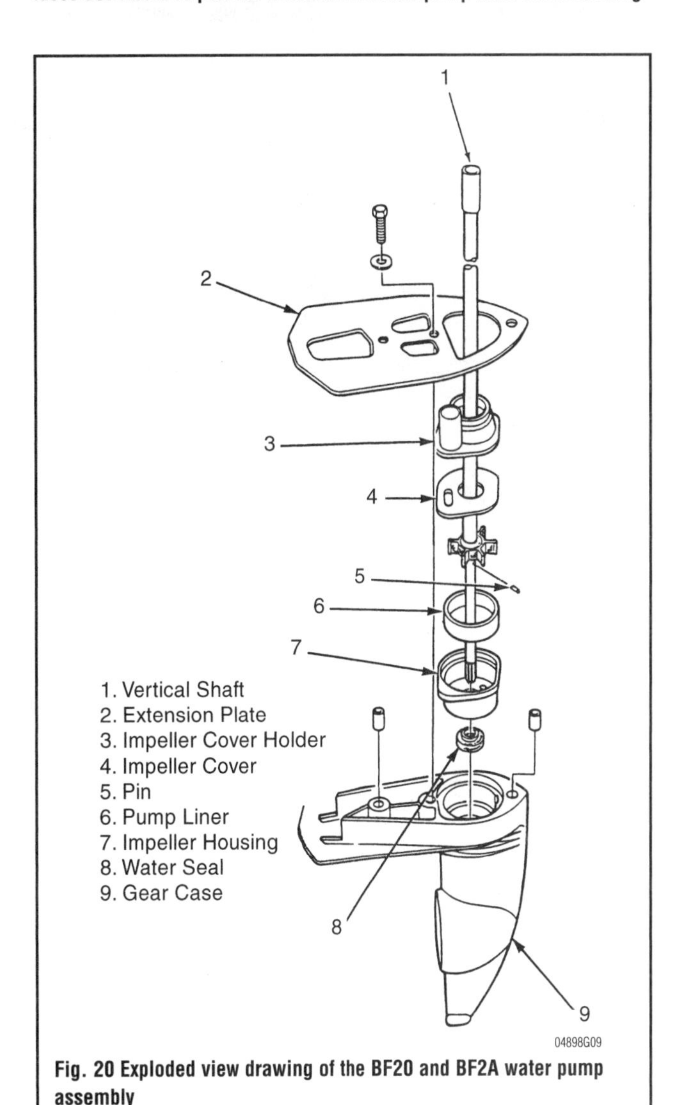

04898G09

Fig. 20 Exploded view drawing of the BF20 and BF2A water pump assembly

1 2

1. Pump Liner
2. Water Pump Impeller

04898G11

Fig. 21 Make sure the impeller vanes are bending in this direction during installation

the lubricant around the Pinion Shaft. However, to install the seal, the water pump cover is turned upside down on the work surface. In this position, the lip of the seal must face upward.

Using the seal installer and handle, install the seal. After installation, pack the seal with water resistant lubricant.

1. Install the impeller housing and pump liner into the gear case.
2. Install the impeller on the pinion shaft with the keyway facing downward.
3. Slide the impeller up the shaft and insert the pin below the impeller.
4. Align the impeller keyway with the pin and slide the impeller down over the pin.
5. Seat the impeller in the pump liner while rotating the impeller and shaft clockwise. The impeller vanes must bend away from the direction of shaft rotation or the impeller will be damaged.
6. Lower the assembled Pinion Shaft into the lower unit but do not mate the cover with the lower unit surface at this time. Leave some space, as shown. The splines on the lower end of the Pinion Shaft will protrude into the lower unit cavity.
7. Slide the thrust washer onto the lower end of the Pinion Shaft. Slide the pinion gear up onto the end of the Pinion Shaft. The splines of the pinion gear will index with the splines of the Pinion Shaft and the gear teeth will mesh with the teeth of the forward gear. Rotating the pinion gear slightly will permit the splines to index and the gears to mesh. Now, comes the hard part. Snap the circlip into the groove on the end of the Pinion Shaft to secure the pinion gear in place. If the first attempt is not successful, try again. Take a break, have a cup of coffee, tea, whatever, then give it another go. With patience, the task can be accomplished.
8. Apply some Loctite, or equivalent, to the extension plate retaining bolt. Tighten the bolt to a torque value of 4.4–8.7 in. lbs. (60–120 kg-cm).

Bearing Carrier

▶ See Figure 22

1. Place the bearing carrier cap on the work surface. Install the oil seal with the lettered side facing out. In this position the seal will prevent water from entering and contaminating the lubricant in the lower unit. Coat the seal lip with water resistant grease.
2. Install the bearing carrier gasket.
3. Install the bearing carrier cap to the lower unit and make sure the cap is seated evenly all the way around.
4. Apply Loctite, or equivalent, to the threads of the bearing carrier attaching bolts. Install and tighten the bolts alternately and evenly to a torque value of 5.8 ft. lbs. (8 Nm).

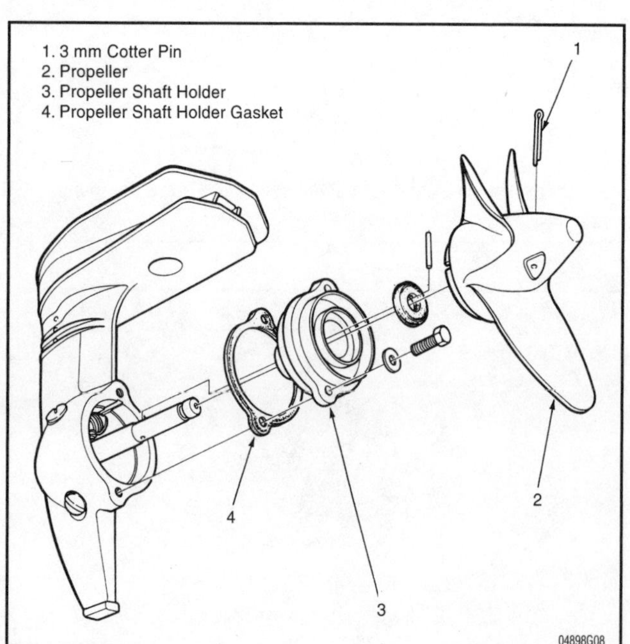

Fig. 22 Exploded view drawing of the BF20 and BF2A gear case assembly

Pinion Shaft

1. On the BF2A/BF20 models, pry the circlip free from the end of the Pinion Shaft with a thin screwdriver. This clip holds the pinion gear onto the lower end of the Pinion Shaft. The clip may not come free on the first try, but have patience and it will come free. With one hand, lift the Pinion Shaft up and out of the lower unit housing and at the same time, with the other hand, catch the pinion gear and SAVE any shim material from behind the gear. The shim material is critical to obtaining the correct backlash during installation. Using the old shim material will save considerable time, especially starting with no shim material.

Except BF20 and BF2A (Reverse Gear)

DISASSEMBLY

Water Pump

Although the various model water pumps have different configurations, they all are secured in similar ways.

BF50 AND BF5A

▶ See Figure 23

1. Remove the retaining bolts.
2. Remove the impeller housing.
3. Remove the impeller gaskets and the impeller cover.
4. Remove the distance spacers.
5. Slide the impeller off the pinion shaft.

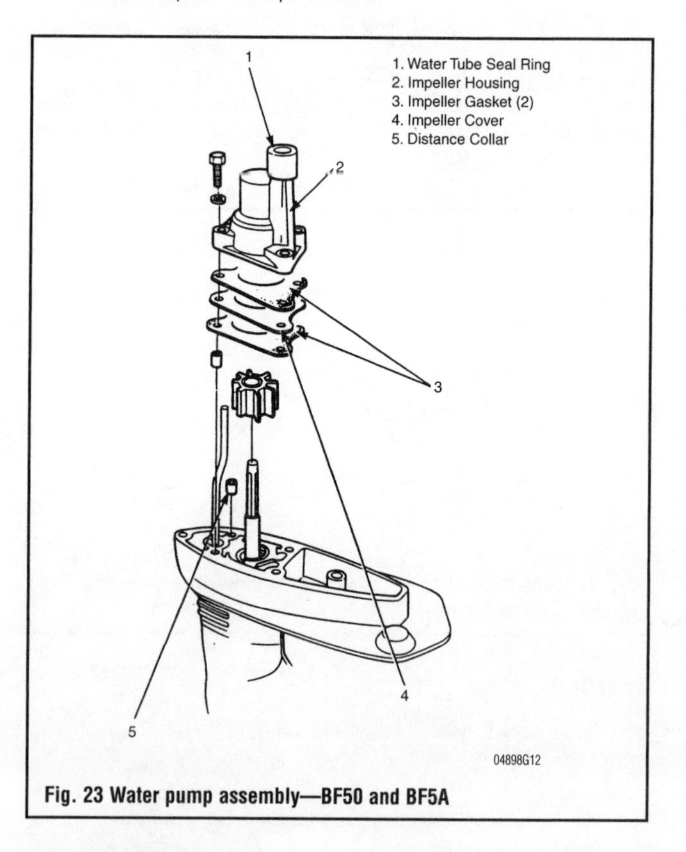

Fig. 23 Water pump assembly—BF50 and BF5A

BF75, BF8A AND BF100

▶ See Figure 24

1. Remove the retaining bolts on the water pump housing and the two retaining bolts on the cover holder.
2. Remove the impeller gaskets and the impeller cover.
3. Remove the distance spacers.
4. Slide the impeller off the pinion shaft.

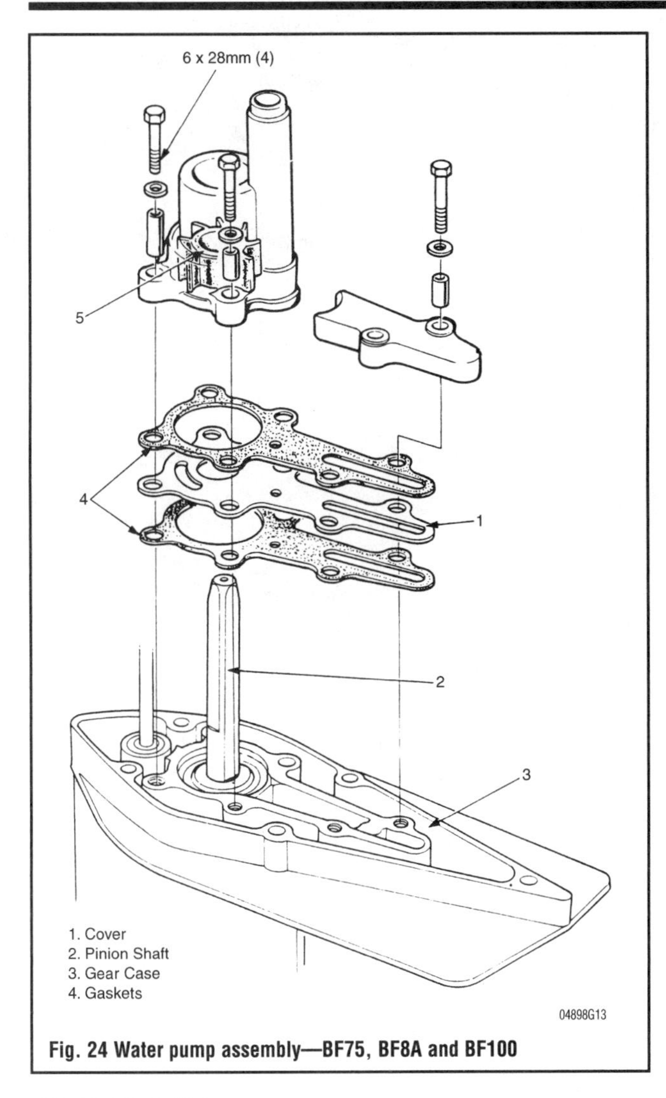

Fig. 24 Water pump assembly—BF75, BF8A and BF100

BF9.9A AND BF15A

➧ See Figure 25

1. Remove the retaining bolts.
2. Remove the impeller housing.

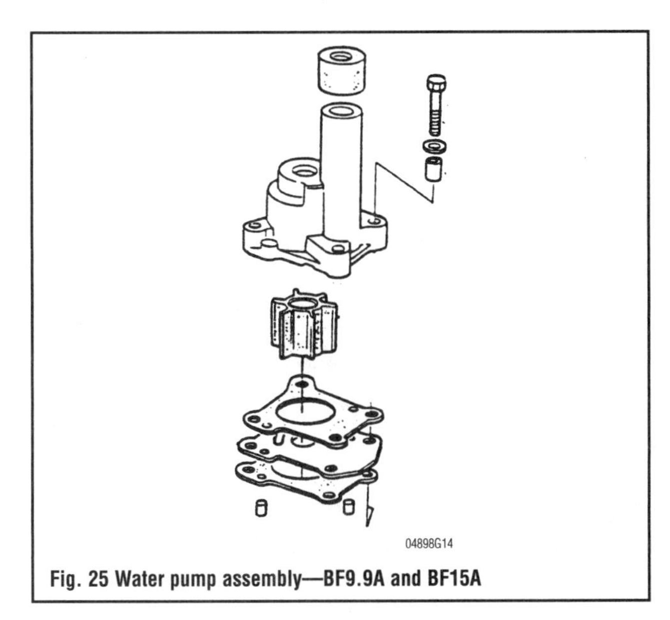

Fig. 25 Water pump assembly—BF9.9A and BF15A

3. Remove the impeller gaskets and the impeller cover.
4. Remove the distance spacers. These collars are not interchangeable and must be installed as shown in the drawing.
5. Slide the impeller off the pinion shaft.

BF25A, BF30A, BF35A, BF40A, BF45A AND BF50A

➧ See Figure 26

1. Remove the retaining bolts on the impeller housing.
2. Remove the impeller housing and water pump O-ring. Discard the O-ring.
3. Remove the water tube seal ring.
4. Remove the pump liner.
5. Remove the impeller. Keep track of the Woodruff key.
6. Remove the impeller cover, impeller gasket "B", water pump base, water pump gasket "A" and the bottom O-ring. Discard the O-ring.
7. Using a commercially available seal remover, pry out the water seals from the water pump base. Discard the seals. Do not reuse.

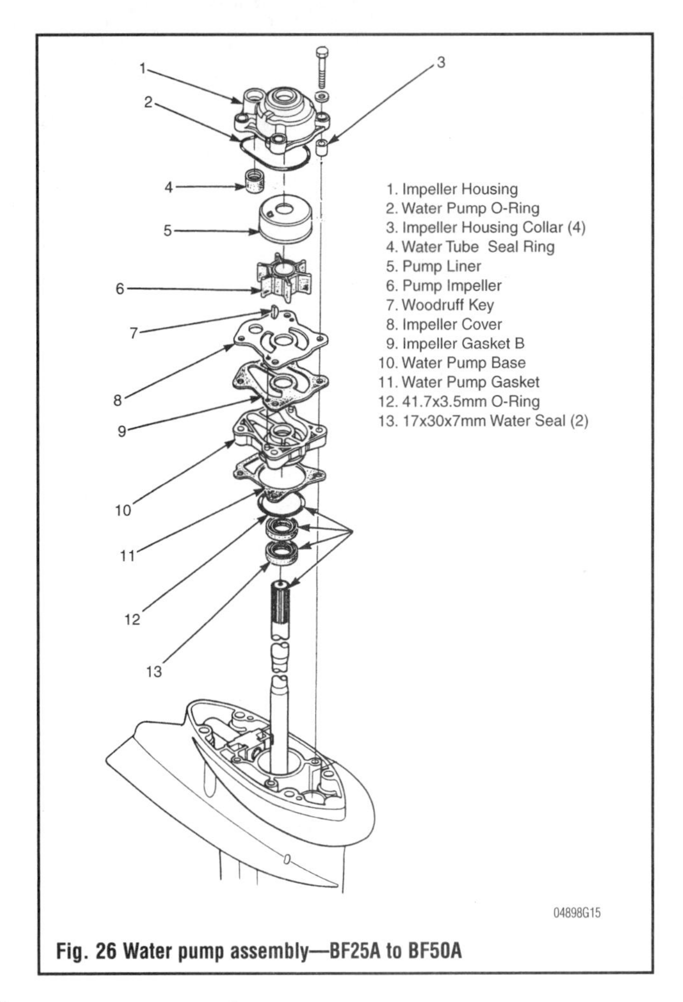

Fig. 26 Water pump assembly—BF25A to BF50A

BF75A AND BF90A

➧ See Figure 27

1. Remove the retaining bolts on the impeller housing.
2. Remove the impeller housing.
3. Remove the upper impeller gasket and discard. Do not reuse the gasket.
4. Remove the impeller. Keep track of the Woodruff key.
5. Remove the impeller cover and the lower impeller gasket. Discard the gasket.
6. Remove the water pump housing retaining bolts and remove the water pump housing and water pump gasket.
7. Using a commercially available seal remover, pry out the upper and lower water seals from the water pump base. Discard the seals. Do not reuse.

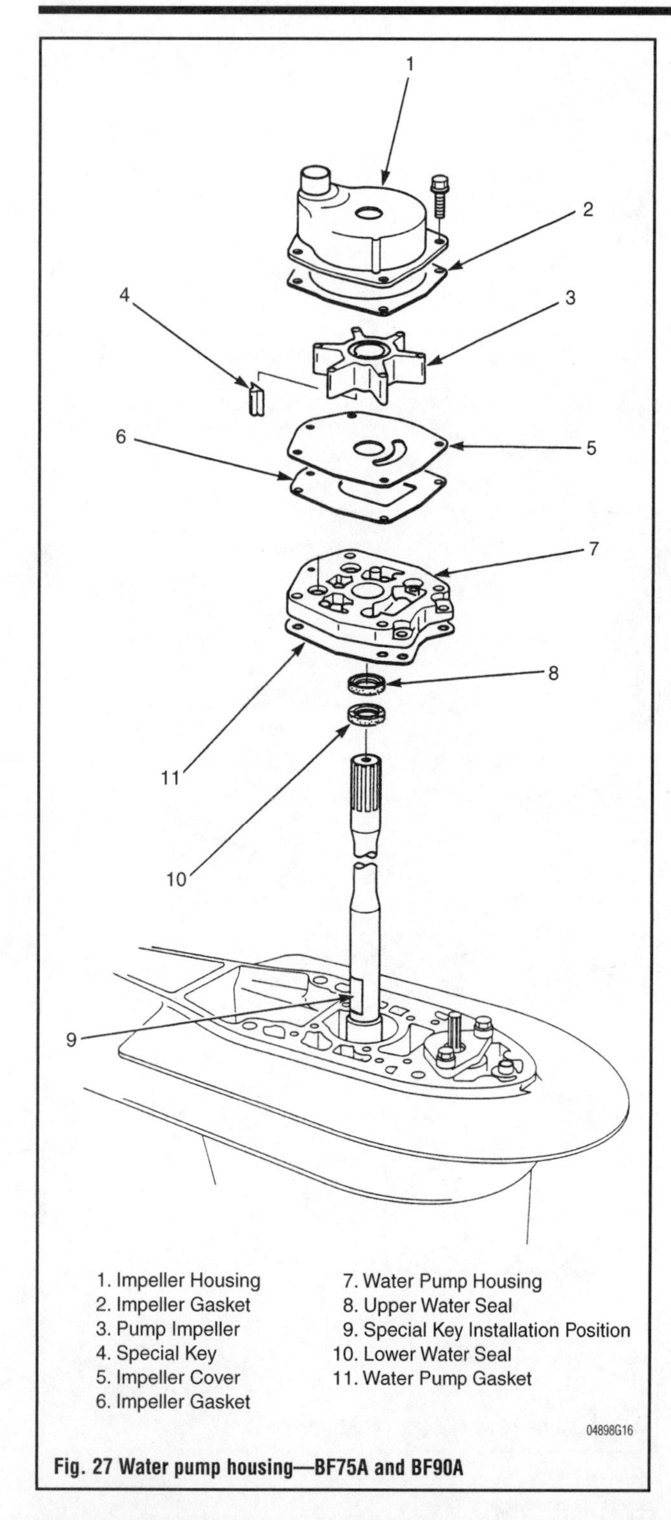

Fig. 27 Water pump housing—BF75A and BF90A

BF115A AND BF130A

➧ See Figure 28

1. Remove the retaining bolts and distance spacers on the impeller housing.
2. Remove the impeller housing and water pump O-ring. Discard the O-ring.
3. Remove the pump liner.
4. Remove the impeller. Keep track of the Woodruff key.
5. Remove the impeller cover and impeller gasket. Discard the gasket.

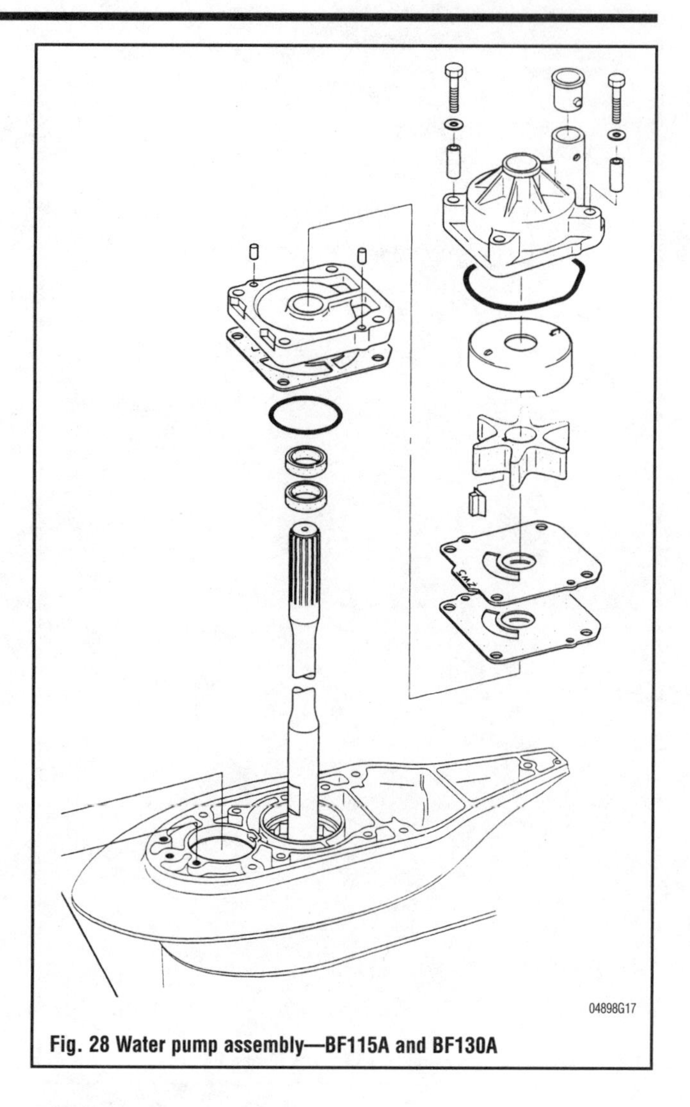

Fig. 28 Water pump assembly—BF115A and BF130A

Bearing Carrier

BF50 AND BF5A

➧ See Figure 29

1. Remove the two bolts securing the cap to the gear case, rotate the cap 90° and, using a rubber or plastic hammer, tap the bearing carrier out of the gear case.

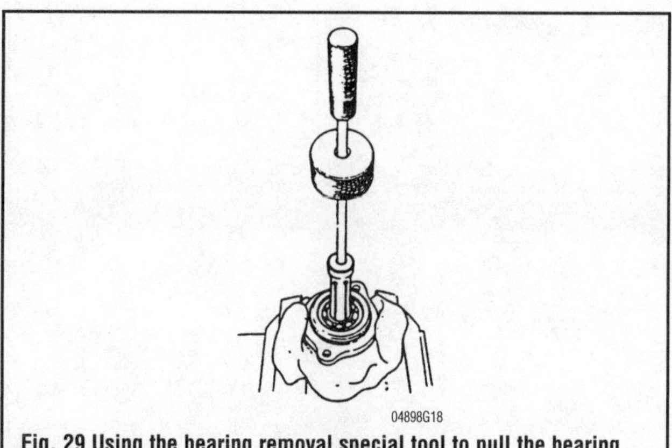

Fig. 29 Using the bearing removal special tool to pull the bearing carrier ball bearing

2. Remove the O-ring and discard.
3. To remove the roller bearing, obtain the following special tools:
- Bearing remover (20 mm)—p/n 07936-3710600
- Bearing remover weight—p/n 07936-3710200
- Bearing remover handle—p/n 07936-3710100
4. Using the special tools, remove the roller bearing from the bearing carrier cap.

BF75, BF8A AND BF100

See Figure 30

1. Remove the two bolts securing the cap to the gear case, rotate the cap 90°and, using a rubber or plastic hammer, tap the bearing carrier out of the gear case.
2. To remove the roller bearing from the bearing carrier, obtain the following special tool:
- Bearing remover—p/n 07945-9350001
3. Using the special tool, remove the roller bearing.

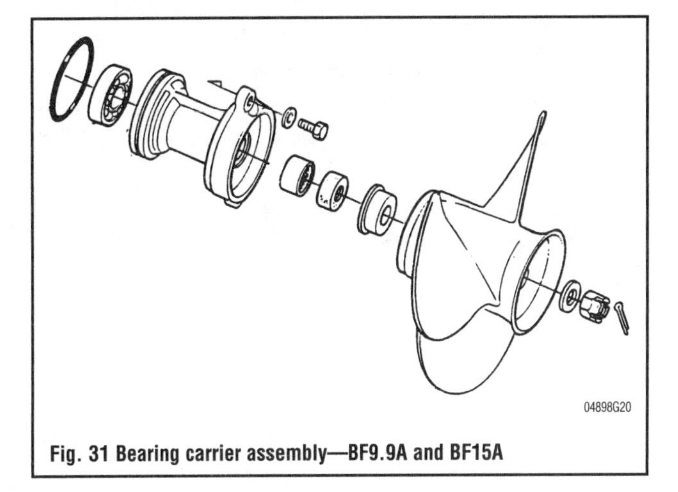

Fig. 31 Bearing carrier assembly—BF9.9A and BF15A

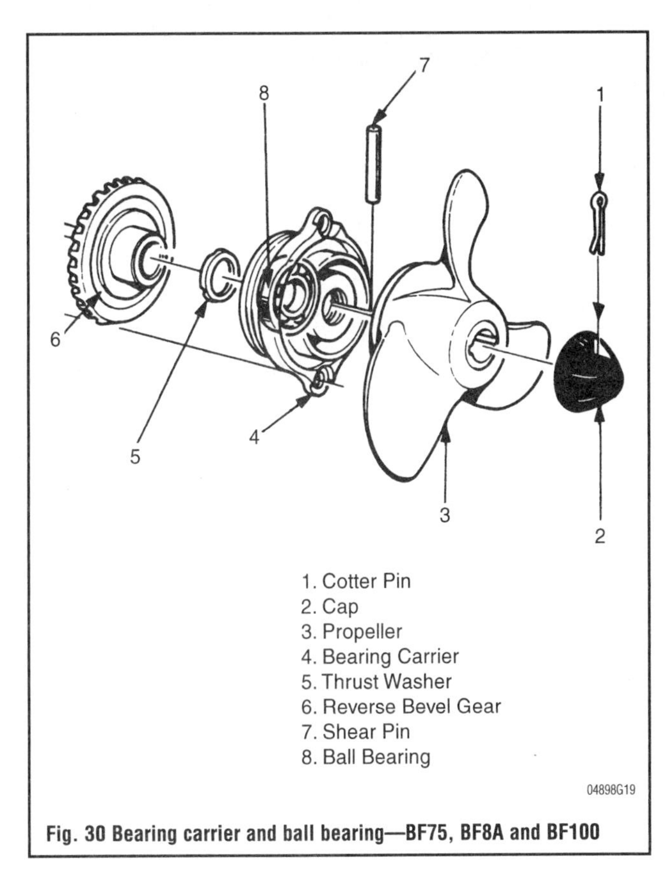

Fig. 30 Bearing carrier and ball bearing—BF75, BF8A and BF100

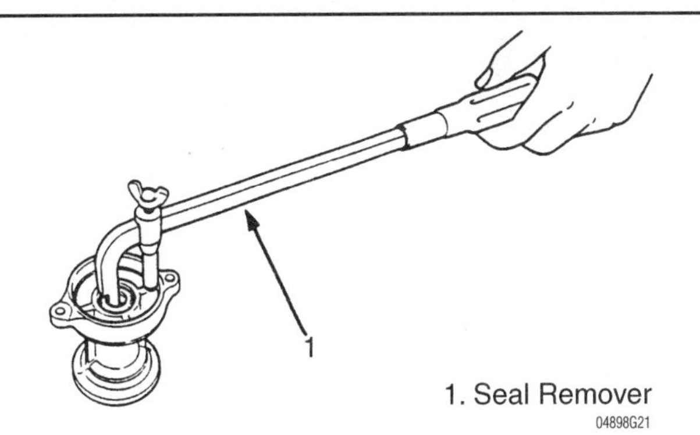

Fig. 32 Using the special tool to remove the bearing carrier water seal—BF9.9A and BF15A

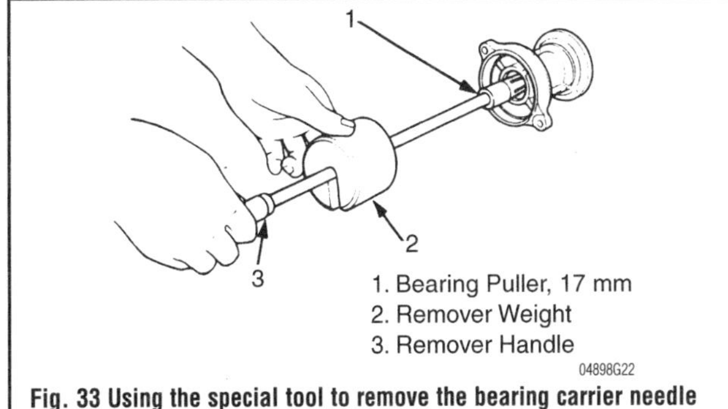

Fig. 33 Using the special tool to remove the bearing carrier needle bearing—BF9.9A and BF15A

BF9.9A AND BF15A

See Figures 31, 32, 33 and 34

1. Remove the two bolts securing the cap to the gear case, rotate the cap 90° and, using a rubber or plastic hammer, tap the bearing carrier out of the gear case.
2. To remove the water seal, obtain the following special tool:
- Seal remover—p/n 07748-0010001
3. Remove the 17 mm water seal out of the bearing carrier with the special tool.
4. To remove the needle bearing, obtain the following special tool:
- Bearing puller (17 mm)—p/n 07936-3710300
5. To remove the ball bearing, obtain the following special tool:
- Adjustable bearing puller (25–40 mm)—p/n 07736-A01000A
6. Using a commercially available slide hammer and the special tool, remove the ball bearing.

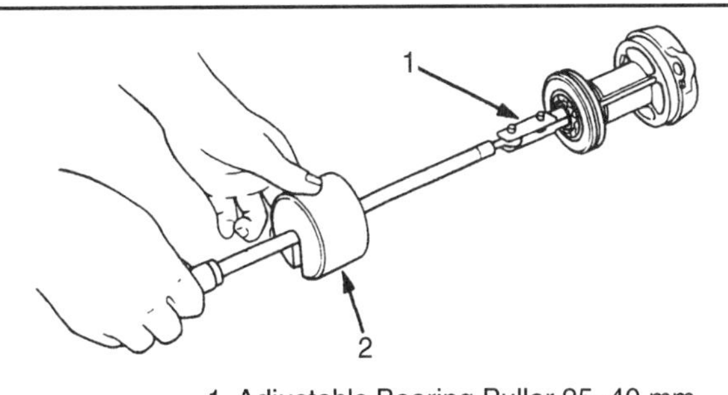

Fig. 34 Using the special tool to remove the bearing carrier ball bearing—BF9.9A and BF15A

BF25A AND BF30A

➧ See Figures 35, 36, 37 and 38

1. Remove the two bolts securing the cap to the gear case, rotate the cap 90°and, using a rubber or plastic hammer, tap the bearing carrier out of the gear case.
2. Remove the reverse bevel gear with a hydraulic press.
3. To remove the radial ball bearing, obtain the following special tools:
   - Remover handle—p/n 07936-3710100
   - Remover weight—p/n 07741-0010201
   - Bearing race puller—p/n 07LPC-ZV30100
4. Use the special tools to remove the bearing from the bearing carrier.
5. To remove the water seal, obtain a commercially available seal removal tool and pry out the seal from the bearing carrier.
6. To remove the needle bearing, obtain the following special tools:
   - Driver shaft—p/n 07946-MJ00100
   - Driver head—p/n 07946-KM40701
7. Using the special tools, drive the needle bearing out from the gear case side of the bearing carrier.

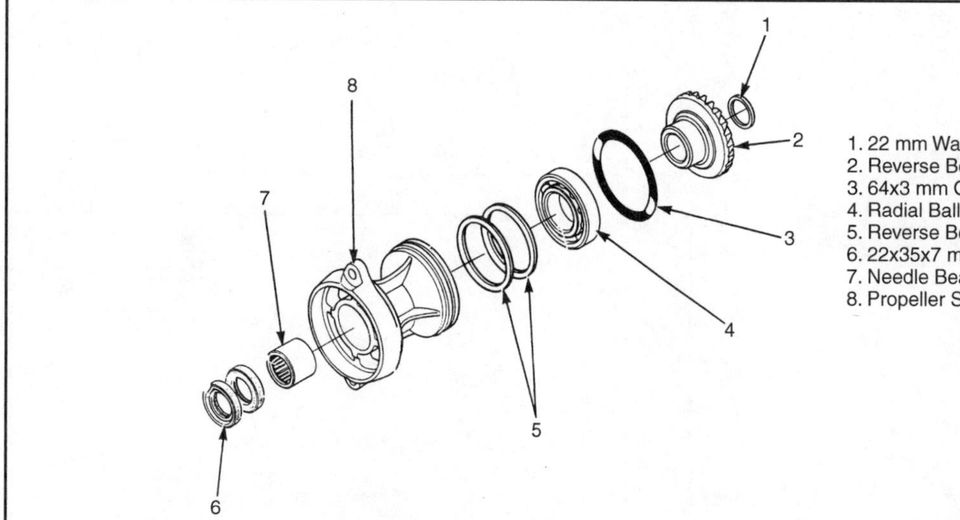

Fig. 35 Bearing carrier assembly—BF25A and BF30A

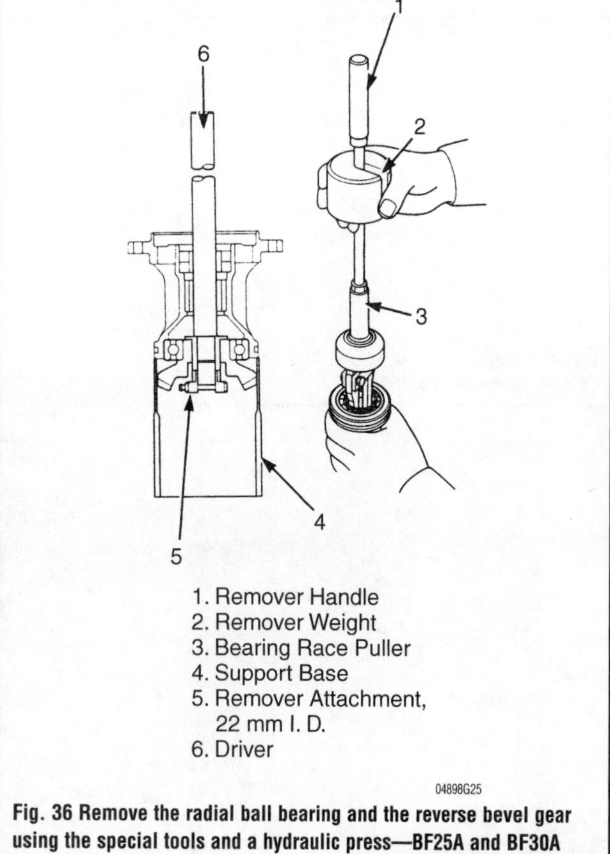

Fig. 36 Remove the radial ball bearing and the reverse bevel gear using the special tools and a hydraulic press—BF25A and BF30A

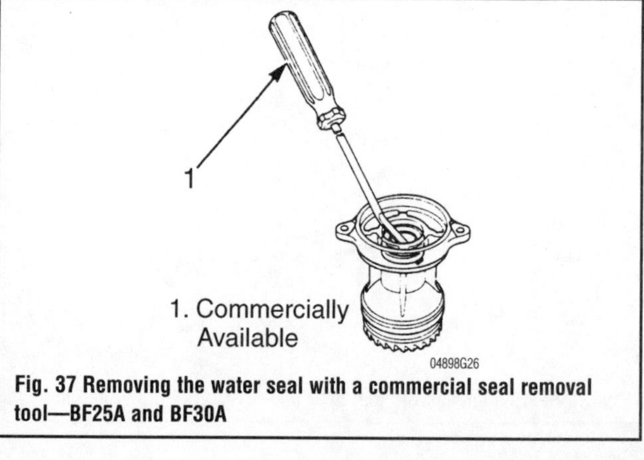

Fig. 37 Removing the water seal with a commercial seal removal tool—BF25A and BF30A

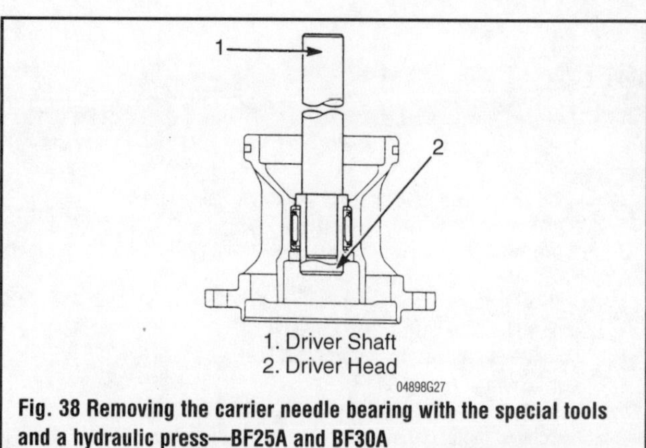

Fig. 38 Removing the carrier needle bearing with the special tools and a hydraulic press—BF25A and BF30A

BF35A, BF40A, BF45A AND BF50A

See Figures 39, 40 and 41

1. To remove the propeller shaft and bearing carrier, obtain the following special tools:
   - End nut wrench—p/n 07LPA-ZV30100
   - Propeller shaft remover—p/n 07LPC-ZV30200
2. Bend the washer tab down and remove the end nut with the end nut wrench.
3. Install the propeller shaft remover and draw out the bearing carrier from the gear case.
4. To remove the radial ball bearing, obtain the following special tools:
   - Bearing remover (35 mm)—p/n 07936-3710400

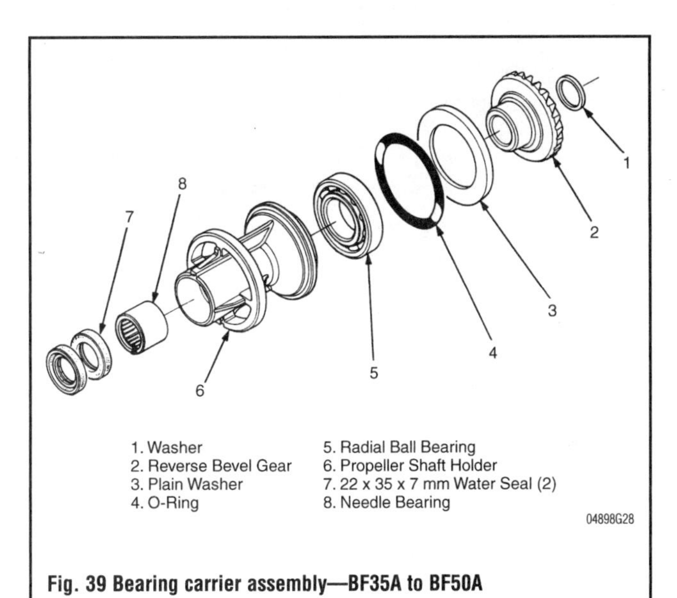

Fig. 39 Bearing carrier assembly—BF35A to BF50A

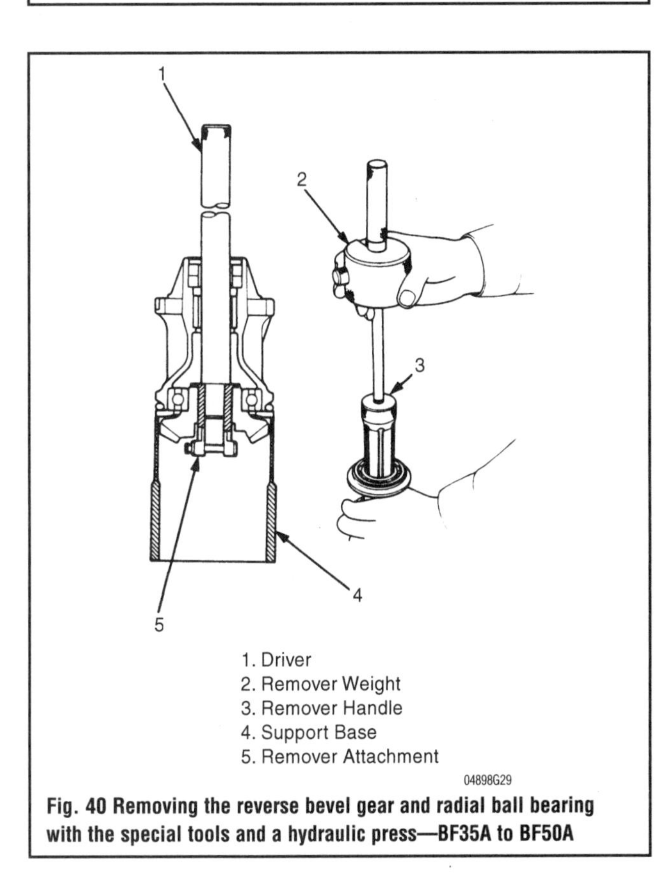

Fig. 40 Removing the reverse bevel gear and radial ball bearing with the special tools and a hydraulic press—BF35A to BF50A

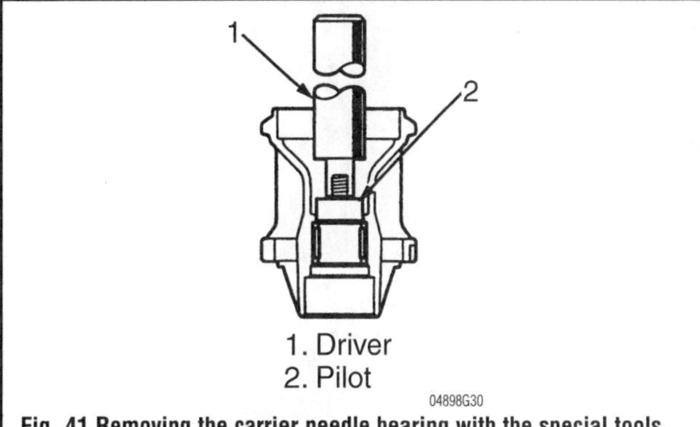

Fig. 41 Removing the carrier needle bearing with the special tools and a hydraulic press—BF35A to BF50A

   - Remover weight—p/n 07936-3710400
   - Remover handle—p/n 07936-3710100
5. Use the slide hammer special tool to remove the bearing from the bearing carrier.
6. To remove the water seal, obtain a commercially available seal remover and pry the seal out of the bearing carrier.
7. To remove the needle bearing, obtain the following special tools:
   - Driver—p/n 07749-0010000
   - Pilot (25 mm)—p/n 07746-0040600
8. Using the special tools, drive the needle bearing out from the gear case side of the bearing carrier.

BF75A AND BF90A

See Figures 42, 43 and 44

1. Adjust the engine until it is in a horizontal position.
2. Set the propeller thrust washer on the propeller shaft.
3. Hook the jaws of a commercially available puller to the inside of the bearing carrier.
4. Tighten the puller and remove the bearing carrier from the gear case.

Remove the propeller shaft assembly from the gear case very carefully. Be careful not to loose the shift slider or the three steel balls.

5. To remove the water seals and needle bearing, obtain the following special tools:
   - Driver—p/n 07949-3710001 (91-37323)
   - Attachment (32 x 35 mm)—p/n 07746-0010100 (91-36569)
   - Pilot (30 mm)—p/n 07746-0040700
6. The inner and outer water seals and the needle bearing must be removed as a set.
7. Use the special tools and a hydraulic press to press out the seals and bearing.
8. To remove the roller bearing, obtain the following special tools:

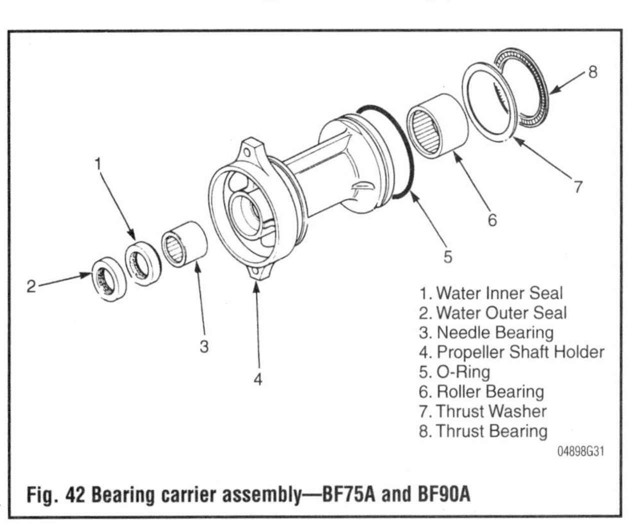

Fig. 42 Bearing carrier assembly—BF75A and BF90A

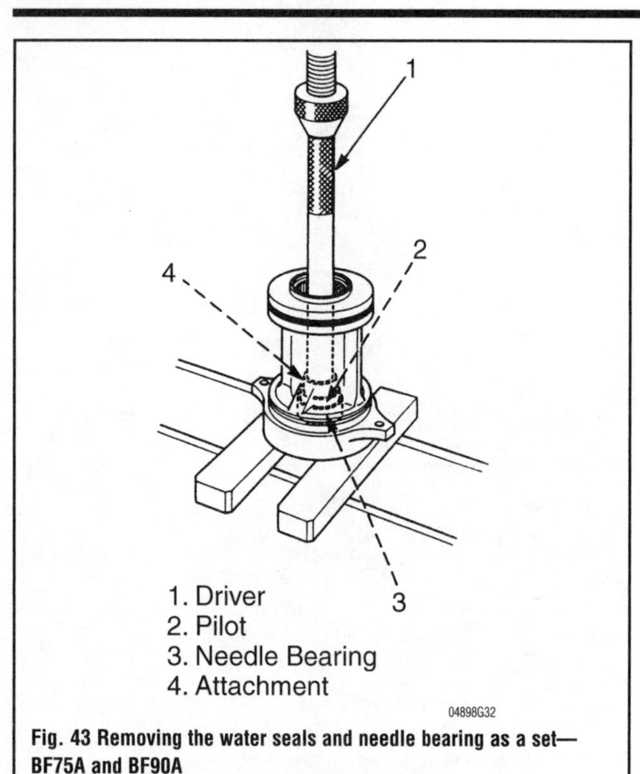

Fig. 43 Removing the water seals and needle bearing as a set—BF75A and BF90A

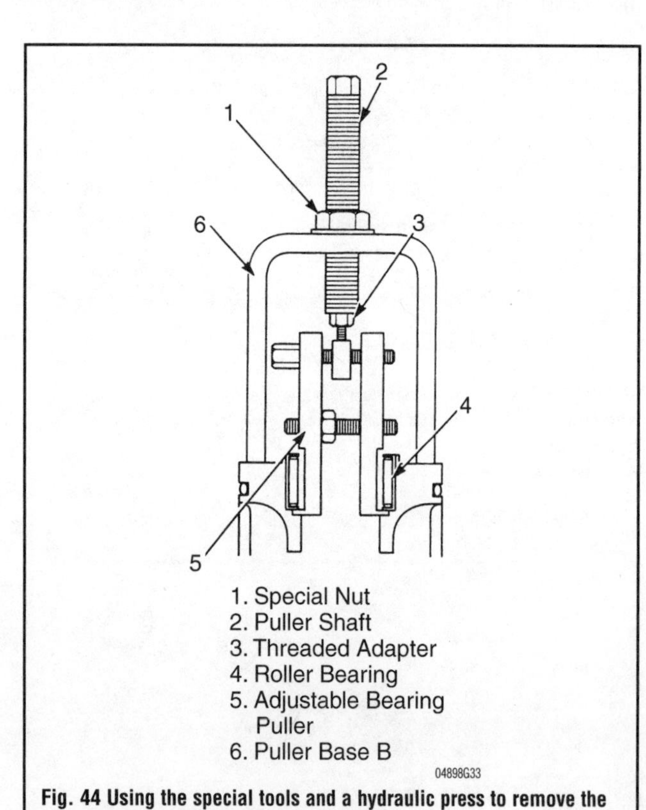

Fig. 44 Using the special tools and a hydraulic press to remove the carrier roller bearing—BF75A and BF90A

- Puller shaft (22 x 1.5 x 240 mm)—p/n 07931-ME4010B
- Special nut—p/n 07931-HB3020A
- Puller base "B"—p/n 07LPC-ZV3020A
- Threaded adapter—p/n 07965-HB3000A
- Adjustable bearing puller (25–40 mm)—p/n 07736-A01000B

9. Install the special tool assembly onto the bearing carrier and draw out the roller bearing from the bearing carrier.

**** CAUTION**

Remove the roller bearing using the special tools only. Do not try to remove the bearing by hitting with a hammer.

BF115A AND BF130A

See Figures 45 thru 55

1. Remove the propeller and gear case assembly.
2. Remove the shift rod.
3. Remove the two bolts securing the bearing carrier to the gear case and turn the bearing carrier 90°and tap out the carrier using a rubber or plastic hammer on the exposed carrier flanges.
4. To remove the two water seals, obtain a commercially available seal removal tool and pry out the old seals and discard.
5. To remove the needle bearing, obtain the following special tools:
   - Bearing remover (30 mm)—p/n 07936-8890300
   - Remover weight—p/n 07741-0010201
   - Remover handle—p/n 07936-3710100
6. Use the special tools to pull the needle bearing out from the bearing carrier.
7. On the LA and XA types, to remove the reverse bevel gear and radial ball bearing from the bearing carrier, obtain the following special tools:
   - Driver —p/n 07749-0010000
   - Attachment (24 x 26 mm)—p/n 07746-0010700
   - Pilot (17 mm)—p/n 07746-0040400
   - Bearing remover—p/n 07HMC-MR70100
8. Using the special tools and a hydraulic press, remove the reverse bevel gear and radial ball bearing as a set.

➡Remove the reverse bevel gear and radial ball bearing with a hydraulic press only. Do not attempt to remove the gear and bearing by hitting with a hammer.

9. To separate the reverse bevel gear and radial ball bearing, obtain the following special tools:
   - Driver—p/n 07749-0010000
   - Attachment (37 x 40 mm)—p/n 07746-0010200
   - Pilot (30 mm)—p/n 07746-0040700
10. Install a commercially available bearing separator between the radial ball bearing and the reverse bevel gear.
11. Remove the reverse bevel gear from the bearing by using a hydraulic press and the special tools.
12. On the LCA and XCA (counter-rotating) types, fabricate a fixture tool according to the dimensions shown in the drawing.

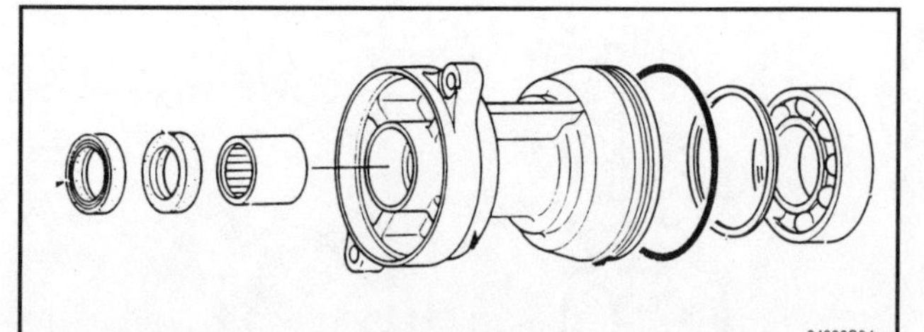

Fig. 45 Bearing carrier assembly (LA / XA type)—BF115A and BF130A

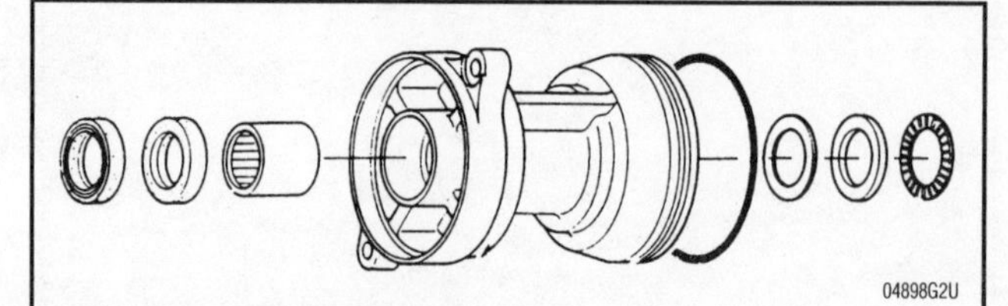

Fig. 46 Bearing carrier assembly (LCA / XCA type)—BF115A and BF130A

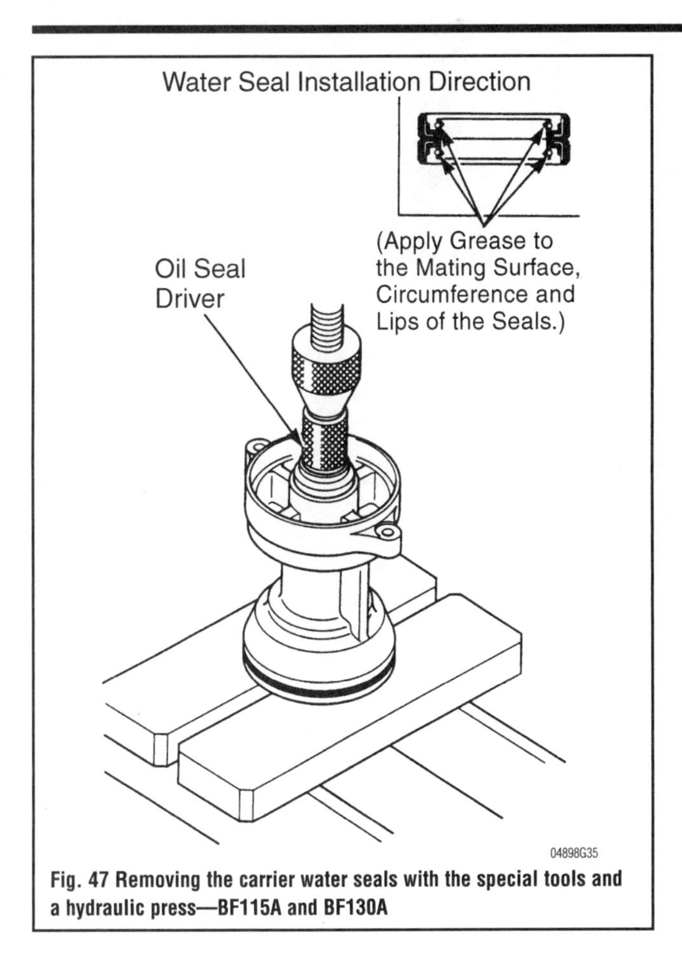

Fig. 47 Removing the carrier water seals with the special tools and a hydraulic press—BF115A and BF130A

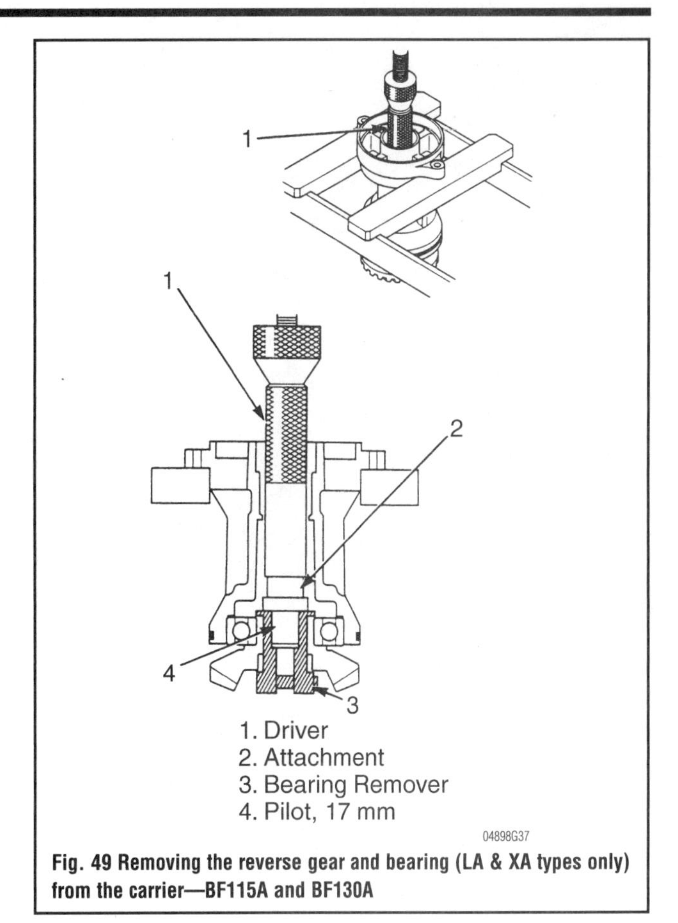

Fig. 49 Removing the reverse gear and bearing (LA & XA types only) from the carrier—BF115A and BF130A

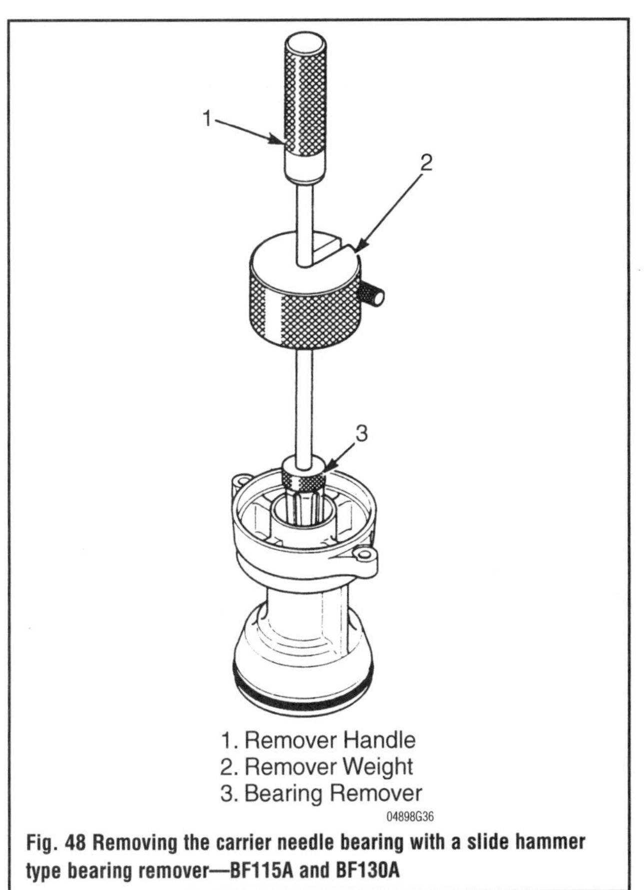

Fig. 48 Removing the carrier needle bearing with a slide hammer type bearing remover—BF115A and BF130A

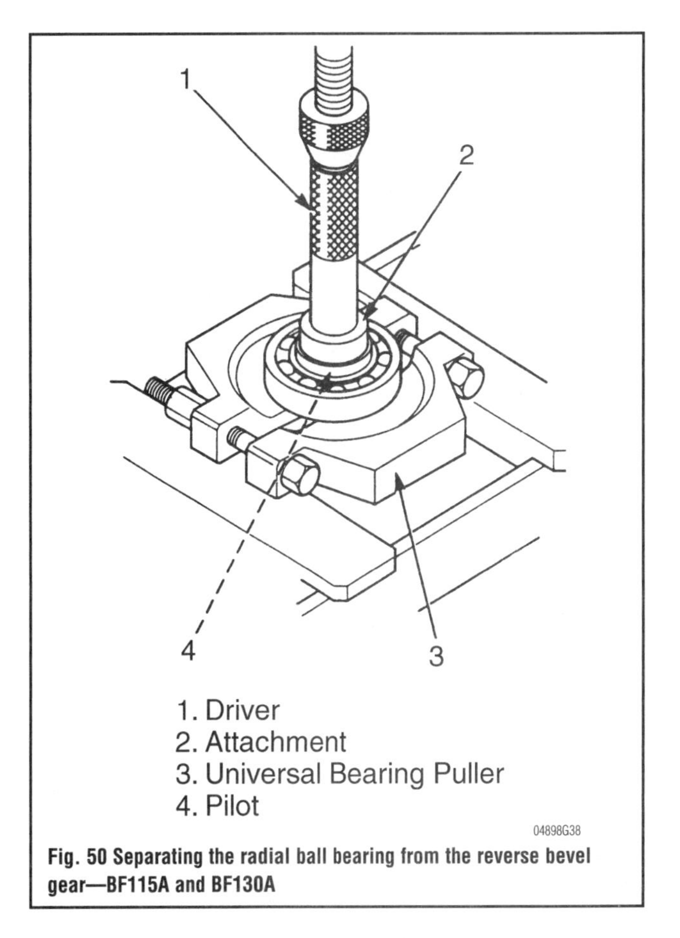

Fig. 50 Separating the radial ball bearing from the reverse bevel gear—BF115A and BF130A

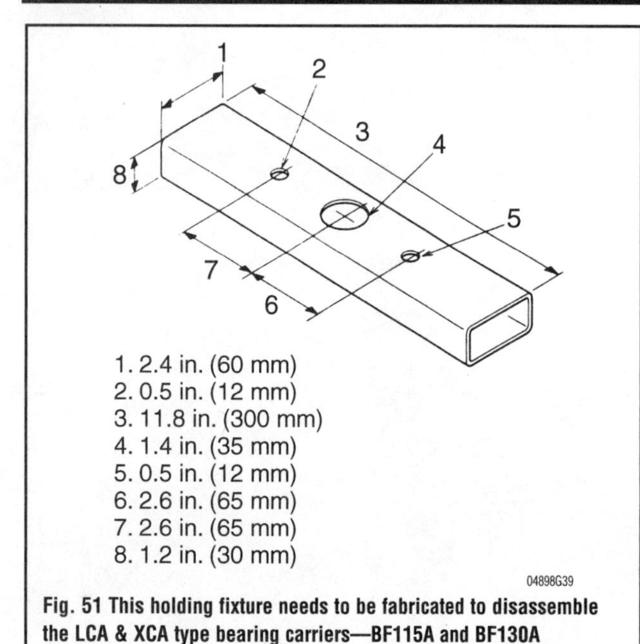

Fig. 51 This holding fixture needs to be fabricated to disassemble the LCA & XCA type bearing carriers—BF115A and BF130A

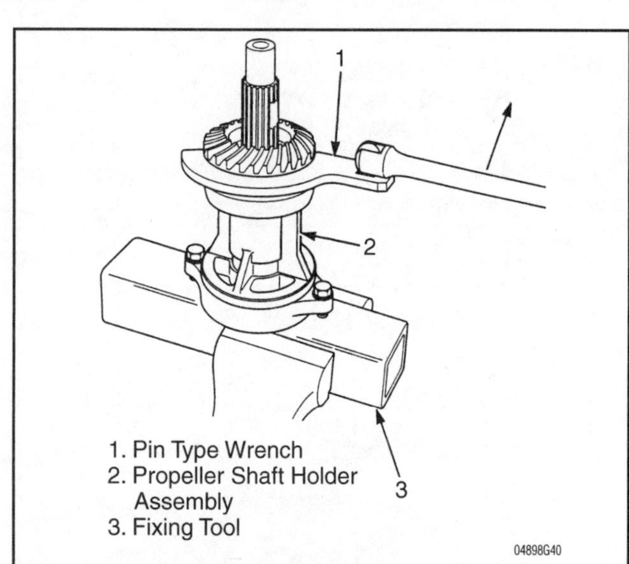

Fig. 52 Using a pin spanner wrench to disassemble the bearing holder assembly—BF115A and BF130A

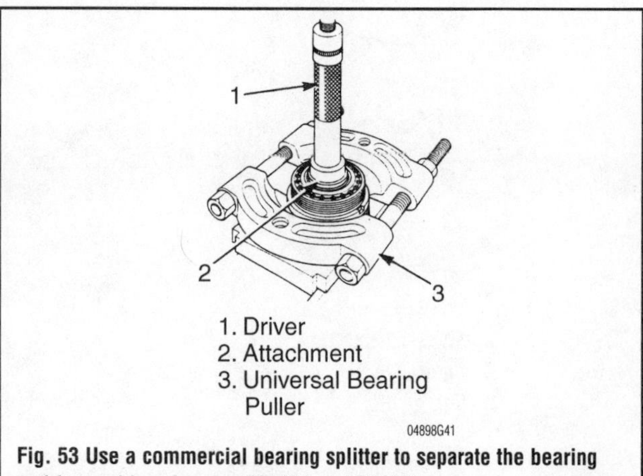

Fig. 53 Use a commercial bearing splitter to separate the bearing and forward bevel gear—BF115A and BF130A

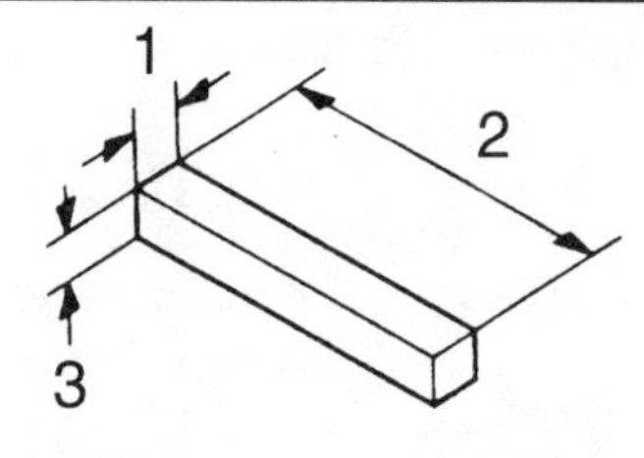

Fig. 54 Fabricate this special tool. . .

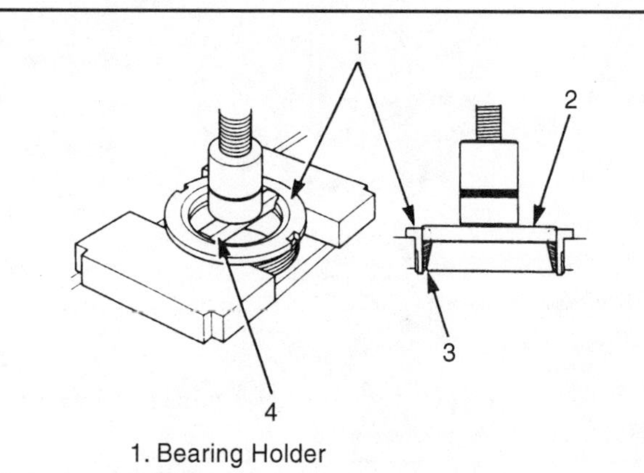

Fig. 55 . . . to remove the tapered outer bearing race from the bearing holder

13. Bolt the bearing carrier into the holding fixture and secure the entire assembly in a vice.
14. Obtain the following special tool:
- Pin type wrench (103 mm)—p/n 07WPA-ZW50100
15. Loosen the bearing holder with the special tool and remove the bearing holder from the propeller shaft.
16. To separate the bearing holder and the forward bevel gear, obtain the following special tools and a commercially available bearing separator:
- Driver—p/n 07749-0010000
- Attachment (42 x 47 mm)—p/n 07746-0010300
17. Using the special tools and a hydraulic press, separate the forward bevel gear and the tapered roller bearing inner race.
18. To remove the tapered roller bearing outer race, fabricate a tool to the specifications shown in the drawing.
19. Set the tool in the groove on the bearing holder. Using the fabricated tool, press out the outer race with a hydraulic press.
20. After removing the outer bearing race, remove the forward bevel gear shim/s.

Propeller Shaft and Clutch shifter

BF50 AND BF5A

➡When removing and installing the propeller shaft, make sure the transmission is in FORWARD gear.

1. Use a 3.0 mm drift and press out the shifter pin. Remove the rod, spring and shifter pin from the end of the propeller shaft. Slide the clutch shifter from the shaft. Observe how the clutch shifter was installed. It must be installed in the same direction.

BF75, BF8A AND BF100

➧ **See Figure 56**

➡**When removing and installing the propeller shaft, make sure the transmission is in FORWARD gear.**

1. Use a drift or obtain the following special tool to push out the shifter pin:
   - Pin flare tool–07968-9350000.
2. After removing the shifter pin, remove the push rod, clutch shifter and spring from the end of the propeller shaft. Slide the clutch shifter from the shaft. Observe how the clutch shifter was installed. It must be installed in the same direction

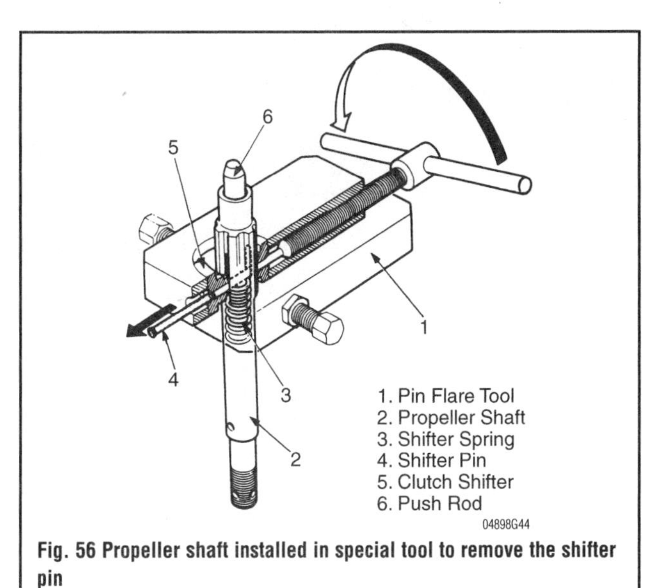

Fig. 56 Propeller shaft installed in special tool to remove the shifter pin

BF9.9A AND BF15A

➧ **See Figures 57 and 58**

➡**When removing and installing the propeller shaft, make sure the transmission is in FORWARD gear.**

1. In order to remove the clutch shifter, obtain the following special tool:
   - Pin flare tool (4 mm)—07968-9350000.
2. Install the propeller shaft in the special tool and align the pin driver of the tool with the shifter pin.
3. Tighten the 2 holding bolts on the tool to hold the clutch shifter.
4. Turn the handle on the tool and push out the shifter pin.
5. Remove the clutch shifter, shift slider and shift spring from the propeller shaft.

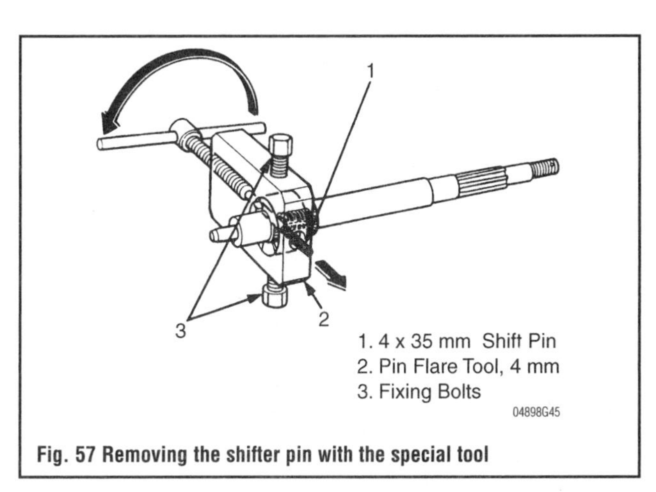

Fig. 57 Removing the shifter pin with the special tool

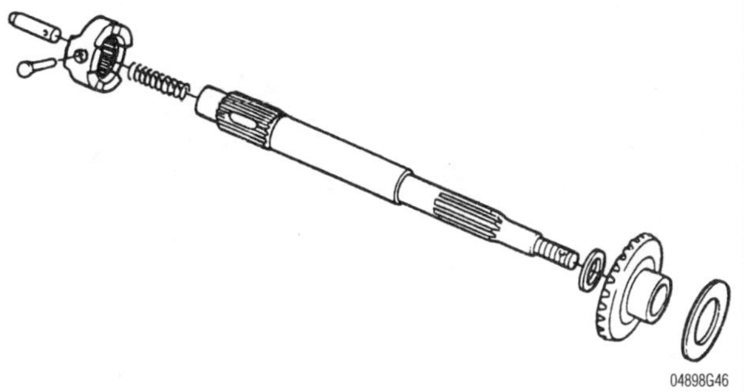

Fig. 58 Propeller shaft and clutch shifter assembly—BF9.9A and BF15A

BF25A, BF30A, BF35A, BF40A, BF45A AND BF50A

➧ **See Figure 59**

1. Use an awl to pry out and remove the cross pin ring.
2. Press the shifter pin against a piece of wood, and press out the shifter pin.
3. Slide the clutch shifter off the propeller shaft.
4. Remove the shift slider, steel ball, shift pin holder and the shift spring from inside the propeller shaft.

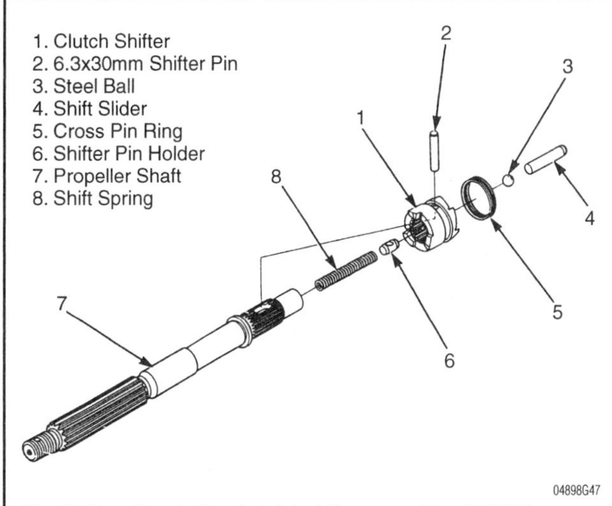

Fig. 59 Propeller shaft and clutch shifter assembly—BF25A to BF50A

BF75A AND BF90A

➧ **See Figures 60 and 61**

1. Use an awl to pry out and remove the cross pin ring.
2. Press the shifter pin against a piece of wood, and press out the shifter pin.
3. Slide the clutch shifter off the propeller shaft.
4. Remove the shift slider, 3 steel balls, shift pin holder and the shift spring from inside the propeller shaft.

BF115A AND BF130A

➧ **See Figures 62, 63 and 64**

1. Pull out on the shift slider carefully (do not let the 2 smaller steel balls pop out) and remove the steel balls.
2. Remove the cross pin ring and press out the shifter pin.
3. Remove the clutch shifter, the 2 larger steel balls and the shift spring.

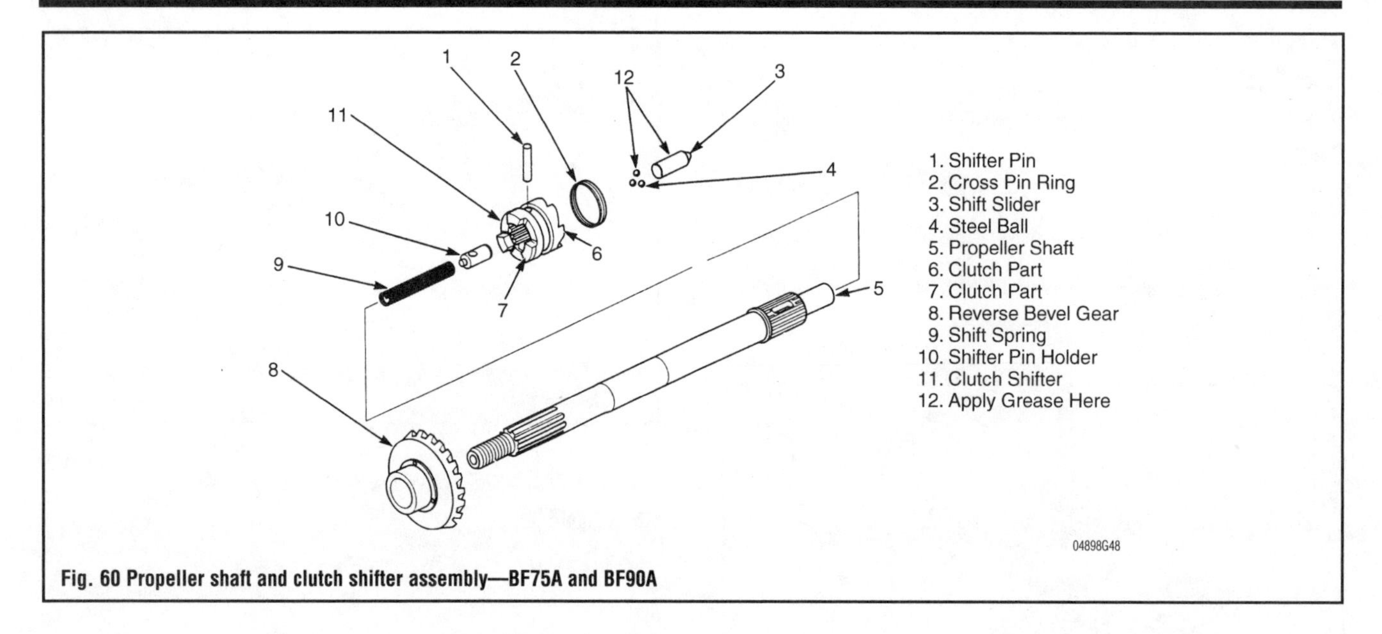

Fig. 60 Propeller shaft and clutch shifter assembly—BF75A and BF90A

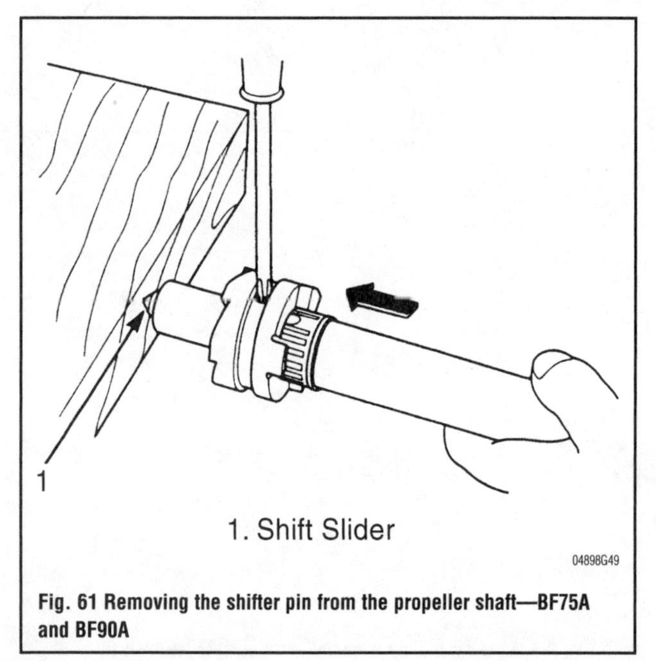

Fig. 61 Removing the shifter pin from the propeller shaft—BF75A and BF90A

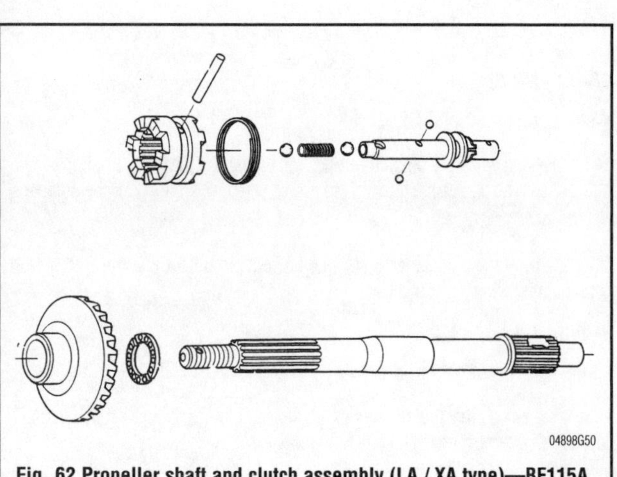

Fig. 62 Propeller shaft and clutch assembly (LA / XA type)—BF115A and BF130A

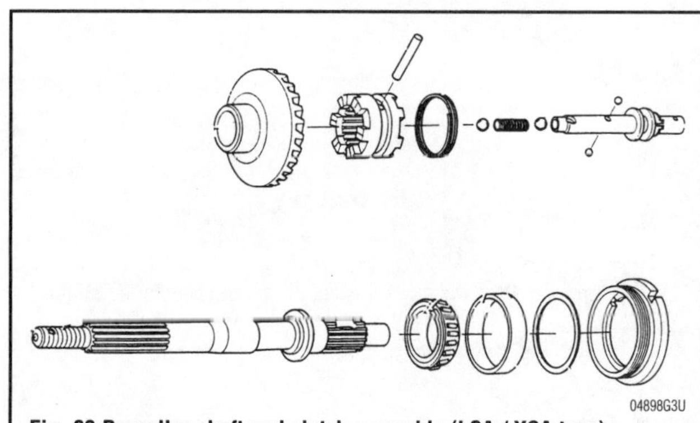

Fig. 63 Propeller shaft and clutch assembly (LCA / XCA type)—BF115A and BF130A

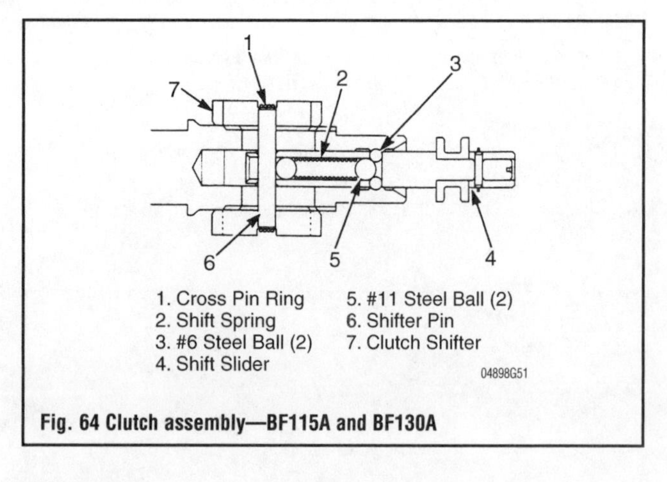

Fig. 64 Clutch assembly—BF115A and BF130A

Pinion Shaft

Before the Pinion Shaft can be removed, the pinion gear at the lower end of the shaft must be removed. The pinion gear may be secured to the pinion shaft with a simple circlip, as on the BF2A and BF20 models, or without any clip or nut as on the BF5A to BF9.9, or with a nut, as on all other models.

➡In most cases, when working with tools, a nut is rotated to remove or install it to a particular bolt, shaft, etc. In the next two steps, the reverse is required because there is no room to move a wrench inside

the lower unit cavity. The nut on the lower end of the Pinion Shaft is held steady and the shaft is rotated until the nut is free.

BF50, BF5A, BF75, BF8A AND BF100

➧ See Figure 65

1. After the bearing carrier and propeller shaft is removed, remove the pinion gear, thrust bearing and thrust washer. The pinion shaft can now be removed from the top of the gear case.

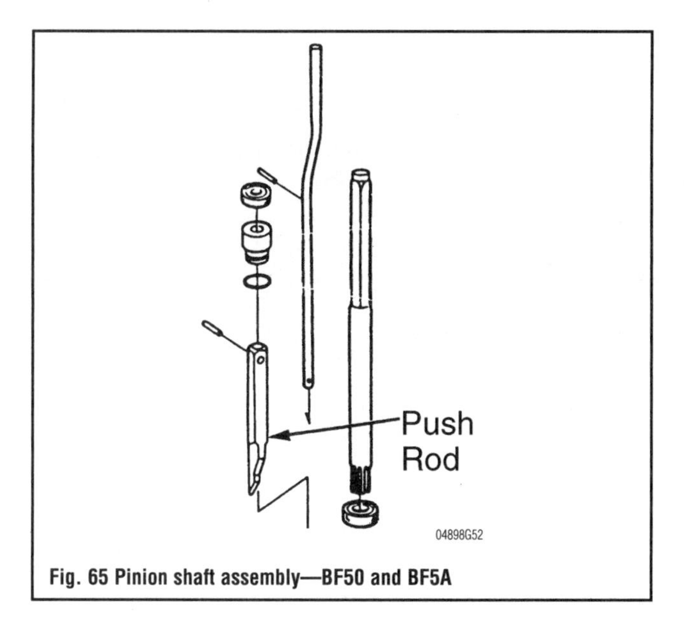

Fig. 65 Pinion shaft assembly—BF50 and BF5A

BF9.9A AND BF15A

1. On the BF9.9A and BF15A engines, hold the nut with a 12 mm socket. Turn the pinion shaft with a 12 mm open end wrench to loosen and remove the pinion nut and remove the shaft from the top.

BF25A TO BF50A

➧ See Figures 66 and 67

On the BF25A to BF50A models, obtain the following special tools:

- Pinion Shaft holding tool—p/n 07LPB-2V30200.

1. Hold the pinion shaft on the upper end with the special tool and remove the 12 mm pinion flange nut. Remove the pinion gear and remove the pinion shaft out through the top of the gear case.

Save any shim material from behind the pinion gear. The shim material is critical to obtaining the correct backlash during installation. Using the old shim material will save considerable time, especially starting with no shim material.

BF75A TO BF130A

➧ See Figure 68

1. On BF75A to BF130A , obtain the following special tool:

- Pinion Shaft holding tool—p/n 07SPB-ZW10200

2. Now, hold the pinion nut steady with the tool and at the same time install the proper Pinion Shaft tool, for the model being serviced, on top of the Pinion Shaft. With both tools in place, one holding the nut and the other on the Pinion Shaft, rotate the Pinion Shaft counterclockwise to break the nut free.
3. Remove the pinion nut and gently pull up on the Pinion Shaft and at the same time rotate the Pinion Shaft. The pinion gear, followed by the shim material (if applicable), washer, thrust bearing and a washer will come free from the lower end of the Pinion Shaft. Pull the Pinion Shaft up out of the lower unit housing. Keep the parts from the lower end of the Pinion Shaft in order, as an aid during assembling.

Save any shim material from behind the pinion gear. The shim material is

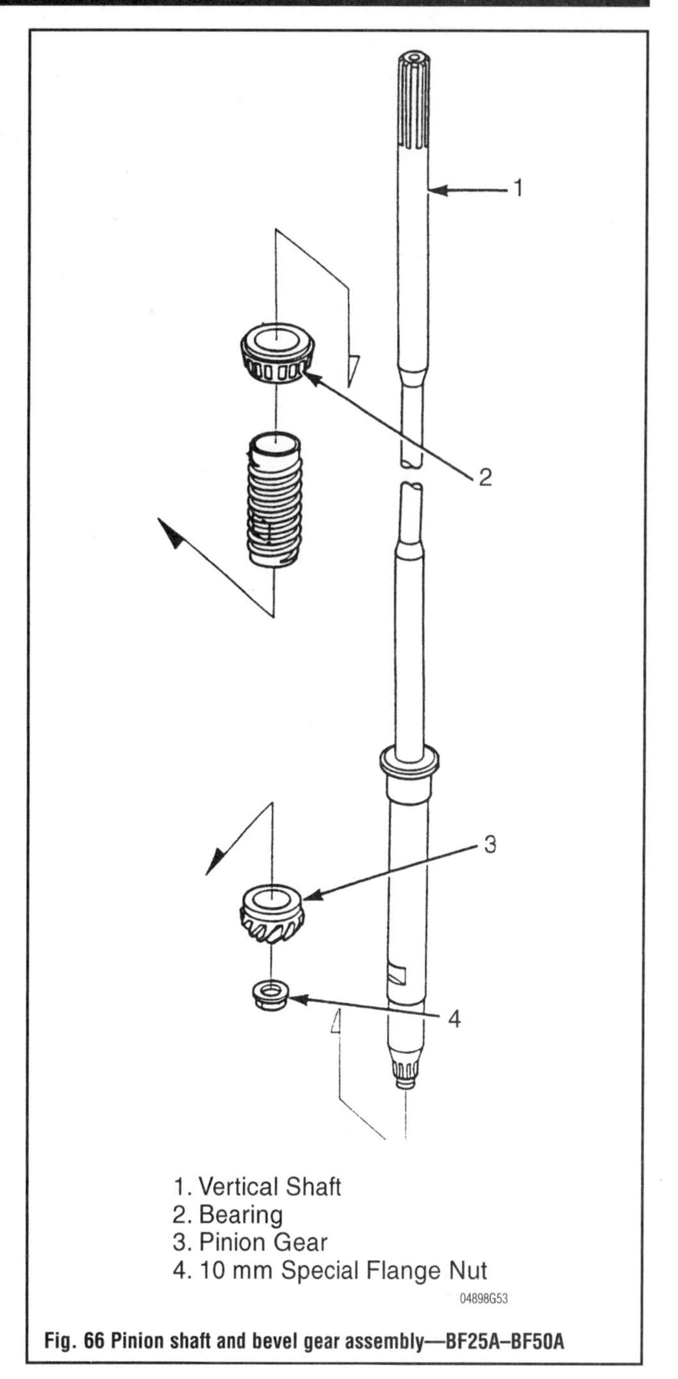

Fig. 66 Pinion shaft and bevel gear assembly—BF25A–BF50A

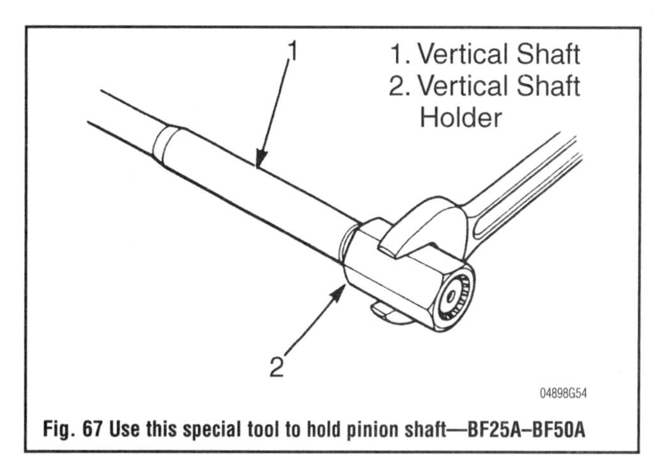

Fig. 67 Use this special tool to hold pinion shaft—BF25A–BF50A

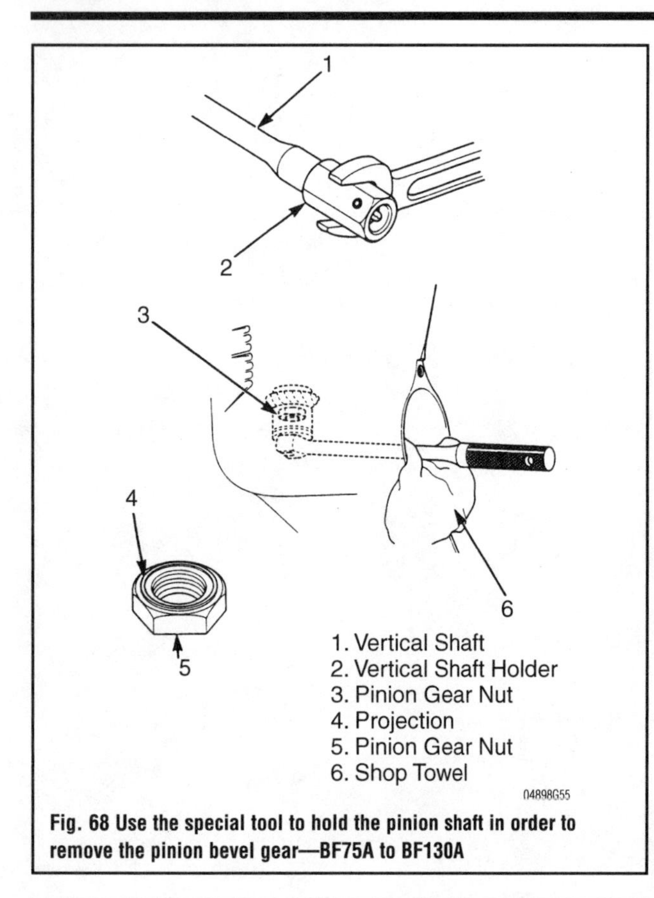

Fig. 68 Use the special tool to hold the pinion shaft in order to remove the pinion bevel gear—BF75A to BF130A

critical to obtaining the correct backlash during installation. Using the old shim material will save considerable time, especially starting with no shim material.

If the tapered roller bearing in the center of the Pinion Shaft is unfit for further service, press the bearing free using the proper size mandrel. Take care not to bend or distort the Pinion Shaft because of its length.

Pinion Shaft Needle Bearing/Tapered Roller Bearing

BF75, BF8A AND BF100

See Figure 69

1. To remove the pinion shaft needle bearing, obtain the following special tool:
   - Bearing remover–07936-9350000
2. Insert the slide hammer into the bearing and turn the handle of the tool clockwise until tight.
3. Slight the weight back against the handle until the bearing comes out of the gear case.

BF9.9A AND BF15A

See Figures 70, 71 and 72

1. To remove the upper water seal, obtain the following special tool:
   - Seal remover—07748-0010001
2. Use the seal remover to pry out the old seal. Discard the seal.
3. To remove the upper ball bearing, obtain the following special tool:
   - Bearing remover (15 mm)—p/n 07936-KC10500
   - Remover weight—p/n 07936-3710200
4. Use the slide hammer to remove the upper Pinion Shaft ball bearing.
5. To remove the lower needle bearing, obtain the following special tool:
   - Bearing remover—07946-MJ00100.
6. Drive the needle bearing out into the gear case.

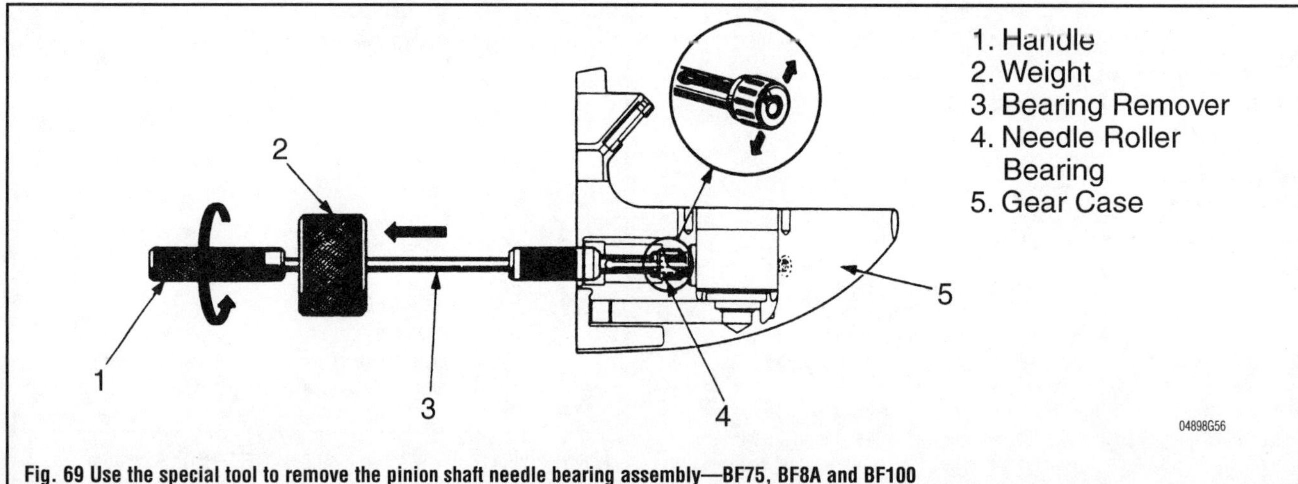

Fig. 69 Use the special tool to remove the pinion shaft needle bearing assembly—BF75, BF8A and BF100

Fig. 70 First remove the seal from the upper gear case. . .

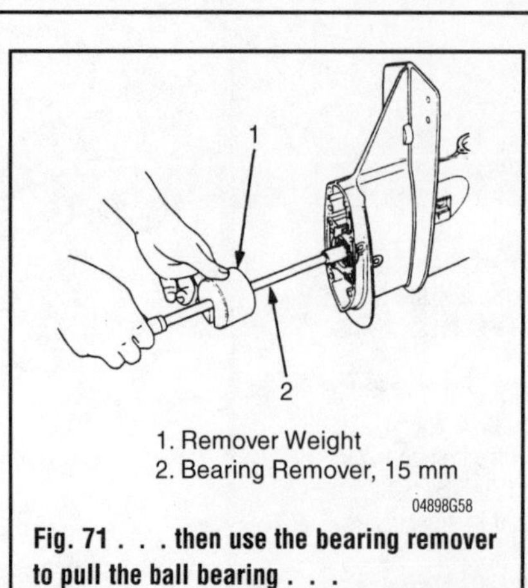

Fig. 71 . . . then use the bearing remover to pull the ball bearing . . .

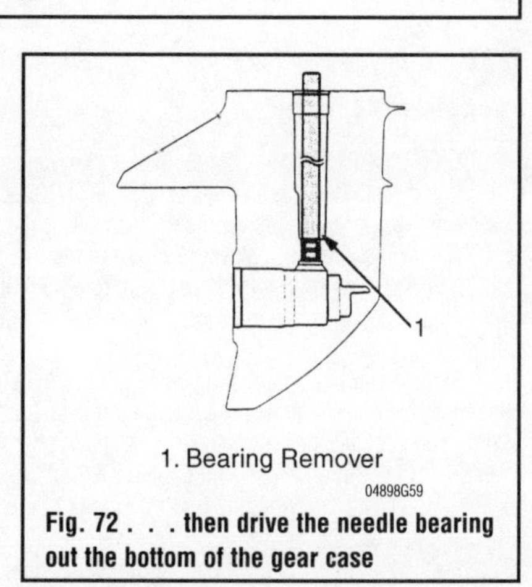

Fig. 72 . . . then drive the needle bearing out the bottom of the gear case

BF25A, BF30A, BF35A, BF40A, BF45A AND BF50A

See Figures 73, 74 and 75

1. To remove the needle bearing, obtain the following special tools:
- Driver—07949-3710001
- Attachment (24 x 26 mm)—07746-0010700
- Pilot (22 mm)—07746-0041000

Before driving out the bearing, mark the level of the gear case on the driver. This way, when you install the new needle bearing, you can tell how far you'll need to drive it in the gear case.

2. Using the special tool, drive out the needle bearing.
3. To remove the tapered bearing (inner race), obtain a commercially available bearing remover.
4. Install the pinion gear 10 mm flange nut on the pinion shaft.
5. Place the tapered bearing in the bearing remover and press the bearing off with a hydraulic press.
6. Remove the oil slinger at this time.

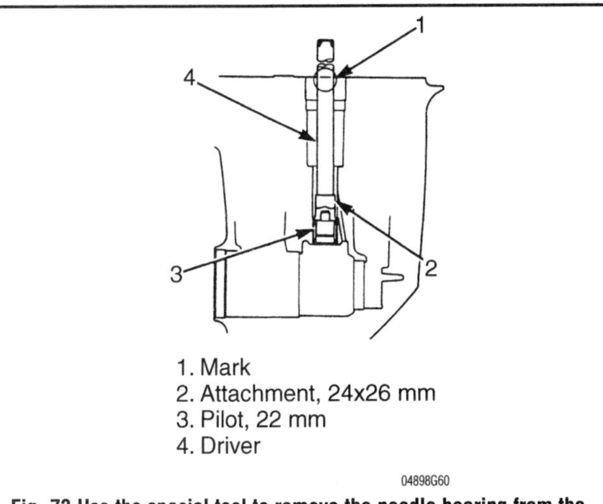

Fig. 73 Use the special tool to remove the needle bearing from the gear case—BF25A, BF30A, BF35A, BF40A, BF45A and BF50A

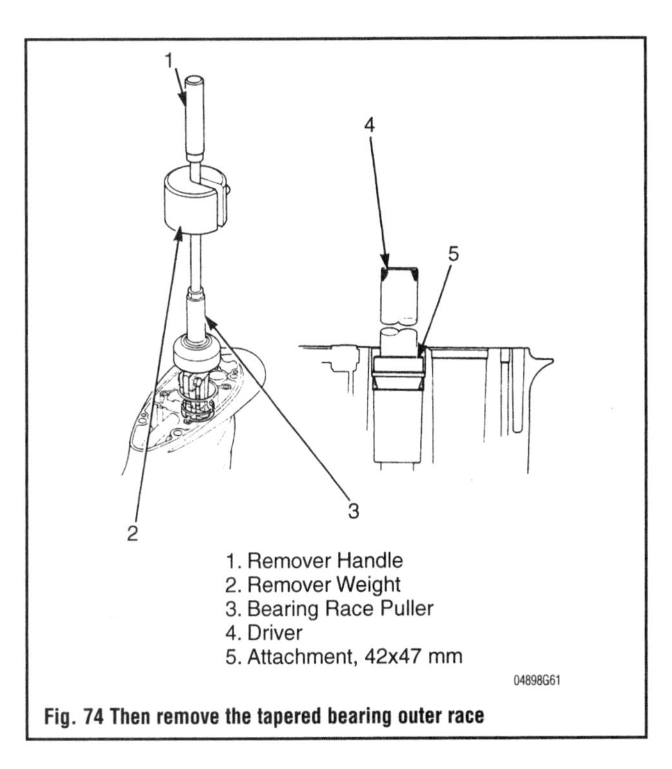

Fig. 74 Then remove the tapered bearing outer race

Press
Press
1
2
3
1. 10 mm Special Flange Nut
2. Stem Race Driver
3. Bearing Remover
04898G62

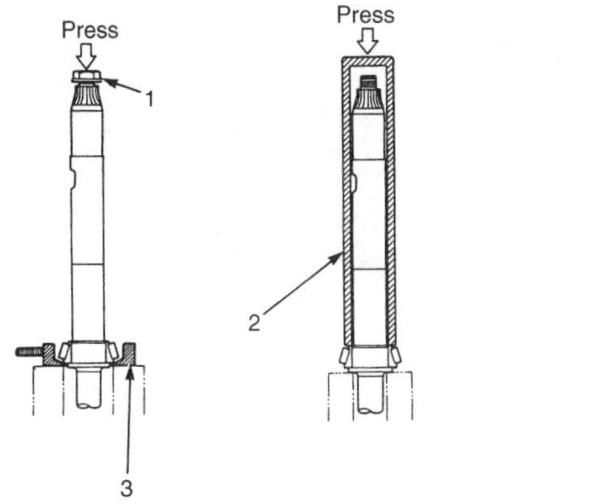

Fig. 75 Use the special tool and hydraulic press to remove the inner race

7. To remove the tapered bearing (outer race), obtain the following special tools:
- Bearing race puller—07LPC-ZV30100
- Remover weight—07741-0010201
- Remover handle—07936-3710100

8. Using the slide hammer and puller, remove the tapered bearing (outer race).

Keep track of all shim/s below the tapered bearing (outer race).

BF75A AND BF90A

See Figures 76, 77, 78 and 79

1. The BF75A and BF90A have an upper needle bearing, upper bearing sleeve and a tapered roller bearing above the pinion gear.

Remove the upper needle bearing first, then the wear sleeve using the special tools listed. If the wear sleeve comes out with the bearing, remove the bearing from inside the sleeve by tapping out the bearing with a drift or punch.

Obtain the following special tools:
- Puller Shaft (22 x 1.25 x 240 mm)—p/n 07931-ME4010B
- Special Nut—p/n 07931-HB3020A
- Puller Base "B"—p/n 07LPC-ZV302A
- Threaded Adapter—p/n 07965-HB3000A
- Adjustable Bearing Puller (25–40 mm)—p/n 07736-A01000B.

To remove the wear sleeve, place a commonly available bearing separator below the wear sleeve and tap out the pinion shaft with a soft hammer (rubber

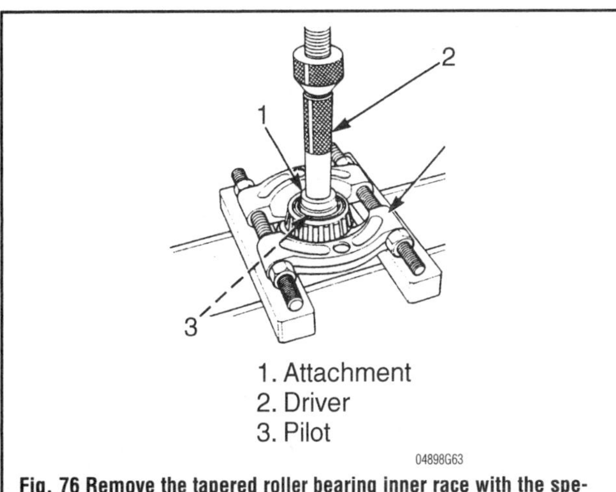

Fig. 76 Remove the tapered roller bearing inner race with the special tools and a press—BF75A and BF90A

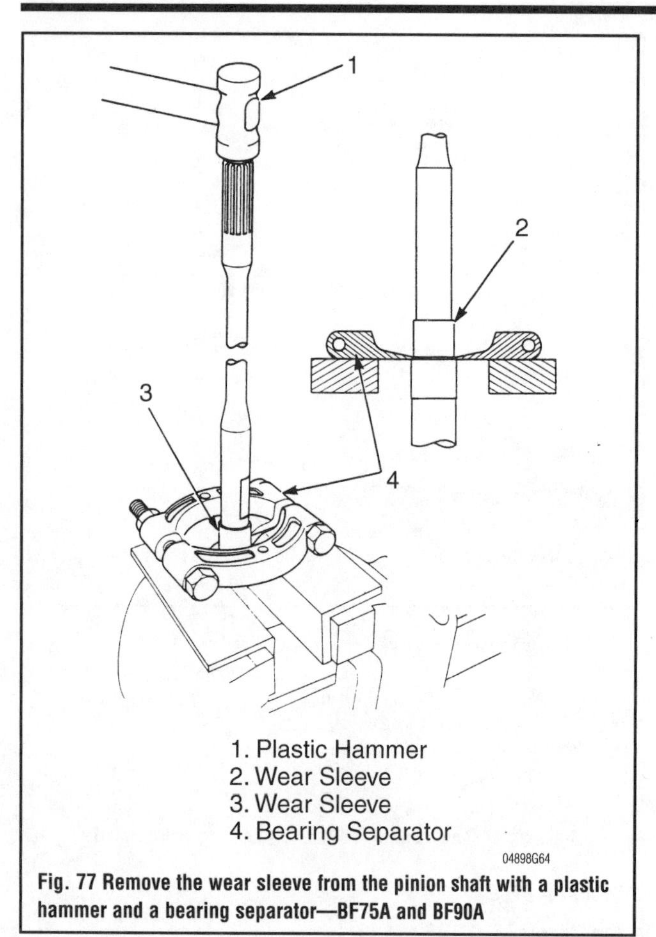

Fig. 77 Remove the wear sleeve from the pinion shaft with a plastic hammer and a bearing separator—BF75A and BF90A

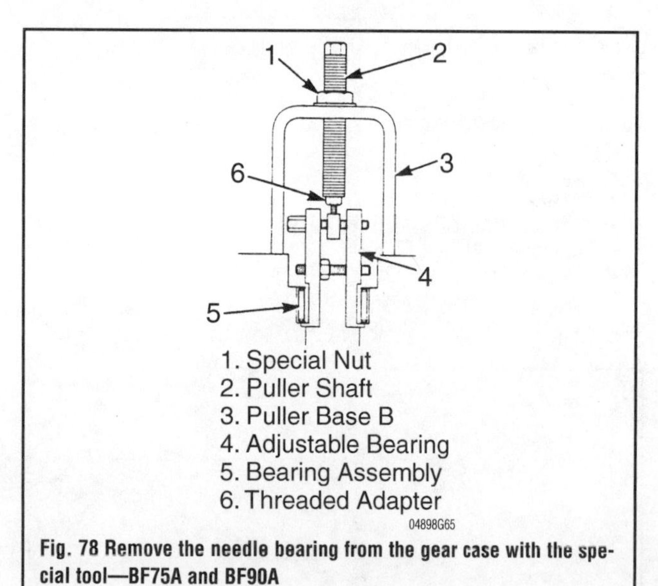

Fig. 78 Remove the needle bearing from the gear case with the special tool—BF75A and BF90A

or plastic) separating the two components. Throw away the used wear sleeve and replace with a new one.

To remove the tapered roller bearing (outer race) and pinion shim, obtain the following special tools:

- Collet Assembly—p/n 07SPC-ZW0021Z (91-13778A1)
- Driver—p/n 07SPC-ZW0022Z (91-13779)

➡The tapered bearing can be removed and installed without removing the bearing assembly and oil slinger. Set a cloth or shop towel underneath the tapered roller bearing and drive out the bearing, being very careful to not damage the gear case. Replace the bearing.

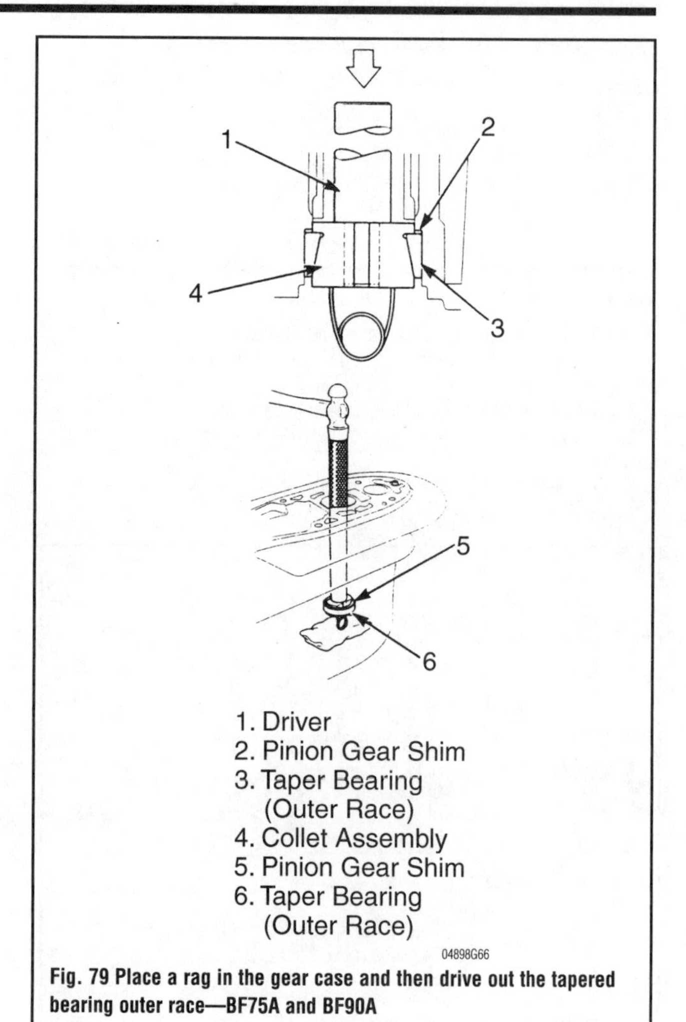

Fig. 79 Place a rag in the gear case and then drive out the tapered bearing outer race—BF75A and BF90A

BF115A AND BF130A

➧ See Figures 80, 81 and 82

To remove the pinion shaft needle bearing and tapered roller bearing, obtain the following special tools:

- Pinion shaft holder—07SPB-ZW10200
- Lock nut wrench (64 mm)—07916-MB00002
- Driver—07949-3710001
- Attachment (32 x 35 mm)—07746-0010100
- Pilot (28 mm)—07746-0041100

1. Hold the pinion shaft with the special tool and remove the pinion gear nut.
2. Remove the lock nut with the special tool.

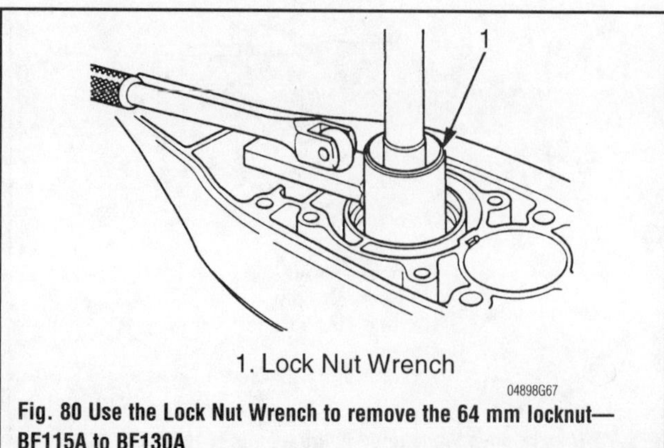

Fig. 80 Use the Lock Nut Wrench to remove the 64 mm locknut—BF115A to BF130A

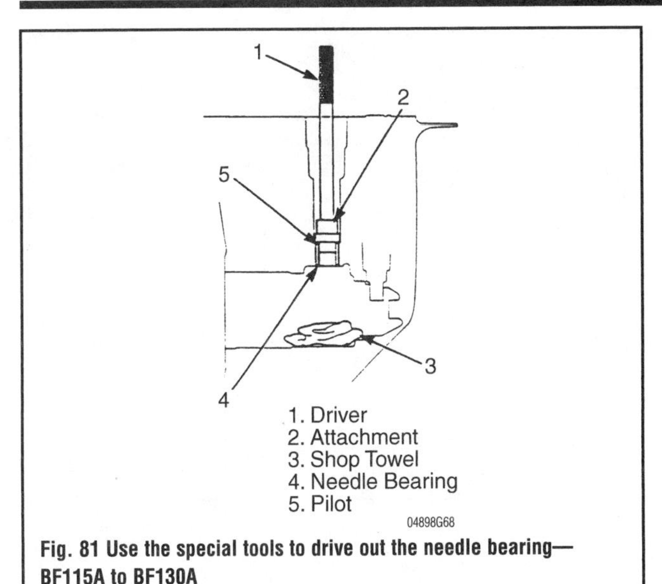

Fig. 81 Use the special tools to drive out the needle bearing—BF115A to BF130A

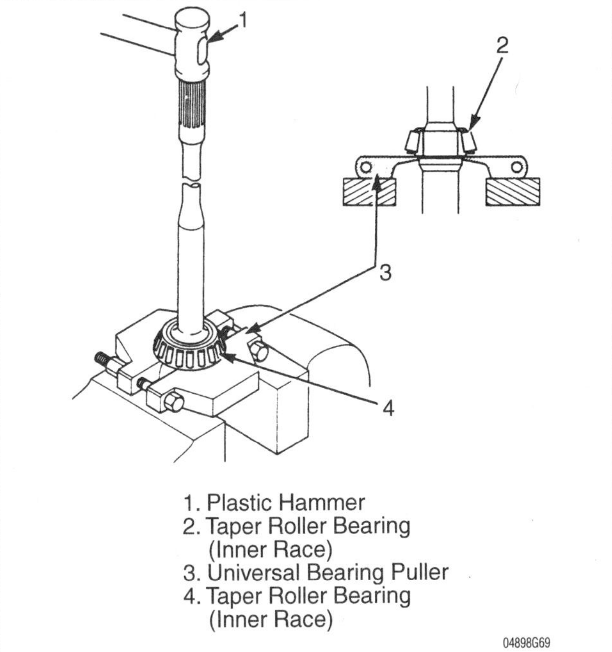

1
2
3
4
1. Plastic Hammer
2. Taper Roller Bearing (Inner Race)
3. Universal Bearing Puller
4. Taper Roller Bearing (Inner Race)
04898G69

Fig. 82 Remove the tapered roller bearing from the pinion shaft using a commercial bearing separator—BF115A to BF130A

3. Remove the pinion shaft from the gear case.
4. Place the gear case in a suitable holding fixture.

➡Before driving out the bearing, mark the level of the gear case on the driver. This way, when you install the new needle bearing, you can tell how far you'll need to drive it in the gear case.

5. Place a shop towel in the gear case and then driver, drive the needle bearing out of the gear case.
6. Using a commercially available bearing separator, drive off the tapered roller bearing inner race using a plastic mallet.

Forward Gear and Bearing

BF20 AND BF2A

1. If the forward ball bearing is no longer fit for service, proceed as follows: Obtain a commercially available slide hammer and jaw attachment and pull the ball bearing set from the lower unit.

BF50 AND BF5A

➧ See Figure 83

After removing the bearing carrier and propeller shaft, remove the pinion gear, pinion washer and forward bevel gear obtain the following special tools:

- Bearing remover (20 mm)—p/n 07936-3710600
- Bearing remover weight—p/n 07936-3710200
- Bearing remover handle—p/n 07936-3710100

1. Using the special tools, remove the forward gear roller ball bearing from the gear case. Make sure to keep track of the gear shim located behind the roller ball bearing.

BF75, BF8A, BF100, 9.9A AND BF15A

➧ See Figure 84

After removing the bearing carrier and propeller shaft, remove the pinion gear, pinion washer and forward bevel gear obtain a commercially available bearing removal tool and remove the forward gear roller bearing from the gear case.

1. Make sure to keep track of the gear shim located behind the roller ball bearing.

BF25A, BF30A, BF35A, BF40A, BF45A AND BF50A

➧ See Figures 85 and 86

1. Remove the forward gear and tapered roller bearing inner race from the gear case.
2. Obtain the following special tools:
   - Bearing race puller—p/n 07LPC-ZV30100
   - Remover weight—p/n 07741-0010201
   - Remover handle—p/n 07936-3710100
3. Using the special tools, pull the forward tapered roller bearing outer race from the gear case.
4. Make sure to keep track of any gear shim/s located behind the roller ball bearing.

To remove the inner race, obtain a commercially available bearing remover and press the gear out of the bearing using a hydraulic press.

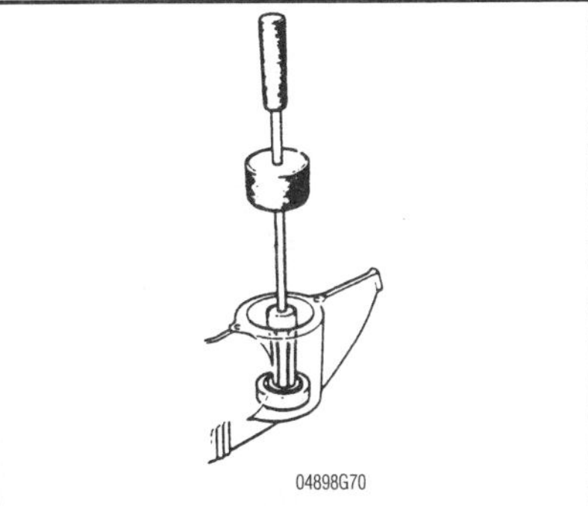

Fig. 83 Use the special tool to remove the forward ball bearing from the gear case—BF50 and BF5A

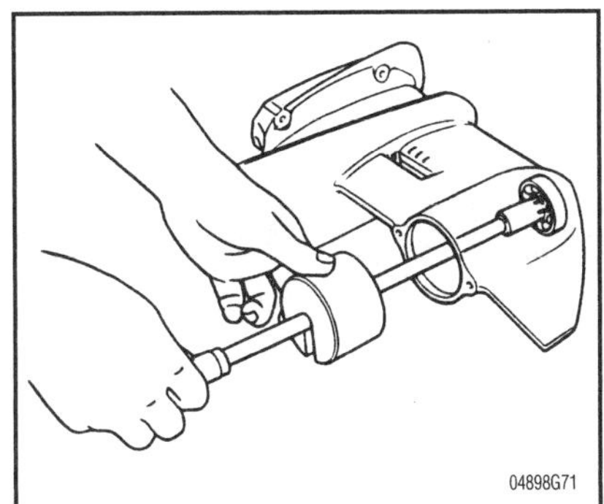

Fig. 84 Use the bearing puller to remove the forward gear ball bearing—BF75, BF8A, BF100, 9.9A and BF15A

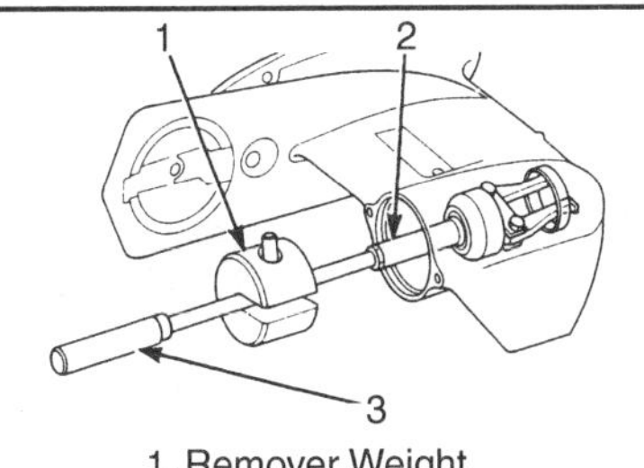

Fig. 85 Removing the forward gear bearing outer race from the gear case—BF25A AND BF30A, BF35A, BF40A and BF45A, BF50A

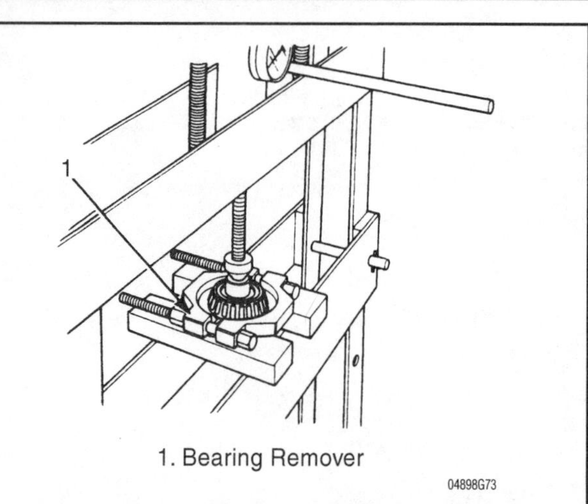

Fig. 86 Using a commercial bearing separator to press off the forward gear bearing inner race—BF25A AND BF30A, BF35A, BF40A and BF45A, BF50A

BF75A AND BF90A

See Figure 87

1. Remove the forward gear and tapered roller bearing inner race from the gear case.
2. Obtain a commercially available bearing puller and remove the tapered roller bearing outer race.

Remove the gear oil drain bolt from the gear case before the removal or installation of the tapered roller bearing. Failure to do this procedure can result in damage to the magnetic end of the drain bolt.

3. Make sure to keep track of any forward gear shims located behind the bearing race.
4. Discard the bearing after removal.
5. Obtain the following special tools to remove the inner race from the forward gear.
   - Driver—p/n 07749-0010000
   - Attachment (37 x 40 mm)—p/n 07746-0010200 (91-37350)
   - Pilot (25 mm)—p/n 07746-0040600
6. Attach the special tools to the bearing.
7. Remove the bearing with a commercially available bearing separator and hydraulic press.

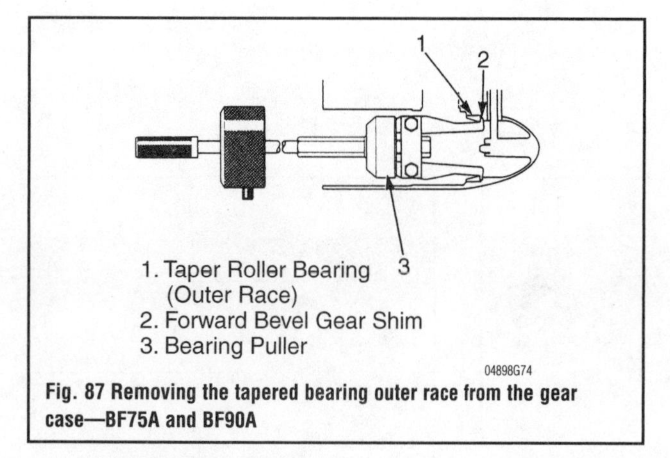

Fig. 87 Removing the tapered bearing outer race from the gear case—BF75A and BF90A

BF115A AND BF130A

See Figures 88, 89, 90 and 91

1. To remove the forward gear tapered bearing outer race from the gear case, obtain the following special tools:

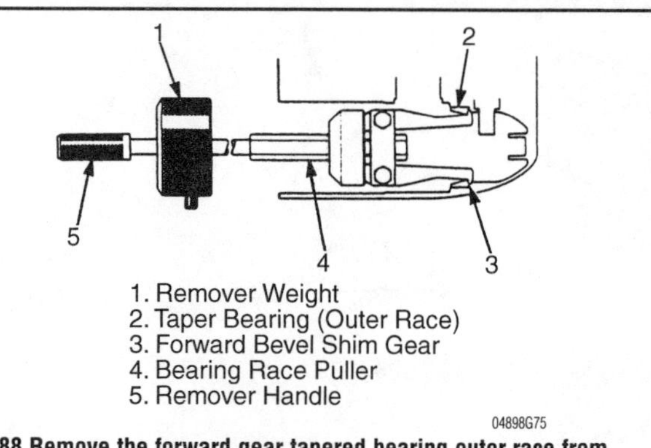

Fig. 88 Remove the forward gear tapered bearing outer race from the gear case—BF115A and BF130A

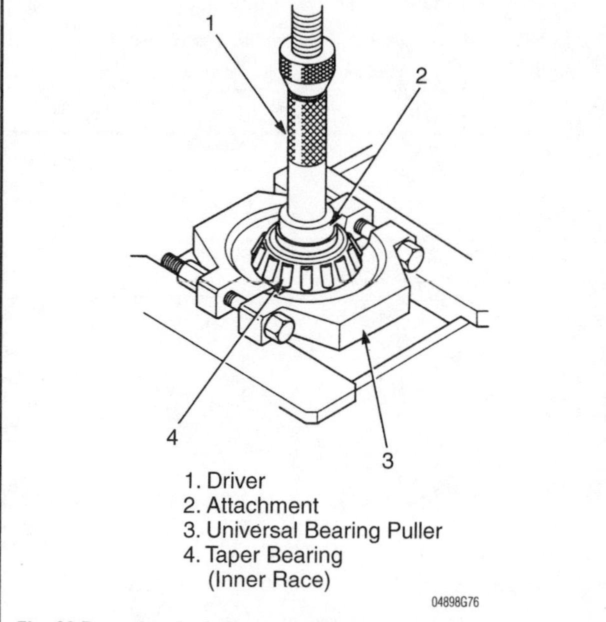

Fig. 89 Removing the inner race with a commercial bearing separator, special tools and a hydraulic press—BF115A and BF130A (LA and XA types only)

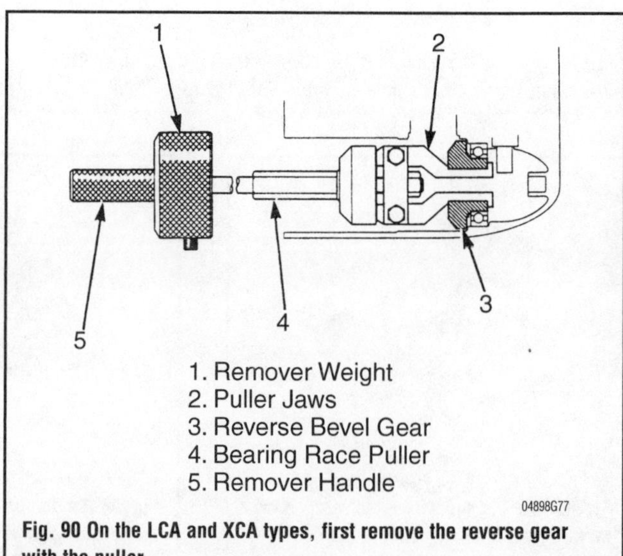

Fig. 90 On the LCA and XCA types, first remove the reverse gear with the puller . . .

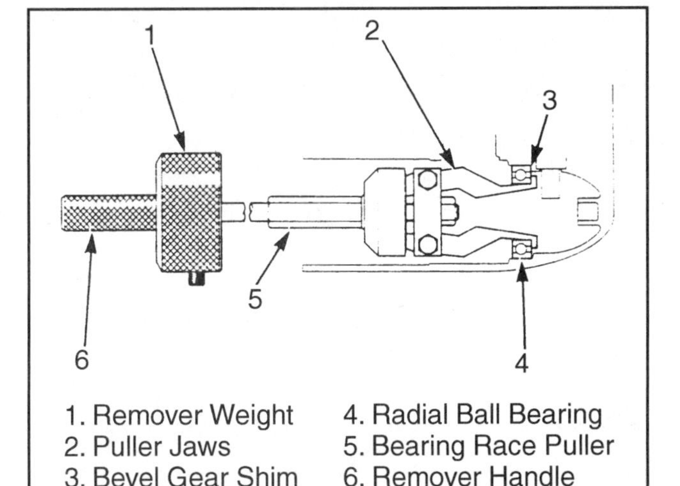

Fig. 91 . . . then remove the radial ball bearing from the gear case

- Bearing race puller—p/n 07LPC-ZV30100
- Remover weight—p/n 07741-0010201
- Remover handle—p/n 07936-3710100

2. Using the special tools, remove the tapered bearing outer race from the gear case.
3. Make sure to keep track of any forward gear shims located behind the bearing race.
4. Obtain the following special tools to separate the forward gear from the tapered roller bearing inner race:

- Driver—p/n 07749-0010000
- Attachment (42 x 47 mm)—p/n 07746-0010300

5. Using a commercially available bearing separator, press apart the gear and bearing using a hydraulic press.

➡On the counter-rotating models (LCA and XCA types) the reverse gear take the place of the forward gear in normal rotation applications.

6. To remove the reverse gear, obtain the following special tools:

- Bearing race puller—p/n 07LPC-ZV30100
- Remover weight—p/n 07741-0010201
- Remover handle—p/n 07936-3710100
- Puller jaws (25 mm)—p/n 07WPC-ZW50100

7. Use the special tool to remove the reverse gear from the gear case.
8. To remove the radial ball bearing, use the same special tools and remove the bearing from the gear case.
9. Again, keep track of any shim/s for use later during reassembly.

Reverse Gear

BF50, BF5A, BF75, BF8A, BF100, BF9.9A AND BF15A

1. Remove the reverse bevel gear from the ball bearing in the bearing carrier.
2. If the gear will not separate from the bearing, a commercially available bearing separator will be needed to remove the reverse bevel gear.

BF25A, BF30A, BF35A, BF40A, BF45A AND BF50A

➧ See Figure 92

1. To separate the reverse bevel gear from the radial ball bearing, obtain the following special tools:

- Driver—07949-3710001.
- Remover attachment (22 mm I.D.)—07GMD-KT70200.
- Support base—07965-SD90100.

2. Using the special tools and a hydraulic press, separate the reverse bevel gear from the radial ball bearing.

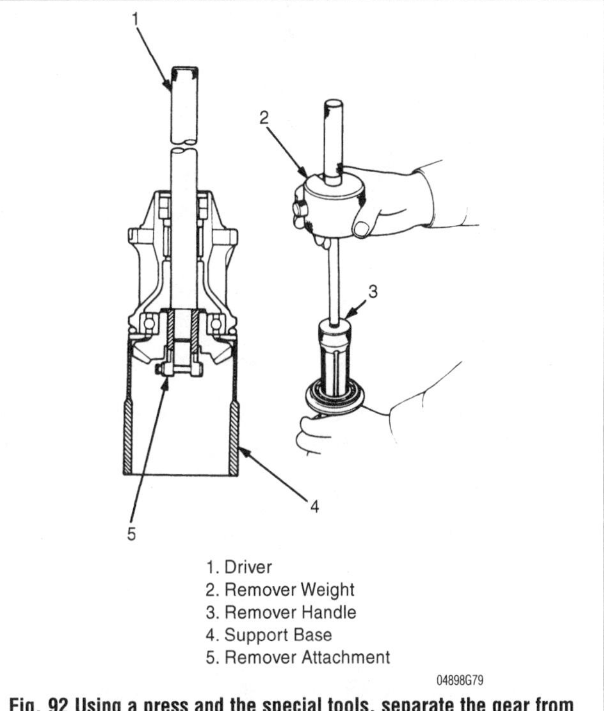

Fig. 92 Using a press and the special tools, separate the gear from the bearing—BF25A, BF30A, BF35A, BF40A, BF45A and BF50A

BF75A AND BF90A

1. After removing the propeller shaft, remove the reverse bevel gear from the bearing carrier.
2. Make sure to keep track of the thrust bearing and thrust washer after removing the reverse gear.

BF115A AND BF130A (LA AND XA TYPES ONLY)

1. Remove the two water seals.
2. Obtain the following special tools to remove the reverse bevel gear:

- Driver—07749-0010000
- Attachment (24 x 26 mm)—07746-0010700
- Pilot (17 mm)—07746-0040400
- Bearing remover—07HMC-MR70100

3. Using the special tools, and a hydraulic press, press out the reverse bevel gear and radial ball bearing together as an assembly.

➡Always use a hydraulic press. Do not attempt to remove the gear and bearing by striking them with a hammer.

4. Obtain the following special tools to separate the gear from the bearing:

- Driver—07749-0010000
- Attachment (37 x 40 mm)—07746-0010200
- Pilot (30 mm)—07746-0040700

5. Install the bearing separator between the reverse bevel gear and the bearing.
6. Using the press and the special tools, separate the reverse bevel gear from the radial ball bearing.

CLEANING & INSPECTING

Good shop practice requires installation of new O-rings and oil seals regardless of their appearance.

Clean all water pump parts with solvent and then dry them with compressed air. Inspect the water pump housing and oil seal housing for cracks and distortion, possibly caused from overheating. Inspect the inner and outer plates and water pump cartridge for grooves and/or rough surfaces. If possible, always install a new water pump impeller while the lower unit is disassembled. A new impeller will ensure extended satisfactory service and give peace of mind to the owner. If the old impeller must be returned to service, never install it in reverse to the original direction of rotation. Installation in reverse will cause premature impeller failure.

If installation of a new impeller is not possible, check the seal surfaces. All must be in good condition to ensure proper pump operation. Check the upper, lower and ends of the impeller vanes for grooves, cracking and wear. Check to be sure that the indexing notch of the impeller hub is intact and will not allow the impeller to slip.

Clean around the Woodruff key or impeller pin. Clean all bearings with solvent, dry them with compressed air and inspect them carefully. Be sure there is no water in the air line. Direct the air stream through the bearing. Never spin a bearing with compressed air. Such action is highly dangerous and may cause the bearing to score from lack of lubrication. After the bearings are clean and dry, lubricate them with Formula 50 oil, or equivalent. Do not lubricate tapered bearing cups until after they have been inspected.

Inspect all ball bearings for roughness, scratches and bearing race side wear. Hold the outer race and work the inner bearing race in-and-out, to check for side wear.

Determine the condition of tapered bearing rollers and inner bearing race, by inspecting the bearing cup for pitting, scoring, grooves, uneven wear, imbedded particles and discoloration caused from overheating. Always replace tapered roller bearings as a set.

Clean the forward gear with solvent and then dry it with compressed air. Inspect the gear teeth for wear. Under normal conditions the gear will show signs of wear but it will be smooth and even.

Clean the bearing carrier or cap with solvent and then dry it with compressed air. Never spin bearings with compressed air. Such action is highly dangerous and may cause the bearing to score from lack of lubrication. Check the gear teeth of the reverse gear for wear. The wear should be smooth and even.

Check the clutch shifters to be sure they are not rounded-off , or chipped. Such damage is usually the result of poor operator habits and is caused by shifting too slowly or shifting while the engine is operating at high rpm. Such damage might also be caused by improper shift rod adjustments.

Rotate the reverse gear and check for catches and roughness. Check the bearing for side wear of the bearing races.

Inspect the roller bearing surface of the propeller shaft. Check the shaft surface for pitting, scoring, grooving, embedded particles, uneven wear and discoloration caused from overheating.

Clean the Pinion Shaft with solvent and then dry it with compressed air. Never spin bearings with compressed air. Such action is dangerous and could damage the bearing. Inspect the bearing for roughness, scratches, or side wear. If the bearing shows signs of such damage, it should be replaced. If the bearing is satisfactory for further service coat it with oil.

Inspect the Pinion Shaft splines for excessive wear. Check the oil seal surfaces above and below the water pump drive pin or Woodruff key area for grooves. Replace the shaft if grooves are discovered.

Inspect the Pinion Shaft-bearing surface above the pinion gear splines for pitting, grooves, scoring, uneven wear, embedded metal particles and discoloration caused by overheating.

Inspect the propeller shaft oil seal surface to be sure it is not pitted, grooved, or scratched. Inspect the roller bearing contact surface on the propeller shaft for pitting, grooves, scoring, uneven wear, embedded metal particles and discoloration caused from overheating.

Inspect the propeller shaft splines for wear and corrosion damage. Check the propeller shaft for straightness.

BF50 AND BF5A

➧ See Figures 93, 94, 95 and 96

1. Measure the pinion shaft O.D. clearance. Standard measurement should be 0.5112 in. (12.984 mm). If the clearance is greater than 0.510 in. (12.95 mm) replace the component.
2. Measure the gear case bushing I.D. Standard measurement should be 0.512 in. (13.00 mm). If the measurement is greater than 0.513 in. (13.04 mm), replace the component.
3. Measure the pinion shaft-to-bushing clearance. The standard measurement should be 0.0006–0.0018 in. (0.016–0.045 mm). If the measurement is greater than 0.004 in. (0.09 mm), replace the component.
4. Measure the propeller shaft O.D. The standard measurement is 0.4718 in. (11.984 mm) for the forward gear part of the shaft, and 0.5112 in. (12.984 mm) for the reverse gear part. If the measurement exceeds 0.470 in. (11.95 mm) for forward and 0.510 (12.95 mm) for reverse, replace the propeller shaft.
5. Measure the bevel gear I.D. The standard measurement is 0.472 in. (12.00 mm) for the forward gear, and 0.512 in. (13.00 mm) for the reverse gear. If the measurement exceeds 0.474 in. (12.04 mm) for forward, and 0.513 in. (13.04 mm) for reverse, replace the bevel gear.

BF75, BF8A AND BF100

1. Measure the propeller shaft O.D. The standard measurement is 0.6682–0.6687 in. (16.973–16.984 mm). If the measurement exceeds 0.667 in. (16.930 mm), replace the propeller shaft.
2. Measure the forward and reverse bevel gear I.D. Standard measurement should be 0.669–0.670 in. (17.00–17.018 mm). If the measurement exceeds 0.672 in (17.06 mm), replace the gear.

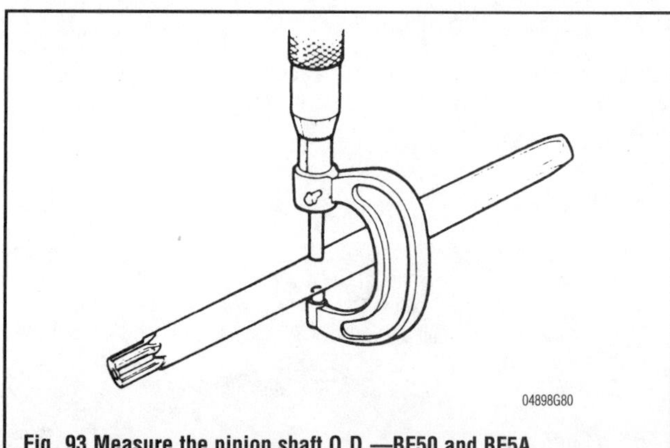

Fig. 93 Measure the pinion shaft O.D.—BF50 and BF5A

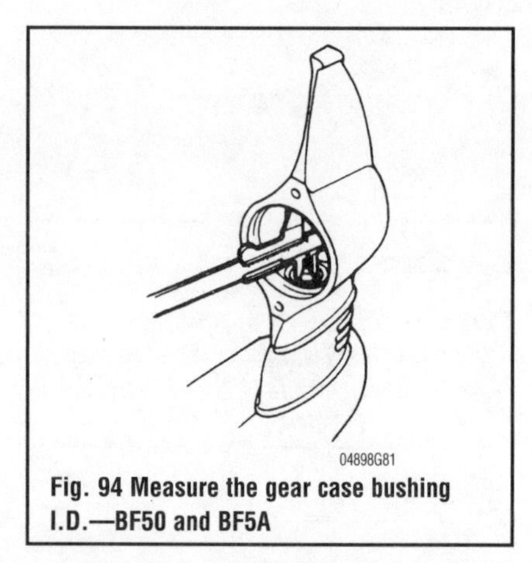

Fig. 94 Measure the gear case bushing I.D.—BF50 and BF5A

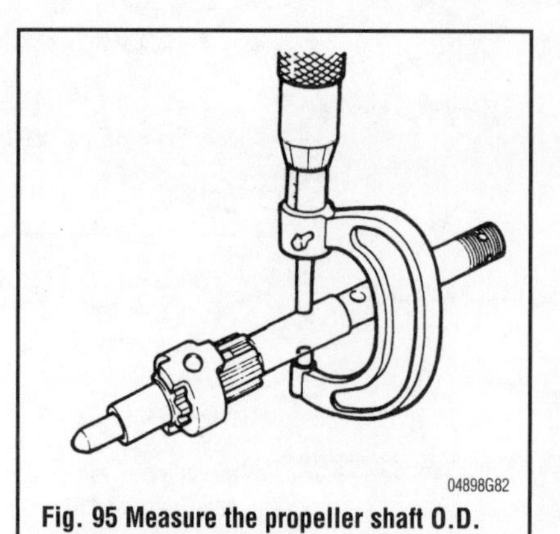

Fig. 95 Measure the propeller shaft O.D. —BF50 and BF5A

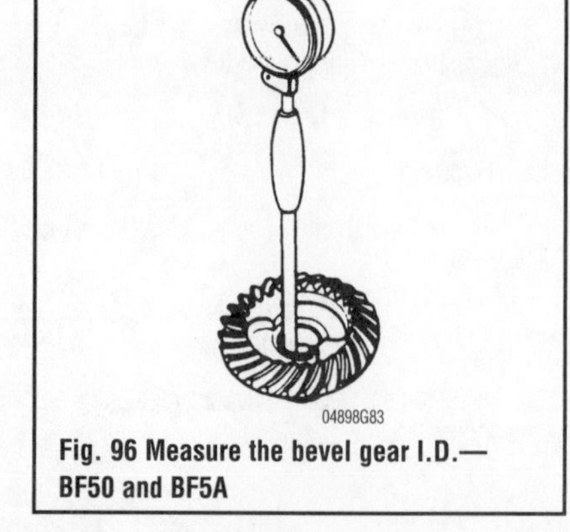

Fig. 96 Measure the bevel gear I.D.—BF50 and BF5A

BF9.9A AND BF15A

➧ See Figures 97 and 98

1. Measure the propeller shaft O.D. (front bevel gear side). The standard measurement is 0.66866 in. (16.984 mm). If the measurement exceeds 0.667732 in. (16.95 mm), replace the propeller shaft.
2. Measure the front bevel gear I.D. The standard measurement is 0.66929 in. (17.000 mm). If the measurement exceeds 0.67086 in. (17.04 mm), replace the gear.
3. Measure the pinion shaft O.D. The standard measurement is 0.5903 in. (14.994 mm). If the measurement exceeds 0.5890 in. (14.96), replace the shaft.

BF25A, BF30A, BF35A, BF40A, BF45A AND BF50A

➧ See Figures 99, 100, 101 and 102

1. Measure the propeller shaft O.D. Standard measurement is 0.7467–0.7472 in. (18.967–18.980 mm) for the front; 0.82–0.83 in. (20.9–21.2 mm) for the rear and 0.8664–0.8669 in. (22.007–22.020 mm) at the needle bearing. The maximum reading for these are: 0.7459 in. (18.946 mm) for the front, 0.821 in. (20.85 mm) for the rear and 0.866 in. (21.99 mm) at the needle bearing. If the measurement exceeds any of these readings, replace the propeller shaft.
2. Measure the reverse bevel gear I.D. The standard measurement is 0.868–0.878 in. (22.05–22.30 mm). If the measurement exceeds 0.880 in. (22.35 mm), replace the gear.
3. Measure the pinion shaft O.D. at the needle bearing. Standard measurement is 0.8747–0.8752 in. (22.217–22.230 mm). If the measurement exceeds 0.8739 in. (22.196 mm), replace the pinion shaft.
4. Measure the forward bevel gear I.D. The standard measurement is 0.7480–0.7489 in. (19.000–19.021 mm). If the measurement exceeds 0.750 in. (19.04 mm), replace the gear.

BF75A AND BF90A

➧ See Figures 103, 104, 105 and 106

1. Measure the propeller shaft O.D.(at the needle bearing). The standard measurement at the forward bevel gear location is 0.9996–1.0010 in. (25.390–25.425 mm) and the measurement at the bearing carrier location is 1.1870–1.1875 in. (30.149–30.162 mm). The maximum measurements are 0.9988 in. (25.369 mm) for the forward bevel gear and 1.1861 in. (30.128 mm) for the bearing carrier. If the measurements are exceeded, replace the propeller shaft.
2. Measure the propeller shaft runout with a dial indicator and V-blocks. Maximum measured runout cannot exceed 0.006 in. (0.15 mm). Replace the propeller shaft if measurement is exceeded.
3. Measure the pinion shaft O.D. at the roller bearing. The standard measurement is 1.1245–1.1250 in. (28.562–28.575 mm). If the maximum wear measurement of 1.1237 in. (28.541 mm) is exceeded, the pinion shaft needs to be replaced.

Perform a gear case pressure test. Assemble the gear case and perform this test before installing the gear case on the engine.

4. Remove the vent screw.
5. Install the joint provided on the gear lubricant oil pump into the vent screw hole.
6. Attach the pressure tester to the joint.
7. Pressurize the gear case to 10–12 psi (69–82 kPa and observe the pressure tester gauge for 5 minutes.
8. Move shift rod B, and rotate the pinion shaft and propeller shaft while the gear case is pressurized to check for leaks. The pressure should not leak.
9. If a pressure drop is noticed, immerse the gear case in water and repressurize to the specified pressure and check for escaping air bubbles.
10. Note the location of the escaping air bubbles and replace the suspected seal.

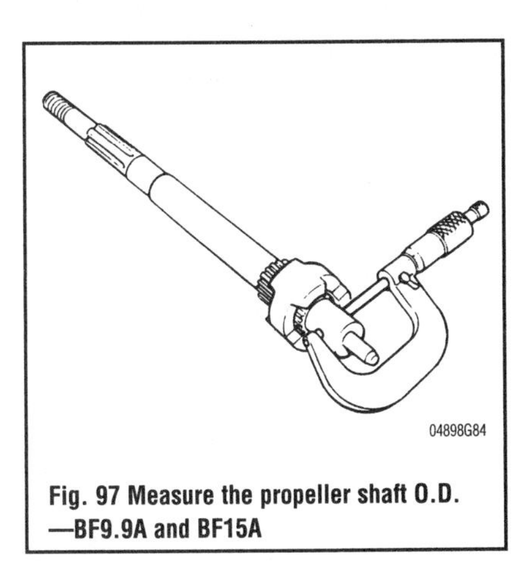

Fig. 97 Measure the propeller shaft O.D. —BF9.9A and BF15A

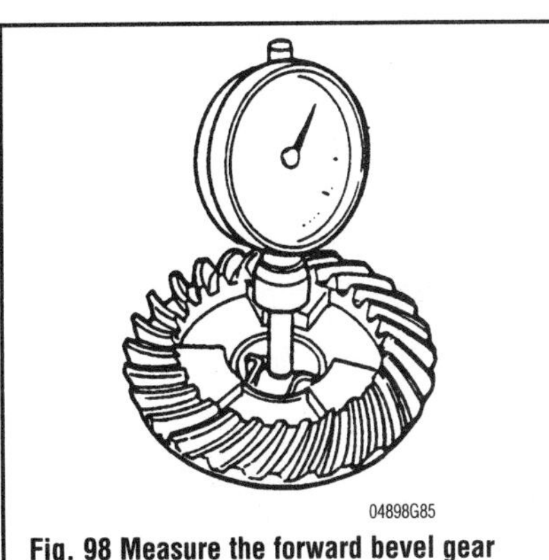

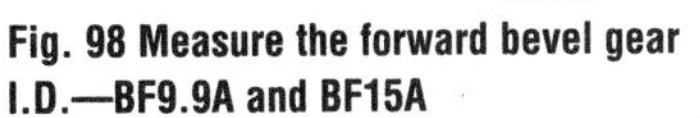

Fig. 98 Measure the forward bevel gear I.D.—BF9.9A and BF15A

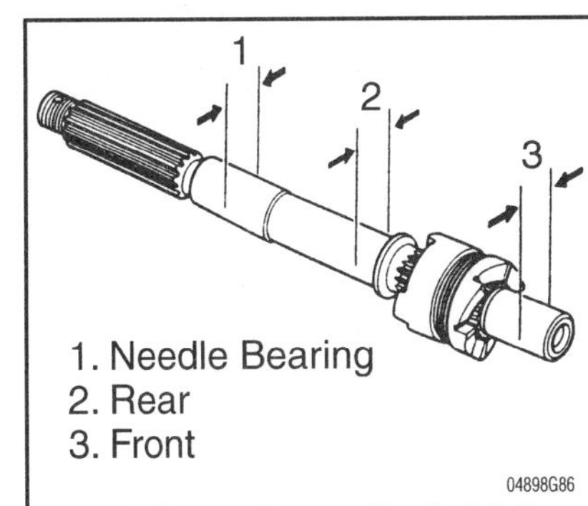

Fig. 99 Measure the propeller shaft O.D. in three locations—BF25A, BF30A, BF35A, BF40A, BF45A and BF50A

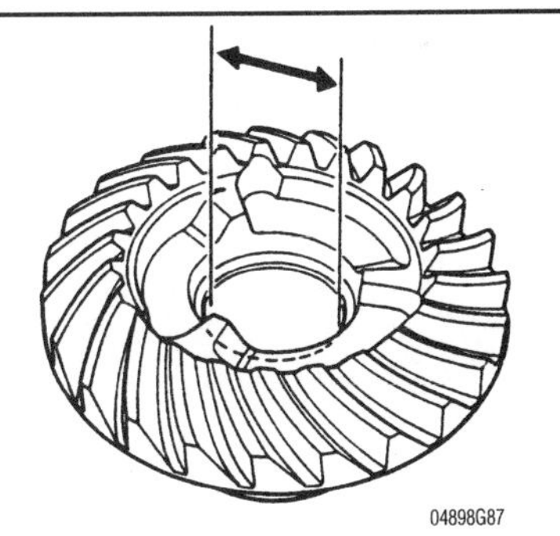

Fig. 100 Measure the reverse bevel gear I.D.—BF25A, BF30A, BF35A, BF40A, BF45A and BF50A

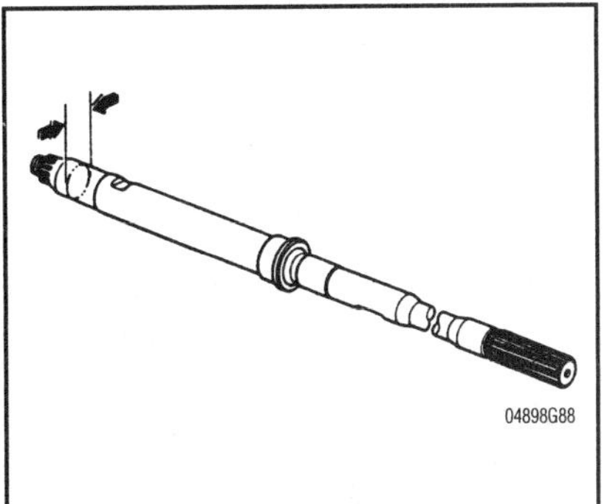

Fig. 101 Measure the pinion shaft O.D. at the needle bearing location—BF25A, BF30A, BF35A, BF40A, BF45A and BF50A

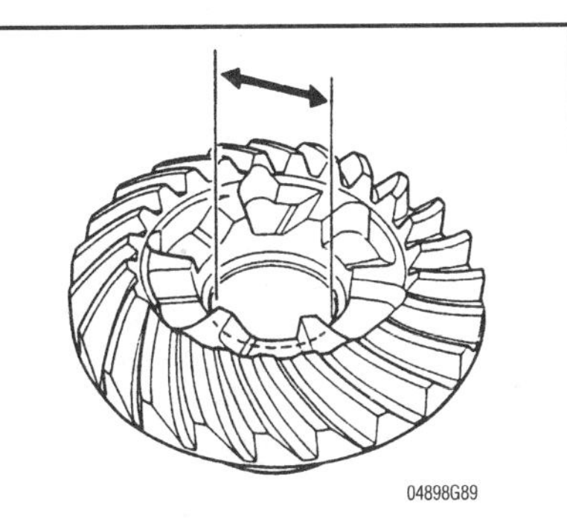

Fig. 102 Measure the forward bevel gear I.D.—BF25A, BF30A, BF35A, BF40A, BF45A and BF50A

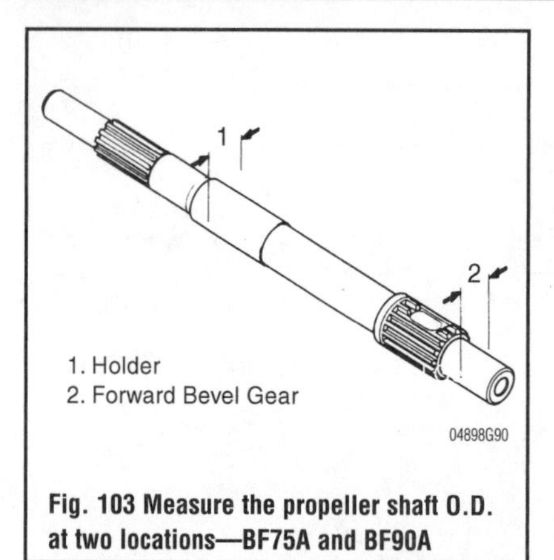

Fig. 103 Measure the propeller shaft O.D. at two locations—BF75A and BF90A

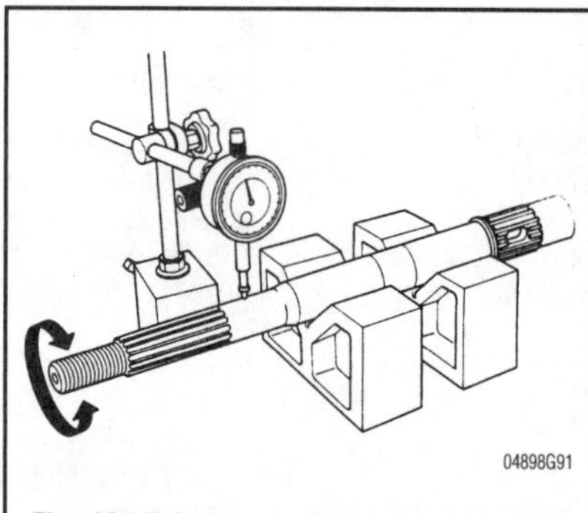

Fig. 104 Using a set of V-blocks and a dial indicator, measure the propeller shaft runout—BF75A and BF90A

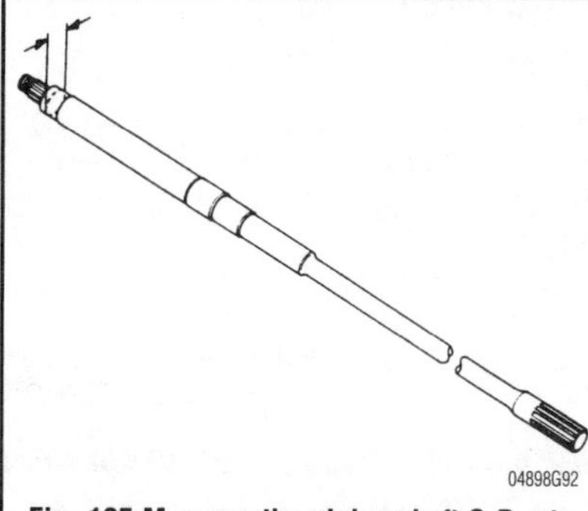

Fig. 105 Measure the pinion shaft O.D. at the roller bearing location—BF75A and BF90A

11. After replacing the affected seal, repressurize the gear case and check for leaks. The gear case should hold pressure for 5 minutes.
12. Remove the attachments and tighten the vent screw to 5.1 ft. lbs. (7 Nm).

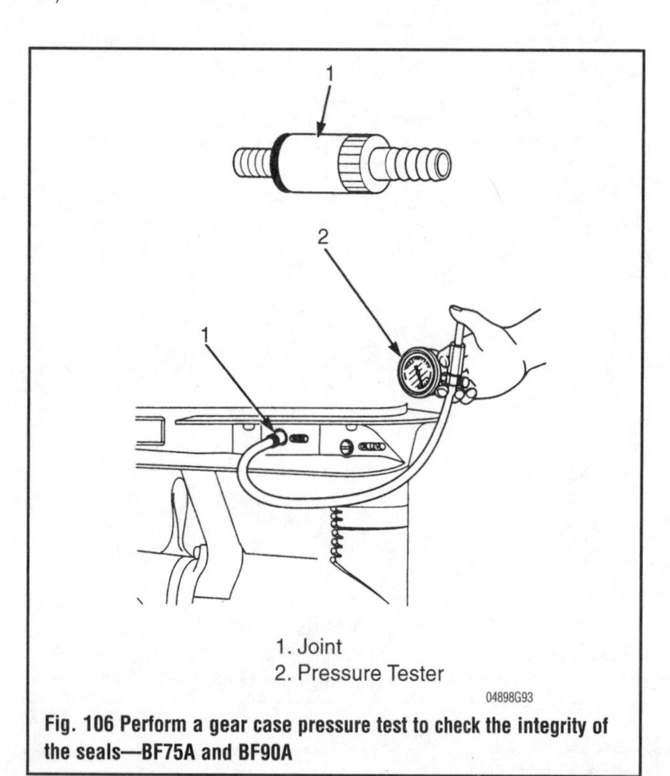

Fig. 106 Perform a gear case pressure test to check the integrity of the seals—BF75A and BF90A

BF115A AND BF130A

See Figures 107, 108, 109 and 110

1. Measure the propeller shaft O.D. On the LA and XA types, the standard measurement at the forward gear location is 0.0837–0.9843 in. (24.987–25.000 mm) and the maximum wear limit is 0.9829 in. (24.966 mm). The standard measurement at the needle bearing location is 1.1814–1.1819 in. (30.007–30.020 mm) and the maximum wear limit is 1.1807 in. (29.990 mm). If the wear limits are exceeded, the propeller shaft needs to be replaced.
2. On the LCA and XCA types, the standard measurement at the reverse gear is 0.9837–0.9843 in. (24.987–25.000 mm) and the service limit is 0.9829 in. (24.966 mm). Thc standard measurement at the needle bearing location is 1.1814–1.1819 in. (30.007–30.020 mm) and the wear limit is 1.1807 in. (29.990 mm). The standard measurement at the forward gear is 1.2954–1.2961 in. (32.904–32.920 mm) and the wear limit is 1.2946 in. (32.883 mm).
3. Measure the forward bevel gear I.D. (LCA and XCA types only). The stan-

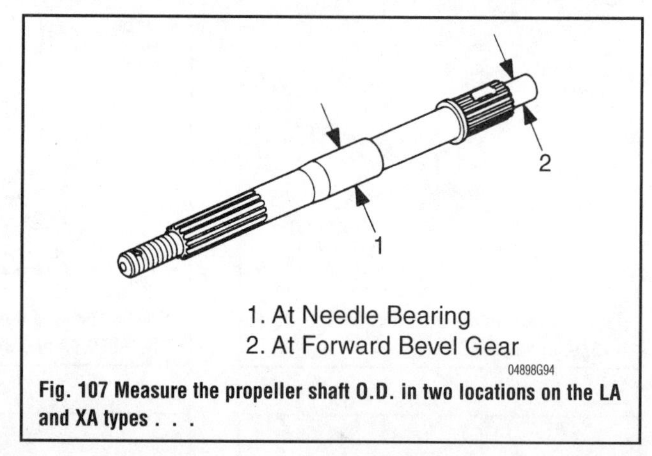

Fig. 107 Measure the propeller shaft O.D. in two locations on the LA and XA types . . .

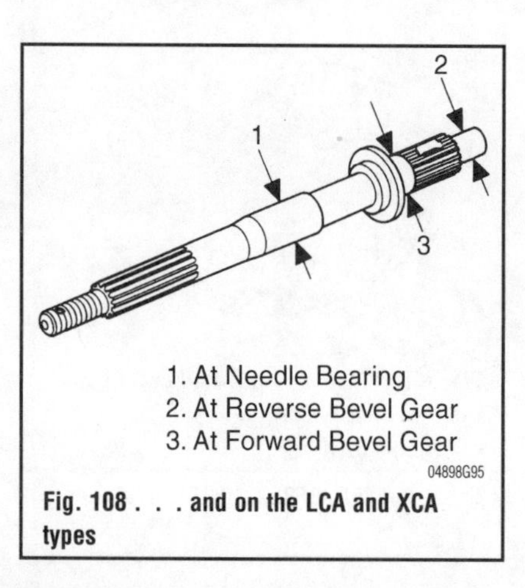

Fig. 108 . . . and on the LCA and XCA types

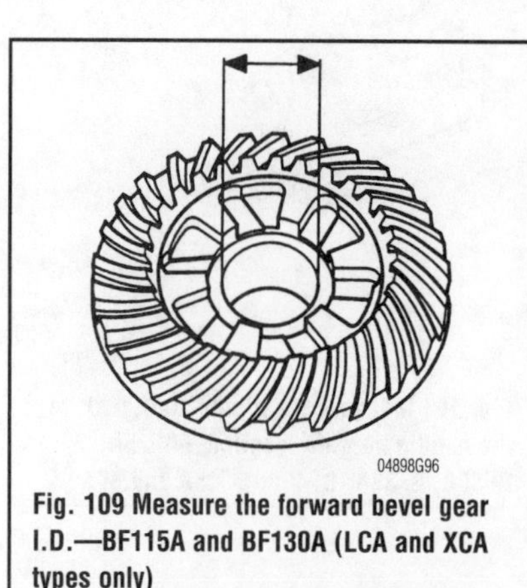

Fig. 109 Measure the forward bevel gear I.D.—BF115A and BF130A (LCA and XCA types only)

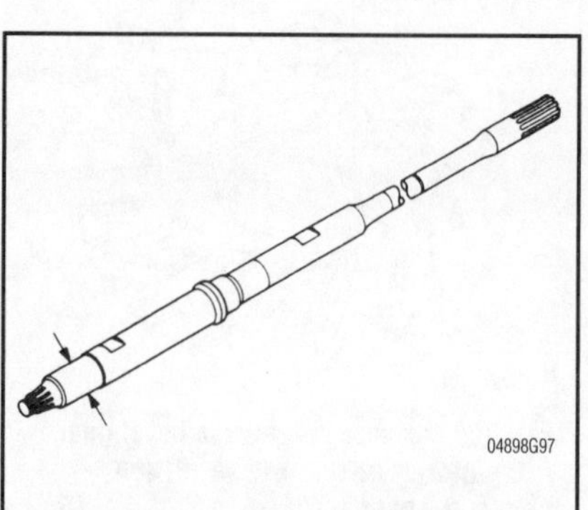

Fig. 110 Measure the pinion shaft OD. at the needle bearing location—BF115A and BF130A

dard measurement is 1.2992–1.3002 in. (33.000–33.025 mm) and the service wear limit is 1.3009 in. (33.044 mm). If the wear limit is exceeded, replace the gear.

4. Measure the pinion shaft O.D. at the needle bearing location. The standard measurement is 1.1246–1.1250 in. (28.566–28.575 mm) and the service wear limit is 1.1238 in. (28.545 mm). Replace the pinion shaft if the wear limit measurement is exceeded.

Inspect the following parts for wear, corrosion, or other signs of damage:

- Shift shaft boot
- Shift shaft retainer
- Shift cam
- Check the circlip to be sure it is not bent or stretched. If the clip is deformed, it must be replaced.
- Clean all parts with solvent and then dry them with compressed air.
- Inspect:
- All bearing bores for loose fitting bearings.
- Gear housing for impact damage.
- Cover nut threads on the BF25A and BF30A for cross threading and corrosion damage.
- Check the pinion nut corners for wear or damage. This nut is a special locknut. Therefore, do not attempt to replace it with a standard nut. Obtain the correct nut from an authorized Honda dealer.

ASSEMBLY

Procedural steps are given to assemble and install virtually all items in the lower unit. However, if certain items, i.e. bearings, bushings, seals, etc. were found fit for further service and were not removed, simply skip the assembly steps involved. Proceed with the required tasks to assemble and install the necessary components.

Propeller Shaft and Clutch shifter

BF5A AND BF50

1. Align the holes in the clutch shifter, propeller shaft and shift slider.
2. Using a 3.0 mm drift, drive a new shifter pin into the holes.
3. After the new pin is centered, flare the ends of the new pin.

BF75, BF8A AND BF100

See Figure 111

1. Install the shift spring in the propeller shaft and align the holes in the clutch shifter, propeller shaft and shift slider.
2. Install the assembly into the special tool (Pin flare tool p/n 07986-9350000).
3. Drive in a new shift pin with the special tool through the hole in the clutch shifter and propeller shaft until it is centered.
4. Shift the assembly in the tool and tighten the bolts so that they will flare the ends of the shift pin.

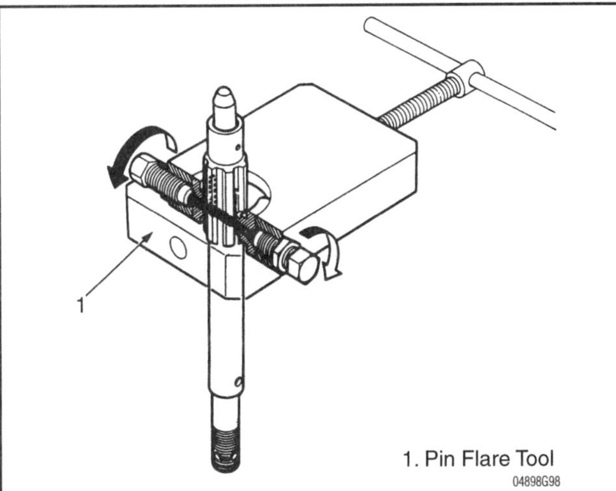

Fig. 111 Use the special tool to flare the shifter pin after installation—BF75, BF8A and BF100

BF9.9A AND BF15A

See Figure 112

1. Install the shift spring in the propeller shaft and align the holes in the clutch shifter, propeller shaft and shift slider.
2. Install the assembly into the special tool (4 mm Pin flare tool).
3. Using the special tool, drive in a new 4 x 35 mm shift pin through the hole in the clutch shifter and propeller shaft until it is centered.
4. Shift the assembly in the tool and tighten the fixing bolts so that they will flare the ends of the shift pin.

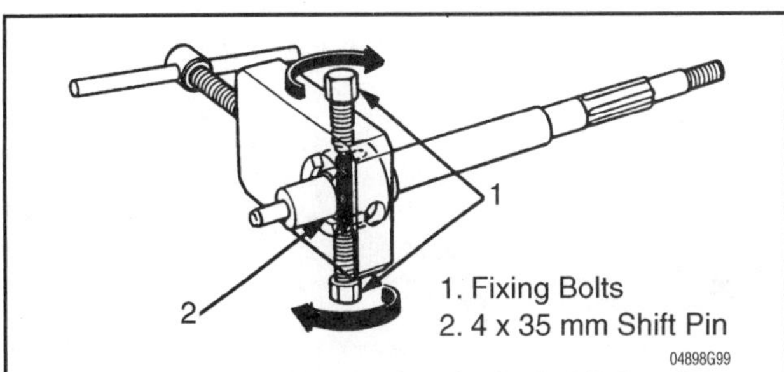

Fig. 112 Use the tool to flare the pin ends after installation of the shifter pin—BF9.9A and BF15A

BF25A, BF30A, BF35A, BF40A, BF45A AND BF50A

See Figures 113 and 114

1. Install the shift spring and shifter pin holder, steel ball and shift slider into the hollow end of the propeller shaft.

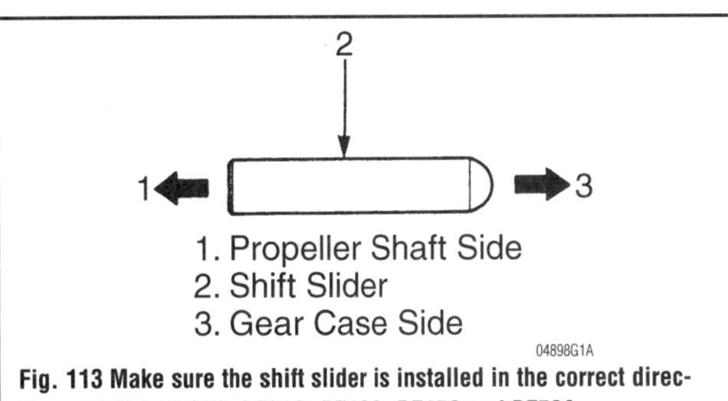

Fig. 113 Make sure the shift slider is installed in the correct direction—BF25A, BF30A, BF35A, BF40A, BF45A and BF50A

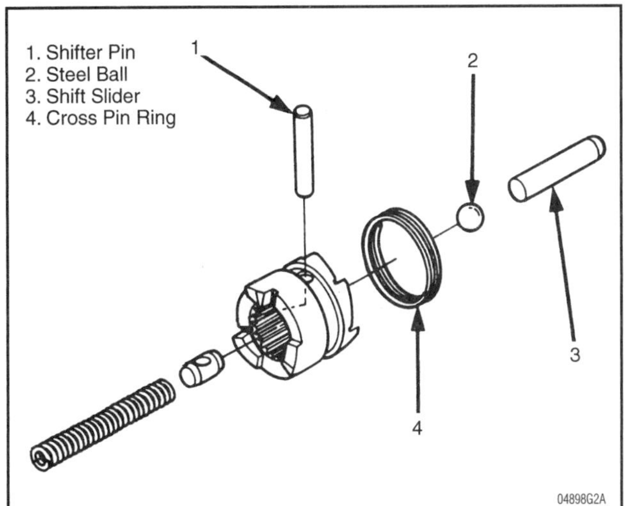

Fig. 114 Clutch shifter assembly—BF25A, BF30A, BF35A, BF40A, BF45A and BF50A

➡Make sure to install the shift slider with the flat end pointing toward the propeller side and the round end pointing toward the gear case.

2. Slide the clutch shifter onto the propeller shaft.

➡Make note of the installation direction. The ramped dogs face toward the gear case and the square dogs face toward the propeller.

3. Press the end of the propeller shaft against a piece of wood to drive in the shift slider until the holes in the clutch shifter and propeller shaft line up. Then press the shifter pin through the clutch shifter and propeller shaft.
4. Install the cross pin ring around the clutch shifter.

BF75A AND BF90A

♦ See Figures 115, 116 and 117

1. Set the shift spring in the propeller shaft. Note the installation direction of the shaft pin holder. The smaller end goes toward the shift spring side and the dished end goes toward the steel ball side.
2. Apply Quicksilver®2-4-C to the entire surface of the three steel balls and shift slider and assemble everything to the propeller shaft.
3. Assemble the clutch shifter to the propeller shaft by aligning the long hole in the propeller shaft with the shifter pin hole. Note the installation direction of the clutch shifter: the ramped dogs go to the shift slider side and the straight dogs go toward the propeller shaft bearing carrier side.
4. Push the shift slider into the propeller shaft against a solid object like a wooden board. While the shift slider is pushed in, align the clutch shifter hole and the hole in the shifter pin holder and install the shifter pin.

BF115A AND BF130A

♦ See Figures 118, 119, 120, 121 and 122

1. On the BF115A and BF130A LCA and XCA models (counter-rotating)follow the following procedure:

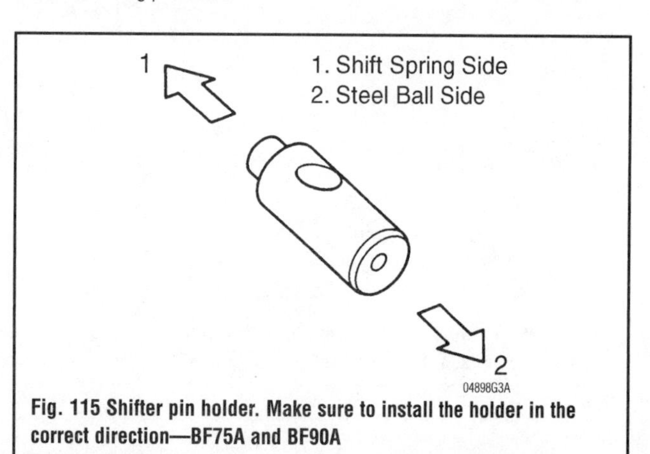

Fig. 115 Shifter pin holder. Make sure to install the holder in the correct direction—BF75A and BF90A

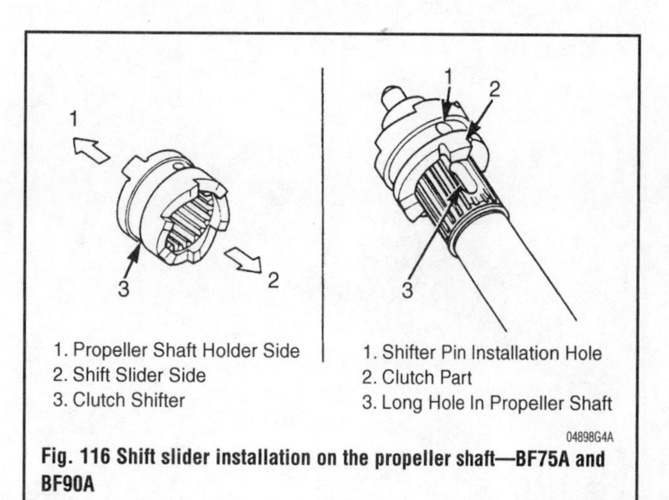

Fig. 116 Shift slider installation on the propeller shaft—BF75A and BF90A

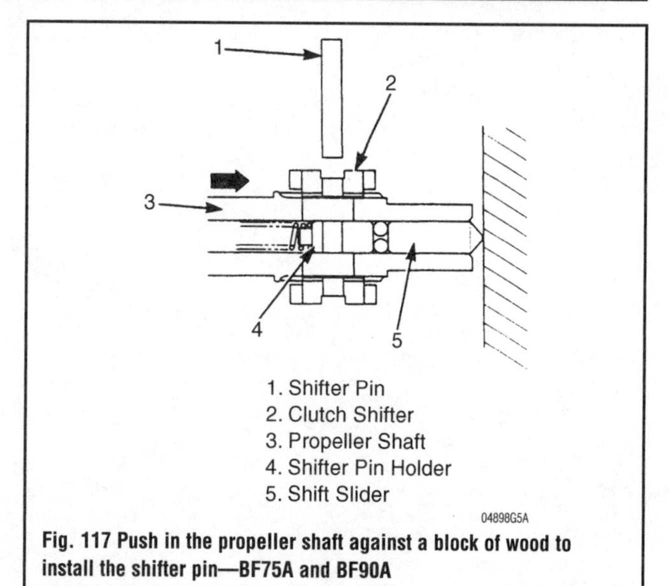

Fig. 117 Push in the propeller shaft against a block of wood to install the shifter pin—BF75A and BF90A

- Be sure the bearing holder assembly, thrust bearing, washer and shim are set on the propeller shaft bearing carrier.
- With the "F" mark on the clutch shifter pointed toward you (toward the forward bevel gear on the LA and XA models or toward the reverse gear on the LCA and XCA models), assemble the clutch shifter and propeller shaft by aligning the long hole in the propeller shaft with the shifter pin hole.
- Install the #11 steel ball, shaft spring and the other #11 steel ball into the shift slider in that order.

2. Making sure to keep the steel balls in the shift slider, align the clutch shifter hole and the hole in the shift slider, and insert the shifter pin.

- Position the propeller shaft upright.
- Push the shift slider up until the holes in the shift slider appear, place the #6 steel balls in each hole and push the shift slider slowly onto the propeller shaft making sure not to let the steel balls fall out of their holes.

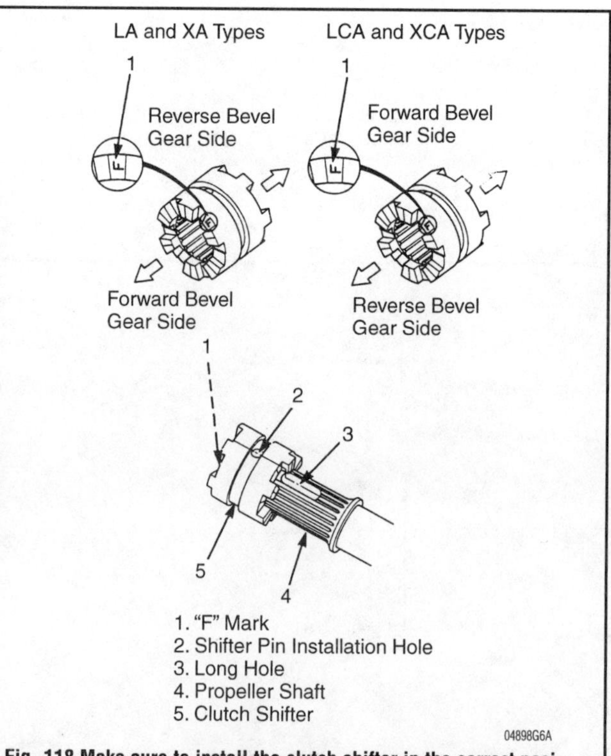

Fig. 118 Make sure to install the clutch shifter in the correct position according to the type of gear case—BF115A and BF130A

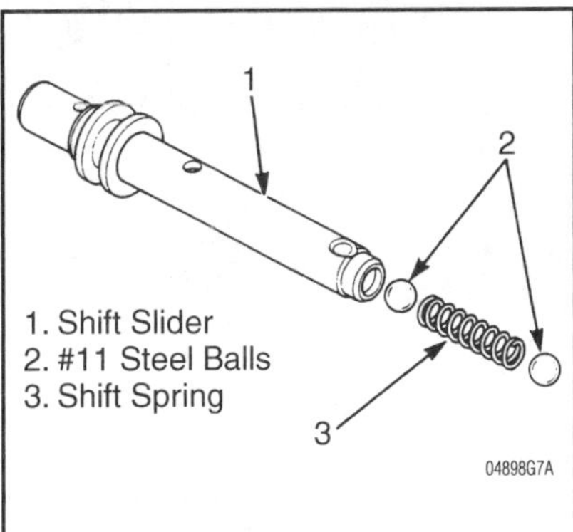

Fig. 119 Exploded view of the shift slider assembly—BF115A and BF130A

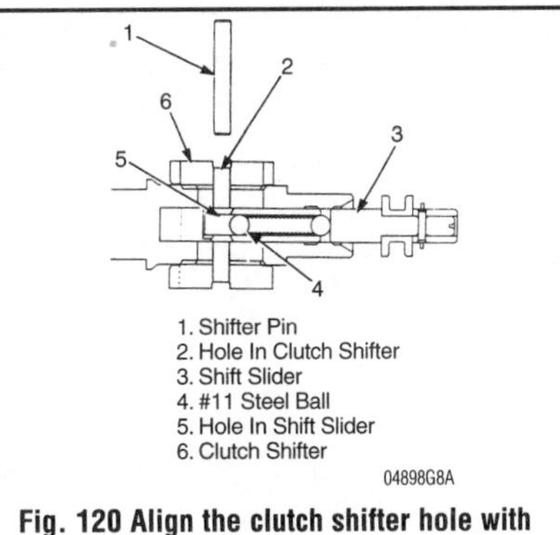

Fig. 120 Align the clutch shifter hole with the hole in the shift slider and install the shifter pin—BF115A and BF130A

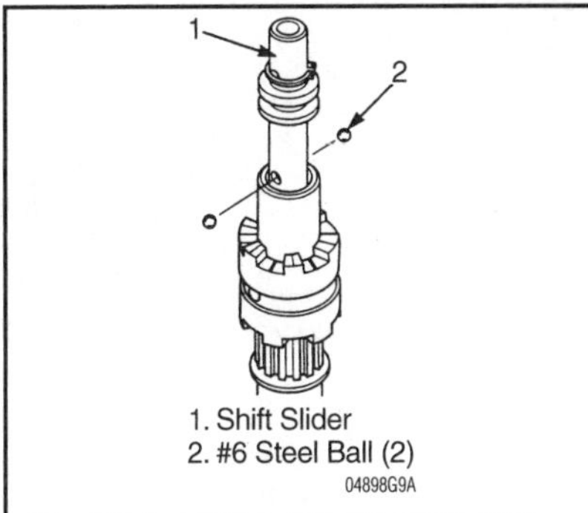

Fig. 121 Install the shift slider and steel balls into the propeller shaft—BF115A and BF130A

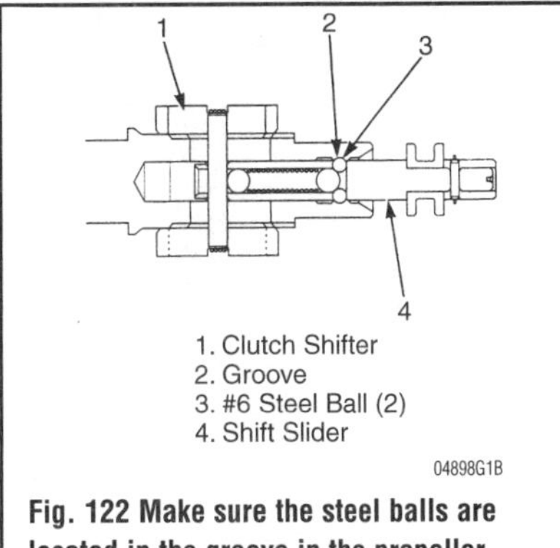

Fig. 122 Make sure the steel balls are located in the groove in the propeller shaft—BF115A and BF130A

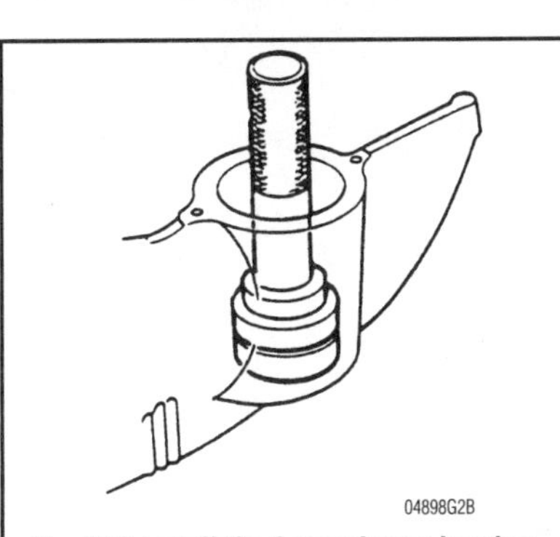

Fig. 123 Install the forward gear bearing into the gear case with the special tools—BF50 and BF5A

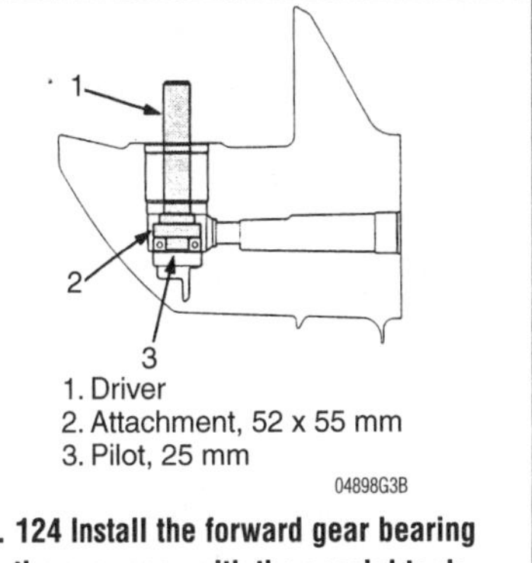

Fig. 124 Install the forward gear bearing into the gear case with the special tools—BF9.9A and BF15A

- Make sure that the #6 steel balls are seated securely in the groove in the propeller shaft.

3. On the most models, press the shift spring down into the propeller shaft. Insert a narrow screwdriver into the slot in the shaft and compress the spring until approximately ½ in. (12mm) is obtained between the top of the slot and the screwdriver.

Hold the spring compressed and at the same time, slide the clutch shifter over the splines of the propeller shaft with the hole in the dog aligned with the slot in the shaft. Double check that the clutch shifter is going back on in the correct direction.

Insert the cross pin into the clutch shifter and through the space held open by the screwdriver. Center the pin and then remove the screwdriver allowing the spring to pop back into place.

Fit the cross pin ring into the groove around the clutch shifter to retain the cross pin in place.

Insert the flat end of the plunger into the propeller shaft, with the rounded end protruding to permit the plunger to slide along the cam of the shift rod.

Forward Gear Ball Bearing

Perform the following steps apply only if the forward gear ball bearing was removed.

BF50 AND BF5A

See Figure 123

1. On BF5A and BF50 models, obtain the following special tools:
- Driver—p/n 07749-0010000
- Attachment (42 mm x 47 mm)—p/n 07746-0010300
- Pilot (20 mm)—p/n 07746-0040500

2. Apply oil to the outer race of the bearing, and then drive the bearing squarely into the gear case using the special tools.

BF75, BF8A AND BF75

1. On the BF75, BF8A and BF75 models, obtain the following special tools:
- Driver—p/n 07749-0010000
- Attachment (6302 bearing assembly)—p/n 07946-9350200

2. Install the correct size gear shim into the gear case.
3. Apply oil to the outer race of the bearing, and drive the bearing squarely into the gear case using the special tools.

BF9.9A AND BF15A

See Figure 124

1. On the BF9.9A and BF15A models, obtain the following special tools:
- Driver—p/n 07749-0010000
- Attachment (52 mm x 55 mm)—p/n 07746-0010400
- Pilot (25 mm)—p/n 07746-0040600

2. Install the correct size gear shim into the gear case.
3. Apply oil to the outer race of the bearing, and drive the bearing squarely into the gear case using the special tools.

BF25A TO BF50A

See Figure 125

1. On the BF25A to BF50A models, obtain the following special tools:
- Driver—p/n 07949-3710001
- Attachment (62 mm x 68 mm)—p/n 07746-0010500.

2. Install the correct size gear shim into the gear case before installing the bearing race
3. Apply oil to the outer race of the bearing and insert the bearing race into the gear case with the larger of the inside diameters of the race facing out. Drive the bearing squarely into the gear case using the special tools.

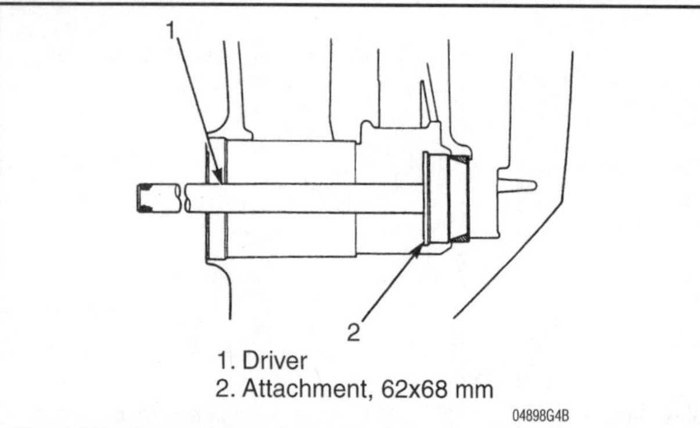

Fig. 125 Install the forward gear bearing outer race in the gear case with the driver—BF25A to BF50A

BF75A AND BF90A

➧ See Figure 126

1. On the BF75A and BF90A models, obtain the following special tool:
• Mandrel—p/n 07SPD-ZW0010Z (91-31106)

➡When the forward shim is missing, damaged or a new gear case is being used, as a temporary measure install a forward bevel gear shim of 0.01 in. (0.25 mm) to get your first backlash measurement.

2. Place the forward bevel gear shim/s into the gear case.
3. Apply Quicksilver® 2-4-C with Teflon® to the circumference of a new taper roller bearing, and set the bearing into the gear case.
4. Remove the propeller shaft assembly from the bearing carrier. Then remove the shift slider, three steel balls, cross pin ring, shifter pin, clutch shifter, shifter pin holder, shift spring thrust bearing and thrust washer. Install the propeller shaft into the bearing carrier.
5. Apply Quicksilver® 2-4-C with Teflon® to the bearing carrier and O-ring.
6. Attach the special tool to the propeller shaft and install the bearing carrier into the gear case.

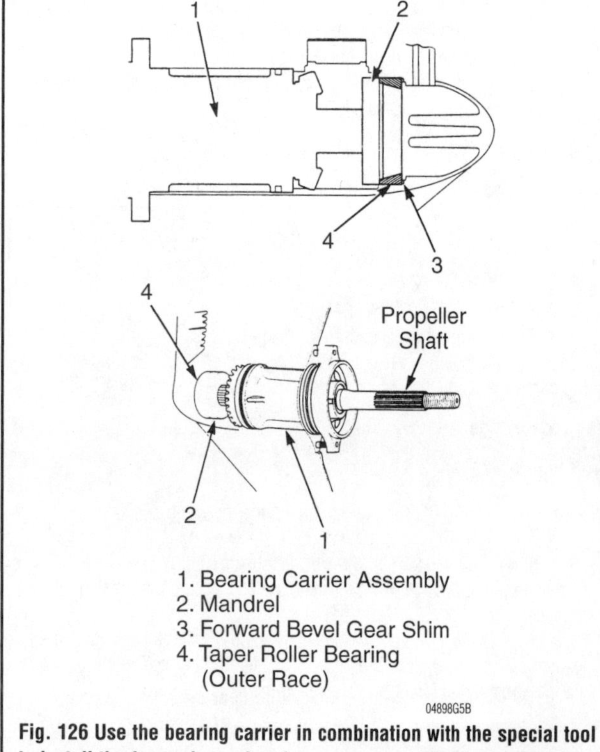

Fig. 126 Use the bearing carrier in combination with the special tool to install the forward gear bearing outer race—BF75A and BF90A

7. Use the bearing carrier and propeller shaft to drive the bearing into place with a brass hammer.
8. After installing the tapered roller bearing, remove the propeller shaft from the bearing carrier.

BF115A AND BF130A

➧ See Figures 127, 128 and 129

1. On the LA and XA models only, obtain the following special tools:
• Driver—p/n 07949-3710001
• Attachment (52 mm x 55 mm)—p/n 07746-0010400
• Pilot (25 mm)—p/n 07746-0040600
• Mandrel—p/n 07SPD-ZW0010Z (91-31106)

Insert the shim material saved during disassembly into the lower unit bearing race cavity. The shim material should give the same amount of backlash between the pinion gear and the forward gear as before disassembling. After applying oil to the circumference of the bearing race, insert the bearing race squarely into the lower unit housing with the larger inside diameter facing out.

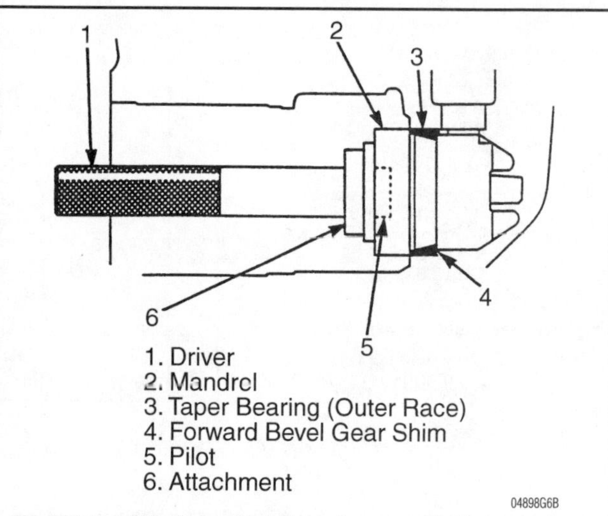

Fig. 127 Install the forward bearing outer race with the bearing driver—BF115A and BF130A

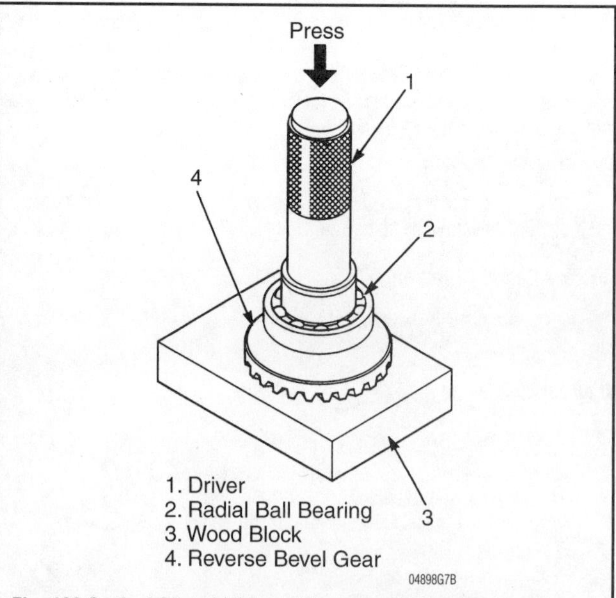

Fig. 128 On the LCA and XCA models, press the bearing into the reverse bevel gear with the driver and hydraulic press—BF115A and BF130A

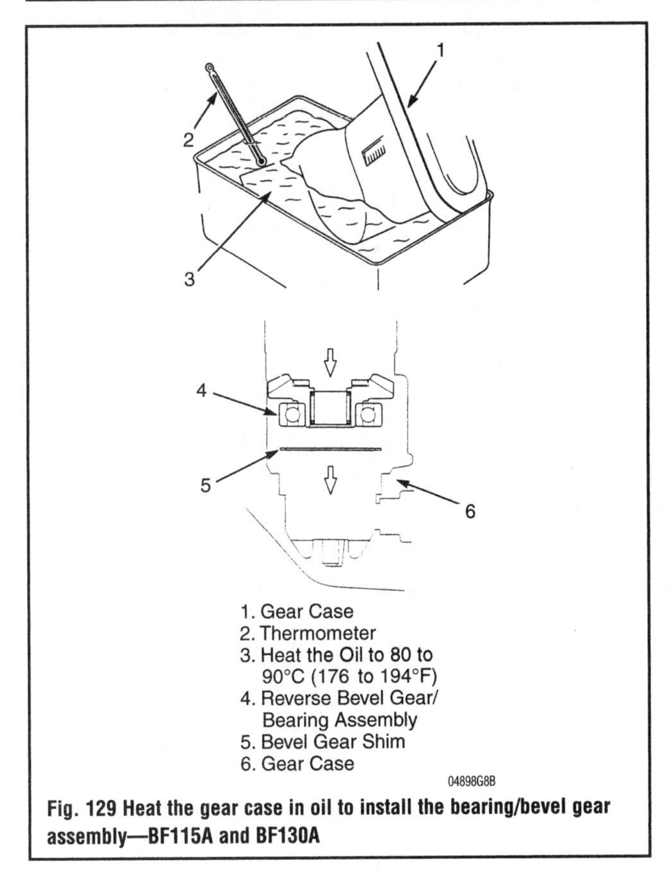

Fig. 129 Heat the gear case in oil to install the bearing/bevel gear assembly—BF115A and BF130A

Then use the appropriate driver for the model unit being serviced and drive the race into the housing until it is fully seated.

2. On the LCA and XCA types, obtain the following special tool:

- Driver (40 mm I.D.)—07746-0030100

3. Place a wood block underneath the reverse bevel gear.
4. Apply oil to the entire surface of the bearing and drive the bearing into the reverse bevel gear using the special tool and a hydraulic press.
5. Soak the entire gear case in a container of oil with the bearing part of the gear case facing down in the container.
6. Heat the oil to 176–194°F (80–90°C).
7. After the bearing part of the gear case becomes hot, remove the gear case from the container and install the bevel gear shim and the reverse bevel gear/ bearing assembly into the gear case very quickly before it cools too much.

Pinion Shaft Needle Bearings And Tapered Bearings

Perform the following steps only if the bushings and/or bearings were removed during disassembly.

BF5A AND BF50

1. The BF5A and BF50 have one bushing for the Pinion Shaft. It is located just above the pinion gear at the lower end.

Obtain a commercially available bearing driver and drive the upper bushing into place from the top until it seats.

BF75, BF8A AND BF100

Use a bearing driver to properly seat the bearing in the gear case.

BF9.9A AND BF15A

See Figures 130 and 131

1. Use the following special tool:

- Bearing remover—p/n 07946-MJ00100

2. To both remove and install the lower needle bearing assembly. Make sure that on assembly the needle bearing does not protrude into the gear case.

To install the upper roller bearing, obtain the following special tools:

- Bearing Driver—p/n 07749-0010000

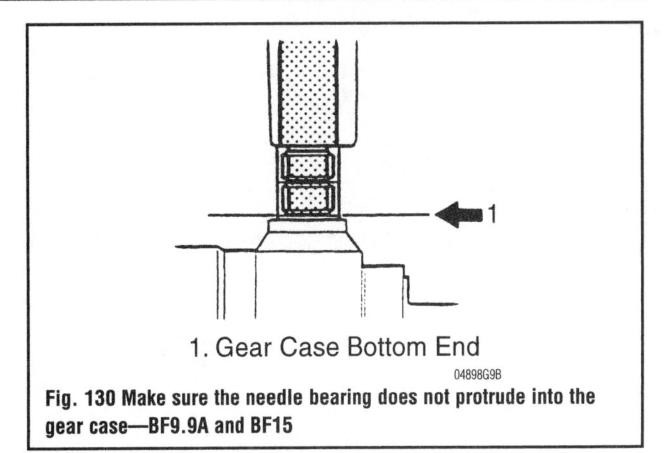

Fig. 130 Make sure the needle bearing does not protrude into the gear case—BF9.9A and BF15

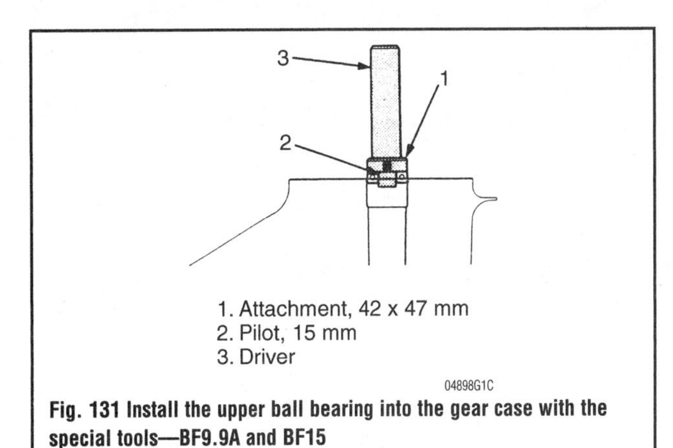

Fig. 131 Install the upper ball bearing into the gear case with the special tools—BF9.9A and BF15

- Pilot (15 mm)—p/n 07746-0040300
- Attachment (42 x 47 mm)—p/n 07746-0010300

3. Drive the bearing in until it seats firmly in the gear case.

BF25A, BF30A, BF35A, BF40A, BF45A AND BF50A

See Figures 132 and 133

To install the needle bearing, obtain the following special tools:

- Bearing Driver—p/n 07949-3710001
- Attachment (24 x 26 mm)—p/n 07746-0010700 and
- Pilot (22 mm)—p/n 07746-0041000.

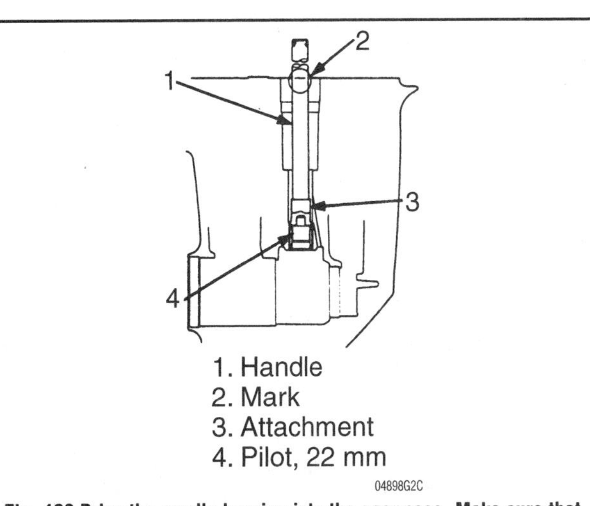

Fig. 132 Drive the needle bearing into the gear case. Make sure that it is flush with the bottom of the gear case—BF25A, BF30A, BF35A, BF40A, BF45A and BF50A

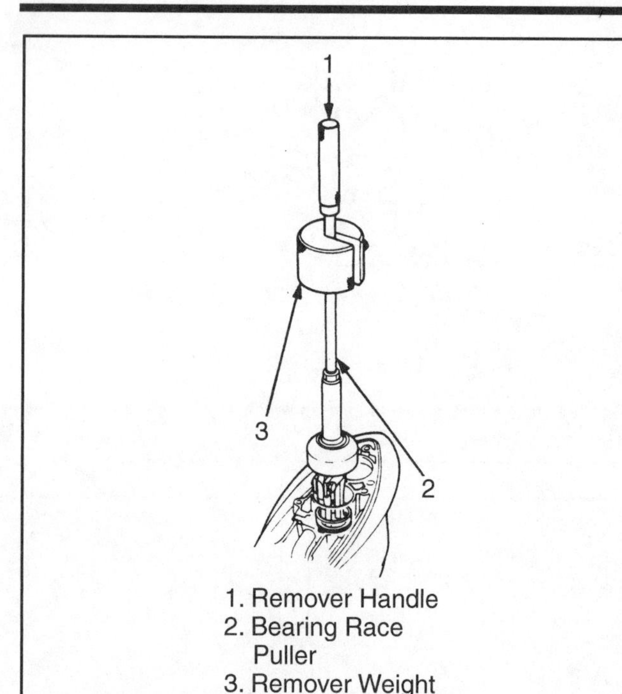

Fig. 133 Use the driver to install the tapered bearing outer race in the upper gear case—BF25A, BF30A, BF35A, BF40A, BF45A and BF50A

To install the upper tapered roller bearing, obtain the following special tools:

- Driver—p/n 07749-0010000
- Attachment (42 x 47 mm)—p/n 07746-0010300

1. Drive the bearing in until it is seated securely.

➡Use the special tool to drive the bearing in with the larger inner diameter side of the bearing facing outward.

BF75A AND BF90A

➧ See Figures 134, 135, 136 and 137

1. To install the wear sleeve, obtain the following special tool:
- Wear Sleeve Installation Tool—p/n 07SPF-ZW0010Z (91-14310A1).

2. Apply a thin coat of Loctite®271 or an equivalent to a new rubber ring and install it on the pinion shaft.

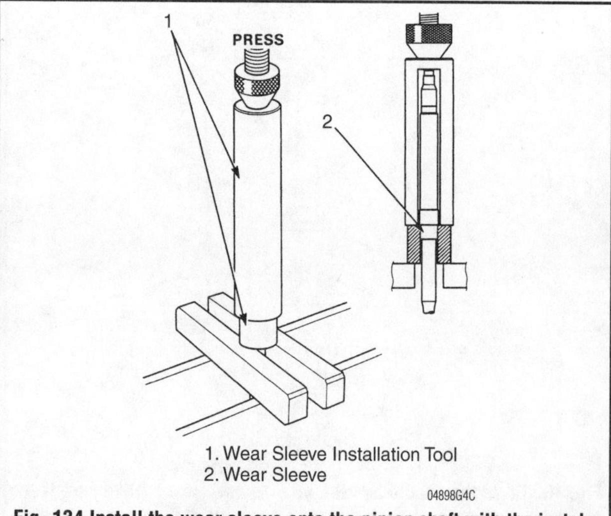

Fig. 134 Install the wear sleeve onto the pinion shaft with the installation tool—BF75A and BF90A

3. Apply a thin coat of Loctite®271 or equivalent to the wear sleeve installation position on the pinion shaft and to the inside of the new wear sleeve.
4. Set the new wear sleeve on the special tool. Then set the pinion shaft in the special tool on which the wear sleeve is mounted and pass the other special tool over the pinion shaft.

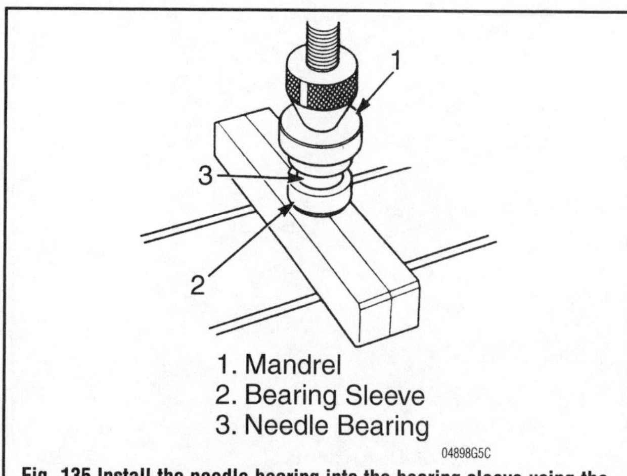

Fig. 135 Install the needle bearing into the bearing sleeve using the mandrel—BF75A and BF90A

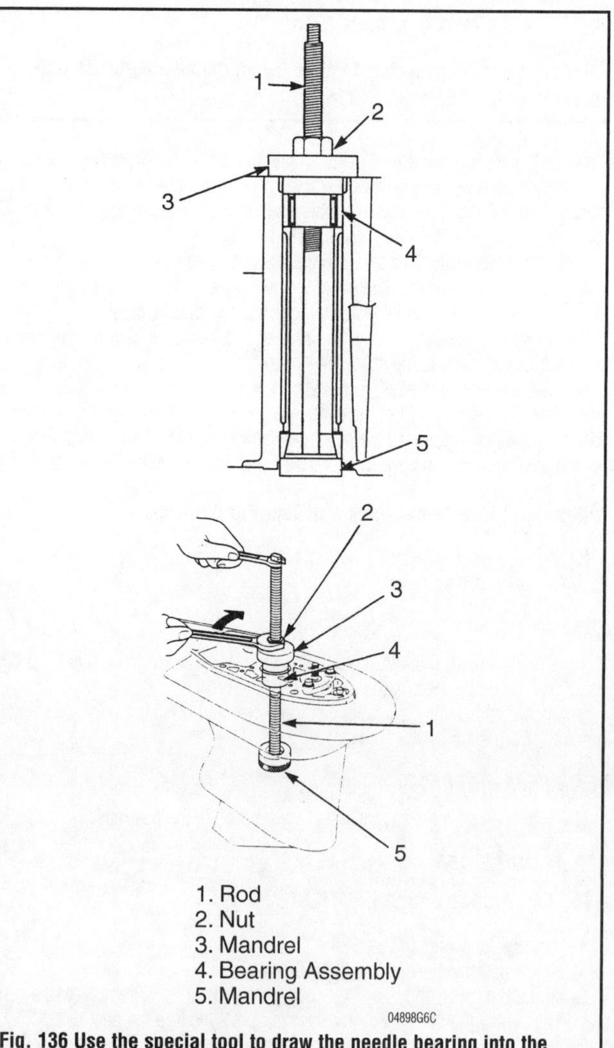

Fig. 136 Use the special tool to draw the needle bearing into the gear case—BF75A and BF90A

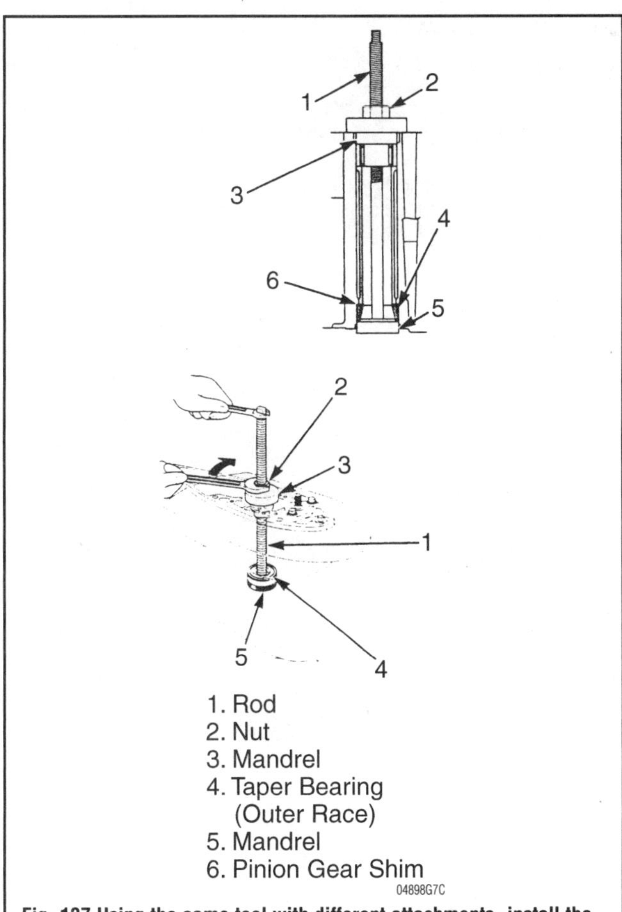

Fig. 137 Using the same tool with different attachments, install the tapered bearing outer race into the gear case—BF75A and BF90A

5. Set both the pinion shaft and the mounted special tool onto a hydraulic press and press the two together until the special tools contact each other. Wipe off the excess Loctite®.

CAUTION

Do not try to install the wear sleeve on the pinion shaft with a hammer. Use a hydraulic press only.

6. To install the needle bearing, obtain the following special tools:
- Mandrel—p/n 07SPD-ZW0072A (91-13781)
- Rod—p/n 07SPD-ZW0050Z (91-31229)
- Nut—p/n 07SPD-ZW0060Z (11-24156)
- Mandrel—p/n 07SPD-ZW0071A (91-13780)
- Mandrel—p/n 07SPD-ZW0072A (91-13781).

7. Apply Quicksilver 2-4-C with Teflon to the outer circumference of a needle bearing and the inner surface of the bearing sleeve.

8. Set the bearing sleeve with the chamfered side facing down and the stamped surface of the needle bearing facing up. Press the needle bearing into the bearing sleeve using the special tool (mandrel 07SPD-ZW0072A).

9. Check that the oil slinger is mounted inside the gear case. If not, install it now.

10. Attach the special tools and the bearing assembly to the gear case. Tighten the nut and draw the bearing assembly into the gear case.

➡The bearing assembly can be mounted on the gear case whether or not the tapered bearing outer race is place.

11. To install the tapered bearing outer race in the gear case, obtain the following special tools:
- Rod—p/n 07SPD-ZW0050Z (91-31229)
- Nut—p/n 07SPD-ZW0060Z (11-24156)
- Mandrel—p/n 07SPD-ZW0071A (91-13780)
- Mandrel—p/n 07SPD-ZW0072A (91-13781)

➡When the pinion gear shim is damaged or missing or a new gear case is being used, adjust the shim as needed.

12. Apply Quicksilver 2-4-C with Teflon®to the outer circumference of a new tapered roller bearing.

13. Attach the special tools, tapered roller bearing and pinion gear shim to the gear case and by tightening the nut on the special tool, draw the bearing and shim into the gear case until they are seated.

BF115A AND BF130A

See Figures 138, 139, and 140

To install the needle bearing and the tapered roller bearing, obtain the following special tools:
- Driver—07949-3710001
- Attachment (32 x 35 mm)—07746-0010100
- Pilot (28 mm)—07746-0041100
- Drive shaft B—07964-MB00200

1. Apply some oil to the outer edge of the new needle bearing.
2. Using the tools, drive the needle bearing into the gear case to the mark on the driver.
3. After installation, check that the edge of the bearing is flush with the bottom of the gear case.
4. To install the pinion shaft tapered roller bearing, install the pinion gear on the shaft and temporally tighten the nut by hand.
5. Apply some oil to the inner wall and roller section of the bearing.
6. Install the pinion gear shim, tapered roller bearing inner race and the special tool (Drive shaft B) on the pinion shaft.
7. Install the pinion shaft upright on the hydraulic press with the pinion gear side up.
8. Press the tapered roller bearing inner race on the pinion shaft using the hydraulic press.

➡Be sure that the ends of the special tool are set securely on the inner part of the tapered roller bearing and on the hydraulic press table.

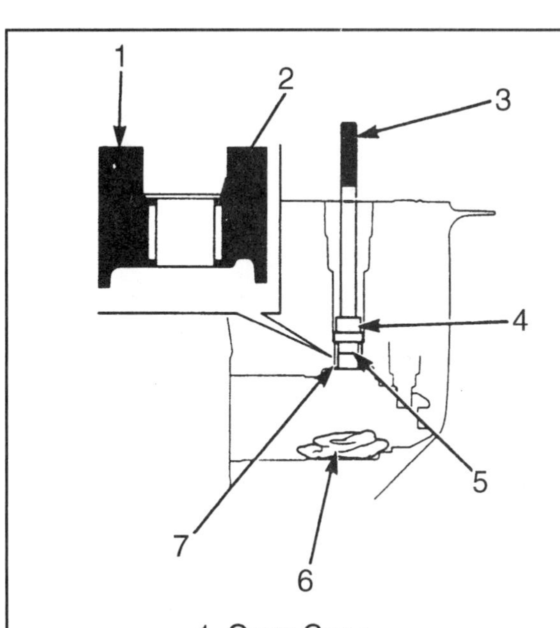

Fig. 138 Use the driver to install the new needle bearing into the gear case—BF115A and BF130A

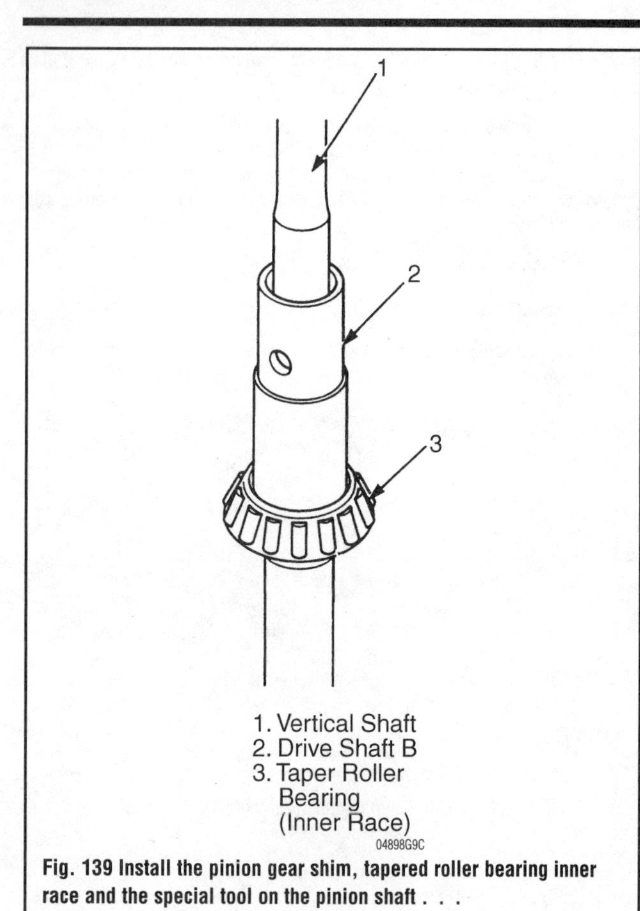

Fig. 139 Install the pinion gear shim, tapered roller bearing inner race and the special tool on the pinion shaft . . .

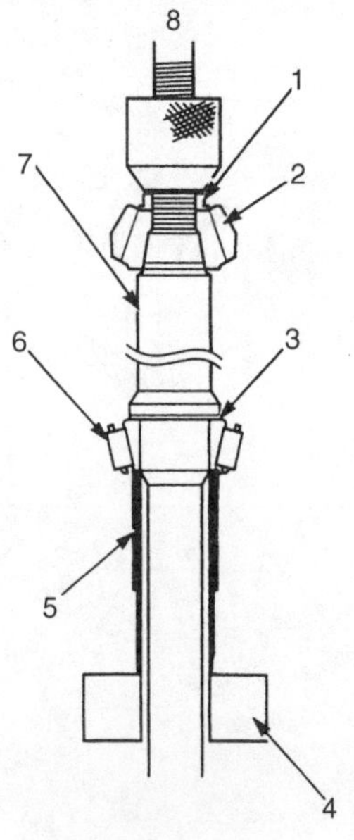

Pinion Gear And Pinion Shaft

BF50 AND BF5A

1. Clean the pinion shaft and pinion gear splines of all grease and dirt.
2. After installing the forward bevel gear, place the pinion washer and pinion bevel gear into the gear case.
3. While holding the pinion bevel gear with one hand, lower the pinion shaft into the gear case and engage the pinion gear splines.

BF75, BF8A AND BF100

1. Clean the pinion shaft and pinion gear splines of all grease and dirt.
2. After installing the forward bevel gear, place the thrust bearing washer, thrust bearing and pinion bevel gear into the gear case.
3. While holding the pinion bevel gear assembly with one hand, lower the pinion shaft into the gear case and engage the pinion gear splines with the other.

BF9.9A AND BF15A

1. Clean the pinion shaft and pinion gear splines of all grease and dirt.
2. After installing the forward bevel gear, place the thrust bearing washer, thrust bearing and pinion bevel gear into the gear case.
3. While holding the pinion bevel gear assembly with one hand, lower the pinion shaft into the gear case and engage the pinion gear splines with the other.
4. Hold the pinion nut with a 12 mm socket and 3/8 breaker bar. Turn the pinion shaft with a 12 mm open end wrench and tighten the pinion nut to 18.8 ft. lbs. (26 Nm).

BF25A AND BF30A

➧ See Figure 141

1. Clean the pinion shaft and pinion gear splines of all grease and dirt.
2. After installing the forward bevel gear, place the pinion bevel gear and 10 mm flange nut into the gear case.
3. While holding the pinion bevel gear assembly with one hand, lower the

1. Pinion Gear Nut
2. Pinion Gear
3. Pinion Gear Shim
4. Hydraulic Press Table
5. Drive Shaft B
6. Taper Roller Bearing (Inner Race)
7. Vertical Shaft
8. Press

04898G1D

Fig. 140 . . . install the entire assembly in a hydraulic press and press the bearing onto the shaft

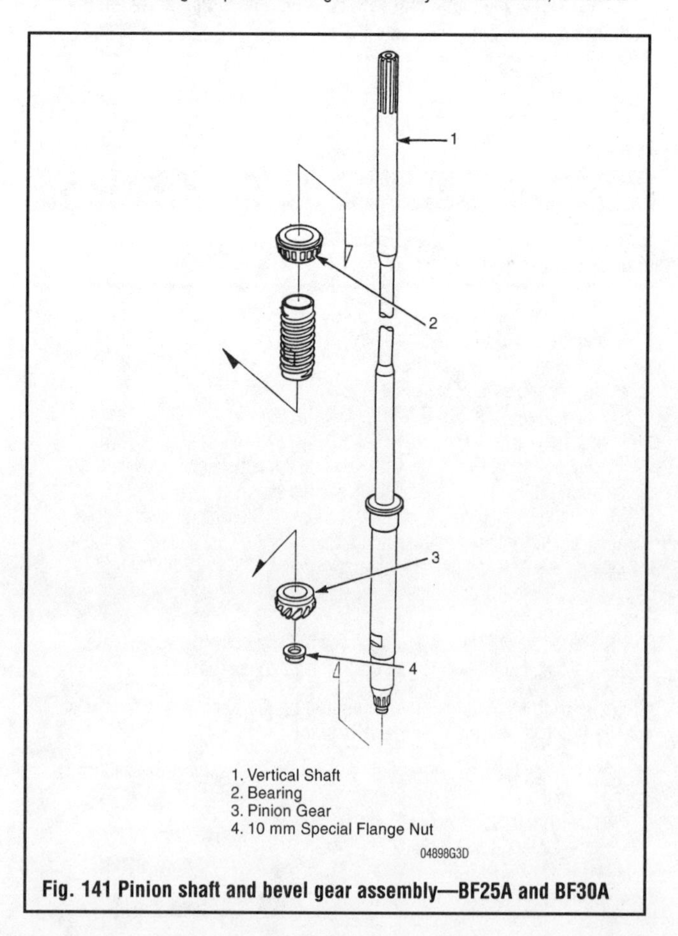

Fig. 141 Pinion shaft and bevel gear assembly—BF25A and BF30A

pinion shaft into the gear case and engage the pinion gear splines with the other.

4. Using the special tool (pinion shaft holder p/n 07LPB-ZV30200) to hold the pinion shaft, tighten the pinion gear nut to 28.9 ft. lbs. (40 Nm).

➡Use shop towels or some kind of padding to protect the inside of the gear case from the breaker bar handle while tightening the pinion nut.

BF35A, BF40A, BF45A AND BF50A

▸ See Figure 142

1. Clean the pinion shaft and pinion gear splines of all grease and dirt.
2. After installing the forward bevel gear, place the pinion bevel gear and 12 mm flange nut into the gear case.
3. While holding the pinion bevel gear assembly with one hand, lower the pinion shaft into the gear case and engage the pinion gear splines with the other.
4. Using the special tool (pinion shaft holder p/n 07LPB-ZV30200) to hold the pinion shaft, tighten the pinion gear nut to 54.2 ft. lbs. (75 Nm).

➡Use shop towels or some kind of padding to protect the inside threads of the gear case from the breaker bar handle while tightening the pinion nut.

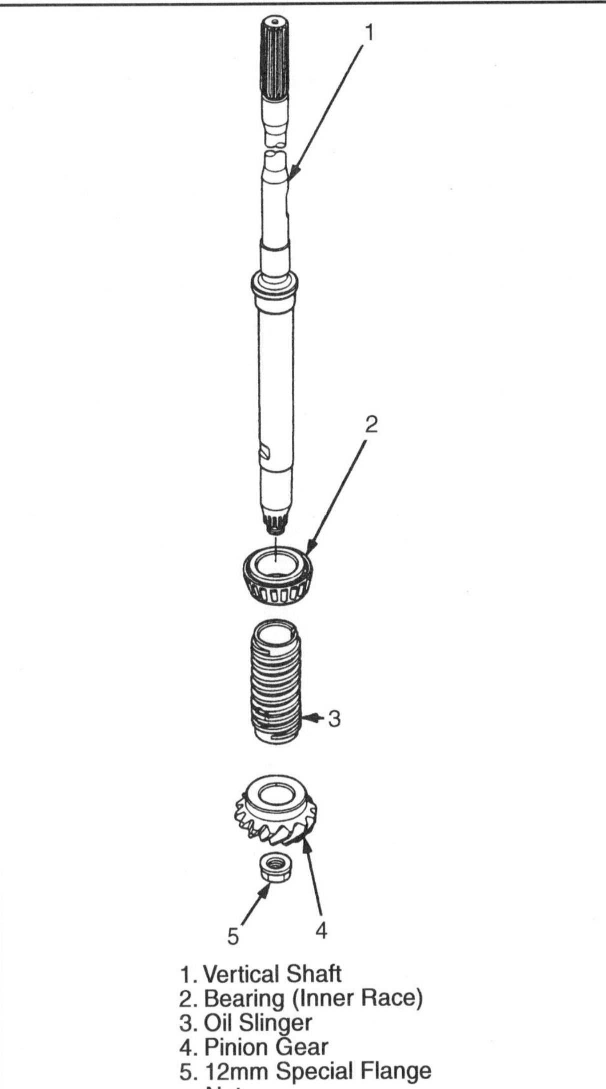

Fig. 142 Pinion shaft and bevel gear assembly—BF35A, BF40A, BF45A and BF50A

BF75A AND BF90A

▸ See Figure 143

1. Clean the pinion shaft and pinion gear splines of all grease and dirt.
2. After installing the forward bevel gear, place the pinion bevel gear and pinion gear nut (with the projection on the pinion nut facing the gear) into the gear case.
3. While holding the pinion bevel gear assembly with one hand, lower the pinion shaft into the gear case and engage the pinion gear splines with the other.
4. Using the special tool (pinion shaft holder p/n 07SPB-ZW10200) to hold the pinion shaft, tighten the pinion gear nut to 70 ft. lbs. (95 Nm).

➡Use shop towels or some kind of padding to protect the inside of the gear case from the breaker bar handle while tightening the pinion nut.

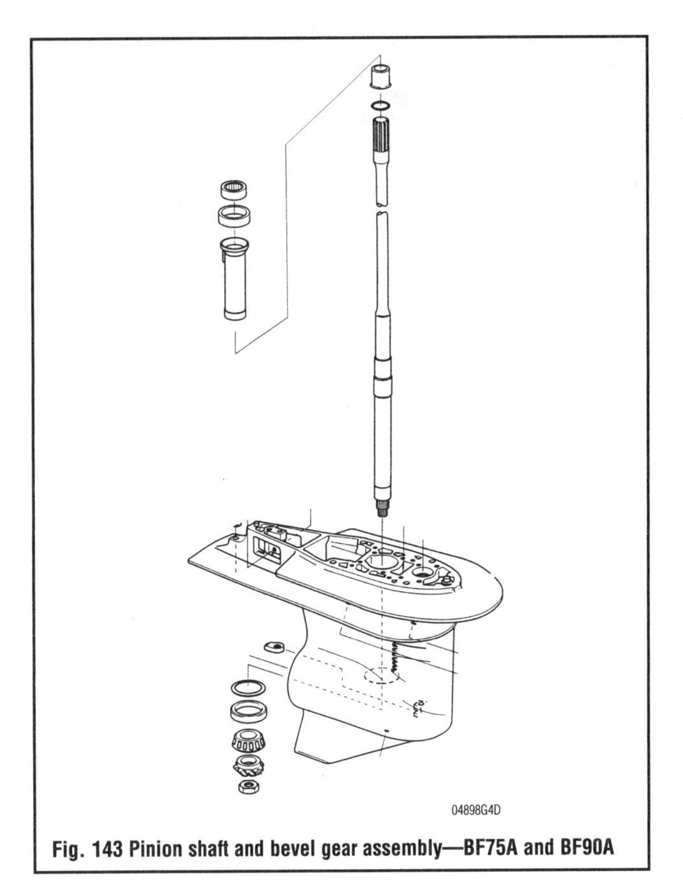

Fig. 143 Pinion shaft and bevel gear assembly—BF75A and BF90A

BF115A AND BF130A

▸ See Figure 144

1. Clean the pinion shaft and pinion gear splines of all grease and dirt.
2. After installing the forward bevel gear, place the pinion bevel gear and pinion gear nut into the gear case.
3. While holding the pinion bevel gear assembly with one hand, lower the pinion shaft into the gear case and engage the pinion gear splines with the other.
4. Using the special tool (pinion shaft holder p/n 07SPB-ZW10200) to hold the pinion shaft, tighten the pinion gear nut to 98 ft. lbs. (132 Nm).

➡Use shop towels or some kind of padding to protect the inside of the gear case from the breaker bar handle while tightening the pinion nut.

5. Install the 64 mm lock nut at the top of the gear case.
6. Using special tool (lock nut wrench-64 mm p/n 07916-MB00002) now tighten the lock nut to 90 ft. lbs. (123 Nm).

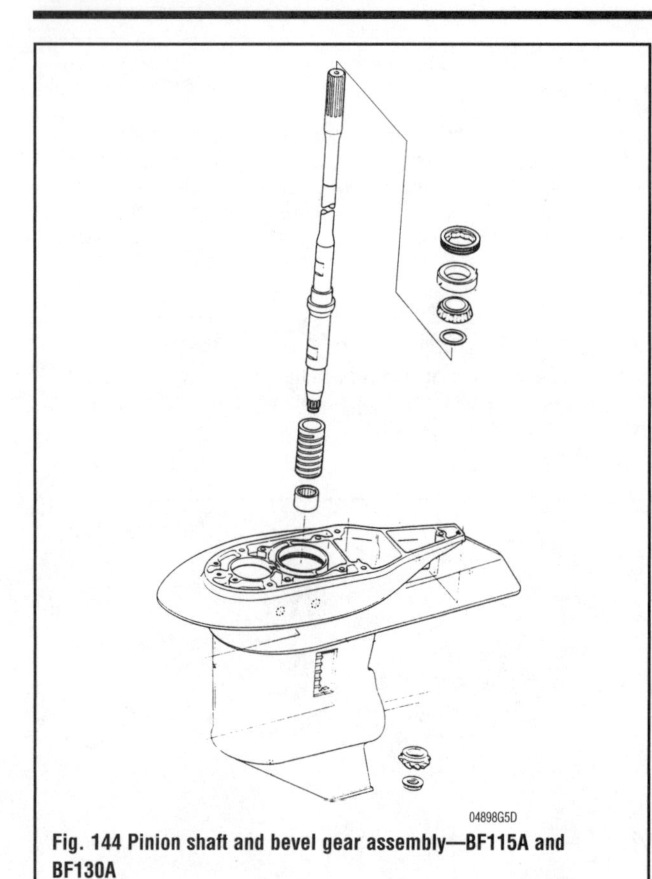
Fig. 144 Pinion shaft and bevel gear assembly—BF115A and BF130A

Bearing Carrier

BF50 AND BF5A

See Figures 145 and 146

1. Install a new O-ring onto the bearing carrier.
2. Obtain the following special tool to install the new 13 mm water seal:
- Attachment (32 x 35 mm)—p/n 07746-0010100
- Driver—p/n 07749-0010000
3. Apply grease to the seal lip.
4. Drive the new water seal into the bearing carrier with the lettering on the seal facing toward the propeller with the special tools.
5. To install the ball bearing, obtain the following special tools:
- Driver—p/n 07749-010000
- Attachment (42 x 47 mm)—p/n 07746-0010300
- Pilot—p/n 07746-0040500
6. Apply engine oil to the outer edge of the bearing.
7. Drive the bearing into the bearing carrier with the special tools until it is seated firmly.

BF75, BF8A AND BF100

1. Obtain the following special tools to install a new water seal and ball bearing:
- Driver—07749-0010000
- Attachment (17 mm)—07946-9350101
2. Install the new water seal first into the carrier. Make sure the seal is seated firmly.
3. Install the bearing next, and again, make sure that it is seated firmly in the carrier.

BF9.9A AND BF15A

See Figures 147, 148 and 149

1. Obtain the following special tool to install a new seal:
- Driver—07749-0010000
- Attachment (32 x 35 mm)—07746-0010100
- Pilot (17 mm)—07746-0040400
2. Coat the outer edge of the new water seal with engine oil and using the special tools, drive the seal firmly into the bearing carrier.
3. Obtain the following special tools to install the needle bearing:
- Driver—07749-0010000
- Attachment (24 x 26 mm)—07746-0010700
- Pilot (17 mm)—07746-0040400
4. Apply engine oil to the outer edge of the bearing and using the special tools, drive the bearing firmly into the bearing carrier.
5. Obtain the following special tools to install the ball bearing:
- Driver—07749-0010000
- Attachment (42 x 47 mm)—07746-0010300
- Pilot (17 mm)—07746-0040600
6. Apply engine oil to the outer edge of the bearing and using the special tools, drive the bearing firmly into the bearing carrier.

BF25A AND BF30A

See Figures 150, 151, 152 and 153

1. To install the reverse bevel gear in to the radial ball bearing, obtain the following special tools:
- Attachment (35 mm I.D.)—07746-0030400
- Attachment (24 x 26 mm)—07746-0010700
2. Using a hydraulic press and the special tools, press the reverse gear into the radial ball bearing.
3. To install the bearing/gear assembly into the bearing carrier, obtain a commercially available bearing remover and a block of wood.
- Install the bearing remover around the radial ball bearing and the wood block on the upper end of the bearing carrier, then use a hydraulic press to press the bearing into the carrier.
4. To install the new water seals, obtain the following special tools:
- Driver—07749-0010000
- Attachment (32 x 35 mm)—07746-0010100
5. Apply oil to the outer edge of the seals and using the special tools, drive each one of the seal firmly into the bearing carrier.

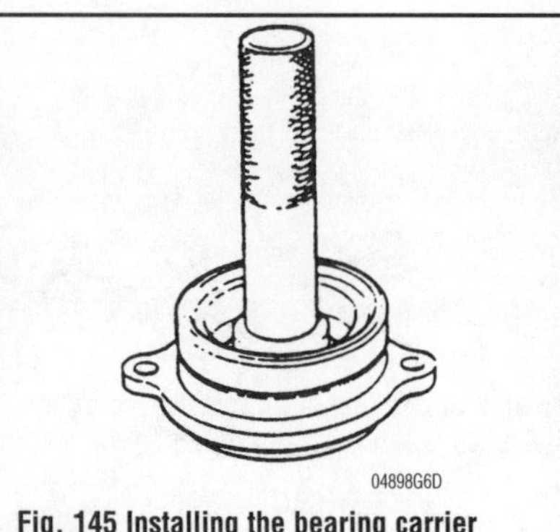
Fig. 145 Installing the bearing carrier water seal with the special tool—BF50 and BF5A

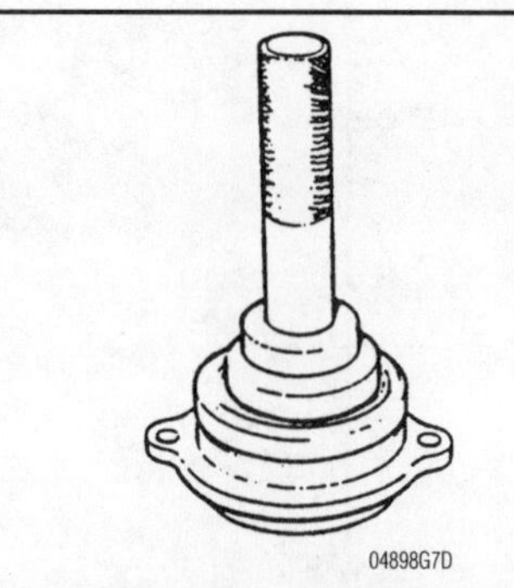
Fig. 146 Installing the bearing carrier ball bearing with the special tools—BF50 and BF5A

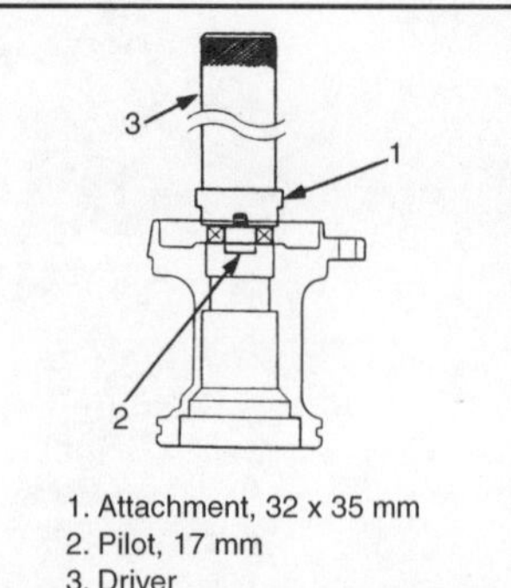

Fig. 147 Install the new water seal into the bearing carrier—BF9.9A and BF15A

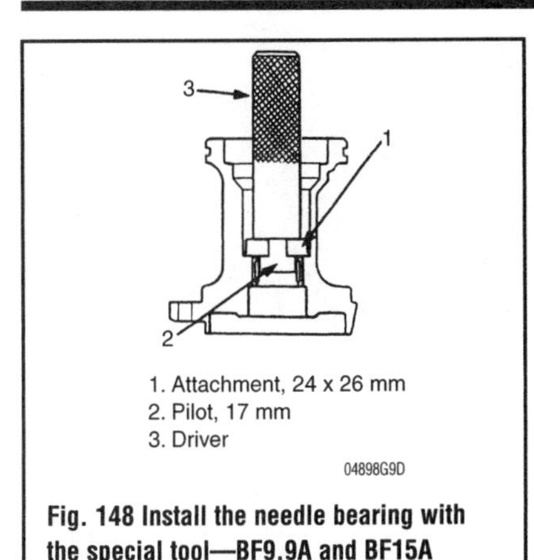

Fig. 148 Install the needle bearing with the special tool—BF9.9A and BF15A

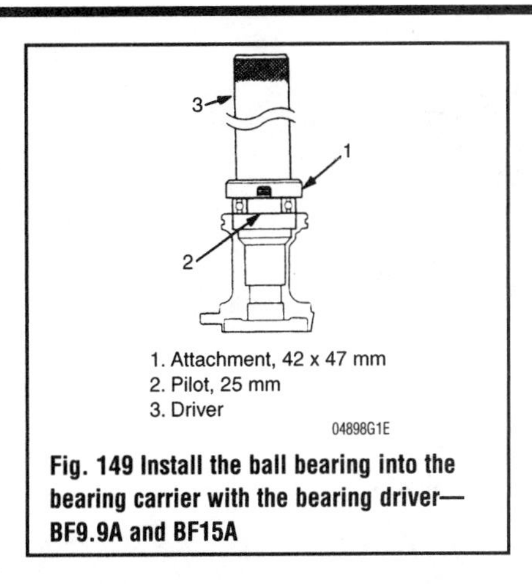

Fig. 149 Install the ball bearing into the bearing carrier with the bearing driver—BF9.9A and BF15A

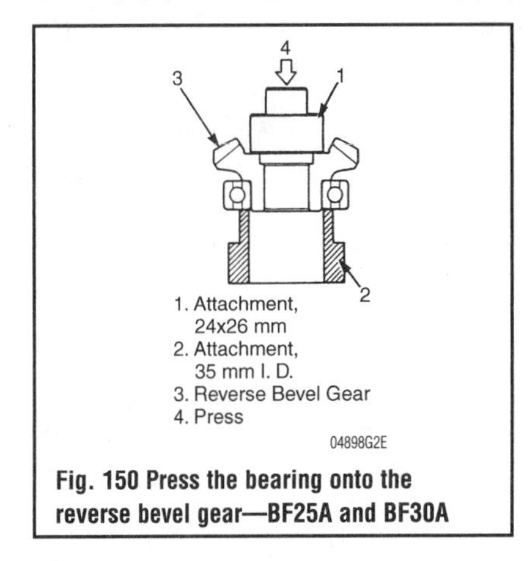

Fig. 150 Press the bearing onto the reverse bevel gear—BF25A and BF30A

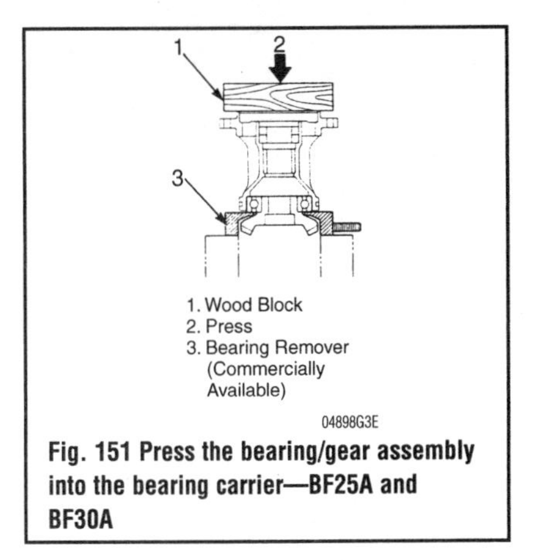

Fig. 151 Press the bearing/gear assembly into the bearing carrier—BF25A and BF30A

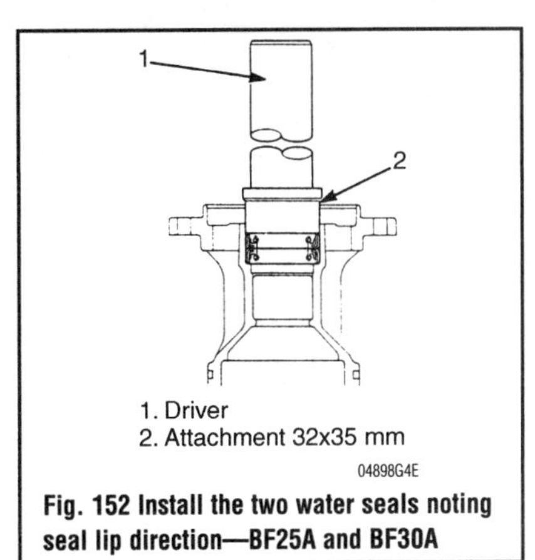

Fig. 152 Install the two water seals noting seal lip direction—BF25A and BF30A

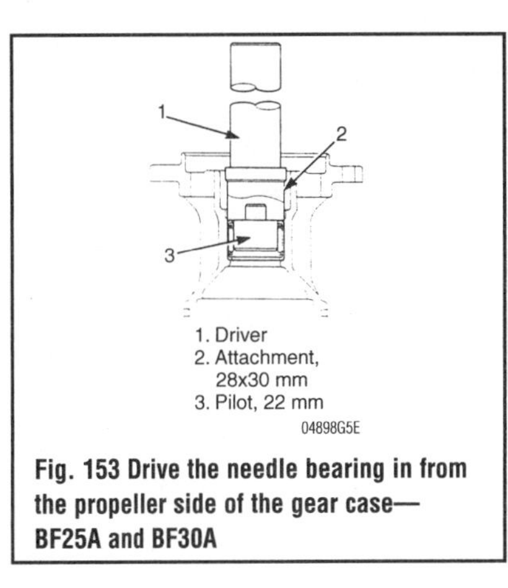

Fig. 153 Drive the needle bearing in from the propeller side of the gear case—BF25A and BF30A

6. Install the seals back to back, with the bottom seal lip facing downward and the top seal lip facing upward.

7. To install the needle bearing, obtain the following special tools:

- Driver—07749-0010000
- Attachment (28 x 30 mm)—07946-1870100
- Pilot (22 mm)—07746-0041000

8. Using the special tools, drive the needle bearing firmly into the bearing carrier from the propeller side.

BF35A, BF40A, BF45A AND BF50A

See Accompanying Illustrations

1. To install the reverse bevel gear in to the radial ball bearing, obtain the following special tools:

- Attachment (35 mm I.D.)—07746-0030400
- Attachment (24 x 26 mm)—07746-0010700
- Support base—07965-SD90100

2. Install the reverse bevel gear and plain washer on the bearing using the special tools. Note the direction of installation of the plain washer.

3. Set the O-ring on the gear assembly and install it into the bearing using the special tools and a hydraulic press. Be very careful to not damage the O-ring during installation.

4. Place a wood block on the one end of the bearing carrier and the special tool support base on the other, then use a hydraulic press to press the bearing/gear assembly into the carrier.

5. To install the new water seals, obtain the following special tools:

- Driver—07749-0010000
- Attachment (32 x 35 mm)—07746-0010100

6. Apply oil to the outer edge of the seals and using the special tools, drive each one of the seal firmly into the bearing carrier.

7. Install the seals back to back, with the bottom seal lip facing downward and the top seal lip facing upward.

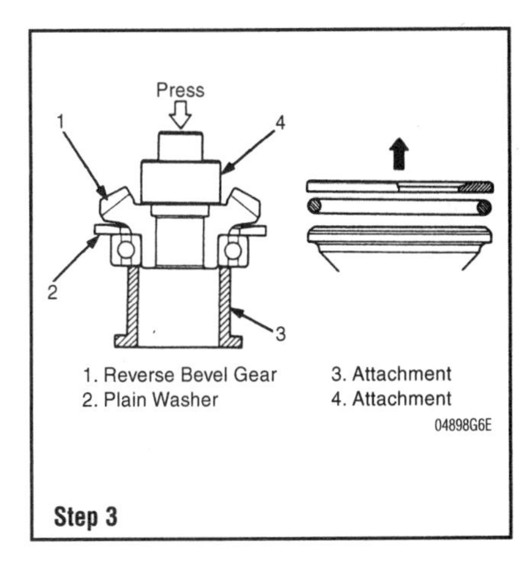

Step 3

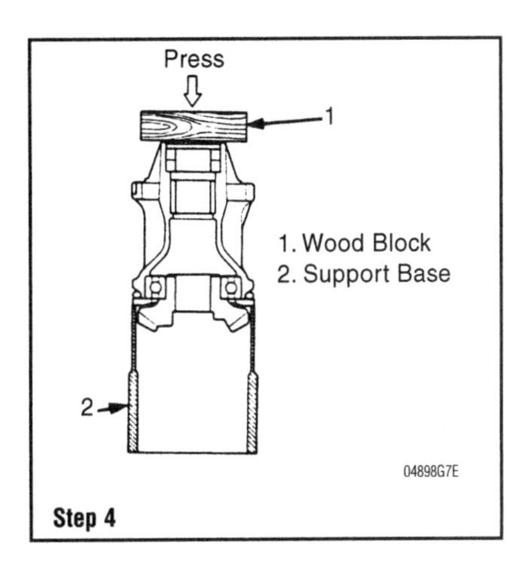

Step 4

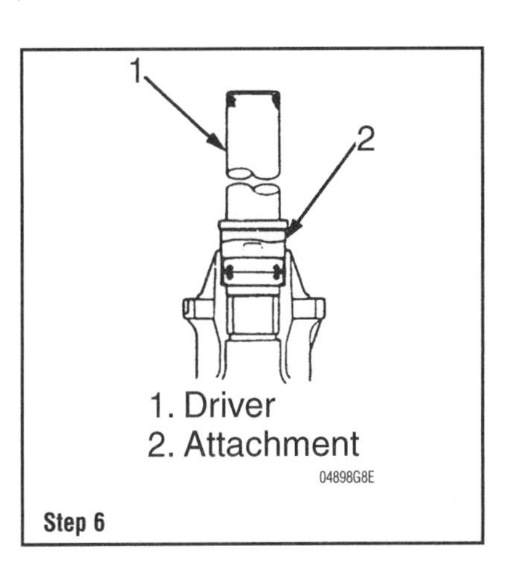

Step 6

8. To install the needle bearing, obtain the following special tools:
- Driver—07749-0010000
- Attachment (28 x 30 mm)—07946-1870100
- Pilot (22 mm)—07746-0041000

9. Using the special tools, drive the needle bearing firmly into the bearing carrier from the propeller side.
10. Install the reverse gear shim/s into the gear case.
11. Install the bearing carrier assembly into the gear case with the "UP" mark facing up.
12. Install the claw washer by aligning the cutout and the tab with the projection and the groove of the gear case.
13. Apply grease to the end nut. Install the gear case end nut into the gear case with the "OFF" mark on the gear case end nut facing out. Turn the end nut a little counterclockwise till the threads take hold and then carefully turn the nut clockwise to tighten the nut. Turn the nut as far as it will go by hand, then tighten it further to 50.6 ft. lbs. (70 Nm) using the special tool.
14. Check to see if the claw washer tab and the end nut groove are aligned. If they aren't, tighten the nut till they are aligned. (Do not exceed 72.3 ft. lbs. (100 Nm).
15. When the washer tab and nut groove are aligned, bend the tab down to stake.
16. Install the propeller thrust washer as shown in the drawing.

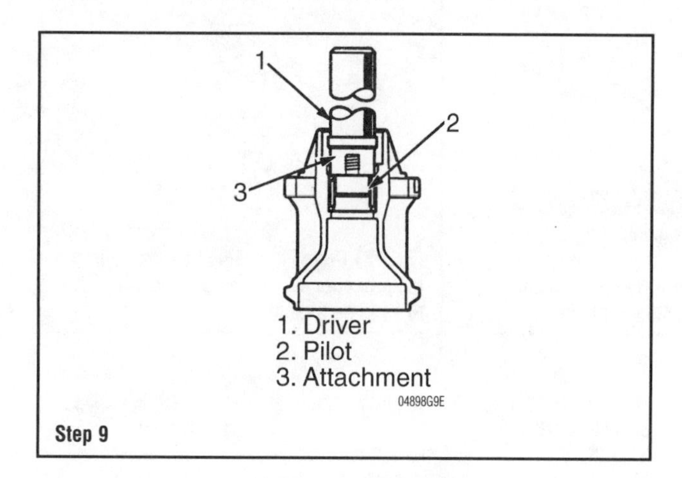

Step 9

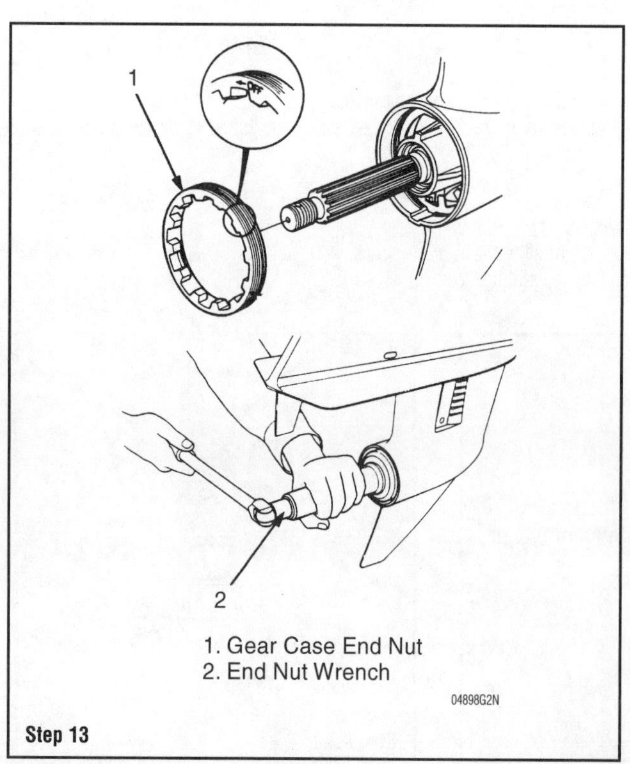

Step 13

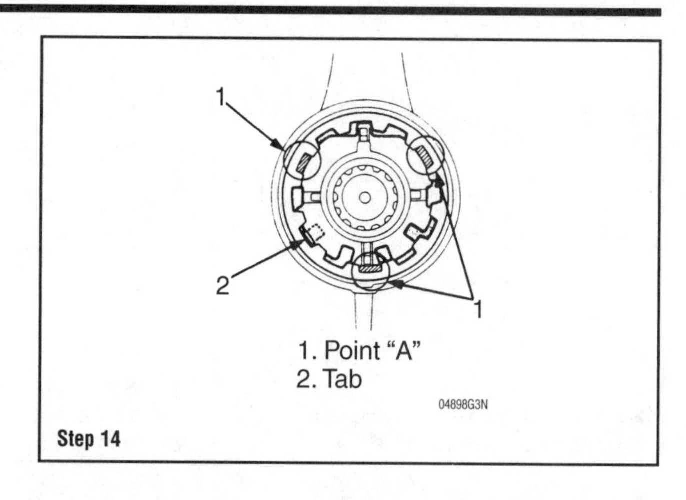

Step 14

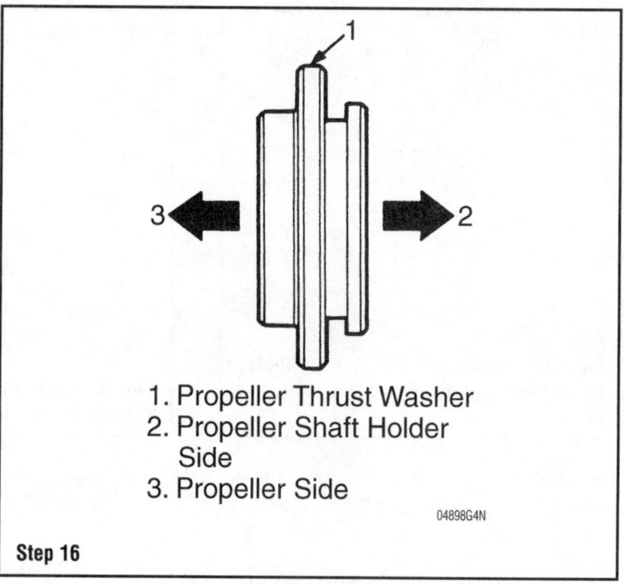

Step 16

BF75A AND BF90A

See Accompanying Illustrations

1. To install the needle bearing, obtain the following special tools:
- Driver—07749-0010000
- Attachment (32 x 35 mm)—07746-0010100 (91-15755)
- Pilot (30 mm)—07746-0040700
- Bearing installation tool—07SPD-ZW0030Z (91-13945)

2. Install the special tool on the installation side of the reverse bevel gear.
3. Apply lubricant (grease or Quicksilver 2-4-C) to the circumference of the new needle bearing.
4. Place the needle bearing in the bearing carrier with the stamp mark on the end of the needle bearing facing up.
5. Using the special tools, press the new needle bearing into the bearing carrier with a hydraulic press.
6. To install the inner water seal, obtain the following special tools:
- Oil seal driver—07SPD-ZW0040Z (91-31108)
- Bearing installation tool—07SPD-ZW0030Z (91-13945)

7. Place the bearing carrier on the special tool to protect the bearing carrier lip.
8. Apply a light coat of grease to the surface of the oil seal driver. Slide the inner water seal onto the end of the oil seal driver that has the longer shoulder.
9. Apply Loctite®271 or an equalivant thread locker around the outer edge of the seal.
10. Place the oil seal driver and seal on the bearing carrier, and using a hydraulic press, press the seal into the bearing carrier.
11. Apply grease to the seal lips.

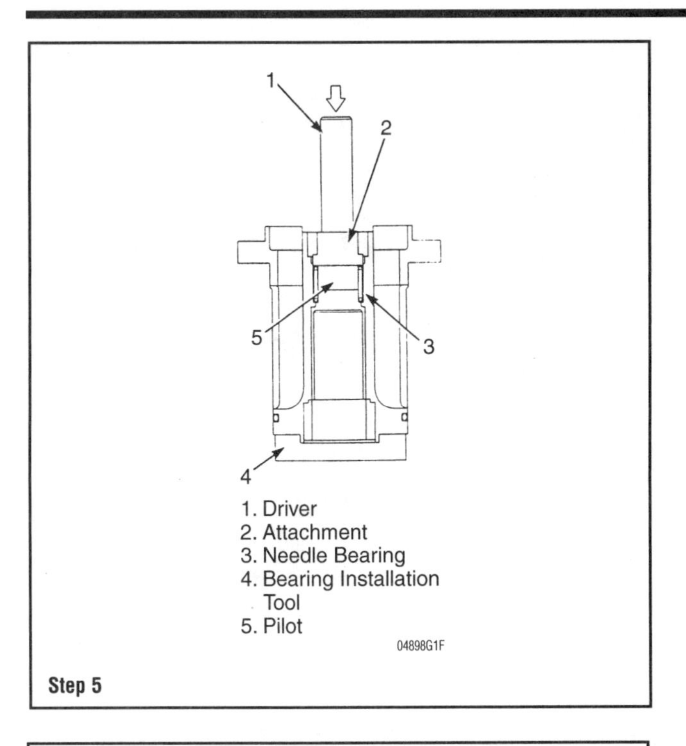

1. Driver
2. Attachment
3. Needle Bearing
4. Bearing Installation Tool
5. Pilot

04898G1F

Step 5

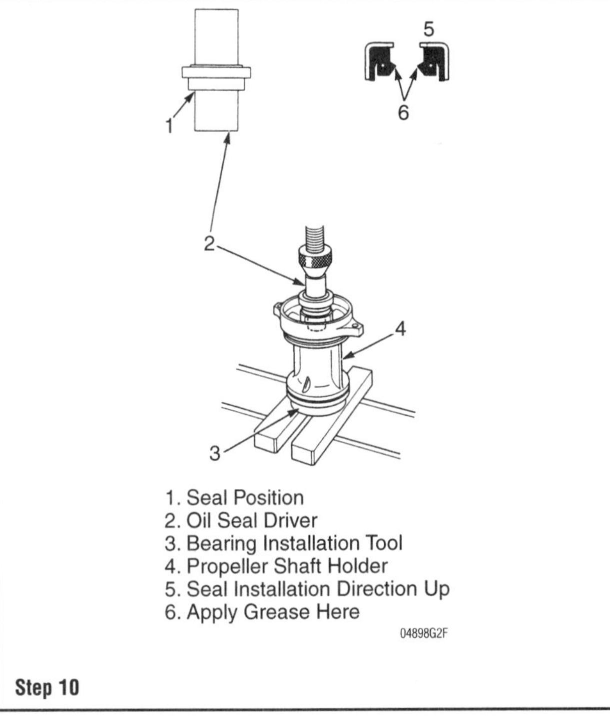

1. Seal Position
2. Oil Seal Driver
3. Bearing Installation Tool
4. Propeller Shaft Holder
5. Seal Installation Direction Up
6. Apply Grease Here

04898G2F

Step 10

12. To install the outer water seal, the procedure and special tools are the same with the following exception:

- Slide the seal onto the end of the special tool which has the shorter shoulder.

➡Make sure the seals are installed in the correct positions.

13. To install the roller bearing, obtain the following special tools:

- Bearing installation tool—07SPD-ZW0030Z (91-13945)

14. Apply grease to the outside edge of the roller bearing.

15. Using the special tool, press the bearing into the bearing carrier with a hydraulic press until it is firmly seated.

16. Prepare a piece of PVC pipe in the following dimensions:

- Length: 6.0 in. (152 mm)
- I.D.: 1.3–1.5 in. (32–38 mm).

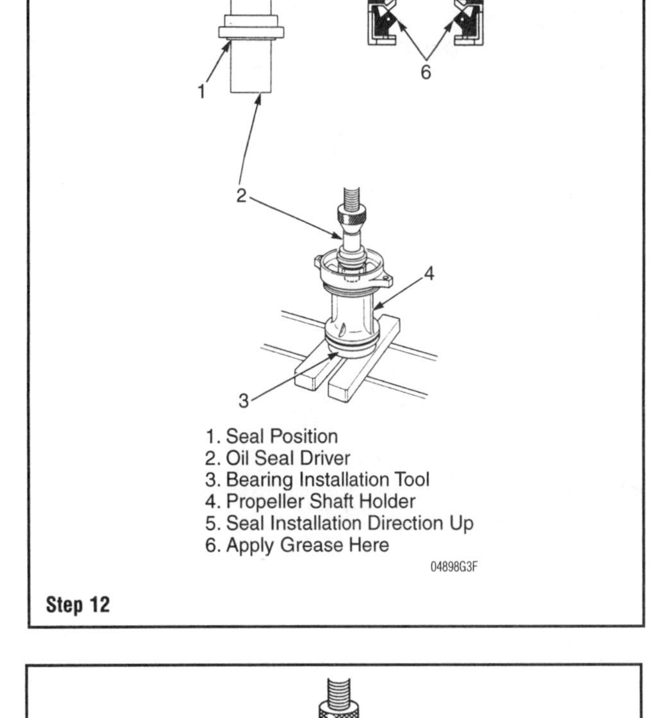

1. Seal Position
2. Oil Seal Driver
3. Bearing Installation Tool
4. Propeller Shaft Holder
5. Seal Installation Direction Up
6. Apply Grease Here

04898G3F

Step 12

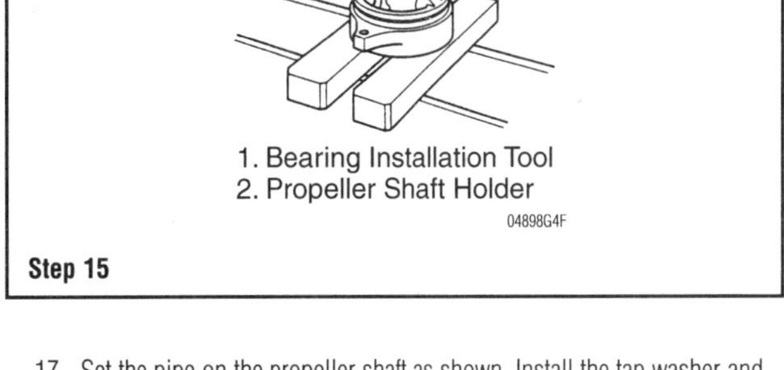

1. Bearing Installation Tool
2. Propeller Shaft Holder

04898G4F

Step 15

17. Set the pipe on the propeller shaft as shown. Install the tap washer and propeller shaft nut on the prop shaft.

18. Apply a light coating of grease to the bearing carrier O-ring and the inside surface of the gear case.

19. Install the bearing carrier into the gear case with the "TOP" mark positioned up.

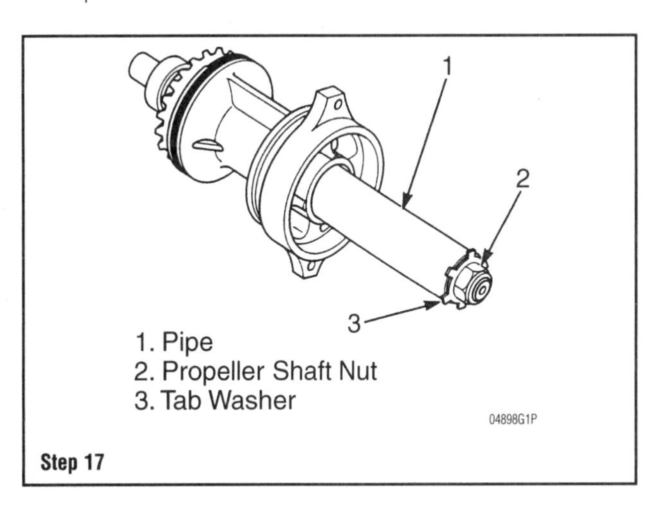

1. Pipe
2. Propeller Shaft Nut
3. Tab Washer

04898G1P

Step 17

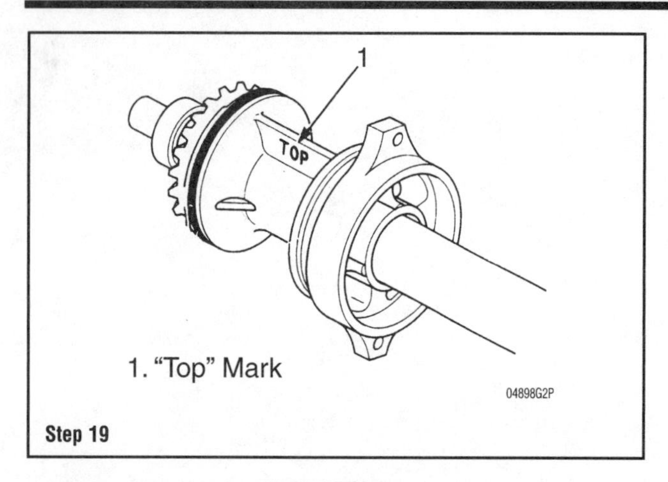

Step 19

20. Install the thrust washer with the wider side of the tapered hole facing the gear case.

➡If the bearing carrier assembly is installed without using a pipe, pull out on the propeller shaft while installing the carrier into the gear case. This will prevent the thrust bearing and thrust washer from falling out of position.

21. Apply Quicksilver 2-4-C with Teflon®to the threads and spline of the propeller shaft. Assemble the propeller shaft nut with a new tab washer and install the propeller shaft nut onto the propeller. Be sure the three tabs of the washer align with the grooves in the propeller hub.
22. Tighten the prop shaft nut to 55 ft. lbs. (75 Nm).
23. Bend the three tabs of the tab washer into the grooves of the prop hub.
24. If the three tabs do not align with the grooves, tighten the nut some more to align them and then bend the tabs.

BF115A AND BF130A

➧ See Figures 154, 155, 156 and 157

1. To install the needle bearing, obtain the following special tools:
- Driver—07749-0010000
- Attachment (37 x 40 mm)—07746-0010200
- Pilot (30 mm)—07746-0040700
2. Apply gear oil to the outer edge of the needle bearing.
3. Using a hydraulic press and the special tools, firmly press in the new needle bearing.

➡Install the needle bearing with the stamp mark at the end of the needle bearing facing the special tool.

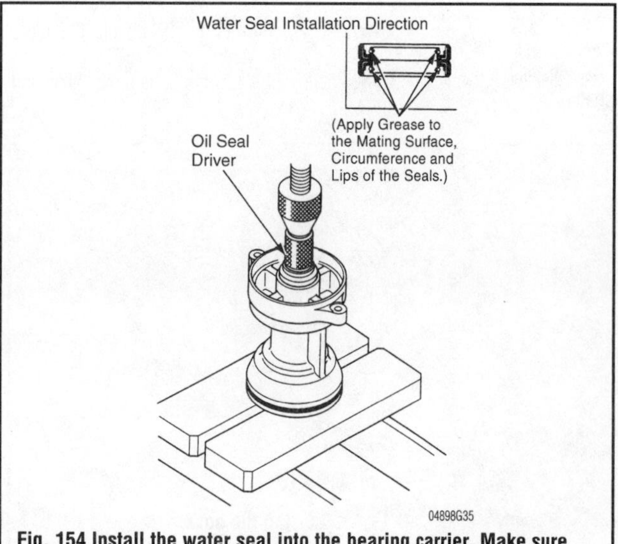

Fig. 154 Install the water seal into the bearing carrier. Make sure the direction of the seals is correct—BF115A and BF130A

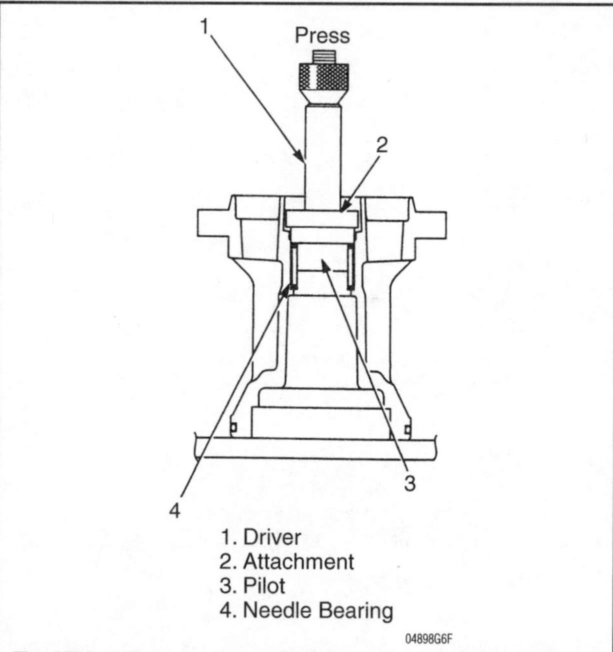

Fig. 155 Install the needle bearing into the bearing carrier. Make sure that stamp mark at the end of the bearing is facing the tool—BF115A and BF130A

4. To install the water seal, obtain the following special tools:
- Oil seal driver—07947-SB00100
5. Apply grease to the outside edge of the new seals.
6. Using a hydraulic press and the special tool, press in the new seals one after another. Make sure to install the seals back to back with the lips facing away from each other.
7. After the seal are installed, apply grease to the seal lips.
8. To install the reverse bevel gear and redial ball bearing (LA and XA types only), obtain the following special tool:
- Driver (40 mm I.D.)—07746-0030100
9. Place a wood block underneath the reverse bevel gear.

➡Check the bearing for play and unusual sounds by turning it by hand. If there is anything out of the ordinary, replace the bearing.

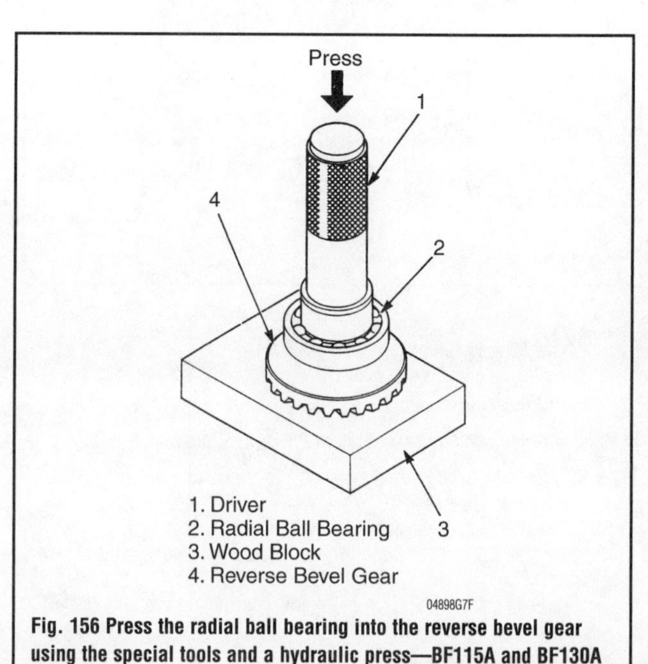

Fig. 156 Press the radial ball bearing into the reverse bevel gear using the special tools and a hydraulic press—BF115A and BF130A

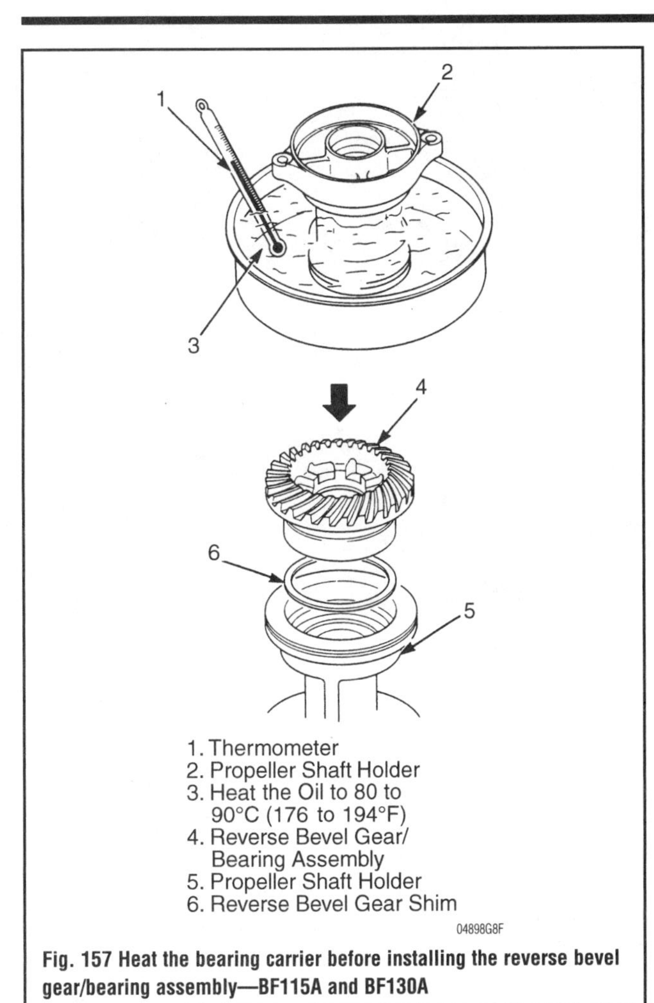

Fig. 157 Heat the bearing carrier before installing the reverse bevel gear/bearing assembly—BF115A and BF130A

10. Apply gear oil to the entire surface of the bearing. Using the special tool and a hydraulic press, press in the bearing until it seats.
11. Remove the 94 x 4.1 mm O-ring from the bearing carrier.
12. Soak the bearing carrier in a container filled with oil. Place the bearing with the bearing side down in the container.
13. Heat the oil to 176°–194°F (80°–90°C).
14. After the entire bearing carrier is heated, remove the bearing carrier from the container, and install the cold reverse bevel gear shim and the reverse gear/bearing assembly into the hot bearing carrier.

CAUTION

Do not heat the oil beyond 194°F (90°C). Always wear gloves when handling hot components during this operation.

15. Install the propeller thrust washer with the wide side of the tapered hole facing the gear case.
16. Install the propeller onto the propeller shaft.
17. Install the washer and the castellated propeller shaft nut. Tighten the nut to 0.7 ft. lbs. (1 Nm). If the cotter pin cannot be installed, turn the nut until it can be installed. Do not exceed 33 ft. lbs. (44 Nm).
18. Do not reuse the cotter pin. Always install a new stainless steel cotter pin and then bend the ends to hold the nut.

Assembling of parts at this time is not to be considered as final. The three gears are coated with the Desenex® powder, or equivalent, to determine a gear pattern. Therefore, the assemblies will be separated to check the pattern. During final installation the two mounting bolts will be coated with Loctite, or equivalent.

If the assembler has omitted the application of Desenex® powder and does not have plans to check the gear pattern, then this step may be considered as the final assembly of the bearing carrier. If such is the case, bend one or more of the lockwasher tabs down over the locknut to secure it in place.

The propeller will be installed after the gear backlash measurements have been made, the water pump installed and the lower unit attached to the intermediate housing.

Water Pump Installation

BF50 AND BF5A

See Figures 158 and 159

1. Install the impeller cover, with the new impeller gaskets on either side of it, onto the gear case. Be sure that the cover is not bent or damaged.
2. Set the distance spacers in place. Make sure the collars are installed in their proper locations.
3. Install the impeller into the impeller housing while turning it counterclockwise.

WARNING

Do not install the impeller any other way or damage to both the impeller and engine will occur.

4. Line up the flats on the impeller and pinion shaft and install the assembly onto the pinion shaft
5. Install a thread locking compound on the retaining bolts and tighten the impeller housing to the gear case.
6. Install the seal ring on the pump discharge port.

BF75, BF8A AND BF100

See Figures 160 and 161

1. Install the impeller cover, with the new impeller gaskets on either side of it, onto the gear case. Be sure that the cover is not bent or damaged.
2. Slide the impeller over the shaft, aligning the flats on the pinion shaft and the impeller.

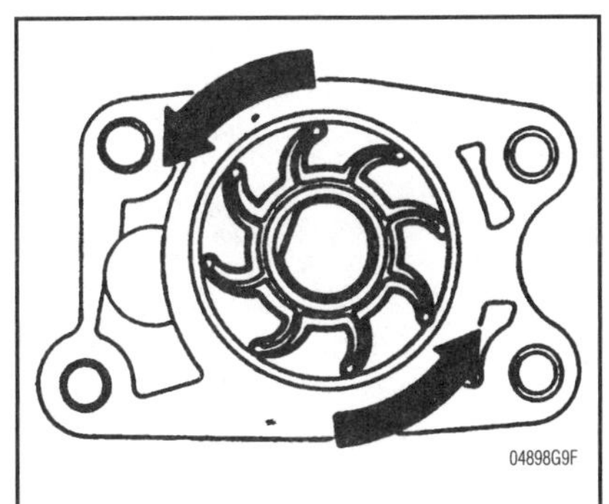

Fig. 158 Install the impeller in the pump housing by turning it counterclockwise—BF50 and BF5A

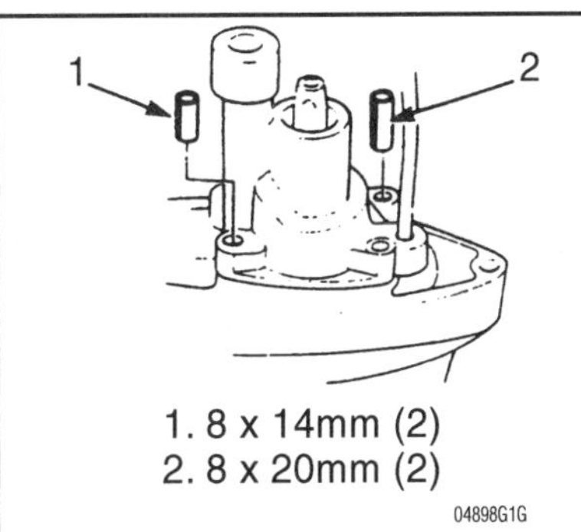

Fig. 159 Make sure the distance spacers are installed in the correct locations. They are not interchangeable—BF50 and BF5A

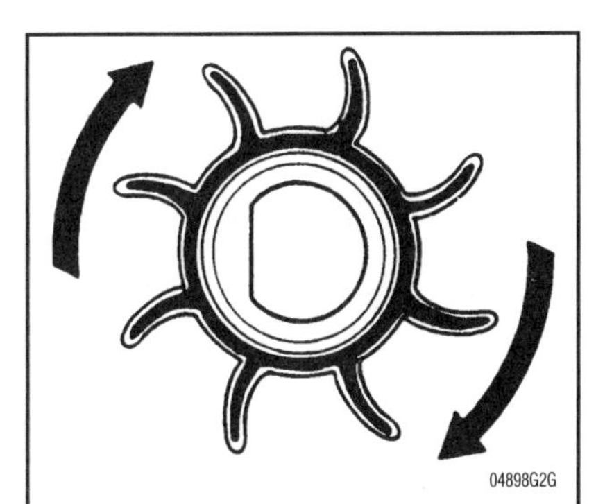

Fig. 160 Install the housing while turning the pinion shaft clockwise—BF75 BF 8A and BF100

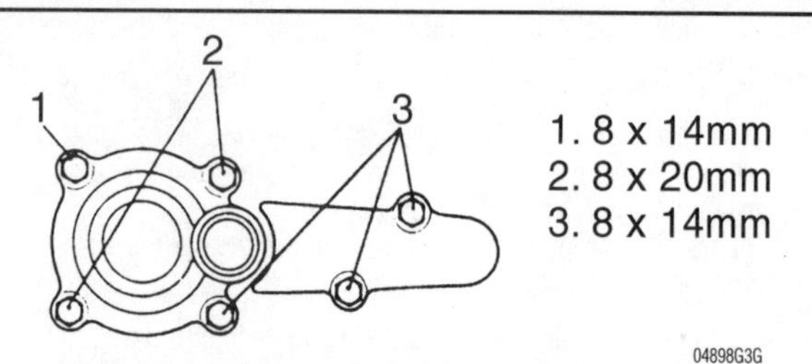

Fig. 161 Make sure the distance spacers are installed in the correct locations. They are not interchangeable—BF75 BF 8A and BF100

3. While turning the pinion shaft clockwise, install the impeller housing over the impeller.
4. Install the distance spacers in their proper locations and tighten the retaining bolts.

BF9.9A AND BF15A

➧ **See Figure 162**

1. Install the impeller cover, with the 2 new impeller gaskets on either side of it, onto the gear case. Be sure that the cover is not bent or damaged.
2. While turning the impeller counterclockwise, install it into the impeller housing.
3. Align up the flats on the pinion shaft and the impeller, and install the assembly onto the pinion shaft.
4. Install the distance spacers, then tighten the retaining bolts.
5. Install the water seal on the discharge port.

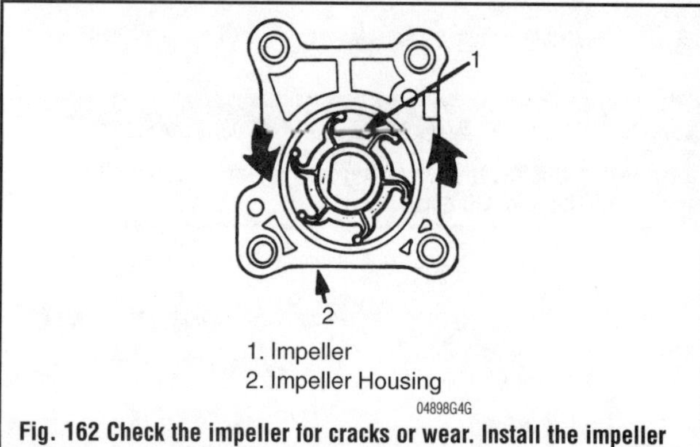

Fig. 162 Check the impeller for cracks or wear. Install the impeller while turning it counterclockwise—BF9.9A and BF15A

BF25A, BF30A, BF35A, BF40A, BF45A AND BF50A

➧ **See Figures 163 and 164**

1. To install the new upper and lower water seals, obtain the following special tools:
   - Driver—07749-0010000
   - Attachment (28 x 30 mm)—07946-1870100
2. Install the 2 new water seals in the direction shown in the drawing. (bf25 pg. 12-9)
3. Install a new bottom O-ring and lower impeller gasket.
4. Install the water pump base.
5. Install the upper impeller gasket.
6. Install the impeller cover.
7. Install the impeller into the pump liner by turning it counterclockwise. Keep the open end of the keyway toward the bottom of the engine.
8. Install the Woodruff key onto the pinion shaft.
9. With the impeller keyway aligned to the Woodruff key, install the pump liner onto the pinion shaft.
10. Install the water tube seal ring, new O-ring, impeller housing collars and impeller housing.
11. Install the retaining bolts and washers and tighten the assembly.

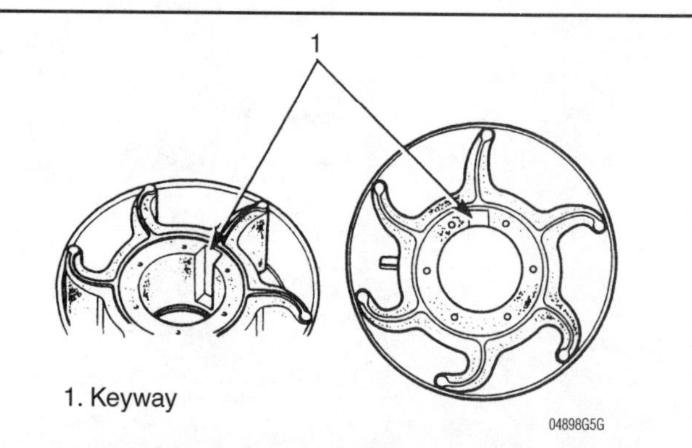

Fig. 163 Check the impeller for cracks and wear. Install the impeller by turning it counterclockwise with the open end of the keyway facing the bottom of the engine—BF25A, BF30A, BF35A, BF40A, BF45A and BF50A

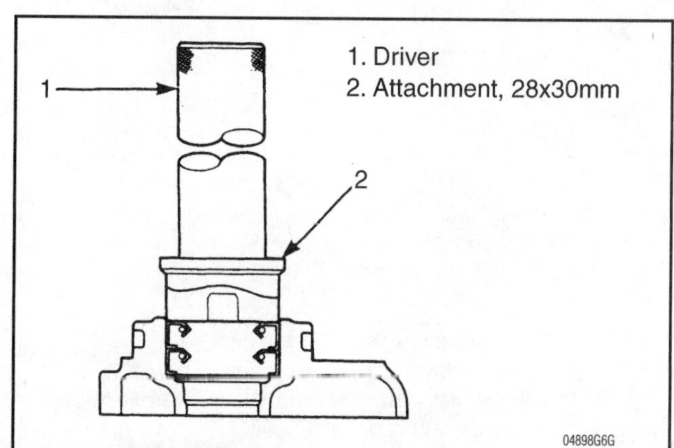

Fig. 164 Install the dual water seals in the water pump base. Make sure the seal are installed in the direction shown in the drawing—BF25A, BF30A, BF35A, BF40A, BF45A and BF50A

BF75A AND BF90A

➧ **See Figures 165, 166 and 167**

1. To install the new upper and lower water seals, obtain the following special tools:
   - Driver—07749-0010000.
   - Attachment (37 x 40 mm)—07946-0010200.
2. Install the new upper water seal first in the water pump housing. Note the direction of installation, the upper seal lip faces down and the lower seal lip faces up.
3. Apply Loctite®271 or equivalent to the circumference of the seal and grease to the seal lips.
4. Using the special tools, drive the seal into the water pump housing until it seats firmly.
5. Install the water pump housing and new water pump gasket. Apply a thread-locking agent and tighten the retaining bolts to 5.1 ft. lbs. (7 Nm)
6. Install a new impeller gasket and the impeller cover.
7. Apply grease to the inside of the impeller housing. While turning the impeller counterclockwise, install the impeller into the housing with the open end of the keyway facing up. Line up the hole in the impeller with the hole in the housing.
8. Slide the impeller housing down the pinion shaft and place the special key into position.
9. Align the keyway in the impeller with the key on the pinion shaft and install the impeller housing by turning the housing clockwise as viewed from the top of the housing.

➡ **After installation, double-check the position of the impeller gasket.**

10. Apply a thread-locking agent and tighten the retaining bolts to 5.1 ft. lbs. (7 Nm)
11. Apply beads of RTV sealant to the mating surface of the water pump housing and the gear case.

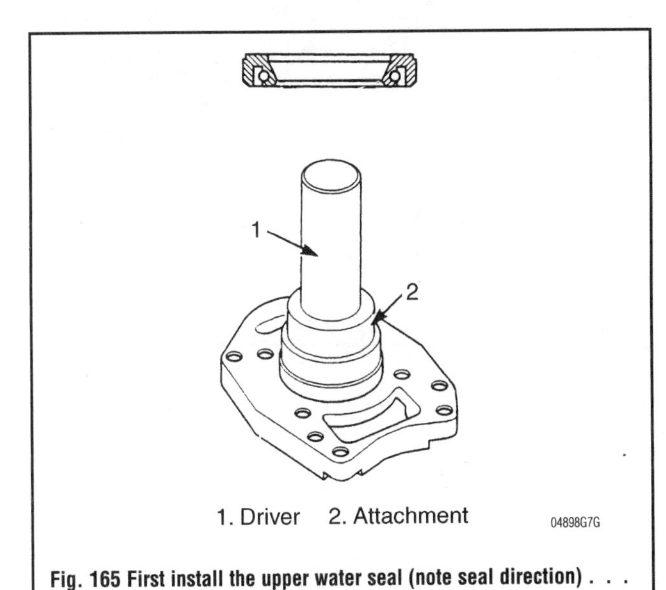

Fig. 165 First install the upper water seal (note seal direction) . . .

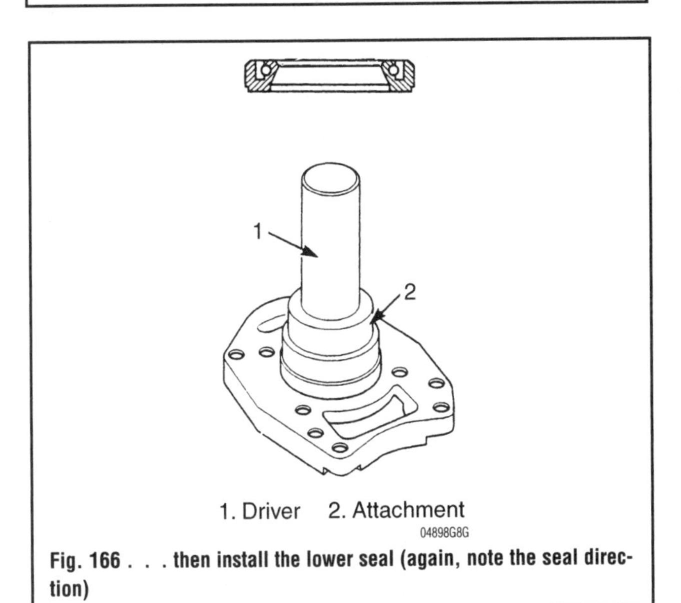

Fig. 166 . . . then install the lower seal (again, note the seal direction)

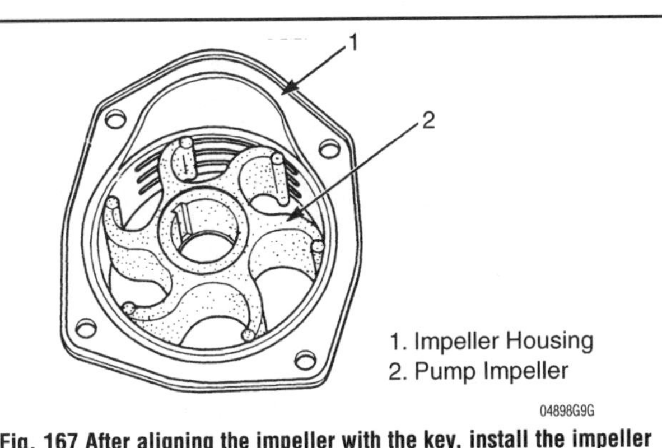

Fig. 167 After aligning the impeller with the key, install the impeller housing on the pump housing by turning the impeller housing clockwise as viewed from the top of the housing—BF75A and BF90A

BF115A AND BF130A

See Figures 168, 169 and 170

1. To install the new upper and lower water seals, obtain the following special tools:
   - Driver—07749-0010000.
   - Attachment (32 x 35 mm)—07746-0010100.
2. Apply grease to the outer circumference and lips of the new seals.
3. Using the special tools, drive the seals into the water pump housing with the lips on both seals facing down.
4. Apply grease to the inside of the pump liner. While turning the impeller counterclockwise, install the impeller into the liner with the open end of the keyway facing up. Line up the hole in the impeller with the hole in the liner.
5. Install the pump liner in the impeller housing by aligning the two projections on the line with the two grooves in the impeller housing. Be sure that the cutout section of the liner is aligned with the open end of the impeller housing.
6. Install the new O-ring and new water pump gasket.
7. Install the water pump housing.
8. Install the new impeller gasket and impeller cover over the water pump housing. Make sure to install the impeller cover with the "ZW5" mark toward the impeller.
9. Slide the impeller housing down the pinion shaft and place the special key into position.
10. Align the keyway in the impeller with the key on the pinion shaft and install the impeller housing by turning the housing clockwise as viewed from the top of the housing.

After installation, double-check the position of the impeller gasket.

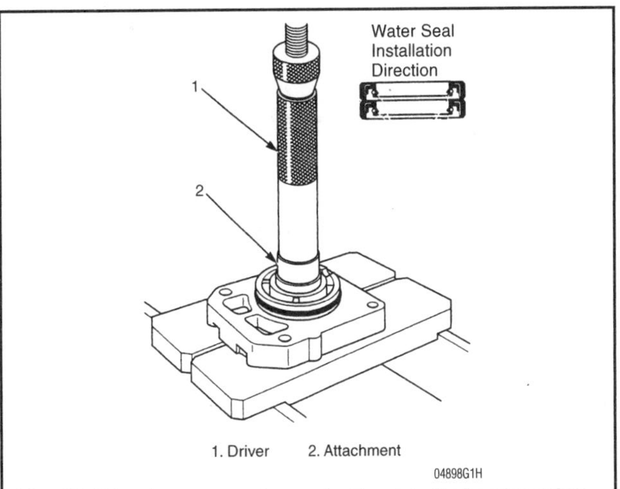

Fig. 168 Drive the water seals one at a time into the housing using the driver. Make sure they are installed in the right direction as shown—BF115A and BF130A

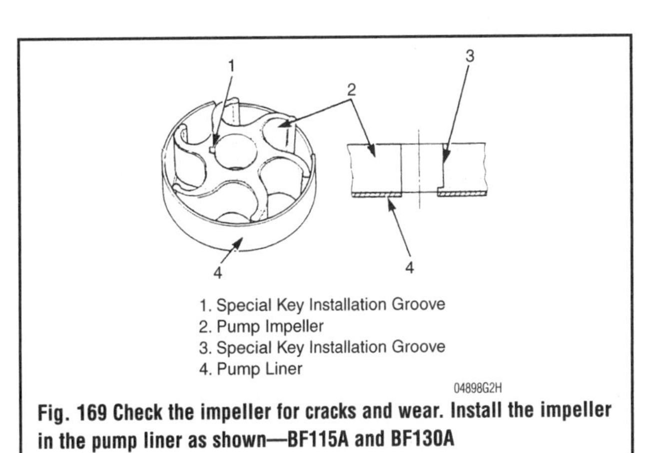

Fig. 169 Check the impeller for cracks and wear. Install the impeller in the pump liner as shown—BF115A and BF130A

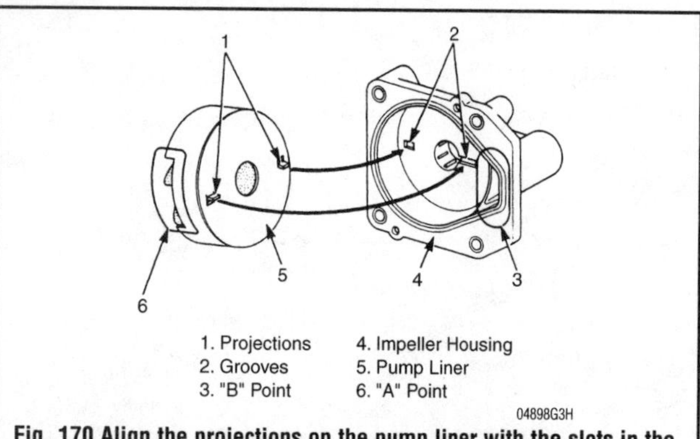

Fig. 170 Align the projections on the pump liner with the slots in the impeller housing as shown—BF115A and BF130A

11. Install the distance spacers and tighten the retaining bolts to 14 ft. lbs. (19.7 Nm).
12. Apply grease to the inside of the seal ring on the pump discharge port.

SHIMMING PROCEDURE

To ensure the proper adjustment of the bevel gear and pinion, different size shims and thrust bearing washers are available. Select the proper shims and washers according to the factory markings inside the gear case. If a new gear case is installed, refer to these codes and select the correct shim and pinion washer.

BF50 and BF5A

➧ See Figures 171, 172 and 173

Select the shims and washers according to the identification marks on the gear case when replacing the gear case.

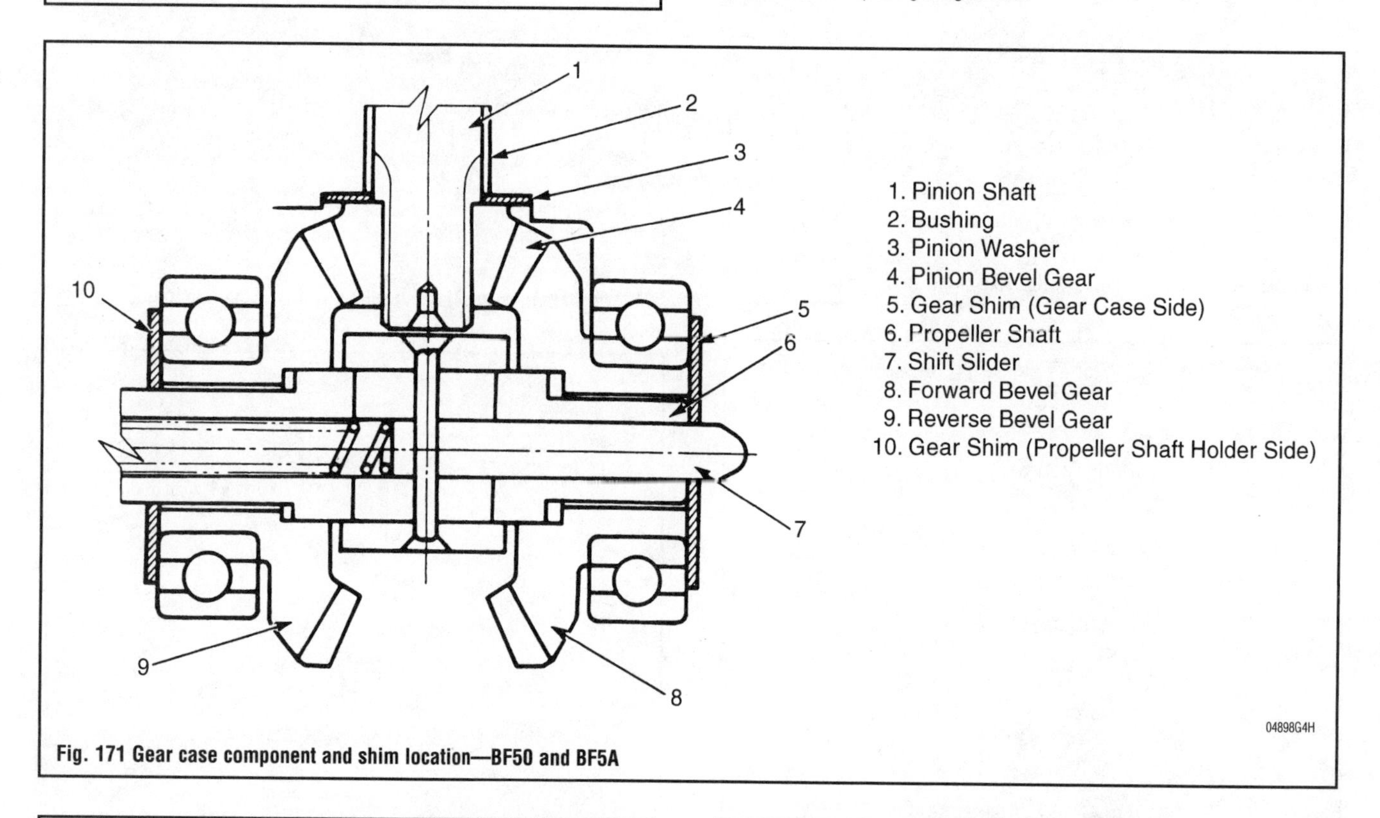

Fig. 171 Gear case component and shim location—BF50 and BF5A

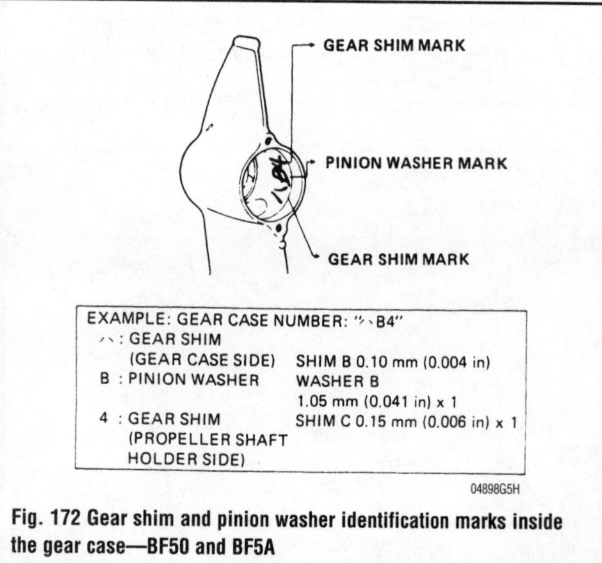

Fig. 172 Gear shim and pinion washer identification marks inside the gear case—BF50 and BF5A

Gear Shim (Gear Case Side)

CODE	SHIM	SHIM Q'TY
イ	NOT NECESSARY	–
ロ	SHIM A 0.05 mm (0.002 in)	1
ハ	SHIM B 0.10 mm (0.004 in)	1
ニ	SHIM C 0.15 mm (0.006 in)	1

Pinion Washer

CODE	WASHER	WASHER Q'TY
A	1.00 mm (0.039 in)	1
B	1.05 mm (0.041 in)	1

Gear Shim (Propeller Shaft Holder Side)

CODE	SHIM	SHIM Q'TY
1	NOT NECESSARY	–
2	SHIM A 0.05 mm (0.002 in)	1
3	SHIM B 0.10 mm (0.004 in)	1
4	SHIM C 0.15 mm (0.006 in)	1

04898G6H

Fig. 173 Gear shim and pinion washer tables—BF50 and BF5A

BF75 and BF8A

See Figures 174, 175 and 176

Select the shims and washers according to the identification marks on the gear case when replacing the gear case.

BF100, BF9.9A and BF15A

See Figures 177, 178 and 179

Select the shims and washers according to the identification marks on the gear case when replacing the gear case.

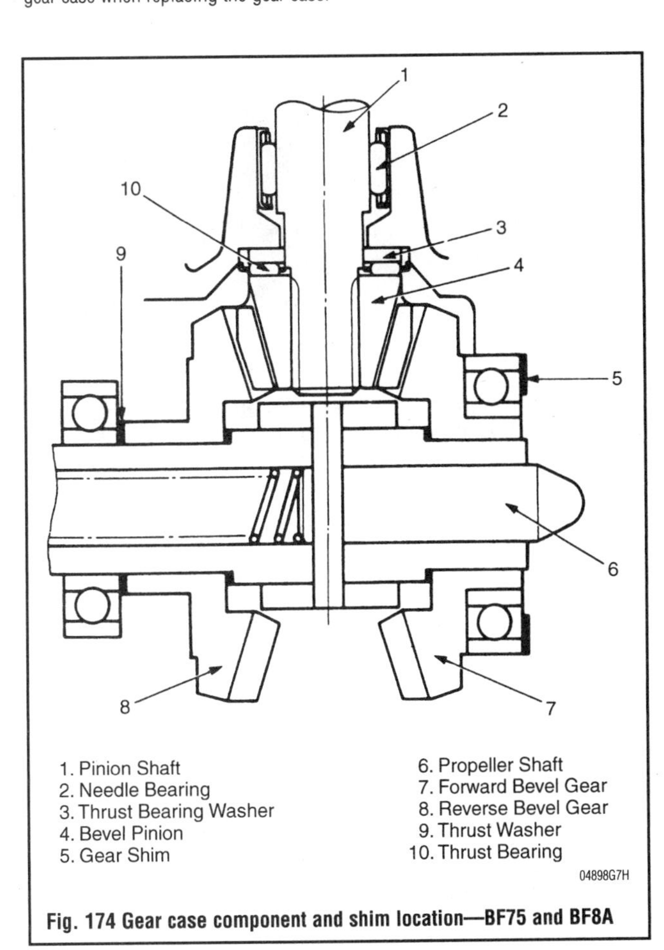

Fig. 174 Gear case component and shim location—BF75 and BF8A

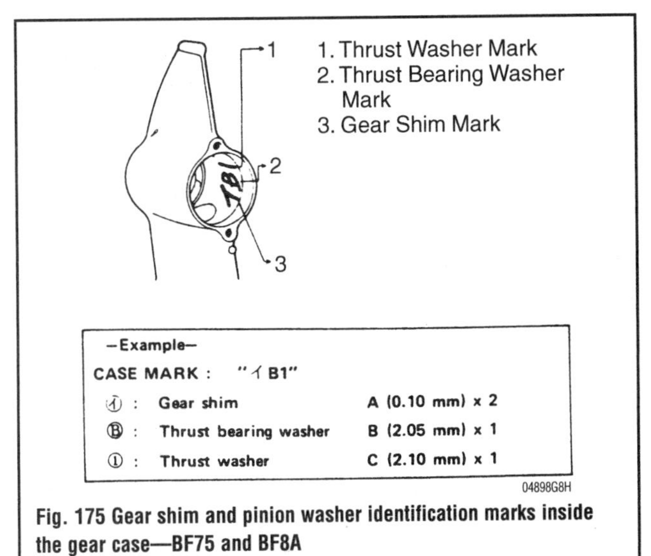

Fig. 175 Gear shim and pinion washer identification marks inside the gear case—BF75 and BF8A

GEAR SHIM

CASE MARK	SHIM	SIZE
イ	A+A	A: 0.10mm(one)
ロ	A+B	B: 0.15mm(one)

THRUST BEARING WASHER

CASE MARK	WASHER	SIZE
A	A	A: 2.00mm(one)
B	B	B: 2.05mm(one)
C	C	C: 2.10mm(one)

THRUST WASHER

CASE MARK	WASHER	SIZE
1	C	C: 2.10mm(one)
2	D	D: 2.15mm(one)

04898G9H

Fig. 176 Gear shim and pinion washer tables—BF75 and BF8A

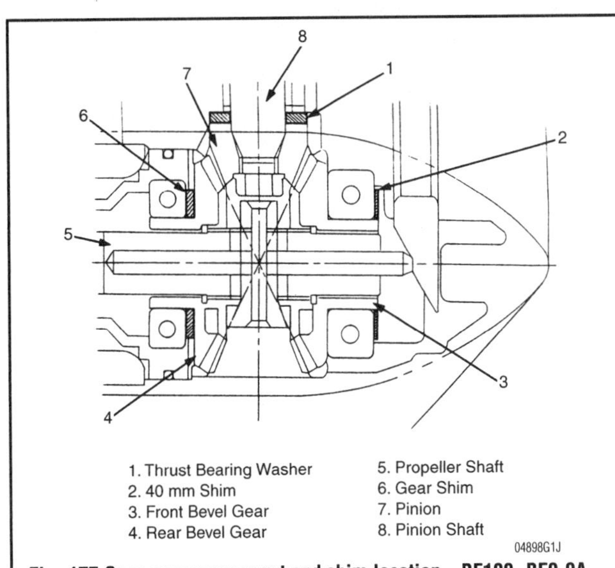

Fig. 177 Gear case component and shim location—BF100, BF9.9A and BF15A

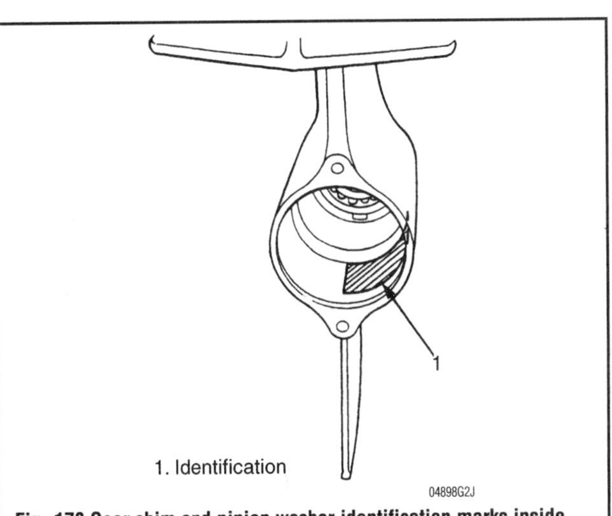

Fig. 178 Gear shim and pinion washer identification marks inside the gear case—BF100, BF9.9A and BF15A

Thrust Bearing Washer

Identification	Size	No. of washer
A	3.00 mm	1
B	3.03 mm	1
C	3.06 mm	1
D	3.09 mm	1

40 mm Shim

Identification	Size	No. of shim
イ	0.10 mm	1
ロ	0.12 mm	1
ハ	0.15 mm	1
ニ	0.08 mm	1
	0.10 mm	1

Gear Shim

Identification	Size	No. of shim
1	0.10 mm	1
2	0.15 mm	1

04898G3J

Fig. 179 Gear shim and pinion washer tables—BF100, BF9.9A and BF15A

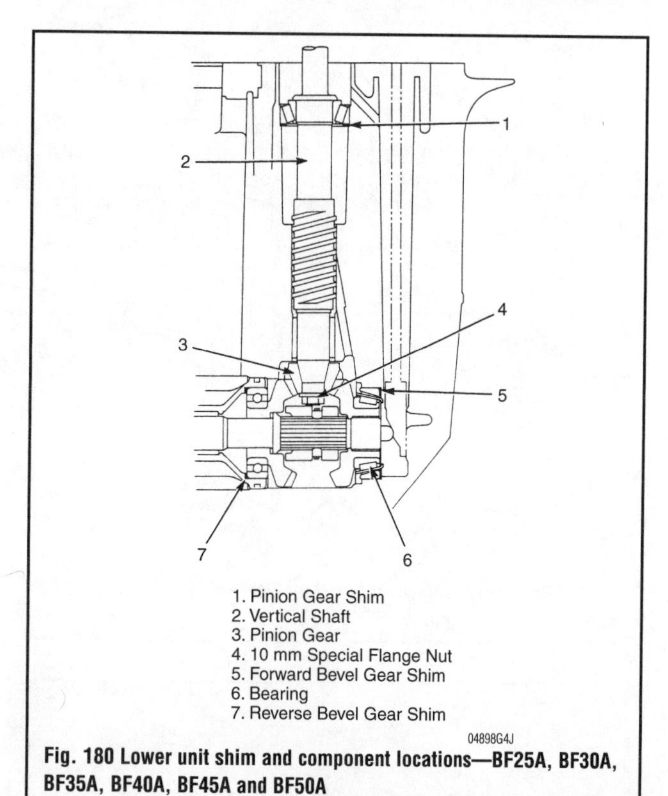

Fig. 180 Lower unit shim and component locations—BF25A, BF30A, BF35A, BF40A, BF45A and BF50A

BF25A, BF30A, BF35A, BF40A, BF45A and BF50A

➧ See Figures 180, 181, 182 and 183

When replacing any parts in the lower unit, refer to the Lower Unit Shimming Guide.

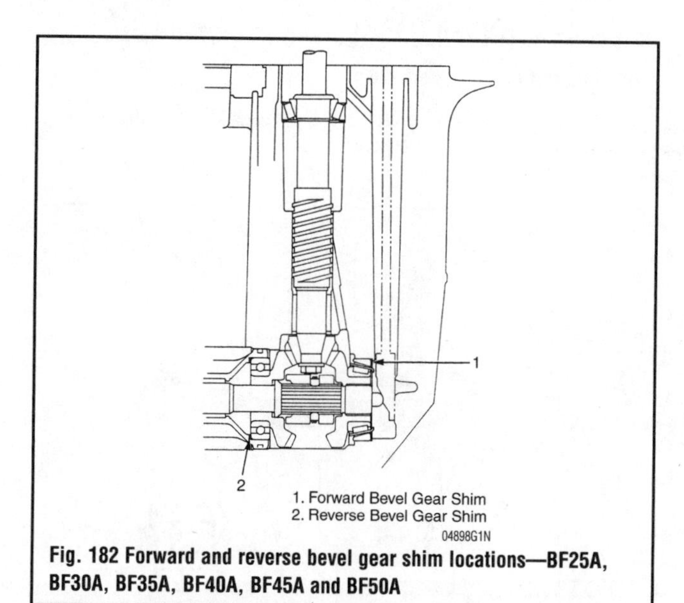

Fig. 182 Forward and reverse bevel gear shim locations—BF25A, BF30A, BF35A, BF40A, BF45A and BF50A

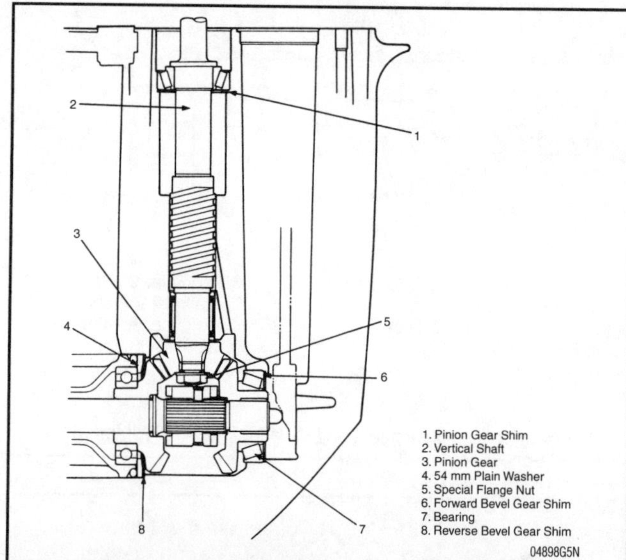

Fig. 183 Gear case shim and component locations—BF25A, BF30A, BF35A, BF40A, BF45A and BF50A

LOWER UNIT SHIMMING GUIDE

Legend	
Operation	Code
Pinion Gear Shim	P
Forward Bevel Gear Shim	F
Reverse Bevel Gear Shim	R
Backlash Check-Adjust	BL

REPLACEMENT PART	SHIMMING OPERATION REQUIRED
Gear Case	P, F, R, and BL
Vertical Shaft, Bearing, and or Pinion Gear	P and BL
Forward Bevel Gear Bearing	F and BL
Forward or Reverse Bevel Gear	BL
Propeller Shaft Holder and or Bearing(s)	BL

04898G5J

Fig. 181 Lower Unit Shimming Guide—BF25A, BF30A, BF35A, BF40A, BF45A and BF50A

PINION GEAR SHIM SELECTION

➧ See Accompanying Illustrations

To make the pinion gear shim selection, you must first obtain the following special tool:

- Pinion shaft gauge—07PPJ-ZV70100

1. Clean and lightly oil the all the components. Install the inner race of the 25 x 47 x 15 mm bearing on the pinion shaft. Place the outer bearing race over the inner race. Position the gauge end of the special tool over the pinion shaft and slide the knurled cap over the pinion shaft.
2. Hold the gauge end of the special tool, then hand tighten the knurled cap onto the gauge end securely. After tightening, there should be no side play at the measuring end.
3. Install the pinion gear on the pinion shaft, aligning the spline. Install and torque the 10 mm flange bolt to 28.9 ft. lbs. (40 Nm) or the 12 mm flange bolt to 54.2 ft. lbs. (75 Nm) depending on your model.
4. Measure the clearance between the bottom of the pinion gear and the special tool. Then obtain the calculation value using the following formula:
   - Clearance—0.5 mm = calculation value
   - Example;
   - When clearance is 0.3 mm.
   - 0.3 mm—0.5 mm = 0.2 mm
   - Therefore, the calculation value is -0.2 mm (-0.008 in.)
5. Cross reference the calculation value and engagement mark located on the bottom water screen section of the gear case.
6. Gear case marks for the BF25A, BF30A models.
7. Gear case mark for the BF35A, BF40A, BF45A, BF50A.
8. Pick the shim from the shim type table.
9. Select the shim of the appropriate thickness from the pinion gear shim selection tables.

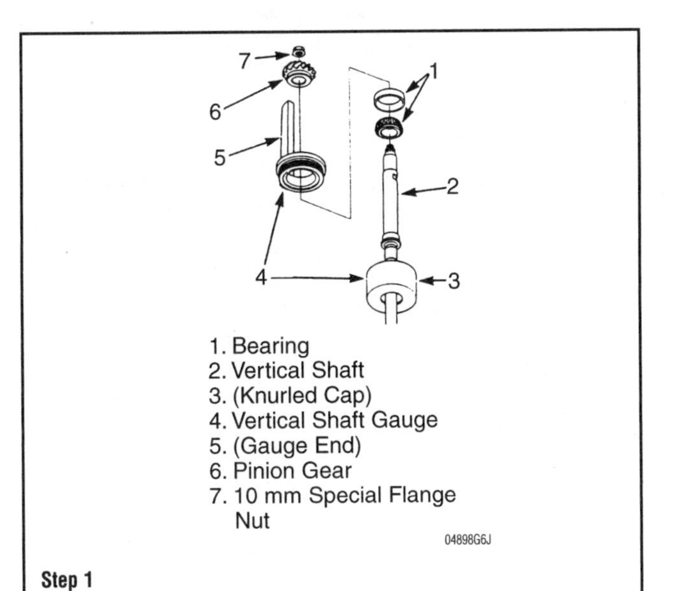

1. Bearing
2. Vertical Shaft
3. (Knurled Cap)
4. Vertical Shaft Gauge
5. (Gauge End)
6. Pinion Gear
7. 10 mm Special Flange Nut

04898G6J

Step 1

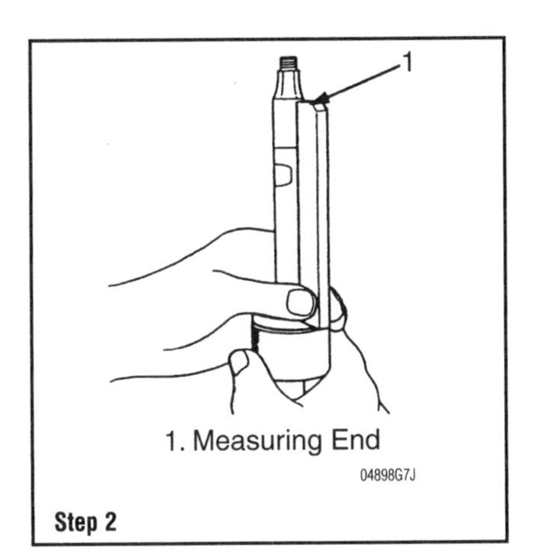

1. Measuring End

04898G7J

Step 2

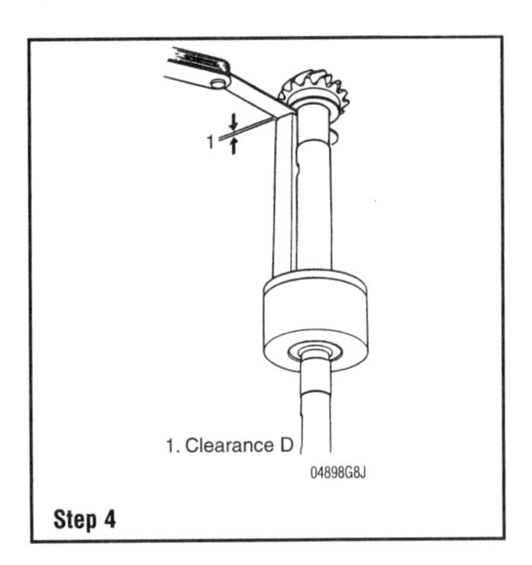

1. Clearance D

04898G8J

Step 4

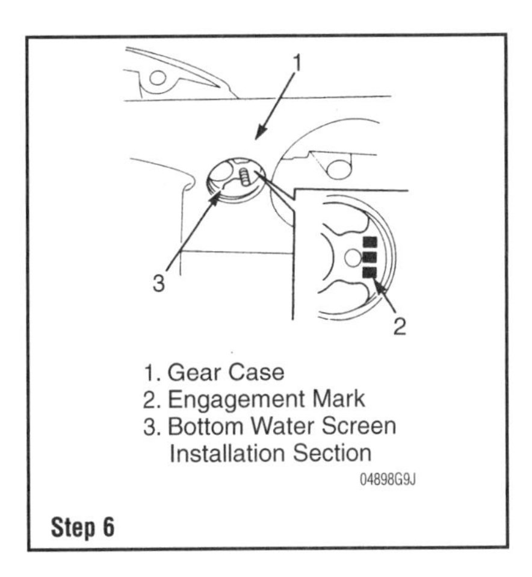

1. Gear Case
2. Engagement Mark
3. Bottom Water Screen Installation Section

04898G9J

Step 6

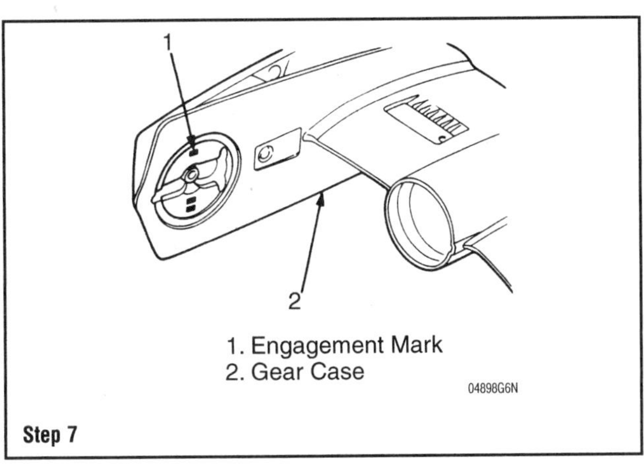

1. Engagement Mark
2. Gear Case

04898G6N

Step 7

Shim Type Table

Parts name	Parts number	Thickness
Gear shim A	90518–ZV5–000	0.10 mm (0.0039 in)
Gear shim B	90519–ZV5–000	0.15 mm (0.0060 in)
Gear shim C	90520–ZV5–000	0.30 mm (0.0118 in)
Gear shim D	90521–ZV5–000	0.50 mm (0.0197 in)

04898G1K

Step 8

Pinion Gear Shim Selection Tables

Unit: mm (in)

Engagement mark on gear case	(+) Calculation value +0.45~0.00 mm (+0.018~0.00 in)								
	0.45~0.40 (0.018~0.016)	0.40~0.35 (0.016~0.014)	0.35~0.30 (0.014~0.012)	0.30~0.25 (0.012~0.010)	0.25~0.20 (0.010~0.008)	0.20~0.15 (0.008~0.006)	0.15~0.10 (0.006~0.004)	0.10~0.05 (0.004~0.002)	0.05~0.00 (0.002~0.000)
A	1.20 (0.047)	1.15 (0.045)	1.10 (0.043)	1.05 (0.041)	1.00 (0.039)	0.95 (0.037)	0.90 (0.035)	0.85 (0.033)	0.80 (0.031)
B	1.15 (0.045)	1.10 (0.043)	1.05 (0.041)	1.00 (0.039)	0.95 (0.037)	0.90 (0.035)	0.85 (0.033)	0.80 (0.031)	0.75 (0.030)
C	1.10 (0.043)	1.05 (0.041)	1.00 (0.039)	0.95 (0.037)	0.90 (0.035)	0.85 (0.033)	0.80 (0.31)	0.75 (0.030)	0.70 (0.028)
D	1.05 (0.041)	1.00 (0.039)	0.95 (0.037)	0.90 (0.035)	0.85 (0.033)	0.80 (0.031)	0.75 (0.030)	0.70 (0.028)	0.65 (0.026)
E	1.00 (0.039)	0.95 (0.037)	0.90 (0.035)	0.85 (0.033)	0.80 (0.031)	0.75 (0.030)	0.70 (0.028)	0.65 (0.026)	0.60 (0.024)
F	0.95 (0.037)	0.90 (0.035)	0.85 (0.033)	0.80 (0.031)	0.75 (0.030)	0.70 (0.028)	0.60 (0.026)	0.60 (0.024)	0.55 (0.022)

Unit:mm (in)

Engagement mark on gear case	(–) Calculation value 0.00~-0.45 mm (0.000~-0.018 in)								
	0.00~-0.05 (0.000~-0.002)	-0.05~-0.10 (-0.002~-0.004)	-0.10~-0.15 (-0.004~-0.006)	-0.15~-0.20 (-0.006~-0.008)	-0.20~-0.25 (-0.008~-0.010)	-0.25~-0.30 (-0.010~-0.012)	-0.30~-0.35 (-0.012~-0.014)	-0.35~-0.40 (-0.014~-0.016)	-0.40~-0.45 (-0.016~-0.018)
A	0.75 (0.030)	0.70 (0.028)	0.65 (0.026)	0.60 (0.024)	0.55 (0.022)	0.50 (0.020)	0.45 (0.018)	0.40 (0.016)	0.35 (0.014)
B	0.70 (0.028)	0.65 (0.026)	0.60 (0.024)	0.55 (0.022)	0.50 (0.020)	0.45 (0.018)	0.40 (0.016)	0.35 (0.014)	0.30 (0.012)
C	0.65 (0.026)	0.60 (0.024)	0.55 (0.022)	0.50 (0.020)	0.45 (0.018)	0.40 (0.016)	0.35 (0.14)	0.30 (0.012)	0.25 (0.010)
D	0.60 (0.024)	0.55 (0.022)	0.50 (0.020)	0.45 (0.018)	0.40 (0.016)	0.35 (0.014)	0.30 (0.012)	0.25 (0.010)	0.20 (0.008)
E	0.55 (0.022)	0.50 (0.020)	0.45 (0.018)	0.40 (0.016)	0.35 (0.014)	0.30 (0.012)	0.25 (0.010)	0.20 (0.008)	0.15 (0.006)
F	0.50 (0.020)	0.45 (0.018)	0.40 (0.016)	0.35 (0.014)	0.30 (0.012)	0.25 (0.010)	0.20 (0.008)	0.15 (0.006)	0.10 (0.004)

- **How to read shim selection table and shim combination:**
  When the engagement mark on the gear case is B and the calculation value is -0.20 mm (-0.008 in), the shim thickness is 0.55 mm (0.022 in).
  After determining the required shim thickness, refer to the above shim type table; the correct combination of gear shims should be A, B and C to bring the total thickness of the shim to 0.55 mm (0.022 in).

Step 9 - NOTE: for 1999 and later BF75A/BF90A motors, please refer to the pinion gear shim selection table in the Supplement

Foward Gear Shim Selection

◆ **See the Accompanying Illustrations**

1. Make your shim selection from the Forward Gear Shim Selection Table

2. To make your forward gear shim selection, you must obtain the following special tool:

- Bearing height gauge-07LPJ-ZV30200

3. Clean the special tool and bearing. Set the bearing inner race on the

- **Forward gear shim selection table**

Unit: mm (in)

Engagement mark on gear case	(+) Calculation value +0.20~0.00 mm (+0.008~0.000 in)				(–) Calculation value 0.00~-0.20 mm (0.000~-0.008 in)			
	0.20~0.15 (0.008~0.006)	0.15~0.10 (0.006~0.004)	0.10~0.05 (0.004~0.002)	0.05~0.00 (0.002~0.000)	0.00~-0.05 (0.000~-0.002)	-0.05~-0.10 (-0.002~-0.004)	-0.10~-0.15 (-0.004~-0.006)	-0.15~-0.20 (-0.006~-0.008)
1	0.40 (0.016)	0.45 (0.018)	0.50 (0.020)	0.55 (0.022)	0.60 (0.024)	0.65 (0.026)	0.70 (0.028)	0.75 (0.030)
2	0.35 (0.014)	0.40 (0.016)	0.45 (0.018)	0.50 (0.020)	0.55 (0.022)	0.60 (0.024)	0.65 (0.026)	0.70 (0.028)
3	0.30 (0.012)	0.35 (0.014)	0.40 (0.016)	0.45 (0.018)	0.50 (0.020)	0.55 (0.022)	0.60 (0.024)	0.65 (0.026)
4	0.25 (0.010)	0.30 (0.012)	0.35 (0.014)	0.40 (0.016)	0.45 (0.018)	0.50 (0.020)	0.55 (0.022)	0.60 (0.024)
5	0.20 (0.008)	0.25 (0.010)	0.30 (0.012)	0.35 (0.014)	0.40 (0.016)	0.45 (0.018)	0.50 (0.020)	0.55 (0.022)
6	0.15 (0.006)	0.20 (0.008)	0.25 (0.010)	0.30 (0.012)	0.35 (0.014)	0.40 (0.016)	0.45 (0.018)	0.50 (0.020)

Step 1 - NOTE: for 1999 and later BF75A/BF90A motors, please refer to the forward gear shim selection table in the Supplement

special tool base, then the outer bearing race, then the top part of the special tool.

4. Rotate the upper part of the special tool 2 full revolutions to center the bearing on the special tool.

5. Now measure the clearance between the upper and lower parts of the special tool with a feeler gauge.

6. Obtain the calculation value by subtracting 0.02 in. (0.5 mm) from the measured clearance.

- Example:
- If the clearance is 0.012 in. (0.3 mm)
- 0.3 – 0.5 = 0.2

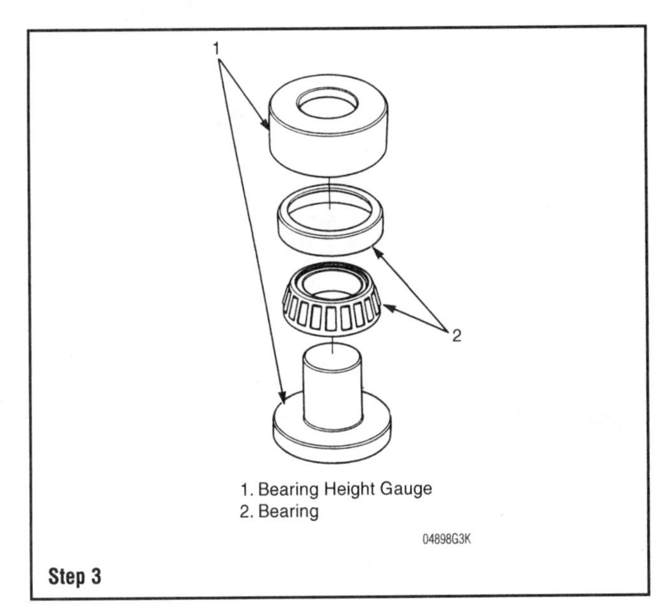

1. Bearing Height Gauge
2. Bearing

04898G3K

Step 3

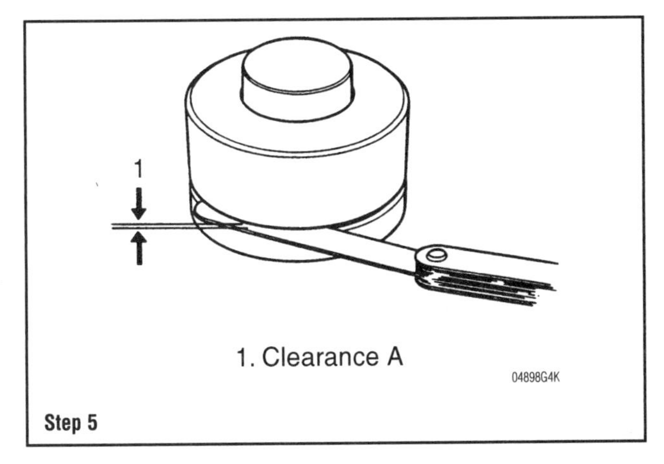

1. Clearance A

04898G4K

Step 5

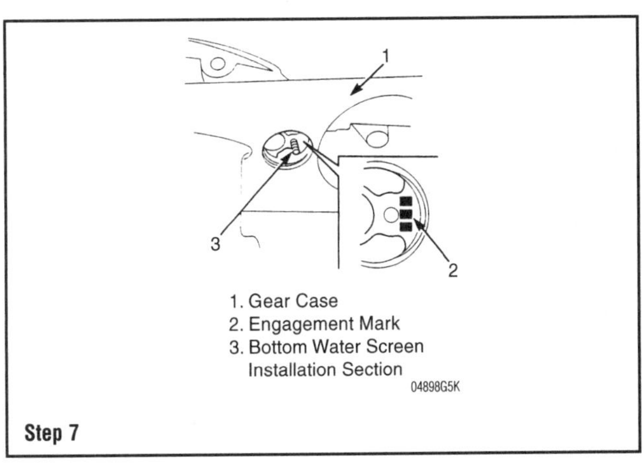

1. Gear Case
2. Engagement Mark
3. Bottom Water Screen Installation Section

04898G5K

Step 7

- Your calculation value is -0.2 mm (-0.008 in.)

7. On the BF25A and BF30A models, cross reference the calculation value and engagement mark located on the bottom water screen section of the gear case.

8. On the BF35A and BF50A models, cross reference the calculation value and engagement mark located on the bottom water screen section of the gear case.

9. Select the appropriate thickness shim from the shim selection table.

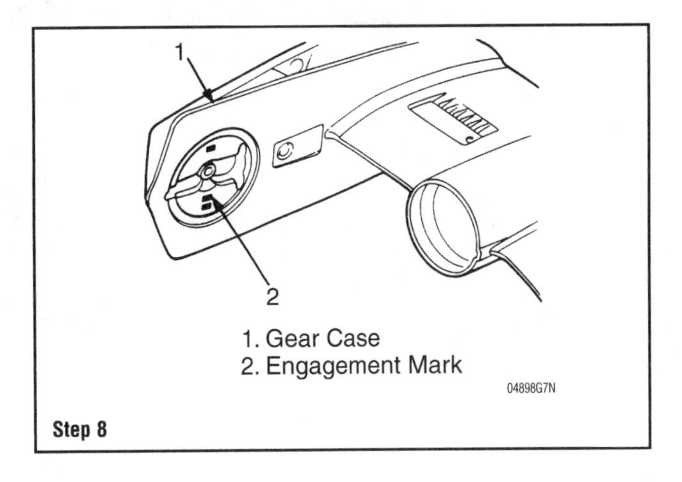

1. Gear Case
2. Engagement Mark

04898G7N

Step 8

Shim Type Table

Parts name	Parts number	Thickness
Forward gear shim A	90528–ZV5–000	0.10 mm (0.0039 in)
Forward gear shim B	90529–ZV5–000	0.15 mm (0.0060 in)
Forward gear shim C	90530–ZV5–000	0.35 mm (0.0118 in)
Forward gear shim D	90531–ZV5–000	0.50 mm (0.0197 in)

04898G2M

Step 9

REVERSE GEAR SHIM SELECTION

See Accompanying Illustrations

1. Make your shim selection from the Reverse Gear Shim Selection Table
2. Obtain the following special tool to measure the reverse gear:

- Propeller shaft gauge—07PPJ-ZV70200

3. Set the special tool on the bearing carrier.
4. Measure the clearance between the bearing carrier and the special tool.
5. Obtain the calculation value using the following formula:

- Clearance R – 0.008 in. (0.2 mm) = calculation value.
- Example:
- When clearance R is 0.012 in. (0.3 mm)
- 0.3 – 0.2 = 0.1
- Your calculation value is 0.1 mm (0.004 in.)

6. Cross reference the calculation value and engagement mark located on the bottom water screen section of the gear case.

7. Pick the appropriate thickness shim from the shim type table. **Perform the following procedure without using the special tool.**

8. Cross reference the engagement mark on the bearing carrier and the engagement mark located on the gear case underneath the bottom water screen.

9. Select the appropriate thickness shim from the reverse gear shim selection table

10. On BF35A to BF50 models, use a micrometer to measure the thickness of the 54 mm plain washer. The calculation value can be obtained by subtracting 0.2 in. (4 mm) from the washer thickness.

Reverse Gear Shim Selection Table

Unit: mm (in)

		(+) Calculation value +0.15~0.00 mm (+0.006~0.000 in)			(–) Calculation value 0.00~-0.15 mm (0.000~-0.006 in)		
		0.15~0.10 (0.006~ 0.004)	0.10~0.05 (0.004~ 0.002)	0.05~0.00 (0.002~ 0.000)	0.00~-0.05 (0.000~ -0.002)	-0.05~-0.10 (-0.002~ -0.004)	-010~-0.15 (-0.004~ -0.006)
Engagement mark on gear case	A	0.35 (0.014)	0.40 (0.016)	0.45 (0.018)	0.50 (0.020)	0.55 (0.022)	0.60 (0.024)
	B	0.30 (0.012)	0.35 (0.014)	0.40 (0.016)	0.45 (0.018)	0.50 (0.020)	0.55 (0.022)
	C	0.25 (0.010)	0.30 (0.012)	0.35 (0.014)	0.40 (0.016)	0.45 (0.018)	0.50 (0.020)
	D	0.20 (0.008)	0.25 (0.010)	0.30 (0.012)	0.35 (0.014)	0.40 (0.016)	0.45 (0.018)
	E	0.15 (0.006)	0.20 (0.008)	0.25 (0.010)	0.30 (0.012)	0.35 (0.014)	0.40 (0.016)
	F	0.10 (0.004)	0.15 (0.006)	0.20 (0.008)	0.25 (0.010)	0.30 (0.012)	0.35 (0.014)

Step 1

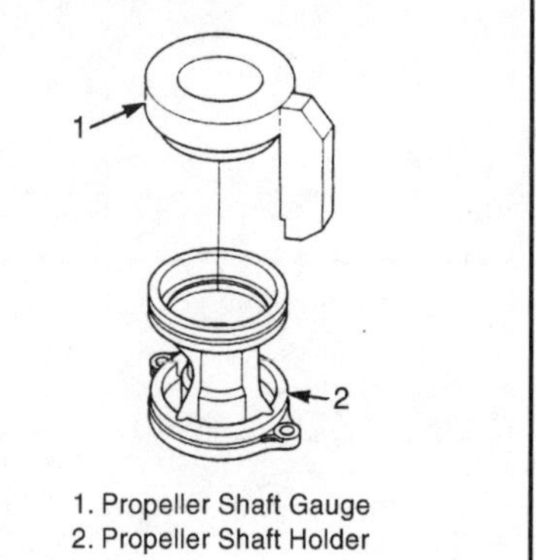

1. Propeller Shaft Gauge
2. Propeller Shaft Holder

Step 3

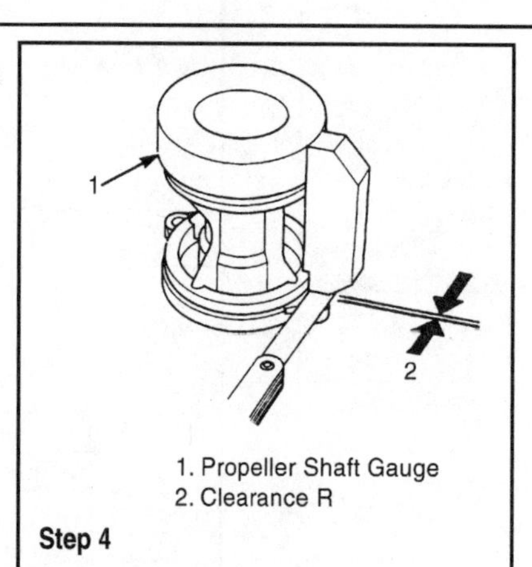

1. Propeller Shaft Gauge
2. Clearance R

Step 4

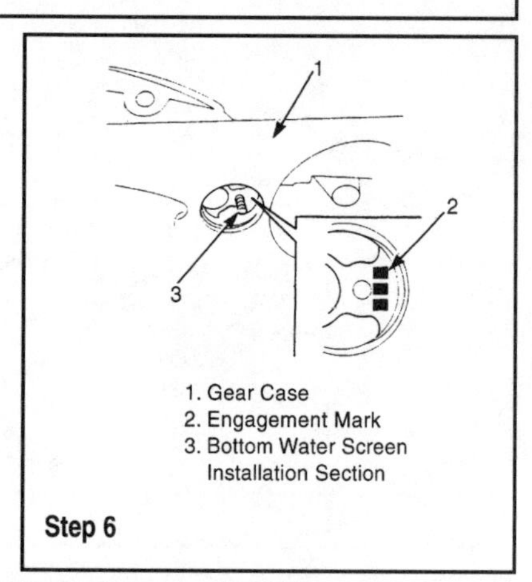

1. Gear Case
2. Engagement Mark
3. Bottom Water Screen Installation Section

Step 6

Shim Type Table

Parts name	Parts number	Thickness
gear shim A	90528-ZV5-000	0.10 mm (0.0039 in)
gear shim B	90529-ZV5-000	0.15 mm (0.0060 in)
gear shim C	90530-ZV5-000	0.35 mm (0.0118 in)
gear shim D	90531-ZV5-000	0.50 mm (0.0197 in)

Step 7

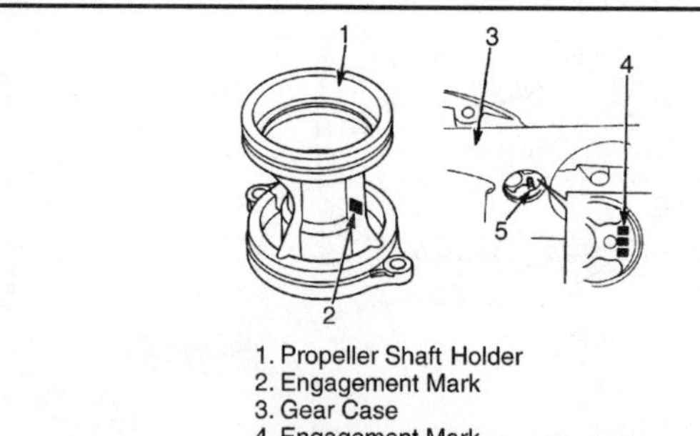

1. Propeller Shaft Holder
2. Engagement Mark
3. Gear Case
4. Engagement Mark
5. Bottom Water Screen Installation Section

Step 8

Reverse Gear Shim Selection Table

Unit: mm (in)

		Engagement mark on propeller shaft holder					
		1	2	3	4	5	6
Engagement mark on gear case	A	0.35 (0.014)	0.40 (0.016)	0.45 (0.018)	0.50 (0.020)	0.55 (0.022)	0.60 (0.024)
	B	0.30 (0.012)	0.35 (0.014)	0.40 (0.016)	0.45 (0.018)	0.50 (0.020)	0.55 (0.022)
	C	0.25 (0.010)	0.30 (0.012)	0.35 (0.014)	0.40 (0.016)	0.45 (0.018)	0.50 (0.020)
	D	0.20 (0.008)	0.25 (0.010)	0.30 (0.012)	0.35 (0.014)	0.40 (0.016)	0.45 (0.018)
	E	0.15 (0.006)	0.20 (0.008)	0.25 (0.010)	0.30 (0.012)	0.35 (0.014)	0.40 (0.016)
	F	0.10 (0.004)	0.15 (0.006)	0.20 (0.008)	0.25 (0.010)	0.30 (0.012)	0.35 (0.014)

Step 9 - NOTE: for 1999 and later BF75A/BF90A motors, please refer to the reverse gear shim selection table in the Supplement

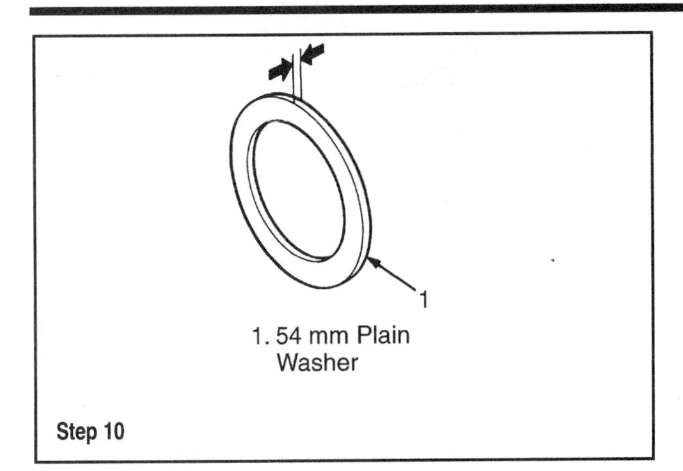

1. 54 mm Plain Washer

Step 10

Shim Type Table

Parts name	Parts number	Thickness
gear shim A	90528-ZV5-000	0.10 mm (0.0039 in)
gear shim B	90529-ZV5-000	0.15 mm (0.0060 in)
gear shim C	90530-ZV5-000	0.35 mm (0.0118 in)
gear shim D	90531-ZV5-000	0.50 mm (0.0197 in)

Step 11

11. Cross reference the washer thickness and engagement mark located on the trim tab section of the gear case, and select the appropriate thickness shim from the shim selection table.

Backlash Adjustment (Forward Gear)

◆ **See Accompanying Illustrations**

1. To check forward gear backlash, obtain the following special tools:
- J Bolt—07LPB-ZV30310
- Flywheel puller—07935-8050003
- Backlash inspection attachment—07MGJ-0010100

2. After you select the appropriate thickness shim, remove the water pump and install the following parts on the gear case.
- Pinion shaft/pinion gear
- Propeller shaft/bearing carrier
- Propeller/bearing carrier assembly

3. Tighten the bearing carrier bolts to the gear case.
4. Hold the propeller shaft securely with the special tool. Tighten the special tool to 3.6 ft. lbs. (5 Nm)
5. Set the other special tool on the pinion shaft and press the tip of the dial indicator gauge to the special tool.

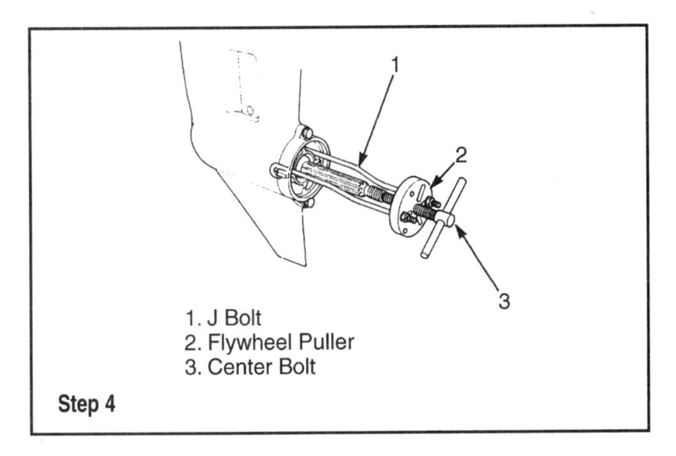

1. J Bolt
2. Flywheel Puller
3. Center Bolt

Step 4

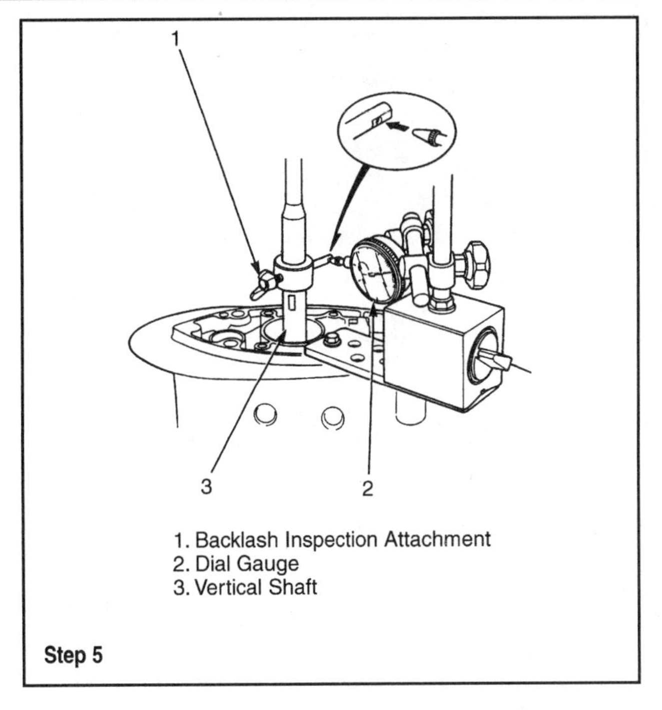

1. Backlash Inspection Attachment
2. Dial Gauge
3. Vertical Shaft

Step 5

6. Turn the pinion shaft back and forth while pressing down with approximately 11 lbs. (5 kg.) of force and the measure the backlash shown on the dial indicator. The standard measurement is 0.0043-0.0134 in. (0.11-0.34mm) for the BF35A—BF50A models and 0.004-0.011 in. (0.10-0.011mm) BF25A—BF30A models
7. If the backlash reading is too large, increase the forward gear shim thickness and recheck the backlash. If the back lash is too small, decrease the shim thickness and recheck the backlash.

Backlash Adjustment (Reverse Gear)

◆ **See Accompanying Illustrations**

1. On BF25A and BF30A, install the propeller and propeller thrust washer onto the propeller shaft and tighten the nut. Specified torque is 0.7 ft. lbs. (1 Nm) and maximum torque is 23.3 ft. lbs. (35 Nm).
2. Install the backlash inspection tool on the pinion shaft and set the dial indicator up to get a reading off it.
3. Turn the pinion shaft back and forth while pressing down with approximately 11 lbs. (5 kg) of force and measure the backlash. The standard measurement is 0.004-0.015 in. (0.10-0.39mm).
4. If the backlash reading is too large, increase the reverse gear shim thickness and recheck the backlash. If the back lash is too small, decrease the shim thickness and recheck the backlash.
5. To measure the reverse gear backlash on BF35A to BF50A, obtain the following special tools:

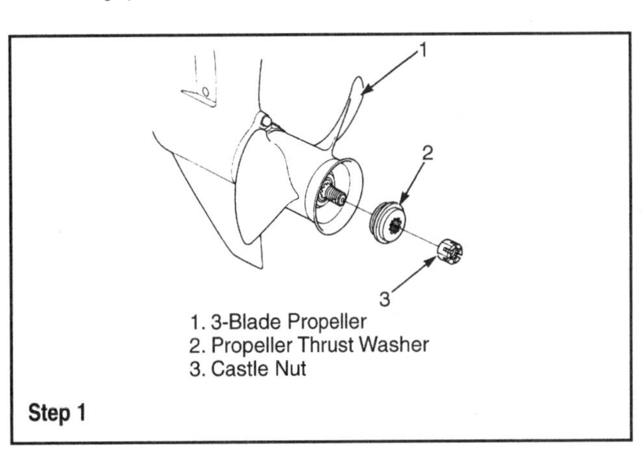

1. 3-Blade Propeller
2. Propeller Thrust Washer
3. Castle Nut

Step 1

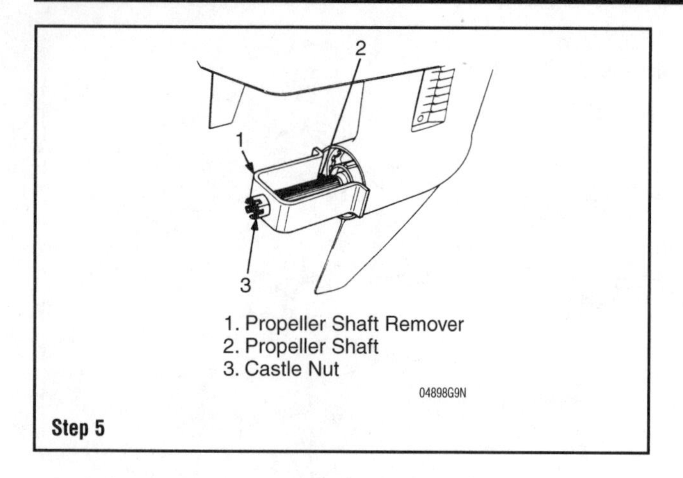

Step 5

- Propeller shaft remover—07LPC-ZV30200
- Backlash inspection attachment—07LPJ—ZV30300
- Ring gear holder—07LPB—ZV30100

6. Turn the pinion shaft back and forth while pressing down on the shaft with approximately 11 lbs. (5 kg) of force and measure the backlash. The standard backlash measurement is 0.0043—0.0134 in. (0.11—0.34 mm).

7. If the backlash reading is too large, increase the forward gear shim thickness and recheck the backlash. If the back lash is too small, decrease the shim thickness and recheck the backlash.

BF75A and BF90A

See Figure 184

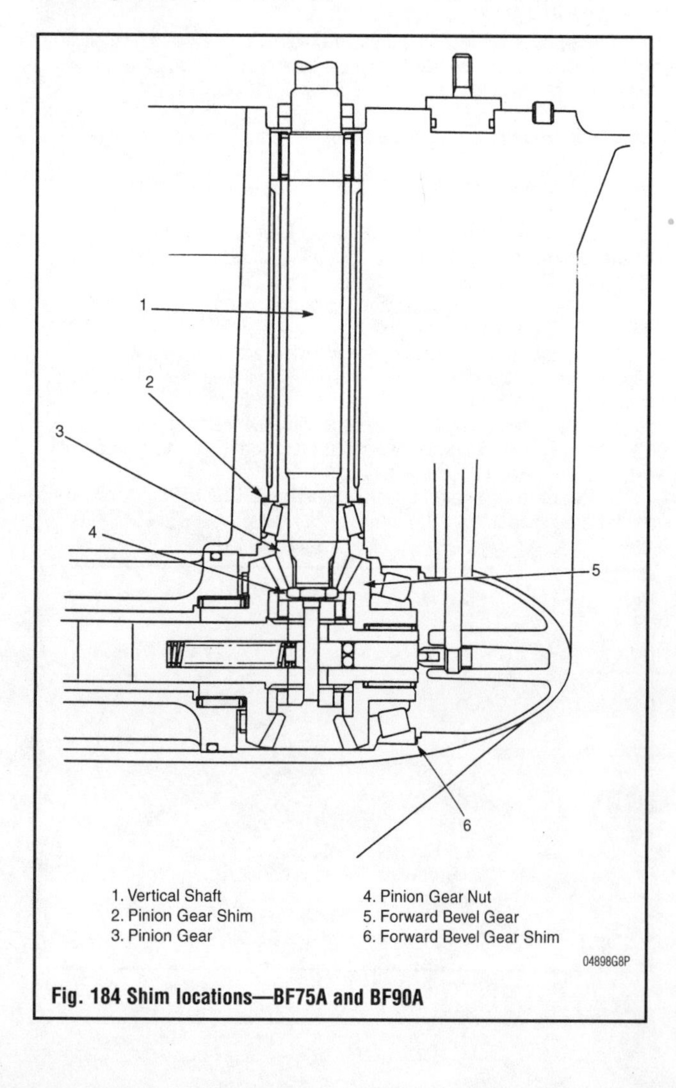

Fig. 184 Shim locations—BF75A and BF90A

PINION GEAR SHIM ADJUSTMENT

See Accompanying Illustrations

1. Obtain the following special tools before performing the pinion gear shim adjustment:
   - Bearing preload bearing/adapter assembly—07SPJ-ZW0011A (91-13948A1)
   - Bearing preload spring—07SPJ-ZW0012A (24-14111)
   - Bearing preload nut—07SPJ-ZW0013A (11-13953)
   - Bearing preload set screws—07SPJ-ZW0015A (10-12575)
   - Backlash indicator attachment (long pinion shaft)—07SPK-ZW10100
   - Sleeve 7/8 in. (22mm)—07SPJ-ZW0016A (23-13946)
   - Bearing preload nut—07SPJ-ZW0013A (11-13953)
   - Bearing preload bolt—07SPJ-ZW0014A (10-12580)
   - Pinion gear locating disc #3—07SPJ-ZW0021A (91-12346-3)
   - Pinion arbor assembly tool—07SPJ-ZW0023A (91-12349A3)
2. Install the all the gear case parts, other than the water pump and bearing carrier assembly, before doing the pinion gear shim adjustment.
3. Position the pinion shaft so that it is in the vertical position.
4. Thread the preload nut against the head of the bearing preload bolt.
5. Back out the bearing preload set screws and slide the special tool component assembly down over the pinion shaft and onto the gear case.
6. Assemble the special tool to preload the pinion shaft during the adjustment procedure.

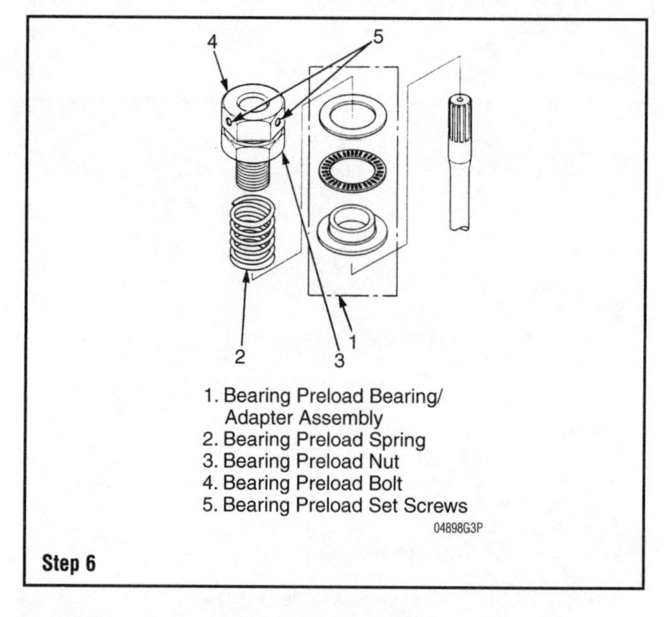

Step 6

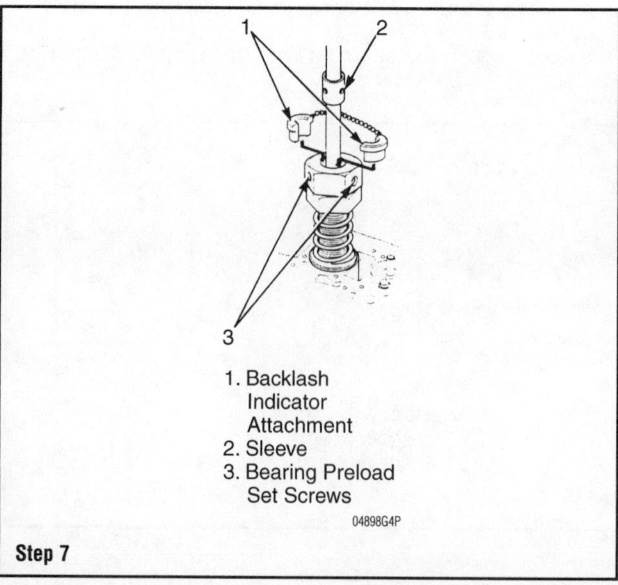

Step 7

7. Based on the O.D. of the pinion shaft, install either the backlash indicator attachment or the ⅞in. (22 mm) sleeve and tighten the bearing preload set screws.

➡Do not press down on the bearing preload bolt when tightening the bearing preload set screws. When using the sleeve be sure the holes align with the bearing preload set screws before tightening them.

8. Measure the clearance between the top of the bearing preload nut and the bottom of the bearing preload bolt.

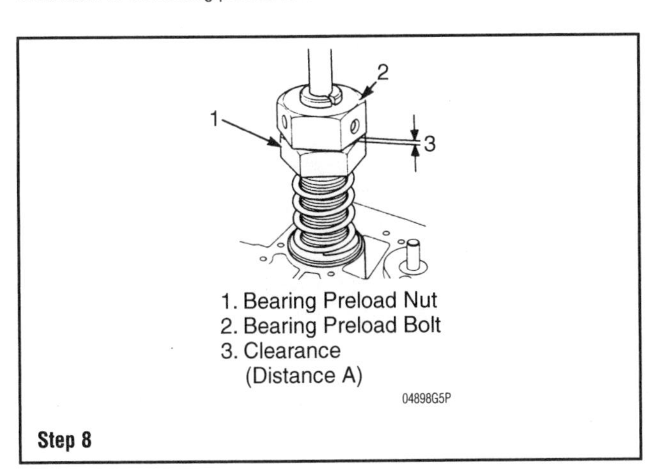

Step 8

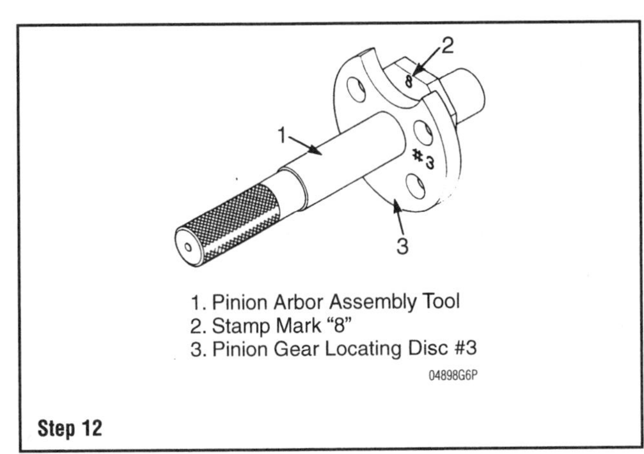

Step 12

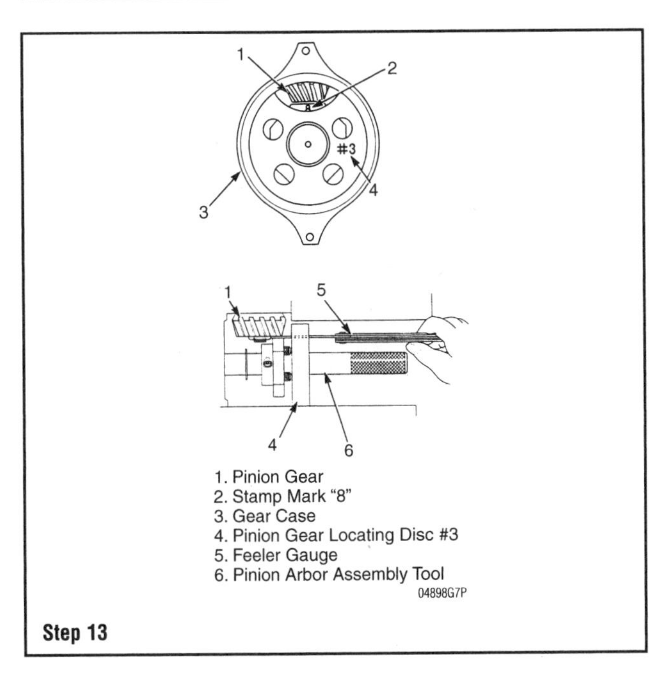

Step 13

9. Use the following formula to calculate the amount to tighten the bearing preload nut.

- Formula: Clearance + 0.96 in. (25 mm) = amount to tighten the bearing preload nut.
- Example: When clearance is 0.04 in. (1 mm). 1 + 25 = 26 mm (1.02 in.) Therefore the amount to tighten the bearing preload nut would be 26 mm (1.02 in.).

10. Tighten the bearing preload nut until the clearance between the top of the bearing preload nut and the bottom of the bearing preload bolt equals the calculated value.

11. After tightening, turn the pinion shaft 5 to 10 revolutions to properly seat the upper pinion shaft tapered roller bearing.

12. Set the special tool so that the "8" stamped on the tool is visible.

13. Insert the special tools in the gear case so that the "8" mark is under the pinion gear, then insert a long feeler gauge between the special tool and the pinion gear. The clearance should read 0.025 in. (0.64 mm).

14. Example:

- When the clearance is 0.021 in. (0.54 mm): 0.54 – 0.64 = -0.1
- Reduce the thickness of the shim by 0.004 in. (0.1 mm)
- When the clearance is 0.029 in. (0.74 mm): 0.74 – 0.64 = 0.1
- Increase the shim thickness by 0.004 in. (0.1 mm)

15. When the clearance is correct, remove the pinion arbor assembly tool and the pinion gear locating disc #3, and adjust the backlash.

➡Do not remove the bearing preload tools from the gear case at this time.

BACKLASH ADJUSTMENT (FORWARD BEVEL GEAR)

➧ See Accompanying Illustrations

The backlash adjustment should be made at the same time as the pinion gear shim adjustment procedure.

1. Obtain the following special tools to perform the forward gear backlash adjustment:

- Backlash indicator attachment—07SPK-ZW10100
- Backlash indicator tool—07SPJ-ZW0030Z (91-78473)

Be sure that the pinion gear shim adjustment special tools are attached to the pinion shaft. If they have been removed, install them as per the previous instructions.

2. Attach the bearing carrier assembly to the gear case, and tighten the self locking nuts to 22 ft. lbs. (30 Nm).

3. Hold the propeller shaft securely with a commercial puller and tighten the puller bolt to 3.6 ft. lbs. (5 Nm).

4. Turn the pinion shaft 5 to 10 turns.

5. Attach the special tool to the pinion shaft and adjust the dial indicator so that its needle is at "line 4" of the backlash indicator attachment.

6. Turn the pinion shaft slightly back and forth and read the dial indicator reading. Measure the backlash at all four points, (by turning the pinion shaft 90°at a time)in the same slight back and forth manner.

➡Do not turn the propeller shaft when turning the pinion shaft.

7. The backlash measurement should be within 0.012–0.019 in. (0.30–0.48 mm). If the backlash is too large, increase the forward bevel gear shim thickness and recheck. If the backlash is too small, reduce the forward bevel gear shim thickness and then recheck the backlash.

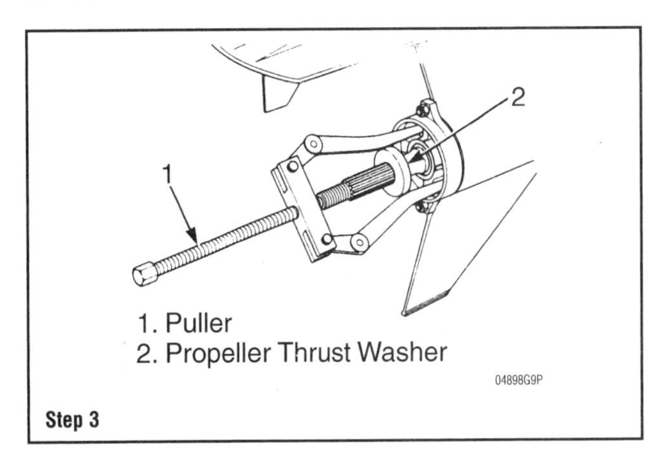

Step 3

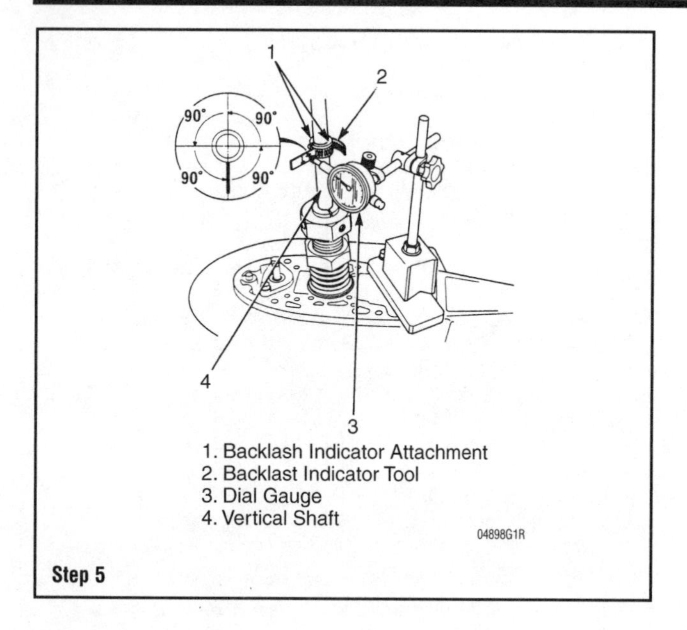

Step 5

BF115A and BF130A

➧ See Figures 185 and 186

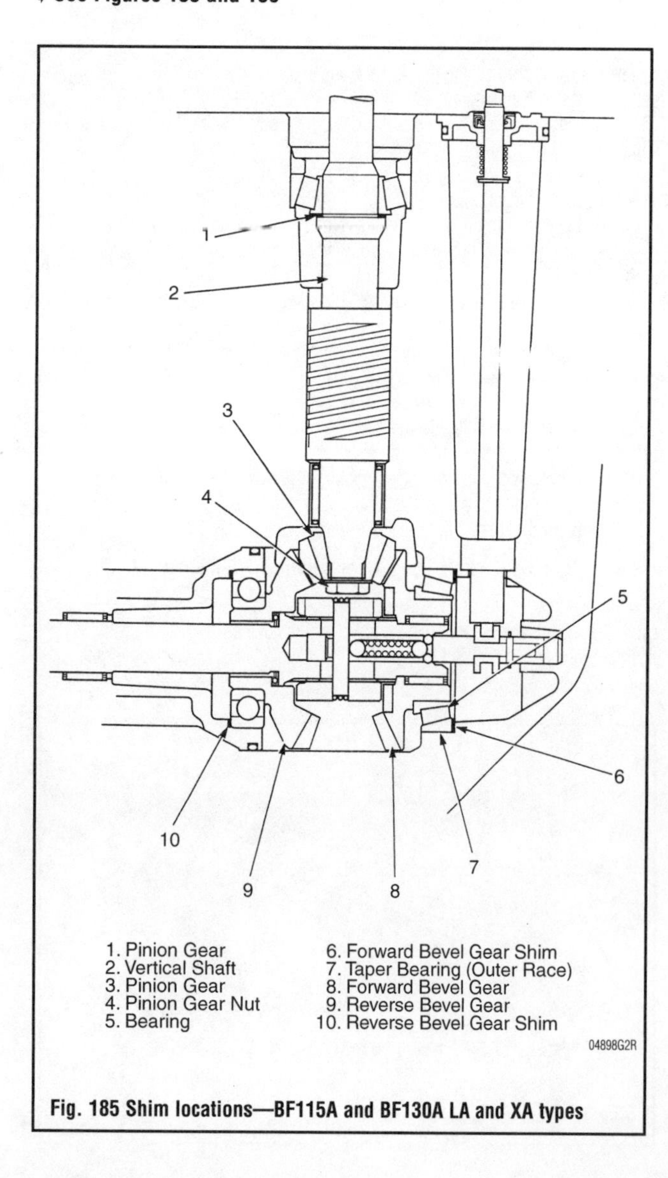

Fig. 185 Shim locations—BF115A and BF130A LA and XA types

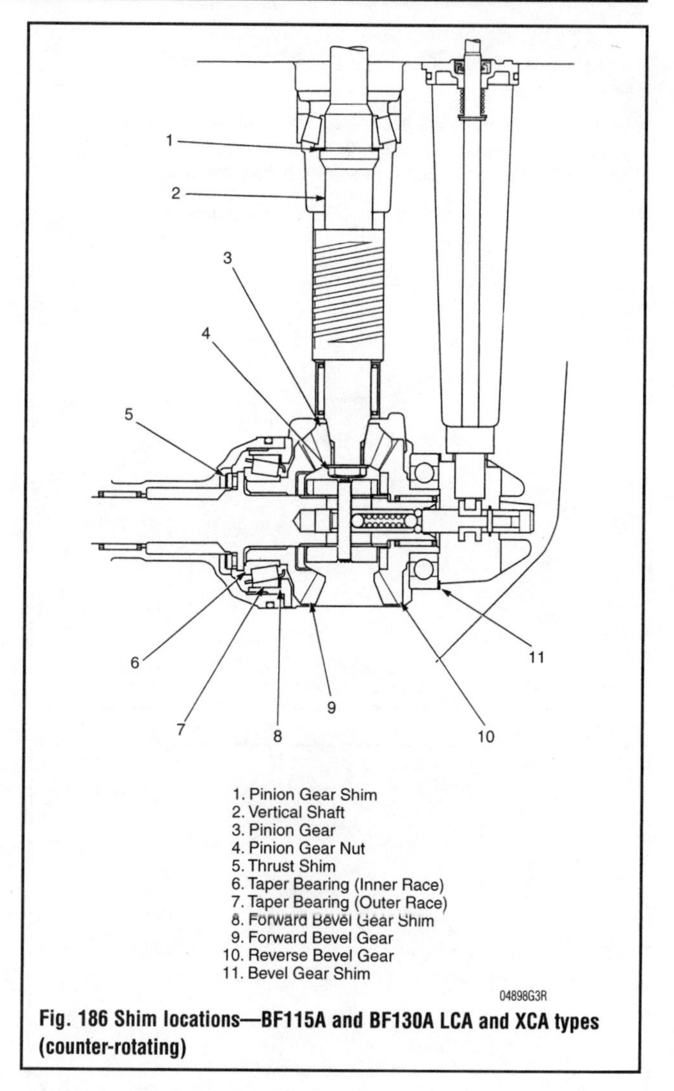

Fig. 186 Shim locations—BF115A and BF130A LCA and XCA types (counter-rotating)

PINION GEAR SHIM ADJUSTMENT

➧ See Accompanying Illustrations

Remove the tapered roller bearing (inner race) if it is still mounted on the pinion shaft.

1. To perform the adjustment, obtain the following special tool:
   - Gauge, adapter (80 mm)—07WPJ-ZW50100
2. Throughly clean the pinion shaft and gear.
3. Install the pinion gear on the pinion shaft and tighten the pinion gear nut to 98 ft. lbs. (132 Nm).

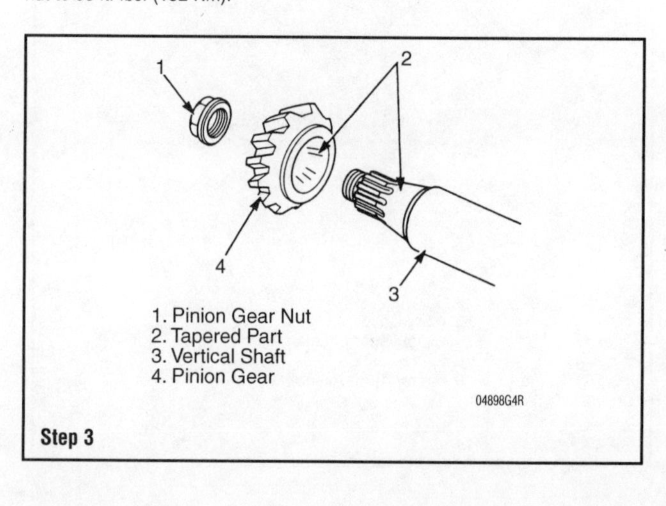

Step 3

➡Do not install the pinion shaft in the gear case. We recommend using the pinion shaft holder on the opposite end of the shaft to tighten the pinion nut correctly.

4. Be sure that the side of the gauge adapter special tool (the side with the stamped tool number) is facing to the opposite side of the pinion gear, and be sure that the tool for the pinion gear nut side is not set on the nut. Tighten the bolts on the tool by hand while pushing both tools toward the pinion gear side of the shaft.

5. Do not confuse the pinion gear side and the tapered roller bearing side of the special tool. Do not score and scratch the opposite side (measurement side) from the side where the tool number is stamped. Do not tighten the bolts with a wrench. The tool must not be loose or wobbly. It must be securely set on the pinion gear.

5. Set both tools on both sides of the projection on the pinion shaft, so that the special tool stamped with the tool number faces to the opposite side from the pinion gear as shown. Tighten the bolts by hand.

6. Align the tool end gap of the pinion gear side with the tool end gap of the taper roller bearing side.
- Do not tighten the bolts with a wrench.
- The tool must not be loose or wobbly. It must be securely set on the pinion gear

7. Hold the pinion shaft upright (with the pinion gear up) and secure the pinion shaft.

8. Install a commercially available depth gauge at the gauge adapter of the pinion gear side. Measure the pinion shaft length (distance D)and record it.

9. Assemble the outer and inner race of the tapered roller bearing.

10. Measure the bearing height from the outer race end to the inner race end and record the measurement.

11. Measure the height of the tapered roller bearing outer race and record the measurement.

12. Calculate the gap between the outer and inner race using the measurements obtained in previous two steps.

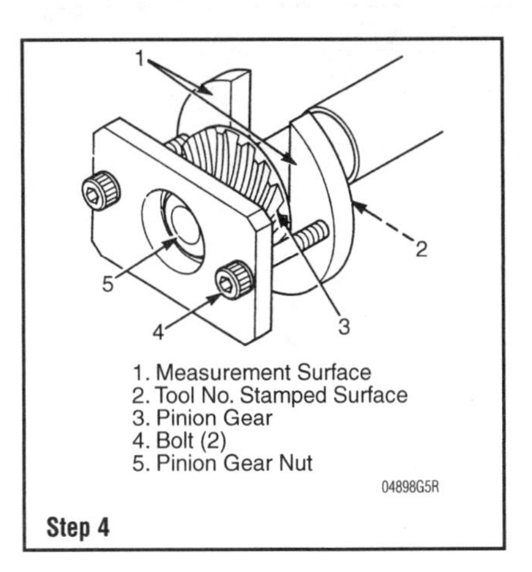

Step 4

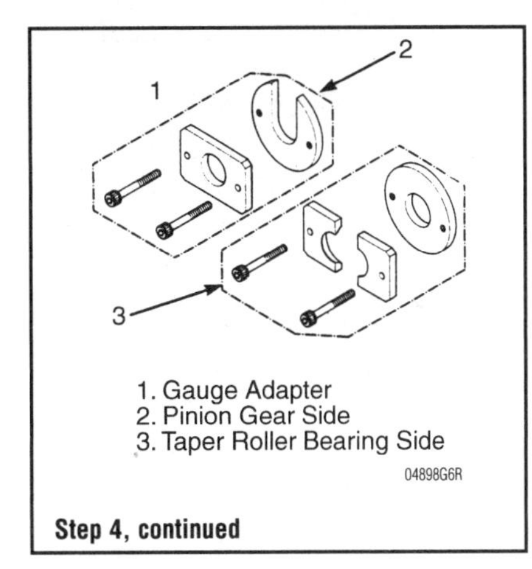

Step 4, continued

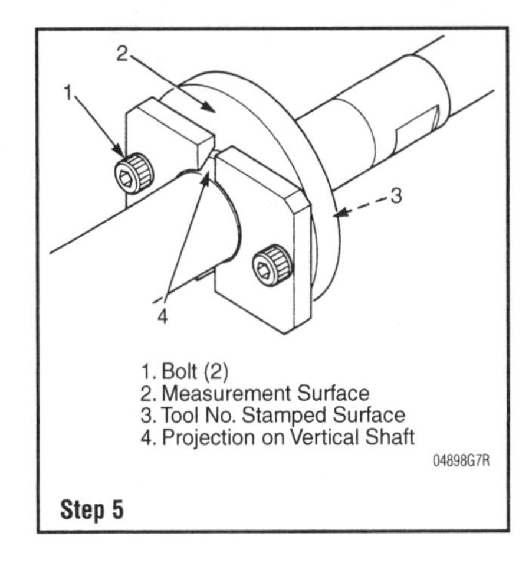

Step 5

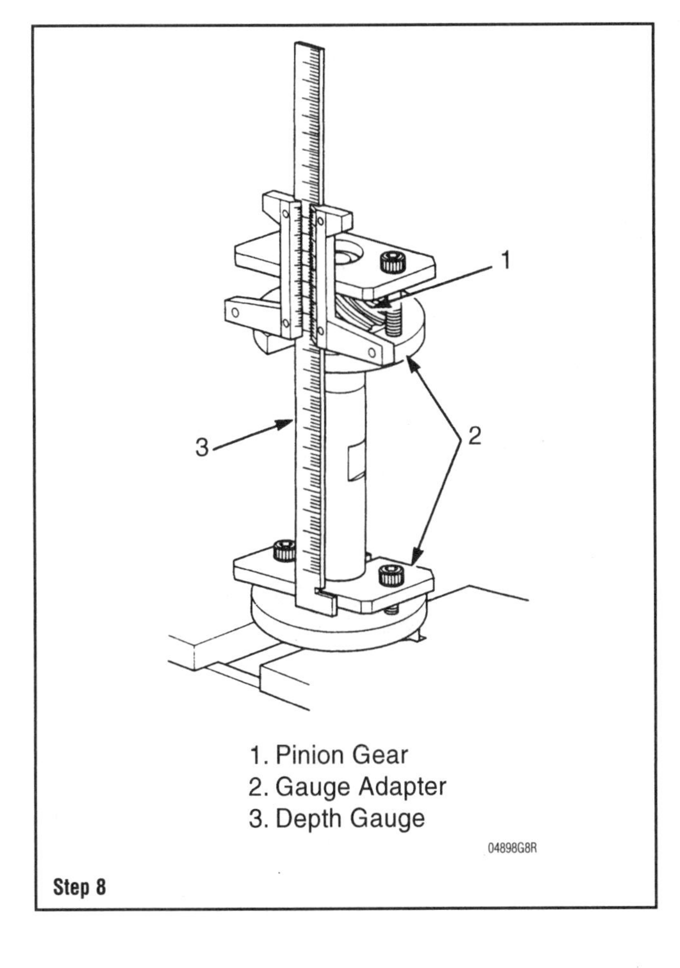

Step 8

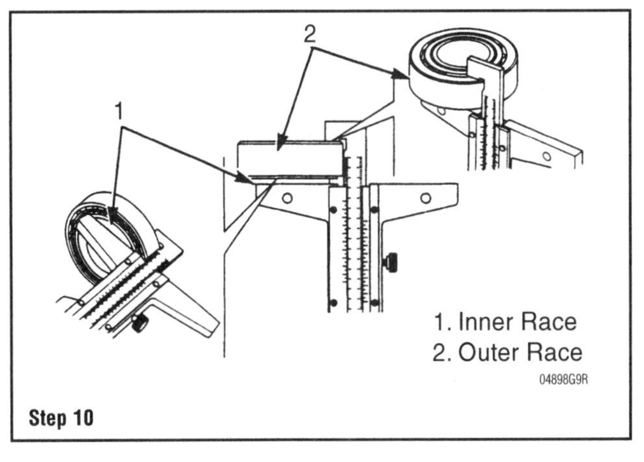

Step 10

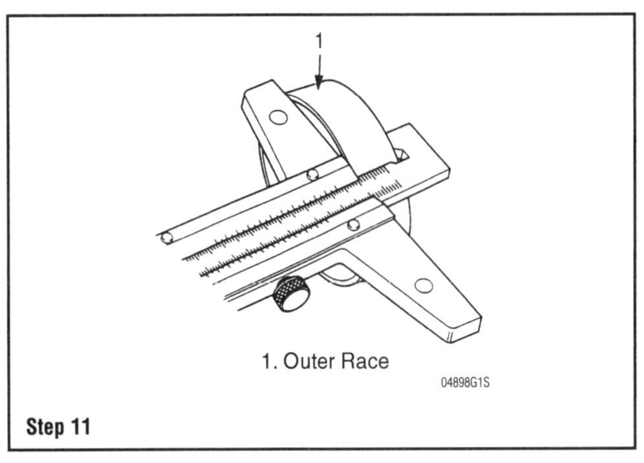

Step 11

- Formula: Bearing height – outer race height = gap (distance E)
- Example: Bearing height from the outer race end to the inner race end is 21.25 mm. Outer race height is 17.0 mm (0.669 in.): 21.25 – 17.00 = 4.25
- The gap distance is 4.25 mm (0.167 in.)

13. Determine the calculation value using the pinion shaft length (distance D), the gap (distance E) and the following formula:

- Pinion shaft length (D) + Gap (E) – 162.95 = calculation value.
- Example: Pinion shaft length (D) is 158.5 mm (6.240 in.), gap (E) is 4.25 mm (0.167 in.).
- 158.5 + 4.25 – 162.95 = -0.20
- The calculation value is -0.20 mm (-0.008 in.)

14. Cross reference the calculation value and the engagement mark located underneath the trim tab on the gear case.

15. Select the correct shim from the Pinion Gear Shim Selection Table—BF115A and BF130A.

16. Select the appropriate thickness shim from the shim type table.

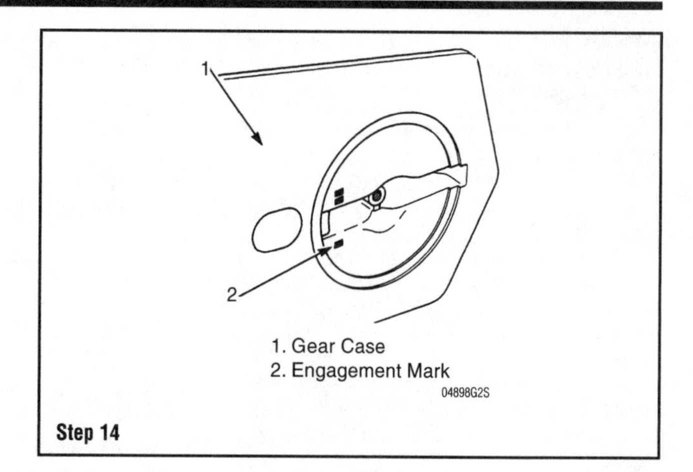

1. Gear Case
2. Engagement Mark

04898G2S

Step 14

Pinion Gear Shim Selection Table

Unit: mm (in)

Engagement mark on the gear case	0.57 (0.022) – 0.55 (0.021)	0.55 (0.021) – 0.50 (0.020)	0.50 (0.020) – 0.45 (0.018)	0.45 (0.018) – 0.40 (0.016)	0.40 (0.016) – 0.35 (0.014)	0.35 (0.014) – 0.30 (0.012)	0.30 (0.012) – 0.25 (0.010)	0.25 (0.010) – 0.20 (0.008)	0.20 (0.008) – 0.15 (0.006)
	Calculation value								
F	0.35 (0.014)	0.40 (0.016)	0.45 (0.018)	0.50 (0.020)	0.55 (0.022)	0.60 (0.024)	0.65 (0.026)	0.70 (0.028)	0.75 (0.030)
E	0.30 (0.012)	0.35 (0.014)	0.40 (0.016)	0.45 (0.018)	0.50 (0.020)	0.55 (0.022)	0.60 (0.024)	0.65 (0.026)	0.70 (0.028)
D	0.25 (0.010)	0.30 (0.012)	0.35 (0.014)	0.40 (0.016)	0.45 (0.018)	0.50 (0.020)	0.55 (0.022)	0.60 (0.024)	0.65 (0.026)
C	0.20 (0.008)	0.25 (0.010)	0.30 (0.012)	0.35 (0.014)	0.40 (0.016)	0.45 (0.018)	0.50 (0.020)	0.55 (0.022)	0.60 (0.024)
B	0.15 (0.006)	0.20 (0.008)	0.25 (0.010)	0.30 (0.012)	0.35 (0.014)	0.40 (0.016)	0.45 (0.018)	0.50 (0.020)	0.55 (0.022)
A	0.10 (0.004)	0.15 (0.006)	0.20 (0.008)	0.25 (0.010)	0.30 (0.012)	0.35 (0.014)	0.40 (0.016)	0.45 (0.018)	0.50 (0.020)

Unit: mm (in)

Engagement mark on the gear case	0.15 (0.006) – 0.10 (0.004)	0.10 (0.004) – 0.05 (0.002)	0.05 (0.002) – 0 (0.000)	0 (0.000) – – 0.05 (– 0.002)	– 0.05 (– 0.002) – – 0.10 (– 0.004)	– 0.10 (– 0.004) – – 0.15 (– 0.006)	– 0.15 (– 0.006) – – 0.20 (– 0.008)	– 0.20 (– 0.008) – – 0.25 (– 0.010)
	Calculation value							
F	0.80 (0.031)	0.85 (0.033)	0.90 (0.035)	0.95 (0.037)	1.00 (0.039)	1.05 (0.041)	1.10 (0.043)	1.15 (0.045)
E	0.75 (0.030)	0.80 (0.031)	0.85 (0.033)	0.90 (0.035)	0.95 (0.037)	1.00 (0.039)	1.05 (0.041)	1.10 (0.043)
D	0.70 (0.028)	0.75 (0.030)	0.80 (0.031)	0.85 (0.033)	0.90 (0.035)	0.95 (0.037)	1.00 (0.039)	1.05 (0.041)
C	0.65 (0.026)	0.70 (0.028)	0.75 (0.030)	0.80 (0.031)	0.85 (0.033)	0.90 (0.035)	0.95 (0.037)	1.00 (0.039)
B	0.60 (0.024)	0.65 (0.026)	0.70 (0.028)	0.75 (0.030)	0.80 (0.031)	0.85 (0.033)	0.90 (0.035)	0.95 (0.037)
A	0.55 (0.022)	0.60 (0.024)	0.65 (0.026)	0.70 (0.028)	0.75 (0.030)	0.80 (0.031)	0.85 (0.033)	0.90 (0.035)

Reading Shim Selection Table

When the engagement mark on the gear case is E and the calculation value is – 0.2 mm (– 0.008 in) or more, the shim thickness is 1.05 mm (0.041 in).
When the calculation value is less than – 0.2 mm (– 0.008 in), the shim thickness is 1.10 mm (0.043 in).

Example 1: Unit: mm (in)

			Calculation value
			– 0.20 mm (– 0.008 in) or above to less than – 0.15 mm (– 0.006 in)
F			
E			1.05 (0.041)

Example 2: Unit: mm (in)

			Calculation value
			– 0.25 mm (– 0.010 in) or above to less than – 0.20 mm (– 0.008 in)
F			
E			1.10 (0.043)

Shim Combination

To obtain 1.05 mm (0.041 in) of shim thickness, combine five gear shim Bs and one gear shim C, or combine three gear shim Bs, one gear shim A and one gear shim D by referring to the shim type table.

04898G4S

Step 15

Shim Type Table

Parts name	Thickness
Pinion gear shim A	0.10 mm (0.0039 in)
Pinion gear shim B	0.15 mm (0.0060 in)
Pinion gear shim C	0.30 mm (0.0118 in)
Pinion gear shim D	0.50 mm (0.0197 in)

04898G3S

Step 16

FORWARD BEVEL GEAR SHIM

See Accompanying Illustrations

1. Assemble the outer and inner race of the tapered bearing.
2. Measure the bearing height (distance F) from the outer race end to the inner race end and record the measurement you get.

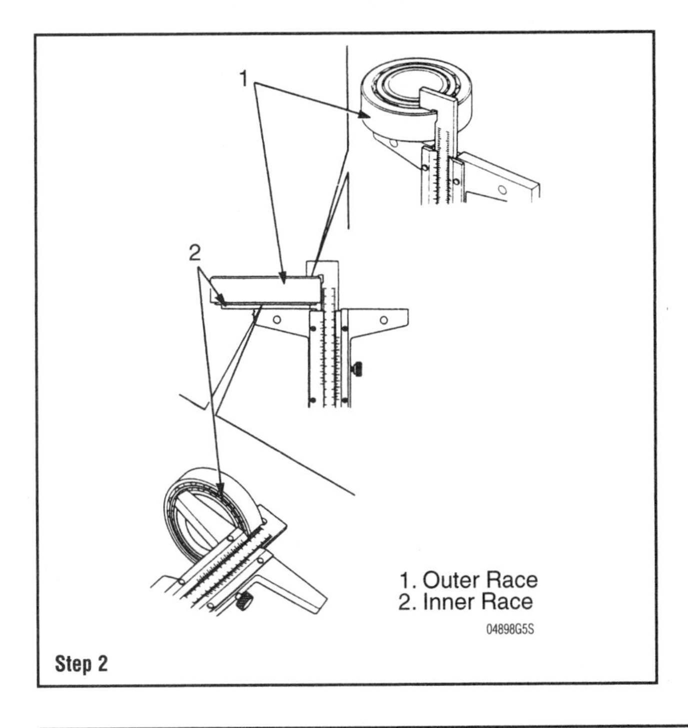

1. Outer Race
2. Inner Race

04898G5S

Step 2

3. Determine the calculation value using the bearing height (distance F) and the following formula:
   - Formula: Bearing height (F) – 21.5 = calculation value
   - Example: When bearing height (F) is 21.55 mm (0.848 in.)
   - 21.55 – 21.5 = 0.05, the calculation value is 0.05 mm (0.002 in.)
4. Cross reference the calculation value and the engagement mark located underneath the trim tab on the gear case.
5. Select the appropriate thickness shim from the shim type table.
6. Select the correct shim thickness from the forward bevel gear shim selection table— BF115A and BF130A LA and XA types

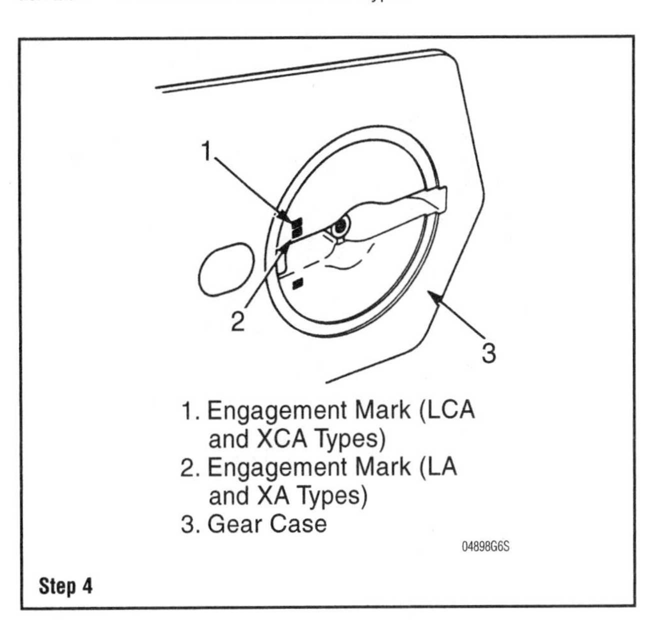

1. Engagement Mark (LCA and XCA Types)
2. Engagement Mark (LA and XA Types)
3. Gear Case

04898G6S

Step 4

Shim Type Table

Parts name	Thickness
Forward bevel gear shim A	0.10 mm (0.0039 in)
Forward bevel gear shim B	0.15 mm (0.0060 in)
Forward bevel gear shim C	0.30 mm (0.0118 in)
Forward bevel gear shim D	0.50 mm (0.0197 in)

04898G7S

Step 5

Forward Bevel Gear Shim Selection Table (LA And XA Types)

Unit: mm (in)

		Calculation value			
		0.20 - 0.15 (0.008 - 0.006)	0.15 - 0.10 (0.006 - 0.004)	0.10 - 0.05 (0.004 - 0.002)	0.05 - 0 (0.002 - 0.000)
Engagement mark on the gear case	1	0.45 (0.018)	0.50 (0.020)	0.55 (0.022)	0.60 (0.024)
	2	0.40 (0.016)	0.45 (0.018)	0.50 (0.020)	0.55 (0.022)
	3	0.35 (0.014)	0.40 (0.016)	0.45 (0.018)	0.50 (0.020)
	4	0.30 (0.012)	0.35 (0.014)	0.40 (0.016)	0.45 (0.018)
	5	0.25 (0.010)	0.30 (0.012)	0.35 (0.014)	0.40 (0.016)
	6	0.20 (0.008)	0.25 (0.010)	0.30 (0.012)	0.35 (0.014)

04898G8S

Step 6

LCA And XCA Types

Unit: mm (in)

Engagement mark on the gear case	Calculation value: 0.32 (0.013) – 0.30 (0.012)	0.30 (0.012) – 0.25 (0.010)	0.25 (0.010) – 0.20 (0.008)	0.20 (0.008) – 0.15 (0.006)	0.15 (0.006) – 0.10 (0.004)	0.10 (0.004) – 0.05 (0.002)	0.05 (0.002) – 0 (0.000)
A	0.30 (0.012)	0.35 (0.014)	0.40 (0.016)	0.45 (0.018)	0.50 (0.020)	0.55 (0.022)	0.60 (0.024)
B	0.25 (0.010)	0.30 (0.012)	0.35 (0.014)	0.40 (0.016)	0.45 (0.018)	0.50 (0.020)	0.55 (0.022)
C	0.20 (0.008)	0.25 (0.010)	0.30 (0.012)	0.35 (0.014)	0.40 (0.016)	0.45 (0.018)	0.50 (0.020)
D	0.15 (0.006)	0.20 (0.008)	0.25 (0.010)	0.30 (0.012)	0.35 (0.014)	0.40 (0.016)	0.45 (0.018)
E	0.10 (0.004)	0.15 (0.006)	0.20 (0.008)	0.25 (0.010)	0.30 (0.012)	0.35 (0.014)	0.40 (0.016)
F	0.05 (0.002)	0.10 (0.004)	0.15 (0.006)	0.20 (0.008)	0.25 (0.010)	0.30 (0.012)	0.35 (0.014)

04898G9S

Step 7

7. Select the correct shim from the forward bevel gear shim selection table— BF115A and BF130A LCA and XCA types

REVERSE BEVEL GEAR SHIM

See Figures 187, 188, 189 and 190

1. Refer to the engagement mark located underneath the trim tab on the gear case and select the appropriate thickness shim from the shim selection table.
 - Example: When the engagement mark on the gear case is C, the appropriate shim thickness should be 0.010 in. (0.25 mm) in the LA and XA types. When the engagement mark on the gear case is 3, the appropriate shim thickness should be 0.022 in. (0.55 mm) in the LCA and XCA types.

Shim Type Table LA And XA Types

Parts name	Thickness
Reverse bevel gear shim A	0.10 mm (0.0039 in)
Reverse bevel gear shim B	0.15 mm (0.0060 in)
Reverse bevel gear shim C	0.30 mm (0.0118 in)
Reverse bevel gear shim D	0.50 mm (0.0197 in)

04898G1T

Fig. 187 Shim type table—BF115A and BF130A LA and XA types

LCA And XCA Types

Parts name	Thickness
Forward bevel gear shim A	0.10 mm (0.0039 in)
Forward bevel gear shim B	0.15 mm (0.0060 in)
Forward bevel gear shim C	0.30 mm (0.0118 in)
Forward bevel gear shim D	0.50 mm (0.0197 in)

04898G2T

Fig. 188 Shim type table—BF115A and BF130A LCA and XCA types

Reverse Bevel Gear Shim Selection Table (LA And XA Types)

Engagement mark on the gear case	Thickness
A	0.15 mm (0.006 in)
B	0.20 mm (0.008 in)
C	0.25 mm (0.010 in)
D	0.30 mm (0.012 in)
E	0.35 mm (0.014 in)
F	0.40 mm (0.016 in)

04898G3T

Fig. 189 Reverse bevel gear shim selection table—BF115A and BF130A LA and XA types

LCA And XCA Types

Engagement mark on the gear case	Thickness
1	0.65 mm (0.026 in)
2	0.60 mm (0.024 in)
3	0.55 mm (0.022 in)
4	0.50 mm (0.020 in)
5	0.45 mm (0.018 in)
6	0.40 mm (0.016 in)

04898G4T

Fig. 190 Reverse bevel gear shim selection table—BF115A and BF130A LCA and XCA types

THRUST SHIM (LCA AND XCA TYPES ONLY)

See Accompanying Illustrations

Remove the tapered bearing inner race if it is mounted on the bearing carrier assembly.

1. Measure the height of the tapered bearing inner race distance (distance J) and record it.
2. Determine the calculation value using the inner race height (J) and the following formula.
 - Formula: Inner race height (J) – 21.5 = tolerance

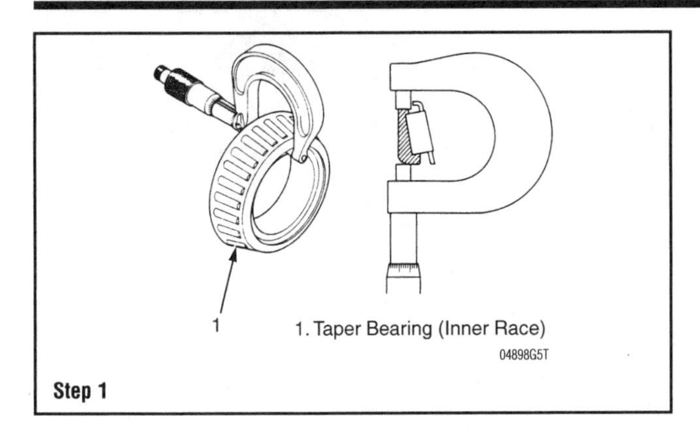

1. Taper Bearing (Inner Race)

04898G5T

Step 1

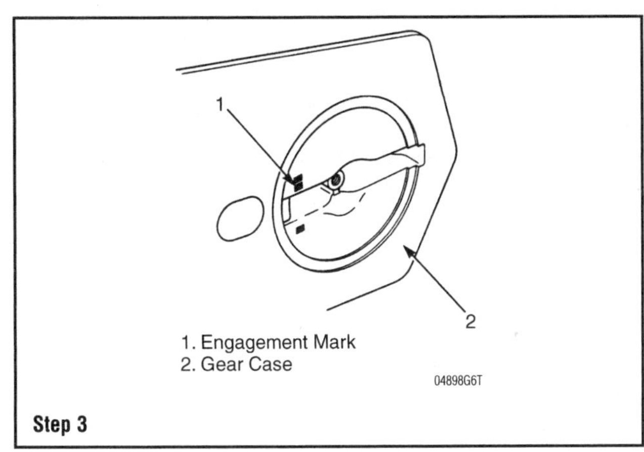

1. Engagement Mark
2. Gear Case

04898G6T

Step 3

Shim Type Table

Parts name	Thickness
Thrust shim A	0.10 mm (0.0039 in)
Thrust shim B	0.15 mm (0.0060 in)
Thrust shim C	0.30 mm (0.0118 in)
Thrust shim D	0.50 mm (0.0197 in)

04898G7T

Step 4

Trush Shim Selection Table

Unit: mm (in)

Engagement mark on the gear case		Calculation value: 0 (0.000) – –0.05 (–0.002)	Calculation value: –0.05 (–0.002) – –0.10 (–0.004)	Calculation value: –0.10 (–0.004) – –0.15 (–0.006)
	F	0.95 (0.037)	1.00 (0.039)	1.05 (0.041)
	E	0.90 (0.035)	0.95 (0.037)	1.00 (0.039)
	D	0.85 (0.033)	0.90 (0.035)	0.95 (0.037)
	C	0.80 (0.031)	0.85 (0.033)	0.90 (0.035)
	B	0.75 (0.030)	0.80 (0.031)	0.85 (0.033)
	A	0.70 (0.028)	0.75 (0.030)	0.80 (0.031)

04898G8T

Step 5

- Example: Inner race height (J) is 21.4 mm (0.84 in.).
- 21.4 – 21.5 = -0.1 mm, the calculation value is -0.1 mm (-0.004 in.)

3. Cross reference the calculation value and the engagement mark located underneath the trim tab on the gear case.
4. Select the appropriate thickness shim from the shim type table.
5. Select the correct shim from the thrust shim selection table— BF115A and BF130A LCA and XCA types only

FORWARD BEVEL GEAR BACKLASH ADJUSTMENT

See Accompanying Illustrations

1. LCA and XCA types: Forward bevel gear backlash adjustment must be made after adjusting the reverse bevel gear backlash.

These procedures are common to both the LA, XA and LCA, XCA types. Backlash adjustment must be made after the adjustment of each gear shim. Install the all the parts except the water pump into the gear case.

2. Obtain the following special tools:

- Bearing preload bearing/adapter assembly—07SPJ-ZW0011A (91-13948A1)
- Bearing preload spring—07SPJ-ZW0012A (24-14111)
- Bearing preload nut—07SPJ-ZW0013A (11-13953)
- Bearing preload bolt—07SPJ-ZW0014A (10-12580)
- Bearing preload set screws—07SPJ-ZW0015A (10-12575)
- Backlash indicator attachment—07WPK-ZW50100
- Backlash indicator tool—07SPJ-ZW0030Z (91-78473)
- Dial indicator adapter kit—07SPJ-ZW0040Z (91-83155)

3. Position the gear case so the pinion shaft is vertical.
4. Thread the bearing preload nut against the head of the bearing preload bolt.
5. Slide the special tool components down the pinion shaft and onto the gear case.
6. Back out the bearing preload set screws so the bearing preload bolt can slide smoothly over the pinion shaft.
7. Based on the O.D of the pinion shaft, install either the backlash indicator attachment or the ⅞in. (22 mm) sleeve, and tighten the bearing preload screws.

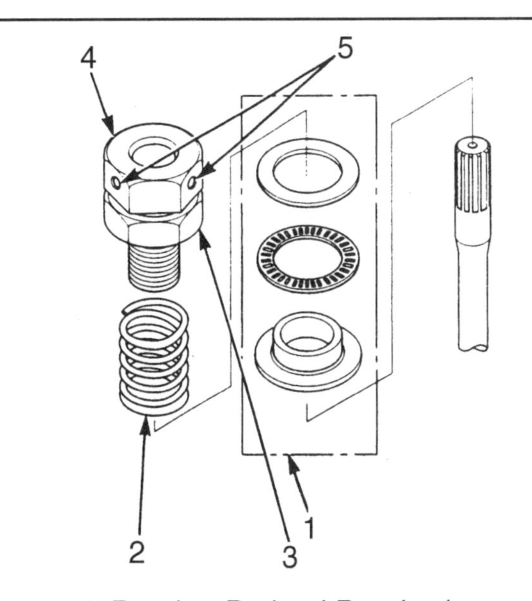

1. Bearing Preload Bearing/ Adapter Assembly
2. Bearing Preload Spring
3. Bearing Preload Nut
4. Bearing Preload Bolt
5. Bearing Preload Set Screws

04898G3P

Step 5

• Do not push down on the bearing preload bolt when tightening the bearing preload screws.
• When using the sleeve, be sure the holes align with the bearing preload set screws before tightening them.

8. Measure the clearance between the top of the bearing preload nut and the bottom of the bearing preload bolt. Calculate the amount to tighten the bearing preload nut using the following formula:
• Formula: Clearance (distance A) + 25 mm (0.96 in.) = amount to tighten the bearing preload nut.

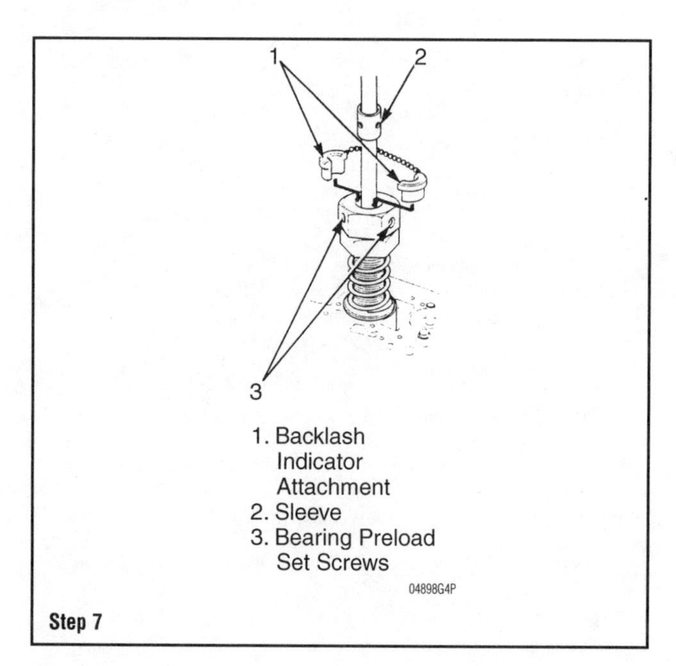

Step 7

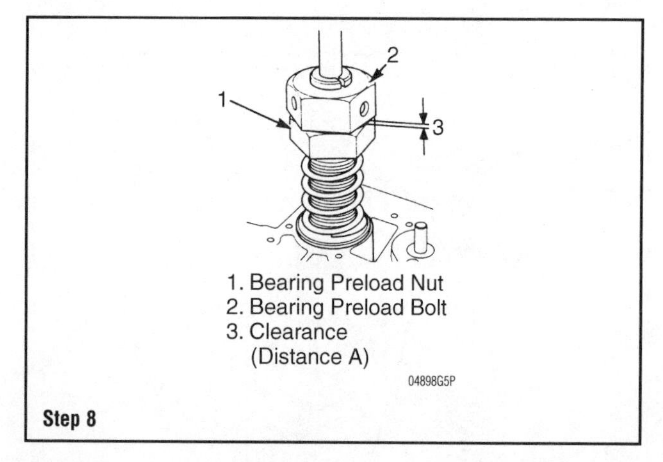

Step 8

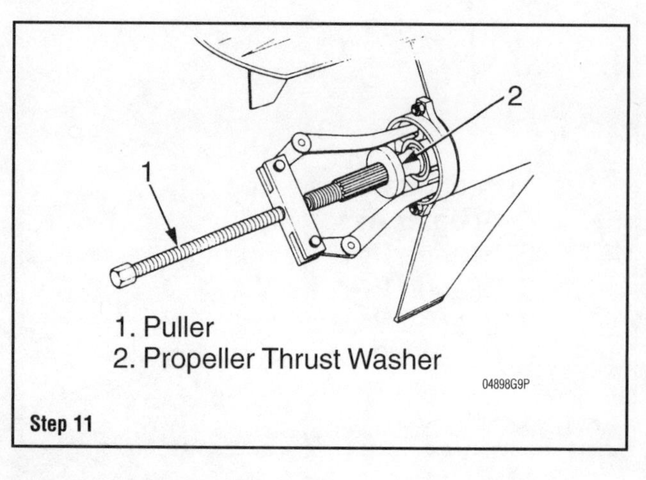

Step 11

• Example: When clearance (A) is 1 mm (0.04 in.). 1 + 25 = 26 mm (1.02 in.). Therefore, the amount to tighten the bearing preload nut would be 26 mm (1.02 in.).

9. Tighten the bearing preload nut until the clearance between the top of the bearing preload nut and the bottom of the bearing preload bolt equals the calculated value.

10. After tightening, turn the pinion shaft about 10 revolutions to properly seat the pinion shaft upper tapered roller bearing.

LA and XA Types Only

11. Hold the propeller shaft securely with a puller and tighten the puller bolt to 3.6 ft. lbs. (5 Nm).

12. Turn the pinion shaft about 10 turns,

13. Attach the backlash indicator attachment tool to the pinion shaft and adjust the dial indicator gauge so that its needle is at "line 2" on the backlash tool. Turn the pinion shaft slightly back and forth and record the backlash reading on the dial indicator.

14. By moving the pinion shaft in 90°increments, measure the backlash at each of the four points. (Do not move the pinion shaft by turning the propeller shaft)

15. Obtain the forward bevel gear backlash using the dial indicator runout and the following formula:
• Formula: Dial indicator runout x 1.03 mm = Backlash
• Example: Dial indicator runout is 0.195 mm (0.0077 in.). 0.195 X 1.03 = 0.20 mm. Therefore the backlash is 0.20 mm (0.008 in.).
• The standard backlash is 0.005–0.011 in. (0.12–0.29 mm).
• If the backlash is too large, increase the forward bevel gear shim thickness and then recheck the backlash.
• If the backlash is too small, lessen the shim thickness and recheck the backlash.

LCA and XCA Types Only

16. Hold the propeller shaft securely with a puller and tighten the puller bolt to 3.6 ft. lbs. (5 Nm).

17. Turn the pinion shaft about 10 turns,

18. Attach the backlash indicator attachment tool to the pinion shaft and adjust the dial indicator gauge so that its needle is at "line 2" on the backlash

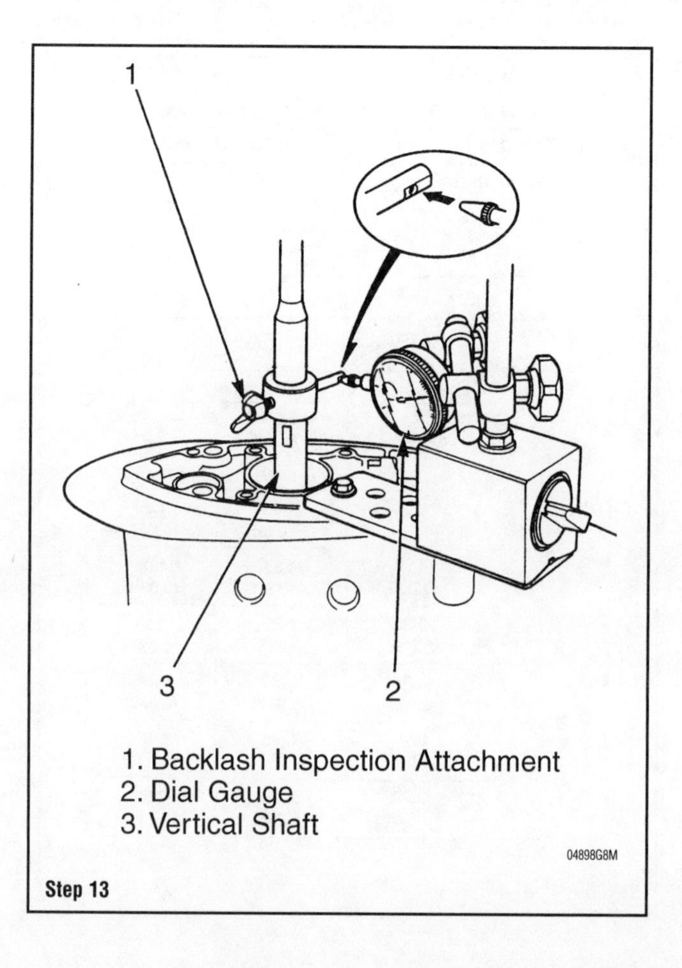

Step 13

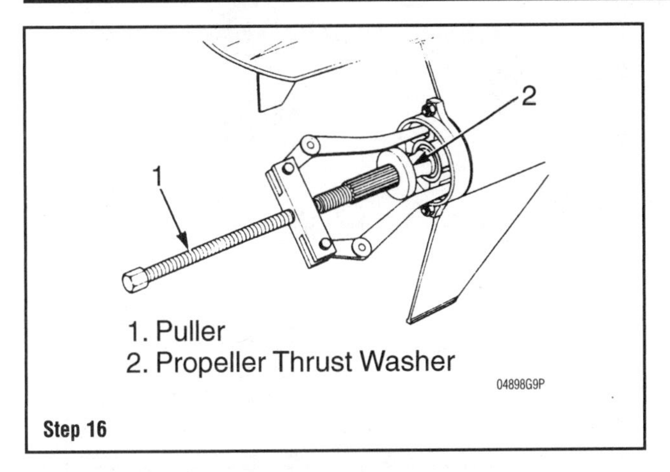

Step 16

tool. Turn the pinion shaft slightly back and forth and record the backlash reading on the dial indicator.

19. By moving the pinion shaft in 90°increments, measure the backlash at each of the four points. (Do not move the pinion shaft by turning the propeller shaft)

20. Obtain the forward bevel gear backlash using the dial indicator runout and the following formula:

- Formula: Dial indicator runout x 1.03 mm = Backlash
- Example: Dial indicator runout is 0.195 mm (0.0077 in.). 0.195 X 1.03 = 0.20 mm. Therefore the backlash is 0.20 mm (0.008 in.).
- The standard backlash is 0.005–0.011 in. (0.12–0.29 mm).
- If the backlash is too large, increase the forward bevel gear shim thickness and then recheck the backlash.
- If the backlash is too small, lessen the shim thickness and recheck the backlash.

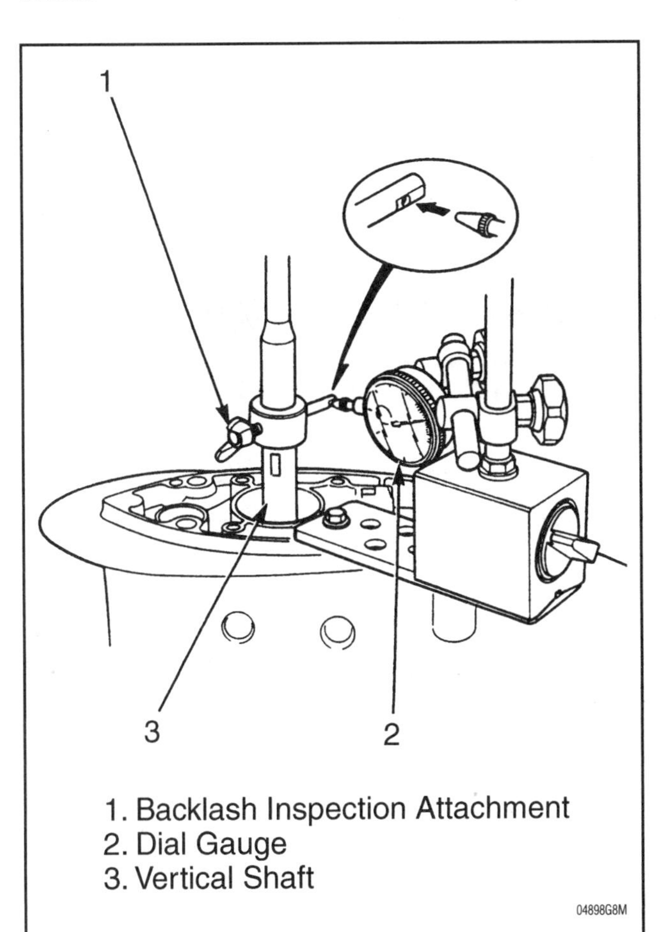

Step 18

REVERSE BEVEL GEAR BACKLASH

See Accompanying Illustrations

- LCA and XCA types: Reverse bevel gear backlash adjustments must be made before adjusting the forward bevel gear backlash.

These procedures are common to LA, XA, LCA and XCA types.

- Reverse bevel gear backlash adjustment should be made after adjusting the forward bevel gear backlash. Be sure that the special tools for the water pump mounting and the pinion shaft are attached to the pinion shaft. If they are removed , install the special tools.

LA and XA Types Only

1. Obtain the following special tools:

- Propeller shaft holder—07TPB-ZW10100
- Backlash indicator attachment—07WPK-ZW50100
- Backlash indicator tool—07SPJ-ZW0030Z (91-78473)
- Dial indicator adapter kit—07SPJ-ZW0040Z (91-83155)

2. Hold the propeller shaft with the special tool.

3. Tighten the propeller nut to 0.7 ft. lbs. (1 Nm). If it is hard to install the cotter pin once the nut has been tightened to specification, tighten the nut additionally until the cotter pin can be inserted.

4. Turn the pinion shaft about 10 turns.

5. Attach the backlash indicator attachment tool to the pinion shaft and adjust the dial indicator gauge so that its needle is at "line 2" on the backlash

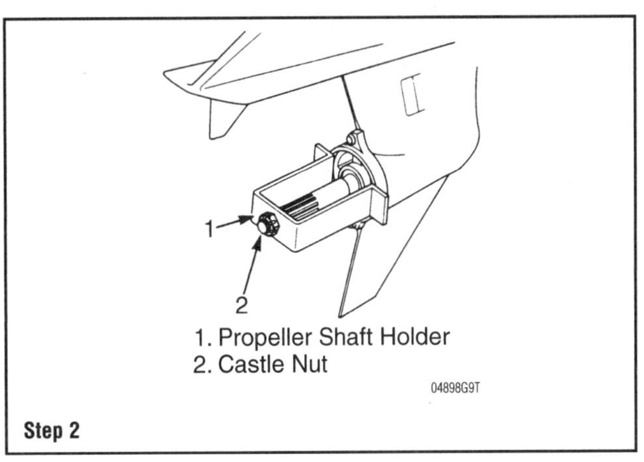

Step 2

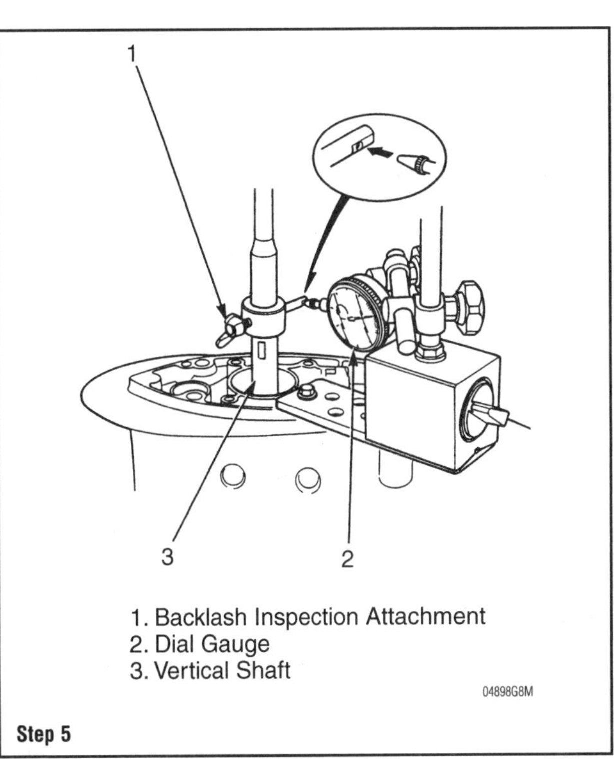

Step 5

tool. Turn the pinion shaft slightly back and forth and record the backlash reading on the dial indicator.

6. By moving the pinion shaft in 90°increments, measure the backlash at each of the four points. (Do not move the pinion shaft by turning the propeller shaft)

7. Obtain the forward bevel gear backlash using the dial indicator runout and the following formula:

- Formula: Dial indicator runout x 1.03 mm = Backlash
- Example: Dial indicator runout is 0.195 mm (0.0077 in.). 0.195 X 1.03 = 0.20 mm. Therefore the backlash is 0.20 mm (0.008 in.).
- The standard backlash is 0.005–0.015 in. (0.12–0.38 mm).
- If the backlash is too large, increase the forward bevel gear shim thickness and then recheck the backlash.
- If the backlash is too small, lessen the shim thickness and recheck the backlash.

LCA and XCA Types Only

8. Obtain the following special tools:

- Propeller shaft holder—07TPB-ZW10100
- Backlash indicator attachment—07WPK-ZW50100
- Backlash indicator tool—07SPJ-ZW0030Z (91-78473)
- Dial indicator adapter kit—07SPJ-ZW0040Z (91-83155)

9. Hold the propeller shaft with the special tool.

10. Tighten the propeller nut to 0.7 ft. lbs. (1 Nm). If it is hard to install the cotter pin once the nut has been tightened to specification, tighten the nut additionally until the cotter pin can be inserted.

11. Turn the pinion shaft about 10 turns.

12. Attach the backlash indicator attachment tool to the pinion shaft and adjust the dial indicator gauge so that its needle is at "line 2" on the backlash tool. Turn the pinion shaft slightly back and forth and record the backlash reading on the dial indicator.

13. By moving the pinion shaft in 90°increments, measure the backlash at each of the four points. (Do not move the pinion shaft by turning the propeller shaft)

14. Obtain the forward bevel gear backlash using the dial indicator runout and the following formula:

- Formula: Dial indicator runout x 1.03 mm = Backlash
- Example: Dial indicator runout is 0.195 mm (0.0077 in.). 0.195 X 1.03 = 0.20 mm. Therefore the backlash is 0.20 mm (0.008 in.).
- The standard backlash is 0.005–0.015 in. (0.12–0.38 mm).
- If the backlash is too large, increase the forward bevel gear shim thickness and then recheck the backlash.
- If the backlash is too small, lessen the shim thickness and recheck the backlash.

THRUST CLEARANCE ADJUSTMENT (LCA AND XCA TYPES ONLY)

See Accompanying Illustration

Adjust the thrust clearance after replacing the forward bevel gear shim.

1. Be sure that the bearing holder assembly is tightened against the bearing carrier to 76 ft. lbs. (103 Nm).

2. Make sure that the shift slider and clutch shifter are not installed.

3. Set the bearing carrier in the holding fixture as shown previously, and tighten the bolts securely.

4. Attach the tip of the dial indicator to the end of the propeller shaft perpendicularly.

5. Move the propeller shaft up and down and read the runout on the dial indicator. The standard runout should be 0.008–0.012 in. (0.2–0.3 mm).

6. If the thrust clearance is larger than specified, increase the thrust shim thickness and recheck the runout. If the thrust clearance is smaller than specified, reduce the shim thickness and recheck the runout.

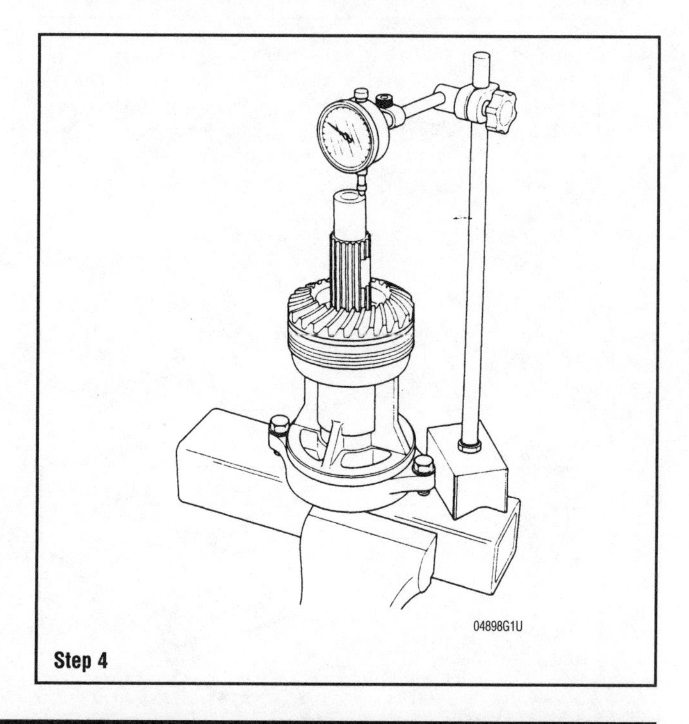

Step 4

JET DRIVE

Description and Operation

The jet drive unit is designed to permit boating in areas prohibited to a boat equipped with a conventional propeller drive system. The housing of the jet drive barely extends below the hull of the boat allowing passage in ankle deep water, white water rapids and over sand bars or in shoal water which would foul a propeller drive.

The jet drive provides reliable propulsion with a minimum of moving parts. Simply stated, water is drawn into the unit through an intake grille by an impeller driven by a driveshaft off the crankshaft of the powerhead. The water is immediately expelled under pressure through an outlet nozzle directed away from the stern of the boat.

As the speed of the boat increases and reaches planing speed, the jet drive discharges water freely into the air and only the intake grille makes contact with the water.

The jet drive is provided with a gate arrangement and linkage to permit the boat to be operated in reverse. When the gate is moved downward over the exhaust nozzle, the pressure stream is reversed by the gate and the boat moves sternward.

Conventional controls are used for powerhead speed, movement of the boat, shifting and power trim and tilt.

Model Identification and Serial Numbers

See Figure 191

A model letter identification is stamped on the rear, port side of the jet drive housing. A serial number for the unit is stamped on the starboard side of the jet drive housing, as indicated in the accompanying illustration.

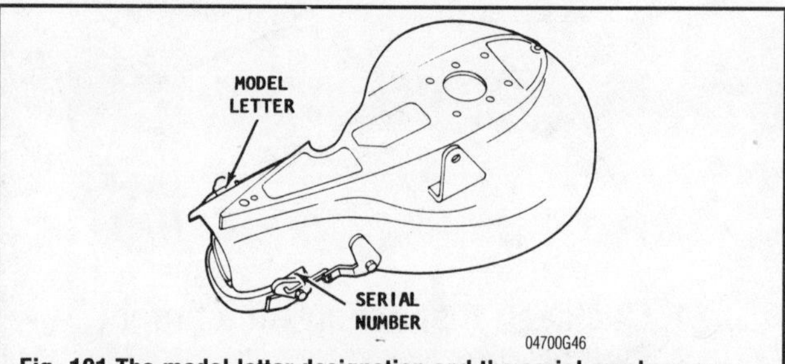

Fig. 191 The model letter designation and the serial numbers are embossed on the jet drive housing

The jet drives are used with the outboard units covered in this manual: Model AI 35R, AI 35LR, AI 35M, AI 35LM, AI 45R, AI 45LR, AI 45M, AI 45LM, AN and AN130. These letters are embossed on the port side of the jet drive housing.

Model AI 35R, AI 35LR, AI 35M AND AI35LM are used on the BF35A and BF40A. The AI 45R, AI 45LR, AI45M and AI 45LM models are used on the BF45A and BF50A. The AN model is used on the BF75A and BF90A and the AN130 model is used on the BF115A and BF130A.

For the most part, jet drive units are identical in design, function and operation. Differences lie in size and securing hardware.

Jet Drive Assembly

REMOVAL & INSTALLATION

See Accompanying Illustrations

1. Remove the two bolts and retainer securing the shift cable to the shift cable support bracket.
2. Remove the locknut, bolt and washer securing the shift cable to the shift arm. Try not to disturb the length of the cable.
3. Remove the six bolts securing the intake grille to the jet casing.
4. Ease the intake grille from the jet drive housing.
5. Pry the tab or tabs of the tabbed washer away from the nut to allow the nut to be removed.
6. Loosen and then remove the nut.
7. Remove the tabbed washer and spacers. Make a careful count of the spacers behind the washer. If the unit is relatively new, there could be as many as eight spacers stacked together. If less than eight spacers are removed from behind the washer, the others will be found behind the jet impeller, which is removed in the following step. A total of eight spacers will be found.
8. Remove the jet impeller from the shaft. If the impeller is frozen to the shaft, obtain a block of wood and a hammer. Tap the impeller in a clockwise direction to release the shear key.
9. Slide the nylon sleeve and shear key free of the driveshaft and any spacers found behind the impeller. Make a note of the number of spacers at both locations—behind the impeller and on top of the impeller, under the nut and tabbed washer.

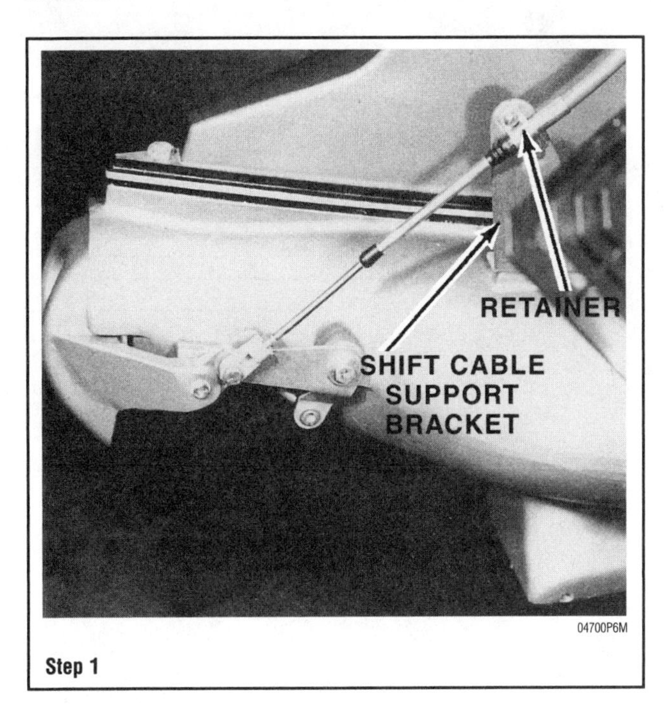

Step 1

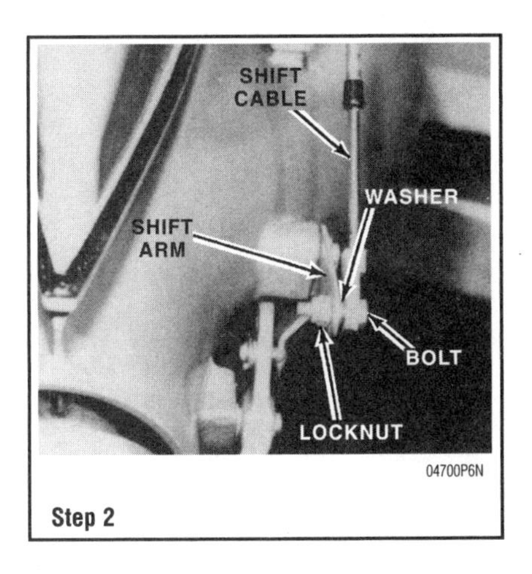

Step 2

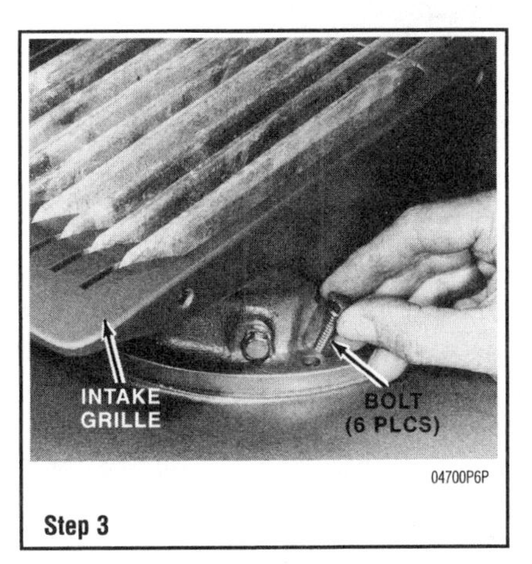

Step 3

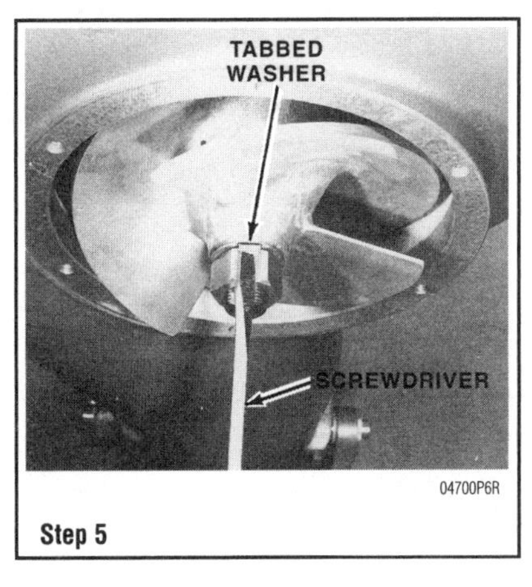

Step 5

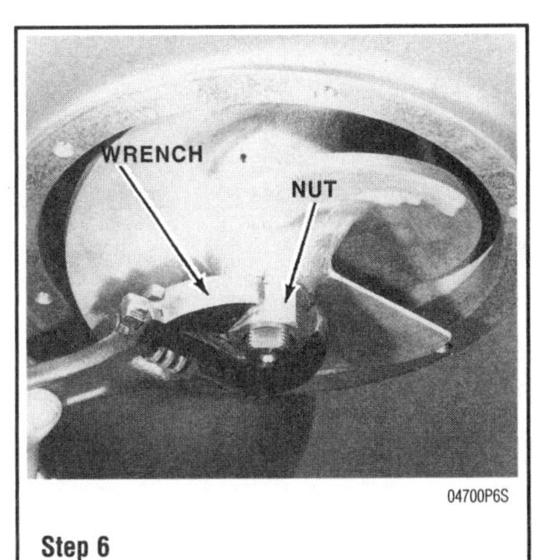

Step 6

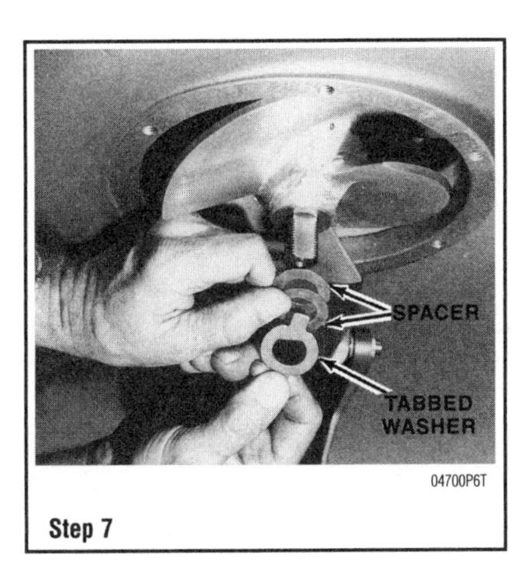

Step 7

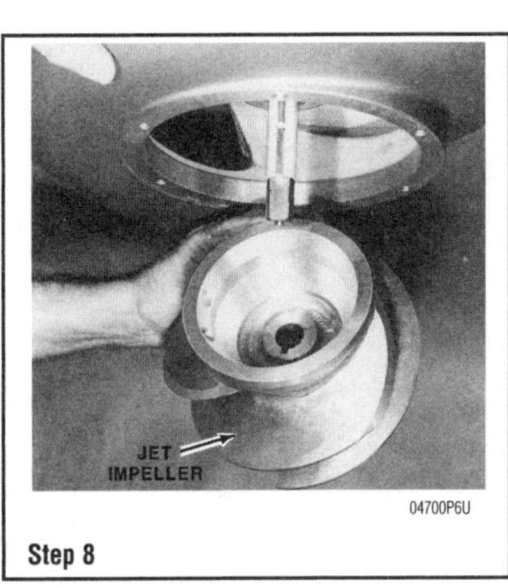

Step 8

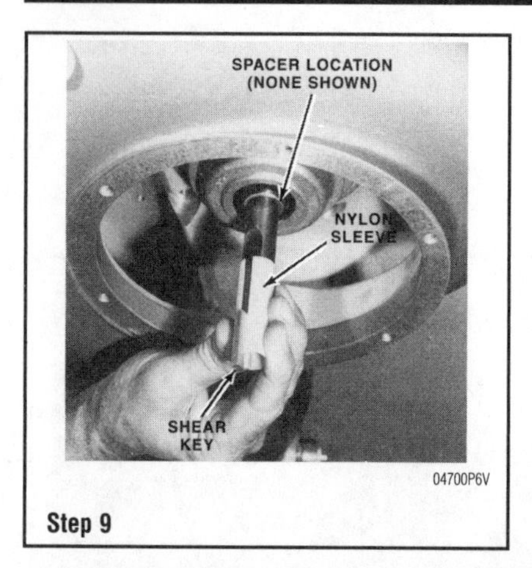

Step 9

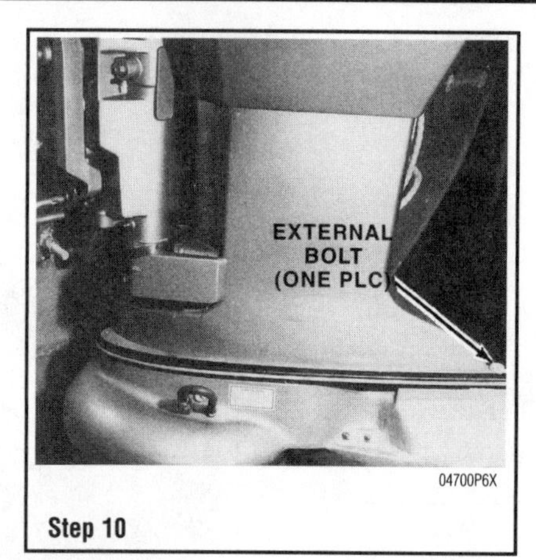

Step 10

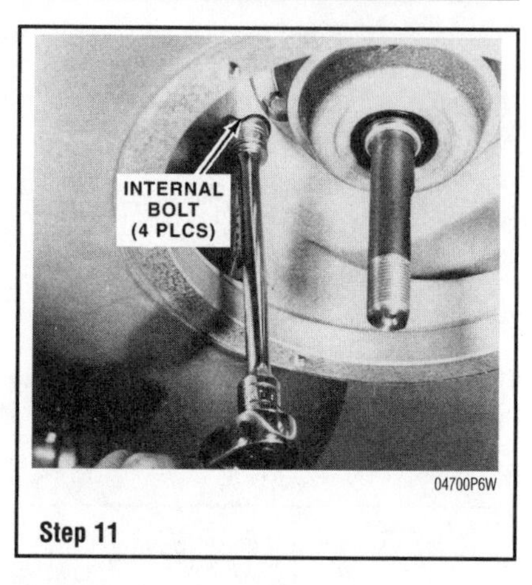

Step 11

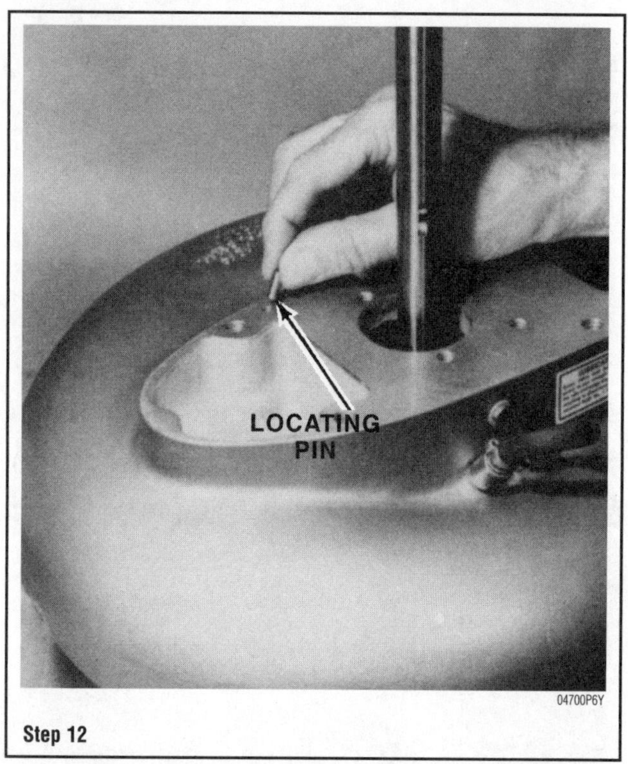

Step 12

10. One external bolt and four internal bolts are used to secure the jet drive to the intermediate housing. The external bolt is located at the aft end of the anti-cavitation plate.

11. The four internal bolts are located inside the jet drive housing, as indicated in the accompanying illustration. Remove the five attaching bolts.

12. Lower the jet drive from the intermediate housing. Remove the locating pin from the forward starboard side (or center forward, depending on the model being serviced) of the upper jet housing.

13. Remove the locating pin from the aft end of the housing. This pin and the one removed in the previous step should be of identical size.

14. Remove the four bolts and washers from the water pump housing.

Pull the water pump housing, the inner cartridge and the water pump impeller, up and free of the driveshaft. Next, remove the outer gasket, the steel plate and the inner gasket.

15. Remove the two small locating pins and lift the aluminum spacer up and free of the drive shaft.

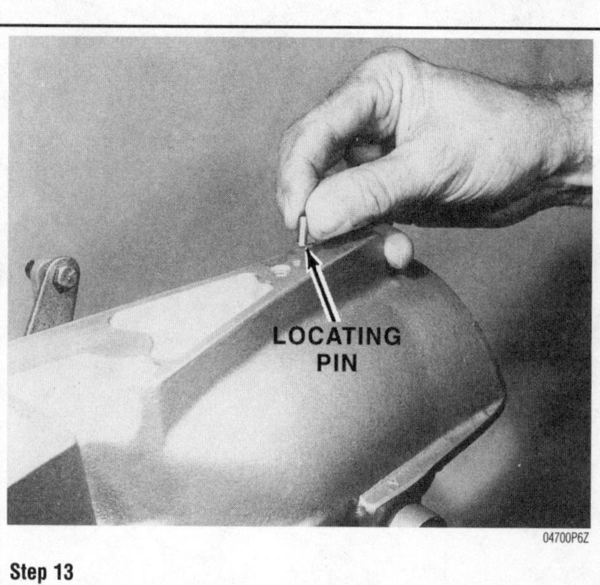

Step 13

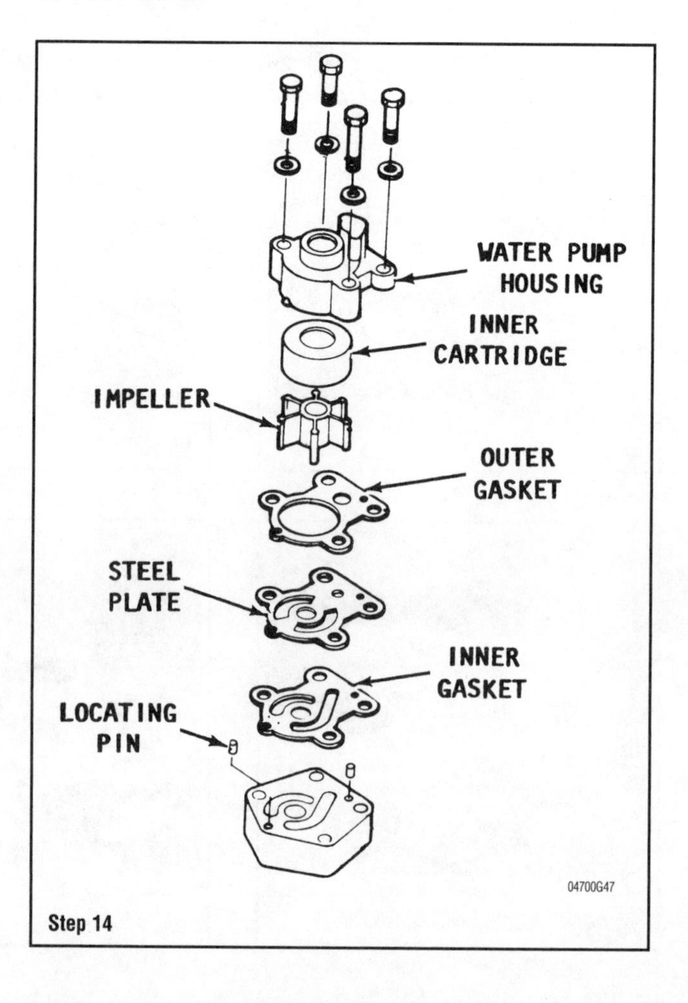

Step 14

Step 15

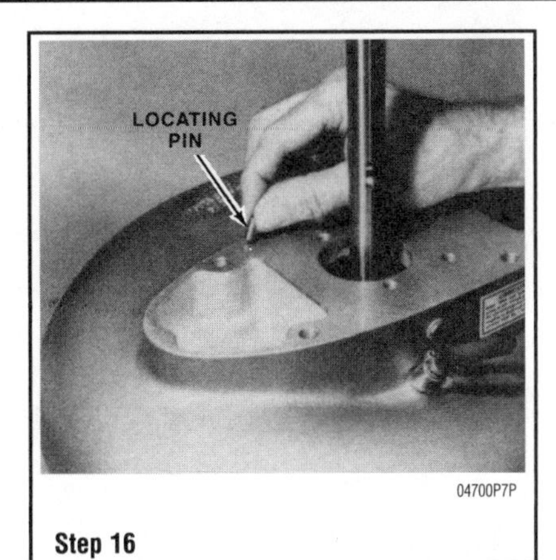

Step 16

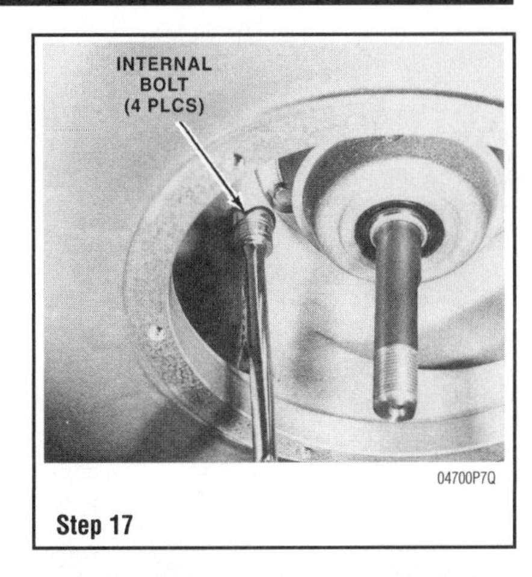

Step 17

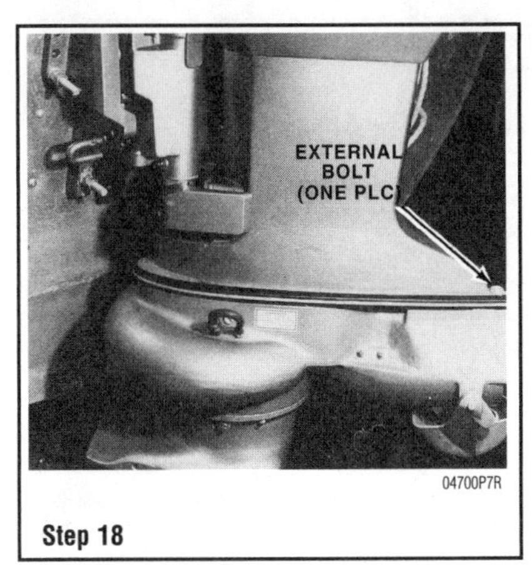

Step 18

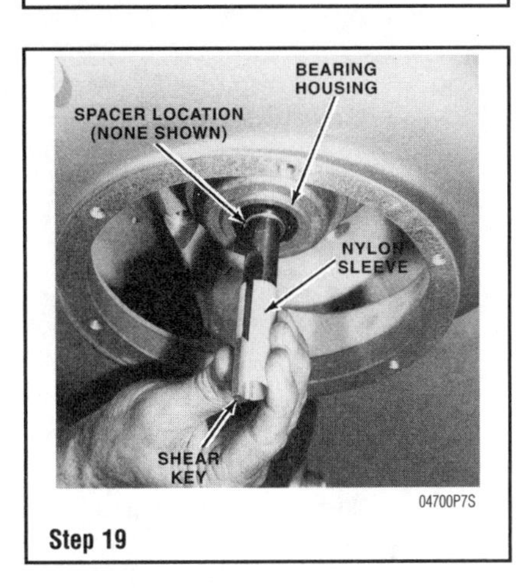

Step 19

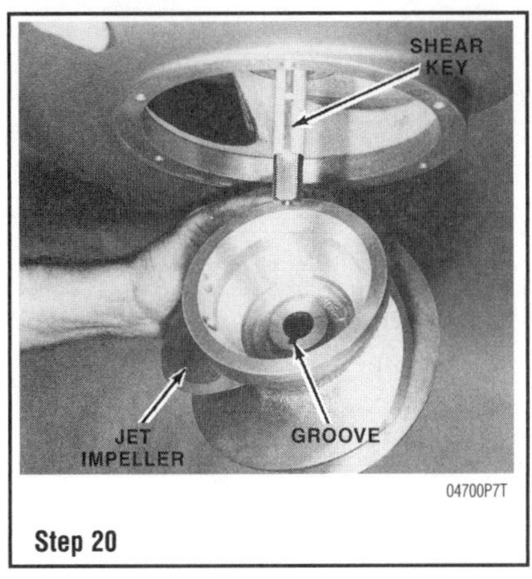

Step 20

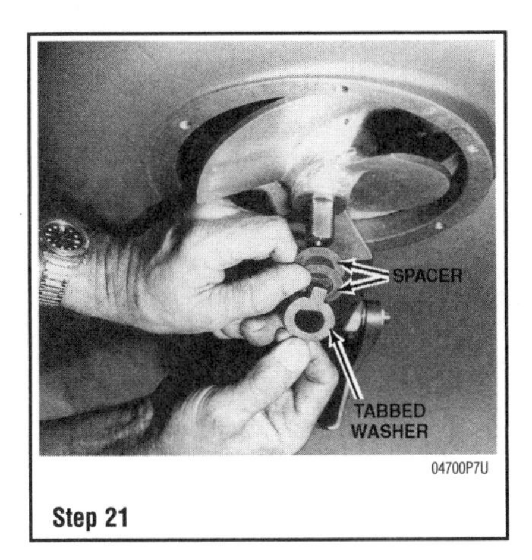

Step 21

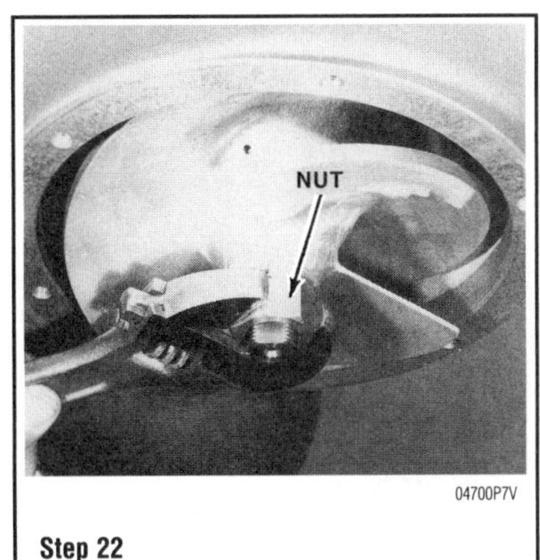

Step 22

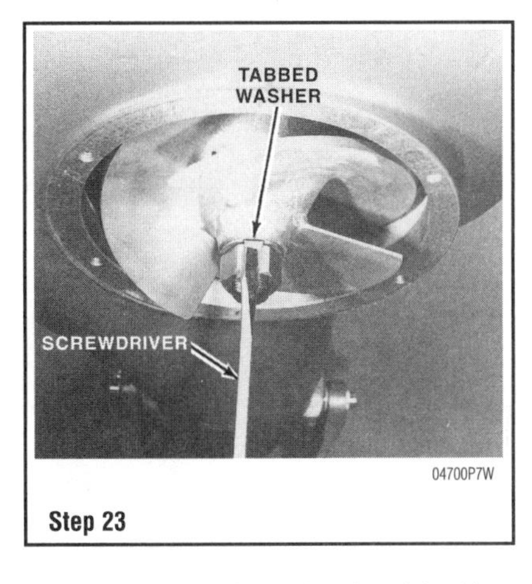

Step 23

Remove the driveshaft and bearing assembly from the housing.

Remove the large thick adaptor plate from the intermediate housing. This plate is secured with seven bolts and lock washers. Lower the adaptor plate from the intermediate housing and remove the two small locating pins, one on the forward port side and another from the last aft hole in the adaptor plate. Both pins are identical in size.

To install:

16. Install the other small locating pin into the forward starboard side (or center forward end, depending on the model being serviced).
17. Raise the jet drive unit up and align it with the intermediate housing, with the small pins indexed into matching holes in the adapter plate. Install the four internal bolts.
18. Install the one external bolt at the aft end of the anti-cavitation plate. Tighten all bolts to a torque value of 11 ft. lbs. (15Nm).
19. Place the required number of spacers up against the bearing housing. Slide the nylon sleeve over the driveshaft and insert the shear key into the slot of the nylon sleeve with the key resting against the flattened portion of the driveshaft.
20. Slide the jet impeller up onto the driveshaft, with the groove in the impeller collar indexing over the shear key.
21. Place the remaining spacers over the driveshaft.
22. Tighten the nut to a torque value of 17 ft. lbs. (23Nm). If neither of the two tabs on the tabbed washer aligns with the sides of the nut, remove the nut and washer. Invert the tabbed washer. Turning the washer over will change the tabs by approximately 15°. Install and tighten the nut to the required torque value. The tabbed washer is designed to align with the nut in one of the two positions described.
23. Bend the tabs up against the nut to prevent the nut from backing off and becoming loose.
24. Install the intake grille onto the jet drive housing with the slots facing aft. Install and tighten the six securing bolts. Tighten ¼ in. bolts to a torque value of 5 ft. lbs. (7Nm). Tighten 5⁄16 in. bolts to 11 ft. lbs. (15Nm).

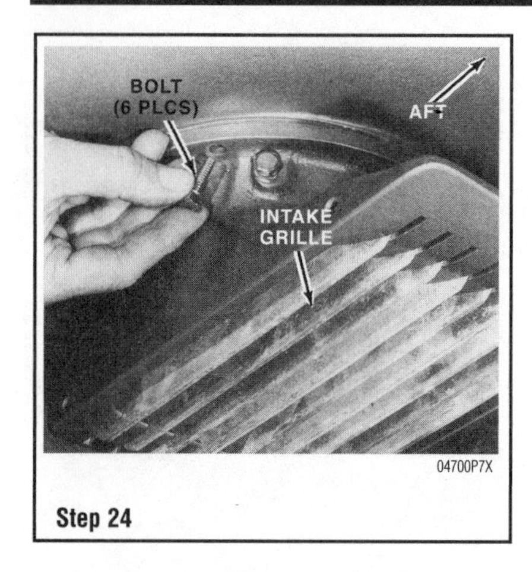

04700P7X
Step 24

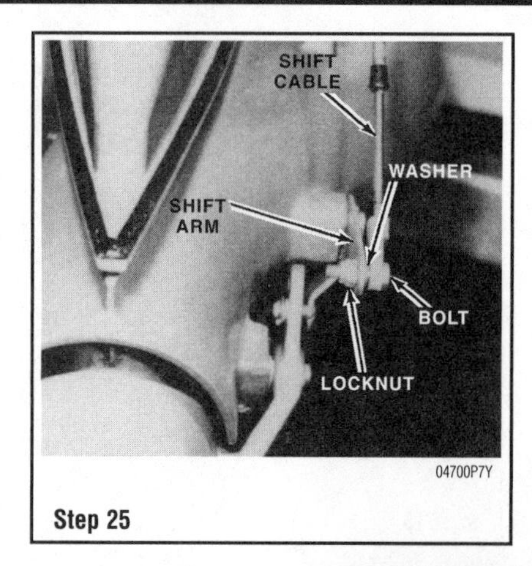

04700P7Y
Step 25

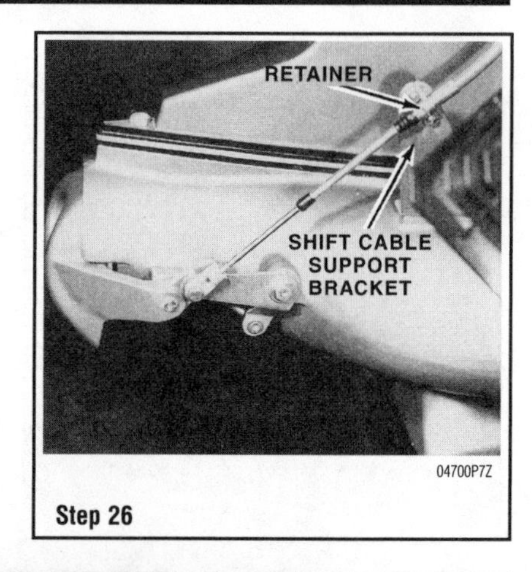

04700P7Z
Step 26

25. Slide the bolt through the end of the shift cable, washer and into the shift arm. Install the locknut onto the bolt and tighten the bolt securely.

26. Install the shift cable against the shift cable support bracket and secure it in place with the two bolts.

ADJUSTMENT

Cable Alignment And Free Play

➧ See Accompanying Illustrations

1. Move the shift lever downward into the forward position. The leaf spring should snap over on top of the lever to lock it in position.

04700P8A
Step 1

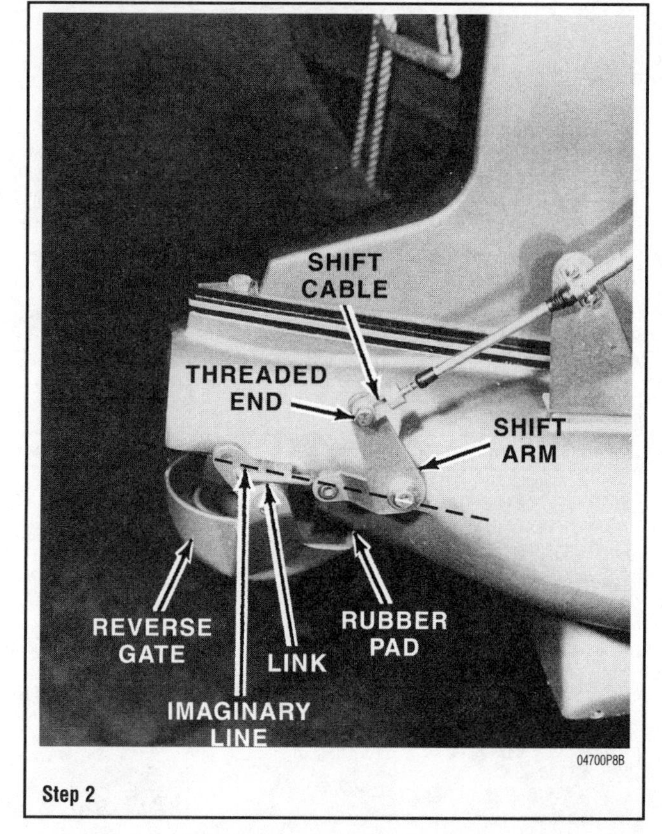

04700P8B
Step 2

2. Remove the locknut, washer and bolt from the threaded end of the shift cable. Push the reverse gate firmly against the rubber pad on the underside of the jet drive housing.

Check to be sure the link between the reverse gate and the shift arm is hooked into the LOWER hole on the gate.

Hold the shift arm up until the link rod and shift arm axis form an imaginary straight line, as indicated in the accompanying illustration. Adjust the length of the shift cable by rotating the threaded end, until the cable can be installed back onto the shift arm without disturbing the imaginary line. Pass the nut through the cable end, washer and shift arm. Install and tighten the locknut.

Neutral Stop Adjustment

➧ See Accompanying Illustrations

In the forward position, the reverse gate is neatly tucked underneath and clear of the exhaust jet stream.

In the reverse position, the gate swings up and blocks the jet stream deflecting the water in a forward direction under the jet housing to move the boat sternward.

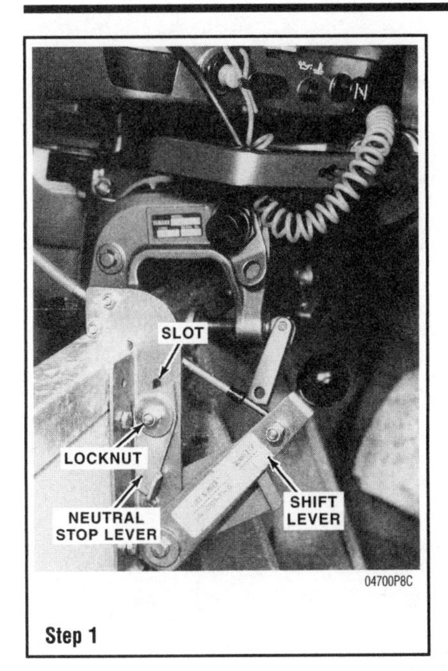

Step 1

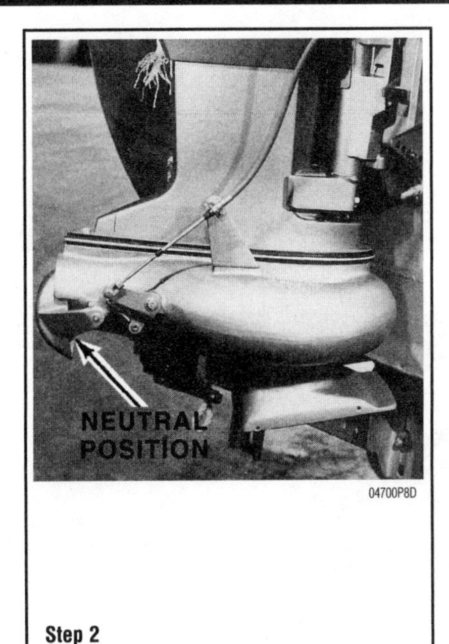

Step 2

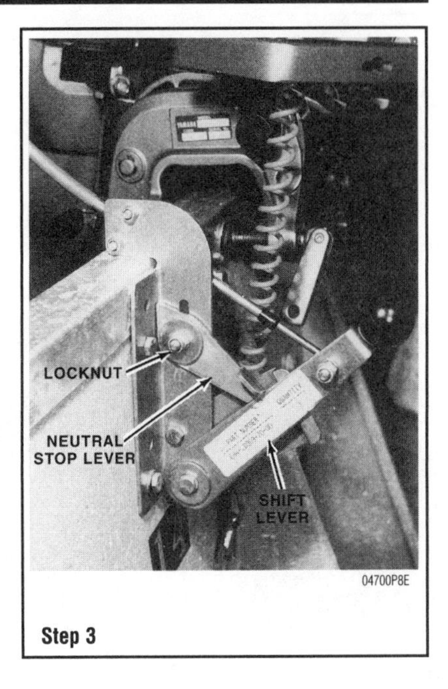

Step 3

In the neutral position, the gate assumes a happy medium—a balance between forward and reverse when the powerhead is operating at IDLE speed. Actually, the gate is deflecting some water to prevent the boat from moving forward, but not enough volume to move the boat sternward.

**** WARNING**

The gate must be properly adjusted for safety of boat and passengers. Improper adjustment could cause the gate to swing up to the reverse position while the boat is moving forward causing serious injury to boat or passengers.

1. Loosen, but do not remove the locknut on the neutral stop lever. Check to be sure the lever will slide up and down along the slot in the shift lever bracket.

➡The following procedure must be performed with the boat and jet drive in a body of water. Only with the boat in the water can a proper jet stream be applied against the gate for adjustment purposes.

**** CAUTION**

Water must circulate through the lower unit to the powerhead anytime the powerhead is operating to prevent damage to the water pump in the lower unit. Just five seconds without water will damage the water pump impeller.

2. Start the powerhead and allow it to operate only at IDLE speed. With the neutral stop lever in the down position, move the shift lever until the jet stream forces on the gate are balanced. Balanced means the water discharged is divided in both directions and the boat moves neither forward nor sternward. The gate is then in the neutral position with the powerhead at idle speed.
3. Move the neutral stop lever up against the shift lever until the stop lever barely makes contact with the shift lever. Tighten the locknut to maintain this new adjusted position. Shut down the powerhead.

➡The reverse gate may not swing to the full up position in reverse gear after the previous steps have been performed. Do not be concerned. This condition is acceptable, because water pressure in reverse will close the gate fully under normal operation.

Trim Adjustment

See Accompanying Illustration

1. During operation, if the boat tends to pull to port or starboard, the flow fins may be adjusted to correct the condition. These fins are located at the top and bottom of the exhaust tube.

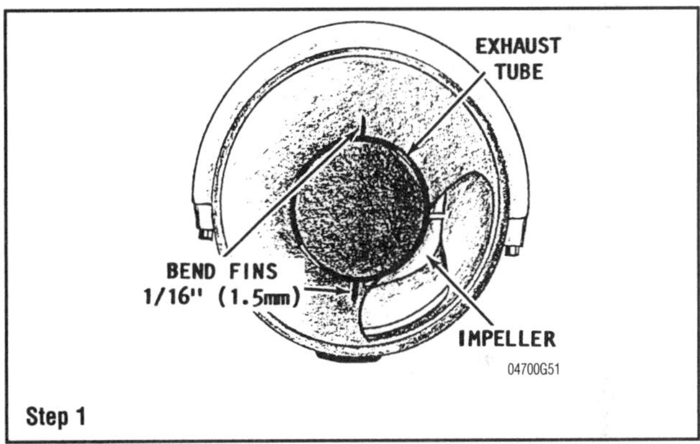

Step 1

If the boat tends to pull to starboard, bend the trailing edge of each fin approximately 1/16 in. (1.5mm) toward the starboard side of the jet drive. Naturally, if the boat tends to pull to port, bend the fins toward the port side.

DISASSEMBLY

1. Remove the locating pin from the forward starboard side (or center forward, depending on the model being serviced) of the upper jet housing.

➡There will be a total of six locating pins to be removed in the following steps. Make careful note of the size and location of each when they are removed, as an assist during assembling.

2. Remove the locating pin from the aft end of the housing. This pin and the one removed in the previous step should be of identical size.
3. Remove the four bolts and washers from the water pump housing.

Pull the water pump housing, the inner cartridge and the water pump impeller, up and free of the driveshaft. Remove the Woodruff key from its recess in the driveshaft. Next, remove the outer gasket, the steel plate and the inner gasket.

4. Remove the two small locating pins and lift the aluminum spacer up and free of the driveshaft.
5. Remove the driveshaft and bearing assembly from the housing.

Remove the large thick adapter plate from the intermediate housing. This plate is secured with seven bolts and lock-washers. Lower the adapter plate from the intermediate housing and remove the two small locating pins, one on the forward port side and another from the last aft hole in the adapter plate. Both pins are identical size.

CLEANING & INSPECTING

See Figures 192 and 193

Wash all parts, except the driveshaft assembly, in solvent and blow them dry with compressed air. Rotate the bearing assembly on the driveshaft to inspect the bearings for rough spots, binding and signs of corrosion or damage.

Saturate a shop towel with solvent and wipe both extensions of the driveshaft.

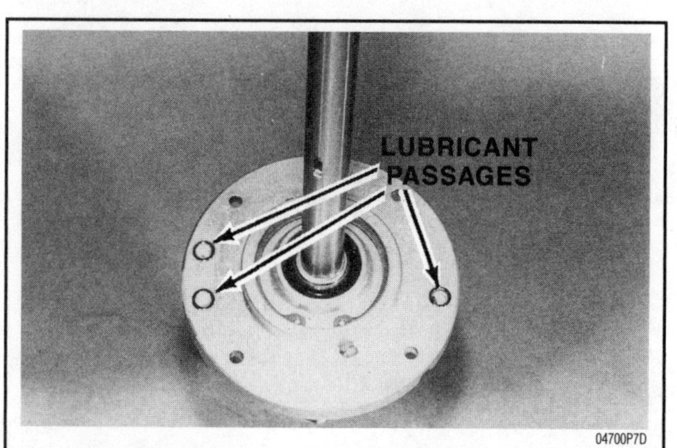

Fig. 192 Take extra precautions to prevent solvent from entering the lubrication passages

Bearing Assembly

See Figure 194

Lightly wipe the exterior of the bearing assembly with the same shop towel. Do not allow solvent to enter the three lubricant passages of the bearing assembly. The best way to clean these passages is not with solvent—because any solvent remaining in the assembly after installation will continue to dissolve good useful lubricant and leave bearings and seals dry. This condition will cause bearings to fail through friction and seals to dry up and shrink—losing their sealing qualities.

The only way to clean and lubricate the bearing assembly is after installation to the jet drive—via the exterior lubrication fitting.

If the old lubricant emerging from the hose coupling is a dark, dirty, gray color, the seals have already broken down and water is attacking the bearings. If such is the case, it is recommended the entire driveshaft bear-ing assembly be taken to the dealer for service of the bearings and seals.

Dismantling Bearing Assembly

A complicated procedure must be followed to dismantle the bearing assembly including torching off the bearing housing. Naturally, excessive heat might ruin the seals and bearings. Therefore, the best recommendation is to leave this part of the service work to the experts at your local Honda dealership.

Driveshaft and Associated Parts

Inspect the threads and splines on the driveshaft for wear, rounded edges, corrosion and damage.

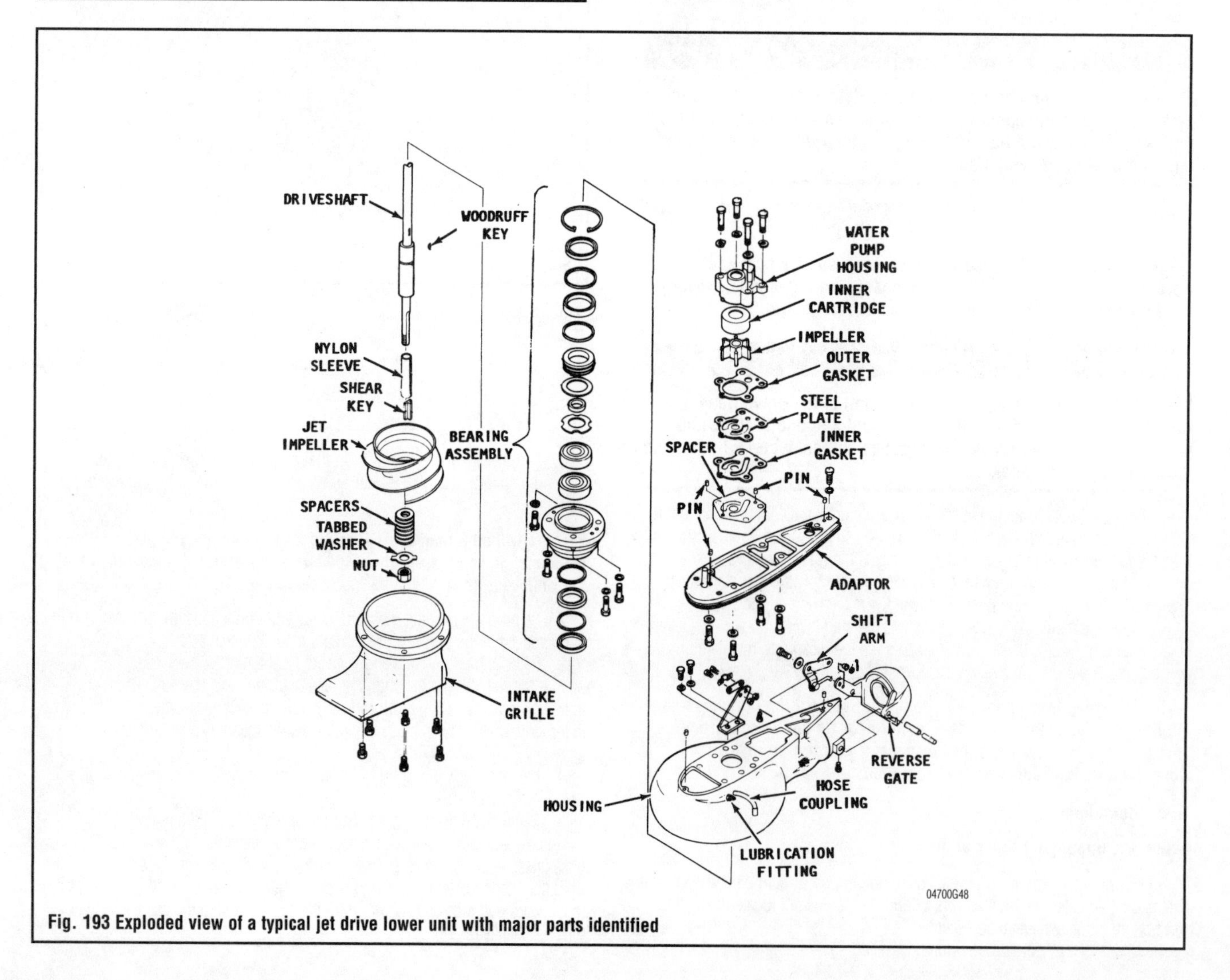

Fig. 193 Exploded view of a typical jet drive lower unit with major parts identified

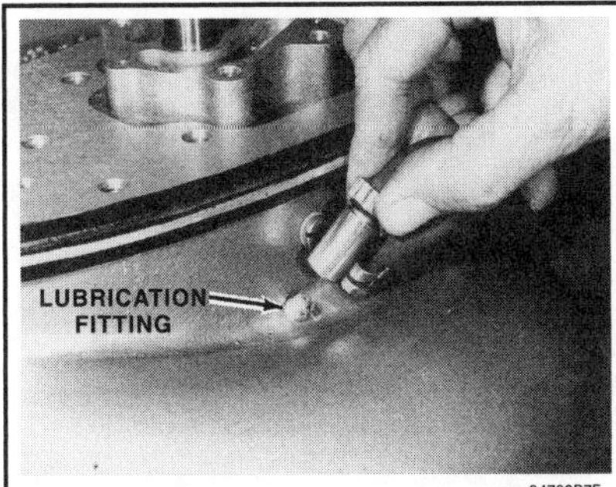

04700P7E

Fig. 194 Cleaning and lubricating the bearing assembly is best accomplished by completely replacing the old lubricant

04700P7F

Fig. 195 The slats of the grille must be carefully inspected and any bent slats straightened for maximum performance of the jet drive

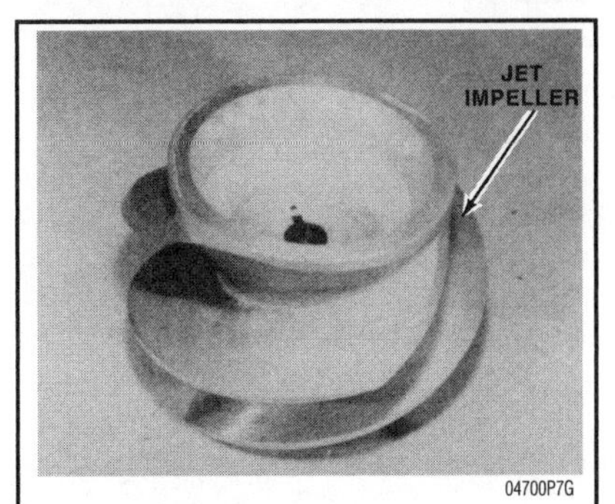

04700P7G

Fig. 196 The edges of the jet impeller should be kept as sharp as possible for maximum jet drive efficiency

Carefully check the driveshaft to verify the shaft is straight and true without any sign of damage.

Inspect the jet drive housing for nicks, dents, corrosion, or other signs of damage. Nicks may be removed with No. 120 and No. 180 emery cloth.

Reverse Gate

Inspect the gate and its pivot points. Check the swinging action to be sure it moves freely the entire distance of travel without binding.

Inspect the slats of the water intake grille for straightness. Straighten any bent slats, if possible. Use the utmost care when prying on any slat, as they tend to break if excessive force is applied. Replace the intake grille if a slat is lost, broken, or bent and cannot be repaired. The slats are spaced evenly and the distance between them is critical, to prevent large objects from passing through and becoming lodged between the jet impeller and the inside wall of the housing.

Jet Impeller

See Figure 195

The jet impeller is a precisely machined and dynamically balanced aluminum spiral. Observe the drilled recesses at exact locations to achieve this delicate balancing. Some of these drilled recesses are clearly shown in the accompanying illustration.

Excessive vibration of the jet drive may be attributed to an out-of-balance condition caused by the jet impeller being struck excessively by rocks, gravel or cavitation burn.

The term cavitation burn is a common expression used throughout the world among people working with pumps, impeller blades and forceful water movement.

Burns on the jet impeller blades are caused by cavitation air bubbles exploding with considerable force against the impeller blades. The edges of the blades may develop small dime size areas resembling a porous sponge, as the aluminum is actually eaten by the condition just described.

Excessive rounding of the jet impeller edges will reduce efficiency and performance. Therefore, the impeller should be inspected at regular intervals.

If rounding is detected, the impeller should be placed on a work bench and the edges restored to as sharp a condition as possible, using a file. Draw the file in only one direction. A back-and-forth motion will not produce a smooth edge. Take care not to nick the smooth surface of the jet impeller. Excessive nicking or pitting will create water turbulence and slow the flow of water through the pump.

Inspect the shear key. A slightly distorted key may be reused although some difficulty may be encountered in assembling the jet drive. A cracked shear key should be discarded and replaced with a new key.

Water Pump

See Figure 196

Clean all water pump parts with solvent and then blow them dry with compressed air. Inspect the water pump housing for cracks and distortion, possibly caused from overheating. Inspect the steel plate, the thick aluminum spacer and the water pump cartridge for grooves and/or rough spots. If possible always install a new water pump impeller while the jet drive is disassembled. A new water pump impeller will ensure extended satisfactory service and give peace of mind to the owner. If the old water pump impeller must be returned to service, never install it in reverse of the original direction of rotation. Installation in reverse will cause premature impeller failure.

If installation of a new water pump impeller is not possible, check the sealing surfaces and be satisfied they are in good condition. Check the upper, lower and ends of the impeller vanes for grooves, cracking and wear. Check to be sure the indexing notch of the impeller hub is intact and will not allow the impeller to slip.

ASSEMBLING

See Accompanying Illustration

Identify the two small locating pins used to index the large thick adapter plate to the intermediate housing. Insert one pin into the last hole aft on the topside of the plate. Insert the other pin into the hole forward toward the port side, as shown.

Lift the plate into place against the intermediate housing with the locating pins indexing with the holes in the intermediate housing. Secure the plate with the five (or seven) bolts.

On the five bolt model, one of the five bolts is shorter than the other four. Install the short bolt in the most aft location.

Tighten the long bolts to a torque value of 22 ft. lbs. (30Nm). Tighten the short bolt to a torque value of 11 ft. lbs. (15Nm).

1. Place the driveshaft bearing assembly into the jet drive housing. Rotate the bearing assembly until all bolt holes align. There is only one correct position.

If installing a new jet impeller, place all eight spacers at the lower or nut end of the impeller and skip the following step.

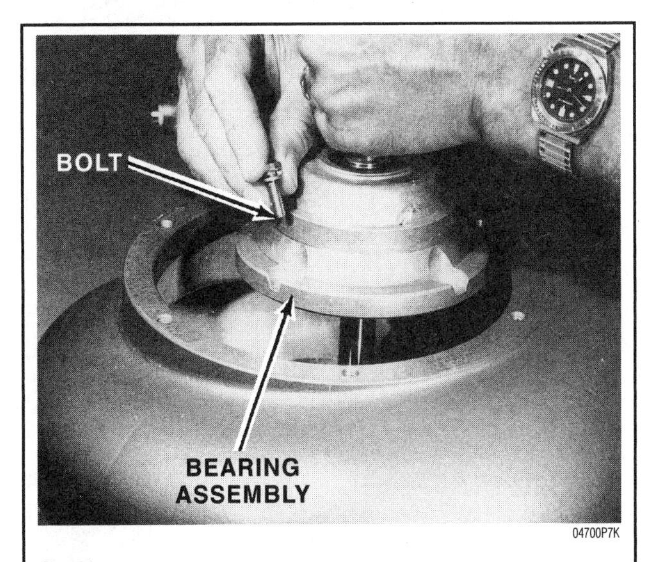

04700P7K

Step 1

Shimming Jet Impeller

➧ See Accompanying Illustrations

1. The clearance between the outer edge of the jet drive impeller and the water intake housing cone wall should be maintained at approximately 1⁄32 in. (0.8mm). This distance can be visually checked by shining a flashlight up through the intake grille and estimating the distance between the impeller and the casing cone, as indicated in the accompanying illustrations. It is not humanly possible to accurately measure this clearance, but by observing closely and estimating the clearance, the results should be fairly accurate.

After continued use, the clearance will increase. The spacers previously removed are used to position the impeller along the driveshaft with a desired clearance of 1⁄32 in. (0.8mm) between the jet impeller and the housing wall.

2. Spacers are used depending on the model being serviced. When new, all spacers are located at the tapered (or nut) end of the impeller. As the clearance increases, the spacers are transferred from the tapered (nut) end and placed at the wide (intermediate housing) end of the jet impeller.

This procedure is best accomplished while the jet drive is removed from the intermediate housing.

Secure the driveshaft with the attaching hardware. Installation of the shear key and nylon sleeve is not vital to this procedure. Place the unit on a convenient work bench. Shine a flashlight through the intake grille into the housing cone and eyeball the clearance between the jet impeller and the cone wall, as indicated in the accompanying line drawing. Move spacers one-at-a-time from the tapered end to the wide end to obtain a satisfactory clearance. Dismantle the driveshaft and note the exact count of spacers at both ends of the bearing assembly. This count will be recalled later during assembly to properly install the jet impeller.

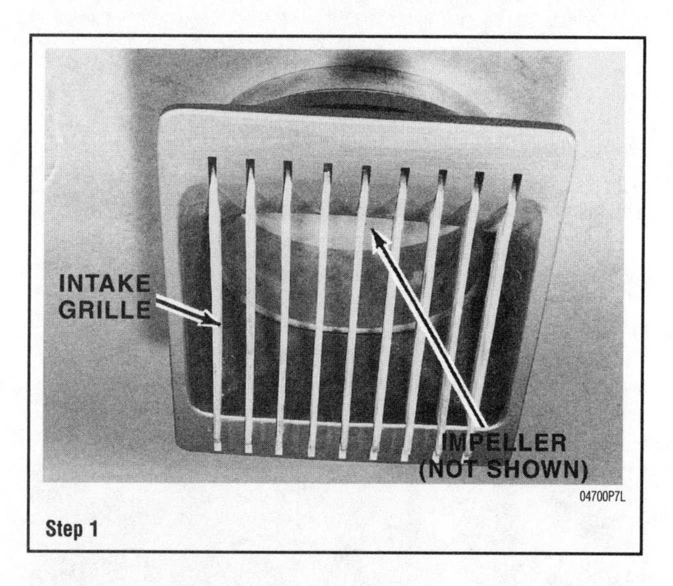

Step 1

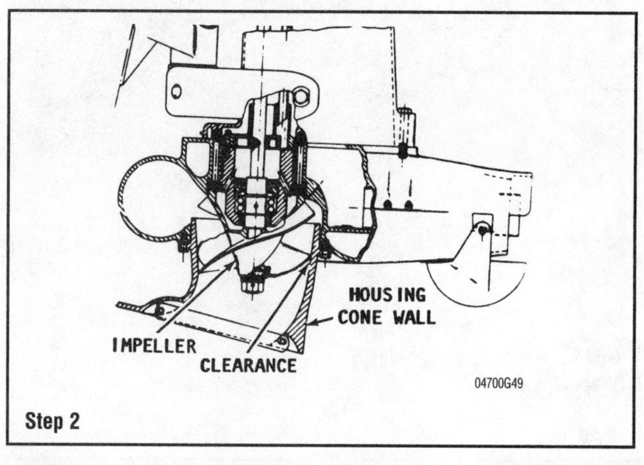

Step 2

Water Pump Assembling

➧ See Accompanying Illustrations

1. Place the aluminum spacer over the driveshaft with the two holes for the indexing pins facing upward. Fit the two locating pins into the holes of the spacer.

➡The manufacturer recommends no sealant be used on either side of the water pump gaskets.

2. Slide the inner water pump gasket (the gasket with two curved openings) over the driveshaft. Position the gasket over the two locating pins. Slide the steel plate down over the driveshaft with the tangs on the plate facing downward and with the holes in the plate indexed over the two locating pins.

Check to be sure the tangs on the plate fit into the two curved openings of the gasket beneath the plate. Now, slide the outer gasket (the gasket with the large center hole) over the driveshaft. Position the gasket over the two locating pins.

Fit the Woodruff key into the driveshaft. Just a dab of grease on the key will help to hold the key in place. Slide the water pump impeller over the driveshaft with the rubber membrane on the top side and the keyway in the impeller indexed over the Woodruff key. Take care not to damage the membrane. Coat the impeller blades with Hondaline Grease or equivalent water resistant lubricant.

Install the insert cartridge, the inner plate and finally the water pump housing over the driveshaft. Rotate the insert cartridge counterclockwise over the impeller to tuck in the impeller vanes. Seat all parts over the two locating pins.

➡On some models, two different length bolts are used at this location.

Tighten the four bolts to a torque value of 11 ft. lbs. (15Nm).

3. Install one of the small locating pins into the aft end of the jet drive housing.

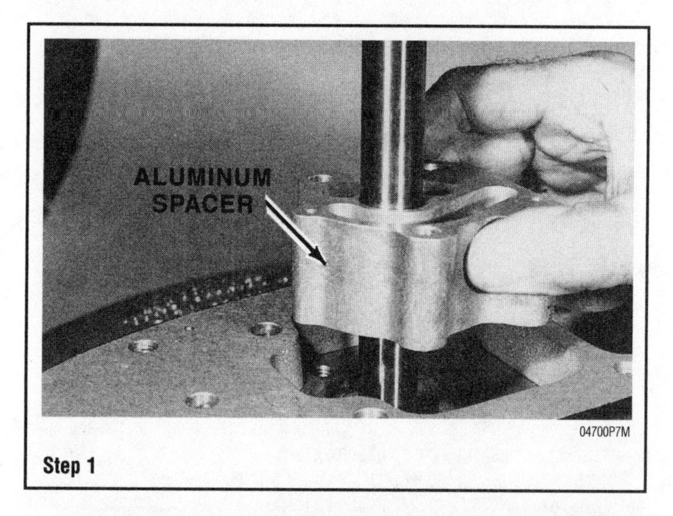

Step 1

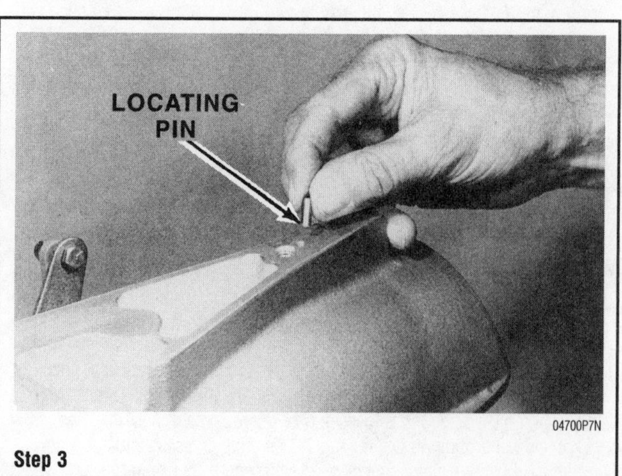

Step 3

MANUAL TILT 9-2
DESCRIPTION AND OPERATION 9-2
SERVICING 9-2
GAS ASSISTED TILT 9-2
DESCRIPTION AND OPERATION 9-2
GAS ASSIST DAMPER 9-3
TESTING 9-3
REMOVAL & INSTALLATION 9-3
OVERHAUL 9-4
POWER TRIM AND TILT 9-4
DESCRIPTION AND OPERATION 9-4
TRIM/TILT MOTOR 9-8
TESTING 9-8
REMOVAL & INSTALLATION 9-8
DISASSEMBLY 9-8
CLEANING & INSPECTION 9-9
ASSEMBLY 9-10
HYDRAULIC UNIT 9-10
CHECKING FLUID LEVEL 9-10
BLEEDING THE SYSTEM 9-10
TESTING 9-10
REMOVAL & INSTALLATION 9-11
DISASSEMBLY 9-14
CLEANING & INSPECTION 9-15
ASSEMBLY 9-15
TRIM/TILT SWITCH 9-16
TRIM ANGLE SENSOR 9-16
TESTING 9-16
REMOVAL & INSTALLATION 9-16

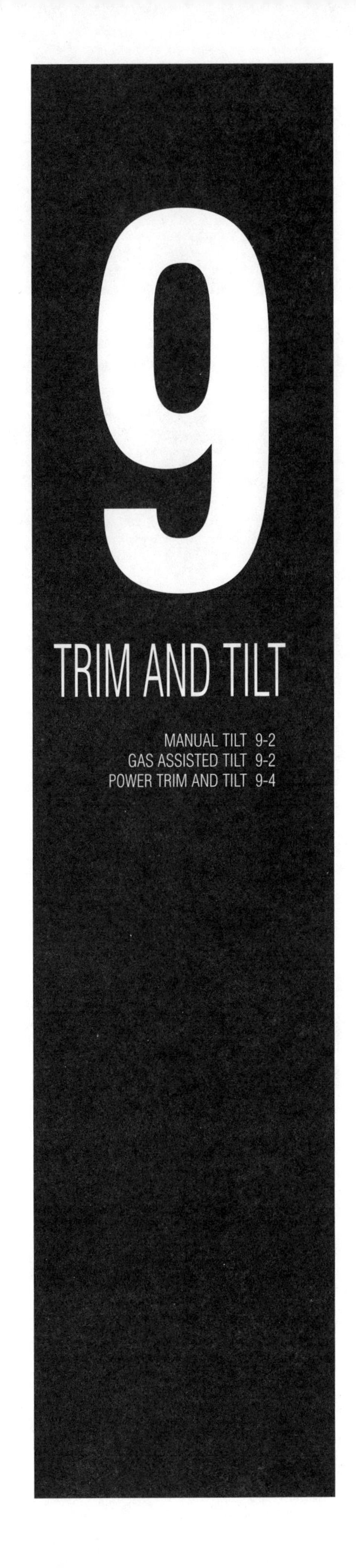

9 TRIM AND TILT

MANUAL TILT 9-2
GAS ASSISTED TILT 9-2
POWER TRIM AND TILT 9-4

MANUAL TILT

Description and Operation

See Figures 1 and 2

All outboard installations are equipped with some means of raising or lowering (pivoting) the outboard for efficient operation under various load, boat design, water conditions, and for trailering to and from the water. By pivoting the outboard, the correct trim angle can be achieved to ensure maximum performance and fuel economy as well as a more comfortable ride for the crew and passengers.

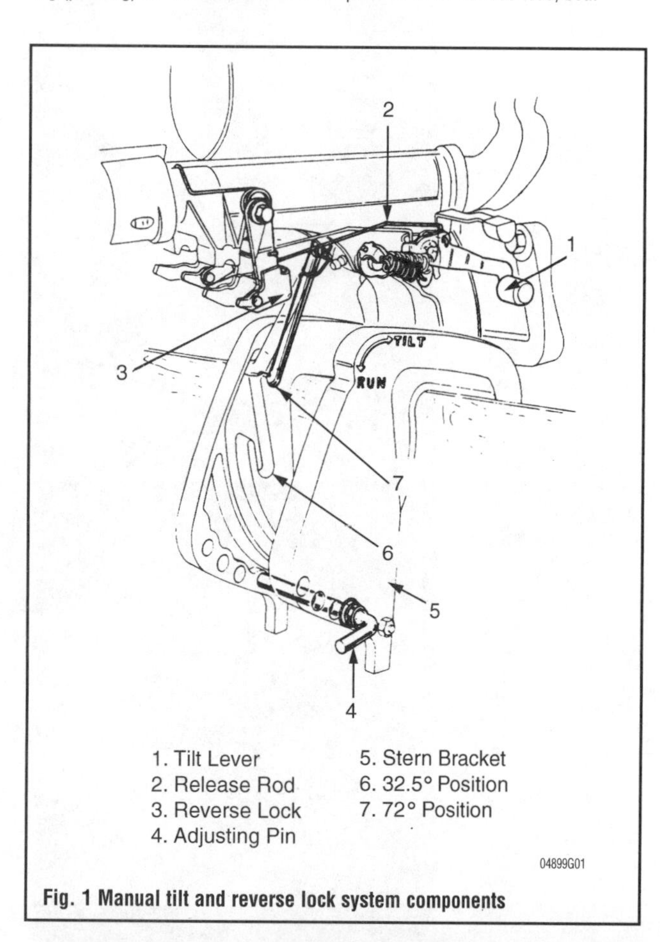

Fig. 1 Manual tilt and reverse lock system components

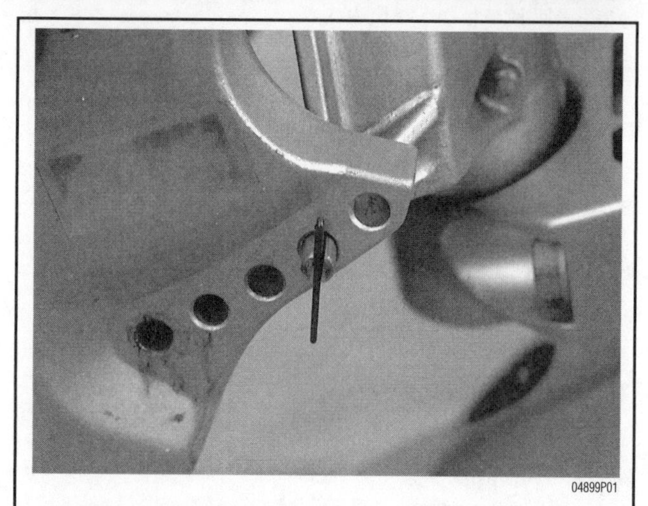

Fig. 2 Adjusting pin on a BF35A outboard, others similar

The manual tilt mechanism is used to tilt the outboard in relation to the stern bracket. To adjust the outboard angle, the tilt lever is moved to the **TILT** position. This disengages the release rod and causes the reverse lock to disengage from the adjusting pin. The unit can be set at varying angles by raising or lowering the outboard. To release the tilt mechanism, return the tilt lever to the **RUN** position, then raise and lower the extension case slightly.

➡The tilt lever should be kept in the RUN position whenever the outboard is operating.

SERVICING

Service procedures for the manual tilt system are confined to general lubrication and inspection. If individual components should wear or break, replacement of the defective components is necessary.

GAS ASSISTED TILT

Description and Operation

See Figures 3 and 4

The gas assisted tilt system consists of a single shock absorber. The shock absorber contains a high pressure gas chamber located in the upper portion of the cylinder bore above the piston assembly. The piston contains a down relief valve and an absorber relief valve. Below the piston assembly, the lower cylinder bore contains an oil chamber. This lower chamber is connected to the upper chamber above the piston by a hydraulic line with a manual check valve.

The manual check valve is activated by the tilt lever, when the lever is rotated from the **LOCK** position to the **FREE** position. The check valve cam rotates and pushes the check valve push rod against the check valve. This action opens the check valve and allows hydraulic fluid to flow from the lower chamber through the hydraulic line, past the open check valve and into the upper chamber. When the outboard unit is tilted up, the volume below the piston decreases, and the volume above the piston increases until the piston has reached the bottom of its stroke. In this position all fluid is contained above the piston.

The tilt lever is then rotated to the **LOCK** position to engage the clamp bracket. When the tilt lever is in this position, the manual valve push rod rests on a flat spot of the manual valve cam and releases pressure on the check valve. Releasing pressure on the check valve closes off the hydraulic line and the flow of hydraulic fluid.

To lower the outboard unit from the full up and locked position, the tilt lever is again rotated from the **LOCK** position to the **FREE** position. The manual check valve cam rotates and pushes the manual check valve push rod against the check valve and opens the valve. As the outboard unit is tilted down, the piston moves up and compresses the fluid in the upper chamber. The fluid pressure overcomes the relief valve spring and opens the relief valve. The valve in the open position permits hydraulic fluid to flow through the piston from the upper chamber to the lower chamber.

➡When the manual check is open, the valve will allow hydraulic fluid to flow in only one direction— from the lower chamber to the upper chamber.

During normal cruising, the tilt lever is set in the **LOCK** position. The manual check valve is closed to prevent the outboard unit from being tilted up by water pressure against the propeller when the unit is in reverse gear. When the unit is in forward gear, the outboard is held in position by the tilt pin through the swivel bracket.

In the event the outboard lower unit should strike an underwater object while the boat is underway, the piston would be forced down. The hydraulic fluid

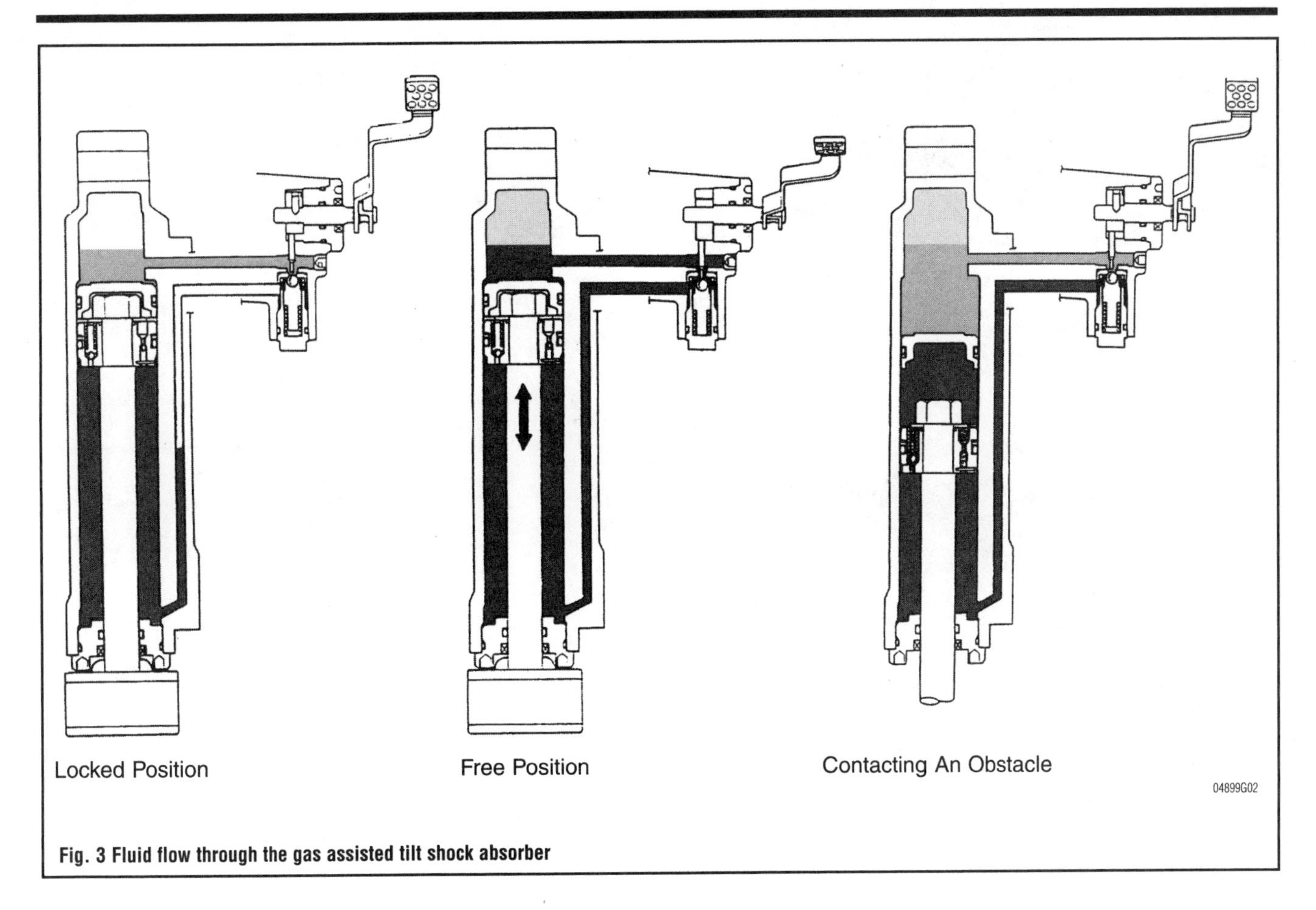

Fig. 3 Fluid flow through the gas assisted tilt shock absorber

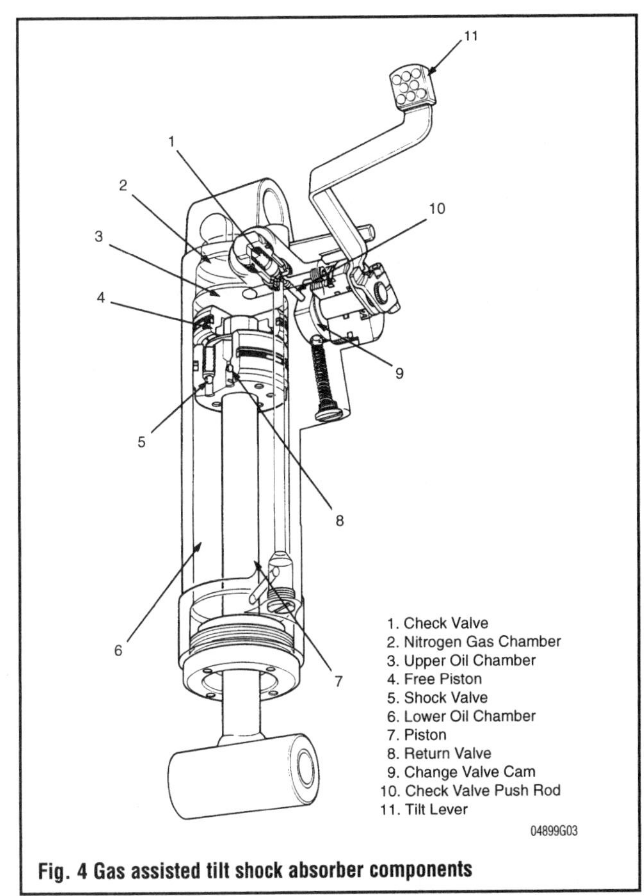

Fig. 4 Gas assisted tilt shock absorber components

below the piston would be under pressure with no escape because the manual check valve is closed. To prevent rupture of the hydraulic line, a safety relief valve is incorporated in the piston. This relief valve permits fluid to pass through the piston from the lower chamber to the upper chamber through the absorber relief valve. After the outboard has passed the obstacle, the fluid returns to the lower chamber through the piston and the down relief valve, because the piston is pushed up.

Gas Assist Damper

TESTING

Testing of the gas assisted damper is a fairly simple operation. Since the operation of the cylinder is solely reliant on the action of the gas pressure on the piston, if the damper fails to provide the appropriate amount of lifting assistance and the steering tube adjustment is not set too tight, then it can be assumed that the damper is faulty.

REMOVAL & INSTALLATION

See Figures 5 and 6

1. Remove the outboard from the boat and lay it on its side.
2. Place the tilt lever in the **FREE** position.
3. Position and support the outboard in its fully tilted position.
4. Remove the steering tube cap and loosen the self locking nut.
5. Remove the E-ring and slide the upper cylinder pin from the mounting frame.
6. Pivot the damper away from the mounting frame taking care to catch the wave washers that fall out.
7. Remove the damper through bolt from the stern bracket and lift out the distance collar.
8. Remove the damper assembly from the stern bracket.
9. Remove the cylinder bushings from the damper.

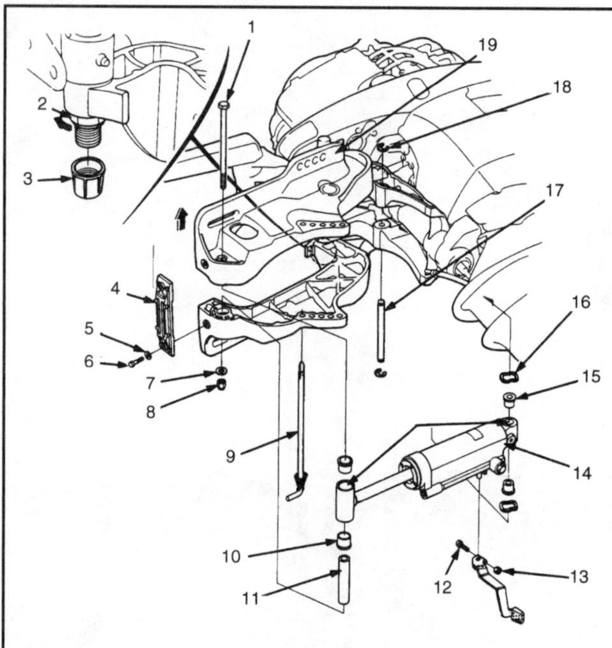

1. Hex Bolt
2. 7/8 UNF Self-Locking Nut
3. Steering Tube Cap
4. Anode
5. Plain Washer
6. Hex Bolt
7. Plain Washer
8. Self-Locking Nut
9. Transom Angle Adjusting Rod
10. Lower Cylinder Bushing
11. Distance Coller
12. Hex Bolt
13. Hex Nut
14. Gas Assist Damper Assembly
15. Upper Cylinder Bushing
16. Valve Washer
17. Upper Cylinder Pin
18. E-Ring
19. Stern Bracket

04899G10

Fig. 5 The gas assist damper is secured between the stern bracket and the mounting frame by the upper cylinder pin and a through bolt

➡Store the gas assisted damper assembly vertically with the upper cylinder bushing upward. Never store the damper horizontally or with the distance collar facing upward.

To install:

10. Lubricate and install the cylinder bushings on the damper.
11. Position the damper assembly on the stern bracket.

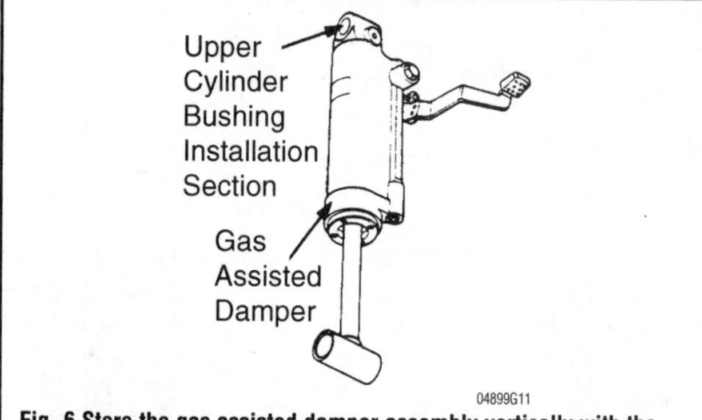

04899G11

Fig. 6 Store the gas assisted damper assembly vertically with the upper cylinder bushing upward

12. Install the damper through bolt and distance collar on the stern bracket. Tighten the through bolt to 25 ft. lbs. (35Nm).
13. Pivot the damper toward the mounting frame.
14. After positioning the wave washers, install the upper cylinder pin on the mounting frame and install the E-ring to secure it.
15. Install the self locking nut and tighten to 25 ft. lbs. (35Nm). Install the steering tube cap.
16. Install the outboard on the boat and test the gas assisted tilt assembly for proper operation.

OVERHAUL

Overhaul procedures for the gas assist damper are confined to removal of the end cap, removing the piston and replacing the O-rings. A spanner wrench is required to remove the end cap. Even with the tool, removal of the end cap is not a simple task. The elements, especially if the unit has been used in a salt water atmosphere, will have their corrosive affect on the threads. The attempt with the special tool to break the end cap loose may very likely elongate the two holes provided for the tool. Once the holes are damaged, all hope of removing the end are lost. The only solution in such a case is to replace the unit.

➡The gas assist damper is filled with high pressure gas and should only be disassembled using the proper tools. Serious damage and personal bodily injury could result if the unit is not properly disassembled.

POWER TRIM AND TILT

Description and Operation

See Figures 7 thru 12

The single and multi-cylinder powered trim/tilt systems consist of a housing with an electric motor, gear driven hydraulic pump, hydraulic reservoir and one or two trim/tilt cylinders. The cylinders perform a double function as trim/tilt cylinders and also as a shock absorbers, should the lower unit strike an underwater object while the boat is underway.

The necessary valves, check valves, relief valves, and hydraulic passageways are incorporated internally and externally for efficient operation. A manual release valve is provided to permit the outboard unit to be raised or lowered should the battery fail to provide the necessary current to the electric motor or if a malfunction should occur in the hydraulic system.

The gear driven pump operates in much the same manner as an oil circulation pump installed on motor vehicles. The gears rotate in either direction, depending on the desired cylinder movement. One side of the pump is considered the suction side, and the other the pressure side, when the gears rotate in a given direction. These sides are reversed, the suction side becomes the pressure side and the pressure side becomes the suction side when gear movement is changed to the opposite direction.

Depending on the model, up to two relays may be used for the electric motor. The relays are usually located at the bottom cowling pan, where they are fairly well protected from moisture.

➡As a convenience, on some models an auxiliary trim/tilt switch is installed on the exterior cowling.

When the up portion of the trim/tilt switch is depressed, the up circuit, through the relay, is closed and the electric motor rotates in a clockwise direction. Pressurized oil from the pump passes through a series of valves to the lower chamber of the trim cylinders, the pistons are extended and the outboard unit is raised. The fluid in the upper chamber of the pistons is routed back to the reservoir as the piston is extended. When the desired position for trim is obtained, the switch on the control handle is released and the outboard is held stationary.

If the trim cylinder pistons should become fully extended, such as in a tilt up situation, fluid pressure in the lower chamber of the trim cylinders increases. This increase in pressure opens an up relief valve and the fluid is routed to the reservoir. The sound of the electric motor and the pump will have a noticeable change.

When the down portion of the trim/tilt switch is depressed, the down circuit, through the relay, is closed and the electric motor rotates in a counterclockwise direction. The pressure side of the pump now becomes the suction side and the original suction side becomes the pressure side. Pressurized oil from the pump

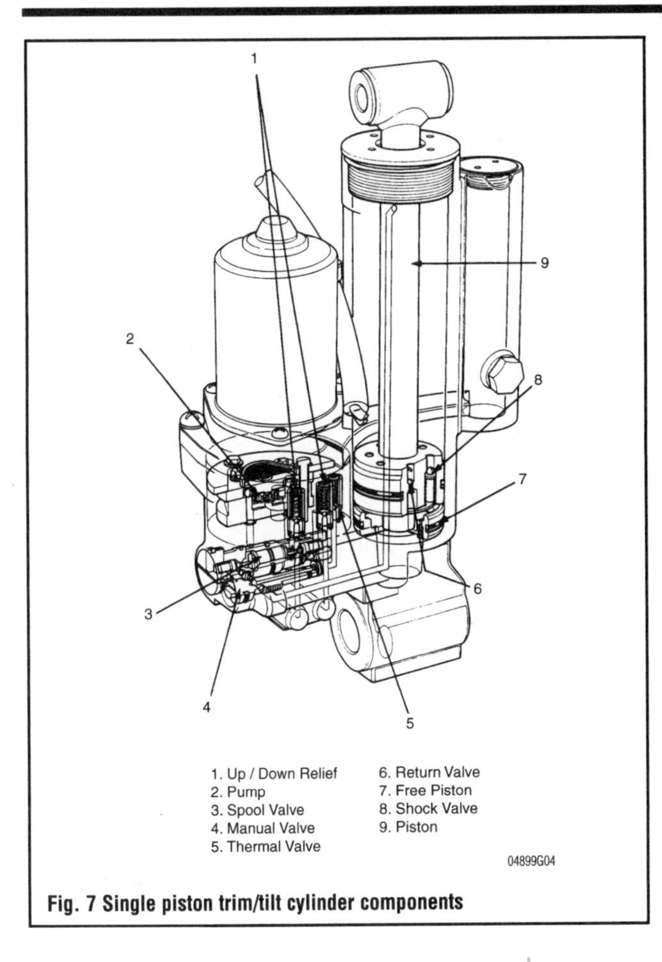

Fig. 7 Single piston trim/tilt cylinder components

passes through a series of valves to the upper chamber of the trim cylinders, the pistons are retracted and the outboard unit is lowered. The fluid in the lower chamber of the pistons is routed back to the reservoir as the retracted is extended. When the desired position for trim is obtained, the switch on the control handle is released and the outboard is held stationary.

If the trim cylinder pistons should become fully retracted, such as in a tilt down situation, fluid pressure in the upper chamber of the trim cylinders increases. This increase in pressure opens an up relief valve and the fluid is routed to the reservoir. The sound of the electric motor and the pump will have a noticeable change.

In the event the outboard lower unit should strike an underwater object while the boat is underway, the tilt piston would be suddenly and forcibly extended, moved upward. For this reason, the lower end of the tilt piston is capped with a free piston. This free piston normally moves up and down with the tilt piston.

The free piston also moves upward but at a much slower rate than the tilt piston. The action of the tilt piston separating from the free piston causes two actions. First, the hydraulic fluid in the upper chamber above the piston is compressed and pressure builds in this area. Second, a vacuum is formed in the area between the tilt piston and the free piston.

This vacuum in the area between the two pistons sucks fluid from the upper chamber. The fluid fills the area slowly and the shock of the lower unit striking the object is absorbed. After the object has been passed the weight of the outboard unit tends to retract the piston. The fluid between the tilt piston and the free piston is compressed and forced through check valves to the reservoir until the free piston reaches its original neutral position.

A manual relief valve, located on the stern bracket, allows easy manual tilt of the outboard should electric power be lost. The valve opens when the screw is turned counterclockwise, allowing fluid to flow through the manual passage. When the relief valve screw is turn fully clockwise, the manual passage is closed and the outboard lock in position.

A thermal valve is used to protect the trim/tilt motor and allow it to maintain a designated trim angle. Oil in the upper chamber is pressurized when force is applied to the outboard from the rear while cruising. Oil is directed through the right side check valve and activates the thermal valve to release oil pressure and lessen the strain on the motor and pump.

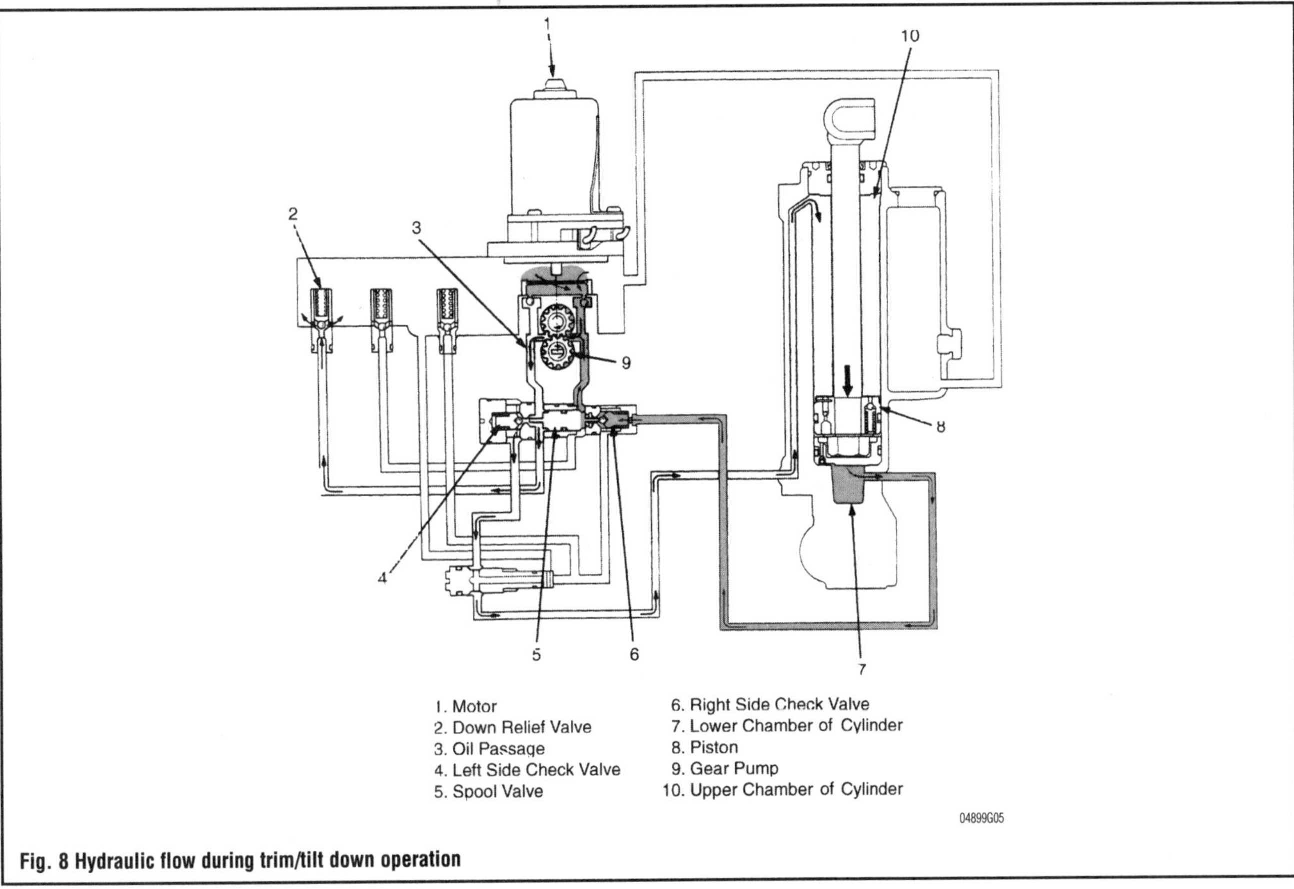

Fig. 8 Hydraulic flow during trim/tilt down operation

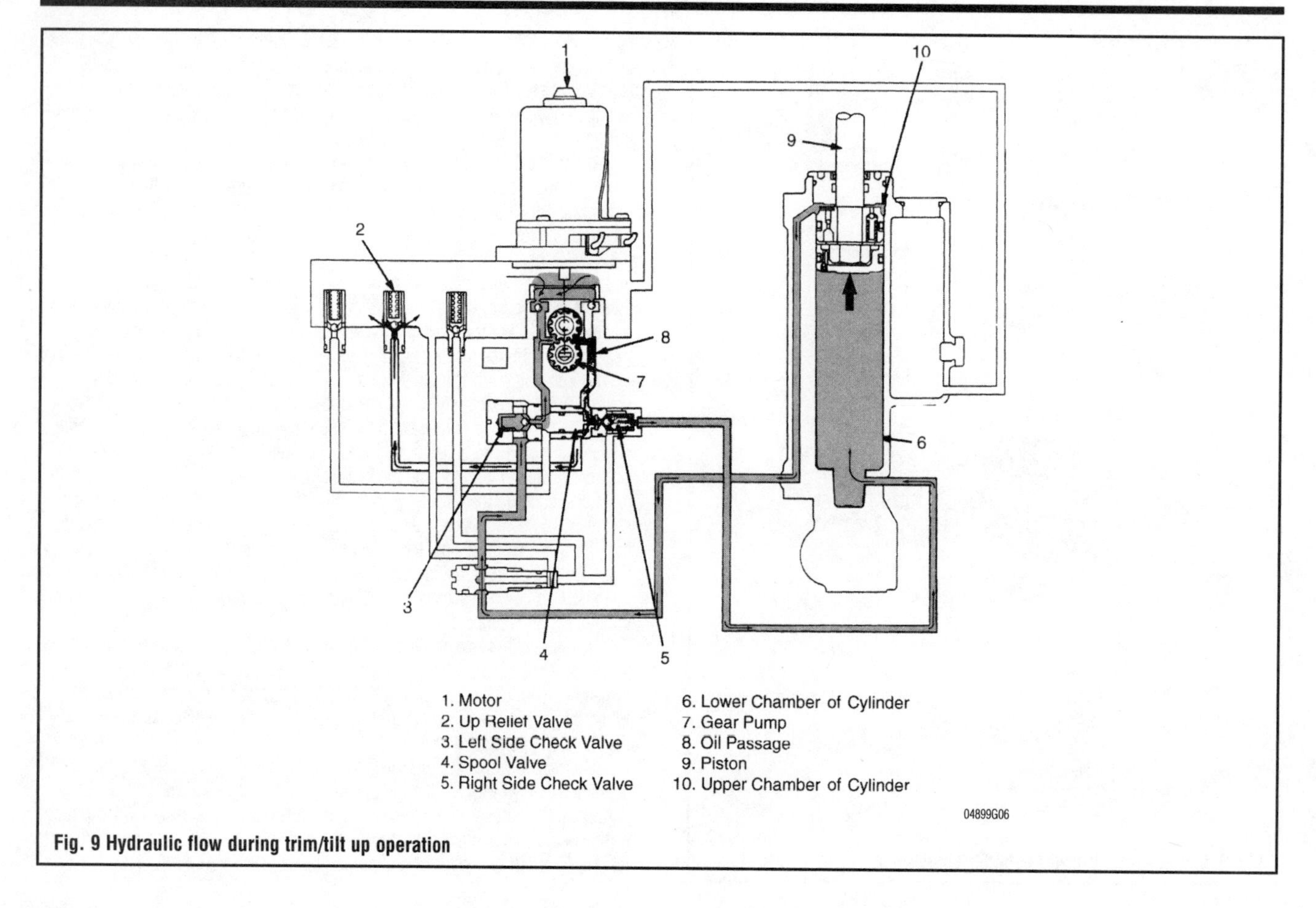

Fig. 9 Hydraulic flow during trim/tilt up operation

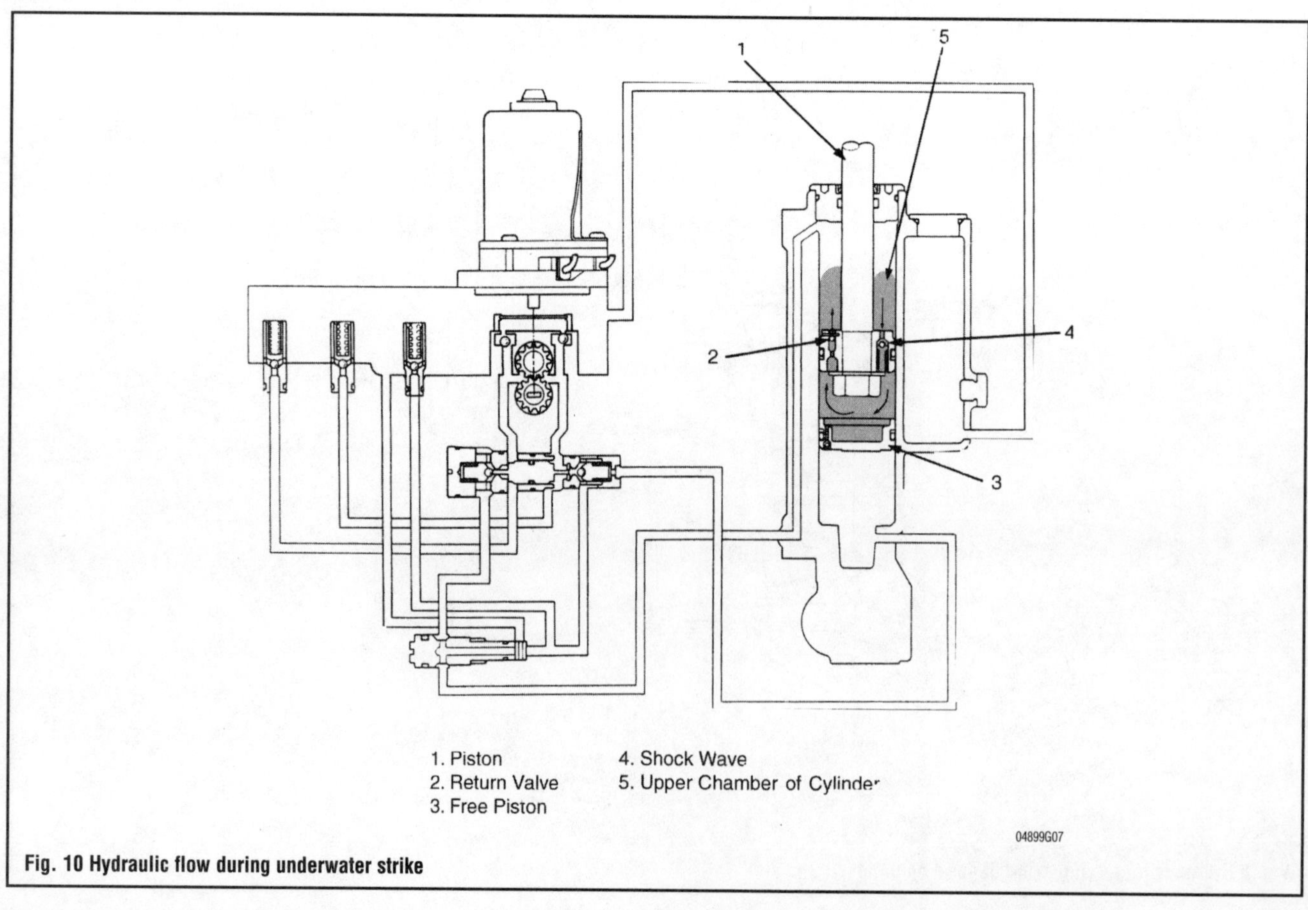

Fig. 10 Hydraulic flow during underwater strike

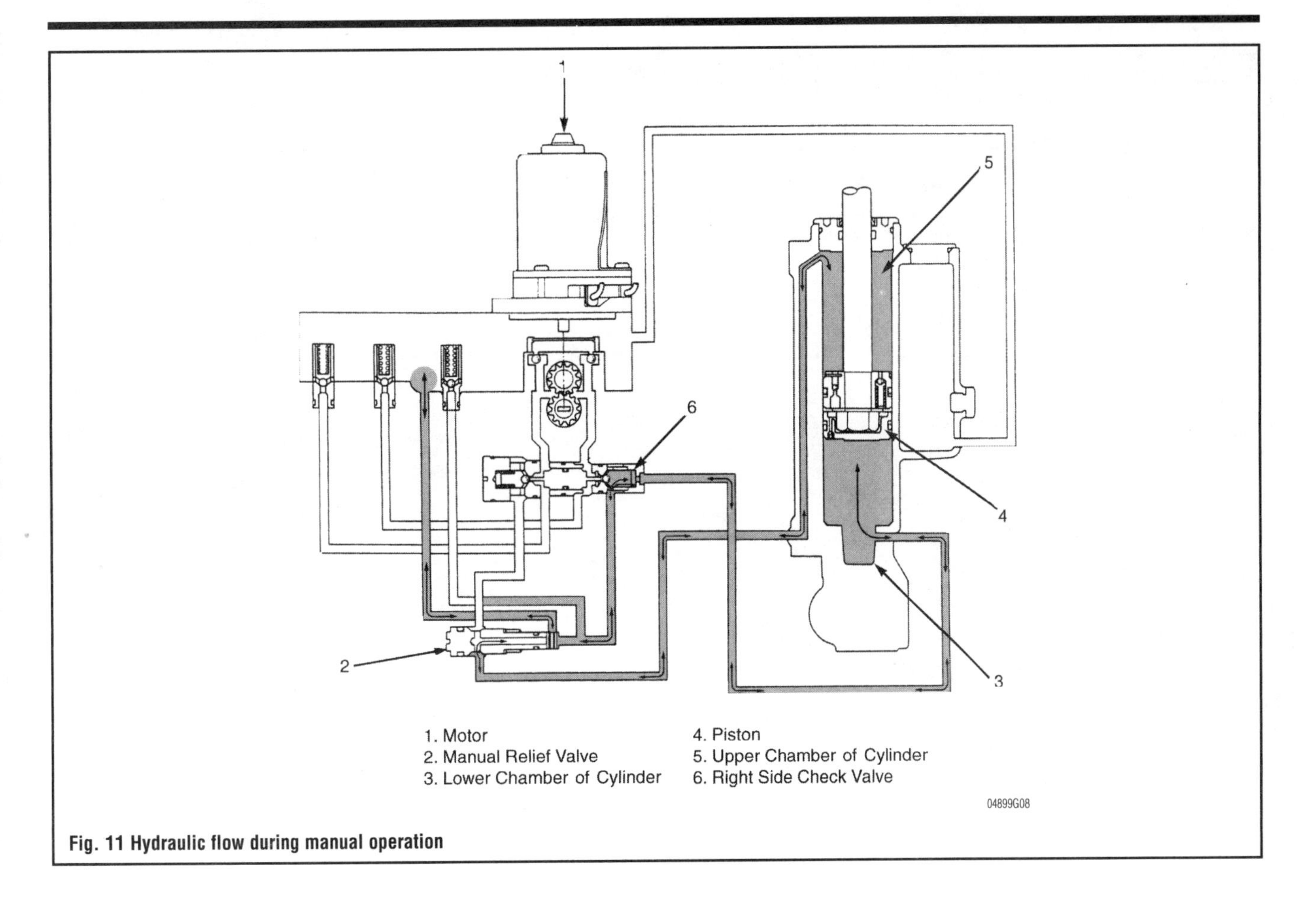

Fig. 11 Hydraulic flow during manual operation

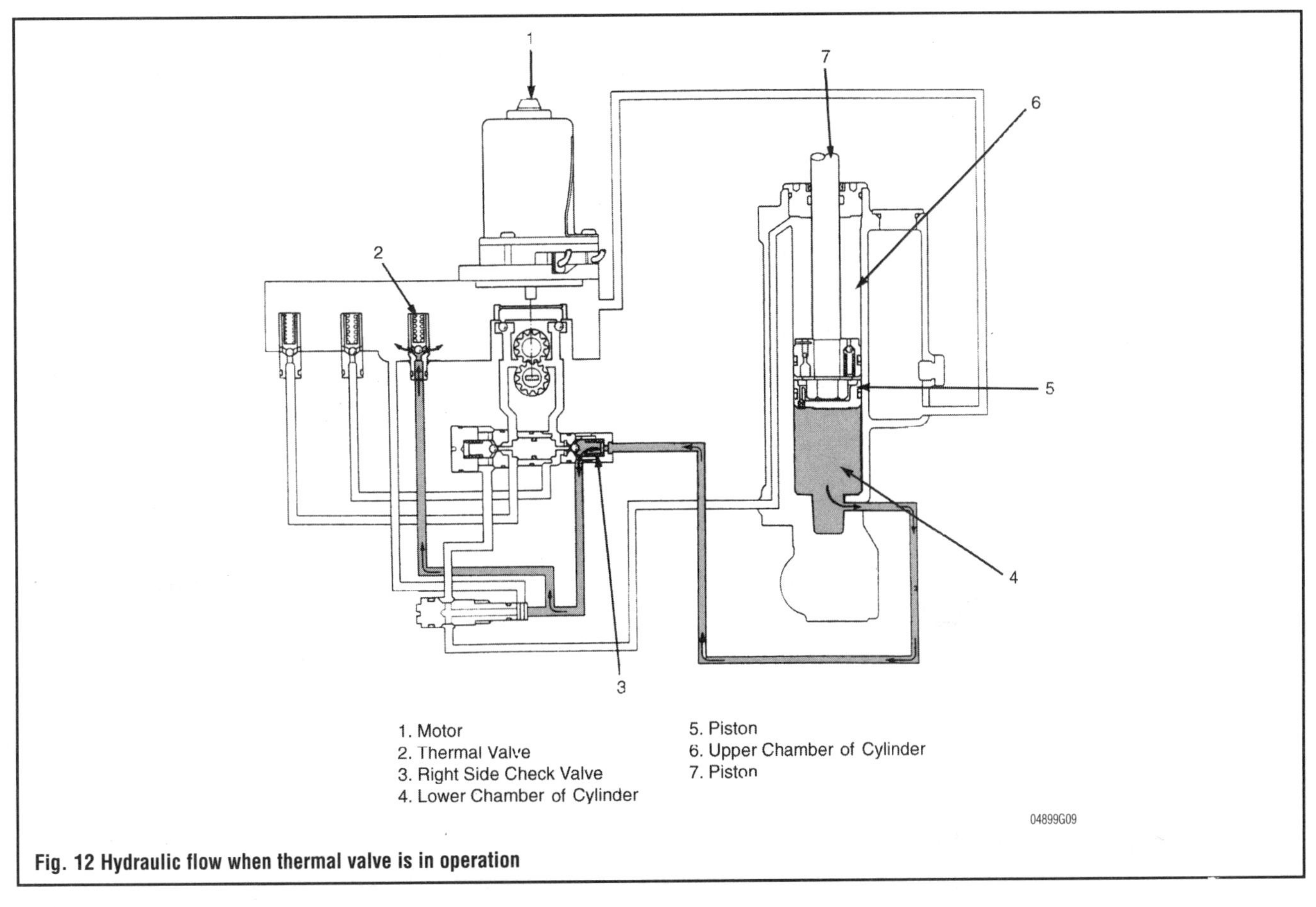

Fig. 12 Hydraulic flow when thermal valve is in operation

Trim/Tilt Motor

TESTING

1. Ensure the manual release valve is in the manual tilt position.
2. Disconnect the trim/tilt motor wiring harness at the quick connect fittings.
3. Using jumper cables, momentarily make contact between the disconnected leads and a fully charged battery.

➡Make the contact only as long as necessary to hear the electric motor rotating.

4. Reverse the leads on the battery posts and again listen for the sound of the motor rotating. The motor should rotate with the leads making contact with the battery in either direction.
5. If the motor operates properly, the problem may be in the trim/tilt switch or associated wiring.
6. If the motor does not operate as specified, it may be faulty.

REMOVAL & INSTALLATION

➧ See Figure 13

1. Remove the hydraulic unit from the outboard.

➡Keep the power trim/tilt damper assembly vertical during service. The upper cylinder bushing should be up and the oil plug on the reservoir should be down.

2. Remove the power tilt motor attaching screws and lift the motor from the trim/tilt assembly.
3. Carefully remove the O-ring and inspect the O-ring groove for damage.
4. Remove the drive joint.

To install:

5. Install the drive joint.
6. Lubricate and install a new O-ring.
7. Position the motor on the trim/tilt assembly and fasten with the attaching screws. Tighten screws securely..
8. Install the hydraulic unit on the outboard.
9. Test the power trim/tilt system for proper operation.

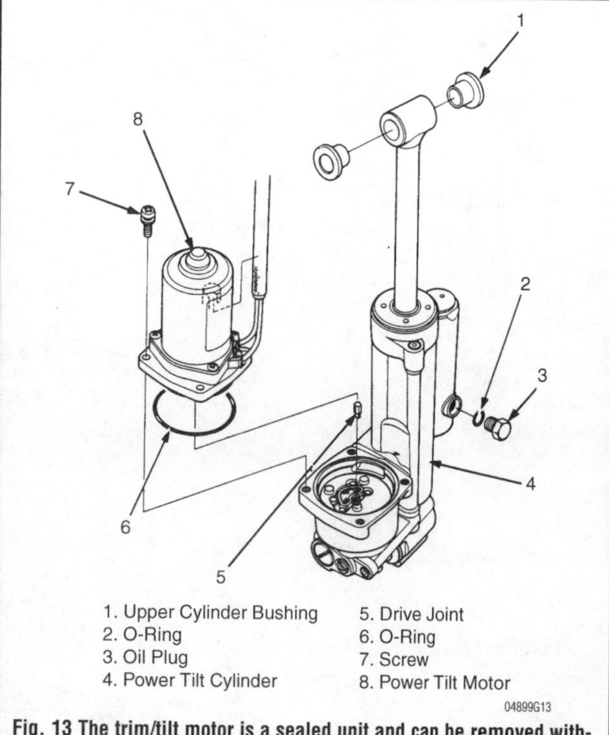

Fig. 13 The trim/tilt motor is a sealed unit and can be removed without draining the hydraulic fluid

DISASSEMBLY

➧ See Figures 14 and 15

1. Remove the screw attaching the wire holder and carefully pull the wires from the motor.
2. Remove the wire grommets from their position in the motor carefully.
3. Remove the screws attaching the front bracket to the yoke assembly.
4. Wrap the armature shaft with a towel and use pliers to pull the armature from the yoke assembly.

➡Push the motor wiring harness toward the yoke assembly while removing the armature assembly.

5. Remove the armature from the front bracket. Once removed take care to keep the commutator clean.
6. Hold the breaker assembly and disconnect the wire terminal.

➡If the breaker assembly is not held during removal of the wire, it can easily break.

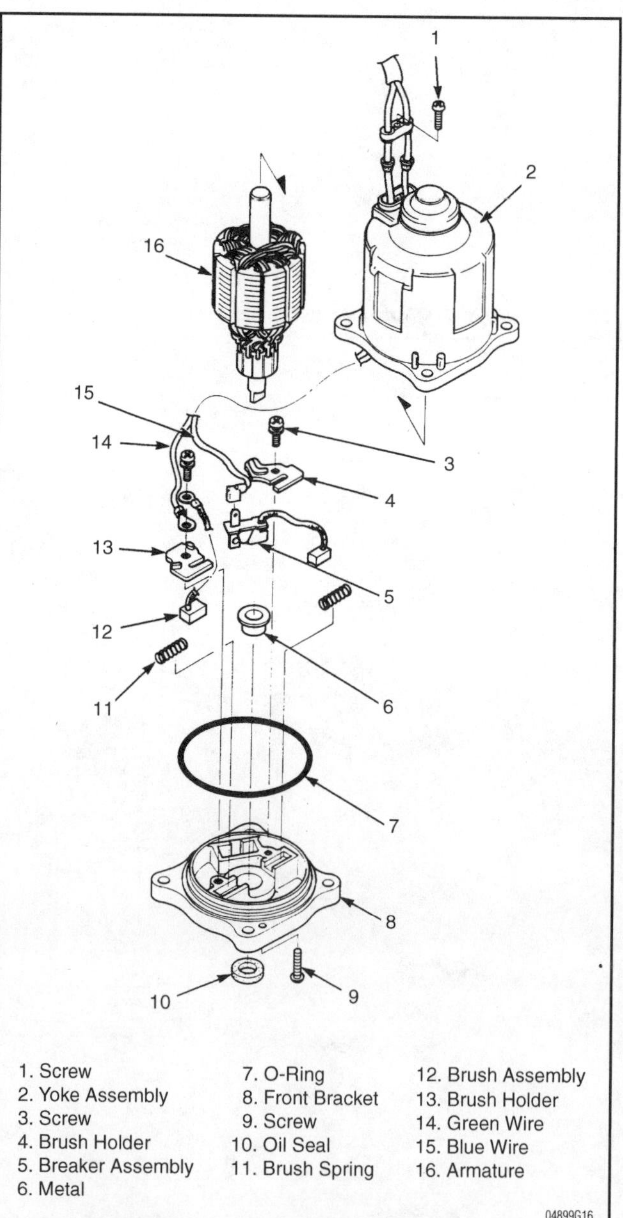

Fig. 14 Exploded view of the trim/tilt motor

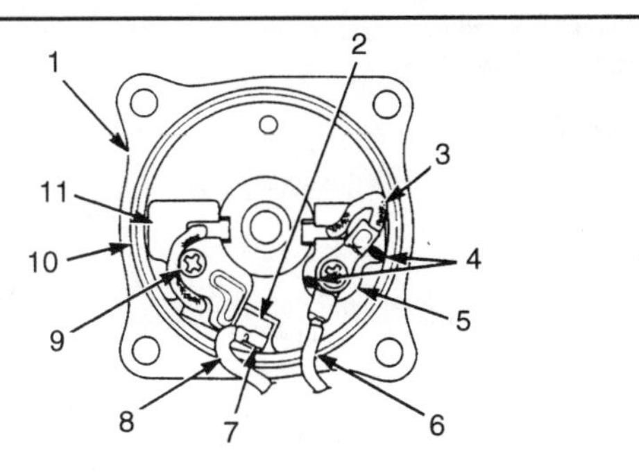

1. Front Bracket
2. Breaker Assembly
3. Brush Assembly
4. Projections
5. Brush Holder
6. Green Wire
7. Terminal
8. Blue Wire
9. Screw
10. O-Ring
11. Brush Holder

04899G17

Fig. 15 Trim/tilt motor components as viewed from the bottom of the yoke assembly

7. Disassemble the brush holders, brush assembly, springs and breaker assembly after remove the attaching screws.

➡Do not touch the bimetal portion of the breaker assembly.

8. Remove the oil seal from the front bracket.

CLEANING & INSPECTION

➧ See Figure 16

1. Clean all components with a electrical contact cleaner and dry using compressed air.
2. Measure brush length using calipers. Standard brush length should be 0.39 in. (9.8mm) and the service limit is .019 in. (4.8mm). If brush length is beyond the service limit, replace the brushes.
3. Check for continuity between the brush and the terminal on the breaker assembly. Replace the breaker assembly if there is no continuity.
4. Inspect the mica depth on the commutator. If the mica depth is less than 0.020 in. (0.5mm) or the grooves are clogged, use a hacksaw blade or small file to deepen the grooves.
5. Check for continuity between each section of the armature. If an open circuit exists between any two segments of the armature it is faulty and should be replaced.

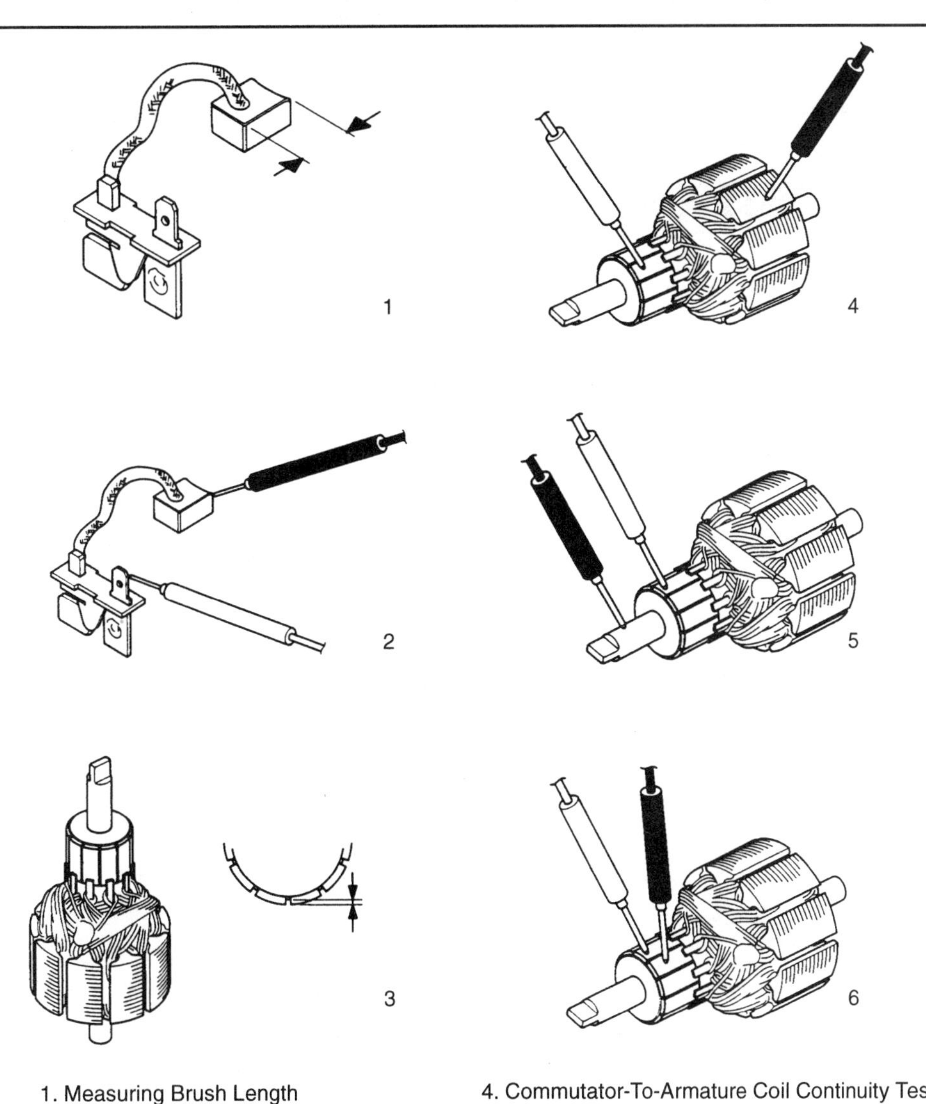

1. Measuring Brush Length
2. Brush-To-Terminal Continuity Test
3. Mica Depth Measurement
4. Commutator-To-Armature Coil Continuity Test
5. Commutator-To-Armature Shaft Continuity Test
6. Section-To-Section Armature Continuity Test

04899G18

Fig. 16 Trim/tilt motor electrical inspection steps

6. Check for continuity between the commutator and armature coil core. If continuity exists the armature is faulty and should be replaced.
7. Check for continuity between the commutator and armature shaft. If continuity exists the armature is faulty and should be replaced.

ASSEMBLY

1. Lubricate and install the O-ring in the front bracket.
2. Install the oil seal and metal on the front bracket.
3. Assemble the brush holders, brush assembly, springs and breaker assembly. Tighten the attaching screws securely.

➡Do not touch the bimetal portion of the breaker assembly.

4. Connect the wire terminals to the breaker assembly.
5. Wrap the armature shaft with a towel and use pliers to install the armature into the yoke assembly.
6. Pull the motor wires through the slot in the yoke while installing the armature.
7. Check the armature for smooth rotation.
8. Install and tighten the screws attaching the front bracket to the yoke assembly.
9. Connect the motor wiring harness to the outboard and test the motor for proper operation. If the motor does not operate properly, the blue wire terminal may not be connected properly.
10. Carefully install the wire grommets and screw the wire holder carefully into the motor. Tighten the wire holder screw to 12 inch lbs. (1.4 Nm). Do not overtighten the screw.
11. Tape the varnished Teflon® tube in position with heat resistant tape so the tube end is 0.5 in. (12mm) from the end of the wire holder.

Hydraulic Unit

CHECKING FLUID LEVEL

➧ See Figure 17

➡Fluid level should be checked with the trim/tilt rods fully extended. If fluid level is tested with the rods compressed, hydraulic fluid will spill from the reservoir when the cap is unscrewed.

1. With the rods fully extended, remove the reservoir cap and visually inspect the fluid level. The hydraulic fluid level should be at the upper level of the filler port. If the level is below specification, add Automatic Transmission Fluid (ATF) until the level is as specified.
2. After adding hydraulic fluid, bleed the system to remove air from the reservoir and fluid passages.

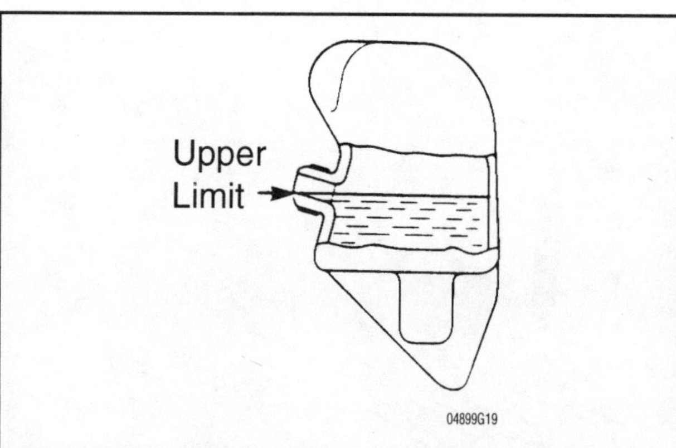

Fig. 17 Fluid level should be at the upper level of the filler port

BLEEDING THE SYSTEM

➡The hydraulic bleeding sequence should be performed whenever the trim/tilt assembly has been removed for service. Bleed the unit as follows: Bleed with unit off outboard, Perform pressure check, Bleed with unit installed on outboard

Unit Off Outboard

1. Connect the power trim/tilt assembly to electrical power and the trim/tilt switch on the boat.
2. Hold the assembly vertically.
3. Press the down side of the trim/tilt switch to compress the rods fully.
4. Press the up side of the trim/tilt switch to extend the rods fully.
5. Remove the reservoir cap with the rods fully extended and fill to the upper limit of the filler port with Automatic Transmission Fluid (ATF). The ATF should be allowed to flow out of the filler port.
6. Once again, press the down side of the trim/tilt switch to compress the rods fully and the up side to extend the rods fully. Perform this sequence several times to fully bleed air from the unit.
7. Ensure the rods fully extend and compress. If the rods do not compress fully, perform the procedure under "Rod Compression Test".

Unit On Outboard

➧ See Figure 18

Perform the following procedure only after completing the "Unit Off Outboard" bleeding procedure.
1. Press the up side of the trim/tilt switch to raise the outboard to the full tilt position.
2. Remove the reservoir cap and check the hydraulic fluid level. Adjust the level as necessary.
3. Lower the outboard slowly to its normal operating position by loosening the manual valve. Tighten the manual valve.
4. Allow the outboard to sit for five minutes or more.
5. Press the up side of the trim/tilt switch to raise the outboard to the full tilt position.
6. Again, allow the outboard to sit for five minutes or more.
7. Check and adjust the fluid level in the reservoir.
8. Repeat the procedure five times.

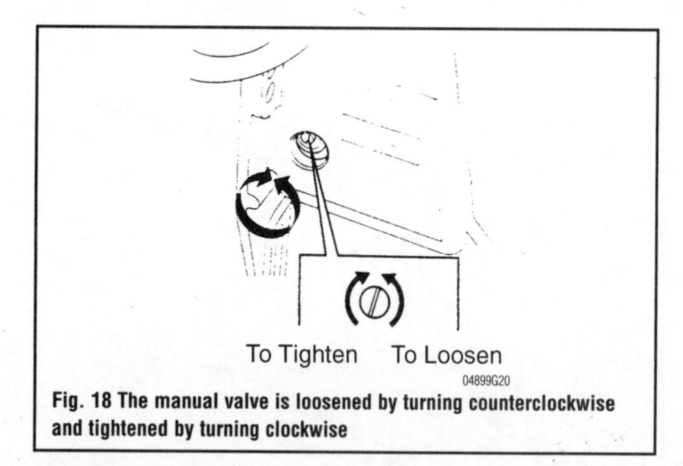

Fig. 18 The manual valve is loosened by turning counterclockwise and tightened by turning clockwise

TESTING

Rod Compression Test

1. Press the up side of the trim/tilt switch to extend the rods fully. Then press the switch again until the hydraulic relief valve makes a blowing sound.
2. Press the down side of the trim/tilt switch and ensure the rods compress fully. If the rods compress fully, perform the pressure test.
3. If the rods do not compress fully, press the down side of the trim/tilt switch while pushing on the trim cylinder rods to assist them in compressing. If the rods compress fully, perform the pressure test.
4. If the rods do not compress fully, loosen the manual valve and compress each rod. Tighten the manual valve.

5. Press the up side of the trim/tilt switch to extend the rods fully, then press the down side of the trim/tilt switch and ensure the rods compress fully. If the rods compress fully, perform the pressure test.
6. If the rods do not compress after performing the above steps, disassemble, reassemble and bleed the power trim/tilt assembly.

Pressure Test

LOWER CHAMBER

See Figures 19 and 20

1. Remove the circlip holding the manual valve in its bore, then remove the manual valve.

➡A small amount of hydraulic fluid may leak out of the manual valve bore so be prepared to catch it with a drip pan.

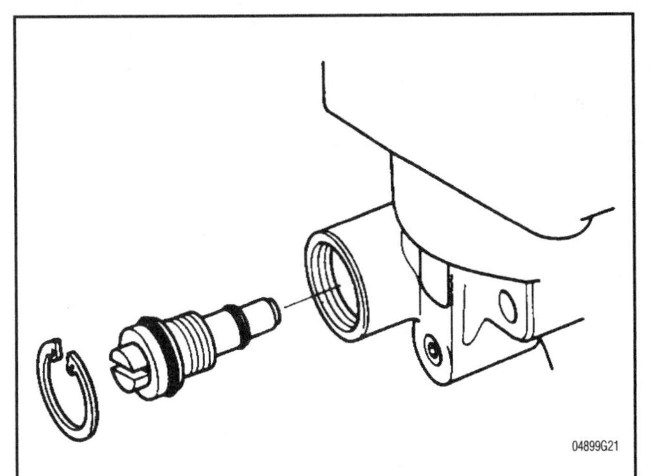

Fig. 19 The manual valve is held in its bore by a circlip. To prevent the valve from leaking, always reassemble using new O-rings

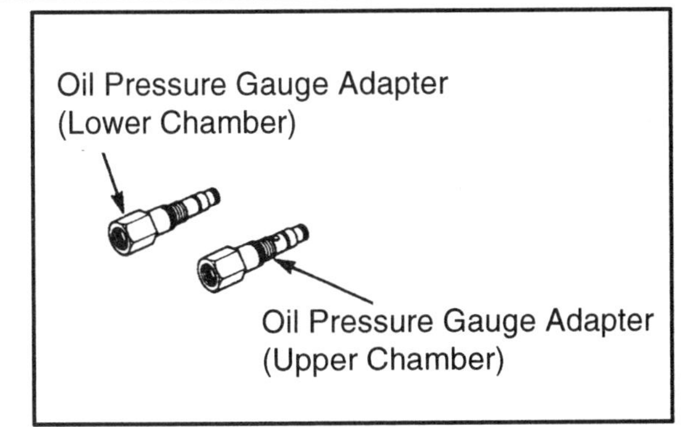

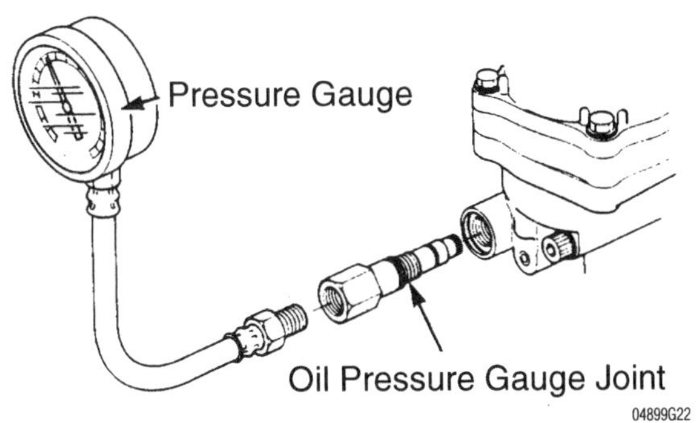

Fig. 20 The oil pressure adapters allow testing of the lower and upper chamber pressure with a standard 0–5000 psi pressure gauge

2. Install and tighten an oil pressure gauge adapter (07SPJ-ZW1020A) or equivalent to the outboard. Tighten the adapter to 6.5 ft. lbs. (9 Nm).
3. Attach a 0–5000 psi pressure gauge to the adapter.
4. Remove the reservoir cap and ensure the fluid level is at the upper limit of the filler port.
5. Press the down side of the trim/tilt switch and ensure the rods compress fully.
6. Press the up side of the trim/tilt switch to extend the rods fully. Measure the lower chamber pressure. Pressure should be between 1280–1707 psi (8826–11768 kPa).
7. If a sharp drop in pressure occurs, check the power tilt motor for a damaged seal. If the pressure is generally lower than specification, check for oil leaks.
8. Remove the adapter and gauge.

UPPER CHAMBER

1. Install and tighten an oil pressure gauge adapter (07SPJ-ZW1010A) or equivalent to the outboard. Tighten the adapter to 6.5 ft. lbs. (9 Nm).
2. Attach a 0–5000 psi pressure gauge to the adapter.
3. Remove the reservoir cap and ensure the fluid level is at the upper limit of the filler port.
4. Press the down side of the trim/tilt switch to compress the rods fully. Measure the upper chamber pressure. Pressure should be between 569–1067 psi (3923–7355 kPa).
5. If the pressure is lower than specification, check for oil leaks.
6. After checking the upper chamber pressure, press the up side of the trim/tilt switch to extend the rods fully.
7. Remove the adapter and gauge.
8. Install the manual valve and circlip.
9. Remove the reservoir cap and ensure the fluid level is at the upper limit of the filler port.

REMOVAL & INSTALLATION

See Figures 21 thru 31

1. Remove the outboard from the boat and lay it on its side.
2. Turn the manual valve screw to allow the outboard to be manually tilted.
3. Position and support the outboard in its fully tilted position.
4. Remove the steering tube cap and loosen the self locking nut to remove the steering tube.
5. Label and disconnect the assembly wiring harnesses and feed them through the hole in the stern bracket.
6. Remove the E-ring and slide the upper cylinder pin from the mounting frame.
7. Pivot the assembly away from the mounting frame taking care to catch the wave washers that fall out.
8. Remove the assembly through bolt from the stern bracket and lift out the distance collar.
9. Remove the assembly from the stern bracket.
10. Remove the upper cylinder bushings.

➡Store the assembly vertically with the upper cylinder bushing upward. Never store the damper horizontally with the distance collar facing upward.

To install:

11. Lubricate and install the cylinder bushings.
12. Position the assembly on the stern bracket.
13. Install the distance collar and through bolt on the stern bracket. Tighten the through bolt to 25 ft. lbs. (35Nm).
14. Pivot the assembly toward the mounting frame.
15. After positioning the wave washers, lubricate and install the upper cylinder pin on the mounting frame. Install the E-ring to secure the pin in place.
16. Feed the assembly wiring harnesses through the hole in the stern bracket and connect them.
17. Install the self locking nut and tighten to 25 ft. lbs. (35Nm). Install the steering tube cap.
18. Install the outboard on the boat and test the power trim/tilt assembly for proper operation.

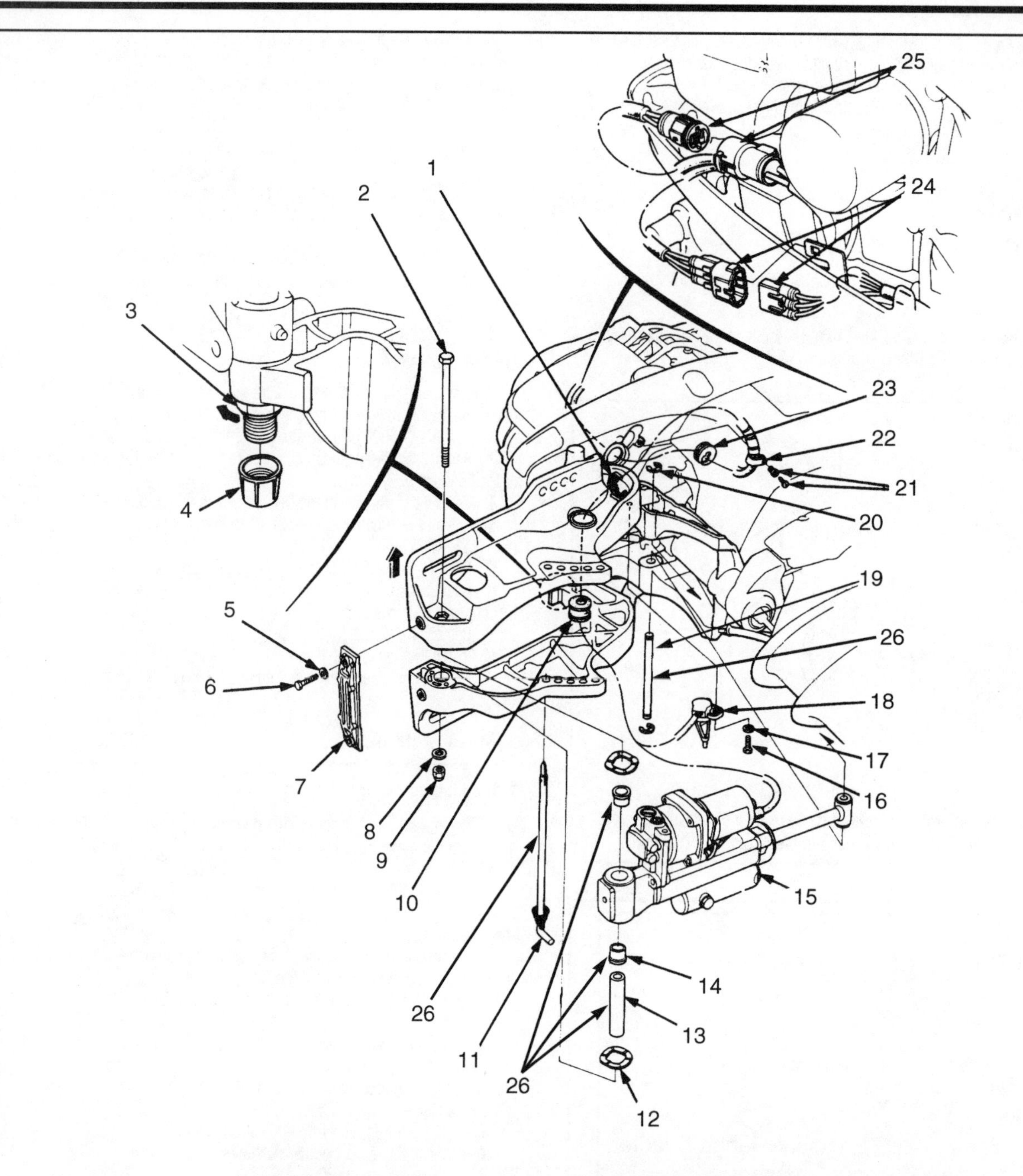

1. Wire Harness Clip
2. Hex Bolt
3. Self Locking Nut
4. Steering Tube Cap
5. Plain Washer
6. Hex. Bolt
7. Anode Metal
8. Plain Washer
9. Self-Locking Nut
10. Motor Cord Bushing
11. Transom Angle Adjusting Rod
12. Wave Washer
13. Distance Collar
14. Lower Cylinder Bushing
15. Power Tilt Assembly
16. Hex. Bolt
17. Plain Washer
18. Trim Angle Sensor
19. Upper Cylinder Pin
20. E-Ring
21. Screw Rivet
22. Clip
23. Under Case Grommet
24. Trim Angle Sensor Connector
25. Power Tilt Motor Connector
26. Apply Grease Here

04899G14

Fig. 21 Exploded view of a power tilt assembly installation

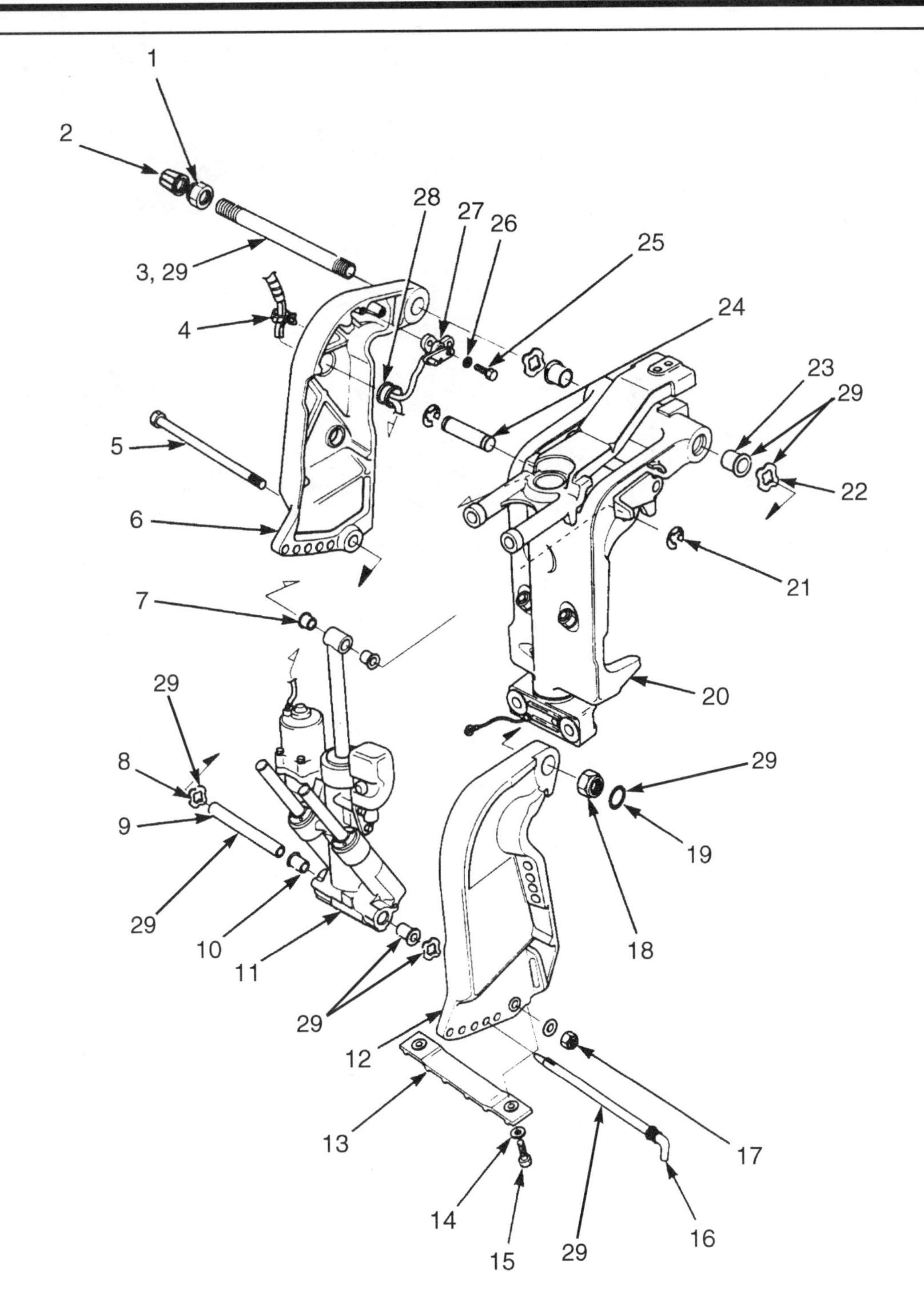

1. Self Locking Nut
2. Tilting Bolt Cap
3. Tilting Shaft
4. Wire Harness Clip
5. Hex Bolt
6. Stern Bracket
7. Upper Cylinder Bushing
8. Wave Washer
9. Lower Cylinder Collar
10. Lower Cylinder Bushing
11. Power Trim Tilt Assembly
12. R. Stern Bracket
13. Anode Metal
14. Washer
15. Hex Bolt
16. Adjusting Rod
17. Self-Locking Nut
18. Self-Locking Nut
19. O-Ring
20. Swivel Case Assembly
21. E-Ring
22. Wave Washer
23. Swivel Case Bushing
24. Upper Cylinder Pin
25. Hex Bolt
26. Washer
27. Trim Angle Sensor
28. Motor Wire Bushing
29. Apply Grease Here

04899G15

Fig. 22 Exploded view of a power trim/tilt assembly installation

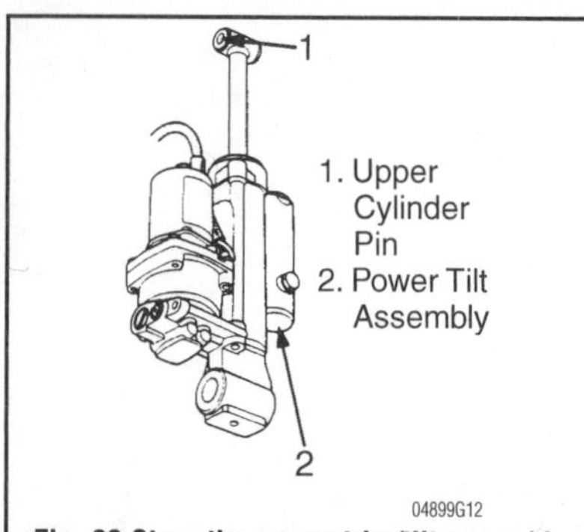

Fig. 23 Store the power trim/tilt assembly vertically with the upper cylinder bushing upward

Fig. 24 Remove the steering tube cap . . .

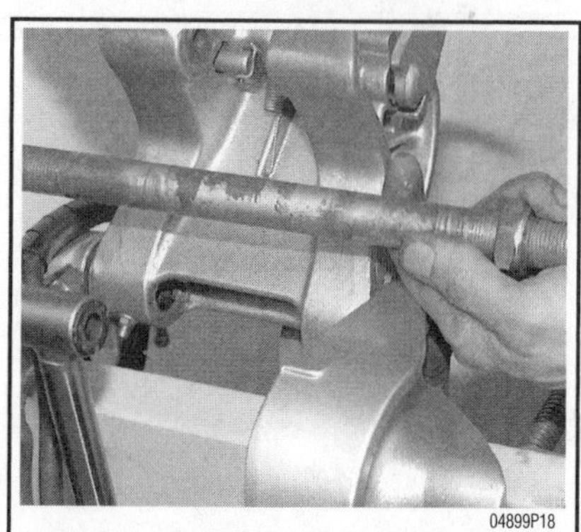
Fig. 25 . . . and loosen the self locking nut to remove the steering tube

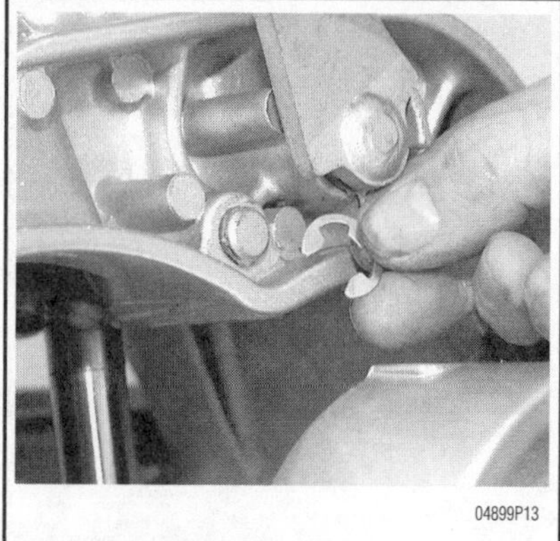
Fig. 26 Remove the E-ring . . .

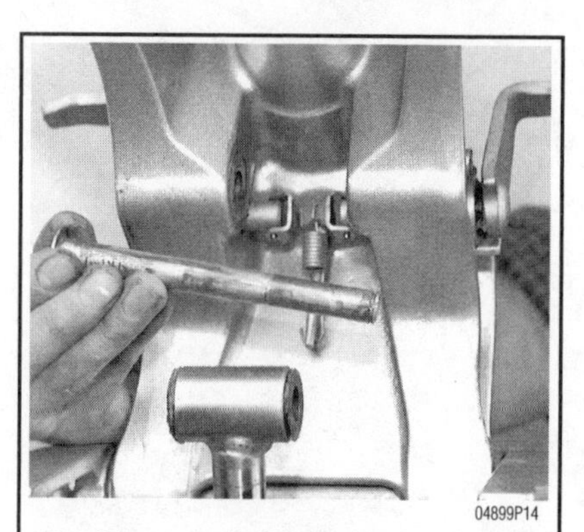
Fig. 27 . . . and slide the upper cylinder pin from the mounting frame

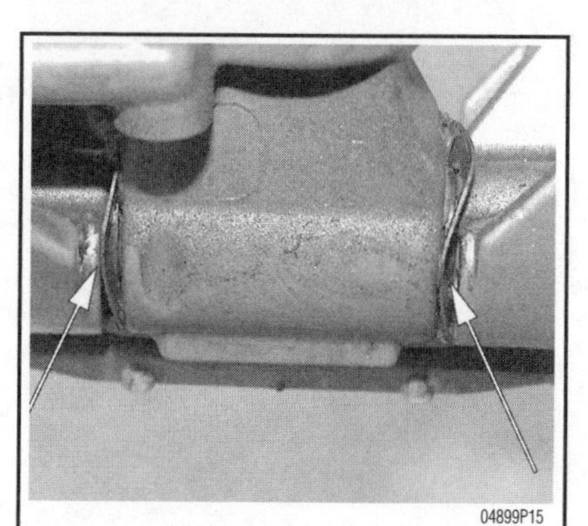
Fig. 28 Take care to catch the wave washers . . .

Fig. 29 . . . as they fall out when the through bolt is removed

Fig. 30 Lower trim/tilt assembly mounting hardware

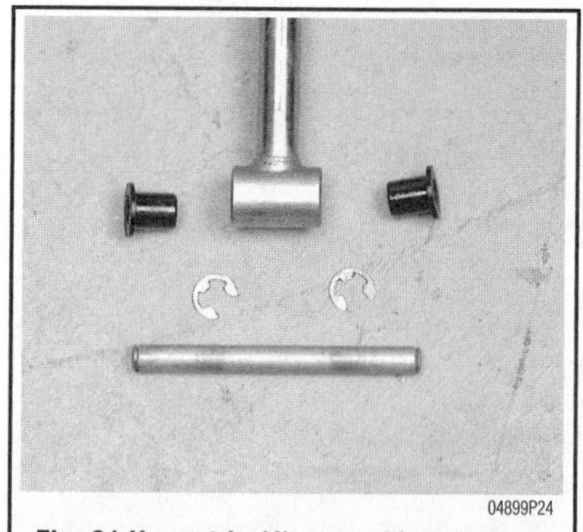
Fig. 31 Upper trim/tilt assembly mounting hardware

DISASSEMBLY

➧ See Figure 32

Hydraulic Pump

1. Place the trim/tilt assembly in a soft jawed vise.

➡Do not tighten the vise excessively as the assembly may be damaged.

2. Remove the oil reservoir cap and drain the reservoir of fluid.
3. Remove the reservoir attaching screws. Remove the reservoir from the power trim/tilt assembly. Discard the O-ring.
4. Remove the motor attaching screws and lift the motor from the power trim/tilt assembly.
5. Remove and discard the power trim/tilt motor O-ring.
6. Remove the filter and drive joint.
7. Remove the pump bolts and lift the pump from the power trim/tilt assembly.
8. Remove and discard the O-rings and orifice collars on the cylinder assembly.
9. Remove the internal circlip and unscrew the manual valve from the pump.

Tilt Cylinder

1. Using a Pin Wrench (07SPA-ZW10100) or equivalent, loosen the cylinder cap.
2. Unscrew the cylinder cap and carefully remove the rod from the cylinder.
3. Remove the piston rod and discard the O-ring on the nut side (bottom) of the piston rod and on the cylinder cap.
4. Remove the free piston.
5. Remove the backup ring and O-ring on the free piston.

Trim Cylinders

1. Using a Pin Wrench (07SPA-ZW10100) or equivalent, loosen the cylinder cap.

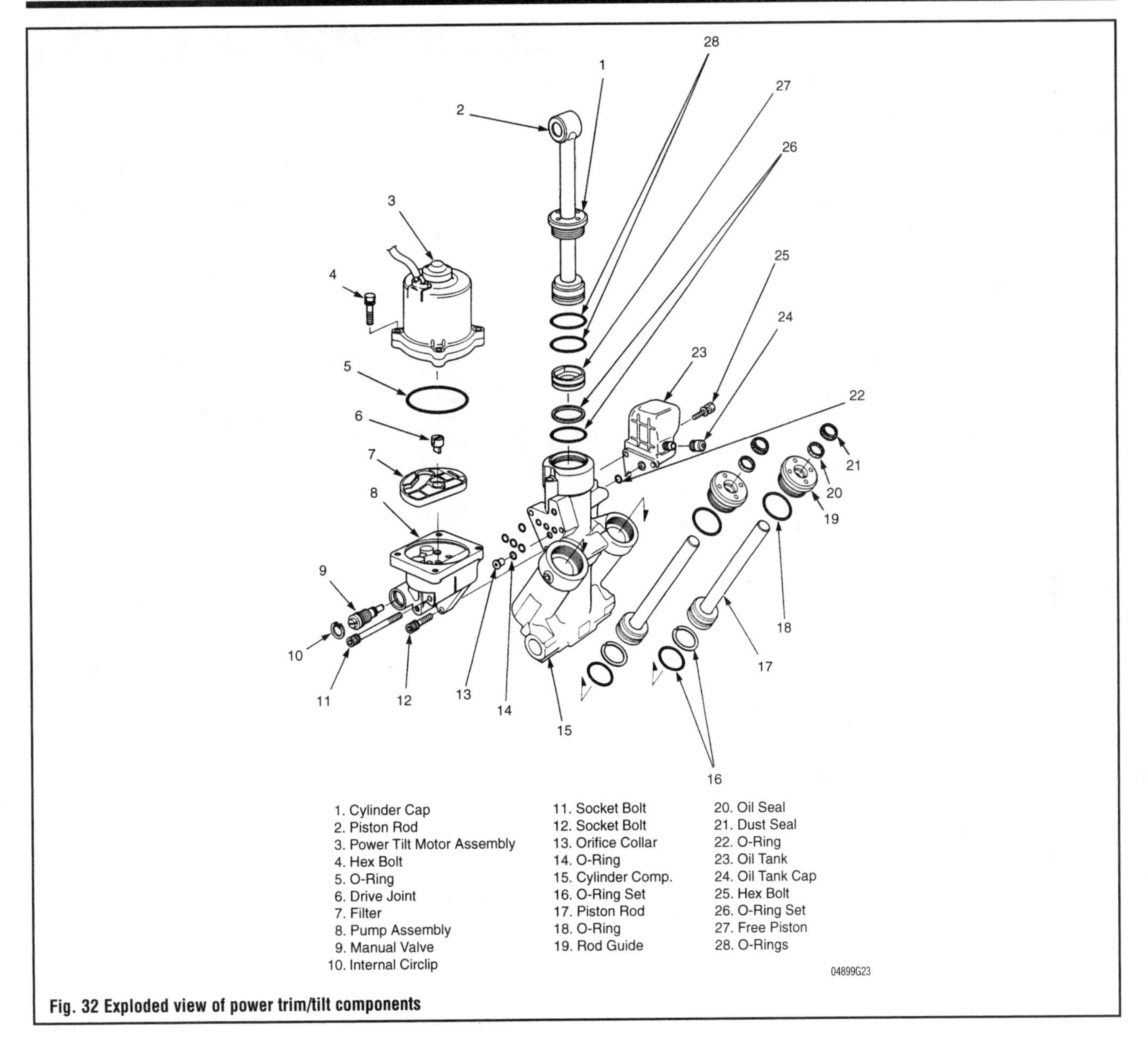

Fig. 32 Exploded view of power trim/tilt components

2. Unscrew the cylinder cap on each rod and carefully remove the rods from the cylinders.
3. Remove and discard the backup ring and O-ring on the piston rod.
4. Remove and discard the O-ring and oil seal on the cylinder cap.
5. Remove and discard the dust seal inside the cylinder.

CLEANING & INSPECTION

Inspect the free piston for damage or wear and replace as necessary.

Inspect the manual valve for damage or wear and replace as necessary.

Inspect and clean the pump filter. The filter may be cleaned using compressed air. Check the filter for damage and replace as necessary.

ASSEMBLY

Tilt Cylinder

1. Place the trim/tilt assembly in a soft jawed vise.

➡Do not tighten the vise excessively as the assembly may be damaged.

2. Pour 1 ounce of automatic transmission fluid into the cylinder.
3. Install a new backup ring and O-ring on the free piston.
4. Lubricate and install the free piston into the cylinder. Push it down until it bottoms out.
5. Fill the cylinder with automatic transmission fluid up to the lower edge of the threaded section on the cylinder.
6. Install a new O-ring on the nut side (bottom) of the piston rod and on the cylinder cap.
7. Place the cylinder cap on the piston rod and install the rod in the cylinder.

➡Do not push the rod down into the cylinder.

8. Using a Pin Wrench (07SPA-ZW10100) or equivalent, tighten the cylinder cap to 119 ft. lbs. (162 Nm).

Trim Cylinders

1. With the trim/tilt assembly in a soft jawed vise, fill both the left and right trim cylinders with automatic transmission up to the upper edge of the threaded section on the cylinder.
2. Install a new dust seal inside the cylinder.
3. Install a new O-ring and oil seal on the cylinder cap.
4. Install a new backup ring and O-ring on the piston rod.

5. Place the cylinder cap on each rod and insert the rods into their respective cylinders.

➡Do not push the rod down into the cylinder.

6. Using a Pin Wrench (07SPA-ZW1020A) or equivalent, tighten the cylinder cap to 58 ft. lbs. (78 Nm).

Hydraulic Pump

1. Install the manual valve and tighten to 2.5 ft. lbs. (3.5 Nm). Install the internal circlip.
2. Install new O-rings and orifice collars on the cylinder assembly and place the pump in position.
3. Install the pump bolts and tighten to 6.4 ft. lbs. (8.3 Nm).
4. Fill the pump with automatic transmission fluid.
5. Install the filter and drive joint. Check the filter for bubbles and remove them by using a plastic squeeze bottle to draw the bubbles from the fluid.

➡Removal of air from the system during assembly is critical to proper system bleeding.

6. After removing the bubbles from the system, refill the pump with automatic transmission fluid.
7. Install a new O-ring on the power trim/tilt motor.
8. Install the motor on the pump by aligning the projection on the motor with the cutout in the drive joint.
9. Install the motor attaching screws and tighten to 3.6 ft. lbs. (5 Nm).
10. Install a new O-ring on the power trim/tilt assembly and position the oil reservoir. Install the attaching screws and tighten to 3.6 ft. lbs. (5 Nm).
11. Remove the oil reservoir cap and fill the reservoir to the upper limit of the filler port.
12. Bleed the air from the power trim/tilt assembly hydraulic system.

Trim/Tilt Switch

Complete diagnosis, testing and servicing procedures for the trim/tilt switch are located in the Remote Control section of this manual.

Trim Angle Sensor

TESTING

1. The trim angle sensor is tested by measuring the resistance between the sensor terminals.
2. On BF35A, BF40A, BF45A and BF50A, resistance between the Orange and Black wires should be 4–6 Kilohms. Resistance between the Yellow/Blue and Black wires should be 2.7–4.3 Kilohms.
3. On all others resistance between the Light Green/Black and Black wires should be 4–6 Kilohms. Resistance between the Yellow/Blue and Black wires should be 2.7–4.3 Kilohms.
4. If resistance is within specification, inspect the wiring harness for damage or shorts.
5. If resistance is not as specified, the sensor may be faulty.

REMOVAL & INSTALLATION

➧ See Figures 33, 34 and 35

1. Raise and support the outboard to the full tilt position.
2. Label and disconnect the tilt angle sensor wiring harness.
3. Remove the tilt angle sensor attaching bolts.
4. Remove the sensor from the stern bracket.

To install:

5. Position the sensor on the stern bracket and tighten the attaching bolts securely.
6. Connect the tilt angle sensor wiring harness.
7. Lower the outboard and check the sensor for proper operation.

Fig. 33 The trim angle sensor is located on the stern bracket

Fig. 34 The trim angle sensor wiring harness runs through a grommet in the stern bracket.

Fig. 35 Trim angle sensor is held in place by two bolts that screw into the stern bracket

REMOTE CONTROL BOX 10-2
DESCRIPTION AND OPERATION 10-2
REMOTE CONTROL BOX ASSEMBLY 10-2
REMOVAL & INSTALLATION 10-2
EXPLODED VIEWS 10-3
CHOKE SWITCH 10-7
TESTING 10-7
REMOVAL & INSTALLATION 10-7
ENGINE STOP SWITCH 10-7
TESTING 10-7
REMOVAL & INSTALLATION 10-8
IGNITION SWITCH 10-8
TESTING 10-8
REMOVAL & INSTALLATION 10-8
INDICATOR LIGHTS 10-8
TESTING 10-8
REMOVAL & INSTALLATION 10-8
SAFETY SWITCH 10-9
TESTING 10-9
REMOVAL & INSTALLATION 10-9
THROTTLE SWITCH 10-9
TESTING 10-9
REMOVAL & INSTALLATION 10-9
TRIM/TILT SWITCH 10-9
TESTING 10-9
REMOVAL & INSTALLATION 10-10
WARNING BUZZER 10-10
TESTING 10-10
REMOVAL & INSTALLATION 10-10
TILLER HANDLE 10-11
DESCRIPTION AND OPERATION 10-11
TROUBLESHOOTING 10-11
TILLER HANDLE 10-12
REMOVAL & INSTALLATION 10-12
THROTTLE CABLE 10-14
REMOVAL & INSTALLATION 10-14
ADJUSTMENT 10-17
ENGINE STOP SWITCH 10-20
TESTING 10-20
REMOVAL & INSTALLATION 10-20
INDICATOR LIGHT 10-20
TESTING 10-20
REMOVAL & INSTALLATION 10-20
STARTER BUTTON 10-20
TESTING 10-20
REMOVAL & INSTALLATION 10-20
TRIM/TILT SWITCH 10-20
TESTING 10-20
REMOVAL & INSTALLATION 10-20

10 REMOTE CONTROL

REMOTE CONTROL BOX 10-2
TILLER HANDLE 10-11

REMOTE CONTROL BOX

Description and Operation

See Figures 1, 2 and 3

The remote control box allows the helmsman to control the throttle operation and shift movements from a location other than where the outboard is mounted. In most cases, the remote control box is mounted approximately halfway forward (midship) on the starboard side of the boat.

The control box usually houses a key switch, engine stop switch, choke switch, neutral safety switch, warning buzzer and the necessary wiring and cable hardware to connect the control box to the outboard unit.

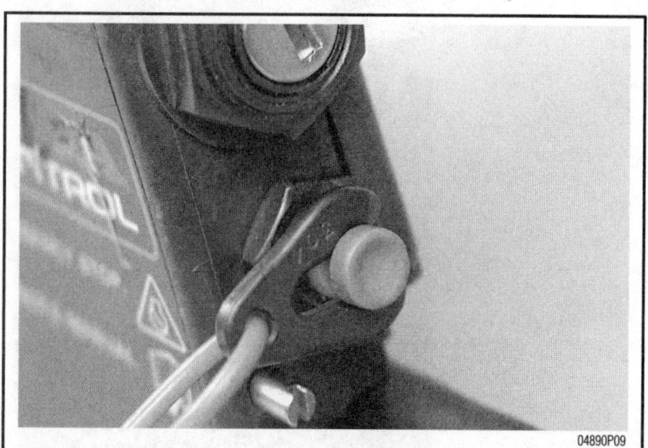

04890P09

Fig. 1 An important part of the remote control box is the emergency tether

Remote Control Box Assembly

REMOVAL & INSTALLATION

See Figures 4, 5, 6 and 7

Side/Panel Mount

1. Remove the bolts/screws securing the control box to the boat.
2. Remove the rear cover of the control box.
3. Label and disconnect the control cables.
4. Label and disconnect the control box electrical harness.
5. Remove the control box from the boat.

To install:

6. Connect the control box electrical harness.
7. Connect the control cables and properly adjust them, as needed.
8. Install the rear cover of the control box, tightening the screws securely.
9. Position the control box in the boat.
10. Install and securely tighten the bolts/screws securing the control box to the boat.

Top Mount

1. Remove the remote control box housing.
2. Remove the bolts/screws securing the control box to the boat.
3. Label and disconnect the control cables.
4. Label and disconnect the control box electrical harness.
5. Remove the control box from the boat.

To install:

6. Connect the control box electrical harness.
7. Connect the control cables and properly adjust them, as needed.

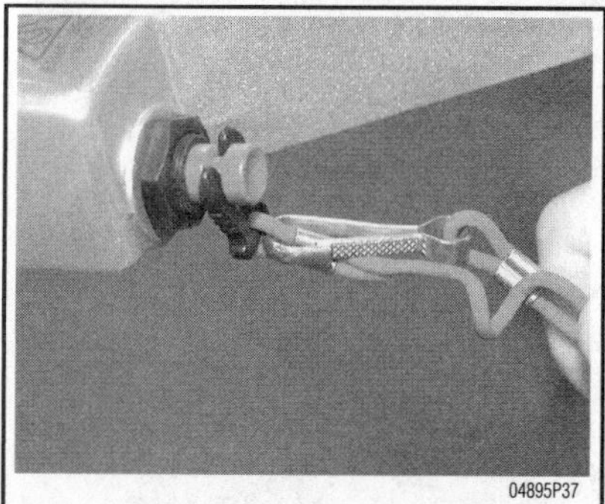

04895P37

Fig. 2 With the emergency switch tether in place, the engine will start and run normally

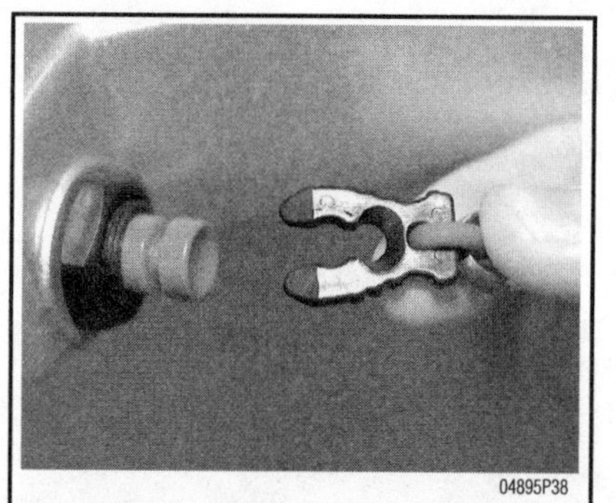

04895P38

Fig. 3 If the operator disengages the tether from the switch, the engine will immediately stall

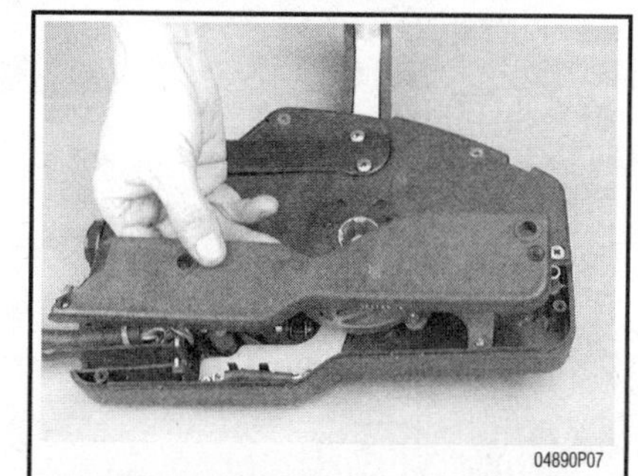

04890P07

Fig. 4 The remote control box has two covers on the rear which can be removed to access components. The first cover conceals the cable assemblies . . .

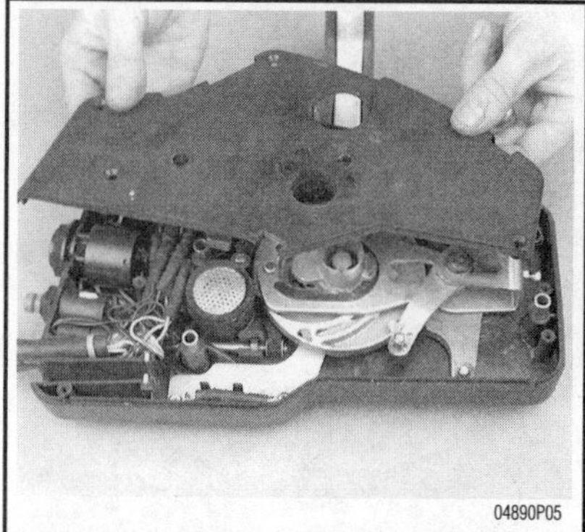

04890P05

Fig. 5 . . . the main cover conceals the majority of the internal components

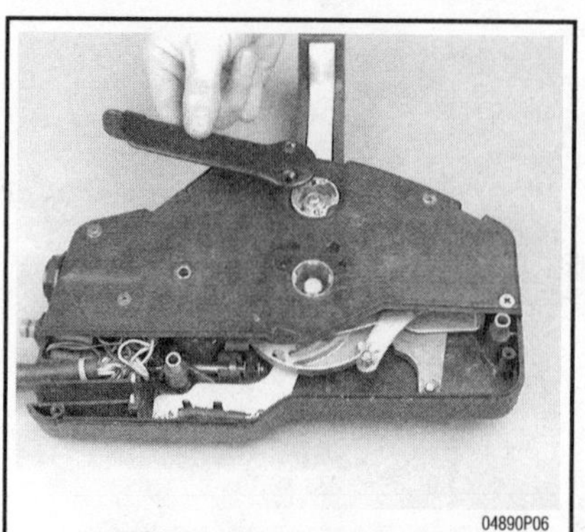

04890P06

Fig. 6 To remove the main cover, the choke switch must be removed first

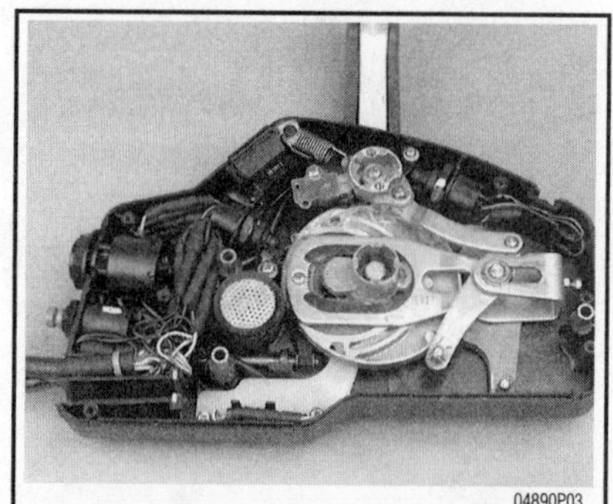

04890P03

Fig. 7 Remote control box with covers removed, allowing full access to all internal components

8. Position the control box in the boat.
9. Install and securely tighten the bolts/screws securing the control box to the boat.
10. Install the remote control box housing.

EXPLODED VIEWS

See Figures 8 thru 20

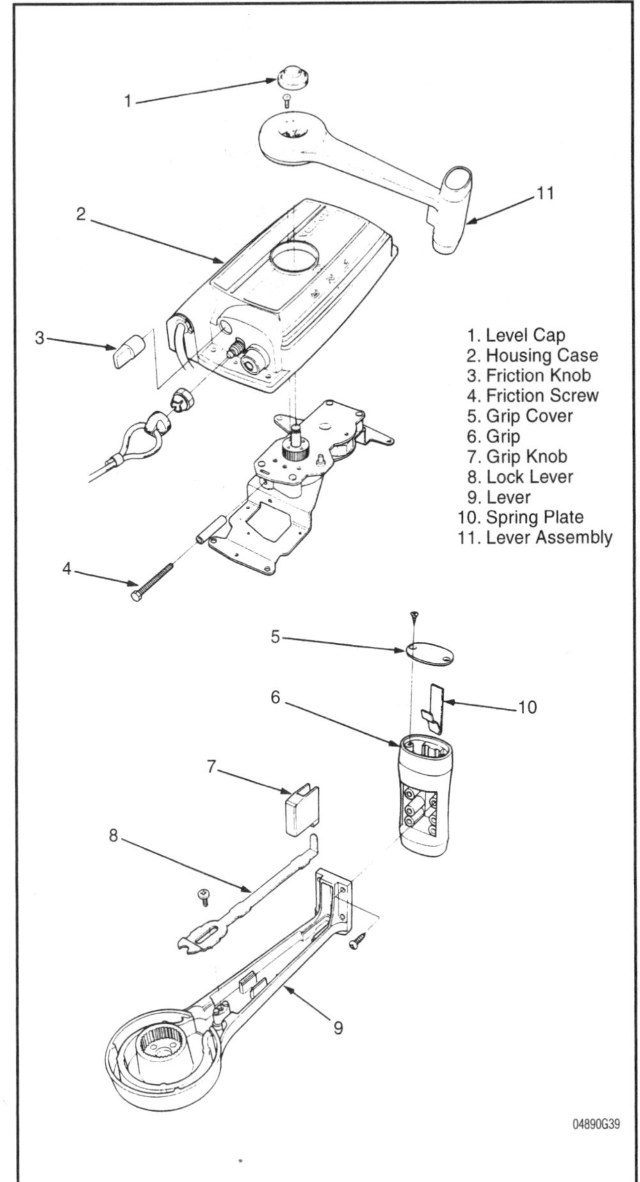

Fig. 8 Side mount remote control box front exploded view—BF9.9A and BF15A

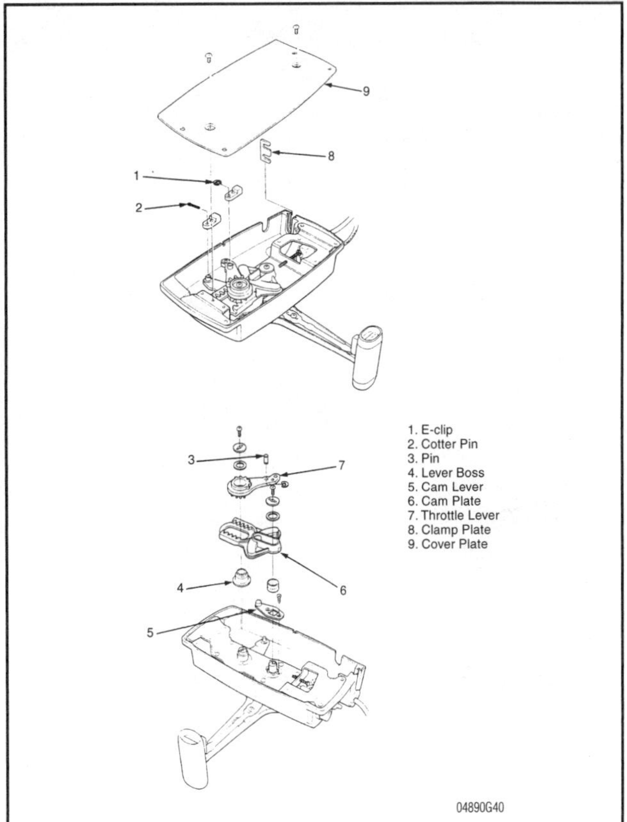

Fig. 9 Side mount remote control box rear exploded view—BF9.9A and BF15A

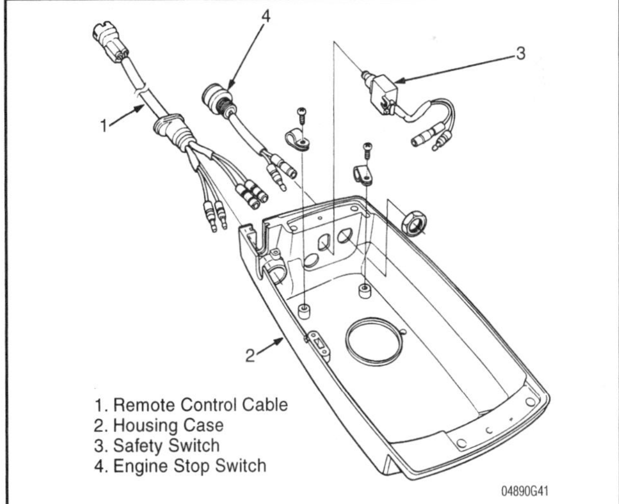

Fig. 10 Side mount remote control box electrical component locations —BF9.9A and BF15A

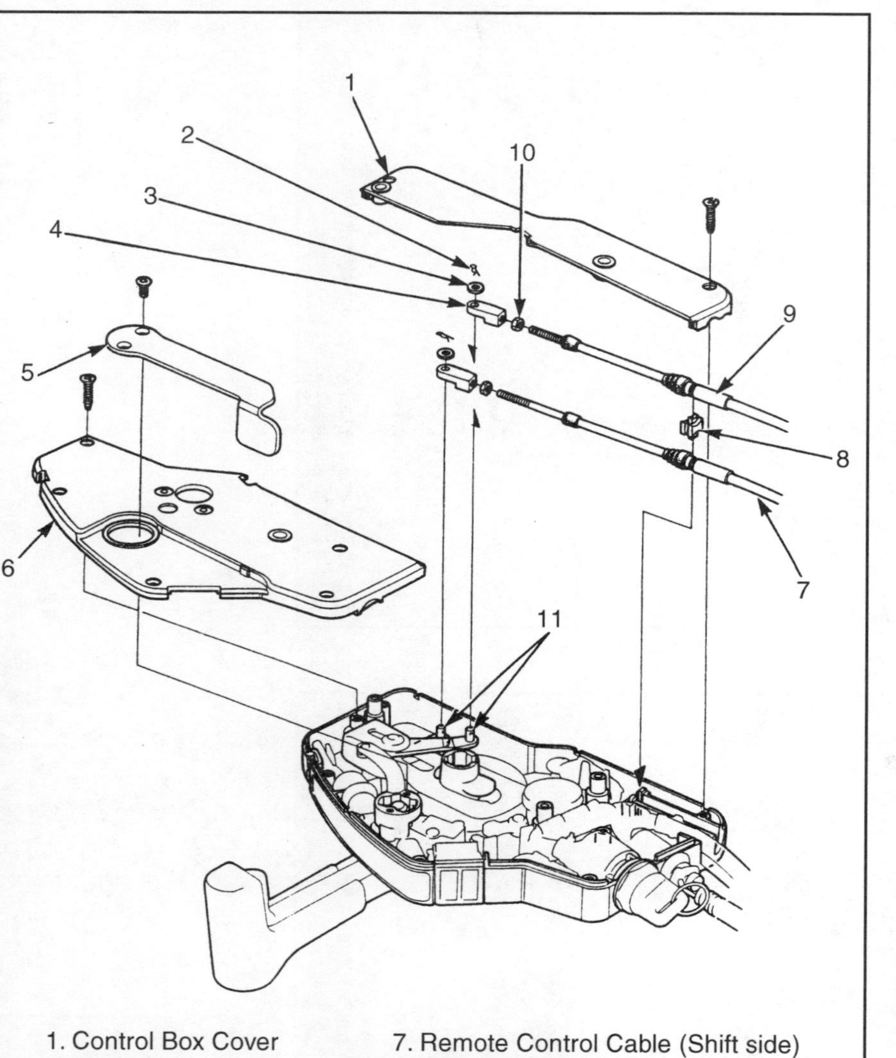

Fig. 11 Side mount remote control box cable detail—except BF9.9A, BF15A, BF75A and BF90A

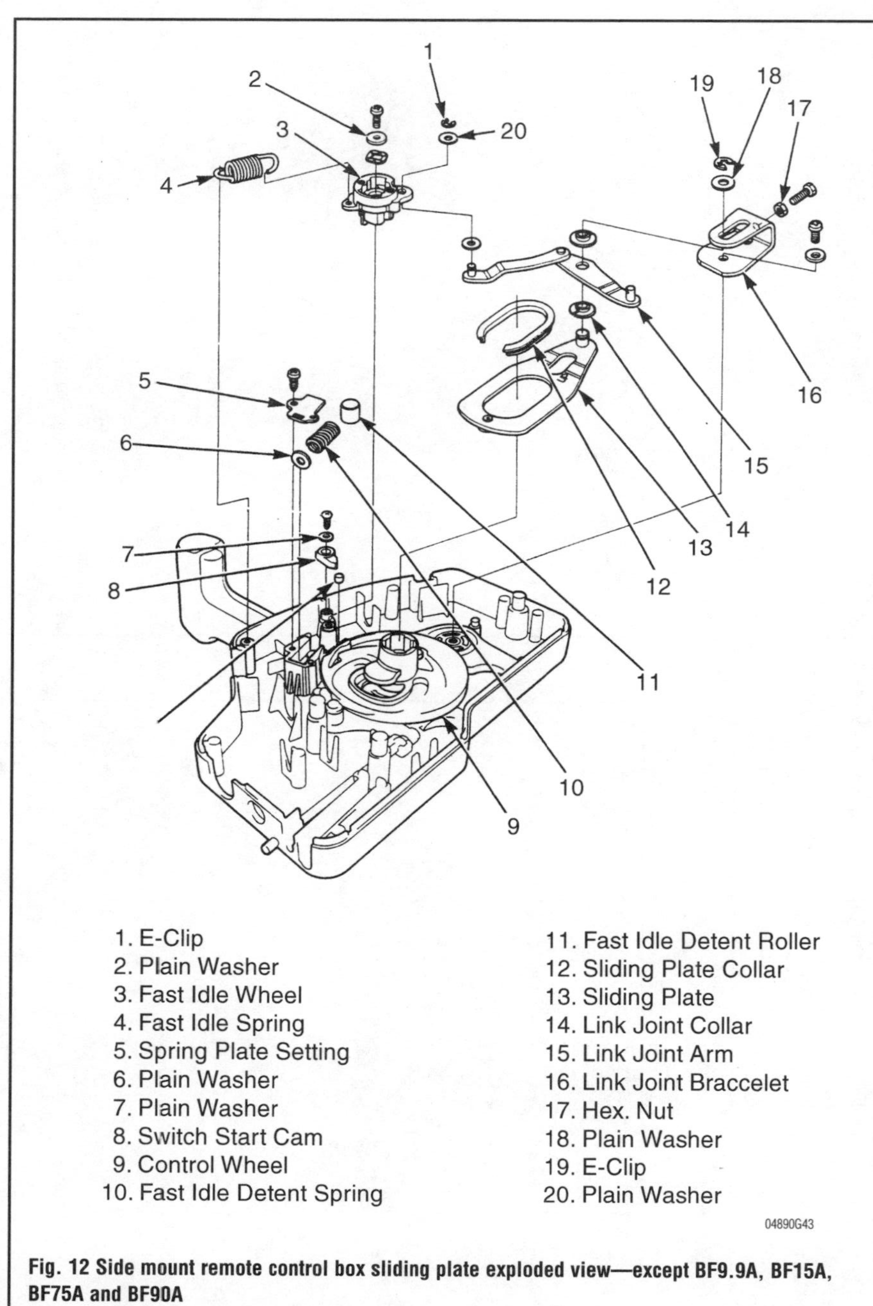

Fig. 12 Side mount remote control box sliding plate exploded view—except BF9.9A, BF15A, BF75A and BF90A

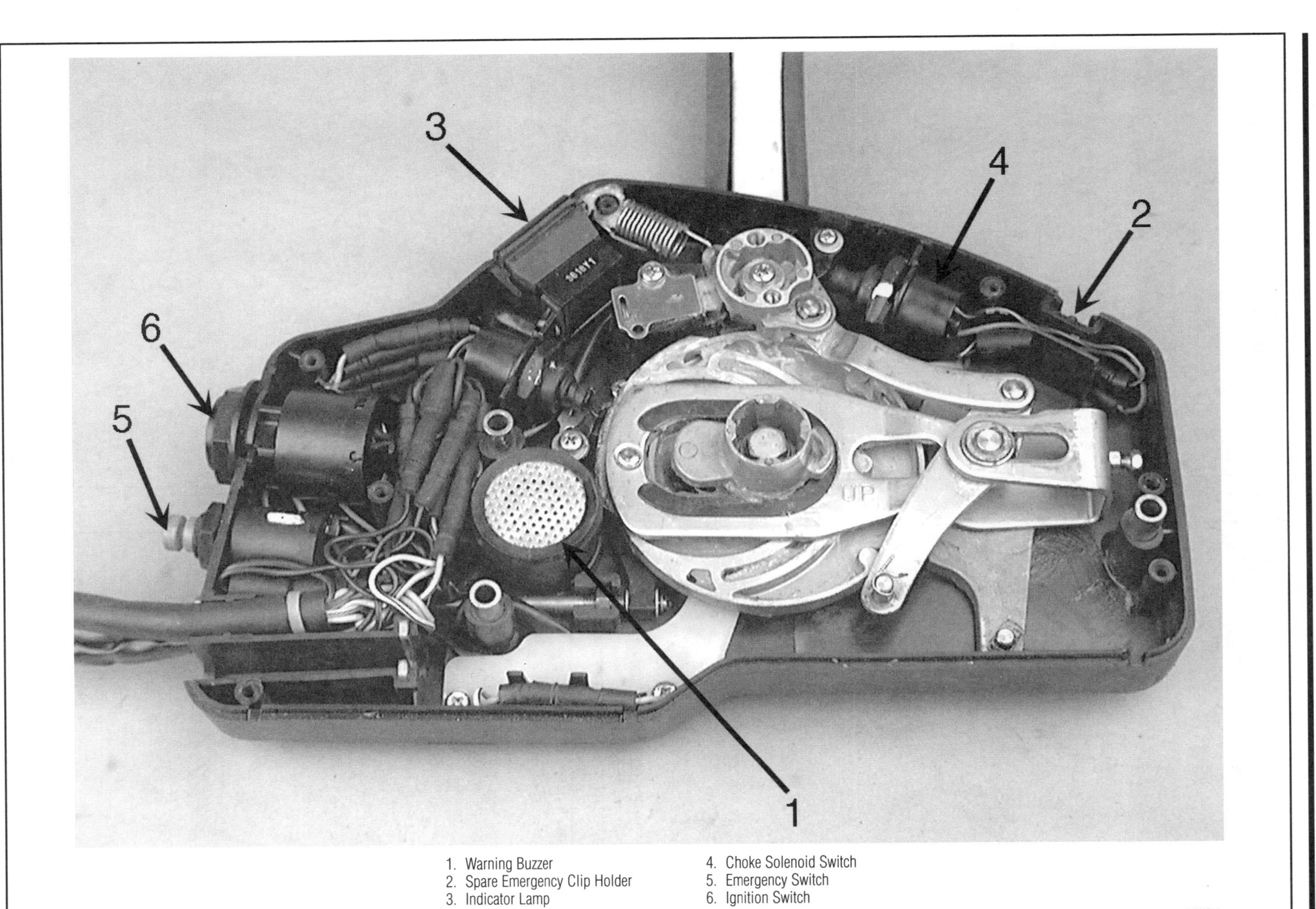

1. Warning Buzzer
2. Spare Emergency Clip Holder
3. Indicator Lamp
4. Choke Solenoid Switch
5. Emergency Switch
6. Ignition Switch

Fig. 13 Side mount remote control box electrical component locations—except BF9.9A, BF15A, BF75A and BF90A

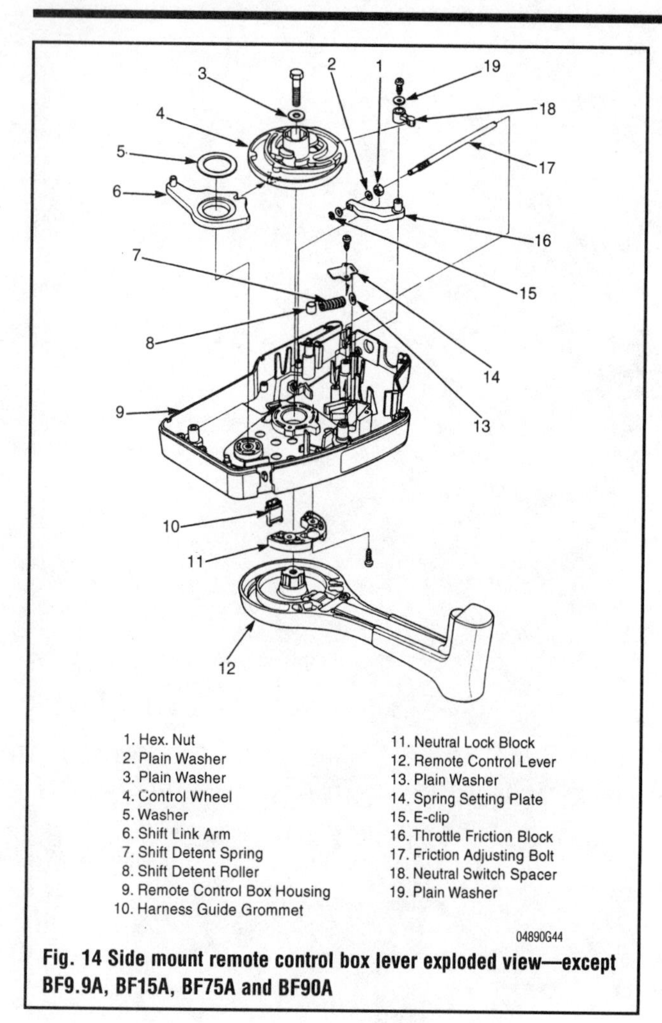

1. Hex. Nut
2. Plain Washer
3. Plain Washer
4. Control Wheel
5. Washer
6. Shift Link Arm
7. Shift Detent Spring
8. Shift Detent Roller
9. Remote Control Box Housing
10. Harness Guide Grommet
11. Neutral Lock Block
12. Remote Control Lever
13. Plain Washer
14. Spring Setting Plate
15. E-clip
16. Throttle Friction Block
17. Friction Adjusting Bolt
18. Neutral Switch Spacer
19. Plain Washer

04890G44

Fig. 14 Side mount remote control box lever exploded view—except BF9.9A, BF15A, BF75A and BF90A

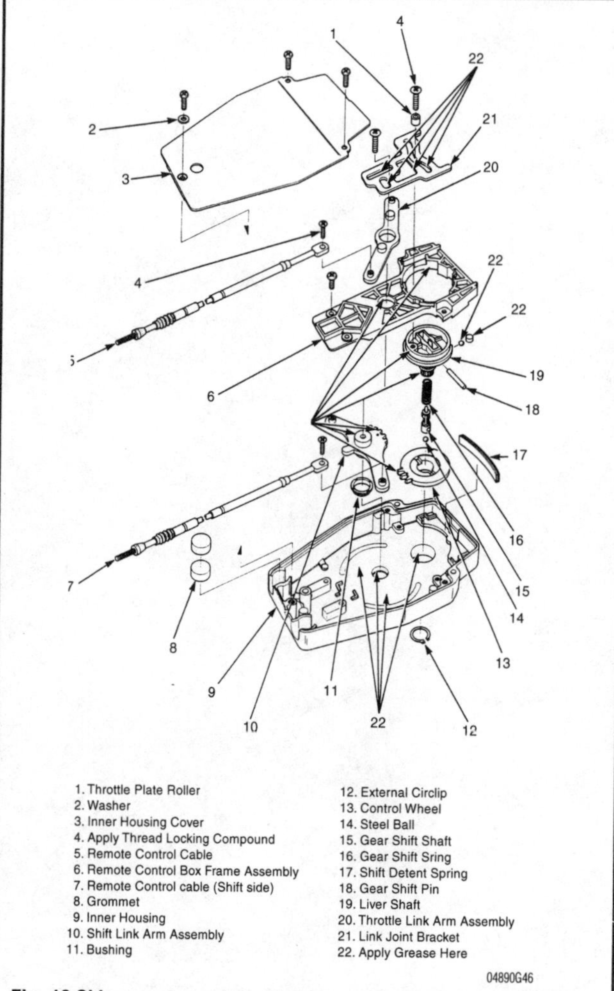

1. Throttle Plate Roller
2. Washer
3. Inner Housing Cover
4. Apply Thread Locking Compound
5. Remote Control Cable
6. Remote Control Box Frame Assembly
7. Remote Control cable (Shift side)
8. Grommet
9. Inner Housing
10. Shift Link Arm Assembly
11. Bushing
12. External Circlip
13. Control Wheel
14. Steel Ball
15. Gear Shift Shaft
16. Gear Shift Sring
17. Shift Detent Spring
18. Gear Shift Pin
19. Liver Shaft
20. Throttle Link Arm Assembly
21. Link Joint Bracket
22. Apply Grease Here

04890G46

Fig. 16 Side mount remote control box rear exploded view—BF75A and BF90A

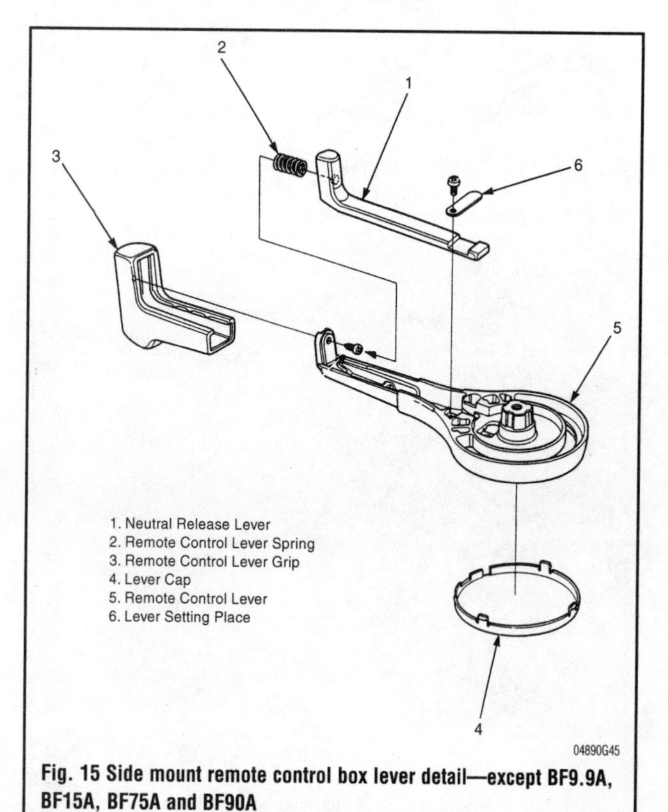

1. Neutral Release Lever
2. Remote Control Lever Spring
3. Remote Control Lever Grip
4. Lever Cap
5. Remote Control Lever
6. Lever Setting Place

04890G45

Fig. 15 Side mount remote control box lever detail—except BF9.9A, BF15A, BF75A and BF90A

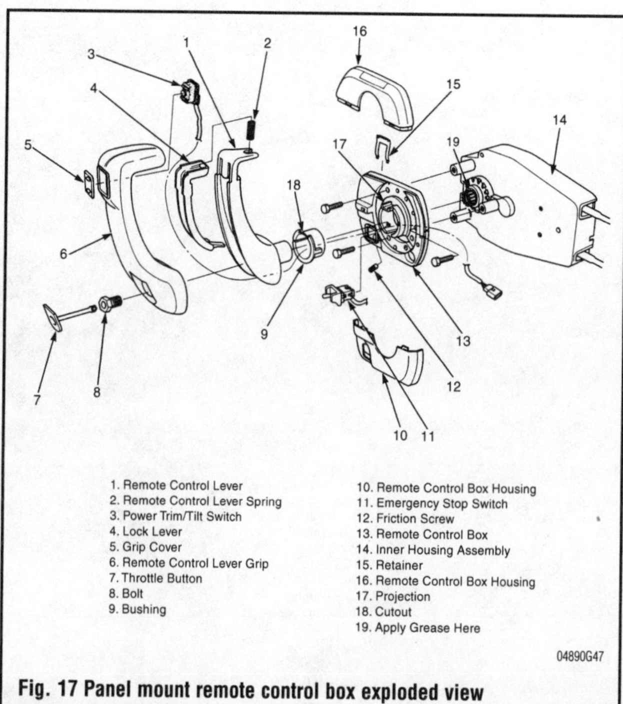

1. Remote Control Lever
2. Remote Control Lever Spring
3. Power Trim/Tilt Switch
4. Lock Lever
5. Grip Cover
6. Remote Control Lever Grip
7. Throttle Button
8. Bolt
9. Bushing
10. Remote Control Box Housing
11. Emergency Stop Switch
12. Friction Screw
13. Remote Control Box
14. Inner Housing Assembly
15. Retainer
16. Remote Control Box Housing
17. Projection
18. Cutout
19. Apply Grease Here

04890G47

Fig. 17 Panel mount remote control box exploded view

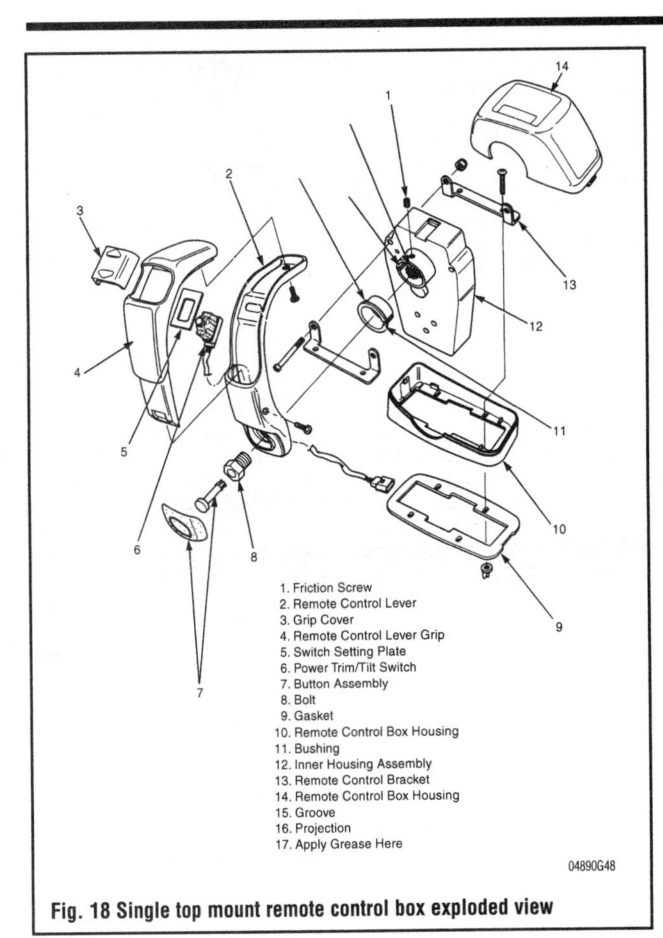

Fig. 18 Single top mount remote control box exploded view

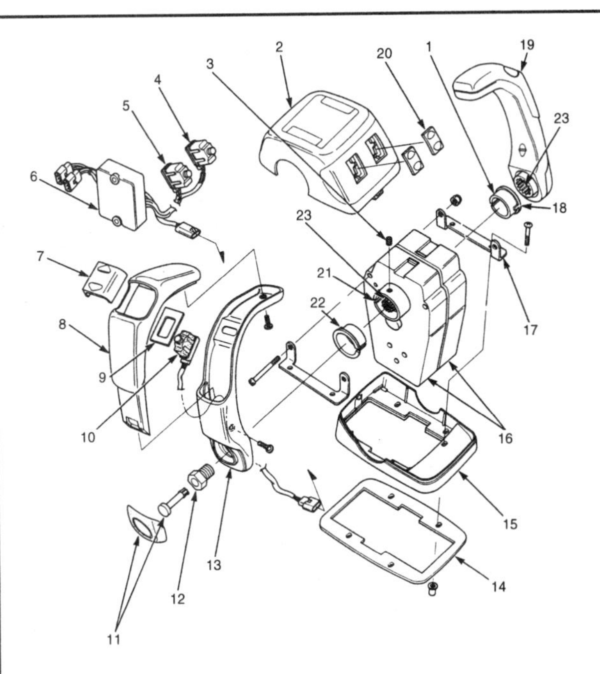

Fig. 19 Dual top mount remote control box exploded view

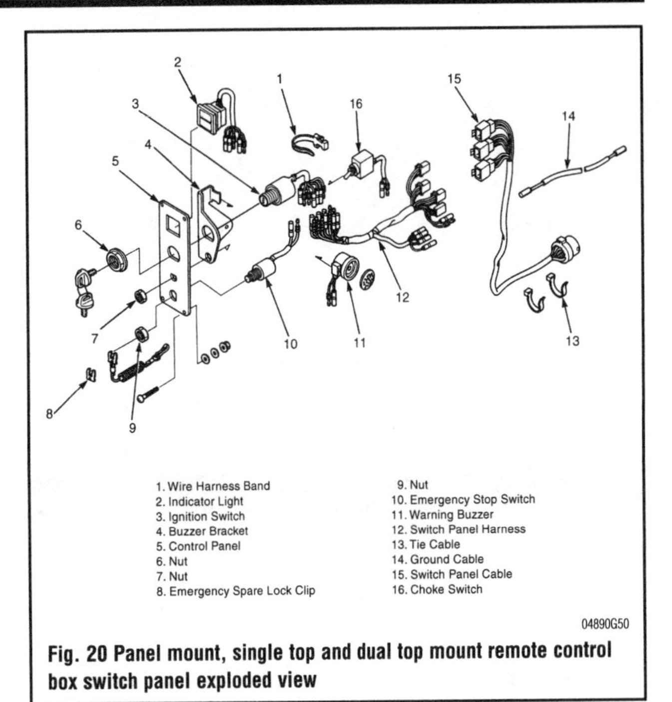

Fig. 20 Panel mount, single top and dual top mount remote control box switch panel exploded view

Choke Switch

TESTING

1. Disconnect the choke switch wiring harness.
2. Connect a multimeter between the switch harness leads.
3. With the switch on, continuity should exist. With the switch off, continuity should not exist.
4. If the switch functions properly, the problem may be in the wiring harness.
5. If the switch does not function as specified there is a short in either the switch or harness and the switch should be replaced.

REMOVAL & INSTALLATION

1. Remove the control box from the side of the boat and open the side covers to allow access to the internal components.
2. Disconnect the engine stop switch wiring harness.
3. Remove any wire straps that connect the switch to the box.
4. Remove the switch attaching nut.
5. Remove the switch from the control box.

To install:

6. Install the switch in the control box.
7. Install the switch attaching nut and tighten securely.
8. Install any wire straps that connect the switch to the control box.
9. Connect the engine stop switch wiring harness.
10. Test the switch for proper operation.
11. Install the control box side covers and mount the control box in the boat.

Engine Stop Switch

TESTING

1. Disconnect the engine stop switch wiring harness.
2. Connect a multimeter between the switch harness leads.
3. With the switch released, continuity should exist. With the switch pressed, continuity should not exist.
4. If the switch functions properly, the problem may be in the wiring harness.
5. If the switch does not function as specified, the switch may be faulty.

REMOVAL & INSTALLATION

1. Remove the control box from the side of the boat and open the side covers to allow access to the internal components.
2. Disconnect the engine stop switch wiring harness.
3. Remove any wire straps that connect the switch to the box.
4. Remove the switch attaching nut.
5. Remove the switch from the control box.

To install:

6. Install the switch in the control box.
7. Install the switch attaching nut and tighten securely.
8. Install any wire straps that connect the switch to the control box.
9. Connect the engine stop switch wiring harness.
10. Test the switch for proper operation.
11. Install the control box side covers and mount the control box in the boat.

Ignition Switch

TESTING

See Figures 21 and 22

1. Disconnect the ignition switch wiring harness.
2. Connect a multimeter between the switch harness leads as noted in the chart.
3. With the switch in the stated positions, check for continuity between the various terminals.

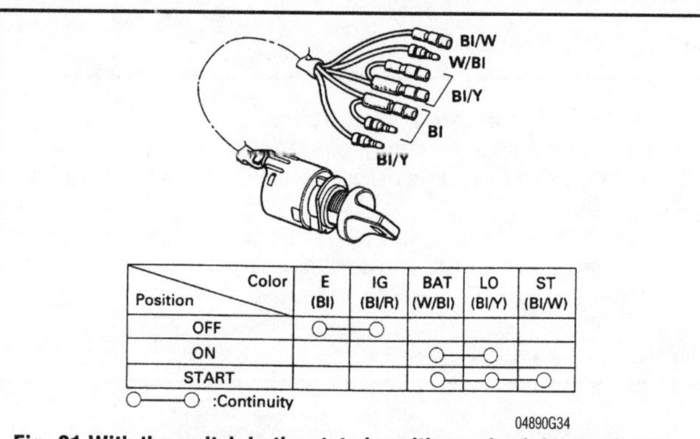

Position \ Color	E (Bl)	IG (Bl/R)	BAT (W/Bl)	LO (Bl/Y)	ST (Bl/W)
OFF	O——	——O			
ON			O——	——O	
START			O——	——O——	——O

O——O :Continuity

04890G34

Fig. 21 With the switch in the stated positions, check for continuity between the ignition switch terminals as illustrated

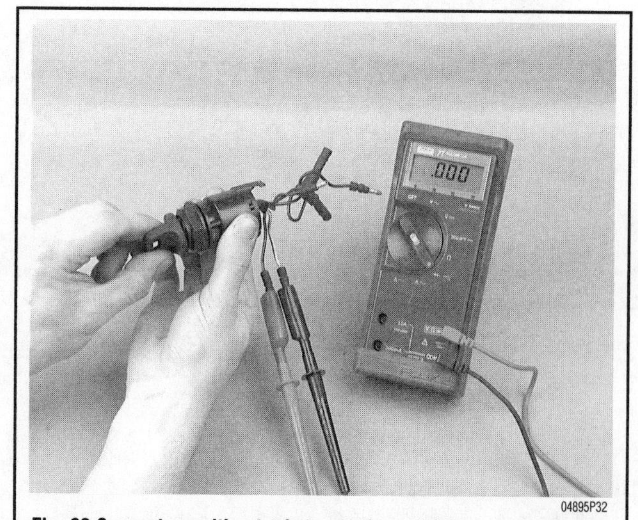

04895P32

Fig. 22 Connect a multimeter between the switch harness leads and check for continuity

4. If the switch functions properly, the problem may be in the wiring harness
5. If the switch does not function as specified, the switch may be faulty.

REMOVAL & INSTALLATION

1. Remove the control box from the side of the boat and open the side covers to allow access to the internal components.
2. Disconnect the engine stop switch wiring harness.
3. Remove any wire straps that connect the switch to the box.
4. Remove the switch attaching nut.
5. Remove the switch from the control box.

To install:

6. Install the switch in the control box.
7. Install the switch attaching nut and tighten securely.
8. Install any wire straps that connect the switch to the control box.
9. Connect the engine stop switch wiring harness.
10. Test the switch for proper operation.
11. Install the control box side covers and mount the control box in the boat.

Indicator Lights

TESTING

See Figure 23

1. Disconnect the indicator light wiring harness.
2. Connect a 12 volt switched power source as noted in the illustration.
3. With switch 1 (SW1) **ON**, the green lamp should illuminate.
4. With switches 1 and 3(SW1, 3) **ON**, the green lamp should not illuminate.
5. If the switch functions properly, the problem may be in the wiring harness
6. With switch 2 (SW2) **ON**, the green lamp should illuminate.
7. If the lights function as stated, there may be a problem in the wiring harness.
8. If the lights do not function as stated, the indicator light assembly may be faulty.

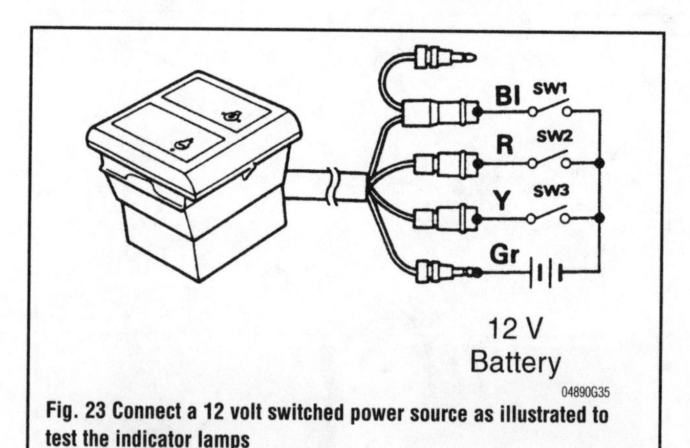

04890G35

Fig. 23 Connect a 12 volt switched power source as illustrated to test the indicator lamps

REMOVAL & INSTALLATION

1. Remove the control box from the side of the boat and open the side covers to allow access to the internal components.
2. Disconnect the light wiring harness.
3. Remove any wire straps that connect the light wiring harness to the box.
4. Remove the interior light bezel.
5. Remove the light assembly from the control box.

To install:

6. Install the light assembly in the control box.
7. Install the interior light bezel.
8. Install any wire straps that connect the light harness to the control box.
9. Connect the light wiring harness.
10. Test the lights for proper operation.
11. Install the control box side covers and mount the control box in the boat.

Safety Switch

TESTING

➧ **See Figure 24**

1. Disconnect the engine stop switch wiring harness.
2. Connect a multimeter between the switch harness leads.
3. With the switch engaged (stop switch lanyard in position), continuity should not exist. With the switch released (stop switch lanyard pulled), continuity should exist.
4. If the switch functions properly, the problem may be in the wiring harness
5. If the switch does not function as specified, the switch may be faulty.

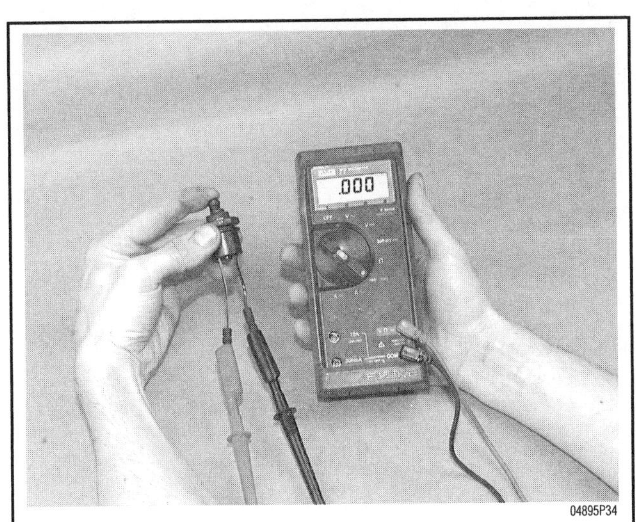

04895P34

Fig. 24 Connect a multimeter between the switch harness leads and check for continuity

REMOVAL & INSTALLATION

➧ **See Figure 25**

1. Remove the control box from the side of the boat and open the side covers to allow access to the internal components.
2. Disconnect the engine stop switch wiring harness.
3. Remove any wire straps that connect the switch to the box.
4. Remove the switch attaching nut.
5. Remove the switch from the control box.

To install:

6. Install the switch in the control box.
7. Install the switch attaching nut and tighten securely.
8. Install any wire straps that connect the switch to the control box.
9. Connect the engine stop switch wiring harness.
10. Test the switch for proper operation.
11. Install the control box side covers and mount the control box in the boat.

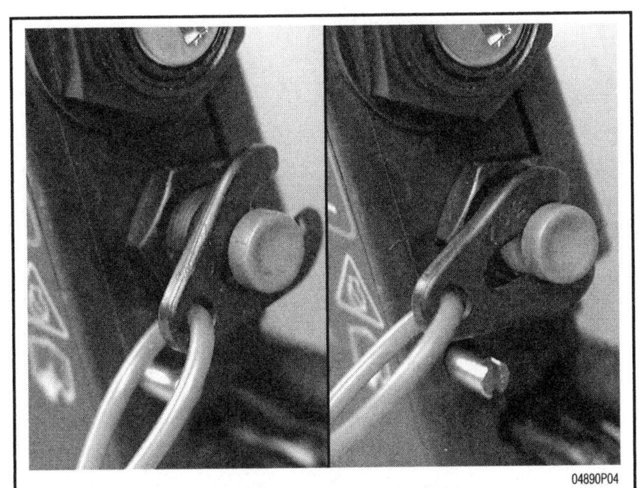

04890P04

Fig. 25 The emergency tether should be fully inserted into the safety switch (left). If the tether is partially inserted (right), the switch will not allow the engine to be started

Throttle Switch

TESTING

1. Disconnect the throttle switch wiring harness.
2. Connect a multimeter between the switch harness leads.
3. With the switch knob pushed, continuity should exist. With the switch knob released, continuity should not exist.
4. If the switch functions properly, the problem may be in the wiring harness
5. If the switch does not function as specified, the switch may be faulty.

REMOVAL & INSTALLATION

1. Remove the control box from the side of the boat and open the side covers to allow access to the internal components.
2. Disconnect the throttle switch wiring harness.
3. Remove any wire straps that connect the switch to the box.
4. Loosen the switch attaching nut.
5. Remove the switch from the control box.

To install:

6. Install the switch in the control box.
7. Tighten the switch attaching nut securely.
8. Install any wire straps that connect the switch to the control box..
9. Connect the throttle switch wiring harness.
10. Test the switch for proper operation.
11. Install the control box side covers and mount the control box in the boat.

Trim/Tilt Switch

TESTING

➧ **See Figures 26, 27 and 28**

1. Disconnect the trim/tilt switch wiring harness.
2. Connect a multimeter between the switch harness terminals as illustrated.

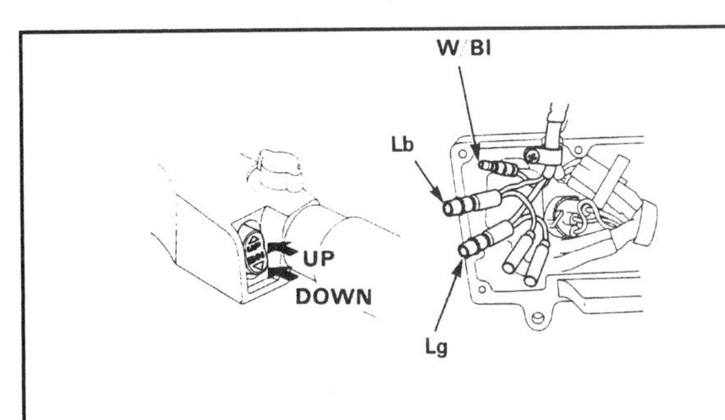

	Lg	W/Bl	Lb
UP	O—	—O	
DOWN		O—	—O

O——O Continuity

04890G36

Fig. 26 With the trim/tilt switch in the illustrated positions, check for continuity between the terminals— side mount remote control box

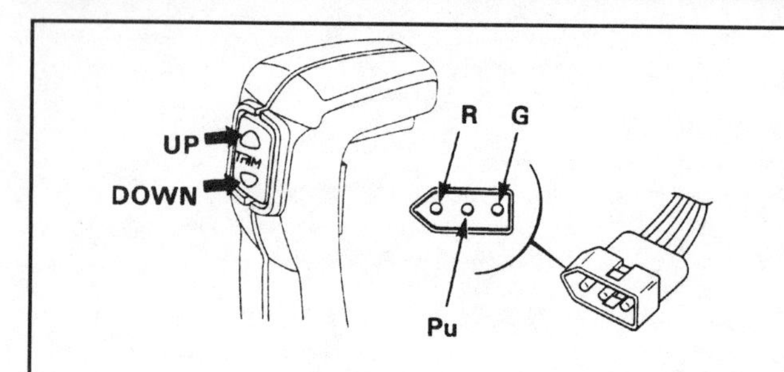

	Pu	R	G
UP	O——	——O	
DOWN		O——	——O

O——O Continuity

04890G37

Fig. 27 With the trim/tilt switch in the illustrated positions, check for continuity between the terminals— panel mount and single top mount remote control boxes

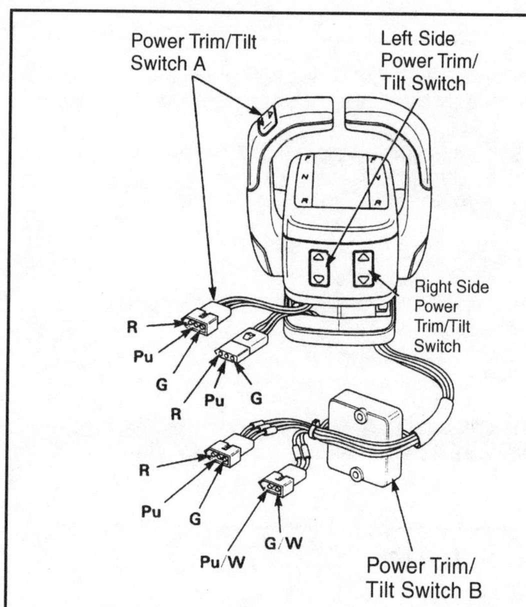

Switch A

	Pu	R	G
UP	O——	——O	
DOWN		O——	——O

Switch B, Right

	Pu	R	G
UP	O——	——O	
DOWN		O——	——O

Switch B, Left

	Pu/W	R	G/W
UP	O——	——O	
DOWN		O——	——O

O——O :Continuity

04890G38

Fig. 28 With the trim/tilt switch in the illustrated positions, check for continuity between the terminals— dual top mount remote control box

3. With the switch in the stated positions, check for continuity between the various terminals.
4. If the switch functions properly, the problem may be in the wiring harness
5. If the switch does not function as specified, the switch may be faulty.

REMOVAL & INSTALLATION

Side Mount

1. Remove the control box from the side of the boat and open the side covers to allow access to the internal components.
2. Remove the bolt and washer that retain the remote control lever to the control box.
3. Remove the remote control lever taking care to not pull on the tilt/trim wiring harness.
4. Label and disconnect the tilt/trim wiring harness.
5. Remove the screw securing the neutral release lever.
6. Remove the neutral release lever and lever spring.
7. Remove the screw securing the control lever grip.
8. Remove the control lever grip from the control lever taking care to not pull on the tilt/trim wiring harness as you thread it though the holes in the control lever.
9. Remove the tilt/trim switch from the lever grip.

To install:

10. Install the tilt/trim switch in the lever grip.
11. Thread the tilt/trim wiring harness through the control lever and install the control lever grip. Secure the grip with the screw.
12. Install the neutral release lever and lever spring, securing them with the setting plate and screw.
13. Connect the tilt/trim wiring harness.
14. Install the remote control lever securing them with the bolt and washer to the remote control box.
15. Install the control box from the side of the boat and open the side covers to allow access to the internal components.

Panel/Top Mount

1. Remove the control box from its attaching point on the boat and open the covers to allow access to the internal components.
2. Remove the bolt that retain the remote control lever to the control box.
3. Remove the remote control lever taking care to not pull on the tilt/trim wiring harness.
4. Label and disconnect the tilt/trim wiring harness.
5. Remove the neutral release lever and lever spring.
6. Remove the tilt/trim switch from the lever grip.

To install:

7. Install the tilt/trim switch in the lever grip.
8. Thread the tilt/trim wiring harness through the control lever and install the control lever grip.
9. Install the neutral release lever and lever spring.
10. Connect the tilt/trim wiring harness.
11. Install the remote control lever, tightening the bolt to 12 ft. lbs. (17 Nm).
12. Install the control box from the side of the boat and open the side covers to allow access to the internal components.

Warning Buzzer

TESTING

1. The warning buzzer is tested by simply connecting it to a 12 volt power source.
2. The buzzer should sound when properly connected to power.
3. If the buzzer functions properly, there may be a problem in the wiring harness.
4. If the buzzer does not perform as stated, it may be faulty.

REMOVAL & INSTALLATION

1. Remove the control box from the side of the boat and open the side covers to allow access to the internal components.

2. Disconnect the buzzer wiring harness.
3. Remove any wire straps that connect the buzzer wiring harness to the box.
4. Remove the buzzer from the control box.

To install:

5. Install the buzzer in the control box.
6. Install any wire straps that connect the buzzer harness to the control box.
7. Connect the buzzer wiring harness.
8. Test the buzzer for proper operation.
9. Install the control box side covers and mount the control box in the boat.

TILLER HANDLE

Description and Operation

➧ **See Figures 29 thru 34**

Steering control for most outboards begins at the tiller handle and ends at the propeller. Tiller steering is the most simple form of small outboard control. All components are mounted directly to the engine and are easily serviceable.

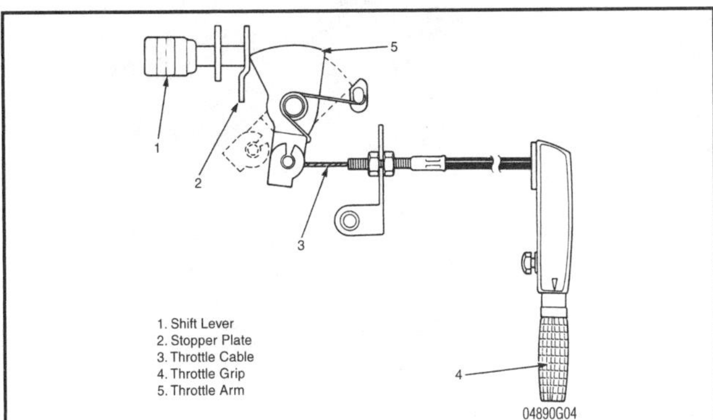

Fig. 29 The throttle stopper system limits throttle opening when the shift lever is in neutral and reverse

Forward

Nuetral

Reverse

04890G05

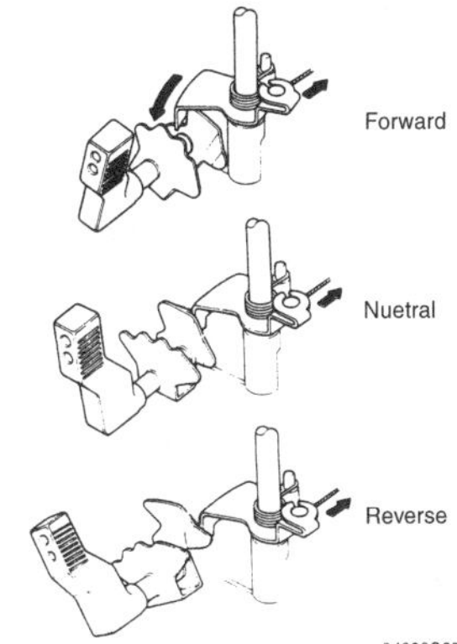

Fig. 30 The throttle stopper system works by blocking the throttle arm using a raised portion of the shift lever

Throttle control is performed via a throttle grip mounted to the tiller arm. As the grip is rotated a cable opens and closes the throttle lever on the engine. An adjustment thumbscrew is usually located near the throttle grip to allow adjustment of the turning resistance. In this way, the operator does not have to keep constant pressure on the grip to maintain engine speed.

An emergency engine stop switch is used on most outboards to prevent the engine from continuing to run without the operator in control. This switch is controlled by a small clip which keeps the switch open during normal engine operation. When the clip is removed, a spring inside the switch closes it and completes a ground connection to stop the engine. The clip is connected to a lanyard that is worn around the helmsman's wrist.

Some tiller systems utilize a throttle stopper system which limits throttle opening when the shift lever is in neutral and reverse. This prevents overrevving the engine under no-load conditions and also limits the engine speed when in reverse.

Troubleshooting

If the tiller steering system seems loose, first check the engine for proper mounting. Ensure the engine is fastened to the transom securely. Next, check the tiller hinge point where it attaches to the engine and tighten the hinge pivot bolt as necessary.

Excessively tight steering that cannot be adjusted using the tension adjustment is usually due to a lack of lubrication. Once the swivel case bushings run

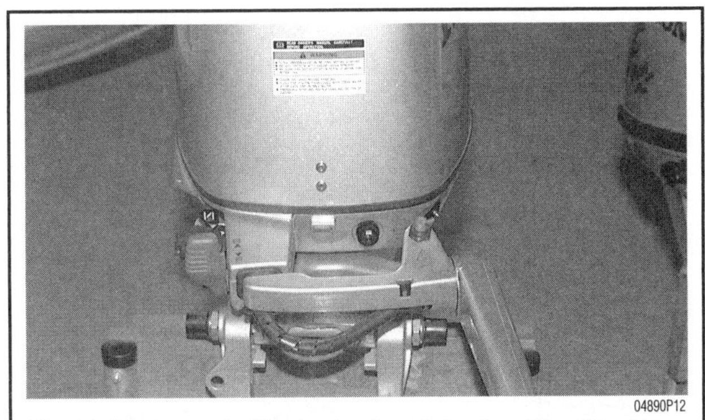

Fig. 31 Steering control for most outboards begins at the tiller handle and ends at the propeller

Fig. 32 The throttle stopper system limits throttle opening when the shift lever is in neutral and reverse

Fig. 33 An adjustment thumbscrew is usually located near the throttle grip to allow adjustment of the turning resistance

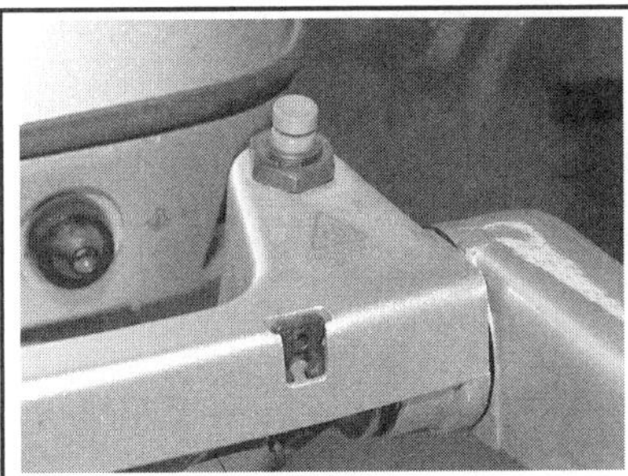

Fig. 34 An emergency engine stop switch is used on most outboards to prevent the engine from continuing to run without the operator in control

dry, the steering shaft will get progressively tighter and eventually seize. This condition is generally caused by a lack of periodic maintenance.

Correct this condition by lubricating the swivel case bushings and working the outboard back and forth to spread the lubricant. However, this may only be a temporary fix. In severe cases, the swivel case bushings may need to be replaced.

Tiller Handle

REMOVAL & INSTALLATION

BF20 and BF2A

See Figure 35

1. Remove the engine cover.
2. Remove the tiller handle pivot bolt and carefully remove the tiller handle.
3. Remove the mounting collar, mounting rubber, distance collar and special washer from the tiller handle.

To install:

4. Install the mounting collar, mounting rubber, distance collar and special washer on the tiller handle.
5. Position the tiller handle on the outboard and install the tiller handle pivot bolt. Tighten the pivot bolt to 15–20 ft. lbs. (20–27 Nm).
6. Check the handle for smooth operation.
7. Install the engine cover.

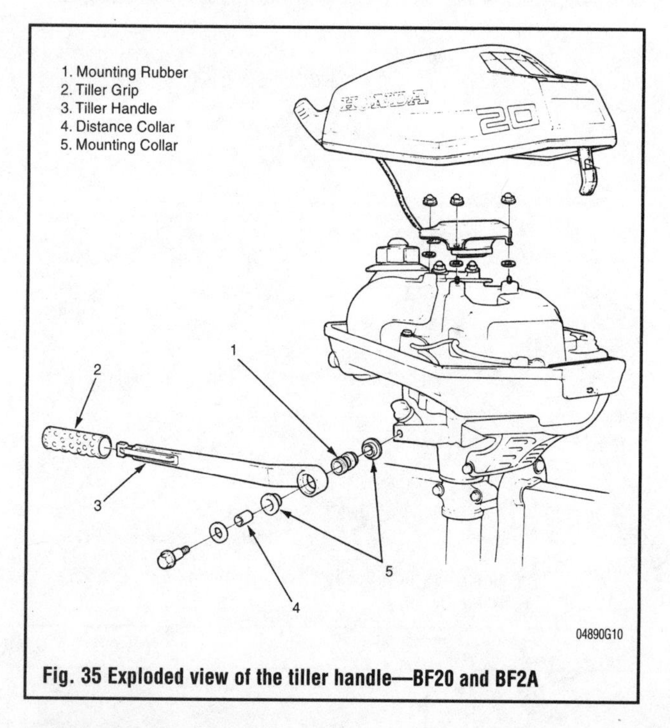

Fig. 35 Exploded view of the tiller handle—BF20 and BF2A

BF50 and BF5A

See Figures 36 and 37

1. Remove the engine cover.
2. Place the throttle grip in the **SLOW** position on the tiller arm.
3. Loosen the locknut and slacken the adjustment on the throttle cable at the throttle lever.
4. Label and disconnect the engine stop switch wiring harness.
5. Remove the tiller handle through bolt.
6. Remove the rubber mounting and handle collar.
7. Slide the tiller handle from the mounting bracket enough to disconnect the throttle cable from the throttle reel.
8. Carefully remove the tiller handle from the bracket.

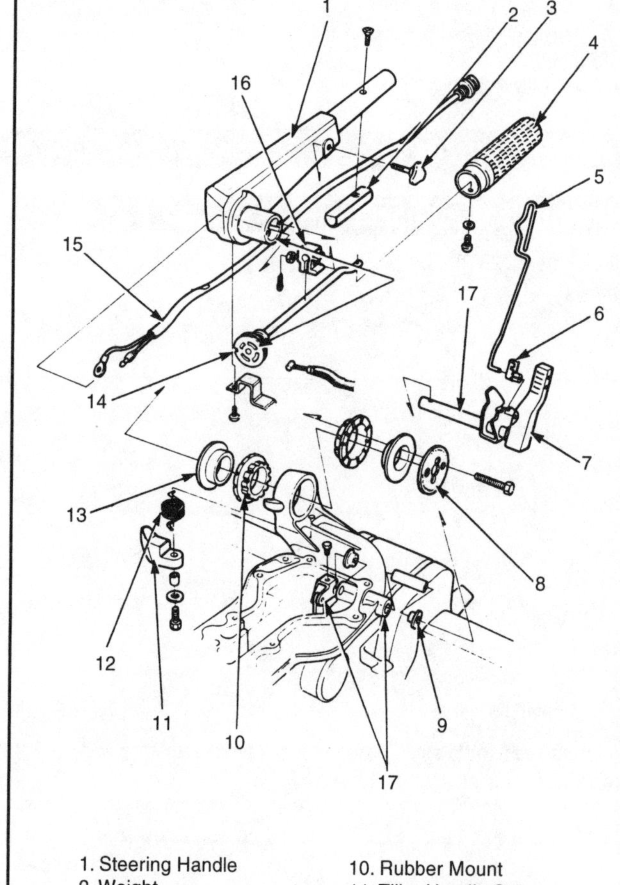

Fig. 36 Exploded view of the tiller handle— BF50 and early model BF5A

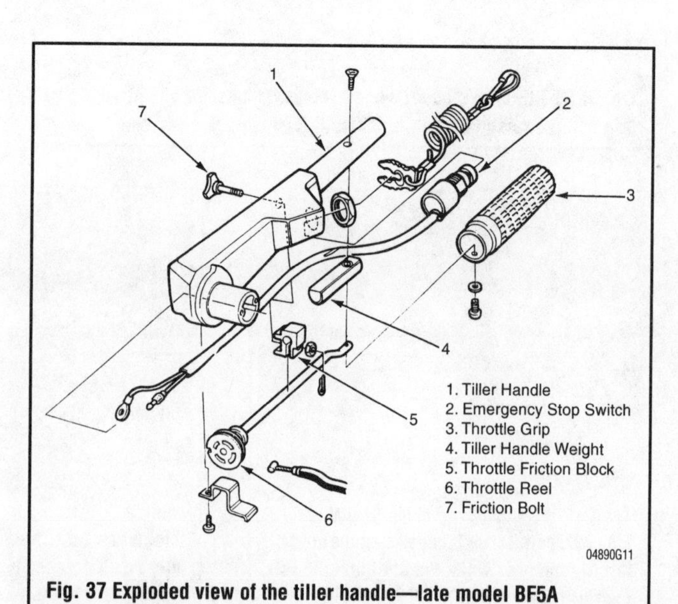

Fig. 37 Exploded view of the tiller handle—late model BF5A

To install:

9. Lubricate the throttle reel where the cable makes contact.
10. Install the throttle cable end on the throttle reel and route the throttle cable around the reel.
11. Lubricate and install the tiller handle into the mounting bracket.
12. Install the rubber mounting and handle collar.
13. Install the tiller handle through bolt and tighten to 6–9 ft. lbs. (8–12 Nm).
14. Connect the engine stop switch wiring harness.
15. Adjust the throttle cable tension and tighten the locknut securely.
16. Check for proper throttle operation.
17. Install the engine cover.

BF75, BF100 and BF8A

➧ See Figure 38

1. Remove the engine cover.
2. Align the word **SHIFT** on the grip pipe cover with the dot on the tiller arm.
3. Loosen the locknut and slacken the adjustment on the throttle cable at the throttle arm.
4. Disconnect the end of the throttle cable from the throttle arm.
5. Remove the grommet which seals the throttle cable to the engine case. This grommet should be replaced if found to be damaged or dry rotted.
6. Remove the tiller handle pivot bolt and carefully remove the tiller handle while feeding the throttle cable through the engine case.
7. Remove the pivot spring from the tiller handle.

To install:

8. Check the pivot spring for adequate tension and replace as necessary. Lubricate and install the spring in the tiller handle.
9. Position the tiller handle on the engine case and install the pivot bolt. Tighten the bolt to 17 ft. lbs. (23 Nm). Ensure the tiller arm moves freely without binding.
10. Install a new grommet on the throttle cable, as necessary and feed the throttle cable through the engine case.
11. Lubricate the end of the throttle cable and position it in the throttle arm.
12. Adjust the throttle cable tension and tighten the locknut securely.
13. Check for proper throttle operation.
14. Install the engine cover.

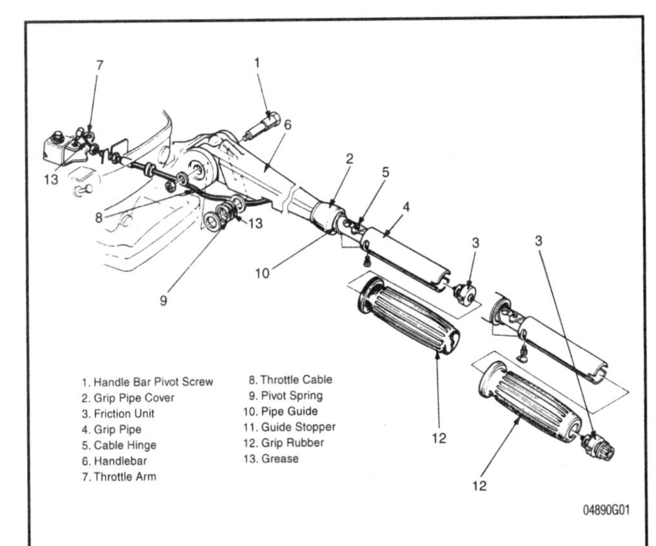

Fig. 38 Exploded view of the tiller handle—BF75, BF100 and BF8A

BF9.9A and BF15A

➧ See Figure 39

1. Remove the engine cover.
2. Place the throttle grip in the **SLOW** or fully closed position.
3. Loosen the locknut and slacken the adjustment on the throttle cable at the throttle lever.
4. Label and disconnect the engine stop switch wiring harness.

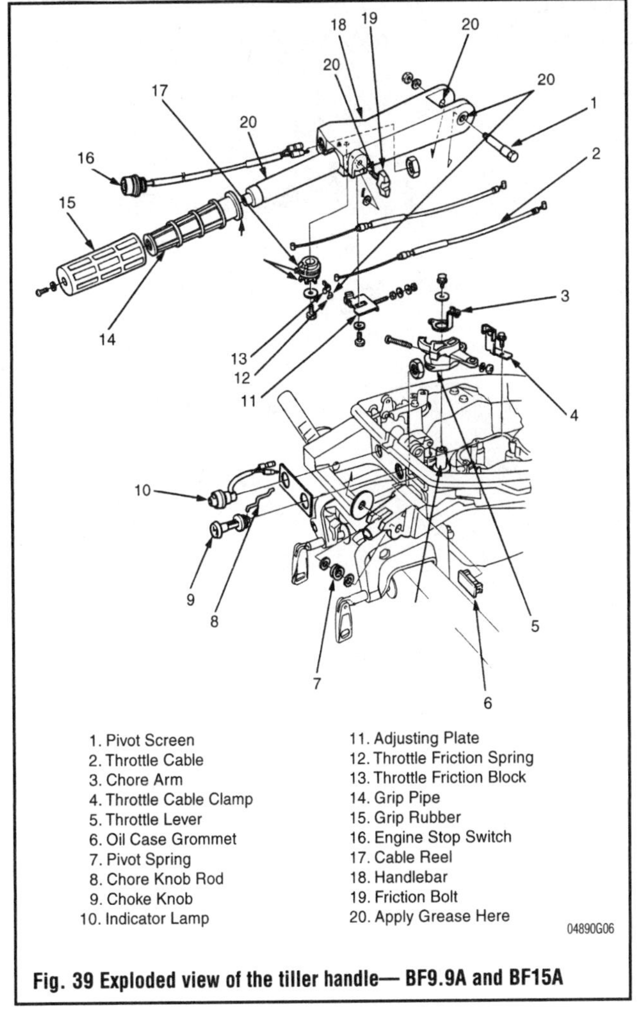

Fig. 39 Exploded view of the tiller handle— BF9.9A and BF15A

5. Disconnect the throttle cables from the throttle lever.
6. Remove the tiller handle through bolt.
7. Carefully remove the tiller handle from the bracket while guiding the throttle cables out of the engine.

To install:

8. Guide the throttle cables into position and install the tiller handle on its bracket.
9. Lubricate and install the tiller handle through bolt. Tighten to 17 ft. lbs. (24 Nm).
10. Connect the throttle cables at the throttle lever and properly adjust the throttle cable.
11. Connect the engine stop switch wiring harness.
12. Check for proper throttle operation.
13. Install the engine cover.

BF35A, BF40A, BF45A and BF50A

➧ See Figure 40

1. Remove the engine cover.
2. Place the throttle grip in the **SLOW** position on the tiller arm.
3. Wrap both ends of one cable with tape to distinguish the cables from each other.
4. Loosen the locknut and slacken the adjustment on the throttle cable at the throttle lever.
5. Remove the throttle reel plate.
6. Slide the throttle reel down enough to disconnect the throttle cables from the throttle reel.
7. Remove the tiller handle through bolt.
8. Remove the rubber mounting and washer.
9. Carefully remove the tiller handle from the bracket.

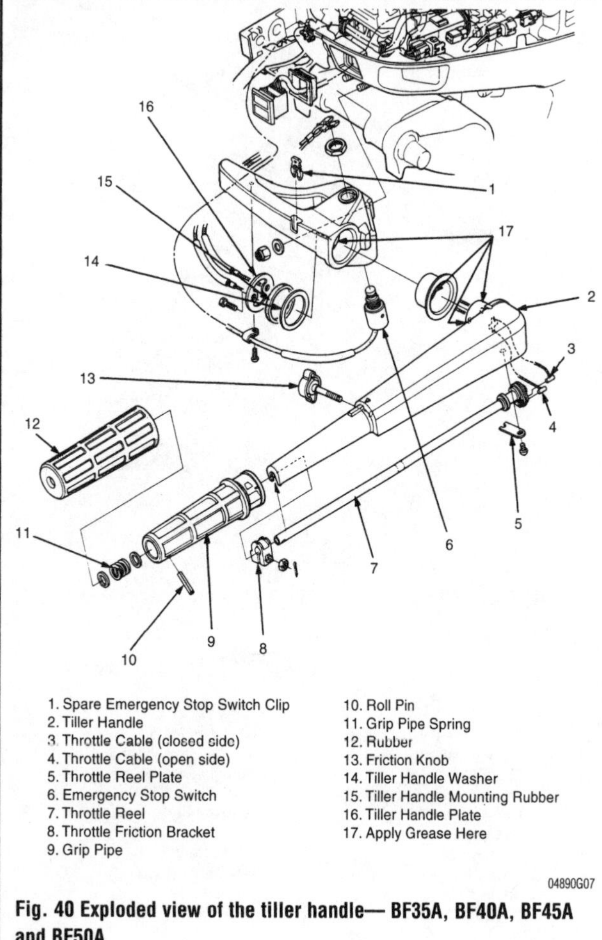

Fig. 40 Exploded view of the tiller handle— BF35A, BF40A, BF45A and BF50A

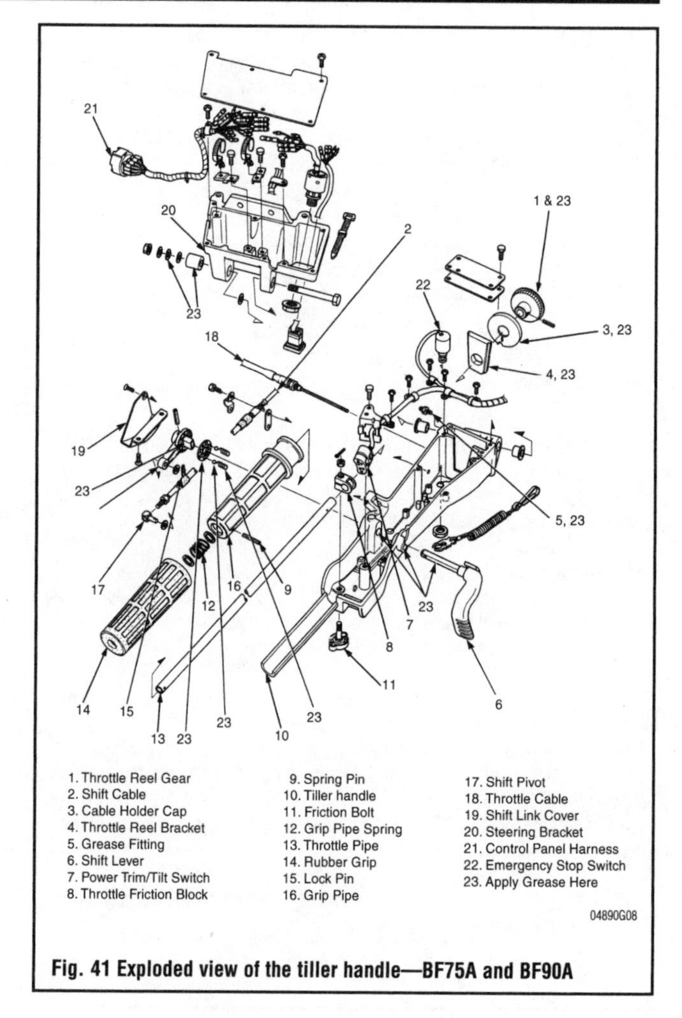

Fig. 41 Exploded view of the tiller handle—BF75A and BF90A

To install:

10. Lubricate and install the tiller handle into the mounting bracket.
11. Install the rubber mounting and handle collar.
12. Install the tiller handle through bolt and tighten to 6–9 ft. lbs. (8–12 Nm).
13. Lubricate the throttle reel where the cable makes contact.
14. Install the throttle cable end on the throttle reel and route the throttle cable around the reel.
15. Install the throttle reel plate.
16. Adjust the throttle cable tension and tighten the locknut securely.
17. Check for proper throttle operation.
18. Install the engine cover.

BF75A and BF90A

▶ See Figure 41

1. Remove the engine cover.
2. Place the throttle grip in the fully closed position.
3. Disconnect the throttle cable from the throttle lever.
4. Pull on the throttle cable inner rod while turning the throttle pipe on the tiller clockwise. Continue turning until the inner cable is released from the cable reel gear.
5. Disconnect the throttle cable from the tiller handle by unscrewing the cable nut.
6. Disconnect the tiller handle electrical connector.
7. Disconnect the shift cable from the tiller handle.
8. Remove the tiller handle through bolt.
9. Remove the tiller handle from the bracket.

To install:

10. Install the tiller handle on the bracket.
11. Install the tiller handle through bolt and tighten to 23 ft. lbs. (31Nm).
12. Connect the shift cable to the tiller handle and properly adjust it.
13. Connect the tiller handle electrical connector.
14. Install the throttle cable using the procedure under throttle cable and properly adjust it using the adjustment procedure under throttle cable.
15. Check for proper throttle operation.
16. Install the engine cover.

Throttle Cable

REMOVAL & INSTALLATION

BF50 and BF5A

▶ See Figures 36 and 37

1. Place the tiller handle in the up position with the throttle grip in the **SLOW** position.
2. Loosen the locknut and slacken the adjustment on the throttle cable.
3. Disconnect the end of the throttle cable from the throttle arm.
4. Remove the screw attaching the throttle reel rod to the throttle grip.
5. Loosen the friction bolt on the throttle friction block.
6. Lower the throttle reel assembly from the tiller handle and disconnect the throttle cable.
7. Remove the throttle cable from the powerhead.

To install:

8. Install the throttle cable into position on the powerhead.
9. Connect the throttle cable end to the throttle reel and route the cable around the reel.
10. Install the throttle reel assembly into the tiller arm and tighten the friction bolt to hold it in place.
11. Install the screw attaching the throttle reel rod to the throttle grip.
12. Connect the end of the throttle cable to the throttle arm.

13. Adjust the throttle cable tension and tighten the locknut securely.
14. Install the engine cover.
15. Check for proper throttle operation.

BF75, BF100 and BF8A

See Figures 42 and 43

1. Remove the engine cover.
2. Align the word **SHIFT** on the grip pipe cover with the dot on the tiller arm.
3. Loosen the locknut and slacken the adjustment on the throttle cable.
4. Disconnect the end of the throttle cable from the throttle arm.
5. Remove the grommet which seals the throttle cable to the engine case. This grommet should be replaced if found to be damaged or dry rotted.
6. Remove the grip rubber by lubricating the inner face and sliding it from the grip pipe.
7. Remove the cable tie from the bottom of the tiller handle.
8. Remove the pipe guide and cable stopper from the tiller handle.
9. Lift the cable from the cable hinge. Once the cable sleeve clears the cable holder, slide the cable from the tiller handle through the slit in the cable holder.

To install:

10. Feed the throttle cable through the tiller handle and connect the end of the cable to the cable holder.
11. Install the pipe guide with the screw hole facing down.
12. Insert the cable stopper in the groove in the cable end and install the hinge.
13. Install the cable tie on the bottom of the tiller handle.
14. Install the grip rubber by lubricating the inner face and sliding it onto the grip pipe.
15. Install a new grommet on the throttle cable, as necessary and feed the throttle cable through the engine case.
16. Lubricate the end of the throttle cable and position it in the throttle arm.

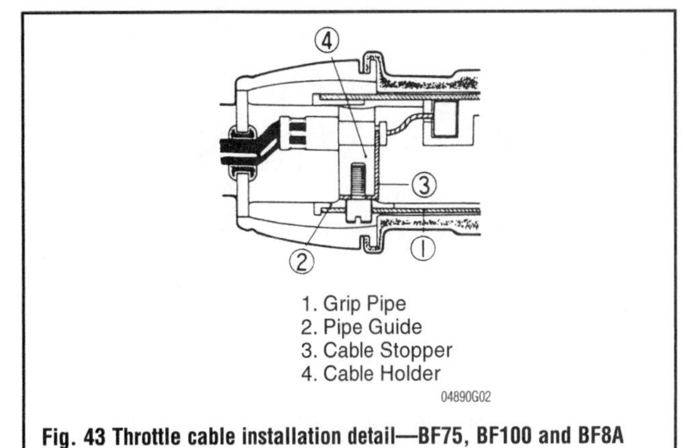

Fig. 43 Throttle cable installation detail—BF75, BF100 and BF8A

17. Adjust the throttle cable tension and tighten the locknut securely.
18. Install the engine cover.
19. Check for proper throttle operation.

BF9.9A and BF15A

See Figures 39 and 44

1. Remove the engine cover.
2. Place the throttle grip in the **SLOW** or fully closed position.
3. Disconnect the throttle cables from the throttle lever.
4. Wrap both ends of one cable with tape to distinguish the cables from each other.
5. Loosen the locknut and slacken the adjustment on the throttle cable.
6. Remove the cable reel through bolt, the washer and the cable reel.
7. Remove the throttle cable adjusting plate, friction block and spring.

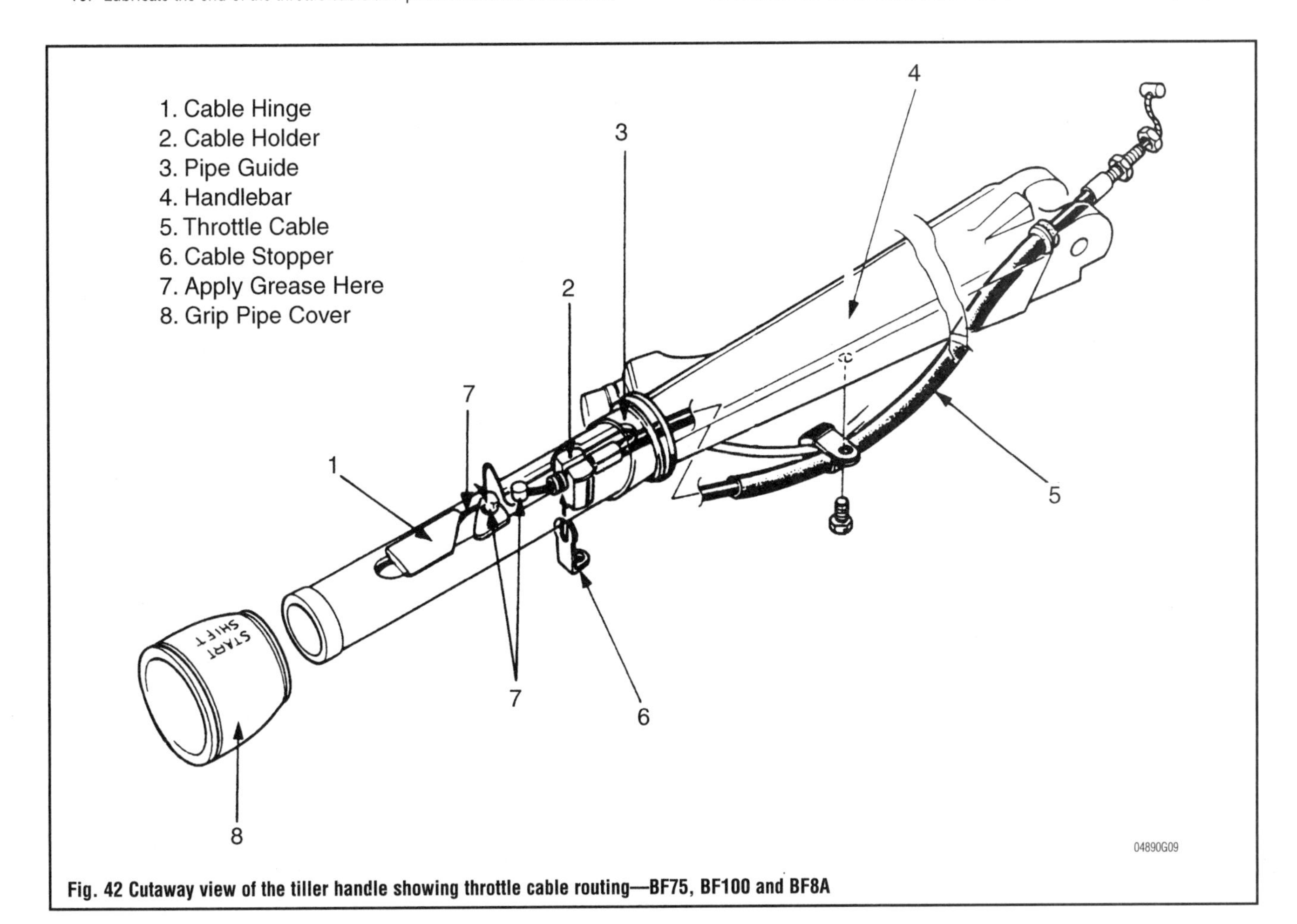

Fig. 42 Cutaway view of the tiller handle showing throttle cable routing—BF75, BF100 and BF8A

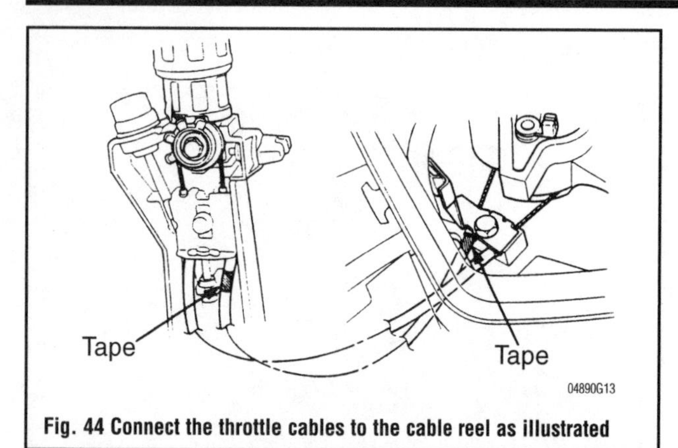

Fig. 44 Connect the throttle cables to the cable reel as illustrated

8. Remove the throttle cable from the tiller handle.
9. Carefully guiding the throttle cables, remove them from the engine.

To install:

10. Carefully guide the throttle cables through the engine and connect them to the throttle lever.
11. Install the throttle cable adjusting plate, friction block and spring.
12. Install the cable reel, washer and the cable reel through bolt.
13. Adjust the throttle cable to specification.
14. Check for proper throttle operation.
15. Install the engine cover.

BF35A, BF40A, BF45A and BF50A

➧ See Figure 40

1. Place the tiller handle in the up position with the throttle grip in the **SLOW** position.
2. Wrap both ends of one cable with tape to distinguish the cables from each other.
3. Loosen the locknut and slacken the adjustment on the throttle cable at the throttle lever.
4. Disconnect the end of the throttle cable from the throttle lever.
5. Remove the throttle reel plate.
6. Slide the throttle reel down enough to disconnect the throttle cables from the throttle reel.
7. Remove the throttle cable from the powerhead.

To install:

8. Install the throttle cable into position on the powerhead.
9. Connect the throttle cable end to the throttle reel and route the cable around the reel.
10. Install the throttle reel assembly into the tiller arm and secure with the throttle reel plate.
11. Install the screw attaching the throttle reel plate and tighten securely.
12. Connect the end of the throttle cable to the throttle lever.
13. Adjust the throttle cable tension and tighten the locknut securely.
14. Install the engine cover.
15. Check for proper throttle operation.

BF75A and BF90A

➧ See Figures 45 thru 51

1. Remove the engine cover.
2. Place the throttle grip in the fully closed position.
3. Disconnect the throttle cable from the throttle lever.
4. Pull on the throttle cable inner rod while turning the throttle pipe on the tiller clockwise. Continue turning until the inner cable is released from the cable reel gear.
5. Disconnect the throttle cable from the tiller handle by unscrewing the cable nut.

To install:

6. Connect the throttle cable outer screw in the tiller handle and tighten to 25 ft. lbs. (34Nm).
7. Pull the inner cable back through the core until it is no longer visible from the throttle cable screw.
8. Lubricate the throttle reel gear and cable holder cap with a marine grade grease.
9. Lubricate the portion of the throttle reel bracket that comes into contact with the throttle gear.
10. Assemble the cable holder cap with the throttle reel bracket by aligning the projection on the cable holder cap with the cutout in the throttle reel bracket.
11. Install the throttle pipe in the throttle reel gear assembly.
12. Install the spring pin in the throttle pipe.
13. Install the throttle friction block on the throttle pipe at a point 6.4 in. (163mm) from the end of the pipe.
14. Install the throttle pipe and reel gear assembly in the tiller handle so that the spring pin holes are aligned.
15. Push in on the throttle cable inner rod while turning the throttle pipe on the tiller counterclockwise. Continue turning until the pivot pin threads on the throttle cable contact the seal on the throttle cable outer shell.
16. Temporarily assemble the grip pipe with the teller handle using a 0.15 in. (4mm) shaft.
17. Insert the shaft through the throttle pipe by aligning the hole in the grip pipe with the holes in the throttle pipe.
18. Place the throttle in the fully closed position. Ensure the projection on the tiller handle contacts the fully closed side stopper on the grip pipe.
19. Measure the length between the inner rod end and the cable outer groove as illustrated. The distance should be 8.1–8.3 in. (205–211mm).
20. If the measurement is less than 8.1 in. (205mm) proceed as follows:
    a. Disassemble the cable assembly and turn the throttle pipe so the spring pin holes are at the 2 o'clock and 8 o'clock positions when viewed from the end of the tiller handle.
    b. Ensure the inner cable is no longer visible at the throttle cable screw.
    c. Push in on the throttle cable inner rod while turning the throttle pipe on the tiller counterclockwise. Continue turning until the pivot pin threads on the throttle cable contact the seal on the throttle cable outer shell.

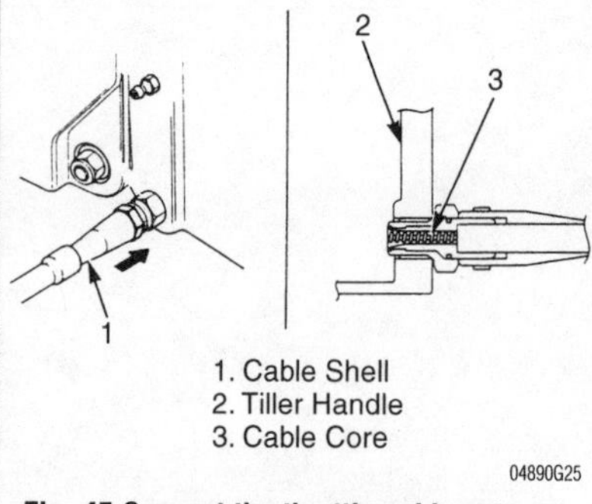

Fig. 45 Connect the throttle cable outer screw to the tiller handle and pull the inner cable core back through the shell until it is no longer visible from the throttle cable screw

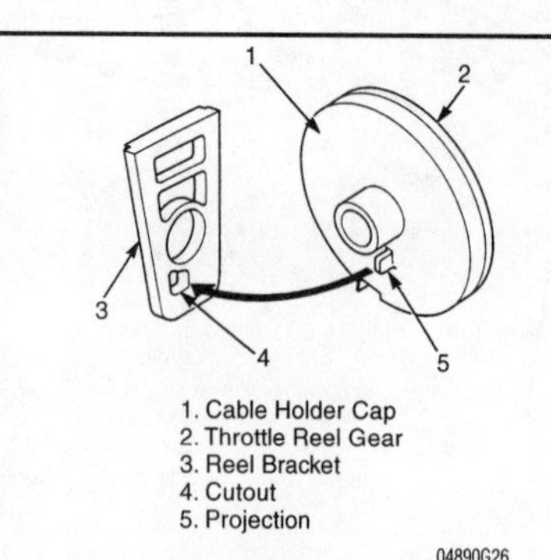

Fig. 46 Assemble the cable holder cap with the throttle reel bracket by aligning the projection on the cable holder cap with the cutout in the throttle reel bracket

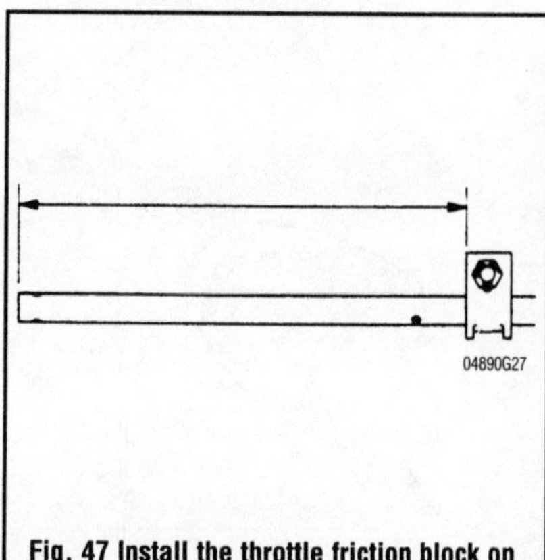

Fig. 47 Install the throttle friction block on the throttle pipe at a point 6.4 in. (163mm) from the end of the pipe

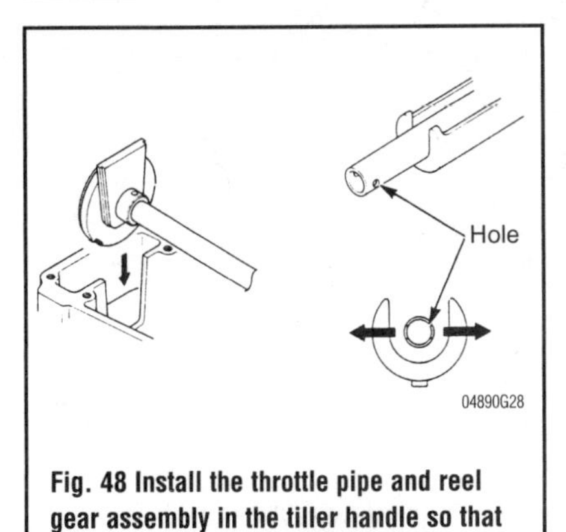

Fig. 48 Install the throttle pipe and reel gear assembly in the tiller handle so that the spring pin holes are aligned

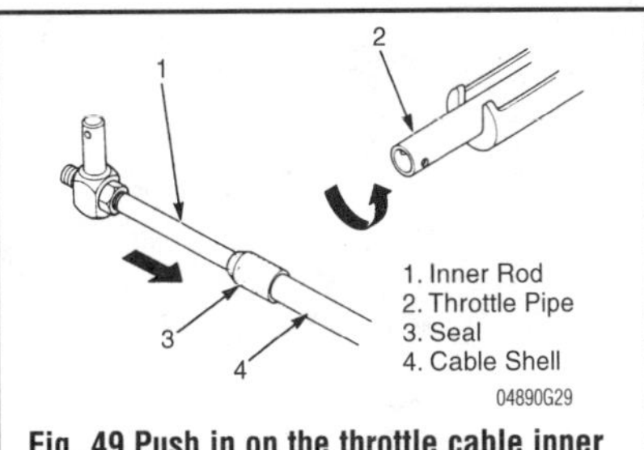

Fig. 49 Push in on the throttle cable inner rod while turning the throttle pipe on the tiller counterclockwise. Continue turning until the pivot pin threads on the throttle cable contact the seal on the throttle cable outer shell

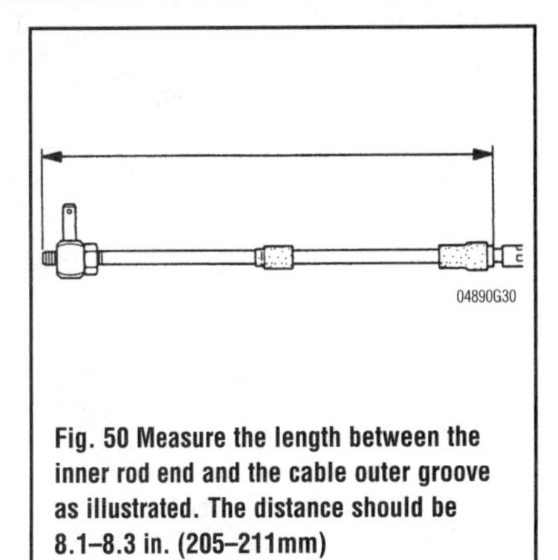

Fig. 50 Measure the length between the inner rod end and the cable outer groove as illustrated. The distance should be 8.1–8.3 in. (205–211mm)

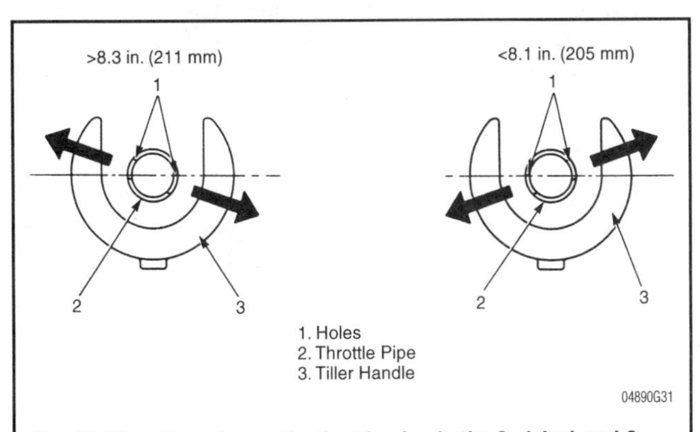

Fig. 51 Place the holes on the throttle pipe in the 2 o'clock and 8 o'clock positions or the 10 o'clock and 4 o'clock positions when viewed from the end of the tiller handle

d. Temporarily assemble the grip pipe with the teller handle using a 0.15 in. (4mm) shaft.

e. Insert the shaft through the throttle pipe by aligning the hole in the grip pipe with the holes in the throttle pipe.

f. Place the throttle in the fully closed position. Ensure the projection on the tiller handle contacts the fully closed side stopper on the grip pipe.

g. Remeasure the length between the inner rod end and the cable outer groove.

21. If the measurement is less than 8.3 in. (211mm) proceed as follows:

a. Disassemble the cable assembly and turn the throttle pipe so the spring pin holes are at the 10 o'clock and 4 o'clock positions when viewed from the end of the tiller handle.

b. Ensure the inner cable is no longer visible at the throttle cable screw.

c. Push in on the throttle cable inner rod while turning the throttle pipe on the tiller counterclockwise. Continue turning until the pivot pin threads on the throttle cable contact the seal on the throttle cable outer shell.

d. Temporarily assemble the grip pipe with the teller handle using a 0.15 in. (4mm) shaft.

e. Insert the shaft through the throttle pipe by aligning the hole in the grip pipe with the holes in the throttle pipe.

f. Place the throttle in the fully closed position. Ensure the projection on the tiller handle contacts the fully closed side stopper on the grip pipe.

g. Remeasure the length between the inner rod end and the cable outer groove.

22. If the measurement is between 8.1–8.3 in. (205–211mm), install the grip pipe on the tiller handle.

23. Place the spring and washers in the grip pipe and push the washer in to the pipe. Secure the assembly to the throttle pipe using a new a spring pin.

24. Install the rubber grip and grease fitting.

25. Install a new reel gear gasket on the tiller handle and install the reel setting plate.

26. Check for proper throttle operation.

27. Install the engine cover.

ADJUSTMENT

BF50 and BF5A

See Figure 52

1. Remove the engine cover.
2. Place the throttle grip in the **SLOW** position.
3. Loosen the locknut on the throttle cable.
4. Adjust the throttle cable so that the distance between the end of the threaded portion of the cable and the cable bracket is 0.33–0.37 in. (8.5–9.5mm).
5. Tighten the locknut securely and check for proper throttle operation.
6. Install the engine cover.

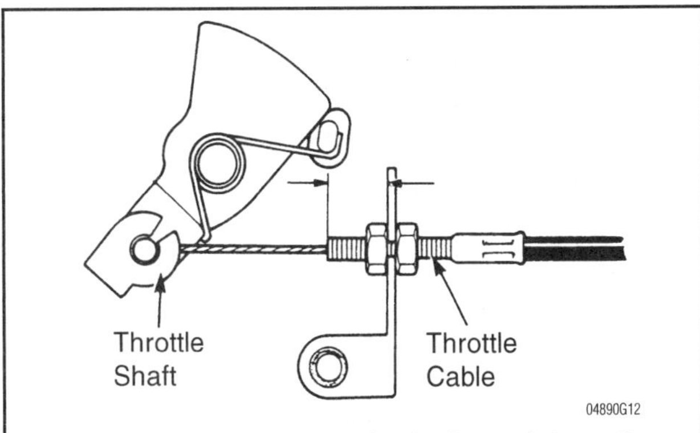

Fig. 52 Adjust the throttle cable so that the distance between the end of the threaded portion of the cable and the cable bracket is 0.33–0.37 in. (8.5–9.5mm)

BF9.9A and BF15A

THROTTLE CABLE

See Figures 53 and 54

1. Remove the engine cover.
2. Place the shift lever in the **NEUTRAL** position.
3. Place the throttle grip in the **SLOW** position and hold the throttle lever fully closed.
4. Loosen the throttle cable locknut and the adjusting plate through bolt.
5. Adjust the throttle cable so there is less than 0.11 in. (3mm) of deflection at the boss of the cable reel.
6. Tighten the locknut to 3.6 ft. lbs. (5Nm) and the adjusting nut to 7.2 ft. lbs. (10 Nm)
7. Check for proper throttle operation.
8. Install the engine cover.

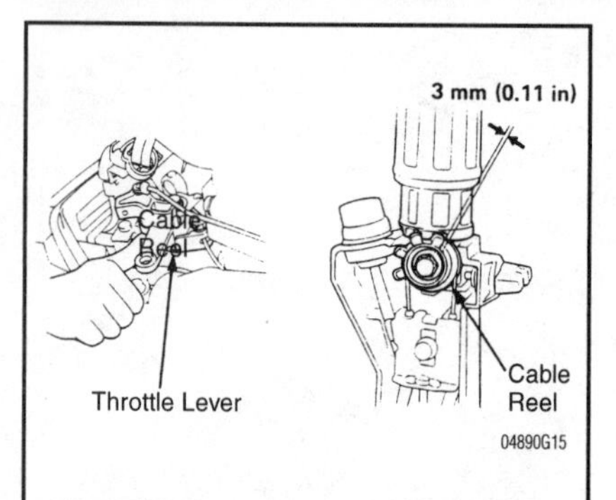

Fig. 53 Adjust the throttle cable so there is less than 0.11 in. (3mm) of deflection at the boss of the cable reel.

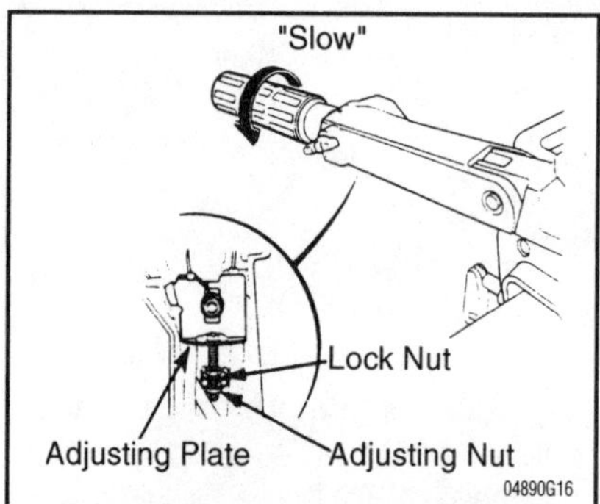

Fig. 54 Loosen the throttle cable locknut and the adjusting plate through bolt prior to adjusting the cable

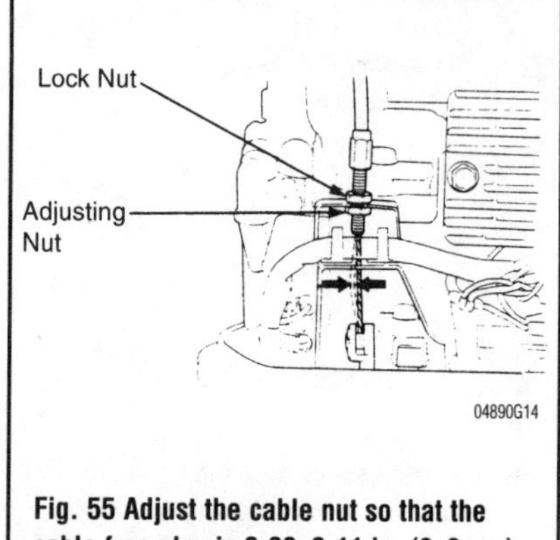

Fig. 55 Adjust the cable nut so that the cable free play is 0.08–0.11 in. (2–3mm).

NEUTRAL START CABLE

See Figure 55

1. Remove the engine cover.
2. Place the shift lever in the **NEUTRAL** position.
3. Loosen the neutral start cable locknut.
4. Adjust the cable nut so that the cable free play is 0.08–0.11 in. (2–3mm).
5. Tighten the locknut to 7.2 ft. lbs. (10 Nm) and check for proper neutral start operation.
6. Install the engine cover.

BF35A, BF40A, BF45A and BF50A

THROTTLE CABLE

See Figure 56

1. Remove the engine cover.
2. Place the shift lever in the **FORWARD** position.
3. Move the throttle grip to the fully open position and confirm the throttle arm contacts the full open position stopper located on the throttle cable bracket.
4. If the throttle arm does not contact the stopper, loosen the open side throttle cable locknut and adjust the cable by turning the adjusting nut.
5. Check for proper throttle operation.
6. Install the engine cover.

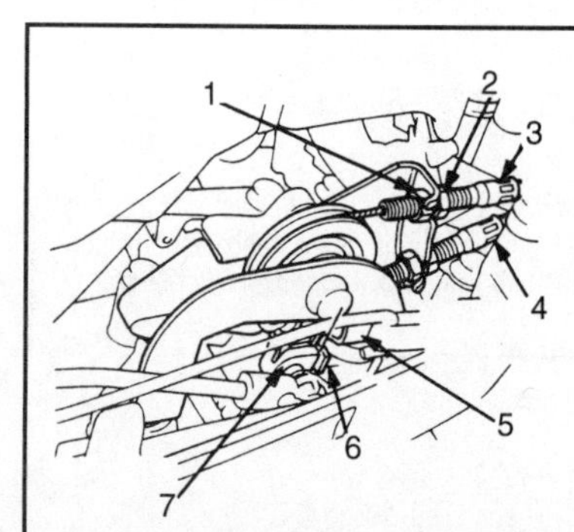

Fig. 56 Move the throttle grip to the fully open position and confirm the throttle arm contacts the full open position stopper located on the throttle cable bracket

THROTTLE ROD

See Figures 57 and 58

1. Remove the engine cover.
2. Place the shift lever in the **FORWARD** position.
3. Move the throttle grip to the fully open position and confirm the throttle arm contacts the full open position stopper located on the throttle cable bracket.
4. Ensure the carburetor throttle lever contacts the full open position stopper on the No. 3 carburetor.
5. If the throttle lever does not contact the stopper, loosen the throttle rod locknut and adjust the throttle rod length by turning the rod pivot.
6. Tighten the locknut securely after adjustment.
7. Ensure the throttle lever moves freely from fully open to fully closed positions.
8. Ensure the throttle lever has a clearance of 0.04 in. (1mm) or less between the lever and the full open position stopper on the No. 3 carburetor. If the clearance is more than specified, readjust the throttle rod pivot.
9. Turn the throttle grip to the fully closed position and align the triangle mark on the tiller handle with the throttle slow mark.
10. Ensure the two reference marks on the throttle cam align with the center

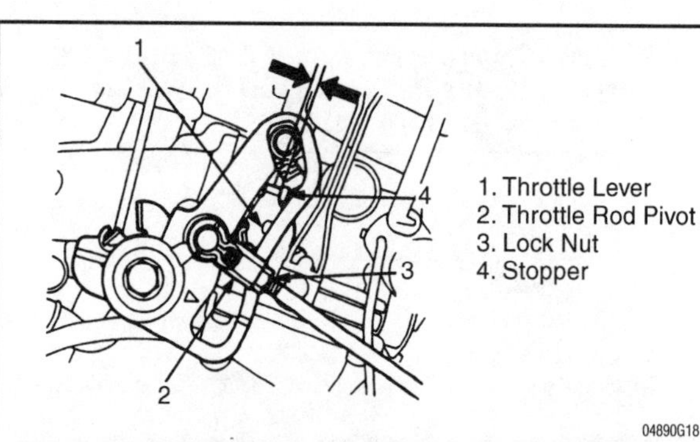

Fig. 57 With the grip at full open throttle, ensure the carburetor throttle lever contacts the full open position stopper. With the throttle closed, ensure the throttle lever has a clearance of 0.04 in. (1mm) or less between the lever and the full open position stopper

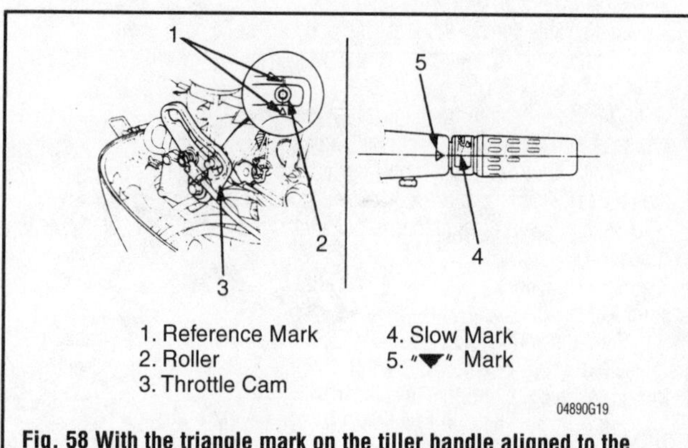

Fig. 58 With the triangle mark on the tiller handle aligned to the slow mark, ensure the two reference marks on the throttle cam align with the center of the roller

of the roller. If they do not align, loosen the locknut on the closed side throttle cable and adjust.

11. Check for proper throttle operation.
12. Install the engine cover.

BF75A and BF90A

THROTTLE CABLE

See Figures 59, 60, 61 and 62

1. With the throttle in the fully closed position, measure the length of the open side and closed side throttle cable at both the throttle arm and the throttle cam. Distances should be as follows:
   - Open side-to-throttle arm—0.5 in. (14mm)
   - Closed side-to-throttle arm—0.6 in. (16mm)
   - Open side-to-throttle cam—1.3 in. (34mm)
   - Closed side-to-throttle arm—1.0 in. (25mm)
2. If lengths are not within specification, adjust the cable by loosening the locknut and adjusting the cable using the adjusting nut.
3. Place the throttle grip in the fully open position and ensure the throttle arm is in contact with the stopper. If the throttle arm is not in contact with the stopper, proceed as follows:
   a. Place the throttle grip in the fully closed position. Loosen the shift pivot locknut and remove the lock pin and washer.
   b. Disconnect the shift pivot from the throttle arm and adjust by turning the pivot.
   c. Place the throttle grip in the fully open position and ensure the throttle arm is in contact with the stopper.

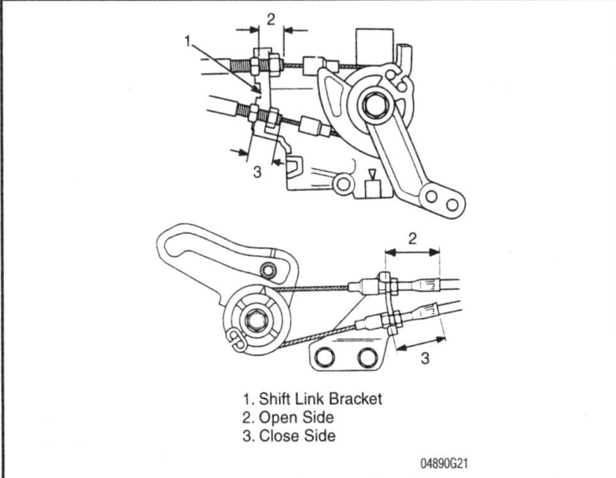

Fig. 59 With the throttle in the fully closed position, measure the length of the open side and closed side throttle cable at both the throttle arm and the throttle cam

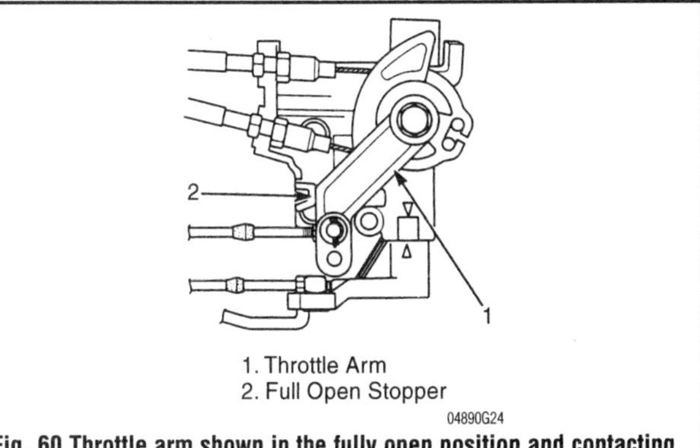

Fig. 60 Throttle arm shown in the fully open position and contacting the stopper

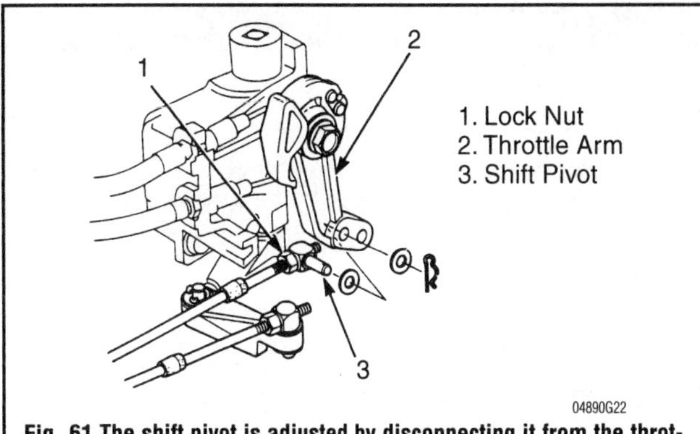

Fig. 61 The shift pivot is adjusted by disconnecting it from the throttle arm and turning the pivot

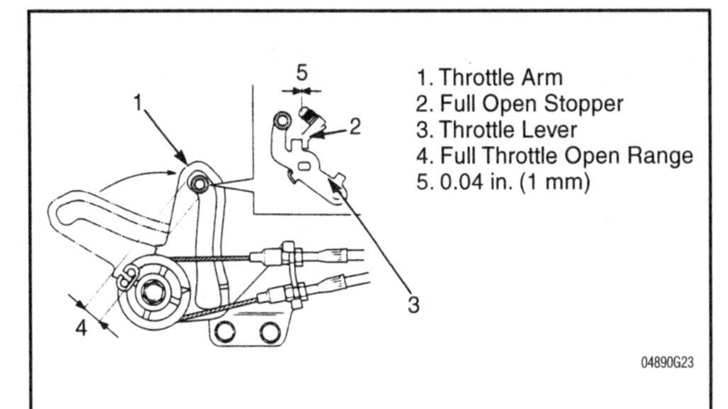

Fig. 62 The clearance between the throttle lever and fully open position stopper should be 0.04 in. (1mm) or less when the throttle arm is in contact with the stopper

   d. Place the throttle grip in the fully closed position and tighten the shift pivot locknut.

4. Place the throttle arm in contact with the fully open position stopper and ensure the clearance between the throttle lever and fully open position stopper is 0.04 in. (1mm) or less. If the clearance is not within specification, proceed as follows:
   a. Place the throttle grip in the fully closed position.
   b. Loosen the adjusting nuts on the open and closed side of the throttle cable.
   c. Place the throttle grip in the fully open position.
   d. Ensure the clearance between the carburetor throttle lever and the fully open position stopper is 0.04in. (1mm) or less. If the clearance is not within specification adjust it by turning the adjusting nut on the open side of the throttle cable.
   e. Place the throttle grip in the fully closed position.
5. Ensure the two throttle cam alignment marks are properly aligned with the center of the throttle cam roller. If they do not align, adjust by turning the adjusting nut at the closed side of the throttle cable.
6. Tighten the adjusting nuts on the open and closed side of the throttle cable securely.

TILLER HANDLE

1. Place the throttle grip in the fully closed position and align the triangle mark on the handle with the throttle mark in the **SLOW** position.
2. The two alignment marks on the throttle cam should align with the center of the throttle cam roller. If the marks do not align, perform the throttle cable adjustment.
3. Place the throttle grip in the fully open position and align the triangle mark on the handle with the throttle mark in the **FAST** position. The throttle arm should be in contact with the fully open position stopper. If the throttle arm is not in contact with the stopper, perform the throttle cable adjustment.

Engine Stop Switch

TESTING

1. Disconnect the engine stop switch wiring harness.
2. Connect a multimeter between the switch harness leads.
3. With the switch engaged (stop switch lanyard pulled), continuity should exist. With the switch released (stop switch lanyard in position), continuity should not exist.
4. If the switch does not function as specified there is a short in either the switch or harness and the switch should be replaced.

REMOVAL & INSTALLATION

1. Disconnect the engine stop switch wiring harness.
2. Remove any wire straps that connect the switch to the tiller handle or bracket.
3. Remove the switch attaching nut.
4. Remove the switch from the tiller handle or bracket.

To install:

5. Install the switch on the tiller handle or bracket.
6. Install the switch attaching nut and tighten securely.
7. Install any wire straps that connect the switch to the tiller handle or bracket.
8. Connect the engine stop switch wiring harness.
9. Test the switch for proper operation.

Indicator Light

TESTING

Testing of the outboard mounted indicator light involves checking all the components around the light and deciding by process of elimination if the light is faulty. If the oil pressure switch and the CDI unit are functioning properly and the light does not turn on when the engine is running, the indicator light may be faulty.

REMOVAL & INSTALLATION

1. Remove the engine cover.
2. Disconnect the light wiring harness.
3. Remove any wire straps that connect the light wiring harness to the case.
4. Remove the light fasteners.
5. Remove the light assembly from the case.

To install:

6. Install the light assembly in the case.
7. Install the light fasteners.
8. Install any wire straps that connect the light harness to the case.
9. Connect the light wiring harness.
10. Test the lights for proper operation.
11. Install the engine cover.

Starter Button

TESTING

1. Disconnect the starter button wiring harness.
2. Connect a multimeter between the switch harness leads.
3. With the starter button pushed, continuity should exist.
4. If the button functions properly, the problem may be in the wiring harness.
5. If the switch does not function as specified, the switch may be faulty.

REMOVAL & INSTALLATION

1. Disconnect the engine stop switch wiring harness.
2. Remove any wire straps that connect the switch to the tiller handle or bracket.
3. Remove the switch attaching nut.
4. Remove the switch from the tiller handle or bracket.

To install:

5. Install the switch on the tiller handle or bracket.
6. Install the switch attaching nut and tighten securely.
7. Install any wire straps that connect the switch to the tiller handle or bracket.
8. Connect the engine stop switch wiring harness.
9. Test the switch for proper operation.

Trim/Tilt Switch

TESTING

➧ See Figure 63

1. Disconnect the trim/tilt switch wiring harness.
2. Connect a multimeter between the switch harness terminals as illustrated.
3. With the switch in the stated positions, check for continuity between the various terminals.
4. If the switch functions properly, the problem may be in the wiring harness
5. If the switch does not function as specified, the switch may be faulty.

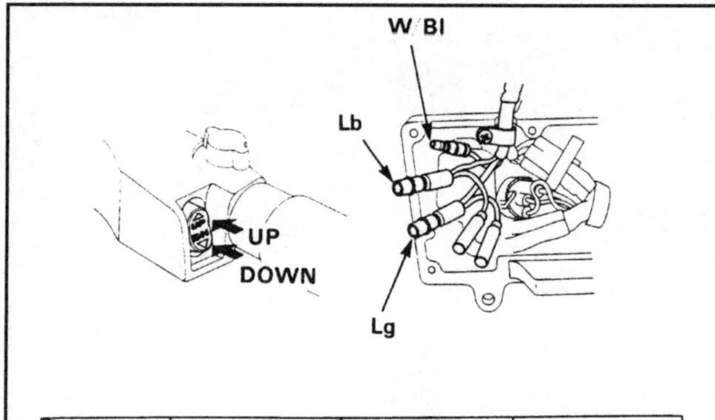

	Lg	W/Bl	Lb
UP	O—	—O	
DOWN		O—	—O

O——O Continuity

04890G36

Fig. 63 With the trim/tilt switch in the illustrated positions, check for continuity between the terminals

REMOVAL & INSTALLATION

1. Disconnect the trim/tilt switch wiring harness.
2. Remove any wire straps that connect the switch to the tiller handle.
3. Remove the switch attaching hardware.
4. Remove the switch from the tiller handle.

To install:

5. Install the switch on the tiller handle.
6. Install the switch attaching hardware and tighten securely.
7. Install any wire straps that connect the switch harness to the tiller handle.
8. Connect the trim/tilt switch wiring harness.
9. Test the switch for proper operation.

HAND REWIND STARTER 11-2
DESCRIPTION AND OPERATION 11-2
TROUBLESHOOTING THE HAND REWIND STARTER 11-2
BF2A AND BF20 11-2
REMOVAL & INSTALLATION 11-2
DISASSEMBLY 11-3
CLEANING & INSPECTION 11-3
ASSEMBLY 11-3
BF5A AND BF50 11-4
REMOVAL & INSTALLATION 11-4
DISASSEMBLY 11-4
CLEANING & INSPECTION 11-4
ASSEMBLY 11-5
BF75, BF100 AND BF8A 11-5
REMOVAL & INSTALLATION 11-5
DISASSEMBLY 11-6
CLEANING & INSPECTION 11-6
ASSEMBLY 11-7
BF9.9A AND BF15A 11-7
REMOVAL & INSTALLATION 11-7
ADJUSTMENT 11-7
DISASSEMBLY 11-7
CLEANING & INSPECTION 11-9
ASSEMBLY 11-9
BF25A AND BF30A 11-9
REMOVAL & INSTALLATION 11-9
ADJUSTMENT 11-10
DISASSEMBLY 11-10
CLEANING & INSPECTION 11-10
ASSEMBLY 11-11

11

HAND REWIND STARTER

HAND REWIND STARTER 11-2

HAND REWIND STARTER

Description and Operation

▸ **See Figures 1 and 2**

The main components of a hand rewind starter (recoil starter) are the cover, rewind spring and pawl arrangement. Pulling the rope rotates the pulley, winds the spring and activates the pawl into engagement with the starter hub at the top of the flywheel. Once the pawl engages the hub, the powerhead is spun as the rope unwinds from the pulley.

0489AP02

Fig. 1 The recoil starter is a compact and easily reparable unit

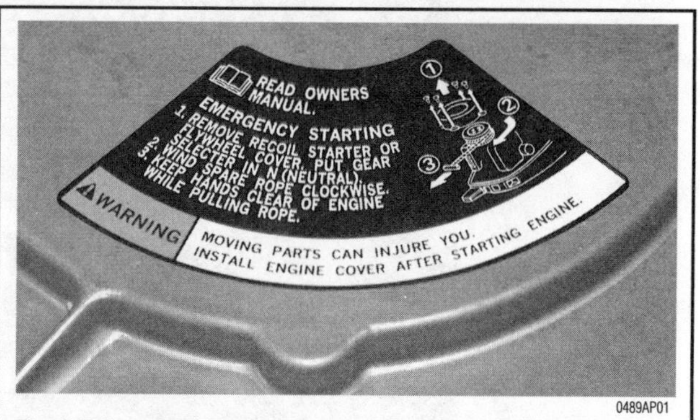

0489AP01

Fig. 2 If the recoil starter malfunctions, in may cases it can be removed and the engine started using an emergency rope

Releasing the rope on rewind starter moves the pawl out of mesh with the hub. The powerful clock-type spring recoils the pulley in the reverse direction to rewind the rope to the original position.

On the BF50 and BF5A , a newer design starter is used. This type employs inertia to slide a gear down a shaft to engage the flywheel as the rope is pulled (much the same as an electric starter does). At the same time, the rewind spring is tightened so there can be a recoil of the starter rope. There is a predetermined ratio between the starter pinion gear and the flywheel that provides maximum cranking speed for easy starts.

Some starters may use a starter interlock system. The hand rewind starter should not function when the shift handle is in any other position than **NEUTRAL**. This prevents starting in gear and possibly throwing the occupants overboard. Always check for the proper function of this system after any repairs.

Troubleshooting The Hand Rewind Starter

Repair on hand rewind starter units is generally confined to rope, pawl and occasionally spring replacement.

** CAUTION

When replacing the recoil starter spring extreme caution must be used. The spring is under tension and can be dangerous if not released properly.

Starters which use friction springs to assist pawl action may suffer from bent springs. This will cause the amount of friction exerted to not be correct and the pawl will not be moved into engagement.

Models equipped with a starter interlock system may experience a no-start condition due to a misadjusted interlock cable. The hand rewind starter should only function when the shift handle is in the **NEUTRAL** position.

BF2A and BF20

REMOVAL & INSTALLATION

▸ **See Figure 3**

1. Disconnect the engine cover fasteners and lift the engine cover off the engine.
2. Untie the rope at the starter knob and remove the starter knob.
3. Remove the recoil starter assembly attaching nuts and lift the assembly from the powerhead.

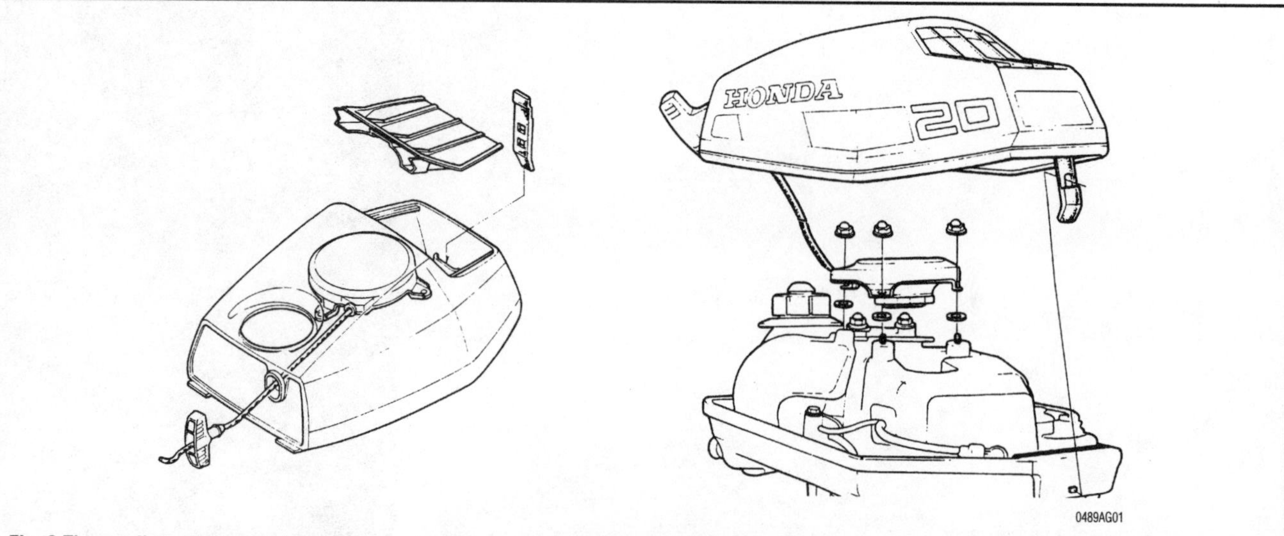

0489AG01

Fig. 3 The recoil starter assembly is attached using three cap nuts—BF2A and BF20

To install:

4. Place the recoil starter assembly on the powerhead and tighten the attaching nuts to 86 inch lbs. (10 Nm).
5. Insert the rope through the starter knob and tie the rope to secure the grip in place.
6. Install the engine cover and connect the cover fasteners.
7. Pull the starter knob several times and check for the proper operation of the ratcheting mechanism.

DISASSEMBLY

◆ **See Figure 4**

■ **Please refer to the Supplement for an exploded view of the starter normally found on BF2D models.**

1. If not already done, untie the rope at the starter knob and remove the starter knob.
2. Carefully loosen the reel cover bolt and remove.
3. Lift off the reel cover, friction spring and ratchet(s).

✳✳ CAUTION

The starter reel spring is under high tension. If the spring should come loose, it may cause serious damage or personal injury. Take all applicable cautions when working with this spring.

■ **It is advisable to wear heavy gloves while removing the spring to prevent your hands from being cut by the sharp spring steel.**

4. Carefully lift off the starter reel, taking extra care to prevent the starter spring from unwinding.
5. Carefully remove the spring from the starter reel.
6. Untie the knot at the starter reel and remove the rope.

CLEANING AND INSPECTION

Clean all components and then blow them dry using compressed air. Remove any trace of corrosion and wipe all metal parts with an oil-dampened cloth to prevent future corrosion.

Inspect the rope. Replace the rope if it appears to be weak or frayed. If the rope is frayed, check the holes through which the rope passes for rough edges or burrs. Remove the rough edges or burrs with a file and polish the surface until it is smooth. Inspect the starter return spring end hooks. Replace the spring if it is weak, corroded or cracked. Inspect the inside surface of the starter recoil case and reel for grooves or roughness. Grooves may cause erratic rewinding of the starter rope.

Inspect and lubricate the ratchet mechanism and check for freedom of movement.

ASSEMBLY

◆ **See Figures 5 and 6**

1. Pass the rope through the hole in the starter reel and tie the end of the rope in a figure eight knot.
2. Wind the rope onto the reel in the direction indicated on the reel.
3. Wedge the end of the rope in the notch on the edge of the reel.
4. Hook the inner end of the spring on the tab in the reel and wind it counterclockwise.

■ **Ensure that the inner end of the spring is hooked on the tab in the case.**

5. Install the starter reel assembly in the starter case. Turn the reel counterclockwise while inserting to ease installation.

✳✳ CAUTION

The starter reel spring is under high tension. If the spring should come loose, it may cause serious damage or personal injury. Take all applicable cautions when working with this spring.

6. Hold the reel and pull the end of the rope out of the case and feed it though the starter knob. Knot the end of the rope to retain it in the starter knob.
7. With a length of rope extending from the starter reel notch, rotate the reel 2 full turns counterclockwise. The bottom of the reel should be facing upward when the reel is rotated.

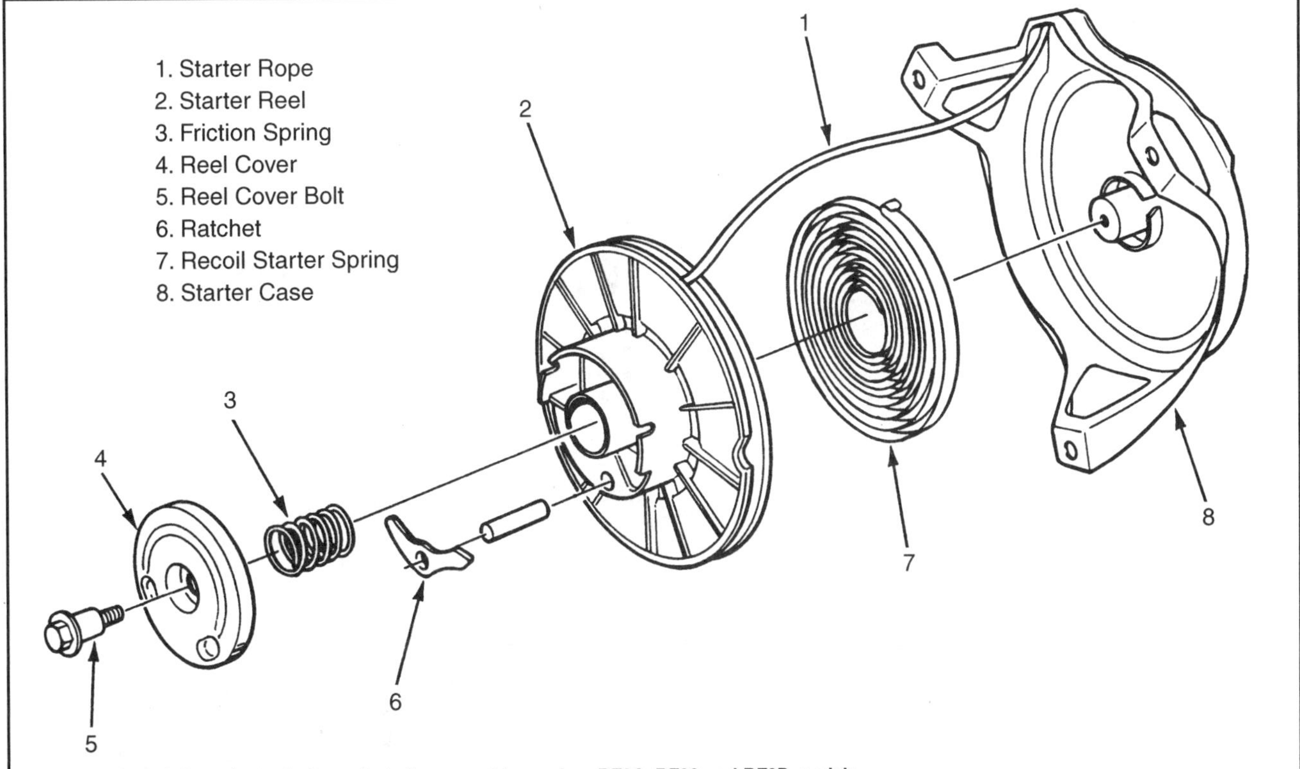

Fig. 4 Exploded view of a typical recoil starter assembly used on BF2A, BF20 and BF2D models (Note: ratchet components may differ slightly on some models)

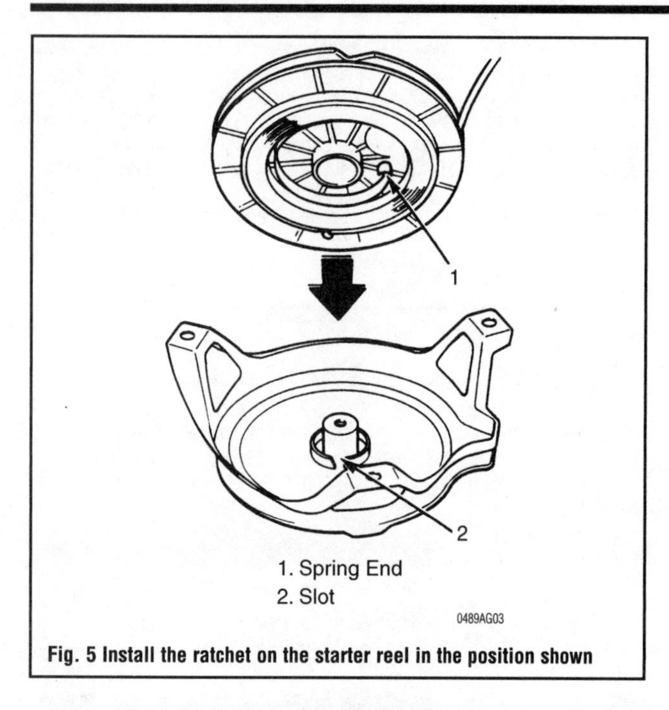

Fig. 5 Install the ratchet on the starter reel in the position shown

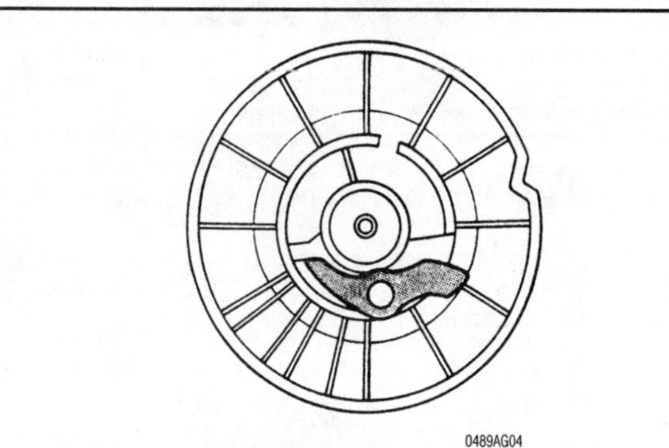

Fig. 6 When installing the starter reel, insert the end of the spring in the slot on the starter case

8. Install the ratchet and friction spring.
9. Mount the reel cover and tighten the cover bolt securely.
10. Pull the starter knob several times and check for the proper operation of the ratcheting mechanism.

BF5A and BF50

REMOVAL & INSTALLATION

➧ **See Figure 7**

1. Disconnect the engine cover fasteners and lift the engine cover off the engine.
2. Remove the recoil starter assembly attaching bolts and lift the assembly from the powerhead.

To install:

3. Place the recoil starter assembly on the powerhead and tighten the attaching nuts to 78 inch lbs. (9Nm).

➡Some of the recoil starter attaching bolts are self tapping. Take care during installation to not overtorque these bolts as they are easily stripped.

4. Install the engine cover and connect the cover fasteners.
5. Pull the starter knob several times and check for the proper operation of the ratcheting mechanism.

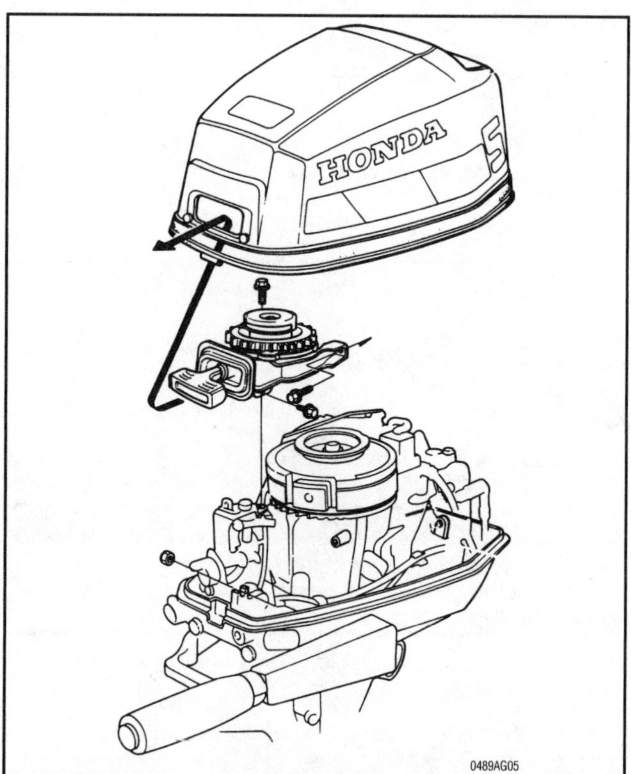

Fig. 7 The recoil starter assembly is attached with both horizontally and vertically orientated fasteners—BF5A and BF50

DISASSEMBLY

➧ **See Figure 8**

1. Remove the friction plate nut and disassemble the friction plate, friction spring, starter gear and runner cushion
2. Remove the starter knob cap. Untie the rope end and remove the starter knob.
3. Pass the rope end through the hole in the starter reel.
4. Slowly release the preload on the return spring and remove the starter reel and spring retainer from the starter case.
5. Untie the knot at the starter reel and remove the rope.

➡It is advisable to wear heavy gloves while removing the spring to prevent your hands from being cut by the sharp spring steel.

⁂ CAUTION

The starter reel spring is under high tension. If the spring should come loose, it may cause serious damage or personal injury. Take all applicable cautions when working with this spring.

6. Carefully remove the return spring from the spring retainer.

CLEANING & INSPECTION

Clean all components and then blow them dry using compressed air. Remove any trace of corrosion and wipe all metal parts with an oil dampened cloth to prevent future corrosion.

Inspect the rope. Replace the rope if it appears to be weak or frayed. If the rope is frayed, check the holes through which the rope passes for rough edges or burrs. Remove the rough edges or burrs with a file and polish the surface until it is smooth. Inspect the starter return spring end hooks. Replace the spring if it is weak, corroded or cracked. Inspect the inside surface of the starter recoil case and reel for grooves or roughness. Grooves may cause erratic rewinding of the starter rope. Proper starter rope length is 43.3–51.1 in. (110–130cm).

Inspect and lubricate the assembly as illustrated, then check for freedom of movement.

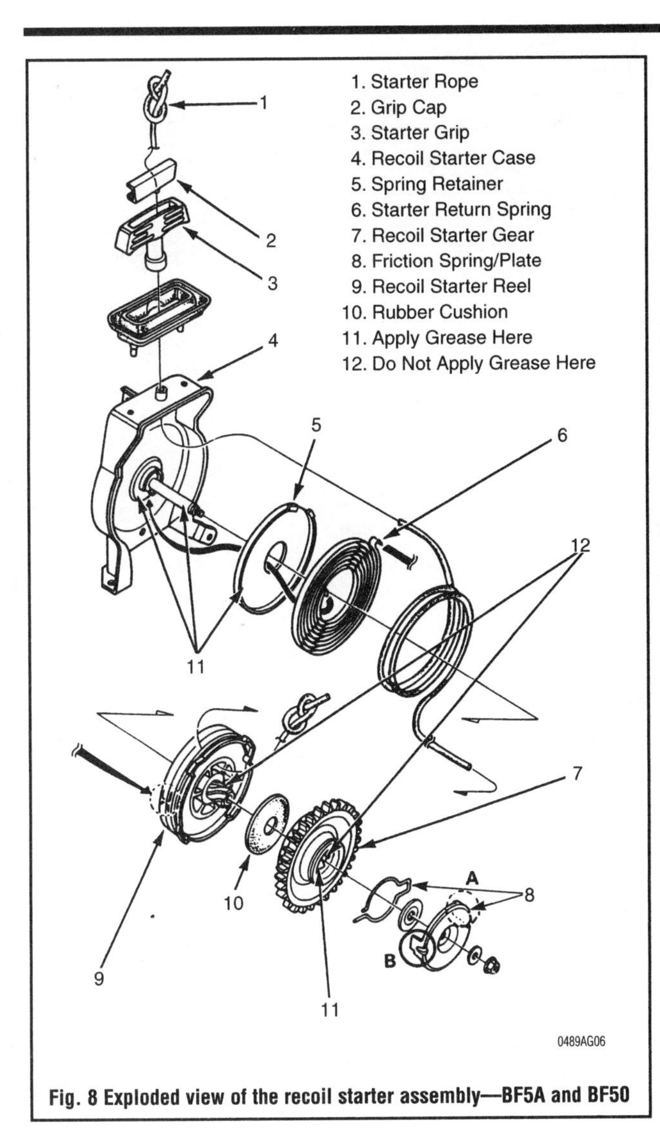

Fig. 8 Exploded view of the recoil starter assembly—BF5A and BF50

ASSEMBLY

See Figure 9

1. Install the return spring on the retainer and then attach it to the starter case by hooking the spring outer end to the center of the groove. Hook the inner end of the spring on the tab of the starter case.
2. Route one end of the starter rope through the hole in the starter reel and tie a knot in the end.

Ensure that the inner end of the spring is hooked on the tab in the case.

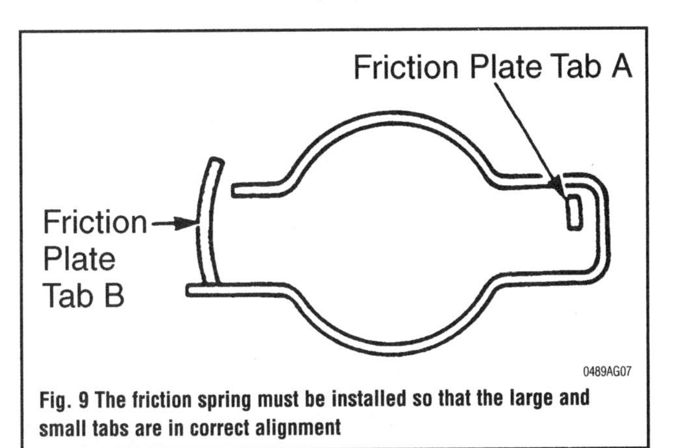

Fig. 9 The friction spring must be installed so that the large and small tabs are in correct alignment

3. Wedge the end of the rope in the notch on the edge of the reel and install the reel aligning the groove in the reel with the spring end.
4. While holding the rope end in the notch on the edge of the reel, rotate the reel counterclockwise 2 turns to preload the return spring.

CAUTION

The starter reel spring is under high tension. If the spring should come loose, it may cause serious damage or personal injury. Take all applicable cautions when working with this spring.

5. Pass the rope end through the hole out of the starter case by holding the starter reel.
6. Insert the rope into the starter knob and tie the rope end. Install the starter knob cap.
7. Install the runner cushion, the starter gear, friction spring and friction plate. Tighten the friction plate nut to 86 inch lbs. (10Nm).
8. Pull the starter knob several times and check for the proper operation mechanism.

BF75, BF100 and BF8A

REMOVAL & INSTALLATION

See Figure 10

1. Disconnect the engine cover fasteners and lift the engine cover off the engine.

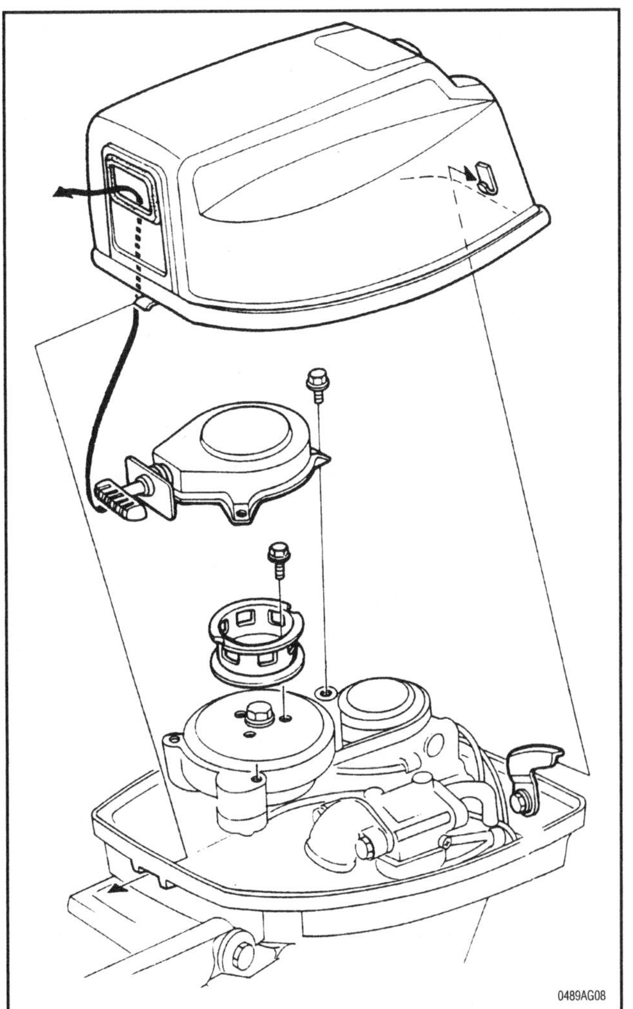

Fig. 10 The recoil starter assembly is attached to the powerhead with three bolts—BF75, BF100 and BF8A

2. Remove the recoil starter assembly attaching bolts and lift the assembly from the powerhead.

To install:

3. Place the recoil starter assembly on the powerhead and tighten the attaching bolts to 86 inch lbs. (10Nm).
4. Install the engine cover and connect the cover fasteners.
5. Pull the starter knob several times and check for the proper operation of the ratcheting mechanism.

DISASSEMBLY

▶ See Figures 11 and 12

CLEANING & INSPECTION

Clean all components and then blow them dry using compressed air. Remove any trace of corrosion and wipe all metal parts with an oil dampened cloth to prevent future corrosion.

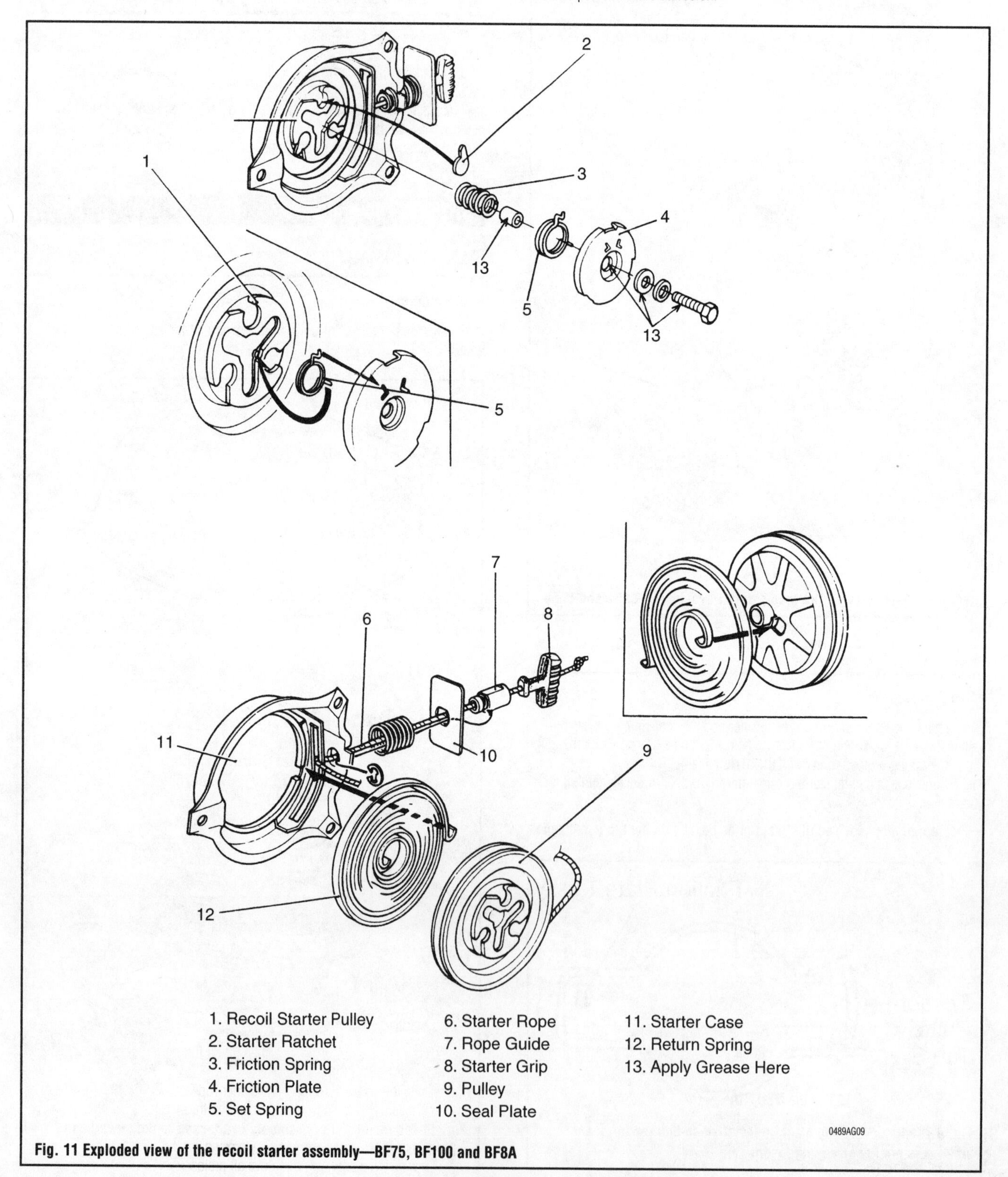

Fig. 11 Exploded view of the recoil starter assembly—BF75, BF100 and BF8A

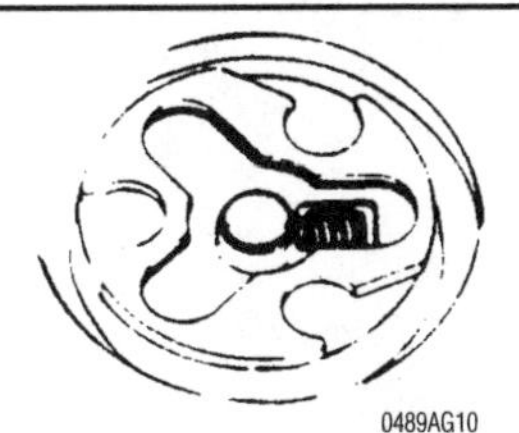

0489AG10

Fig. 12 Slip the inner end of the return spring onto the projection on the back of the pulley as illustrated

Inspect the rope. Replace the rope if it appears to be weak or frayed. If the rope is frayed, check the holes through which the rope passes for rough edges or burrs. Remove the rough edges or burrs with a file and polish the surface until it is smooth. Inspect the starter return spring end hooks. Replace the spring if it is weak, corroded or cracked. Inspect the inside surface of the starter recoil case and reel for grooves or roughness. Grooves may cause erratic rewinding of the starter rope.

Inspect and lubricate the friction clutch components and check for freedom of movement.

ASSEMBLY

1. Route one end of the starter rope through the hole in the starter pulley and tie a knot in the end. Secure the rope with the tab on the pulley.
2. Wind the rope around the pulley and install the pulley in the case.
3. Install the springs, ratchets, and friction plate in their proper order.
4. Install the securing bolt and tighten to 86 inch lbs. (10Nm).
5. While holding the tope end in the notch on the pulley, rotate the pulley 3–4 turns counterclockwise to reload the spring.
6. Attach a spring scale to the tope end and measure the force necessary to pull the rope out. If resistance is greater than 2.9–5.2 lbs. (1.3–2.3 kg), loosen the securing bolt and turn the pulley to obtain a proper resistance.
7. Slip the rope guide, seal plate and grip over the rope and tie a knot in the end of the rope.
8. Secure the rope guide with the E-ring.
9. Pull the starter knob several times and check for the proper operation of the ratcheting mechanism.

BF9.9A and BF15A

REMOVAL & INSTALLATION

See Figure 13

1. Disconnect the engine cover fasteners and lift the engine cover off the engine.
2. Place the lower unit in the **NEUTRAL** position.
3. Loosen the neutral starting cable locknut and remove the cable from its bracket.
4. Remove the recoil starter assembly attaching bolts and lift the assembly from the powerhead.

To install:

5. Place the recoil starter assembly on the powerhead and tighten the attaching bolts to 86 inch lbs. (10Nm).
6. Connect the neutral starting cable its bracket and adjust it to specification.
7. Install the engine cover and connect the cover fasteners.
8. Pull the starter knob several times and check for the proper operation of the ratcheting mechanism.

ADJUSTMENT

The neutral starting cable should be adjusted anytime it is removed for servicing.

1. Place the lower unit in the **NEUTRAL** position.
2. Loosen the cable locknut and turn the adjusting nut so that the cable freeplay is 0.08–0.11 in. (2–3mm).
3. Tighten the locknut to 86 inch lbs. (10Nm)

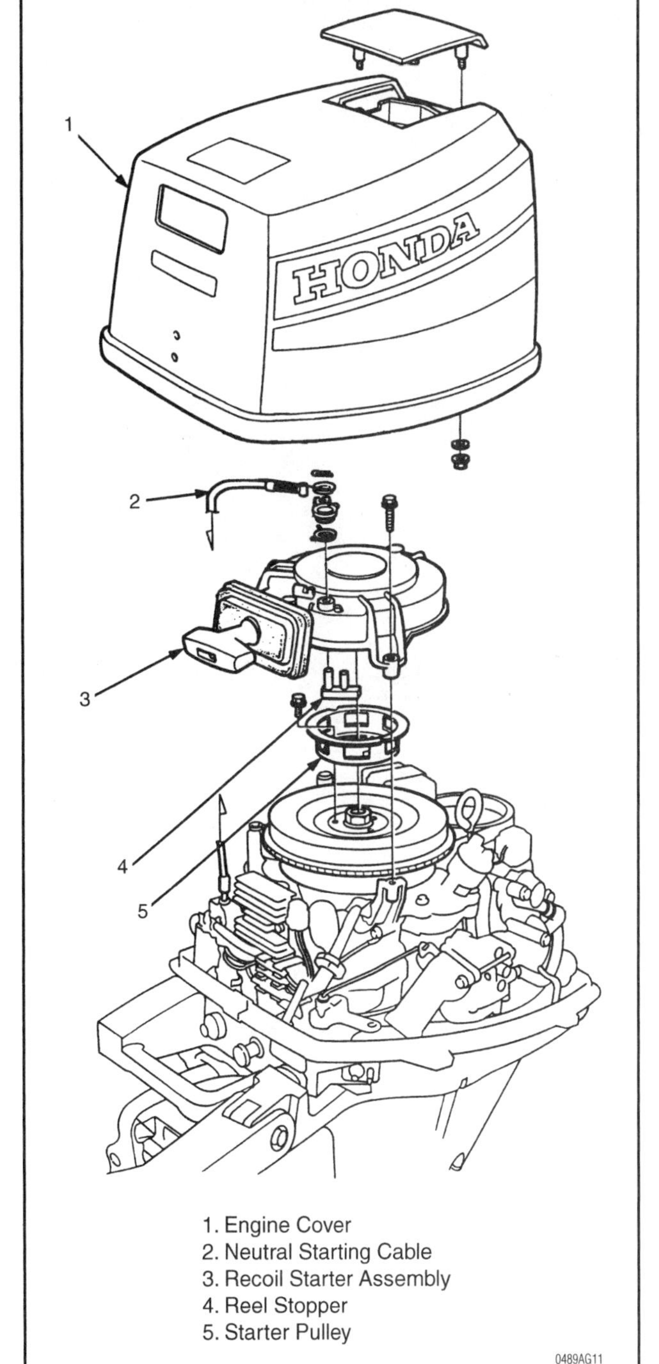

0489AG11

Fig. 13 Some engines utilize a shift interlock system which prevents starting the powerhead in gear—BF9.9A and BF15A

4. Ensure the proper function of the neutral starting system. The system should not allow the engine to be started in any other position but **NEUTRAL**.

DISASSEMBLY

See Figures 14, 15 and 16

1. Remove the roller guide bolt, then lift off the starter roller, roller guide, thrust washer and spring washer.
2. Untie the knot in the end of the starter rope and remove the recoil starter knob and recoil seal. Pass the starter rope end through the rope hole in the starter case.
3. Remove the friction plate bolt and spring washer.
4. Unwind the starter reel to remove all preload.

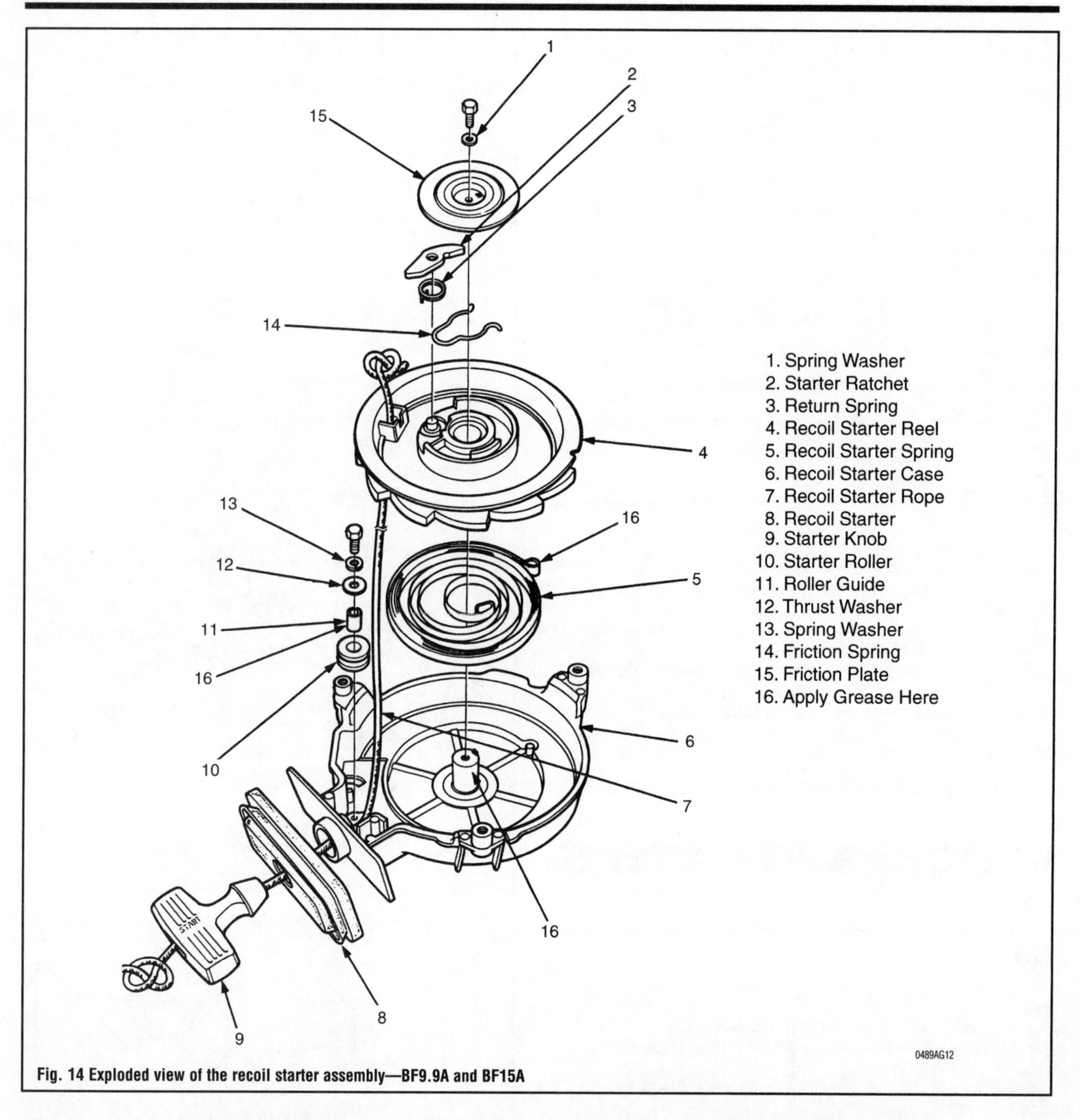

Fig. 14 Exploded view of the recoil starter assembly—BF9.9A and BF15A

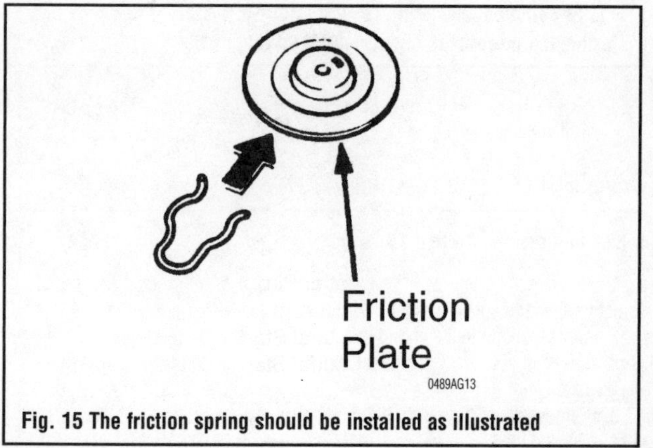

Fig. 15 The friction spring should be installed as illustrated

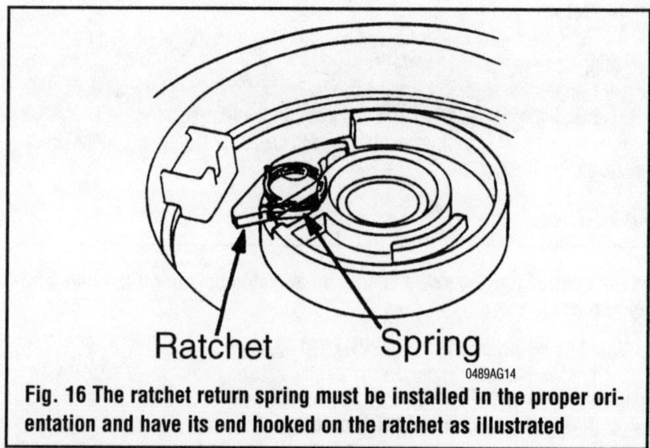

Fig. 16 The ratchet return spring must be installed in the proper orientation and have its end hooked on the ratchet as illustrated

CAUTION

The starter reel spring is under high tension. If the spring should come loose, it may cause serious damage or personal injury. Take all applicable cautions when working with this spring.

5. Remove the ratchet and return spring from the reel.
6. Pull the rope end out of the cutout in the reel.
7. Carefully lift the starter reel from the starter case, containing the starter spring as it unwinds.

➡It is advisable to wear heavy gloves while removing the spring to prevent your hands from being cut by the sharp spring steel.

CLEANING & INSPECTION

Clean all components and then blow them dry using compressed air. Remove any trace of corrosion and wipe all metal parts with an oil dampened cloth to prevent future corrosion.

Inspect the rope. Replace the rope if it appears to be weak or frayed. If the rope is frayed, check the holes through which the rope passes for rough edges or burrs. Remove the rough edges or burrs with a file and polish the surface until it is smooth. Inspect the starter return spring end hooks. Replace the spring if it is weak, corroded or cracked. Inspect the inside surface of the starter recoil case and reel for grooves or roughness. Grooves may cause erratic rewinding of the starter rope.

Inspect and lubricate the ratchet mechanism and check for freedom of movement.

ASSEMBLY

1. Route on end of the starter rope through the hole in the starter reel and tie a knot in the end.
2. Wind the starter rope on the reel in a counterclockwise direction.
3. Hook the starter spring end on the starter case tab and wind the spring so that it is installed in the case securely.

➡Ensure that the end of the spring is securely hooked on the tab in the case.

CAUTION

The starter reel spring is under high tension. If the spring should come loose, it may cause serious damage or personal injury. Take all applicable cautions when working with this spring.

4. Align the hook on the recoil starter spring with the hook on the recoil starter reel and install the reel in the starter case.
5. Pull the rope end out of the cutout in the reel
6. Install the return spring and ratchet on the reel.
7. Install the friction plate on the reel and align the hole in the friction plate with the projection on the starter reel.
8. Install the spring washer and tighten the friction plate attaching bolt to 86 inch lbs. (10Nm).
9. While holding the rope end in the notched portion of the reel, rotate the reel in a clockwise direction to preload the spring.
10. Pass the starter rope end through the rope hole in the starter case. Install the recoil seal and starter knob and tie a knot in the end of the rope.
11. Install the starter roller, roller guide, thrust washer and spring washer. Tighten the roller guide bolt securely.
12. Pull the starter knob several times and check for the proper operation of the ratcheting mechanism.

BF25A and BF30A

REMOVAL & INSTALLATION

▶ **See Figures 17 and 18**

1. Loosen the neutral starting cable locknuts and slide the cable from its bracket.
2. Disconnect the neutral starting cable from the stopper arm.
3. Remove the recoil starter attaching bolts and lift the recoil starter from the powerhead.

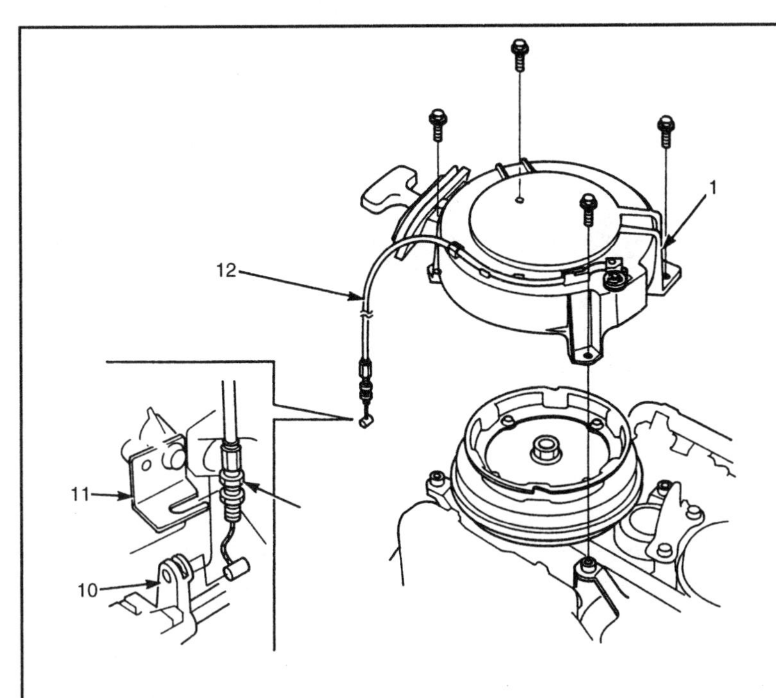

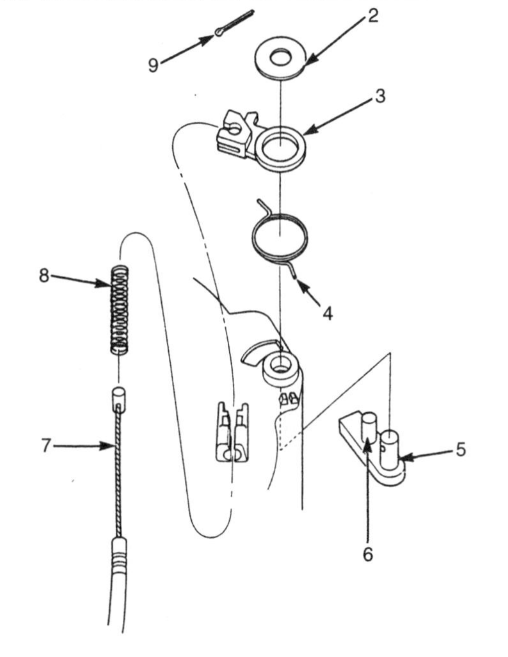

1. Recoil Starter
2. Washer
3. Stopper Arm
4. 'A' Side
5. Reel Stopper
6. Projection
7. Neutral Starter Cable
8. Neutral Starter Cable Spring
9. Cotter Pin
10. Shift Arm
11. Neutral Starter Cable Bracket
12. Neutral Starter Cable

0489AG15

Fig. 17 Removal and installation detail—BF25A and BF30A

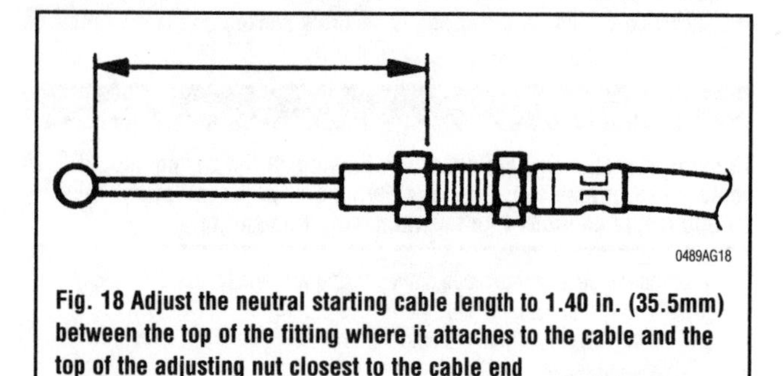

Fig. 18 Adjust the neutral starting cable length to 1.40 in. (35.5mm) between the top of the fitting where it attaches to the cable and the top of the adjusting nut closest to the cable end

To install:

4. Connect the neutral starting cable on the stopper arm.
5. Install the recoil starter on the powerhead and tighten the attaching bolts to 86 inch lbs. (10Nm).
6. Adjust the neutral starting cable length to 1.40 in. (35.5mm) between the top of the fitting where it attaches to the cable and the top of the adjusting nut closest to the cable end.

➡When installing the cable, take care not to change the cable length.

7. Connect the cable to the shift arm and insert it into the bracket. Tighten the locknut securely.
8. With the lower unit in **NEUTRAL**, pull the starter knob several times and check for the proper operation of the ratcheting mechanism.

ADJUSTMENT

See Figures 19 and 20

➡The neutral starting cable should be adjusted anytime it is removed for servicing.

1. Check to see if the cutout in the stopper arm is aligning with the **I** mark on the starter case.
2. If the cutout does not align properly, readjust the neutral starting cable length until the cutout properly aligns.
3. With the lower unit in either **FORWARD** or **REVERSE** positions, pull the starter knob slowly to check for proper operation of the neutral starting mechanism.
4. With the lower unit in **NEUTRAL**, pull the starter knob slowly to see if the engine can be started.

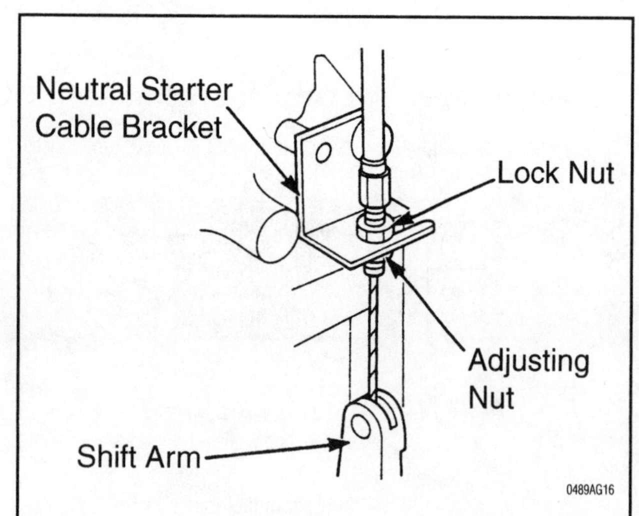

Fig. 19 The neutral starter cable is attached to the shift arm at one end . . .

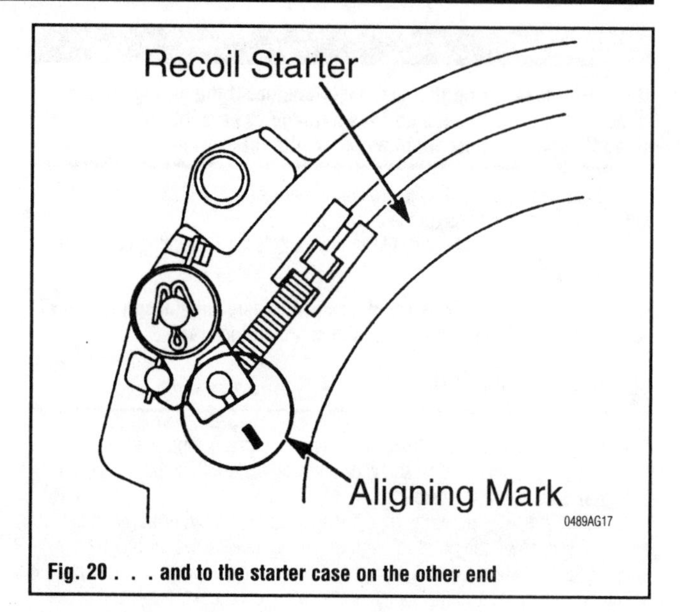

Fig. 20 . . . and to the starter case on the other end

DISASSEMBLY

See Figure 21

1. Disconnect the neutral starting cable from the recoil starter.
2. Remove the cotter pin, washer, stopper arm, stopper spring and reel stopper from the recoil starter.
3. Remove the roller guide through bolt, washer, roller guide and starter roller from the case.
4. Untie the knot in the end of the rope and pass the starter rope through the starter knob, recoil seal and the hole in the case.
5. Remove the recoil starter seal noting its orientation.
6. If not already done, rotate the starter reel clockwise to release the preload on the spring.
7. Remove the friction plate though bolt, friction plate, friction spring and spacer.

➡It is advisable to wear heavy gloves while removing the spring to prevent your hands from being cut by the sharp spring steel.

8. Remove the starter reel taking extra care to contain the starter spring.

**** CAUTION**

The starter reel spring is under high tension. If the spring should come loose, it may cause serious damage or personal injury. Take all applicable cautions when working with this spring.

9. Unwind the rope on the reel and untie the rope end to release it from the reel.
10. Remove the starter ratchet, E-ring, starter guide and return spring from the starter reel.

CLEANING & INSPECTION

Clean all components and then blow them dry using compressed air. Remove any trace of corrosion and wipe all metal parts with an oil dampened cloth to prevent future corrosion.

Inspect the rope. Replace the rope if it appears to be weak or frayed. If the rope is frayed, check the holes through which the rope passes for rough edges or burrs. Remove the rough edges or burrs with a file and polish the surface until it is smooth. Inspect the starter return spring end hooks. Replace the spring if it is weak, corroded or cracked. Inspect the inside surface of the starter

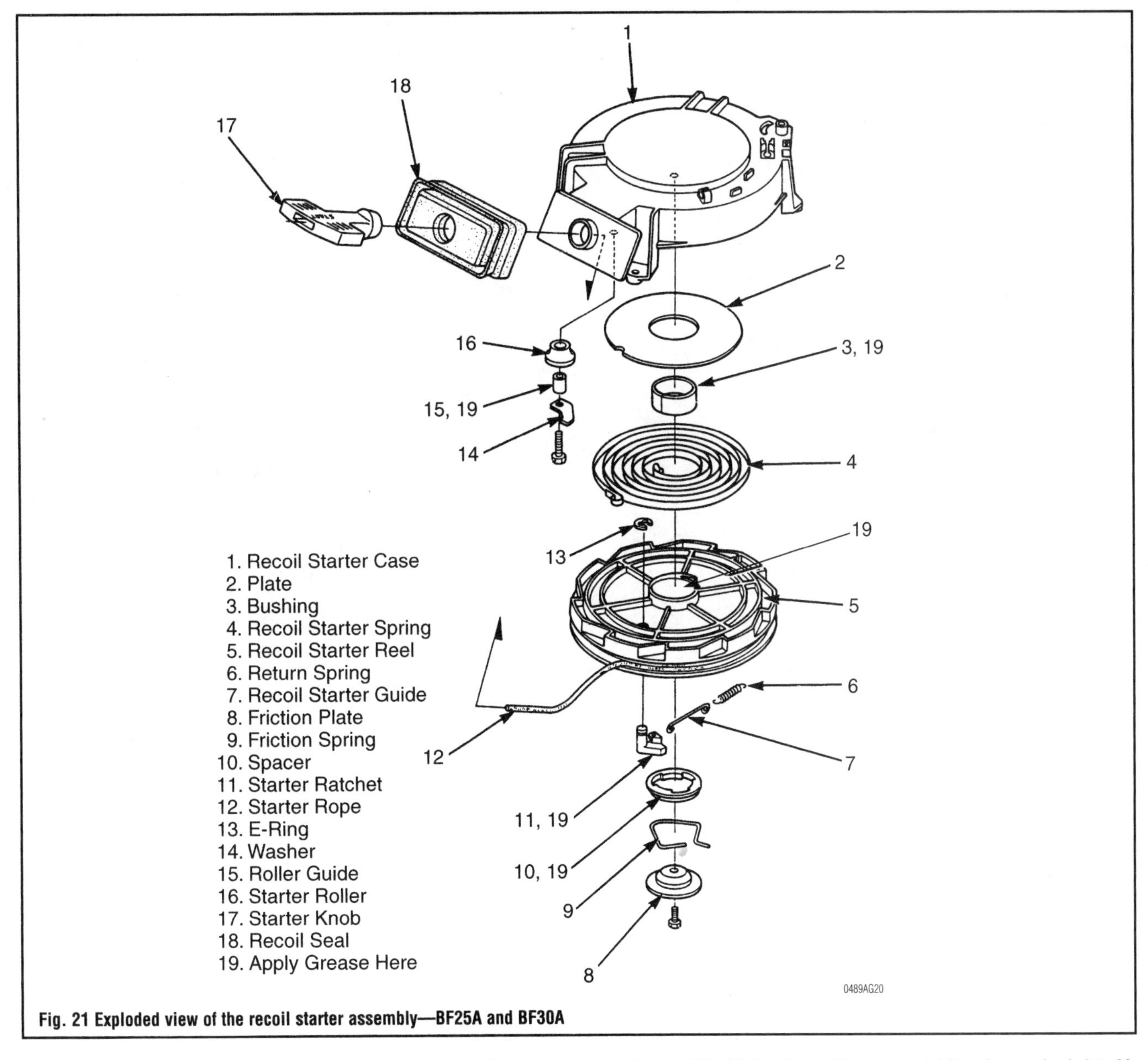

Fig. 21 Exploded view of the recoil starter assembly—BF25A and BF30A

recoil case and reel for grooves or roughness. Grooves may cause erratic rewinding of the starter rope.

Inspect and lubricate the ratchet mechanism and check for freedom of movement.

ASSEMBLY

➧ See Figure 22

1. Install the starter ratchet, E-ring, starter guide and return spring on the starter reel.
2. Route one end of the starter rope through the hole in the starter reel and tie a knot in the end.
3. Wind the rope on the reel in a counterclockwise direction.
4. Set the recoil starter spring end on the starter case tab and wind the spring so that it is installed in the case securely.
5. Lubricate the reel bushing and install it in the starter case.
6. Align the hook on the starter spring with the hook on the recoil starter reel and install the reel in the case.
7. Pull the end of the rope out of the cutout in the reel.
8. Install the spacer on the reel bushing and set the friction spring on the spacer.

➡Place the end of the friction spring at the end of the recoil starter guide.

9. Install the friction plate on the spacer and tighten the securing bolt to 86 inch lbs. (10Nm).
10. While holding the rope end in the cutout, rotate the reel 5 turns counterclockwise to preload the spring.

⁂ CAUTION

The starter spring is under high tension. If the spring should come loose, it may cause serious damage or personal injury. Take all applicable cautions when working with this spring.

11. Install the recoil starter seal to the starter case.
12. Pass the starter rope through the hole in the case, the recoil seal and the starter knob and tie a knot in the end of the rope to secure it.
13. Install the starter roller, roller guide and washer. Tighten the roller guide through bolt to 86 inch lbs. (10Nm).

➡Ensure the starter rope is securely set in the groove of the starter roller.

14. Pull the starter knob several times and check for the proper operation of the ratcheting mechanism.
15. Install the reel stopper, stopper spring, stopper arm and washer on the case and secure them with a new cotter pin.

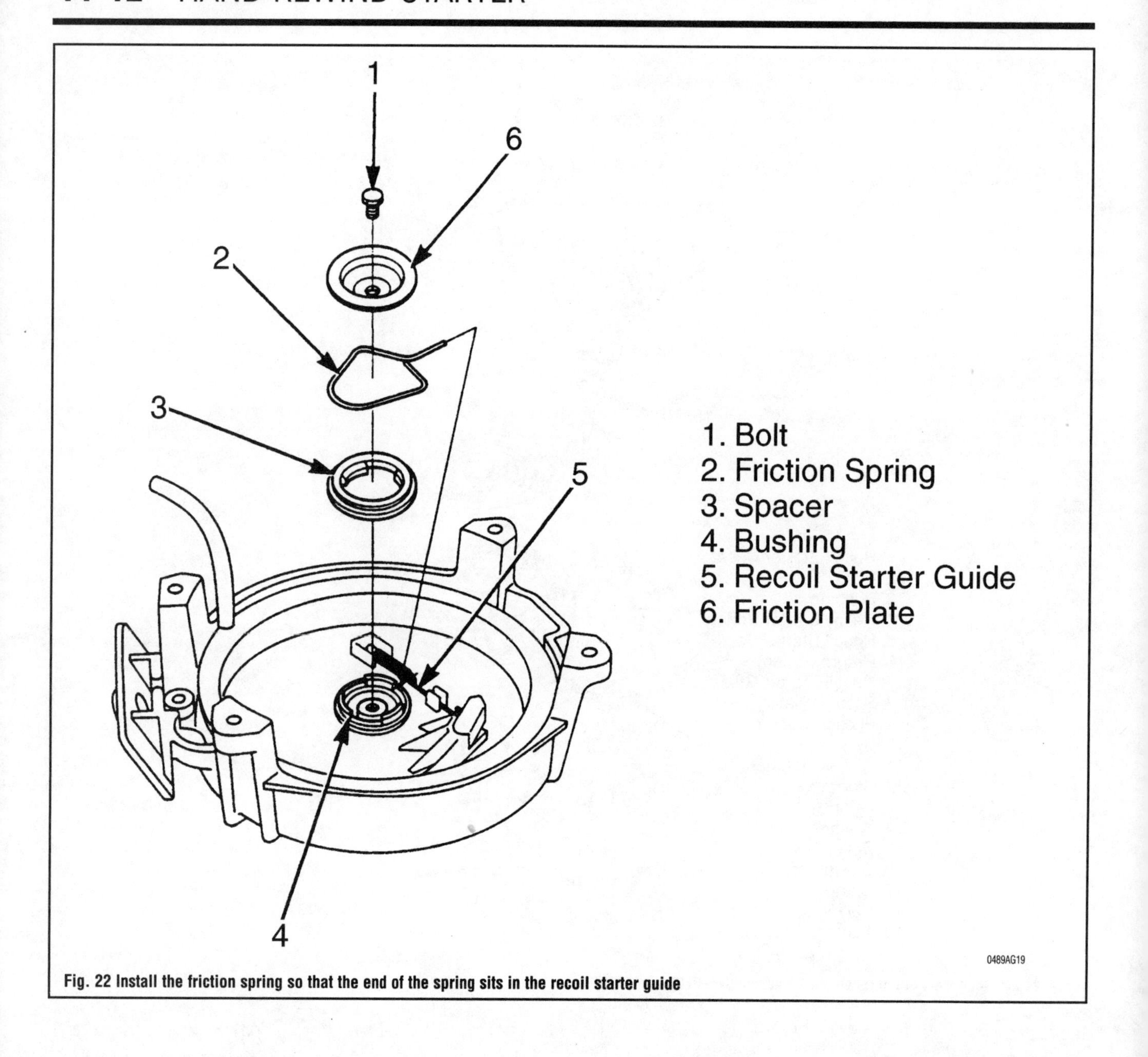

Fig. 22 Install the friction spring so that the end of the spring sits in the recoil starter guide

GLOSSARY

Understanding your mechanic is as important as understanding your marine engine. Most boaters know about their boats, but many boaters have difficulty understanding marine terminology. Talking the language of boats makes it easier to effectively communicate with professional mechanics. It isn't necessary (or recommended) that you diagnose the problem for him, but it will save him time, and you money, if you can accurately describe what is happening. It will also help you to know why your boat does what it is doing, and what repairs were made.

AFTER TOP DEAD CENTER (ATDC): The point after the piston reaches the top of its travel on the compression stroke.

AIR CLEANER: An assembly consisting of a housing, filter and any connecting ductwork. The filter element is made up of a porous paper or a wire mesh screening, and is designed to prevent airborne particles from entering the engine. Also see Intake Silencer.

AIR/FUEL RATIO: The ratio of air-to-fuel, by weight, drawn into the engine.

ALTERNATING CURRENT (AC): Electric current that flows first in one direction, then in the opposite direction, continually reversing flow.

ALTERNATOR: A device which produces AC (alternating current) which is converted to DC (direct current) to charge the battery.

AMMETER: An instrument, calibrated in amperes, used to measure the flow of an electrical current in a circuit. Ammeters are always connected in series with the circuit being tested.

AMP/HR. RATING (BATTERY): Measurement of the ability of a battery to deliver a stated amount of current for a stated period of time. The higher the amp/hr. rating, the better the battery.

AMPERE: The rate of flow of electrical current present when one volt of electrical pressure is applied against one ohm of electrical resistance.

ANTIFREEZE: A substance (ethylene or propylene glycol) added to the coolant to prevent freezing in cold weather.

ARMATURE: A laminated, soft iron core wrapped by a wire that converts electrical energy to mechanical energy as in a motor or relay. When rotated in a magnetic field, it changes mechanical energy into electrical energy as in a generator.

ATDC: After Top Dead Center.

ATMOSPHERIC PRESSURE: The pressure on the Earth's surface caused by the weight of the air in the atmosphere. At sea level, this pressure is 14.7 psi at 32°F (101 kPa at 0°C).

ATOMIZATION: The breaking down of a liquid into a fine mist that can be suspended in air.

AXIAL PLAY: Movement parallel to a shaft or bearing bore.

BACKFIRE: The sudden combustion of gases in the intake or exhaust system that results in a loud explosion.

BACKLASH: The clearance or play between two parts, such as meshed gears.

BALL BEARING: A bearing made up of hardened inner and outer races between which hardened steel balls roll.

BATTERY: A direct current electrical storage unit, consisting of the basic active materials of lead and sulphuric acid, which converts chemical energy into electrical energy. Used to provide current for the operation of the starter as well as other equipment, such as the radio, lighting, etc.

BEARING: A friction reducing, supportive device usually located between a stationary part and a moving part.

BEFORE TOP DEAD CENTER (BTDC): The point just before the piston reaches the top of its travel on the compression stroke.

BLOCK: See Engine Block.

BLOW-BY: Combustion gases, composed of water vapor and unburned fuel, that leak past the piston rings into the crankcase during normal engine operation. These gases are removed by the evacuation system to prevent the buildup of harmful acids in the crankcase.

BORE: Diameter of a cylinder.

BTDC: Before Top Dead Center.

BUSHING: A liner, usually removable, for a bearing; an anti-friction liner used in place of a bearing.

CAMSHAFT: A shaft in the engine on which are the lobes (cams) which operate the valves. The camshaft is driven by the crankshaft, via a belt, chain or gears, at one half the crankshaft speed.

CARBON MONOXIDE (CO): A colorless, odorless gas given off as a normal byproduct of combustion. It is poisonous and extremely dangerous in confined areas, building up slowly to toxic levels without warning if adequate ventilation is not available.

CHECK VALVE: Any one-way valve installed to permit the flow of air, fuel or vacuum in one direction only.

CIRCLIP: A split steel snapring that fits into a groove to hold various parts in place.

CIRCUIT BREAKER: A switch which protects an electrical circuit from overload by opening the circuit when the current flow exceeds a pre-determined level. Some circuit breakers must be reset manually, while most reset automatically.

CIRCUIT: Any unbroken path through which an electrical current can flow. Also used to describe fuel flow in some instances.

COMBUSTION CHAMBER: The part of the engine in the cylinder head where combustion takes place.

COMPRESSION CHECK: A test involving cranking the engine with a special high pressure gauge connected to an individual cylinder. Individual cylinder pressure as well as pressure variance across cylinders is used to determine general operating condition of the engine.

COMPRESSION RATIO: The ratio of the volume between the piston and cylinder head when the piston is at the bottom of its stroke (bottom dead center) and when the piston is at the top of its stroke (top dead center).

CONDUCTOR: Any material through which an electrical current can be transmitted easily.

CONNECTING ROD: The connecting link between the crankshaft and piston.

CONTINUITY: Continuous or complete circuit. Can be checked with an ohmmeter.

COOLANT: Mixture of water and anti-freeze circulated through the engine to carry off heat produced by the engine.

CRANKCASE: The lower part of an engine in which the crankshaft and related parts operate.

CRANKSHAFT: Engine component (connected to pistons by connecting rods) which converts the reciprocating (up and down) motion of pistons to rotary motion used to turn the driveshaft.

CYLINDER BLOCK: See engine block.

CYLINDER HEAD: The detachable portion of the engine, usually fastened to the top of the cylinder block and containing all or most of the combustion chambers. On overhead valve engines, it contains the valves and their operating parts. On overhead cam engines, it contains the camshaft as well.

CYLINDER: In an engine, the round hole in the engine block in which the piston(s) ride.

DETONATION: An unwanted explosion of the air/fuel mixture in the combustion chamber caused by excess heat and compression, advanced timing, or an overly lean mixture. Also referred to as "ping".

DIAPHRAGM: A thin, flexible wall separating two cavities, such as in a vacuum advance unit.

DIESELING: The engine continues to run after the it is shut off; caused by fuel continuing to be burned in the combustion chamber.

DIGITAL VOLT OHMMETER: An electronic diagnostic tool used to measure voltage, ohms and amps as well as several other functions, with the readings displayed on a digital screen in tenths, hundredths and thousandths.

DIODE: An electrical device that will allow current to flow in one direction only.

DIRECT CURRENT (DC): Electrical current that flows in one direction only.

DISPLACEMENT: The total volume of air that is displaced by all pistons as the engine turns through one complete revolution.

DOUBLE OVERHEAD CAMSHAFT: The engine utilizes two camshafts mounted in one cylinder head. One camshaft operates the exhaust valves, while the other operates the intake valves.

DVOM: Digital volt ohmmeter

ELECTROLYTE: A solution of water and sulfuric acid used to activate the battery. Electrolyte is extremely corrosive.

END-PLAY: The measured amount of axial movement in a shaft.

ENGINE: The primary motor or power apparatus of a vessel, which converts fuel into mechanical energy.

ENGINE BLOCK: The basic engine casting containing the cylinders, the crankshaft main bearings, as well as machined surfaces for the mounting of other components such as the cylinder head, oil pan, transmission, etc.

ETHYLENE GLYCOL: The base substance of antifreeze.

EXHAUST MANIFOLD: A set of cast passages or pipes which conduct exhaust gases from the engine.

FEELER GAUGE: A blade, usually metal, of precisely predetermined thickness, used to measure the clearance between two parts.

FIRING ORDER: The order in which combustion occurs in the cylinders of an engine.

FLAME FRONT: The term used to describe certain aspects of the fuel explosion in the cylinders. The flame front should move in a controlled pattern across the cylinder, rather than simply exploding immediately.

FLAT SPOT: A point during acceleration when the engine seems to lose power for an instant.

FLYWHEEL: A heavy disc of metal attached to the rear of the crankshaft. It smoothes the firing impulses of the engine and keeps the crankshaft turning during periods when no firing takes place. The starter also engages the flywheel to start the engine.

FOOT POUND (ft. lbs. or sometimes, ft. lb.): The amount of energy or work needed to raise an item weighing one pound, a distance of one foot.

FUEL FILTER: A component of the fuel system containing a porous paper element used to prevent any impurities from entering the engine through the fuel system. It usually takes the form of a canister-like housing, mounted in-line with the fuel hose, located anywhere on a vessel between the fuel tank and engine.

FUEL INJECTION: A system that sprays fuel into the cylinder through nozzles. The amount of fuel can be more precisely controlled with fuel injection.

FUSE: A protective device in a circuit which prevents circuit overload by breaking the circuit when a specific amperage is present. The device is constructed around a strip or wire of a lower amperage rating than the circuit it is designed to protect. When an amperage higher than that stamped on the fuse is present in the circuit, the strip or wire melts, opening the circuit.

FUSIBLE LINK: A piece of wire in a wiring harness that performs the same job as a fuse. If overloaded, the fusible link will melt and interrupt the circuit.

HORSEPOWER: A measurement of the amount of work; one horsepower is the amount of work necessary to lift 33,000 lbs. one foot in one minute. Brake horsepower (bhp) is the horsepower delivered by an engine on a dynamometer. Net horsepower is the power remaining (measured at the flywheel of the engine) that can be used to power the vessel after power is consumed through friction and running the engine accessories (water pump, alternator, fan etc.)

HYDROCARBON (HC): Any chemical compound made up of hydrogen and carbon. A major pollutant formed by the engine as a by-product of combustion.

HYDROMETER: An instrument used to measure the specific gravity of a solution.

INCH POUND (inch lbs.; sometimes in. lb. or in. lbs.): One twelfth of a foot pound.

INJECTOR: A device which receives metered fuel under relatively low pressure and is activated to inject the fuel into the engine under relatively high pressure at a predetermined time.

INTAKE MANIFOLD: A casting of passages or pipes used to conduct air or a fuel/air mixture to the cylinders.

INTAKE SILENCER: An assembly consisting of a housing, and sometimes a filter. The filter element is made up of a porous paper or a wire mesh screening, and is designed to prevent airborne particles from entering the engine. Also see Air Cleaner.

JOURNAL: The bearing surface within which a shaft operates.

JUMPER CABLES: Two heavy duty wires with large alligator clips used to provide power from a charged battery to a discharged battery.

JUMP START: Utilizing one sufficiently charged battery to start the engine of another vessel with a discharged battery by the use of jumper cables.

KNOCK: Noise which results from the spontaneous ignition of a portion of the air-fuel mixture in the engine cylinder.

LITHIUM-BASE GREASE: Bearing grease using lithium as a base. Not compatible with sodium-base grease.

LOCK RING: See Circlip or Snapring

MANIFOLD VACUUM: Low pressure in an engine intake manifold formed just below the throttle plates. Manifold vacuum is highest at idle and drops under acceleration.

MANIFOLD: A casting of passages or set of pipes which connect the cylinders to an inlet or outlet source.

MISFIRE: Condition occurring when the fuel mixture in a cylinder fails to ignite, causing the engine to run roughly.

MULTI-WEIGHT: Type of oil that provides adequate lubrication at both high and low temperatures.

NEEDLE BEARING: A bearing which consists of a number (usually a large number) of long, thin rollers.

NITROGEN OXIDE (NOx): One of the three basic pollutants found in the exhaust emission of an internal combustion engine. The amount of NOx usually varies in an inverse proportion to the amount of HC and CO.

OEM: Original Equipment Manufactured. OEM equipment is that furnished standard by the manufacturer.

OHM: The unit used to measure the resistance of conductor-to-electrical flow. One ohm is the amount of resistance that limits current flow to one ampere in a circuit with one volt of pressure.

OHMMETER: An instrument used for measuring the resistance, in ohms, in an electrical circuit.

OXIDES OF NITROGEN: See nitrogen oxide (NOx).

PING: A metallic rattling sound produced by the engine during acceleration. It is usually due to incorrect timing or a poor grade of fuel.

PISTON RING: An open-ended ring which fits into a groove on the outer diameter of the piston. Its chief function is to form a seal between the piston and cylinder wall. Most pistons have three rings: two for compression sealing; one for oil sealing.

POLARITY: Indication (positive or negative) of the two poles of a battery.

POWERTRAIN: See Drivetrain.

PPM: Parts per million; unit used to measure exhaust emissions.

PREIGNITION: Early ignition of fuel in the cylinder, sometimes due to glowing carbon deposits in the combustion chamber.

PRELOAD: A predetermined load placed on a bearing during assembly or by adjustment.

PRESS FIT: The mating of two parts under pressure, due to the inner diameter of one being smaller than the outer diameter of the other, or vice versa; an interference fit.

PSI: Pounds per square inch; a measurement of pressure.

PUSHROD: A steel rod between the hydraulic valve lifter and the valve rocker arm in overhead valve (OHV) engines.

RACE: The surface on the inner or outer ring of a bearing on which the balls, needles or rollers move.

RADIATOR: Part of the cooling system for some water-cooled engines. Through the radiator, excess combustion heat is dissipated into the atmosphere through forced convection using a water and glycol based mixture that circulates through, and cools, the engine.

REAR MAIN OIL SEAL: A synthetic or rope-type seal that prevents oil from leaking out of the engine past the rear main crankshaft bearing.

RECTIFIER: A device (used primarily in alternators) that permits electrical current to flow in one direction only.

REGULATOR: A device which maintains the amperage and/or voltage levels of a circuit at predetermined values.

RELAY: A switch which automatically opens and/or closes a circuit.

RESISTANCE: The opposition to the flow of current through a circuit or electrical device, and is measured in ohms. Resistance is equal to the voltage divided by the amperage.

RESISTOR: A device, usually made of wire, which offers a preset amount of resistance in an electrical circuit.

ROLLER BEARING: A bearing made up of hardened inner and outer races between which hardened steel rollers move.

RPM: Revolutions per minute (usually indicates engine speed).

RUN-ON: Condition when the engine continues to run, even when the key is turned off. See dieseling.

SENDING UNIT: A mechanical, electrical, hydraulic or electromagnetic device which transmits information to a gauge.

SENSOR: Any device designed to measure engine operating conditions or ambient pressures and temperatures. Usually electronic in nature and designed to send a voltage signal to an on-board computer, some sensors may operate as a simple on/off switch or they may provide a variable voltage signal (like a potentiometer) as conditions or measured parameters change.

SHIM: Spacers of precise, predetermined thickness used between parts to establish a proper working relationship.

SHORT CIRCUIT: An electrical malfunction where current takes the path of least resistance to ground (usually through damaged insulation). Current flow is excessive from low resistance resulting in a blown fuse.

SLUDGE: Thick, black deposits in engine formed from dirt, oil, water, etc. It is usually formed in engines when oil changes are neglected.

SNAP RING: A circular retaining clip used inside or outside a shaft or part to secure a shaft, such as a floating wrist pin.

SOLENOID: An electrically operated, magnetic switching device.

SPECIFIC GRAVITY (BATTERY): The relative weight of liquid (battery electrolyte) as compared to the weight of an equal volume of water.

SPLINES: Ridges machined or cast onto the outer diameter of a shaft or inner diameter of a bore to enable parts to mate without rotation.

STARTER: A high-torque electric motor used for the purpose of starting the engine, typically through a high ratio geared drive connected to the flywheel ring gear.

STROKE: The distance the piston travels from bottom dead center to top dead center.

TACHOMETER: A device used to measure the rotary speed of an engine, shaft, gear, etc., usually in rotations per minute.

TDC: Top dead center. The exact top of the piston's stroke.

THERMOSTAT: A valve, located in the cooling system of an engine, which is closed when cold and opens gradually in response to engine heating, controlling the temperature of the coolant and rate of coolant flow.

TOP DEAD CENTER (TDC): The point at which the piston reaches the top of its travel on the compression stroke.

TORQUE: Measurement of turning or twisting force, expressed as foot-pounds or inch-pounds.

TUNE-UP: A regular maintenance function, usually associated with the replacement and adjustment of parts and components in the electrical and fuel systems of a engine for the purpose of attaining optimum performance.

VACUUM GAUGE: An instrument used to measure the presence of vacuum in a chamber.

VISCOSITY: The ability of a fluid to flow. The lower the viscosity rating, the easier the fluid will flow. 10 weight motor oil will flow much easier than 40 weight motor oil.

VOLT: Unit used to measure the force or pressure of electricity. It is defined as the pressure

VOLTAGE REGULATOR: A device that controls the current output of the alternator or generator.

VOLTMETER: An instrument used for measuring electrical force in units called volts. Voltmeters are always connected parallel with the circuit being tested.

WATER PUMP: A component of the cooling system that circulates the coolant under pressure.

AIR CLEANER 3-13
ALTERNATOR (STATOR) 5-38
REMOVING & INSTALLATION 5-39
TESTING 5-38
AVOIDING THE MOST COMMON MISTAKES 1-3
AVOIDING TROUBLE 1-2
BASIC ELECTRICAL THEORY 5-2
HOW DOES ELECTRICITY WORK: THE WATER ANALOGY 5-2
OHM'S LAW 5-2
BATTERY (BOAT MAINTENANCE) 3-23
BATTERY AND CHARGING SAFETY PRECAUTIONS 3-24
BATTERY CHARGERS 3-25
BATTERY TERMINALS 3-24
CHECKING SPECIFIC GRAVITY 3-23
CLEANING 3-23
REPLACING BATTERY CABLES 3-25
BATTERY (CHARGING CIRCUIT) 5-40
BATTERY & CHARGING SAFETY PRECAUTIONS 5-42
BATTERY CABLES 5-43
BATTERY CHARGERS 5-43
BATTERY CONSTRUCTION 5-40
BATTERY LOCATION 5-41
BATTERY RATINGS 5-40
BATTERY SERVICE 5-41
BATTERY STORAGE 5-43
BATTERY TERMINALS 5-42
MARINE BATTERIES 5-40
BOAT MAINTENANCE 3-21
BOATING SAFETY 1-3
BOLTS, NUTS AND OTHER THREADED RETAINERS 2-12
BREAKER POINTS 5-9
GAP CHECK 5-9
REMOVAL & INSTALLATION 5-9
BREAKER POINTS IGNITION (MAGNETO IGNITION) 5-7
BREAKER POINTS IGNITION OVERHAUL 5-11
ASSEMBLY 5-12
CLEANING & INSPECTION 5-11
DISASSEMBLY 5-11
BUY OR REBUILD? 7-32
CAMSHAFT, BEARINGS AND LIFTERS 7-25
REMOVAL & INSTALLATION 7-25
CAN YOU DO IT? 1-2
CAPACITIES 3-11
CAPACITOR DISCHARGE IGNITION (CDI) SYSTEM 5-12
CARBURETION 4-4
CARBURETOR FLOAT HEIGHTS 4-13
CARBURETOR IDENTIFICATION 4-4
CARBURETOR PILOT SCREW SPECIFICATIONS 4-13
CARBURETORS 4-6
ASSEMBLY 4-11
CLEANING & INSPECTION 4-11
DESCRIPTION AND OPERATION 4-6
DISASSEMBLY 4-9
REMOVAL & INSTALLATION 4-8
CDI TEST CHARTS 5-26
CDI UNIT 5-25
DESCRIPTION & OPERATION 5-25
REMOVAL & INSTALLATION 5-30
TESTING 5-26
CHARGE COIL (BREAKER POINTS IGNITION) 5-10
DESCRIPTION & OPERATION 5-10
REMOVAL & INSTALLATION 5-10
TESTING 5-10
CHARGE COIL (CAPACITOR DISCHARGE IGNITION SYSTEM) 5-18
DESCRIPTION & OPERATION 5-18
REMOVAL & INSTALLATION 5-19
TESTING 5-18
CHARGING CIRCUIT 5-35
CHEMICALS 2-3
CLEANERS 2-4

LUBRICANTS & PENETRANTS 2-3
SEALANTS 2-3
CHOKE SWITCH 10-7
REMOVAL & INSTALLATION 10-7
TESTING 10-7
COMBUSTION 4-3
ABNORMAL COMBUSTION 4-3
FACTORS AFFECTING COMBUSTION 4-3
COMPONENT LOCATIONS (IGNITION & ELECTRICAL) 5-48
COMPRESSION CHECK 3-26
CHECKING COMPRESSION 3-26
LOW COMPRESSION 3-26
CONDENSER 5-10
DESCRIPTION & OPERATION 5-10
REMOVAL & INSTALLATION 5-10
TESTING 5-10
CONVERSION FACTORS 2-14
COOLING SYSTEM 6-9
COURTESY MARINE EXAMINATIONS 1-11
CYLINDER HEAD (ENGINE MECHANICAL) 7-11
REMOVAL & INSTALLATION 7-11
CYLINDER HEAD (ENGINE RECONDITIONING) 7-34
ASSEMBLY 7-38
DISASSEMBLY 7-34
INSPECTION 7-35
REFINISHING & REPAIRING 7-37
DESCRIPTION AND OPERATION (CAPACITOR DISCHARGE IGNITION SYSTEM) 5-12
FOUR-CYLINDER CARBURETED IGNITION 5-13
FOUR-CYLINDER EFI IGNITION 5-13
SINGLE-CYLINDER IGNITION 5-12
TWO & THREE-CYLINDER IGNITION 5-13
DESCRIPTION AND OPERATION (CHARGING CIRCUIT) 5-35
DESCRIPTION AND OPERATION (COOLING SYSTEM) 6-9
OVERHEAT WARNING SYSTEM 6-9
THERMOSTAT & PRESSURE RELIEF 6-9
WATER PUMP 6-9
DESCRIPTION AND OPERATION (ELECTRONIC IGNITION) 5-34
DESCRIPTION AND OPERATION (GAS ASSISTED TILT) 9-2
DESCRIPTION AND OPERATION (HAND REWIND STARTER) 11-2
DESCRIPTION AND OPERATION (JET DRIVE) 8-68
DESCRIPTION AND OPERATION (LUBRICATION SYSTEM) 6-2
FORCED LUBRICATION 6-2
SPLASH LUBRICATION 6-2
DESCRIPTION AND OPERATION (MANUAL TILT) 9-2
SERVICING 9-2
DESCRIPTION AND OPERATION (POWER TRIM AND TILT) 9-4
DESCRIPTION AND OPERATION (REMOTE CONTROL BOX) 10-2
DESCRIPTION AND OPERATION (STARTING CIRCUIT) 5-43
DESCRIPTION AND OPERATION (TILLER HANDLE) 10-11
DETERMINING ENGINE CONDITION 7-31
COMPRESSION TEST 7-31
OIL PRESSURE TEST 7-32
DIAGNOSTIC TROUBLE CODE CHART 4-23
DIAGNOSTIC TROUBLE CODES 4-22
CLEARING 4-24
READING 4-23
SERVICE PRECAUTIONS 4-22
DIRECTIONS AND LOCATIONS 1-2
DO'S 1-11
DON'TS 1-11
DRAINING THE FUEL SYSTEM 4-4
ELECTRICAL COMPONENTS 5-2
CONNECTORS 5-4
GROUND 5-3
LOAD 5-3
POWER SOURCE 5-2
PROTECTIVE DEVICES 5-3
SWITCHES & RELAYS 5-3
WIRING & HARNESSES 5-4
ELECTRICAL SYSTEM PRECAUTIONS 5-7
ELECTRONIC IGNITION 5-34
ENGINE BLOCK 7-39
ASSEMBLY 7-42
DISASSEMBLY 7-39
GENERAL INFORMATION 7-39
INSPECTION 7-40
REFINISHING 7-42
ENGINE COOLANT TEMPERATURE (ECT) 4-27
DESCRIPTION & INSTALLATION 4-27
REMOVAL & INSTALLATION 4-27
TESTING 4-27
ENGINE COVER 3-7
REMOVAL & INSTALLATION 3-7
ENGINE MAINTENANCE 3-2
ENGINE MECHANICAL 7-2
ENGINE OIL 3-7
ENGINE OIL RECOMMENDATIONS 3-8
OIL & FILTER CHANGE 3-10
OIL LEVEL CHECK 3-8
ENGINE OVERHAUL TIPS 7-32
CLEANING 7-33
OVERHAUL TIPS 7-32
REPAIRING DAMAGED THREADS 7-33
TOOLS 7-32
ENGINE PREPARATION 7-34
ENGINE REBUILDING SPECIFICATIONS 7-46
ENGINE RECONDITIONING 7-31
ENGINE START-UP AND BREAK-IN 7-45
BREAKING IT IN 7-45
KEEP IT MAINTAINED 7-45
STARTING THE ENGINE 7-45
ENGINE STOP SWITCH (REMOTE CONTROL SWITCH) 10-7
REMOVAL & INSTALLATION 10-8
TESTING 10-7
ENGINE STOP SWITCH (TILLER HANDLE) 10-20
REMOVAL & INSTALLATION 10-20
TESTING 10-20
EQUIPMENT NOT REQUIRED BUT RECOMMENDED 1-10
ANCHORS 1-10
BAILING DEVICES 1-10
FIRST AID KIT 1-10
SECOND MEANS OF PROPULSION 1-10
TOOLS AND SPARE PARTS 1-10
VHF-FM RADIO 1-10
FASTENERS, MEASUREMENTS AND CONVERSIONS 2-12
FIBERGLASS HULLS 3-21
FLYWHEEL 7-29
REMOVAL & INSTALLATION 7-29
FUEL 4-2
ALCOHOL-BLENDED FUELS 4-3
HIGH ALTITUDE OPERATION 4-3
OCTANE RATING 4-2
RECOMMENDATIONS 4-2
THE BOTTOM LINE WITH FUELS 4-3
VAPOR PRESSURE AND ADDITIVES 4-2
FUEL FILTER 3-13
RELIEVING FUEL SYSTEM PRESSURE 3-14
REMOVAL & INSTALLATION 3-14
FUEL INJECTION 4-19
FUEL INJECTION BASICS 4-19
FUEL INJECTORS 4-28
DESCRIPTION & OPERATION 4-28
REMOVAL & INSTALLATION 4-28
TESTING 4-28

FUEL LINES (CARBURETION) 4-15
GENERAL INFORMATION 4-15
REPLACEMENT & ROUTING 4-15
FUEL LINES (FUEL INJECTION) 4-32
COMMON PROBLEMS 4-34
GENERAL INFORMATION 4-32
REPLACEMENT & ROUTING 4-33
FUEL PRESSURE REGULATOR 4-32
DESCRIPTION & OPERATION 4-32
INSPECTION 4-32
REMOVAL & INSTALLATION 4-32
FUEL PUMP (CARBURETION) 4-14
ASSEMBLY 4-15
CLEANING & INSPECTION 4-15
DESCRIPTION & OPERATION 4-14
DISASSEMBLY 4-14
FUEL PRESSURE CHECK 4-14
REMOVAL & INSTALLATION 4-14
FUEL PUMP (FUEL INJECTION) 4-31
DESCRIPTION & OPERATION 4-31
FUEL PRESSURE CHECK 4-31
OVERHAUL 4-32
REMOVAL & INSTALLATION 4-31
FUEL SYSTEM BASICS 4-2
FUEL SYSTEM SERVICE 4-3
FUEL/WATER SEPARATOR 3-15
GAS ASSIST DAMPER 9-3
OVERHAUL 9-4
REMOVAL & INSTALLATION 9-3
TESTING 9-3
GAS ASSISTED TILT 9-2
GENERAL ENGINE SPECIFICATIONS 3-3
GENERAL INFORMATION (LOWER UNIT) 8-2
HAND REWIND STARTER 11-2
BF2A AND BF20 11-2
ASSEMBLY 11-3
CLEANING & INSPECTION 11-3
DISASSEMBLY 11-3
REMOVAL & INSTALLATION 11-2
BF5A AND BF50 11-4
ASSEMBLY 11-5
CLEANING & INSPECTION 11-4
DISASSEMBLY 11-4
REMOVAL & INSTALLATION 11-4
BF9.9A AND BF15A 11-7
ADJUSTMENT 11-7
ASSEMBLY 11-9
CLEANING &INSPECTION 11-9
DISASSEMBLY 11-7
REMOVAL & INSTALLATION 11-7
BF25A AND BF30A 11-9
ADJUSTMENT 11-10
ASSEMBLY 11-11
CLEANING & INSPECTION 11-10
DISASSEMBLY 11-10
REMOVAL & INSTALLATION 11-9
BF75, BF100 AND BF8A 11-5
ASSEMBLY 11-7
CLEANING & INSPECTION 11-6
DISASSEMBLY 11-6
REMOVAL & INSTALLATION 11-5
HAND TOOLS 2-4
ELECTRONIC TOOLS 2-9
GAUGES 2-9
HAMMERS 2-8
OTHER COMMON TOOLS 2-8
PLIERS 2-7
SCREWDRIVERS 2-8
SOCKET SETS 2-4
SPECIAL TOOLS 2-8
WRENCHES 2-6
HOW TO USE THIS MANUAL 1-2
HYDRAULIC UNIT 9-10
ASSEMBLY 9-15
BLEEDING THE SYSTEM 9-10
CHECKING FLUID LEVEL 9-10
CLEANING & INSPECTION 9-15
DISASSEMBLY 9-14
REMOVAL & INSTALLATION 9-11
TESTING 9-10
IGNITION AND ELECTRICAL WIRING DIAGRAMS 5-52
IGNITION COIL (BREAKER POINTS IGNITION) 5-11
DESCRIPTION & OPERATION 5-11
REMOVAL & INSTALLATION 5-11
TESTING 5-11
IGNITION COILS (CAPACITOR DISCHARGE IGNITION SYSTEM) 5-20
DESCRIPTION & OPERATION 5-20
REMOVAL & INSTALLATION 5-22
TESTING 5-20
IGNITION SWITCH 10-8
REMOVAL & INSTALLATION 10-8
TESTING 10-8
IGNITION SYSTEM 3-36
INDICATOR LIGHT (TILLER HANDLE) 10-20
REMOVAL & INSTALLATION 10-20
TESTING 10-20
INDICATOR LIGHTS (REMOTE CONTROL BOX) 10-8
REMOVAL & INSTALLATION 10-8
TESTING 10-8
INSIDE THE BOAT 3-21
INTAKE AIR TEMPERATURE SENSOR (IAT) 4-27
DESCRIPTION & OPERATION 4-27
REMOVAL & INSTALLATION 4-27
TESTING 4-27
INTAKE MANIFOLD 7-9
REMOVAL & INSTALLATION 7-9
INTRODUCTION (TUNE-UP) 3-25
JET DRIVE 8-68
JET DRIVE ASSEMBLY 8-69
ADJUSTMENT 8-72
ASSEMBLING 8-75
CLEANING & INSPECTING 8-74
DISASSEMBLY 8-74
REMOVAL & INSTALLATION 8-69
LIGHTING COIL 5-20
DESCRIPTION & OPERATION 5-20
REMOVAL & INSTALLATION 5-20
TESTING 5-20
LOWER UNIT 8-2
LOWER UNIT (ENGINE MAINTENANCE) 3-12
DRAINING AND FILLING 3-12
LOWER UNIT OVERHAUL 8-10
LOWER UNIT—NO REVERSE GEAR 8-5
ASSEMBLY 8-11
CLEANING & INSPECTION 8-10
DISASSEMBLY 8-10
REMOVAL & INSTALLATION 8-5
LOWER UNIT—WITH REVERSE GEAR 8-8
ASSEMBLY 8-33
CLEANING & INSPECTING 8-29
DISASSEMBLY 8-12
REMOVAL & INSTALLATION 8-8
SHIMMING PROCEDURE 8-50
LUBRICATION SYSTEM 6-2
MAINTENANCE INTERVALS 3-6

MAINTENANCE OR REPAIR? 1-2
MANUAL TILT 9-2
MEASURING TOOLS 2-10
  DEPTH GAUGES 2-12
  DIAL INDICATORS 2-11
  MICROMETERS & CALIPERS 2-11
  TELESCOPING GAUGES 2-12
METRIC BOLTS 2-15
MODEL IDENTIFICATION AND SERIAL NUMBERS 8-68
OIL PAN 7-16
  REMOVAL & INSTALLATION 7-16
OIL PRESSURE SWITCH 6-6
  REMOVAL & INSTALLATION 6-7
  TESTING 6-6
OIL PRESSURE WARNING LAMP 6-7
  REMOVAL & INSTALLATION 6-7
  TESTING 6-7
OIL PUMP 6-3
  OVERHAUL 6-4
  REMOVAL & INSTALLATION 6-3
OVERHEAT SENSOR 4-28
  DESCRIPTION & OPERATION 4-28
  REMOVAL & INSTALLATION 4-28
  TESTING 4-28
OVERHEAT THERMOSWITCH 6-16
  REMOVAL & INSTALLATION 6-16
  TESTING 6-16
OVERHEAT WARNING LAMP 6-15
  REMOVAL & INSTALLATION 6-16
  TESTING 6-15
POWER TRIM AND TILT 9-4
PRIMARY COIL 5-11
  DESCRIPTION & OPERATION 5-11
  REMOVAL & INSTALLATION 5-11
  TESTING 5-11
PROFESSIONAL HELP 1-2
PROPELLER (ENGINE MAINTENANCE) 3-19
PROPELLER (LOWER UNIT) 8-4
  REMOVAL & INSTALLATION 8-4
PULSAR COIL 5-16
  DESCRIPTION & OPERATION 5-16
  REMOVAL & INSTALLATION 5-17
  TESTING 5-16
PURCHASING PARTS 1-3
RECTIFIER 5-31
  REMOVAL & INSTALLATION 5-32
  TESTING 5-31
REGULATIONS FOR YOUR BOAT 1-3
  CAPACITY INFORMATION 1-4
  CERTIFICATE OF COMPLIANCE 1-4
  DOCUMENTING OF VESSELS 1-4
  HULL IDENTIFICATION NUMBER 1-4
  LENGTH OF BOATS 1-4
  NUMBERING OF VESSELS 1-4
  REGISTRATION OF BOATS 1-4
  SALES AND TRANSFERS 1-4
  VENTILATION 1-4
  VENTILATION SYSTEMS 1-5
REGULATOR (CAPACITOR DISCHARGE IGNITION SYSTEM) 5-33
  DESCRIPTION & OPERATION 5-33
  REMOVAL & INSTALLATION 5-33
  TESTING 5-33
REGULATOR/RECTIFIER (CAPACITOR DISCHARGE IGNITION SYSTEM) 5-33
  REMOVAL & INSTALLATION 5-34
  TESTING 5-33
REMOTE CONTROL BOX 10-2
REMOTE CONTROL BOX ASSEMBLY 10-2
  EXPLODED VIEWS 10-3
  REMOVAL & INSTALLATION 10-2
REQUIRED SAFETY EQUIPMENT 1-5
  FIRE EXTINGUISHERS 1-5
  PERSONAL FLOTATION DEVICES 1-6
  SOUND PRODUCING DEVICES 1-8
  TYPES OF FIRES 1-5
  VISUAL DISTRESS SIGNALS 1-8
  WARNING SYSTEM 1-6
ROCKER ARM/SHAFTS 7-5
  REMOVAL & INSTALLATION 7-5
SAE BOLTS 2-16
SAFETY IN SERVICE 1-11
SAFETY SWITCH 10-9
  REMOVAL & INSTALLATION 10-9
  TESTING 10-9
SAFETY TOOLS 2-2
  EYE AND EAR PROTECTION 2-2
  WORK CLOTHES 2-3
  WORK GLOVES 2-2
SELF DIAGNOSIS 5-34
SERIAL NUMBER IDENTIFICATION 3-2
SHIFTING PRINCIPLES 8-2
  COUNTERROTATING UNIT 8-3
  STANDARD ROTATING UNIT 8-2
SINGLE PHASE CHARGING SYSTEM 5-36
  DESCRIPTION AND OPERATION 5-36
SPARK PLUG WIRES 3-30
  CHECKING RESISTANCE 3-30
  REMOVAL & INSTALLATION 3-30
  TESTING 3-30
SPARK PLUGS 3-27
  INSPECTION & GAPPING 3-28
  READING SPARK PLUGS 3-28
  REMOVAL & INSTALLATION 3-27
  SPARK PLUG HEAT RANGE 3-27
  SPARK PLUG SERVICE 3-27
SPECIFICATIONS CHARTS
  CAPACITIES 3-11
  CARBURETOR FLOAT HEIGHTS 4-13
  CARBURETOR PILOT SCREW SPECIFICATIONS 4-13
  CDI TEST CHARTS 5-26
  CONVERSION FACTORS 2-14
  DIAGNOSTIC TROUBLE CODE CHART 4-23
  ENGINE REBUILDING SPECIFICATIONS 7-46
  GENERAL ENGINE SPECIFICATIONS 3-3
  MAINTENANCE INTERVALS 3-6
  METRIC BOLTS 2-15
  SAE BOLTS 2-16
  TUNEUP SPECIFICATIONS 3-26
  USING A VACUUM GAUGE 2-10
  VOLTAGE REGULATOR RECTIFER 5-32
SPRING COMMISSIONING CHECKLIST 3-38
STANDARD AND METRIC MEASUREMENTS 2-13
STARTER BUTTON 10-20
  REMOVAL & INSTALLATION 10-20
  TESTING 10-20
STARTER MOTOR 5-44
  DESCRIPTION & OPERATION 5-44
  REMOVAL & INSTALLATION 5-45
  TESTING 5-44
STARTER MOTOR SOLENOID SWITCH 5-46
  DESCRIPTION AND OPERATION 5-46
  REMOVAL & INSTALLATION 5-47
  TESTING 5-46
STARTING CIRCUIT 5-43

SYSTEM TESTING (BREAKER POINTS IGNITION) 5-8
SPARK CHECK 5-8
SYSTEM TESTING (CAPACITOR DISCHARGE IGNITION SYSTEM) 5-16
PROCEDURE 5-16
TEST EQUIPMENT 5-4
JUMPER WIRES 5-4
MULTIMETERS 5-5
TEST LIGHTS 5-5
TESTING 5-6
OPEN CIRCUITS 5-7
RESISTANCE 5-6
SHORT CIRCUITS 5-7
VOLTAGE 5-6
VOLTAGE DROP 5-6
THE FOUR-STROKE CYCLE 7-3
COMPRESSION STROKE 7-3
EXHAUST STROKE 7-3
INTAKE STROKE 7-3
POWER STROKE 7-3
THREE-PHASE CHARGING SYSTEM 5-36
DESCRIPTION AND OPERATION 5-36
PRECAUTIONS 5-37
SERVICING 5-37
TROUBLESHOOTING 5-37
THROTTLE CABLE 10-14
ADJUSTMENT 10-17
REMOVAL & INSTALLATION 10-14
THROTTLE POSITION SENSOR (TPS) 4-26
DESCRIPTION & OPERATION 4-26
REMOVAL & INSTALLATION 4-26
TESTING 4-26
THROTTLE SWITCH 10-9
REMOVAL & INSTALLATION 10-9
TESTING 10-9
TILLER HANDLE 10-11
TILLER HANDLE 10-12
REMOVAL & INSTALLATION 10-12
TIMING AND SYNCHRONIZATION 3-36
CARBURETOR SYNCHRONIZATION 3-37
IDLE SPEED ADJUSTMENT 3-37
PREPARATION 3-36
SYNCHRONIZATION 3-36
TIMING 3-36
TIMING BELT (ENGINE MAINTENANCE) 3-15
ADJUSTMENT 3-16
INSPECTION 3-15
TIMING BELT (ENGINE MECHANICAL) 7-19
ADJUSTMENT 7-22
INSPECTION 7-19
REMOVAL & INSTALLATION 7-20
TIMING BELT COVER 7-19
REMOVAL & INSTALLATION 7-19
TOOLS 2-4
TOOLS AND EQUIPMENT 2-2
TORQUE (FASTENERS, MEASUREMENTS AND CONVERSIONS) 2-13
TRIM ANGLE SENSOR 9-16
REMOVAL & INSTALLATION 9-16
TESTING 9-16
TRIM TABS, ANODES AND LEAD WIRES 3-22
TRIM, TILT & PIVOT POINTS 3-19
INSPECTION AND LUBRICATION 3-19
TRIM/TILT MOTOR 9-8
ASSEMBLY 9-10
CLEANING & INSPECTION 9-9
DISASSEMBLY 9-8
REMOVAL & INSTALLATION 9-8
TESTING 9-8
TRIM/TILT SWITCH (POWER TRIM AND TILT) 9-16
TRIM/TILT SWITCH (REMOTE CONTROL BOX) 10-9
REMOVAL & INSTALLATION 10-10
TESTING 10-9
TRIM/TILT SWITCH (TILLER HANDLE) 10-20
REMOVAL & INSTALLATION 10-20
TESTING 10-20
TROUBLESHOOTING (TILLER HANDLE) 10-11
TROUBLESHOOTING ELECTRICAL SYSTEMS 5-6
TROUBLESHOOTING THE CARBURETED FUEL SYSTEM 4-4
COMMON PROBLEMS 4-5
COMMON SYMPTOMS 4-5
LOGICAL TROUBLESHOOTING 4-4
TROUBLESHOOTING THE CHARGING SYSTEM 5-38
TROUBLESHOOTING THE COOLING SYSTEM 6-10
TROUBLESHOOTING THE FUEL INJECTION SYSTEM 4-20
COMMON PROBLEMS 4-22
DIAGNOSIS BY SYMPTOM 4-21
LOGICAL TROUBLESHOOTING 4-20
PRELIMINARY INSPECTION 4-20
TROUBLESHOOTING THE HAND REWIND STARTER 11-2
TROUBLESHOOTING THE LOWER UNIT 8-4
TROUBLESHOOTING THE LUBRICATION SYSTEM 6-2
TROUBLESHOOTING THE STARTING SYSTEM 5-44
TUNE-UP 3-25
TUNE-UP SEQUENCE 3-25
TUNEUP SPECIFICATIONS 3-26
UNDERSTANDING AND TROUBLESHOOTING ELECTRICAL SYSTEMS 5-2
USING A VACUUM GAUGE 2-10
VALVE (ROCKER) COVER 7-3
REMOVAL & INSTALLATION 7-3
VALVE CLEARANCE 3-30
ADJUSTMENT 3-30
VAPOR SEPARATOR 4-29
CLEANING & INSPECTION 4-30
DESCRIPTION & OPERATION 4-29
REMOVAL & INSTALLATION 4-30
VOLTAGE REGULATOR RECTIFER 5-32
WARNING BUZZER (COOLING SYSTEM) 6-16
REMOVAL & INSTALLATION 6-16
TESTING 6-16
WARNING BUZZER (LUBRICATION SYSTEM) 6-8
REMOVAL & INSTALLATION 6-8
TESTING 6-8
WARNING BUZZER (REMOTE CONTROL BOX) 10-10
REMOVAL & INSTALLATION 10-10
TESTING 10-10
WATER PUMP 6-10
REMOVAL & INSTALLATION 6-10
WHERE TO BEGIN 1-2
WINTER STORAGE CHECKLIST 3-38
WIRE AND CONNECTOR REPAIR 5-7

SUPPLEMENT—BF2D,BF75A AND BF90A 12-2
BF2D SUPPLEMENTAL INFORMATION 12-2
POWERHEAD REMOVAL & INSTALLATION 12-2
VALVE CLEARANCE ADJUSTMENT 12-3
IGNITION COIL TESTING ART 12-4
SWIVEL CASE LINER
REMOVAL & INSTALLATION 12-4
VERTICAL SHAFT BUSHING
REMOVAL & INSTALLATION 12-4
WATER SEAL REMOVAL & INSTALLATION 12-7
HAND REWIND STARTER 12-8
ENGINE REBUILDING-BF2D 12-9
BF75A/BF90A SUPPLEMENTAL INFORMATION
(1999 AND LATER MODELS) 12-10
REVISED GEAR/SHIM SELECTION CHARTS-
LOWER UNIT OVERHAUL 12-10

12

SUPPLEMENTAL INFORMATION

SUPPLEMENT—BF2D,BF75A
AND BF90A 12-2

SUPPLEMENT—BF2D, BF75A AND BF90A

■ **Throughout this manual you will find references to Honda models with the suffix "A" (BF#A). Procedures for later model suffix "D" (BF#D) motors are the same, unless noted or unless listed here. For instance, most procedures on the BF2A apply to the BF2D, unless otherwise noted. Further, there are a few production changes to the BF75A/BF90A that are also noted in this section.**

BF2D Supplemental Information

POWERHEAD REMOVAL & INSTALLATION

◆ **See Figures 1 thru 5**

Powerhead removal and installation is necessary for a large variety of maintenance procedures on this tiny powerhead. It is a relatively straightforward procedure that requires the replacement of the air guide plate (see the accompanying illustration). It allows access to items such as the valve cover, the clutch lining and assembly.

1. Remove the starter rope from the grip, but to prevent the rope from being pulled into the hand rewind starter housing tie a knot in the rope.
2. Remove the engine cover.
3. For safety and to prevent a mess, drain the fuel tank into a suitable container using a siphon.
4. Remove the 2 cap nuts, then lift the fuel tank up for access to the fuel hose. Squeeze the clips and disconnect the fuel hose, then remove the tank completely from the powerhead.
5. Remove the spark plug cap and disconnect the engine switch wire.
6. Loosen the 3 cap nuts and remove the hand rewind starter assembly.
7. Remove the engine cooling fan cover from the powerhead.
8. Push inward on the choke knob and hold while disconnecting the choke rod from the back of the knob shaft.
9. Disconnect the choke rod and throttle wire from the carburetor.
10. Remove the two 16mm flange bolts securing the exhaust pipe to the engine. Disconnect the exhaust pipe from the engine, then remove and discard the air guide plate.
11. On the same side as the exhaust pipe, loosen and remove the 10mm flange bolt securing the engine stop switch wiring ground terminal.
12. Remove the 4 large powerhead mounting bolts and washers threaded upward from underneath the cowling.
13. Carefully lift the powerhead straight up and off the housing.

■ **The clutch lining is attached to the bottom of the powerhead. Be sure to keep this lining and the outer clutch surface in the cowling clean and free of oil or grease during service and installation.**

To Install:

14. Carefully lower the powerhead straight downward onto the cowling, inserting the clutch lining into the cowling.
15. Install the 4 powerhead mounting bolts and tighten to 9 ft. lbs. (13 Nm).
16. Install the engine stop switch ground wire and secure using the retaining bolt.
17. Position a NEW air guide plate assembly, then connect the exhaust pipe and secure using the 2 retaining bolts. Tighten the bolt to 4.3 ft. lbs. (5.9 Nm).
18. Reconnect the throttle wire and choke rod to the carburetor, then hold the choke button inward while you reconnect the rod to the button shaft.
19. Install the fan cover, followed by the hand rewind starter assembly and secure the 3 capnuts.
20. Reconnect the engine switch wire and the spark plug cap.
21. Reconnect the fuel line and position the fuel tank to the powerhead. Secure by tightening the 2 capnuts to 5.8 ft. lbs. (7.8 Nm).
22. Refill the fuel tank and verify that there are no leaks.
23. Install the engine cover and reinstall the starter grip.

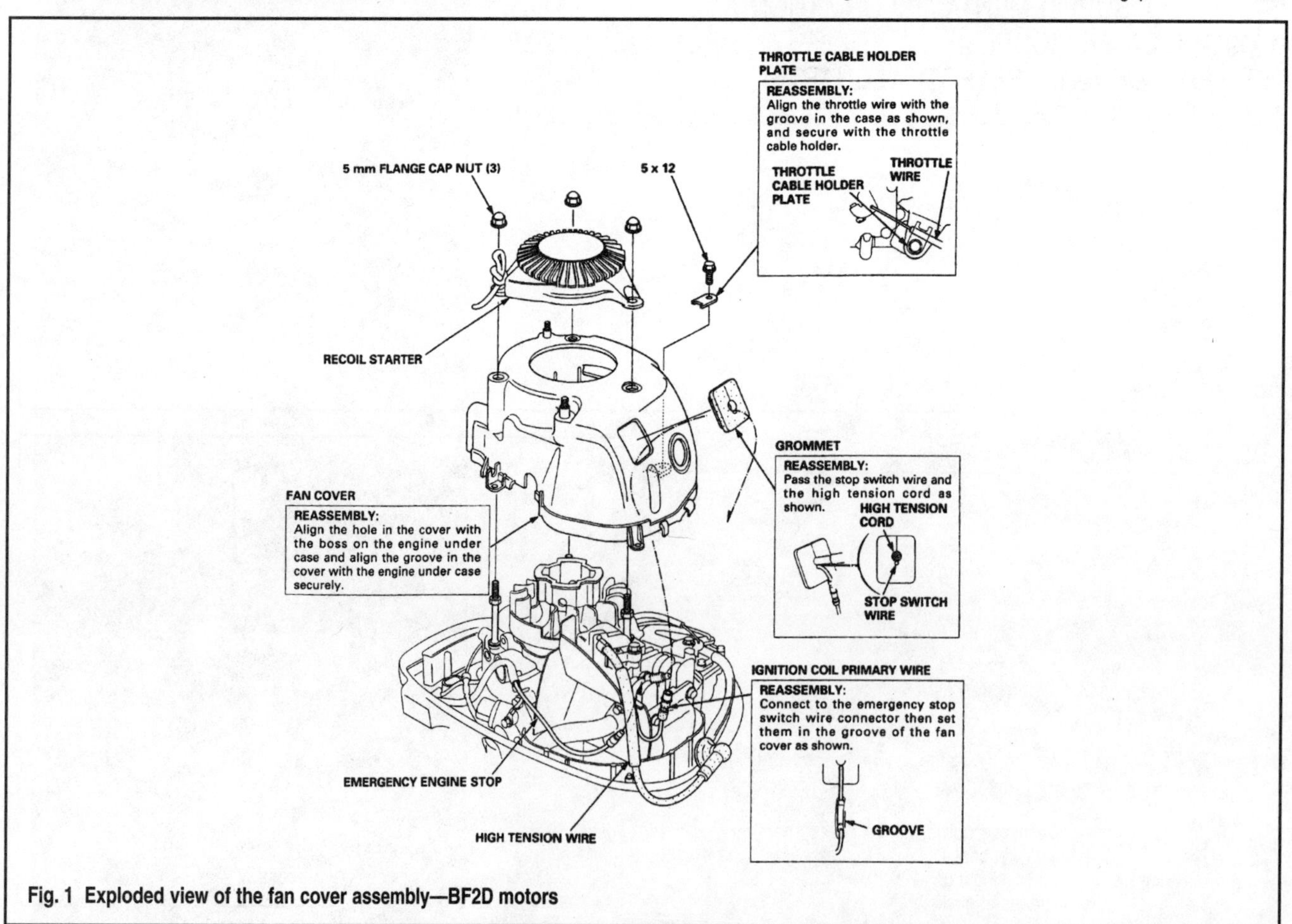

Fig. 1 Exploded view of the fan cover assembly—BF2D motors

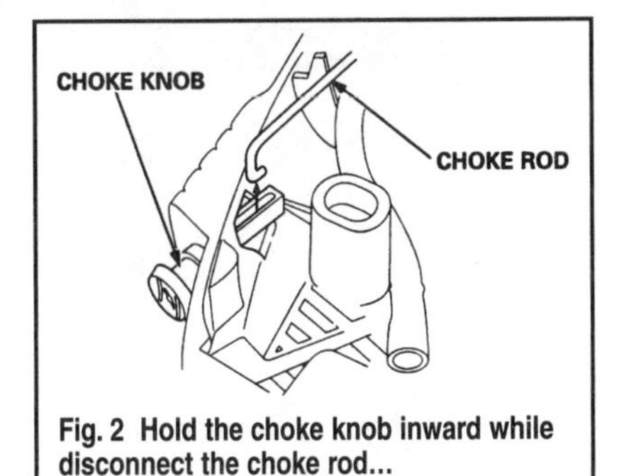

Fig. 2 Hold the choke knob inward while disconnect the choke rod...

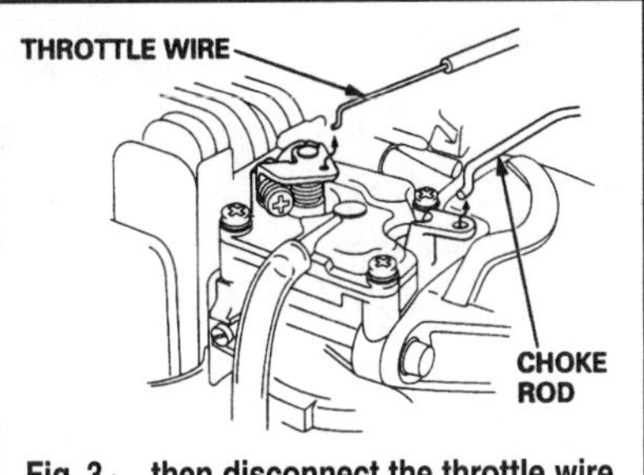

Fig. 3 ...then disconnect the throttle wire and choke rod at the carb

EMERGENCY ENGINE STOP SWITCH WIRE
AIR GUIDE PLATE
EXHAUST PIPE
5 x 10
5 x 16 (2)

Fig. 4 The air guide plate should not be reused

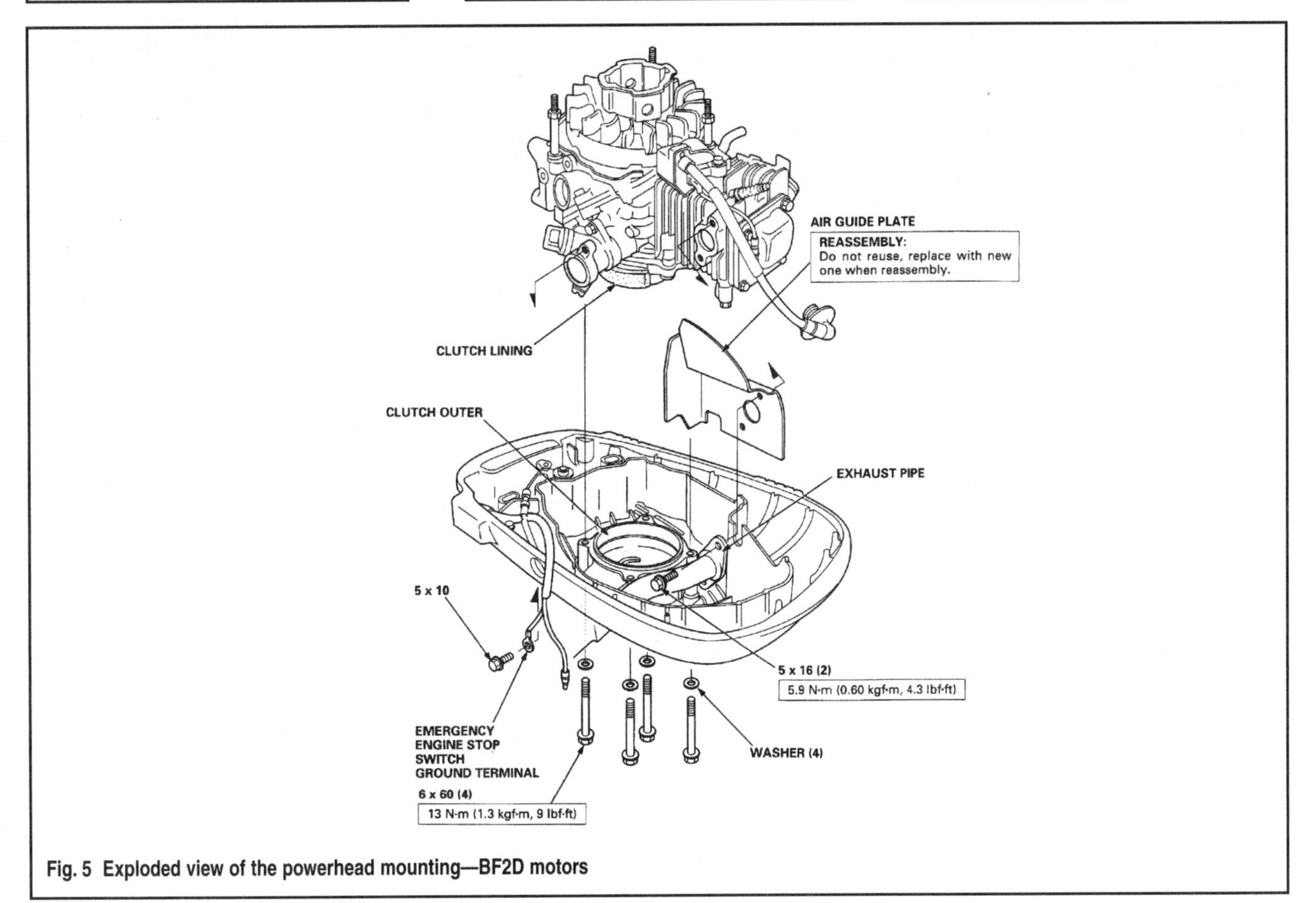

Fig. 5 Exploded view of the powerhead mounting—BF2D motors

VALVE CLEARANCE ADJUSTMENT

◆ **See Figures 6 thru 10**

Valve clearance checking and adjustment must occur with the engine completely cold. The best way to do this is to let the motor sit overnight before checking. Unfortunately, the positioning of the valve cover makes access virtually impossible without first removing the powerhead from the cowling.

1. Remove the Powerhead for access, as detailed earlier in this section.
2. Loosen and remove the four 12mm flange bolts securing the valve cover to the powerhead.

■ **Have a shop rag handy for the next step as a small amount of engine oil will be released when the cover gasket seal is broken.**

3. Using a small prytool, gently pry outward at the four corners of the valve cover until you break the gasket seal. DO NOT use excessive force as you could distort the valve cover making it unsuitable for reuse. Remove the cover and carefully remove all traces of gasket material/sealant.

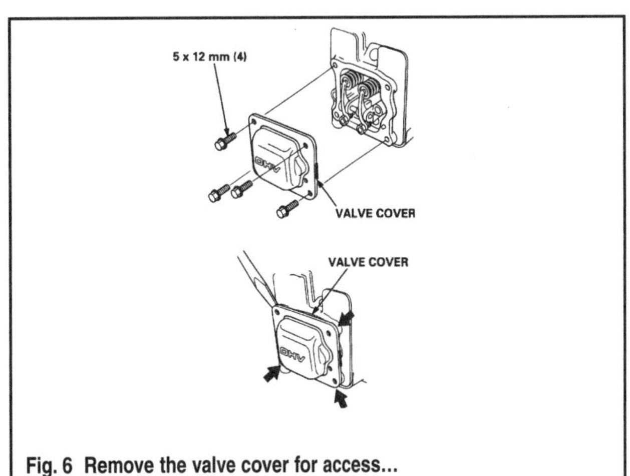

Fig. 6 Remove the valve cover for access...

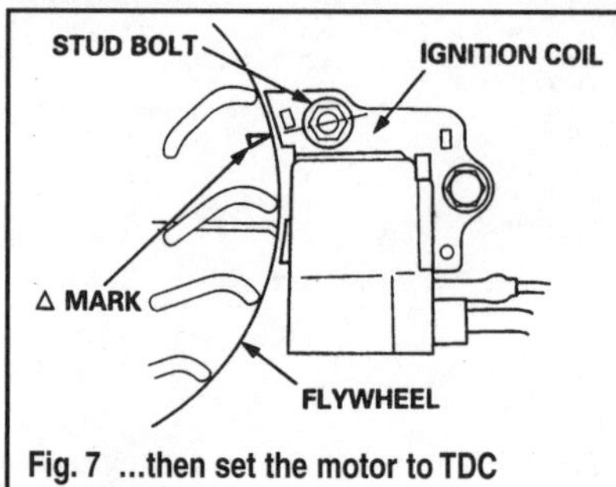

Fig. 7 ...then set the motor to TDC compression

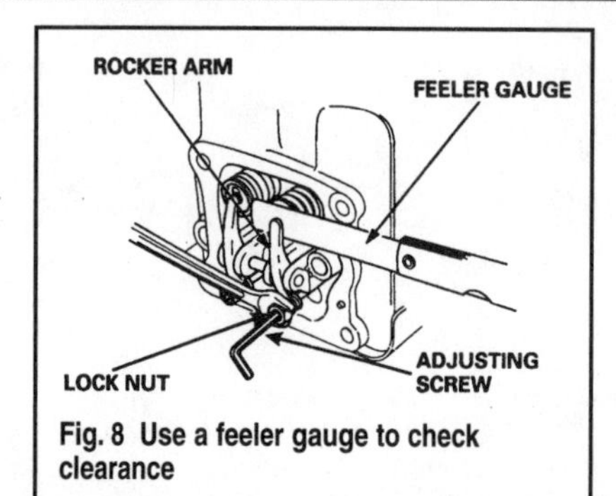

Fig. 8 Use a feeler gauge to check clearance

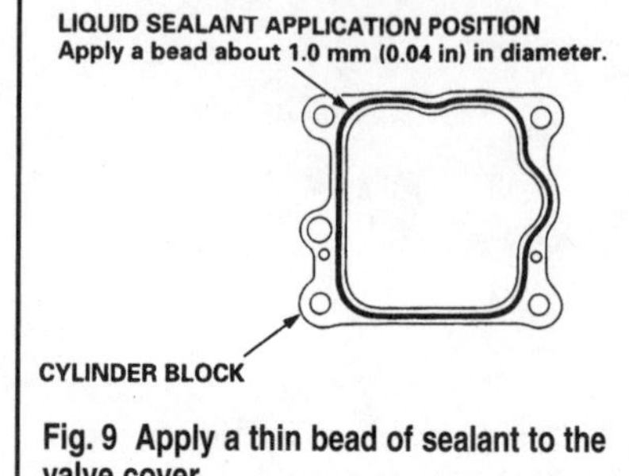

Fig. 9 Apply a thin bead of sealant to the valve cover

4. Slowly turn the flywheel clockwise until the piston reaches TDC of the compression stroke (as noted by watching for the triangular mark on the flywheel to align with the center of the ignition coil mounting stud).

■ **You can also watch the movement of the valves as the flywheel is rotated. If the exhaust valve opens as the mark aligns with the coil stud, then the engine is at TDC of the exhaust stroke (a full turn of the crankshaft away from TDC of the compression stroke). If this is the case, rotate the crankshaft one full turn and watch as the flywheel mark approaches alignment again. This time both valves should be closed at TDC!**

5. The intake valve clearance should be 0.0024-0.0039 in. (0.06-0.10mm), while the exhaust valve clearance should be 0.0035-0.0051 in. (0.09-0.13mm). Gently insert a feeler gauge between the rocker arm and the valve in order to measure the gap. When using feeler gauges, remember that the proper sized gauge will pass through the gap with a slight drag, while the next size down should pass freely and the next size up should not fit.
6. If adjustment is necessary, loosen the lock nut using an open end wrench, then turn the valve adjuster screw inward or outward using a hex key until the proper gap is obtained. DO NOT tighten the screw down against the gauge with any force, as you will only succeed in damaging the feeler gauge and you won't set the right gap in any case.
7. Once you've got the gap set correctly, HOLD the adjuster screw from turning while you tighten the locknut securely. Recheck the valve clearance to make sure the adjuster did not move while tightening the locknut.
8. Use a suitable solvent or degreaser to remove all traces of oil from the gasket mating surface of the cylinder head and valve cover.
9. Apply a light coat (about a 0.04 in. / 1mm diameter bead) of Three Bond 1207B or an equivalent sealant along the inner perimeter of the valve cover.

■ **To make sure there is a proper seal, install the valve cover within 3 minutes of sealant application.**

10. Install the valve cover to the cylinder head and thread the 4 retaining bolts by hand. Tighten the bolts using 2-3 steps of a criss-crossing pattern to 4.3 ft. lbs. (5.9 Nm).
11. WAIT for at least 20 minutes before adding oil to or starting the motor to give the sealant additional time to fully cure.
12. Install the powerhead assembly, as detailed earlier in this section.

IGNITION COIL TESTING ART

◆ **See Figures 11 and 12**

SWIVEL CASE LINER REMOVAL & INSTALLATION

◆ **See Figures 13 and 14**

At least every 3 years Honda recommends replacing the swivel case liner.

1. Remove the Powerhead for access, as detailed earlier in this section.
2. Remove the steering handle from the outboard.
3. Carefully remove the exhaust pipe from the outboard.
4. Loosen and remove the 3 bolts threaded upward from underneath the engine cowling. Remove the cowling and the clutch housing from the outboard.
5. Loosen and remove the 2 bolts securing the swivel case cap to the stern bracket. Remove the cap and carefully separate the extension housing from the bracket.
6. Remove the 2 halves of the swivel case liner and discard.
7. If you are also replacing the Vertical Shaft Bushing, refer to that procedure at this time to remove and replace the bushing before installation.

To Install:

8. Apply a light coating of marine grade grease to the halves of the swivel case liner, and to the friction surfaces of the stern bracket.
9. Install the liner halves making sure the hole in the liner aligns with the friction block.
10. Install the swivel case cap and tighten the retaining bolts to 17 ft. lbs. (24 Nm).
11. Install the engine cowling and clutch housing, then tighten the 3 retaining bolts to 9 ft. lbs. (13 Nm).
12. Before installing the exhaust pipe, GENTLY tab it with a plastic or rubber mallet to help dislodge carbon deposits. Apply a light coating of oil to the bottom of the pipe (where it inserts into the vertical shaft bushing), then gently install the pipe into the outboard, making sure not to cock or damage the vertical shaft bushing.
13. Install the tiller handle.
14. Install the powerhead assembly, as detailed earlier in this section.

VERTICAL SHAFT BUSHING REMOVAL & INSTALLATION

◆ **See Figures 15 and 16**

At least every 3 years Honda recommends replacing the vertical shaft bushing. The bushing is mounted in the bottom of the extension/swivel housing, just above the gearcase (which must be removed for access). Although it is possible to replace the bushing with the powerhead installed, it is probably much easier to replace it while the powerhead is removed and the extension housing is separated from the stern bracket. In this way you won't have to worry about the exhaust pipe making bushing installation more difficult. If necessary, refer to the Swivel Case Liner procedure earlier in this section for more details.

1. Remove the gearcase from the extension housing for access to the bushing.
2. Carefully pull the bushing from the bottom of the housing. If the powerhead and exhaust pipe have been removed from the top of the housing you may be able to use a driver to carefully push the bushing down and out of the housing.

To Install:

3. Carefully clean and degrease the interior of the extension case (where the bushing is installed) and the outer surface of the vertical shaft bushing.
4. Apply a light coating of water to the outer surface of the bushing, then insert the bushing into the bottom of the extension housing (with the water pipe hole on L models positioned toward the starboard side of the housing). Carefully seat the bushing squarely into the housing until it stops, about 1.34-1.42 in. (34-36mm) from the bottom of the housing.
5. Apply a light coating of marine grade grease to the bushing.
6. If removed, install the exhaust pipe and powerhead. For details, refer to the Swivel Case Liner procedure in this section.
7. Install the gearcase assembly.

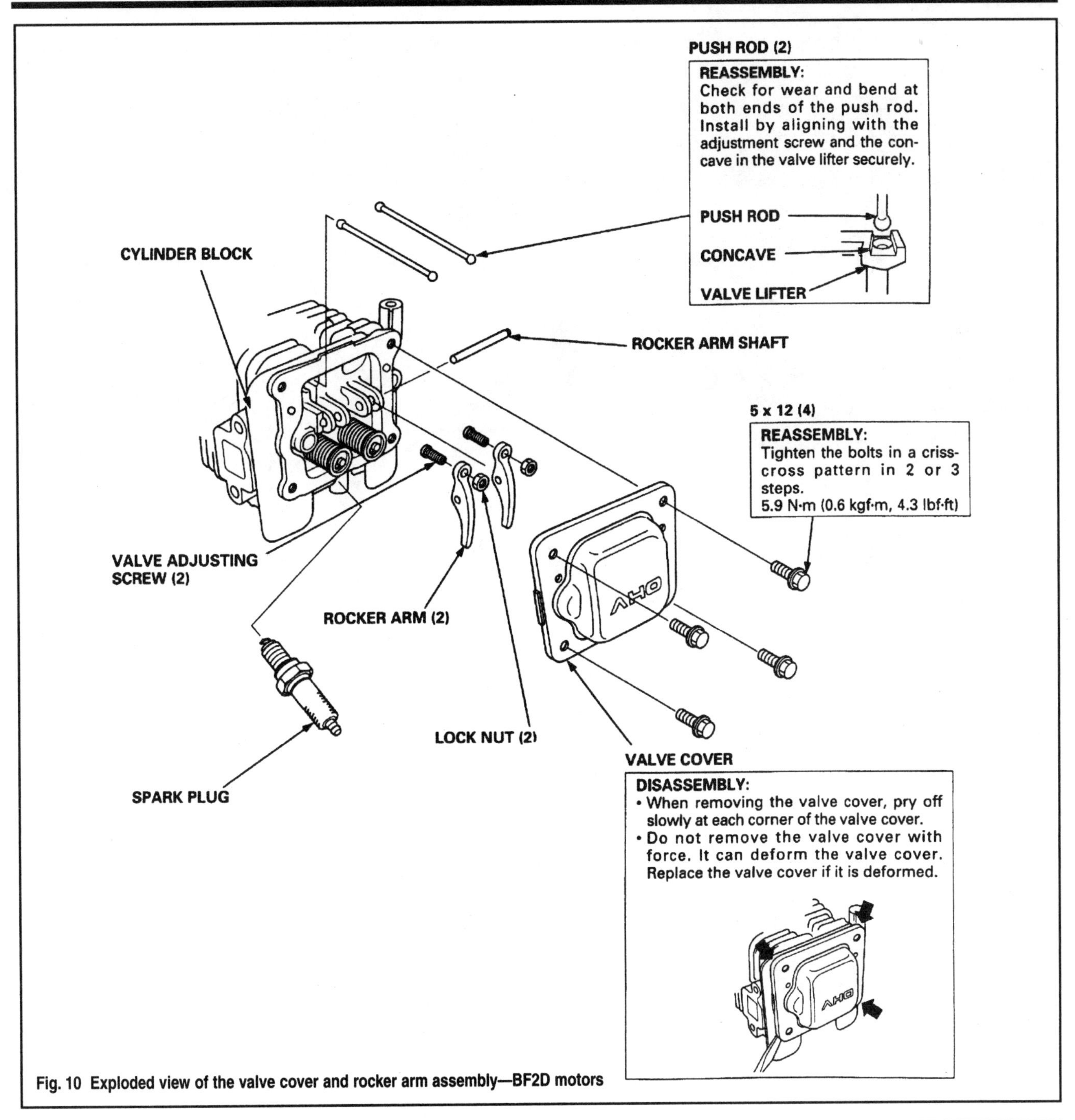

Fig. 10 Exploded view of the valve cover and rocker arm assembly—BF2D motors

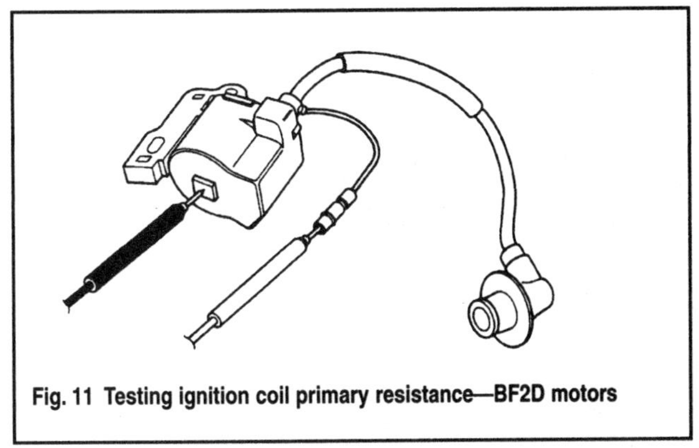

Fig. 11 Testing ignition coil primary resistance—BF2D motors

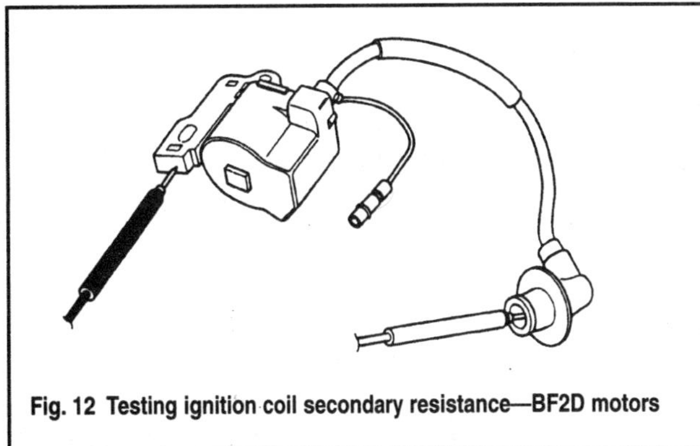

Fig. 12 Testing ignition coil secondary resistance—BF2D motors

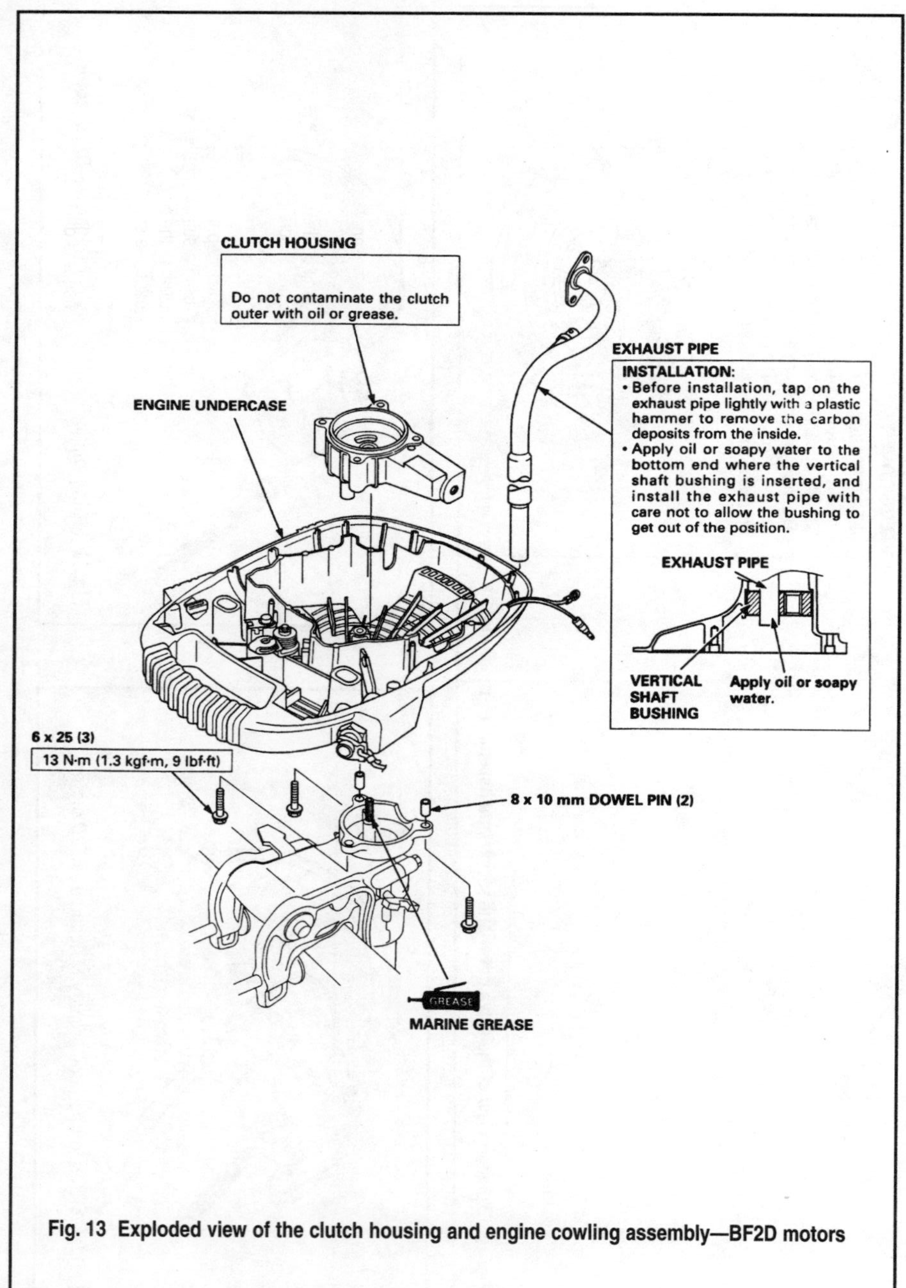

Fig. 13 Exploded view of the clutch housing and engine cowling assembly—BF2D motors

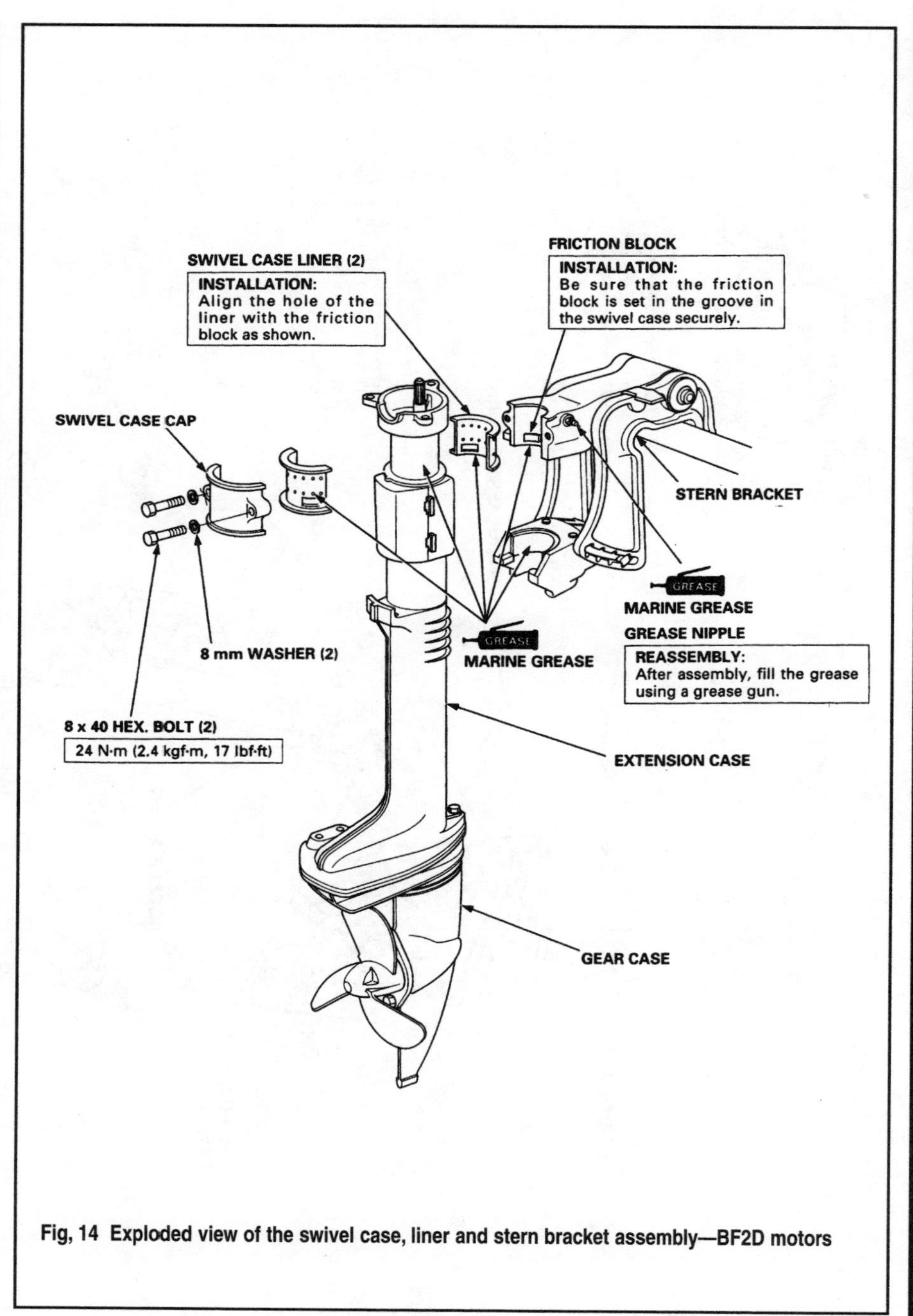

Fig, 14 Exploded view of the swivel case, liner and stern bracket assembly—BF2D motors

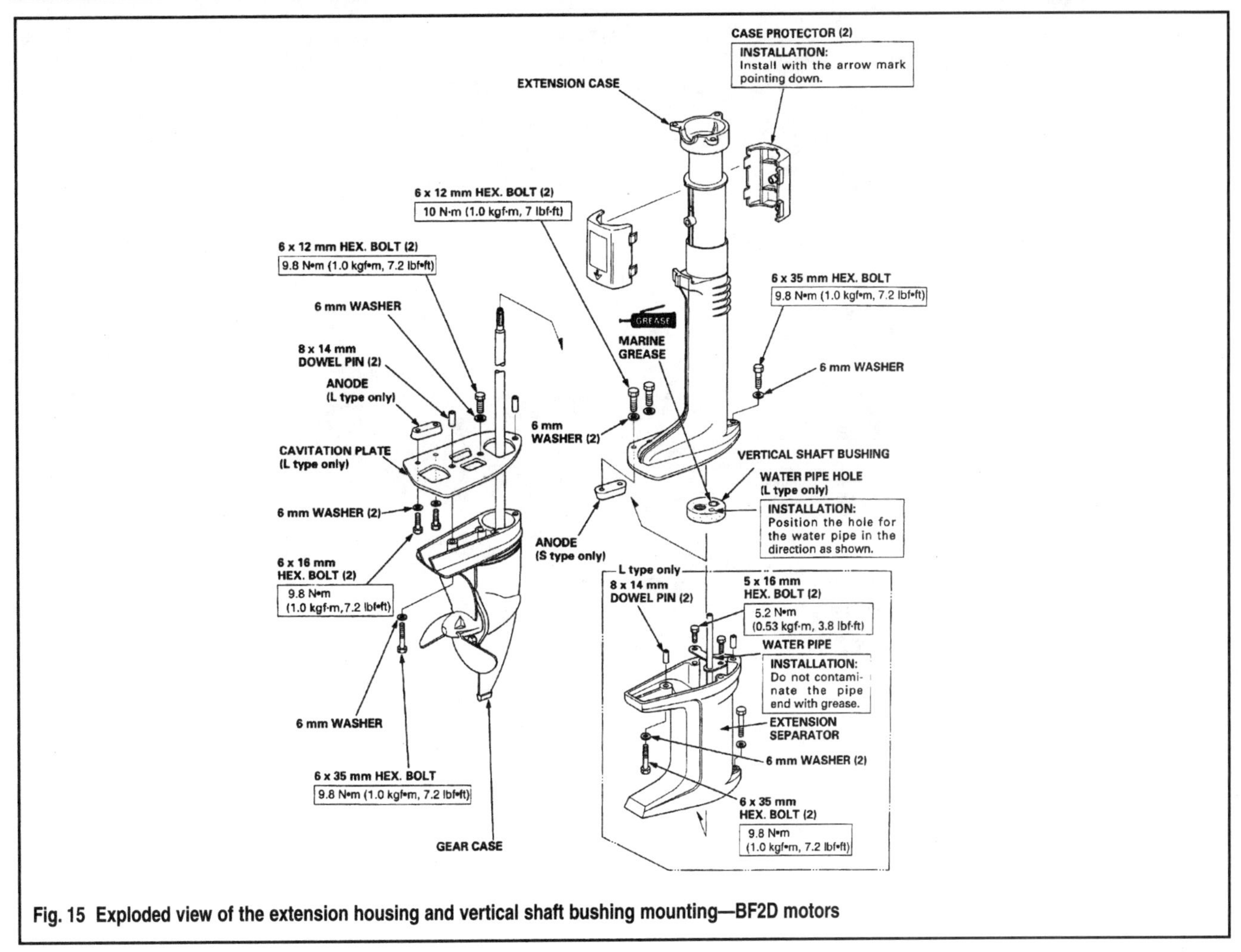

Fig. 15 Exploded view of the extension housing and vertical shaft bushing mounting—BF2D motors

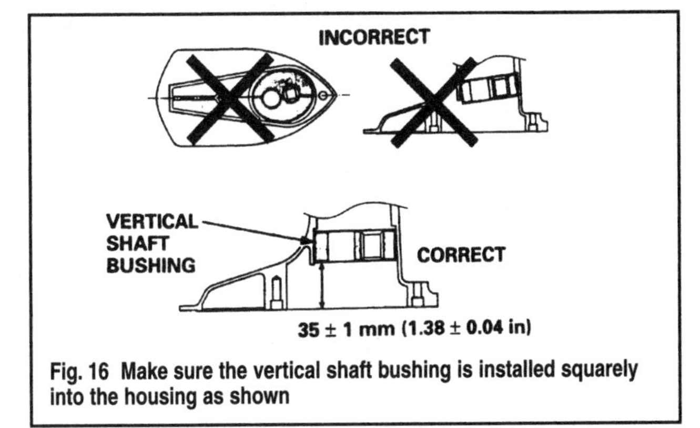

Fig. 16 Make sure the vertical shaft bushing is installed squarely into the housing as shown

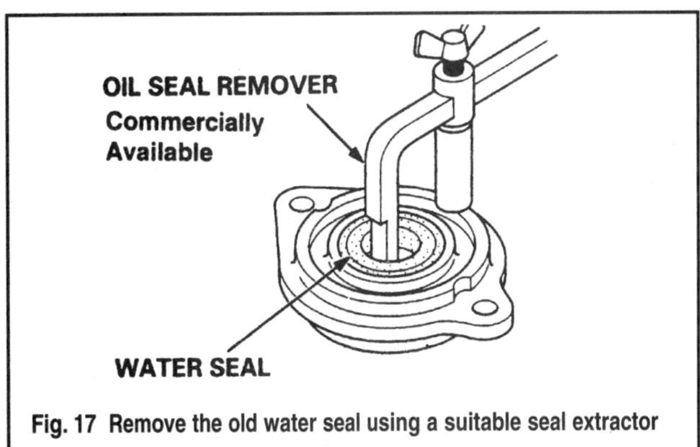

Fig. 17 Remove the old water seal using a suitable seal extractor

WATER SEAL REMOVAL & INSTALLATION

◆ **See Figures 17 and 18**

At least every 3 years Honda recommends replacing the gearcase water seal. This can easily be done with the gearcase still installed on the outboard. The outboard should be tilted fully up to keep gearcase oil from escaping OR if you are doing this during an annual service, simply drain the gearcase before starting the procedure and refill it with fresh oil afterward. The choice is yours.

1. Remove the propeller from the gearcase.
2. Loosen the 2 bolts securing the propeller shaft seal holder to the gearcase.
3. Remove the propeller shaft seal holder and gasket. Discard the old gasket.
4. Using a seal extractor, carefully remove the old water seal from the propeller side of the housing. BE SURE not to scratch and damage the housing.

To Install:

5. Using a suitably sized driver carefully tap the new seal into position in the housing, then pack the seal lips using a suitable marine grade grease.
6. Install the propeller shaft seal holder to the gearcase using a new gasket and taking great care not to damage the seal with the propeller shaft during installation.
7. Install the retaining bolts and tighten to 7.2 ft. lbs. (9.8 Nm).
8. Install the propeller. If the gearcase was drained, be sure to properly refill it with fresh gear oil.

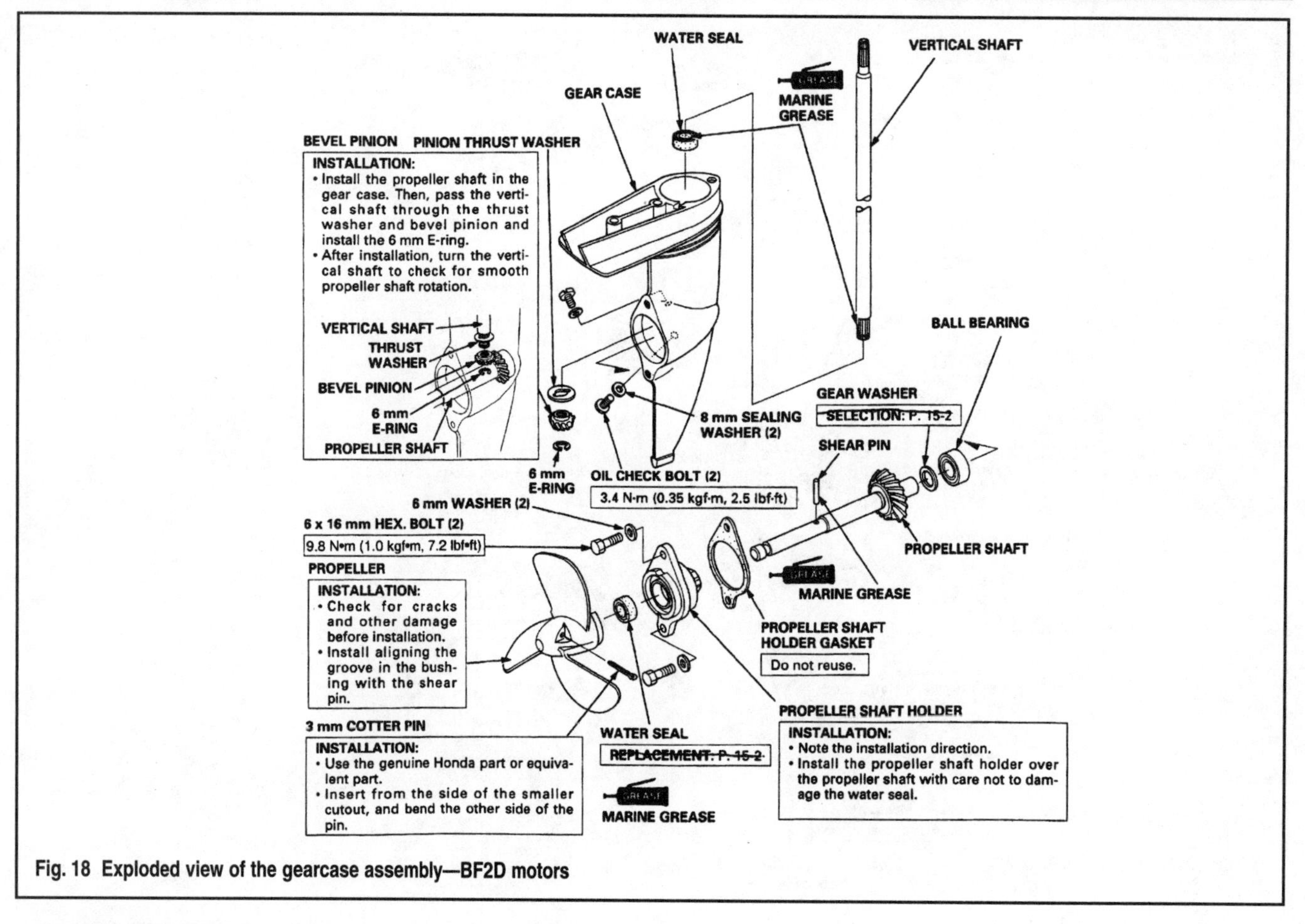

Fig. 18 Exploded view of the gearcase assembly—BF2D motors

HAND REWIND STARTER

◆ See Figure 19

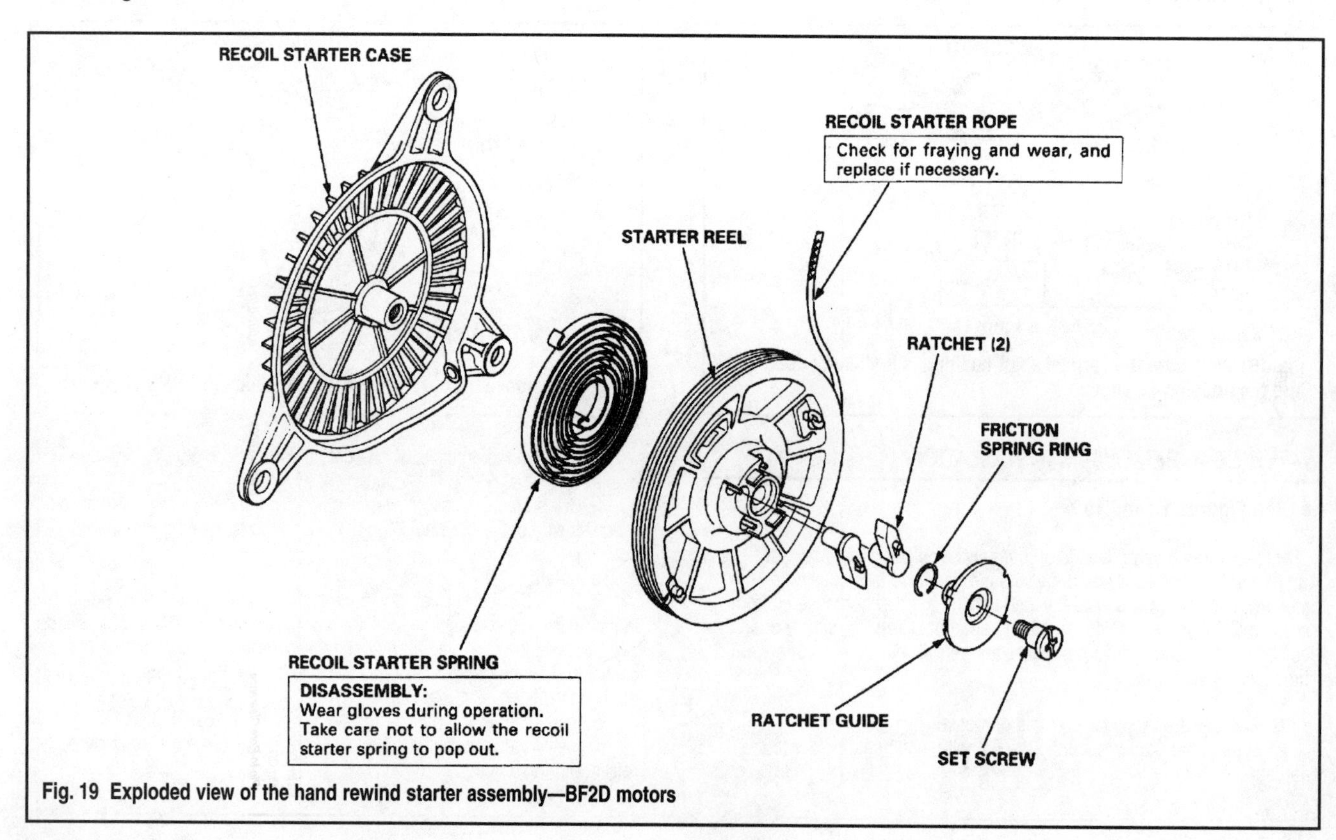

Fig. 19 Exploded view of the hand rewind starter assembly—BF2D motors

Engine Rebuilding—BF2D

Engine Rebuilding Specifications—BF2D

Component	Standard		Service Limit	
	Standard (in.)	Metric (mm)	Standard (in.)	Metric (mm)
Engine				
Type	4-Stroke, Overhead Valve, 1-Cylinder			
Fuel Consumption	420 g/Psh			
Cooling System	Forced Air Cooling			
Ignition System	Transistorized Magneto			
Carburetor	Float Type, Horizontal Butterfly Valve			
Lubrication System	Spalsh-Type			
Starting System	Recoil Starter			
Stopping System	Ground Primary Circuit			
Fuel	Regular Grade Automobile Gasoline			
Exhaust System	Underwater Exhaust System			
Cylinder Compression	128 psi @ 1000 rpm			
Lower Unit				
Gear Ratio	2.4:1			
Propeller				
Number of Blades	3			
Diameter	7.25	184		
Pitch	4.75	120		
Propeller Rotation	Clockwise (Viewed From Rear)			
Valve				
Stem OD				
Intake	0.1563-0.1569	3.970-3.985	0.1535	3.980
Exhaust	0.1549-0.1555	3.935-3.950	0.1528	3.880
Guide ID	0.1575-0.1582	4.000-4.018	0.1598	4.060
Valve Clearance				
Intake	0.0024-0.0039	0.06-0.10	n/a	n/a
Exhaust	0.0035-0.0051	0.09-0.13	n/a	n/a
Spring Free Length	0.93	23.7	0.90	22.8
Valve Lifter				
Bearing ID	0.1970-0.1978	5.005-5.025	0.1988	5.050
Roller OD	0.1965-0.1969	4.990-5.000	0.1949	4.950
Cylinder				
Sleeve ID	1.7717-1.7722	45.000-45.015	1.7756	45.100
Piston				
Skirt OD	1.770-1.771	44.97-44.99	1.768	44.90
Piston-to-Cylinder Clearance	0.0004-0.0019	0.010-0.045	0.0047	0.120
Pin Bore ID	0.3938-0.3940	10.002-10.008	0.3957	10.05
Piston Pin				
Pin OD	0.3935-0.3937	9.994-10.000	0.3917	9.95
Pin-to-Bore Clearance	0.0001-0.0006	0.002-0.014	0.0039	0.10

Engine Rebuilding Specifications—BF2D

Component	Standard		Service Limit	
	Standard (in.)	Metric (mm)	Standard (in.)	Metric (mm)
Piston Ring				
Width				
Top/Second	0.038-0.039	0.97-0.99	0.0362	0.920
Side Clearance				
Top/Second	0.0006-0.0020	0.015-0.050	0.0047	0.120
Piston Ring				
End Gap				
Top	0.0039-0.0098	0.100-0.250	0.0236	0.600
Second	0.0098-0.0157	0.250-0.400	0.0236	0.600
Connecting Rod				
Small End ID	0.3939-0.3944	10.006-10.017	0.3957	10.05
Big End Oil Clearance	0.0006-0.0015	0.016-0.038	0.0039	0.10
Big End Axial (Side) Clarance	0.004-0.024	0.10-0.60	0.031	0.80
Big End ID	0.5906-0.5910	15.000-15.011	0.5921	15.040
Crankshaft				
Crank Pin OD	0.5895-0.5899	14.973-14.984	0.5882	14.940
Camshaft				
Cam Lift	1.1013	27.972	1.0619	26.970
Bearing ID	0.1976-0.1988	5.020-5.050	0.20008	5.100
Roller OD	0.1965-0.1969	4.990-5.000	0.1949	4.950
Crankcase- Side Cover				
Camshaft Bearing - ID	0.1970-0.1978	5.005-5.023	0.1988	5.050
Valve Lifter Roller Bearing - ID	0.1970-0.1978	5.005-5.023	0.1988	5.050
Crankcase- Cylinder Block				
Camshaft Roller Bearing - ID	0.1970-0.1978	5.005-5.023	0.1988	5.050
Valve Lifter Roller Bearing - ID	0.1970-0.1978	5.005-5.023	0.1988	5.050
Rocker Arm Roller Bearing - ID	0.1575-0.1582	4.000-4.018	0.1594	4.050
Rocker Arm				
Bearing - ID	0.1577-0.1585	4.005-4.025	0.1594	4.050
Roller - ID	0.1571-0.1575	3.990-4.000	0.1555	3.950
Clutch				
Lining Thickness	0.08	2.0	0.04	1.0
Outer Clutch - ID	3.071-3.081	78.00-78.25	30.09	78.5
Propeller Shaft				
OD at Bevel Gear (Holder)	0.4320-0.4324	10.973-10.984	0.4303	10.93

Engine Rebuilding Specifications—BF2D

Component	Standard		Service Limit	
	Standard (in.)	Metric (mm)	Standard (in.)	Metric (mm)
Propeller Shaft Holder				
Shaft Bore ID	0.4331-0.4338	11.000-11.018	0.4354	11.06
Shaft-to-Bore Clearance	0.0006-0.0018	0.016-0.045	-	-
Vertical Shaft				
OD at Gear Case	0.432-0.433	10.97-10.99	0.430	10.93
Vertical Bushing OD	0.432-0.433	10.97-10.99	0.430	10.93
Gear Case				
Vertical Shaft Bore ID	0.4331-0.4338	11.000-11.018	0.4354	11.06
Case-to-Vertical Shaft Clearance	0.0004-0.0019	0.010-0.048	-	-
Vertical Shaft Bushing				
Vertical Shaft Bore ID	0.439-0.441	11.15-11.20	0.461	11.70
Bushing-to-Vertical Shaft Clearance	0.006-0.009	0.16-0.23	-	-

BF75A/BF90A Supplemental Information (1999 and Later Models)

REVISED GEAR/SHIM SELECTION CHARTS—LOWER UNIT OVERHAUL

◆ See Figures 20, 21 and 22

*** Reverse bevel gear shim selection table**

Engagement mark on the gear case	Thickness
A	0.15 mm (0.006 in)
B	0.20 mm (0.008 in)
C	0.25 mm (0.010 in)
D	0.30 mm (0.012 in)
E	0.35 mm (0.014 in)
F	0.40 mm (0.016 in)

Fig. 20 Reverse gear shim selection table—1999 and later BF75A and BF90A

• Pinion gear shim selection table

Unit: mm (in)

Engagement mark on the gear case	Calculation value								
	0.57 (0.022) – 0.55 (0.021)	0.55 (0.021) – 0.50 (0.020)	0.50 (0.020) – 0.45 (0.018)	0.45 (0.018) – 0.40 (0.016)	0.40 (0.016) – 0.35 (0.014)	0.35 (0.014) – 0.30 (0.012)	0.30 (0.012) – 0.25 (0.010)	0.25 (0.010) – 0.20 (0.008)	0.20 (0.008) – 0.15 (0.006)
F	0.35 (0.014)	0.40 (0.016)	0.45 (0.018)	0.50 (0.020)	0.55 (0.022)	0.60 (0.024)	0.65 (0.026)	0.70 (0.028)	0.75 (0.030)
E	0.30 (0.012)	0.35 (0.014)	0.40 (0.016)	0.45 (0.018)	0.50 (0.020)	0.55 (0.022)	0.60 (0.024)	0.65 (0.026)	0.70 (0.028)
D	0.25 (0.010)	0.30 (0.012)	0.35 (0.014)	0.40 (0.016)	0.45 (0.018)	0.50 (0.020)	0.55 (0.022)	0.60 (0.024)	0.65 (0.026)
C	0.20 (0.008)	0.25 (0.010)	0.30 (0.012)	0.35 (0.014)	0.40 (0.016)	0.45 (0.018)	0.50 (0.020)	0.55 (0.022)	0.60 (0.024)
B	0.15 (0.006)	0.20 (0.008)	0.25 (0.010)	0.30 (0.012)	0.35 (0.014)	0.40 (0.016)	0.45 (0.018)	0.50 (0.020)	0.55 (0.022)
A	0.10 (0.004)	0.15 (0.006)	0.20 (0.008)	0.25 (0.010)	0.30 (0.012)	0.35 (0.014)	0.40 (0.016)	0.45 (0.018)	0.50 (0.020)

Unit: mm (in)

Engagement mark on the gear case	Calculation value								
	0.15 (0.006) – 0.10 (0.004)	0.10 (0.004) – 0.05 (0.002)	0.05 (0.002) – 0 (0.000)	0 (0.000) – – 0.05 (– 0.002)	– 0.05 (– 0.002) – – 0.10 (– 0.004)	– 0.10 (– 0.004) – – 0.15 (– 0.006)	– 0.15 (– 0.006) – – 0.20 (– 0.008)	– 0.20 (– 0.008) – – 0.25 (– 0.010)
F	0.80 (0.031)	0.85 (0.033)	0.90 (0.035)	0.95 (0.037)	1.00 (0.039)	1.05 (0.041)	1.10 (0.043)	1.15 (0.045)
E	0.75 (0.030)	0.80 (0.031)	0.85 (0.033)	0.90 (0.035)	0.95 (0.037)	1.00 (0.039)	1.05 (0.041)	1.10 (0.043)
D	0.70 (0.028)	0.75 (0.030)	0.80 (0.031)	0.85 (0.033)	0.90 (0.035)	0.95 (0.037)	1.00 (0.039)	1.05 (0.041)
C	0.65 (0.026)	0.70 (0.028)	0.75 (0.030)	0.80 (0.031)	0.85 (0.033)	0.90 (0.035)	0.95 (0.037)	1.00 (0.039)
B	0.60 (0.024)	0.65 (0.026)	0.70 (0.028)	0.75 (0.030)	0.80 (0.031)	0.85 (0.033)	0.90 (0.035)	0.95 (0.037)
A	0.55 (0.022)	0.60 (0.024)	0.65 (0.026)	0.70 (0.028)	0.75 (0.030)	0.80 (0.031)	0.85 (0.033)	0.90 (0.035)

• How to read shim selection table

When the engagement mark on the gear case is E and the calculation value is – 0.2 mm (– 0.008 in) or more, the shim thickness is 1.05 mm (0.041 in).

When the calculation value is less than – 0.2 mm (– 0.008 in), the shim thickness is 1.10 mm (0.043 in).

Fig. 21 Pinion gear shim selection table—1999 and later BF75A and BF90A

• Forward bevel gear shim selection table

Unit: mm (in)

Engagement mark on the gear case	Calculation value			
	0.20 - 0.15 (0.008 - 0.006)	0.15 - 0.10 (0.006 - 0.004)	0.10 - 0.05 (0.004 - 0.002)	0.05 - 0 (0.002 - 0.000)
1	0.45 (0.018)	0.50 (0.020)	0.55 (0.022)	0.60 (0.024)
2	0.40 (0.016)	0.45 (0.018)	0.50 (0.020)	0.55 (0.022)
3	0.35 (0.014)	0.40 (0.016)	0.45 (0.018)	0.50 (0.020)
4	0.30 (0.012)	0.35 (0.014)	0.40 (0.016)	0.45 (0.018)
5	0.25 (0.010)	0.30 (0.012)	0.35 (0.014)	0.40 (0.016)
6	0.20 (0.008)	0.25 (0.010)	0.30 (0.012)	0.35 (0.014)

Fig. 22 Forward gear shim selection table—1999 and later BF75A and BF90A